REMEDIATION ENGINEERING of CONTAMINATED SOILS

Environmental Science and Pollution Control Series

1. Toxic Metal Chemistry in Marine Environments, *Muhammad Sadiq*
2. Handbook of Polymer Degradation, *edited by S. Halim Hamid, Mohamed B. Amin, and Ali G. Maadhah*
3. Unit Processes in Drinking Water Treatment, *Willy J. Masschelein*
4. Groundwater Contamination and Analysis at Hazardous Waste Sites, *edited by Suzanne Lesage and Richard E. Jackson*
5. Plastics Waste Management: Disposal, Recycling, and Reuse, *edited by Nabil Mustafa*
6. Hazardous Waste Site Soil Remediation: Theory and Application of Innovative Technologies, *edited by David J. Wilson and Ann N. Clarke*
7. Process Engineering for Pollution Control and Waste Minimization, *edited by Donald L. Wise and Debra J. Trantolo*
8. Remediation of Hazardous Waste Contaminated Soils, *edited by Donald L. Wise and Debra J. Trantolo*
9. Water Contamination and Health: Integration of Exposure Assessment, Toxicology, and Risk Assessment, *edited by Rhoda G. M. Wang*
10. Pollution Control in Fertilizer Production, *edited by Charles A. Hodge and Neculai N. Popovici*
11. Groundwater Contamination and Control, *edited by Uri Zoller*
12. Toxic Properties of Pesticides, *Nicholas P. Cheremisinoff and John A. King*
13. Combustion and Incineration Processes: Applications in Environmental Engineering, Second Edition, Revised and Expanded, *Walter R. Niessen*
14. Hazardous Chemicals in the Polymer Industry, *Nicholas P. Cheremisinoff*
15. Handbook of Highly Toxic Materials Handling and Management, *edited by Stanley S. Grossel and Daniel A. Crow*
16. Separation Processes in Waste Minimization, *Robert B. Long*
17. Handbook of Pollution and Hazardous Materials Compliance: A Sourcebook for Environmental Managers, *Nicholas P. Cheremisinoff and Nadelyn Graffia*
18. Biosolids Treatment and Management, *Mark J. Girovich*
19. Biological Wastewater Treatment: Second Edition, Revised and Expanded, *C. P. Leslie Grady, Jr., Glen T. Daigger, and Henry C. Lim*
20. Separation Methods for Waste and Environmental Applications, *Jack S. Watson*
21. Handbook of Polymer Degradation, Second Edition, Revised and Expanded, *S. Halim Hamid*
22. Bioremediation of Contaminated Soils, *edited by Donald L. Wise, Debra J. Trantolo, Edward J. Cichon, Hilary I. Inyang, and Ulrich Stottmeister*
23. Remediation Engineering of Contaminated Soils, *edited by Donald L. Wise, Debra J. Trantolo, Edward J. Cichon, Hilary I. Inyang, and Ulrich Stottmeister*

Additional Volumes in Preparation

REMEDIATION ENGINEERING of CONTAMINATED SOILS

edited by

Donald L. Wise
Debra J. Trantolo
Cambridge Scientific, Inc.
Cambridge, Massachusetts

Edward J. Cichon
Ionics, Incorporated
Watertown, Massachusetts

Hilary I. Inyang
University of Massachusetts
Lowell, Massachusetts

Ulrich Stottmeister
University of Leipzig
Leipzig, Germany

CRC Press
Taylor & Francis Group
Boca Raton   London   New York

CRC Press is an imprint of the
Taylor & Francis Group, an **informa** business

CRC Press
Taylor & Francis Group
6000 Broken Sound Parkway NW, Suite 300
Boca Raton, FL 33487-2742

© 2000 by Taylor & Francis Group, LLC
CRC Press is an imprint of Taylor & Francis Group, an Informa business

No claim to original U.S. Government works

This book contains information obtained from authentic and highly regarded sources. Reasonable efforts have been made to publish reliable data and information, but the author and publisher cannot assume responsibility for the validity of all materials or the consequences of their use. The authors and publishers have attempted to trace the copyright holders of all material reproduced in this publication and apologize to copyright holders if permission to publish in this form has not been obtained. If any copyright material has not been acknowledged please write and let us know so we may rectify in any future reprint.

Except as permitted under U.S. Copyright Law, no part of this book may be reprinted, reproduced, transmitted, or utilized in any form by any electronic, mechanical, or other means, now known or hereafter invented, including photocopying, microfilming, and recording, or in any information storage or retrieval system, without written permission from the publishers.

For permission to photocopy or use material electronically from this work, please access www.copyright.com (http://www.copyright.com/) or contact the Copyright Clearance Center, Inc. (CCC), 222 Rosewood Drive, Danvers, MA 01923, 978-750-8400. CCC is a not-for-profit organization that provides licenses and registration for a variety of users. For organizations that have been granted a photocopy license by the CCC, a separate system of payment has been arranged.

Trademark Notice: Product or corporate names may be trademarks or registered trademarks, and are used only for identification and explanation without intent to infringe.

Visit the Taylor & Francis Web site at
http://www.taylorandfrancis.com

and the CRC Press Web site at
http://www.crcpress.com

Preface

Remediation Engineering of Contaminated Soils presents updated discussion of the topics covered in *Remediation of Hazardous Waste Contaminated Soils* (Marcel Dekker, Inc., 1994), as well as new work in this important global focus on technology. We are well aware of the advantages of a clean environment. Nonetheless, our environment is still the victim of industrial growth and defense-related activities. There is now international recognition of the problems of environmental damage, especially to soil. The damage to our environment substantially affects our health and welfare. It is a credit to our human spirit that we remain optimistic and share an enthusiasm about environmental issues.

It is a matter of international concern that the number of recognized waste sites continues to grow daily. Past and continuing abuses exact high cleanup tolls in terms of our technical, financial, and social resources. Environmental remediation is thus a rapid growth area ripe for technological applications and innovations. Whether driven by government mandate, social responsibility, economics, or other forces, remediation of hazardous wastes is a necessary exercise in our present environmentally active climate and it is international in scope.

This book provides an up-to-date source of technical information relating to current and potential remediation practices. Over 120 recognized experts provide an in-depth treatment of this rapidly growing field, which draws its resources from many disciplines and countries. We have deliberately solicited input from governmental, industrial, and academic specialists to ensure a multidimensional/multinational presentation of the hazardous waste remediation schemes that are shaping our environmental outlook.

The book is divided into four parts. It begins with the presentation of general engineering issues and the regulatory, ethical, and technical framework within which these processes are managed. The text then introduces specific case studies in hydrocarbon remediation that offer a concise view of the many technological approaches possible in remediation. The following parts balance presentations of traditional and emerging technologies in remediation, with special attention to exciting developments in bioremediation. Throughout, we have attempted to provide a sense that although the now global scope of hazardous waste remediation is immense, it is not overwhelming.

We trust this volume will contribute to this important, global field and emphasize the need for continued progress. One way to better our environment is to eliminate some of the burdens of our wasteful past. Potentially great benefits await us if we can develop economical, effective, and efficient solutions to our waste problems. All readers of this text will contribute something to the environment of tomorrow.

Donald L. Wise
Debra J. Trantolo
Edward J. Cichon
Hilary I. Inyang
Ulrich Stottmeister

Contents

Preface   *iii*

Part 1: Engineering Issues in Waste Remediation

1. International Perspectives on Contaminated Land   1
   Stephen C. James and Walter W. Kovalick, Jr.

2. Methods of Analysis of Contaminant Migration in Barrier Materials   63
   Hilary I. Inyang, John L. Daniels, and Calvin C. Chien

3. Design Considerations for Hazardous Waste Landfills   83
   M. A. Gabr

4. Basics and Applications of Electrokinetic Remediation   95
   Akram N. Alshawabkeh and Ray Mark Bricka

5. Measurement of Pollutant Distribution, Toxicity, and Effectiveness of Emerging Soil Remediation Techniques   113
   K. Bundy, F. Mowat, P. Taverna, and M. Shettlemore

6. Theoretical Basis for the Simulation of Electrokinetic Remediation   155
   Akram N. Alshawabkeh and Christian J. McGrath

7. Conditions for the Use of Innovative Remediation Techniques Set by New Policy   173
   Cors van den Brink, Timo J. Heimovaara, and Mari P. J. C. Marinussen

8. Cone Penetrometer Technology for Groundwater Remediation Assessment in the Vadose Zone   183
   John F. Peters, Daniel A. Leavell, and Landris T. Lee

9. Evaluation of the Adequacy of Hazardous Chemical Site Remediation by Landfilling   193
   G. Fred Lee and Anne Jones-Lee

10. Bioavailability: The Major Challenge (Constraint) to Bioremediation of
    Organically Contaminated Soils 217
    A. L. Juhasz, M. Megharaj, and R. Naidu

Part 2: Case Studies in Hydrocarbon Remediation

11. Modeling Extractive Washing of Hydrophobic Organic
    Chemical-Contaminated Soils 243
    Gerald R. Eykholt and Walid K. Al-Fayyoumi

12. Air Sparging, Soil Vapor Extraction, and Bioremediation in
    Low-Permeability Soils 261
    C. Marjorie Aelion and Brian C. Kirtland

13. Intensification of Removal of Hydrocarbon Contamination from Water
    and Soil Using Oxygen Transferors 279
    Tamara V. Sakhno, Viktor M. Emelyanov, and Viktor M. Kurashov

14. Hydrocarbon Uptake and Utilization by *Streptomyces* Strains 289
    *György Barabás, András Penyige, István Szabó, György Vargha,
    Sándor Damjanovich, János Matkó, János Szöllősi, Samir S. Radwan,
    Anita Mátyus, and Tadashi Hirano*

15. Enhanced Naphthalene Bioavailability in a Liquid–Liquid Biphasic System 309
    J. A. G. F. Menaia, M. Rosário Freixo, and F. M. Gírio

16. Intensive Treatment of Mineral Oil-Contaminated Drilling Cuttings 321
    A. Noke, R. A. Müller, and Ulrich Stottmeister

17. Ecological Consequences of Enhanced UV Radiation on the Phenolic
    Content of *Brassica oleracea*: A Review 329
    Jeffrey M. Lynch and Alicja M. Zobel

18. RC1 Consortium for Soil Decontamination: Its Preparation and Use 357
    Dana M. V. Horáková and Miroslav Nemec

19. Distribution and Fate of Organic and Inorganic Contaminants in a River
    Floodplain—Results of a Case Study on the River Elbe, Germany 373
    *Kurt Friese, Guenter Miehlich, Barbara Witter, Werner Brack, Olaf Buettner,
    Alexander Groengroeft, Frank Krueger, Maritta Kunert, Holger Rupp,
    René Schwartz, Andrea van der Veen, and Dieter W. Zachmann*

20. Formation, Long-Term Stability, and Fate of Nonextractable ^{14}C-PAH
    Residues in Contaminated Soils 427
    Annette Eschenbach, Reinhard Wienberg, and Bernd Mahro

21. Industrial Trickling Bed Biofilters for Abatement of Volatile Organic
    Compounds from Air Emissions 447
    *Vladimir O. Popov, Alexey M. Bezborodov, Phillip Cross, and
    Adrian Murphy*

Contents

Part 3: Traditional Soil-Specific Technologies

22. Air Stripping and Soil Vapor Extraction as Site Remediation Measures 473
    Constantine J. Gregory and Frederic C. Blanc

23. Structural and Functional Aspects of Artificial Soils Construction 489
    Natalia Zaimenko, T. Cherevchenko, and G. Rusin

24. Soil Trace Metal Remediation Using Solid Additives: Thermodynamic and Physical-Chemical Aspects 505
    Isabelle Yousfi and Alain Bermond

25. Surfactants-Enhanced Photolysis of Polychlorinated Biphenyls (PCBs): A Solution for Remediation of PCB Contaminated Soils 523
    Zhou Shi, Mriganka M. Ghosh, and Kevin G. Robinson

26. Decontamination of Heavy-Metal-Laden Sludges and Soils Using a New Ion-Exchange Process 541
    Sukalyan Sengupta and Arup K. SenGupta

27. Cost-Effective Method of Determining Soil Respiration in Contaminated and Uncontaminated Soils for Scientific and Routine Analysis 573
    Markus Robertz, Thomas Muckenheim, Susanne Eckl, and Leslie Webb

28. Remediation of Soils Contaminated with Chromium Due to Tannery Wastes Disposal 583
    K. Ramasamy, S. Mahimairaja, and R. Naidu

29. Aqueous Solvent Removal of Contaminants from Soils 617
    James C. O'Shaughnessy and Frederic C. Blanc

30. Relationships Between Heavy-Metal Mobility in Soils and Sediments and the Presence of Organic Wastes 639
    Luis Madrid, Encarnación Díaz-Barrientos, and María Bejarano

31. Field Techniques for Sampling and Measurement of Soil Gas Constituents at Contaminated Sites 657
    Alan J. Lutenegger and Don J. DeGroot

32. Speciation of Heavy Metals: An Approach for Remediation of Contaminated Soils 693
    G. S. R. Krishnamurti

33. Arsenic-Contaminated Soils: I. Risk Assessment 715
    A. Brandstetter, E. Lombi, W. W. Wenzel, and D. C. Adriano

34. Arsenic-Contaminated Soils: II. Remedial Action 739
    E. Lombi, W. W. Wenzel, and D. C. Adriano

Part 4: Developing Remediation Technologies

35. Advanced Risk-Based Biodegradation Study Using Environmental Information System and the Holistic Macroengineering Approach 759
    Stergios Dendrou, Basile Dendrou, and Mehmet Tumay

36. Application of Carbon Dioxide in Remediation of Contaminants: A New Approach 829
    Katta J. Reddy

37. TORBED® Process Reactor Technologies: A Novel Treatment Approach for Contaminated Soils 839
    C. E. Dodson and R. G. W. Laughlin

38. Dispersing by Chemical Reactions Remediation Technology 849
    F. Bölsing

39. Processing of Vegetable Raw Material and Its Waste by Energy-Saving, Nature-Preserving Technology to Obtain Carbohydrate-Protein Fodder 931
    I. Shakir, V. Panfilov, D. Kulinenkov, and M. Manakov

40. Natural Attenuation of Explosives 949
    Maurice V. Cattaneo, Judith C. Pennington, James M. Brannon, Douglas Gunnison, Danny W. Harrelson, and Mansour Zakikhani

41. Municipal Solid Waste Generation and Management in Caracas and Metropolitan Area, Venezuela 971
    Adriana Diaz-Triana

Index *981*

REMEDIATION ENGINEERING of CONTAMINATED SOILS

1
International Perspectives on Contaminated Land

Stephen C. James
U.S. Environmental Protection Agency, Office of Research and Development, National Risk Management Research Laboratory, Cincinnati, Ohio

Walter W. Kovalick, Jr.
U.S. Environmental Protection Agency, Office of Solid Waste and Emergency Response, Technology Information Office, Washington, DC

I. INTRODUCTION

In 1980, a NATO/CCMS pilot study addressing contaminated land and groundwater was formed. In 1986, 1992, and 1998, the pilot study was renewed by participating countries because of the continued interest in this topic. In this initial pilot study (1980–1985), contaminated land was defined for the purposes of the study as land that contains substances that, when present in sufficient quantities or concentrations, are likely to cause harm, directly or indirectly, to man, to the environment, or on occasion to other targets. The emphasis is on the presence of potentially harmful contaminants rather than on past use. This definition embraces both old industrial sites that have become contaminated owing to their former usage, and hazardous waste problem sites or uncontrolled hazardous waste sites. It also implies that a problem is only defined to exist following a proper site investigation and after the evaluation of all available information specific to the site and taking into account the intended use. It is generally agreed that there are three main stages to dealing with contaminated sites:

Identification
Assessment of hazard
Remedial action

With subsequent phases of the pilot study starting in 1986, the focus has been on technologies and policy updates provided by participating countries. This chapter addresses both and reports on recent information provided to the pilot study by participating countries. Currently the following countries participate in the pilot study: Australia, Austria, Belgium, Canada, Czech Republic, Denmark, Finland, France, Germany, Greece, Hungary, Ireland, Japan, New Zealand, Norway, Poland, Portugal, Slovenia, Sweden, Switzerland, The Netherlands, Turkey, United Kingdom, and the

United States. References provided at the end of the chapter make available a wealth of information on the pilot study which can be accessed through several Internet sites.

II. CONTAMINATED LAND POLICY ISSUES

The following summarizes contaminated land policy issues from selected countries.

A. Australia

Australia is a federation of six states and two territories, with a federal government overseeing specific areas of national policy including trade and commerce, external affairs, and defense. The federal government does not have direct legislative powers on environmental policy but works with state and territory governments to implement a national strategy. The key national bodies which provide technical guidance for environment and health are the Australian and New Zealand Environment and Conservation Council (ANZECC) and the National Health and Medical Research Council (NHMRC), respectively.

In September 1995, the state, territory, and federal governments established the National Environment Protection Council (NEPC) with the aim of ensuring greater regulatory consistency in implementing national standards for environmental protection. The council consists of one minister (not necessary the Environment Minister) from each state and territory government and is currently chaired by the Commonwealth Minister for the Environment. The council's principal role is to develop National Environmental Protection Measures (NEPMs) in the form of standards, protocols, goals, or guidelines. Since each state and territory government implements its own regulatory framework, this has lead to many differences between them concerning their arrangements for contaminated sites.

Although the NEPC is considering the future development of national assessment procedures for contaminated sites, present guidance for the assessment and management of contaminated sites has been issued by both the ANZECC and NHMRC. First issued in January 1992, the main purpose of the ANZECC/NHMRC guidelines is to provide a framework for the proper assessment, remediation, and management of contaminated sites. In addition, they are intended to ensure that a consistent standard of site assessment and subsequent management is implemented for all contaminated sites in Australia. The guidelines have no formal legal status. The framework outlined in the guideline document covers such issues as site identification, initial evaluation, community consultation, occupational health and safety, sampling and chemical analysis, development of site-specific guidelines, and remedial strategy options. The guidelines define the goals of site management to be

> Ensuring a site is acceptable and safe for long-term continuation of its existing use
> Minimizing environmental and health risks on- and off-site
> Maximizing the potential options for future site use

The guidelines do not recommend remediation of contaminated sites regardless of site-specific circumstances, since treatment may be either technically difficult to achieve or outweighed by social, economic, and environmental considerations. A site-specific assessment which recommends a restricted site use is the preferred option can be acceptable.

The guidelines are an evolving document which is currently undergoing a major revision by the state, territory, and federal environmental agencies. A major aim of revising them is to ensure that the document is more responsive to Australian environmental conditions and better reflects the needs of the community and associated industrial sectors through:

> Development of a national health and environmental risk assessment framework
> Development of national soil quality objectives
> Preparation of technical, specific issue guidance for regulatory agencies

In June 1993, ANZECC released a discussion paper for public comment with the aim of developing a consistent, national approach for addressing issues associated with the financial liability of contaminated sites. After submissions to the discussion paper were received, an ANZECC Position Paper entitled "Financial Liability for Contaminated Site Remediation" was released in April 1994 to address and respond to particular public comments. The Position Paper identified the general principles which ANZECC believed should underpin a liability scheme for contaminated-site remediation. These principles included:

> The polluter pays, except where the polluter or original site owner is insolvent or unknown, in which case the current site owner/occupier should be liable.
> A distinction is made between risk and nonrisk contaminated sites, and the government should intervene only in cases of the former.
> Strict liability to comply with a direction to take remedial action should apply (although a statutory right to recover costs from the polluter should also exist).
> Governments (state, federal, etc.) should assume responsibility for remedial action in the case of orphan sites.

ANZECC recognizes that regulatory responsibility lies with individual state and territory governments and therefore each government may adopt the principles which fit within their own administrative and legal framework.

B. Austria

The Government of Austria operates using a federal system with a national government supporting the work of nine provincial governments. In July 1989, Austria introduced the Federal Clean-up Law (ALSAG), which provides the legislative structure for management of contaminated sites. ALSAG set out a management framework entitled the National Clean-up Program, which includes the following aspects:

> A national uniform structure of registration and assessment for contaminated sites
> The creation of public funds to support site management
> A mechanism for the national distribution of public funds to stimulate voluntary activities at contaminated sites (Guidelines for Funding)
> Description of the responsibilities for different regulatory authorities operating within the program
> Statement of liabilities and mechanisms for enforcement (referring to related environmental laws)

The National Clean-up Program is coordinated by the Federal Ministry of the Environment, with the provincial governments being responsible for the identification of potentially contaminated sites and for enforcing compulsory remedial actions by potential responsible parties in urgent cases. The information provided by the provinces is transferred to the Federal Environment Agency (UBA), which carries out registration, assessment, and prioritization of sites according to their level of risk. Financing of remedial actions occurs via a levy on certain waste types, with the Federal Ministry of Finance being responsible for its collection. Initial estimates of the required funding for site management are of the order of $1 billion over the next decade. However, this has been revised upwards in light of experiences in the first years of the National Clean-up Program. It is proposed to raise levies by as much as 500% to account for increased cost estimates and a significant shortfall in tax collection efficiency. The procedures for collection will also be strengthened. Eighty percent of funds generated by levies was intended to finance remedial actions, with 20% for additional risk assessments. The Guidelines for Funding, which came into force in July 1991, specify the required conditions for applicants and are intended to encourage voluntary actions. Every application should include an evaluation of possible treatment options based on ecological, economic, and technical considerations. The guidelines aim to enhance cooperation between potential responsible partners and public authorities, and to encourage the selection of the most effective remedial solutions.

By May 1996, 93 applications for financial support had been received for support, totaling $430 million. The share of agreed public funds used to support site management is related to the degree of negligence of the potential responsible party. In the past ALSAG has generally covered 70–90% of the overall costs for remediation, but this has had to be reduced in accordance with Austria's membership in the European Union, which began on January 1, 1995.

By May 1996, 24,155 potentially contaminated sites had been identified, of which 1807 were included on the national register according to one of four categories of estimated risk level. Ninety-four percent of registered sites are classified as municipal landfills, whereas only 6% are industrial sites. It is suspected that the number of potentially contaminated industrial sites will increase as a program for their specific identification is fully implemented. Of registered sites, 105 have been classed as contaminated in terms of ALSAG. Remedial activity has been undertaken at 38 of them. At present the total number of contaminated sites in Austria is estimated to be 5000–10,000.

By 1994, after five years of experience with the National Clean-up Program, several problems in the framework were identified:

> Enforcement of ALSAG has been weakened because of the complexity of different regulations and regulatory agencies in the management process.
> Revenue generated by the waste levy has been only 33% of the amount predicted, due to ineffective payment controls at the Federal Ministry of the Environment.
> The value for remedial action support is currently twice as large as the incoming revenue.

It is expected that legislation to amend the National Clean-up Program to address these weaknesses will be implemented in the short term. This may include changes to enforcement legislation, simplification of responsibilities, restructuring of the priority-setting system, and improvements to industrial cooperation.

Austria has a new landfill ordinance. This new law specifies parameters related to the quality of the waste to be deposited in landfills in terms of limit values for total pollutant content and related contaminant constituent values in waste samples. In addition, this law specifies four types of landfills—landfills for excavated soils, landfills for demolition waste, landfills for residual materials, and solid or mass waste landfills—with each type of landfill legally accepting only certain types of waste. The following aspects of landfill operation are defined in the new landfill ordinance: waste acceptance inspections; waste generator identity checks; waste sample collection and analysis; special provisions for solidified wastes; precipitation runoff and management; and requirements for bottom-sealing systems (i.e., landfill liner systems). This new law calls for the predisposal segregation of waste streams as well as improved pretreatment of waste prior to disposal.

C. Belgium

Interest in contaminated soil and groundwater has been growing in Belgium since the early 1980s, when a few highly publicized cases came to the fore—for example, the presence of severe arsenic contamination at an abandoned factory at Bocholt in Flanders.

Belgium is a federal state composed of three communities (Flemish, French, and German) and three regions (the Flemish Region, the Brussels Capital Region, and the Walloon Region). The federal government delegates part of its legislative and executive power to the communities and the regions. Communities have responsibility for such matters as culture, education, health, and other social affairs. The regions have responsibility for territorial issues, including the environment, employment, transport infrastructure, and the production and supply of materials including energy and water. The legal framework for policy on contaminated land issues for each region is summarized below.

At the end of February 1995 the Flemish government ratified a new Decree concerning the management of contaminated sites. This law covered five key issues:

An inventory of contaminated land
The difference between historical and new soil contamination
The difference between duty and liability for decontamination
The soil decontamination procedure
The transfer of land

In the Flemish Region, regulatory authorities are now empowered to investigate sites at the time of property transfer or during the closure of certain installations to determine the extent of any contamination which may be present. This information is stored on a regional database under the administration of OVAM (the Public Waste Agency of Flanders), with the dual purposes of supporting policy decision making and as a mechanism to heighten the awareness of potential site purchasers to contamination. In addition to creating an inventory, the new legislation defined contamination as either historical or new, depending on whether pollution occurred before or after the new Decree came into force. It is a requirement that new contamination be remediated as soon as the soil standards for contaminants in soil and groundwater are exceeded. The decision to remediate historical contamination is based on a site-specific risk assessment approach and, due to the limited financial resources available, will be subject to prioritization by the government. In either case, where remediation is required, this should be to background levels wherever possible.

According to the Decree, responsibility for remediation rests with the user of the land where pollution enters the soil. Therefore, in the case where site contamination has arisen from migration of pollutants from off-site, no responsibility rests with the site operator. Further, where the operator can prove that either he did not cause the pollution or could have known about the pollution when purchasing the site, he is exempt from an obligation to remediate. The liability for new contamination is in accordance with the "polluter pays" principle. In the case of historic contamination the owner/user/polluter can be held liable under the rules adopted by the Act of Waste Management (1981).

The Decree outlined the soil decontamination procedure, which involves four phases:

Exploratory investigations (similar to a desk study to determine any reason why this site may be contaminated)
Descriptive soil examination (a site investigation to determine the extent and quantity of any contamination present)
Soil remediation plans (design of a remedial strategy)
Soil decontamination operations (implementation of remediation)

In accordance with the new legislation, a soil register has been created. The Flemish authorities proceed with a systematic examination of potentially polluted areas mainly on three occasions:

At the time of property transfer
At the closure of licensed installations
Whenever the license (authorization) has to be renewed

Considering the varying delays for industrial license renewals, a special soil control obligation has been introduced in the general authorization procedure, so the ultimate deadline seems to be the year 2003 (with intermediate deadlines in 1999 and 2001). By that time, all industrial sites in use should have been checked, and reauthorized or compelled to consider clean-up measures (to be implemented before 2006). The information on soil pollution is compiled in the soil register under the administration of OVAM. This register serves as a database for policy decisions and also as an instrument to protect and inform all potential land purchasers. A soil certificate is requested for all sorts of property transfers. This system has increased the number of voluntary investigations, and sometimes induces voluntary remediations, in order to avoid being listed as contaminated in the register.

The Flemish legislation lays a special responsibility on registered soil decontamination experts. These are the responsible body for soil examination, under the supervision of OVAM, which selects them according to expertise criteria and which controls their work.

According to OVAM, at the end of February 1998 there were 5528 potentially contaminated sites in different parcels of land listed in the soil register. As of the same date, about 8000 parcels of land were mentioned as contaminated in the register. Remediation programs are launched for about 70 sites. Registered soil decontamination experts have to develop and carry out those programs, according to the procedure and soil standards. They will also have to control the final result of the clean-up, under the supervision of OVAM.

In the Brussels Capital Region, the development of a soil protection policy is mainly the responsibility of the Minister of Housing, Environment, Nature Conservation, Water Policy, Historical Monuments, and Sites, together with the Brussels Institute for Environmental Management (BIM/IBGE). Duties for soil protection are delegated to BIM/IBGE in accordance with the Royal Decree of 1989. Existing legislation for soil protection is limited: According to the National Act on Protection of Groundwater (1971) and to the Ordinance on Waste Prevention and Management (1991), any producer or owner of waste has to remove that waste without causing damage to soil, flora, fauna, air, and water, and at no risk for the environment and public health. Therefore the "polluter pays" principle applies in the region. Where liable parties cannot be found or do not comply with a requirement to take remedial measures, then the government may act to protect human health and the environment.

No registration system is known at this moment. A first investigations/mapping strategy is in preparation.

In the Walloon Region the responsibility for soil protection is shared among a number of regional ministries and semipublic institutions. A distinction is made between land contaminated due to abandoned industrial operations and those contaminated as a result of waste disposal practices, since they are regulated by different authorities. Legislation has been enacted for each type of contaminated site, including the Walloon Town and Country Planning Code (1978, amended 1995) and the Walloon Act on Waste Management (1985).

It has been estimated that Walloonia has 22,240 acres of derelict land, of which more than 50% are related to coal mining. It was for this reason that the Town and Country Planning Code (1978) contained specific regulations concerning site reclamation and alternative economic uses. The principle of these regulations is that the owner of any derelict land is obligated to bring derelict land back into good condition. The Walloon Region provides assistance to site owners/operators to meet these regulations in the form of a subsidy or loan:

> If the owner is a public body, then regional support will meet 50% of the site purchase cost and 100% of the remedial management cost.
> If the owner is in the private sector, then the region will provide interest-free loans.

The 1978 regulations did not include a requirement to investigate and/or control chemical contamination of the soil. However, following concerns raised internationally over the problems of contaminated land, the issue of soil contamination was added to the more general derelict land policy. A public body responsible for the reclamation of contaminated sites, called SPAQUE (Société publique d'aide à la qualité de l'environnement), was created. The main mission of SPAQUE is to assess, set priorities, and reclaim former waste disposal sites. As of November 1995, SPAQUE was responsible for 17 priority sites, of which 4 are subject to ongoing remedial action.

A registration system has existed since 1978 for industrial derelict land and brownfield sites, based on specific town and country planning legislation aimed at the redevelopment of those sites. In 1989, a special program, entrusted to the Brussels University, was launched by the Town and Country Planning Administration to assess the risk of contamination on all registered sites. It is based on preliminary assessments and includes a four-level risk ladder. The resulting database serves for policy decisions, to select priorities for detailed site investigations, and for remediation plans if proven necessary. A more elaborate hazard ranking system has been developed recently for dumping sites by the SPAQUE under the supervision of the Walloon Waste Office. The ranking is performed on the basis of a checklist considering source, vectors, and risk groups.

An estimation of about 5000 potentially contaminated sites is currently mentioned. Of these, 2200 industrial derelict sites are already registered and classified in the Town and Country Planning database. Among the sites presenting a high risk factor, about 90 have been submitted to detailed investigations (as of February 1998). A dozen are now benefitting from remediation programs. For sites presenting lower risk factors, detailed investigations are ordered only when a redevelopment strategy is planned, whether by a public or a private operator. In addition, SPAQUE assessed 17 heavily polluted priority sites, among former dumping and deposit sites. Four of them are in the remediation process.

Since the adoption of the Flemish Decree on soil remediation, there has been a growing recognition of soil and groundwater contamination issues in Belgium. The implementation and the first results of the Flemish Decree are generally considered satisfactory by public authorities, and this stimulates the two other regions, Brussels and Walloonia, to define their own policies. However, these policies might be based on rather different legal schemes, clean-up guidelines, and soil criteria. For instance, whether these criteria should be compulsory or subject to site-specific interpretation is a matter of debate in the two regions. At the same time, in the private sector, the big companies are preparing the groundwork for the future legal impositions. Their main question is to what extent it will be possible to adopt different strategies and levels of soil protection in the three Belgian regions. More generally, the two main problems to be tackled in the near future will probably be

> The lack of resources of many liable parties for the clean-up of historical pollution
> The cost efficiency and environmental merit of the remediation programs, whether funded by public or private money

D. Canada

Canada is a federal state consisting of a central federal government and 10 provincial and 2 territorial governments. Although there is a Canadian Environmental Protection Act, which covers most environmental issues, the provincial and territorial governments are the primary legislative authorities for the environment. The levels of government work extensively together, in particular through the Canadian Council of Ministers for the Environment (CCME).

In October 1989 the CCME initiated the National Contaminated Sites Remediation Program (NCSRP) in order to provide a consistent national approach for the classification and remediation of polluted sites. The program's guiding principle is that the polluter pays. NCSRP is administered through bilateral agreements between the federal government and the provincial/territorial governments and focuses on:

> Identification, assessment, and remediation of all contaminated sites that form a (potential) risk to human health or the environment
> Providing joint government funding for high-priority orphan sites
> Stimulation of the development and demonstration of new and innovative remediation technologies

Two other objectives of the program are to

> Define clearly the liabilities of those responsible for contaminating a site, in an effort to encourage pollution prevention
> To communicate with stakeholders

A budget of $150 million was allocated for orphan site remediation for a period of five years from 1989. By the end of fiscal year 1994–1995, work had commenced on seven new high-risk contaminated orphan sites and had been completed at 11 sites. The total number of orphan sites being addressed by the NCSRP is now 45, at a total cost to March 1995 of nearly $50 million.

As custodian for nearly 40% of Canada's land mass, the federal government has a significant responsibility to ensure that contaminated sites within its remit are managed effectively. Federal Crown land is put to both public and private-sector uses, with the responsibility for each site resting with the government departments that control its use. In April 1990, through the NCSRP, Environment Canada committed a further $20 million over five years to assist federal government departments in identifying, assessing, and remediating high-risk contaminated sites within their jurisdictions. Funding is provided on a 50–50 cost-shared basis between Environment Canada and the custodian department. As of the end of March 1995, a total of 325 federal sites had been investigated, with remediation being initiated at 14 sites.

A component of the NCSRP is the Development and Demonstration of Site Remediation Technology (DESRT) program, which has a budget of $35 million. The DESRT program encourages the development and testing of new methods for assessing and treating contaminated sites to eventually bring these new methods to commercial viability. The total number of DESRT projects at the end of the 1994–1995 fiscal year stood at 50, with a total of $20 million having been spent by the end of March 1995 (of which $10 million has come from the private sector). The private-sector contributions are negotiated on an individual project basis, depending largely on the nature of the proposed project. It is reported that five companies which have demonstrated remediation technologies under the DESRT program have been successful at marketing them for commercial use.

In addition to the financial resources created under the NCSRP, a general framework was implemented for developing tools for the consistent identification, assessment, and remediation of contaminated sites. It includes:

- A National Classification System for Contaminated Sites. This is a technical system designed for screening of contaminated sites to identify priority by assessing relative risks, and to evaluate the need for further measures (site characterization, risk assessment, or remediation).
- The Interim Canadian Environmental Quality Criteria for Contaminated Sites. These include assessment and remediation criteria for evaluating the extent of contamination and the need for remediation. Accommodation for various land uses is provided.
- Remediation objectives. These are site-specific treatment goals, which may be either criteria-based or risk-based.

To ensure that the updated criteria, which will replace the interim criteria, are derived on a satisfactory scientific basis, a Protocol for the Derivation of Ecological Effects-Based and Human Health-Based Soil Quality Criteria for Contaminated Sites is under development. The Guidance Manual for Developing Site-Specific Soil Quality Remediation Objectives sets out the circumstances under which the criteria-based or risk-based approach should be used to establish remediation objectives.

Outside the DESRT program, the application of innovative technologies in the wider NCSRP orphan site program is limited to approximately 36% of remedial work for sites where a strategy had been decided upon by the end of fiscal year 1994–1995. In over 60% of projects, on- and off-site disposal and/or containment measures are reported as the preferred or selected option. In the case of two sites it was concluded that no further remedial action was required.

The past year has been one of very great change within the Canadian government in the area of site remediation. The Departments of National Defense (DND) and of

Fisheries and Oceans (DFO) have been active in assessment and remediation of their contaminated sites. Environment Canada has carried out a number of site assessments at its properties. These will be followed by further assessments in the coming year and, potentially, some initial site remediation. The Treasury Board, a central government agency, has given all departments until March 31, 1999, to come up with an approximation of the financial liability which they may be facing as a result of contaminated sites.

E. Czech Republic

The Czech Constitution was established on December 16, 1992. The Act on the Environment No. 17/1992 was adopted from federal law, like many other legal provisions. A new Environmental Policy was adopted in August 1995. Numerous existing laws dated prior to the division of Czechoslovakia have been gradually updated or replaced. The systematic approach for polluted soil remediation started in connection with the ecological damages assessment of former Soviet military bases in the former Czechoslovakia in 1990. The second, more significant wave of soil pollution assessment is in progress now, in connection with property transfers in privatization processes.

There are three sources of financing for environmental projects in the Czech Republic: (a) the State Budget, (b) the State Environmental Fund (created mostly by pollution levies, e.g., for emissions, for waste disposal), and (c) the National Property Fund (created by money from privatization). Legislation which is important for soil and groundwater pollution has been enacted in association with the privatization of former state property. Act No. 92/1991 on Conditions for the Transfer of State Property to Other Persons and connecting Methods of Assessing Environmental Liabilities of Companies for the Preparation of Privatization Projects (Methodological Instructions of the Ministry for Administration of National Property and Its Privatization of the Czech Republic and of the Ministry of Environment of the Czech Republic of May 18, 1992) introduced guideline limit levels for soil, soil gas, and groundwater. These limits have been implemented according to the future use of contaminated sites based on the environmental assessment of property to be privatized. Resolutions of the Government of the Czech Republic No. 455/1992 and No. 123/93 describe and limit the degree to which the state retains liabilities for past environmental damage (items of damage were contamination of groundwater, contamination of soil, and landfills of harmful wastes). Resolution of the Government of the Czech Republic No. 810/1997 changed the above-mentioned resolution when extending the environmental damages to be remediated in the sense of this document by contamination of constructions and their parts. Other legislation on soil quality presents Act No. 13/1994 concerning agricultural soil quality. Limits for discharging pumped out and treated water in the process of remediation are controlled by Order No. 171/1992 on standards of admissible levels of water pollution.

The Czech Army, of course, follows the above-mentioned laws and guidelines. There are additional special regulations issued by Ministry of Defense. There are three types of contaminated sites in the Czech republic:

> Former Soviet Army bases (Soviet occupation lasted from 1968 to 1991). Contaminated sites were evaluated by investigation carried out by environmental companies in 1990 and 1991. Contaminated sites were considered as those where concentration of pollutants more-or-less exceeded Dutch C limits or Czechoslovak Drinking Water Standards when contaminated groundwater

was used for drinking water supply or due to historical use. If necessary, a new risk assessment may be carried out to distinguish which localities are (still) contaminated and/or are to be remediated.

Czech army bases have no special or legal definition in the Czech Republic; nevertheless, the initial assessment procedures differed in the past from those mentioned below.

Other sites are considered contaminated when concentration of pollutants in soils, rocks, groundwater, wastes, soil gas, and buildings are dangerous to the environment when deemed so by the Czech Environmental Inspectorate, mostly based on a risk assessment.

The highest environmental authority is the Ministry of Environment. Each district office has an environmental authority with responsibility for administrative tasks related to the environment. The Czech Environmental Inspectorate enforces the environmental laws through its 42 branches. There are separate divisions for water, air, wastes, forestry, and natural protection.

Concerning the registration of contaminated sites, major problems include:

The most widespread pollutants are oil products (gasoline, diesel fuel, kerosene, lubricating and heating oils), chlorinated aliphatic hydrocarbons (DCE, TCE, PCE), heavy metals (As, Cr, Cu, Pb, Hg, Cd, etc.), phenols, and cyanides.

The most harmful and/or untreatable are organic refractants (PCBs, PAHs, wood-preserving agents, tars), radioactive materials, poisons, and combat chemicals.

There are hundreds of illegal waste disposal sites, a lot of them without any legal or known owner or user.

Registration of former waste disposal sites was slowed down due to the lack of money. Only about 40 districts (50%) have gotten them registered.

There is no common and complete registration of all former and recently contaminated sites.

Background information on site registration includes:

Former Soviet Army bases have been completely registered, and their records of remedial progress are at the Ministry of Environment.

Contaminated sites of the Czech Army are registered by the Ministry of Defence and its regional branches. A new concept of areal registration of contaminated and potentially contaminated sites with the help of the global information system (GIS) has been started.

Registration of waste dumps and landfills with the help of questionnaires was started by one district office in 1995.

Registration of contaminated sites including waste dumps and landfills has been organized by the Ministry of the Environment as a research project (10 districts are involved in the first phase).

There are 60 contaminated former Soviet Army bases in the Czech Republic. Remediation will be accomplished at most of them by 2000. The clean-up of the largest sites, Mladá-Milovice and Ralsko-Hradèany, will last until 2006 and 2008, respectively. Registration of former Soviet Army bases is administered by the Ministry of the Environment.

The eight largest contaminated Czech military sites being returned to civilian use will be remediated by 2005. The five other largest contaminated Czech military sites that will be used by the Czech Army also will be remediated by 2005. Some 300 small contaminated military sites also exist that will be environmentally assessed, and some of them will be remediated step by step. The administration of contaminated sites is done by regional offices of the Ministry of Defense. The National Property Fund administered (and guaranteed remediation) of 300 contaminated sites between 1991 and 1997.

About 100 contaminated sites from the former privatization phase are neither registered nor guaranteed for remediation. About 3000 former illegal municipal or industrial waste dumps, without known ownership, exist throughout the country. A number of them contain harmful wastes and may pose a threat to the environment. Their registration will be carried out by district offices, but only half of these have been done.

The estimated number of sites for future remediation are

Czech military sites: 300
Contaminated sites guaranteed by the National Property Fund: 300–500
Illegal waste dumps without known owner or user: 200

Factors influencing use of remedial methods in the Czech Republic include hydrogeological and physical properties of soil and rocks, chemical-physical properties of pollutants, target concentration of pollutants, amount of contaminated soils, infrastructure of contaminated site (buildings and roads which cannot be destroyed or removed), time and money.

Legislation is under development because to date there have been objections to the Waste Act. The main problems of contaminated sites, more technical than financial, are related to the environmental issues of privatization projects. In particular, there are some discrepancies between planned remedial goals on the one hand and time and economic limits on the other. Both technical and legislative/financial difficulties are related to environmental assessments and the eventual remediation of orphan former waste dumps and contaminated industrial sites from the first phase of privatization. Remediation of fissured aquifers contaminated with chlorinated organics and refractants will pose a significant task for the future. One way to tackle remediation will be through research grants; another way may be the demonstration of up-to-date or developing technologies.

F. Denmark

In Denmark, the national agency responsible for management of contaminated sites is the Environmental Protection Agency (Miljøstyrelsen, DEPA). However, the responsibility for practical implementation of management strategies and remedial work falls on local government (i.e., regional or county-level administration). The DEPA approves the design of remedial actions nationally and compiles a revised national priority list each year based on data supplied by local governments. The DEPA provides national guidelines and sets up regulations.

Contaminated land is addressed mainly in two pieces of legislation:

Environmental Protection Act of 1974, amended in 1992, which covers sites contaminated after 1972–1974
Danish Contaminated Sites Act 1982, revised in 1990, which covers sites contaminated before 1972–1974

A central principle in the Environmental Protection Act is that the polluter should pay. However, according to the statute of limitation of the High Court, the time for bringing such actions expires 20 years after the polluter ceased his activities. Related to the Contaminated Sites Act, a national program on contaminated sites has been in operation since 1983. The main objective of this program is whenever possible to make the site fit for reuse without restrictions.

A Statutory Order and an accompanying guideline on registration and deregistration of contaminated sites was introduced in 1993. The order set out the procedural requirements for registration, with the need for technical documentation emphasized.

In 1994, the Minister for Environment and Energy set up a Contaminated Land Committee. In March 1996, the committee submitted a report on contaminated land, with a proposal for a revised act on contaminated sites. A revised proposal was set out in a public hearing on February 6, 1998. A revised proposal was expected to be presented to Parliament in 1998 and includes a proposal for inclusion of all contaminated land in an act on soil contamination. The intention of the proposal is to expand the regulation from a single-source view into an expanded source view and provide legislation covering all aspects of soil contamination, also including management of soil excavated and transported from one place to another. The five main topics are

> The conclusions to be drawn from the fact that extended contaminated areas, as well as contaminated sites, are in existence
> New concept for mapping
> How to secure groundwater and population within a short time span
> How to secure environmentally safe handling of contaminated soil
> How to accomplish the "polluter pays" principle

According to the proposal, the polluter pays principle will be the basic principle. No distinction will be made between contamination taking place before and after the mid-1970s. All contamination taking place before the new act comes into force will be subject to the same regulation. However, according to the proposal, there will be a difference for contamination taking place before and after the date of enforcement of the new legislation. Whereas contamination taking place after the enforcement of the new act will be subject to strict liability, the applicability of the polluter pays principle will regard contamination that has taken place before the enforcement only if it can be proved that the polluter was acting in bad faith at the time the polluting activity was taking place.

Polluted sites are registered by the Official Property Registry. A condition for registration is that there is evidence that the site is contaminated to a level which presents risks to human health and/or the environment through site use and/or water pollution. After registration, any change in land use must be authorized in advance by the regional authorities. A prioritization system ranks registered sites according to the actual risks related to the present site use. Sites where there is no conflict with the current land use will not be remediated within 10–15 years. Groundwater protection has a very high priority in Denmark because 99% of the drinking water comes from groundwater sources. The Ministry of the Environment is preparing a Groundwater Strategy to speed up the identification and protection of good-quality groundwater areas.

Under the Contaminated Sites Act, approximately 2800 sites were included in the national register by January 1994. Of these sites, 53% are waste disposal sites, 36% are industrial sites, 5% are gasworks sites, and 4% are petrol (gasoline) stations. The total number of waste disposal sites is expected to reach 11,000 nationally.

A budget of $80 million was allocated for registration and management of contaminated sites over the period 1990–1993. A report covering the period 1990–1993 concluded that the total expected cost of cleaning up all identified waste disposal sites is $4 billion, including voluntary actions. It is estimated that remedial actions will take between 30 and 50 years to complete. Approximately half of the remediation projects in Denmark are voluntary, conducted under the guidance and approval of the county authorities. Licenses are normally given on condition of limited remedial action. The severity of licence conditions depends on the sensitivity of the proposed future use. Sites that are only partly remediated remain on the register with a licence for restricted land use.

Other sites which require remediation include about 10,000 sites occupied or formerly occupied by gas stations. In 1992 the petroleum and gas industry reached an agreement with regulatory authorities to remediate old gas station sites under the organization of an operating company. By the middle of 1993 management had been initiated at 140 sites, with remedial actions (mostly by excavation) completed at 66 of them. The cost of remediation is expected to be $7–15 million annually over a 10-year period.

In March 1997, 3650 contaminated sites were registered. The total number is expected to be 11,000. The total number of new and older contaminated sites is estimated to be about 14,000–15,000 sites. About 4500 sites are assigned to areas with vulnerable water use (threat to drinking water) and about 2500 sites are assigned to vulnerable land use (residential areas, playgrounds, etc.). In 1998, the total budget for registered contaminated sites was $45 million. From 1993 to the spring of 1997, 4400 preliminary investigations were carried out. The total number of preliminary investigations for registered sites referring to the Contaminated Sites Act is estimated to 9000–10,000. The investigation usually involves two or three soil samples and possibly one or two groundwater samples. Of the 3650 sites registered up to 1997, there have been remedial activities on 800 sites. In 1994 the Ministry of the Environment appointed a national commission (the Contaminated Land Committee) to review the contaminated land situation and recommend a revised approach for its management. The commission was to consider a proposal for new legislation to produce an integral Contaminated Land Act and to regulate the excavation of soil for reuse or disposal.

Excavation and off-site treatment or disposal are the most commonly used techniques in Denmark for remedial actions. Emerging technologies are seldom used in publically financed actions. Several reasons for this were reported:

> Under the deregistration procedures for sensitive land uses such as for housing, removal from the register can take place only when remedial objectives are met. These targets are strict and can be confidently met only by excavation and removal of soil. The threat of economic blight plays an important role in technology selection.
> No general criteria for soil remediation and for disposal of treated soil are available. Landfill disposal of contaminated soils is in many cases cheaper than alternative technologies.
> The average area of individual contaminated sites in Denmark is small, which makes the implementation of in-situ remediation technically difficult and costly.

Emerging technologies have been more frequently applied in voluntary clean-ups where there is no requirement for deregistration. The aim of most voluntary actions is to obtain a site licence for a particular land use when the developer/owner can show that there is no conflict between planned use and residual contamination. The site remains on the register. Example technologies which have been used in Denmark include:

> Soil vapor extraction
> Ex-situ biological methods, such as land farming and composting, for treating petroleum oils and BTEX compounds
> In-situ approaches such as soil flushing and bioremediation

Looking back over the past 15 years, Danish policy seems to be changing from a very idealistic approach (clean-up of all contaminated sites, making them, wherever possible, clean/uncontaminated) to a more pragmatic approach. One of the reasons is the realization of the extent of the problem, as well as better understanding of the phenomenon. This does not mean that contamination/pollution should be accepted, especially not recent pollution, but it does mean that there might be some sites where Denmark achieves the most environmental benefit for the least money by making sure that there will be no exposure from the site.

Many countries have technology programs that aim to reduce barriers by providing, for example, reliable data on performance and cost, translation of laboratory-scale technologies to technologies implemented at full scale, improved investor confidence in new technologies, and by dealing with technical barriers. Reducing other barriers, and thereby providing for a wider use of the various treatment technologies, will also provide the technology programs with such initiatives as guidelines for target criteria for remedial actions (including acceptable residual concentrations left after treatment is ended), documentation of remediation, and affect policies such as landfill disposal of contaminated soil. In relation to technology choices, it seems that long-term remedial activities are given high priority, meaning that short-term solutions (such as solidification) will become less attractive.

In-situ activities will continue and will be expanded even more than at present. Implementation and prevalence of new technologies will increase. An example is reactive permeable walls. In the few years since its introduction, many countries have either implemented this technology, scaled up from laboratory to field test to full scale, or are already running full-scale projects (helped, of course, because it is a simple technology, which is very suitable for field tests).

Methods for treatment of soil contaminated with heavy metals seem to cause problems, and so does handling remediation of composite contamination. Aspects such as costs of operation, monitoring, life-cycle analysis, sustainability, etc., will also have considerably weight.

From being a engineering phenomenon, remedial activities will become multidisciplinary, involving different scientific specialties; already biology and chemistry are involved, but in the future even more interdisciplinary themes will be necessary. However, no matter what the future may bring, excavation will always be necessary.

G. Finland

In the Report to Parliament on Environmental Protection given by the Council of State on May 31, 1988, the following statement on the policy on contaminated sites in Finland is summarized:

Studies will be made of contaminated land areas, and steps will be taken as necessary to clean them up systematically. The most urgent reclamation work will be investigated as soon as the need for it is established. The following measures will be necessary to achieve this objective: appropriate administrative arrangements must be made by the environmental authorities; techniques must be investigated and organized; and, where necessary, work must begin on revisions to legislation. As far as possible, the Polluter Pays Principle will be held to in meeting costs.

Finnish legislation of primary importance in connection with soil contamination includes, first, waste management legislation, and second, public health legislation and water legislation. There is no separate act concerning soil protection or remediation of contaminated soil.

The Waste Management Act (January 1994) is the main legislative remedial tool for soil cleaning on older sites (created prior to January 1, 1994). According to this act, contaminated soil is defined as waste, and the responsible parties are defined as the polluter, owner, or occupier of the property. In such cases where it is impossible to find a property owner or occupier, the municipality has the responsibility for risk assessment and remediation.

Soil conservation has been included in the waste legislation as a separate chapter. In the Waste Act, new soil polluting activities will be prohibited, or the soil must be returned to its original state so that it can be used for any purpose for which it could have been used without the contamination. The Waste Act will also gives the owner of property a responsibility to find out the state and contamination of the property and to transfer this information to the buyer. The act will also enable the State Council to give detailed regulations concerning soil contamination.

There is also other legislation for public health, air pollution, construction, and neighbor relations. Water legislation prohibits pollution of ground and surface waters. The Construction Act requires that land areas prejudicial to health may not be built on. The existence of soil contamination must be known whenever land use is planned.

The Environmental Damage Act came into effect in June 1996. It is applied to new environmental damage and is based on the polluter pays principle. There is also a need for a complementary scheme for secondary compensation of damages. The secondary compensation is based on compulsory insurance (new law in 1998).

The Ministry of the Environment is the national environmental authority. It formulates environmental policies and does strategic planning. It is also responsible for preparing legislation, setting binding standards, and allocating public funding. The regional environmental administration is proved by 13 regional environmental centers. They are responsible for data collection and allocation of public funding. They create and run clean-up programs, give permits for all clean-up work, and make plans for remediation (called state waste management works). Municipal boards for environmental protection supervise local environmental affairs.

The Finnish Environment Institute conducts environmental R&D. It provides independent expertise for the identification, assessment, clean-up, and control of chemically contaminated environments including contaminated soil.

At the moment, urgent remedial works will be carried out without waiting for the priority plans of contaminated soil remediation. Contaminated soil areas will be cleaned until the year 2015.

The private sector is carrying out clean-up actions on its own contaminated sites. Restoration of old, abandoned sites is funded by budgeted financing and carried on by authorities. There is a state waste management system that makes it possible for the state to participate in or finance (on average by 50%) remedial action in cooperation with the municipalities.

The oil sector, municipalities, and the state made an agreement in 1996 for 10 years, dealing with clean-up of contaminated gasoline stations which will be or have already been closed. The program is based on an agreement among the Finnish Petroleum Federation, the Ministry of the Environment, and the Association of Finnish Localities. It is being funded mainly by oil industry marketing companies, as well as the Oil Pollution Compensation Fund, operating in conjunction with the Ministry of the Environment, to which oil companies pay a fee levied according to their oil imports into Finland. The oil sector will pay the clean-up costs for their own stations, and the Oil Damage Fund will pay for actions needed on abandoned gasoline stations.

Soil contamination will to be taken into account in land use plans and land use changes to a greater extent. At the moment there are no brownfields in Finland.

To date, common remediation methods used in Finland are composting, stabilization, purification of the pore air of soils, containment, incineration at high or low temperature, disposal of slightly contaminated soils at landfills, and also groundwater treatment. Soil washing techniques and biological in-situ methods have not been applied at full scale.

Research and development work is being carried out with composting of soil contaminated by oil and chlorinated phenols, purification of soil contaminated by dioxins and furans (containment, incineration, biological decomposition), use of fungi for aerobic decomposition, use of plants (vegetation) in the clean-up of polluted soil and biological filters for the gases, consolidation and stabilization techniques, as well as biological in-situ treatment of soil and groundwater.

Soil protection, including soil contamination as well as other conservation and degradation processes, is now viewed in general terms. The condition, loading, and protection of soil in Finland has been investigated. Major and irreversible degradation of soil has been avoided so far. The next step will be a target program for soil protection. Also, local and regional soil protection projects have been discussed.

The Finnish legislation is at the moment the powerful instrument for remedial actions of contaminated soil, and the Council of State will issue more detailed regulations concerning the implementation of provisions on soil contamination. Under consideration is a draft to target and limit values of concentrations of harmful substances in soil.

The cooperation between central and regional authorities and the private sector is ongoing in the oil sector, as mentioned above, but also in such others as the forest sector, which is cleaning up old sawmill sites contaminated with organochlorinated compounds. The private sector is carrying on the clean-up actions on its own contaminated sites. Restoration of old abandoned sites is funded by budgeted financing and carried on by authorities. Financial responsibility can also be apportioned between state and private parties if necessary, for example, in cases where the requirement to restore the site would be unreasonable severe.

H. France

At the present time, a polluted site is defined as a site generating a risk—either actual or potential—for human health or the environment related to the pollution of one of the media, resulting from past or present activities. In practice, polluted sites are industrial sites, active or inactive, waste sites, and accidental pollution sites. The French approach to dealing with polluted sites is basically connected with the legislation on the environmental management of industrial installations (IC law) and to a more limited degree on the management of waste (waste law). This means that there is a no specific legislation relative to soil protection or polluted sites. Although the development of such legislation has been considered, it seems that it will probably not happen in the short or middle term and that the existing approach will continue. Basically, legal and administrative action is based on the polluter pays principle, the polluter being, according to the IC law, the operator of the installation at the origin of the pollution.

In this view, the IC law will be applied under the technical directives issued by the Ministry of the Environment to organize the management of polluted sites. These technical directives are related to technical guides developed at the present time.

A first technical guide was issued in 1996 and 1997 (revised) to organize the preliminary evaluation and priority ranking of suspected polluted sites. The proposed preliminary evaluation includes two steps:

Step A: A documentary study (a historical review and a vulnerability study) based on available and accessible data, and completed with a site visit. The historical review includes a description of the sequences of activities that have taken place in the course of time, their precise locations, and any associated environmental practices that may have been carried out. The vulnerability study includes an investigation of the parameters (geology, etc.) that could have relevance for the fate and transport of the contaminants and the potential targets (housing, drinking water supply, etc.) likely to be affected. During the site visit the data deriving from the documentation study should be verified and additional data acquired. An evaluation and identification of existing and potential impacts takes place and a further investigation program is prepared.

Step B: The simplified risk assessment (SRA) includes the collection of data that were not available in the previous study but are conditional for the simplified risk assessment. The SRA demands an understanding of the contamination's spatial distribution and transport mechanisms, the identification of possible hazards, and descriptions of possible rehabilitation methods. At this stage it is necessary to develop some field investigation in order to acquire the data to make this understanding possible.

Based on the results of the preliminary evaluation, a simplified risk assessment is conducted according to a scoring system. The site in question is classified into one of three groups:

Sites needing further investigation and detailed risk assessment
Sites for which monitoring systems should be applied
Sites that can be used for specific purposes without further investigations or implementation of measures

At the present time, another methodology guide is in preparation, under the responsibility of the Ministry of the Environment, in cooperation with a national working group. This guide will define the objectives and contents of the impact study (detailed investigations) and detailed risk assessment.

For the sites where the preliminary investigation concludes that the pollution and risks are serious, the realization of the impact study and risk assessment will give the basis to determine the clean-up objectives and to select the remedial options.

Although France was probably one of the first countries to conduct an inventory of polluted sites, limited attention had been given to the problems of land pollution until the beginning of the 1990s. Apart from the initial national surveys on contaminated sites conducted in 1978, new activities have been taken recently. At the national level, since 1993 the national register has been managed by the Ministry of the Environment. Listed in this register are the sites that are known by the local authorities and can be considered polluted. These sites are listed in a computerized database, and reports are issued by the Ministry periodically, to inform the public of the situation. A publication of this register was issued in December 1994 and included 669 sites. Another one based on the situation of December 1996 was issued in December 1997, with 896 polluted sites, plus 125 sites already restored without any limitation of use.

The inventory has two specific components:

> Historical inventories initiated at the regional level, based on consideration of local industrial history, in order to discover, in connection with the existence of past polluting industrial activities, the places where pollution can be suspected. These inventories are based mainly on consideration of the archives and indicate suspected sites (or potentially polluted sites). At the present time about half of the departments located in 17 regions have initiated such inventories. It is expected that about 200,000–300,000 suspected locations will be reported at the end of these studies for the whole national territory, among which some thousand will require corrective action.
>
> Evaluation of active industrial sites (including industrial waste treatment and disposal sites). A preliminary classification indicates that some 1500–2000 sites assigned priority I will be listed for further investigations.

According to data collected in the national register published in December 1997, the techniques used for the clean-up of polluted soils in sites where a rehabilitation project has been carried out are as follows:

Landfilling, 44
On-site isolation, 60
Stabilization, 12
Natural attenuation, 15
Biotreatment, 29
Soil washing, 10
Thermal/incineration, 29
Other, 33

In more than one-third of these cases, combinations of techniques have been used.

For the first cases of rehabilitation during the 1980s and at the beginning of the 1990s, the techniques used were mostly containment and treatment or disposal at approval sites. It soon appeared that waste treatment plants (incineration) were often technically inappropriate and very expensive and, because of recent regulations inducing restriction of use and technical constraints, landfilling has become more and more difficult and costly. These circumstances create a positive evolution for the use of specific soil treatment techniques.

Containment remains one of the most frequently used techniques, mainly in cases where no treatment technique can be technically or economically applied. The techniques that have been and are still the most frequently used to clean soils are microbiological degradation and soil venting.

Biodegradation is generally carried out on site by the mean of composting or bio-piles. Contaminants degraded are petroleum compounds, light and heavy oils, and even polyaromatic hydrocarbons. Soil venting addresses volatile hydrocarbons and chlorinated solvents in the unsaturated zone. It is sometimes associated with in-situ biodegradation (bio-venting). To remediate the saturated levels (groundwater) venting is combined with air sparging.

More recently, new treatment capabilities have been made available either by technology development or by technology transfer. The techniques concerned are soil washing (solvent washing) and thermal desorption. At the present time, four thermal treatment installations, with various levels of performance (quantity and complexity of pollution that can be treated), have been made available in France, and a fifth is under consideration.

I. Germany

According to the German Constitution, the 16 federal states are responsible for contaminated site remediation. For the enforcement of contaminated site remediation, which includes registration, assessment, and remediation in general, the federal states have enacted specific legislation. In the framework of the federal state's laws regulating contaminated site remediation, different criteria and values are currently used. Together with these regulations, more than 35 different lists containing values such as soil screening, action, and clean-up values exist all over the country. As these values differ more or less from each other depending on different derivation criteria; harmonization and standardization is still urgently needed.

Therefore, the federal government submitted the Federal Act on Soil Protection and Remediation of Contaminated Sites (Bundes-Bodenschutzgesetz) to the Parliament in 1996. In February 1998 the Federal Soil Protection Act was passed by the Parliament and the Federal Council. The act came into force on March 1st, 1999.

The Federal Soil Protection Act (SPA) includes precaution issues as well as remediation of contaminated soils and sites. The main purpose of the SPA is to protect against harmful changes in the soil. Harmful changes in the soil exist when soil functions are impaired, and when this becomes dangerous, leads to adverse effects for individuals or the general public. The definitions of the SPA include natural soil functions and functions of soil utilization.

The following basic duties guarantee that the soil as a living basis for human beings, animals, plants, and soil organisms will be maintained and secured for future utilization:

> Preventive duties exist that the soil is not demanded too much in its ecological efficiency by material and physical influences.
> Existing harmful changes in the soil which cause dangers to human beings and the environment have to be remedied. The duty for remediation also includes groundwater pollution caused by the contaminated soil.
> Site owners are obliged to take care that no hazards are caused by conditions.
> Everyone has to behave in such a manner that harmful changes in the soil do not occur.

> Contaminated sites (CS) are defined as

> Closed-down disposal facilities or other estates on which wastes have been treated, stored, or disposed (abandoned waste disposal sites, AWDS).
> Estates of closed-down facilities and other estates on which environmentally hazardous substances have been handled (abandoned industrial sites, AIS).

For the remediation of contaminated sites, the following are a substantial part of the SPA:

- Federal states' (Länder) authorities are responsible for registration, investigation, and assessment of suspected contaminated sites (SCS).
- Under certain conditions, authorities may require remedial investigations and a remedial plan by those who are obliged for remediation.
- The remedial plan should be straightforward, even in cases of serious and complex contaminated sites, in order to gain the acceptance of the necessary remedial measures by the affected persons.
- The remedial plan should cover a summary of the risk assessment and the remedial investigations as well as the remedial goals and the remedial measures.
- By regulation, the remedial plan is prepared by an expert.
- In the cases of CS and SCS, responsible persons are obliged to announce these sites and to carry out self-control measures; the authorities are responsible for supervision.
- Together with the remedial plan, the regulated person can submit a public contract for the remedial measures.
- To enhance the approval procedure the official obligation of the remedial plan as well as the official order for remediation concentrates all necessary permissions from other laws.

The registration of SCS, which is carried out by the Länder, is focused on the registration of AWDS and AIS. As a result of a nationwide survey in 1997, more than 190,000 SCS were registered, nearly 90,000 are AWDS, and more than 100,000 are AIS. The registration is not complete. The estimate is that more than 240,000 SCS will be registered in the future.

The end of the Cold War, the dissolution of the former Soviet Union, the withdrawal of the West Group of the former Soviet Troops (WGT) from East Germany, and the significant decrease in the number of active military personnel and installations of Allied and German forces results in around 1,235,500 acres of former military land which has been returned to civil control and reuse. Until 1995, of the 1026 West Group of the former Soviet Troops (WGT) bases with a total area of 633,000 acres, 33,738 suspected contaminated sites have been registered. According to a preliminary assessment 12% require immediate action, 32% require further medium-term investigation, and 56% are not environmentally relevant. Most of the sites were handed over to the Länder in East Germany.

Former armament production sites are sites which were contaminated during World War I and II by ammunition production facilities, depots, delaboration works, and storage of chemical warfare agents. As a result a nationwide inventory, 3240 suspected former armament production sites were registered. The assessment of these sites, which is conducted by the Länder, is not finished yet.

As the federal states (Länder) and the local communities are responsible for the remediation, there is no nationwide overview on technologies used or available. In 1996 an evaluation for the Federal State of Northrhine-Westphalia on applied technologies indicated that there is a significant trend toward containment techniques. Of 660 applied remedial measures at 498 industrial sites, 50% were excavation measures with subsequent disposal of the soil, 32% were decontamination measures including pumping and treating of groundwater and soil vapor extraction techniques, and 18% were containment measures. Actual soil clean-up technologies (thermal treatment, soil washing, biological treatment) constituted 5%. Due to the economic situation of the communities there is a common trend to use low-cost technologies (containment, disposal) rather than more expensive decontamination technologies.

Contaminated soils are mostly treated off-site in stationary treatment facilities. As of October 1997, 107 soil treatment facilities with a treatment capacity of 3.7 million ton per year (t/y) are available in Germany:

4 thermal treatment facilities (total capacity 168,000 t/y)
24 soil washing facilities (total capacity 1.5 million t/y)
81 biological treatment facilities (total capacity 2.0 million t/y)

In 1996 the total soil treatment capacity was about 3.5 million t/y; the average capacity used was about 64%. Fifty-six percent of the soil was treated by biological techniques, 39% by soil washing, and only 5% by thermal treatment.

J. Greece

Greece has taken positive legislative action for the protection of public health and the environment and has included in its national legislation since 1986 the basic environmental law 1650/86, which covers all environmental fields and aspects. In this law, specific provisions are included, referring to soil protection from the disposal of municipal and industrial wastes, as well as from the excessive use of fertilizers and pesticides. More specific legislation, concerning some aspects of contaminated land (soil and underground water) is included mainly in Joint Ministerial Decisions (JMD) addressing groundwater, soil waste management, hazardous waste management, and agricultural contamination. Also, according to the legislation, the person or party responsible for waste disposal is charged with the cost of disposal site rehabilitation or reclamation, but in cases of orphan sites the cost is covered by public resources.

No official survey or registry, or official guidelines, exist in Greece regarding contaminated land. The land contaminated by industrial activities is rather limited, because of the lack of important heavy industry. Suspected sites are the industrial areas of Athens (Thriassion Pedion, west of Athens; Oinofyta, north of Athens), Thessaloniki, Volos, and Kavala. Moreover, redundant or operating polymetallic sulfide mines are suspected sites. This includes the redundant mines in Lavrion, where heavy metal pollution has been documented, and other sites, such as Thassos and Ermioni, where no studies have been done. In the operating mines in northern Greece, an extensive rehabilitation plan is underway.

Concerning landfills, the first inventory carried out in 1988 revealed that 1500 sites were operating with some rules, while 3500 sites were operating without any environment protection measures. Since 1990, all new sanitary landfill sites should follow the procedure defined in the JMD. Local authorities are responsible for municipal and household waste management.

The waste disposal must be performed under control according to the environment terms defined by the competent authorities and under continuous monitoring. Secondary landfill sites should be rehabilitated at the end of the operation, and the local authorities are responsible for the restoration costs. Redundant landfill sites, in which operations stopped before 1990, cause serious pollution problems.

The hazardous waste produced in Greece is estimated to be around 450,000 t/y. The disposal of hazardous solid waste and sludge is done either in common landfill sites or in specific sites under control. Co-disposal is applied for those industrial wastes and sludges which have compositions similar to household wastes. Dangerous industrial wastes are disposed of according to their origin and grade of risk posed. PCBs, cyanide wastes, pesticides, etc., either are stored in a safe place or are exported for thermal decomposition according to the existing legislation.

The main contaminated areas in Greece include:

- The greater Lavrion area, 60 km SE of Athens. Intensive polymetallic sulfide mining and smelting activities practiced for over 3000 years resulted in extensive contamination of land and groundwater by heavy metals. Mining and smelting activities stopped in 1988. Urban expansion has led to changes in land use from industrial to residential, recreational, and, to a lesser degree, agricultural. An extended program is underway to define the pollution and develop remedial action.
- Thriassion Pedion, 20 km W of Athens. This is the main industrial area in Athens, with major industries (refineries, steel plants, shipbuilding, clement plants), as well as minor ones. Contamination of land and groundwater has been determined at various sites. The sea bottom sediments in the Gulf of Eleusis are also suspected to be contaminated.
- Ano Liossia Landfill. Studies have been completed concerning biogas composition and groundwater contamination by leachates.

For the other industrial areas (Volos, Thessaloniki, Kavala), a limited amount of data exists.

Three rehabilitation projects on existing landfill sites are currently being undertaken:

- Ano Liossia Landfill site, Athens. This is the main municipal landfill of the Athens area and lies close to the Thriassion Pedion industrial area of Athens. Leachates from this landfill seriously pollute the groundwater, which eventually ends up in the Gulf of Eleusis. The rehabilitation studies have been completed and the work is underway.
- The environmental impact assessment studies have been completed for the now-closed Schistos landfill, which was serving the Piraeus area and stopped operating in 1992.
- Taragades landfill site of Thessaloniki. At this site the rehabilitation work has already started, with the installation of pipe system for the collection of leachate and biogas.

Relevant action is also planned or underway for other minor landfills or uncontrolled dumping sites in Greece. A major project is underway for site selection and installation of two modern sanitary landfills for municipal waste to serve the greater Athens area. The main problem encountered is public perception and acceptance of the proposed sites.

Two projects are underway for site selection and construction of plants for the controlled disposal of hazardous wastes from northern and southern Greece, respectively. In addition, the installation of a treatment plant for hazardous liquid wastes and sludges produced from industries of Attica and Viotia Prefecture is under study. The major problem being faced here is, again, public perception and acceptance of the proposed sites.

An extensive rehabilitation project of a sulfidic tailings dump has been completed in Lavrion, involving the addition of ground limestone to neutralize the acid generation potential, followed by an earth cover to isolate the toxic tailings from the environment and establish an esthetic vegetative cover. More rehabilitation action is being undertaken on laboratory and demonstrate scales, involving soil rehabilitation using chemical fixation as well as soil washing-leaching technologies. Extensive rehabilitation projects are being carried out in the Kassandra mines, northern Greece, by the mine owner. (TVX HELLAS), involving reactive sulfide tailings and acid mine drainage. The work involve chemical stabilization of the material, as well as collection and treatment of the contaminated mine waters as well as leachates.

Land contamination in Greece is related to industrial activities as well as municipal landfills. No specific legislation or guidelines regarding soil quality standards exist. Registration of land affected by landfills has been done, but no registry exists for industrially contaminated sites. Research carried out by universities and research organizations has identified a number of industrially contaminated sites and studied technological solutions for rehabilitation. The research is funded through EC, state, and municipal grants. Municipal landfill rehabilitation has attracted considerable interest and many remediation projects are currently underway. However, very little attention has been given to industrially contaminated land. Although research has been carried out, application of the remediation solutions to full scale is extremely expensive and has not been practiced to any considerable extent.

K. Hungary

Economic growth, especially vigorous industrial development, took place in Hungary without the constraints of strong environmental protection regulations up to the end of the 1970s and the beginning of the 1980s. Although legal regulations concerning environmental protection later caught up with contemporary requirements, compliance with the regulations fell far short of the theoretical strictness of the limits and other regulations for a decade or so. In the midst of the economic difficulties of the time, only insubstantial amounts could be spent on environmental protection. This situation has led to a gradual accumulation of nondegradable and slowly degrading pollutants in groundwater and the soil.

There are approximately 10,000 polluted areas in Hungary where clean-up should have been an imperative for years, even decades. The environmental protection authorities, local governments, and possibly other organizations have information (which is far from complete) concerning only a fraction of these. The pollution of the soil and groundwater is obviously less perceptible than smoke-emitting factory chimneys, petrochemical city smog, or dead fish in oil-stained rivers, but their harmful effects can be felt. At best, we have to pay for the cost of replacing polluted water utility wells (it must be noted that more than 90% of public water consumption in Hungary comes from groundwater); at worst, human health can be threatened by the consumption of polluted water or garden-grown vegetables or even by recreation in polluted areas. Nevertheless, pollutants do get washed into surface waters, and decomposing hazardous wastes do threaten the environment through the air. This long-term environmental damage constitutes one of the factors of environmental pollution that has an unfavorable effect on public health and, ultimately, life expectancy. With most long-term damage, the person (legal entity) responsible for the pollution cannot be compelled to clean up. In these cases, the coordinated use of government funds is necessary in order to clean up.

In practice, most countries today tend toward using limits in the first phase of uncovering pollution and individual risk analyses in detailed investigations. Although there are many methods of financing, government budgets dominate most of the time. The differences lie in the kinds of income schemes that provide coverage. These include product charges, waste taxes, and environmental protection contributions and fines, though on occasion there are no special sources and the public bears the full burden through the budget. The principle of polluter pays, which is widely accepted in environmental protection, is least applicable here, since it is often impossible to prove the responsibility of the polluters.

The government's 1991 short- and medium-term action plan, which identified the tasks of surveying, uncovering, and terminating accumulated environmental pollution, can be considered as the starting point for the Remediation Program. The same plan deals with solutions to the environmental problems presented by abandoned Soviet barracks and training grounds. Owing to the lack of funds, only the latter task could be started before 1995, under the technical direction of the Ministry for Environment and Regional Policy and the Environmental Management Institute.

The experiences obtained in the course of privatization (many foreign investors were concerned about the risk of inherited environmental damage connected to properties), the revival of the real estate market, the experiences acquired as a result of the upsurge in bankruptcies and liquidation, and, hopefully, developing public participation in environmental protection, all helped provide justification for the Ministry for Environment and Regional Policy's original initiative. Therefore, the government launched the National Environmental Remediation Program in 1996 in order to assess polluted areas, uncover damage that falls within the scope of the government's responsibility, and eliminate the damage. In 1997, Parliament approved the National Environmental Program, which contains the Remediation Program.

The new environmental protection law stipulates that if no other person can be made responsible, it is the task of the government to eliminate the consequences of significant environmental damage. Under certain conditions, the law stipulates the joint and several responsibility of the polluter and the owner of the area in which the activity causing the pollution is, or was, pursued. This provision will, in the long run, increase the chance of having the responsible persons, not the government, pay for eliminating environmental damage. If, therefore, the polluter has no funds, the pollution must be considered a government responsibility and the damage to the given area must be eliminated within the framework of the Remediation Program. Naturally, it is also the responsibility of the government to clean up long-term environmental damage caused by government budget agencies.

The purpose of the Remediation Program is to terminate the harmful and hazardous effects of long-term environmental pollution that falls within the scope of the government's responsibility. In order to achieve this, the first step that needs to be taken is a comprehensive survey of long-term environmental damage (sources of pollution and polluted areas).

The remediation concept extends over the entire process. Prior to the Remediation Program, the environmental protection authorities started to survey the entire country in 1995 for pollution whose clean-up was particularly important. As a result, approximately 200 areas were registered; 86% of the soil and groundwater and a lesser degree of the air and surface waters were at risk.

The Remediation Program incorporates three distinct groups of activities. The general tasks include operating, managing, and coordinating the program. The recurrent tasks include compiling the annual priority lists and announcing tenders for companies that undertake to search out and clean up pollution (usually by means of the public procurement procedure in accordance with the size of the project). Strategic tasks include research and technical development that meets the program's needs and the creation of a basis for developing legal, technical, and economic regulations. For public acceptance of the program, the development of a communication strategy and public relations, the organization of educational programs, and the editing and publishing of technical publications are indispensable. The development of a two-tier (central and regional) information technology system is considered a general task.

In 1996, 16% of the program funds was used for the performance of general tasks. The assessment of sources of pollution and polluted areas is the most important of the national tasks. This includes the registration of long-term pollution in the property register (in accordance with uniform nationwide procedures), central operation of the monitoring system, and development of groundwater monitoring and the Soil Protection Monitoring (TIM) system.

Another national task entails the development of so-called subprograms for remediation projects for which government organizations bear statutory (or contractual) obligations. Six percent of the program funds were used in 1996 for carrying out national tasks. Individual tasks include the investigation of damage and remediation projects for which the government is responsible, as well as local monitoring, both according to schedules that comply with priorities.

There are also close ties with the National Environmental Health Action Plan (NEKAP), whose purpose is to survey and rank the most important environmental hygiene problems and study possible solutions at the national, regional, and local levels. The National Environmental Health Action Plan has also established a common database for the polluted areas, and it rates planned interventions.

The National Environmental Health Action Plan rates pollution by evaluating the environmental hygiene risks and considering local characteristics and possibilities. In some sample areas, highly detailed analyses (e.g., environmental epidemiological investigations) are made in order to calculate risks. The findings of the National Environmental Health Action Plan provide a reasonably good basis for making comprehensive priority calculations, especially from the perspective of the Remediation Program.

The first two years of operation (1996 and 1997) can be considered the program's short-term phase. The development of the research, information technology, regulatory, and monitoring systems started in the period during which the program was established and its methodology created.

The program's medium-term phase (1998–2002) has five principal aspects:

> The information technology background and research and technical development will continue to be emphasized in the course of carrying out the general tasks. These include, for example, compiling and publishing a list of the most suitable modern technologies for clean-up projects as wells as developing methodologies for risk evaluation and cost/benefit calculations.
> Important national tasks include a comprehensive assessment of the actual and potential sources of pollution that cause long-term environmental damage, as well as registration of the findings by developing an information system for inventorying polluted sites which is integrated into the Environmental Information System. The most comprehensive version of the Remediation Priorities List was prepared on the basis of this for 2002.
> The individual remediation tasks entail the investigation and clean-up of pollution for which the government is responsible in accordance with the schedule determined by the priorities. In this, diagnostic or partial investigations can be car-

ried out in approximately 200 areas. Accordingly, we can anticipate that emergency measures will be needed in approximately 50 cases. The annual number of clean-ups after fact finding will gradually increase. It is possible to estimate approximately 50–80 remediation projects for the period leading up to 2002.

Most of the pollution that falls within the scope of the government's responsibility must be cleaned up within the framework of the program. According to the plans, State Privatization Agency (ÁPV Rt.) will be in charge of two specialized subprograms. ÁPV Rt.'s clean-up tasks are aimed not only at existing government properties, but at the properties that ÁPV Rt. has already sold and on which it has assumed environmental protection guarantees on the basis of contracts of sale or the law. The government clean-up of former Soviet properties is carried out within the framework of one of the subprograms. The other subprogram is the so-called corporate privatization subprogram, which also incorporates environmental protection guarantees that were made mostly on the basis of individual decisions in the course of sales negotiations

In the future, the government should not be held responsible for new long-term environmental damage, and the means of prevention must be created and developed in the program's medium-term phase. Prevention can best be served by the introduction of regulators, which have already been specified in the environmental protection law; these include environmental liability insurance and collateral requirements in proportion to anticipated environmental expenses.

The Remediation Program is coordinated by the Ministry for Environment and Regional Policy with the participation of the ministries and professional and scientific organizations concerned. The program is operated by the Remediation Program Office, which was developed within the Environmental Management Institution, with the participation of the environmental protection authorities. The office's activities are supervised by the Assistant Undersecretary of State for the Ministry for Environment and Regional Policy. A team of professionals assigned by the various departments of the ministry assist the Assistant Undersecretary of State in his duties. The Ministry of Environment and Regional Policy makes regular reports to the government concerning the program and the manner in which it is being implemented.

In 1996, the Remediation Program Office announced open tenders for the diagnostic investigation of 15 areas and separate tenders for emergency measures in the case of eight of these areas. Nearly 100 offers were received for the public procurement announcements. The emergency measures were, with two exceptions, completed by the end of 1996, while the Remediation Program Office concluded contracts with the winners who bid for the diagnostic investigations at the beginning of 1997. Most of the investigations were completed by June 1997. Most of the program's first (rapid) responses were aimed at neutralizing the pollution that was left mainly by companies which had been terminated or liquidated. In summary, the Ministry for Environment and Regional Policy's remediation project was launched in 17 areas in the second half of 1997. Investigations will begin in nine of these areas, and emergency measures are necessary in four areas. With four exceptions, the remediation projects begun in the previous year are continuing, with detailed investigation of the work, supplementary emergency measures, and/or clean-up.

L. Netherlands

According to present estimates, the application of the multi-functionality approach to the estimated 110,000 seriously contaminated sites would have incurred costs of around $50 billion. The Netherlands is now spending about $0.5 billion per annum, which equals the sum that was initially thought to be sufficient to resolve the entire problem. At this pace, however, it would take about 100 years to conclude the operation. In the meantime, soil contamination would hamper construction and redevelopment essential to economic and social development, and dispersal of contaminants in the groundwater keeps making the problem even bigger. For this reason another policy is needed. Recently, a document passed Parliament containing a new policy on soil remediation. The new approach abandons the strict requirement for contamination to be removed to the maximum extent, and instead permits clean-up on the basis of suitability for use. At the same time, the government proposed other changes to soil protection legislation, including greater devolution of responsibility for clean-up to local authorities and the creation of more stimulating instruments. Basically, the policy has switched from a sectoral to an integrated approach. This means that the market has to play a more prominent role and take more of the financial burden. Soil contamination should not be treated only as an environmental problem. The soil contamination policy should also be geared to other social activities such as spatial planning and social and economic development and vice versa.

The strategy is

To protect clean soil
To optimize use of contaminated soil
To improve the quality of contaminated soil where necessary
To monitor soil quality

This new approach will encourage the development and application of new technology, resulting in a more cost-effective organization of the actual clean-up. These measures taken together are expected to cut costs by 30–50%. In this approach, remediation is part of a comprehensive policy regarding soil contamination. Prevention, use, treatment of excavated soil, reuse of excavated soil (for example, as building material), monitoring of soil quality, and remediation have to be geared to each other in a more sophisticated manner. This internal integration is being promoted under the concept of active soil management. To stimulate market investment, a different approach to government funding has been announced. Taxpayers' money will be used in such a way that it evokes private investment. This will be done by improving the existing financial instruments and by the creation of a private-sector fund for contaminated land. The legal instruments will be made more effective. The discretion of provinces and municipalities will be further enlarged to create the flexibility which is needed to initiate and stimulate measures that are best suited to the local situation (tailor-made solutions). In September 1997, Parliament accepted the new policy.

With these measures, the Dutch government wants to achieve ambitious goals:

Each year almost four times as many sites will have to be remediated as is the case now.
Presuming that the costs will be reduced by 30–50% this requires a duplication of the total annual expenditure on soil remediation.

In order to monitor the results of these efforts and to make information on soil quality accessible to the general public (for example, potential buyers) and to authorities (for example, planning authorities), we want to have a system of soil quality maps covering the whole country in 2005.

Based on the Soil Protection Act, there are two driving forces to investigate soil quality:

Anyone intending to excavate or move soil for building activities has to report the quality of the soil to provincial authorities.
Companies that do not investigate soil quality voluntarily might be required to do so.

Based on these activities, a lot of seriously contaminated sites have been identified. These numbers have increased enormously since the first case at Lekkerkerk.

In the policy on contaminated land, three phases are recognized. In the first phase, restoration was the aim of the technology. In the second phase, control of the spreading was added; and in the third phase, control of risks has become the aim of technology. In the first phase, the treatment technology for contaminated excavated soil was developed. This was mainly physicochemical technology that was originally applied in mining and road building, such as particle classification (soil washing) and thermal treatment. In the next phase, containment was added to these technologies. The main containment technologies are the isolation of a site by a nonpermeable wall and pump-and-treat. In the latest phase, in-situ technologies are developed, especially in-situ bioremediation.

The Netherlands policy was changed drastically in 1997. This has resulted in an increasing demand for knowledge of new approaches and new technologies. Therefore, research efforts will be continued and increased in the coming years.

M. New Zealand

The issues of contaminated land have only recently come to prominence in New Zealand. Since 1991, problems associated with contaminated sites and their management have become an important part of the environmental agenda in New Zealand, for both regulatory agencies and industry. Environmental management in New Zealand is divided among three tiers of government: national, regional, and city or district councils.

At the national level, the Ministry for the Environment was established to advise the central government. It has two key roles:

To provide policy advice to the Minister and the government that promotes sustainable management of the environment
To enable the implementation of sustainable management through the administration of environmental statutes, advocacy, education, and advice

In addition, the Ministry also has the ability to develop and promote national environmental standards and to develop policy and legislation on environmental issues. The management of government-owned contaminated sites is the responsibility of individual government agencies. The Department of Survey and Land Information is the responsible agent where the government is the liable party on private land.

New Zealand has 14 regional councils and two unitary authorities. These councils are the primary environmental regulatory agencies, with comprehensive responsibilities for management of discharges to air, water, and land. All tiers of local government draw their authority from laws passed by the central government. In the management of contaminated land, the principal roles of regional councils are to:

> Gather and maintain information on potential and actual contaminated sites
> Undertake environmental monitoring
> Set priorities for specific sites according to their likely adverse effect on the environment (including human health)
> Regulate site investigation, risk assessment, and remedial actions where required
> Develop effective management strategies for existing contaminated sites in the region and to prevent new contamination
> Ensure public awareness of the issues and maintain an information support system for all parties affected by contaminated sites

The city or district councils are responsible for land use planning and control. They address site-specific land use controls, issue consents for building construction, and provide a wide range of other services, including:

> Administration of building controls and restrictions related to sites
> Maintenance of land information memoranda relating to the physical characteristics and chemical contamination on each parcel of land

To date, contaminated sites have been largely the responsibility of the regional councils and the major urban city councils. These regulatory agencies have been most affected by the discovery of contaminated sites and have undertaken major initiatives to investigate and characterize such sites.

The legal framework for contaminated land in New Zealand is based on the Resources Management Act (RMA 1991), which integrated a large number of previous environmental statutes into a single, all-encompassing law. The aim of the act is to promote sustainable management of natural and physical resources and represents a fundamental shift in focus from previous legislation, which emphasized planning activities, to regulating the effects of such activities. The RMA sets out a framework of policies and plans, providing processes and mechanisms to allocate resources and to grant consents to minimize adverse environmental effects from industrial activities. There are restrictions on discharge of waste to land, water, and air, with enforcement procedures and liability provisions which are particularly relevant to contaminated sites:

> Liability is to be assigned to the owner/occupier of the site or the polluter.
> Any party may initiate legal action to enforce environmental improvement, with liability determined by the acting party.
> Increased fines and imprisonment (up to two years) for actions in breach of these regulations.
> Extension of liability of site pollution to company executives.

These latter provisions have significantly affected the policies of major companies in New Zealand, which are reportedly now carrying out considerable work to assess their own environmental practices and potential liabilities. However, prosecutions are seen as a regulatory action of last resort, and fines are generally low in comparison to the overall cost of contaminated site management.

Other relevant legislation includes the Health Act and the Toxic Substances Act, which contain provisions which could be used to manage contaminated sites. It is unlikely, however, that they will be used for this purpose, since the major initiatives have been taken by the environmental regional agencies rather than the health authorities. New Zealand law does not provide a statutory definition of contaminated land, nor are there are statutory threshold levels to define whether a site is contaminated by a specified substance.

A broad review of potentially contaminated sites was commissioned in 1992 to identify suspected contaminated sites from historical records of past industrial use. The likelihood and severity of site contamination for each type of industrial site identified was then determined from international experiences of similar industries. Sites were categorized into three risk classes:

> High-risk sites are expected to pose an actual or imminent significant adverse effect to human health and/or the environment.
> Moderate-risk sites have soil contamination levels above ANZEEC "B" values.
> Low-risk sites are considered to be contaminated but not a long-term risk unless land use changes to a more sensitive application.

A total of 7800 suspected contaminated sites were identified by this review, with gasoline stations (2616), engine works (878), and landfill sites (716) being the largest categories. The study estimated that 23% of these sites were high risk, with a time frame for their remediation being around 5–10 years at a cost of $330 million. The treatment cost of moderate-risk sites was estimated to exceed $537 million and take between 20 and 30 years to complete.

Regional councils are currently undertaking a more detailed survey of their regions to assess the validity of the initial study. This has generally resulted in an increase in the number of potentially contaminated sites expected to be identified. The number of small operations with potential contamination has increased significantly, with key activities being the use of underground storage tanks in association with gas stations, stationary boilers, and metal finishing operations. This has not resulted in an increase in expected remedial costs, however, since the number of large sites requiring remediation has changed little from original estimates.

New Zealand is a party to the Australian and New Zealand Environmental and Conservation Council (ANZEEC), which in conjunction with the National Health and Medical Research Council (NHMRC) has developed guidelines for the assessment and management of contaminated sites. These guidelines have been adopted as policy and are considered best practice within New Zealand. These ANZEEC guidelines are outlined in more detail in Section II.A.

Key features of the ANZEEC Guidelines applying to New Zealand in particular are

> Definition of contaminated land
> Provision of a methodology for site assessment
> Provision of a methodology for setting site-specific treatment criteria
> Reference to the old Dutch A, B, C values where no Australian or New Zealand-based value exists

In addition to the ANZEEC Guidelines, the Ministry for the Environment, in conjunction with the Timber Industry Federation, has developed guidelines and standards to address the specific problems of organochlorine chemicals in estuarine sediments. Such contamination is derived largely from the use of preservatives by the timber industry. Guidelines have been prepared which consider the most important and widespread treatment chemical, including pentachlorophenol and compounds of boron, copper, chromium, and arsenic. They have been established within the overall framework of the ANZEEC Guidelines and are consistent with its approach. The major components of the timber industry guidelines are

Sampling strategies and chemical analysis protocols
Soil acceptance criteria
Surface and groundwater acceptance criteria
Landfill disposal

Additional guidelines under development include:

Management of gasworks sites
Management of hydrocarbon-contaminated sites (especially gas stations) in conjunction with the oil industry

Recent developments in New Zealand policy have aimed to address three key areas of uncertainty in the initial 1991 legislation:

Liability
Funding of remedial actions for orphan sites
The role of the regulatory agencies and their interaction at the district, regional, and national levels

There is considerable uncertainty under current legislation over who is liable for sites contaminated prior to 1991. This is due for the most part to the legislation not being tested under case law. To resolve this uncertainty the government has agreed to the development of legislation which will be clear regarding its retrospective intent. As a consequence there is an opportunity to establish a clear liability regime for contaminated sites which incorporates the following components:

The polluter, owner, occupier of a site will all be potentially liable.
An innocent landowner legal defense.
A secured lender defense.

Although government policy in waste management is to promote the polluter pays principle, this has not yet been incorporated into statute. Therefore, policy has yet to be established on an appropriate regulatory framework to fully address liability at historically polluted sites.

At present, government has no specific mechanism to raise funds for management of contaminated sites. In the past, government has appropriated money on an as-required basis according to the nature of previous government responsibilities and activities at the site. Funding of the management of orphan sites is a policy issue which has not been fully addressed to date. The size of the orphan site problem has yet to be fully determined, although it is expected that this can be estimated once the national review of suspected sites is completed. It is anticipated that the funding of orphan site management will focus on minimizing litigation costs; creating incentives for treatment; and keeping overall costs low, especially the funding requirement from taxpayers.

There is also uncertainty over the interaction between local and national government in policy on contaminated land, for example, whether regional registers of contaminated sites should conform to a national standard.

N. Norway

The main law regulating clean-up of contaminated land in Norway is the Pollution Control Act of 1981. The polluter pays principle forms an important basis of the Pollution Control Act. If the original polluter can no longer be identified or held responsible, the current landowner may be held liable for investigations and remedial actions. Regulation of contaminated land in Norway under the Pollution Control Act is the responsibility of the Norwegian Pollution Control Authority (NPCA). While almost all sites are regulated directly by the national agency, a few cases are left to regional authorities (counties). The Planning and Building Act, however, requires that local authorities consider possible soil contamination before a new construction project or land development is licensed. During recent years national authorities have encouraged municipalities to use this law in their regulatory work and hence contribute to a reduction in the number of construction projects which have to be stopped temporarily due to the discovery of soil contamination.

Contaminated land is generally accepted as a local environmental problem. Therefore the national and regional authorities are considering whether regulation of contaminated land should be the responsibility of the counties, or alternatively, how and to what extent counties should be involved. Clean-up of contaminated sites is at present regulated through permits/licenses under the Pollution Control Act. As the Norwegian procedures for licensing clean-up and remedial actions are complicated and time-consuming, the NPCA is preparing a General Regulation for Contaminated Sites. This will allow private and public companies to conduct clean-up programs for their sites without detailed permits or licenses from the authorities, saving time and costs. Norway has developed a decision model consisting of a two-tiered system for regulation of contaminated sites. Generic target values are developed for most sensitive land use. For other sites or when target values are exceeded, a system of site-specific risk assessment is applied. The target values are based on data from other countries. Improving the target values and development of a systematic approach for risk assessment were issues of high priority in the NPCA in 1997–1998. This was part of the decision model for contaminated sites in Norway, which was revised in 1998.

Norway has decided not to apply the principle of multifunctionality as the basis for remediation. Because clean-up goals are adjusted to actual or potential land use, site-specific information regarding level of contamination, remedial measures, and land use restrictions should be kept for future generations. Therefore, it is important that results from regulation of contaminated land are included in the land use planning system.

Contaminated land in Norway is considered an important source of contamination of rivers, lakes, and fjords. More than 85% of the Norwegian water supply is based on surface water, and consequently groundwater contamination has been of less concern in Norway compared to many other countries. Potential impact from industry, contaminated sediments, and landfills on the marine environment is of greater concern. In some fjords, recommendations of reduced intake of seafood is recommended, due to pollutants such as heavy metals, PCBs, PAHs, or dioxins. During the years from 1989 to 1991, a national survey of landfills and contaminated sites was carried out in Norway. Approximately 2100 possibly contaminated sites were registered. The total number includes municipal and industrial landfills, industrial sites, gasworks, military sites, and sites from World War II. In 1992, the NPCA presented an action plan for contaminated sites. A status and revised plan was presented with the national budget from the government in 1996. New contaminated sites have continuously been discovered through land development or construction activities.

In 1997 the NPCA decided to produce an annual status report to the public with an overview of contaminated sites and status of remediation. One annual report will satisfy the need for information in the public (media, NGOs, politicians, etc). The status for 1997 shows that more than 3350 contaminated sites are now registered in Norway. About

150 of these are given high priority and about 600 additional sites need to be investigated. Of these 750 sites, investigation has started on about 350 and in 250 sites remediation is going on or finished. The remaining 2600 sites are given low priority with the recent land use. When redevelopment or construction work is planned for these sites, necessary investigations and measures must be considered. A GIS database has been developed by the NPCA to keep track of all registered sites and any investigation or remedial action carried out at the different sites. Information from the database will be used for reporting and by the NPCA in general, by the counties, and by municipalities for their planning purposes.

Recent market research on treatment technologies for contaminated land in Norway (November 1997) shows that the following technologies are commercially available through Norwegian companies: bio-venting, vacuum extraction, air sparging, pump-and-treat, bio-piles, land farming, soil washing, solidification/stabilization, and incineration. In-situ and ex-situ bioremediation technologies are conducted mainly by contractors. Five to 10 consulting companies have experience with these technologies. In addition to the contractors, about three to five companies have specialized in treatment of contaminated soil in Norway as they major activity. They have so far concentrated on solidification/stabilization, soil washing, land farming, and partly incineration. Few sites are in the remediation phase so far, and easy access to and low prices on landfills are major reasons for the limited development and accessibility of treatment technologies on the market. The NPCA has started projects on national and local scales to develop guidelines for management of excavated contaminated soil. The guidelines will be administrative tools for local, regional, and national authorities and support the existing legislation on contaminated land. A more predictable assessment by the authorities is of great importance for society.

The Norwegian Pollution Control Authority gives priority to the following issues:

- Transfer of responsibility, competence, and resources to county or regional authorities for the regulation of contaminated sites
- Preparation of a General Regulation for Contaminated Sites, which will allow private and public companies to conduct clean-up programs for their sites without detailed permits or licenses from the authorities, saving time and costs
- Development of an improved decision model for regulation of contaminated land, including target values for sensitive land use and a systematic approach for site-specific risk assessment
- Annual public status reports with an overview of number of sites and status of remediation
- Development of guidelines for management of excavated contaminated soil

O. Slovenia

Recently, a few changes have occurred in the field of environmental protection in Slovenia. The ministry responsible for the environment has decided to expand its activities from the strictly legislative to more practical areas. This means that it has begun to carry out certain activities in order finally to break the deadlock on environmental protection, although it is currently limited to certain areas only. In many other areas with acute problems the state has still not decided whether to become practically involved, in spite of pressure from the public, the media, and local communities. Selected legislation which has come into effect in recent years includes the Law on Environmental Protection, adopted in June 1993, and the following regulations based on the law:

Decree on the Input of Toxic Substances and Plant Nutrients into the Soil (December 14, 1996)

Decree on the Emission of Substances into the Atmosphere from Waste Incinerators and During the Combined Incineration of Waste (December 10, 1996)

Decree on the Export, Import, and Transit of Waste (August 10, 1996)

Decree on the Management of Infectious Wastes Which Appear in the Performance of Health Care Activities (October 7, 1994)

Decree on the Management of Wastes Which Appear in the Performance of Health Care Activities (June 3, 1995)

With these documents, the state has taken legal and, in certain cases, practical environmental protection measures. In 1997, an environmental inspection body began operating in Slovenia; its duties include the monitoring and registering of all events and activities connected with environmental pollution. All hazardous wastes in factories in the process of privatization have now been registered. This was done because wastes of this type have been lying around factory estates for years, and also in order to ensure that the new owners would provide for the proper processing of these wastes, as this activity will now be monitored by the inspection body. There are large amounts of these wastes: according to data published in the media so far, at least 250 factories have considerable amounts of hazardous waste, which are mainly stored simply in yards or in improvised shelters. According to some sources, the actual amounts of these wastes far exceed those recorded by the inspectors. They can be found not only on factory estates but also on illegal dumps, and were often simply buried at different locations. Slovene experts estimate that about 30,000 tons or more of hazardous waste have been produced by industry in recent decades, which cannot be processed or incinerated since Slovenia does not have the proper tools and technologies. In this area the state has done virtually nothing in recent years, although state institutions and the ministry responsible for the environment have in their possession data on the amounts of these wastes and also on their locations. At present, Slovenia has no strategy of waste treatment at a state level.

The monitoring of industrial waste waters has been conducted for one year beginning in 1997. High fines are prescribed for violators and for those who do not arrange measurements with the authorized institutions. The measurements have already yielded some results, particularly in changing the attitude of polluters toward their waste waters. Corrective measures which are being introduced in certain factories and the construction of water purification plants are the most significant steps in this field. In addition to the above problems and certain activities which are already underway in order to solve or at least mitigate them, a number of other burning environmental issues exist in Slovenia today which will have to be tackled in the coming years. The most pressing problems are probably those of waste treatment strategy, both municipal and industrial, the protection of soil, rivers, lakes and the sea, and the protection of the atmosphere in certain areas.

P. Sweden

Sweden is still suffering from a lack of legislative support. There is no special legislation covering remediation-related issues yet. There is the Environmental Protection Act from 1996, which was not drafted to take into account remediation problems. The legislation is both unclear and incomplete concerning remediation. Due to the recently announced ruling of the Supreme Administrative Court, the possibilities of placing demands on companies that have closed down have been limited to the period after 1989. This means that about 75% of the remedial cost must be covered by society. The need for new legislation has been obvious for some time, and in 1996 the Swedish EPA submitted a proposal for new legislation on remediation to the government. This legislation has been incorporated in the new Environmental Code, which is currently under consideration and will come into effect on January 1, 1999.

The purpose of this new legislation is to clarify liability and give the authorities greater opportunity to promote, control, and steer remedial action. With this new code, it will be possible to place demands on companies from 1969 onwards. This new legislation also introduces official registration of confirmed contaminated sites.

In 1997 the Swedish EPA presented guideline values for 36 contaminants in contaminated soil. Guidelines for the remediation of gas stations, including guideline values for soil and groundwater, are under consideration. The Swedish definition of a contaminated site is a site, deposit, land, groundwater, or sediment which has been contaminated, intentionally or unintentionally, by industry or some other activity. The definition of contaminated is that the levels of contamination apparently exceed the local/regional background values. The new Environmental Code will give the authorities quite a different role in remedial work. It will make it possible to take a more active role and force private companies to take greater responsibility for their actions than they do at present.

So far about 3000 potential sites in Sweden have been identified. The total number of contaminated sites is estimated to be 10,000 sites. Due to the industrial structure, sites with metallic contaminants dominate; mines with acid mine drainage are the heaviest and most costly remedial problem. Other problems are caused by metalworks, iron and steel works and surface plating facilities. Second, there is a group of industries that use complex mixtures of metals and persistent organic substances such as chloralkali (mercury and dioxins/furans). These include gasworks, the pulp and paper industry (mercury and PCBs), and wood preservation plants (CCA, Cu, PAHs, PCP, and dioxins/furans). Third, the petroleum industry, with oil refineries, oil depots, and gas stations, represents the largest group by number but also causes problems which are easiest to solve.

Today there is an informal registration of identified, suspected sites at the Swedish EPA. This register is not official and is open only to environmental authorities. A more developed and regionally based computer system at the county administrative boards (CABs) will replace this first database. The Swedish EPA is responsible for the development of this regionally based site registration data system, in order to ensure that the regional registers are consistent. The purpose of this database is to provide a basis for regional planning and prioritization of inventories, investigations, and remedial work, as well as serve as a support in the ongoing work on licensing and supervision. With the new Environmental Code, the CABs will be authorized to decide which sites can, with certainty, be classified as contaminated in an official register. General criteria for this registration will be regulated by law. This registration can, in certain cases, lead to land use restrictions, obligation to report certain kinds of activities (such as excavation) at the site to the municipality, etc. This information will also be entered into the national land register. The CABs will also be given the right to decide if and when such a classification should be annulled.

The EPA's policy in the context of remediation is to choose long-term solutions that, if possible, solve the problem once and for all. That means, in the first instance, selecting methods which destroy the contaminant through biodegradation or combustion. When this is not possible, as in the case of metals, for example, methods should be used where the contaminant is concentrated/collected for further treatment and/or landfilling. Concentration methods include, for example, soil washing, soil venting, and thermal desorption. Only in the last instance should methods such as containment, immobilization, and landfilling of untreated residues be selected. This is an application of the BAT (best available technology) principle in the remedial field.

The second principle that concerns the choice of technology is the ecocycle principle. Site remediation has to do with the rational management of land and water resources. Methods which enable land and soil to be reused are given higher priority than methods which involve excavation and removal of waste as well as landfilling. Landfilling, encapsulation, and incineration are still the dominant remediation measures in Sweden. During 1997 two rather large sites, both of which were former wood preservation plants, were successfully remediated using soil washing. The trend is that some kind of treatment is becoming more and more common. In particular, biological methods such as composting and in-situ methods such as vapor extraction and bioventing are becoming more and more frequent.

The state of the art in Sweden is as follows:

Soil washing: There are three pilot plants and two full-scale plants in Sweden. In addition, three more full-scale plants are planned.

Thermal desorption: Two pilot plants have been tested, and one full-scale plant is under construction.

Composting: There are a great number of companies dealing with uncontrolled composting, in the open air without evaporation or leaching control. In controlled composting, two companies are working with some kind of on-site static, encapsulated compost.

In-situ methods such as soil vapor extraction, bioventing, and air sparging are used by one company, mostly for remediating gas stations.

Finally, the problem in Sweden is that there are still only a small number of remediations being carried out. Despite the fact that quite a lot of companies are interested in working in this field, the market is still very small. One bright spot is the initiative from the Swedish Petrol Institute to get the petroleum companies to form an environmental commission to clean up petrol stations which have closed. The work will be financed by a marginal increase in petrol prices. The aim is that 6000 petrol stations will be remediated within a 10-year period. This will surely increase the demands for remedial work and make the market larger, at least for biological methods and in-situ methods such as vapor extraction and bioventing.

Another positive development is the government's investment in building a new ecological society. Together with housing, energy, and transportation, remedial action is one of the sectors where money will be spent. A total of $700 million will be spent over three years. Local authorities will present plans to the government, which will prioritize and allocate the funds. The Swedish EPA is not much involved in these decisions, and funds for long-term plans have been cut to a minimum. This is a general trend in Sweden. The environmental authorities get less and less money and temporary organizations, often run by politicians, are formed to administrate regular authority work on an ad-hoc basis. Based on rather few remediations, the conclusion is that biological treatment, such as composting, should be used if there is an easily degradable organic contamination as at petrol stations, oil depots, and refineries. In-situ methods such as vapor extraction, bioventing, and air sparging are also useful in some of these cases. These methods are rather cheap. Composting could be used for lighter PAH, but if there are 4–6 ringed PAHs or PCPs, a bioslurry reactor is needed. There are many sites with mixed contaminants, metals and organics. Soil washing is a very useful technology in Sweden for this. The two full-scale remediations in 1997 worked out very well. Concerning thermal treatment, there are no full-scale treatment facilities yet, but tests show that it could be useful for PAHs, mercury, dioxins, etc.

Metals and complex mixtures of metals and persistent organics are the dominating problem in Sweden. Acid mine drainage is the major, and most costly, remedial problem. The lack of technology has been a great problem, but in the last few years there has been a change for the better. The interest from treatment companies has increased, and today there are around 15 companies active on the market. Some of these are developing their technology from the beginning, others are seeking collaboration with companies in other countries, such as the Netherlands or Germany. The lack of legislative support and of governmental long-term funding make the market unsure. The financial sector's increasing awareness makes it more and more difficult to avoid these problems, giving companies the incentive to clean up voluntarily. Let us hoped that the new Environmental Code and the remedial programs for gas stations will help the market survive until the remedial program can get more stable financing.

Q. Switzerland

The population of Switzerland is about 7 million, living in an area of 41,000 s km^2. Outside the sparsely populated mountainous region, which comprises about 60% of the country's surface, most people live or work in the urban areas of the lowland. The country's political structure is federalist, organized and divided into 23 states, called cantons. These cantons are very different in terms of surface area and population, as well as economy, industrialization, and scientific background. Industrial waste, waste management, and environmental impacts also vary considerably. The first steps toward a systematic assessment and remediation of contaminated sites were made by local authorities in 1985. Today about 75% of the estimated 50,000 suspected sites are registered by the cantons and the Federal Department of Defense. The Cantons are responsible for entering the sites contaminated with waste in a register, differentiating among landfill, industrial, and accident sites. The registration of industrial sites, which is carried out according to the branch of industry concerned, is difficult. Sites should not be put on the register if they are not polluted with waste. In total we can reckon on more than 3000 contaminated sites that will have to be remediated in the next 20–25 years. Up to 200 contaminated sites have been remediated to date. Based on current experience, 5–10% of the polluted sites (about 3000) need to be remediated. The overall remediation costs for these contaminated sites are estimated by the Federal Agency at over $3.6 billion.

Regulations for management of contaminated sites were established in the 1995 revised law relating to the Protection of the Environment (LPE). This amendment to the Law of 1983 relating to the Protection of the Environment, regulates the management of contaminated sites for the first time in Swiss environmental legislation, in the following three articles:

> Registration and remediation: Obligation to register landfills and other sites polluted by waste (contaminated sites) in a register open to public; obligation to remediate polluted sites, if they result in harmful effects or cause a nuisance to the environment or if there is a danger that such effects may arise (contaminated sites).

Regulation of financing: Polluter pays principle; the owner of a site is excepted if he or she could not have had any knowledge of the contamination, did not stand to gain from the contamination, or will not stand to gain from the remediation. The authorities rule on the division of the costs if people with an obligation to remediate so require.

Levy to fund remediation: Levy on landfills of up to 20% of the average costs in order to finance remediation projects, where the polluter cannot be identified or cannot pay, or where domestic waste is to be remediated.

Based on the revised law relating to the protection of the environment, the Ministry of Environment, Traffic, Energy, and Communication planned to put into force the ordinance on contaminated sites relating to the remediation of contaminated sites during 1998. This ordinance has the following objectives:

Stop emissions at source: The remediation criteria are not based on the pollution itself, but on the emissions from it that lead to unacceptable levels in waters, air, or soil; decontamination, containment, and use restrictions for the soil are all therefore acceptable as remediation measures.

Cooperation between polluters and authorities: Authorities and polluters may carry on working as long as possible under agreements, instead of needing a ruling; agreements among branches of industry should be encouraged.

Legal equality through harmonized criteria (e.g., 72 intervention values, remediation targets, leaching tests) and uniform requirements for the elaboration and management of registers, planning and execution of investigations, as well as monitoring and remediation projects.

Prevention against new risks: Building activities on polluted sites are permitted only if it can be proved the site does not need remediation, if the project does not hinder future remediation, or if it will be remediated in the course of the project; containment measures have to be effective long-term, controllable, reparable, and financially guaranteed.

In the revised LPE, the section on remediation of contaminated sites and of the financing has its own regulations due to its importance. The LPE gives the Federal Council the authority to introduce a tax to finance remediations. The tax should be levied on the deposition of wastes; the rate is limited to a maximum of 20% of average deposition costs in Switzerland. The revenue is expressly related to this purpose and flows to the cantons (if they fulfill certain conditions), which must in turn find the finance to remediate contaminated sites. The amount of the compensation is limited to 40% of the countable remediation costs; at least 60% of the remediation costs must be borne by the cantons.

The issuing of federal regulations to cover financial cooperation in the remediation of contaminated sites is justified because for many sites the polluter is no longer identifiable, or is unable to pay. In these cases the costs of remediation, insofar as they cannot be passed on to the proprietor, will be carried by the cantons and thus by public taxes. This ordinance will enable the cantons to receive financial support from the Confederation. Furthermore, this fiscal instrument should offer an incentive for the quick and environmentally sound remediation of contaminated sites. It is the Confederation's aim that contaminated sites, which represent a severe potential danger, should not only be investigated but should also be rapidly remediated. Remediations should be provoked by the actual danger to the environment, and not just be development, building plans, or the presence of adequate sources of money.

In the approximately 200 remediations carried out to date, traditional methods of remediation were predominantly used. These are primarily:

> Excavation of the contaminated material and treatment in a soil-washing facility or disposal in a landfill
> Securing of the site (e.g., sealing of surfaces, barrier walls)

New and innovative remediation technology, particularly in-situ measures, are still not completely accepted. Efforts are especially necessary in this area, to which the authorities can contribute. With the ordinance on contaminated sites in force, it will be possible to keep a register of remediations carried out in Switzerland and to keep more comprehensive information on individual cases than was formerly possible.

Current Federal policy on the treatment of contaminated sites is oriented primarily according to the following important principles:

> Uniform goals for the treatment of contaminated sites should be valid throughout Switzerland.
> The authorities work with those directly affected, especially with industry.
> The contaminated sites should be treated according to objective urgency (danger to the environment).
> Remediations should be carried out quickly, with realistic solutions (principle of commensurability); the search for perfect solutions, and thus leaving the problem for future generations, should be avoided.
> The requirements of remediation should be set, as far as possible, according to the environmental situation at the time.
> The remediation should guarantee that illegal effects are permanently halted and that the measures are sustainable overall.
> Contaminated sites are to be decontaminated where possible and to be secured as a secondary priority.
> Future contaminated sites should be avoided through consistent implementation of precautionary environmental regulations.
> Industrial and commercial contaminated sites are to be remediated as far as possible for future use. Brownfields, and their subsequent replacement with greenfields, are to be avoided.

The legislator has the difficult task of issuing regulations with which environmentally legitimate treatment of contaminated sites is possible and on the other hand ensuring that these regulations are acceptable to the population and those affected by a remediation. The registration of sites contaminated with waste is valuable. On the other hand, there is still great necessity to investigate the sites and their possible remediations, which could in some cases be very cost-intensive. Prerequisites must be created so that investigation and, if necessary, remediation can be carried out, not just where there are plans for construction, but also where it is necessary for purely environmental reasons. We hope that the planned ordinance on the financing of the remediation of contaminated sites will offer significant support to this.

R. Turkey

A Ministry for the Environment was established in Turkey in 1991. This ministry holds power of veto over government projects instigated by other ministries including the Ministry of Housing and Public Works and the Ministry of Tourism and Transport.

In 1992 the Turkish Government enacted the Solid Waste Control Act, which aimed to regulate the landfill disposal of industrial and municipal waste. The act prohibits dumping of hazardous solid wastes in the environment and sets limit values for contaminants in stored and disposed wastes. It controls the use of sewage sludge in agriculture and defines terms used in waste management practice. In addition, it prohibits co-disposal of dangerous industrial and medical wastes with municipal solid. Considerable effort has taken place to implement these measures in the major industrial cities of Istanbul, Izmit, and Izmir. Regular waste collection and controlled disposal has been implemented for both municipal and industrial wastes.

Under the Solid Waste Project (SWP), all but one of these illegal dumps has been closed. The SWP work program has been supported by a grant from Germany. Remedial work has started at Ümraniye-Hekimbasi where, in 1993, 27 people died as a result of a landslide of illegally dumped waste caused by a methane gas explosion. This environmental disaster raised public awareness on the dangerous location of these uncontrolled dump sites, often near inhabited areas. The first stages of remediation involve construction of stone walls capable of preventing the waste mass moving farther along the valley. Eventually this waste will be collected and transported for further treatment and proper disposal.

There is growing recognition of soil and groundwater pollution problems in Turkey since the enforcement of the regulation of the Control of Hazardous Wastes in August 1995. The main purpose of the regulation is to provide a legal framework for the management of hazardous wastes throughout the nation. It basically regulates prevention of direct or indirect release of hazardous wastes that can be harmful to human health and the environment, control of production, transportation and exports, technical and administrative standards for construction and operation of disposal sites, waste recycle, treatment, minimization at the source, and related legal and punitive responsibilities. The regulation is applicable not only to hazardous wastes to be generated in the future, but also concerns existing hazardous wastes and their safe disposal in compliance with the current regulation within three years.

The Control of Hazardous Wastes regulation does not explicitly define the concept of contaminated sites. Rather, it defines what a hazardous waste is and provides lists categorizing hazardous wastes based on their sources, chemical compositions, and accepted disposal techniques. Thus, any site contaminated with or subjected to any of these categorized hazardous wastes can implicitly be defined as a contaminated site. However, difficulties arise from the lack of information for most of chemicals in these lists regarding specific maximum concentration levels (MCLs) or remedial action levels. Currently, identification of any contaminated site is not based on a systematic approach. These sites are mostly identified after some potential environmental problem becomes obvious and public as a result of the efforts of local authorities or concerned citizens. However, some current policy developments by the Ministry of Environment can make the identification of contaminated sites somewhat more systematic.

In this new policy development, the waste management commission, an administrative body proposed by the Control of Hazardous Wastes regulation, initiates preparation of industrial waste inventory on a regional basis. Waste inventory is planned to be achieved by requiring all the industry to fill out annual waste declaration forms revealing the type, amount, composition, and current disposal practice for their wastes. This way, it is expected that waste generation activities and pollution potentials of industries can be monitored, regionally effective waste reutilization and recycling programs can be implemented, and regional needs for the type and capacity of waste disposal facilities can be identified. In response to such efforts, an integrated waste management facility, including a landfill and incineration unit for disposal of industrial wastes, is becoming operational at full scale in the heavily industrialized Izmit region. Another policy development related to identification of contaminated sites is the work progressing toward the preparation of a Soil Pollution Control regulation. It is expected that this regulation will clarify the existing confusion over the remedial action and clean-up levels and set a guideline for the selection of appropriate clean-up technologies for various different types of contaminated soil sites.

Currently, there are no reliable and comprehensive case study-based statistics or data on remedial methods and technologies used for a clean-up of soil and groundwater in Turkey. Regulatory aspects of acceptable remedial methods and technologies are provided by the Control of Hazardous Wastes regulation, which specifies acceptable remedial and/or disposal methods for a given type of contaminant group. In the Control of Hazardous Wastes regulation, acceptable methods for a large number of contaminant group is given as physical, chemical, and biological treatment, without stating the specific name of the method. However, it clearly states that use of remedial technologies is a must for wastes containing a large group of contaminants. Currently, there is no official knowledge regarding the widespread past use of particular technologies for soil and groundwater clean-up in Turkey. Most probably the remedial technologies that will be used for the Beykan, Incirlik, and COPR Dump sites are going to be the first site-specific examples and set precedence, in terms of both cost and performance, for clean-up of other similar sites.

There is a pressing need for research and development of soil and groundwater clean-up technologies in Turkey. There has been a significant increase in the number of soil and groundwater remediation research projects supported financially by the Turkish State Planning Organization. Among this group, a project will be initiated on the performance assessment of solidification/stabilization technology for remediation of a large waste group (e.g., soils, mining waste, and paper and pulp industry sludge) containing organic contaminants and heavy metals. The main purpose of this project is to investigate the reliability of this technology for remediation of certain waste groups and provide technical and economical guidance for field-scale applications. Another component of this project is to emphasize consideration of the risk-based corrective action (RBCA) approach in the application of regulatory process for site-specific cases. Considering the high cost of subsurface remediation problems, the RBCA approach will offer significant savings compared to the current regulatory approach based on a fixed clean-up level.

There is a growing recognition of soil and groundwater degradation problems in Turkey. Because the enforcement of hazardous waste regulations is relatively new, some difficulties in the identification of soil and groundwater contamination sites remain unresolved. Recent regulatory efforts are helpful for identification of those sites contaminated as result of past activities. In the near future a considerable increase in the number of registered contaminated sites is expected. Turkey presently relies heavily on surface water resources to satisfy water supply demands, mainly because of the relative abundance of surface water resources. Groundwater constitutes a relatively small component of total available resources (10%), but it represents a significant portion (27%) of total water withdrawal. However, due to growing water demand parallel to rapid population and industrial growth, an increasing demand for food production, urban expansion, and accelerated degradation of surface water quality, protection of clean groundwater resources as well as remediation of contaminated soil and groundwater sites are becoming environmental issues of high priority. The sustainable development of groundwater resources requires proper treatment of municipal and industrial wastes. Groundwater is the major source of drinking water supply and needs to be fully protected and allocated only for high-quality uses. Although legislation on groundwater exists, its protection appears to be neglected at least in certain areas. With the spread of irrigation practices, the pollution threat to groundwater is also increasing. To date, unsatisfactory efforts has been made to protect groundwater from the increasing variety of potential pollution sources, such as agricultural chemicals, septic tanks, and waste dumps. The control of soil and groundwater contamination is essential to Turkey's ongoing reliance on groundwater resources for potable water.

The management of hazardous wastes in Turkey is inadequate to ensure proper handling and treatment. Industrial waste, particularly hazardous waste, has grown proportionately with industrial production. Treatment facilities are minimal and disposal is usually haphazard. Wastes pose serious dangers for soil and groundwater and in some cases for public health. The legal gap has to a certain extent has been filled with the regulation of the Control of Hazardous Wastes. Minimization of the generation and availability of facilities for proper storage and disposal of hazardous wastes has been embodied in this Turkish regulation. The policies are being strengthened by the application of such mechanisms of industrial waste management as the full implementation of environmental impact assessment for new proposals, the requirement that waste management programs be prepared and implemented by existing industries, and the encouragement of waste reuse.

S. United Kingdom

In 1995 the UK introduced a major piece of legislation which significantly affects the way issues of contaminated land are dealt with. This legislation followed a wide-ranging review and consultation exercise of contaminated land and liabilities which resulted in the publication by the Department of the Environment and Welsh Office of the document framework for contaminated land in November 1994. The Environment Act (1995), and specifically Section 57, inserted a new piece of law into the Environmental Protection Act (1990) concerning contaminated land. Section 57 created a new regime for the control of environmental problems associated with contaminated land in accordance with the objective of the framework document which sought to establish a modern specific contaminated land power. Now, for the first time in UK law, there will be a specific definition of contaminated land and dedicated procedures for its control.

The legal regime will implement the suitable-for-use approach, which requires regulatory action only where necessary to deal with unacceptable risks to human health or the environment, taking into account the use of the land in question and its environmental setting. Although the contaminated land regime is new, its overall structure and the nature of its controls are broadly similar to more general powers under statutory nuisance. These other powers will cease to apply to contaminated land.

The primary regulatory role under the new regime will rest with local government, through the borough and district councils. This reflects their existing powers under statute, and will complement their roles as planning authorities. Their role will be to:

Inspect their areas in order to identify contaminated land
Consult on what remediation might be required in any individual case

Require remediation to take place through issuing of a formal Remediation Notice if necessary, and with powers to act in default

Record information about remediation carried out under the regime

The identification of any contaminated land will be based on a process of risk assessment. For any land to be identified, the local authority will be required to have found contaminating substances in the land and to have established that they are likely to cause harm to particular targets. Under the liability provisions in the legislation, the responsibility for paying for remedial actions follows the polluter pays principle. Therefore a person who caused or knowingly permitted the contamination will be liable. However, if the polluter cannot be found, then liability passes to the current owner or occupier of the site except in the case of water pollution.

Section 57 also provided for the creation of the Environment Agency on April 1, 1996, which is an integration of the functions of the National Rivers Authority, Her Majesty's Inspectorate of Pollution, and the local waste regulation authorities. The agency will build on the role of its predecessor's by protecting and enhancing the environment in line with the government's commitment to sustainable development.

The agency has four principal roles with respect to contaminated land:

To provide site-specific guidance on remediation requirements
To act as the regulator for a defined category of special sites
To compile a national report on contaminated land
To sponsor technical research while acting as a center of expertise

The primary legislation sets the principles which will govern the new regime. Within this framework a package of statutory guidance will provide the detailed parameters to support professional and technical judgment in specific cases. This itself will be supported by technical guidance from the Department of the Environment's research program. This will include standard procedures and soil guideline values in an advisory, not statutory role. Secondary legislation, in the form of regulations, will also be issued to cover the detailed procedural matters.

The government announced on December 22, 1997, that it had concluded that Part IIA of the Environmental Protection Act 1990 (inserted by Section 57 of the Environment Act 1995 and passed by the previous government) sets out, in principle, the right framework for controlling land that in its current use poses health or environmental dangers. Part IIA is modeled on the existing statutory nuisance provisions and will replace them with respect to contaminated land. The timetable for the implementation of Part IIA will be decided by the government after it has concluded its present Comprehensive Spending Review. Local authorities will cause their areas to be inspected in order to identify contaminated land, and they will ensure that appropriate remediation takes place when they identify such land.

Contaminated land is identified on the basis of risk assessment. Land is contaminated land only where it appears to the authority, by reason of the substances in, on or under the land, that:

Significant harm is being caused or there is a significant possibility of such harm being caused, or
Pollution of controlled waters is being, or is likely to be caused.

Where necessary, authorities will ensure that appropriate remediation is undertaken by serving a Remediation Notice. Such a notice is served on any person who caused or knowingly permitted the substances causing the land to be contaminated to be present. If no such person can be found, the notice has to be served on the owner or occupier of the land. The provisions allow for the apportionment of liability where there is more than on polluter. Failure to comply with a Remediation Notice is an offence. However, a person who is the owner or occupier of the land cannot be required, under this legislation, to carry out remediation which is only needed to deal with water pollution. This is dealt with by separate legislation to cover the protection of water resources.

In some circumstances the authority can carry out the remediation itself and recover its costs from the persons or persons liable. In setting any remediation requirements, an authority has to have regard to the costs which are likely to be involved and to the seriousness of the relevant harm or water pollution. The authority also has to consider whether the person liable for carrying out the remediation might suffer financial hardship if he did the work. If so, the cost to him is waived or reduced, and the cost is met by the local authority. The Environment Agencies are responsible for dealing with special sites.

Recent estimates of the extent of contaminated land in the United Kingdom indicate an area of 123,550–617,750 acres. The amounts of government money spent yearly on redevelopment of contaminated land are approximately $350 million.

The Royal Commission on Environmental Pollution (RCEP) has recently published its findings on a wide-ranging review of soils and soil use in the UK. The report emphasized the pressures on soils in the UK as a result of factors including:

The great intensification of agriculture in the last hundred years
Continuing demand for land for building, particularly in areas with fertile soils
Increasing amounts of sewage sludge (for which disposal at sea was banned after 1998)
Large quantities of other wastes spread onto land or buried in the ground
Pollutants reaching soils from the atmosphere
Contamination from industry

The central recommendations of the report were that:

The government should draw up and implement a soil protection policy for the UK which takes full account for long-term environmental considerations
The newly formed Environment Agency must take a genuinely integrated view of the environment and give proper attention to safeguarding and remediating soil

The Royal Commission stressed the need to reduce pressure on greenfield sites by increasing the recycling of derelict and contaminated land. The Royal Commission asked the government to consider promoting land banks of sites for redevelopment which have been treated to remove contamination. They suggested that after a site had been remediated to the standards set by the local regulatory authorities, the landowner should be absolved from further liabilities. The government is expected to respond to the Royal Commission's report in due course.

In May 1995 leading members of UK industry, with the support of the Department of the Environment, formed the Soil and Groundwater Technology Association (SAGTA). The primary aim of this new association is to share expertise and experience of the technical aspects of dealing with contaminated land and to promote the development of the most effective management methodologies and remedial technologies. The SAGTA will pay particular attention to the introduction of best practice to prevent future land contamination and to deal with historic problems. In February 1996, SAGTA held a major meeting with the general participation of the academic research community. Its objectives were to:

> Bring together research performers from many scientific and engineering disciplines with interests in contaminated land research
> Provide problem holders with a clear understanding of the current state of the art in contaminated land management
> Explain the needs of problem holders to research performers and facilitators
> Stimulate the development of multidisciplinary collaborative links among researchers, funders, and problem holders.

General conclusions from the meeting included:

> A requirement for research and development investment in many of the areas identified by the four workshops: site assessment, risk assessment, control measures, and measurement and monitoring. Several important issues which were identified included validation, fundamental understanding of mechanisms affecting the accessibility and availability of contaminants, and dealing with problems of heterogeneity.
> Recommendations were made for the development of centers of excellence; establishment of larger, better targeted projects; and in particular, for setting up of field-scale sites for research investigations and technology demonstrations.

The Environment Agency carries out a significant program of research into contaminated land, which was inherited from the Department of the Environment, Transport and the Regions. The program focuses on the production of best-practice guidance to support the proposed new regulatory regime. Specifically, the Research Program:

> Develops current scientific knowledge on risk assessment and risk management of contaminated land (with particular emphasis on issues of sustainability and a consideration of the costs and benefits)
> Develops procedures for the effective delivery of regulatory activities in land contamination
> Reviews and identifies information needs for the preparation of a report on the state of contaminated land in England and Wales

Collaboration with other organizations is sought, where appropriate, to achieve the program objectives. The Research Program also takes account of work in other countries and of possibilities for the UK to influence and contribute to important international developments in this area.

T. United States

Three different federal programs provide the authority to respond to threatened releases of hazardous substances that endanger public health or the environment. (a) In response to growing concern about contaminated sites, Congress passed the Comprehensive Environmental Response, Compensation, and Liability Act (CERCLA) in 1980. Commonly known as Superfund, the program under this law is the central focus of federal efforts to clean up releases of hazardous substances at abandoned or uncontrolled hazardous waste sites. The program is funded, in part, by a trust fund based on taxes on petroleum and other basic organic and inorganic chemicals. (b) The second program is directed at corrective action at currently operating industrial facilities. This program is authorized by the Resource Conservation and Recovery Act of 1980 (RCRA) and its subsequent amendments. This law also regulates the generation, treatment, storage, and disposal of hazardous waste at industrial facilities. RCRA corrective action sites tend to have the same general types of waste as Superfund sites, and environmental problems are generally less severe than at Superfund sites, although some RCRA facilities have corrective action problems that could equal or exceed those of many Superfund sites. (c) The third clean-up program, also authorized by the RCRA, addresses contamination resulting from leaks and spills (primarily petroleum products) from underground storage tanks (USTs). This law has compelled clean-up activities at many UST sites. By the end of 1996, over 300,000 confirmed releases had been reported, over 250,000 clean-ups initiated, and over 150,000 clean-ups completed.

Each clean-up program has a formal process for identifying, characterizing, and cleaning up contaminated sites. These processes generally involve joint implementation with state agencies and the involvement of various groups, such as local government agencies, local residents, businesses, and environmental public interest groups. Superfund is administered by Environmental Protection Agency (EPA) and the states under the authority of the CERCLA. The procedures for implementing the provisions of CERCLA substantially affect those used by other federal and state clean-up programs. These procedures are spelled out in the National Oil and Hazardous Substances Pollution Contingency Plan, commonly referred to as the National Contingency Plan (NCP). The NCP outlines the steps that the EPA and other federal agencies must follow in responding to releases of hazardous substances or oil into the environment. Although the terminology may differ from one program to another, each follows a process more or less similar to this one. Thus, in addition to comprising a defined single program, activities in the Superfund program substantially influence the implementation of the other remediation programs. The RCRA assigns the responsibility for corrective action to facility owners and operators and authorizes the EPA to oversee corrective action. Unlike Superfund, the RCRA responsibility is delegated to the states. As of the end of 1996, the EPA had authorized 32 states and territories to implement RCRA corrective action. The processes for characterizing and remediating RCRA corrective action sites are analogous to those used for Superfund sites, although the specific terminology and details differ. The UST regulations require tank owners to monitor the status of their facilities and immediately report leaks or spills to the regulatory authority, which usually is the state. Clean-up requirements generally are similar to those under RCRA corrective action and are entirely overseen by state agencies.

The nature and scope of remediation policies are driven largely by federal and state requirements and public and private expenditures. A number of legislative and regulatory initiatives may affect the operation of the Superfund, RCRA corrective action, and UST programs. For example, some of the proposed changes to Superfund would require consideration of land use in setting clear-up standards, emphasize the treatment and disposal of only highly contaminated and highly mobile media, limit the addition of new sites to the Superfund remediation program, and change the liability aspect of CERCLA to reduce the cost and time needed to assign the liability for a clear-up project. Some of these changes have already being implemented, to some extent, under EPA administrative reforms. Congress and EPA also are considering proposals to revise the RCRA to exempt wastes from remediation activities from certain hazardous waste management requirements, streamline the permitting process, and modify land disposal restrictions. There is widespread and growing interest in using risk assessment to determine clean-up priorities, as may be done under the Risk Based Corrective Action initiative in the

UST program. There is also increasing interest in the issue of bioavailability of contaminants as an alternative to chemical concentrations alone to set clean-up standards. Much scientific work and consensus building has yet to be completed on this issue. Finally, the brownfields policy initiative has become prominent at the federal and state levels. This concept uses economic redevelopment as the driving force for site clean-up and is gaining widespread acceptance.

Almost half a million sites with potential contamination have been reported to state or federal authorities over the past 15 years. Of these, about 217,000 still require remediation for which contracts have not been issued. Almost 300,000 other sites were either cleaned up or were found to require no further action. Regulatory authorities have identified most of the contaminated sites. Nevertheless, new ones continue to be reported each year, but at a declining rate. The data on number of sites come from disparate sources because these sites are not all registered in one data repository. The EPA maintains detailed data on Superfund sites and summary information for RCRA corrective action and UST sites. The states and other federal agencies generally maintain separate records of the sites for which they are responsible. It is estimated that the cost of remediating the 217,000 sites will be about $187 billion in 1996 dollars, and that it will take at least several decades to completely remediate all the identified sites. Legislative, regulatory, and programmatic changes may alter the nature and sequence of clean-up work done at Superfund, Department of Defense, and Department of Energy sites. If some of the current proposals become law, more emphasis may be placed on cleaning up the most severely contaminated areas on a site, making government properties available for economic reuse, increased consideration of future land use in remedy selection, and more explicit consideration of cost and performance in remedy decisions.

After a significant increase in the selection of newer treatment technologies—such as SVE, thermal desorption, and bioremediation—in the early 1990s, the selection of innovative technologies leveled off or decreased, and the selection of containment became more common. Nevertheless, treatment remedies still are more common. New technologies offer the potential to be more cost-effective than conventional approaches. In-situ technologies, in particular, are in large demand because they are usually less expensive and more acceptable than above-ground options. New technology development programs emphasize in-situ technologies, in particular bioremediation and enhancements to SVE. Although metals are common at most sites, alternatives to treat metals are limited. Government and corporate owners of contaminated sites have targeted several technologies to treat metals in soil for further development, including electrokinetics and phytoremediation. While groundwater is contaminated at more than 70% of the sites, not all of these sites will be actively remediated. Available technology cannot always meet the desired clean-up goals for a site, because the methods leave residual aquifer contamination, known as nonaqueous-phase liquids (NAPLs). The most frequently used method for groundwater remediation at Superfund sites is conventional pump-and-treat technology. In-situ treatment technologies, primarily bioremediation and air sparging, have been selected at only 6% of Superfund groundwater treatment sites, most of which also are using pump-and-treat. New management approaches recently receiving more attention include treatment walls, electrokinetics, use of surfactants and co-solvents, hydraulic and pneumatic fracturing, and selective application of natural attenuation. If more effective in-situ groundwater technologies were available, a larger portion of contaminated groundwater sites could be fully remediated.

III. TECHNOLOGIES AND ISSUES

Since 1980, the pilot study has examined numerous issues and technologies. The most recent group of technologies to be report on was at the end of the Phase II study in 1997. At that time, there were 52 active projects in the Pilot Study. Summary information on each project is provided in the final report and other pilot study reports. These are listed in the References. The project summaries provide a technical abstract, which summarizes the project's progress and results, but is not a critical review of the project. The summaries also provide the name of a technical contact for further information.

While the objective of the pilot study was to evaluate applications of particular technologies, a large proportion of the projects involved more than one technology. Some involved the use of integrated treatment systems combining more than one technology, and others involved the application of more than one technology to deal with separate aspect of site contamination. Other projects concerned theoretical studies, strategic scientific studies, or large-scale remediation projects for which the remediation strategy had yet to be developed.

During the Phase II study, types of technologies addressed were

Biological: 24 projects including bioventing, biopiles, slurry reactors, white rot fungi, etc.
Physical-chemical: 29 projects including soil vapor extraction, soil washing, solvent extraction, and ultraviolet treatment.
Chemical: 4 projects including photochemical oxidation, ozone treatment, sorption, and leaching
Thermal: 5 projects including thermal desorption, incineration, and thermal vitrification
Stabilization solidification: 2 projects including chemical fixation, and grouting
Other projects including site characterizations, free-product recovery, etc.

There were 23 projects that relied upon a single technology, 19 that used integrated technologies, 7 mixed technologies, and 3 that did not involve treatment. Typical combinations were soil vapor extraction with in-situ biotreatment, soil washing followed by biotreatment, and soil washing followed by thermal treatment

Forty of the 52 projects were concerned only with the treatment of organic contaminants, including polycyclic aromatic hydrocarbons, polychlorinated biphenyls, and BTEX compounds (benzene, toluene, ethylbenzene, and xylenes). Six projects dealt exclusively with metals, and six dealt with both inorganic and organic contaminants. One project focused on remediation of inorganic sulfates and cyanides.

A. In-Situ Technologies

Environmental remediation technologies can be broadly divided into two categories: ex situ and in situ. Ex-situ technologies treat contaminated materials after gross removal and transport of contaminated media to the treatment facility. Actual treatment often occurs on-site—reducing costs, risks, and administrative burden incurred with hazardous material portage. In contrast in-situ technologies apply the remediation process directly to the contaminants, with little or no gross movement of hazardous material.

The cost of environmental remediation is, to a large degree, directly proportional to the amount of material handled in the process. When large masses of earth or water are removed and cleaned, costs are incurred for both the physical handling of the material—large fractions of which may be uncontaminated—and for application of the treatment process in order to ensure complete decontamination. Deep contamination can involve extensive excavation of uncontaminated overburden. In addition to direct costs, excavation of contaminated soil is often impractical due the presence of overlying structures.

In-situ processes attempt either to destroy the contaminants where they are found or, at the very least, to remove the contaminants from the contaminated matrix. Postextraction physical separations are avoided or minimized, even if destruction or recovery is necessary. The technical challenge common to all of these processes involves moving mass to some desired area, moving reagents (oxygen, nutrients, oxidants, etc.) to the contaminants, or moving the contamination to some subsurface treatment zone. In general, in-situ processes require less capital outlay than ex-situ treatments. Material handling requirements are lower, transportation costs are avoided, and postprocess treatment (e.g., landfilling) is avoided. In-situ processes are also less invasive, which is often the reason for their use, as in the case of treatment under a building. On the other hand, in-situ treatments, especially biotreatments, are generally slower and require longer implementation. In many circumstances, such as in the sale of property, the need to act quickly can outweigh the lower capital costs.

In-situ strategies frequently use biological processes to destroy contaminants. Bioremediation uses microorganisms to transform the hazardous organic contaminants into harmless products, such as carbon dioxide and water. Microorganisms require mineral nutrients and a carbon and energy source (food) to carry out these biodegradation processes. Ideally, the target contaminant will be the food source, and sometimes a treatment process can be designed around fortuitous incidental biochemical reactions. Microbes also require a terminal electron acceptor to complete the circuit of reactions by which they survive. The most familiar electron acceptor is oxygen, but certain other oxidized ionic species, such as nitrate, sulfate, or ferrous iron, can support bacterial growth. Several other factors (e.g., temperature and pH) affect the efficiency of these processes. Degradation capabilities of microorganisms have been used for decades to treat municipal and industrial wastes. Recent advances in biotechnology allow these processes to be applied to hazardous chemicals in situ.

In general, petroleum hydrocarbons can serve as primary growth substrates for bacteria. The ease of biodegradability of a hydrocarbon is inversely proportional to its molecular weight and complexity. Short-chain aliphatic hydrocarbons and simple aromatic molecules are fairly readily consumed, while large, polycyclic aromatic hydrocarbons (PAHs) are more recalcitrant. Synthetic organic compounds, such as chlorinated solvents [tetrachloroethene (PCE), trichloroethene (TCE), carbon tetrachloride, etc,], are much more resistant to biodegradation. Chlorinated solvents cannot serve as growth substrates for most microorganism, but can nonetheless be degraded or transformed by populations that grow on other substrates. TCE, for example, can be transformed and even mineralized by a variety of microorganisms growing on different organic compounds, including methane, phenol, toluene, propane, methanol, and n-butane. The current challenge for bioremediation is the encouragement of microorganisms to degrade these manufactured compounds.

Bioremediation can be very effective for removing contaminants that serve as growth substrates, particularly if low concentrations of the contaminants are present in an appropriate environment. Bioremediation can provide a cost effective alternative to traditional technologies (e.g., air stripping, carbon sorption, and excavation) for a wide range of natural organic compounds, such as motor or jet fuel. Biological treatment offers a permanent and often less expensive solution than strictly physical treatments, because microorganisms convert toxic organic compounds to environmentally benign products. However, bioremediation is no panacea. In-situ bioremediation systems are often integrated with other remediation technologies to effect total clean-up.

Physical processes will also be considered as in situ for purposes for this study if the intent of the process is to physically remove only the contaminant from the contaminated media. As an example, air sparging is intended to remove volatile organic compounds (VOCs) from groundwater. This judgement could be debated, as such a process still merely transfers contamination from water-saturated soil to air, which often still requires postreatment. However, as opposed to the pumping and treating of groundwater, air sparging promises several advantages in material handling, as well as certain challenges associated with the physical transfer of matter at the contaminated area, and it is appropriate to discuss these processes here.

Knowledge of contaminant location and physical state in the subsurface is critical in implementing in-situ remediation techniques. In the vadose or unsaturated zone, contamination may exist as a vapor phase, adsorbed to particles, dissolved in the thin film of water surrounding soil particles, or as a nonaqueous-phase liquid (NAPL). Contamination in the saturated zone might consist of residual saturation or material trapped within the soil matrix, matter sorbed to solids, a pool of NAPL, or dissolved material in the groundwater. Each situation can pose unique challenges to the remediation engineer.

Each of the technologies examined by the Phase II Pilot Study offers innovations over more traditional, mass-intensive approaches to remediation. As they are implemented, a greater understanding of the dynamic interaction of contamination with the subsurface is gained. Comparing and contrasting the results of these demonstrations suggests further innovations, as well as contextual evaluation of the technologies themselves.

The most successful in-situ technologies were those directed toward simple hydrocarbons in the vadose zone, i.e., some form of bioventing or land farming. Complicating factors involve mass transfer and bioavailability because tight clayey soils or tarry deposits prevent the intimate contact of microbes, oxygen, water contaminants, and other nutrients necessary to carry out the destruction of the contaminant. Cold temperatures pose no great threat to implementing bioremediation—provided one is willing to heat the soil or wait longer for success.

SVE has also come into its own during this period, with several enhancements being demonstrated.

Extension of bioventing and SVE through pneumatic fracturing of tight soils is notable.

Aquifer stripping entered the remediation scene during this Pilot Study period and met with mixed success. These technologies appear to be most effective in relatively homogeneous aquifers contaminated with highly volatile contaminants. Microbial filters, as demonstrated to the Pilot Study, show promise in chlorinated hydrocarbon contamination, but still require development. Electroosmotic transport of contaminants through treatment zones is also promising.

Attempts to extend bioremediation to PAHs have met with mixed success, due to the recalcitrance of the substrate. Land farming was somewhat more successful than bioventing toward these contaminants.

B. Physical-Chemical Treatment

1. Typical Soil Washing

As a pretreatment for excavated material, soil washing exploits the fact that contaminants are often preferentially adsorbed to the fine particles. This approach relies on physical processes to separate a small volume of contaminated material from the bulk of relatively uncontaminated material. Current commercial soil washing processes remove mainly fine fractions (<0.063 mm) containing the highest concentrations of contaminants. The remaining coarse fraction (>0.063 mm) is relatively clean. The clean material is often reused as inert fill.

Separating the contaminated fines often results in lower costs for overall treatment, because it is only this smaller volume of contaminated material, not all the original material, that requires further treatment or disposal. The contaminated material is shipped to a controlled landfill or treated further by a variety of processes that destroy, immobilize, or recycle the contaminants.

Soil washing is economically feasible only if the volume reduction is large enough to provide financial benefits. An extensive review of commercial and pilot-scale soil washing systems showed that soil washing is most effective on soils containing less than 30–35% clay and slit, i.e., particles smaller than 0.063 mm. At higher percentages of these fines, the volume reduction is not large enough for the process to be economical. In addition, difficulties arise in handling these materials, separating the contaminated and uncontaminated materials, and handling the products.

Despite these problems for soils rich in fines, soil washing may be seen as a cost-effective treatment if the technology overcomes difficulties presented by the high levels of silt and clay. Possible solutions include enhancing soil separation techniques and developing processes to treat the fine fractions downstream.

2. Soil Washing Combined with Other Technologies

With some soils, physical treatment alone will not reduce the absolute concentration of contaminants to acceptably low levels. For soils containing more than 30% clay and silt by weight, physical pretreatment could reduce the volume of contaminated material requiring downstream treatment. Pretreatment may also present the separated contaminant concentrate in a form suitable for the downstream process, e.g., bioslurry, solvent extraction, and vacuum distillation. Case studies in this chapter illustrate soil washing combined with biological or physical-chemical treatment such as vacuum distillation, photooxidation, biodegradation, and chemical dehalogenation.

3. Physical-Chemical Treatment (No Soil Washing)

Some of the downstream processes mentioned above may be used directly on contaminated material that was not previously washed. Some of the case studies in the chapter examine processes such as

Solvent extraction and treatment of extracts by stabilization (for heavy metals)
Leaching and treatment of leachate
In-situ electroosmosis and adsorption

Like soil washing, these conventional technologies experience difficulties in treating contaminated materials where the fines exceed 30–35%. When the levels of fines are this high, methods such as thermal treatment and solvent extraction become more expensive, while others such as biological treatment take a longer time.

4. *Photooxidation Treatment*

Photooxidation treatment does not transfer or concentrate contaminants that may require further treatment or costly disposal. Treated water can be disposed on-site or off-site. Options for on-site disposal include groundwater recharge or temporary on-site storage for sanitary use. Off-site disposal options include discharge into surface water bodies, storm sewers, and sanitary sewers. Depending on permit requirements, discharged water may have to be adjusted for pH. Factors influencing the applicability of photooxidation can be grouped into four categories: site characteristics, influent characteristics, operating parameters, and maintenance requirements.

C. Biological Processes

Biological processes for the remediation of contaminated land depend on one or more of four basic processes: (a) biodegradation; (b) biological transformation to a less toxic form (e.g., for metals); (c) biological accumulation into biomass; or, conversely, (d) mobilization of contaminants for downstream recovery. In general, established commercial processes are limited to those based on biodegradation.

The vast majority of practical biological treatments exploit degradation and are variously described as bioremediation, bioreclamation, biotreatment, or biorestoration. Contaminated sites are also commonly revegetated to improve their stability and esthetic appeal and to reduce windblow of contaminated dust.

Concerns about current biological processes include:

- Their susceptibility to inhibition by toxic contaminants (e.g., for example heavy metals), although some biodegradation processes appear quite robust
- The low biodegradability and or bioavailability of some common organic pollutants, found with other more degradable contaminants (current research includes the use of chemical pretreatments to enhance biodegradability and treatment of an increasing number of compounds is found to be feasible, for example, chlorinated solvents)
- Residual concentrations of contaminants after treatment, whose environmental significance is not known
- The mobilization and release of potentially toxic, partially degraded contaminants from in-situ treatments

Biodegradation describes the decomposition of an organic compound into smaller chemical subunits through the action of organisms. Both aerobic and anaerobic degradation pathways exist, although there are some differences in the types of compound that will degrade under aerobic and anaerobic conditions. Principally, soil microorganisms (bacteria, fungi, and actinomycetes) are responsible for bioremediation processes, but some researchers are interested in prospects for plants and algae. Plants may be of more immediate use in the accumulation of contaminants or as a means of stimulating soil microbial activity. These approaches are known collectively as phytoremediation and are regarded as an important emerging technology for future research.

Completely degraded compounds are said to be mineralized, and the end products of the aerobic degradation of chlorinated hydrocarbon might be carbon dioxide, water, and chloride ions. Biotransformation may be of use in biological treatment of contaminated soil, but has not been exploited. It has the drawback that further transformations could regenerate toxic forms. In addition, biotransformation can be accompanied by an enhancement in toxicity.

Biodegradation may proceed via enzymic activity on compounds adsorbed into cells or through the activity of extracelluar enzymes active outside the confines of the cell. Cells also use enzymes to generate free radicals or peroxide ions that attack organic compounds, particularly insoluble compounds. In many cases, organic compounds do not readily enter microbial cells, since the compounds are either sorbed to soil surfaces, are too large, or are physically incapable of being sorbed into cells.

Bioavailability is regarded as one of the key limiting factors for bioremediation. More complex compounds may not be completely degradable by single organisms, but are degraded by consortia of organisms, or in some cases may not be completely degradable in any circumstance. Some organic compounds may be coincidentally degraded as a result of microbial activity against other substrates, a process called co-metabolism. An example of this is the use of methane oxidation to degrade some chlorinated solvents. There are a number of organic compounds, such as tetrachloroethene (PCE), whose degradation is not energetically favorable to microorganisms. However, in some cases, under anaerobic conditions, these compounds may be biodegraded. The compound does not serve as an energy source or carbon source, but is used as an electron acceptor, i.e., it is reduced during the conversion of other organic materials.

It is likely that several of these processes may occur simultaneously in practical bioremediation treatments; however, some techniques are designed to capitalize on particular microbial processes, such as the use of fungal lignase systems to degrade recalcitrant organic contaminants such as pentachlorophenol (PCP) and polycyclic aromatic hydrocarbons (PAHs).

Inorganic compounds may also be changed by microorganisms, either by direct metabolism (as in the oxidation of sulfur or the methylation of mercury) or indirectly through the release of ligands or acids. These processes may mobilize inorganic contaminants such as heavy metals. There may be potential applications for mobilization as a means of stripping inorganic contaminants from soils, and several laboratory- and pilot-scale initiatives based on microbial mobilization are underway. Arsenic and some heavy metals may be converted into volatile methylated forms by microbial activity (also referred to as biotransformation). However, the toxicity of the methylated compounds may raise serious issues of operational safety and environmental emissions from such an approach.

D. Thermal Treatment

Three main types of thermal treatment can be identified for contaminated soils, sediments, sludges, filter cakes (e.g., from soil washing), and similar materials:

Thermal desorption, in which contaminants are removed from the feedstocks at relatively low temperatures and then destroyed or collected from the gas stream in a subsequent stage

Incineration (thermal destruction), in which contaminants are destroyed at high temperature

Vitrification, in which very high operating temperatures destroy some contaminants and trap others in a glassy product.

The thermal process projects reviewed involved either thermal desorption or incineration, and the introductory sections that follow concentrate on these forms of treatment. In practice, there is no clear technical distinction between thermal desorption and incineration, since thermal desorption of contaminants occurs during incineration of soils or other solids, and partial combustion of desorbed organic compounds often occurs within a desorber unit or downstream in a fume incinerator, depending on the design.

Existing industrial thermal processors, such as cement kilns and coal-fired boilers, are also sometimes used for organic-rich residues. Other thermal processes can be used for specific contaminants; for example, retorts have been specified by the U.S. EPA for the treatment of mercury-contaminated soil.

Thermal desorption methods physically separate volatile and semivolatile contaminants from soils, sludges, and sediments. They do not generally result in a high degree of thermal decomposition of contaminants, although temperature variations between different systems may allow for some localized oxidation or pyrolysis. The thermal desorption unit is only one part of a treatment train; some pretreatment of feedstocks and posttreatment of treated soil or separated contaminants is usually required.

Efficient separation can occur at temperatures of up to 600°C, although temperatures may reach 900°C during the primary stage in some specialized systems. In practice, many systems operate at relatively low solids temperatures; even polychlorinated biphenyls (PCBs) can be removed at 450–500°C. An important design parameter is the length of time that soils remain at the target temperature. Separated contaminants, water vapor, and particulates must be collected and treated. Typically, this is done using conventional methods of condensation, adsorption, incineration, filtration, etc. The methods are selected according to the nature and concentration of contaminants, regulatory regime, and economics of the system employed. It may be possible to recover separated contaminants for reuse. Thermal desorption systems that employ combustion or other oxidation processes for treating the off-gas can accomplish the same goal is incineration—i.e., destruction of contaminants.

Incineration (thermal destruction) destroys contaminants at high temperature (800–1200°C). Specifically, incineration is a high-temperature oxidation reaction between combustible substances and oxygen under controlled conditions of retention time, temperature, and turbulence within a single- or multiple-stage combustion chamber. Although organic contaminants are destroyed in the process, air pollution control equipment must be provided to collect and treat combustion products, particulates, and volatile metals present in exhaust gases. Incineration of soils and sediments involves volatilization and desorption of water and organic contaminants (and some inorganic contaminants), as in thermal desorption. A secondary combustion chamber to complete oxidation of the volatilized materials is generally required.

The high temperatures used during incineration have implications for the reuse of the treated soil due to changes to the physical, chemical, and biological properties of the material. Changes in soil texture, together with the loss of natural organic constituents, reduce the ability of treated material to support vegetation and may affect engineering properties. The loss of soil structure and organic content also may increase the leachability of any heavy metals remaining in the treated product. Further treatment, e.g., stabilization/solidification, may therefore be required before the material is acceptable for reuse.

A wide range of incineration techniques has been developed for the treatment of contaminated soils, sediments, and sludges, including direct-fired rotary kilns, fluidized beds, and infrared belt conveyor systems.

Vitrification destroys contaminants by oxidation and thermal decomposition and immobilizes residual contaminants in a vitreous product. The advantages of vitrification over other thermal treatment processes are that it produces fewer air emissions and a solid residue with favorable leaching characteristics.

Vitrification systems consist of a melter, heat recovery system, air pollution control system, and storage and handling for feedstock and raw materials. There are various configurations for melters, some of which are multichamber and others of which use mechanical agitation. Energy requirements are significant where feedstocks have a high mineral content. The most common melters are heated by electrical currents passed through the melt mixture from electrodes. Variations between melters include the method of introducing feed, the degree and type of mixing, electrode design, and the means of achieving complete combustion of organic compounds. More recent melter designs utilize alternative methods of introducing heat and have different heat and mass transfer characteristics.

Typical melt temperatures are about 1500°C. Sufficient glass-forming material (silicate) must be present to produce a proper melt that will result in a durable vitrified product. This may require the addition of fluxing agents. Molten product is continuously drawn off the melter, either into containers for cooling, solidification, and handling, or through some type of cooling process to produce granular solids. Emissions of the more volatile metals is a potential concern, and air pollution control systems must be highly efficient. Process residues include glass/vitrified waste, molten metal (not produced as a separate phase in most processes), scrubbing and cooling liquors, and off-gases.

Commercial vitrification systems have been developed for the treatment of contaminated soils and sediments in the United States, where a number of vendors have field-, bench-, and pilot-tested the technology. Most of these systems were modifications of different types of glass-making furnaces, and development was directed initially toward radioactive or other highly hazardous solid wastes. One commercial facility in the United States was used to treat organic wastes for several years.

E. Solidification/Stabilization

Stabilization/solidification methods (sometimes called immobilization methods) change the physical state of a contaminated material, such as solidifying a contaminated sludge. In addition, chemical stabilization can reduce the availability of contaminants to potential targets, usually by containment within a solid product of low permeability.

Stabilization involves adding chemicals to the contaminated material to produce more chemically stable constituents, for example, the formation of virtually insoluble metal hydroxides. Stabilization may not result in an improvement in the physical characteristics of the material. For instance, the material may remain as a relatively mobile sludge, but the stabilization process will have reduced the toxicity or mobility of the hazardous constituents within it.

Solidification involves adding reagents to the contaminated material to reduce the material's fluidity or friability and to prevent access by external mobilizing agents, such as wind or water, to the contaminants contained in the solid product. Solidification does not necessarily require that chemical reactions occur between contaminants and the solidification agent, although such reactions may take place depending on the nature of the reagent.

In practice, many commercial systems and applications involve a combination of stabilization and solidification processes. Solidification follows stabilization to reduce exposure of the stabilized material to the environment through, for example, formation of a monolithic mass of low permeability.

Although volatile constituents may be driven off (because heat is often generated) and some hydrolysis of chlorinated organic compounds may occur during the application of some processes, the destruction or removal of contaminants is not the objective of stabilization/solidification.

Contaminants may become available once again if the physical or chemical nature of the treated product alters in response to changes in the external environment, such as exposure to an acidic discharge or leachate or physical breakdown of a compacted soil mass due to freezing and thawing. Solidified products also may be subject to internal degradation reactions over time (e.g., the oxidation of sulfides to form expansive sulfates). Key points for selecting a stabilization/solidification method are therefore:

- Its ability to achieve and retain the desired physical properties, chemically stabilize or permanently bind contaminants, and contain (physically entrap) contaminants over the long term
- The methods to be used to determine treatability and short- and long-term performance

The effectiveness of stabilization/solidification methods depends on:

- Proper characterization of the material to be treated so that the most appropriate formulation can be selected
- Effective contact between the contaminants and treatment reagents—for many systems this can be achieved by ensuring a high degree of chemical and physical consistency of the feedback, and the use of appropriate mixing equipment
- Control over external factors, such as temperature, humidity, and amount of mixing after gel formation, since these affect the setting and strength development processes and the long-term durability of the product
- Absence of substances that inhibit the stabilization/solidification process and development of the required physical characteristics, or pretreatment to render such substances harmless, such as by sorbent addition

Treatability studies are always required to establish anticipated effectiveness and materials handling requirements.

Because most stabilization/solidification methods involve the addition of solid reagents to the contaminated material, some increase in the final volume of the treated product can be expected. Increases in the range 30–130% are typical. A major advantage of most stabilization/solidification methods is that they improve the handling characteristics of sludges and other high-water-content materials, and may confer additional structural strength on contaminated material. Sometimes these are the primary reasons for their use, and they may be an important consideration for contaminated sites undergoing redevelopment. Stabilization/solidification methods are readily applied on-site using mobile mixing and blending equipment.

Most commercial stabilization/solidification systems are derived from established hazardous waste treatment techniques and use relatively simple equipment and conventional reagents (binders) to immobilize the contaminants. Systems may be classified according to the primary stabilization agent used: cement-based, pozzolanic-based, silicate-based, thermoplastic-based, or polymer-based systems. In practice, a combination of these reagents may be employed. The formulations actually used on a commercial basis are often proprietary in nature.

In the United States, stabilization/solidification is considered an established technology for the treatment of certain inorganic forms of contamination, and long-term monitoring data are available on the performance of solidified wastes in the field. Stabilization/solidification techniques have also been used in Europe for the treatment of hazardous waste. However, doubts remain over their long-term performance. These doubts arise from the chemical and physical nature of the processes themselves and from observed deficiencies in the quality of application in the field (and in fixed plant). In addition, methods of testing and predicting performance are not well developed and are the subject of continuing debate. Proprietary formulations for the treatment of organic contaminants prior to the use of conventional binders are available in the United States, but practical experience in their application is limited.

F. Integration of Technologies

Experience shows that contaminated sites frequently cannot be remediated by a single technology. Complex contamination problems require the combination of different technologies for either different contaminated areas and media or for a specific medium exhibiting complex contamination. However, remediation strategies employing a single treatment technology and those employing a combination of technologies are commonly more complicated and more expensive than is removal to a landfill. Consequently, not only technical effectiveness (in terms of ability to achieve remediation objectives, time requirements, potential environmental impacts, and cost), but also political (policy) factors will influence the choice of strategy. These policy considerations may lead to adoption of complex treatment systems that otherwise might be rejected in terms of short-term costs alone (assuming comparable technical effectiveness).

The term "integrated" refers to approaches involving process integration where two or more technologies are used simultaneously or in series to treat a specific problem. The term "mixed" refers to projects involving two or more technologies to treat different contaminated areas or media at a site as part of an overall remedial strategy.

Effective and efficient treatment of environmental contamination requires tailor-made solutions meeting the specific requirements of the media to be treated, of the contaminants to be removed or destroyed, and of the policy framework within which the project is to be implemented. In many projects, the contamination problem can be adequately addressed by applying a single treatment technology. However, problems may occur due to one or more technical or organizational factors:

- Difficult-to-treat media, such as soil with high proportion of fine-grained material; mixed solids (soil, ashes, slags, brick, debris, concrete, plastics, wood, etc.); solids with a high proportion of organic matter (e.g., peat); low-permeability soils and sediments; fine or uneven distribution of contaminants (i.e., giving rise to low bioavailability); or large volumes of contaminated material requiring treatment
- Contaminants that are difficult to treat due, for example, to physical properties (e.g., low solubility); chemical properties (e.g., not biodegradable); presence of complex contaminant mixtures (e.g., metals and organic compounds)

Frequently, combinations of technologies can overcome these limitations. In many cases, efforts are made to reduce the amount of material requiring expensive treatment by separating fractions of materials that can be reused without further treatment or with limited effort. In other cases, materials or contaminants are difficult to treat, which means that the limitations of a single technology are evident early in the development of the remediation strategy. In these cases, means of modifying the material's physical and chemical conditions have to be identified and evaluated in order to allow treatment at all, or to optimize cost and results.

The organizational factors mentioned above are commonly reduced to the policy requirement to avoid generating secondary wastes that would have to be landfilled.

Technology integration can be generally classified into methods involving

> Separation of fractions for volume reduction or to apply different downstream treatments
> Increasing the availability of contaminants for treatment by mobilizing of contaminants in the medium to be treated; modifying the chemical or physical properties of contaminants; or employing treatment trains for sequential removal/treatment of different types of contaminants

In practice, a combination of these options may be employed in an integrated treatment system to deal with particularly complex contamination.

Separation of different fractions may be carried out to reduce the volume to be treated using a more expensive technology or to be treated at all, or to separate out fractions that need to be treated differently. Separation is usually achieved through dry physical separation (e.g., crushing or sieving) or wet physical separation (e.g., soil washing or other wet mechanical separation processes or flotation with or without chemical pretreatment). Besides separating out the more contaminated fine-grained concentrate, creating a clean coarse fraction may be intended when applying wet mechanical separation processes. One group of integration options is based on the principle of washing off the contaminants and the (highly) contaminated fine particles from the surfaces of the coarser particles, leaving the coarse fraction relatively clean. A second group of options uses physical techniques to separate fractions by exploiting in the physical properties (e.g., specific gravity or surface hydrophobicity).

Mobilization of contaminants may be achieved by altering the medium to be treated. Examples include:

> In-situ methods, such as through fracturing, steam injection, air sparging, or soil flushing with agents
> Ex-situ methods, such as crushing clay clumps

Mobilization may also be achieved by modifying the contaminants to increase availability to microbial degradation by concentrating or pretreating the contaminants (e.g., partial oxidation of organics). Also, treating combinations of different contaminants (e.g., organics and metals) usually will require application of different processes in sequence.

IV. CONCLUSIONS

The major Pilot Study conclusions were:

> Involvement of more countries led to better and wider awareness of the problems posed by contaminated land. A total of 23 countries were involved in the Pilot Study, and 14 contributed projects. The increased number of countries participating in Phase II undoubtedly increased the overall value and impact

of the technology exchange process. Although more countries took part largely as recipients of technology information rather than as contributors, the representatives of these countries brought new insights and priorities to the Pilot Study and were able to make valuable contributions to the discussions.

Presentation of additional full-scale experiences was helpful. A number of the participants found the case studies involving full-scale remediation, as opposed to demonstration-, pilot-, or bench-scale studies, to be particularly valuable.

In a number of countries, remediation strategies are moving from technology-intensive treatment processes to greater recognition of land use management and extensive approaches, such as natural attenuation. Further research into these approaches is needed. It was apparent during the Pilot Study that there is increasing interest in land use management and extensive approaches to remediation. While these approaches can be as effective as more intensive methods, they may take longer to complete. Extensive approaches demand fewer resources and are less costly; thus, they can be viewed as more sustainable. Extensive remediation options, which are less dependent on technology and energy inputs, etc., and are likely to have less impact on other aspects of economic activity, are needed. Cost considerations are particularly appropriate for less developed countries having pollution problems that would be prohibitively expensive to treat using technology-intensive processes. Consideration of the overall potential environmental, social, and economic impacts of planned remedial actions are of increasing interest and the subject of formal study in some countries. It should be noted that taking these issues into account can both increase the level of clean-up required (e.g., in situations where fit for current or immediate future use is a prime criterion) and lessen the level of clean-up required (e.g., in a situation where multifunctional land use might otherwise be required or where groundwater has no economic value).

The intended future use of a site is increasingly a determining factor when setting clean-up objectives and selecting a remediation strategy. This conclusion reflects the convergence in thinking between those countries that have always seen land use as an important factor and those that have tended to set clean-up requirements regardless of the future use of the land. Consideration of future land use can lead to a better allocation of scarce resources.

All remediation activities require proper operation and management. The success of field demonstration and treatability studies is not enough to ensure success of the remediation activities. The overall effectiveness of a remediation scheme, which may include many interrelated elements of civil engineering works and soil and groundwater clean-up technologies, will depend heavily on the care with which the individual technologies are operated in the field. Site and operating conditions may change over time, and skilled people are needed to adjust technologies to these changes or discontinue them if they do not meet expectations. Similarly, a strong quality assurance program needs to be in place, and activities must be carried out by a dedicated and effective management team.

Whenever possible, the wider environmental impacts of a chosen remedial strategy should be considered during remedial selection. Short-term performance goals should not be the sole factor in technology selection when developing a remediation strategy. Some remediation strategies may only be effective over a longer time frame, but may have lower environmental impacts during implementation (e.g., reduced traffic, lower emissions, and lower energy requirements).

ACKNOWLEDGMENTS

The authors would like to thank the many Pilot Study participants and authors that have contributed over the past 18 years:

Initial Study (1980–1984). Author: Michael A. Smith (UK).

Phase I Study (1986–1991). Authors: Thomas Dahl (US), Merten Hinsenveld (Netherlands), Stephen James (US), Norma Lewis (US), Donald Sanning (US), James Schmidt (Canada), Sjef Staps (Netherlands), and Robert Olfenbuttel (US).

Phase II Study (1992–1997). Authors: Michael A. Smith (UK), Mark Smith (US), Cathy Vogel (US), Alison Thomas (US), Alex Lye (Canada), Robert Booth (Canada), Paul Bardos (UK), Diane Dopkin (US), Kai Steffens (Germany), Walter Kovalick, Jr. (US), Diane Dopkin (US), and Stephen James (US).

Annual Reports: John Moerlins (US), Jay Bassin (US), and Diane Dopkin (US).

Participants: Per Antonsen (Norway), Resat Apak (Turkey), Stephen Aston (UK), Nora Auer (Austria), Erik Backland (Sweden), Paul Bardos (UK), Jay Bassin (US), Paul Beam (US), Robert Bell (UK), James Berg (US), Mats Bergstrom (Sweden), Bjorn Bjornstad (Norway), Robert Booth (Canada), Jan Bovendeur (Netherlands), Harald Burmeier (Germany), Teresa Chambino (Portugal), David Cooper (Canada), Thomas Dahl (US), Judith Denner (UK), Ludek Domaci (Czech Republic), Diane Dopkin (US), Branko Druzina (Slovenia), Louise Emmett (Australia), Erol Ercag (Turkey), Marco Estrala (Portugal), Volker Franzius (Germany), Rene Goubier (France), Domenic Grasso (US), Bernard Hammer (Switzerland), Mary Harris (UK), Ingrid Hasselsten (Sweden), George Hill (Canada), Merten Hinsenveld (Netherlands), Ian Hosking (Australia), Semund Haukland (Norway), Stephen James (US), Walter Kovalick, Jr. (US), Edda Kasamas (Austria), Harald Kasamas (Austria), Lisa Keller (Canada), John Kingscott (US), Antonios Kontopoulos (Greece), Michael Kosakowski (US), Ian Lambert (Australia), Tomas Lederer (Czech Republic), Eric Lightner (US), Edward Marchand (US), Ewa Marchwinska (Poland), Ian Martin (UK), Jacqueline Miller (Belgium), Jan Mrkos (Czech Republic), Corneliu Negulescu (Romania), Robert Olfenbuttel (US), Anna Orlova (Russian Federation), Michael Pearl (UK), Carlos de Miguel Perales (Spain), Robert Reiniger (Hungary), Peter Richter (Hungary), Raymond Salter (New Zealand), Ari Seppanen (Finland), Robert Siegrist (US), Inge-Marie Skovgard (Denmark), Michael Smith, Esther Soczo (Netherlands), (UK), Marek Stanzel (Czech Republic), Kai Steffens (Germany), Hans-Joachim Stietzel (Germany), Neel Strobaek (Netherlands), Jan Svoma (Czech Republic), Alison Thomas (US), Robert Thomas (Australia), Kahraman Unlu (Turkey), Johan van Veen (Netherlands), Pal Varga (Hungary), John Vijgen (Netherlands), Wilma Visser (Netherlands), Cathy Vogel (US), Harry Whittaker (Canada), and Urs Ziegler.

Special recognition goes to John Moerlins from Florida State University for organizing the meetings and producing many of the annual reports. Without his assistance, the Pilot Study would certainly not have made the accomplishments to date.

Another special recognition goes to Dr. Deniz Beten and Mrs. Martine Deweer from the NATO/CCMS Program in Brussels, Belgium. Without their support, the Pilot Study would not been able to expand and include emerging countries where remedial technologies are needed.

REFERENCES

1. U.S. Environmental Protection Agency. 1998. NATO/CCMS Pilot Study: Evaluation of demonstrated and emerging technologies for treatment and clean up of contaminated land and groundwater, Phase II final report, EPA/542/R-98/001a.
2. U.S. Environmental Protection Agency. 1998. NATO/CCMS Pilot Study: Evaluation of demonstrated and emerging technologies for treatment and clean up of contaminated land and groundwater, Phase II overview report, EPA/542/R-98/001b.
3. U.S. Environmental Protection Agency. 1998. NATO/CCMS Pilot Study: Evaluation of demonstrated and emerging technologies for treatment and clean up of contaminated land and groundwater, Phase II project summaries, EPA/542/R-98/001c.
4. U.S. Environmental Protection Agency. 1995. NATO/CCMS Pilot Study: Evaluation of demonstrated and emerging technologies for treatment and clean-up of contaminated land and groundwater (Phase II): Interim status report, EPA/542/R-95/006.
5. U.S. Environmental Protection Agency. 1993. NATO/CCMS Pilot Study: Demonstration of remedial action technologies for contaminated land and groundwater (Phase I), final report, volume 1, EPA/600/R-93/012a.
6. U.S. Environmental Protection Agency. 1993. NATO/CCMS Pilot Study: Demonstration of remedial action technologies for contaminated land and groundwater (Phase I), final report, volume 2, part 1, EPA/600/R-93/012b.
7. U.S. Environmental Protection Agency. 1993. NATO/CCMS Pilot Study: Demonstration of remedial action technologies for contaminated land and groundwater (Phase I), final report, volume 2, part 2, EPA/600/R-93/012c.
8. Smith, M. A. (ed.). 1985. *Contaminated Land: Reclamation and Treatment*. Plenum Publishers, London, UK.
9. U.S. Environmental Protection Agency. 1998. NATO/CCMS Pilot Study: Evaluation of demonstrated and emerging technologies for treatment and clean up of contaminated land and groundwater, Phase III 1998 annual report, EPA/542/R-98/002.
10. U.S. Environmental Protection Agency. 1998. NATO/CCMS Pilot Study: Evaluation of demonstrated and emerging technologies for treatment and clean up of contaminated land and groundwater, Phase III special session report—Treatment walls and permeable reactive barriers, EPA/542/R-98/003.
11. Martin, I., and Bardos, R. P. 1996. A review of full scale treatment technologies for the remediation of contaminated soil. (Final report for Royal Commission on Environmental Pollution, October 1995.) EPP Publications, Richmond, Surrey, UK.

Note: Publications pertaining to the Pilot Study are available from the following Internet sites: www.nato.int/ccms and www.clu-in.com

2
Methods of Analysis of Contaminant Migration in Barrier Materials

Hilary I. Inyang and John L. Daniels
University of Massachusetts, Lowell, Massachusetts

Calvin C. Chien
DuPont Corporate Remediation, Wilmington, Delaware

I. INTRODUCTION

Several techniques that are complementary can be used to assess the extent of contaminant transport in barrier systems. Due to the long time interval it takes for a contaminant to travel through even small distances in low ($< 10^{-5}$ cm/s)-permeability materials, several numerical formulations, most of which are based on solutions for the one-dimensional advection-dispersion equation, have been developed for use in scaling contaminant concentrations in time and space within a barrier. Examples are the use of solutions provided by (1–4).

The most common laboratory-based experimental methods as batch and column tests in which the barrier materials and targeted contaminants are often used as described below. Field tests usually involve drilling through emplaced barriers to extract samples for analyses or direct field testing to assess barrier integrity. The latter approach may also fit within a monitoring scheme for an in-situ containment system. These techniques are discussed and illustrated as follows.

II. NUMERICAL APPROACHES

Contaminant retardation in barrier materials is often described by the one-dimensional advection-dispersion equation. It may be expressed in the following partial differential form that accounts for contaminant retardation and decay:

$$\frac{\partial C}{\partial t} = D_1 \cdot \frac{\partial^2 C}{\partial L^2} - V \cdot \frac{\partial C}{\partial L} + \left(\frac{\rho_b}{n}\right) \cdot \frac{\partial S}{\partial t} - \lambda C \tag{1}$$

where

C = solute concentration (ML^{-3})

t = elapsed time (T)
D_l = coefficient of hydrodynamic dispersion (L^2T^{-1})
L = curvilinear length in the direction of flow (L)
V = average linear groundwater velocity (LT^{-1})
ρ_b = bulk density of the medium (ML^{-3})
n = porosity of the medium (dimensionless)
S = mass of chemical retained on solid phase (M)
λ = decay constant (T^{-1})

Boundary conditions and integral approximations are applied to solve the differential equation. Considering a soil column of length x and analyzing the concentration as a function of time provides two situations. At zero time, and any distance along the length of the specimen, the concentration C is assumed to equal zero. At zero length and any time, the concentration C equals the initial concentration C_0. Stated another way:

$$C(x, 0) = 0 \qquad (2)$$

$$C(0, t) = C_0 \qquad (3)$$

The following equation can then be derived, presented in a shortened form (5):

$$C_{x,t} = 0.5 \cdot C_0 \cdot \text{erfc}\left[\frac{R \cdot x - V \cdot t}{2 \cdot (D_1 \cdot t \cdot R)^{1/2}}\right] \qquad (4)$$

where

$C_{x,t}$ = aqueous concentration at position x and time t (M/L^3)
C_0 = initial aqueous concentration (M/L^3)
R = retardation of solute (dimensionless)
x = position along flow path (L)

The complementary error function, erfc, has been implemented to approximate the integration of Eq. (1) (3). Its properties are such that $\text{erfc}(x) = 1 - \text{erf}(x)$, $\text{erf}(-x) = -\text{erf}(x)$, $\text{erf}(0) = 0$, and $\text{erf}(\infty) = 1$. There are extensive tables revealing the value of the error function and its complement (6). The term R represents retardation. Physically, R represents the ratio of seepage velocity to the rate of contaminant migration. It can be derived using a linear sorption isotherm and mass balance approach to yield (7)

$$R = \frac{V}{V_c} = 1 + \left(\frac{\rho_b}{n}\right)K_d \qquad (5)$$

where

V_e = solute velocity (L/T)
K_d = distribution of contaminant between solid and liquid phase (L^3/M)

The determination of K_d involves batch testing and is discussed in conjunction with experimental analyses in the next section. Often, contaminants are removed by the material through which contaminated water flows, and R is therefore greater than 1. Other chemicals (e.g., anionic constituents) are repulsed by the surrounding media and this renders R less than unity.

Analysis of contaminant retardation by barrier materials requires the modification of the general equation. In the field, barrier materials often experience low hydraulic gradients. Therefore, the advection component of the equation is of less relevance with

Analysis of Contaminant Migration

respect to the molecular diffusion that is occuring. To illustrate the relative contribution of advective and diffusive transport, the Peclet number is analyzed (8). It is given by

$$P_L = \frac{V \cdot d}{D_1} \tag{6}$$

where

P_L = Peclet number (dimensionless)
d = average particle diameter or other characteristic length (L)

Peclet numbers exceeding 50 indicate an advection-dominated system, whereas values less than 1 imply diffusion. Barrier materials maintain low Peclet numbers.

Many other solutions to the advection-dispersion equation exist. Another form, presented in (3) and applied to clay barriers where advection in negligible (9), is given by

$$C = C_0 \cdot \left\{ 1 - \frac{z}{L} - \frac{2}{\pi} \sum_{n=1}^{\infty} \frac{1}{n} \sin\left(\frac{n\pi z}{L}\right) \exp\left[\frac{-(D_1/R)n^2\pi^2 t}{L^2}\right] \right\} \tag{7}$$

where

z = position along specimen (L)
n = expansion series variable (dimensionless)

In addition to the conditions applied in Eq. (4), another boundary condition was used to derive this solution. The concentration at the effluent must be flushed to zero. In other terms,

$$C(L, t) = 0 \tag{8}$$

The term L refers to the total length of the specimen, while z refers to the location at which the concentration is being determined. Using the same boundary conditions as Eq. (7), the following solution has been derived and applied to containment systems (4):

$$C = C_0 \cdot \left[\exp\left(\frac{xV}{2D_1}\right) \left(\frac{\sinh\{[(L-x)V]/2D_1\}}{\sinh(LV/2D_1)}\right) + \frac{2\pi}{L^2} \exp\left(\frac{xV}{2D_1}\right) \sum_{m=1}^{\infty} \frac{(-1)^m m \sin(m\pi x/L)}{(V^2/4D_1^2) + (m\pi^2/L^2)} \right.$$
$$\left. \cdot \exp\left[-\left(\frac{V^2}{4D_1} + \frac{D_1 m\pi^2}{L^2}\right)\left(\frac{t}{R}\right)\right] \right] \tag{9}$$

where

x = position along specimen (L)
m = expansion series variable (dimensionless)

Care must be exercised, however, in applying the correct solution to the relevant situation. For example, many barrier materials are tested in laboratories at hydraulic gradients in excess of what is experienced in the field. Additionally, many solutions assume isotropic, saturated conditions. Additionally, the use of a distribution coefficient implies that local equilibrium applies, although this is not always the case. Under these circumstances, effort must be placed into establishing the kinetics of the targeted contaminants.

With the continued improvement of computing power and associated software, a number of robust programs have been developed for the modeling of contaminant transport through barriers. By interfacing the fundamental equations described above with graphical interfaces, two- and three-dimensional views of contaminant plumes can easily be obtained. Examples of such software include MIGRATE and POLLUTE from GAEA Software Engineering Group, and Visual MODFLOW from Waterlow Hydrogeologic.

III. EXPERIMENTAL METHODS

Measurements of transport-related parameters have been made using batch techniques, cation exchange, and column flow through experiments. These tests produce useful results but they differ in the degree to which they simulate sorption conditions in real barriers. Batch-test results may be considered as upper-bound estimates of contaminant sorption because in real systems the degree of access of contaminant-bearing fluids to adsorption sites may not be as high. Column permeation tests incorporate transport phenomena which are relevant to real systems, but at rates that may be too high. Furthermore, boundary conditions and the limited size of samples in column tests may influence collected data undesirably. Some investigators have compared batch and column test results for contaminant sorption on the same soils under similar pH conditions. In studies of the adsorption of potassium by three Delaware soils, it was found that sorption equilibrium was reached much more quickly in batch tests than in column tests (10). This result was attributed to the greater ease with which soil particles can be disaggregated to increase available surface area for sorption in batch tests.

A. Batch Adsorption Test Protocols

In physical experiments, sorption isotherms determined represent the totality of phenomena that may collectively result in the establishment of thermodynamic equilibrium in the solid/fluid mixture. Essentially, measurements are conducted using the following test protocol to calculate the distribution coefficient:

$$K_d = \frac{C_s}{C_a} \tag{10}$$

where

C_s = concentration of solute sorbed to solid phase (MM^{-1})
C_a = concentration of solute in aqueous phase (ML^{-1})

Detailed information on the test protocols is provided in (11). The laboratory ware needed for this test include plastic bottles, centrifuge tubes, open dishes, pipets, and graduated cylinders. The following equipment is needed:

Filtering apparatus or centrifuge capable of attaining 1400g acceleration level
Laboratory shaker/rotator
Environmental monitoring instruments: pH meter, electrodes for Eh measurement, and thermometer
Balance for mass determination
Analytical instrumentation for contaminant concentration measurements in the fluid phase

First, the sample of the soil and/or additive to be used in the batch test is characterized. Characterization may involve measurement of the particle size distribution and/or specific surface area. For particles within the size range common in soils, fine sieve

Analysis of Contaminant Migration

sizes may be used. Particles in the micrometer range may require the use of a particle size or surface area analyzer.

A solution containing the initial concentration, C_0, of the solute is prepared, and a known volume, V_0, of the solution is transferred into a container. A known mass of the sorbent, M_s, is added to the solution. The mixture is shaken and allowed to equilibrate. The path to equilibration is tracked through measurements of the changed in the liquid-phase concentration of the solute at specific time intervals. Environmental conditions (e.g., pH, Eh) are also monitored.

The sorbent is separated from the solution by centrifuging the mixture. An aliquot of the supernatant is sampled for measurement of the final "equilibrium" concentration, C_f, of the solute in the liquid phase. From these measurements, the concentration of the solute on the adsorbent can be computed:

$$S = \frac{V_0(C_0 - C_f)}{M_s} \tag{11}$$

where

S = amount of solute adsorbed by the soil matrix (MM^{-1})
V_0 = volume of the original solution (L^3)
C_0 = solute concentration in the original solution (ML^{-3})
C_f = final concentration of the solute in the equilibrated liquid (ML^{-3})
M_s = mass of the adsorbent (M)

This test is repeated at a number of initial concentration C_0, and a plot of S (vertical axis) and C_f (horizontal axis) is developed. The slope of this plot, as indicated in Eq. (10), is the value of K_d at that solution condition. The chemistry of the initial solution can be altered such that K_d values can be obtained for a range of solution conditions.

Potential problems have been pointed out in (12) and (13) with the use of batch sorption tests to determine K_d:

Vigorous agitation can result in sorbent particle breakdown, such that higher reaction rates are observed.
Desorption of some chemical species may influence the course of reactions.
Some particles may not settle before the separation of the supernatant and could raise the measured concentration of solutes in the fluid phase.
Equilibrium may not be attained for some reactions before the separation of adsorbent from the supernatant.

It has also been shown that ineffective separation of colloidal particles from the supernatant prior to concentration measurements may result in high values of C_f, thereby causing underestimation of K_d (14).

B. Cation-Exchange Test Protocols

In cation-exchange tests, the objective is to measure the capacity of the solid material to give up its exchangeable cations for others present in solution. Cation exchange is an important sorption mechanism, especially for alkali metals and alkaline earth elements. However, it does not accurately explain the interaction of several transition metals with minerals (13). The latter is better explained in terms of the adsorption processes. The negatively charged surfaces of clay minerals attract cations but still maintain a constant total charge in the case of ion exchangers. The total cation-exchange capacity (CEC) of clay minerals has been subdivided into two components: a permanent charge CEC that results from isomorphous substitution within clay mineral structures; and a pH-dependent CEC (15). In the CEC test, the amount of cations that can neutralize the charge on the surface of a sorbent is measured in units of milliequivalent per 100 g of the sorbent. Two sets of equipment are needed for this test:

An apparatus for saturating the sample
Analytical equipment for NH_4^+

The saturation equipment comprises a mechanical extractor, fluid extraction syringes, 1/8–1/4-in.-diameter rubber tubing for connecting syringe barrels; analytical filter pulp, polyethylene bottles (about 25 mL), reciprocating shaker, tubes, and a centrifuge. If the ammonium displacement–flow injection analysis is chosen, a flow injection analyzer (FIA) is required. The following reagents are required in the saturation procedure: acetic acid (CH_3COOH), concentrated ammonium hydroxide (NH_4OH), ammonium acetate ($NH_4C_2H_3O_2$ or NH_4OAc), ammonium chloride (NH_4Cl), 95% ethanol (CH_3CH_2OH), and Nessler's reagent.

When ammonium acetate is used as the saturating solution at pH 7, a theoretical estimate of the maximum (total) CEC is obtained. In acid soils (such as acid sulfate soils that are commonly used in clayey barriers in the eastern United States; see (16), the adsorption of NH_4^+ onto otherwise neutral sites may result in overestimation of CEC. The use of 1.0 N NH_4Cl can buffer the soil, thereby reducing this effect (15). When NH_4Cl is used instead of NH_4OAc, the CEC determined is termed the effective CEC. In the general protocol, the soil or additive is saturated with the NH_4^+-containing solution. NH_4^+ then occupies sorption sites on the sorbent. Ethanol is used to rinse off excess NH_4^+, and the content of NH_4^+ in the adsorbent is determined through titration or ammonium displacement-flow injection methods. Minerals such as biotite, vermiculite, and muscovite contain K^+ and NH_4^+ in their interlayers that are not readily exchangeable. More details on these protocols, including the role of micaceous clay minerals in CEC determinations, is provided in (15).

C. Column Test Protocols

It has been noted that the number of published column flow experiments that address sorption kinetics is limited (12). Flow techniques for sorption measurements have the following advantages over batch methods: the contaminant transport situation in the subsurface is simulated more appropriately; the need to separate supernatant from sorbent is nonexistent; information on the permeability of the soil medium can be obtained: flow rates through the soil can be adjusted; and it is not necessary to conduct tests at several influent concentrations. However, column systems have some disadvantages too, the primary ones being that experiments may take a long time to complete, especially when the materials in the column are highly compacted; and the degree of dispersion of the particles of materials in the column can affect the sorption parameters measured. These factors have been discussed in greater detail elsewhere (17,18). Some additional factors that pertain to column flow-through methods are briefly described below.

1. Hydraulic Gradient and Flow Rate Effect Sorption

In column tests, the applied gradient influences the flow rate of the permeant and may influence the interactions of contaminants with the sorbent. In general, the higher the hydraulic gradient, the higher is the flow rate. Test results are somewhat conflicting on the effects of permeant flow rates on contaminant sorption. Flow rate experiments were conducted in columns of three natural soils, revealing that Zn, Cu, and Cr were sorbed in greater quantities by the soils at higher flow rates (19). Other tests, (20), conducted on Cd adsorption on a volcanic soil from Pope Ridge, near Ardenvoir, Washington, shows similar results. The sorption rate increased from 0.07×10^{-3} s^{-1} at static conditions to 0.4×10^{-3} s^{-1} at a permeant flow rate of 2.7 cm/h. In their study of nine subsoils that represent seven of the 10 major soil orders collected from several locations in the United States, (21) found that the attenuations of Al, Be, Cr(VI), and Fe(II) contained in solid waste leachate were significantly affected by leachate flow rates through soils and that flow rate did not influence the attenuation of Cd, Ni, and Zn. In contrast to the findings reported in (19), (21) also found that slower flow rates favored the attenuation of Al, Be, Cr(VI), and Fe(II).

2. Hydraulic Gradient and Flow Rate Effect on Soil Texture

The retardation coefficient is a function of both K_d (a sorption parameter) and textural parameters: porosity, bulk density, and their surrogates. Due to the relatively low effective porosities of slurry backfill mixes, it is somewhat desirable to use higher hydraulic gradients to lessen the permeation time of samples. When the retardation coefficient exceeds 10, column tests may take too much time to be practical (8,13).

Unfortunately, the use of high hydraulic gradients may change the texture, and, hence, the transport-related properties of the barrier. This can lead to imprecision in the measured values of R, the retardation coefficient. It is noted that in the case of slurry walls, minimal hydraulic gradients are usually realized in the field (generally, between 1 and 5, sometimes up to 10), while the gradients used in laboratory tests generally range from 20 to 50 (22). An evaluation of the permeability of samples of silty clay soil from central New Jersey at a confining pressure of 55 psi (379 kPa), driving pressure of 40 psi (275 kPa), back pressure of 30 psi (207 kPa), and hydraulic gradient of 100, using a flexible wall permeameter (23), found that fines migrated from the upper part of samples downward. This displacement and accumulation of fines at the bottom can result in a decrease in permeability (24–26). The exit of fines through the bottom of the sample, which is more representative of field conditions, would result in a permeability increase. Excessively high gradients can also cause seepage-induced consolidation (27), and soil dispersion due to permeant chemistry (28).

3. Column Testing Details

a. Constant Versus Falling-Head Permeameter. Constant-head permeameters are not commonly used in fine-grained soil permeability measurements. The relative imperviousness of barrier materials implies that permeation times would be excessive. The advantage of the falling-head setup is that small flows can be measured easily (24). Furthermore, flow can be enhanced by the use of air pressure to increase the head on the tested sample. It was found that the constant-head permeability test values had a tendency to be slightly higher than falling-head values for the same clayey soil (29).

b. Type of Permeameter Chamber. The columns used in contaminant permeation/hydraulic conductivity tests fall into four categories: fabricated (custom-made) columns; compaction molds, consolidation cells, and flexible-wall permeameters. A comprehensive analysis of the relative advantages and disadvantages of the aforementioned permeameters has been given in (30). The compaction mold permeameter is simple and economical for practical applications but may not be precise enough for research-type applications: it is difficult to saturate the sample completely; stress cannot be controlled; and the permeant can migrate through channels at the specimen/sidewall interface. No vertical stress is applied to compaction mold samples. This may apply more to cover material than for liners or slurry walls at depth.

The small sample size (diameter 5–8 cm and height 13–25 mm) used in consolidation cell permeameters allows faster permeation rates. Vertical stress can be applied to simulate overburden stress. It is relatively simple to trim the test sample. Testing proceeds by loading the sample to the selected stress level, and hydraulically connecting the sample at the bottom with the permeant. It is implied that it may still be difficult to ensure sample saturation (30).

Specially fabricated columns have been used in permeation experiments by several investigators, exemplified by (12). Most of these columns are relatively simple in configuration and do not generally allow the application of stresses. This is very common in investigations that are focused primarily on contaminant retardation rather than on the hydraulic conductivity of the tested sample. These columns operate essentially as rigid-wall permeameters.

In flexible-wall permeameters, the specimen is bounded on the sides by a membrane to which pressure is applied. This arrangement makes sidewall leakage negligible. The application of backpressure usually saturates the test specimen such that gases can be eliminated. Sample deformation can be tracked and stresses controlled during testing. There is usually a concern with chemical attack of the flexible membrane. Concentrated organic chemicals pose a greater threat to flexible membranes than inorganic constituents. Figure 1 shows a schematic of a flexible-wall permeameter interfaced with a data acquisition system. As shown in Fig. 2, laboratory data generated from tests on soil-bentonite indicate that the rigid- and flexible-wall permeameters produced reasonably similar data (31). Hydraulic conductivity tests were conducted on highly

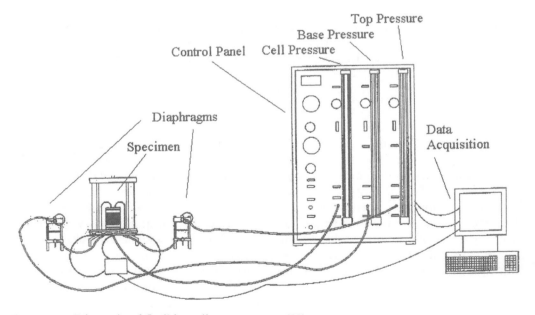

Figure 1 Schematic of flexible wall permeameter (31).

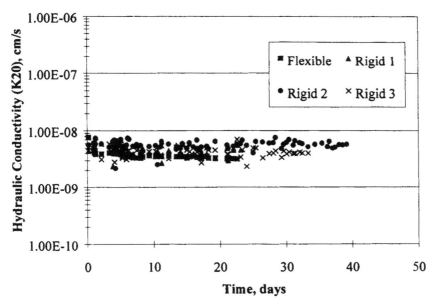

Figure 2 Hydraulic conductivity of a soil–bentonite mixture in both rigid and flexible wall configurations (31).

plastic clay from Houston, Texas, using both flexible-wall and fixed-wall permeameters (32). For a maximum clod size of 19 mm, molding water content of 12%, and dry unit weight of approximately 14.5 kN/m^3, the hydraulic conductivities measured with the flexible- and fixed-wall permeameters were 8.5×10^{-4} cm/s and 4.1×10^{-4} cm/s, respectively.

The flexible-wall permeability test is the most amenable to the control of test parameters. It is likely to produce the most precise results when slurry backfill or linear material is tested. The barrier material is carefully placed into a flexible membrane that is sealed at the base with O-rings. A top platen is placed on top of the specimen and the upper portion of the membrane is sealed around it with more O-rings.

Specimen dimensions are measured and the cell is assembled. The chamber is connected to the control panel and de-aired tap water or other appropriate solution is filled into the cell and pore-water lines. Backpressure is then applied to saturate the specimen. The specimen is consolidated at the appropriate pressure to simulate its likely condition in the field, and flow is started with entry at the bottom at the selected hydraulic gradient. Inflow and outflow volumes through the sample are monitored until both are equilibrated for 5 consecutive days. Furthermore, a minimum of one pore volume of water should be allowed to flow through the specimen prior to the introduction of contaminated permeant. Inflow and outflow of the new permeant is monitored again until about two pore volumes exit the specimen. The effluent from the specimen is monitored periodically so that breakthrough calculations can be made. The specimen is then removed for measurement of final weight, moisture content, and other characteristics.

Another, separate category of column tests involving chemical migration is diffusion measurement. In the initial stages of barrier service, diffusion is most likely the most relevant transport mechanism for contaminants because the rates of transport through materials with low effective porosity are higher for diffusion than for advection. A number of investigators (33–35) have discussed diffusion measurement techniques and protocols.

IV. FIELD ASSESSMENT OF CONTAMINANT TRANSPORT IN BARRIERS

In the field, contaminants are unlikely to migrate in a straight, uniform front, due to the heterogeneity of the transport media. For emplaced barrier materials, this heterogeneity can stem from poor construction quality, natural variability due to mix design, and environmental stresses which degrade the material nonuniformly. Therefore, after the service time period of interest, it may be necessary to assess contaminant transport within the barrier in order to evaluate potential in-situ release rates of contaminants from the containment system. Approaches to such an in-situ assessment fall into two major categories, sample retrieval for testing and in-situ monitoring of contaminants using sensors and other techniques.

A. Retrieval and Testing of Samples

The challenge in retrieving and testing samples is to do so without creating flaws that could serve as conduits for contaminant transport. Nevertheless, this approach offers the potential for analyses of barrier samples to determine contaminant concentrations in portions that represent specific locations within the barrier matrix. An example of such a scheme is illustrated in Fig. 3. Figure 3 represents the sampling scheme for sample retrieval from a slurry wall and surrounding soil.

Contact with a slurry wall can be made underground if a hole is drilled at an angle toward the wall. Delineation of the wall location is based on the assumption that samples of the slurry wall would exhibit different physical and chemical properties from those of the surrounding soil medium. The slurry wall mix (backfill) is designed to attain a low permeability and may be contaminated to various levels along its width, depending on the rate of contaminant transport from the source. Bentonite is often added to clayey soils to form the backfill. With the addition of bentonite, the backfill textured is altered, making the wall discernable from surrounding soils. This textural difference is observable through fines content and mineralogy. The increase in the soil fines results in changes in soil parameters which are discernable through conventional geotechnical testing (e.g., grain size distribution and Atterberg limits testing). Soil mineralogy and contamination level can then be identified through advanced methods such as X-ray diffraction and X-ray fluorescence as described by (36).

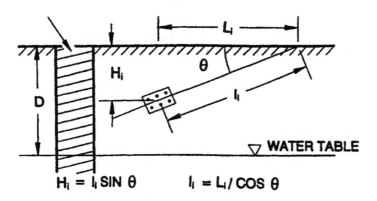

Figure 3 Potential orientation of drill holes for retrieving subsurface samples.

The horizontal distance between the drill entry point and the approximate external boundary of the slurry wall as indicated by a geophysical technique such as ground penetrating radar (GPR) is selected such that the drill hole does not penetrate the water table. The drill angle for both in the example is 45°, measured from the horizontal as illustrated in Fig. 3. Drilling can proceed toward the slurry wall from locations outside the containment system. Samples can be retrieved at various points along the inclined borehole (known as I_i). It is possible to establish the corresponding horizontal distance toward the wall through the use of simple trigonometric relationships as shown in Fig. 3.

B. In-Situ Monitoring of Contaminants

Traditionally, monitoring wells have been used to assess the effectiveness of barrier systems. Through periodic sampling, contaminant concentrations in well water from wells just outside the barrier system provide information that is modestly useful for indirect estimation of the transport rates of contaminants through the containment system. Such an approach is useful for the evaluation of the functional effectiveness of the barrier system but is incapable of yielding meaningful data on contaminant concentrations at various locations within the barrier. Another problem with monitoring wells is the uncertainties associated with estimates of the water-table elevation when a barrier is bounded on both sides by the wells. As illustrated in Fig. 4, depending on hydrogeological conditions, any of the orientations labeled 1, 2, or 3 could plausibly be regarded as the correct water-table elevation. This uncertainty could influence the precision of estimates of contaminant transport rates across the barrier.

C. Geophysical Monitoring Techniques

The movement of moisture and contaminants can be tracked within and outside barriers to assess the effectiveness of a containment system. The direct objective can be one or a combination of the following:

> Successive measurements at a point to determine the arrival time of moisture and contaminants at the particular point following the construction of a barrier system
> Successive measurements at various points to estimate the transport of moisture and contaminants through barriers and the surrounding geomedia
> Direct determination of the boundaries of a contaminated area around a waste containment facility
> Direct measurements of contaminant flow through velocities at a point
> Measurement of moisture content or contaminant concentrations at a point

Geophysical methods represent an effective means of assessing barrier performance in the field, although they differ in terms of their cost effectiveness when they are used to acquire data on the parameters mentioned above. Generally, they are largely unsuitable for point measurements but are of high utility when the objective is to get information on an area or volume of geomedia. Geophysical methods are based on the existence of physical and chemical property contrasts between different portions of the ground. The sensing instrument does not need to be in contact with all the locations tested. Thus, in a sense, all geophysical methods can be categorized as remote sensing methods.

It is not technically desirable or cost-effective to use geophysical techniques to sense moisture and contaminant existence within barrier components such as covers, liners,

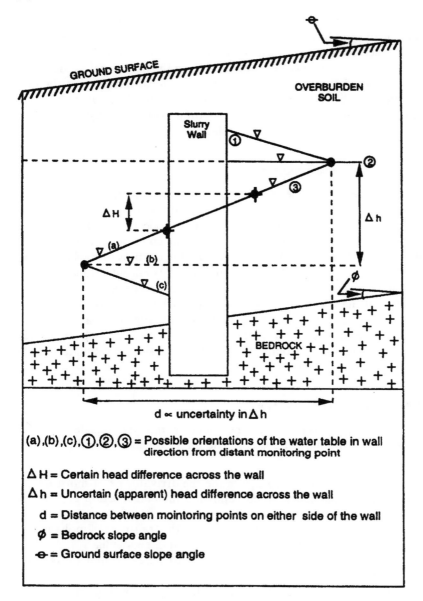

Figure 4 Schematic illustration of the effects of monitoring point displacement on the uncertainty of head loss estimate across a slurry wall.

slurry walls, cemented barriers, and drainage layers. Usually, the direction of concern with regard to the migration of liquid substances in barriers is downward. This implies that even workable geophysical methods would have to be employed in the downhole mode, a system that cannot be effectively implemented within or across barrier layers. Otherwise, there is the problem of inadequate resolution. Each barrier layer seldom exceeds a thickness of 1.5 m. Geophysical methods do not have an adequately low-resolution regime to allow the delineation of moisture and contaminant distribution zones within such a thickness.

Geophysical methods are useful for estimating the sizes and locations of contaminated zones around a waste containment structure. Measurements can be done successfully over a given time interval in order to determine the rate of increase in the size of the contaminated zone. This approach also provides a basis for assessing the effectiveness of the waste containment system. Within this context, the most promising geophysical techniques are ground-penetrating radar and electromagnetic resistivity methods.

1. Ground-Penetrating Radar

Ground-penetrating radar (GPR) uses electromagnetic waves to penetrate the ground and delineate differences in pore-water content and quality, soil texture, and soil density. Usually, wave frequencies the range from 80 to 1000 MHz (broad band) are used. Upon contact with materials of different properties, a fraction of the wave energy is reflected back to an antenna located on the ground surface and the remaining fraction travels deeper into the ground. As illustrated in Fig. 5, reflected electromagnetic pulses are received by recorders and converted to plots of amplitude versus wave travel time. Generally, the electromagnetic wave velocity can be computed as follows:

$$V = \frac{C}{D^{0.5}} \quad (12)$$

where

V = electromagnetic wave velocity (LT^{-1})
C = velocity of light in a vacuum (LT^{-1})
d = relative dielectric permittivity of the transport medium (dimensionless)

$$V = \lambda f \quad (13)$$

where

λ = selected wavelength (L)
f = selected frequency (T^{-1})

The units for Eqs. (12) and (13) should be selected to match the depth and resolution of interest. Soils with high conductivity of electromagnetic waves dissipate radar energy quickly, so depth penetration is limited. This is the case with clays and other fine-grained saturated soils. In general, penetration depths of up to 10 m are common. However, 5–10% (by weight) of montmorillonitic clay can reduce the penetration depth to about 1 m (37). Conductivity is directly proportional to the concentration of dissolved solids in the soil pore solution. In field applications, the receiving antenna can be towed around a contaminated site at speeds up to 8 km/h. In Port Washington, New York, GPR was used at the Roslyn/Beacon Hill Landfill site to optimize the selection of monitoring well locations (38).

2. Electromagnetic (EM) Resistivity

The electromagnetic (EM) resistivity method is also based on the electrical conductivity contrasts that usually exist between zones of different physical and chemical properties in the ground. It is not necessary to use electrodes embedded in the ground to transmit and receive current. Electromagnetic phenomena are generated in the media investigated. This is an effective means of delineating contaminated zones beneath covered areas such

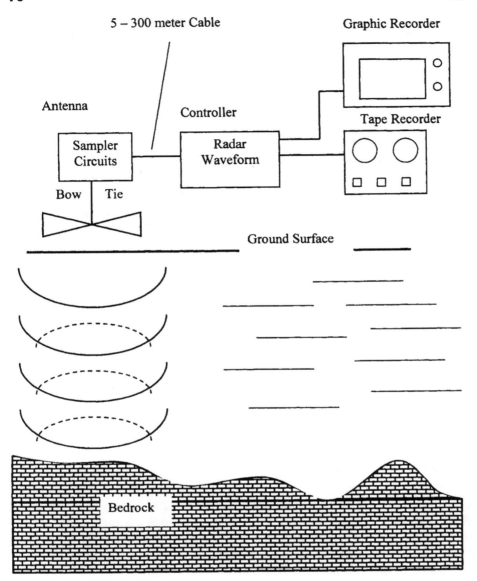

Figure 5 Sketch of a ground-penetrating radar (GPR) system.

as structural foundations and paved areas because there is no requirement for the use of direct-contact electrodes.

The principle of inducement of electromagnetic field in geomedia for resistivity measurements is illustrated in Fig. 6. A transmitter coil is held or fixed on or close to the ground surface. An alternating current is applied to the terminals of the coil to induce the flow of a current. This phenomenon generates an alternating magnetic field, which in turn causes electric currents to permeate the earth. Within the earth, a secondary magnetic field is induced. The secondary and primary magnetic fields are detected by a receiving coil placed near the transmitting coil. For a fixed intercoil spacing and operating frequency, the magnitude of the secondary magnetic field (or ratio of the secondary to primary magnetic

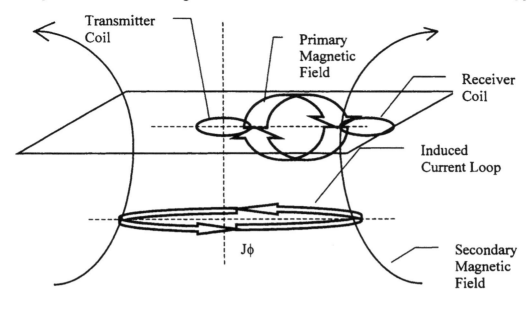

Figure 6 Illustration of induced electromagnetic fields.

fields) is directly proportional to the conductivity of the ground. Resistivity can be computed from conductivity data. In a geomedium, the solid particles act as insulators while soil moisture acts largely as the conductor of the electric current. Furthermore, the conductivity of the pore fluid is proportional to the concentration of ions within the fluid. Thus, these measurements can be used to delineate contaminated zones around waste containment barriers. If metallic objectives are present near the surveyed site, the readings recorded may be negatively affected. To effective depth of penetration can be as high as 60 m. This method is often called the frequency-domain electromagnetic (FDEM) method.

There are two other variations of the electromagnetic induction method. The decay rate of magnetic field after the transmission is turned off can be measured. An eddy current flows through the ground at successively greater depths. The data obtained are interpretated to obtain resistivity variation with depth. This is called time-domain electromagnetic (TDEM) survey. It has been applied to trace the migration of brines from an oilfield brine pit in southwestern Texas (39). Penetration depths of up to 500 m are estimated by (39). The third type of EM measurement involves the use of very low frequencies (15–25 kHz). In this method, ground contact is required for the potential electrodes. It is very suitable for investigating relatively shallow contaminant plumes (20–50 m).

D. Electrical Methods

Electrical methods can be used for detecting contaminant and moisture contents in both barriers and subsuface materials. With successive measurements at a point or concurrent measurements at various points, they can be used to estimate the rate of contaminant migration. Moisture flow rates can also be measured directly at a point through the

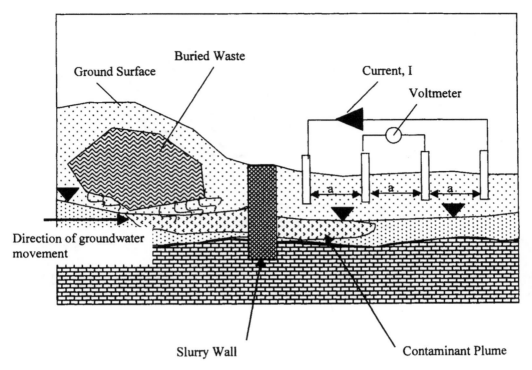

Figure 7 Sketch of Wenner resistivity array for delineating contaminant plumes behind a subsurface barrier.

use of electrothermal systems. Another important electrical method is the direct-current (DC) electrical resistivity method, which can be used to delineate contaminated zones. Its utility is described below.

1. Direct-Current (DC) Resistivity Method

Except for the means by which an electric current is generated, the DC resistivity method is similar to the EM resistivity method. Both are based on contrasts in the electrical conductivity or resistivity of geomedia with different textures, moisture contents, and contamination levels. In DC resistivity measurements, voltage is applied across a pair of electrodes that are embedded in the soil. Directly, this causes an electric current to flow through the soil. A pair of receiving electrodes are used to record the incoming voltage (from the soil). The Wenner array illustrated in Fig. 7 is commonly used in geoenvironmental site characterization. The depth of penetration of current is approximated by the interelectrode spacing (40). The resistivity is computed as follows:

$$p = 2\pi a \left(\frac{V}{I}\right) \quad (14)$$

where

p = resistivity of the soil (ohm-m)
a = interelectrode spacing (m)
I = current flow through outer electrodes (amps)
V = Voltage across inner electrodes (volts)

Following a system called the profiling method, horizontal variations in electrical resistivity of the ground can be tracked by moving the setup laterally from location to location on the ground surface. Thus, the horizontal extent of contaminant plumes can be mapped. The spacing of electrodes can be increased progressively to reach deeper locations in the ground. This is referred to as the sounding method and can provide information on material variability with depth. Usually, apparent resistivity values obtained from soundings are plotted as function of the separation distance between electrodes. Detailed information on computations relevant to the Wenner electrode configuration as well as other configurations is provided in (41,42). Effective ground penetration depths of several hundred meters can be attained if powerful equipment and adequate electrode spacing is used.

The electrical resistivity method was used to delineate the contaminated zone around a concrete barrier for impoundments of toxic wastes in the Jurupa Mountains, Riverside, California. It was reported that impoundments containing about 140 million liters of industrial solvents, acids, heavy metals, and organic residues were flooded by heavy rains in 1969 (43). In 1972, contaminants were discovered in a monitoring well located 1100 m downstream from the site. The DC electrical resistivity method was used to delineate the plume.

E. Electrochemical Methods

Electrochemical methods are the newest techniques for detecting contaminants in barriers and geomedia around barriers. They are based on the sensitivity of the detector to moisture variations or specific chemicals at the sensor locations. They are contact systems. Thus, the contaminants of concern must travel to the sensing points before detection can occur. There is the attendant risk that some of the sensing points will be short-circuited. The closer the spacing among the sensors, the lower the probability of short-circuiting. Chemical sensing cables have also been developed for multipoint detection of contaminants. Some of the cables can be looped for increased areal coverage.

Electrical moisture and contaminant sensors can be installed at the following points within and around barriers:

Under clay barriers of impoundments, landfills, and waste piles
Under joints in a Portland cement concrete cover
Below and at the potential water table on the downgradient and upgradient sides of slurry walls and sheet pile walls
Under the settled slurry blanket of confined disposal facilities for dredged sediments

1. Pressure Sensors in Unsaturated Media

Some moisture sensors operate on the basis of humidity and pore pressure changes in the soil. Psychrometers, tensiometers, and neutron probes have been effective in detecting moisture in soils. Their probe points can be embedded at various depths in barriers such as landfill cover and slurry wall sections above the water table. In other examples, the objective might be to monitor the potential increase in moisture under a building foundation or similar structure. There is a risk that such moisture may be contaminated. The monitoring system also allows a continuous assessment of the infiltration of moisture into the drainage system above the isolated waste.

V. CONCLUSIONS

Contaminant migration in barrier materials can be assessed through numerical methods, laboratory experimentation, field studies, or a combination of these. Additionally, today there are ample computing resources with which to approach complicated modeling scenarios. The optimal approach depends on the nature of the target contaminant, site conditions, and financial constraints. A more in-depth discussion on the fundamental processes that affect barrier performance may be found in (45,46).

REFERENCES

1. Carslaw, H. S., and Jaeger, J. C. 1959. *Conduction of Heat in Solids*. Oxford, London.
2. Bear, J. 1979. *Hydraulics of Groundwater*. McGraw-Hill, New York.
3. Crank, J. 1975. *The Mathematics of Diffusion*. Oxford, London.
4. Rabideau, A. J. Contaminant transport modeling. In: R. A. Rumer and J. K. Mitchell (eds.), *Assessment of Barrier Containment Technologies. A Comprehensive Treatment for Environmental Remediation Applications*, pp. 247–299. U.S. Department of Commerce, National Technical Information Service, Springfield, VA.
5. Ogata, A., and Banks, R. B. 1961. A solution of the differential equation of longitudinal dispersion in porous media. *U.S. Geological Survey Professional Paper 411-A*.
6. Inyang, H. I., and Tumay, M. T. 1995. Containment systems for contaminants in the subsurface. In: *Encyclopedia of Environmental Control Technology*. Gulf Publishing, New York.
7. Bouwer, H. 1991. Simple derivation of the retardation equation and application to preferential flow and macrodispersion. *Ground Water*, 29(1):41–46.
8. Freeze, R. A. and Cherry, J. A. 1979. *Groundwater*. Prentice-Hall, Englewood Cliffs, NJ.
9. Johnson, R. L., Cherry, J. A., and Pankow, J. F. 1989. Diffusive contaminant transport in natural clay: A field example and implications for clay-lined waste disposal sites. *Environ. Sci. Technol.*, 23(3):340–349.
10. Sparks, D. 1986. *Soil Physical Chemistry*. CRC Press, Boca Raton, FL.
11. ASTM. 1993. Standard test method for distribution ratios by the short-term batch method, ASTM D-4319-93. American Society of Testing and Materials, Philadelphia.
12. Heijn, E. C., Pierce, J. J., and Sperry, J. M. 1995. Pore water velocity influence on zinc sorption by clay-bearing sand. *Waste Management Res.*, 13:451–465.
13. U.S. EPA. 1991. Site characterization for subsurface remediation. Seminar publication. Center for Environmental Research Information, U.S. EPA, Washington, DC.
14. Gshwend, P. M., and Wu, S. C. 1984. On the constancy of sediment-water partition coefficients of hydrophobic organic pollutants. *Environ. Sci. Technol.*, 19:90–96.
15. U.S. EPA. 1987. Analytical methods manual for the direct/delayed response project soil survey, EPA/600/8-87/020. Office of Acid Deposition, Environmental Monitoring Quality Assurance, U.S. EPA, Washington, DC.
16. Kargbo, D. M., Fanning, D. S., Inyang, H. I., and Duell, R. W. 1993. The environmental significance of acid sulphate clays as waste covers. *Environ. Geol.*, 22:218–226.
17. Sparks, D. 1989. *Kinetics of Soil Chemical Processes*. Academic Press, San Diego, CA.
18. Brusseau, M. L., Jessup, R. E., and Rao, P. S. 1990. Sorption kinetics of organic chemicals: Evaluation of gas-purge and miscible-displacement techniques. *Environ. Sci. Technol.*, 24:727–735.
19. Wentnick, G. R., and Etzel, J. E. 1976. Removal of metal ions by soil. *J. Water Pollution Control Fed.*
20. Akratanakul, S., Boersma, L., and Klock, G. O. 1983. Sorption processes in soils as influenced by pore water velocity: 2. Experimental results. *Soil Sci.*, 135(6):331–341.

21. Aleshi, B. A., Fuller, W. H., and Boyle, M. V. 1980. Effect of leachate flow rate on metal migration through soil. *J. Environ. Qual.*, 9(1):119–126.
22. Tallard, G. 1984. Slurry trenches for containing hazardous waste. *Civil Eng.*, February.
23. Korfiatis, A. M., Rabah, N., and Lekmine, D. 1987. Permeability of compacted clay liners in laboratory scale models. In: *Proc. ASCE Specialty Conf. on Geotechnical Practice for Waste Disposal*, Ann Arbor, MI, pp. 611–624.
24. Olson, R. E., and Daniel, D. E. 1981. Measurement of the hydraulic conductivity of fine-grained soils. In: T. F. Zimmie and C. O. Riggs (eds.), ASTM STP 746, pp. 18–64. American Society for Testing and Materials, Philadelphia.
25. Leonards, G. A., Huang, A. B., and Ramos, J. 1991. Piping and erosion tests at Conner Run Dam. *ASCE J. Geotech. Eng.*, 117(1):108–117.
26. Goldenburg, L. C., Hutcheon, I., Wardlaw, N., and Melloul, A. J. 1993. Rearrangement of fine particles in porous media causing reduction of permeability and formation of preferred pathways of flow: Experimental findings and conceptual model. *Transport in Porous Media*, 13:221–237.
27. Fox, P. J. 1996. Analysis of hydraulic gradient effects for laboratory hydraulic conductivity testing. *ASTM Geotech. Testing J.*, 19(2):181–190.
28. Dunn, R. J., and Mitchell, J. K. 1984. Fluid conductivity testing of fine-grained soil. *ASCE J. Geotech. Eng.*, 110(1):1648–1665.
29. McIntyre, D. S., Cunningham, R. B., Vatanakul, V. I., and Stewart, G. A. 1979. Measuring the hydraulic conductivity in clay soils: methods, techniques and errors. *Soil Sci.*, 128(3):171.
30. Bowders, J. J., Daniel, D. E., Broderick, G. P., and Lijestrand, H. M. 1986. Methods for testing the compatibility of clay liners with landfill leachate. In: *Proc. Fourth Symp. on Hazardous and Industrial Solid Waste Testing*, ASTM STP 886, pp. 233–250. ASTM, Philadelphia.
31. Daniels, J. L. 1998. Textural and mineralogical controls on heavy metal attenuation in barrier materials. M.S. thesis, University of Massachusetts, Lowell, MA.
32. Benson, C. H., and Daniel, D. E. 1990. Influence of clods on hydraulic conductivity of compacted clay. *ASCE J. Geotech. Eng.*, 116(8):1231–1248.
33. Jahnke, F. M., and Radke, C. J. 1987. Electrolyte diffusion in compacted montmorillonite engineered barriers. In: C. Tsang. (ed.), *Coupled Processes Associated with Nuclear Waste Repositories*, pp. 287–297. Academic Press, Orlando, FL.
34. Cheung, S. C. H., and Gray, M. N. 1988. Mechanism of ionic diffusion in dense bentonite. In: *Scientific Basis for Nuclear Waste Management/Proc. Material Research Society Symp. XII*, pp. 677–681. Berlin, Germany.
35. Shackelford, C. D. 1991. Laboratory diffusion testing for waste disposal—A review. *J. Contaminant Hydrol.*, 7:177–217.
36. Inyang, H. I., Fang, H. Y., Choquette, M. R., and Iskandar, A. 1998. Clay barriers, chemical and mineralogical analyses. In: R. A. Meyers (ed.), *Encyclopedia of Environmental Analysis and Remediation*, pp. 1158–1165. Wiley, New York.
37. Walther, E. G., Pitchford, A. M., and Olhoeft, G. R. 1986. A strategy for detecting subsurface organic contaminants. In: *Proc. NWWA/API Conf. on Petroleum Hydrocarbons and Organic Chemicals in Ground Water: Prevention, Detection and Restorations*, Houston, TX, pp. 357–381.
38. Kardos, J. J., and Ennis, G. B. 1993. Subsurface interface radar as an investigative tool. *Public Works*, September, pp. 82–83.
39. Hoekstra, P., Lahti, R., Hild, J., Bates, C. R., and Phillips, D. 1992. Case histories of shallow time domain electromagnetics in environmental site assessment. *Ground Water Management Res.*, Fall, pp. 110–117.
40. McNeil, J. D. 1982. Electromagnetic resistivity mapping of contaminant plumes. In: *Proc. National Conf. on Management of Uncontrolled Hazardous Waste Sites*, Washington, DC, pp. 1–6.
41. Van Zijl, J. S. V. 1978. On the uses and abuses of the electrical resistivity method. *Bull. Assoc. Eng. Geol.*, 15(1):85–111.

42. U.S.A.C.E. 1979. *Geophysical Exploration: Engineering and Design*, Engineering Manual EM 1110-1-1802. U.S.A.C.E., Washington, DC.
43. Stierman, D. J. 1984. Electrical methods of detecting contaminated groundwater at the Stringfellow waste disposal, Riverside County, CA. *Environ. Geol. Water Sci.*, 6(1):11–20.
44. Evans, R. B. 1982. Currently available geophysical methods for use in hazardous waste site investigations. In: *Proc. American Chemical Society Symp. on Risk Assessment at Hazardous Waste Sites*, Las Vegas, NV, pp. 94–115.
45. Inyang, H. I., Parikh, J. M., and Iskandar, A. 1998. Waste containment barriers. In: R. A. Meyers (ed.), *Encyclopedia of Environmental Analysis and Remediation*, pp. 5131–5141. Wiley, New York.
46. Fang, H. Y., Daniels, J. L., and Inyang, H. I. 1997. Enviro-geotechnical considerations in waste containment system design and analysis. In: *Proc. International Containment Technology Conf.*, St. Petersburg, FL, pp. 414–420.

3
Design Considerations for Hazardous Waste Landfills

M. A. Gabr
North Carolina State University, Raleigh, North Carolina

I. INTRODUCTION

The use of disposal as a remedial alternative for hazardous waste is one of the common techniques implemented in practice. Land disposal, including landfills, deep-well injection, surface impoundments, and land farming, accommodate 12.3% of the total amount of hazardous waste reported by Resource Conservation and Recovery Act (RCRA) large-quantity generators (LQGs). As of 1995, the amount of hazardous waste reported by LQGs was of the order of 214 million tons (1). Approximately 1 million tons of this waste was landfilled (1). RCRA prohibits the disposal of hazardous waste-containing free liquids in hazardous waste landfills. Free liquids are defined as those that readily separate from the solid portion of waste under ambient temperature and pressure (40 CFR §260.10). A typical breakdown of the constituents forming "hazardous waste" is shown in Fig. 1.

The identification, permitting, and design of disposal sites often represent the most controversial part of utilizing landfills as a remediation measure. Consequently, RCRA was enacted in 1976 as an amendment to the Solid Waste Disposal Act. This amendment was aimed directly at regulating the disposal of municipal and industrial solid wastes. Owners of facilities that treat, store, and/or dispose of hazardous waste must obtain a permit under RCRA Subtitle C. Owners of facilities that store or dispose of solid waste must obtain a permit under the enacted RCRA Subtitle D.

Permitting under the RCRA is a complicated process that has to satisfy several environmental, socioeconomic, regulatory, and engineering criteria. In general, each state maintains its own permitting program. It is usually the case that a state's permitting program is consistent with the national standards but encompasses a broader and more stringent scope. In general, an application for permitting a storage and disposal landfill must meet minimum design guidelines, described in the Hazardous and Solid Waste Amendment (HSWA), and consists of two parts. The first part includes mainly abstract information about the facility and the applicants, and the second part includes detailed engineering analyses and design.

An overview of the current state of practice for designing hazardous waste landfills in compliance with the RCRA is presented in this chapter. An example using a proposed

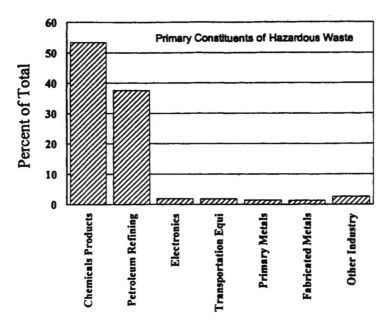

Figure 1 Percent components of hazardous waste materials (after Ref. 2).

hazardous waste landfill is used to identify the engineering analyses and design usually performed to prepare a permit application. Key design issues are presented and discussed.

II. DESIGN CONSIDERATIONS

As RCRA regulations prescribe performance rather than design standards (3), the design of a landfill is often a challenging task which encompasses applications from most disciplines of civil engineering. In general, such a design involves expertise in areas such as geotechnical and environmental engineering, mechanics of materials, transportation engineering, hydraulics and hydrology, structural engineering, and construction management. Figure 2 illustrates the major engineering components and activities that makes the design of a landfill a multidisciplinary endeavor.

III. DESIGN OBJECTIVES

In addition to waste containment, landfills are structures that are constructed for control of hydraulic heads, both externally and internally. External head control implies the routing and management of stormwater such that a minimum amount of water reaches the waste. Internal head control implies the drainage and collection of the generated leachate such that its migration into the environment is minimized.

The siting of a landfill has to be suited for the geological conditions of the selected site and has to indulge the socioeconomic conditions of the community in which it

Design for Hazardous Waste Landfills

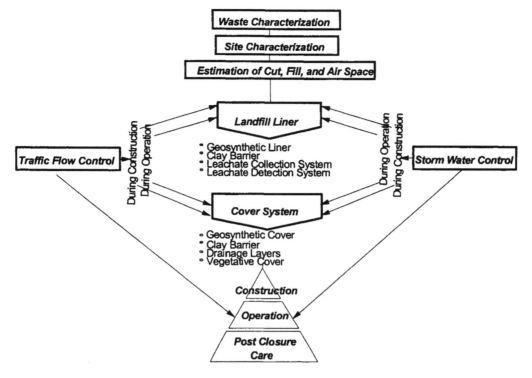

Figure 2 Major design components for an RCRA Part B landfill.

is being constructed. The design of a landfill has to be environmentally sound, with built-in safety against liner/cover failure and leachate migration. The air space has to be adequate such that the volume and type of waste are handled throughout the life of the landfill.

IV. EXAMPLE CASE

The landfill to be used as an example to illustrate the current state of practice is proposed for construction for the disposal of refinery ash. The landfill site encompasses challenging geological formations of the mantled Karst terrane zones. These zones, usually overlain by unconsolidated sediments, are susceptible to solution weathering and therefore the development of sinkholes.

The landfill area is approximately 100 acres. The waste stream to be landfilled consists mainly of stabilized fly ash and bottom ash generated from incineration of sludge from a petroleum refining process. The design life of the landfill is 22 years, with a total capacity of approximately 23.6 million cubic feet.

The landfill was designed using a modular scheme. As shown in Fig. 3, the landfill area was divided into 11 rectangular cells, with each cell divided into two modules. The storage capacity for each module was selected to accommodate a disposal volume of one year period. Construction of the landfill using a modular scheme allows for

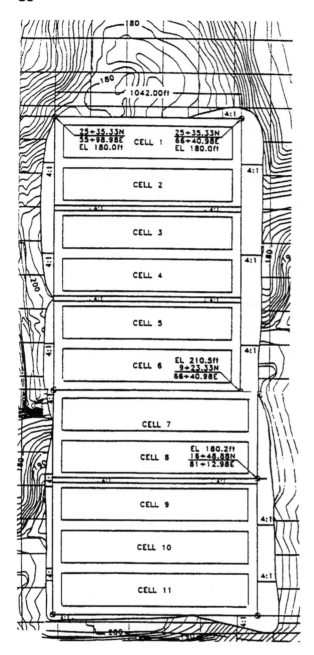

Figure 3 Layout of the landfill and compatible siting with surrounding topography.

the active disposal of waste without the need to wait for the construction of the full landfill area. Such a development scheme, however, presents the designer and the contractor with the challenge of synchronizing the construction and disposal activities to accomplish safe operation.

V. CUT, FILL, AND AIR SPACE

One of the most important steps in designing a landfill is to ensure that the proposed layout will provide the required capacity. Computation of cut and fill to achieve the landfill base grades is usually a trial and error process. The use of a computer-aided drafting (CAD) system supplemented by earthwork-calculation software is highly recommended for this purpose. Vargas and Porter (4) presented a case study where the automated design on a basic CAD system proved to be advantageous. The computerization of this step allowed the optimization of the landfill layout whereby the optimal amounts of cut and fill were achieved while at the same time arriving at the required landfill air space.

The estimation of the available air space to satisfy the waste storage requirement is always a formidable task. This is because an assumption regarding the depth below existing grades and the magnitude of the side slopes for the liner and the cover systems has to be made. The established slope magnitudes using this assumption also have to satisfy the requirements of slope and deformational stability. For optimum design, an iterative scheme can be employed. For the example case presented here, a computerized workbench was employed whereby the design was conducted using an integrated CAD/geotechnical analysis. The CAD system was used to generate cross-section information digitally, for direct input into the geotechnical analysis, which fed back into the CAD system for revised cross sections to accommodate stability requirements.

VI. WASTE CHARACTERIZATION

The classification of waste is usually a straightforward task, since it is established according to the regulatory framework of land disposal restrictions. However, a challenging task is the evaluation of its engineering and physical properties. Waste properties needed for the design include unit weight (γ), shear strength parameters defined in terms of cohesion (c) and friction angle (ϕ), and elastic properties (E, ν). These properties have a direct impact on the design of the cover system since the integrity of the cover will depend on the deformation characteristics of the stored waste. Also, predisposal storage and transportation activities are influenced by these waste properties, although they are often overlooked.

VII. SITE CHARACTERIZATION

The objective of the site characterization is to estimate the subsurface stratification, the engineering properties of the different soil and rock layers, and the groundwater conditions. Figure 4 presents the major components that need to be evaluated for site characterization. It should be mentioned that not all of the in-situ and laboratory tests outlined in Fig. 4 are normally performed at a given site. Depending on the site conditions and the type of subsurface soils, a site-specific laboratory and field testing program is developed for a given landfill. However, it is the author's experience that investing funds to gain a comprehensive database about the site soils and groundwater conditions always results in a "smart" design that is sound and cost-effective, with considerable return on the investment.

In cases where geological irregularities are expected, additional testing may be required. For the example presented here, the subsurface profile contained soluble rock

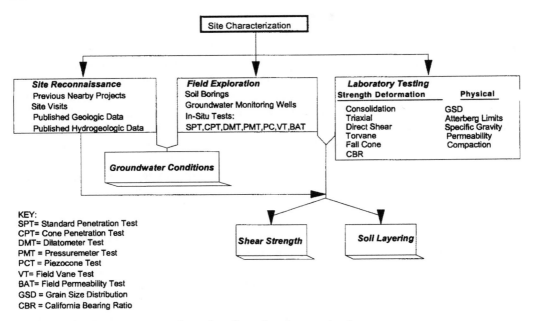

Figure 4 Major components for subsurface site characterization.

and an additional investigation program that included seismic and resistivity testing on a grid configuration was conducted to characterize any sinkhole formation. Geophysical data often prove valuable in these situations.

Once all the data are collected, a project database must be developed for the management of the large quantity of data from the different site investigation studies. The structure of the computerized database must allow flexibility in cross-referencing information. Such flexibility is important especially when several firms are involved in the projects. Maintaining a computerized database for a given landfill is also valuable for planned and unforeseen project modifications and expansions.

VIII. LINER CONFIGURATION AND DESIGN

Section 3004(o) of the RCRA requires that hazardous waste landfills have two or more liners, a leak detection system, and a leachate collection and removal system (LCRS) above and between the liners.

The establishment of liner configuration is usually a straightforward task, since regulatory framework has to be followed. However, the design of the different components of the liner system is rather challenging. Figure 5 shows a typical cross section of the liner section used for the example case. Design manuals by the U.S. EPA (5) and by Richardson and Koerner (6), and a textbook by Koerner (7) cover the design procedures for the different geosynthetic components of the liner system. While these references provide the tools to perform such a design, the engineer still bears the responsibility for applying these methods correctly, as well as for proper characterization of the material properties.

Manufacturers' literature which includes data on the physical and engineering properties of the synthetic material are available. However, it should be noted that most of the physical and engineering properties of synthetic materials reported in

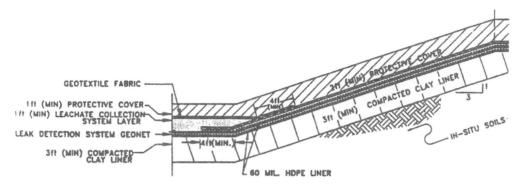

Figure 5 Configuration of liner system.

manufacturer's literature are from short-term testing programs. In general, the properties of the geosynthetic material are time-dependent, and deterioration with time due to several factors should be anticipated. A state-of-the-art review of long-term tensile strength of geosynthetics has been presented (8). The importance of accounting for factors such as chemical aging, creep, and biological degradation on the performance of geosynthetics was emphasized. Failure cases due to inaccurate estimation of geosynthetic material properties have been reported in the literature (9,10).

In addition to the tests that are usually performed on synthetic material as a part of the compatibility and quality control/assurance plans, additional performance testing should be performed to justify the design parameters. Such a testing program should be viewed in the same manner as, for example, conducting a testing program for the evaluation of the in-situ soil properties.

The design of the liner synthetic components must address both short-term and long-term situations A general design objective should be to subject the geosynthetic material to the minimal tensile strains under long-term and sustained loading conditions. Short-term design conditions include accounting for loads and stresses due to construction equipment and construction procedures. Long-term design conditions include accounting for loads and stresses due to waste fill, time-rate deformation of the supporting soil profile fluctuations in the groundwater table, and future land use.

IX. LEACHATE COLLECTION SYSTEM

The design of a leachate collection system includes specifying the size of a sump area, the grain size gradation and thickness of leachate collection and detection layers, and sizing the subdrainage pipe network. Factors that influence the design of a leachate collection system include the permeability ratio of the liner material to the leachate collection material, the slope of the collection system, and the hydraulic head on the liner system, which is specified not to exceed 12 in. at any given time.

Before the design can be accomplished, the quantity of flow reaching the collection system has to be estimated. The computer program HELP (11) is widely used for this purpose. The magnitude and duration of the design storm used in HELP analysis usually varies from state to state. However, the leachate collection system should be designed to handle stormwater reaching the landfill excavation during construction and collect leachate generated during operation and postclosure. Most often the sizing of a leachate collection system is governed by the conditions during operation (landfilling), since once the cover system is installed the quantity of leachate is significantly reduced. A minimum liner slope of 1–3%, toward the sump area, has to be maintained to ensure the flow of leachate under gravity.

The design of a leachate collection system also includes the hydraulic and structural design of a piping network. Usually the hydraulic capacity of 6-in. to 12-in. pipes is adequate to handle leachate flow in most landfills. The challenge is in selecting an appropriate pipe stiffness to satisfy structural stability during construction and operation. The selected pipe cross section should be flexible enough to avoid brittle failure under applied loads. At the same time, high deformation magnitudes due to high flexibility should be avoided so that the pipes' function is not hindered.

In addition, and as shown in Fig. 5, the use of a sand layer in a leachate collection system on the side slopes should be avoided. Leachate flow velocities and associated seepage forces can cause migration of fines and erosion of this layer. Instead, it is recommended that the sand layer be replaced by a geonet drainage layer. The transmissivity, or the drainage capacity, of the geonet has to be equivalent to that of the sand layer, and several layers of geonet can be used to accomplish this purpose. The change in the transmissivity of the geonet(s) as a function of the applied waste loads should be considered, however, as described by Koerner (7).

X. SETTLEMENT AND STABILITY ANALYSES

The design of the liner system is highly dependent on the amount of settlement and differential settlement to be developed over the lifetime of the landfill. The differential settlement affects directly the amount of tensile strains generated in the liner system components. Also, differential settlements will induce a change in the base slopes and may adversely affect the flow of leachate. When the landfill is constructed on a clay profile, a major settlement component that is usually overlooked is the secondary compression. In highly organic or soft marine clays, the magnitude of the secondary settlement is significant. The secondary settlement is considered a long-term settlement, and the majority of its magnitude will take place after closure of the landfill. Accounting for this settlement component assists in planning the postclosure care and avoiding unpleasant surprises.

Stability analysis is also critical for a sound design of the side slopes of the liner and cover systems. Richardson and Reynolds (12) present case study in which landslide occurred due to pore pressure buildup in the supporting marine clay layer. Another famous case is the slope failure at the Kettleman City hazardous waste treatment and storage facility (13). For the example described in this chapter, stability analysis was conducted using both short-term and long-term shear strength parameters. Specification of the side slope magnitudes for the liner and cover systems was based on long-term conditions which were estimated to be the most critical.

XI. COVER CONFIGURATION AND DESIGN

The main function of the cover system is to isolate the environment from the waste and minimize the amount of stormwater infiltration. Similar to the liner system, the components of a cover system are prescribed by regulations. Figure 6 shows the cover

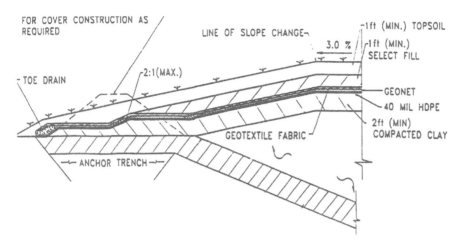

Figure 6 Configuration of cover system.

system that was designed for the example case discussed here. The slopes of the cover system were designed such that surface drainage was promoted with minimum erosion. Also esthetics and the possibility of future land use were taken into consideration. The cover system is not usually subjected to high structural loads and in many cases its geosynthetic components will not require anchoring. Subsidence and stability conditions should, however, be fully analyzed to ensure cover integrity. Similar to the liner system, soil drainage layers on the slopes were replaced by geosynthetic drainage layers. Such a replacement is important especially if the cover slope exceeds 4 horizontal to 1 vertical (4h : 1v).

A discussion on the suitability of different cover systems for landfills has been presented (14). It was concluded that the components of a cover system should not necessarily encompass a compacted clay layer but rather vary according to the weather and geological conditions of the landfill site. Such a conclusion was based on the intolerance of compacted clay liners to relatively large differential settlements, desiccation, and freeze/thaw cycles. While alternatives to the compacted clay layer were presented, a compacted clay cover is still required by the regulations.

XII. CONSTRUCTION, OPERATION, AND POSTCLOSURE

Construction and operation of the landfill are integral elements of the landfill design. The construction sequence of a given cell affects the stability of the liner system. Once the synthetic liner is placed, it is usually the case that a soil buttress will be needed at the bottom of the side excavation before placement of the protective soil layer on the side slope can proceed. Also, the sequence of filling during operation is important to maintain waste and cover stability. The designer should be familiar with the construction equipment that will be used, as well as specify the construction sequence to avoid adverse stability conditions.

Stormwater management and traffic flow plans have to be developed to address conditions during construction, operation, and postclosure. Detailed design of the hydraulic structures and conveyance of stormwater to management ponds was performed as a part of the permit in the example case. Traffic flow operation as well as detailed cross sections of access roads were also developed. As part of the traffic flow plans, geometric design of turns and road superelevation were presented.

XIII. PERMIT APPLICATION: DRAWINGS AND SUBMISSION

The design part of a permit application for landfills can be divided into four main categories:

1. Landfill layout and details
2. Construction and operation
3. Stormwater management
4. Closure and postclosure

A typical permit application will contain 50–100 drawings and several volumes of narrative, technical specifications, quality control/assurance plans, and financial assurance. The permit process usually consists of the following steps (15).

1. Submitting the permit application
2. Permit review
3. Development of a draft permit
4. Public comments
5. Development of a final permit

A groundwater monitoring program is an important aspect of permitting hazardous waste facilities. Groundwater monitoring requirements of 40 CFR Part 264, Subpart F, specify that owners/operators of permitted hazardous waste landfills need to sample the groundwater at least semiannually for permit-specific indicator parameters and waste constituents. This sampling program is used to monitor for statistically significant evidence of liner leakage. Statistical procedures written into permits to comply with the groundwater monitoring requirements usually involve verification resampling and retesting procedures as a means of checking "false negatives," error rates, and improve statistical analysis.

XIV. SUMMARY AND CONCLUSIONS

Landfill permitting is a complicated process that has to satisfy several environmental, socioeconomic, regulatory, and engineering criteria. An overview of the state of practice for the design of hazardous waste landfills and permit application has been presented in this chapter.

Development of a permit application for a landfill necessitates the assembly of a design team with expertise in almost all areas of civil engineering. While the regulations prescribe performance standards, the design needs to utilize evolving and innovative technologies. The challenge is to embrace these innovative technologies while constructing cost-effective and safe landfills for containment of environmental hazards.

REFERENCES

1. U.S. EPA. 1997. Executive summary: The national biennial RCRA hazardous waste report.

2. U.S. EPA. 1996. RCRA environmental indicators progress report: 1995 update. Office of Solid Waste, June.
3. Landreth, R. E. 1990. Landfill containment systems regulations, waste containment systems, *ASCE Geotechnical Special Publication No. 26*, pp. 1–13.
4. Vargas, J. C., and Porter, D. 1992. Landfills: Anatomy of automated design. *Civil Eng.*, March, pp. 52–55.
5. U.S. EPA. 1989. Requirement for hazardous waste landfill design, construction and closure, Seminar Publication, EPA/625/4-89/022.
6. Richardson, G. N., and Koerner, R. M. 1987. Geosynthetic design guidance for hazardous waste landfill cells and surface impoundments, EPA/600/2-87/097.
7. Koerner, R. M. 1990. *Designing with Geosynthetics*, 2nd ed. Prentice-Hall, Englewood Cliffs, NJ.
8. Allen, T. M. 1991. Determination of long-term tensile strength of geosynthetics: A state-of-the-art review. *Geosynthetics '91 Conf.*, Atlanta, GA, pp. 351–375.
9. Peggs, I. D., Winfree, J. P., and Giroud, J. P. 1991. A shattered geomembrane liner case history: Investigation and remediation. *Geosynthetics '91 Conf.*, Atlanta, GA, pp. 495–505.
10. Mitchell, J. K., Seed, R., and Seed, H. 1990. Kettleman Hills waste landfill slope failure. I: Liner-system properties. *J. Geotech. Eng.*, V 116(4):647–668.
11. Schroeder, P. R., Morgan, J. M., Walski, T. M., and Gibson, A. C. 1988. The hydrologic evaluation of landfill performance (HELP) model. Version 2, EPA 600/2-87-049 and EPA 600/2-87-050.
12. Richardson, G. N., and Reynolds, R. G. 1992. Geosynthetic considerations in a landfill on compressible clays. *Geosynthetics '91 Conf.*, Atlanta, GA, pp. 507–516.
13. Mitchell, J. K. 1989. Failed expectations. The 1989 Woodward Lecture, Woodward-Clyde Consultants Silver Anniversary Symposium.
14. Koerner, R. M., and Daniel, D. E. 1992. Better cover-ups. *Civil Eng.*, May, pp. 55–57.
15. Clay, D. R., U.S. EPA. 1990. RCRA orientation manual 1990 edition, Office of Solid Waste Report No. EPA/530/SW-90/036.

APPENDIX: RCRA ACRONYMS

```
BDAT   = best demonstrated available technology
CMS    = corrective measures study
HHW    = household hazardous waste
MSW    = municipal solid waste
NCAPS  = National Corrective Action Prioritization System
NPDES  = National Pollutant Discharge Elimination System
POTW   = Publicly owned treatment works
RCRA   = Resources Conservation and Recovery Act
RCRIS  = Resource Conservation and Recovery Information System
RFA    = RCRA facility assessment
RFI    = RCRA facility investigation
SIC    = Standard Industrial Classification
TSD    = treatment, storage, and disposal
TSDF   = treatment, storage, and disposal facility
```

4
Basics and Applications of Electrokinetic Remediation

Akram N. Alshawabkeh
Northeastern University, Boston, Massachusetts

Ray Mark Bricka
U.S. Army Engineer Research & Development Center, Vicksburg, Mississippi

I. INTRODUCTION

The use of electric fields has emerged as an innovative method for in-situ restoration of contaminated hazardous waste sites. Direct currents (DC) are applied across electrodes inserted in the soil to generate an electric field for mobilization and extraction of contaminants and for biogeochemical modifications of polluted soils and slurries. The driving mechanisms for this technique, known as electrokinetic remediation, are transport under electric fields (in particular, electroosmosis and ionic migration) coupled with electrolysis and geochemical reactions. Extraction and removal are generally achieved by electrodeposition, precipitation, or ion exchange (for heavy metals), and collection and treatment of organics in external systems. Contaminants that could be treated by electric field applications include inorganic, organic, and radioactive compounds that are charged (ionic) or noncharged (polar and nonpolar). This chapter provides a review of the fundamentals and applications of electrokinetic remediation. The chapter describes the general electrokinetic and transport phenomena in soil under electric fields, followed by identification of electrolysis and geochemical reactions associated with application of electric fields in soils. Finally, current technology status and considerations for practical implementation of the technology are presented.

II. ELECTROKINETIC PHENOMENA IN SOILS

Application of electric fields (and/or hydraulic gradients) in fine-grained soils (silt and clay) results in electrokinetic phenomena that influence transport of water, charge, and mass. Electrokinetics is generally defined as the physicochemical transport of charge, action of charged particles, and effects of applied electric potentials on formation and fluid transport in porous media. Therefore, the term "electrokinetics" refers to coupled fluid and charge (electric current) transport. Electrokinetic phenomena in soils are developed mainly because discrete clay particles have a negative surface charge that influences and controls the particle environment. The soil particle surface charge can be developed in different ways, including the presence of broken bonds and due to isomorphous substitution (1). The net negative charge on the clay particle surfaces requires an excess positive charge (or exchangeable cations) distributed in the fluid zone adjacent to the clay surface forming the diffuse double layer. The quantity of these exchangeable cations required to balance the charge deficiency of clay is termed the cation-exchange capacity (CEC) and is expressed in milliequivalents per 100 g of dry clay. Several theories have been proposed for modeling charge distribution adjacent to clay surface. The Gouy-Chapman diffuse double-layer theory has been widely accepted and applied to describe clay behavior. A detailed description of the diffuse double-layer theories for a single flat plate is found in (1–3).

The presence of the diffuse double layer causes several electrokinetic phenomena in soils, which may result from either the movement of different phases with respect to each other, including transport of charge, or the movement of different phases relative to each other, due to the application of electric field. Electrokinetic phenomena identified in soils include (1) electroosmosis, electrophoresis, streaming potential, and sedimentation potential. Electroosmosis is defined as fluid movement with respect to a solid wall as a result of an applied electric potential gradient. In other words, if the soil is placed between two electrodes in a fluid, the fluid will move from one side to the other when an electromotive force is applied. Electrophoresis is the movement of charged particles (e.g., clay particles or microorganisms) suspended in a liquid due to application of an electric potential gradient. Streaming potential is the reverse of electroosmosis. It defines the generation of an electric potential difference due to fluid flow is soils. Sedimentation (or migration) potential, known as Dorn effect (4), is an electric potential generated by the movement of particles suspended in a liquid. The electrokinetic phenomena that affect electrokinetic remediation are electroosmosis and electrophoresis. The other two phenomena (streaming and sedimentation potentials) are not related directly to electrokinetic soil remediation.

Under certain conditions (such as the presence of appropriate minerals, high water content, and low ionic strength of the pore fluid), electroosmosis will have a significant role in electrokinetic soil remediation. Several theories are established to describe and evaluate water flow by electroosmosis, including Helmholtz-Smoluchowski theory, Schmid theory, the Spiegler friction model, and ion hydration theory. Descriptions of these theories are given by Gray and Mitchell (5) and by Mitchell (1). The Helmholtz-Smoluchowski model is the most common theoretical description of electroosmosis and is based on the assumption of fluid transport in the soil pores due to transport of the excess positive charge in the diffuse double layer toward the cathode. The rate of electroosmotic flow is controlled by the coefficient of electroosmotic permeability of the soil (k_e), which is a measure of the fluid flux per unit area of the soil per unit electric gradient. The value of k_e is assumed to be a function of the so-called zeta potential of the soil–pore fluid interface (which describes the electrostatic potential resulting from the soil surface charge), the viscosity of the pore fluid, soil porosity, and soil electrical permittivity. Vane and Zang (6) investigated the effect of pore fluid properties on electroosmotic permeability. The results displayed that the effect of pH on zeta potential and electroosmostic flow vary significantly depending on the mineral type. Lockhart (7) demonstrated that high electrolyte concentration in the pore fluid causes strong electrolyte polarization, which limits electroosmotic flow. At a certain pH value and pore-fluid ionic strength, the soil surface charge could drop to zero, rendering a zero zeta potential or

Basics and Applications of Electrokinetic Remediation

what is called the isoelectric point (8). Negative surface charge of clay particles (negative zeta potential) causes electroosmosis to occur from anode to cathode, while positive surface charge causes electroosmosis to occur from cathode to anode (9,10). The electroosmotic flow can be virtually eliminated at the isoelectric point.

III. ION MIGRATION

Electric currents occur in soils due to ion migration, which is the transport of charged ions in the pore fluid toward the electrode that is opposite in polarity. Ionic mobility is the term used to describe the rate of migration of a specific ion under a unit electric field. A similar term is used in soils, but to account for soil porosity and tortuosity, the term is modified to "effective" ionic mobility. Rates of contaminant extraction and removal from soils by electric fields are dependent on the values of the effective ionic mobilities of contaminants. Heavy-metal ionic mobilities at infinite dilution are in the range of 10^{-4} cm^2/Vs. Accounting for soil porosity and tortuosity, the effective ionic mobilities are in the range of 10^{-4} to 10^{-5} cm^2/Vs. Accordingly, the rate of heavy-metals transport in clayey soil is about a few centimeters per day under a unit electric gradient (1 V/cm). As a result of ion migration in the soil pores, cations are collected at the cathode and anions at the anode.

In summary, application of electric gradients in soil will result in two significant transport mechanisms; electroosmosis and ion migration (Fig. 1). Electroosmosis draws contaminants with the flowing water under electric fields. Ion migration transports ions to the electrode opposite in polarity under electric fields. Electroosmosis and any other hydraulic flow will usually carry all types of solutes from one location to another, depending on flow direction. However, ion migration separates negatively and positively charged ions and causes their migration to opposite electrodes. Consequently, hydraulic flow might enhance the migration of certain ions but retard migration of other ions (with

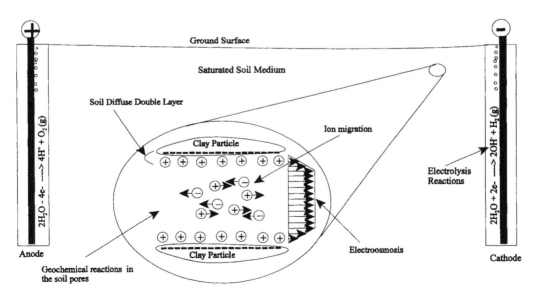

Figure 1 Schematic diagram showing the processes involved during application of electric fields in soils.

opposite charge). The relative contribution of electroosmosis and migration to ion transport under electric fields varies for different soil types, water contents, types of ion, pore fluid chemistry, and boundary conditions.

IV. ELECTROLYSIS REACTIONS

Application of direct electric current through electrodes immersed in water induces electrolysis reactions at the electrodes (Fig. 1). Oxidation of water at the anode generates an acid front, while reduction at the cathode produces a base front as described by the following electrolysis reactions:

$$2H_2O - 4e^- \longrightarrow O_2^- + 4H^+ \quad \text{(anode)}$$
$$4H_2O + 4e^- \longrightarrow 2H_2^- + 4OH^- \quad \text{(cathode)}$$

The prevailing of electrolysis reactions at the electrodes depends on the availability of chemical species and the electrochemical potentials of these reactions. Although some secondary reactions might be favored at the cathode because of their lower electrochemical potential, the water reduction half-reaction (H_2O/H_2) is dominant at early stages of the process. Within the first few days of processing, electrolysis reactions drops the pH at the anode to below 2 and increases it at the cathode to above 10, depending on the total current applied (11–13). Studies (10,14–16) showed that while acid production enhances the process, development of a high pH zone at the cathode adversely affects extraction of heavy metals from soils.

V. SOIL pH AND GEOCHEMICAL REACTIONS

While the acid generated at the anode advances through the soil toward the cathode by ionic migration and electroosmosis, the base developed at the cathode initially advances toward the anode by diffusion and ionic migration. However, the counterflow due to electroosmosis retards the back-diffusion and migration of the base front. The advance of this front is slower than the advance of the acid front because of the counteracting electroosmotic flow and also because the ionic mobility of H^+ is about 1.76 times that of OH^-. As a consequence, the acid front dominates the chemistry across the specimen except from small sections close to the cathode (12,15,17,18).

Geochemical reactions in the soil pores significantly affect electrokinetic remediation and can enhance or retard the process. Precipitation and sorption of heavy metals prevent their transport and thus limit extraction. Complexation could reverse the charge of the ion and might reverse the direction of migration. These geochemical reactions, including precipitation/dissolution, sorption, redox, and complexation reactions, are highly dependent on the pH condition generated by the process. The advance of the acid front from anode toward the cathode assists in desorption and dissolution of metal precipitates. However, formation of the high-pH zone near the cathode results in immobilization to precipitation of metal hydroxides. Limitations of electrokinetic remediation due to catholyte high pH required utilizing innovative methods to enhance the technique and prevent immobilization of metals close to the cathode.

VI. ELECTROLYTE ENHANCEMENT

Several procedures have been proposed to enhance electrokinetic remediation. Some of these procedures attempt to control production of hydroxyl ions at the cathode. Other procedures attempt to enhance complexation of heavy metals to enhance extraction at the anode. Some of these procedures are presented.

A. Catholyte Neutralization

One way of controlling the catholyte pH is to neutralize the hydroxyl ions produced by electrolysis using weak acids or catholyte rinsing. The advantages of using weak acids include (a) they form soluble metal salts; (b) their low solubility and migration rates will not increase the electric conductivity of the soil; and (c) they are biodegradable and, if properly selected, environmentally safe. However, improper selection of some acids may pose a health hazard. For example, the use of hydrochloric acid may pose an environmental concern or a health hazard because (a) it may increase the chloride concentration in the groundwater; (b) it may promote the formation of some insoluble chloride salts, e.g., lead chloride; and (c) if chloride ions reach the anode compartment, chloride gas will be generated by electrolysis.

Rødsand et al. (19) and Puppala et al. (20) demonstrated that neutralization of the cathode reaction by acetic acid can enhance electrokinetic extraction of lead. Hicks and Tondorf (21) indicated that development of a pH front could cause isoelectric focusing, which retards ions transport under electric fields. They showed that this problem can be prevented simply by rinsing away the hydroxyl ions generated at the cathode. They demonstrated 95% zinc removal from kaolinite samples by using the catholyte rinsing procedure.

B. Ion-Selective Membranes

Another procedure to control hydroxyl ions and enhance metals transport toward the cathode is the use of membranes. Ion-selective membranes, which are impermeable to hydroxyl ions, could be used to separate the catholyte from the soil and thus prevent or minimize the transport of hydroxyl ions into the soil. These membranes are insoluble in most solvents and chemically resistant to strong oxidizing agents and strong bases. Rødsand et al. (19) and Puppala et al. (20) showed that this technique has limited success when compared to catholyte neutralization. The reason is that heavy metals accumulate and precipitate on these membranes, resulting in a significant increase in the electrical resistivity of the membrane. Unless these membranes are continuously rinsed and cleaned, the energy cost of this technique will increase substantially.

C. Chelating or Complexing Agents

An acid front may not develop in soils of high buffer capacity and in soils that produce reverse electroosmosis, i.e., water movement from the cathode toward the anode (22). Reddy et al. (23) showed that soils that contain high-carbonate buffers, such as glacial till, hinder the development and advance of the acid front. On the other hand, acid advance in soils with low buffering capacity may cause uncontrolled dissolution of soil minerals, resulting in an excessive release of soil constituents, such as Al and Si. Under these circumstances, it is necessary to use enhancement agents to solubilize the contaminants without acidification. Chelating or complexing agents, such as citric acid and ethylenediaminetetraacetic acid (EDTA), have been demonstrated to be feasible for the extraction of different types of metal contaminants from fine-grained soils. The enhancement agents should form charged soluble complexes with the metal contaminants. Cox et al. (24) demonstrated the feasibility of using iodine/iodide lixivant to remediate mercury-contaminated soil. The use of EDTA as an enhancement agent has also been demonstrated for the removal of lead from kaolinite (22) and lead from sand (25).

D. Enhancement of Anolyte pH

While acidification of the soil causes dissolution of the soil minerals, it also increases the ionic strength and electric conductivity of the soil, which may hinder contaminant transport (13). If hydrogen ion generation and supply at the anode is not controlled, most of the energy may be consumed by generation and migration of this ion between the electrodes rather than the transport of charged contaminants. Therefore, it may be necessary to neutralize the anode reaction and/or control acid production and introduction into the soil mass.

VII. PRACTICAL CONSIDERATIONS

Although significant research was conducted in the 1980s and 1990s on electrokinetic remediation, one can argue that the fundamentals of the process are not yet well understood. The reason is that the objective of most of these studies was to demonstrate the feasibility of the remediation process and not to investigate the fundamental physiochemical and geochemical processes. For example, rates of electroosmosis and ionic migration in soils under heterogenous, anisotropic, and partially saturated soils are not easy to establish. Even in homogenous and saturated conditions, the complex geochemical reactions make it difficult to predict transport rates. In most cases these reactions (e.g., sorption) are assumed to be instantaneous, whereas in fact they are time dependent and the rate of these reactions can and will significantly affect time and energy requirements. In any case, a discussion is provided for practical applications of the process based. Few studies have looked into practical considerations for field implementations. Schultz (26) provided economic modeling and calculations of optimum spacings, time, and energy requirements of one-dimensional field applications based on electroosmotic transport. Some of the prominent practical considerations were also discussed by Alshawabkeh et al. (27).

A. Soil Type

Bench-scale and pilot-scale tests indicated that the technology can be successful in clayey to fine sandy soils. However, contaminant transport rates and the efficiency of the process depend heavily on soil type, mineral composition, and pore fluid conditions. Reddy et al. (23) demonstrated that presence of iron oxides in glacial till creates complex geochemical conditions that retard Cr(VI) transport. On the other hand, the same study showed that presence of iron oxides in kaolinite and Na-montmorillonite did not seem to affect Cr(VI) extraction significantly. Pamukcu and Wittle (28) and Wittle and Pamukcu (29) demonstrated removal of Cd^{2+}, Co^{2+}, Ni^{2+}, and Sr^{2+} from different soil types at variable efficiencies. The results showed that kaolinite, among different types of soils, had the highest removal efficiency, followed by sand with 10% Na-montmorillonite, while Na-montmorillonite showed the lowest removal efficiency. The results indicated that soils of high water content, high degree of saturation, low ionic strength, and low activity (soil activity describes soil plasticity and equals plasticity index divided by percent fines, clay, and silt in the soil) provide the most favorable conditions for transport of contaminants by electroosmotic advection and ionic migration. Highly plastic soils, such as illite, montmorillonite, or soils that exhibit high acid/base buffer capacity, require excessive acid and/or enhancement agents to desorb and solubilize contaminants before they can be transported through the subsurface and removed (30), thus requiring excessive energy.

Hydraulic conductivities of different soil types can vary many orders of magnitude within a heterogeneous deposit. For a contaminated soil deposit containing interlayers of sand and clay, typical values of hydraulic conductivities of these soils are 1×10^{-3} and 1×10^{-9} m/s, respectively. If pump-and-treat remediation techniques are used to remediate such a heterogeneous deposit, most of the fluid flow induced will occur in the sandy layer and the clayey layer will be practically untreated. On the other hand, electric conductivities of these soils are within an order of magnitude (1). As a result, the electric field strengths in the different soil layers will be similar when an externally electric potential is applied across the deposit. Similar ionic migration rates of contaminants can be generated in different soil layers within the heterogeneous deposit, resulting in a more homogeneous clean-up. The ability to remove contaminants uniformly from a heterogeneous natural deposit is another distinct advantage of the electrokinetic remediation technology. A combination of hydraulic and electric gradients may also enhance the process. Runnells and Wahli (31) showed the use of ion migration combined with soil washing for removal of Cu^{2+} and SO_4^- from fine sands. A field study reported by Banerjee et al. (32) also investigated the feasibility of using electrokinetics in conjunction with pumping to decontaminate a site from chromium. While soil chromium profile were not evaluated in this study, the results showed increase in effluent chromium concentrations.

B. Contaminant Types and Concentrations

Studies have showed that removal of heavy metals, radionuclides, and selected organics by electrokinetics is feasible. Hamed (33) and Hamed et al. (14) demonstrated electrokinetic remediation using kaolinite samples mixed with Pb^{2+} at various concentrations below and above the soil cation-exchange capacity. The process removed 75–95% of lead at concentrations of up to 1500 mg/kg, at reported energy expenditure of 29–60 kWh/m^3. However, since no enhancement procedure was used, most of the removed lead was found deposited at sections close the cathode. Acar et al. (34) demonstrated 90–95% removal of Cd^{2+} from kaolinite specimens with initial concentration of 99–114 mg/kg. Other laboratory studies reported by Runnels and Larson (35), Lageman et al. (36), Eykholt (9) and Acar et al. (37) further substantiate the applicability of the technique to a wide range of heavy metals in soils. The process can potentially remove radionuclides from clayey soil samples (38). Bench-scale tests demonstrated that uranium at 1000 pCi/g of activity was efficiently removed from kaolinite. A yellow uranium hydroxide precipitate was found in sections close to the cathode. Enhanced electrokinetic processing showed that 0.05 M acetic acid was enough to neutralize the cathode reaction and overcome uranium precipitation in the soil. Other radionuclides such as thorium and radium showed limited removal (39). In the case of thorium, it was postulated that precipitation of these radionuclides at their hydroxide solubility limits in the cathode region formed a gel that prevented additional transport and extraction. Limited removal of radium is believed to be either due to precipitation of radium sulfate or because radium binds strongly to the soil minerals, causing its immobilization (39).

Lageman et al. (36) showed that the process can move a mixture of different contaminants in soil simultaneously. Lageman (40) reported 73% removal of Pb at a concentration of 9000 mg/kg from fine argillaceous sand, 90% removal of As at 300 mg/kg from clay, and varying removal rates ranging between 50% and 91% of Cr, Ni, Pb, Hg, Cu, and Zn from fine argillaceous sand. Cd, Cu, Pb, Ni, Zn, Cr, Hg, and As at concentrations of 10–173 mg/kg also were removed from a river sludge at efficiencies of 50–71%. The energy expenditures ranged between 60 and 220 kWh/m^3 of soil processed. Therefore, one can conclude that the type of contaminant does not pose a significant limitation on the technology provided it does not exist in an immobile form, e.g., sorbed on the soil particle surface or precipitated in the soil pores.

Regarding contaminant concentrations, existing experimental data indicate that removal of Cu(II) of concentration up to 10,000 mg/kg of soil and Pb(II) concentration up to 5000 mg/kg are possible. Acar and Alshawabkeh (41) demonstrated extraction of lead at 5300 mg/kg from pilot-scale kaolinite samples. However, high ion concentrations in the pore fluid increase the electrical conductivity of the soil and thus reduce the efficiency of electroosmotic fluid flow (5,7). Moreover, the strength of electric field applied may have to be reduced to prevent excessive power consumption and heat generation during the process. The results of Hamed (33), Pamukcu and Wittle (28), and Wittle and Pamukcu (29) demonstrated that lower initial concentrations of cadmium result in higher electroosmotic efficiency; however, removal efficiencies were higher for samples with higher initial concentrations. Alshawabkeh et al. (30) investigated electrokinetic extraction of heavy metals from clay samples retrieved from a contaminated U.S. Army ammunition site. The soil contained cations at the following concentrations: calcium, 19,670 mg/kg; iron, 11,840 mg/kg; copper, 10,940 mg/kg; chromium, 9930 mg/kg; zinc, 6330 mg/kg; and lead, 1990 mg/kg. The high calcium concentration hindered extraction of the metals. However, the results showed that metals with higher initial concentration, less sorption affinities, higher solubilities, and higher ionic mobilities are transported and extracted more quickly than other metals.

Electrokinetic remediation may also effective for the removal of organic pollutants such as phenol, gasoline hydrocarbons, and TCE from contaminated soils. Successful application of the process has been demonstrated for extraction of the BTEX (benzene, toluene, ethylene, and *m*-xylene) compounds and trichloroethylene from kaolinite specimens at concentrations below the solubility limit of these compounds (42,43). High degrees of removal of phenol and acetic acid (up to 94%) were also achieved by the process (44,45). Acar et al. (39) reported removal of phenol from saturated kaolinite by the technique. Two pore volumes were sufficient to remove 85–95% of phenol at an energy expenditure of 19–39 kWh/m^3. Wittle and Pamukcu (29) investigated the feasibility of removal of organics from different synthetic soil types. Tests were conducted on kaolinite, Na-montmorillonite, and sand samples mixed with different organics. Their results showed the transport and migration of acetic acid and acetone toward the cathode. Samples mixed with hexachlorobenzene and phenol are reported to show accumulation at the center of each samples. The results of some of these experiments were inconclusive, either because contaminant concentrations were below detection limits or because the samples were processed for only 24 h, which might not be sufficient to demonstrate feasibility in electrokinetic soil remediation. Recently, the U.S. Department of Energy (DOE), the U.S. Environmental Protection Agency (EPA), Monsanto, General Electric, and Dupont have also applied electric fields for electroosmotic extraction using layered horizontal electrodes in what is called the "Lasagna^{TM}" process.

Ho et al. (46) reported successful extraction of TCE from a site in Paducah, Kentucky, using the Lasagna^{TM} process. They also reported 98% removal efficiency of p-nitrophenol, as a model organic compound, from soil in a pilot study. Although removal of free-phase nonpolar organics is questionable, Mitchell (47) stated that this could be possible if the organic was present as small bubbles (emulsions) that could be swept along with the water moving by electroosmosis. Acar et al. (37) stated that unenhanced electrokinetic remediation of kaolinite samples loaded with up to 1000 mg/kg hexachlorobutadiene has been unsuccessful. However, Acar et al. (37) reported that hexachlorobutadiene transport was detected only when surfactants were used.

C. Voltage and Current Levels

Electric current intensities used in most reported studies are in the order of a few amperes per square meter. Although high current levels generate more acid which may solubilize more metals, this will also increase the total ionic concentration and decrease the overall electroosmotic flow. Selection of the most appropriate current density and voltage gradients depends on the soil electrochemical properties, especially electric conductivity. Soils with higher electric conductivities require more charge and higher currents than lower conductivity soils. A voltage gradient of the order of 100 V/m can be used as an estimate for initial processing. Increasing the current densities (or voltage gradients) will increase transport rates under ionic migration. However, increasing current densities will increase energy expenditures and cost of the process. An optimum current density or voltage gradient could be approximated based on soil properties, electrode spacing, and time requirements of the process. A procedure is provided in this chapter for approximating the current density.

D. Electrode Configuration and Time Requirements

Electric currents could be applied in soils to generate one-dimensional (1D) or two-dimensional electric fields. 1D electric fields could be generated by electrode sheets with specific spacing between the anodes and cathodes. Using sheets requires trenches and is not expected to be cost-effective in most cases. Placing electrodes in bore holes is expected to provide an optimum and cost-effective method for in-situ application of electric fields. However, problems associated with bore-hole configuration include development of inactive (dead zones) electric field spots between the treatment zones.

Several factors will affect electrode configuration, spacing, and time requirements for electrokinetic remediation. These factors include (a) location and size of any electrical dead zones, (b) number and costs of electrodes per unit area to be treated, and (c) time requirements of the remediation process. Large electrode spacings reduce the number of boreholes and installation costs but increase the processing time and operation costs. It will be necessary to optimize these variables prior to selecting a configuration and spacing arrangement.

In general, the processing time required is a function of the rate of transport and electrode spacing. As electroosmotic advection and ionic migration are the prominent transport mechanisms, hydrodynamic dispersion and retardation can be neglected in a preliminary analysis. Accordingly, rate of species transport under an electric field is given by

$$v = (u^* + k_e)\nabla(-\phi) \quad (1)$$

where v = rate of species transport (or velocity) assuming the soil is a homogeneous medium (m/s); u^* = effective ionic mobility (m²/V·s); k_e = coefficient of electroosmotic conductivity (m²/V·s), and ϕ is the electric potential (V). If the spacing between electrodes of opposite polarity is chosen to be L, the time (T) required for remediation can be estimated by dividing the spacing (L) over the velocity of species transport (v), i.e.

$$T = \frac{L}{(u^* + k_e)\nabla(-\phi)} \quad (2)$$

where T is the time required for clean-up. However, Eq. (2) represents a simplified estimation of time requirements, where the contaminant of interest is assumed to be readily available for transport in the soil pore fluid. This is probably the exception rather than the rule in field implementation of electrokinetic remediation. Heavy metals are usually either sorbed on the soil particle surface or precipitated in the soil pore. Therefore, their transport is retarded. A delay factor similar to the retardation factor in advection-dispersion contaminant transport can be introduced to account for the extra time required for acid transport, metal desorption and dissolution, etc. Therefore, the modified transport equation correcting for ion retardation is

$$T = \frac{R_t L}{(u^* + k_e)\nabla(-\phi)} \quad (3)$$

where R_t = delaying factor (dimensionless). The value of R_t depends on soil type, pH, and type of contaminant and is determined through experimentation. Sorption retardation factor can be used as an initial estimate of R_t and equals unity for nonreactive contaminants which are readily available for transport. If enhancement agents are used to solubilize heavy metals, this factor should be modified accordingly. Alshawabkeh et al. (27) defined a new parameter (β) to estimate the reactive transport rate of a species relative to the electric conductivity of a medium, i.e.,

$$\beta = \frac{u^* + k_e}{R_t \sigma^*} \quad (4)$$

where σ^* is the effective electric conductivity of the soil medium (siemens/m or s/m). Accordingly, the time required for remediation in 1D applications is given by

$$T = \frac{1}{\beta} \frac{L}{\sigma^* \nabla(-\phi)} \quad (5)$$

Equation (5) indicates that for a given electric field strength, the time required for remediation is linearly related to the spacing between the electrodes.

E. Energy Expenditure

Total energy required to treat a unit volume of contaminated soil depends on many factors, including soil properties, contaminant properties, and electrode configuration and spacing. If the electrical conductivity of the contaminated soil is assumed to be constant throughout the process as an approximation, the energy expenditure per unit volume of contaminated soil is given by the following equation:

$$W = \frac{\phi_{max} I_d T}{L} \tag{6}$$

where W = energy expenditure per unit volume of soil (J/m^3) and I_d = current density per unit cross area (A/m^2). Substituting the equation that describes time requirements for remediation [Eq. (5)] into Eq. (6) leads to

$$W = \frac{\phi_{max}}{\beta} \tag{7}$$

As described by Eq. (4), β is a lumped property of the contaminant and the soil. It represents the rate of transport of a specific species per unit electric current density. It is evident from Eq. (7) that the energy expenditure depends on both the soil and contaminant characteristics. A high coefficient of electroosmotic conductivity, k_e, and/or a high ionic mobility of the contaminant, u^*, increase the value of β and reduce energy expenditure W. High ionic strength of the pore fluid increases the electrical conductivity of the soil, reduces the value of β, and thus increases energy expenditure W. An increase in the delaying factor R_t also increases energy expenditure. Typical values of β for contaminated fine-grained soils are estimated to be in the range of 1×10^{-9} to 1×10^{-6} m^3/C.

VIII. COSTS

The total costs for full-scale in-situ implementation of electrokinetic remediation can be divided into five major components (27): (a) cost of electric energy, (b) cost for fabrication and installation of electrodes and system, (c) cost of enhancement agents, if necessary, (d) costs of any posttreatment, if necessary, and (e) fixed costs. Impacts of electrode configuration and spacing on these cost components are addressed separately.

A. Electric Energy Cost

Equation (7) provides an estimate for energy expenditure per unit volume of the soil treated. Thus, the electric energy cost of the treatment process can be estimated by

$$C_{energy} = \frac{C_2 \phi_{max}}{3,600,000 \times \beta} = \frac{C_2 (\nabla \phi L)}{3,600,000 \times \beta} \tag{8}$$

where C_{energy} = electric energy cost per unit volume of soil treated (\$/m^3); C_2 = electric energy cost (\$/kWh). It is noted that the equations are provided as a function of the voltage gradient and not the maximum voltage applied. The reason is that constant voltage gradient could be used in different applications and arrangement, while the maximum voltage will be dependent on the spacing between the electrodes.

B. Costs for Fabrication and Installation of Electrodes

If acid production at the anode is not controlled, then inert electrodes, such as graphite, coated titanium, should be used to prevent dissolution of the electrode and generation of undesirable corrosion products. Acid production at the anode doesn't preclude the use of sacrificial anode, but proper design and maintenance issues must be considered. In contrast, any conductive materials that do not corrode in a neutral or basic environment can be used as cathode. Important considerations for the choice of electrode material are (27) (a) electrical conduction properties of the material, (b) availability of the material, (c) ease of fabrication to the form required for the process, (d) ease of installation in the field, and (e) material, fabrication, and installation costs. The electrodes can be installed horizontally or vertically. The costs of each electrode depend on the material used, complexity of installation, and dimensions. The number of electrodes per unit volume of soil to be treated depends on electrode configuration and spacing. The installation costs depend on the method of installation, depth of the electrodes to be installed, and number of electrodes to be installed. The total electrode costs per unit volume of soil to be treated can be calculated by evaluating the number of electrodes per unit cell of an area of L^2, i.e.,

$$C_{\text{electrode}} = C_1 \frac{F_1}{L^2} \tag{9}$$

where $C_{\text{electrode}}$ = electrode costs per unit volume of soil to be treated, C_1 = cost of an electrode to be installed per unit depth, and F_1 is the number of electrodes per the area (L^2). Values for F_1 could be easily calculated based on configurations and spacings (27). C_1 includes the unit costs for material and fabrication of the electrode, drilling and preparation of the borehole, and placement of the electrode.

C. Cost for Enhancement Agent

Cost of chemical amendments could be a significant component of the total cost of electrokinetic remediation. Various chemicals could be used for different enhancement schemes, such as neutralizing the pH conditions or enhancing solubility of target contaminants. In this case, we will consider the cost of chemicals required for neutralizing the catholyte pH, which depends on electric current applied and is given by the following equation:

$$C_{n-\text{chemical}} = C_3 \frac{I_d}{L} \frac{M_W}{\alpha F} T \tag{10}$$

where $C_{n-\text{chemical}}$ is the cost of chemicals required to neutralize electrolytes per unit soil volume ($/L^3$), C_3 is the cost of the chemical agent ($/M); M_W is the molecular weight of the neutralizing chemical, α is a factor depending on the stoichiometry of the neutralizing reaction (dimensionless), and F is Faraday's constant (96,485 C/mol-electron). Substituting the time required for remediation results in the following equation:

$$C_{n-\text{chemical}} = \frac{C_3}{\beta} \frac{M_W}{\alpha F} \tag{11}$$

Equation (11) shows that chemicals cost is independent of electric current or spacing and is dependent on soil characteristics. This is due to the fact that electric current and electrode spacings affect time requirements. For example, increasing the current decreases the time required for remediation, such that the same total charge is introduced for any electric current value.

D. Cost of Posttreatment

Posttreatment costs should also be considered if effluent treatment is required. These costs are highly site and contaminant specific. An estimate of effluent treatment costs could be evaluated per unit volume of the soil for 1D case as follows:

$$C_{posttreat} = C_4 \frac{k_e \nabla(-\phi)}{(L/n)} T \tag{12}$$

where $C_{posttreat}$ is the posttreatment cost per unit volume of the soil ($/L^3$), n is the soil porosity and C_4 is the cost of treatment per unit volume of the electrolyte (effluent) collected ($/L^3$). Substituting for the value of T (time required for remediation), then effluent treatment cost is given by

$$C_{posttreat} = C_4 \frac{nk_e}{\beta \sigma^*} = C_4 \frac{nR_t k_e}{u^* + k_e} \tag{13}$$

Derivations for radial transport posttreatment costs also result in Eq. (13). Volume and cost of effluent treatment depends on the ratio of transport under electroosmosis relative to total transport rate. In order to minimize the volume collected, it is necessary to maximize transport by ionic migration and minimize transport by electroosmosis. If contaminant transport occurs only due to migration, then this cost component will be limited to the cost of frequent treatment of the electrolyte in the electrode well. However, if electroosmosis is the only mechanism used for contaminant transport (e.g., for noncharged contaminants), then cost of treatment will be equal to $(C_4 n R_t)$, which indicates that the cost depends on the number of pore volumes required for remediation. If the contaminant is readily available for transport, then $R_t = 1$ and one pore volume is enough for remediation. However, if extraction is retarded due to geochemical reactions, then it is obvious that the pore volumes required will increase depending on the value of R_t. Sometimes catholyte recycling is used, which will add another component that should be considered for evaluation of total volume of water collected.

E. Other Fixed and Variable Costs

Other costs for full-scale implementation include mobilization and demobilization costs of various equipment, site preparation, security, progress monitoring, insurance, labor, contingency, and miscellaneous expenses. The equipment will not be consumed in a particular project. However, there are capital, depreciation, or rental costs involved. These cost components will be divided into fixed (e.g., mobilization and demobilization) and variable (e.g., monitoring, insurance, rentals) components. Variable costs are simply evaluated by multiply cost rate by the total time required for remediation, i.e.,

$$C_{variable} = \frac{C_5}{\beta} \frac{L}{\sigma^*(-\nabla \phi)} \tag{14}$$

where $C_{variable}$ is the total variable cost per unit soil volume ($/L^3$) and C_5 is the variable cost rate per unit soil volume ($/L^3 T$). C_5 is evaluated by estimating the variable daily cost (for monitoring, insurance, rentals, etc.) and dividing by the total volume of site. C_5 is highly dependent on the size of the site and decreases as volume of contaminated soil increases.

F. Total Costs

The total costs per unit volume of soil to be treated are thus given by

$$C_{total} = C_{electrode} + C_{energy} + C_{chemical} + C_{posttreat} + C_{fixed} + C_{variable} \tag{15}$$

where C_{total} = total costs per unit volume of soil to be treated ($/L^3$), and C_{fixed} = fixed costs per unit volume of soil to be treated ($/L^3$). Cost evaluation indicates that electrode configuration will affect electrode, energy, and variable costs. Other costs (chemicals, fixed, and posttreatment) are independent of electrode configuration and spacing.

IX. OPTIMUM ELECTRODE SPACINGS

Assuming posttreatment, chemicals, and fixed costs are independent of electrode spacing, the optimum electrode spacing can be obtained for 1D applications by equating the partial derivative of C_{total} with respect to L to zero, which renders the following equation:

$$L_{optimum}^3 = \frac{7,200,000 \beta C_1 F_1 \nabla \phi \sigma^*}{3,600,000 C_5 + C_2 (\nabla \phi)^2 \sigma^*} \quad (16)$$

where $L_{optimum}$ = optimum electrode spacing (L). Equation (16) provides an estimate for the optimum electrode spacing that minimizes the total costs of 1D applications as a function of the properties of the contaminated soil and electric field strength. The optimum spacing is also dependent on electrode costs (C_1), energy costs (C_2), variable costs (C_5), and electrode configuration (F_1). Equation (16) demonstrates the impact of the cost ratios C_2/C_1 (energy cost/electrodes cost) and C_5/C_1 (variable cost/electrodes cost) on optimum electrode spacings.

In some cases, time requirements might be the limiting factor, where remediation needs to be completed within a specific time period. For such cases, one can substitute the values of $\nabla \phi$ from Eq. (5) in Eq. (16), so that the optimum electrode spacing could be evaluated based on time requirement by the following equation for 1D applications:

$$L_{optimum}^2 = \frac{7,200,000 \beta^2 C_1 F_1 \sigma^* T}{3,600,000 \sigma^* \beta^2 T^2 C_5 + C_2 L_{optimum}^2} \quad (17)$$

Thus, the procedure for design of field implementation is to decide whether the limiting factor is time or energy.

X. EXAMPLE

An example (27) is given on the estimation of total costs per unit volume of contaminated soil to be treated by electrokinetics. In this example, a contaminated area of 50 m × 100 m in plan is assumed. The depth of contamination is assumed to be 5 m. The soil is a saturated silty clay. The electrical conductivity and coefficient of electroosmotic conductivity of the soil are determined to be 0.02 S/m and 1×10^{-9} m^2/V-s. A heavy-metal contaminant is assumed with effective ionic mobility of 5×10^{-9} m^2/V-s. The value of the delaying factor R_t is dependent on soil sorption capacity, soil acid/base buffer capacity, and pore fluid chemistry. It is taken to be 3 in this example. The value of β is thus calculated to be 1×10^{-7} m^3/C.

The mobilization cost of a drilling rig and the labor cost of a two-man operating crew are taken to be $1000/day to evaluate C_1. For boreholes to be drilled without installation of casing and sampling, a continuous flight auger can achieve approximately 65 m/day. Therefore, the drilling cost is estimated to be $15/linear meter. Costs for fabrication and installation of electrodes are approximately $5/linear meter, as the electrodes are reusable. Therefore, C_1 is taken to $20/linear meter. The electricity cost C_2 is assumed

$0.04/kWh. Cost of enhancement reagent (molecular weight assumed 100 g/mole) is taken as \$2/kg. Variable cost rate is calculated based on two-man crew, 14 h/week, at a rate of 25 \$/labor-h. Accordingly, the variable rate, C_5, will be around 0.001 \$/m^3 h, including 30% increase for insurance. Total cost of other components, including posttreatment and fixed costs, is taken as \$25/m^3. If 1D electrode configuration is used with spacing between electrodes of the same polarity equals one-third the anode–cathode spacing, then the factor F_1 in Eq. (9) equals 3.

The two limiting factors, i.e., processing time and electric voltage, are considered. If time available is taken to be the limiting factor and that the remediation has to be finished in six months, Eq. (17) yields an optimum electrode spacing of 2.2 m. The electric field strength required is determined by Eq. (5) to be 71 V/m. The electric voltage across electrodes of opposite electrodes is thus 156 V. The electrode costs per unit volume of soil are given by Eq. (9) to be \$12.4/m^3. The energy cost per unit volume of soil is given by Eq. (8) to be \$17.3/m^3. If chemical costs, posttreatment costs, and fixed costs are estimated to be \$30/m^3, the total costs are \$70/m^3. Total costs of the project will thus be \$1,750,000.

If the electric voltage available is the limiting factor, the limiting voltage should be used in Eq. (16) to obtain the optimum electrode spacing. The processing time required, energy expenditure, and total costs per unit volume of contaminated soil of the remediation process can then be estimated similarly. If the limiting voltage is taken to be 100 V in this example, the optimum electrode spacing given by Eq. (16) is 2.5 m. The electrode costs are \$9.6/m^3 as given by Eq. (9). The energy cost per unit volume of soil is \$11.1/m^3. The total costs per unit volume of soil are \$51/m^3. Total costs of the project are thus reduced to \$1,275,000. However, it will take one year to remediate the site as estimated by Eq. (5).

REFERENCES

1. Mitchell, J. K. 1993. *Fundamentals of Soil Behavior*, 2nd ed. Wiley, New York.
2. Hunter, R. J. 1981. *Zeta Potential in Colloid Science*. Academic Press, New York.
3. Stumm, W. 1992. *Chemistry of the Solid–Water Interface, Processes at the Mineral–Water and Particle–Water Interface in Natural Systems*. Wiley-Interscience, New York.
4. Kruyt, H. R. 1952. *Colloid Science (I): Irreversible Systems*, Elsevier.
5. Gray, D. H., and Mitchell, J. K. 1967. Fundamental aspects of electro-osmosis in soils. *J. Soil Mech. Found. Div., Proc. ASCE*, 93(6):209–236.
6. Vane, M. L., and Zang, G. M. 1997. Effect of aqueous phase properties on clay particle zeta potential and electroosmostic permeability: Implications for electrokinetic remediation processes. Electrochemical Decontamination of Soil and Water, Special Issue of *J. Hazardous Mater.*, 55(1–3):1–22.
7. Lockhart, N. C. 1983. Electro-osmotic dewatering of clays I, II, and II. *Colloids Surfaces*, 6:238–269.
8. Lorenz, P. B. 1969. Surface conductance and electrokinetic properties of kaolinite beds. *Clays and Clay Minerals*, 17:223–231.
9. Eykholt, G. R. 1992. Driving and complicating features of the electrokinetic treatment of contaminated soils. PhD thesis, Department of Civil Engineering University of Texas at Austin.
10. Eykholt, G. R., and Daniel, D. E. 1994. Impact of system chemistry on electroosmosis in contaminated soil. *J. Geotech. Eng.*, 120(5):797–815.

11. Acar, Y. B., Gale, R. J., Putnam, G., and Hamed, J. 1989. Electrochemical processing of soils: Its potential use in environmental geotechnology and significance of pH gradients. In: *2nd Int. Symp. Environmental Geotechnology*, Shanghai, China, May 14–17, Vol. 1, pp. 25–38. Envo Publishing, Bethlehem, PA.
12. Acar, Y. B., Gale, R. J., Putnam, G. A, Hamed, J., and Wong, R. L. 1990. Electrochemical processing of soils: Theory of pH gradient development by diffusion, migration, and linear convection. *J. Environ. Sci. Health*, A25(6):687–714.
13. Acar, Y. B., and Alshawabkeh, A. 1993. Principles of electrokinetic remediation. *Environ. Sci. Technol.*, 27(13):2638–2647.
14. Hamed, J., Acar, Y. B., and Gale, R. J. 1991. Pb(II) removal from kaolinite using electrokinetics. *J. Geotech. Eng.*, 117(2):241–271.
15. Probstein, R. F., and Hicks, R. E. 1993. Removal of contaminants from soil by electric fields, *Science*, 260:498–503.
16. Yeung, A. T., and Datla, S. 1995. Fundamental formulation of electrokinetic extraction of contaminants from soil. *Can. Geotech. J.*, 32(4):569–583.
17. Alshawabkeh, A. N., and Acar, Y. B. 1992. Removal of contaminants from soils by electrokinetics: A theoretical treatise. *J. Environ. Sci. Health*, A27(7):1835–1861.
18. Acar, Y. B., and Alshawabkeh, A. N. 1994. Modeling conduction phenomena in soils under an electric current. *Proc. XIII Int. Conf. on Soil Mechanics and Foundation Engineering* (ICSMFE), New Delhi, India.
19. Rødsand, T., Acar, Y. B., and Breedveld, G. 1995. Electrokinetic extraction of lead from spiked Norwegian marine clay. In: *Characterization, Containment, Remediation, and Performance in Environmental Geotechnics*, Geotech. Spec. Publ. No. 46, vol. 2, pp. 1518–1534. ASCE, New York.
20. Puppala, S., Alshawabkeh, A. N., Acar, Y. B., Gale, R. J., and Bricka, R. M. 1997. Enhanced electrokinetic remediation of high sorption capacity soils. Electrochemical Decontamination of Soil and water, Special Issue of *J. Hazardous Mater.*, 55(1–3):203–220.
21. Hicks, R. E., and Tondorf, S. 1994. Electrorestoration of metal contaminated soils. *Environ. Sci. Technol.*, 28(12):2203–2210.
22. Yeung, A. T., Hsu, C., and Menon, R. M. 1996. EDTA-enhanced electrokinetic extraction of lead. *J. Geotech. Eng.*, 122(8):666–673.
23. Reddy, K. R., Parupudi, U, S., Devulapalli, S. N., and Xu, C. Y. 1997. Effect of soil composition on removal of chromium by electrokinetics. Electrochemical Decontamination of Soil and Water, Special Issue of *J. Hazardous Mater.*, 55(1–3):135–158.
24. Cox, C. D., Shoesmith, M. A., and Ghosh, M. M. 1996. Electrokinetic remediation of mercury-contaminated soils using iodine/iodide lixivant. *Environ. Sci. Technol.*, 30(6):1933–1938.
25. Wong, J. S., Hicks, R. E., and Probstein, R. F. 1997. EDTA-enhanced electroremediation of metal contaminated soils. Special Issue on Electrochemical Decontamination of Soil and Water, *J. Hazardous Mater.*, 55(1–3):61–80.
26. Schultz, D. S. 1997. Electroosmosis is technology for soil remediation: Laboratory results, field trial and economic modeling. Electrochemical Decontamination of Soil and Water, Special Issue of *J. Hazardous Mater.*, 55(1–3):81–92.
27. Alshawabkeh, A. N., Yeung, A., and Bricka, R. M. 1999. Practical aspects of in situ electrokinetic remediation. *J. Environ. Eng.*, 125(1):27–35.
28. Pamukcu, S., and Wittle, J. K. 1992. Electrokinetic removal of selected heavy metals from soil. *Environ. Prog.*, 11(4):241–250.
29. Wittle, J. K., and Pamukcu, S. 1993. Electrokinetic treatment of contaminated soils, sludges, and lagoons. Final Report, Contract No. 02112406, DOE/CH-9206, Argonne National Laboratory, Chicago.
30. Alshawabkeh, A. N., Puppala, S. K., Acar, Y. B., Gale, R. J., and Bricka, R. M. 1997. Effect of solubility on enhanced electrokinetic extraction of metals. In situ Remediation of the Geoenvironment (In Situ Remediation '97), Minneapolis, MN, Oct. 5–8.

31. Runnells, D. D., and Wahli, C. 1993. In situ electromigration as a method for removing sulfate, metals, and other contaminants from ground water. *Ground Water Monitor. Remediation*, 13(1):121–129.
32. Banarjee, S., Horng, J., Ferguson, J., and Nelson, P. 1990. Field scale feasibility of electrokinetic remediation. Report presented to U.S. EPA, Land Pollution Control Division, PREL, CR 811762-01.
33. Hamed, J. 1990. Decontamination of soil using electro-osmosis. Ph.D. thesis, Louisiana State University.
34. Acar, Y. B., Hamed, J. T., Alshawabkeh, A., and Gale, R. J. 1994. Cd(II) removal from saturated kaolinite by application of electrical current. *Géotechnique*, 44(3):239–254.
35. Runnels, D. D., and Larson, J. L. 1986. A laboratory study of electromigration as a possible field technique for the removal of contaminants from ground water. *Ground Water Monitor. Rev.*, Summer, pp. 81–91.
36. Lageman, R., Pool, W., and Seffinga, G. 1989. Electro-reclamation: theory and practice, *Chem. Ind.*, 18:585–590.
37. Acar, Y. B., Alshawabkeh, A., and Gale, R. J. 1993. Fundamentals of extracting species from soils by electrokinetics. *Waste Management*, 13(2):141–151.
38. Ugaz, A., Puppala, S., Gale, R. J., and Acar, Y. B. 1994. Electrokinetic soil processing: Complicating features of electrokinetic remediation of soils and slurries: Saturation effects and the role of the cathode electrolysis. *Chem. Eng. Commun.*, 129:183–200.
39. Acar, Y. B., Li, H., and Gale, R. J. 1992. Phenol removal from kaolinite by electrokinetics. *J. Geotech. Eng.*, 118(11):1837–1852.
40. Lageman, R. 1993. Electro reclamation: application in the Netherlands. *Environ. Sci. Technol.*, 27(13):2638–2647.
41. Acar, Y. B., and Alshawabkeh, A. N. 1996. Electrokinetic remediation: I. Pilot-scale tests with lead-spiked kaolinite. *J. Geotech. Eng.*, 122(3):173–185.
42. Bruell, C. J., Segall, B. A., and Walsh, M. T. 1992. Electroosmotic removal of gasoline hydrocarbons and TCE from clay. *J. Environ. Eng.*, 118(1):68–83.
43. Segall, B. A., and Bruell, C. J. 1992. Electroosmotic contaminant removal processes. *J. Environ. Eng.*, 118(1):84–100.
44. Shapiro, A. P., Renaud, P. C., and Probstein, R. F. 1989. Preliminary studies on the removal of chemical species from saturated porous media by electroosmosis. *PCH PhysicoChemical Hydrodynam.*, 11(5/6):785–802.
45. Shapiro, A. P., and Probstein, R. F. 1993. Removal of contaminants from saturated clay by electroosmosis. *Environ. Sci. Technol.*, 27(2):283–291.
46. Ho, S. V., Athmer, C. J., Sheridan, P. W., and Shapiro, A. P. 1997. Scale-up aspects of the LasagnaÔ process for in situ soil decontamination. Electrochemical Decontamination of Soil and Water, Special Issue of *J. Hazardous Mater.* 55(1–3):39–60.
47. Mitchell, J. K. 1991. Conduction phenomena: From theory to geotechnical practice. *Géotechnique* 41(3):299–340.
48. Denisov, G., Hicks, R. E., and Probstein, R. F. 1996. On the kinetics of charged contaminant removal from soils using electric fields. *J. Colloid Interface Sci.*, 178(1):309–323.
49. Pamukcu, S., Khan, L., and Fang, H. 1990. Zinc detoxification of soils by electroosmosis. Electro-kinetic Phenomena in Soils, Transportation Research Record, TRB, Washington, DC.
50. West, L. J., and Stewart, D. I. 1995. Effect of zeta potential on soil electrokinesis. In: *Characterization, Containment, Remediation, and Performance in Environmental Geotechnics*, Geotech. Spec. Publ. No. 46, Vol. 2, pp. 1535–1549. ASCE, New York.
51. Yeung, A. T., Hsu, C., and Menon, R. M. 1997. Physicochemical soil-contaminant interactions during electrokinetic extraction. Electrochemical Decontamination of Soil and Water, Special Issue of *J. Hazardous Mater.*, 55(1–3):221–238.

5
Measurement of Pollutant Distribution, Toxicity, and Effectiveness of Emerging Soil Remediation Techniques

K. Bundy, F. Mowat, P. Taverna, and M. Shettlemore
Tulane University, New Orleans, Louisiana

I. INTRODUCTION

Environmental pollution from industrial, energy, and defense-related operations is widespread in many countries and results in contamination of the land, water, and air with which all living organisms interact, often producing toxic responses. For example, approximately 16,000 U.S. Department of Defense (DoD) installations are currently thought to be contaminated (1). Those with the worst problems (in terms of contaminant ubiquity and human health hazards) appear on the National Priorities List (NPL). Developing valid methods for recording the extent of pollution is central to assessing environmental damage and remediation efforts. This development is often difficult, however, due to speciation effects and distribution of contaminants among multiple phases. In addition, contaminants are often present in the environment as mixtures of unknown composition, making the task of assessing hazards even more difficult.

For the reasons given above, ecosystem damage and potential human health threats from a myriad of contaminants are creating increasing concern worldwide. Complexities exist both for identifying severity of hazards posed by contamination at given sites as well as determining efficient means for remediation. Not surprisingly, given the complexity mentioned, many approaches have been developed, or are emerging, to both detect pollutants and eliminate, or reduce, their harm through remediation. To remediate waterways and wetlands with complicated pollution patterns (including those polluted by heavy metals, organic compounds, or both), a number of strategies can be employed. Some of these are based on traditional techniques, while others use newly emerging technologies, such as extraction treatment, electrokinetics, zero-valent metals, and phytoremediation.

Despite the diverse, multifactorial nature of these problems, there is also a degree of commonality among them, and certain questions and concerns are universally applicable. This chapter explores such themes and presents various strategies we have used to address these issues. It focuses on chemical analysis and toxicity assessment of heavy metals and organics in soil and sediment relevant to defense and energy installations, but other types of phases and environments are considered.

II. BACKGROUND

Identifying polluted sites and gauging the extent of and hazards associated with the contamination (through fieldwork and laboratory studies) are essential first steps for a rational approach to remediation. We have recently developed an approach for on-site measurements using field-deployable Microtox, DeltaTox, and cyclic voltammetry (CV) sensor technology. These techniques are based, respectively, on bioluminescent bacterial toxicity assays and electrochemical principles. The CV test is capable of measuring concentrations of metals and selected organics, as well as metal speciation, an important issue since various chemical species may exhibit differential toxicity. We use a dual approach of measuring both concentration and toxicity, allowing identification of possible synergistic or antagonistic interactions among pollutants in complex mixtures. Besides being useful for identifying polluted areas and assessing their status, these same techniques are equally applicable to monitoring the effectiveness of clean-up efforts and helping to guide remediation planning strategies.

A. Contaminants of Interest

Industrial, defense, and energy operations involve a broad spectrum of contaminants. To consider just one specific example, some of the major types at the DoD NPL sites are heavy metals, pesticides, phenols, volatile and semivolatile organic compounds (VOCs and SVOCs), polycyclic aromatic hydrocarbons (PAHs), petroleum-related compounds, explosives, phosphates, and nitrates (2,3). Many of these interact with receptors in the immune, nervous, and endocrine (INE) systems, resulting in various toxic and systemic effects. A literature review (4–10) has identified 124 toxicants affecting one or more of these systems, including environmental estrogens, polyaromatic hydrocarbons, organophosphorous compounds, organohalogens, chlorinated hydrocarbons, aromatic carbamates, organotins, and heavy metals. Of these, 75 are endocrine disrupters, 69 are neurotoxins, and 24 are immunotoxins. Of these toxicants, 36 act on at least two systems. Some toxicants affecting all three systems include the heavy metals lead, cadmium, arsenic, and mercury, and the organics aldrin, dieldrin, hexachlorobenzene (HCB), DDT, polychlorinated biphenols (PCBs), methoxychlor, tetrachlorodibenzodioxin (TCDD), Mirex, and pentachlorophenol (PCP). Similarly, many substances are known to be cytokine disrupters and immunotoxins, such as vanadium, lead oxide, sulfuric acid, methylmercury chloride, tobacco, benzo(a)pyrene, and tetrandine (11–17).

Considering the specific case of heavy metals, the American Toxic Substance and Disease Registry and the U.S. Environmental Protection Agency compile a listing of the nation's 20 most threatening chemical pollutants (18). The three most hazardous substances are metals: arsenic (1st place), lead (2nd), and metallic mercury (3rd). Other metal hazards are cadmium (ranked 7th) and hexavalent chromium (16th). The hazards posed by these and other metallic pollutants to the natural environment of the United States makes combating metal contamination one of the country's top environmental priorities. Similarly, five of the organic species mentioned above are identified on the ATSDR/EPA list (ranked 6th, 8th, 10th, 13th, and 18th). Louisiana, a state where heavy industries related to energy production and manufacturing (petroleum, chemical, etc.) and commercial fishing coexist, is an area where such problems are particularly acute. Several areas, including the Mississippi River, the Atchafalaya Basin, and Lake Pontchartrain, are prime examples. Clearly, there is an urgent need for sensors that can reliably measure concentrations and effects of organic and metal pollutants affecting human health and/or causing environmental damage to air, water, soil, and biota.

B. Speciation and Risk Assessment Issues

A great many factors influence risks posed by environmentally released chemicals to ecosystems and human health. For example, consider the case of heavy metals. Heavy metals often exist in chemical forms differing widely in structure, solubility, and oxidation state, depending on characteristics of the environment such as pH, oxygen content, and salinity (19). These environmental parameters can often undergo substantial fluctuations in their levels. For example, in the Gulf coast off Louisiana, oxygen concentration can show a wide seasonal variation in which 30-fold changes in concentration can occur (20). Thus, there clearly is a possibility that various chemical forms of a single metal can be present in a particular waterway. These chemical forms may also differ substantially in toxicity. For instance, chromium can exist in a trivalent form, Cr^{3+}, which is an essential trace element with low toxicity for occupational and environmental exposure. Alternatively, Cr^{6+} has been implicated in allergic hypersensitivity reactions and carcinogenesis (21). It has been suggested that photochemical reactions can convert labile Cr^{3+}, including dissolved and colloidal forms, to Cr^{6+} in aerobic surface waters (22,23) indicating that even natural processes such as sunlight may serve to amplify or alter the toxic effects of certain metals. Besides chromium, other heavy metals known to have chemical species-dependent toxicity include arsenic ($+3$ and $+5$) and copper ($+1$ and $+2$). Knowledge of the actual metallic species (and their concentrations) in fresh and salt water would therefore dramatically reduce uncertainties associated with assessing heavy-metal risks to human health, food supplies, ecosystems, and commercial fishing industries. Thus, for the most meaningful results, not only must analytical methods identify a particular element's presence but also its specific chemical form to determine the associated risk. This issue may be particularly acute where different forms of pollutants or chemical agents work through various INE receptors, for example, due to extremely specific interactions between receptors and target molecules.

III. MATERIALS AND METHODS

A. Basic Principles of Chemical Concentration Measurement

Most metal assays of environments such as those mentioned above have been concerned with determining the total amount of metal present, as most chemical analytical techniques, such as atomic absorption spectroscopy (AAS) and inductively plasma atomic emission spectroscopy (ICP-AES), do not differentiate between different heavy-metal oxidation states. Polarography is a technique that can differentiate between various chemical forms and is therefore very useful for speciation studies. Polarography is actually a family of methods, one of which (cyclic voltammetry) forms the basis for our field sensor.

1. Polarographic Techniques

To consider the basic principles involved in more detail, the polarographic method is an electroanalytical technique based on the following approach. As employed in a laboratory setting for analyzing water samples, the potential (E) of a mercury drop substrate (relative to a reference electrode) is changed, and the substance of interest is electrochemically reduced at the substrate surface. Suppose the substance to be assayed is a metal ion M^{j+}. Then the electrochemical reduction reaction can be represented as

$$M^{j+} + (j-i)e^- \longrightarrow M^{i+} \qquad (1)$$

where M^{i+} is the reduced form of the metal (which could represent an ion of lower oxidation state or the zero-valent metallic form). The water sample to be analyzed is placed in a special solution termed the supporting electrolyte. The supporting electrolyte contains an excess of ionic charge carriers, which ensures that the ions reduced at the substrate arrive at the surface by a process of diffusional mass transport. In electrochemical terminology, the reaction is said to be occurring under diffusion control. For diffusion controlled conditions, the largest reduction current possible, I_L, when the concentration of M^{j+} is C_b (the unknown concentration to be determined), is given by

$$I_L = \frac{A_{\text{sub}} D C_b (j-i) F}{\Delta} \qquad (2)$$

where A_{sub} is the surface area of the reduction substrate, D is the diffusion coefficient in the supporting electrolyte of the ion being reduced, $(j-i)$ is the number of electrons involved in the reduction reaction, F is Faraday's constant, and Δ is the thickness of the diffusion layer. Conditions in the polarographic cell (the container in which the measurement is conducted) can be controlled so that all of these terms are constant. Thus, from Eq. (2), the limiting current I_L is directly proportional to bulk concentration C_b.

Since currents can be measured to quite low (nanoampere) levels, so too can concentrations. Sensitivity in the parts-per-billion range is achievable with polarography under laboratory conditions (24). Instead of using Eq. (2) directly to determine concentration, generally standards containing known amounts of the ion to be assayed are measured to obtain a calibration curve from which the concentration corresponding to the current measured from the unknown sample can be determined. A number of techniques make use of this general approach. These differ from each other depending on the nature of the reduction substrate and the time variation of the applied potential.

When a slowly time-varying potential is applied as a linear ramp to the mercury drop (the DC polarographic technique), the relationship between potential E and reduction current I is given by the Heyrovsky-Ilkovic equation:

$$E = E_{1/2} + \left[\frac{2.3 RT}{(j-i)F}\right] \log\left[\frac{(I_L - I)}{I}\right] \qquad (3)$$

where R is the gas constant, T is the absolute temperature, and $E_{1/2}$ is a unique potential (for a given reaction and supporting electrolyte) termed the half-wave potential. For certain polarographic techniques a current peak proportional to C_b is present at a potential of $E_{1/2}$. In a given supporting electrolyte, the half-wave potential is unique for each element and its different valence states and chemical forms. Observation of a current peak at a specific half-wave potential therefore identifies the chemical species producing the current. Thus, this method, in addition to being useful for its ability to measure concentrations at trace levels, is very useful in speciation studies to identify specific chemical forms present.

Although polarography as described above is mostly a laboratory tool, this basic approach has been used to some extent to make field sensors for environmental monitoring. For example, this method is commonly used for sensing dissolved oxygen in waterways. Over 20 years ago, a portable custom electronic circuit that performed polarographic nitrate analyses was described (25), but no field applications were reported. Also, polarographic analyzers (called voltammetric analyzers or electrochemical transducers in the air pollution field) have been used routinely for source monitoring of air quality mandated by the Clean Air Act (26), often in connection with semipermeable membranes. Both portable and continuous monitoring versions are available for monitoring of various gases.

2. Cyclic Voltammetry

Field sensing of heavy metals using polarographically based methods has been less widely investigated for environmental monitoring. This is a research area we have been involved in for the past several years and have reported previously on the use of these technologies (in some cases coupled with Microtox) in environmental science studies of water, sediment, soil, and animal tissues (27–37). We have found it most beneficial to use an electroanalytical method closely related to polarography, cyclic voltammetry (CV), for environmental field testing.

CV is capable of detecting concentration levels in the parts-per-billion (ppb) range under favorable conditions. Though the analyte for CV is often a specific heavy-metal oxidation state, organic compounds can also be analyzed in this manner (38,39). Here, a triangle-wave potential is applied and the resulting current is measured. A plot of electrical current as a function of electrode potential exhibits peaks from electrochemical oxidation and reduction reactions that are proportional to concentration. CV is not only useful in speciation studies, it can differentiate ionically dissolved from complexed metal forms under optimal circumstances. Thus CV can be used in studies where the interaction of inorganic and/or organic ligands is of interest, such as the effects of humic substances on metal speciation.

3. Backup Methods

Although polarographic and CV techniques are generally highly effective, there are some disadvantages associated with such methods. These include difficulties in performing simultaneous multielement analysis and measuring in complex media due to (a) interferences from other ions present with similar half-wave potentials, (b) the need for multiple supporting electrolytes, and (c) the possibility of electrode fouling. These difficulties also potentially apply to our CV field electrodes. For backup metal concentration measurement capability, we use colorimetry (40,41) or selective ion electrodes (SIEs) (42,43).

Colorimetry can be used to measure concentrations of a variety of heavy-metal contaminants in water or acid-digested sediment and soil samples. In a colorimetric assay a beam of light of given wavelength is passed through a test sample which has been treated with a chromagenic reagent specific for the ion being tested. The concentration of the substance of interest is proportional to the amount of light absorbed by the solution. As discussed later, this method shows promise for development as a field method.

SIEs can also be used to detect ionic concentrations in treated water, sediment, and soil specimens. The concentration measurement is based on a Nernstian electrode response:

$$E_m = E_0 + S \log(C_x + C_b) \tag{4}$$

where E_m is the measured potential of the SIE, S and E_0 are constants, C_x is the concentration of interest, and C_b is a factor related to blank correction. Generally, detection limits are not as low for SIEs compared to CV and colorimetry, but they can often be used to detect metals at levels found in the environment.

B. Basic Principles of Toxicity Assessment

The Microtox method employs a biosensor based on optical transduction principles. The assay uses a nonpathogenic bioluminescent marine bacterium, *Vibrio fisheri* (formerly known as *Photobacterium phosphoreum*). In the presence of a toxicant, the health and viability of the organisms are compromised, and they accordingly exhibit diminished light output to a degree proportional to toxicant concentration. Photometric methods are used to convert the light output into an appropriate electrical signal for analysis. In an actual measurement, a series of 2:1 dilutions from an initial concentration value is tested. The toxicant concentration required to extinguish 50% of the light output relative to a control is known as the effective concentration 50%, or EC_{50}, which is a quantitative measure of a substance's toxicity. The EC_{50} value generated using Microtox can be expressed in terms of percent concentration. This represents the percentage of the initial concentration of the test sample resulting in half the light being extinguished from baseline levels. For example, an EC_{50} of 40% measured for a 1000 ppm test sample corresponds to 400 ppm. Note that small EC_{50} values therefore indicate high levels of toxicity. A Microtox raw data sample is shown in Fig. 1. An alternative unit for the EC_{50} value, most useful in laboratory testing of one component, can be given in terms of a direct concentration value (e.g., milligrams per liter).

The DeltaTox PS1 system is a simple, rapid, portable test method for toxicity screening and monitoring of environmental samples, and can be operated at ambient temperatures of 10–28°C. It uses a highly sensitive photometer to measure light attenuation of *Vibrio fisheri* in comparison to the light output of a control not exposed to toxicants and is based on the same optical principals as Microtox. However, the percent light loss determined using DeltaTox is only an indication of the qualitative toxicity of the sample, where large light losses indicate high levels of toxicity. This device cannot obtain EC_{50} values and is thus most useful as a screening device. Water and sediment/soil samples can be assayed with both methods.

The feasibility of studying metals using Microtox has been extensively investigated. When Microtox is compared to more traditional bioassay organisms, overall agreement varies from 72% to 100%, 80% to 96%, and 74% to 97% for fathead minnows, daphnids, and rainbow trout assays, respectively (44–47). Differences in agreement between assays arise from the differing types of chemical tested and on the water quality variables of the sample. It should be noted that interspecies variability (i.e., between differing species of fish and between different types of daphnids) is significant and may account for some of the variability.

The literature indicates that Microtox has been used successfully to monitor many organics (48), although some compounds with high ($>10^5$) octanol–water partition coefficients (K_{ow}), such as PCBs, and highly halogenated compounds may be problem-

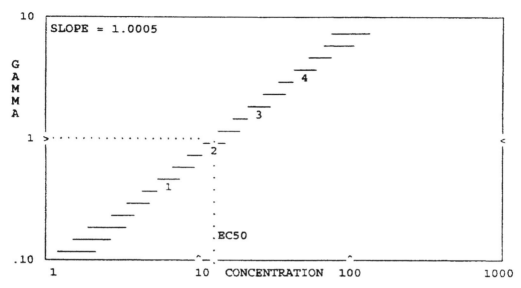

Figure 1 Microtox raw data sample. The gamma value represents the ratio of light lost to light remaining at time t for a particular sample concentration.

atical (49). K_{ow} represents the influence of hydrophobicity on solubility and cell wall/membrane permeation of organic compounds (49,50). This indicates that, due to certain anatomical and/or metabolic factors, the bacteria may have different reactions to certain organics when compared to other toxicity assays that show success with high-K_{ow} compounds. In other words, certain organic compounds may interact with bacterial cell membranes at faster rates, resulting in a high sensitivity to such compounds compared to other contaminant types.

We have found K_{ow} values for 26 of the compounds in our INE toxicant database mentioned above. Of the 26, only 3 are problematic for Microtox tests. An additional concern is that for solid-phase measurements, reliable Microtox tests often require toxicants to be in a liquid carrier. However, the extractant solution may have its own toxicity, which can confound the measurement. Literature demonstrates that, depending on chemical properties of the organics involved, acetone, ethanol, and methanol may be optimum solvents for extracting maximum toxicity, while contributing minimal amounts on their own (51,52).

C. Remediation Techniques

The above methods have been implemented successfully at Louisiana sites of interest to characterize the extent and potential effects of pollution using appropriate chemical and toxicological field testing. Together they form an efficient, powerful approach for assessing pollutant status. This knowledge is essential when considering potential risk and clean-up strategies, since complex issues of speciation and presence of contaminant mixtures may dictate the necessary remedial action. Several novel remediation techniques have recently been developed in an attempt to accommodate these and other complicating issues.

Some important remediation technologies that have emerged in recent years are extraction, electrokinetics, zero-valent metals, and phytoremediation. The effectiveness of all of these can be gauged by Microtox/DeltaTox/CV monitoring. The following briefly describes the principles on which these are based.

1. Extraction

Extraction is an emerging technology which has several variants (e.g., soil flushing, heap pile leaching, and soil washing). It is used most often for heavy-metal remediation and involves exposing a contaminated solid phase to an extractant (typically an acidic, chelating, and/or complexing solution). Results of batch extraction tests conducted at the U.S. Army Corps of Engineers Waterways Experiment Station in Vicksburg, Mississippi, have shown that up to 97% of soil metal contamination can be removed in this manner (53), although results can vary depending on soil type. In pilot-scale testing, an EPA study has shown that hydrochloric acid extraction can lower TCLP (Toxic Characteristics Leaching Protocol) metals below RCRA (Resource Conservation and Recovery Act) limits (54).

2. Electrokinetics

As with polarography and CV, the electrokinetics (EK) method works on electrochemical principles (55). It involves placement of electrodes in soil and application of a voltage across them. The resulting current creates movement of contaminants through electroosmotic and electromigration mechanisms. If controlled correctly, EK can change soil pH to values which accelerate metal extraction. Cadmium levels have been shown to be reduced by 90% after EK treatment (53). Uranium and lead concentrations have been diminished by over 80%, and trivalent chromium levels have been lowered by 50–66% in soil when current densities of 1 mA/cm^2 or less have been applied (56).

3. Zero-Valent Metals

The zero-valent metal strategy is also based on electrochemistry and involves a corrosion (oxidation) reaction of a metal that is relatively benign to the environment such as iron (57,58). The half-reactions for iron corroding in water are

Oxidation: $Fe \rightarrow Fe^{2+} + 2e^-$
Reduction: $O_2 + 4H^+ + 4e^- \rightarrow 2H_2O$ or
$O_2 + 2H_2O + 4e^- \rightarrow 4OH^-$ or
$2H^+ + 2e^- \rightarrow H_2$

The particular reduction reaction depends on pH and the degree to which the solution is aerobic. An example where zero-valent iron treatment would reduce environmental hazards is

$$Pb^{2+} + SO_4^{2-} + 8H^+ + 4Fe^0 \rightarrow PbS + 4Fe^{2+} + 4H_2O$$

Here, due to iron corrosion in an acidic, sulfate-laden solution, soluble lead ions have been combined into insoluble sulfides which can then be separated out. Cadmium, mercury, and many other heavy metals can also be treated using this strategy (59,60). Another mechanism of toxicity reduction involves species conversion. Hexavalent chromium can be converted to the trivalent form using zero-valent iron (61), thereby reducing the metal's overall toxicity. Inorganic nitrate can also be reduced (62,63). This is important because nitrate contamination is one of the most widespread pollution problems worldwide, especially in agricultural areas where fertilizers are used.

Toxic organic molecules also can be broken down in this manner (57,63–65). For example, trichloroethylene, perchloroethylene, carbon tetrachloride, and a whole host of other chlorinated hydrocarbons, indicated generically below as RCl, can be modified by dechlorination reactions:

$$RCl + Fe^0 + H^+ \longrightarrow RH + Fe^{2+} + Cl^-$$

The degradation products are either less harmful or are more easily broken down by further treatment than is the parent compound. Other types of organics amenable to zero-valent iron treatment include dye effluent from textile mills and pesticides such as toxaphene (63). Aromatic nitro compounds, $ArNO_2$, used in munitions, can be broken down as follows (57):

$$ArNO_2 + 3Fe^0 + 6H^+ \longrightarrow ArNH_2 + 3Fe^{2+} + 2H_2O$$

4. Phytoremediation

Phytoremediation is an emerging technology that uses plants to treat soils contaminated with heavy metals. The plants can be used either (a) to extract metal from soil and accumulate it into their roots or shoots or (b) immobilize metals in soil adjacent to the roots (66). Plants used for phytoremediation which can take up significant amounts of metals compared to their biomass are known as hyperaccumulators (67). Soil amendments such as EDTA can be used to chelate metal ions in order to increase metal accumulation (68). Hyperaccumulators have been identified for several different metals including lead, which is of special interest for DoD sites. A lead hyperaccumulator is defined as a plant where a concentration of at least 1000 µg/g has been recorded in the dry matter of any above-ground tissue in at least one specimen growing in its natural habitat (67). Five known lead hyperaccumulators exist, including *Brassica juncea* (Indian mustard) and *Minuartia verna* (Carophyllaceae) (69,70). Indian mustard has been used successfully to remediate a former battery recycling factory located in New Jersey (71), as well as a prior industrial site in Massachusetts where children suffered from lead poisoning (72).

All of these remediation methods have advantages and drawbacks (53). For example, phytoremediation works only for shallow contamination, but it can be very cost-effective. Extraction, zero-valent iron, and electrokinetics all work with a variety of substances and thus can treat pollutant plumes composed of mixtures of various contaminants. For EK, as with most treatment methods, the effectiveness varies with the types of soils and contaminants involved. With some of the remediation strategies there are issues to be considered related to proper disposal of reclaimed wastes. For example, the metal residue in extraction can be highly concentrated. If recycling is not possible, it must be disposed of in a secure landfill.

D. Field Testing Issues

As part of the effort to combat pollution and to judge the efficacy of remediation processes, means for accurately measuring and monitoring concentrations of heavy metals and organic compounds in the environment are essential. Although such measurements can be conducted by harvesting samples in the field for later laboratory analysis (and certainly there is much to learn from such studies), in many situations a better, more rapid approach for surveying pollutant distributions (when feasible) is to measure pollutant concentrations directly in the field. A main rationale for field sensors is increased ease, speed, and cost effectiveness of determining contaminant toxicity, particularly with regard to identification of hot spots. In addition, this type of measurement minimizes artifacts associated with sample transport and storage, as there may be considerable delay between sample harvesting and laboratory analysis. A mobile field sensor capable of monitoring pollutant levels in near real time would overcome these problems. Hexavalent chromium, for example, according to EPA guidelines, must be assayed within 24 h of sampling (because of the possibility of conversion to other chemical forms) for the measurement to be considered valid. Time constraints are all the more important if the polluted site is remote from the analytical laboratory.

A further improvement beyond the use of mobile field instrumentation is to measure metal concentrations using in-situ field sensors. Various additional technologies could be particularly helpful here. For example, acquired data could be telemetered back to a central location in near real time for display on a Web site. Such sensors could have a one-month data storage capability in the event that telemetering capability is interrupted and repair of the transmitting equipment is delayed due to weather or other conditions. A reasonable goal would be to have an in-situ sensor that functions for 30 days between routine maintenance visits to the site, and 60 days as a worst case if adverse weather or other factors prevent a routine maintenance stop. Such contaminant detection devices could prove to be extremely useful in a variety of ways. First, they could monitor isolated pollutant overload releases from point sources (e.g., due to an industrial effluent emission or chemical spill) as well as normal metal or organic concentrations over time due to point or distributed pollution sources. Since these techniques could be broadly applicable to all types of waters (rivers, lakes, wetlands, bays, estuaries, coastal waters, offshore oceans, etc.), they would be extremely valuable for aquatic research projects. For example, waterways can be very dynamic, with extremely variable depths, flow rates, and other physical/chemical characteristics. Pollutant levels therefore fluctuate dramatically as well. Thus, at any given time, spatial spot sampling may fail to identify polluted "hot spots," or conversely may overestimate contaminant levels. In addition to sudden metal burden overload events possibly related to industrial activities or catastrophic weather conditions, time variations in pollutant concentrations also occur due to daily, monthly, and seasonal changes. Looking at temporal effects over these different time scales is necessary to understand pollutant dispersal, because processes that govern the fate of contaminant distribution have different time constants that influence them (73). Another important related issue is the influence of weather patterns on pollutant levels. Wind, wave, and precipitation patterns may play a role in solubility of contaminants from sediments and/or their deposition from atmospheric sources into water. Thus, sampling and measurement in a fashion that is temporally restricted may result in the missing of pollution phenomena that depend on weather patterns and seasonal changes. These dynamic conditions are of particular concern for many polluted waters in Louisiana (74).

Field sensors could be used in contaminant monitoring and surveillance programs that are of interest to many state and federal agencies, as well as to companies that must comply with environmental pollution laws and regulations. These devices could also be used to monitor the effectiveness of environmental clean-up efforts at a wide variety of Department of Energy (DoE), DoD, Superfund, and other sites. In-situ sensors, besides helping to monitor the effectiveness of solutions to present-day pollution problems, could aid in anticipating future environmental problems. Such sensors clearly could help government agencies concerned with the status of the environment in establishing their immediate and long-term policies and decision-making priorities on a sound scientific basis.

1. Equipment Considerations

As indicated above, direct field sensing can be an efficient, cost-effective strategy for surveying and monitoring pollution and remediation, particularly in situations where such sensing devices have been validated in prior field studies. Several difficulties exist when considering the specific adaptation of CV/Microtox/DeltaTox to field use, however.

Different challenges occur in polarographic measurements, even under laboratory conditions, depending on the materials being analyzed (e.g., aqueous phases, soil/sediment, vapors, animal tissue, etc.). Though Microtox is not used for tissue analysis, it can be applied to the other media, and a similar comment can be made. Operation in the field poses an additional set of challenges for both techniques, some of which are common to both methods and some of which are unique to each. For field use such sensors should be "hardened" and simplified, as described below.

a. Common Difficulties and Their Solutions. Field measurements with any technique generally need to be done independent of line power. This is not a problem with DeltaTox, which is battery powered, but it is a concern for CV and Microtox. Protection of equipment from effects of weather is a concern with all techniques. For testing in waterways, vibrational interference from the sampling boat may also present difficulties. There is a need for portable data acquisition and storage equipment. Appropriate extraction methods for removal of contaminants of interest from carrier matrices to which they are bound, while posing no harm to the environment or to the personnel handling them, is an important issue for all three testing techniques that is discussed more completely later. It should also be noted that methods that are very effective under laboratory conditions may not be feasible in the field.

To solve the power supply problem we use a battery and a DC/AC inverter which can maintain Microtox and our CV sensor for up to 8 h. Polyethylene wrap is employed to protect the field measurement equipment from acute effects of weather. For use in waterways such as Bayou Trepagnier (BT), where we have conducted measurements reported in the Results section, foam insulation minimizes vibrations from the boat used and maintains their level under 0.5 GRMS. A laptop computer with measurement software (described in more detail below) allows for rapid data acquisition and portability.

b. Challenges Unique to Microtox/DeltaTox Field Testing and Their Solution. The Microtox reagent, *Vibrio fisheri*, must be refrigerated and stored between -20 and $-25°C$ to remain viable. This is possible in the laboratory, but temperature control without the use of a refrigerator is needed for up to 8 h in the field. We have been able to achieve this using a portable cooler filled with dry ice and insulation to protect the bacteria from thermal shock.

c. Challenges Unique to Cyclic Voltammetric Field Testing and Their Solution. For our CV field work we use graphite electrodes (GEs) as the oxidation/reduction (redox) reaction substrate. These are made from high-surface-area pitch-based graphite yarn and replace the Hg drops used in polarography (to avoid accidental spilling of this toxic metal into the environment). In the laboratory polarographic technique described earlier, nitrogen purges the polarographic cell of oxygen, an electroactive substance whose reduction current can sometimes interfere with measurements of heavy metals, and blankets the cell while the measurement is being made. The graphite field sensor, however, appears to be effective when the cell is exposed to the ambient atmosphere using conventional supporting electrolytes.

In laboratory tests using this GE/CV technique to assay aqueous samples, we have found the detection limits for Pb^{2+}, Cu^{2+}, Fe^{2+}, and Hg^{2+} to be 3, 5, 5, and 0.1–0.5 ppm, respectively. These limits are substantially above those for a mercury electrode, but still much lower than values that often characterize sites polluted by heavy metals. Although the data given later focus on lead, we are optimistic that this method is adaptable to analysis of additional heavy metals under field conditions.

The durability of GEs is an important question, particularly for in-situ CV sensors. Maintaining cleanliness and preventing fouling of electrodes by organics are important concerns for any field testing operations. The durability of GEs in aqueous environments has been verified in the laboratory in tests involving over 1300 h of exposure to salt solutions, which provides reason for optimism that GEs can provide the basis for a durable in-situ monitoring system with a reasonable maintenance interval. The resistance of the graphite electrode to fouling by organics has been checked in short-term albumin adsorption tests, and the electrode was still functional after 1 h. This test represented conditions more severe than would generally be encountered in field environmental testing. The influence of humic substances and other native organic compounds on short-term resistance to fouling as well as on long-term performance and durability of GEs should be checked in future experiments, to better understand the potential of GEs for environmental monitoring.

2. Equipment Configuration for Field Measurements

Our CV/Microtox/DeltaTox field measurement system configuration is shown in Fig. 2. This apparatus has been used successfully to conduct various field tests in Bayous Trepagnier, Saint John, and Traverse, as described in more detail later in this chapter. A Dell Latitude LM notebook computer configured with appropriate software and a

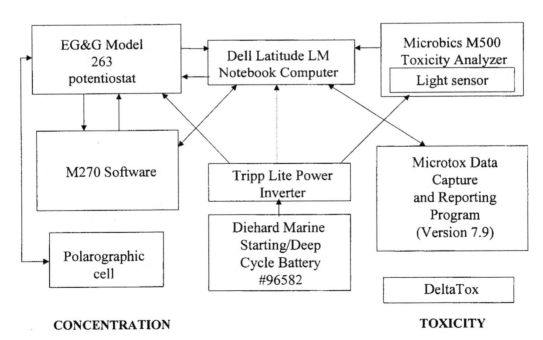

Figure 2 Instrumentation for field toxicity and chemical concentration assays.

GPIB interface runs an EG&G model 263 potentiostat for CV measurements and a Microbics M500 toxicity analyzer for the Microtox assay. It appears that this technology may be extremely cost effective. For example, Microtox measurements can be performed in the field at a supply cost of $5–$15 with about 0.5 h of labor. CV requires about 1.75 h. In either case, costs are much lower than they would be to send samples to an analytical laboratory or even to harvest them for later analysis at a dedicated laboratory.

A requirement for our field concentration detection system is robust operation in a wide range of environments that avoids interference from matrix components in extraction solutions. As indicated previously, we use selective ion electrodes (SIEs) and colorimetric techniques as backup methods for CV. Although these may be prone to their own interference (since all three methods are based on different principles), it is felt that at least one of the three should allow valid measurements in most situations. So far we have focused on using these backup methods to measure lead in soil samples, due to this heavy metal's importance as a munitions component, a common pollutant (particularly in our local environment), and a neuro-, immune, and endocrine system toxicant.

We can supplement the test configuration of Fig. 2 with a colorimetric sensor (the Hach DR/2010 spectrophotometer). This instrument is ideal for both field and laboratory use since it is portable, operates from either battery or line power, and is capable of data logging. In addition, we use an Orion model 801 microprocessor ionanalyzer for lead SIE testing.

3. Specimen Preparation/Methods for Field and Laboratory Analysis of Field Samples

In our studies a number of different methods are used for sample preparation and extraction of the substances of interest, depending on whether concentration or toxicity assays are being conducted, whether measurements are being made in the field or back in the laboratory, whether organics or heavy metals are being assayed, and on other factors.

A modified basic solid-phase test (mBSPT) based on protocols devised by the Microtox manufacturer and others (75,76) is employed when assaying solid samples. This test exposes the bacteria to sediments to obtain a direct measure of toxicity. The advantage of the mBSPT is that, contrary to most solid-phase tests, the organism is exposed to the sediment surface, as opposed to an aqueous extract or filtrate of the sample. In this procedure (see Fig. 3), a sample in a cuvette is allowed to settle for 15 min and then mixed by shaking. Measurements are taken before and after mixing to account for effects of turbidity due to solid particles.

For CV measurement of sediments and soils in the field, a salt extraction procedure based on ion-exchange processes is used. Here 1 g of sediment is combined with 12 mL of saturated (20%) NaCl solution. The mixture is agitated for 5 min at ambient temperature and the supernatant is obtained using a syringe filter (fitted with 5- and 0.45-μm Gelman Acrodisk filters). Although we have used this method of extraction for Microtox also, the modified basic solid-phase test described previously was seen to give better results because it exposes bacteria directly to sediment particles and does not require osmotic adjustment (with consequent sample dilution).

Under laboratory conditions we have found various methods to be useful for extraction of metals from sediments. For polarographic testing, the method we often use is ASTM D3974-81, "Extraction of Trace Elements from Sediments." This technique is capable of partial extraction of metals from soils, bottom sediments, suspended

Bacteria exposed directly to sediment particles:

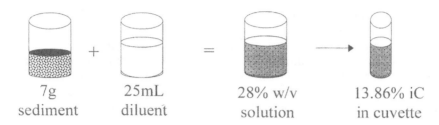

Figure 3 Modified basic solid-phase test (mBSPT). Sample in cuvette settles for 15 min and is then mixed by shaking. Measurements taken before and after mixing account for turbidity (iC = initial concentration).

sediments, and waterborne material and can be used to determine the labile or easily extractable metal concentration. It consists of hot acid digestion with HCl and HNO_3 until 10–15% of the original volume is left. The protocols described above remove metals in the bulk sediment without accounting for how they are bound in the sample. For a more complete assessment of the partitioning of metals into differing sediment or soil moieties, we employ the "Sequential Extraction Procedure for the Speciation of Particulate Trace Metals" (SEQ) (77). This method is used to identify the metal contents of different sediment fractions (exchangeable, bound to carbonates, bound to Fe-Mn oxides, bound to organic matter, and residual) into which metals are partitioned, to obtain a better idea of their availability.

When assays of organic compounds are needed, specimens are sent to an analytical laboratory at Tulane (the Coordinated Instrumentation Facility, CIF). The CIF laboratory analyzes the organic content using accelerated solvent extraction with dichloromethane and acetone followed by hexane exchange and GC/MS measurement. In this manner 65 organic compounds (including 25 PAHs) can be detected, including 18 compounds found in Bayou Trepagnier (see the Test Sites section) that represent most of the major classes of organic pollutants in this waterway. To complement the organic analysis of these specimens, we have the CIF perform inorganic analyses for a suite of 10 heavy metals, including As, Cd, Cr, Cu, Pb, and Zn. For inorganic analyses, microwave digestion is performed according to U.S. EPA method SW-846 3052, "Microwave Assisted Acid Digestion of Siliceous and Organically Based Matrices." This technique is used for digestion of siliceous and other complex matrices including soils, sediments, and sludges. Figure 4 provides an overview of the different extraction and analytical methods used in the research reported here.

4. Comparability of Field and Laboratory Toxicity Data

To compare field and laboratory measurements taken using the techniques described, split samples of sediment from the test sites were either analyzed directly in the field or else back at the laboratory using DeltaTox or Microtox. Based on comparison of 13 DeltaTox laboratory values from BSJ and BT sediment with 40 field measurements and 16 Microtox laboratory values with 17 field measurements, we found that the absolute value of the average discrepancy between laboratory and field tests results (expressed as a percentage of the laboratory toxicity value) was 29.5% for Microtox and 40.4% for DeltaTox. Since

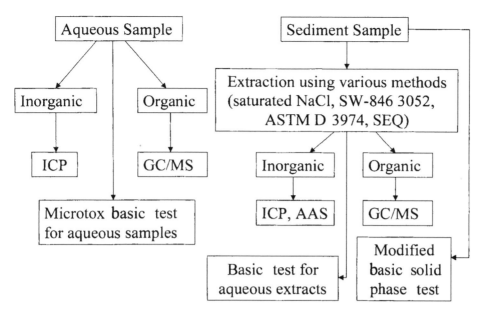

Figure 4 Methods of specimen preparation and analysis used for toxicity and concentration determination under laboratory and field conditions.

there was no systematic pattern of difference between laboratory and field results and since the deviations observed are comparable to the scatter in field test results themselves (as given later in the Results section), it appears that there is good comparability between the field and laboratory toxicity assessments.

5. Test Sites

To validate the field sensors and methods described previously, we have conducted a series of measurements at sites in our local area. Because of our interests in measuring both concentration and toxicity and our concern with synergistic/antagonistic interactions, we conducted tests at sites where there were mixtures of pollutants present that included both heavy-metal and organic compounds. Based on review of a database (78) compiled by Tulane's Center for Bioenvironmental Research (CBR), we began our field measurement program in Bayou Trepagnier (BT), which is part of the LaBranche wetlands [see Fig. 5, which contains maps of the test areas (79,80)]. Review of the CBR data showed that lead levels ranged from 10 to 3500 ppm in bottom sediments and from 29 to 76,000 ppm in spoil banks. Significant levels of chromium and zinc also had been observed, and detectable amounts of copper, nickel, arsenic, and cadmium were present. All of these metals are readily detectable by CV in the laboratory, so this location was thought to be a good starting point for our field testing. Regarding organic contaminants in BT resulting from petrochemical industry activities, polyaromatic hydrocarbons (including naphthalenes, phenanthrenes, and pyrenes), dibenzothiophenes, and other organics have been detected. We selected four sites for conducting tests that (a) had high concentrations of metals and organics and (b) collectively showed a wide metal/organic concentration ratio range. These sites are indicated by stars in Fig. 5. Although our detection methods as well as the remediation methods described previously

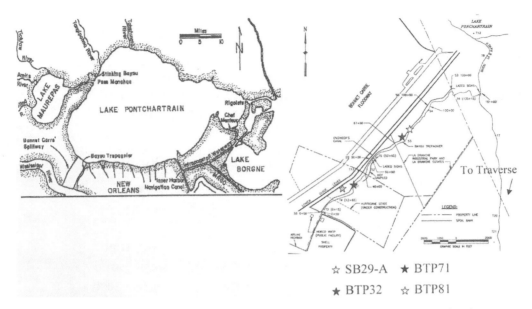

Figure 5 Sites for CV/Microtox/DeltaTox field testing in the LaBranche Wetlands, an area polluted by petrochemical activity located 22 miles west of New Orleans adjacent to the Bonne Carre spillway. Sites where measurements have been performed are indicated by stars. The maps shown are from Refs. 79 and 80.

have broader application, our Microtox/DeltaTox/CV work has focused mainly on sediments and spoilbanks. To further validate these field sensing methods and examine the generality of their usefulness, we selected an urban area where we performed field measurements. The test location was Bayou St. John (BSJ), a waterway located in mid-city New Orleans that is polluted with trace metals (primarily lead, zinc, and cadmium) resulting from effects of traffic and urbanization. This waterway is a natural channel of brackish water that flows northward through the city into Lake Pontchartrain. We sampled a total of four sites in BSJ: Orleans, Dumaine, Esplanade, and Maribeau Avenues.

We have also obtained samples of soils from Keesler Air Force Base and a New Orleans Coast Guard (NOCG) facility in a suburban location, mainly for the purpose of testing our backup methods under realistic conditions in the laboratory.

E. Theoretical Concerns

Since substances can pose health threats that vary widely depending on their particular forms, speciation associated with these substances is often of major concern, as pointed out previously. Toxicity and bioavailability of heavy metals depend strongly on the chemical forms in which they are found. The oxidation states present, as well as whether metals are ionically dissolved, complexed, sorbed to colloidal particles, or precipitated as various salts, markedly affect the hazards that they pose. In turn, these chemical forms will be strongly dependent on various environmental and water quality variables such as temperature, salinity, conductivity, turbidity, pH, and oxygen concentration, as mentioned previously. Various theoretical approaches exist, using principles of thermodynamics and kinetics, to predict speciation, such as Pourbaix diagrams, distribution of species diagrams, and pE–pH diagrams, as well as numerous equilibrium constants governing solubility, dissociation, chelation, sorption, ion exchange, etc. (81,82). Computer software in this area is highly developed, e.g., HSC Chemistry and MINTEQA2, and can model quite involved aquatic chemical systems. However, when environmental chemistry is complex, as is very often the case, it is prudent also to employ direct experimental means to determine speciation, such as polarographic methods and related techniques, supplemented by colorimetry and selective ion electrodes as discussed above.

Even the most extensive chemical analysis of a site, including speciation and partition of pollutants among various phases and moieties of interest, cannot generally give a complete picture of hazards present. Toxicity hazards are affected both by bioavailability and possibilities of synergistic or antagonistic interactions among pollutants in mixtures. To assess toxicity, here too there are helpful theoretical tools, such as QSARs (quantitative structure–activity relationships) (83–86), based on physicochemical properties such as K_{ow}, refractivity, molecular weights, etc. Unfortunately, these are not always reliable. For example, QSARs for mutagenic activity of PAHs favor the 4-, 5-, and 6-ring compounds, rather than the smaller (or larger) ones (87). However, studies with petroleum indicate that the naphthalenes and other mono- and diaromatic compounds are usually responsible for acute toxicity to organisms, with LC_{50} values in the low microgram/liter level (87,88). Experimental approaches to toxicity assessment such as Microtox and DeltaTox also yield valuable information, particular in situations as above, for example, where QSARs give misleading results.

A problem where theoretical analysis is particularly helpful involves identification of synergistic and antagonistic interaction in mixtures of hazardous substances. Synergism refers to a situation where the combined action of two (or more) toxicants in a mixture creates more toxicity than that expected based on additive behavior of the individual components. Antagonism, on the other hand, refers to a situation where the mixture is less toxic than expected based on additive behavior. We have previously presented a theoretical approach to study such interactions (89,90) based on differences between measured Microtox EC_{50} values and those predicted assuming two (or more) chemicals with known individual EC_{50}'s are acting independently. In brief, the basic ideas behind the theory are as follows. The model's main parameter is Γ_i [see Eq. (5)], where I_{oi} represents the initial light reading and I_{ti} represents the final light reading:

$$\Gamma_i = \frac{I_{oi} - I_{ti}}{I_{ti}} \tag{5}$$

Thus Γ_i represents the ratio of light lost to light remaining. Γ_i is related to concentration using the fit equation given by Microtox:

$$\log C_i = m_i \log \Gamma_i + \log EC_{50i} \tag{6}$$

where C_i is the concentration of toxicant i and m_i is the slope (see the Microtox raw data sample in Fig. 1 for a graphical example). If Eq. (6) is solved for Γ_i for each component of a three-component mixture, and the values then are substituted into Eq. (7),

$$\Gamma_{tot} = \Gamma_1 + \Gamma_2 + \Gamma_3 + \Gamma_1\Gamma_2 + \Gamma_2\Gamma_3 + \Gamma_1\Gamma_3 + \Gamma_1\Gamma_2\Gamma_3 \tag{7}$$

and the proportionality constants α and β (representing the ratio of C_1 to C_2 and C_1 to C_3, respectively) are introduced, Γ_{tot} can be expressed as a function of C_1. This equation may then be solved graphically to identify the mixture EC_{50}, as shown in Fig. 6 for a hypothetical case. The concentration of the primary component, C_1', representative of the

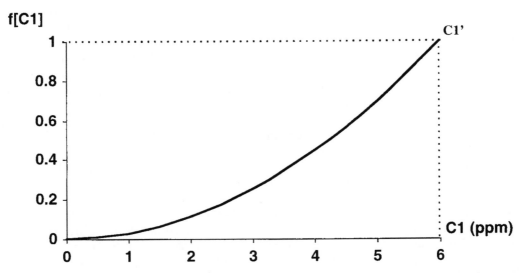

Figure 6 Determination of C_1'. The solution occurs for the value of C_1 where $f(C_1)$ equals unity.

mixture EC_{50}, is where Γ_{tot} equals unity on the graph in Fig. 6 (i.e., the point where light lost equals light remaining). This concentration may then be converted to the mixture EC_{50} by dividing C_1' by its proportion, P_1, in the mixture:

$$(EC_{50})_{\text{mixture}} = \frac{C_1'}{P_1} \tag{8}$$

In principle, this procedure can be extended to a situation involving an N-component mixture.

IV. RESULTS

We have conducted an extensive field measurement test series (91–93) focusing on sediment samples comparing DeltaTox and Microtox at four BT sites (indicated as SB29-A, BTP32, BTP71, and BTP81 in Fig. 5), a Bayou Traverse control site (BTV8), and the four urban BSJ locations. Three trips have been taken to Bayou Trepagnier and four to Bayou St. John. CV lead concentration tests have also been made on these trips to the field. In addition, samples harvested have been taken back to the laboratory for metal and organic analysis as described previously (ICP or AAS for 10 heavy-metal and GC/MS for 65 organic concentrations). The purpose of the chemical analyses was to see if correlation between concentration and toxicity could be observed in samples from the same sites, taken at the same time as the toxicity tests, as well as to compare with field CV tests.

A. Chemical Analyses and Toxicity Measurements of Bayou Trepagnier Samples

Concentrations of the five most abundant nonferrous heavy metals (besides aluminum) at the BT and Bayou Traverse sites (lead, chromium, zinc, copper, and cadmium) are given in Fig. 7. The distribution of each metal in the LaBranche wetlands is very inhomogeneous,

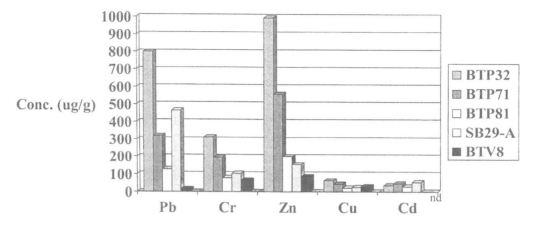

Figure 7 Concentration of the most abundant metals found in Bayou Trepagnier sediments. Values measured with ICP-AES. Extraction by EPA SW-846 3052. Values are averages of six replicate measurements, CV < 3.68% [where CV = 100% × (standard deviation/mean)]; nd = not detectable.

with the three most common metals (Pb, Cr, and Zn) showing almost an order of magnitude difference between the most and least polluted locations. The concentrations of heavy metals are quite elevated in Bayou Trepagnier, with maximum concentrations close to 1000 ppm being observed in these tests. Figure 8 shows the concentration analyses for the three most abundant classes of organic compounds in Bayou Trepagnier: polyaromatic hydrocarbons, phenols, and total organic matter (defined as the first two fractions plus all other organic compounds present in the 65-compound group analyzed). Again, these materials are rather inhomogeneously distributed, although they are present at much lower concentrations (3–36 ppm total) than those characterizing the heavy metals in BT.

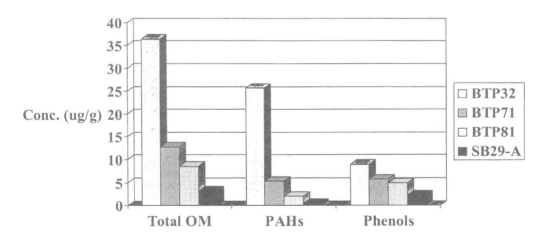

Figure 8 Concentrations of various classes of organic materials found in Bayou Trepagnier sediments. Values measured with GC/MS. OM = organic matter.

Table 1 Microtox EC_{50} Measurement of Field Samples Using Various Extraction Methods

Site	mBSPT reading EC_{50} (%)	NaCl extract reading (%)	D 3974 extract reading (%)[a]
BTP32	0.450 ± 0.070 ($n=4$)	0.396 ± 0.376 ($n=2$)	8.6768 ± 3.05 ($n=4$)
BTP71	0.148 ± 0.001 ($n=3$)	0.500 ± 0.239 ($n=2$)	7.376 ± 1.767 ($n=3$)
BTP81	0.140 ± 0.038 ($n=2$)	1.012 ($n=1$)	8.446 ± 2.194 ($n=4$)
SB29-A	1.560 ± 0.135 ($n=3$)	0.632 ($n=1$)	5.965 ± 1.112 ($n=5$)

[a] D 3974 measurements made in the laboratory from samples taken from Bayou Trepagnier. Other measurements made in the field.

Table 2 Comparison of Microtox and DeltaTox with Concentration in Bayou Trepagnier Sediments

Site	Microtox mBSPT EC_{50} (%)	DeltaTox mBSPT (%)	Toxicity ranking	Concentration (μg/g) of metal/organics
BTP81	0.140 ± 0.038	70.2 ± 11.67	1	457.545/ 12.733
BTP71	0.148 ± 0.001	67.2 ± 11.5	2	1160.545/ 12.733
BTP32	0.450 ± 0.070	55.8 ± 6.419	3	2201.210/ 36.3705
SB29-A	1.560 ± 0.135	53.5 ± 6.45	4	798.021/ 3.19397

At the onset of our field measurement studies it became clear that Microtox readings would be sensitive to the type of method used to extract pollutants from sediment samples. Table 1 shows results of toxicity measurements made using the mBSPT method and NaCl extraction in the field, compared with D 3974 extraction in the laboratory, on split samples. It is interesting to note that the different methods rank the toxicity of the four sites differently and that the apparently chemically less aggressive methods (mBSPT and NaCl) extract more toxicity as seen by the lower EC_{50} values.

Another important issue related to our field testing approach is the correlation between Microtox and DeltaTox measurements. A direct comparison between the two bioassays for measurements in BT is given in Table 2, where the mBSPT test was used. Here and in what follows, the sample size for toxicity measurements is $n=5$, unless otherwise noted. Both types of toxicity measurements agree in terms of rank order, which demonstrates the soundness of our approach to use DeltaTox for rapid screening and hotspot identification. Note that the Microtox EC_{50} value declines as toxicity increases (here, % indicates percent of original sample concentration resulting in 50% light attenuation), while the DeltaTox value (a direct reading of the percentage of light attenuation caused by the sample) increases as toxicity is greater. The R^2 value for linear regression comparison of Microtox and DeltaTox measurements is 0.65. Metal and organic concentrations determined by the CIF (as shown in Figs. 7 and 8) also are provided in Table 2. These concentrations represent total metal (Pb + Zn + Cr + Cu + Cd) and total organic matter (defined as PAHs + phenols + others).

The toxicity values do not seem to be closely correlated with the measured concentrations. In other words, high concentrations do not necessarily mean high toxicity. It appears that there are several possible causes for these disparities. First, Microtox and DeltaTox are probably most affected by and sensitive to the labile and bioavailable portion of the most toxic anthropogenic pollutants present, as compared to the more tightly bound fractions (which could be a prominent portion of the materials measured in the CIF concentration profiles of samples where microwave digestion extracts total pollutants). Second, toxicity bioassays may be affected by synergistic or antagonistic interaction among pollutants, making them more or less toxic than expected based on additive, independent behavior of toxicant concentrations. Third, it is possible that the bacterial assays are sensitive to pollutants not measured in the analytical work.

B. Chemical Analyses and Toxicity Measurements of Bayou Saint John Samples

Table 3 presents results of tests similar to those shown in Table 2, but for Bayou St. John. Here metal concentrations were determined by ICP measurements using the CIF procedures. Concentration measurements, in Bayou Saint John were quite heterogeneous as was the case for BT. For example, total metal content at BSJ was in the 150–2719 ppm range, while at BT the range was 458–2201 ppm. Only PAHs were measured in BSJ, and these were found at much lower levels than in BT (0.062–0.14 ppm compared to 0.5–25 ppm).

A consistent toxicity rank-order correlation between the two bioassays is observed for the BSJ data as well. As the compositions at the two bayous are quite different, our experimental approach has some generality in terms of its applicability to differing environments since we have had success in measuring toxicity at both locations. We have particular interest in applying it at firing ranges where spent small-arms munitions have accumulated. Toxicity readings from Microtox, however, generally indicate that the BT sediment is more toxic than that from BSJ, while the DeltaTox measurements seem to indicate the opposite. We hope to identify the reason for this discrepancy in the near future.

Table 3 Comparison of Microtox and DeltaTox Field Measurements with Metal Concentration in Bayou St. John Sediments

Site	Microtox mBSPT EC_{50} (%)	DeltaTox mBSPT (%)	Total metal concentration (µg/g)
BSJ-O	0.023 ± 0.012 (n = 3)	99.75 ± 0.5 (n = 4)	2719.40
BSJ-D	0.058 ± 0.005 (n = 2)	98.5 ± 1.00 (n = 6)	2196.07
BSJ-E	1.966 ± 0.028 (n = 5)	95.6 ± 1.14 (n = 5)	243.40 (PAH = 140 ng/g)
BSJ-M	3.78 ± 0.193 (n = 5)	68.57 ± 4.20 (n = 7)	149.90 (PAH = 62 ng/g)

C. Sequential Extraction Studies

As described previously, comparing the concentration and toxicity results raised the possibility that the pollutants are sequestered in multiple compartments varying in bioavailability. We explored the metal partitioning question, using the sequential extraction method described by Tessier et al. (77) to examine metal distribution in two BT sites. The results of this work are presented in Figs. 9 and 10. In these figures the metals are separated into five fractions, indicated as F1, exchangeable; F2, bound

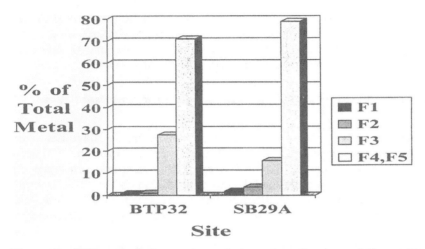

Figure 9 SEQ analysis for total metals in various fractions of Bayou Trepagnier sediments. F1 = exchangeable, F2 = bound to carbonates, F3 = Fe-Mn oxides, F4 = bound to organic matter, and F5 = bound to residuals (sand, silts, clays).

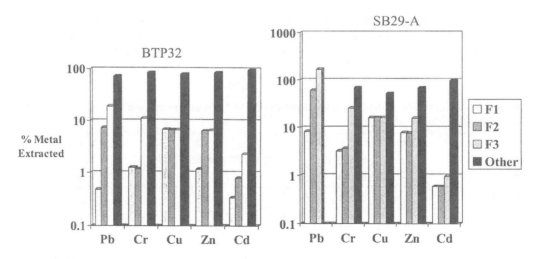

Figure 10 SEQ analysis for selected individual metals in various fractions of Bayou Trepagnier sediments. F1 = exchangeable, F2 = bound to carbonates, F3 = Fe-Mn oxides, other = F4 (bound to organic matter) + F5 (bound to sand, silt, clay residuals). Note log scale. Average of six replicates (CV < 16.6%).

to carbonates; F3, bound to Fe-Mn oxides; F4, bound to organic matter; and F5, residuals (bound to sand, silt, and clay). Of these, F1 through F4 are considered to be bioavailable under conditions that can be found in the environment.

Considering total metals, there seems to be a significant proportion in the last two fractions, between 70% and 80% (see Fig. 9). Individual metals differ in this regard, however (see Fig. 10). For certain metals, the first three fractions appear to be important. For example, a significant proportion of Pb (26 or 100%) and Cu (20 or 48%) is found in F1, F2, and F3 at the two sites. These data confirm that the metals present in BT are dispersed in a complex fashion in a range of fractions varying in bioavailability, which could account for at least a portion of the discrepancy between toxicity readings and concentration measurements.

D. Synergistic and Antagonistic Interactions

As a further probe into the origin of the discrepancy between concentration and toxicity measurements, calculations were performed using the mathematical model we have developed regarding identification of synergistic and antagonistic interactions among pollutants (89,90). A parameter termed the "toxicity index," TI, was defined for a particular contaminant as the quotient of its sediment concentration in BT and its EC_{50} value in order to determine the primary component affecting toxicity. Large TI values therefore indicate a higher potential contribution to overall toxicity. The EC_{50} values used in these TI calculations are based on measurements we have made for various metals (see Fig. 11).

A similar database for organics was more difficult to develop. We have tested several specific organics (with log K_{ow} below or near 5) in our laboratory with Microtox. Other EC_{50} values are available from the literature (51,52,94,95). These measured EC_{50} values and those from the literature are given in Table 4. The organics we have tested include ethanol (EtOH), p,p'-DDD, hexachlorobenzene (HCB), naphthalene, and phenol. EtOH

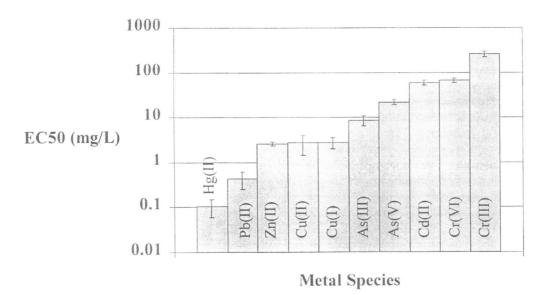

Figure 11 Microtox EC_{50} (15-min exposure) values for various metals. Note log scale. Values determined in laboratory from standard stock solutions, average of at least five replicates.

Table 4 Laboratory and Literature Derived EC_{50} Values of Selected Organic Compounds

Compound	Laboratory value (mg/L)	Literature value (mg/L)	Reference
Ethanol	Not toxic	34,948	DeZwart and Sloof, 1983
p,p'-DDD	30.80 ± 4.10	7.07–13.8[a]	Kaiser and Palabrica, 1991
Napthalene	0.3235 ± 0.033	0.929	Kaiser and Ribo, 1988
HCB	0.846	[b]	Kaiser and Ribo, 1988
Phenol	14.08 ± 6.45	12–26	Microbics Corp., 1995

[a] Published EC_{50} for p,p'-DDT, the more toxic parent compound of p,p'-DDD.
[b] Published EC_{50} values for benzene and 1-, 2-, 3-, and 4-chlorobenzenes of 103, 11.8–20.0, 4.24–10.2, 2.5–3.97, and 0.985–1.88, respectively, indicate that the laboratory-derived value is in the correct range.

Table 5 Toxicity Indices for Metal and PAH Species Found in Bayou Trepagnier

Species	Toxicity Index (TI)
Pb	301–1873
Zn	59–380
Cu	7–23
Fluoranthene	1–22.8
Phenanthrene	1–14.5
Cr	0.3–5
Cd	0.5–0.9
Benzo[g,h,i]perylene	0.6–0.8
Benzo[a]pyrene	0.5–8.4
Nickel	0.29–0.62
2-methylnapthalene	0.15–4.5
Pyrene	0.1–9.5
Cobalt	0.093–0.18
Napthalene	0.03–0.4
Arsenic	0

was used as the solvent for several organics due to their minimal solubility in water, and since it contributes minimally to overall toxicity ($EC_{50} > 34,000$ mg/L). Additional motivation for measuring these substances was that they are either neurotoxic, immunotoxic, or disrupt the endocrine system (or a combination of all three) and are present at several local wetland areas. The toxicity of these (from most to least toxic) based on 5-min EC_{50} values is as follows: naphthalene > HCB > phenol > p,p'-DDD > EtOH. This ranking appears to makes sense based on TLV-TWA (threshold limit value-time weighted average) values quoted for these compounds in their MSDSs. There is two orders of magnitude difference in TLV-TWA limits for the test compounds: naphthalene's TLV is 10 ppm, while ethanol has a threshold limit of 1000 ppm.

The TI values for various pollutants in Bayou Trepagnier are given in Table 5. For metals, Pb (TI = 301–1873) and Zn (TI = 59–380) exert the most toxicity at all Bayou Trepagnier sites tested. To make the calculations with the model tractable (considering the many components present in BT), the top two TI metals (Pb and Zn) were combined with the top TI ranking PAH into a three-component model. PAH toxicity was

Table 6 Mathematical Prediction of Pb + Zn + PAH Synergistic/Antagonistic Interaction for Various Metal : PAH Ratios for Bayou Trepagnier Field Sites

Site	Predicted (mg/L)	Actual (mg/L)	Metal : PAH[a]	Pb + Zn : PAH	Int[b]
BTP32	0.926–0.938	8.18–10.02	86 : 1	70 : 1	ANT
BTP71	1.111–1.115	1.32–1.726	217 : 1	164 : 1	ADD/ANT?
BTP81	1.046–1.049	0.464–0.643	222 : 1	160 : 1	SYN/ADD?
SB29-A	0.567	9.64–12.46	1467 : 1	1136 : 1	ANT

[a] Metal : PAH indicates total concentration of species.
[b] Int = interaction, ADD = additive, SYN = synergistic, ANT = antagonistic.

site-dependent with either 2-methylnaphthalene (TI = 0.15–4.5), pyrene (TI = 0.1–9.5), or benzo[g,h,i]perylene (TI = 0.6–0.8) dominating. The (Pb + Zn)/(top PAH) concentration ratio is similar to the total metal/total PAH ratio, so it was thought that this ratio would be a good marker for the possibility of metal/organic synergistic or antagonistic interactions. To study the possible influence of additional metals, EC_{50} calculations using the theoretical model described were performed for Pb + Zn and Pb + Zn + a tertiary metal (Cu, the third-ranking metal according to the TI scale). The effect of the tertiary metal did not appear to affect mixture EC_{50}, which was adequately predicted using just Pb and Zn. Thus, two metals appear to be sufficient to model the metal-ion contribution to overall toxicity.

Table 6 gives the two concentration ratios mentioned above along with the predicted and measured EC_{50} values at the four BT test sites and the type of interaction observed. The actual measured EC_{50} values given in milligrams/liter in Table 6 were calculated from the raw percentage values measured by Microtox and the concentrations of chemicals at the sites. The lower value of the measured range is a sum of concentrations (Pb + Zn + 1st PAH), while the higher value is total metal + total PAH. The PAH component used in the predicted value calculations was either the highest or second highest TI ranking PAH. This resulted in a range of predicted values. It can be seen from Table 6 that choice of PAH did not appear to have a great effect on overall predicted toxicity, since the predicted range is so narrow. Interactions were classified as ADD (predicted = actual), SYN (predicted > actual) and ANT (predicted < actual).

The two sites where the concentration ratios were below 100 or above 1100 were found to have antagonistic interactions (lower toxicity than expected). At the two sites where the concentration ratios were intermediate (160–225), additive interactions could not be ruled out. The data presented here include only interactions between metals and PAHs. So far our calculations are not definitive enough to rule out other possible interactions, such as synergistic/antagonistic interactions among metals (which are present in higher concentration than organics), between organic species, and between metals and other organic constituents. Compounds not included in the metal/organic analyses may also be influencing toxicity. These possibilities will be investigated in future work. Although we cannot say that these results are unaffected by these factors, they suggest that complex interactions probably occur at these sites. These findings (a) definitely underscore the inadequacy of using only concentration measurements as a marker for toxicity and (b) validate our approach of using both CV and Microtox/DeltaTox measurements to get a refined picture of pollution status.

E. Cyclic Voltammetry Testing Using Graphite Electrodes

1. Heavy Metals

Using the CV/GE method under field conditions has proven to be more challenging to implement than was the case for Microtox and DeltaTox. We have had some success in this regard, however. Figure 12 shows a field cyclic voltammogram taken of a spoilbank sample from Bayou Trepagnier (transect 29-A). Both a reduction and oxidation peak for Pb^{2+} are observable, with the oxidation peak being more sharply defined, though smaller. The current measured from the baseline for the oxidation peak was 9.01 µA. This current corresponds to 7.1 ppm Pb^{2+} in the polarographic cell. The sample measured was an extract after saturated NaCl treatment. The standard addition method was used for calibration. When dilution factors were taken into account, this corresponded to 70.0 ppm in the sediment. This concentration value is clearly much lower than the amount shown in Figure 7. It appears that, though salt extraction is fairly effective with regard to extracting toxicity, only a small fraction of the metal is solubilized in this manner. When compared to SW 846 3052 and ASTM D 3974, NaCl extraction removes less than 0.8% of total metals, so the CV field testing is presumably measuring only a small fraction of the metal of interest (the most bioavailable portion).

To further investigate the CV sensor's performance for lead detection, salt extraction was used in field deployment of our CV sensor at BSJ sites. However, lead levels appear to be below CV detection limits. This suggests the need for more aggressive extractant solutions, less dilution, and treatments to lower the detection threshold of the GE's. To investigate the first point, samples from BSJ at the corner of Esplanade Avenue were collected, and ASTM D 3974 extractions were performed in the laboratory. The specimens were then CV tested using GE's and the field portable instrumentation shown in Fig. 2. The observed concentration was 375.5 mg Pb^{2+}/kg sediment, comparable to

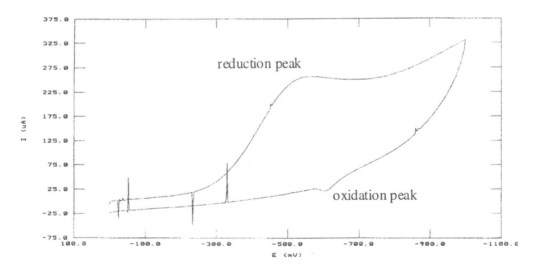

Figure 12 Cyclic voltammogram for lead concentration measurement made at a spoilbank in Bayou Trepagnier. NaCl extraction. Oxidation current peak measured from background level, 9 µA, corresponds to 70 ppm Pb^{2+} in sediment.

the 541 ppm value reported by Mielke et al. (96) using total microwave digestion and ICP. Since ASTM D 3974 would not be practical to implement in the field, other BSJ samples were subjected to aggressive cold acid extraction procedures in the laboratory. These tests showed that between 43 ppm (50% H_2SO_4) and 978 ppm (50% HNO_3) of sediment lead could be detected. Since these extraction methods could be used in the field with appropriate precautions, in future testing we plan to investigate their usefulness for CV measurements at BT and BSJ sites.

2. Organics

Besides heavy metals, several organic compounds are known to be electrochemically reducible/oxidizable (63), making them potential candidates amenable to CV analysis. PAHs known to undergo redox processes include naphthalene, phenanthrene, and acenaphthalene. These and other electrochemically active compounds, including quinones, carbon tetrachloride, and nitro compounds, can also be detected using Microtox, indicating that the dual Microtox/CV approach may be feasible. We have thus also been interested in developing CV protocols for assaying toxic organics. Besides measurement of traditional industrial pollutants, these methods may possibly be extended to detection of organic compounds of defense interest, such as phosphorous-containing nerve gases. These deadly agents block nervous system receptors and, for example, interfere with acetylcholine function. They cause severe, even fatal, reactions in small doses. Literature has shown, though, that these components are similar in chemical structure to organophosphorous insecticides (97), compounds that are more amenable to analysis in the laboratory.

One such substance is Malathion. It is hypothesized in the literature that the insecticide Malathion can be reduced electrochemically by transforming ester groups to alcohols (39). Since ester group reduction had been measured previously using voltammetric methods, it appears that there is reason for optimism that Malathion concentration could be monitored with CV. Since ester reductions occur at high half-wave potentials, nonaqueous solvents were thought to be appropriate candidates for the supporting electrolyte. We investigated dimethylformamide (DMF), which has been used to monitor ester reductions of other compounds previously, and detected Malathion with CV in acidified DMF. Since an aqueous medium is more convenient, a sulfuric acid supporting electrolyte was also investigated. Detection limits of 10 ppm were found with each. Much smaller detection limits were found using sulfuric or tartaric acid supporting electrolytes to test additional organic compounds. These data are summarized in Table 7.

Table 7 Organic Compounds Detected with CV Using GE

Compound	Supporting Electrolyte	Detection Limit
Acetone	1 N H_2SO_4	< 50 ppb
Malathion	Dimethyl formamide or 1 N H_2SO_4	10 ppm
Naphthalene	0.1 M tartaric acid, pH 9 or 1 N H_2SO_4	< 50 ppb
Benzene	1 N H_2SO_4	< 50 ppb

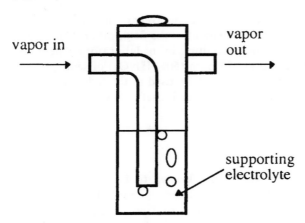

Figure 13 Vapor trapping apparatus based on a modified Greenburg-Smith impinger which can be used for CV measurements of airborne samples.

3. Airborne Pollutants

Another important extension of field CV methods is the potential detection of vapor and aerosol concentrations. Adaptation of air scrubbing technology using vapor trapping equipment appears to be feasible for this purpose. Here, known volumes of air containing dispersed liquid drops or gases of interest can be passed through a trapping solution (usually at lower than ambient temperature) to retain the substances of interest in dissolved form. This solution is then amenable to analysis by CV/Microtox techniques. We have conducted preliminary experiments along these lines using acetone. Due to its high vapor pressure and relatively low detection threshold using GE/CV techniques (see Table 7), it is an excellent compound to use for such studies. A modified Greenburg-Smith impinger (see Fig. 13), where acetone was bubbled through a trapping solution, was used in this work. This impinger can be used to collect particles 0.5–25 μm in size or organic vapors in atmospheric or industrial settings, which then may be used in CV or toxicity measurements. In this experiment, a syringe was employed to pull 60 mL of acetone vapor through 9 mL of 1.0 N H_2SO_4. Approximately 52.71 ppm acetone was detected by CV. Using published values for acetone's vapor pressure (98), this corresponds to a 1.4% trapping efficiency. We plan to repeat these experiments using bacterial toxicity assays with water (since H_2SO_4 contributes a toxic effect) as the trapping solution in the impinger to see if measurable light loss can be seen. If so, this would demonstrate the utility of this method of aerosol monitoring, since many industrial pollutants and agents of defense interest are significantly more toxic than acetone.

F. Testing with Colorimetric and Selective Ion Electrode Backup Methods

A lead kit based on the U.S. EPA dithizone method for colorimetric testing has shown that 100 ppb of Pb in water is detectable, and we have started to use these methods to measure environmental soil and sediment samples. To provide a comparative baseline, polarography and metal extraction using ASTM D 3974 were used to investigate specimens from Bayou Trepagnier site 32 and soils from firing ranges polluted with heavy metals. We have analyzed soil from Keesler Air Force Base and a suburban New Orleans Coast Guard (NOCG) facility. Polarographic lead concentrations have been observed to be

337.5 ppm at Keesler, 1430 ppm at NOCG, and 197.0 ppm at the BT site (about double the value given in Fig. 7). The colorimeter was used to measure soil from NOCG, and 681 ppm of lead was found. The values determined with the lead SIE were lower than observed with polarography or colorimetry: 165 ppm at BT site 32, 300 ppm at Keesler, and 64.4 ppm at NOCG.

V. DISCUSSION

A. Status of Field Microtox/DeltaTox/CV Testing

The DeltaTox/Microtox/CV approach to field testing appears to be successful, rapid, and cost-effective for analyzing metal and organic pollutants, particularly when these are present as complex mixtures. Use of these methods may lead to an increased understanding of potential hazards in Louisiana posed by heavy-metal and organic components and their interactions, and can be extended for use in sites worldwide. We envision development of versatile sensors that can be applied to a broad band of toxicants in water, air, sediments, and soils introduced into the environment from industrial, defense, and energy operations. Ultimately, we hope that elements of these technologies can be incorporated into cone penetrometer systems and/or can serve as the basis for in-situ sensors.

B. Future Directions for Microtox/DeltaTox/CV Testing

While these sensors appear to be useful measurement tools, their sphere of application can be extended beyond areas previously discussed. One of the first areas we plan to explore is the extension of the synergism/antagonism studies to consider more than three components, by which we hope to model mixture interactions more accurately. Another area that is important to investigate is the use of the bacterial toxicity assays in salt-water environments. We have conducted preliminary Microtox measurements of saline solutions, starting with lead, zinc, and arsenic, since the first two metals form chloride-based complexes, while As does not (99,100). The initial results indicate that the degree of salinity is an important consideration, since salt can create hormetic effects leading to difficulties in test interpretation.

An intriguing possibility for extending CV field sensor technology (which at present is essentially land-based) would be to use it in a mode where the sensing element contacts ocean water directly. In effect the device would use the salt water as its own supporting electrolyte. In prior testing of hexavalent chromium for research related to the corrosion of surgical implant alloys, we have shown that a potassium chloride electrolyte at near-neutral pH can be used to perform laboratory polarographic measurements (101). Although the detection limits were roughly double that for the optimal supporting electrolyte (0.1 M tartaric acid adjusted to pH 9 with ammonium hydroxide), they still were low enough (36 ppb) to allow metal concentrations of environmental relevance to be made. Since the supporting electrolyte used for those tests was a saline solution similar to salt water, there is reason for optimism that direct sea contact sensing is feasible. As a means for making a direct sea water contact sensor based on graphite more sensitive, anodic stripping voltammetry (operated in a differential pulse mode) might be useful for lowering detection thresholds. In this approach, a DC potential could be applied for a relatively long period of time to deposit the metal of interest on the electrode surface, after which an anodic potential measurement scan could be employed to oxidize the deposited metal, resulting in a detectable oxidation current proportional to concentration. Other materials besides graphite (such as platinum and gold) could also be used for the sensing element and might have higher sensitivity.

Use of our field sensors to survey other sites on land (such as defense-related sites) is also of significant interest. Firing ranges, used for sport and military purposes, polluted with lead and its degradation products released from corrosion of spent small-arms munitions (102), would be an obvious additional application for the CV sensor, as mentioned previously. Toxicity assays would be useful to prioritize these sites to identify those most in need of remediation.

We plan to extend our backup SIE and colorimeter capabilities as well. The latter method appears to be particularly promising. Colorimetric kits to measure many different heavy metals and chemical compounds are available, including kits with reagents practical for field deployment. These include arsenic, cadmium, hexavalent chromium, copper, nickel, and zinc. Since detection limits for these metals are reported to be about 10 ppb, it appears that in many cases colorimetry could be quite competitive with CV, or superior to it, for field use.

C. Implications for Remediation

Another logical use for the CV/Microtox/DeltaTox approach described is evaluation of remediation studies at the bench-top scale and ultimately in the field as well. Microtox and CV sensors operating in laboratory, field-deployable, and/or in-situ modes could monitor reduction in toxicity and in levels of various heavy-metal and organic species following remediation treatments.

In terms of fieldwork, for example, it would be of interest to harvest indigenous plants at BT spoilbank locations, BSJ, the NOCG firing range, and other nearby sites heavily contaminated with lead. The plants could be taken to the laboratory and analyzed for the lead content in the leaves, shoots, roots, and soil surrounding roots to determine whether any lead-tolerant plants show potential as hyperaccumulators. If any are found, these could be planted in soil plots, and after a given time CV/DeltaTox/Microtox could be used to determine the success of lead removal.

A number of bench-top remediation pilot studies using the four remediation technologies considered previously could be monitored with CV/Microtox/DeltaTox techniques. Toxicity (as measured by Microtox) and heavy-metal and organic-compound levels of the soils before and after remediation could be determined using the techniques described earlier in this chapter. For instance, the influence of moisture, as well as other relevant soil variables (such as pH and oxygen content) could be examined. Certain remediation strategies are influenced by additional parameters. For example, electrode placement is of primary concern for EK. Lighting and nutrient levels are relevant for phytoremediation (31). CV/Microtox/DeltaTox monitoring could be used to study these variables as well as to assist in optimizing pilot-study design regarding variables such as Fe^0 to soil ratios (for zero-valent metal treatment), potentials and current densities for EK reclamation, ionic strengths for extraction, etc. A particular issue to be considered in using the bacterial assays for monitoring extraction treatment is how to completely wash extraction solutions from the soil so that residuals do not cause artifacts in Microtox/DeltaTox tests.

Mechanistic studies of these remediation processes could also be conducted and monitored with CV/Microtox/DeltaTox. Mechanistic investigation would be particularly appropriate for zero-valent iron treatment and EK reclamation. The compounds and metals present on iron particles could be investigated using scanning electron microscopy (SEM), energy dispersive analysis via X-rays (EDAX), and X-ray diffraction to determine their structure and composition. Theoretical and experimental characterization of the remediation process should also be conducted based on thermodynamic, kinetic, and electrochemical principles using Pourbaix diagrams (for example, to consider the influence of soil pH on EK remediation), potentiodynamic polarization, AC impedance measurements, and other tools (81,103,104). Soil characterization in terms of particle size, resistivity, organic content, moisture level, and particulate and pore water chemistry could also be performed to aid in understanding of other factors that could influence remediation processes.

Besides assisting pilot remediation tests and mechanistic investigations, CV/Microtox/DeltaTox monitoring would help efforts to optimize process parameters and protocols for the four remediation methods. For EK treatment for example, the most effective combination of reclamation time, applied potential, current density, electrode placement, and additions of depolarizing and chelating agents could be identified for particular sites. For soil extraction treatment, column or pile design, aspects of extractant chemistry (concentration, pH, ionic strength, etc.), and treatment time could be investigated. For phytoremediation, soil amendments such as sulfate and phosphate fertilizers, pH adjusters, and chelating agents to solubilize metals to aid in their uptake by roots could be studied. Optimization of (a) the ratio of iron to soil and (b) solution chemistry (including pH and aeration) could be investigated for zero-valent iron treatment. Also, the influence of the form of iron on the effectiveness of remediation could be studied, the optimum specific surface area could be determined, and use of other metals to couple galvanically with iron to increase its rate of corrosion (57) could be explored. Combining various remediation methods is also worthy of study. For example, EK could be used to transport contaminants to areas where phytoremediation could be performed. Zero-valent iron treatment could be combined with EK in a manner in which the iron could be polarized to corrode at a higher rate, leading to faster pollutant breakdown. Additional zero-valent metals besides iron could be further researched to assess their potential for remediation. For example, zero-valent metals that readily corrode, such as magnesium and tin, have already been examined for this purpose (53).

D. New Field Sensing Approaches

Microtox/DeltaTox and CV methods have many actual and potential arenas of application, as has been discussed at length previously. In some situations, though, these techniques have limitations related to interference, insensitivity, and/or nonspecificity. Such disadvantages may possibly be overcome using other types of biosensors. This approach may be particularly appropriate for neuro-, immuno-, and endocrine system toxin assessment. Employing more than one environmental biosensor concept in principle adds versatility, reliability, and cost effectiveness to field measurements.

Biosensor devices usually yield digital signals proportional to concentration of a specific analyte or group of analytes and incorporate biological materials (tissue, microorganisms, organelles, cell receptors, enzymes, antibodies, nucleic acids, etc.), biologically derived materials, or biomimics coupled to optical, thermometric, electrochemical, piezoelectric, or magnetic transducers (105). Approximately 27 discrete biosensor types have been described in a comprehensive review (106). These devices are used in medicine, industrial bioprocess monitoring, defense applications, and recently, environmental monitoring. Some biosensor systems work in an automated and unattended remote sensing mode, while others comprise analytical test systems used under field conditions, in closed quarters such as submarines, or as probes to monitor drugs and metabolites in body fluids. Receptor biosensors (more generally, affinity biosensors) are employed as single-use disposable devices to measure hormones, steroids, drugs, microbial toxins, cancer markers, and viruses in micromolar to picomolar ranges. Such biosensors are potentially useful because biomolecule/ligand interactions are extremely specific, thus allowing very low concentrations to be detected even in the presence of much larger amounts of other contaminants. Some difficulties currently presenting drawbacks for use of enzyme-, natural receptor-, or antibody-based biosensors in unattended in-situ environmental toxicant monitoring include instability or unavailability of suitable proteins and, for receptors and antibodies, essentially irreversible binding, necessitating continual supply of reagents for sequential measurements (106).

A usable biosensor must demonstrate (a) ability to detect toxicant bound to target molecules, (b) specificity (overcoming potential difficulties regarding interferences and matrix effects), (c) low detection thresholds, and (d) durability for repeated use. Also, appropriate positive and negative controls must be identified, and the detection limits and dynamic range must be determined. As target goals for sensor performance, the requirements delineated by Rogers and Lin (107) for biosensors in environmental field applications are a good starting point: an easily transportable, battery-operated device that can be carried by one person, is capable of performing an assay in under 1 h for under $15 per analysis, and has parts-per-million to parts-per-billion sensitivity with a dynamic range of at least two orders of magnitude. It should also have high specificity, be easy to operate, and require minimal preparation of test samples.

Presently available environmental biosensor technology suffers drawbacks associated with retention of molecular biofunctionality and limited ability to sense airborne contaminants. We believe that such problems can be lessened using hydrogels, strongly hydrophilic polymeric materials first developed for biomedical purposes, with biomolecules in active form incorporated into their structure. If perfected, such hydrogel (HG) technology could be applied to quantitative field monitoring of a wide range of hormone- and cytokine-disrupting agents, as well as immune, nervous, and endocrine system toxicants.

To describe this biosensor concept further, a hydrogel is a material consisting of an extremely hydrophilic polymer. When hydrated, such materials can have water contents exceeding 75% (108). The water infiltrates a continuous network of open porosity in the polymer, and its percentage can be controlled to a desired value depending on the polymer, cross-linking agents, and the manner of synthesis. Besides percent water, hydrogel pore size can also be controlled during fabrication. Presently the most common applications of hydrogels are biomaterials used in ophthalmology and controlled-release drug delivery. Although they have been employed in other industrial applications to a limited extent, they have rarely been used for environmental purposes. However, the attributes of hydrogels described above may make them attractive for environmental biosensor applications. Because of their high water content, hydrogels should be suited for (a) detection of airborne vapors, droplets, and reactive particles that dissolve or can be suspended in water and (b) uses where aqueous-phase contaminants are monitored. Specificity for pollutants in small amounts, in the midst of a complicated pollutant mixture where other components are present in much higher concentration, is a prime potential advantage of the HG biosensor concept, since conventional sensors may be more prone to masking interferences. To consider a specific example, the insulin receptor can bind its target protein specifically when there is a 100,000-fold excess of unrelated proteins

(109). To further ensure adequate specificity of the HG sensors, pore size can be controlled to pass molecules of interest and to exclude larger ones that would elicit spurious signals (were the receptor to be activated by more than one ligand).

To consider the concept of such environmental biosensors in more detail, receptors are made of proteins that have binding sites with a high affinity for a particular signaling substance (e.g., a hormone, pheromone, or neurotransmitter) (109). There are three basic types: ligand-gated receptors, voltage-gated ion channels and receptors, and G-protein-coupled receptors. Although many receptors are present on cell surfaces (e.g., those interacting with cytokines), some reside on the nuclear membrane (e.g., the estrogen receptor). Some receptors that may be useful in environmental biosensors are the gamma-aminobutyric acid (GABA) receptor which binds to cyclodienes, pyrethroids, bicyclophosphates, and orthocarboxylates; the muscarinic receptor which binds orthophosphate insecticides; and the aryl hydrocarbon receptor that binds with high affinity to dioxins and PAHs (107,110). Although receptors (as well as enzymes and some antibodies) used in environmental biosensors show high affinity to their respective analytes, they typically bind to a closely related group of molecules. They can therefore also be used in devices that screen for the presence of closely related environmental pollutants [e.g., organophosphorous (OP) insecticides, triazine herbicides, and BTX (benzene, toluene, and xylene)] (107).

Obtaining reliable sources of receptors is often problematic, since specific target molecules for toxins may not be available or identifiable, or cannot be isolated in stable form. Either tissues rich in the desired receptors or cultured cells can serve as the raw material for obtaining receptors. Isolation and purification protocols for various receptors are well developed, including those for the adrenergic receptors (109,111,112), the nicotinic acetylcholine receptor (nAChR) from the electric organ of the ray *Torpedo california* (112–114), the glutamate ion channel (115), muscarinic cholinergic receptors (111), olfactory receptors for the bullfrog *Rana catesbeiana* (116), dopaminergic receptors (111,117), estrogen receptors (118), and receptors for IgE, insulin, thyrotropin, and asialoglycoproteins (112).

Among many potential applications, such biosensors in an in-situ mode could continuously monitor changes in concentrations of pollutants. This sensor type could form the core of an early-warning system for detecting such agents. These systems could acquire data continuously and store it for later retrieval during periodic trips to the field site, as previously pointed out. Ultimately, in-situ field sensor designs could be devised where data can be telemetered back to a central location in near real time for display on a World Wide Web site. Probably the best strategy for a continuous monitoring system would be to use enzyme-based electrodes (so-called catalytic biosensors) (119). These can detect millimolar to micromolar concentrations of the analyte of interest and are usable for continuous monitoring over 1- to 60-day periods (106,120). A particularly notable feature of enzyme electrodes is that they can be renewed and regenerated by appropriate chemical treatments. Another particular advantage of enzyme-based biosensors is the range of transducable components (protons, ions, heat, light, electrons, and mass) that can be exchanged as part of their catalytic mechanism. The catalytic activity is controlled by pH, ionic strength, temperature, and the presence of cofactors (106). Several types of catalytic biosensors have been described (121–125). Various enzymes attached to cellulose nitrate membranes (121) or glassy carbon electrodes covered by a nylon membrane (124) have been used to detect OPs. Iwuoha and Smyth (126) have developed electrodes for detection of phenols, organic peroxides, and pesticides in organic solutions using horseradish peroxidase, tyrosinase, and glucose oxidase attached to various substrates. These enzyme-based sensors use amperometric (i.e., polarographic) means for detection.

Biosensors of the various types described above would appear to have a bright future for field testing and for evaluation of remediation processes. They could prove to be valuable supplements to the Microtox/DeltaTox/CV approach described in this chapter.

VI. CONCLUSION

This chapter describes a combination of various techniques for concentration measurement and toxicity assessment that have been adapted for cost-effective field deployment. In the first category, these include cyclic voltammetry (CV), colorimetry, and selective ion electrodes. Microtox and DeltaTox assays are used in the second area. This combination of methods can monitor heavy metals, organics, and mixtures of these. In this investigation, soils, sediments, and aqueous phases have been examined with this technology, supplemented where appropriate by laboratory analyses and theoretical calculations to examine synergistic or antagonistic interactions among toxicants. Besides detection of harmful chemical agents, this technology can help to guide remediation efforts.

Our methods have been tested extensively in the field and validated at sites in the New Orleans area and compare well with duplicate samples tested under laboratory conditions. Specifically, the LaBranche wetlands (Bayou Trepagnier and Bayou Traverse) and Bayou St. John, an urban location, have been studied. The challenges posed by field testing have been overcome through development of appropriate portable field-deployable equipment and a field kit containing auxilliary supplies for phase separation, contaminant extraction, and safe disposal of spent reagents. Laboratory analyses have supplemented this field testing effort. In terms of the bacterial toxicity assays, DeltaTox is capable of pinpointing areas of high and low pollutant toxicity, while Microtox makes EC_{50} measurements, which can then be related to concentration data. Site toxicity rankings are well correlated between these bioassays. This combination appears to be successful for rapid on-site identification and quantification of pollutant toxicity. CV can be used to detect lead under field conditions and also shows promise for measurement of certain organic compounds.

Besides developing field techniques, we have increased understanding of the distribution of pollutants in these contaminated sites. Judging by their abundance and toxicity, our testing indicates that Pb^{2+} and Zn^{2+} appear to be the primary heavy metals contaminating Bayou Trepagnier sediments, while PAHs are the largest fraction of organic pollutants. Bayou Saint John appears to have a high level of metal pollution also, but organic levels are rather low. Sequential extraction analysis indicates that heavy metals in BT are partitioned primarily in the residual fraction (bound to sands, silts, and clays) or bound to organic matter, indicating that most metal contaminants are unavailable for interaction under normal environmental conditions. Pb is most easily extracted (26–100%) from the more bioavailable fractions of sediments (exchangeable, bound to carbonates, or bound to Fe-Mn oxides), with Cu following closely behind (20–48%). These results are consistent with our finding that Bayou Trepagnier contaminant concentration measurements do not necessarily agree with toxicity rank order from the bacterial assays. This implies that Microtox field tests are sensitive to only a fraction (the more bioavailable portion) of total contaminant concentration (which includes both available and more tightly bound contaminants). In addition, we have studied patterns of synergism and antagonism between metal and organic moieties in Bayou Trepagnier. Antagonistic interactions appear to be involved in the discrepancy between toxicity and concentration measurements, although mechanistic understanding of these differences is presently unknown. These interactions appear to be dependent on the metal : PAH ratio in sediments tested, with low and high ratios indicating less toxicity than expected based on additive behavior.

There are a number of research areas for the CV/Microtox/DeltaTox methods that we wish to explore in the future, such as the extension of the CV sensor to detect airborne pollutants and to in-situ configurations. Testing at other sites heavily contaminated with lead, (e.g., firing ranges and battery factories, is another application that could be easily implemented given the success of these methods in measuring this metal. Extension of these methods to other important types of sites, such as marine environments, should also be investigated. A main motivation for interest in this technology is its potential for monitoring efficiency (regarding reduction in concentration of pollutants and toxicity) of efforts to remediate sites using emerging technologies. In these new studies we hope to incorporate biosensors based on polymeric hydrogels into our test battery of field deployable environmental monitoring methods, allowing us to obtain a clearer understanding of pollutant status.

ACKNOWLEDGMENTS

Funding from the Tulane/Xavier Center for Bioenvironmental Research (CBR) DoD DSWA Receptor-Based Hazard Monitoring program (grant no. DSWA 01-96-01-0004) and CBR DoD "Laboratory and Field Methodology for Speciation Studies and Toxicity Assessment of Complex Heavy Metal Mixtures" project (grant no. 93DNA-2) is gratefully acknowledged.

The assistance of Steve Adams (of the CBR) for collecting soil, sediment, and water samples from Bayou Trepagnier and for providing sampling equipment is acknowledged. We would also like to thank Dr. Mark Bricka of the Environmental Laboratory of the U.S. Army Corps of Engineers Waterways Experiment Station in Vicksburg, Mississippi for providing soil samples from defense installations.

REFERENCES

1. The Tri-Service Environmental Quality R&D Strategic Plan.
2. http://www.ndcee.ctc.com.
3. http://www.acq.osd.mil/ens.
4. Isaacson, R. L., and Jensen, K. F. 1994. *The Vulnerable Brain and Environmental Risks: Toxins in the Air and Water, 3*. Plenum Press, New York.
5. Fawell, J. K., and Hunt, S. 1988. *Environmental Toxicology: Organic Pollutants*. Ellis Horwood Series in Water and Wastewater Technology, Ellis Horwood, Chichester, England.
6. U.S. EPA Office of Water Health Advisories. 1993. *Health Advisories for Drinking Water Contaminants*. Lewis Publishers, Boca Raton, FL.
7. Safe Drinking Water Committee, Board of Environmental Studies and Toxicology. 1989. *Drinking Water and Health*. Selected Issues in Risk Assessment, 9. National Academy Press, Washington, DC.
8. U.S. EPA Office of Drinking Water Health Advisories. 1991. *Drinking Water Health Advisory: Volatile Organic Carbons*. Lewis Publishers, Ann Arbor, Michigan.

9. Manno, J., Myers, S., Riedel, D., and Trembley, N. (eds.). 1995. Effects of the Great Lakes Basin environmental contaminants on human health. State of the Great Lakes Ecosystem Conference, EPA 905-r-95-013 (http://epasaver.ciesin.org:7777/gl ... onpo/ndata/solec/health/health.html).
10. Keith, L. H. 1997. *Environmental Endocrine Disruptors: A Handbook of Property Data*. Wiley, New York.
11. Cohen, M. D., Parsons, E., Schlesinger, R. B., and Zelikoff, J. T. 1993. Immunotoxicity of in vitro vanadium exposure: Effects on interleukin-1 tumor necrosis factor, and prostaglandin E2 production by macrophages. *Int. J. Immunopharmacol. Immunotoxicol.*, 15:437–446.
12. Zelikoff, J. T., Parsons, E., and Schlesinger, R. B. 1993. Immunomodulating activity of inhaled particulate lead oxide disrupts pulmonary macrophage-mediated functions important for host defense and tumor surveillance in the lung. *Environ. Res.* 62:207–222.
13. Zelikoff, J. T., Sisco, M., Yang, Z., Cohen, M. D., and Schlesinger, R. B. 1994. Immunotoxicity of sulfuric acid aerosol: Effects on pulmonary macrophage effector and functional activities critical for maintaining host resistance against infectious diseases. *Toxicology*, 92:269–286.
14. Ortega, H., Salvaggio, J., and Lopez, M. 1995. In-vitro effect of methylmercury chloride on lymphocyte proliferative response and cytokine production. *J. Allergy Clin. Immunol.*, 95(1):212.
15. Ortega, H., Lopez, M., and Salvaggio, J. E. 1994. Effect of methylmercury compounds on lymphocyte proliferation and IL-6 production. Presented at the 5th Biennial Meeting of the Transpacific Allergy and Immunology Society, Lanai, HI.
16. Kang, J. H. H., Lewis, D. M., Castranova, V., Rojanasakul, Y., Banks, D. E., Ma, J. Y. C., and Ma, J. K. H. 1992. Inhibitory action of tetrandine on macrophage production on interleukin-1 (IL-1)-like activity and thymocyte proliferation. *Exp. Lung Res.*, 18:715–729.
17. Salvaggio, J. E. Alterations in immunoregulation by environmental agents. Project description 1992-4 P42 ES05946, Tulane University.
18. Top 20 Hazardous Substances—ATSDR/EPA Priority List for 1997, http://atsdr1.atsdr.cdc.gov:8080/cxcx3.html.
19. Evans, L. J. 1989. Chemistry of metal retention by soils. *Environ. Sci. Technol.*, 23:1046–1056.
20. Rabelais, N. N., Turner, R. E., Wiseman, Jr., W. J., and Boesch, D. F. 1991. A brief summary of hypoxia on the northern Gulf of Mexico continental shelf: 1985–1988. In: *Modern and Ancient Continental Shelf Anoxia*, Tyson, R. V., and Pearson, T. H. (eds.), *Geol. Soc. Spec. Publ.* 58, pp. 35–47, The Geol. Soc., London.
21. Williams, D. F. 1981. Toxicity of implanted metals. In: *Fundamental Aspects of Biocompatibility*, Williams, D. F. (ed.) vol. 2, pp. 45–61. CRC Press, Boca Raton, FL.
22. Pettine, M., and Millero, F. J. 1990. Chromium speciation in seawater: The probable role of hydrogen peroxide. *Limnol. Oceanogr.*, 35:730–736.
23. Government of Canada. 1994. Chromium and Its Compounds—Priority Substances List Assessment Report. Beauregard Printers, Ottawa, Canada, p. 10.
24. Bond, A. M. 1980. *Modern Polarographic Methods in Analytical Chemistry*. Marcel Dekker, New York.
25. Young, R. L., Spell, J. E., Siu, H. M., and Phillip, R. H. 1975. Determination of nitrate in water samples using a portable polarographic instrument. *Environ. Sci. Technol.*, 9(12):1075–1077.
26. Prasad, A. Meeting the demands of tougher air pollution regulations. *Instrument. Control Syst., Sensors*: Part 8, 64(11):74.
27. Bundy, K. J., and Berzins, D. 1998. Differential pulse polarographic analysis of lead and chromium content in Louisiana waters. *Environ. Geochem. Health*, 20:45–51.
28. Berzins, D., Bundy, K. J., and Chan P. 1994. Polarographic trace level analysis can be applied to the detection of environmental contaminants. In: R. Cothern (ed.), *Trace Substances, Environment, and Health*, pp. 63–72. Science Reviews, Northwood, U.K.

29. Bundy K. J., and Berzins, D. 1994. Heavy metal concentration in Louisiana waterways, sediments, and biota. 15th Annual Meeting Soc. Environ. Tox. and Chem., Denver, CO, Oct. 30–Nov. 3, Abstract Book, Abstract 249, p. 45.
30. Bollinger, J. E., Bundy, K., Anderson, M. B., Millet, L., Jolibois, L., Chen, H., Kamath, B., and George, W. J. 1997. Bioaccumulation in red swamp crayfish (*Procambarus clarkii*). *J. Hazardous Mater.*, 54:1–13.
31. Bundy, K., Berzins, D., and Millet, L. 1996. Heavy metal speciation and uptake in crayfish and tadpoles. *Proc. HSRC/WERC Joint Conf. on the Environment*, Albuquerque, NM, May 21–23, pp. 21–34.
32. Bundy, K. J., and Berzins, D. 1995. Lead bioaccumulation in tadpoles from sediment and water. V. M. Goldschmidt Conference, State College, PA, May 24–26, Abstract Book, p. 34.
33. Bundy, K., Millet, L., Bollinger, J., and Anderson, M. 1996. Speciation of chromium in crayfish. Presented at Experimental Biology 96 (annual FASEB meeting), Washington, DC, Apr. 14–17, Abstract No. 1007, published in *FASEB J.*
34. Bollinger, J. E., Bundy, K., Anderson, M., Millet, L., Preslan, J. E., and George, W. J. 1996. Bioaccumulation of chromium in red swamp crayfish (*Procambarus clarkii*). Presented at Experimental Biology 96 (annual FASEB meeting), Washington, DC, Apr. 14–17, Abstract No. 1006, published in *FASEB J.*
35. Bundy, K. J., and Mowat, F. 1996. Speciation and toxicity studies of complex heavy metal mixtures. *Proc. HSRC/WERC Joint Conf. on the Environment*, Albuquerque, NM, May 21–23, pp. 35–47.
36. Bundy, K. J., and Mowat, F. 1996. Speciation, complexation, and sorption effects on toxicity of heavy metal mixtures in water and sediment. Abstract Book, 17th Annual Meeting Soc. Environ. Tox. and Chem., Washington, DC, Nov. 17–21, Abstract 152, pp. 28–29.
37. Bundy, K., Berzins, D., and Taverna, P. 1996. Development of polarographic field sensors for heavy metal detection. *Proc. HSRC/WERC Joint Conf. on the Environment*, Albuquerque, NM, May 21–23, pp. 176–185.
38. EG&G Princeton Applied Research. 1982. Application Note F-1: Polarographic determination of formaldehyde and other aldehydes.
39. Lund, H., and Baizer, M. M. (eds.). 1991. *Organic Electrochemistry: An Introduction and a Guide*. Marcel Dekker, New York.
40. Charlot, G. 1964. *Colorimetric Determination of Elements*. Elsevier, Amsterdam.
41. Hach Co. 1996. *Spectrophotometer Handbook*, DR/2010.
42. Rieger, P. H. 1994. *Electrochemistry*, 2nd ed. Chapman & Hall, New York.
43. Laboratory Products Group. 1984. *Model 901 Microprocessor Ionanalyzer Instruction Manual*, Orion Research.
44. Chang, J. C., Taylor, P. B., and Leach, F. R. 1981. Use of the Microtox assay system for environmental samples. *Bull. Environ. Contam. Toxicol.*, 26:150–156.
45. Indorato, A. M., Snyder, K. B., and Usinowicz, P. B. 1984. Toxicity screening using the Microtox analyzer. In: D. Liu and B. J. Dutka (eds.) *Toxicity Procedures Using Bacterial Systems*, pp. 37–54. Marcel Dekker, New York.
46. Firth, B. K., and Backman, C. J. 1990. A comparison of Microtox testing with rainbow trout acute and *Ceriodaphnia* chronic bioassays using pulp and paper mill wastewaters. TAPPI, 1990 Environmental Conference, Apr. 9–10.
47. Ribo, J. M., and Kaiser, K. L. E. 1983. Effects of selected chemicals to photoluminescent bacteria and their correlations with acute and sublethal effects on other organisms. *Chemosphere* 12(11/12):1421–1442.
48. Kaiser, K. L. E., and Esterby, S. R. 1991. Regression and cluster analysis of the acute toxicity of 267 chemicals to six species of biota and the octanol/water partition coefficient. *Sci. Tot. Environment*, 109/110:499–514.
49. Hermens, J., et al. 1985. Quantitative structure–activity relationships and mixture toxicity of organic chemicals in *Photobacterium phosphoreum*: The Microtox test. *Ecotoxicol. Environ. Safety*, 9:17–25.

50. Ribo, J. M., and Kaiser, K. L. E. 1984. Toxicities of chloranilines to *Photobacterium phosphoreum* and their correlations with effects on other organisms and structural parameters. In K. L. E. Kaiser (ed.), *QSAR in Environmental Toxicology*, pp. 319–336. D. Reider Publishers, Dordrecht, Holland.
51. Dezwart, D., and Sloof, W. 1983. The Microtox as an alternative assay in the acute toxicity assessment of water pollutants. *Aquatic Toxicol.*, 4:129–138.
52. Kaiser, K. L. E., and Ribo, J. M. 1988. *Photobacterium phosphoreum* toxicity bioassay. II. toxicity data compilation. *Toxicol. Assess.*, 3:195–237.
53. Bricka, R. M. 1997. An overview of remediation technologies for treating military heavy metal contaminated soils. 4th Int. Conf. on Biogeochemistry of Trace Elements, Berkeley, CA, June 23–26.
54. Paff, S. W., Bosilovich, B. E., and Kardos, N. J. 1994. Acid extraction treatment system for treatment of metal contaminated soils, EPA/540/SR-94/513, Aug., pp. 1–6.
55. Acar, Y. B., et al. 1995. Electrokinetic remediation: basics and technology status. *J. Hazardous Mater.*, 40:117–137.
56. Puppala, S., Ozsu, E. E., and Gale, R. J. 1997. Effects of electrolyte enhancement to accelerate metal migration for in situ electrokinetic remediation processes. 4th Int. Conf. on Biogeochemistry of Trace Elements, Berkeley, CA, June 23–26.
57. Tratnyek, P. G. 1996. Putting corrosion to use. Remediating contaminated groundwater with zero-valent metals. *Chem. Ind.*, July 1, pp. 499–503.
58. Weber, E. J. 1996. Iron-mediated reductive transformations: Investigation of reaction mechanism. *Environ. Sci. Technol.*, 30(2):716–719.
59. Cantrell, K. J., Kaplan, D. I., and Wietsma, T. W. 1995. Zero-valent iron for the in situ remediation of selected metals in groundwater. *J. Hazardous Mater.*, 42(2):201–212.
60. Gould, J. P., Khudenko, B. M., and Wiedeman, H. F. 1986. Kinetics and yield of the magnesium cementation of cadmium. In: *Metals Speciation, Separation and Recovery*, Patterson, J. W., and Passino, R. (eds.), pp. 175–190. Lewis, Chelsea, MI.
61. Gould, J. P. 1982. The kinetics of hexavalent chromium reduction by metallic iron. *Water Res.*, 16:871–877.
62. Young, G. K., Bungay, H. R., Brown, M., and Parsons, W. A. 1964. Chemical reduction of nitrate in water. *J. Water Pollution Control Fed.*, 36(3, Pt. 1):395–398.
63. Wilson, E. K. 1995. Zero-valent metals provide possible solution to groundwater problems. *Chem. Eng. News*, July 3, 73:19–22.
64. Burris, D. R., Campbell, T. J., and Manoranjan, V. S. 1995. Sorption of trichloroethylene and tetrachloroethylene in a batch reactive metallic iron-water system. *Environ. Sci. Technol.*, 29(11):2850–2855.
65. Helland, B. R., Alvarez, P. J. J. and Schnoor, J. L. 1995. Reductive dechlorination of carbon tetrachloride with elemental iron. *J. Hazardous Mater.*, 41:205–216.
66. Gupta, S., Wenger, K., and Krebs, R. 1997. In situ gentle remediation and stabilization approaches for heavy-metal-contaminated soils. In: I. K. Iskandar et al. (eds.), *Proc. 4th Int. Conf. on Biogeochemistry of Trace Elements*, Berkeley, CA, June 23–26, pp. 639–640.
67. Watanabe, M. E. 1997. Phytoremediation on the brink of commercialization. *Environ. Sci. Technol.*, 31(4):182A–186A.
68. Huang, J. W., et al. 1997. Phytoremediation of lead-contaminated soils. Role of synthetic chelates in lead phytoextraction. *Environ. Sci. Technol.*, 31(3):800–805.
69. Blaylock, M. J., et al. 1997. Enhanced accumulation of Pb in Indian mustard by soil-applied chelating agents. *Environ. Sci. Technol.*, 31(3):860–865.
70. Streit, B., and Strumm, W. 1993. Chemical properties of metals. In: B. Markert (ed.), *Plants as Biomonitors*, Weinheim. VCH, p. 57.
71. Adler, T. 1996. Botanical cleanup crews using plants to tackle polluted water and soil. *Sci. News*, July 20, 150:42–43.

72. Blaylock, M. J., et al. 1997. Field demonstrations of phytoremediation of lead-contaminated soils. In: Iskandar, I. K. et al. (eds.) *Proc. 4th Int. Conf. on Biogeochemistry of Trace Elements*, Berkeley, CA, June 23–26, pp. 629–630.
73. Eadie, B. J., and Robbins, J. A. 1987. The role of particulate matter contaminants in Great Lakes. In: R. A. Hites and S. J. Eisenreich (eds.), *Sources and Fates of Aquatic Pollutants*, pp. 319–364. Advances in Chemistry Series 216, American Chemical Society, Washington, DC.
74. Rabalais, N. N., McKee, B. A., Reed, D. J., and Means, J. C. 1992. Fate and effects of nearshore discharges of federal OCS produced waters. In: J. P. Ray and F. R. Engelhart (eds.), *Produced Water*. Plenum Press, New York.
75. Azur Environmental. 1997. *DeltaTox PS1 Portable System*. Azur, Carlsbad, CA.
76. True, C. J., and Heyward, A. A. 1990. Relationships between Microtox test results, extraction methods, and physical and chemical compositions of marine sediment samples. *Toxicol. Assess.*, 5:29–45.
77. Tessier, A., Campbell, P. G. C., and Bisson, M. 1979. Sequential procedures for the speciation of particulate trace metals. *Anal. Chem.*, 51(7):844–851.
78. http://www.tulane.edu~dmc_spot/PPT/maps.
79. Flowers, G. C., Koplitz, L. V., and McPherson, G. L. 1994. Chemical stability of heavy metals in the bottom sediments of Bayou Trepagnier. Report submitted to Tulane/Xavier DoE/EM Project by the Natural and Active Chemical Remediation Cluster.
80. EA Engineering, Science and Technology, Inc. 1996. Feasibility study of remediation alternatives for Bayou Trepagnier. EA Project #11468.40.
81. Pourbaix, M. 1973. *Lectures on Electrochemical Corrosion*. Plenum Press, New York.
82. Manahan, S. E. 1994. *Environmental Chemistry*, 6th ed. Lewis Publishers, Boca Raton, FL.
83. Blum, D., and Speece, R. 1990. Determining chemical toxicity to aquatic species—The use of QSAR's and surrogate organisms. *Environ. Sci. Technol.*, 24:284–293.
84. Devillers, J., Chambon, P., Zakarya, D., and Chastrette, M. 1986. A new approach to ecotoxicological QSAR studies. *Chemosphere*, 15:993–1002.
85. Kaiser, K. L. E., and Ribo, J. M. 1985. QSAR of chlorinated aromatic compounds. In: M. Tichy (ed.), *QSAR in Toxicology and Xenobiochemistry*, pp. 27–38. Elsevier, Amsterdam.
86. Koch, R. 1984. Quantitative structure–activity relationships in ecotoxicology: Possibilities and limits. In: K. L. E. Kaiser (ed.), *QSAR in Environmental Toxicology*, pp. 207–222. D. Reidel, Dordrecht, Holland.
87. National Research Council of Canada. 1983. *Polycyclic Aromatic Hydrocarbons in the Aquatic Environment: Formation, Sources, Fate, and Effects on Aquatic Biota*, No. 18981, National Academy Press, Washington, DC.
88. Nagpal, N. K. 1994. Development of water quality and sediment criteria for PAH's to protect aquatic life. In: D. D. MacKinley (ed.), *High Performance Fish: Proc. Int. Fish Physiology Symp.*, Vancouver, BC, July 16–21, pp. 193–198.
89. Shettlemore, M., Bundy, K., Mowat, F., and Greene, M. 1997. Bioluminescent bacterial assays of implant corrosion product toxicity. In: J. Bumgardner and A. Puckett (eds.), *Proc. 16th Southern Biomedical Engineering Conf.*, Biloxi, MS, Apr. 4–6, pp. 190–193.
90. Shettlemore, M. G., and Bundy, K. J. 1999. Toxicity measurement of orthopaedic implant alloy degradation products using a bioluminescent bacterial assay. *J. Biomed. Mater Res.*, 45: 395–403.
91. Bundy, K. J., Mowat, F., Taverna, P., and Shettlemore, M. 1998. Toxicity assays and chemical analysis under field conditions. In: *Abstract Book 8th Annual SETAC-Europe Meeting*, Bordeaux, France, Apr. 14–18, p. 47.
92. Mowat, F., Shettlemore, M., and Bundy, K. J. 1998. Toxicity assessment directly in the field. In: *Abstract Book 19th Annual SETAC Meeting*, Charlotte, NC, Nov. 15–19, p. 205.
93. Taverna, P. J., Mowat, F. S., and Bundy, K. J. 1998. Environmental field assessment in Louisiana waterways. *Proc. Third Environmental State of the State Conf.*, Nov. 19, Baton Rouge, LA, pp. 36–37.

94. Kaiser, K., and Palabrica, V. 1991. *Photobacterium phosphoreum* toxicity data index. *Water Pollution Res. J. Can.*, 26:361–431.
95. Microbics Corporation. 1995. *Microtox Acute Toxicity Basic Test Procedures*. Microbics Corp., Carlsbad, CA.
96. Gonzalez, C., Smith, M. K., Mielke, H. W., and Kale, F. P. 1997. Trace metals in sediment, soils and water of urban Bayou Saint John and rural Jean Lafitte National Park, Louisiana. Abstract No. 414. *Fund. Appl. Toxicol.*, 36(1, Pt. 2):278.
97. Corbridge, D. E. C. 1990. *Phosphorus: An Outline of Its Chemistry, Biochemistry, and Technology*. Elsevier, Oxford, U.K.
98. Budvari, S. (ed.), 1989. *The Merck Index*. Merck & Co., Rahway, NJ.
99. Nriagu, J. O. 1978. *The Biogeochemistry of Lead in the Environment. Part A. Ecological Cycles*. Elsevier/North Holland and Biomedical Press, New York.
100. Benko, V. 1987. Arsenic. In: L. Fishbein, A. Furst, and M. A. Mehlman (eds.), *Advances in Modern Environmental Toxicology*, Vol. 2, pp. 1–30. Princeton Scientific, Princeton, NJ.
101. Bundy, K. J., and Chan, P. 1991. Polarographic analysis of chromium concentration levels in blood. 10th Southern Biomedical Engineering Conf., Atlanta, Oct. 18–21, Digest of Papers, pp. 54–57.
102. Bundy, K., Bricka, M., and Morales, A. 1996. Environmental factors affecting corrosion of munitions. In: *Abstract Book, 17th Annual Meeting Soc. Environ. Toxicol. and Chem.*, Washington, DC, Nov. 17–21, Abstract PO525, pp. 217–218.
103. Kumar, P. B. A., et al. 1995. Phytoextraction: The use of plants to remove heavy metals from soils. *Environ. Sci. Technol.*, 29:1232–1238.
104. Bundy, K. J. 1994. Corrosion and other electrochemical aspects of biomaterials. *Crit. Rev. Biomed. Eng.*, 22(3/4):139–251.
105. Lowe, C. R., Yon Hin, B. F. Y., Cullen, D. C., Evans, S. E., Stephens, L. D. G., and Maynard, P. 1990. Biosensors. *J. Chromatogr.*, 510:347–354.
106. Paddle, B. M. 1996. Biosensors for chemical and biological agents of defence interest. *Biosensors & Bioelectronics*, 11(11):1079–1113.
107. Rogers, K. R., and Lin, J. N. 1992. Biosensors for environmental monitoring. *Biosensors & Biolectronics*, 7:317–321.
108. Refojo, M. F. 1996. Ophthalmologic applications. In: B. D. Ratner, et al. (eds.), *Biomaterials Science—An Introduction to Materials in Medicine*, pp. 328–334. Academic Press. San Diego, CA.
109. Darnell, J., Lodish, H., and Baltimore, D. 1990. *Molecular Cell Biology*, pp. 710–752. Scientific American Books, New York.
110. Timbrell. J. A. 1991. *Principles of Biochemical Toxicology*, 2nd ed. Taylor & Francis, London.
111. Venter, J. C. 1984. Evolution and structure of neurotransmitter receptors. In: *Monoclonal and Anti-Idiotypic Antibodies: Probes for Receptor Structure and Function Analysis*, J. C. Venter, C. M. Fraser, and J. Lindstrom, (eds.), pp. 117–139. Alan R. Liss, New York.
112. Klausner, R. D., van Renswoude, J., Blumenthal, R., and Rivnay, B. 1984. Reconstitution of membrane receptors. In: *Molecular and Chemical Characterization of Membrane Receptors*, J. C. Venter and L. C. Harrison, (eds.), pp. 209–239. Alan R. Liss, New York.
113. Rogers, K. R., Valdes, J. J., and Eldefrawi, M. E. 1991. Effects of receptor concentration, media pH and storage on nicotinic receptor-transmitted signal in a fiber-optic biosensor. *Biosensors & Bioelectronics*, 6:1–8.
114. Miller, C. (ed.). *Ion Channel Reconstitution*, pp. 186–197. Plenum Press, New York.
115. Uto, M., Michaelis, E. K., Hu, I. F., Umezawa, Y., and Kuwana, T. 1990. Biosensor development with a glutamate ion-channel reconstituted in a lipid bilayer. *Anal. Sci.*, 6:221–225.
116. Labarca, P., Simon, S. A., and Anholt, R. R. H. 1988. Activation by odorants of a multistate cation channel from olfactory cilia. *Proc. Natl. Acad. Sci. USA.*, 85:944–947.
117. Strange, P. G. 1987. Isolation and molecular characterisation of dopamine receptors. In: *Dopamine Receptors*, I. Creese and C. M. Fraser, (eds.), pp. 29–43. Alan R. Liss, New York.

118. Toft, D., and Gorski, J. 1966. A receptor molecule for estrogens: Isolation from the rat uterus and preliminary characterisation. *Proc. Natl. Acad. Sci. USA*, 55:1574–1581.
119. Griffiths, D., and Hall, G. 1993. Biosensors—What real progress is being made? *Trends Biotechnol.*, 11:122–130.
120. Rechnitz, G. A., and Ho, M. Y. 1990. Biosensors based on cell and tissue material. *J. Biotechnol.*, 15:201–218.
121. Marty, J. L., Sode, K., and Karube, I. 1992. Biosensor for detection of organophosphate and carbamate insecticides. *Electronalysis*, 4:249–252.
122. Palleschi, G., Bernabei, M., Cremsini, C., and Macini, M. 1992. Determination of organophosphorous insecticides with a choline electrochemical biosensor. *Sensors & Actuators B*, 7:513–517.
123. Nyamsi Hendji, A. M., Jaffrezic-Renault, N., Martelet, C., Clechet, P., Shul'ga, A. A., Strikha, V. I., Netchiporuk, L. I., Soldatkin, A. P., and Wlodarski, W. B. 1993. Sensitive detection of pesticides using differention ISFET-based system with immobilized cholinesterases. *Anal. Chim. Acta*, 281:3–11.
124. La Rosa, C., Pariente, F., Hernandez, L., and Lorenzo, E. 1995. Amperometric flow-through biosensor from the determination of pesticides. *Anal. Chim. Acta*, 308:129–136.
125. Skladal, P. 1992. Detection of organophosphate and carbamate pesticides using disposable biosensors based on chemically modified electrodes and immobilized chloinesterase. *Anal. Chim. Acta*, 269:281–287.
126. Iwuoha, E. I., and Lyons, M. E. G. 1997. Organic phase enzyme electrodes: Kinetics and analytical applications. *Biosensors & Bioelectronics*, 12(1):53–75.

6
Theoretical Basis for the Simulation of Electrokinetic Remediation

Akram N. Alshawabkeh
Department of Civil and Environmental Engineering, Northeastern University
Boston, Massachusetts

Christian J. McGrath
U.S. Army Engineer Research & Development Center, Waterways Experiment Station
Vicksburg, Mississippi

I. INTRODUCTION

Electrokinetic soil remediation employs electric currents through electrodes inserted in the contaminated soil to produce an electric field. Ambient or introduced solutes migrate in response to the imposed electric field by electroosmosis and ionic migration. Electroosmosis mobilizes the pore fluid to flush solutes, usually from the anode (+ve) toward the cathode (−ve), while ionic migration effectively separates anionic (−ve) and cationic (+ve) species, drawing them to the anode and cathode, respectively. This transport, coupled with geochemical reactions such as sorption, precipitation, and dissolution, are the fundamental mechanisms of electrokinetic remediation. Contaminant extraction and removal are accomplished by electrodeposition, precipitation, or ion exchange, either at the electrodes or in an external extraction system. The major advantages of the technology include: (a) it can be implemented in situ with minimal disruption, (b) it is well suited for fine-grained, heterogeneous media, where other techniques such as pump-and-treat can be ineffective, and (c) accelerated rates of contaminant transport and extraction may be obtained. Electrokinetic extraction of metals and radionuclides from soils has been investigated by bench-scale tests, pilot-scale tests, and limited field applications (1–11). Recent developments of the process are summarized in a special issue of the *Journal of Hazardous Materials* on "Electrochemical Decontamination of Soil and Water" (12). Principles and applications of the electrokinetic remediation are also presented in Chapter 4.

Theoretical understanding and simulation of the technique demand a grasp of the mathematical formulation of transport processes, which are controlled by such variables as electrolysis reactions at the electrodes, pH and soil-surface chemistry, equilibrium chemistry of the aqueous system, electrochemistry of the contaminants, and geotechnical/hydrological characteristics of the porous medium. The complexity of transport processes necessitates simplifying assumptions which allow adequate numerical simulation. The following assumptions are employed in the theoretical development presented in this chapter: (a) the soil medium is isotropic and saturated; (b) the porous medium is a solid framework of ion-exchange surfaces with the pore space occupied by chemically reactive species in aqueous solution; (c) all fluxes are linear homogeneous functions of all driving forces (or potential gradients); (d) isothermal conditions prevail (coupled heat transfer is neglected); (e) all the applied voltage is effective in fluid and charge transport; (f) electrophoresis is negligible; (g) the chemicoosmotic coupling is negligible; (h) soil particles are treated as electrically nonconductive (insulators); (i) surface conductance and streaming potential are negligible.

Based on the identified assumptions, theoretical formulations are provided for transport mechanisms under electric fields. Contaminant transport mechanisms include hydraulic or fluid flow, species or mass transport, and charge transport. Following is a description of the theoretical basis of these transport mechanisms.

II. FLUID FLUX

Fluid flux results from application of a hydraulic gradient (Darcy's law) and/or an electric gradient (electroosmosis). Fluid flux per unit area of the porous medium due to hydraulic and electric gradients, $J_w(L^3L^{-2}T^{-1})$, is given by

$$J_w = k_h \nabla(-h) + k_e \nabla(-\Phi) \tag{1}$$

where k_h is the coefficient of hydraulic conductivity (LT^{-1}), k_e is the coefficient of electroosmotic permeability $(L^2V^{-1}T^{-1})$, h is the hydraulic head (L), and Φ is the electric potential (V). Extensive research has been carried out on the hydraulic conductivity of fine-grained soils with a relatively good understanding of the fundamental factors affecting its value (13–17). These studies indicate that microstructure and fabric are the factors that highly influence fluid transport in fine-grained deposits. Dispersed microstructure generally result in lower hydraulic conductivity than flocculated microstructure. Other factors that affect k_h include soil porosity and pore size distribution. Presence of uniformly distributed fine-size pores results in lower hydraulic conductivity, while presence of few macropores results in higher hydraulic conductivity even if soil porosity is the same in both cases.

Electrokinetic soil remediation induces changes in the pore fluid chemistry, diffuse double layer, soil fabric, and consequently the hydraulic conductivity. Furthermore, electroosmotic consolidation is expected to take place and influence the hydraulic conductivity. In attempting to provide a mathematical formulation for electrokinetic soil remediation, hydraulic conductivities are generally assumed to be constant because (a) there is no clear mathematical formalism that can describe the effect of pore fluid chemistry on soil fabric and consequently the hydraulic conductivity, and (b) the uncertainties in evaluating the hydraulic conductivities are more significant than the changes expected in its values.

The Helmholtz-Smouluchowski theory for electroosmosis is the most commonly adopted description of fluid transport through soils due to electrical gradients. Similar to hydraulic conductivity, this theory introduces the coefficient of electroosmotic permeability, k_e, as the volume rate of fluid flowing through a unit cross-sectional area due to a unit electrical gradient. The value of k_e is described as a function of zeta potential, viscosity of the pore fluid, porosity, and electrical permittivity of the soil medium. When the soil pores are treated as capillary tubes, the coefficient of electroosmotic permeability is given by

$$k_e = \frac{\varepsilon \zeta}{\eta} n \qquad (2)$$

where ε is the permittivity of the medium (farad L^{-1}), ζ is the zeta potential (V), n is the porosity dimensionless), and η is the viscosity (FTL^{-2}).

Contribution of each component due to hydraulic and electric gradients may be described by the ratio of the coefficient of electroosmotic permeability relative to hydraulic conductivity (k_e/k_h). Soil type, microstructure, and pore fluid conditions are the factors that affect this ratio. In course-grained soils this ratio is very small and goes to zero due to the negligible electroosmotic flow and relatively high hydraulic conductivities ($>10^{-3}$ cm/s) of such soils. On the other hand, in soft, fine-grained soils the ratio k_e/k_h becomes significant, as k_e is usually of the order of 10^{-5} cm^2/V-s while k_h is less than 10^{-5} cm/s (10^{-7} cm/s for clayey soils).

III. MASS FLUX

Mass flux of different chemical species relative to pore fluid is a result of multiple coupled potential gradients. Electromigrational mass flux is mass transport of charged species to the electrode opposite in polarity due to an electric potential gradient.

Hydrodynamic dispersion is a result of two basic phenomena: mechanical dispersion and molecular diffusion. While mechanical dispersion occurs as a result of scale dependent velocity variation within the porous medium, molecular diffusion is mass transport due to the random migration of the molecules which macroscopically fit a Fickian model. Mechanical dispersion is a significant mechanism in contaminant transport in groundwater (18,19) because of the relatively high hydraulic conductivity and advective hydraulic flow in such deposits (higher than 10^{-5} cm/s). On the other hand, molecular diffusion is the primary process that controls hydrodynamic dispersion in clay deposits due to the low advective hydraulic flow in these deposits. Therefore, only molecular diffusion is considered in mass transport under concentration gradients.

Total mass flux of dissolved species also includes the advective component due to species transport by the flowing fluid. The total mass transport of chemical species per unit cross-sectional in a saturated soil medium under hydraulic, electric, and chemical concentration gradients is described by

$$J_i = D_i^* \nabla(-c_i) + c_i(u_i^* + k_e)\nabla(-\Phi) + c_i k_h \nabla(-h) \qquad (3)$$

where J_i(ML^{-2}T^{-1}) is the total mass flux of the ith chemical species per unit cross-sectional area of the porous medium, c_i(ML^{-3}) is the molar concentration of the ith chemical species, D_i^*(L^{2}T^{-1}) is the effective diffusion coefficient of the ith chemical species, and u_i^*(L^{2}TV^{-1}) is the effective ionic mobility of the ith species. Transport mechanisms included in Eq. (3) are diffusion, ion migration, electroosmotic advection, and hydraulic advection. A schematic of mass transport of cationic and anionic species is provided in Fig. 1. Transport profiles in Fig. 1 are based on the assumptions that water advection components (electroosmosis and hydraulic) act from anode to cathode. The advective flow enhances transport of cationic species, which migrates from anode to cathode, and retards transport of anionic species, which migrates from cathode to anode.

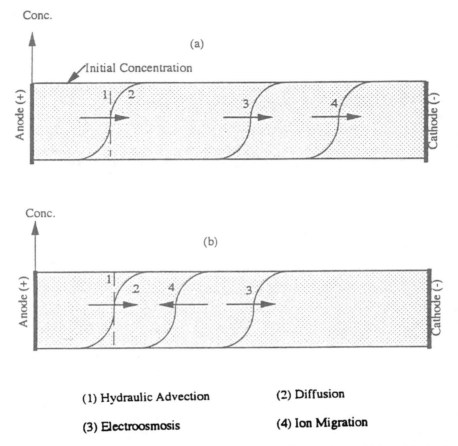

Figure 1 Schematic showing concentration profiles in transport of (a) positively charged ions and (b) negatively charged ions.

The effective diffusion coefficient in the porous medium, D_i^*, is related to the respective diffusion coefficient in free solution, D_i, by (19–21)

$$D_i^* = D_i \tau n \qquad (4)$$

where τ (dimensionless) is an empirical coefficient accounting for the tortuosity of the medium. Values of τ span a wide range for different saturated and unsaturated soils. Experiments are often necessary to determine its value for a specific soil type. Shackelford and Daniel (21) summarize reported τ values for different soil types. These values are as low as 0.01 and as high as 0.84, mostly ranging between 0.2 and 0.5. Diffusion coefficients for different ions at infinite dilution are available in most electrochemistry handbooks and references. These values represent the maximum values attained under ideal conditions. Many factors might affect the molecular diffusion coefficient, such as the electroneutrality requirement, complexation, solubility, concentration, and electrolyte strength (22). Shackelford and Daniel (21) investigated the effective diffusion coefficients, D^*, of different inorganic chemicals in compacted clay. Generally, changing molding water content in compaction tests results in significant changes in the compacted soil microstructure (23). According to results of Shackelford and Daniel (21), the change in compacted clay microstructure due to different molding water contents will have little effect on the effective diffusion coefficient of different chemicals.

The effective ionic mobility, u_i^*, defines the velocity of the ion in soil pores under unit electric field. There is no method yet devised to measure the effective ionic mobility (24); however, u_i^* can be theoretically estimated by assuming that the Nernst-Townsend-Einstein relation between D_i, the molecular diffusion coefficient, and u_i, holds for ions in the pore fluid of soils (25):

$$u_i^* = n\tau u_i = \frac{D_i^* z_i F}{RT} \tag{5}$$

where u_i is the ionic mobility of species i at infinite dilution, z_i is the charge of the ith species, F is Faraday's constant (96,485 C/mol electrons), R is the universal gas constant (8.3144 J/K-mol), and T is the absolute temperature. Note that each u_i in this case has a value and a sign that reflects the charge of species i (i.e., the ionic mobilities and effective ionic mobilities, u_i and u_i^*, will have negative values for anions and positive values for cations). The signs are included in the ionic mobilities of cations and anions to simplify the mathematical equations.

Equation (3) demonstrates that electrical gradient affects two mass transport mechanisms, ion migration and electroosmotic advection. The relative contribution of these components to species transport depends on soil and contaminant characteristics. Both mechanisms require a high degree of saturation. However, electroosmosis requires specific conditions, including the presence of clay minerals. It is also affected by pore-fluid ionic strength and pH. Acidic pore fluid conditions could reverse the charge on the clay mineral surface, and consequently the electroosmostic flow (26–29). At specific pH, the soil could have a zero net charge [pzc: point of zero charge, (28)] and electroosmosis could cease. Ionic migration on the other hand, occurs in all soil types, including course-grained soils (sand and gravel). Acar and Alshawabkeh (8) compared the role of electroosmosis and ionic migration for different conditions. The results showed that in most cases ionic migration was more significant than electroosmosis. Best conditions for electroosmosis can result in a flow of the order of 10^{-5} cm/s per unit voltage gradient (V/cm). Average effective ionic migration rates of different ions are usually of the order of 10^{-5} cm/s, but also could be of the order of 10^{-4} cm/s.

IV. CHARGE FLUX

Applying a DC current through a soil–water–electrolyte medium generates an electric field that causes charge transport. It is assumed that the soil pore fluid has a relatively high ionic strength, which makes the contribution of the free pore fluid dominate the other charge transport mechanisms. Therefore, the contribution of the soil solids and the diffuse double-layer ions on charge transport is neglected. The simplest form of electrical conductance of the soil is governed by Ohm's law describing the current density (charge transport) in the pore fluid due to electrical gradients,

$$I = \sigma^* \nabla(-\Phi) \tag{6}$$

where I is the electric current density ($CL^{-2}T^{-1}$), and σ^* is the effective electrical conductivity of the soil. However, charge flux has another component due to ion diffusion.

Therefore, the total charge flux can be evaluated by using Faraday's law for equivalence of mass flux and charge flux,

$$I = \sum_{j=1}^{N} z_j F J_j \tag{7}$$

substituting Eq. (3) into Eq. (7), the total charge flux is given by

$$I = F \sum_{j=1}^{N} z_j D_j^* \nabla(-c_j) + \sigma^* \nabla(-\Phi) \tag{8}$$

where the effective electrical conductivity, σ^*, of the soil bulk due to charge flux is given by

$$\sigma^* = \sum_{i=1}^{N} F z_i u_i^* c_i \tag{9}$$

It should be noted that the advective components (electroosmosis and hydraulic advection) of species mass flux will not result in any charge flux due to preservation of the electrical neutrality of free pore fluid, given by

$$\sum_{j=1}^{N} c_j z_j = 0 \tag{10}$$

V. CONSERVATION OF MASS AND CHARGE

To evaluate changes per unit pore volume of the soil, it is necessary to apply the conservation equations to transport equations. Conservation of mass and charge in a unit volume of the soil under the set of assumptions employed require that

$$\frac{\partial \varepsilon_v}{\partial t} = m_v \gamma_w \frac{\partial h}{\partial t} = -\nabla \cdot J_w \tag{11}$$

$$\frac{\partial n c_i}{\partial t} = -\nabla \cdot J_i + n R_i \qquad i = 1, 2, \ldots, N \tag{12}$$

$$\frac{\partial T_e}{\partial t} = C_p \frac{\partial \Phi}{\partial t} = -\nabla I \tag{13}$$

where ε_v is the volumetric strain of the soil mass, m_v is the coefficient of volume compressibility of the soil (L^2F^{-1}); T_e is the volumetric charge density of the soil medium (CL^{-3}); C_p is the electrical capacitance per unit volume (FL^{-3}); and $R_i(ML^{-3}T^{-1})$ is the production/consumption rate of the ith aqueous chemical species per unit fluid volume due to geochemical reactions. The partial differential equation (PDE) given by Eq. (11) describes change in hydraulic head due to soil volume change (consolidation). This equation becomes significant in the case where hydraulic gradients are used to enhance transport under electric gradients. Significant changes in the hydraulic head and development of negative pore water pressure (suction) could occur due to nonuniform pH and voltage gradient distributions. Equation (12) describes transient reactive transport of chemical species i under hydraulic, electrical and chemical concentration gradients. For zero electrical gradients, this equation will reduce to the classical advective-diffusive equation. The last PDE of the described system, Eq. (13), describes conservation of charge in the porous medium. The equations describes the rate of change in the electrical potential required to maintain electrical neutrality of the medium. For zero net change in charge, one should assume zero electric capacitance (C_p) of the soil.

VI. GEOCHEMICAL REACTIONS

Reactive transport of charged species, described by Eq. (12), is controlled by rates of geochemical reactions (R_i). Electrolysis reactions at the electrodes usually produce substantial compositional changes, such as extreme pH at the boundaries (unless amendments are used). The chemistry boundary conditions results in a complex system of geochemical reactions that include sorption (surface complexation and ion exchange), redox, and precipitation-dissolution reactions. The term R_i can then be expanded to account for each reaction type, e.g.,

$$R_i = R_i^s + R_i^{aq} + R_i^p \tag{14}$$

where R_i^s is a term for sorption, similarly R_i^{aq} is a term for aqueous reactions, and R_i^p is for precipitation/dissolution reactions. Two approaches have been developed and used to describe chemical reactions, the instantaneous equilibrium approach and the kinetics approach. In instantaneous equilibrium reactions, species concentrations reach equilibrium instantaneously; whereas in the kinetic reactions approach, concentrations in solution vary with time until they reach equilibrium. For several species, chemical reactions, e.g., dissolution, sorption, and redox, have been found to vary with time before reaching equilibrium. The kinetics approach is expected to be more realistic for modeling these reactions. However, one could assume that chemical reactions reach equilibrium in a very short time (relative to transport time) and thus use an equilibrium model. Geochemical models (kinetics or equilibrium) exist that could be applied in the description of the geochemical reactions induced by electric currents. An overview of these reactions follows.

A. Sorption Reaction

The following general equation describes the sorption rate term:

$$R_i^s = -\frac{\rho}{n}\frac{\partial s_i}{\partial t} = -\frac{\rho}{n}\frac{\partial s_i}{\partial c_i}\frac{\partial c_i}{\partial t} \quad i = 1, 2, \ldots, N \tag{15}$$

where ρ is the bulk dry density of the soil and s_i is the adsorbed concentration of the component j per unit mass of the soil solids (MM^{-1}). The reversible term ($\partial s_i/\partial t$) is often used to describe the sorption rate. The equilibrium partitioning between the adsorbed phase and the aqueous phase of the chemical components are commonly measured under controlled temperature and pressure, and the resulting correlations of s_i versus c_i are called adsorption isotherms. Several isotherm (equilibrium) models (linear, Freundlich, and Langmuir models) have been used to describe sorption of heavy metals on soils. Assuming instantaneous equilibrium in sorption reactions and linear isotherms,

$$\frac{\partial s_i}{\partial c_i} = K_{di} \tag{16}$$

where K_{di} is called the distribution coefficient, K_d, of species i (L^3/M). A retardation factor, R_{di}, may be used to describe the inhibitory effect of adsorption on transport. For a linear isotherm the retardation factor is,

$$R_{di} = 1 + \frac{\rho K_{di}}{n} \tag{17}$$

The retardation factor of species i, R_{di}, defines the relative rate of transport of a nonsorbing species to that of a sorped species. For a nonsorbing species, $R_{di} = 1$.

Simple isotherm sorption models ignore the potential effects of variations in pH, solute composition and ionic strength, redox potential, and processes such as competitive adsorption. Alternative, more robust (and complicated) sorption models include ion or ligand exchange, mass action models, and surface complexation models, as described by Langmuir (30), Kirkner and Reeves (31), Yeh and Tripathi (32), Davis and Kent (33), Stumm and Morgan (34), and Bethke (35).

B. Aqueous Reactions

In aqueous-phase reactions, any complex j is the product of Nc reactant components, i.e.,

$$\sum_{i=1}^{N_c} a_{ji} \overline{c_i} \leftrightharpoons \overline{x_j} \quad j = 1, \ldots, N_x \tag{18}$$

where $\overline{c_i}$ is the chemical formula for component i, $\overline{x_j}$ is the chemical formula for the complex j, and a_{ji} is the stoichiometric ratio in complex j of component i. The law of mass action implies that

$$x_j = K_j^{eq} \prod_{i=1}^{N_c} c_i^{a_{ji}} \quad j = 1, \ldots, N_x \tag{19}$$

where K_j^{aq} is the equilibrium constant for aqueous reaction j. From Eq. (18), the rate of accumulation of component i due to aqueous reaction j, R_{ji}^{aq}, is

$$R_{ji}^{aq} = -a_{ji} R_j \tag{20}$$

The total rate of accumulation of component i due to all aqueous reactions is

$$R_i^{aq} = \sum_{j=1}^{N_x} R_{ji}^{aq} = -\sum_{j=1}^{N_x} a_{ji} R_j \tag{21}$$

C. Precipitation/Dissolution Reactions

It is necessary to account for the precipitation/dissolution reactions in the formulation of mass transport equations. In precipitation reactions, the chemical components are assumed to be composed of reactants,

$$\overline{p_j} \leftrightharpoons \sum_{i=1}^{N_c} b_{ji} \overline{c_i} \quad j = 1, \ldots, N_p \tag{22}$$

where $\overline{p_j}$ is the chemical formula for precipitate j, b_{ji} is the stoichiometric coefficient in precipitate j for component i, and N_p is the number of precipitates for component i. The production of the precipitate will not occur until the solution is saturated and any kinetic barriers are overcome. Therefore, assuming equilibrium, the law of mass action is written as

$$K_j^{sp} \geq \prod_{i=1}^{N_c} c_i^{b_{ji}} \qquad j = 1, \ldots, N_p \tag{23}$$

where K_j^{sp} in the solubility-product equilibrium constant for precipitate j. By the same rationalization of previous formulations, the total rate of production of component i due to precipitation/dissolution reactions, R_i^p, is

$$R_i^p = \sum_{j=1}^{N_p} R_{ij}^p = -\sum_{j=1}^{N_p} b_{ji} R_j^p \tag{24}$$

where R_j^p is the rate of production of precipitate j.

VII. GENERAL SYSTEM FOR MODELING TRANSPORT

The theoretical formalism results in a mathematical system of equations describing transient multicomponent reactive transport under hydraulic, electric, and chemical gradients. The resulting system consists of partial differential equations for transport and algebraic equations for geochemical reactions. The transport PDEs are divided into three types. The first consists of one equation that describes transient fluid flow. The second consists of N number of equations that describe reactive transport of N species. The third type is described by one equation for charge transport.

The equation describing fluid flow is given by Eq. (11). Substituting fluid flux equation [Eq. (1)] in Eq. (11) results in

$$\frac{\partial h}{\partial t} = c_v \nabla^2 h + \frac{k_e}{m_v \gamma_w} \nabla^2 \Phi \tag{25}$$

Equation (25) is necessary to describe changes in the hydraulic head across the soil, which affect the advective component of mass transport. The PDEs transient reactive for mass transport are derived by substituting Eq. (3), which describes mass flux, into Eq. (12), i.e.,

$$\frac{\partial n c_i}{\partial t} = D_i^* \nabla^2 c_i + \nabla(c_i[(u_i^* + k_e)\nabla\Phi + k_h \nabla h]) + n R_i \tag{26}$$

where $i = 1, 2, \ldots, N$. Note that for the case of nonreactive solute transport ($R_i = 0$), steady-state fluid flux ($\partial h/\partial x = $ const.) and no electrical gradient, Eq. (26) becomes

$$\frac{\partial n c_i}{\partial t} = D_i^* \nabla^2 c_i - v \nabla c_i \tag{27}$$

where $v = -k_h \tilde{N} h$. Equation (27) is the advective-diffusive solute transport equation used widely to describe nonreactive solute transport.

Changes in the electric potential distribution across the soil as a result of changes in the geochemistry is formulated by substituting the charge flux equation in the charge conservation equation [Eq. (8) in Eq. (13)],

$$C_p \frac{\partial \Phi}{\partial t} = F \sum_{j=1}^{N} z_j D_j^* \nabla^2 c_j + \nabla(\sigma^* \nabla \Phi) \tag{28}$$

The total number of differential equations described for this system is $N+2$, which include N equations for mass transport, one equation for charge conservation, and one equation for fluid flow. The unknowns described in this system are N species concentrations, c_i, one electric potential F, one hydraulic potential h, and N unknowns for the rate of i chemical reactions. Therefore $2n+2$ unknowns are described by $N+2$ differential equations. The other N equations required for this system are the mass balance equations for the chemical reactions.

VIII. PRESERVATION OF ELECTRICAL NEUTRALITY

The following derivation is used to demonstrate that the change transport equation preserves the electrical neutrality of the porous medium. For a unit volume of the soil, the rate of change in the electric charge equals the total rate of change of chemical species concentrations multiplied by their charge and Faraday's constant, i.e.,

$$\text{rate of change in electric charge} = \sum_{j=1}^{N} z_j F \frac{\partial n c_j}{\partial t} \tag{29}$$

Preservation of electrical neutrality requires that the total change in electric charge per unit volume equals zero.

$$\sum_{j=1}^{N} z_j F \frac{\partial n c_j}{\partial t} = -\sum_{j=1}^{N} z_j F \nabla \cdot J_j + n \sum_{j=1}^{N} z_j F R_j \tag{30}$$

Substituting Eq. (3) in Eq. (30),

$$\sum_{j=1}^{N} z_j R_j = 0 \tag{31}$$

The total rate of change of all chemical species due to chemical reactions times their charge is zero. In other words, consider the following chemical reaction:

$$A \rightleftharpoons m B^{+l} + l D^{-m} \tag{32}$$

One mole of A will produce m moles of B^{+l} and l moles of D^{+m}. The total change in B^{+l} concentration times its charge is m moles of $B^{+l} \times (+l) = ml$. The total change in D^{+m} concentration times its charge is l moles of $D^{+m} \times (-m) = -ml$. Therefore, the total change in each species molar concentration times its charge is $ml - ml = 0$. Substituting Eq. (3) for mass flux and Eq. (31) in Eq. (30),

$$\sum_{j=1}^{N} z_j F \frac{\partial n c_j}{\partial t} = \sum_{j=1}^{N} z_j F D_j^* \nabla^2 c_j + \sum_{j=1}^{N} z_j F \nabla (c_j [(u_j^* + k_e) \nabla \Phi + k_h \nabla h]) \tag{33}$$

Simplifying and substituting $\sum z_j c_j = 0$,

$$\sum_{j=1}^{N} F z_j \frac{\partial n c_j}{\partial t} = \sum_{j=1}^{N} F z_j D_j^* \nabla^2 c_j + \sum_{j=1}^{N} F z_j \nabla (c_j u_j^* \nabla \Phi) \tag{34}$$

Equation (28) already describes the right-hand side of Eq. (34) to be equal to zero when

$C_p = 0$ (no charge capacitance). Therefore, the left-hand side of Eq. (34) should be zero when soil has zero charge capacitance,

$$\sum_{j=1}^{N} z_j F \frac{\partial n c_j}{\partial t} = 0 \tag{35}$$

IX. BOUNDARY CONDITIONS

Boundary conditions are required for hydraulic head, N chemical species concentrations, and electric potential. Usually, hydraulic heads are controlled either to provide constant head difference or constant flow rates. Thus, hydraulic head boundary conditions are easily identified in practice. In most cases, zero head difference is applied between the cathodes and anodes. Boundary conditions for this case are

$$h|_{S1} = h|_{S2} = 0 \tag{36}$$

where $S1$ is the boundary surface at the anode and $S2$ is the boundary surface at the cathode.

Boundary conditions for charge conservation equation are developed from the current value at the boundary. Two types of boundary conditions can be applied: constant voltage or constant current. If the voltage difference is kept constant between the anode and cathode (current density changes depending on electric conductivity), then the following boundary conditions are used:

$$\Phi_{S1} = \Phi_{max} \qquad \Phi_{S2} = 0 \tag{37}$$

However, if the current is maintained constant (voltage changes depending on electric conductivity), then the following boundary conditions are used:

$$\left[F \sum_{j=1}^{N} z_j D_j^* \nabla c_j - \sigma^* \nabla \Phi \right]\bigg|_{S1} = I \tag{38}$$

$$\Phi|_{S2} = 0$$

Identifying boundary conditions for the partial differential equations describing species transport are calculated based on electrolysis reactions at the electrodes. When inert electrodes are used in groundwater, oxidation of water at the anode generates an acid front, while reduction at the cathode produces a base front by the following electrolysis reactions:

$$\begin{aligned} 2H_2O - 4e^- &\to 4H^+ + O_2 \quad \text{anode} \\ 2H_2O + 2e^- &\to 2OH^- + H_2 \quad \text{cathode} \end{aligned} \tag{39}$$

While these electrolysis reactions are the most likely to occur (if no amendments are used), the prevailing electrolysis reactions at the electrodes depend on the availability of other species, pH, and the electrochemical potentials of their reactions. In the presence of multiple species at the cathode or anode, electrolysis reactions will depend on the electrochemical potential of each reaction. It is necessary to evaluate these potentials for different species. In general, the overall electrolysis (cell) reaction comprises two independent half-reactions at the anode and at the cathode. Consider the half-cell reaction

$$v_o O + n e^- \to v R \tag{40}$$

where O is the oxidized form, R is the reduced form, and v_o and v_r are the stoichiometric coefficients. From basics of thermodynamics, the free energy of this cell reaction is given by

$$\Delta G = \Delta G^0 + RT \ln \frac{(R)^{v_r}}{(O)^{v_o}} \tag{41}$$

where G is the Gibbs free energy and parentheses represent activities. Since $\Delta G = -nFE$,

$$E = E^0 + \frac{RT}{nF} \ln \frac{(O)^{v_o}}{(R)^{v_r}} \tag{42}$$

Equation (42) is called the Nernst equation (36), and it provides the potential of the O/R electrode versus the natural hydrogen electrode (NHE) as a function of the activities of O and R. Equation (42) is useful in identification of the type of electrolysis reactions expected at the anode and at the cathode in the presence of multiple species. The electrolysis reaction that has the highest positive E will occur at the cathode, while the electrolysis reaction with the most negative E will occur at the anode. As a result of these electrolysis reactions, the chemistry at the electrodes will undergo continuous changes. These changes either enhance the electrokinetic process (e.g., generation of acid at the anode) or retard the process (e.g., generation of the base at the cathode). Chemical reagents are introduced at the electrodes to enhance electrokinetic process.

Once the types of electrolysis reactions are identified, the boundary conditions can be evaluated from mass equilibrium in electrodes compartments. The rate of concentration change of a specific species i in the electrode compartment will be equal to the net mass flow rate of the species. Figure 2 displays a schematic of the anode well showing the mass fluxes of a specific species, i. The in and out fluxes of species i in this well are (a) generation or electroplating at the electrode due to electrolysis reactions, (b) in/out flux at the soil boundary due to transport, (c) injection of any enhancement agent in the compartment, (d) extraction of the anolyte from the electrode well, and (e) chemical reactions of species i. Accordingly, the rate of change of species i concentration in the anode well is given by

$$V^a \frac{\partial c_i^a}{\partial t} = \oint_{S_A} J_i dS - q^E c_i^a + q^A C_i^A \pm R_i^A V^a \pm R_i^{elect} V^a \tag{43}$$

where V^a is the volume of water in the anode well, c_i^a is the concentration of species i in the anode well, J_i is the flux of species i in/out of the anode/soil boundary, S_A is the boundary surface area of the soil/anode well, q^E is the flow rate of fluid extracted from the anode, q^A is the flow rate of the (enhancement) fluid injected into the anode well, C_i^A is the concentration of species i in the enhancement fluid, R_i^A is the production/sink rate of species i in the anode well due to chemical reactions, and R_i^{elect} is the production/sink rate of species i due to electrolysis reactions at the electrode. A set of differential equations in the form of Eq. (11) can be developed for all species. A similar set of equations can be developed at the cathode. Solving Eq. 43 should define the flux of species i at the boundaries. However, the boundary equations will need to be solved simultaneously with the transport equations through the soil, since the concentration in the anode is another dependent variable.

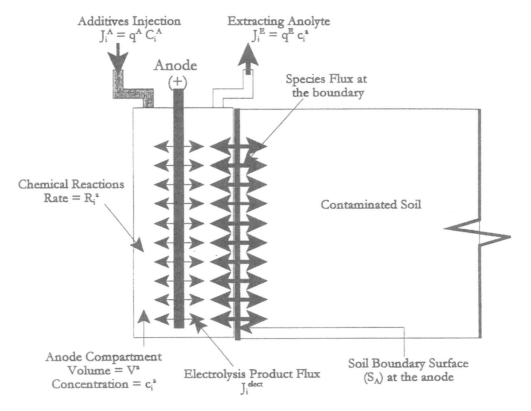

Figure 2 Schematic showing electrochemical processes at the anode (defines the boundary conditions for transport).

X. NUMERICAL STRATEGIES

Three common approaches to the numerical solution for the developed system of differential and algebraic equations are:

1. *Differential and algebraic equations approach (DAE)*. This approach consists of providing a solution to the mixed differential and algebraic equations in which the transport equations and chemical equilibrium reactions are solved simultaneously, as a system (37,38).
2. *Direct substitution approach (DSA)*. This approach consists of direct substitution of the algebraic chemical equilibrium equations into the differential transport equations to form a highly nonlinear system of partial differential equations (39–42).
3. *Sequential iteration approach (SIA)*. This approach consists of iterating between the sequentially solved differential and algebraic equations (43–45).

XI. ELECTROKINETIC REMEDIATION MODELS

Researchers have attempted to model contaminant transport under electrical gradients. Shapiro et al. (3) and Shapiro and Probstein (46) describe a 1-D model accounting for species transport under electric fields. The model accounts for ion diffusion, migration, and electroosmotic advection in predicting species transport rate. The model solves the charge flux equation, Eq. (8), in order to evaluate the nonlinear electric field distribution. The model assumes incompressible soil medium and thus constant hydraulic head distribution. Water electrolysis reactions are used to calculate constant flux boundary conditions for hydrogen ions at the anode and hydroxyl ions at the cathode. A steady-state electroosmotic flux is calculated by averaging the electrical gradient and zeta potential across the soil sample. The results are compared with experiments for the case of acetic acid extraction with constant voltage at the boundaries. Species incorporated in the code include acetate, hydrogen, hydroxyl, sodium, and calcium ions. Geochemical reactions included are first-order sorption and water and acetic acid dissociation. Numerical solution is achieved using the finite-element method in the spatial domain and Adams-Bashforth integration in time. Comparisons show good agreement in one case of acetic acid removal from a 0.4-m-long kaolinite sample.

Jacobs et al. (47) followed the model described by Shapiro and Probstein to predict 1-D transport of zinc under electric fields. The model uses an averages electroosmotic flow rate across the soil (i.e., electroosmotic flow is assumed to be independent of location). The model accounts for zinc precipitation and dissolution reactions and demonstrates the role of background ion concentrations on the process. Jacobs and Probstein (48) further modified the code to model 2-D species transport under electric fields. They applied the 2-D code for the case of electroosmotic extraction of phenol from kaolinite. The model solves three PDEs for transport of phenol, sodium ion, and chloride ion. Hydrogen and hydroxyl ion concentrations are calculated using the zero net charge equation and the water equilibrium equation. The model demonstrate 2-D phenol between one anode and one cathode. Limited geochemical reactions are incorporated (water and phenol dissociation), due to the complex 2-D simulation of the process.

Mitchell and Yeung (49) propose a model in a study of the feasibility of using electrical gradients to retard or stop migration of contaminants across earthen barriers. Principles of irreversible thermodynamics are employed and 1-D model is developed for transport of contaminants across the liner. An integral finite-difference method is used to solve the problem, and the model reasonably predicts the transport of sodium and chloride ions across the liner. Geochemical reactions were not incorporated in this model. Eykholt (50) attempted to model the pH distribution during the process using the mass conservation equation accompanied by empirical relations to account for nonlinearities in the parameters controlling the process. One transport differential equation is formed assuming that hydrogen and hydroxyl ions have the same diffusion coefficients and ionic mobilities. In this model, the development of negative pore-water pressure is modeled using the modified Smoluchoweski equation of Anderson and Idol (51), and the complexity in electrical potential distribution is modeled using proposed empirical relations. Haran et al. (52) presented a 1-D model for extraction of hexavalent chromium from soils using electric fields. The model accounts for transport of H^+, OH^-, CrO_4^{2-}, K^+, Na^+, and SO_4^-. Geochemical reactions included sorption (described by a retardation coefficient) and water equilibrium.

Acar et al. (1) and Acar et al. (53) present a 1-D model to estimate pH distribution during electrokinetic soil processing. The model demonstrates the impact of electrolysis reactions on pH distribution during electrokinetic remediation. Alshawabkeh and Acar (54) describe a modified formulation and present a system of differential/algebraic equations for the process and accounting for the chemical reactions of adsorption/desorption, precipitation/dissolution, and acid/base reactions. Acar and

Alshawabkeh (55) model the change in soil and effluent pH during electrokinetic soil processing. Two transient transport equations for hydrogen and hydroxyl ions are used together with water autoionization equation. This approach assume linear electric and hydraulic gradients throughout the process and disregards the coupling of these components. Alshawabkeh and Acar (11) and Acar and Alshawabkeh (56) enhance the model and modify the code for stimulating reactive extraction of heavy metals by electric fields. The model predicts reactive transport of hydrogen, lead, hydroxyl, and nitrate ions. The charge conversation equation is used to solve for electric field distribution, and the electroosmotic consolidation equation is used for predicting hydraulic head profile. The model accounts for lead hydroxide precipitation-dissolution, lead sorption (assuming linear pH dependent isotherm), and water equilibrium reactions.

REFERENCES

1. Acar, Y. B., Gale, R., Putnam, G., Hamed, J., and Juran, I. 1988. Determination of pH gradients in electrochemical processing of soils. Report presented to the Board of Regents of Louisiana, Civil Engineering Department, Louisiana State University.
2. Lageman, R., Wieberen, P., and Seffinga, G. 1989. Electro-reclamation: Theory and Practice. *Chem. Ind.* (Lond.), 9:585–590.
3. Shapiro, A. P., Renauld, P., and Probstein, R. 1989. Preliminary studies on the removal of chemical species from saturated porous media by electro-osmosis. *Physiocochem. Hydrodynam.*, 11(5/6):785–802.
4. Hamed, J., Acar, Y. B., and Gale, R. 1991. Pb(II) removal from kaolinite by electroosmosis. *J. Geotech. Eng.*, 117(2):241–271.
5. Probstein, R. F., and Hicks, R. E. 1993. Removal of contaminants from soils by electric fields. *Science* 260:498–504.
6. Runnels, D. D., and Wahli, C. 1993. In situ electromigration as a method for removing sulfate metals, and other contaminants from groundwater. *Groundwater Monitor. Rev.*, Winter: 121–129.
7. Lageman, R. 1993. Electro-reclamation. *J. Environ. Sci. Technol.*, 27(13):2648–2650.
8. Acar, Y. B., and Alshawabkeh, A. N. 1993. Principles of electrokinetic remediation. *Environ. Sci. Technol.*, 27(13):2638–2647.
9. Acar, Y. B., Alshawabkeh, A. N., and Gale, R. J. 1993. Fundamentals aspects of extracting species from soils by electrokinetics. *Waste Manage.*, 12(3):1410–1421.
10. Acar, Y. B., Hamed, J., Alshawabkeh, A., and Gale, R. 1994. Cd(II) removal from saturated kaolinite by application of electrical current. *Geotechnique*, 44(3):239–254.
11. Alshawabkeh, A. N., and Acar, Y. B. 1996. Electrokinetic remediation: II. Theory. *J. Geotech. Eng.*, 122(3):186–196.
12. Acar, Y. B., and Alshawabkeh, A. N. 1997. *Electrochemical Decontamination of Soil and Water*, Special Issue of *J. Hazardous Mater.*, 55:(1–3).
13. Mitchell, J. K. 1956. The fabric of natural clays and its relation to engineering properties. *Proc. Highway Res. Board*, 35:693–713.
14. Olson, R. E., and Daniel, D. E. 1981. Measurements of the hydraulic conductivity of fine-grained soils, ASTM STP 746. ASTM, Philadelphia.
15. Boynton, S. S., and Daniel, D. E. 1985. Hydraulic conductivity tests on compacted clay. *J. Geotech. Eng.*, 111(4):465–478.
16. Acar, Y. B., and Olivieri, I. 1989. Pore fluid effect on the fabric and hydraulic conductivity of laboratory compacted clay. *Transport. Res. Rec.*, no. 1219:144–159.
17. Daniel, D. E. 1989. In situ hydraulic conductivity tests for compacted clays. *J. Geotech. Eng.*, 115(9):1205–1226.
18. Perkins, T. K., and Johnston, O. C. 1963. A review of diffusion and dispersion in porous media. *J. Soc. Petroleum Eng.*, 19:70–84.

19. Bear, J. 1972. *Dynamics of Fluids in Porous Media*. American Elsevier, New York.
20. Gilham, R. W., and Cherry, J. A. 1982. Contaminant migration in saturated unconsolidated geologic deposits. *Geol. Soc. Am. Spec. Paper 189*, pp. 31–61.
21. Shackelford, C. D., and Daniel D. E. 1991. Diffusion in saturated soil (I): Background. *J. Geotech. Eng.*, 117(3):467–484.
22. Shackelford, C. D. 1991. Diffusion of contaminants through waste containment barriers. *Transport. Res. Record*, no. 1219, pp. 169–182.
23. Mitchell, J. K. 1993. *Fundamentals of Soil Behavior*. Wiley, New York.
24. Koryta, J. 1982. *Ions, Electrodes, and Membranes*. Wiley, New York.
25. Holmes, P. J. 1962. *The Electrochemistry of Semiconductors*. Academic Press, London.
26. Lorenz, P. B. 1969. Surface conductance and electrokinetic properties of kaolinite beds. *Clays and Clay Minerals*, 17:223–231.
27. Hunter, R. J. 1981. *Zeta Potential in Colloid Science*. Academic Press, London.
28. Stumm, W. 1992. *Chemistry of the Solid–Water Interface, Processes at the Mineral–Water and Particle–Water Interface in Natural Systems*. Wiley-Interscience, New York.
29. Eykholt, G. R., and Daniel, D. E. 1994. Impact of system chemistry on electroosmosis in contaminated soil. *J. Geotech. Eng.*, 120(5):797–815.
30. Langmuir, D. 1997. *Aqueous Environmental Geochemistry*, pp. 1–600. Prentice-Hall, Upper Saddle River, NJ.
31. Kirkner, D. J., and Reeves, M. 1988. Multicomponent mass transport with homogeneous and hetrogeneous chemical reactions: Effect of the chemistry on the choice of numerical algorithm, 1, Theory. *Water Resources Res.*, 24(1):1719–1729.
32. Yeh, G. T., and Tripathi, V. S. 1989. A critical evaluation of recent developments in hydrogeochemical transport models of reactive multicomponent components. *Water Resources Res.*, 25(1):93–108.
33. Davis, J. A., and Kent, D. B. 1990. Surface complexation modeling in aqueous geochemistry. In M. F. Hochella and A. F. White (eds), *Mineral-Water Interface Geochemistry. Reviews in Mineralogy*, 23:177–260.
34. Stumm, W., and Morgan, J. J. 1995. *Aquatic Chemistry: Chemical Equilibria and Rates in Natural Waters*, 3rd ed. Wiley, New York.
35. Bethke, C. M. 1996. *Geochemical Reaction Modeling: Concepts and Applications*. Oxford University Press, New York.
36. Bard, A. J., and Faulkner, L. R. 1980. *Electrochemical Methods: Fundamentals and Applications*. Wiley, New York.
37. Miller, C. W., and Benson, L. V. 1983. Simulation of solute transport in a chemically reactive heterogeneous system: Model development and application. *Water Resources Res.*, 19(2):381–391.
38. Lichtner, P. C. 1985. Continuum model for simultaneous chemical reactions and mass transport in hydrothermal systems. *Geochem. Cosmochim. Acta*, no. 49:779–800.
39. Vallocchi, A. J., Street, R. L., and Roberts, P. V. 1981. Transport of ion-exchange solutes in groundwater: Chromatographic theory and field simulations. *J. Water Resources Res.*, 17(5):1517–1527.
40. Jennings, A. A., Kirkner, D. J., and Theis, L. L. 1982. Multicomponent equilibrium chemistry in groundwater quality models. *Water Resources Res.*, 18(4):1089–1096.
41. Rubin, J. 1983. Transport of reaction solute in porous media; relation between mathematical nature of problem formulation and chemical nature of reactions. *J. Water Resources Res.*, 19(5):1231–1252.
42. Lewis, F. M., Voss, C. I., and Rubin, J. 1987. Solute transport with equilibrium aqueous complexation and either sorption or ion exchange; simulation methodology and applications. *J. Hydrol.*, 90:81–115.
43. Kirkner, D. J., Theis, I. L., and Jennings, A. A. 1984. Multicomponent solute transport with sorption and soluble complexation. *Adv. Water Resources*, no. 7, pp. 120–125.

44. Kirkner, D. J., Theis, I. L., and Jennings, A. A. 1985. Multicomponent mass transport with chemical interaction kinetics. *J. Hydrol.*, 76:107–117.
45. Yeh, G. T., and Tripathi, V. S. 1991. A model for simulating transport of reactive multi-species components: Model development and demonstration. *Water Resources Res.*, 27(12):3075–3094.
46. Shapiro, A. P., and Probstein, R. F. 1993. Removal of contaminants from saturated clay by electroosmosis. *Environ. Sci. Technol.*, 27(2):283–291.
47. Jacobs, R. A., Sengun M. Z., Hicks, R. E., and Probstein, R. F. 1994. Model and experiments on soil remediation by electric fields. *J. Environ. Sci. Health*, A29(9):1933–1955.
48. Jacobs, R. A., and Probstein, R. F. 1996. Two-dimensional modeling of electromigration. *AICHE J.*, 42(6):1685–1696.
49. Mitchell, J. K., and Yeung, T. C. 1991. Electro-kinetic flow barriers in compacted clay. *Transport. Res. Record*, no. 1288, Soils Geology and Foundations, Geotechnical Engineering 1990, pp. 1–10.
50. Eykholt, G. R. 1992. Driving and complicating features of the electrokinetic treatment of contaminated soils. PhD. dissertation, University of Texas at Austin.
51. Anderson, J. L., and Idol, W. K. 1986. Electroosmosis through pores with non uniformly charged walls. *Chem. Eng. Commun.*, 38:93–106.
52. Haran, B. S., Popov, B. N., Zheng, G., and White, R. E. 1997. Mathematical modeling of hexavalent chromium decontamination from low surface charged soil. *Electrochemical Decontamination of Soil and Water*, Special Issue of *J. Hazardous Mater.*, 55(1–3):93–108.
53. Acar, Y. B., Gale, R. J., Putnam, G., and Hamed, J. 1989. Electrochemical processing of soils: Its potential use in environmental geotechnology and significance of pH gradients. 2nd *Int. Symp. on Environmental Geotechnology*, Shanghai, China, May 14–17, vol. 1, pp. 25–38, Envo Publishing, Bethlehem, PA.
54. Alshawabkeh, A. N., and Acar, Y. B. 1992. Removal of contaminants from soils by electrokinetics: A theoretical treatise. *J. Environ. Sci. Health*, A27(7):1835–1861.
55. Acar, Y. B., and Alshawabkeh, A. N. 1994. Modeling conduction phenomena in soils under an electric current. *Proc. XIII Int. Conf. on Soil Mechanics and Foundation Engineering* (ICSMFE), New Delhi, India, Jan. 1994, vol. 2, pp. 662–669.
56. Acar, Y. B., and Alshawabkeh, A. N. 1996. Electrokinetic remediation: I. Pilot-scale tests with lead spiked kaolinite. *J. Geotech. Eng.*, 122(3):173–185.

7
Conditions for the Use of Innovative Remediation Techniques Set by New Policy

Cors van den Brink, Timo J. Heimovaara, and Mari P. J. C. Marinussen
IWACO BV, 's-Hertogenbosch, The Netherlands

I. INTRODUCTION

Cleaning the approximately 60,000 contaminated sites in The Netherlands according to the current goals—which requires the entire removal of the contaminants—has been estimated to cost approximately $50 billion US. At present approximately $0.5 billion US is spent on soil remediation every year.

Aiming to improve the soil remediation effort and to limit the expense of soil remediation, the Dutch policy on soil remediation is being reconsidered (1). One of the important changes in the Dutch policy is that the *strict and uniform goals* of soil remediation of the former policy are replaced by a *range of acceptable goals*. In the former policy the goals of every soil remediation was a potentially multifunctional use of the soil (and site) after remediation. According to the new policy, it may be sufficient to take into account the intended use and function of soil and groundwater. Hence, it may be acceptable that certain amounts of contaminants remain in the soil after remediation. As a result, the goals of soil remediation are variable and ought to be viewed on a regional scale.

The range of acceptable goals is determined by the *strategic goal* and the *minimum operational goal* (MOG) (Fig. 1; Ref. 2). The strategic goal requires the entire removal of pollutants or a removal until the remaining concentration does not exceed a certain natural background level. However, financial and technical constraints often require flexibility with respect to the remediation goal at any specific site. To prevent that the remaining pollution levels do not pose risks or damage to the environment or public health, this flexibility is limited.

The MOG procedure aims that a site-specific soil quality will be reached after soil remediation. The question of what pollution level is acceptable, however, has to be discussed. The MOG will be defined with respect to the use and function of the soil and groundwater (e.g., recreation, agriculture, urban area).

Both the strategic goal and the MOG can be established nationwide. In the case of MOG this means that a nationwide map with region-specific quality standards must be developed.

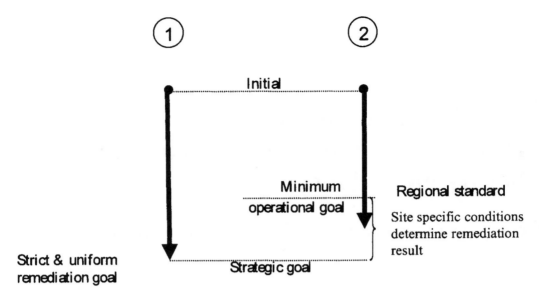

Figure 1 Relationship between strategic goal, regional minimum operational goal (MOG), and site-specific conditions as actors in determining the remediation goals.

This flexibility in remediation goals can be worked out in two ways:

With a (pre)fixed goal, based on the function and use of soil and groundwater in a certain region (e.g., groundwater protection area, urban area, industrial area); the background concentration of heavy metals could be the strategic goal in inner-city neighborhoods, whereas the total removal of mobile pollutants is required within groundwater protection zones.

An end point in soil concentrations based on negotiations between all parties involved and the pollution situation of a certain region.

Allowing for a certain amount of flexibility in remediation goals offers the opportunity to use innovative remediation techniques that can be tailor-made to fit the remediation goal. It is also possible to make a trade-off between the different aspects in the remediation operation (time versus costs, risk reduction versus spreading, etc.). In this article one of the opportunities that arise as a result of the new Dutch policy will be demonstrated. We present a case study in which the remediation must fulfill a regional goal: protection of a drinking-water well. This case illustrates the need for a cyclic remediation approach in which the remediation goals are determined on a regional level based on the "common interest" of all parties involved.

II. GROUNDWATER REMEDIATION IN REGIONAL PERSPECTIVE

A study has been performed to set up a remediation approach on a regional scale within the boundaries of the new policy. In order to demonstrate the impact of the modified policy we present both the situation under the former policy and the situation under the new policy.

A. Background of the Situation of Emmen

In the town of Emmen (The Netherlands), groundwater is abstracted for public water supply (Fig. 2). The well field is located on a topographically higher area, which is part of the Drents Plateau, a regional infiltration area with several brook valleys on the slopes of the plateau. The soils are sands. The abstraction volume is 5 million m^3/yr. The well field of Emmen can be considered vulnerable for activities inside the recharge area resulting in diffuse or point pollution. The groundwater protection area protects groundwater with travel times—starting at the soil surface—of less than 100 years, which is approximately 82% of the abstracted groundwater. In addition, 60% of the abstracted volume has a spreading in travel time of only 50 years (3).

The well field of Emmen is threatened by the occurrence of several sites polluted by chlorinated aliphatic hydrocarbons (CAH). For two sites the extent of the pollution is known in detail, at other sites the extent of the pollution is still unknown. A remediation operation carried out for more than 10 years at one of the "known sites" (De Weiert) led to the conclusion that the total removal of CAH was technically not feasible (4). At the other "known site" (VINEX-locatie) the remediation operation is about to start, but still requires a remediation goal.

In addition to the known sites, several unknown sites with polluted soil occur in the groundwater protection area, most of them polluted with CAH or BTEX. An overview of the pollution sites is given in Fig. 2.

B. Analysis of the Current Situation Under the Former Policy

Under the former policy, each of the polluted sites was treated as an individual case. The reason for this was the ambition of the Dutch government to clean up the soil nationwide to a multifunctional level. This was thought to be feasible both technically and financially, because it was supposed that the total number of polluted sites in The Netherlands was limited to a small number.

A result of treating pollution sites as individual cases was that a site was investigated as long as necessary to choose the remediation approach and consequently clean the site. The urgency of the remediation of the site was determined by a risk analysis, which included the regional context of the site. The remediation approach and goal of the individual site were also partly determined by the regional context of the site. A problem was that no remediation goal for the region was available. The chosen remediation goal was mostly a complete removal of the pollutants. The conceptual model is illustrated by Fig. 3.

In case the situation consists of only a single polluted site, the conceptual model presented covers the whole pollution problem. However, in the case of multiple sites, because sites are treated as individual cases, this may end up in a situation in which in the remediation effort (budget and personnel) is focused on one or two sites. At these sites (almost) all pollution is removed (removal of the source) to meet the strategic remediation goal. At the same time, other pollution sources which are "still on the list" may pollute the path (groundwater flow system) or even the object (well field). In the situation of the well field at Emmen, the result was that substantial effort was put into two sites (De Weiert and the VINEX-locatie), without even having an overview of the total impact from all pollution sites on the well field. In addition, the financial feasibility of removing (almost) all pollution from all polluted sites is low.

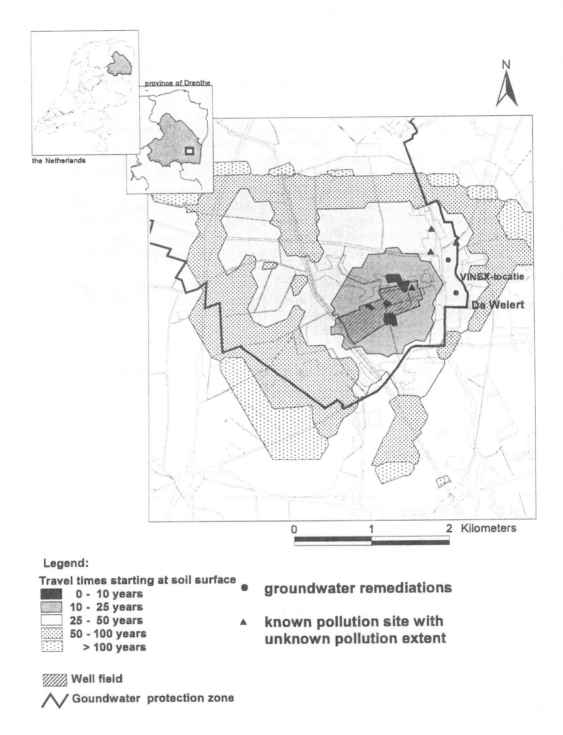

Figure 2 Overview of the (known and unknown) pollution sites within the groundwater protection zone of the well field of Emmen.

Use of Innovative Remediation Techniques

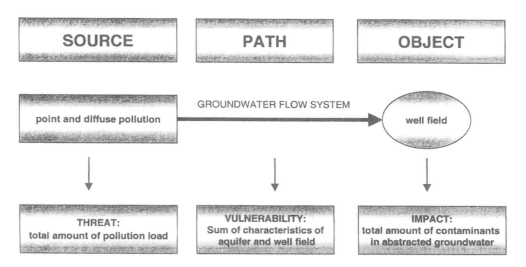

Figure 3 The SOURCE–PATH–OBJECT concept within the context of groundwater protection.

In summary, the former Dutch policy resulted in the following situation at the well field of Emmen:

An unsuccessful groundwater remediation (De Weiert) that has lasted for over 10 years without meeting the remediation goal.

A polluted site (VINEX-locatie) for which the remediation operation is about to start but for which no remediation goal and approach could be set because of the experience at De Weiert.

A number of CAH pollution sites within the groundwater protection area which are not even investigated because the input of budget and personnel was used for the two other sites (De Weiert and VINEX-locatie).

Because each pollution site was treated as an individual case, each pollution site was investigated by a separate project team, in which different people represented all parties involved.

C. Framework of the Regional Remediation Approach Under the New Policy

Within the framework of the new policy, the authorities responsible for the soil remediation wondered whether a remediation concept could be developed based on the interests of all involved parties (5). These parties included the provincial authorities responsible for the soil protection and soil remediation, the regional inspection, the water company, and the local authorities. The parties involved all have their own interest in the function and use of the contaminated area and thus in the soil remediation. Hence the first step in setting a regional goal is to formulate the common interest of these parties. All parties involved share the interest of protection of public health. This leads to the conclusion that the pollution situation should be treated at the level of the well field (regional scale). The specific individual remediation goals of the individual sites are of minor importance. The regional approach changes the way pollution sites are handled. The effect of adopting a regional approach is illustrated in Fig. 4.

The new concept for the Emmen well field consists of three "lines of defense":

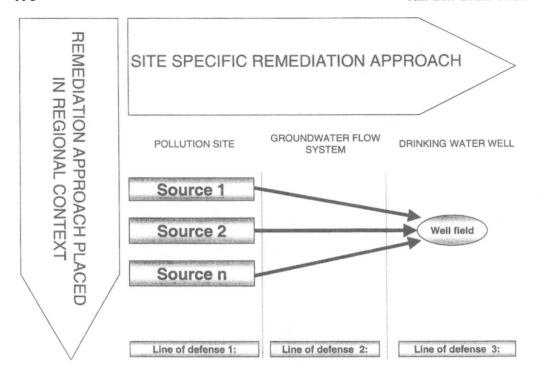

Figure 4 Schematized outline of groundwater remediation in regional perspective. Within this regional approach, first an overview of all pollution sources is obtained and remediation measures are focused on the removal of pollution from the bulk zone (line of defense 1). In addition, spreading is prevented by intercepting the pollution plumes as close as possible to the pollution or by natural attenuation processes (line of defense 2). If the pollution reaches the well field, the water will be treated or the well field has to be reallocated or closed (line of defense 3).

Line 1: Obtain an overview of all pollution sites and remove as much pollution as necessary in order to prevent spreading.
Line 2: Intercept the plumes from the sites as close as possible to the pollution sources.
Line 3: Treat the polluted groundwater at the well field or reallocate or close the well field.

The idea of the lines of defense is that line 2 becomes relevant only after line 1 has to be given up. This means that, for each line of defense, criteria must be established in order to choose between "defend" and "surrender."

Line of defense 1: The primary goal of this line of defense is twofold. First, we want to obtain an overview of all polluted sites within the region; and second, we want to prevent the spreading of pollution from these sites. Several approaches are available, such as removing the bulk of the pollution from the source zone or controlling the spreading of the pollution on a local scale. Of course, the best approach from a sustainability point of view is removal of the bulk of the pollution, as was the case under the former policy. The main difference is that all activities related to the soil investigation and remediation are not serving a site-specific but a regional goal. In this way decisions can be made on the optimal application of funds and personal in order to protect the well field.

Line of defense 2: In the case of the well field of Emmen monitoring shows that the aquifer has not yet been polluted. However, if it turns out that the number of pollution sources is too high or if the removal of pollutants is (technically or financially) not feasible, one is forced to take measures to prevent pollution from affecting the well field. This can be achieved by monitoring the groundwater quality at the pollution sources. If the (remaining) pollution at the sites results in a pollution plume, this plume should be intercepted as close as possible to the pollution source. Because most pollution sources are known, the monitoring can be optimized. If all pollution sources are known and all monitoring data are analyzed, the regional remediation plan (including the goals of the individual sites) must be optimized. The following questions are relevant:

> Should the effort be focused on the (additional) removal of pollution sources, and if so, which sources?
> Should the effort be focused on the interception of the pollution plumes, and if so, which plumes?
> Should the effort be focused on line of defense 3, which means that the total amount of plumes moving toward the well field is too big to intercept?
> Do natural processes, such as natural attenuation (NA), occur, and if so, are these processes protective for the well field?

Line of defense 3: If it is not possible to prevent pollution plumes from reaching the well field, measures focusing on removing the pollution at the water production facility or reallocating or even closing of the well field have to be considered. These include, e.g., water treatment but also selective pumping.

D. Implication

As a result of the new policy, all polluted sites in the region are treated in a cyclic manner. During the process an insight into the total pollution situation threatening the well field evolves. This not only changes the remediation approach but also the remediation goals set for the individual sites and, most probably, the order in which the sites will be remediated. The measures at the well field of Emmen will be focused on a combination of defense lines 1 and 2, because at the moment not all pollution sources are known. The total removal of all pollutants is technically and financially not feasible, and up to now no pollutant has been observed in the monitoring system.

E. Action Plan

Based on the strategy described in the previous section, the parties involved in this process agreed on the following action plan:

1. The local and regional authorities are responsible for an overview of the total pollution problem.
2. With respect to the two known polluted sites the strategy is

    De Weiert: Maintaining the current status quo. This means that the remaining concentration of approximately 130 µg/L PER will be monitored. This rather high value occurs only in a very limited spot because almost all pollution is removed (6). In case spreading occurs, the plume should be intercepted.

VINEX-locatie: Removal of the bulk of the pollution up to a minimum operational goal of 30 µg/L and the installation of a monitoring system to check whether spreading takes place. This MOG corresponds to the maximum acceptable concentration, based on toxicological considerations. If spreading occurs, the plume should be intercepted.

The remediation goal may be updated after new information comes available.

3. The local authorities, inspectors, and the water company judge the remediation goals (which allow more remaining pollution than the strict and uniform goals of the old policy) within the (regional) context of this pollution situation.
4. After obtaining the overview of all polluted sites, the monitoring system of the provincial authority (which monitors at each pollution site) and the water company (which monitors the quality of the aquifer) should be integrated and adjusted.
5. The financial and legal impact of the remaining pollution in the soil at the polluted sites will be investigated.
6. After 6 months, all parties involved will jointly evaluate the action plan and decide whether the new information requires modification of the action plan.

III. SUMMARY

With respect to the situation under the old policy of strict and uniform goals, major differences of the new policy in remediation approach and remediation goals can be identified:

A regional goal (protection of the well field) is set with all parties involved. This goal overrules the remediation goals of the individual pollution sites as set under the old policy. Within the regional goal of protection of the well field, the minimum operational goal should be met at the specific sites.

The remediation approach is a cyclic process that includes all polluted sites that potentially endanger the well field. In this process the approach and goals at individual sites may be adjusted after evaluating new information (on the extent of as yet unknown polluted sites or the quality of the aquifer, etc.).

The remediation goals mentioned in the action plan (MOG) are less strict than the strategic goals in the old policy. However, the removal of the bulk of the pollution at all sites that potentially endanger the well field removes substantially more pollution from the soil (and aquifer). In addition, the spreading from all polluted sites will be monitored. As a result, adjusting the remediation goals for individual sites within a regional context substantially increases the total regional remediation efficiency.

REFERENCES

1. Kabinetsstandpunt Bodemsanering. 1997. (Point of view of the cabinet/government with respect to the soil remediation.)
2. Beinat, E., van den Brink, C., Koolenbrander, J., in't Veld, M., and Vermeulen, H. 1998. Cost effectiveness of soil remediation. A comparative system for mobile pollutants (Kosteneffectiviteit van bodemsanering. Een afwegingssystematiek voor verontreinigingen in het mobiele regime).

3. van den Brink, C., and Ouboter, P. S. H. 1998. Groundwater pollution problems of the groundwater protection area "Emmen," Results workshop (Grondwaterverontreinigingsproblematiek grondwaterbeschermingsgebied Emmen, Verslag workshop). Report 2244690.
4. van den Brink, C., Nipshagen, A. A. M., and Blommers, J. 1997. Evaluation of the groundwater remediation "De Weiert, Emmen" (Evaluatie van de grondwatersanering De Weiert, Emmen). Report 2231760.
5. van den Brink, C., and Ouboter, P. S. H. 1998. Groundwater pollution problems of the groundwater protection area "Emmen," Results covenant (Grondwaterverontreinigingsproblematiek grondwaterbeschermingsgebied Emmen, Verslag afspraken). Report 2246480.
6. van den Brink, C., Ouboter, P. S. H., and van den Haspel, J. 1997. Groundwater remediations in regional context (Grondwatersaneringen in gebiedsgericht perspectief). Report 2231570.

8
Cone Penetrometer Technology for Groundwater Remediation Assessment in the Vadose Zone

John F. Peters, Daniel A. Leavell, and Landris T. Lee
U.S. Army Engineer Waterways Experiment Station, Vicksburg, Mississippi

I. INTRODUCTION

Well-engineered groundwater remediation requires accurate and timely information on the distribution of contaminants for both design and implementation phases. This chapter describes tools now available, the rational for their design, and recommendations for their use.

A. Observation Method

Geotechnical engineers have long faced the problem of uncertainty in subsurface conditions. Added to this uncertainty is the difficulty in predicting behavior of geologic materials even when conditions are known. The observational method, championed by Peck (1), allows the engineer to incorporate design modifications in the field in response to observations from a well-designed instrumentation program. Groundwater remediation schemes share the same uncertainty as traditional civil engineering and can benefit from an observational approach. Traditional monitoring wells provide high-quality air and water samples but are expansive to install and maintain. Also, monitoring wells are static systems, with placements that must be determined prior to initiation of work. In contrast, observational procedures for groundwater remediation can benefit from a dynamic approach that evolves as performance data become available. Often, lower accuracy can be accepted, provided data on plume movement, changing water levels, and evolution of water chemistry can be obtained in a flexible and timely manner.

B. Role of the Cone Penetrometer

A key advantage for remediation is that cone penetrometer investigations can be readily extended as needed to update assessment of groundwater conditions. The cone penetrometer is an established geotechnical tool that allows rapid assessments of subsurface conditions. The cone has been used for some 70 years to evaluate soil stratigraphy and estimate shear strength. Developmental research on the cone particularly flourished in the 1970s and 1980s as a cost-effective procedure for geotechnical investigations. This research elevated cone technology to be on par with traditional drilling and sampling technology as a reliable indicator of subsurface conditions. The past decade has seen the capabilities of the cone extended far beyond that of a sounding probe to that of a screening and sampling tool. The cone can be used in tandem with conventional sampling methods by identifying optimal locations for monitoring wells. Thus, the number of conventional monitoring wells, water and air samples, and analytical laboratory tests required to characterize and monitor clean-up activities can be reduced. In many cases, the cone can provide data that rival the accuracy of those obtained from monitoring wells.

II. CONE PENETROMETER FOR GROUNDWATER MONITORING

Research and developments in sampling technology through use of the cone offer a variety of tools having accuracy and cost (2,3) that are appropriate for both screening-level surveys and detailed chemical analyses. The U.S. Army Engineer Waterways Experiment Station (WES), under the sponsorship of the Army Environmental Center (AEC), initiated the development of the Site Characterization and Analysis Penetrometer System (SCAPS). SCAPS development was in response to the need to provide the Department of Defense (DoD) with a rapid and cost-effective means to characterize soil conditions at DoD sites undergoing installation restoration (clean-up). WES partnered with the U.S. Naval Command, Control and Ocean Surveillance Center, and the U.S. Air Force Armstrong Laboratory to accelerate and coordinate the Tri-Service SCAPS technology development, demonstration, and technology transition under the sponsorship of the Strategic Environmental Research and Development Program (SERDP). The Department of Energy has partnered with WES via interagency agreement to receive SCAPS technology. The Environmental Protection Agency (EPA) has joined with the Tri-Service SCAPS developers to conduct validation studies that will lead to regulatory acceptance of SCAPS contaminant sensing and sampling technologies.

A. The SCAPS System

The SCAPS platform consists of an 18,200-kg truck (Fig. 1) equipped with vertical hydraulic rams that are used to force a cone penetrometer into the ground (Fig. 2), instrumentation for data acquisition and reduction (Fig. 3), and shelter for the crew. The cone is advanced at a speed of 2 cm/s to depths of approximately 50 m in normally consolidated fine-grained soils when using a 100-m umbilical cable (25 m when using 50-m umbilical cables). During a vertical push, data are continuously collected and recorded with 2-cm spatial resolution. The truck contains separate enclosed compartments for data acquisition/processing and the hydraulic ram/rod handling room. Each compartment is temperature controlled and monitored for air quality. As for the original geotechnical cone, SCAPS multisensor penetrometer probes are equipped to measure tip and sleeve resistances simultaneously to determine soil stratigraphy, layer boundaries, and soil type. Contaminant-specific sensor data can be obtained from a suite of samplers and instrumentation to determine the presence of pollutants in each soil strata. The SCAPS data acquisition room contains a real-time data acquisition and processing computer system; electronic signal processing equipment; and a networked postprocessing computer system

Cone Penetrometer Technology

Figure 1 SCAPS truck.

Figure 2 Hydraulic ram/drill rod handling room.

for three-dimensional visualization of soil stratigraphy and contaminant plumes. A mobile laboratory truck, equipped with field-portable ion-trap mass spectrometer and/or gas chromatography equipment, accompanies SCAPS for near-real-time analytical analysis of analyte vapor samples collected by SCAPS in situ samplers.

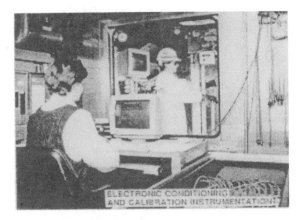

Figure 3 Data acquisition room.

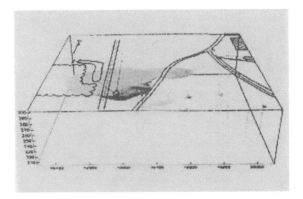

Figure 4 A computer-generated three-dimensional rendering of a contamination plume.

B. Site Characterization and Data Visualization

The use of the SCAPS as a site characterization tool is similar to that of the conventional civil engineering application of the cone penetrometer (4). The resistance at the cone tip and frictional resistance can be used to interpret soil types (5) through a computerized classification system that provides a digital database for visualization.

A vital part of site characterization is the ability to visualize data in a three-dimensional representation. Data obtained from the cone can be represented with traditional linear logs, which permits detailed analysis of results from a particular probe sounding. In addition, data can be depicted in three-dimensional renderings that aid in developing interpretations of stratigraphic facies and contaminant distribution (Fig. 4).

C. Fluorescence Measurements

The seminal development for the SCAPS system was fluorescence sensors for detection of organic contaminants and byproducts of petroleum, oil, and lubricants (POL). The laser-induced fluorometry (LIF) POL sensor uses ultraviolet laser energy to induce fluorescence in POL contaminants. The laser, mounted in the SCAPS truck, is linked via fiber-optic cables to a sampling "window" mounted on the side of a penetrometer probe. Laser energy emitted through the window causes fluorescence in adjacent POL-contaminated media. The fluorescent energy is returned to the surface via fiber-optic cables for real-time spectral data acquisition and processing (spectral analysis) in the SCAPS truck. The SCAPS LIF POL sensor has undergone numerous successful field investigations at various government facilities to determine soil classification, stratigraphy, and POL contaminant data (6–11). The SCAPS LIF POL sensor is currently undergoing EPA demonstration/validation investigations and has been licensed to private industry for commercialization. See http://www.epa.gov/etv/02/frwextoc.htm for information on recent technology demonstration plan of the laser-induced fluorometry/cone penetrometer at the hydrocarbon national test site at the Naval Construction Battalion Center Port Hueneme, California.

D. Sample Recovery

An early limitation of the SCAPS system was the inability to obtain samples for accurate analyses. The primary difficulty was obtaining access to the soil outside the cone through a system that had the following attributes: (a) sufficiently robust to withstand the large stresses imposed on the cone, (b) mechanically simple to be operable under field conditions, (c) (preferably) capable of obtaining several samples from a single push that are free from cross-contamination, (d) capable of integrating the cone tip resistance and sleeve friction needed to interpret soil strata, and (e) compatible with a grouting system. Sampling technology available in the early stages of SCAPS development were simple and relatively robust but lacked the ability for multiple samples and integration with cone tip resistance and sleeve friction measurements.

A system that was designed to address all of these issues was the multiport sampler (MPS) described by Leavell and Lee (12). Figure 5 is a photograph of the MPS showing an assembled probe and an exploded view of its components. The MPS consists of

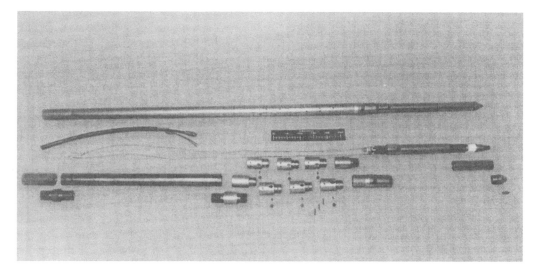

Figure 5 Photograph of the MPS probe with an exploded view of its components.

vertically stacked sampling modules that are independently operated from the surface and allows collection of multiple vapor samples during a single penetration. In the initial design, up to 12 modules could be stacked, depending on the purpose of the sampling. For site exploration it was found that the ideal number of ports was six. This number allowed for a smaller umbilical cable (the grout tube is evenly surrounded by the sampling tubes and instrumentation cables). Each sampling port is connected to the surface through 1.6-mm stainless steel tubing. During the advancement of the MPS, each sampling port is pressurized with nitrogen to $1000 \, kN/m^2$. The MPS is advanced to the desired sampling depth, a module is selectively opened, and analyte is drawn through a sidewall port. If additional sampling ports are needed, each sampling port can be closed and reused. Because stainless steel tubing is used there is little to no memory of previously sampled contaminants. Sampling is conducted by either bringing the analyte/carrier gas to the surface via tubing for analysis using field-portable analytical equipment or storing the analyte in probe-mounted ion traps that are analyzed after the MPS is brought to the surface. In addition to the sampling modules, the MPS is capable of real-time soil layer (stratigraphy) mapping and grouting through the probe tip as the MPS is retracted to prevent cross-layer contamination. The MPS has been used successfully to sample vapors from soils contaminated with chlorinated organic solvents to determine the relative concentration of the contaminants in different soil strata. MPS technology is available via license through the USAE-WES. The key purpose of the MPS system is to provide access to the soil from multiple ports. In principle, the MPS could be used to sample fluids below the water table, provided suitable collection systems can be designed. Another potential use is as rapidly deployed monitoring sites. For example, a multiple-depth monitoring station could be created from a string of several MPS modules separated by spacers to allow their placement at predetermined depths. Such a monitoring station could thus be deployed from a single push.

E. Volatile Organic Compound Sensors

The Hydrosparge VOC sensing system consists of a direct-push groundwater-sampling device coupled to an in-situ sparge device interfaced to an ion-trap mass spectrometer. A commercially available direct-push groundwater-sampling tool, Hydropunch,[tm] is pushed to the direct depth. A temporary screen is opened and essentially a temporary monitoring well is developed to provide access to groundwater. An in-situ sparge device, developed by Oak Ridge National Laboratory, uses a helium gas flow to strip VOCs from the groundwater. The VOCs are then returned to the surface via a sampling tube and analyzed in real time by an onboard field-portable ion-trap mass spectrometer or similar detection system.

The SCAPS Thermal Desorption VOC Sampler combines thermal desorption and cone penetrometer technologies to provide a means for real-time detection and mapping of solvent and hydrocarbon contamination in both the vadose and saturated zones. In operation, the thermal desorption VOC sampler is pushed to a desired depth, an interior rod retracts the penetrometer tip, and a known volume of soil is collected in a sample chamber. While in the sample chamber, heat is applied around the soil sample to purge contaminant vapors. Volatilized compounds are transferred to the surface via carrier gas, where they are trapped on tenax, desorbed, and analyzed using a field-portable gas chromatograph and/or an ion-trap mass spectrometer. The soil sample is then expelled and the cone penetrometer pushed to a new depth, where the process is repeated. Alternatively, the sampler may be used as a vapor sampler in the vadose zone by applying

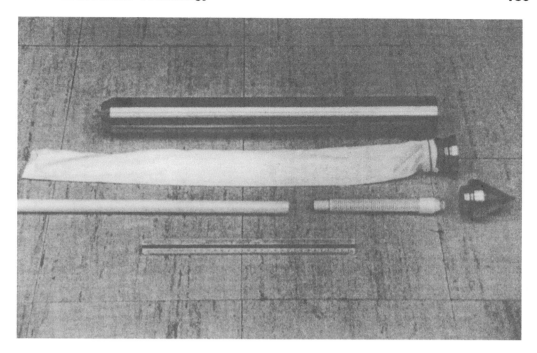

Figure 6 Exploded view of the MicroWell.

a vacuum to the transfer line to draw soil vapors to the surface, where they are trapped and analyzed.

F. MicroWell

The MicroWell was designed and developed to monitor and obtain samples of subterranean water contamination. Figure 6 shows an exploded view of the MicroWell. A preassembled well casing, well screen with filter sand, and clay seal are placed insider a penetrometer rod. The MicroWell is pushed to depth and the penetrometer rod retracted, leaving an installed MicroWell complete with sand filter and clay seal. A vacuum pump is used to develop the well prior to sampling.

A MicroWell installation conducted using a single penetration contributes to a faster and more economical well installation during the site characterization and screening phase. Contaminant plume monitoring and water sampling are obtained with minimal labor, equipment, and exposure to hazardous conditions. Because a screen length of 15–25 cm is used, specific strata can be isolated for examination. Currently, the maximum installation depth is 25 cm. Also, its application is limited to normally consolidated soils and it is not appropriate for highly consolidated or cemented materials, cobbles, or rocky strata.

G. Grouting and Cleaning Technique

A primary concern with all invasive exploration techniques is the potential for creating avenues for cross-layer contamination among strata intersected by a boring or probe.

Figure 7 Grout trailer with steam cleaner.

Although the cone is less intrusive that other techniques, suitable grouting techniques present a difficult technical challenge because the grout must be introduced at the base of the probe as it is retracted. The problem becomes particularly severe as more sophisticated instrumentation is added to the cone package. However, most monitoring and sampling packages described above provide access to the cone's sacrificial tip for a grout tube. For those that do not, a grouting module was developed that can be incorporated into the probe behind the sampling package. A trailer-mounted grout pumping system accompanies the SCAPS truck (Fig. 7). This system is attached to a specially designed grouting system that has been incorporated into each SCAPS probe to facilitate backfilling the hole with grout as the penetrometer push rods and probe are retracted. The SCAPS truck is also equipped with a specially designed steam cleaning system mounted beneath the truck rod handling room that removes soil and contaminants that may adhere to the push rods and probe during retraction. The contaminated effluent is collected for proper disposal. Grouting techniques and materials are described by Wright (13), Bean et al. (14), and Lee (15).

III. CONCLUDING REMARKS

Cone penetrometer technology has grown to a powerful investigative tool that is particularly well suited for planning and monitoring groundwater remediation activities. The development of the SCAPS system by the Tri-Services offers a suite of methods that can be used to devise contaminant- and site-specific strategies for observational-based design and implementation. The Tri-Services operate four Army and three Navy SCAPS vehicles. The Army maintains the original SCAPS truck at WES for research, development, and demonstration/validation purposes. The Corps of Engineers (COE) Kansas City, Savannah, and Tulsa Districts operate three SCAPS trucks for operational site characterization and monitoring field investigations at government facilities. The Air Force conducts SCAPS work under contract to the COE and private contractors. The Department of Energy has SCAPS technology through interagency agreement with WES. SCAPS technologies are now found in the private sector as a result of licensing agreements, cooperative research and development agreements, and technology reinvestment programs. Information on SCAPS, and the cone penetrometer in general, is extensive. A recent inquiry found over 1000 citations on the World Wide Web for cone penetrometers, including technical data on SCAPS applications. Specific information about the Tri-Services program can be found at http://www.wes.army.mil/el/scaps.html.

ACKNOWLEDGMENTS

The design and development of the equipment sited in this chapter was performed at WES. The AEC and SERDP sponsored the equipment development and field evaluation. Permission to publish this information was granted by the Chief of Engineers.

REFERENCES

1. Peck, R. B. 1969. Advantages and limitations of the observational method in applied soil mechanics. *Geotechnique*, 19(2):171–187.
2. Booth, S. R., Durepo, C. J., and Temet, D. L. 1993. Cost effectiveness of the cone penetrometer technology. Los Alamos Natl. Lab. Rep. LA-UR-93-3383.
3. Schroeder, J. D., Booth, S. R., and Trocki, L. K., 1991. Cost effectiveness of the Site Characterization and Analysis Penetrometer System. Los Alamos Natl. Lab. Rep. LA-UR-91-4016.
4. Cooper, S. S., Douglas, D. H., Sharp, M. K., Olsen, R. A., Comes, G. D., and Malone, P. G. 1993. Initial field trials of the Site Characterization and Analysis Penetrometer System (SCAPS). Tech. Rep. GL-93-30, U.S. Army Engineer Waterways Experiment Station, Vicksburg, MS.
5. Olsen, R. S., and Malone, P. G. 1988. Soil classification and site characterization using the cone penetrometer test. *Proc. First Int. Symp. on Penetration Testing (ISOPT-1)*.
6. Sharp, M. K., and Kala, R. 1992. Use of the Site Characterization and Analysis Penetration System at Grandville, Michigan, Superfund Site. Tech. Rep. GL-92-38, U.S. Army Engineer Waterways Experiment Station, Vicksburg, MS.
7. Sharp, M. K., Olsen, R. S., and Kala, R. 1992. Field evaluation of the Site Characterization and Analysis Penetrometer System at Philadelphia Naval Shipyard Philadelphia, Pennsylvania. Tech. Rep. GL-92-39, U.S. Army Engineer Waterways Experiment Station, Vicksburg, MS.
8. Koester, J. P., Lee, L. T., Olsen, R. S., Douglas, D. H., Comes, G. D., Cooper, S. S., and Powell, J. F. 1993. Field trials of the Site Characterization and Analysis Penetrometer System at the Savannah River Site (SRS). Tech Rep. GL-93-16, U.S. Army Engineer Waterways Experiment Station, Vicksburg, MS.
9. Lee, L. T., and Powell, J. F. 1993. Use of the Site Characterization and Analysis Penetrometer System at the Walnut Creek Watershed, Ames, Iowa. Misc. Paper GL-93-12, U.S. Army Engineer Waterways Experiment Station, Vicksburg, MS.
10. Lee, L. T., Davis, W. M., Goodson, R. A., Powell, J. F., and Register, B. A. 1994. Site Characterization and Analysis Penetrometer System (SCAPS) field investigation at the Sierra Army Depot, California. Tech. Rep. GL-94-4, U.S. Army Engineer Waterways Experiment Station, Vicksburg, MS.
11. Davis, W. M., Lee, L. T., and Powell, J. F. 1994. Site Characterization and Analysis Penetrometer System (SCAPS) field investigation at the Building 4020 Site, Aberdeen Proving Ground, Maryland. Tech. Rep. EL-96-17, U.S. Army Engineer Waterways Experiment Station, Vicksburg, MS.
12. Leavell, D. A., and Lee, L. T. 1995. Design, development, and operation of the multiport sampler. Tech. Rep. GL-95-16, U.S. Army Engineer Waterways Experiment Station, Vicksburg, MS.
13. Wright, T. D. 1993. Environmental assessment of selected cone penetrometer grouts and a tracer. Misc. Paper IRRP-93-1, U.S. Army Engineer Waterways Experiment Station, Vicksburg, MS.

14. Bean, D. L., Green, B. H., Walley, D. M., Malone, P. G., and Lee, L. T. 1995. Selection of materials and techniques for usage in sealing geotechnical investigation holes. Tech. Rep. SL-95-4, U.S. Army Engineer Waterways Experiment Station, Vicksburg, MS.
15. Lee, L. T. 1996. Cone penetrometer grouting evaluation. Tech. Rep. GL-96-33, U.S. Army Engineer Waterways Experiment Station, Vicksburg, MS.

9
Evaluation of the Adequacy of Hazardous Chemical Site Remediation by Landfilling

G. Fred Lee and Anne Jones-Lee
G. Fred Lee & Associates, El Macero, California

I. INTRODUCTION

Hundreds of millions to ultimately billions of dollars are being spent across the country in "Superfund" and Resources Conservation and Recovery Act (RCRA) hazardous chemical site investigation and remediation. In the past few years, there has been a considerable emphasis on redeveloping so-called brownfield properties as part of this effort. This redevelopment will result in the public and/or the environment being exposed to hazardous/deleterious chemicals left at the site as part of site remediation/closure. While initially in the Superfund program the primary focus of remediation was clean-up to background, today the emphasis is changing to the use of remediation technologies that leave waste residues at the hazardous chemical site. In general, it is being assumed that an on-site waste management area that is covered with a landfill that meets minimum RCRA specifications, on-site RCRA landfills, solidified waste residues, "naturally attenuated" groundwater pollution plumes, and other on-site remediation technologies that leave appreciable quantities of hazardous or deleterious chemicals at the site upon site closure will, after a 30-year period, require no further monitoring and/or maintenance of the waste residues left at the site. However, a critical review of the adequacy of such remediation technology shows that in many instances the residues will be a threat to public health and the environment for a long, if not infinite, period of time.

Typically, some type of use-deed restrictions are placed on future uses of the remediated property, which presents at least a superficial appearance that future users of the property and nearby properties will not experience problems due to the waste residues present at the site upon closure. The adequate implementation of the use-deed restrictions, however, by public agencies and future property owners/users is tenuous at best. At this time, as part of closure of a hazardous chemical site, inadequate attention is being given to ensuring that the principal responsible parties (PRPs), as well as any new owners or users of the "remediated" hazardous chemical site property, will adequately monitor and maintain the waste residues left at the site upon closure for as long as the waste components represent a threat to public health, groundwater resources, and/or the environment.

This chapter reviews a number of issues that need to be addressed in evaluating the adequacy of hazardous chemical site remediation relative to long-term liabilities and potential public health and environmental impacts associated with residual waste/chemical components left at the "closed" site. Particular attention is given to:

- The adequacy of using RCRA landfill-type covers to reduce infiltration of moisture into the waste that leads to the transport of pollutants to the underlying groundwater system
- The adequacy of U.S. Environmental Protection Agency (EPA) Subtitle D and C RCRA landfills in protecting groundwaters from pollution by landfill leachate for as long as the residual wastes at the site will be a threat
- The adequacy of current waste characterization procedures in determining the potential for residual waste components to endanger public health, groundwater resources, and the environment for as long as the waste residues at the site will be a threat
- The potential problems with the use of natural attenuation as an approach for remediating contaminated groundwaters
- The issues that need to be addressed in the redevelopment of brownfield properties to protect public health, the environment, groundwater resources, and the financial interests of the property redevelopers/purchasers/users.

While the discussion of examples of problem areas given are presented in a generic form, they are based on the authors' experience on these topics and the literature.

II. LIABILITY FOR SITE CLEAN-UP

One of the commonly made statements by PRPs for hazardous chemical sites, such as areas where chemicals have been inadequately or improperly stored, or were inappropriately managed through landfilling or waste disposal ponds, is that those who deposited chemicals or wastes in these areas were following regulatory requirements of the time when the environmental contamination first occurred. The implication is that they should not be responsible for having to clean up the contaminated soils or groundwaters, since they met regulatory requirements. A critical review of what was known or should have been known by PRPs about the potential hazards associated with depositing wastes on the ground, in lagoons, or in sanitary landfills shows that, since the 1950s, there is substantial literature demonstrating that the management of chemical wastes by means that now lead to inadequately closed RCRA sites and/or "Superfund" sites would lead to groundwater pollution. The issue often becomes one of whether the PRPs knowingly managed hazardous chemicals and hazardous chemical wastes in accord with both regulatory and appropriate professional expertise that was available at the time that the hazardous chemicals/wastes were deposited at the location that ultimately led to the hazardous chemical site that requires remediation.

III. ADEQUACY OF REGULATIONS RELATIVE TO PROTECTING PUBLIC HEALTH AND THE ENVIRONMENT

Those familiar with how regulatory agencies develop regulations relative to protection of public health and the environment know that, with few exceptions, the adoption and appropriate implementation of regulations lags behind the professional recognition of the need for regulations, often by decades or more. This situation has existed throughout time, where usually the only way that regulations are adopted is through the occurrence of a major crisis. An example of this kind of situation is the regulation of waterborne pathogens in domestic water supplies. It has been well known in the professional literature since the mid-1980s that the current treatment of domestic water supplies for consumption has been inadequate to prevent significant human disease and deaths associated with the consumption of treated water that meets water-supply public health standards. It took a Milwaukee *Cryptosporidium* outbreak, where 400,000 people became ill and 100 people died, to get the national and state regulatory agencies to begin to meaningfully address what was a well-known problem by those familiar with the topic area well before the Milwaukee incident occurred.

In the mid-1980s, the Center for Disease Control published information which indicated that approximately 1000 people per year in the United States die from consuming water from what is accepted as adequately treated domestic water supplies. Millions of U.S. residents become ill every year because of inadequately disinfected water supplies that meet the fecal coliform standard. This is not the situation where there is inadequate treatment, but one of inadequate technical procedures and incorporation of basic information into the regulatory process. It has been known since the 1940s that the approach used for evaluating the sanitary quality of domestic water supplies was inadequate for enteroviruses and parasitic protozoan cyst-forming organisms, such as *Cryptosporidium*. The fecal coliform standard that has been and continues to be used was known since the 1940s work of a number of individuals, principally Dr. Chang at Harvard University, to be inadequate for determining the safety of a water that contains protozoan cyst-forming parasitic organisms such as those that cause amebic dysentery, *Cryptosporidosis*, and giardiosis. These and other disease-causing organisms of this type are not killed by conventional water supply or wastewater chlorination practices. The conventional domestic water supply and wastewater treatment/chlorination is designed primarily for the control of organisms that respond to disinfection like fecal coliforms, such as the salmonella-type organisms that cause bacterial enteric diseases. Even today, while it is now well known that the fecal coliform standard is inadequate to protect domestic water supply consumers from disease, the regulatory agencies at the federal and state level have still not established monitoring programs for enteroviruses and protozoan parasites as part of the routine monitoring of the safety of a treated domestic water supply.

Lee and Jones (1–3) have discussed how inadequate development and implementation of regulations that are being used as requirements for remediation for hazardous chemical sites create conditions where inadequate long-term protection of public health and/or the environment is occurring. The major problems that exist are associated with failure to establish well-defined, appropriately funded evaluations of the adequacy of site remediation and monitoring relative to existing and new information that develops on the potential hazards of residual chemicals left at the site as part of remediation. This situation will cause many so-called remediated Superfund/RCRA sites to require additional remediation at some time in the future. Because of inadequate long-term funding to address these issues, it is likely that both the public and the environment will be unnecessarily exposed to hazardous or deleterious conditions associated with current remediation approaches for hazardous chemical sites.

IV. INADEQUATE APPROACHES FOR LANDFILLING OF WASTES

The regulation of chemicals in waste with respect to the potential to pollute groundwaters has been and continues to be woefully inadequate. Beginning in the 1950s, there were a number of reports in the professional literature about groundwater pollution by various types of wastewater lagoons, sanitary landfills, and other areas where hazardous or deleterious chemicals or wastes are located. In 1959 the American Society of Civil Engineers (4) issued a landfill design manual which discussed the potential for sanitary landfill waste to pollute groundwaters with waste-derived constituents. It was known by many professionals in the 1950s that the approach that was being used for managing industrial and municipal solid wastes could readily lead to groundwater pollution, rendering the groundwaters unusable for domestic purposes. This situation was not begun to be addressed, however, until the mid-1970s, with the passage of the Resource Conservation Recovery Act (RCRA). Even today, the U.S. EPA and state regulatory agencies are still allowing the development of municipal and industrial so-called nonhazardous and hazardous waste landfills that, at best, will only postpone the problem for a short period of time compared to the time that the wastes in the landfill will be a threat when groundwater pollution occurs.

Lee and Jones-Lee (5) have recently summarized the significant technical problems associated with today's minimum Subtitle D landfills in protecting groundwaters from pollution by landfill leachate for as long as the wastes in the landfill will be a threat. Basically, a minimum Subtitle D "dry tomb" of landfill, which uses a single composite liner consisting of a 60-mil HDPE plastic sheeting layer in contact with 2 ft of compacted clay with a permeability at the time of development of less than 1×10^{-7} cm/s, will prevent groundwater pollution for a small period of time relative to the time that the waste in the landfill will be a threat. The landfill liner system allowed in Subtitle D landfills will be a significant barrier to leachate passing through the liner into the underlying groundwater system for those landfills hydraulically connected to usable groundwaters for a short period of time relative to the time the waste in the landfill will be a threat. The HDPE liner, if it is properly constructed and no holes are punched into it at the time of waste deposition, will be effective in collecting leachate generated in the wastes through infiltration of precipitation until the HDPE begins to disintegrate. While no one can reliably predict how long HDPE layers of the type being used in municipal landfills will last, it is certain that they will not be effective barriers for leachate migration through them for as long as the wastes in the landfill will be a threat.

Eventually, through breakdown of the polymeric HDPE, the leachate will come in contact with the underlying clay layer and start to pass through it. If the original 1×10^{-7} cm/s permeability still exists at that time, then the leachate will pass through this layer in about 25 years. There are, however, a wide variety of well-known mechanisms by which the permeability of the clay layer may increase significantly over time, the most important of which are desiccation cracks (5,6). Clay layers are compacted with a certain optimum moisture content. Over time, the moisture will migrate out of the clay, causing the permeability to increase significantly, likely leading to cracking. Therefore, holes that develop in the HDPE layer, if they intersect cracks in the clay layer, could result in quite rapid transport of large amounts of leachate through the liner system into the underlying groundwater system.

Another current example of inadequate regulation implementation that has applicability to some of the technologies that are used to remediate/close hazardous chemical sites occurs with the monitoring of liner leakage from solid waste landfills. In 1988

Evaluation of Remediation by Landfilling

the U.S. EPA (7), in its promulgation of Subtitle D municipal solid waste regulations, stated:

> First, even the best liner and leachate collection system will ultimately fail due to natural deterioration, and recent improvements in MSWLF [municipal solid waste landfill] containment technologies suggest that releases may be delayed by many decades at some landfills.

The U.S. EPA *Criteria for Municipal Solid Waste Landfills*, released in July 1988 (8) stated:

> Once the unit is closed, the bottom layer of the landfill will deteriorate over time and, consequently, will not prevent leachate transport out of the unit.

The information available today, 10 years later, strongly supports the EPA's position of the inevitable failure of the landfill liner system. The agency, however, in promulgating the final version of Subtitle D regulations in 1991 (9), established groundwater monitoring requirements stipulating that:

> The design must ensure that the concentration values listed in Table 1 of this section not be exceeded in the uppermost aquifer at the relevant point of compliance...

and specifying that

> (a) A ground-water monitoring system must be installed that consists of a sufficient number of wells, installed at appropriate locations and depths, to yield ground-water samples from the uppermost aquifer (as defined in §258.2) that: (b) Represent the quality of ground water passing the relevant point of compliance.... (c) The sampling procedures and frequency must be protective of human health and the environment.

In support of this requirement, the agency required a three-phase monitoring program which consisted of detection monitoring where groundwater monitoring wells are to be placed at the point of compliance for the monitoring of the landfill in sufficient number and characteristics to detect leachate-polluted groundwaters in order to protect human health and the environment.

When leachate is detected by the monitoring wells, the landfill owner/operator is required to determine the extent of groundwater pollution that has occured. This requires a site-specific investigation where a number of additional monitoring wells are constructed to determine the characteristics of the pollution plume. After defining the magnitude of the pollution plume, the Subtitle D regulations require that the pollution be cleaned up to the extent possible, typically through a pump-and-treat operation, where the polluted groundwaters are pumped to the surface and treated before discharge to surface waters or reinjected into the groundwater system. Verification monitoring is required to confirm that the pollution has been cleaned up and that no further pollution occurs.

While the Subtitle D groundwater monitoring system and the associated remediation requirements should in principle be effective in preventing off-site groundwater pollution by landfill leachate, in practice, as being implemented today in many states across the United States, the Subtitle D monitoring system has a low probability of detecting groundwater pollution at the point of compliance for groundwater monitoring before significant off-site/adjacent property groudwater pollution occurs. Subtitle D landfill siting requirements do not preclude placing a Subtitle D landfill immediately next to the adjacent property owner's property line. The monitoring requirements established that the point of compliance for groundwater monitoring shall be no more than 150 m from the downgradient edge of the waste management unit, i.e., where wastes are deposited,

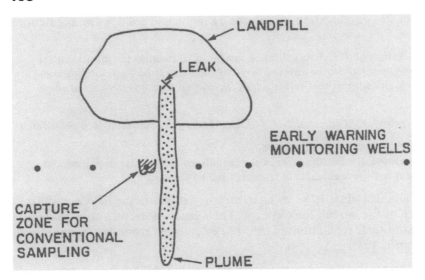

Figure 1 Pattern of landfill leakage—groundwater contamination from lined landfills. (After Cherry [11].)

and must be on the landfill owner's property. For many landfills, there are a few hundred yards of bufferlands between the edge of the landfill and adjacent properties. This means that, within a short time, once the leachate plumes pass the point of compliance, the polluted groundwaters will be trespassing under adjacent properties.

Lee and Jones-Lee (10) have recently reviewed the 1990 work of Cherry (11) and others on the adequacy of monitoring groundwater pollution at lined landfills. They point out that the work of Cherry discusses the unreliability of the typical groundwater monitoring systems that are being used for lined landfills. Figure 1, adopted from Cherry (11), shows the situation that will typically occur, where the initial leakage through the flexible membrane liner and underlying clay layer will produce fingerlike plumes of leachate which can be a few meters in width at the point of compliance for groundwater monitoring. Typically, groundwater monitoring wells are located hundreds to a thousand or so feet apart at the point of compliance. This means that a 3-m-wide plume at the point of compliance has a low probability of being detected by typical groundwater monitoring wells, since each well has a zone of capture of about 1 ft associated with the sampling event.

Basically, today's groundwater monitoring systems for Subtitle D and C landfills evolved from the groundwater monitoring systems that were used for unlined landfills, which leaked leachate into the underlying groundwaters across the entire bottom of the landfill. For that type of landfill, it was relatively easy to locate monitoring wells that would readily detect leachate-polluted groundwaters. However, with lined landfills, where the initial leakage would be through holes, rips, tears, or points of deterioration of the liner which produce fingerlike plumes of leachate of limited lateral dimensions at the point of development and limited lateral spread along the groundwater flow path downgradient from the landfill, the monitoring system for unlined landfills is highly unreliable for detecting leachate-polluted groundwaters before widespread off-site pollution occurs.

In order for today's Subtitle D landfills to be effectively monitored to comply with Subtitle D regulations for detecting leachate-polluted groundwaters for many types of geologic formations, groundwater monitoring wells would have to be placed about 10 ft apart. There are some types of geologic formations, such as fractured rock or cavernous limestone, where groundwater monitoring by vertical monitoring wells at the point of compliance is highly unreliable. Haitjema (12) states:

> An extreme example of Equation (I) (aquifer heterogeneity) is flow through fractured rock. The design of monitoring well systems in such an environment is a nightmare and usually not more than a blind gamble. ... Monitoring wells in the regional aquifer are unreliable detectors of local leaks in a landfill.

There are alternative monitoring approaches that could be used to more reliably detect when the composite liner for a Subtitle D landfill fails to prevent leachate from migrating through the liner that could pollute groundwaters. As discussed by Lee and Jones-Lee in 1998 (10), the state of Michigan adopted a double composite-lined landfill for municipal solid wastes, where the lower composite liner is a leak-detection system for the upper composite liner. This design is the same as that used for a double composite-lined Subtitle D landfill as well as the EPA's Subtitle C landfills for hazardous waste, where there is a leak detection layer of highly permeable materials that is designed to take leachate that passes through the upper composite liner to a sump where it can be sampled and removed.

In accord with the Michigan regulations, whenever leachate is detected in the leak detection system between the two composite liners, action must be taken to stop leachate generation within the landfill. As discussed in the next section, this can be done through improving the quality of the landfill cover. If that approach is not taken, it is only a matter of time until leachate passes through the lower composite liner and pollutes groundwater. The double composite-lined system also will ultimately fail to prevent groundwater pollution, since the lower composite liner is subject to the same types of failure mechanisms as the upper liner.

The double composite-lined landfill, operated where the lower composite liner is a leak-detection system for the upper liner and where action is taken to stop the leakage into the leak detection system between the two liners shortly after it is detected, provides a far more reliable groundwater monitoring system than is typically used today. The key to the operation of this system, however, is the availability of funds to continue to remove leachate in the leachate collection system and to operate the leak-detection system between the two composite liners for as long as the wastes in the landfill will be a threat. For funding planning purposes, it should be assumed that the wastes in almost all landfills will be a threat, effectively, forever. Therefore, there should be an adequate funding mechanism to achieve in perpetuity funds that could be used to monitor the landfill for liner leakage.

Hickman (13,14) and Lee and Jones-Lee (10) have come to the conclusion that the only reliable funding mechanism to ensure that funds will be available when needed for plausible worst-case scenario failures of a landfill waste management system is a dedicated trust which is established at the time that the wastes are placed in the landfill. Lee and Jones-Lee (15) have discussed the long-term liabilities associated with hazardous waste landfills, and point out that there is inadequate assurance of long-term funding being developed today as part of permitting Subtitle C landfills to ensure that the funds that will eventually be needed to remediate these landfills when the liner and other components of the containment system ultimately fail will, in fact, be available. For municipal solid waste landfills or hazardous waste landfills, the trust can be developed from part of the disposal fees. For hazardous chemical site remediation, the trust should be established at the time of closure of the site from funds provided by the PRPs. It is important not to rely on the various financial instruments that the EPA and states allow for providing financial assurance associated with Subtitle D and C landfills. Many of these are unreliable in assuring that funds will be available in perpetuity, i.e., for as long as the wastes are a threat. A dedicated trust, however, where the only way that the funds can be used is for meeting monitoring, maintenance, and remediation needs associated with the landfill, is a significantly more reliable funding mechanism than many of the approaches allowed today by the EPA and the states in closing hazardous chemical sites that allow residual waste components/chemicals to be left at the site in on-site landfills or under RCRA covers of a waste deposition area.

Eight states or parts of states now require double composite-lined landfills for municipal solid waste management, and all Subtitle C landfills are double composite-lined, but neither the EPA nor the state regulations require the development of a reliable funding mechanism to ensure with a high degree of reliability that funds will be available to implement the double composite-lined landfill leak-detection system and remediation approach for as long as the wastes in the landfill will be a threat.

A dedicated trust established to address plausible worst-case scenario failures over the time that the wastes in the landfill will be a threat should also be required for minimum Subtitle D landfills. However, for a single composite-lined landfill which relies on vertical monitoring wells at the point of compliance for detection of groundwater pollution, the magnitude of the trust needs to be considerably larger to eventually address the groundwater pollution plume that will occur. For Subtitle C or for double composite-lined landfills, the trusts could be somewhat smaller, since there is a high probability of detecting leakage through the upper composite liner before groundwater pollution occurs. It is important, however, that the closure plan for a landfill or waste management unit clearly define that effective action must be taken to stop leachate from entering the leak detection system between the two composite liners, or else even the double composite-lined landfill system will eventually pollute groundwaters.

Lee and Jones-Lee (16,17) have discussed the significant technical problems with Subtitle C hazardous waste landfills as they are being permitted today. Basically, they are the same problems as Subtitle D landfills, with the exception that there is an opportunity for Subtitle C landfills to provide for higher degrees of protection because of a double composite liner and the fact that many of the hazardous wastes are treated before being placed in the landfill to reduce the potential for constituent migration. This treatment, however, is not adequate to produce residues in a Subtitle C landfill that represent no significant threat to public health, groundwater resources, or the environment. The treatment reduces the hazards to groundwater pollution and public health but does not eliminate them.

V. PERIOD OF TIME THAT WASTES IN LANDFILLS OR CONTAMINATED SOILS ARE LIKELY TO BE A THREAT

For most situations where remediation of a hazardous chemical site is needed, the wastes have already been in the soil/landfill area for considerable periods of time. Any readily decomposable/leachable components have likely been removed from the wastes, with the result that less readily leachable, more resistant to transformation residues are present. Under these conditions, there will be few situations where a hazardous chemical site remediation should be planned for anything less than an infinite period of time. The

30-year postclosure care period and site-remediation monitoring period that is specified in RCRA and CERCLA represents an infinitesimally small part of the real time over which the waste residues at remediated-closed hazardous chemical sites or at a landfill will be a threat to public health, the environment, and groundwater resources.

The 30-year postclosure care period arose out of a significant error made by Congress in the 1970s, when Congress received inappropriate advice from individuals on how long wastes in a classic sanitary, unlined landfill were a threat. Thirty years was the period during which classic sanitary landfills typically produced landfill gas. Those advising Congress did not understand that landfill gas production, which is often stated to be landfill "stabilization," does not address the leaching properties of the wastes. As discussed by Lee and Jones-Lee (18), Belevi and Baccini (19) have conducted investigations that show that classic sanitary landfills in Switzerland would be expected to produce leachate that would contain hazardous chemicals, such as lead. From their review, it is concluded that lead above drinking-water MCLs could be leached from a classic sanitary landfill for over 2000 years. According to Freeze and Cherry (20), Roman Empire landfills are still producing leachate approximately 2000 years after they were developed.

These periods of time represent the fermentation and leaching characteristics of wastes in classic sanitary landfills in relatively wet climates, where there is appreciable infiltration of precipitation into the wastes that promotes landfill gas production through bacterial fermentation processes and leaching of the wastes. Lee and Jones-Lee (18) have discussed that the classic sanitary landfill groundwater pollution and landfill gas production characteristics are significantly different than today's Subtitle D and C landfills, where there is an attempt to create a "dry tomb" of plastic sheeting-entombed waste. As long as the tomb's (plastic) integrity is maintained, and the wastes are in fact kept dry, then no landfill gas production or leaching of the wastes will occur, since both of these processes are dependent on moisture. Eventually, however, under current landfill cover regulatory requirements, today's landfill covers will fail to keep the wastes dry; it is only a matter of time until a closed landfill or waste management area that stopped producing leachate at the time of a cover installation starts to produce leachate again when the integrity of the cover is no longer maintained. This issue is discussed further in a separate section of this chapter. Therefore, for today's dry tomb-type landfills, where there is an attempt to isolate the wastes from moisture which can provide treatment of the wastes, the period of time that waste monitoring, management, and maintenance activities must be conducted should be based on for as long as the wastes in the landfill are a threat. The state of California Water Resources Control Board Chapter 15 regulations adopted this approach in 1984. However, the State Water Resources Control Board and regional water quality control boards have continued to permit municipal solid waste landfills, largely ignoring this requirement of the regulations.

VI. LANDFILL AND WASTE MANAGEMENT AREA COVERS AS A SITE REMEDIATION TECHNOLOGY

One of the commonly considered methods of remediation of hazardous chemical sites is to develop an RCRA or a sometimes "less than RCRA" landfill-type cover over the waste-containing area, such as a burial pit, former lagoon, old landfill, etc. It is reasoned that if the supply of moisture to the landfill through precipitation can be stopped, then it should be possible to stop the leaching of the waste components that are leading to groundwater pollution. The same type of reasoning is used for on-site Subtitle D for nonhazardous waste components of a hazardous chemical site and Subtitle C for hazardous waste components of a hazardous chemical site remediation program where on-site landfills are constructed and the wastes present at the site are managed in the landfills. Such landfills suffer from all of the same characteristics as the classic Subtitle D and C landfills.

There is widespread recognition that landfill covers for Subtitle C and D landfills have limited periods of time over which they can be expected to function effectively in preventing moisture from entering a landfill and generating leachate. Lee and Jones-Lee (21) have discussed the inadequacies of today's landfill covers in keeping wastes dry in a dry tomb-type landfill for as long as the wastes represent a threat. While landfill owners and operators and some regulatory agency staff, as part of permitting landfills, will claim that the integrity of a landfill cover can be maintained through visual inspection and maintenance of the cover, the facts are that often such claims are superficial and unreliable. The basic problem with visual inspection of the cover, which can be readily accomplished by walking over the surface, is that it only detects superficial problems with the soil cover layer, such as major cracks caused by erosion, differential settling, etc. It can also detect major areas of landfill gas release since typically the release of landfill gas through uncontrolled conditions of the cover leads to the killing of the vegetative layer where the gas release occurs.

It is not possible to determine the integrity of the low-permeability layer in the landfill cover, since this layer, whether it is compacted clay, an HDPE layer, or combination of the two, is buried below several feet of topsoil and a drainage layer. The net result is that for minimum Subtitle D landfills, there can readily be cracks, holes, rips, tears, etc., in the key layer of the cover which is designed to keep the wastes dry that will not likely be detected for a minimum Subtitle D landfill until off-site pollution of a neighboring property's well occurs. For double composite-lined Subtitle D or C landfills, it will be possible to detect the failure of the cover to keep moisture out through detecting leachate in the leak-detection system between the two composite liners.

The situation of special concern is a low rate of leakage through the landfill cover that does not produce sufficient leachate to cause leachate generation to occur in sufficient amounts to lead to the need for leachate collection systems to operate effectively and remove the leachate. As discussed by Lee and Jones (6), however, leachate collection and removal systems are well known to clog, due to buildup of chemical and biological precipitates and the accumulation of fine particulates in the leachate collection system. Therefore, especially in older systems, the detection of leachate in the leachate collection system is not necessarily a reliable indicator of leachate production in a landfill. There can readily be leachate production that would pass through the deteriorated liner system that would not be detected in the leachate collection system.

Lee and Jones-Lee (5,21–23) have discussed how a hazardous chemical site or a Subtitle C or D landfill can be covered using a leak-detectable cover that has a high degree of reliability if the system is operated and maintained as designed, preventing moisture from entering or passing through the cover into the underlying wastes. As they discussed, several commercial systems are available. One is based on the Robertson vacuum system (24). Another is based on the electrical discharge system (25,26). Both indicate the location where the cover's low-permeability layer no longer maintains its integrity and therefore moisture could pass through the cover into the underlying wastes. While a number of other kinds of landfill covers are being developed, none of them thus far has the ability to provide a high degree of assurance that moisture will not enter the underlying landfill wastes or chemical management area and leach the chemicals, transporting them into the groundwater system.

In order to make a leak-detectable cover system work to keep wastes dry for as long as the waste components represent a threat to public health, groundwater resources, and/or the environment, it will be necessary to have a highly reliable funding mechanism to ensure that funds will, in fact, be available for as long as the wastes in the area under the cover are a threat. Again, the reliable funding approach for this type of situation is a dedicated trust which is of sufficient magnitude to cover plausible worst-case failure scenarios for as long as the waste residues represent a threat. Recently, there have been increasing efforts on the part of public agencies who have gained the support of the EPA to relax long-term funding requirements associated with postclosure care activities for landfills or other waste management units. This approach is strongly contrary to the future generations' interests and, at best, only passes the inevitable costs of monitoring and maintenance and eventual remediation on to future generations.

It is time to stop passing on the real costs of landfills, waste-covered areas, and clean-up of hazardous chemical sites to future generations and start using the technologies and funding mechanisms that are readily available to ensure that existing Superfund, closed RCRA facilities, and/or other sites or Subtitle C or D landfills do not become future Superfund sites. Today's hazardous chemical remediation approaches often virtually ensure that there will not be adequate funding available to address what can be readily expected to be significant problems in the future associated with how hazardous chemical sites and landfills are being closed today. Often, legitimate concerns can be raised about the financial stability of private companies in the waste management business and their being able to meet the long-term liability that is being developed in the closure of landfills and Superfund/hazardous chemical sites. Hickman (13,14) has pointed out that public agencies, such as county boards of supervisors, are no better, and in fact may be less reliable than some private interests in meeting the financial obligations associated with funding closure and postclosure activities on former landfills and waste management areas. The future funding of operation and maintenance of leak-detectable covers cannot be left to the whims of public agencies or private entities. It should be addressed in a reliable way as part of closure of the site, as one of the components of site closure costs.

Overall, on-site Subtitle C or D landfills or RCRA or less than RCRA covers for waste management areas, as being developed today, are not reliable approaches for managing hazardous or deleterious chemicals for as long as these chemicals will represent a threat to public health, groundwater resources, and/or the environment. It is possible through alternative approaches, using available technology, to develop on-site landfills or waste-covered areas that will have a high reliability of protecting the resources of a region from further pollution by releases from the waste management area. The key components are adequate and reliable groundwater monitoring to detect failure of the waste/hazardous chemical management system. All on-site waste management areas where there is a potential for leaching of constituents by precipitation that infiltrates the waste area should include closure with a leak-detectable cover and a dedicated trust of sufficient magnitude to operate and maintain this cover for as long as the wastes or residual chemicals represent a threat. For planning purposes, the funding needs should be considered to be for an infinite period of time.

VII. MONITORING OF CAPPED WASTE MANAGEMENT UNITS

Two kinds of situations can readily occur that are associated with hazardous chemical site remediation. One of these is where the chemicals of concern are distributed fairly evenly over the site and therefore, under unmanaged conditions, precipitation infiltrates through the waste management area and leaches some of the chemicals, carrying them into the underlying groundwater system. The other situation, which is likely more common, is that the hazardous chemicals are not evenly distributed across the site, but there are pockets of more hazardous chemicals or certain types of hazardous chemicals in localized areas. While the overall site may have generated a fairly large plume of contaminated groundwaters, there may be one or more smaller, more intense plumes due to the disposal of a particular type of chemical at a certain location within the overall site. An example of this type of situation occurred at the University of California, Davis, national Superfund site, where the university attempted to manage its own campus wastes by disposal in shallow pits located in campus landfills. Certain pits within a part of the landfill received certain types of wastes over considerable periods of time. An example was the waste chloroform, apparently developed from veterinary medicine, that was dumped into one pit in the corner of a landfill. This practice resulted in a chloroform plume over 1 mile with concentrations well above those that are considered acceptable risk for unchlorinated domestic water supplies. This occurred at two different campus landfill chloroform dumping areas. This kind of situation points to the importance of evaluating the potential hazards of a hazardous chemical site to contain small or limited dimension plumes that carry high concentrations of constituents for considerable distances down the groundwater gradient.

Once the site is remediated through capping, the plume characteristics will likely change significantly, since no longer will there be the potential for generalized pollution. The pollution plumes that arise from a capped landfill or waste management area will, at least initially and for considerable periods of time, produce limited-dimension plumes that will not likely be detected by broad-brush-type monitoring systems where a few downgradient wells are used to characterize groundwater quality. Basically, the characteristics of the waste deposition area and the characteristics of the groundwater hydrologic regime should be understood sufficiently well to develop a site-specific monitoring program that will reliably detect the groundwater pollution plumes that were generated prior to site remediation and those that will be generated after remediation. Because of the change in the character of the infiltration and the associated plume characteristics, far more monitoring wells will likely be needed to monitor failure of the remediation approach system to prevent further groundwater pollution by waste-derived constituents than were needed for the initial characterization of the site.

VIII. ADEQUACY OF SOLID WASTE HAZARD CLASSIFICATION

One of the frequently explored and sometimes used procedures for remediation of hazardous chemical sites, such as Superfund sites, is the addition of reagents to the waste-containing soils or waste residues to solidify or stabilize the constituents in the treated wastes. These so-called stabilized/solidified wastes are typically placed in an on-site or off-site landfill. A variety of solidification/stabilization reagents/materials are used for this purpose, ranging from lime/cement-based materials through various petroleum derivatives, plastics, etc. The Association for Environmental Engineering

Evaluation of Remediation by Landfilling

Professors (AEEP) is publishing a review of many of the procedures that can be used for this purpose, which discusses some of the potential benefits associated with stabilization/fixation of various types of wastes with various reagents/materials. The focus of the AEEP review is on short-term compliance with the U.S. EPA Toxicity Characteristic Leaching Procedure (TCLP). This section reviews the appropriateness of using the TCLP-based short-term evaluations of the adequacy of stabilization/fixation as a measure of the threat associated with long-term potential problems for public health, groundwater resources, and the environment by the chemical constituents in the stabilized/fixed waste. It also addresses the reliability of the TCLP procedure to properly characterize the hazards that contaminants in soils and wastes represent to public health and the environment.

A. Reliability of TCLP

The Toxicity Characteristic Leaching Procedure (TCLP) evolved from the U.S. EPA Exaction Procedure toxicity test that was specified in RCRA. Basically, this test was contrived by the EPA as a political test designed to minimize the size of the hazardous waste stream that would have to be managed as part of implementing the initial RCRA requirements for managing hazardous waste. The test was not then, nor is the TCLP which evolved from the EP-tox test now, a reliable assessment of the potential for chemical constituents present in a waste that has been stabilized and/or fixed to cause public health, groundwater quality, or environmental problems. The conditions of the EP-tox/TCLP test were selected arbitrarily, without reference to the physical conditions of the EP-tox/TCLP test such as solid–liquid ratio, leaching times, physical degree of dispersion of the waste, leaching solution characteristics, etc. As discussed by Lee and Jones-Lee (27,29), many of the conditions of the hazardous waste leaching test were selected arbitrarily and bear no relation to the real world, in which the leaching of constituents present in a soil or waste or fixed by various stabilization/fixation reagents would take place. The test is alleged by the EPA to simulate the conditions of leaching that would occur if the materials were placed in a municipal solid waste landfill. The only characteristic of the TCLP that begins to resemble a municipal solid waste leaching environment is the use of a dilute acetic acid solution to establish a pH during the leaching test similar to that encountered in some municipal solid waste landfills.

At the time of development of the EP-tox test, Lee and Jones (27) provided a detailed discussion of the unreliability of this test as a true measure of the leaching potential of constituents in waste or contaminated soils with reference to their potential impact on public health, groundwater resources, and/or the environment. They report on a number of significant problems with the test that still have not been adequately or reliably addressed today.

The EP-tox test was patterned somewhat arbitrarily after the elutriate test that was developed by the U.S. Army Corps of Engineers for leaching of dredge sediments to determine what might be released when dredged sediments are dumped in open waters. Lee and Jones (28) have reviewed the development and evaluation of the elutriate test that is used by the U.S. EPA and Army Corps of Engineers to evaluate the potential for chemical constituents in contaminated dredged sediments to be leached to the water column during open-water disposal of the dredged sediments. The 1 : 4 solid-to-liquid ratio used in the elutriate tests was selected based on the typical pumping ratio that is used in hydraulically dredging sediments. The 1 : 4 solid-to-liquid ratio in the EP-tox and now TCLP test has no technical foundation; it is completely arbitrary. Leaching is, for many substances, influenced by liquid–solid ratios.

In the elutriate test, the sediments are slurried to resemble what might happen under worst-case conditions where release of constituents associated with the interstitial water in the dredged sediments, as well as the actual release upon any leaching that would occur in the receiving waters for the dredged sediment disposal, is assessed. In the EP-tox/TCLP tests the wastes are ground to a fine powder. Such an approach can significantly distort the release of constituents in a leaching test. For some constituents in some types of solids, the grinding will expose additional waste components that would not be leached otherwise. In other situations, the grinding exposes surfaces which could serve to take up the constituents of concern in the wastes and therefore actually reduces the amount of release that would occur compared to that which occur in the landfill.

In a landfill the liquid-to-solid ratio can vary from unsaturated conditions where the transport through the wastes occurs as a thin film of moisture moving on the surface of the particles with no fluid between the waste particles to a saturated condition where the wastes are fully saturated with moisture. The 1-h period of leaching in the elutriate test resembles the period of time over which leaching would typically occur in an open-water disposal of hydraulically dredged sediments. The period of leaching in the EP tox/TCLP test has no relationship to the leaching time over which there is concern in a municipal landfill.

The EP-tox and now TCLP tests were designed for a specific purpose, namely, to determine whether a waste placed in a classic municipal solid waste landfill could leach sufficient materials to produce a leachate that represented a significant threat to groundwater quality for use of the groundwaters as a domestic water supply. Any waste that leaches more than an arbitrarily defined amount (100 times the drinking-water MC) is classified as a hazardous waste and must be managed in a hazardous waste landfill. Waste that leaches less than this amount could be placed in a municipal solid waste landfill. This test was never designed to be used as it is widely used today, to determine whether a contaminated soil or other medium is hazardous or not and therefore requires remediation. Lee and Jones (29) discuss the appropriate approach that should be used to determine whether constituents at a particular location represent a significant hazard to public health and the environment to require remediation. This involves a site-specific hazard assessment approach where the leaching characteristics of the materials under the conditions that will be encountered at a particular location are evaluated and the transport/fate of the leached materials is determined to assess whether concentrations above critical levels for various beneficial uses of a water receiving the materials of interest occurs.

In the elutriate test, the receiving waters for the dredged sediment disposal are used to leach the dredged sediments, since this is where leaching will occur. In the EP-tox/TCLP test a dilute acetic acid solution is used. While at one time leachate produced in the classic sanitary landfill would tend to have the pH of dilute acetic acid solution, today that situation is changing. First, the classic sanitary landfills had a certain moisture content dependent on the amount of infiltration into the wastes through the landfill cover. The landfill covers were not designed and constructed to keep moisture out. Moisture that did not run off from the landfill cover surface penetrated into the waste. Today, upon closure of the landfill and the application of a low-permeability cover, the moisture supplied to the waste is shut off or significantly reduced, and to the extent that the cover maintains its integrity, the wastes will be kept dry. This is in accord with the dry tomb-type landfilling that was adopted in the United States for hazardous waste in the mid-1980s and municipal solid wastes in the early 1990s. Eventually, because of the inability to maintain the low-permeability layer within the cover of a dry tomb-type landfill, appreciable moisture can enter the wastes again. The amount of moisture entering the wastes will likely be less than that in the classic sanitary landfill. The situation therefore could become one of not having as low a pH in the leachate generated within today's and tomorrow's dry tomb landfills as occurred in the classic sanitary landfill.

Another factor that influences the pH in a municipal solid waste landfill leachate is the fact that green wastes such as yard wastes are now, as the result of recycling efforts, being diverted from municipal solid waste landfills. As a result, the readily fermentable components of the municipal solid waste stream, which are the precursors of the acid that decreases the pH in leachate, are no longer being added to the landfill to the same degree that they were a few years ago. Recently, the California Environmental Protection Agency Department of Toxic Substances Control (DTSC) (30) has conducted some fairly comprehensive studies on the characteristics of leachates from a variety of landfills in California, in which it was found that the pH in these landfills was not 5 or so as is typically assumed, but rather around 7 or 8. Changing the pH from 4 or 5 to 7 or 8 significantly changes the leaching characteristics for many constituents.

While the TCLP is now being used to determine the leaching characteristics of waste components in a hazardous chemical on-site or off-site landfill, the use of this test for that purpose is technically invalid from several aspects, since the test conditions are not designed to mimic in any way the leaching characteristics that would occur in a hazardous chemical site-type landfill, where there may be little or no municipal solid waste components which determine the characteristics of the leaching solution. Many of the hazardous chemical site on-site landfills will have low fermentable organic content. As a result, the pH of the leaching solution and its complexing tendencies will be significantly different from those that would occur in a municipal solid waste landfill or those of the TCLP test.

The California EPA DTSC studies (30) on the leaching of various types of industrial solid wastes which were potentially hazardous wastes included leaching with acetic acid in the typical TCLP procedure and with citric acid under the California Waste Extraction Test (WET). It was found, as expected, that for most metals the citric acid, because of its stronger complexing tendencies for metals, tended to leach more heavy metals from the wastes than the TCLP. Also used in this study was actual leachate from several municipal solid waste landfills. It was found that the leaching characteristics of the most aggressive of the municipal solid waste landfill leachates that were tested were similar to the leaching characteristics of the TCLP for many of the heavy metals investigated. A number of leachates, however, leached significantly less heavy metals than the TCLP test, indicating that TCLP tends to overestimate the amount of leachable heavy metals in many landfill settings.

B. Definition of Excessive Leaching

As part of developing the EP-tox political test for classifying wastes as hazardous or nonhazardous, the U.S. EPA arbitrarily, without a technical foundation, decided that a 100-fold leaching above the drinking-water MCL (maximum contaminant level–standard) represented an excessive leaching of constituents that, when present in municipal solid waste landfill leachate, should cause the wastes to be placed in a so-called hazardous waste Subtitle C landfill. The agency has described this 100-fold factor between the drinking-water MCL and the allowable leaching in the EP-tox/TCLP test as the "attenuation" factor that typically occurs in municipal solid waste landfills between when leachate leaves the landfill and when the groundwaters which contain the leachate in a diluted form could be used for domestic water supply purposes. The EPA, as part of promulgating the TCLP test as a revision of the EP-tox test, initially proposed abandoning the use of the arbitrary 100-fold "attenuation" factor in favor of a site-specific evaluation that would consider the characteristics of the aquifer system into which the leachate is being added. This was along the lines of those recommended by Lee and Jones (27,28). However, in adopting the final TCLP regulations, without providing any technical justification, the EPA abandoned what could have become a technically valid approach, if appropriately implemented, of site-specific evaluation of attenuation factors, and reverted to the technically invalid approach of using a 100-fold "attenuation" factor.

Following this action, several years ago, the authors attempted, through contacts with EPA administration, to ascertain any technical basis for this action. They were informed that a group within the agency had done some groundwater transport modeling which justified the 100-fold "attenuation" factor. When an attempt was made to obtain a copy of the models and reports on this, the authors were told that there were no reports covering this work. This was then followed up by a request to contact the people doing the work, and the authors were told by the EPA administration member responsible for this area that those individuals had left the agency and had left no records of their work with the agency.

This situation does not give a lot of confidence to the idea of the agency ever having any technical justification for the 100-fold "attenuation" factor. It was a political decision without technical merit. A critical examination of this issue shows that there can readily be conditions within the landfill and in aquifer systems where 100-fold attenuation is underprotective of domestic water supply wells located near the landfill or overprotective of the groundwaters used for domestic purposes. The amount and location of deposition of wastes within a municipal landfill that is of concern because of the leaching of potentially hazardous constituents will influence the concentration of the constituents of concern in the groundwaters underlying the landfill. Further, the characteristics of the aquifer through precipitation/solid formation, sorption, dilution, biological transformations, etc., will influence the transport of waste-derived constituents and their transformation products.

In addition, certain kinds of aquifers, such as fractured rock and cavernous limestone, where there can be substantial transport of leachate with limited dilution along fractures or in solution channels, can lead to concentrations of leachate in groundwaters intercepted by a well where there is certainly less than 100-fold attenuation between the point of release of the leachate from the landfill and the point of interception of the leachate-polluted groundwaters by a domestic water supply well. The deposition of wastes in an on-site landfill from a hazardous chemical site can readily result in parts of the wastes that are placed in the landfill having significantly different characteristics, therefore causing leachate generated in those areas to have different characteristics than other parts of the landfill which receive wastes with different characteristics.

It can be concluded that the 100-fold "attenuation" factor typically used to determine whether a waste is hazardous or not is unreliable in characterizing whether constituents in a waste that is placed in a municipal landfill or a hazardous waste landfill represent significant threats to public health, groundwater resources, or the environment. The same, even to a greater extent, can be said about the use of this factor to characterize the potential for leaching constituents from a hazardous chemical, on-site remediation landfill. The appropriate approach in evaluating the potential for constituents in an on-site remediation landfill is to conduct detailed site-specific leaching and leachate transport characteristic evaluations to predict, prior to the construction of the landfill, whether constituents in the on-site landfill could likely lead to groundwater pollution with the inevitable failure of a dry tomb-type landfill of either Subtitle D or C characteristics, as well as a waste management area which is covered with a RCRA landfill cover. If it appears feasible to manage the hazardous chemical site wastes in an on-site landfill, then a comprehensive groundwater monitoring and landfill gas monitoring program should be implemented, operated, and maintained for as long as the constituents in the landfill represent a threat to public health, groundwater resources, and/or the environment.

This threat should be evaluated not just for the "priority pollutants," as is typically done today, focusing on a limited number of potentially hazardous constituents compared to the vast arena of hazardous constituents present in most landfill leachates and hazardous chemical site wastes, but should also include consideration of unknown hazardous constituents which are a threat to public health through drinking water consumption as well as the so-called nonhazardous components which are not a threat to health, but are a threat to the quality of the groundwater used for domestic purposes. Constituents that cause taste and odors or high dissolved solids which lead to either scaling or corrosion in municipal water supply systems and in residences/commercial establishments are as important to the public as the hazardous constituents which receive the focus of attention. Tastes and odors or unusable waters because of esthetic or other characteristics typically require abandonment of the domestic water supply well and the development of alternative water supplies. While this "musical wells" approach is possible in some areas where the total demand for water supplies is less than the available supply, there are areas, especially in the arid West as well as the more humid East, where the total water supply, especially in drought years, is inadequate to meet the needs of the population in the region. The destruction of a groundwater resource by pollution by landfill leachate is strongly contrary to the public's interests.

As discussed herein, it is essential that a dedicated trust be developed by private and public entities or responsible parties to ensure that reliable groundwater monitoring will be conducted to detect with a high degree of certainty any subsequent releases of constituents from the on-site landfill or covered waste management area for as long as the wastes represent a threat. The magnitude of this trust should be sufficient so that if those responsible for managing the on-site wastes cannot terminate the migration of constituents from the landfill, then the wastes in the on-site waste management unit will have to be removed (mined). Failure to provide this level of protection will mean that future generations will face the same problems as the current generation of inadequately managed hazardous waste.

In summary, the current TCLP test which is widely used to characterize the leaching potential of constituents in a waste for deposition in a municipal solid waste Subtitle D landfill, a hazardous waste Subtitle C landfill, or an on-site hazardous chemical site remediation landfill, which could be C or D depending on the materials placed in the landfill, is not a reliable test to assess the potential mobility of the constituents in the landfill. This test should not be used for anything other than a political test to determine whether wastes can be placed in a municipal solid waste landfill or a hazardous solid waste landfill. The results of the TCLP test bear no relationship to reliably assessing the potential threat that constituents in the waste represent to public health, groundwater resources, or the environment.

IX. NATURAL ATTENUATION

"Natural attenuation" for remediation of groundwater pollution is becoming an increasingly popular approach toward "remediation" of hazardous chemical sites. The basic premise of this approach is that if the source of the constituents that is causing groundwater pollution is removed and the pollution plume has apparently stabilized, then, rather than trying to pump and treat the polluted groundwaters for their remediation, the plume should be allowed to attenuate naturally. While this approach saves PRPs considerable funds, it, in general, must be more carefully implemented than is being done in some areas today.

It should be remembered that in most cases the pollution of groundwaters is the result of the PRP failing to follow common sense and/or good business practice with respect to leaking tanks, which is one of the major causes of groundwater pollution. In most cases, it should have been obvious to those responsible for management of inventories of tanks of stored materials that the tanks were leaking, as the result of more material being purchased than was actually used or sold. Further, in the case of pollution by landfills or waste disposal facilities, the potential to pollute groundwaters by such facilities has been well understood since the 1950s. While regulatory agencies then, and unfortunately in some cases, still today, are not taking the action required by regulations to ensure that groundwater quality is protected with a high degree of certainty, the manager of the waste should have been aware and acted on controlling the disposal of wastes in landfills, pits, etc., so that groundwater pollution would not occur or would have a low probability of occurring. One of the basic problems is that companies, commercial establishments, public agencies, etc., have for years provided inadequate funds to hire appropriately trained individuals who would become familiar with and understand the literature that has been available since the 1950s on the potential to cause groundwater pollution. Further, there has been and continues to be a chronic problem where the management of the industrial or commercial facilities or agencies opt for a short-term economic gain for reduced spending compared to the funds that should be spent to properly manage wastes so they do not cause long-term problems of groundwater pollution.

Basically, the long-term costs of pump-and-treat or other active groundwater remediation approaches represent part of the costs that will have to be borne by industry, commercial establishments, public agencies that manage wastes, etc., arising out of the cheaper-than-initial-cost approach that was used at the time of waste deposition. The public and future generations are entitled to clean-up of polluted groundwaters to the maximum extent practicable to ensure that no further spread of pollutants will occur which would further damage the groundwater resources, health, welfare, and interests of those within the sphere of influence of the polluted groundwater. If this can be achieved through a properly developed, implemented, and most important, monitored natural attenuation approach, then natural attenuation is a viable option that should be considered.

Natural attenuation arises from a variety of factors which limit the size of a groundwater pollution plume. These factors include the dilution of this plume through lateral and vertical mixing with unpolluted groundwaters. One of the sources of dilution, especially of the upper parts of the plume, is the infiltration of precipitation along the plume path, which adds additional water to the aquifer and tends to dilute the plume. Further, there are several chemical and biochemical reactions (precipitation, sorption, oxidation reduction, volatilization, biochemical transformations) which for some constituents tend

Evaluation of Remediation by Landfilling

to remove one or more constituents from the plume. Not all these reactions, however, especially chemical and biochemical transformations, lead to nonhazardous transformation products.

One of the areas in which there has been considerable interest in natural attenuation of polluted groundwaters is associated with the pollution of groundwaters by gasoline from service stations. However, it is inappropriate to assume, as has been done by Lawrence Livermore National Laboratory (LLNL) and the State of California Water Resources Control Board staff, that removal of benzene, xylene, and toluene from a gasoline plume leads to conversion to carbon dioxide and water. There could readily be a variety of transformation products formed which are hazardous or otherwise deleterious to groundwater quality, public health, and/or the environment where the groundwater has become part of the surface water system.

One of the areas of particular concern that is often ignored is the presence of unidentified hazardous or deleterious chemicals in groundwater plumes. At this time, only from about 50 to a couple of hundred chemical constituents are routinely monitored/regulated in groundwater pollution investigation and remediation programs. There are over 75,000 chemicals in use today, with about 1000 new chemicals being developed each year. Further, there is a large, unknown number of transformation products from chemicals that could readily be present in groundwater plumes arising from pollution of groundwaters by various types of wastes, waste spill materials, or tank or transmission line leaks. Frequently, measurement of total organic carbon or dissolved organic carbon in a plume and/or waste shows that the concentrations of total organics far exceed the identified organic components. This can lead to a significant potential for there being unknown/unidentified constituents in the groundwater plume that can be hazardous/deleterious to public health, groundwater quality, and the environment.

With increasing frequency today, risk-based approaches are being used to establish clean-up objectives for groundwater systems and soils. These approaches are typically significantly deficient when these objectives are applied to situations where there are complex mixtures of chemicals in the source or where the source chemicals can be transformed into a variety of hazardous and deleterious chemicals within the soil aquifer system. An appropriately conducted risk-based approach for establishing clean-up objectives for hazardous chemical mixture-polluted groundwaters or soils is to establish a surrogate hazard level for an unknown/unidentified constituent(s) that could be present and be transported in the groundwater plume at the same rate as water. It is suggested that the surrogate be assigned a hazard concentration MCL of 0.01 times that of vinyl chloride and that it is assumed to move as readily in groundwater systems as vinyl chloride. This would make the critical MCL for the unknown constituents about 0.01 mg/L.

There is a situation where vinyl chloride does not persist in groundwaters as long as other constituents. This occurs with municipal landfill systems where the groundwater plume, which originally contained TCE in the landfill, disposed of as a waste from industrial or residential/commercial use, converts to vinyl chloride in the landfill environment. At the leading edge of the groundwater plume from municipal solid waste there is a conversion from anaerobic to aerobic conditions. Under these conditions, there is the potential for methane that could be present in the groundwaters where the conversion of methane to CO_2 under aerobic conditions leads to the removal of vinyl chloride to the groundwater plume.

A prime example of the significant errors that can occur is the LLNL's assessment of the persistence of gasoline derivatives—benzene, xylene, and toluene—where LLNL ignored the fact that there are other well-known constituents in gasoline that can be polluting groundwaters as well. For example, MTBE has been found to be a common component of gasoline that is far more resistant to biotransformations than benzene, toluene, and xylene and moves as readily as water in groundwater systems. It is now known that MTBE is a widespread common pollutant of groundwaters associated with gasoline leaks that is a significant problem to groundwater quality through imparting severe taste and odor problems.

Another example of how assuming that only those constituents that are on the U.S. EPA MCL list or the Priority Pollutants list represent the only hazards in a polluted groundwater plume is provided by a situation that occurred in the Sacramento, California, area with the pollution of groundwaters by Aerojet Corp. in connection with its rocket engine testing. Aerojet created a very large plume that is polluting groundwaters with trichloroethylene (TCE) over a substantial area. For a number of years, the approach toward remediation was to pump the polluted groundwaters, airstrip the TCE, and then reinject the pumped and treated groundwaters downgradient from where they were pumped. About two years ago it was discovered that Aerojet polluted the groundwaters not only with TCE but also with perchlorate, which has been found to be highly hazardous to human health. Perchlorate is not on the U.S. EPA drinking-water MCL list; it is not normally analyzed for in any groundwater pollution studies. The Aerojet-created plume is affecting a number of domestic water supply wells and is threatening many others. Other exotic chemicals associated with Aerojet's mismanagement of its wastes are also being found in this plume.

One of the key aspects of any natural attenuation program is appropriate monitoring of the pollutant plume. A detailed, comprehensive monitoring program should be undertaken as part of any natural attenuation program to assure that the plume is, in fact, being naturally attenuated and that no new situations have developed which would cause the behavior of the plume to change significantly. An area of concern is where new off-site production wells cause the groundwater flow pattern to change, which in turn alters the natural attenuation plume.

The monitoring program should be designed to be operated for as long as any residual potentially hazardous constituents are present in the groundwaters. This will likely require that a dedicated trust be established to ensure that funds are available to continue the monitoring program for as long as there are constituents in the groundwater that represent a potential threat. Initially, the monitoring program should be conducted quarterly for at least two to three years. This would provide information on seasonal changes in the characteristics of the plume, which is especially important where there are significant changes in the water table elevation and where there are large production wells in the vicinity of the plume that could influence its behavior and characteristics. If, after two to three years, depending on the overall plume setting, it is possible to predict reliably the next set of analytical results for sampling the monitoring wells around and within the plume, then the frequency of monitoring can be cut back to semiannually. The approaches that are being used by some regulatory agencies of monitoring once every year to once every five years are totally inappropriate. There is an insufficient database generated from this approach to determine whether changes in plume characteristics are occurring.

Every five years the overall character of the groundwater monitoring program should be subjected to a public review with appropriate public notice and a hearing where the PRP presents a comprehensive report of the results of the monitoring report during the past five years. This report should include a listing of parameters being monitored, with a critical review of the appropriateness of this list considering what is known at that time about constituents of concern. This review should be designed to bring up to date the monitoring program relative to new information that has developed during the past five years on the potential hazards of chemicals present in the plume to public health, groundwater resources, and the environment. It is important to emphasize that, in addition to considering hazardous chemicals which are a threat to human health, consideration must also be given to chemicals that are detrimental to the use of the groundwaters for domestic water supply purposes, including tastes and odors, increased TDS, increased hardness, constituents that cause staining, etc. The monitoring program can be terminated once it has been established after five years of monitoring that there is no residual evidence for any constituent associated with the plume.

X. CONCLUSIONS

The current trend of remediating hazardous chemical sites, in which potentially hazardous chemical constituent residues are left at the site as part of site closure, is typically not being adequately evaluated with respect to the long-term hazards that the waste residues represent to public health and the environment. Of particular concern is the use of on-site landfills or waste management units covered with a RCRA landfill cover to prevent further groundwater pollution by waste-derived constituents for as long as wastes in the landfill or covered waste management will be a threat. Current landfilling approaches allowed by regulatory agencies under RCRA Subtitle C and D, at best, only postpone when further groundwater pollution will occur.

While this approach enables hazardous chemical sites to be cleaned up at initially reduced cost, ultimately the true costs of using landfills or RCRA covered waste management units will have to be borne by future generations. Because of inadequate funding for long-term costs as part of current hazardous waste site remediated closure approaches, future generations not only will have to pay for the cost of remediation, but also will be exposed to the hazards of the residual chemicals at inadequately remediated hazardous chemical sites.

The current regulatory approach for classification of a waste as hazardous involving the use of the TCLP is significantly deficient in properly assessing the real hazards that constituents in wastes represent to public health and the environment. Site-specific evaluations of leachability, transport, and transformations of waste-derived constituents should be used to evaluate the hazards that constituents in wastes represent to public health and the environment.

Many of the approaches being used today in support of natural attenuation as a remediation approach for polluted groundwaters are deficient in properly evaluating and managing the hazards that constituents in a groundwater plume arising from a hazardous chemical site represent to public health and the environment. Far more comprehensive evaluations of constituents of concern in groundwater plumes should be made as part of developing a natural attenuation approach for remediation of contaminated groundwaters. Also, the PRPs for a groundwater pollution plume should be required to reliably monitor the groundwater plume for as long as the plume contains constituents that are a hazard or are detrimental to the use of the groundwater for domestic water supply purposes.

By far the greatest deficiency in hazardous chemical site remediation is the failure to require that the PRPs develop a dedicated trust fund as part of site closure that is sufficient to address all plausible worst-case scenario failures that could occur at the site for as long as the wastes remaining at the site of closure will be a threat to public health and the environment. For most hazardous chemical sites, the planning period during which the trust fund will be needed should be considered to be infinite and should be of sufficient magnitude to monitor and maintain the site in perpetuity for as long as the waste residues and their transformation products represent a threat, and should be of sufficient magnitude to exhume all waste residues and contaminated soils from the site should this prove to be needed to protect public health and the environment.

REFERENCES

1. Lee, G. F., and Jones-Lee, A. 1994. Does meeting cleanup standards mean protection of public health and the environment? *Superfund XV Conf. Proc.*, pp. 531–540. Hazardous Materials Controls Resources Institute, Rockville, MD.
2. Lee, G. F., and Jones, R. A. 1991. Redevelopment of remediated superfund sites: Problems with current approaches in providing long-term public health protection. *Proc. Environmental Engineering 1991 Specialty Conference*, pp. 505–510. ASCE, New York.
3. Lee, G. F., and Jones, R. A. 1991. Evaluation of adequacy of site remediation for redevelopment: Site assessment at remediated-redeveloped "Superfund" sites. *Proc. 1991 Environmental Site Assessments Case Studies and Strategies*: The Conference, pp. 823–837. Association of Ground Water Scientists and Engineers—NWWA, Dublin, OH.
4. American Society of Civil Engineers. 1959. Sanitary landfill. Report Committee on Sanitary Landfill Practice of the Sanitary Engineering Division of the American Society of Civil Engineers. New York: ASCE.
5. Lee, G. F., and Jones-Lee, A. 1998. Assessing the potential of minimum Subtitle D lined landfills to pollute: Alternative landfilling approaches. *Proc. Air and Waste Management Association 91st Annual Meeting*, San Diego, CA. Available on CD-ROM as paper 98-WA71.04(A46); also available at http://members.aol.com/gfredlee/gfl.htm.
6. Lee. G. F., and Jones, R. A. 1992.Municipal solid waste management in lined, "dry tomb" landfills: A technologically flawed approach for protection of groundwater quality. Report of G. Fred Lee & Associates, El Macero, CA.
7. U.S. EPA. 1988. Solid waste disposal facility criteria; Proposed rule. *Fed. Reg.*, 53(168):33314–33422, 40 CFR Parts 257 and 258. Washington DC: U.S. EPA.
8. U.S. EPA. 1988. *Criteria for Municipal Solid Waste Landfills*. Washington, DC: U.S. EPA.
9. U.S. EPA. 1991. Solid waste disposal facility criteria: Final rule, Part II. *Fed. Reg.*, 40 CFR Parts 257 and 258. Washington, DC: U.S. EPA.
10. Lee, G. F., and Jones-Lee, A. 1998. Deficiencies in Subtitle D monitoring for liner failure and groundwater pollution. *Proc. Natl. Monitoring Conf., "Monitoring: Critical Foundations to Protect Our Waters,"* Reno, NV.
11. Cherry, J. A. 1990. Groundwater monitoring: Some deficiencies and opportunities. In: *Hazardous Waste Site Investigations: Towards Better Decisions, Proc. 10th Oak Ridge Natl. Lab. Life Sciences Symp.* Gatlinburg, TN: Lewis Publishers.
12. Haitjema, H. 1991. Ground water hydraulics considerations regarding landfills. *Water Res. Bull.*, 27(5):791–796.
13. Hickman, L. 1992. Financial assurance—Will the check bounce? *Municipal Solid Waste News*.
14. Hickman, L. 1995. Ticking time bombs? *Municipal Solid Waste News*.
15. Lee, G. F., and Jones-Lee, A. 1996. Permitting of new hazardous waste landfills and landfill expansions: A summary of public health, groundwater resource and environmental issues. Report of G. Fred Lee & Associates, El Macero, CA.
16. Lee, G. F., and Jones-Lee, A. 1992. Municipal landfill post-closure care funding: The "30-year post-closure care" myth. Report of G. Fred Lee & Associates, El Macero, CA.

17. Lee, G. F., and Jones-Lee, A. 1996. Superfund site remediation by on-site RCRA landfills: Inadequacies in providing groundwater quality protection. *Proc. Environmental Industry Association's Superfund/Hazwaste Management West Conference*, Las Vegas, NV, pp. 311–329.
18. Lee, G. F., and Jones-Lee, A. 1994. Landfilling of solid & hazardous waste: Facing long-term liability. In: *Proc. 1994 Federal Environmental Restoration III & Minimization II Conference*, pp. 1610–1618. Rockville, MD: Hazardous Materials Control Resources Institute.
19. Belevi, H., and Baccini, P. 1989. Water and element fluxes from sanitary landfills. In: *Sanitary Landfilling: Processes, Technology and Environmental Impact*, pp. 391–397. San Diego, CA: Academic Press.
20. Freeze, R. A., and Cherry, J. A. 1992. *Groundwater*. Englewood Cliffs, NJ: Prentice-Hall.
21. Lee, G. F., and Jones-Lee, A. 1995. Overview of landfill post closure issues. Presented at American Society of Civil Engineers Convention session devoted to Landfill Closures—Environmental Protection and Land Recovery, San Diego, CA.
22. Lee. G. F., and Jones-Lee, A. 1997. Hazardous chemical site remediation through capping: Problems with long-term protection. *Remediation*, 7(4):51–57.
23. Lee, G. F. 1997. Redevelopment of brownfield properties: Future property owners/users proceed with your eyes open. *Environ. Prog.*, 16(4):W3–W4.
24. Robertson, A. 1990. The "Robertson barrier liner": A testable double liner system. Vancouver, B. C., Canada: Robertson Barrier System Corp.
25. Nosko, V., and Andrezal, T. 1993. Electrical damage detection system in industrial and municipal landfills. *Geocontinue*, 93:691–695.
26. GSE Lining Technologies Inc. 1994. *Sensor Damage Detector Systems*. Houston, TX: GSE.
27. Lee, G. F., and Jones, R. A. 1981. Application of site-specific hazard assessment testing to solid wastes. In: Hazardous Solid Waste Testing: First Conf., STP 760, pp. 331–344. Philadelphia: ASTM.
28. Lee, G. F., and Jones, R. A. 1992. Water quality aspects of dredging and dredged sediment disposal. In *Handbook of Dredging Engineering*, pp. 9–23 to 9–59. New York: McGraw-Hill.
29. Lee, G. F., and Jones, R. A. 1982. A risk assessment approach for evaluating the environmental significance of contaminants in solid wastes. In: *Environmental Risk Analysis for Chemicals* pp. 529–549. New York: Van Nostrand.
30. California EPA Department of Toxic Substances Control. 1996. *The RSU Extraction Test project*. Sacramento, CA: DTSC.

10
Bioavailability: The Major Challenge (Constraint) to Bioremediation of Organically Contaminated Soils

A. L. Juhasz, M. Megharaj, and R. Naidu
CSIRO Land and Water, Adelaide, South Australia, Australia

I. INTRODUCTION

Bioremediation of organically contaminated soils is currently regarded as one of the most successful techniques for remediating some contaminated soils. The technique is based on optimization of biological processes to remediate or to minimize the concentrations of hazardous pollutants at contaminated sites. The underlying basis of bioremediation of organic pollutants is the detoxification or mineralization of the contaminating species to CO_2 and H_2O. Therefore, it makes an attractive, environmentally friendly, and relatively cost-effective alternative to conventional physicochemical techniques, which rely mainly on incineration, volatilization, or immobilization of the contaminant. However, the success of bioremediation is determined by the metabolic potential of microorganisms to detoxify or utilize the contaminant, provided the contaminant is accessible for microbial attack and other environmental factors are optimal. The process is therefore dependent on both accessibility and bioavailability of the contaminants to microbes. There is, however, considerable controversy in the literature as to what constitutes the bioavailable fraction, including the definition itself and the methods of its measurements. For instance, while microbiologists often regard the concentration that causes ecotoxicity to microorganisms as the bioavailable fraction, plant scientists regard the plant-available fraction as the bioavailable fraction. Consequently, terms such as "bioavailable," "phytoavailable," and "available" are in use. The concept of bioavailability is applicable to all organisms regardless of their nature, including microorganisms, plants, animals, and humans.

Bioavailability refers to the availability of a substrate to an organism. In terms of bioremediation, bioavailability refers to the extent to which contaminant is available for biological conversion. This is a function of many factors, involving the biological system, physicochemical properties of the contaminant, and environmental factors. The fundamental question that poses the concept of bioavailability is: Is bioavailability the major constraint to the bioremediation of contaminants? And if so, can bioavailability be increased to enhance microbial degradation of the otherwise "unavailable" con-

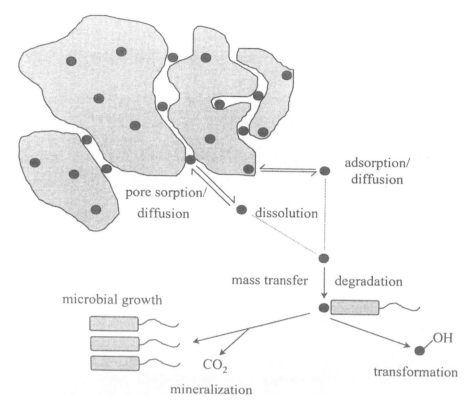

Figure 1 Schematic diagram illustrating the physical, chemical, and biological factors influencing the bioavailability of a contaminant in the environment.

taminants in the soil? A number of physical, chemical, and biological factors can influence the bioavailability of a contaminant (Fig. 1).

In this chapter, we discuss the fundamental principles of bioavailability, its relation to toxicity, and the potential role of bioavailability in the bioremediation process.

II. BIOAVAILABILITY

Bioavailability is of paramount importance because it frequently accounts for the persistence of biodegradable contaminants that are otherwise assumed to be easily degraded. It may also limit the remediation of the polluted sites (3). Hence, prior to implementing the bioremediation process, it is important to understand the factors that control the bioavailability of contaminants.

Following entry into the soil environment, contaminants rapidly bind to mineral and organic matter via a combination of physical, chemical, and biological processes. The ability of the soil to release (desorb) contaminants controls its susceptibility to microbial degradation, thereby influencing the effectiveness of the bioremediation process. Therefore, desorption is considered to be an important process controlling contaminant persistence in soil. Bioavailability is also limited by the transport of the contaminant to the cell in the case of soil aggregates, i.e., diffusion of the contaminant out of a soil

Bioavailability: The Major Challenge

aggregate to a cell attached to the external surface of the aggregate. Thus the unavailability of the organic contaminant could be due to a combination of one of more of the following:

Sorption to soils (i.e., low desorption)
Accessibility
Partitioning into nonaqueous-phase liquids (NAPLS).
Entrapment within the physical matrix of the soil.
Chemical structure of the contaminant (95).

Numerous researchers [e.g., Yaron et al. (95)] have demonstrated a close link between bioavailability and the rate of biodegradation of organic contaminants by microorganisms. This suggests that all factors that control the binding, release, and accessibility of contaminants are important for enhancing the effectiveness of the bioremediation process. However, there is a dearth of information pertaining to the mechanisms of contaminant attachment and its "ageing" in the soil and the resulting effects on bioavailability and consequent toxicity to organisms. Examples of some long-term issues that need to be considered include:

Bioavailability and toxicity of treated residues in soil
Standardized clean-up assessment and toxicity testing protocols
Postbioremediation monitoring requirements to assess remobilisation of contaminants
Achievement of environmentally acceptable pollutant end points in soil

Alexander (5) has suggested several possible reasons for the lack of biodegradation of a pollutant that is biodegradable; among these, bioavailability is only one factor that is of paramount importance for bioremediation. These factors are:

The concentration of toxic substances may be so high at the site so as the microbial proliferation and metabolism is prevented.
One or more nutrients needed for microbial growth is limiting.
The substrate may be present in very low concentration, not enough for replication of the organisms containing the catabolic enzymes.
The substrate may not be in a form that is readily available for microorganisms.

Where these parameters are not limiting, bioavailability could be the rate determining factor for bioremediation of organically contaminated soils.

III. BIOAVAILABILITY AND TOXICITY

Contaminant toxicity to microorganisms has often been regarded as a direct measure of bioavailability (73). Toxicity tests in conjunction with chemical analyses and field surveys reveal the linkage between site contamination and its adverse ecological impacts (64). Also, to assess the efficacy of bioremediation, both chemical and toxicological assays are required. Thus, toxicity testing should be an integral part of the bioremediation program, since a reduction in toxicity is considered to be a very important and necessary characteristic of bioremediation process. Here the problem is which toxicity tests are suitable for use with soils that are naturally heterogeneous and may contain a mixture of organic and inorganic pollutants and a variety of living organisms both prokaryotic and eukaryotic. To date, there is no ideal test suitable for heterogeneous chemical mixtures in the soil, and there is a need for developing new assays or modifying existing ones to this end.

Toxicological assays also reveal information about the synergistic interactions of two or more contaminants present as mixtures in soil, which cannot be measured by chemical assays alone. Despite the importance of toxicological assays, only a few bioremediation studies have attempted to include such assays (21,51,89). Moreover, these were based on only one or two assays. Since no single species is consistently sensitive to all toxicants, it is important to include a range of assays on a variety of organisms. Thus a series of toxicological assays is required.

The ideal toxicological assay should be simple to use, rapid, economical, sensitive, easily reproducible, amenable to statistical analysis, predictable, ecologically relevant, and satisfactory from scientific and regulatory points of view. A variety of toxicity tests involving soil microorganisms, invertebrates, vertebrates, or plants may be used with soil samples. In general, these tests measure the toxicity by directly exposing the test organisms to the samples collected from the contaminated site, or indirectly, by exposing to eluates (filtered aqueous extract of the contaminant soil, which generally contains the water-soluble portion of the contaminants) or leachates from the site samples. Where bioremediation (and not phytoremediation) is the major issue, microbial bioassays are used to assess the resilience of microbes to the contaminants and the ultimate success of the remediation process. However, plant establishment, seedling germination, earthworms, etc., have also been used to assess contaminant toxicity in soils. While some of these tests may yield information that is useful for the remediation process, often the success of tests was assessed using freshly contaminated soils in which the bioavailability of the contaminant may be much higher than in long-term-contaminated soils. Despite these limitations, an attempt is made to summarize some of the toxicity tests that are currently being used by remediators (Table 1).

A. Microbial Bioassays

The physiological functioning of soil microorganisms forms the basis of microbial bioassays used to study the toxicity of contaminants. "Short-term" microbial assays are simple, rapid, and can be relatively cost-effective. The bacterial luminescence test involving *Photobacterium phosphoreum*, that is, the reduction of light emission by the bacterium due to the presence of toxicant, has been used as a measure of toxicity (1,14,72). Possible genotoxic effects induced by pollutants can be measured by using microbial-based mutagenic assays. *Salmonella typhimurium* strains TA98 and TA100 (Salmonella/microsome assay) and *Escherichia coli* PQ37 have been used for SOS Chromotest (43).

B. Bioluminescence Reporter System

Bioluminescence is a natural phenomenon associated with several organisms, mostly fireflies (*Photinus pyralis*) and the marine bacteria *Photobacterium* and *Vibrio*. The bacterial luciferase (*lux*) genes (bioluminescence genes) are well characterized in terms of their expression and regulation (61). The *lux* genes can be inserted into foreign DNA under the control of specific regulatory promoters and the bacterial activity can be monitored by measuring the light produced from such systems.

Recently, the bioluminescent bacterial reporter system has been linked to naphthalene degradation in *Pseudomonas fluorescens* HK44 (76). This system has been used to assess the bioavailability of naphthalene. When naphthalene is bioavailable, luminescence is produced and the amount of light emitted is directly proportional to the concentration of naphthalene in solution (42,76).

C. Microalgal and Cyanobacterial Growth Inhibition Tests

Photosynthetic microorganisms, such as microalgae and cyanobacteria, are ubiquitous in soils, and their contribution to soil carbon, nitrogen (particularly heterocystous cyanobacteria), and soil structure are well known. Since microalgae resemble higher plants in terms of their cellular organization (whereas cyanobacteria are prokaryotes), development of bioassays with these organisms are valuable indicators of toxic effects to fundamentally different cells. Differential sensitivity of these organisms to various pollutants (monocrotophos, quinalphos, nitrophenols, pentachlorophenol) has been demonstrated (56–59). These investigators have showed that two successive applications of quinalphos, even at close to field application rates, were significantly toxic to native algal populations in black vertisol under nonflooded conditions (57). In addition, a single application of quinalphos and two successive applications of monocrotophos were significantly toxic to algae and cyanobacteria under flooded conditions, as evidenced by a decrease in the population size and changes in the species composition (59). Increasing concentrations of priority pollutants (0.5–5.0 kg/ha) such as p-nitrophenol (PNP), m-nitrophenol (MNP), and 2,4-dinitrophenol (DNP) resulted in a gradual increase in toxicity to microalgae and cyanobacterial populations in black vertisol from cotton fields (58). Recently, Megharaj et al. (56) have demonstrated that an estimation of microalgae and cyanobacteria proved to be an important criterion for detection of pentachlorophenol pollution in a soil from a former timber processing facility. These investigators have further suggested that the alteration in species composition of algae can serve as a useful bioindicator of pollution. In addition, Miller et al. (63) have showed that algae are more sensitive for the measurement of toxicity than Microtox (*Photobacterium fischerii*), DO depletion rate, seed germination, and earthworms tests when the organisms were exposed to the water-soluble fraction of hazardous waste site-containing heavy metals and insecticides. Most algal bioassays are based on aquatic species, although soil organisms are more relevant and preferable.

D. Earthworm and Invertebrate Tests

Earthworms and other terrestrial invertebrates can serve as useful targets to examine the adverse effects of pollutants in toxicological assays and are relevant for soils because of their essential function in the soil ecosystem. Toxicity end points are growth, survival, reproductive success, and behavioral changes. Earthworms are useful to assess the acute toxic effects of the pollutants (66), and these tests can be used in a wide range of soil habitats containing a variety of contaminants. These tests are cost-effective and can be conducted rapidly (16,24,86).

E. Vertebrate Tests

Vertebrate toxicity tests are useful to assess acute, subacute, and chronic toxicity of pollutants by examining the growth, survival (e.g., amphibians such as frog), reproductive success, and body burdens of test organisms (e.g., amphibians, small mammals, and

Table 1 Toxicity Tests Currently Being Used to Assess the Toxicity of Contaminated Soil and Water

Test	Description of test	Duration	Advantages/disadvantages	Where used	References
Algae	Change in the species composition of algae is monitored. Soil treated with test chemical is incubated and algal populations determined by MPN method and qualitative changes in species composition is made on the positive soil dilutions.	7–21 days	Both quantitative and qualitative changes in algal populations can be observed at the same time. Alteration in species composition of algae is more sensitive than the estimation of total population density. However, development of resistant species by selection of persistent contaminant may cause alteration of species composition and at times may result in higher total population densities.	Toxicity of pesticides	58–61
Algae	Growth inhibition in liquid cultures using soil elutriates.	1 – <7 days	Aqueous elutriates of soil may underestimate the potential bioavailable fraction of sorbed compounds; nevertheless, it is found to be a sensitive assay.	Pesticides and heavy metals	63
Microtox	A marine bacterium, *Vibrio frischeri*, is added to polluted soil, incubated, and then extracted from soil. Light emission of extracted bacteria is quantified in buffer.	<0.5 h	The bacterium used in this test is not a soil organism. The extraction efficiencies of bacteria from soil varies. This test may be used as a screening tool.	Pesticide and heavy metals	63
Mutagenicity test	The number of revertants of the bacterial strain, *Salmonella typhimurium*, is determined on a histidine-deficient medium in the presence of test substance.	<7 days	Extraction of the test substance is crucial since solvent extracts have been shown to induce stronger response compared to the aqueous leachates. Mutagenicity is often detected in unpolluted soils as well (48). Ecological relevance of this method is hard to assess.	Organics	21,48

Test	Procedure	Duration	Comments	Contaminants	Refs
Earthworm	Mortality rate/LC_{50} of the earthworms is determined. *Eisenia fetida* is exposed to test substances mixed through an artificially composed medium (70% quartz sand, 20% Kaolin clay, 10% sphagnum peat, pH 6.0).	14 days	Earthworm tests have often been used for a variety of contaminants and are internationally accepted. Bioaccumulation of contaminants can be assessed by analyzing the soil samples and earthworms containing nonlethal concentration of the test chemical.	Metals, pesticides and various organics	16,24,66,88
Collembola, bioassay	Uses the same artificial medium as for earthworms. This test is based on the incubation of 10 juvenile *Collembola, Folsomia candida* sp. in a glass vessels containing 30 g of artificial soil and determining the number of animals and their offsprings.	28 days	This test is simple and very easy to carry out. The disadvantage of this test is that reproduction cannot be observed directly and cannot be distinguished from juvenile mortality.	Pesticides	44,65
Seedling emergence	Seeds are exposed to different concentrations of contaminant, mixed with the soil/artificial soil and maintained at 85% moisture holding capacity for specified time. The number of seeds that germinate are counted.	5 days	This test accounts for direct toxicity of soil rather than the water-soluble fraction.	Organics and inorganics	36
Root elongation	This test evaluates the effect of water soluble components of soil elutriates and is carried out in the absence of soil on a moistened filter paper.	5 days	This test is more sensitive compared to seed germination. Needs much caution when extrapolating to soil condition.	Organics and inorganics	36,90

birds). Feeding trials are especially useful for determining the entry of contaminants into the foodweb which cause adverse effects on animal and human health (8–10,74).

F. Plant Tests

Plant toxicity tests include seed germination tests, seedling growth tests, root elongation tests, and chlorophyll fluorescence assays (62,90). These tests are relatively inexpensive, simple, and can easily be applied in both laboratory and field conditions. The assays reveal the effect of pollutants on different stages of plant development, since these stages of plant development are influenced differently by pollutants. For example, seed germination is less sensitive than young roots and seedling growth. Hence any inhibition of seed germination can reveal a potential adverse effect of pollutants at later stages of the plant development. However, extrapolation of laboratory test results to field conditions needs much caution, and environmental conditions (temperature, moisture etc.) should be taken into account. More details about these tests can be obtained from Refs. 15 and 36 (earthworm survival tests); Refs. 53 and 84 (amphibian tests); and Refs. 52, 83, and 84 (seed germination tests).

The main advantage of using bioassays is that they provide a direct indication of the toxicity of a particular soil, integrating the effect of all the contaminants present in it. The disadvantage of these assays is that the specific pollutant causing the toxic effect cannot be identified. Nevertheless, these bioassays are highly useful in estimating the bioavailability of pollutants to the organisms.

G. Uptake of Contaminants by Soil Organisms

The uptake of contaminants by soil organisms depends primarily on soil and environmental properties, characteristics of the contaminant, and the physiology of the organism. The uptake of the contaminant relates to its bioavailability, i.e., the fraction of the contaminant that may be taken up, which varies widely among soil organisms. In relation to the availability to the organism, the amount of contaminant present can be divided into three pools:

> The available portion that can be taken up directly
> The portion that cannot be taken up directly but is in kinetic equilibrium with the available pool
> The portion that cannot be taken up even in the long term

Often the contaminant that is directly available is defined as the intensity factor, and the portion that may be available in the long term as the quantity factor. It is the ratio of the intensity to quantity factor (K_d) that determines the rate of replenishment of the available portion:

> Intensity $\Leftrightarrow$ Quantity
>
> $$K_d = \frac{Q}{I}$$

By manipulating the properties of the soil, K_d and hence the intensity factor may be altered to suit the bioremediation process. Although biological availability of the contaminant can be determined by using various extraction procedures (aqueous, hydrogen peroxide, and solvents), it appears to be difficult to explain bioavailability based solely on this.

Often the extractants used to assess bioavailability alter the composition of soil solution and the surface chemical properties of soils such that they no longer reflect the characteristics typical of that under field condition. Moreover, biological activity can cause speciation of the contaminant, thereby changing both its bioavailability and toxicity. For example, earthworms can have an impact on the speciation of metals due to their actions in improving aeration in soil. Likewise, plant root exudates can influence the contaminant present in the vicinity of roots. Thus bioavailability is a kinetic process and depends on uptake processes, not just on the concentration in the available pool. It has to be viewed with time.

The uptake efficiency of a contaminant varies among different organisms and species. Literature on the uptake of organic contaminants by terrestrial organisms is limited compared to the data on heavy metals. Few available reports indicate the accumulation of persistent organochlorines such as DDT and chlorophenols by earthworms. Polycyclic aromatic hydrocarbons (PAHs) were taken up by invertebrates such as isopods and earthworms sampled from a forest close to a blast furnace plant (85). These investigators found that the bioconcentration factor (BCF) for isopods decreased with log K_{ow} of PAH molecules, whereas in the case of earthworms it was independent of the polarity of PAH molecules. This suggests the presence of a substance-dependent metabolism of PAHs in isopods but not in earthworms (85). Fuchs et al. (32) noticed the accumulation of high levels of DDT in earthworms from apple and pear orchards, which was attributed to the previous use of this pesticide. When two populations of earthworms, *Aporrectodea tuberculata* (Eisen), one from chlorophenol-contaminated and the other from uncontaminated soils, were exposed to acute, toxic, and sublethal concentrations of pentachlorophenol, both populations showed similar uptake capabilities (39).

For several soil organisms, the available pool can be related to the contaminant concentration in the soil pore water. This has been demonstrated in the case of earthworms (86). The distribution of contaminant can be viewed as a consequence of the equilibrium partitioning process between soil, pore water, and the organism. For organic chemicals, often this can be explained based on the sorption coefficient K_d (a function of organic matter content and the contaminant's hydrophobicity). The concentration of the contaminant in the organism (bioconcentration factor, BCF) is expressed relative to the contaminant concentration in the soil, which may be affected by the amount and nature of clay, organic matter, iron, aluminum hydroxides, pH, redox potential, and cation-exchange capacity.

IV. FACTORS AFFECTING MASS TRANSFER

The microbial degradation of organic pollutants is dependent on the rate of uptake and metabolism of the contaminants by the cells and by the rate of transfer of the contaminant to the cell. The bioavailability of a contaminant can be determined by the rate of mass transfer relative to the intrinsic activity of the microbial cells (12). As discussed above, often bioremediation failures, both in situ and ex situ, can be linked to poor bioavailability of the target compound. The limited availability of the contaminant for microbial degradation may be associated with processes such as multiphase partitioning, speciation, rate of desorption, and mass transfer, as well as pollutant and matrix properties and characteristics of the microorganisms. Singly or in combination, these factors can restrict the contaminants' bioavailability, which in turn limits degradation rates and the extent of, bioremediation.

Numerous researchers have demonstrated that the bioavailability of organic compounds can appreciably affect the bioremediation of the contaminated matrix. Weissenfels et al. (91) showed that PAH contaminants in coking plant soil were unable to be degraded even after inoculation of the soil with bacteria known to be effective in degrading PAHs. However, rapid degradation of PAHs was observed after the PAHs were extracted from the coking plant soil and reapplied into the extracted soil material. Erickson et al. (29) observed similar results when investigating the bioremediation of PAH-contaminated soil from a manufacturing gas plant. Incubation of soil with the indigenous microflora (for three months), at different temperatures, soil moistures, or nutrient conditions, did not result in a reduction in PAH concentration. Augmentation of soils with organisms known to be capable of degrading PAHs did not stimulate PAH loss in the manufacturing gas plant soil. Erickson et al. (29) concluded that the failure to observe PAH loss in any of the soils may be due to the toxicity of the soil or the poor availability of the PAHs to the soil microorganisms. However, when naphthalene or phenanthrene was spiked into augmented or nonaugmented soils, rapid degradation of these compounds occurred. It also appeared that background naphthelene and phenanthrene from the contaminated soil remained undegraded at the conclusion of the incubation period. These results led to the conclusion that the soils were not toxic to the indigenous or augmented microorganisms, but the PAHs within the soil were bound in such a way that made them unavailable for degradation.

Similar results have been reported for soils contaminated with chlorophenols (75) and sediments contaminated with α-hexachlorocyclohexane (11). These observations indicate that the residence time of contaminants in soils or sediments can influence the bioavailability of the compounds. The decrease in bioavailability, as a function of time, is often referred to as ageing (Fig. 2) or weathering and may occur due to the compounds' incorporation into organic matter by chemical oxidations, diffusion into small soil pores, and absorption onto organic matter, or by sequestering into nonaqueous-phase liquids

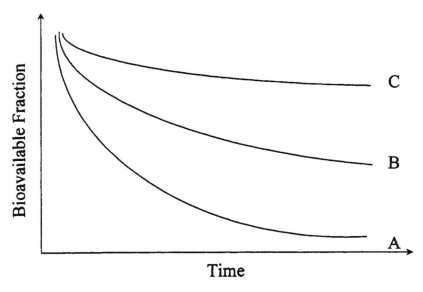

Figure 2 Effect of time on contaminant bioavailability in soils with varying organic matter content that increases in the order A>B>C.

(NAPLs) with high resistance toward NAPL–water mass transfer. The term "ageing" does not include reactions that alter the structure of the molecule, such as polymerization or covalent binding to humic substances (41). The following sections of this chapter will discuss the factors that limit the bioavailability of organic contaminants to microbial degradation.

V. EFFECT OF SORPTION AND MATRIX COMPOSITION

The sorption of hydrophobic contaminants to natural solids affects the transport, degradation, and biological activity of organic compounds in the environment (68). Studies have shown that the microbiol mineralization of high-molecular-weight PAHs (compounds containing four or more fused benzene rings) decreases with increasing residence time in soils (29,91). This has been attributed to association of the compounds to soil organic matter (SOM) (41). The rate and extent of PAH degradation may be reduced by sorption and partitioning processes due to the slowing of PAH desorption from SOM into the soil aqueous phase (38). As most evidence indicates that biotransformation of PAHs occurs in the aqueous phase (94), sorption or partitioning of compounds to humic material or the clay fraction of the soil reduces the concentration of the compound available in the aqueous phase (13,93) and ultimately reduces the degradation rate. Complete degradation of PAHs may be inhibited by nonsorptive processes by slowing desorption of PAH transformation products (38).

Although the nature of sorptive processes is not fully understood, a number of mechanisms such as through van der Waals forces (disperson/London forces) and chemical binding, cation bridging, hydrogen bonding, ion exchange, covalent bonding, or ligand exchange have been identified (46,50). A number of factors influence sorption of organic compounds. These include the type and concentration of solutes in the surrounding solution (2) the type and quantity of clay minerals, the amount and nature of organic matter in the soil, pH, temperature, and the specific compound involved. Many of the major processes of concern to sorption take place at the surfaces of the clay minerals and humic materials (3). The extent of contaminant retention is also directly correlated with the octanol–water partition coefficients, expressed as K_{ow}, and the percentage of organic carbon in the soils: the more organic matter is present in the solid phase, the more hydrophobic compounds sorb (23,86). However, given that the nature of organic matter is so diverse, it would not be reasonable to consider that organic matter uniformly adsorbs the contaminant and hence the nature of organic matter is more valuable to accurately predict the absorption characteristic of a particular compound (23).

Ortega-Calvo et al. (67) demonstrated that phenanthrene bioavailability in soils was affected by its sorption to organic matter and clays. In addition, the bioavailability of phenanthrene was also affected by interactions of phenanthrene with humic fractions–clay complexes. Although the indigenous microbiol population was capable of mineralizing [9-^{14}C]phenanthrene, a reduction in phenanthrene mineralization was observed in soils containing higher organic matter contents: after 120 days incubation, 19% of phenanthrene was mineralized in soils containing 0.8% organic matter, compared to 5.1% in soils containing 36.9% organic matter. Phenanthrene mineralization was also found to decrease with increasing clay contents. In soils containing 5.6% clay, 11.4% of phenanthrene was mineralized after 120 days, compared to 1.3% in soils containing 22.4% clay. Ortega-Calvo et al. (67) explained the differences in phenanthrene mineralization in terms of sorption of the compound to the soil organic matter. Similar results have been observed by other researchers (55,91).

In addition to sorption, tortuosity influences the bioavailability of soil-borne contaminants. Tortuosity refers to the diffusion path of molecules due to the presence of other particles. In soils containing high clay contents, tortuosity as well as sorption may contribute to decreased bioavailability. As microorganisms are barely mobile in soil, acquisition of contaminants by microorganisms is dependent on the flow of nutrients through the soil. With increasing clay contents, tortuosity increases, resulting in reduced flow and nutrient/contaminant delivery. Ortega-Calvo et al. (67) attributed the differences in the mineralization of phenanthrene in soils, similar in organic matter but different clay contents, by the increase in tortuosity caused by higher clay contents. Phenanthrene mineralization was approximately ninefold greater after 120 days in soils containing 5.6% clay compared to those containing 22.4% clay. Presumably, accessibility of the contaminant to the microorganisms caused the reduction in phenanthrene biodegradation.

In addition, soil structure and composition, e.g., grain size, porosity, aggregation, surface chemistry, mineralogy, etc., will influence sorption characteristics and the bioavailability of contaminants (4). Microbiol access to a contaminant may be restricted due to the soil pore size. Most soil microorganisms range in size from 0.5 to 0.8 μm and occupy pores with a mean diameter of 2 μm. Over 50% of the total pore volume of a soil may consist of pores with diameters of less than 2 μm, and therefore up to 50% of the soil pores are not accessible to microorganisms (40). As such, the accumulation of contaminants in fissures and cavities renders the compounds inaccessible to microorganisms. Microbial accessibility to a contaminant may be increased by physical dispersion of the soil matrix. Rasiah et al. (70) increased the degradation of oily waste organics in soils from land treatment farms by ultrasonic treatment. Soils were ultrasonically dispersed at increasing energy levels (0–30 kJ/kg), incubated for up to 12 weeks, and carbon mineralization rates (CMR) determined at weekly intervals. Rasiah et al. (70) observed that CMR from the dispersed soils increased with increasing energy levels: the CMR from the oily waste treated soil dispersed at 30 kJ/kg was 710% higher compared to that from the untreated soil. Significant increases in CMR with increasing energy levels indicated that physical dispersion gradually increased the availability of the oily waste organics for biodegradation.

Along with the factors reported above, numerous investigators have reported enhanced degradation of organic contaminants by composting contaminated soils with a suitable substrate such as green waste (20,71). Although composting has been reported to enhance biodegradation, the underlying reasons for the enhanced bioremediation of the contaminants is not apparent from the studies reported in the literature. In order to assess the effect of composts on DDT-contaminated soils, we incubated soils with a range of composing material. After 40 days of anaerobic incubation we measured both DDT and its metabolites and the dissolved organic carbon (DOC) content of the soil aqueous phase. The results in Fig. 3 show that increasing the rate of application of composting material increases the DOC content of the soil aqueous phase, and this also corresponds with increased DDT (DDT+DDD+DDE) concentration. Although no effort was made to estimate the changes in microbial population in the incubated soils, our preliminary study suggests that one of the factors enhancing bioremediation of DDT was presumably DOC-induced DDT release in the contaminated soils. Further detailed study on the effect of composting and the nature of composting material on soil solution chemistry and DDT compounds is in progress in our laboratory.

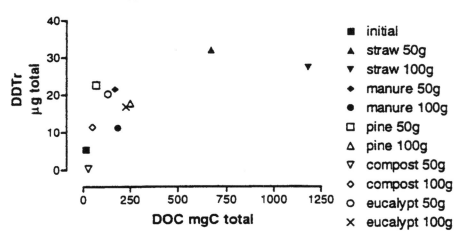

Figure 3 Relationship between dissolved organic carbon and DDT in soil solution in soils incubated with varying amounts of different composting material under oxygen limiting conditions. (From Naidu and Smith, unpublished.)

VI. MULTIPHASE PARTITIONING

The availability of a contaminant to microbial attack may be affected by the presence of nonaqueous-phase liquids (26,34,69). Hydrophobic compounds have been shown to be sequestered into NAPLs, and this can affect the rate of contaminant degradation (26). A decrease in the rate of degradation may occur due to a reduction in the concentration of the compound in the aqueous phase or by the slow partitioning of the compound from the NAPL into the aqueous phase. At low concentrations of phenanthrene [0.6–20 µg/mL of di-2-ethylhexyl phthalate (DEHP)], Efroymson and Alexander (27), observed that the concentration of phenanthrene in the aqueous phase at equilibrium was less than 1 ng/mL. Mineralization of phenanthrene by a phenanthrene-enriched mixed culture was limited due to the availability of the growth substrate (phenanthrene). The existence of a threshold concentration below which little or no degradation occurs has been observed by other researchers.

The presence of NAPLs may also affect the rate of contaminant degradation through preferential utilization of the NAPL by the microorganisms, through toxic effects exerted by the NAPL, or through depletion of essential nutrients in the NAPL region. Nutrient limitation may suppress biodegradation of its constituents. In addition, the composition of the NAPL can influence the rate of partitioning to the aqueous phase. For example, it has been shown that phenanthrene partitions more rapidly and to a greater extent from heptamethylnonane to water than from DEHP (27).

VII. CHARACTERISTICS OF THE MICROORGANISMS

Catabolic enzymes or microbial transporters involved in the degradation of organic contaminants may be regulated by the concentration of the contaminant. In some instances, enzymes are only synthesized in response to a contaminant concentration above a threshold level. Spain and van Veld (79) tested the adaptation period of microorganisms in eco-cores on the Escambia River near Pensacola, Florida, to *p*-nitrophenol. Results obtained with different concentrations of *p*-nitrophenol indicated that there was a concentration threshold of *p*-nitrophenol (10 mg/mL) below which there was no detectable adaptation of the community. Rebber (71) demonstrated that enzymes for the catabolism of 3- and 4-chlorobenzoate by *Acinetobacter calcoaceticus* were induced at substrate concentrations above 1 µM. In addition, microorganisms may need a continuous flux of substrate to keep enzyme synthesis turned on. Microbial degradation of a contaminant may proceed only if suitable organisms, with the appropriate metabolic capabilities, are present. Significant degradation will occur only if these organisms are present in relatively high numbers. Growth of the specific degraders will be favored in areas close to the contamination, where the flux of substrate is sufficient to provide a nutritional advantage (40). This is affected by the spatial distribution of the microorganisms. As soil microorganisms occur mainly in the attached state and in many cases the pollutant is nonrandomly distributed, substrate availability may be limited, increasing acclimation periods and slowing microbial growth. Where microorganisms are in close proximity to the contaminant, the diffusion of the contaminant may be driven by the affinity of the microorganisms by reducing the local concentration of the substrate.

Contaminant availability to microorganisms with the appropriate metabolic capabilities may result in growth of the organisms. Growth will continue until the microbial population reaches a density at which substrate utilization fulfills only its maintenance needs. Theoretical considerations have shown that a drop in the substrate flux below the maintenance rate will result in a threshold concentration below which the substrate is not degraded further (40).

Adaptation of the organisms to contaminated environment may also play an important role in survival of that organism. For example, a bacterial strain, *Sphingomonas* sp RW1, which survived poorly in a dioxin/dibenzofuran-contaminated soil, survived well and degraded the compounds upon its adaptation to that soil (60).

VII. CURRENT APPROACHES TO OVERCOME BIOAVAILABILITY PROBLEMS

The central role of bioavailability in the biological remediation of organically contaminated soils is well evident from the approach that many researchers have adopted to enhance the bioremediation process. Central to this approach is the inclusion of steps that enhance the bioremediation process through increased bioavailability of the contaminant. Given that most contaminated sites consist of a heterogeneous mixture of inorganic and organic contaminants, clean-up of these sites may involve bioremediation through enhanced bioavailability of organic contaminants followed by immobilization of toxic metal contaminants. Some of the processes used to enhance bioavailability are briefly discussed in the following sections.

A. Surfactants

The application of surfactants to contaminated soils has been used as a treatment strategy for increasing the mass transfer of hydrophobic organic contaminants (5,22,33). Surfactants are amphiphilic molecules that contain polar and nonpolar structural moieties

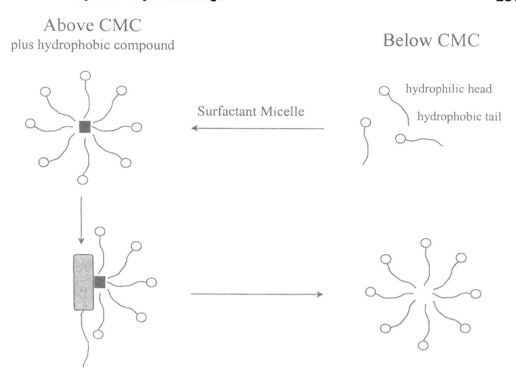

Figure 4 Schematic representation of the solubilization of the hydrophobic compound and its enhanced availability to microorganisms in the soil solution (Adapted from Ref. 37.)

(92). They tend to concentrate at surfaces and interfaces, thereby decreasing surface and interfacial tension. At low concentrations, surfactant molecules exist in monomeric forms, but when a specific thresholds (known as the critical micelle concentration, CMC) is exceeded, micelle aggregates form in the water phase, similar to the soap/detergent molecule (Fig. 4). Solubilization of hydrophobic contaminants is attributed to incorporation of the molecule into the hydrophobic core of micelles in solution (Fig. 4) (37). Surfactant solubilization of hydrophobic contaminants generally starts at the CMC and then is a linear function of surfactant concentration (Fig. 5). In an aqueous system, the extent to which a solute will concentrate in a micelle can be related to the octanol–water partitioning coefficient (K_{ow}) of the solute: the larger the K_{ow}, the greater is its affinity to concentrate in the micelle (92). Edwards et al. (25) observed that the solubilization of naphthalene, phenanthrene, and pyrene in four nonionic surfactants correlated with PAH octanol–water partitioning coefficients. In addition, solubilization of PAHs increased linearly with surfactant dose at aqueous concentrations above the CMC. In soil applications, surfactants can enhance the rate of mass transfer from solid and sorbed phases by increasing the rates of dissolution and desorption of the compounds.

Surfactants have been shown to enhance desorption and solubilisation of PAHs with appreciable desorption in excess of the CMC (93). Thibault et al. (80) demonstrated that Witconol SN70, a nonionic surfactant, was effective in partitioning pyrene into the

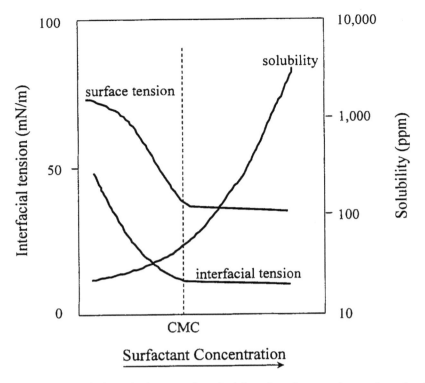

Figure 5 Relationship between interfacial and surface tension and contaminant solubilization with increasing surfactant concentrations.

hydrophobic core of the micelle. The enhanced solubilization of pyrene resulted in increased pyrene degradation rates by the inoculated pyrene-degrading microorganisms. Phenanthrene degradation rates in mineral soils by indigenous microorganisms were also enhanced in the presence of the nonionic alcohol ethoxylate surfactant Alfonic 810-60 (5). Nearly 50% of the added phenanthrene was mineralized after 495 h in the presence of the surfactant, compared to only 4.8% in the absence of the surfactant. Tiehm et al. (82) also observed surfactant-enhanced mobilization and biodegradation of PAHs in manufacturing gas plant soils.

Some researchers have also observed water solubility and degradation enhancements of organic compounds by some surfactants below the CMC (5,6,47). Aronstein et al. (15) observed enhanced rates of degradation of naphthalene and phenanthrene in the presence of some nonionic surfactants at applications below their CMC. Significant solubility enhancements of DDT in Triton and Brij 35 surfactants were observed by Kile and Chiou (47) below their CMC. Kile and Chiou (47) attributed this phenomenon to a partitioning-like interaction with the nonpolar content of the dilute surfactant, which is similar to the enhancement effect caused by dissolved humic materials (18,19). Furthermore, Aronstein and Alexander (7) suggested that the use of low concentrations of surfactants may be useful for in-situ bioremediation of sites contaminated with hydrophobic pollutants and also in preventing the movement of parent contaminants to groundwater.

Although enhanced degradation of organic contaminants has been demonstrated by the addition of surfactants to contaminated soils, some inhibitory effects have also been observed. For example, Graves and Leavitt (35) found that a surfactant amendment (a blend of ethoxylated fatty acids, Adsee, 799) did not enhance the biodegradation of petroleum hydrocarbons; instead, the surfactant served as a primary carbon source to the soil microorganisms. Some surfactants may exhibit toxic effects to microorganisms at concentrations above their CMC through the detergent-induced disruption of microbial cell membranes. For instance, Volkering et al. (88) noticed the detrimental effects of nonionic surfactants to bacteria. Kawai et al. (45) also noted the growth inhibition by Triton X-100 of four strains of bacteria while investigating the biodegradation of polypropylene glycol. However, it has been recently reported that nonionic surfactants with a high hydrophilicity obtained by a long ethoxylate chain were nontoxic to PAH-degrading bacteria (81). In the study by Tiehm (81), the effect of nine surfactants on the degradation of polycyclic aromatic hydrocarbons showed that some surfactants were toxic to microorganisms, some served as preferential carbon sources, while others enhanced the degradation of model aromatic compounds.

Although surfactants in general aid in increasing the bioavailability of hydrophobic contaminants, a careful consideration of their use is necessary. The use of nonionic surfactants above their critical micelle concentration may inhibit the biodegradation process due to extensive micellation resulting in a decrease in equilibrium of free-phase (bioavailable) aromatic hydrocarbon (78). This may result in the further contamination of soil or ground water. In addition, the application of a surfactant may result in the solubilization of the compound in the surfactant micelle; however, a proportion of the partitioned compounds may not be directly available for microbial degradation. Hence, it is important to choose the appropriate surfactant and concentration for the remediation strategy. Some of the limiting factors in the use of surfactants to increase contaminant bioavailability include the cost of the surfactants and the application of the surfactant to the soil system. In addition, increasing the bioavailability of a contaminant above the "threshold" concentration for microorganisms may retard the bioremediation process due to toxic effects of the contaminant. However, limited research has addressed this issue.

B. Co-solvents

Water-miscible polar solvents can significantly increase the solubility of hydrophobic compounds present as solutes within tars or compounds sorbed onto soil (54). The co-solvency power is proportional to the logarithm of the solute's octanol–water partitioning coefficient. As with surfactants, the larger the K_{ow} of a compound, the greater is the enhancement of solubility in a particular co-solvent.

Fan and Jafvert (30) also observed that for solute–solvent–co-solvent systems (i.e., PAH–water–alcohol), the solubility of PAHs (phenanthrene, pyrene, and perylene) increased with increasing alcohol volume fractions. However, the proportion of co-solvent added will influence microbial activity (i.e., high concentrations of co-solvent may inhibit microorganisms) and also the economics of the remediation strategy.

Enhanced bioremediation of phenanthrene has been observed in laboratory-scale biphasic culture systems. Kohler et al. (49) reported that the rate of phenanthrene degradation by *Pseudomonas aeruginosa* AK1 in the presence of 2,2,4,4,6,8,8-heptamethylnonane (HMN) was governed mainly by the HMN/water interface area, whereas with crystalline phenanthrene, the rate of degradation was correlated directly to the particle surface area. Field et al. (31) observed that when acetone was added as the co-solvent to increase the solubility of anthracene, each 10% increment in solvent concentration correlated to a 6.4-fold increase in the soluble anthracene concentration. However, acetone concentrations of 45–58% (v/v) reduced manganese-dependent peroxidase activities of *Beijernickia* strain BOS55 and *Phanerochaete chrysosporium* to 20% of their normal activity, and lignin peroxidase activity was two to three times less tolerant compared to manganese peroxidase (31).

C. Thermal Enhancements

Currently, thermal enhancements, i.e., soil heating, have been used as a remediation technology for the removal of some organic contaminants from soil. A number of heating processes may be used, such as radiofrequency and electrical resistance heating, steam injection, or hot-water injection. Availability of the contaminant is increased by an increase in the contaminant's vapor pressure and diffusivity, an increase in the permeability of the soil with the release of water vapor and contaminant, and a decrease in the viscosity of the contaminant, which improves mobility. Radiofrequency, electrical resistance, or steam-injection heating systems are not applicable for a bioremediation strategy, as soil temperatures may reach 100°C or above, therefore destroying microorganisms. The injection of moderately hot water (50°C) into a contaminant zone may reduce the viscosity of contaminants, increase the solubility of free-phase organics, and result in enhanced contaminant degradation by microbial processes. In many cases, residues collected from thermal treatments (i.e., pump-and-treat processes) may be treated by bioremediation.

D. Hydraulic/Pneumatic Fracturing

Hydraulic and pneumatic fracturing are enhancement technologies designed to increase the effectiveness of many in-situ remediation and extraction processes by increasing the contact between contaminants adsorbed onto soil particles and microorganisms or extraction medium. Pressurized water (hydraulic fracturing) or air (pneumatic fracturing) is injected through wells to develop cracks in low-permeability soils and sediments. The enhanced fracturing increases the permeability of the soil to liquids and vapors which may result in increased contaminant removal by vapor extraction, bioremediation, or thermal treatment. Venkatraman et al. (87) demonstrated the effectiveness of integrating pneumatic fracturing and in-situ bioremediation for the treatment of gasoline-contaminated, low-permeability soil. A pneumatic fracturing system was used to increase the availability of the contaminants, enhance air flow and transport rates, as well as to deliver nutrient directly to the indigenous microbial population. Over a 50-week period, 79% of soil-phase benzene, toluene, and xylenes were removed and, based on mass balance calculation, 85% of the total reduction was attributed to microbial degradation.

E. Electrokinetics

Heavy metals, radionuclide, and organic contaminants may be separated from soil, sludges, and sediments by electrokinetics. The technology involves the supply of a low, intense, direct current across a pair of electrodes that have been placed on either side of the contaminated matrix mass. The electrical current causes electroosmosis and ion migration, which results in desorption of the contaminant from the soil surface and transportation to the corresponding electrode depending on the charge of the contaminant. Surfactants or complexing agents may be added to increase the solubility and movement of the contaminant. The volumetric rate of transport of the contaminant may be increased 50–60 times (17). Electrokinetics may also be used to supply nutrients for bioremediation. Elektorowicz and Boeva (28) demonstrated the effectiveness of this technology for delivering uniform nitrogen concentrations in fine soils and at depth.

IX. CONCLUSION

Although bioremediation of organically contaminated sites is currently regarded as one of the most successful remediation strategies, the process is often limited by the fraction of contaminant available for degradation. It is not clear, however, whether the limiting factor is the lack of the availability of contaminant per se or the toxic effect of contaminant on the microbes or a combination of both. While some effort has been directed toward assessing the bioavailable fraction in soils, the lack of appropriate methods for assessing the dynamic nature of the processes controlling in-situ bioavailability has limited the applicability of bioremediation technique to field remediation of a wide range of organic contaminants.

The availability of a contaminant in soil is not a function of its measured concentration but depends on other factors such as the soil physicochemical characteristics, nature of the contaminant, duration of contact (i.e., ageing), and nature of organisms. It is surprising to note, however, that while many of these properties have been manipulated to optimize the bioremediation process, the accessibility of contaminants to microbes and soil amendments has received little attention. Given that the amount of clay, organic matter, and silt and sand have all been shown to influence biodegradation of contaminants, it may seem prudent to access whether tortuosity could be one of the key factors controlling contaminant bioavailability in soils.

Although research conducted in recent years has contributed substantially to our understanding of the processes that result in sequestration and release of contaminants within the soil, the effect of treatments (e.g., surfactants, etc.) on contaminant availability, mobility, and soil toxicity, further detailed research is needed to better understand the above processes.

Along with the issues raised above, it is imperative that future research on bioremediation consider the following:

- Development of inexpensive, rapid, and simple soil-based toxicity assays. Although a range of bioassays for toxicity testing of contaminated soils is available, toxicity tests suitable for terrestrial environments are still underdeveloped compared to aquatic toxicity tests. There is an urgent need to develop standardized protocols based on soil organisms for terrestrial ecotoxicity testing. For example, for toxicity tests in which artificial soil is used, it is important to devise the test so that it reflects the organic matter and clay contents of real soil. Such toxicity tests are important as they reflect the tolerance of microbes to contaminants.
- Processes involved in contaminant sequestration and its ageing in soil, its availability, and ecotoxicity need to be studied in order to establish the major factors affecting contaminant bioavailability.

Distinguish between true and false degradation/remediation of contaminant, i.e., develop methods to validate the data obtained in biodegradation experiments to show that disappearance of contaminant is due to microbial degradation and not due to volatilization or any physical process and there is reduction in the toxicity.

As bioremediation studies progress from laboratory to field scale, the question of low bioavailability assumes utmost importance and hence, in order to develop efficient bioremediation technologies understanding the concept of bioavailability and the ways and means to enhance the bioavailability of contaminants is crucial. Therefore further research is warranted in this direction.

REFERENCES

1. Ahn, B. K., and Morrison, G. 1991. Soil toxicity screening test with *Photobacterium* activity in environmental site assessment. EcoTech, Inc., Irving, TX.
2. Alexander, M. 1985. Biodegradation of organic chemicals. *Environ. Sci. Technol.*, 18:106–111.
3. Alexander, M. 1994. In: *Biodegradation and Bioremediation.* San Diego, CA: Academic Press.
4. Apitz, S. E., and Meyers-Schulte, K. J. 1996. Effects of substrate mineralogy on the biodegradability of fuel components. *Environ. Toxicol. Chem.*, 15:1883–1893.
5. Aronstein, B. N., Calvillo, Y. M., and Alexander, M. 1991. Effects of surfactants at low concentrations on the desorption and biodegradation of sorbed aromatic compounds in soil. *Environ. Sci. Technol.*, 25:1728–1731.
6. Aronstein, B. N., and Alexander, M. 1992. Surfactants at low concentrations stimulate biodegradation of sorbed hydrocarbons in samples of aquifer sands and soil slurries. *Environ. Toxicol. Chem.*, 11:1227–1233.
7. Aronstein, B. N., and Alexander, M. 1993. Effect of a non-ionic surfactant added to the soil surface on the biodegradation of aromatic hydrocarbons within the soil. *Appl. Microbiol. Biotechnol.*, 39:386–390.
8. American Society for Testing and Materials. 1992. Standard test method for estimating acute oral toxicity in rats, ASTM E1163. In: *Annual Book of ASTM Standards: Pesticides; Resource Recovery; Hazardous Substances and Oil Spill Responses; Waste Management; Biological Effects*, vol. 11.04, pp. 770–775. Philadelphia, PA: ASTM.
9. American Society for Testing and Materials. 1992. Standard guide for conducting the frog embryo teratogenesis assay—*Xenopus* (Fetax), ASTM E1439. In: *Annual Book of ASTM Standards; Waste Management; Biological Effects*, vol. 1 11.04, pp. 1199–1209. Philadelphia, PA: ASTM.
10. American Society for Testing and Materials. 1992. Standard test method for efficacy of acute mammalian predacides, ASTM E552. In: *Annual Book of ASTM Standards: Pesticides; Resource Recovery; Hazardous Substances and Oil Spill Responses; Waste Management; Biological Effects*, vol. 11.04, pp. 235–238. Philadelphia, PA: ASTM.
11. Beurskens, J. E. M., Dekker, C. G. C., Jonkhoff, J., and Pompstra, L. 1993. Microbial dechlorination of hexachlorobenzene in a sedimentation area of the Rhine river. *Biogeochemistry*, 19:61–81.
12. Bosma, T. N. P., Middeldorp, P. J. M., Schraa, G., and Zehnder, A. J. B. 1997. Mass transfer limitation of biotransformation: Quantifying bioavailability. *Environ. Sci. Technol.*, 31:248–252.
13. Bouwer, E., Durant, N., Wilson, L., Zhang, W., and Cunningham, A. 1994. Degradation of xenobiotic compounds in situ: Capabilities and limitations. *FEMS Microbiol. Rev.*, 15:307–317.

14. Bulich, A. A. 1986. Bioluminiscence assays. In: G. Bitton and B. J. Dutka (eds.), *Toxicity Testing Using Microorganisms*, vol. 1, pp. 57–74. Boca Raton, FL: CRC Press.
15. Callahan, C. A., Menzie, C. A., Birmester, D. E., Wilborn, D. C., and Ernst, T. 1991. On-site methods for assessing chemical impact on the soil environment using earthworms: A case study at the Baird and McGuire Superfund site, Holbrook, MA. *Environ. Toxicol. Chem.*, 10:817–826.
16. Callahan, C. A., Russell, L. K., and Peterson, S. H. 1985. A comparison of three earthworm bioassay procedures for the assessment of environmental samples containing hazardous wastes. *Biol. Fertil. Soils*, 1:195–200.
17. Chilingar, G. V., Loo, W. W., Khilyuk, L. F., and Katz, S. A. 1997. Electrobioremediation of soils contaminated with hydrocarbons and metals: Progress report. *Energy Sources*, 19:129–146.
18. Chiou, C. T., Kile, D. E., Brinton, T. I., Malcolm, R. L., and Leenheer, J. A. 1987. A comparison of water solubility enhancements of organic solutes by aquatic humic materials and commercial humic acids. *Environ. Sci. Technol.*, 21:1231–1234.
19. Chiou, C. T., Malcolm, R. L., Brinton, T. I., and Kile, D. E. 1986. Water solubility enhancement of some organic pollutants and pesticides by dissolved humic and fulvic acids. *Environ. Sci. Technol.*, 20:502–508.
20. Crawford, S. L., Johnson, G. E., and Goetz, F. E. 1993. The potential for bioremediation of soils containing PAHs by composing. *Compost Sci. Utilisation*, 1:41–47.
21. Donnelly, K. C., Brown, K. W., Anderson, C. S., Thomas, J. C., and Scott, B. R. 1991. Bacterial mutagenicity and acute toxicity of solvent and aqueous extracts of soil samples from an abandoned chemical manufacturing site. *Environ. Toxicol. Chem.*, 10:1123–1131.
22. Ducreux, J., Baviere, M., Seabra, P., Razakarisoa, O., Shafer, G., and Arnaud, C. 1995. Surfactant-aided recovery/in situ bioremediation for oil-contaminated sites. In: R. E. Hinchee, J. A. Kittel, and H. J. Reisinger (eds.), *Applied Bioremediation of Petroleum Hydrocarbon* pp. 435–443. Columbus, OH: Battelle Press.
23. Dzombrak, D. A., and Luthy, R. G. 1984. Estimating adsorption of polycyclic aromatic hydrocarbons in soils. *Soil Sci.*, 137:292–308.
24. Edwards C. A. 1983. Report of the second stage in development of a standardised laboratory method for assessing the toxicity of chemical substances to earthworms. Report to the Commission of the European Communities, Rothamsted Experimental Station, Harpendon Herts. England.
25. Edwards, D. A. Luthy, R. G., and Liu, Z. 1991. Solubilisation of polycyclic aromatic hydrocarbons in micellar nonionic surfactant solutions. *Environ. Sci. Technol.*, 25:127–133.
26. Efroymson, R. A., and Alexander, A. 1994. Role of partitioning in biodegradation of phenanthrene dissolved in nonaqueous-phase liquids. *Environ. Sci. Technol.*, 28:1172–1179.
27. Efroymson, R. A., and Alexander, M. 1995. Reduced mineralisation of low concentrations of phenanthrene because of sequestering in nonaqueous-phase liquids. *Environ. Sci. Technol.*, 29:515–521.
28. Elektorowicz, M., and Boeva, V. 1996. Electrokinetic supply of nutrients in soil bioremediation. *Environ. Technol.*, 17:1339–1349.
29. Erickson, D. C., Loehr, R. C., and Neuhauser, E. F. 1993. PAH loss during bioremediation of manufactured gas plant site soil. *Water Res.* 27:911–919.
30. Fan, C., and Jafvert, C. T. 1997. Margules equations applied to PAH solubilities in alcohol-water mixtures. *Environ. Sci. Technol.* 31:3516–3522.
31. Field, J. A., Vledder, R. H., Vanzeist, J. G., and Rulkens, W. H. 1996. The tolerance of lignin peroxidase and manganese-dependent peroxidase to miscible solvents and the in vitro oxidation of anthracene in solvent-water mixtures. *Enzyme Microbial Technol.* 18:300–308.
32. Fuchs, P., Ma, W.-C., and Smies, M. 1985. Bioaccumulatie van milieucontaminaten in terristrische voedselketens. *Vakblad Biologie*, 65:75–80.

33. Ghosh, M. M., Yeom, I. T., Shi, Z. Cox, C. D., and Robinson, K. G. 1995. Surfactant-enhanced bioremediation of PAH- and PCB-contaminated soils. In: R. E. Hinchee, C. M. Vogel, and F. J. Brockman, (eds). *Microbial Processes for Bioremediation*, pp. 15–23, Columbus, OH: Battelle Press.
34. Ghoshal, S., and Luthy, R. G. 1996. Bioavailability of hydrophobic organic compounds from nonaqueous-phase liquids: The biodegradation of naphthalene from coal tar. *Environ. Toxicol. Chem.* 15:1894–1900.
35. Graves, D., and Leavitt, M. 1991. Petroleum biodegradation in soil: The effect of direct application of surfactants. *Remediation*, 147–166.
36. Green, J. C., Bartels, C. L., Warren-Hicks, W. J., Parkhurst, B. R., Peterson, S. A., and Miller, W. E. 1988. Protocols for short-term toxicity screening of hazardous waste sites, EPA/600/3-88/029. Corvallis, OR: U.S. Environmental Protection Agency.
37. Guha, S., and Jaffe, P. R. 1996. Bioavailability of hydrophobic compounds partitioned into the micellar phase of nonionic surfactants. *Environ. Sci. Technol.* 30:1382–1391.
38. Guthrie, E. A., and Pfaender, F. K. 1998. Reduced pyrene bioavailability in microbially active soil. *Environ. Sci. Technol.* 32:501–508.
39. Haimi, J., and Pavola, S. 1998. Responses of two earthworm populations with different exposure histories to chlorophenol contamination. *Environ. Toxicol. Chem.* 17:1114–1117.
40. Harm, H., and Bosma, T. N. P. 1997. Mass transfer limitation of microbial growth and pollutant degradation. *J. Ind. Microbiol. Biotechnol.* 18:97–105.
41. Hatzinger, P. B., and Alexander, A. 1995. Effect of aging of chemicals in soil on their biodegradability and extractability. *Environ. Sci. Technol.* 29:537–545.
42. Heitzer, A., Webb, O. F., Thonnard, J. E., and Sayler, G. S. 1992. Specific and quantitative assessment of naphthalene and salicylate bioavailability using bioluminiscent catabolic reporter bacterium. *Appl. Environ. Microbiol.* 58:1938–1946.
43. Helma, C., Mersch-Sundermann, V., Houk, V. S., Glasbrenner, U., Klein, C., Wenquing, L., Kasssie, F., Schulte-Hermann, R., and Knasmuller, S. 1996. Comparative evaluation of four bacterial assays for the detection of genotoxic effects in the dissolved water phases of aqueous matrices. *Environ. Sci. Technol.* 30:897–907.
44. ISO. 1991. Soil quality-effects of soil pollutants on *Collembola*: Determination of the inhibition of reproduction, 190/SC.4/W6 2N 34.
45. Kawai, F., Hanada, K., Yoshiki, T., and Ogata, K. 1997. Bacterial degradation of a water-insoluble polymer (polypropylene glycol). *J. Fermentation Technol.* 55:89–96.
46. Khan, S. U., and Ivarson, K. C. 1982. Release of soil bound (nonextractable) residues by various physiological groups of microorganisms. *J. Environ. Sci. Health B*, 17:737–749.
47. Kile, D. E., and Chiou, C. T. 1989. Water solubility enhancements of DDT and trichlorobenzene by some surfactants below and above the critical micelle concentration. *Environ. Sci. Technol.* 23:832–838.
48. Kool, H. J., van Kreyl, C. F., and Persad, S. 1989. Mutagenic activity in groundwater in relation to mobilization of organic mutagens in soil. *Sci. Total Environ.* 84:185–199.
49. Kohler, A., Schuttoff, M., Bryniok, D., and Knackmuss, H.-J. 1994. Enhanced biodegradation of phenanthrene in a biphasic culture system. *Biodegradation*, 5:93–103.
50. Koskinen, W. C., and Harper S. S. 1990. The retention processes: mechanisms. In: H. H. Chang (ed.), *Pesticides in the Soil Environment: Processes, Impacts, and Modelling*, pp. 51–77. Madisin, WI: Soil Science Society of America.
51. Leung, K. T., Errampalli, D., Cassidy, M., Lee, H., and Trevors, J. T., 1997. A case study of bioremediation of polluted soil: Biodegradation and toxicity of chlorophenols in soil. In: J. D. van Elsas, J. T. Trevors, and E. M. M. Wellington (eds.), *Modern Soil Microbiology*, pp. 577–605. New York: Marcel Dekker.

52. Linder, G., Greene, J. C., Ratsch, H., Nwosu, J., Smith, S., and Wilborn, D. 1990. Seed germination and root elongation toxicity tests in hazardous waste site evaluation: Methods, development and applications. In: W. Wang, J. W. Gorsuch, and W. R. Lower (eds.), *Plants for Toxicity Assessment*, ASTM STP 1091, pp. 177–187. Philadelphia, PA: American Society for Testing and Materials.
53. Linder, G., Wyant, J., Meganck, R., and Williams, B. 1991. Evaluating amphibian responses in wetlands impacted by mining activities in the Western United States. In: R. D. Comer, P. R. Davis, S. Q. Foster, C. V. Grant, S. Rush, O. Thorne, and J. Todd. (eds.), *Issues and Technology in the Management of Impacted Wildlife* pp. 17–25. Boulder, CO: Thorne Ecological Institute.
54. Luthy, R. G., Dzombak, D. A., Peters, C. A., Roy, S. B., Ramaswami, A., Nakles, D. V., and Nott, B. R. 1994. Remediating tar-contaminated soil at manufactured gas plant sites. *Environ. Sci. Technol.*, 28:266A–276A.
55. Manilal, B., and Alexander, M. 1991. Factors affecting the microbial degradation of phenanthrene in soil. *Appl. Microbiol. Biotechnol.* 35:401–405.
56. Megharaj, M., Singleton, I., and McClure, N. C. 1998. Effect of pentachlorophenol pollution towards microalgae and microbial activities in soil from a former timber processing facility. *Bull. Environ. Contam. Toxicol.* 61:108–115.
57. Megharaj, M., Venkateswarlu, K., and Rao, A. S. 1986. Effect of monocrotophos and quinalphos on soil algae. *Environ. Pollution*, A40:121–126.
58. Megharaj, M., Venkateswarlu, K., and Rao, A. S. 1986. The toxicity of phenolic compounds to soil algal population and to *Chlorella vulgaris* and *Nostoc linckia*. *Plant and Soil*, 96:197–203.
59. Megharaj, M., Venkateswarlu, K., and Rao, A. S. 1988. Microbial degradation and algal toxicity of monocrotophos and quinalphos in flooded soil. *Chemosphere*, 17:1033–1039.
60. Megharaj, M., Wittich, R.-M., Blasco, R., Pieper, D. H., and Timmis, K. N. 1997. Superior survival and degradation of dibenzo-*p*-dioxin and dibenzofuran in soil by soil-adapted *Sphingomonas* sp. Strain RW1. *Appl. Microbiol. Biotechnol.* 48:109–114.
61. Meighen, E. A. 1991. Molecular biology of bacterial bioluminiscence. *Microbial. Rev.* 55:123–142.
62. Miles, D. 1990. The role of chlorophyll fluorescence as a bioassay for assessment of toxicity in plants. In: W. Wang, J. W. Gorsuch, and W. R. Lower (eds.), *Plants for Toxicity Assessment*, ASTM STP 1091, pp. 297–307. Philadelphia, PA: American Society for Testing and Materials.
63. Miller, W. E., Peterson, S. A., Greene, J. C., and Callahan, C. A. 1985. Comparative toxicology of laboratory organisms for assessing hazardous waste sites. *J. Environ. Qual.* 14:569–574.
64. Miller, W. E. 1988. Protocols for short-term toxicity screening of hazardous waste sites, EPA/600/3-88/029. USEPA, Corvallis, OR: U.S. Environmental Protection Agency.
65. Mola, L., Sabatini, M. A., Fratello, B., and Bertolani, R. 1987. Effects of atrazine on two species of *Collembola* (Onychiuridae) in laboratory tests. *Pedobiologia* (Berl.), 42:241–251.
66. OECD. 1984. Guideline for testing of chemicals. No. 207. Earthworm acute toxicity tests. Adopted 4 April 1984.
67. Ortega-Calvo, J. J., Lahlou, M., and Saiz-Jimenez, C. 1997. Effect of organic matter and clays on the biodegradation of phenanthrene in soils. *Int. Biodeterior. Biodegrad.*, 40:101–106.
68. Pignatello, J. J., and Xing, B. 1996. Mechanisms of slow sorption of organic chemicals to natural particles. *Environ. Sci. Technol.* 30:1–11.
69. Ramaswami, A., and Luthy, R. G. 1997. Mass transfer and bioavailability of PAH compounds in coal tar NAPL-slurry system. 1. Model development. *Environ. Sci. Technol.* 31:2260–2267.
70. Rasiah, V., Voroney, R. P., and Kachanoski, R. G. 1992. Bioavailability of stabilised oily waste organics in ultrasonified soil aggregates. *Water, Air and Soil Pollution*, 63:179–186.
71. Rebber, H. H. 1982. Inducibility of benzoate oxidising cell activities in *Acinetobacter calcoaceticus* strain BS5 by chlorobenzoates as influenced by the position of chlorine atoms and the inducer concentrations. *Eur. J. Appl. Microbiol. Biotechnol.*, 15:138–140.
72. Ribo, J. M., and Kaiser, K. L. E. 1987, Photobacterium phosphoreum toxicity bioassay. 1. Test procedures and applications. *Toxicity Assessment*, 2:305–323.

73. Ronday, R., van Kammen-Polman, A. M. M., and Dekker, A. 1997. Persistence and toxicological effects of pesticides in top soil: Use of the equilibrium partitioning theory. *Environ. Toxicol. Chem.*, 16:601–607.
74. Rose, N., and Friedman, H. (eds.). 1976. *Manual of Clinical Immunology*. Washington, DC: American Society of Microbiology.
75. Salkinoja-Salonen, M. S., Middeldorp, P. J. M., Briglia, M., Valo, R. J., Haggblom, M. M., McBain, A., and Apajalahti, J. H. A. 1989. Cleanup of old industrial sites. In: D. Kamely, A. Chakrabarty, and G. Omenn (eds.), *Advances in Applied Biotechnology*, pp. 347–365. Houston, TX: Gulf Publishing Company.
76. Sayler, G. S. 1990. Rapid, sensitive, bioluminiscent reporter technology for naphthalene exposure and biodegradation. *Science*, 249:778–781.
77. Singleton, I., McClure, N. C., Bentham, R., Xie, P., Kantachote, D., Megharaj, M., Dandie, C., Franco, C. M., Oades, J. M. and Naidu, R. 1998. Bioremediation of organochlorine-contaminated soil in South Australia: A Collaborative venture. In: Kennedy et al. (eds.), *Seeking Agricultural Produce Free of Pesticide Residues*, Proc. Int. Workshop held in Yogyakarta, Indonesia, 17–19 February 1998, ACIAR Proceedings No. 85:334–337.
78. Smith, M. J., Lethbridge, G., and Burns R. G. 1997. Bioavailability and biodegradation of polycyclic aromatic hydrocarbons in soils. *FEMS Microbiol. Lett.* 152:141–147.
79. Spain, J. C., and van Veld, P. A. 1983. Adaptation of natural microbial communities to degradation of xenobiotic compounds: Effects of concentration, exposure time, inoculum and chemical structure. *Appl. Environ. Microbiol.*, 45:428–435.
80. Thibault, S. L., Anderson, M., and Frankenberger, W. T., Jr. 1996. Influence of surfactants on pyrene desorption and degradation in soil. *Appl. Environ. Microbiol.*, 62:283–287.
81. Tiehm, A. 1994. Degradation of polycyclic aromatic hydrocarbons in the presence of synthetic surfactants. *Appl. Environ. Microbiol.*, 60:258–263.
82. Tiehm, A., Stieber, M., Werner, P., and Frimmel, F. H. 1997. Surfactant-enhanced mobilisation and biodegradation of polycyclic aromatic hydrocarbons in manufactured gas plant soil. *Environ. Sci. Technol.*, 31:2570–2576.
83. U.S. Environmental Protection Agency. 1989. Ecological assessment of hazardous waste sites: A field and laboratory reference document, EPA/600-//3-89/013, Corvallis, OR: U.S. Environmental Protection Agency, Office of Research and Development.
84. U.S. Environmental Protection Agency. 1992. Evaluation of terrestrial indicators for use in ecological assessments at hazardous waste sites, EPA/600/R92/183. Washington, DC: U.S. Environmental Protection Agency, Office of Research and Development.
85. van Brummelen, T. C., Verweij, R. A., Wedzinga, S. A., and van Gestel, C. A. M. 1996. Polycyclic aromatic hydrocarbons in earthworms and isopods from contaminated forest soils. *Chemosphere*, 32:315–341.
86. van Gestel, C. A. M., and Ma., W. 1988. Toxicity and bioaccumulation of chlorophenols in earthworms, in relation to bioavailability in soil. *Ecotoxicol. Environ. Safety*, 15:289–297.
87. Venkatraman, S. N., Schuring, J. R., Boland, T. M., Bossert, I. D., and Kosson, D. S. 1998. Application of pneumatic fracturing of enhance in situ bioremediation. *J. Soil Contam.*, 7:143–162.
88. Volkering, F., Breure, A. M., van Andel, J. G., and Rulkens, W. H. 1995. Influence of nonionic surfactants on bioavailability and biodegradation of polycyclic aromatic hydrocarbons. *Appl. Environ. Microbiol.*, 61:1699–1705.
89. Wang, X, and Bartha, R. 1990. Effects of bioremediation on residues, activity and toxicity in soil contaminated by fuel spills. *Soil Biol. Biochem.*, 22:501–505.
90. Wang, W. 1991. Literature review on higher plants for toxicity testing. *Water, Air, Soil Pollution*, 59:381–400.
91. Weissenfels, W. D., Klewer, H.-J., and Langhoff, J. 1992. Adsorption of polycyclic aromatic hydrocarbons (PAHs) by soil particles: Influence on biodegradability and biotoxicity. *Appl. Microbiol. Biotechnol.* 36:689–698.

92. West, C. C., and Harwell, J. H. 1992. Surfactants and subsurface remediation. *Environ. Sci. Technol.*, 26:2324–2330.
93. Wilson, S. C., and Jones, K. C. 1993. Bioremediation of soils contaminated with polynuclear aromatic hydrocarbons (PAHs): A review. *Environ. Pollution*, 88:229–249.
94. Wodzinski, R. S., and Bertolini, D. 1972. Physical state in which naphthalene and biphenyl are utilised by bacteria. *Appl. Microbiol.*, 23:1077–1081.
95. Yaron, B., Calvet, R., and Prost, R. 1996. *Soil Pollution: Processes and Dynamics*. Berlin: Springer-Verlag.

11
Modeling Extractive Washing of Hydrophobic Organic Chemical-Contaminated Soils

Gerald R. Eykholt and Walid K. Al-Fayyoumi
University of Wisconsin, Madison, Wisconsin

I. INTRODUCTION

In a recent review article, Luthy et al. (1) address the lack of fundamental understanding of the behavior of geosorbents (soils and sediments) and the chemical interactions of hydrophobic organic chemicals (HOCs). The claim is made that, while macroscopic assessments of sorption/desorption isotherms and mass-transfer rates are useful, a greater fundamental understanding of geosorbents at the microscale is needed to advance the science of HOC fate, transport, and risk assessment. Luthy et al. (1) cited the rich literature on the characteristics of HOC partitioning by geosorbents.

On another front affecting soil and sediment remediation options, high concentrations of HOCs are often left in place in river and harbor sediments, at industrial sites, and in combined disposal facilities because of the high costs of treatment (2). For instance, polychlorinated biphenyl (PCB) contamination in the Great Lakes region is extensive and well documented. Due to the relatively high chemical inertness and low solubilities, PCBs have not decayed appreciably and have accumulated within biota, sediments, and associated organic matter. While corrective actions have been initiated at some sites, there are several primary areas of concern (AOCs) with heavily contaminated sediments located at harbors and river mouths. Corrective actions are difficult and costly for several reasons:

Conventional dredge and disposal techniques not allowed
Lack of adequate, off-the-shelf remediation technologies
Few technologies demonstrated at scale that meet regulatory targets
Poor economic forecasting tools available
Diverse site and contamination conditions

Keillor (3) reports that an average cost of dredging and disposal was $12/m^3 (1991 dollars) for 28 sites managed by the U.S. Army Corps of Engineers. While maintenance dredging may cost $2.50/m^3 to $7.50/m^3, full remediation costs for contaminated sediments may range from $26/m^3 up to $1300/m^3 depending on the site, contaminants, and technology chosen (4). The main differences in these costs are the costs associated with treatment, from physical processing and conditioning of the sediments, extraction, to destruction and/or disposal. The wide variance and uncertainty in treatment cost and performance estimates (in addition to other factors) has stalled the rate of clean-up.

Of the two main problems, the problem of understanding the complex behavior of geosorbents and the problem affecting removal and treatment of large deposits, the latter seems to be more pressing. On the basis of risk management, finding more efficient ways to limit exposure via remediation may be better than having a better fundamental understanding of the sediment–contaminant interactions. The more fundamental studies have driven more thorough, risked-based clean-up decisions, but there has been less impact on clean-up techniques and operations once a clean-up decision is made. This is especially true for complex sediments with a variety of contaminants and solid debris.

Less research effort has been devoted to the operations level (cost-effective materials processing) than on understanding secondary effects at the microscale. The same tools and models developed for more fundamental understanding may not be practical at the process scale. Granted, fundamental understanding of geosorbents and HOC interactions has intrinsic value and should enable the field to develop more accurate models. However, the more fundamental models also require more sophisticated materials and chemical characterization for the models to be useful at the operations level.

Currently, the Wisconsin Department of Natural Resources is overseeing two sediment dredging and remediation demonstration projects along the Fox River, Wisconsin. Approximately 12,000 yd^3 of PCB-contaminated sediment will be dredged, dewatered, and disposed of in a Wisconsin municipal landfill. While there is little sophistication in the process, a great deal of engineering effort is placed on safe dredging, efficient dewatering, and characterization of the processed sediments prior to landfilling. The applied aspects of consolidation and strength behavior, leachability, and wastewater treatment drive the costs of the process. While more fundamental models and sediment characterization have been used to determine the risks associated with the sediments, these models are generally inconsistent with the engineering tasks. Very similar engineering problems are found at confined disposal facilities (CDFs). The diminishing capacity has forced the U.S. Army Corps of Engineers to find ways to remove clean deposits in the CDF and minimize the amount of sediment requiring storage. Although the treatment processes being pursued are primarily physical (soil washing), there is a critical need in the next decade to provide methods for low-cost screening of physical and chemical properties to support decisions on removal options. This problem was also found when integrating the modeling of ex-situ contaminant remediation processes into the recently developed sediment remediation framework model, REMSIM (5,6). The ex-situ remediation processes of volume reduction (soil washing), thermal extraction, and solvent extraction were modeled and cost estimates were made. The level of detail that could be expected from existing sediment databases or from user input was a significant limitation in general, essentially prohibiting the use of more fundamental extraction models. However, there was also a limitation with regard to predictions of physical processing, and the treatment costs were, in some cases, more sensitive to uncertainties from assumptions made for the physical processing (7).

Ideally, a great portion of the contaminated sediments would not be contaminated or could be treated quickly with a physical process to greatly reduce the volume of material requiring more costly treatment. Allen (8) discussed the feasibility of various mineral processes for pretreatment of several Great Lakes sediments, and the summary is presented in Table 1. Here, feasibility refers to the ability of the process to separate contaminated

Table 1 Feasibility of Mineral Processing Pretreatment of Contaminated River Sediments from Great Lakes Areas of Concern[a]

Operation/River	Ashtabula	Buffalo	Grand Calumet	Saginaw
Grain size separation	U	L	L	F
Froth flotation	L, M	L, M	L, M	L
Attrition scrubbing	U	U	U	L, M
Density separations	U	L	L	F
Magnetic separations	U	U	L, M	U

[a] F = feasible; L = limited applications; U = unlikely; M = more study recommended.
Source: From Ref. 8.

materials from cleaner sediments (volume reduction). Unfortunately, there are few situations that are clearly feasible and several others that have an uncertain or limited feasibility.

One important consideration that was not addressed by the study was the possibility of enhancement to the subsequent treatment processes. For instance, if attrition scrubbing causes uniform mixing of the sediment and enables carbon-rich phases to sorb more contamination, the leachability of the sediment may be greatly reduced. Alternatively, if attrition scrubbing increased the bioavailablity, a more efficient posttreatment process may be possible. It is important to recognize that most of the work associated with physical treatment (soil washing) has not addressed the enhancements to posttreatment by other technologies, but has focused on the primary issue of volume reduction. Most of the work has also been focused on total concentrations and has not addressed the issues of leachability dynamics rigorously. Like the cases of solidification/stabilization strategies, leachability studies are typically inconclusive unless the test reveals a significant difference in PCB levels found in the treated and untreated specimens.

Another complexity is mixed wastes. Soils and sediments contaminated with organic contaminants and radioactive compounds need to be treated at several U.S. Department of Energy sites. Finding suitable extraction or other remediation technologies requires a careful investigation of materials processing and separations. In general, ex-situ treatments involving significant levels of physical processing of contaminated soils are not as popular as containment or in-situ remediation strategies. One of the factors limiting remediation of contaminated soils is the extensive scale and capital expense required of the physical processing equipment.

The purpose of this chapter is to establish a simple strategy for evaluating extractive washing of soil through the introduction of a new soil washing and separations model. While the complexity of contaminant desorption from geosorbents is an advanced topic, the approach taken here is to develop a mathematical basis for evaluating washing and separations for the soil washing process. The washing model presented is novel, yet simplistic enough that it can be easily integrated with physical separation models to evaluate the feasibility of soil washing. While this chapter provides no new washing data, several guidelines for experimental work are included.

II. BACKGROUND: EXTRACTIVE, EX-SITU SOIL WASHING

If extraction agents are used in soil washing, the process may be called extractive soil washing and the removal function is a combination of volume reduction and extraction.

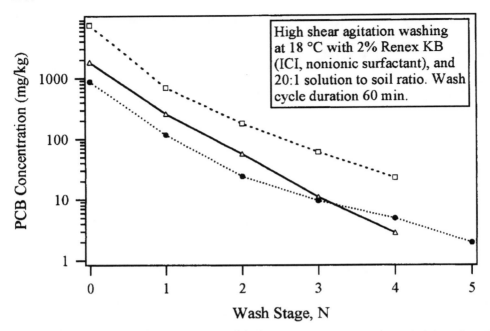

Figure 1 Demonstration of the effectiveness of surfactant soil washing for the cleansing of several highly PCB-contaminated soils from an industrial site. Samples contained high levels of Aroclor 1260 and transformer oil and were washed with a nonionic surfactant, Renex KB. (Courtesy H. Krabbenhoft, GE Company, Corporate R&D Center, Schenectady NY.)

As an example, scientists at the GE Corporate R&D Center extracted PCBs (A1260) from a highly-contaminated industrial soil (approx. 1000 mg/kg total PCBs as A1260). For gentle agitation, a surfactant mass concentration of 1% (Renex KB, ICI), and a 3 : 1 solution to soil ratio (w/w), three or more wash stages were required to reduce the PCB concentration to 100 mg/kg (9). Subsequent wash stages were not effective, indicating a significant resistant fraction. Upon further testing at a higher shear rate, a higher surfactant concentration, and a higher solution-to-soil ratio, washing was more successful, leading to concentrations below 10 mg/kg after several wash stages. These results are shown in Fig. 1. This demonstrates that attrition and other physical pretreatments may play a substantial role in extraction processes.

The staged washing process can be simulated with a simple two-parameter model. A resistant fraction, C_{rf}, can be considered not to play a role in the partitioning, and the soil concentration after N washing stages, C_N, can be shown to be

$$C_N = C_{rf} + \frac{C_0 - C_{rf}}{(1 + K^*)^N} \quad (1)$$

where C_0 is the initial concentration and K^* is a dimensionless "washing power" dependent on the solution volume (V_L) to soil mass (M_s) ratio and the effective, linear partition coefficient (K)

$$K^* = \frac{V_L}{KM_S} = \frac{w}{K} \quad (2)$$

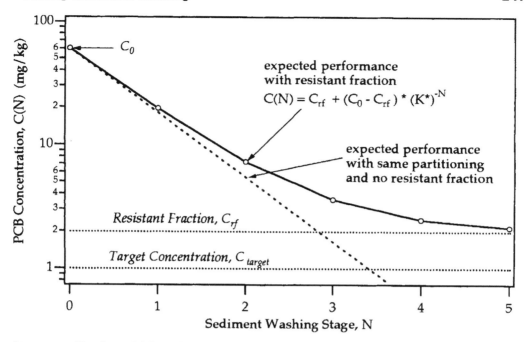

Figure 2 Simple model for sediment washing with resistant fraction. Model assumes equilibrium partitioning at each stage and complete solid–liquid separation between stages.

The solution-to-soil ratio (w) and K can vary depending on the volume and strength of the solvent. Also, the initial liquid-phase concentration is zero, and perfect solid–liquid separation between stages is assumed. In the case of Eq. (1), K has units of (L/kg) and is operational ratio of the concentration of the contaminants in the nonresistant soil fraction (mg/kg) versus the concentration in the liquid phase (mg/L). An example of this model is shown in Fig. 2.

For this simple model, the rate of adsorption and desorption from the resistant fraction is assumed to be negligible. For other cases with highly contaminated soils, one phase may be readily extracted into the liquid, leaving behind another phase that has a higher effective partition coefficient. An example indicating this behavior is shown in Fig. 3. It is likely that an oil phase containing high PCB concentrations and relatively low K (2.4 L/kg) was dominant for the first two washing stages. For later stages, for which the oil phase has been largely removed, a second phase with a much higher K (18 L/kg) may have controlled the extraction. In this case, the improvement with increasing numbers of stages indicated that contaminants in the more resistant fraction could be extracted to solution over the washing stage.

While concurrent, multistage washing tests provide useful information, there may be other tests that are less intensive and more descriptive for the engineering behavior of extractive soil washing operations at scale. The main reason for this is developed in light of the practical application of volume reduction processes as part of extractive soil washing operations. In many cases, contaminated soils and sediments contain mass fractions that represent a substantial portion of the contaminant mass, but are resistant to washing ("hard" organic fractions) and cause operational problems (fines). In some cases, some ferromagnetic materials representing oily cuttings may be present as a small mass percentage, represent a significant portion of the contamination, and are easily

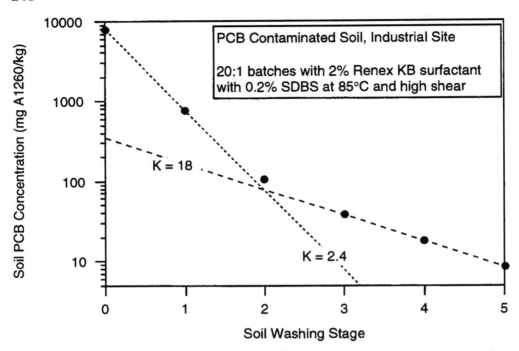

Figure 3 Demonstration of surfactant washing of a heavily contaminated soil, indicating suitability of two-region washing model. (Courtesy H. Krabbenhoft, GE Company, Corporate R&D Center, Schenectady NY.)

removed by magnetic separation. The remaining soil or sediment mass fraction (coarse fraction) may require mild extractive washing or agitation to remove adsorbed organic matter and fines. If the contaminant partitioning, extraction mass transfer characteristics, and contaminant mass distribution of the main soil fractions are characterized initially, a better overall washing strategy should be obtained. For soil washing operations based on the volume reduction, soils and sediments with a large mass fraction of fine (>45%) are considered difficult to wash (10). This is because the fines cause many operational problems and represent a primary waste fraction—material that must be removed from the bulk soil, dewatered, and disposed or treated with another process. Costs for residuals management typically range from $200 to 450/dry ton, so a high mass fraction of fines usually infers that volume reduction will be too expensive.

Another important limitation is that extractive soil washing creates large volumes of water to treat, and there is need for surfactant recycle strategies [see Ang and Abdul (11) and Clarke et al. (12)]. Eykholt and Mackenzie (13) have demonstrated the performance of several polymers that can selectively sorb PCBs and oils in surfactant streams. There has been significant activity in surfactant remediation strategies recently, but "factors affecting the kinetics of surfactant solubilization of PCBs in heterogeneous systems need to be understood to improve process efficiency and chemical use" (2).

The general criteria regarding percent fines effectively restricts a large fraction of contaminated soils from the soil washing remediation option. This is unfortunate, because mild extractive washing of many HOC-contaminated sediments and soil may greatly reduce the leachability of the treated materials. Even if the total contamination level classifies the soil or sediment as a hazardous waste, the effective bioavailability and risk from the material may be reduced greatly. If extractive washing is to be considered more thoroughly, even amidst the difficulties associated with costs and regulatory classification, models describing the washing behavior of several main fractions of the soil should be considered.

III. DESCRIPTION OF STAGED, EXTRACTIVE WASHING MODEL

Modeling strategies that consider rate-limited desorption, multiple sites for sorption, incomplete solid–liquid separation, and countercurrent extractive washing have been addressed by Fayyoumi (14). In order to describe the modeling tasks, three issues will be discussed below: multifraction equilibrium washing, rate-limited and multisize washing, and physical separations modeling.

A. Multiple-Fraction Equilibrium Washing

If all soil or sediment fractions i are combined into one wash stage and equilibrium is achieved, the resulting contaminant concentration on the sediment fraction can be shown to be

$$C_{i,\text{eq}} = \frac{K_i(\bar{C}_0 + wC_0^L)}{w + \bar{K}} \tag{3}$$

where the overbar signifies a mass average (i.e., $\bar{K}$ represents $\Sigma K_i M_i / \Sigma M_i$), and $\bar{C}_0$ and C_0^L are the initial concentrations in the bulk sediment and liquid phases, and w is the liquid-to-solid ratio. The effective partitioning factor for each size fraction K_i may be estimated from the total organic carbon concentration from each fraction and a global partition factor for organic carbon, K_{oc}:

$$K_i = f_{\text{oc},i} K_{\text{oc}} \tag{4}$$

where $f_{\text{oc},i}$ represents the mass fraction of organic carbon in the sediment mass fraction.

An example calculation for multsize, equilibrium washing is shown in Table 2 and Figure 4. These calculations reveal the utility of combining volume reduction and extractive washing. It is also important to note that fractions with high partition coefficients may cause "resistant fraction" behavior, similar to the staged-extraction resistance shown in Figs. 2 and 3.

B. Rate-Limited Extraction for One Size Fraction

The kinetics of contaminant extraction in each sediment fraction may follow a film-mass transfer model trend. The dynamics of mass transfer are described by the differential equation

$$\frac{dC_i}{dt} = k_i(C_i - C_i^*) \tag{5}$$

where k_i is the film mass transfer rate constant and C^* represents the solid-phase (interface) concentration in local equilibrium with the bulk liquid-phase concentration ($C^* = K_i C_L$). Considering mass balance between the sediment and liquid fraction, a simplified equation can be derived:

$$\frac{dC_i}{dt} = k_i^*(C_i - C_{i,\text{eq}}) \tag{6}$$

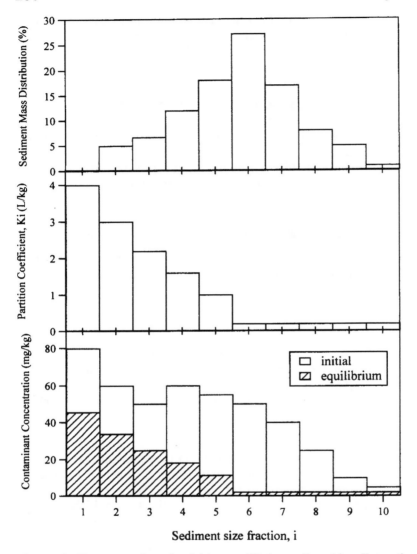

Figure 4 Demonstration of multisize, equilibrium soil washing. Extractions into the liquid phase and separations of the poorly washed sediment fractions may extend the opportunities of conventional soil washing (volume reduction). Additional information provided by Table 2.

where the effective mass transfer coefficient k^* for the size fragment is dependent on a more fundamental rate constant, the liquid-to-solid ratio, and the partition coefficient:

$$k_i^* = k_i\left(1 + \frac{w}{K_i}\right) \quad (7)$$

If the size fraction is extracted independently of other fractions Eq. (6) can be integrated using the equilibrium concentration,

$$C_{i,\text{eq}} = \frac{C_{i,0}}{1 + w/K_i} \quad (8)$$

Modeling Extractive Washing

Table 2 Example Case for One-Stage Equilibrium Washing for Multiple Size Fractions Containing Different TOC Fractions

One-Stage Equilibrium Washing			Many Size Fractions		
Liquid volume (L) = 400 K_{oc} (L/kg) = 20				Liquid concentration (mg/L) Initial = 2.00 Final = 11.37	
Size fraction	Sediment mass (kg)	Organic carbon fraction	K_i (L/kg)	Concentration on solids (mg/kg) Initial	Final
1	0.2	0.20	4.0	80	45.5
2	5.0	0.15	3.0	60	34.1
3	6.7	0.11	2.2	50	25.0
4	12.0	0.08	1.6	60	18.2
5	18.0	0.05	1.0	55	11.4
6	27.1	0.01	0.2	50	2.3
7	17.0	0.01	0.2	40	2.3
8	8.0	0.01	0.2	25	2.3
9	5.0	0.01	0.2	10	2.3
10	1.0	0.01	0.2	5	2.3
Sediment mass (kg) 100.0			Avg. K 0.79	Avg. init. C_s 46.5	Equil. C_s 9.02

to obtain:

$$C_i(t) = C_{i,0} e^{-k_i^* t} + C_{i,eq}(1 - e^{-k_i^* t}) \tag{9}$$

This general equation can be used to estimate the extraction rate constant and effective partition factor from experimental extraction measurements performed on one size fraction.

C. Rate-Limited Extraction for Multisize Washing

Once rate constant and equilibrium parameters are gathered for each major sediment fraction, the dynamics of the extraction process can be modeled using the same fundamental mass transfer equation [Eq. (5)]. However, the pseudo-concentration C^* for each sediment fraction will vary in a more complex fashion.

For this case, the dynamics of desorption are modeled numerically. The contaminant concentration in each size fraction will approach the equilibrium value at different rates, and the rate of mass accumulated in the liquid phase will be affected by the desorption of every phase. Unlike the case for the simple film mass transfer model with one fraction, the desorption rate of a given fraction will depend on the equilibrium condition and the rate of accumulation into the liquid phase. A mass balance can be written for each fraction, and the average concentration on the soil can be incorporated to yield the difference equation

$$\frac{dC_i}{dt} = -k_i \left[C_i - \frac{K_i}{\bar{K}} \bar{C}_{eq} \left(1 + \frac{\bar{K}}{w}\right) + \frac{K_i \bar{C}}{w} \right] \tag{10}$$

With each fraction considered, and without simplification, the system of differential equations can be solved numerically with a Newton-Raphson technique (14).

Under the idealized case that the extraction rate constants k_i are the same for each size fraction, an analytical solution can be derived. Under this case, the dynamics of the average concentration can be tracked with the differential equation

$$\frac{d\bar{C}}{dt} = -k_i\left(1 + \frac{\bar{K}}{w}\right)(\bar{C} - \bar{C}_{eq}) \tag{11}$$

This equation is integrated to obtain the average concentration in the sediment with time [similar form to Eq. (9)]. Then the average concentration can be incorporated into the general difference equation [Eq. (10)], with the following result:

$$C_i(t) = C_{i,eq} + \left(C_{i,0} - \frac{K_i}{K}\bar{C}_0\right)e^{-kt} - \left(C_{i,eq} - \frac{K_i}{K}\bar{C}_0\right)e^{-\bar{k}^*t} \tag{12}$$

where

$$\bar{k}^* = k\left(1 + \frac{\bar{K}}{w}\right) \tag{13}$$

While it is an unlikely condition that the film mass transfer rate constant is identical for every sediment fraction, the analytical solution provided by Eq. (12) can be used to test the accuracy of the numerical solution. Example simulations using the analytical solution and input from Table 3 are shown in Fig. 5. Agreement between the model and analytical solution for this case was excellent. Although the fundamental mass transfer dynamics of each segment is parameterized by the simple, film mass transfer model, the behavior of the mixture indicates more complex, nonlinear trends.

D. Modeling Physical Separation Processes

Wills (15), Svarovsky (16), and others have presented useful models for simulating physical separations of coagulation/flocculation, sedimentation, filtration, as well as the performance-specific process units of hydrocyclones, filter presses, centrifuges, and other equipment. Witt (17) and, later, Vaidya (7) modeled physical separations to simulate conventional volume reduction processes, and these models were then implemented within the sediment remediation simulator REMSIM (5). Particle classification by size can be handled through use of a grade-efficiency function, $G(x_i)$. The grade efficiency function expressed the probability of separating a particle with a specific particle diameter x_i to the product (or underflow) stream (7). For a hydrocyclone,

$$G(x_i) = \frac{1 - Q_u/Q}{1 + (x_i/x_{50})^\lambda} + \frac{Q_u}{Q} \tag{14}$$

where Q_u/Q is the underflow to feed volumetric flow rate ratio, x_{50} is deemed the "cut diameter," and λ is a coefficient related to the sharpness index,

$$\lambda = \frac{-2.1973}{S} \tag{15}$$

The sharpness index S can be defined as the ratio of x_{25} and x_{75}, the particle diameters corresponding to $G(x_i)$ of 0.25 and 0.75, respectively, when Q_u/Q is ignored. For perfect separation at the cut diameter, $S=1$. For a hydrocyclone, separation varies with hydrocyclone diameter, percent solids in the feed, solids density, and pressure, but

Modeling Extractive Washing

Table 3 Input for Analytical Solution Check of Multisize, Rate-Limited Washing of Contaminated Soil (liquid-to-solid ratio, $w = 2$ L/kg)

Material [i]	Sediment mass (kg)	$K[i]$ (L/kg)	Initial $C[i]$ (mg/kg)	$k[i]$ (1/h)	Equilibrium $C[i]$ (mg/kg)
Clay [1]	20	2.2	100	0.05	16.41
Silt [2]	30	1.2	80	0.05	8.95
Coarse [3]	48	0.8	20	0.05	5.97
Sorbent [4]	2	200	0	0.05	1492
Mass average		5.18	53.6	0.05	38.67

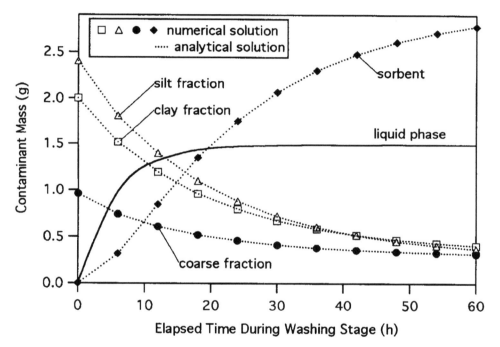

Figure 5 Comparison of analytical and numerical solutions for multisize, rate-limited washing of contaminated sediment. Key model parameters are summarized in Table 3.

Q_u/Q typically ranges between 0.5 and 0.3 and S ranges between 0.3 and 0.6. For this chapter, default values for x_{50}, Q_u/Q, and S were 75 m, 0.15, and 0.4, respectively. A lower-than-usual value of the underflow ratio was selected because a batch-mode operation was considered, leading to some additional dewatering of the underflow solids. More details on sediment classification are discussed by Vaidya (7). More sophisticated models of hydrocyclone operation are available (18), but the above empirical model was deemed adequate for the simulation set.

E. Modeling Wash Stages and Countercurrent Separations

The countercurrent, rate-limited washing model was developed with the following tasks in mind:

1. Solve for both the steady-state and unsteady-state problems of the countercurrent washing and separation.
2. Accept contaminated sediments composed of variable particle sizes with their own partition factor and sorption-desorption rates.
3. Perform multistage washing and separation processes with knowledge of a prescribed grade efficiency function.
4. Deal with washing stages consisting of a number of washing batches.

The basic assumptions of this model may be deduced from the fundamental mechanism of the film mass transfer model. The following assumptions apply to this model:

1. One solute (contaminant) is present in the solution.
2. Partitioning is linear and reversible, and desorption kinetics follow a film model.
3. The sorbing surfaces between the solid and the liquid phases are always available and do not change.
4. The aqueous phase in the washing batch reactor is well mixed as to provide a uniform distribution of the contents.
5. The sediment material in each particle size of the sediment mass is conserved.
6. The grade efficiency function is identical for each batch and is prescribed.
7. No contaminant mass transfer from the liquid to the bulk solids occurs during the separation process.

Once the sediment, sorbent, and liquid report to the same batch, Eq. (10) is solved using a Newton-Raphson procedure [MNEWT.C, (19)]. This gives the time-dependent concentrations in the batch at regular intervals over the duration of the batch. If equilibrium washing is invoked, Eq. (3) is used. Then, the sediment fractions report to the underflow and overflow streams according to the prescribed grade efficiency function for hydrocyclone separation [described above, Eq. (14)]. New batches are composed and the process continues until near steady-state operation is achieved. For most runs, between 12 and 20 cycles of countercurrent washing was required to keep cycle-to-cycle variations below 5% on a contaminant mass basis.

Users of the model have the potential to simulate a great variety of cases, through varying washing stages, partition factors, mass transfers rates, number of size fractions, and the grade efficiency function. However, a major limitation was that no real case was tested, so experimental verification and optimization could not be demonstrated. Instead, a small set of modeling realizations are presented. The default configuration was a countercurrent treatment scheme consisting of three batches, two sediment mass fractions, a sorbent mass fraction, and a liquid : solid feed ratio $w = 3.81$ L/kg. Default parameters for the model are summarized in Table 4.

Table 4 Default Input Data for Countercurrent Washing Simulations (liquid-to-solid ratio, $w = 3.81$ L/kg)

Material [i]	Sediment mass $M_s[i]$ (kg)	$K[i]$ (L/kg)	Initial $C_s[i]$ (mg/kg)	$k[i]$ (1/h)	Grade curve $G[x_i]$
Coarse [1]	100.	2.	200.	0.10	0.93
Fine [2]	100.	2.	500.	0.05	0.61
Sorbent [3]	10.	40.	0.	0.30	0.00

In order to summarize the performance of the washing operation, two types of contaminant removal efficiency were calculated for each case. The total contaminant removal efficiency, ε_{CR}, is defined as the percent contaminant mass removal from the feed solids to the overflow solids, sorbent, and liquid, as well as contaminant mass to the underflow liquid. The mass in the underflow liquid is considered recoverable upon mild rinsing and dewatering. This includes the effects of volume reduction (overflow solids) and extraction (mass transfer to liquid and sorbent). The extraction efficiency, $\varepsilon_{extr.}$, is defined as the percent mass removal from the feed solids to the overflow liquid and sorbent, as well as the mass in the underflow liquid. Also, to indicate the general degree of volume reduction, a sediment yield ratio was calculated. The sediment yield, Y, is defined as the mass ratio of sediments reporting to the process underflow to the feed sediments.

Further details of the model, including source code for the model in the C programming language, are presented by Fayyoumi (14). Verification steps included testing the model against analytical solutions for multisize washing [Eq. (12)], comparing multistage equilibrium washing against spreadsheet solutions, accounting mass balance errors, and other methods. Relative mass balance errors were reported for every case and were generally less than 2%.

IV. RESULTS OF SIMULTATIONS

Equilibrium and rate-limited, countercurrent washing processes were evaluated for a contaminated sediment containing two main size fractions, with default characteristics as reported in Table 4. The contaminated sediment fractions included a coarse fraction and fine fraction. The fine fraction was set to be more heavily contaminated, slower to desorb, and more difficult to retain in the underflow. Three countercurrent batches were considered for every case. Several default parameters were varied to investigate the importance of a sorbent, resistant sediment fractions, and physical separations on contaminant removal efficiency (ε_{CR}), extraction efficiency ($\varepsilon_{extr.}$), and sediment yield (Y).

The default case presents a sediment with two size fractions of the same partition factor, although there are differences in the initial concentration, the rate constant for desorption, and the grade efficiency function. The performance of the washing scheme for the equilibrium case with sorbent is summarized in Fig. 6. The scheme is fairly effective, removing most of the contaminant mass to the overflow liquid and sorbent. The sorbent enhanced the performance of the wash in a mild manner, since the removal efficiency ε_{CR} was 0.98 in the case with sorbent and 0.94 without. Similarly, $\varepsilon_{extr.}$ was 0.88 in the case with sorbent and 0.79 without. While a small mass fraction of sorbent is used, it is important to note that it is separated with 100% efficiency to the overflow and has a moderately high partition coefficient. Selective removal of the sorbent is feasible, since the size, shape, density, and magnetic properties of the sorbent can be selected to be quite different from the sediment.

The ratio of liquid feed to solids feed for the process is approximately 3.8 L/kg. It is important to note that, with no efforts to control the liquid–solid ratio in each batch, a steady-state liquid–solid ratio is nevertheless achieved in each batch. The model could be modified to use make-up water such that the liquid–solid ratio was fixed.

Since the default conditions indicate a highly effective, extractive washing scheme is possible, one pressing question deals with the processing rate. The key variable is the duration of the washing cycle for each batch. The desorption rate constants are selected arbitrarily, and long batch periods are required to obtain effective washing. The effect

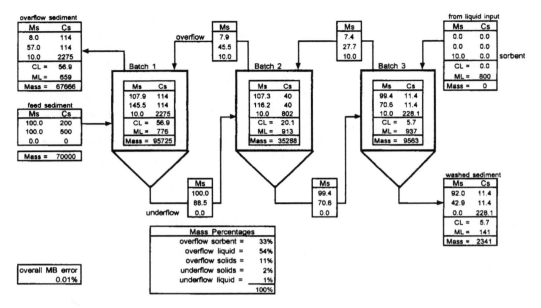

Figure 6 Summary of performance of countercurrent washing of contaminated sediment under default conditions (see Table 4). Result from equilibrium wash with sorbent, $K[i] = \{2, 2, 40\}$, and $G[i] = \{0.926, 0.608, 0\}$.

of batch time on contaminant removal efficiency is shown in Fig. 7. Contaminant removal efficiency increases only mildly with batch time after 12 h, but the effect on extraction efficiency is more significant. The volume reduction function is somewhat independent of batch time, and yet a significant portion of the contaminant mass removal is limited by the slow rates for desorption. The effect of the sorbent is mild for any fixed batch time. However, this effect could be magnified with more sorbent, greater partitioning to the sorbent, and faster sorption mass transfer. Another effect of the sorbent is less apparent. For a given, target removal efficiency, sorbent-aided washing reduced the required batch time significantly. For instance, for a target $\varepsilon_{extr.}$ of 0.70, the required batch time drops from approximately 18 h to 12 h in the presence of the sorbent.

It is difficult to select the ideal cycle time, as this would depend on the economics of processing versus storage, and the relative difficulty of solid–liquid separation and wastewater treatment. However, the model may quite useful for designing pilot experiments and for optimization exercises once more information is gathered.

The sediment yield Y for the default case is 0.67, so approximately 33% of the sediment is removed to overflow. However, the role of extractive washing is still important for this stream, since the average contaminant concentration on the sediment is reduced significantly from the feed stream. Other strategies for posttreatment of this material may increase yield.

Grade efficiency function and sorbent strength changes for more difficult cases. A more difficult washing case is presented if the partitioning factor for the fine sediment fraction, K[2], was increased from 2 to 25 L/kg. Typical responses for this problem include increasing the level of volume reduction and increasing the sorbent strength. Several cases are presented in Table 5 for which equilibrium washing is considered. Without either action (case B), ε_{CR} dropped from 0.98 to 0.80 and $\varepsilon_{extr.}$ dropped from 0.88 to 0.35 from the default case (case A).

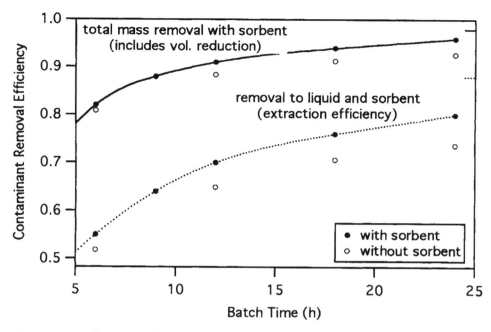

Figure 7 Performance of countercurrent washing scheme as a function of batch time, with and without sorbent. Default conditions are summarized in Table 4.

Since $\varepsilon_{extr.}$ is low, a logical step would be to increase the extent of volume reduction by reducing the grade efficiency function for the fines fraction (G[2] reduced from 0.61 to 0.30). This effect could be accomplished by changing the hydrocyclone diameter. The improvement is significant, since the drop in ε_{CR} due to the difficulty in washing is recovered. However, the sediment yield has dropped from 0.67 to 0.48. The effectiveness of a staged washing strategy is quite limited by the washing difficulty, and the change in the grade efficiency function results in a drastic reduction in yield and solids loadings.

Another corrective action is to increase the strength of the sorbent, instead of changing the grade efficiency (case D). If a stronger sorbent is used (K[3] = 500 L/kg), ε_{CR} is restored. In contrast to the volume reduction strategy, $\varepsilon_{extr.}$ and the sediment yield

Table 5 Changes in Removal Efficiency and Sediment Yield on Countercurrent, Equilibrium Wash Scheme: Effects of Partitioning, Solid–Liquid Separation Grade Efficiency, and Sorbent Strength[a]

Case	A	B	C	D	E	F
K[1]	2	2	2	2	2	2
K[2]	2	25	25	25	25	25
K[3]	40	40	40	500	500	500
G[2]	0.61	0.61	0.30	0.61	0.30	0.80
ε_{CR}	0.98	0.80	0.98	0.98	0.999	0.96
$\varepsilon_{extr.}$	0.88	0.35	0.31	0.79	0.70	0.86
Yield, Y	0.67	0.67	0.48	0.67	0.48	0.84

[a] All other conditions summarized in Table 4.

Table 6 Recommendations for Difficult Washing Conditions by Fayyoumi (14)

Problem	Options	Other issues
Poor extraction due to high partition coefficient	More effective extractants Test volume reduction	Separating agents and sorbent
High percent fines cause great difficulties in interstage separations	Reduce number of stages by providing more extractive washing and sorbent	Fines have high K_p
Slow desorption from matrix slows down process throughput	Precondition sediment for long period, then test Test volume reduction Use stronger solvent, or attrition scrubbing More time/reduce number of stages	Stronger solvent yields more difficult wastewater treatment

are still high. If the sorbent strength is limited, the mass of sorbent could be increased to gain the same general effect. However, the sorption kinetics and impacts on the required batch time should also be considered.

If both actions are pursued, volume reduction and an increased sorbent strength (case E), the removal efficiency is nearly 100% but, ε_{extr} and the sediment yield are lower than case D. It may be impractical to implement volume reduction with three countercurrent batches, and the reductions in ε_{extr} follow from a reduced level of contact of contaminated soil and the sorbent in the batches. In fact, decreasing the amount of volume reduction (case F) may result in an acceptable level of ε_{CR}, and increased level of ε_{extr} and an increased soil yield. Therefore, extractive washing may still be possible for more resistant soils without the need for increased volume reduction. If extractive washing is to function properly amidst volume reduction, the goal should be to maintain a relatively high sediment yield if possible.

A variety of related simulations are discussed by Fayyoumi (14). Although more sediment fractions were considered, similar trends were found. A summary of recommendations from Fayyoumi (14) is presented in Table 6. The use of sorbent was found to reduce the number of countercurrent stages required for the same removal efficiency. For cases in which solid–liquid separation is difficult, use of a sorbent as the main phase in countercurrent flow may also reduce the demand for dewatering between stages. While the time required for washing in each batch may be long, the benefit from extractive washing without severe requirements for dewatering may allow sediments containing a large fraction of fines to be washed more feasibly. Further experimental work, washing modeling, and cost modeling are needed to thoroughly address the complex trade-offs experienced in the countercurrent washing of contaminated sediments.

V. CONCLUSIONS

The desorption and separations behavior of contaminated sediments is complex, involving multiple sorbent phases, contaminants, and irreversible processes. A model based on simplifying assumptions is presented to simulate the performance of countercurrent separations from an operational perspective. The main principle is to segregate the sediment into dominant mass fractions. The dynamics of extractive washing of each segment are assessed independently using a film mass transfer model. This model addresses the readily extracted portion, but can also be used to approximate more complex dynamics. The characteristics of a sorbent can also be assessed in a similar manner. Then, a multisize, dynamic washing model can be implemented to simulate the performance of a countercurrent washing scheme. The combined functions of volume reduction and extractive washing are investigated, such that an effective treatment strategy for the material is identified.

Several key trends were identified with a small subset of modeling realizations. First, using a sorbent within the washing process was an effective way to enhance extraction efficiency and to shorten the batch time required for a given removal efficiency. However, the trade-off between the usefulness of the sorbent and the extent of volume reduction was evident. As the role of volume reduction was increased, the extraction efficiency decreased due to the reduced exposure of the extractive phases to contaminated sediments. Increasing the sorbent strength and mass of sorbent was shown to be an effective way of compensating for an increase in the washing resistance of a particular sediment fraction. Volume reduction is expected to be most useful when the rate of extraction is very low, such that the sorbent or liquid are not effective extractants over the batch cycle.

While the model is useful, the value of the model ultimately relies on its integration with experimental work, costs modeling, and pilot demonstrations. However, the sediment characterization does not necessarily need to be extensive. In fact, the primary objective is to develop testing strategies that can be reproduced and scaled up without extensive characterization. Additional modeling may be needed to address the complexity of the soil washing process, especially those factors that most significantly effect processing rates and costs. Further experimental and modeling work is needed to more thoroughly address the complex trade-offs experienced in the countercurrent washing of contaminated sediments.

ACKNOWLEDGMENTS

Partial research support for this project was made possible through funding by the Great Lakes Protection Fund (GLPF) for a research project coordinated by Dr. Anders Andren and Phil Keillor of The Wisconsin Sea Grant Institute. Additional funding was made available from the National Science Foundation, through a CAREER award to Gerald Eykholt (Award No. 9625087). The soil washing data used in the chapter was made available by Dr. Herm Krabbenhoft of the GE Company Corporate R&D Center, Schenectady, New York.

REFERENCES

1. Luthy, R. G., Aiken, G. R., Brusseau, M. L., et al. 1997. Sequestration of hydrophobic organic chemicals by geosorbents. *Environ. Sci. Technol.*, 31(12):3341–3347.
2. National Research Council. 1997, *Innovations in Ground Water and Soil Cleanup: From Concept to Commercialization*. Washington, DC: National Academy Press.
3. Keillor, J. P. 1993. Obstacles to the remediation of contaminated soils and sediments in North America art reasonable cost. In: *Proc. CATS II Congress: Characterization and Treatment of Contaminated Dredged Material*, Antwerp, Belgium, Technological Institute of the Royal Flemish Society of Engineers (University of Wisconsin Sea Grant Reprint WISCU-W-93-001).

4. ARCS. 1993. The Assessment and Remediation of Contaminated Sediments (ARCS) Program. Remediation Guidance Document, U.S. Environmental Protection Agency, Great Lakes National Program Office, EPA Contract 68-C2-0134.
5. Andren, A. 1997. Cost and benefits of cleaning up contaminated sediments in Great Lakes Areas of Concern. Final report to The Great Lakes Protection Fund and U.S. Environmental Protection Agency, Great Lakes National Program Office. Madison, WI: University of Wisconsin Sea Grant Institute.
6. Keillor, J. P. 1996. Creating an economic decision framework for estimating the benefits and costs of sediment remediation. In: G. De Schutter and R. Vanbrabant (eds.), *CATS III Congress*, Oostende, Belgium (University of Wisconsin Sea Grant Reprint WISCU-R-96-017).
7. Vaidya, A. 1996. Modeling of ex situ sediment remediation technologies. M.S. Thesis, Department of Civil and Environmental Engineering, University of Wisconsin—Madison.
8. Allen, J. P. 1994. Mineral processing pretreatment of contaminated sediments, EPA 905-R94-022. U.S. Environmental Protection Agency ARCs Program Report.
9. Krabbenhoft, H. O. 1994. Remediation of PCB-contaminated soil—Part 3. Surfactant-assisted washing of PCB-containing soil, Report #94CRD104. Schenectady, NY: GE Company, Corporate Research and Development Center.
10. U.S. Environmental Protection Agency. 1990. *Engineering Bulletin: Soil Washing*, EPA/540/2-90/017. Cincinnati, OH: U.S. EPA.
11. Ang, C. C., and Abdul, A. S. 1994. Evaluation of an ultrafiltration method for surfactant recovery and reuse during in situ washing of contaminated sites: laboratory and field studies. *Ground Water Monitor. Remed.*, 14(3):160–171.
12. Clarke, A. N., Oma, K. H., Megehee, M. M., Mutch, R. D., Jr, and Wilson, D. J. 1994. Surfactant-enhanced in-situ remediation: Current and future techniques. In: G. W. Gee and N. R. Wing (eds.), *In-Situ Remediation: Scientific Basis for Current and Future Technologies: Thirty-Third Hanford Symposium on Health and the Environment, Part 1*, pp. 965–987. Columbus, OH: Battelle Press.
13. Eykholt, G. R., and Mackenzie, P. D. 1993. Organic Sponge Engineering Program: 1. Overview of column experiments, Report #93CRD026. Schenectady, NY: GE Company, Corporate Research and Development Center.
14. Fayyoumi, W. 1997. Modeling of countercurrent washing and separation of contaminated sediments. M.S. thesis, Department of Civil and Environmental Engineering, University of Wisconsin—Madison.
15. Wills, B. A. 1992. *Mineral Processing Technology*, 5th ed. Pergammon Press.
16. Svarovsky, L. 1981. *Solid-Liquid Separation*, 2nd ed. London: Butterworths.
17. Witt, C. N. 1996. Investigation of mass balance modeling of ex situ harbor sediment remediation. M.S. Thesis, Department of Civil and Environmental Engineering, University of Wisconsin—Madison.
18. Svarovsky, L., and Thew, M. T. 1992. *Hydrocyclones: Analysis and Applications*, Dordrecht: Kluwer.
19. Press, W. H., Flannery, B. P., Teukolsky, S. A., and Vetterling, W. T. 1988. *Numerical Recipes in C*. Cambridge: Cambridge University Press.

12
Air Sparging, Soil Vapor Extraction, and Bioremediation in Low-Permeability Soils

C. Marjorie Aelion and Brian C. Kirtland
University of South Carolina, Columbia, South Carolina

I. INTRODUCTION

A major problem facing public health and the environment is the release of petroleum from leaking underground storage tanks (UST) into soil and its migration into groundwater aquifers. Aquifers are a vital natural resource supplying more than 50% of the U.S. population with water for domestic use (1). Public health may be threatened with exposure to groundwater containing petroleum due to its toxic constituents, particularly benzene, which is classified as a human carcinogen, but also by less toxic constituents including toluene, ethyl benzene, and xylene (BTEX). The U.S. Environmental Protection Agency (EPA) estimated that 100,000–400,000 USTs are currently leaking (2), mainly due to corrosion of the older bare-steel tanks (1). The potential threat exists for substantial increases in contaminant release, with several million UST systems in use in the United States containing petroleum or hazardous chemicals (1). It is estimated that service station gasoline tanks alone could be losing 11 million gal/year of product (3). The potential threat of leaking USTs was illustrated from a recent groundwater contamination survey in South Carolina which showed a dramatic increase in groundwater-contaminated sites due to leaking USTs. The number of reported groundwater-contaminated sites in South Carolina rose from 60 in 1980 to 3619 in 1998 (4). The sharp increase coincided with the enactment of UST control legislation requiring increased groundwater monitoring efforts. Of the reported 3619 contaminated groundwater sites, 3028 were attributed to leaking USTs (4). Petroleum products accounted for more than 88% of all chemical contaminants reported at these contaminated sites (4), and air sparging/soil vapor extraction (AS/SVE) is well suited to address this type of contamination.

A. Definition of AS/SVE

AS/SVE physically removes volatile organic contaminants from soil and groundwater and stimulates aerobic biological remediation of both volatile and less volatile contaminants

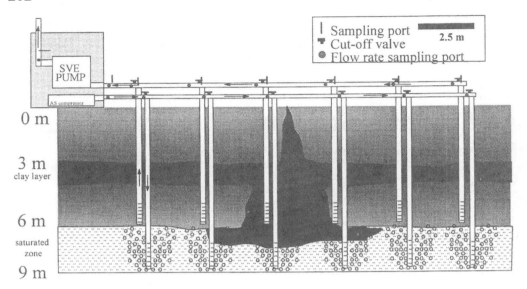

Figure 1 Subsurface diagram of the air sparging/soil vapor extraction system and location of cutoff valves and sampling ports.

by supplying oxygen to the subsurface. Air sparging (AS) involves injecting atmospheric air into the aquifer to induce mass transfer of volatile organic chemicals to the vapor phase and mass transfer of oxygen to the aqueous phase (Fig. 1). The injected air migrates either as bubbles, in discrete channels, or as a combination of both bubbles and channels through the contaminant plume as it flows upward through the saturated zone and into the vadose zone. In soils containing relatively large particles (approximately 4 mm in diameter) such as gravel, air travels as bubbles; while in smaller particles (~ 0.75 mm) air travels in channels (5–7). The injected air volatilizes the contaminants in the flow channels and transports them to the vadose zone, where they are either biodegraded or removed by SVE. The injected air may induce bulk water flow toward air channels due to water evaporation. Mathematical models have indicated that this bulk water flow is a significant contributor to contaminant removal in the saturated zone (8).

Often soil vapor extraction (SVE) is coupled with AS to enhance remedial efforts in the vadose zone and to prevent migration of contaminant vapors off-site. Large quantities of contaminants can be removed in this fashion. Through a manifold system, the location of these contaminants can be redirected from being dispersed in the subsurface environment, to being concentrated in a single outflow exhaust pipe which can be controlled readily. A vacuum is applied to the manifold system and the contaminants are vaporized and sucked from the SVE wells into the manifold system to the exhaust system. As vapor is removed by the system, vapor recharge occurs from surrounding soil vapor. If this vapor is uncontaminated, then a dilution effect also occurs.

B. Classes of Contaminants and Site Conditions for Which AS/SVE Is Appropriate

AS/SVE is best suited for remediating volatile contaminants from homogeneous sandy soils in which air permeability is high and thus a large area of the contaminant plume

Air Sparging, Soil Vapor Extraction and Bioremediation

will be affected by the injected air. AS/SVE is best suited for removing nonaqueous-phase liquids (NAPLs) that are soluble in water and volatile, as indicated by their vapor pressure and Henry's law constant. Henry's law defines the relation between the equilibrium concentration of gas dissolved in solution and the partial pressure of the gas (9). SVE is most suited for removing volatile NAPLs trapped between soil particles. A contaminant is considered appropriate for AS/SVE if it possesses a vapor pressure greater than 5 mm Hg and a Henry's constant greater than 10^{-5} atm m^3 mol^{-1} (10). Modeling studies indicated that AS is applicable to contaminants with high solubilities, even if their Henry's law constants are relatively low (e.g., methyl *t*-butyl ether) (8).

Silt and clay sediments are not considered appropriate for AS/SVE. The low permeability characteristics of clayey soil inhibit air flow through the subsurface, thus lowering contaminant removal efficiencies by reducing mass exchange rates of volatile contaminants to the vapor phase. Air flow patterns are affected by soil permeability, hydraulic conductivity, and soil structure. Less permeable sediments may cause the formation of distinct air flow channels up to the unsaturated zone producing poor air distribution (11). Silty and clayey sediments generally require higher air injection pressures than sandy sediments to achieve air flow through the saturated zone. Excessive pressure can destroy the soil formation and promote soil fracturing, which reduces the effectiveness of AS/SVE (12). Laboratory studies have demonstrated that this channeling or "fracturing" may be enhanced in fine-grain, low-permeability sediments, thus reducing the effectiveness of AS/SVE at such sites (5). Johnson et al. (13) and Loden (14) suggested a hydraulic conductivity and air permeability limit of 10^{-3} cm/s, and 10^{-9}–10^{-8} cm^2, respectively, for effective use of AS/SVE.

C. Biological Remediation

Biological remediation methods rely on subsurface microbes to degrade contaminants, and include bioventing, biosparging, bioaugmentation, and intrinsic bioremediation. The end products of aerobic hydrocarbon mineralization are microbial biomass, CO_2, and water. The main advantages of in-situ bioremediation are that (a) it may be carried out without removing the hydrocarbon-impacted soil and groundwater; (b) it may be more cost-effective compared to conventional treatment (15); and (c) microbial mineralization destroys the contamination, thus providing a permanent solution to the problem (Table 1).

Two essential criteria for effective microbial degradation of subsurface contaminants are that microorganisms with the ability to mineralize the contaminants of interest must be present in sufficient numbers, and the subsurface geology must have a relatively large hydraulic conductivity in order to transport essential nutrients and electron acceptors through the aquifer. Geologic formations with a hydraulic conductivity value greater than 10^{-4} cm/s are considered to be the most treatable using bioremediation methods (16).

Although hydrocarbons may be degraded both aerobically or anaerobically, aerobic degradation generally is considered the most efficient and rapid reaction because it requires less free energy for initiation and produces more energy per reaction (17). Aerobic biodegradation is usually limited by the amount of O_2 that can be transported to microorganisms in the subsurface. Physical remediation technologies commonly enhance bioremediation efforts by supplying oxygen to the subsurface, thus providing an adequate electron acceptor to stimulate aerobic biodegradation. Oxygen has a saturation concentration of 8 mg/L in the water, depending on the water's temperature. Using atmospheric

Table 1 Advantages and Disadvantages of Bioremediation as a Treatment Technology (57)

Advantages
1. Can treat hydrocarbons and other organic compounds, especially water-soluble pollutants and low levels of other compounds that would otherwise be difficult to remediate.
2. Does not usually generate waste products and typically results in complete degradation of the contaminants.
3. Utilizes the indigenous microbial flora and does not introduce potentially harmful microbial organisms.
4. Fast, safe, and generally economical.
5. Treatment moves with the groundwater.
6. Good for short-term treatment of organic contaminated groundwater.

Disadvantages
1. Can be inhibited by heavy metals and some organic chemicals.
2. Bacteria can plug the soil and reduce circulation.
3. Introduction of nutrients could adversely affect nearby surface waters.
4. Nutrient residues and chemical intermediates may cause taste and odor problems.
5. Labor and maintenance requirements may be high, especially for long-term treatment.
6. May not work for aquifers with low permeabilities that do not permit adequate circulation of nutrients.
7. Long-term effects are unknown.

air, in which O_2 is present at 20.9%, as an oxygen source is an effective and cost-efficient method to oxygenate the subsurface. Following addition of atmospheric air, the vadose zone also should be oxygenated at this same concentration of 20.9%.

Biodegradation stimulated by AS generally is thought to be limited by the rate of oxygen transfer from the gaseous to the aqueous phase, and diffusion within the aquifer system. Thus biodegradation and physical removal are both diffusion limited processes (18). Based on this diffusion limitation concept, biodegradation rates in low-permeability, clayey soil should be slower than those measured in sandy sediments and may render bioremediation efforts ineffective. Laboratory and in-situ studies have shown the capability of microbes to degrade hydrocarbon contamination in sandy soils aerobically (19–21), but in-situ rates of biodegradation in clayey soil are not well documented. In-situ petroleum biodegradation studies have suggested that biodegradation may be occurring, but limited in clayey soils (22), and possibly may be compound-specific (21).

II. TECHNICAL DESCRIPTION OF AS/SVE SYSTEM

A. Construction/System Design

AS/SVE is commonly used in petroleum hydrocarbon clean-up and therefore the risk of the presence of free product is high. Some considerations in AS/SVE design therefore include precluding displacing and mobilizing hazardous constituents of the contaminant plume, the vapors, the free-product phase, or the dissolved-phase contaminants (23). To be efficient in contaminant removal, the free product must be removed and properly disposed of prior to running an AS/SVE system. Although SVE may address this contamination, volatilizing it and passing it through an AS/SVE system can contribute significantly to air pollution, escalate costs associated with treatment of the off-gas, and greatly extend the operational time required to meet remediation goals.

Site assessment is one of the most important steps to remediation system design. First, the horizontal and vertical extent, and contaminant concentrations associated with the subsurface, must be determined. Second, a decision must be made as to which part of the plume and at what concentrations the AS/SVE system is to be directed. The SVE wells are designed to be vacuum-tight and screened over the appropriate depth corresponding to the contamination measured. Both vertical and horizontal wells have been used in AS/SVE systems. Horizontal wells may be more difficult to install, particularly as depth increases. However, the area over which vapors can be collected increases significantly compared to that of vertical wells.

In order to capture the vapors from the site, the radius of influence of the SVE wells must be calculated and the appropriate spacing between wells established (13). Soil permeability, porosity, moisture content, stratigraphy, and depth to groundwater are factors which contribute to the radius of influence (25). Thus, if these can be determined at a site, even a site with low permeability, the appropriate number and distribution of SVE wells can be estimated. The cost of the system may increase as the number of wells increases. If too few wells are used, areas which are not affected directly by the AS/SVE system will occur. If too many wells are selected, the cost and maintenance of the system increase. Vapor flow models exist to determine design parameters (26) for various geological systems, from clays and silts, to sands, gravel, fractured rock, and karst (25).

The efficiency of the mass transfer associated with the AS portion of AS/SVE is affected by the bubble size of the injected oxygen. Mathematical models of AS indicated that smaller bubbles with a greater surface-area-to-volume ratio will increase mass transfer from the liquid to the vapor phase (27). Air diffusers can be used to select bubble size. Typical air flow rates are 3 to 10 standard cfm per sparge point (23). Reported values for the radius of influence for AS have ranged from 0 to 2 m, to 8 to 10 m (12), based on information from the American Petroleum Institute, in Situ Air Sparging database.

B. Air Quality Control of Emissions

A successful AS/SVE system produces large concentrations of volatile organic contaminants in the air flow returning to the land surface. Soil venting of these off-gases is allowed in certain states if an air permit is received and the concentrations vented are below those limits designated in the air permit. In this case dilution into the atmosphere is the treatment system.

In other cases local, state, or federal regulations require the treatment of those off-gases. Often treatment of the off-gases itself produces a waste which must be disposed of properly. In many cases traditional technologies such as granular activated carbon are used to treat off-gases. Adsorption of the contaminants occurs onto the activated carbon until all active sites are used. The carbon must be removed, disposed of, and replaced, or regenerated by desorption. Regeneration on-site is preferable. Extracted chemicals must again be properly disposed of or treated. Regeneration time for, and costs associated with the carbon, can be estimated from the gas loading, which is a function of the gas velocity, the stripping factor for the particular contaminants, and the influent concentration to the exhaust gas, similar to those calculations for air stripping towers (28).

Other treatment systems are based on thermal processes, such as incineration or catalytic oxidation of the off-gases. Although used less frequently than carbon adsorption, catalytic oxidation destroys the contaminants, and thus has a limited cost associated with the generation of wastes from the treatment of the off-gases. However, the catalysis requires an energy input and potential reduction in efficiency over time. Also, the catalyst may become poisoned and need to be regenerated or replaced. Removal of the off-gas must be monitored to assure that the system is truly capturing and not releasing the contaminants into the atmosphere. Also, as with incinerators, thermal oxidation of chlorinated compounds produces unwanted products including HCl and potentially products of incomplete combustion (PICs).

Biological treatment of gaseous waste streams is currently under development. In this case the gases are passed through a biological filter and contaminants are destroyed. Although theoretically this process would allow for the destruction of the contaminants without the need for subsequent disposal of the treated waste, few sites are in operation which use this option and it has not been thoroughly evaluated in field-scale AS/SVE systems.

III. MONITORING AS/SVE

A. Physical Removal

The major factors which can be manipulated in the operation of AS/SVE are air flow rates or gas injection pressure, and vacuum rates imposed. The major equipment therefore is an air compressor, normally an oil-free air compressor, a manifold directed to one or more air sparging wells, and a vacuum pump. Pressures, moisture content, and vacuum flow rates should be monitored. Exhaust samples and flow rates should be measured daily to obtain petroleum mass removal rates. Hydrocarbon removal rates are calculated based on the ideal gas law at a temperature of 25°C and pressure of 1 atm using the equation:

$$R = \frac{Q * (\text{ppm}_{HC}) * \text{MW}_{HC} * 10^{-6} * 1 \text{ atm}}{0.0821 \text{ L-atm/K-mol} * 298 \text{ K}} * 1.44 \tag{1}$$

where R is the mass removed (kg/day); Q is the measured flow rate in the exhaust stack (L/min); ppm_{HC} is the hydrocarbon concentration (ppm); MW_{HC} is the average molecular weight of the more volatile fraction of weathered gasoline, estimated to be 111 g/mol (13); K is temperature in ° kelvin; and 1.44 is a conversion factor for grams to kilograms and minutes to days.

Equation (1) can be simplified by substituting 111 g/mol as the model hydrocarbon and correcting for alternative pressures (atm) and temperatures (K) by

$$R = Q * (\text{ppm}_{HC}) * 1.95 \times 10^{-3} * \frac{P}{T} \tag{2}$$

where T is the exhaust temperature (K); and P is the pressure (atm) is the exhaust stack.

Exhaust stack samples may be collected and analyzed for total petroleum hydrocarbons using gas chromatography (GC) or hand meters. Hand meters can save significant time and effort. Hand meters may also be applied in estimating biological contributions by monitoring CO_2 and O_2 in the exhaust gas. Studies have shown an excellent correlation between hand meter and GC values of CO_2, and an adequate correlation between hand meter and GC values of O_2 (29).

B. Biological Removal

Monitoring petroleum hydrocarbon mineralization and demonstrating that microbial remediation is occuring in the subsurface are vital parts of assessing the success of a remediation program. It may be difficult to show that the disappearance of hydrocarbons is attributable to microbial degradation instead of physical processes such as sorption, volatilization, transformation, or dilution. Providing evidence of microbial degradation under field conditions is commonly achieved by monitoring soil gas CO_2, O_2, and contaminant concentrations over time (30–32). An increase in CO_2 in conjunction with a decrease in O_2 and contaminant concentrations over time has been used to estimate rates of aerobic petroleum biodegradation for bioremediation, bioventing, and soil vapor extraction (22, 30, 33–35).

Hinchee and Ong (30) developed an in-situ test method to estimate the aerobic biodegradation rates of hydrocarbons in contaminated soil. The test consisted of injecting atmospheric air and a tracer gas (helium) into the vadose zone to oxygenate the subsurface, then monitoring CO_2 and O_2 concentrations over time after the air was turned off. The injected helium (1–2%) was measured over time to ensure that the air injected into the vadose zone was the same gas being monitored over time. Hinchee (36) suggested that O_2 utilization rates may provide a more accurate estimate of contaminant biodegradation rates than CO_2 production rates because hydrocarbons may be only partially mineralized to intermediate by-products rather than to CO_2, and CO_2 may be converted to bicarbonate in alkaline soils. Both of these underestimate biodegradation rates, thus calculated respiration rates based on O_2 utilization, may provide a quick and effective technique to determine if a site is suitable for applied bioremediation technologies, such as bioventing. O_2 utilization rates between 1.2% and 24% per day have been reported in petroleum biodegradation field studies, and 1.2% per day has been suggested as the rate limit for contaminant biodegradation for effective implementation of enhanced bioremediation (36).

Even though soil gas monitoring is commonly incorporated into a bioremediation program, problems may be encountered with its use. A disadvantage of soil gas monitoring is that fluctuations in CO_2 concentrations may require many data points over time to show that microbial mineralization is occurring in the subsurface. It may be difficult or impossible to differentiate microbially derived CO_2 from geochemically or plant-derived CO_2 soil gas if CO_2 and O_2 concentrations are near background or at atmospheric concentrations. Under natural conditions, soil CO_2 concentrations vary over a wide range [approximately 0.35–3.5% (v/v)] (37). Therefore, elevated CO_2 concentrations in the vadose zone do not necessarily indicate contaminant mineralization.

Petroleum biodegradation studies are necessary at AS/SVE sites in order to understand the ultimate fate of the contaminant in the subsurface and to evaluate the effectiveness of microbial remediation efforts compared to physical removal. At most petroleum hydrocarbon sites where AS/SVE is employed, volatilization accounts for a significantly greater fraction of hydrocarbon removal than biological processes during short-term operation (weeks/months) (8, 18, 24). Biodegradation becomes relatively more important in long-term system operations, during which it may contribute significantly to the removal of residual or less volatile NAPLs (18).

IV. IMPACT OF SYSTEM OPERATION ON EFFICIENCY OF AS/SVE

A. Removal Efficiencies

The depth to groundwater and water table fluctuations may have a major impact on AS/SVE's effectiveness. One strategy of SVE is to place SVE wells as close to the water table as possible, which normally coincides with the greatest contaminant concentrations. AS/SVE studies in low-permeability soils have shown that when the groundwater table rises due to increased precipitation and infiltration, petroleum mass removal rates decrease due to the reduction in the length of well screen in the vadose zone and the saturation of sediments at the previously exposed groundwater table–vadose zone interface (38). Studies have shown the benefits to SVE of lowering water table levels to expose the usually highly contaminated capillary fringe (39). Higher removal rates would be expected when groundwater levels are lower because the highly contaminated capillary fringe, also called a vertical smear zone of residual hydrocarbons (25,40), would be exposed to AS/SVE. Laboratory infiltration experiments using toluene in heterogeneous unsaturated sediments also suggested that significant amounts of contaminant would be entrapped within low-permeability lenses as the result of capillary force (41).

Soil type directly affects the air flow and ultimately the degree to which contaminants are removed from the subsurface. Laboratory studies of AS using five soil types (fine sand to fine gravel) found that when the effective grain size was <0.2 mm, the time for complete benzene removal increased dramatically (6). Air injected into the fine gravel traveled in bubble form and was distributed uniformly allowing effective volatilization of benzene. In contrast, higher injection pressures were necessary for fine sand. Air traveled in small channels, and benzene had to migrate via diffusion to these channels to be removed. Laboratory studies found that the time required for complete toluene removal increased eight times in fine sand compared to fine gravel (7). Results also showed that volatilization was the dominant removal process of toluene in gravelly soil and removal rates increased with increased flow rates (380 to 960 mL/min), but the benefit of rates higher than 960 mL/min was limited.

B. Continuous Versus Pulsed Operation

Strategies for operating AS/SVE systems include continuous operation or intermittent operation termed "pulsing." Pulsed AS/SVE has been suggested to enhance mixing in the subsurface, thereby providing enhanced oxygenation and volatilization of dissolved-phase and NAPL contaminants while conserving operational costs (11). Modeling studies of pulsed SVE suggested that when mass transport is limiting, one-fifth or less volume of air flow may be used and still sustain comparable clean-up times while maximizing savings (42). AS is known to cause groundwater mounding during the first minutes or hours of operation (depending on sediment grain size), due to the displacement of water by the injected air. Conversely, there is a collapse of the air channels when AS stops, which causes a temporary depression in the aquifer. Pulsed AS uses this mounding and depression of an aquifer to induce bulk groundwater mixing, which redistributes dissolved-contaminants relative to the air channels and thus may yield a larger radius of influence and reduce diffusion limitations (18). Continuous AS may induce groundwater mixing by the frictional drag from flowing air, physical displacement of groundwater forming channels, capillary interaction of air and water, thermal convection, migration of fine materials resulting in redirection of airflow, and evaporative loss of water in the air stream and the resulting groundwater inflow to maintain water balance (43). The magnitude of groundwater mixing resulting from these processes, however, is suggested to be greater during pulsed than continuous operation (43).

Several studies have measured enhanced contaminant removal efficiencies of pulsed system operation compared to continuous operation in sandy sediments (6,43,44). Clayton et al. (44) measured the effects of groundwater mixing from pulsed AS by measuring saturated-zone moisture content, and changes in dissolved oxygen and mass removal rates. Based on a pulsing frequency of 12–24 h over a 16-week period, there was a three- to fivefold increase in mass removal rates going from continuous operation to pulsed AS, from 20 kg/day to 35 kg/day. Payne et al. (43) compared mass removal of trichloroethylene (TCE) in groundwater during continuous and pulsed AS. They measured the percent change from initial TCE concentrations in groundwater at 1.5- and 3.0-m radial distance from the AS well. During the 2-week study, the AS system was pulsed using a time schedule of 14 min on and 14 min off. Results showed a slight, but insignificant, increase in TCE concentrations at the 1.5- and 3.0-m radius during continuous AS operation, and a significant decrease at 1.5 m during pulsed AS operation. Reddy and Adams (6) investigated pulsed (2- and 6-h intervals) and continuous AS in coarse and fine sand columns in the laboratory. No benefit was observed in the coarse sand, but in fine sand pulsed AS removed greater concentrations of contaminants and greatly increased residual contaminant removal. Pulsed AS in the fine sand caused relocation of the air channels, which aided in distributing air to the contaminants in these areas. These three studies suggest that pulsed AS operation may induce groundwater mixing and contaminant volatilization, while minimizing displacement of the contaminant plume by reducing vacuum-induced fluid transport, thereby producing greater mass removal rates in sandy aquifers.

A small number of studies on AS/SVE has been carried out in low-permeability sediments relative to the number of carried out in sandy sediments. Studies of pulsed AS/SVE in low-permeability sediments were carried out in the Appalachian Piedmont at a site where sediment permeability increased with depth due to increasing sand and gravel fractions with depth (22,38,45). For this AS/SVE site, 8-h or 24-h operational periods did not produce significantly higher mass extraction rates compared to continuously operating the system 24 h per day for 44 days. However, pulsed operation did improve extraction rates compared to the last 15 days of continuous operation, during which removal rates decreased significantly (38). Also, the contaminant removal rates were vulnerable to water table fluctuations because most of the contamination was located at the water table–vadose zone interface. These periods of inefficiency may have been due to changes in water table level, sediment moisture content, mass transfer limitations for O_2, or contaminant or air channeling. Results at this site suggested that when mass extraction rates fell below approximately 18 kg/day, it was more efficient to operate the system on a pulsed schedule of either 8 h every other day or 24 h every other day. This study demonstrated the importance of closely monitoring AS/SVE mass extraction rates over time in order to better manage periods of inefficient operation.

V. AS/SVE FIELD STUDIES

A. Use of AS/SVE for Addressing Groundwater Contamination

Many AS/SVE field studies have demonstrated the effectiveness of physically removing contaminants from the subsurface. Fields projects carried out using AS/SVE are often not scientifically detailed in the research literature. Malot (25,40) and Marley et al. (23) have summarized several AS, SVE, or AS/SVE studies from the United States (Table 2).

Pijls et al. (34) investigated AS/SVE at a gasoline station in sediment consisting of fine sand and gravel. Within two years of system operation, 4000 kg of hydrocarbons were removed and the unsaturated zone hydrocarbon levels decreased from 10,000 mg/kg

Table 2 Comparison of Several AS/SVE Remedial Systems[a]

Site (Ref.)	System	Chemical	System operation	Depth	Soil type	Amount/rate removed
Belleview (40)	SVE	Gasoline	NS	6.0 m	Clayey sand	136–430 kg/day, 9,979 kg, 95.9–99.7%
Verona (25)	SVE	PCE, TCE, Solvents	NS	7.6 m	Slit sand	1996 kg/day
Barceloneta (25)	SVE	Carbon tetrachloride	1.0 atm	91 m	Clay	113 kg/day, 99.98%
Connecticut (23)	AS/SVE	VOC/TCE	AS: 1–4 atm SVE: 0.08–0.3 m^3/min	4.6 m bwt	Fine-grained material	1.8 kg VOC, 30 days
Rhode Island (23)	AS/SVE	BTEX	AS: 0.4–0.5 atm SVE: 0.06–0.17 m^3/min	6.1 m bwt	Fine to very fine sand	2.3–4.5 kg, 60 days
Massachusetts (23)	AS/SVE	Gasoline	AS: 0.3–0.4 atm SVE: 0.08–0.14 m^3/min, intermittent	15–19 ft	Medium to fine sands	254–280 L AS/SVE, 120 days; 2271 L SVE, 357 days
Ohio (58)	MRVS	VOC	SVE: 40 m^3/min	4.6–6.7 m	Silty clay	88–98%
New Jersey (56)	SVE PFE	TCE, BTEX	AS: 0.75 atm SVE:0.12 m^3/min	2.0–3.6 m	Glacial till	4.5 × 10^{-6} kg/min TCE
South Carolina (22)	AS/SVE	Gasoline	AS: 0.2 atm SVE: 3.4 m^3/min	6.9 m	Clayey sand	0.1 kg^3/min, 21 days
North Carolina (59)	AS/SVE	JP-4	NS	1.0–3.5 m	NS	3 kg/day, 1700 kg, 55% TPH; 98% benzene
Kwinana, Australia (24)	AS/SVE	Gasoline	AS: 35 kPa, 0.3 m^3/min SVE: 1.08 m^3/min	6.2–7.2 m	Medium to fine sand	0.2 kg/day, 0.150 kg TPH bio; 1.090 kg volatilized, 6 days

[a] MRVS = mixed-region vapor stripping; PFE = pneumatic fracturing extraction; NS = not stated; bwt = below water table.

to 100–260 mg/kg. The concentration of volatile hydrocarbons in the soil vapor decreased from 160 g/m^3 to 0.5 g/m^3. Pijls et al. (34) calculated a petroleum mass removal rate of 5.5 kg/day.

Mehran (39) examined AS/SVE in a fine-grain sand aquifer contaminated with gasoline. After two years of operation, approximately 5900 kg of total petroleum hydrocarbons (TPH) were removed and benzene concentrations is groundwater were reduced from 4800 mg/L to 0.3 mg/L. Mehran (39) calculated a petroleum mass removal rate of 8.9 kg/day, which was similar to that calculated by Pijls et al. (34) of 5.5 kg/day.

Another recent study using SVE reported considerably greater petroleum mass removal efficiency compared to the above-mentioned studies. Hinchee et al. (46) removed 11,300 kg of JP-4 jet fuel over 15 months using only soil venting in sandy sediments. Hinchee et al.'s (46) initial exhaust rate was 44 m^3/h and was increased to 2500 m^3/h, which is relatively high and may have resulted in the elevated petroleum mass removal rates.

Modeling and field studies of AS/SVE have suggested that biodegradation of petroleum hydrocarbons was significantly less than physical removal (24,46–48). Assessment of the biological removal is more difficult to estimate than the physical removal and therefore less information is available. Thornton and Wootan (47) measured CO_2 levels in an AS/SVE exhaust stack to estimate the microbial contribution to petroleum removal in the subsurface during 11 days of remedial efforts. Assuming that all CO_2 produced was from the mineralization of gasoline, they estimated that 2% of the gasoline mass removed was from biodegradation based on the complete mineralization of hexane to CO_2 and water. Hinchee et al. (46) removed 11,300 kg of JP-4 jet fuel with AS/SVE. They estimated that 2100–2200 kg of the JP-4 removed were microbially mineralized, based on O_2 and CO_2 measurements over 15 months of remedial efforts. The initial fraction of JP-4 biodegraded dropped rapidly in the first 30 days from 30% to a steady-state value of 15%.

Petroleum volatilization and biodegradation in groundwater were critically assessed at a field site in Kwinana, Australia (24). The gasoline-contaminated site contained coarse, medium, and fine sands. Based on an O_2 utilization rate of at least 10 mg/L-day, 150–210 g of TPH were biodegraded, compared to 1090 g volatilized during the 6-day AS/SVE test. Johnston et al.'s (24) O_2 utilization rate was comparable to Pijls et al.'s (34) rate of 12 mg/L-day in groundwater with the addition of nutrients. Johnston et al.'s (24) estimate of biodegradation contributing approximately 16% of total mass removal is in agreement with predictions from modeling studies of AS/SVE (48).

Bioventing is essentially analogous to conventional SVE but utilizes strategic well placement and lower flow rates to maximize the vapor retention time within the soil to encourage microbial degradation of the hydrocarbon vapors, and reduce or remove the need to treat off-gases. Dupont (49) used a combination of conventional SVE and bioventing to remediate a JP-4 jet fuel spill. The air permeability at this site was measured at 2.2×10^{-6} cm^2. During a high-rate venting period (2100 m^3/h), the volatilization rate of petroleum hydrocarbons ranged from 90 to 180 kg/day. Low venting rates (490–970 m^3/h) produced petroleum removal rates less than 9 kg/day. Over approximately two years of system operation, an estimated 53,650 kg of JP-4 were removed from the vadose zone through volatilization and 42,150 kg were estimated to have been microbially mineralized. This corresponded to a 56–44% ratio of volatilization to biodegradation. Billings et al. (50) used lower AS injection rates to enhance aerobic biodegradation (termed biosparging) and, based on CO_2 measurements, attributed approximately 75% of the total hydrocarbons removed to microbial mineralization.

This wide range of the reported biological versus physical removal (2–75%) at AS/SVE sites may be due to variations in the AS/SVE systems employed, such as various operational strategies implemented at each site, and most important, the duration of remediation efforts. Also, site-specific variations may account for differences, such as the diversity of microbial populations, and subsurface geology at each site which may influence air flow, moisture content, contaminant bioavailability, and thus the rates of petroleum biodegradation.

B. Use of AS/SVE in Low-Permeability Sediments

Several field studies of AS/SVE in low-permeability sediment have measured comparable petroleum mass removal rates to those conducted in sandy soils, and have effectively lowered contaminant levels in the subsurface (22,38,45,51,52). Clodfelter (51) employed AS/SVE in East Texas to remediate clayey sediments with a measured hydraulic conductivity of 1.5×10^{-4} cm/s which is beyond the recommended limit (10^{-3} cm/s). He reported initial groundwater BTEX levels between 60 and 34 ppm. After three months of AS/SVE operation, nondetectable BTEX groundwater concentrations were reported.

Ghandehari et al. (52) used AS/SVE to clean up a petroleum spill from an UST in Charlotte, NC, where low-permeability, saprolitic soils of the Piedmont are dominant. The aquifer's hydraulic conductivity was measured at 1.04×10^{-5} cm/s, which is typical of clayey sediments in the Piedmont. TPH concentrations measured in sediments ranged from 24 to 54 ppm, and 147 kg of BTEX were estimated to be at the site before commencement of AS/SVE. Horizontal wells were used in conjunction with vertical wells to increase the screened well surface-to-aquifer ratio. A 11.2-kW vacuum pump generated exhaust flows between 340 and 425 m^3/h. All monitored groundwater and vadose zone wells showed a decrease in BTEX levels, with an average concentration of 15,000 ppb to nondetectable concentrations following 11 months of AS/SVE operation (53).

Several studies of AS/SVE effectiveness in low-permeability soil of the Piedmont region have been conducted at a field site in Columbia, SC. Widdowson et al. (45) conducted a pilot study of SVE using a 5.22-kW vacuum pump and produced hydrocarbon mass extraction rates ranging from 22 to 68 kg/day with the system operating from 6–8 h/day. Flow rates in the SVE wells ranged from 0.65 to 1.7 m^3/min over a wellhead vacuum varying between 152 and 305 mmHg. Groundwater BTEX and TPH concentrations showed no discernible decrease during the first several months of the pilot study, suggesting that the majority of the contamination removed was from the vadose zone. Aelion et al. (22) conducted further studies with this system and found that initial extraction rates were two to three times higher than subsequent samples. This demonstrated that total mass extraction rates may be grossly miscalculated if only a few sampling points are used and data are extrapolated over a large time period. Depleted O_2 and elevated CO_2 concentrations measured in the deepest sediment vapor probe, located at 6.7 m, suggested that petroleum biodegradation was occurring, but that the greatest decreases of vadose zone BTEX levels were measured during AS/SVE. In a later study at this site, Kirtland and Aelion (38) operated the AS/SVE system continuously for 44 days and removed 608 kg of hydrocarbons for an average extraction rate of 14.3 ± 13.1 kg/day. BTEX concentrations in the vadose zone were reduced from 5 ppm to 1 ppm. The average SVE exhaust flow rate was 257 ± 5.4 m^3/h. These studies at this site showed that AS/SVE was an effective method for removing petroleum hydrocarbons from the subsurface and BTEX concentrations can be reduced in the vadose zone.

C. Pneumatic and Hydraulic Fracturing in Low-Permeability Formations

SVE is generally not amenable to sites consisting of igneous rock, shale, clay, dense silt, and glacial till. However, new techniques such as pneumatic and hydraulic fracturing can increase the number of air pathways in the subsurface and allow SVE to be an effective remedial technique. Pneumatic fracturing extraction (PFE) involves injecting air at high pressure into the geologic formation to create a network of air channels (Fig. 2A). Specifically, bursts of compressed air are injected at 500 psig for 10–20 s into narrow boreholes to induce fracturing of the formation. Pneumatic fracturing has also been employed in low-permeability soils to enhance air flow rates and transport rates of soil amendments to microorganisms to enhance in-situ biodegradation (54). Hydraulic fracturing is widely used in the petroleum industry and involves the injection of a highly viscous liquid–sand slurry (guar gum gel and coarse-grained sand) or water in the subsurface at pressures that induce sediment fracturing (Fig. 2B). Enzymes in the mixture degrade the viscous liquid and it is pumped out of the subsurface, leaving sand to support the open fissures.

Field demonstrations of PFE and hydraulic fracturing were performed in low-permeability sediments. PFE was demonstrated at a site consisting of glacial till and shale which was contaminated with chlorinated and petroleum hydrocarbons (55,56) (Table 2). PFE increased the flow rate in SVE wells between 400% and 700% and increased mass removal by approximately 675%. TCE was removed at a rate of 4.5 mg/min during the 225-day test.

A field site for hydraulic fracturing was located on Oak Brook, IL, and was contaminated with TCE and other VOCs (56). The Oak Brook site consisted of clayey glacial drift interbedded with sand deposits and had air permeabilities ranging between 4×10^{-6} and 7×10^{-7} cm/s. Two fractured wells were compared to a conventional unfractured well during a 160-day SVE test. Vapor discharge in the fractured wells averaged 13–20

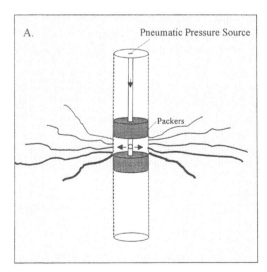
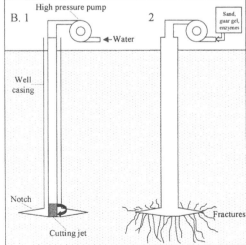

Figure 2 (A) Subsurface diagram of pneumatic fracturing extraction (PFE). (B) Hydraulic fracturing. (Modified from Ref. 56.)

times higher than the conventional well, with flow rates between 404.7 and 967.9 L/min. The mass recovered in the fractured wells was 7–14 times greater than that recovered in the conventional wells.

VI. CONCLUSION

Many field studies have demonstrated that AS/SVE is effective in removing petroleum hydrocarbons from both sandy sediments and low-permeability sediments at hydraulic conductivities lower than those previously recommended. Studies that used AS/SVE to remediate UST petroleum spills in low-permeability, clayey soil reported a large decrease in BTEX concentrations from soil and in many cases from groundwater, and had comparable mass extraction rates compared to other studies conducted in more permeable, sandy soils (22,38,45,51,52). Physical removal processes, however, become less efficient with time and may have limited success in removing petroleum contamination from clayey sediments. It is important to monitor the efficiency of the AS/SVE system and have flexibility to determine what type of operational program is appropriate as conditions change over time. Altering the vacuum, air flow rates, and selecting pulsing versus continuous operation may be necessary to maximize physical and biological removal of contaminants.

Petroleum clean-up operations may physically remove contaminants from the subsurface, impede further migration, or enhance microbial degradation of the subsurface contaminants. The percentage of total contaminant removal represented by the microbial component has varied in extent at several field studies. Microbial removal increases relative to physical removal as the physical removal efficiency decreases with time, and as flow rates are reduced. Selecting the proper combination of physical and biological removal technologies can improve overall contaminant removal. Physical removal may be reduced by design in cases in which physical removal has become less effective over time, exhaust gas concentrations are close to the permitted level, or in clayey, low-permeability sediments in which AS/SVE is not able to reach homogeneously throughout the subsurface. However, more field studies are necessary to evaluate the capability and limitations of AS/SVE technologies in clayey soils and sediments, as well as to develop easily implemented technological modifications to address the special challenges associated with low-permeability sediments.

ACKNOWLEDGMENTS

This research was funded by the National Science Foundation (93-50314), the South Carolina Hazardous Waste Management Research Fund, and the South Carolina Department of Health and Environmental Control.

REFERENCES

1. Blackman, C. W. 1996. *Basic Hazardous Waste Management*, 2nd ed., p. 293. Boca Raton, FL: Lewis Publishers.
2. U.S. Environmental Protection Agency, 1994. *UST Program Facts Cleaning up Releases*, EPA 510-F-94-006. Washington, DC: Office of Solid Waste and Emergency Response.

3. Riser-Roberts, E. 1992. Bioremediation of Petroleum Contaminated Sites, p. 197. Boca Raton, FL: C. K. Smoley.
4. Devlin, R., and Bucklin, C. 1998. South Carolina Groundwater Contamination Inventory 1998, p. 183. South Carolina Department of Health and Environmental Control, Water Monitoring Assessment & Protection Division, Columbia.
5. Ji, W., Dahmani, A., Ahlfeld, D., Lin, J. D., and Hill, E. 1993. Laboratory study of air sparging: air flow visualization. *Ground Water Monitor Rev.*, 13(4):141–145.
6. Reddy, K. R., and Adams, J. A. 1998. System effects on benzene removal from saturated soils and ground water using air sparging. *J. Environ. Eng.*, 124(3):288–299.
7. Semer, R., and Reddy, K. R. 1998. Mechanisms controlling toluene removal from saturated soils during in situ air sparging. *J. Hazard. Mater.*, 57:209–230.
8. Johnson, P. C. 1998. Assessment of the contributions of volatilization and biodegradation to in situ air sparging performance. *Environ. Sci. Technol.*, 32:276–281.
9. Davis, M. L., and Cornwell, D. A. 1991. *Introduction to Environmental Engineering*, 2nd ed., p. 822. New York: McGraw-Hill.
10. Brown, R., Herman, C., and Henry, E. 1991. The use of aeration in environmental clean-ups. In: *Proc. Haztech International Pittsburgh Waste Conf.*, Pittsburgh, PA, May 14–16, pp. 2A.1–2A.42.
11. Johnson, R. L., Johnson, P. C., McWhorter, B. B., Hinchee, R. E., and Goodman, I. 1993. An overview of in situ air sparging. *Ground Water Monitor Rev.*, 13(4):127–135.
12. Marley, M. C., Bruell, C. J., and Hopkins, H. H. 1995. Air sparging technology: A practice update. In: R. E. Hinchee, R. N. Miller, and P. C. Johnson (eds.), *In Situ Aeration: Air Sparging, Bioventing, and Related Remediation Processes*, pp. 31–37. Columbus, OH: Battelle Press.
13. Johnson, P. C., Stanely, C. C., Kemblowski, M. W., Byers, D. L., and Colthart, J. D. 1990. A practical approach to the design, operation, and monitoring of in situ soil-venting systems. *Ground Water Monitor Rev.*, 10(2):159–178.
14. Loden, M. E. 1992. *A Technology Assessment of Soil Vapor Extraction and Air Sparging*, EPA/600/R-92/173. Washington, DC: U.S. Environmental Protection Agency, Office of Research and Development.
15. Davis, K. L., Reed, G. D., and Walter, L. 1995. A comparative cost analysis of petroleum remediation technologies. In: R. E. Hinchee, J. A. Kittel, and H. J. Reisinger (eds.), *Applied Bioremediation of Petroleum Hydrocarbons*, pp. 73–80. Columbus, OH: Battelle Press.
16. Bedient, P. B., and Rifai, H. S. 1992. Ground-water contaminant modeling for bioremediation—a review. *J. Hazard. Mater.*, 32(2–3):225–243.
17. Cookman, J. T., Jr. 1995. *Bioremediation Engineering Design & Application*. New York: McGraw-Hill.
18. Boersma, P. M., Piontek, K. R., and Newman, P. A. B. 1995. Sparging effectiveness for groundwater restoration. In: R. E. Hinchee, R. N. Miller, and P. C. Johnson (eds.), *In Situ Aeration: Air Sparging, Bioventing, and Related Remediation Processes*, pp. 39–46. Columbus, OH: Battelle Press.
19. Atlas, R. M. 1978. Microorganisms and petroleum pollutants. *BioScience*, 28:387–391.
20. Fogel, S., Findlay, M., and Moore, A. 1990. Enhanced bioremediation techniques for in situ and onsite treatment of petroleum contaminated soils and groundwater. In C. Bell (ed.), *Petroleum Contaminated Soils*, pp. 201–209. Chelsea, MI: Lewis Publishers, Inc.
21. Aelion, C. M., and Bradley, P. M. 1991. Aerobic biodegradation potential of subsurface microorganisms from a jet fuel-contaminated aquifer. *Appl. Environ. Microbiol.*, 57(1):57–63.
22. Aelion, C. M., Widdowson, M. A., Ray, R. P., Reeves, H. W., and Shaw, J. N. 1995. Biodegradation, vapor extraction, and air sparging in low-permeability soils. In: R. E. Hinchee, R. N. Miller, and P. C. Johnson (eds.), *In Situ Aeration: Air Sparging, Bioventing, and Related Remediation Processes*, pp. 127–134. Columbus, OH: Battelle Press.

23. Marley, M. C., Hazebrouck, D. J., and Walsh, M. T. 1992. The application of in situ air sparging as an innovative soils and ground water remediation technology. *Ground Water Monitor. Rev.*, 12:137–145.
24. Johnston, C. D., Rayner, J. L., Patterson, B. M., and Davis, G. B. 1998. Volatilisation and biodegradation during air sparging of dissolved BTEX-contaminated groundwater. *J. Contamin. Hydrol.*, 33:377–404.
25. Malot, J. J. 1990. Vacuum extraction: Effective cleanup of soils and groundwater. In: H. M. Freeman (ed.), *Innovative Hazardous Waste Treatment Technology Series, Volume 2, Chemical Processes*, pp. 33–43. Lancaster, PA: Technomic.
26. Sawyer, C. S., and Kamakoti, M. 1998. Optimal flow rates and well locations for soil vapor extraction design. *J. Contamin. Hydrol.*, 32:63–76.
27. Wilson, D. J., Gómez-Lahoz, C., and Rodriguez-Maroto, J. M. 1994. Groundwater cleanup by in-situ sparging. VIII. Effect of air channeling on dissolved volatile organic compounds removal efficiency. *Separation Sci. Technol.*, 29(18):2387–2418.
28. Dvorak, B. I., Lawler, D. F., Speitel, G. E., Jr., Jones, D. L., and Broadway, D. A. 1993. Selecting among physical/chemical processes for removing synthetic organics from water. *Water Environ. Res.*, 65:827–838.
29. Aelion, C. M., Shaw, N. J., Ray, R. P., Widdowson, M. A., and Reeves, H. W. 1996. Simplified methods for monitoring petroleum-contaminated ground water and soil vapor. *J. Soil Contamin.*, 5(3):225–241.
30. Hinchee, R. E., and Ong, S. K. 1992. A rapid in situ respiration test for measuring aerobic biodegradation rates of hydrocarbons in soil. *Air Waste Manage. Assoc.*, 42(10):1305–1312.
31. Wood, B. D., Keller, C. K., and Johnstone, D. L. 1993. In situ measurement of microbial activity and controls on microbial CO_2 production in the unsaturated zone. *Water Resources Res.*, 29(3):647–659.
32. Hickey, W. J. 1995. In situ respirometry: Field methods and implication for hydrocarbon biodegradation in subsurface soils. *J. Environ. Qual.*, 24:583–588.
33. Hinchee, R. E., and Arthur, M. 1991. Bench scale studies of the soil aeration process for bioremediation of petroleum hydrocarbons. *Appl. Biochem. Biotechnol.*, 28:901–906.
34. Pijls, C. G. J. M., Urlings, L. G. C. M., van Vree, H. B. R. J., and Spuij, F. 1994. Applications of in situ soil vapor extraction and air injection. In: R. E. Hinchee (ed.), *Air Sparging for Site Remediation*, pp. 128–136. Boca Raton, FL: Lewis Publishers.
35. Martinson, M., Linck, J., Manz, C., and Petrofske, T. 1995. Optimized air sparging coupled with soil vapor extraction to remediate groundwater. In: R. E. Hinchee, R. N. Miller, and P. C. Johnson (eds.), *In Situ Aeration: Air Sparging, Bioventing, and Related Remediation Processes*, pp. 203–214. Columbus, OH: Battelle Press.
36. Hinchee, R. E. 1992. Test plan and technical protocol for a field treatability test for bioventing. AFCEE, D301.2: T 28.
37. Hem, J. D. 1985. Study and interpretation of the chemical characteristics of natural water. *U.S. Geol. Survey Water Supply Paper 2254*, p. 262.
38. Kirtland, B. C., and Aelion, C. M., 1999. Petroleum mass removal from low permeability sediment using air sparging/soil vapor extraction: Impact of continuous or pulsed operation. *J. Contamin. Hydrol.* In press.
39. Mehran, M. 1995. Combined effects of water table drawdown, vapor extraction and air sparging on soil and groundwater remediation. In: *Emerging Technologies in Hazardous Waste Management VI*, pp. 998–1001. Washington, DC: American Chemical Society.
40. Malot, J. J. 1990. Cleanup of a gasoline contaminated site using vacuum extraction technology. In: E. J. Calabrese and P. T. Kostecki (eds.), *Petroleum Contaminated Soils*, Vol. 2, pp. 283–301. Chelsea, MI: Lewis Publishers.
41. Smith, J. L., Reible, D. D., Koo, Y. S., and Cheah, E. P. S. 1996. Vacuum extraction of a nonaqueous phase residual in a heterogeneous vadose zone. *J. Hazard. Mater.*, 49:247–265.

42. Gomez-Lahoz, C., Rodriguez-Maroto, J. M., Wilson, D. J., and Tamamushi, K. 1994. Soil clean up by in situ aeration. XV. Effects of variable air flow rates in diffusion-limited operation. *Separation Sci. Technol.*, 29(8):943–969.
43. Payne, F. C., Blaske, A. R., and VanHouten, G. A. 1995. Air sparging and bioremediation: The case for in situ mixing. In: R. E. Hinchee (ed.), *In Situ Aeration: Air Sparging, Bioventing, and Related Remediation Processes*, pp. 177–184. Columbus, OH: Battelle Press.
44. Clayton, W. S., Brown, R. A., and Bass, D. H. 1995. Air sparging and bioremediation: The case for in situ mixing. In: R. E. Hinchee (ed.), *In Situ Aeration: Air Sparging, Bioventing, and Related Remediation Processes*, pp. 75–86. Columbus, OH: Battelle Press.
45. Widdowson, M. A., Aelion, C. M., Ray, R. P., and Reeves, H. W. 1995. Soil vapor extraction pilot study at a Piedmont UST site. In: R. E. Hinchee, R. N. Miller, and P. C. Johnson (eds.), *In Situ Aeration, Air Sparging, Bioventing, and Related Remediation Processes*, pp. 455–461. Columbus, OH: Battelle Press.
46. Hinchee, R. E., Downey, D. C., Dupont, R. R., Aggarwal, P. K., and Miller, R. N. 1991. Enhancing biodegradation of petroleum hydrocarbons through soil venting. *J. Hazard. Mater.*, 27:315–325.
47. Thornton, J. S., and Wootan, W. L. 1982. Venting for the removal of hydrocarbon vapors from gasoline contaminated soil. *J. Environ. Sci. Health*, A17(1):31–44.
48. Campagnolo, J. F., and Akgerman, A. 1995. Modeling of soil vapor extraction systems. 2. Biodegradation aspects of soil vapor extraction. *Waste Manage.*, 15(5–6):391–397.
49. Dupont, R. R., 1993. Fundamentals of bioventing applied to fuel contaminated sites. *Environ. Prog.*, 12(1):45–53.
50. Billings, J. F., Cooley, A. I., and Billings, G. K. 1994. Microbial and carbon dioxide aspects of operating air-sparging sites. In: R. E. Hinchee, R. N. Miller, and P. C. Johnson (eds.), *In Situ Aeration: Air Sparging, Bioventing, and Related Remediation Processes*, pp. 112–119. Columbus, OH: Battelle Press.
51. Clodfelter, C. L. 1992. Use of air sparging systems for petroleum hydrocarbon remediation in clayey East Texas aquifers. *Third Int. Conf. Groundwater Quality Research*, SRC, Dallas, TX, pp. 283–284.
52. Ghandehari, M. H., Kelly, B. J., Holt, A. W., Hines, S. M., and Doesburg, J. M. 1994. In situ remediation of groundwater using horizontal well air injection. NWWA/API Conf. on Petroleum Hydrocarbons and Organic Chemicals in Ground Water: Prevention, Detection, and Restoration, Houston, TX.
53. Clark, A. 1995. The factors controlling the application of air sparging to remediate petroleum hydrocarbon contamination. MERM thesis, University of South Carolina, Columbia.
54. Venkatraman, S. N., Schuring, J. R., Boland, T. M., Bossert, I. D., and Kosson, D. S. 1998. Application of pneumatic fracturing to enhance in situ bioremediation. *J. Soil Contamin.*, 7(2):143–162.
55. Frank, U. 1994. United-States-Environmental Protection-Agency Superfund innovative technology evaluation of pneumatic fracturing extraction (sm). *J. Air Waste Manage. Assoc.*, 44(10):1219–1223.
56. Frank, U., and Barkley, N. 1995. Remediation of low permeability subsurface formations by fracturing enhancement of soil vapor extraction. *J. Hazard. Mater.*, 40:191–201.
57. Lee, M. D., Thomas, J. M., Borden, R. C., Bedient, P. B., Ward, C. H., and Wilson, J. T. 1988. Biorestoration of aguifers contaminated with organic compounds. *CRC Crit. Rev. Environ. Control*, 18(1):29–89.
58. Siegrist, R. L., West, O. R., Morris, M. I., Pickering, D. A., Greene, D. W., Muhr, C. A., Davenport, D. D., and Gierke, J. S. 1995. *In situ* mixed region vapor stripping in low-permeability media. 2. Full-scale field experiments. *Environ. Sci. Technol.*, 29:2198–2207.
59. Cho, J. S., DiGiulio, D. C., Wilson, J. T., and Vardy, J. A. 1997. In situ air injection, soil vacuum extraction and enhanced biodegradation: A case study in a JP-4 jet fuel contaminated site. *Environ. Prog.*, 16(1):35–42.

13
Intensification of Removal of Hydrocarbon Contamination from Water and Soil Using Oxygen Transferors

Tamara V. Sakhno, Viktor M. Emelyanov, and Viktor M. Kurashov
Kazan State Technological University, Kazan, Russia

I. INTRODUCTION

A quite extensive information testifying that the oil and oil products contamination leads to the disruption of natural biocenoses, deterioration of agrophysical and agrochemical soil properties, the fall in productivity of crops, worsening of the ecological situation in oil extracting regions has been accumulated by now. The natural processes of self-purification and restoration of soil spoiled by oil and oil products proceed slowly especially with the high level of contamination.

A major part of available nowadays physical-chemical technologies of the oil contaminated soil remediation in homeland and world practice are as a rule, labor-consuming and rather expensive. In this connection the biotechnological methods building around the capability of microorganisms to adopt oil hydrocarbons as the only source of energy, are the most perspective methods among the ecologically and economically justified ones. The natural rate of microbiological oil decontamination in water and in soil, however, is rather small and takes a great amount of time. The effectiveness of biodegradation by microorganisms is often limited by such factors as an access of oxygen and nutriments, an accumulation of metabolism products, etc. Various stimulating substances are used in order to diminish an influence of the limiting factors on the growth and productivity of microorganisms [1,2].

The Proxanole surface active substance (SAS) consisting of ethylene and propylene oxides block copolymers is very perspective in this regard [3]. Proxanoles fall into the group of unionogenic SAS. Unionogenic SAS unlike other sorts of SAS, do not dissociate on ions being dissolved, and their solubility are determined by the presence of hydrophobic part of the molecule [4]. Unionogenic SAS are the most perspective and advanced group of substances. These compounds have low hydrophilicity, high chemical stability and minimal toxic activity. The presence of hydrophobic ethylene oxide chain in unionogenic SAS is responsible for the appearance of some characteristic properties that can not be found in other kind of SAS. When ethylene oxide groups of SAS react with water, hydrogen bonds form. An attachment of water molecule to oxyethylated compounds makes them soluble in water [5]. Proxanoles are SAS with hydrophobic radical in the middle of a molecule and hydrophilic ones at the ends of it.

Nowadays proxanoles are widely used in medicine as a component of gas-transferring medium (artificial blood) [6]. Proxanole as a biologically active substance was studied on isolated heart [7].

As proxanoles are applied to the vital biological systems, they must be easily soluble in water and of minimal danger. The most appropriate substances found on evidence derived from biological and physical-chemical tests are proxanoles containing no more than 20% of oxypropylene blocks [3]. It has been found that the best proxanoles over the all parameters are 168 and 268 proxanoles synthesized from the available monomers with the molecular weight of 7200 and 5700, respectively.

The investigations of the culture of cells using the foreign analogue of proxanole Pluronic-68 have a contradictory character demonstrating both a presence and an absence of activating effect of Pluronic-68 on the growth productivity of cells [8–10]. Probably, it is connected with the use of different concentrations of Pluronic-68 as well as with the carrying out experiments under optimal laboratory conditions where the effect of limiting factors manifest itself poorly.

II. EXPERIMENTAL PROGRAM

A. Methods and Materials

The subject of our investigation is the oil-degrading bacterial culture of *Pseudomonas putida-36*. Biopreparations Proxanole-168 and Proxanole-268 were used as SAS. All SAS were injected before sowing the culture as a fresh sterile 1% water solution in amount of 0.01% from the weight of the cultural medium.

The bacterial culture *Pseudomonas putida-36* was grown in an F04 fermenter with a volume of 20 dm^3, temperature of medium 30–32°C and pH 7–7.2 with the supply of sterile air to the medium. An intensity of aeration was selected so as the lowering of concentration of oxygen [O] dissolved in water (owing to oxygen consumption by the culture) would have gone rather quickly, and the retarding of cultural growth would occur due to a lack of oxygen instead of running out one or another component of nutritive medium or accumulating the matabolites. During the bacterial cultivation in the fermenter, the oxygen concentration in the medium was detected by polarographic method with closed platinum Klarck-type electrode using the LP-7 polarograph. Concentration of dissolved oxygen was expressed in percent from the total saturation of the medium by oxygen in laboratory conditions, i.e. at a temperature of 30–32°C, at atmospheric pressure (760 mm Hg), and aeration of medium by air (dissolved oxygen has $\sim$160 mm Hg). The nutritive medium consisted of glucose, maize extract, $Na_2HPO_4 \cdot 12H_2O$, $(NH_4)_2SO_4$, K_2SO_4, and $MgSO_4$.

Crude oil of different concentrations (0.8 g/L and 2 g/L) was injected to the nutritive medium before sowing the culture in the fermenter in the model experiment. During the whole cultivation period the oil concentration in the medium was kept constant within 0.5–1% by systematic oil adding. Samples were taken from the fermenter every hour, and optical density was measured spectrophotometrically in 1 cm cuvettes and wavelength of 540 nm. The pH value during fermentation was maintained in the range of 7–7.2 by means of 20% NaOH solution or 20% H_3PO_4.

Biodegradation in the model experiment was run for 18 h.

The destructive capability of the microorganisms was checked by the weight method through the residual contents of oil and oil products in the medium after processing three times by hexane, or carbon tetrachloride, or chloroform [11].

Experiments for cleaning oil contaminated soil were carried out in the territory of Vostochno-Suleevskii Romashkinskii deposit of Tatarstan Republic.

Three plots with the area of pollution equal to 100% were chosen. The cause of this pollution was a break in the oil pipeline. The first plot had chernozem soil with oil concentration of 24 g/kg, the depth of penetration 15 cm, the second plot had a chernozem soil with oil concentration of 62 g/kg, and the depth of penetration 20 cm. The third one had sand soil with oil concentration of 50 g/kg, and the depth of penetration 30 cm. Every plot was divided into 9 microplots each 2×2 m in size, located 10 m from each other. Four microplots were reference ones without SAS being added. Two of them were not treated with biopreparation Putidoil. Another three mocroplots were treated with Proxanole-168, the rest three of them being treated with Proxanole-268. All SASs were inserted as fresh-made 1% water solutions at a rate of 0.1 g per square meter of soil.

Model and reference plots were fenced with viniplast frames 65 cm high placed to a depth of 40 cm. Plots were ploughed beforehand. The biopreparation Putidoil on the base of *Pseudomonas putida-36* in amount of 1.5 g was dispersed per square meter as an aqueous solution. After that a mineral feeding was performed at a rate of 0.8 g of ammonium phosphate plus KNO_3 per square meter of soil. The soil was made friable to a depth of oil penetration once a week. Humidity was created by adding fresh water up to soil's hydrocapacity. Ten samples (10 g each) were withdrawn from the upper layer of soil every day for 70 days, they were thoroughly averaged out and were subjected to a study of the biodegradation of oil spills through residual content of oil in the soil by means of the weight method.

Electrophoretic mobility (EM) of cells was detected with the help of automatic microscope Parmoguant-2 in manually operated registration of microorganisms moving in a right-angled cell. Electrophoretical mobility was studied in the nutritive medium for the periodic growth of Pseudomonas putida-36 culture.

B. An Influence of Oxygen Transferors Upon the Growth of *Pseudomonas putida-36* Bacteria

The dependence of the kinetic curves of biomass growth along with concentration of oxygen dissolved in the medium [O] upon saturation was obtained as a result of the experiments in the fermenter. Critical concentrations of dissolved oxygen $[O]_{cr.g.}$ for the growth of *Pseudomonas putida-36* in optimal and extreme conditions were determined as well.

An increase of substrate concentration from 0.8 g/L to 2 g/L of the periodic culture *Pseudomonas putida-36* in the medium with oil as a source of carbon and energy leads to the retard of the growth. It can justify the inhibiting effect of the high oil concentrations on microorganisms. The curves 1 and 3 for growth in Fig. 1 show it apparently.

In the optimal conditions with oil concentration of 0.8 g/L, the retard of *Pseudomonas putida-36* growth begins when concentration of dissolved oxygen falls to 30% of oxygen saturation of the medium (curves 1 and 2 in Fig. 1). An exponential phase begins after 1.5 h. The maximal specific growth rate of microorganisms *Pseudomonas putida-36* is as much as $0.54\,h^{-1}$. The maximal biomass concentration comprises 39 g/L after 8.5 h.

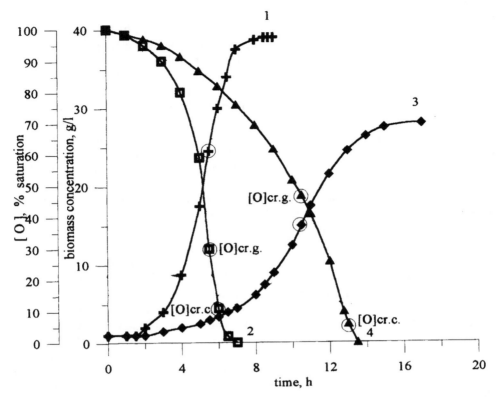

Figure 1 Influence of the dissolved oxygen concentration [O] on the growth of *Pseudomonas putida-36*. The oil concentration 0.8g/L: (1), the growth of *Pseudomonas putida-36*; (2), the concentration of the dissolved oxygen. The oil concentration 2 g/L: (3), the growth of *Pseudomonas putida-36*; and (4), the concentration of the dissolved oxygen.

In extreme conditions when oil concentration [C_{oil}] is 2 g/L, the retard of the growth of *Pseudomonas putida-36* culture begins when the concentration of dissolved oxygen falls to 47% of oxygen saturation of the medium (curves 1 and 2 in Fig. 1). An exponential phase begins after 4 h. The maximal specific growth rate of microorganisms *Pseudomonas putida-36* reduces to $0.29\,h^{-1}$. The maximal biomass concentration reaches 28 g/L only after 18 h. Thus, the inhibiting action of oil lowers the rate of the oxygen consumption, the maximal specific growth rate, the final biomass concentration, and extends the fermentation time.

Injecting of the oxygen transferors Proxanole 168 (P168) and Proxanole 268 (P268) into the cultural medium under optimal conditions does not affect the growth of the biomass and the oxygen consumption.

However, adding the oxygen transferors P168 and P268 under extreme conditions give a positive impact to the growth of the biomass and the oxygen consumption (Fig. 2).

An addition of the oxygen transferors P168 and P268 into the cultural medium with oil concentration of 2 g/L leads to the shortening of lag-phase (an exponential phase begins after 2–3 h), accelerating the maximal speed of growth to $0.38\,h^{-1}$ and $0.46\,h^{-1}$, enlarging the maximal *Pseudomonas putida-36* biomass concentration to 37 g/L after 13.5 h, and to 38 g/L after 10 h, respectively (curves 1 and 3 in Fig. 2). The critical oxygen

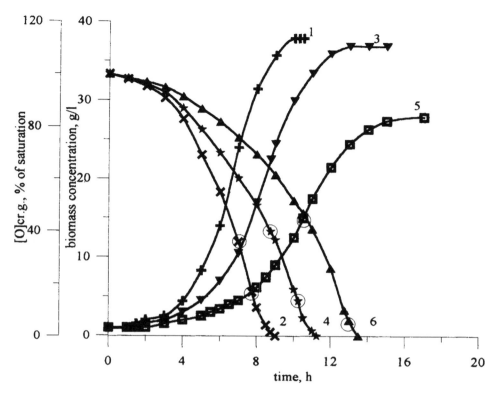

Figure 2 Dependence of an influence of Proxanole-168 and Proxanole-268 oxygen transferors on the growth of *Pseudomonas putida-36* and dissolved oxygen concentration upon the saturation [O]. In the presence of P168: (1), the curve of *Pseudomonas putida-36*; (2), the concentration of the dissolved oxygen. In the presence of P268: (3), the curve of *Pseudomonas putida-36*; (4), the concentration of the dissolved oxygen; (5), the curve of *Pseudomonas putida-36* growth; and (6), the concentration of the dissolved oxygen.

concentration for the cultural growth using P168 and P268 makes 40% and 36% of saturation, respectively (curves 2 and 4 in Fig. 2).

In all experiments the critical oxygen concentration for culture growth lies in the range of 30–50% of saturation. It is easy to see that initially the rate of oxygen decreasing was accelerated in the growing culture, but then it began to retard (curves 2, 4 in Fig. 1; and 2, 4, 6 in Fig. 2 bend to the right side). Let us denote the concentration of oxygen in the medium beyond which the speed of oxygen consumption in the growing culture lowers, by $[O]_{cr.c.}$ (a crtitical concentration for oxygen consumption).

When curves 2 and 4 in Fig. 1 are compared, one can see that at an excess of the substrate (curve 4 in Fig. 1), the critical concentration of oxygen consumption falls to 65% of dissolved oxygen saturation.

The oxygen transferors P168 and P268 being added (Fig. 2), the critical concentration of oxygen consumption increases to, respectively, 18% and 14% of dissolved oxygen saturation.

Comparing the curves 1, 2, 5 in Fig. 1 with the curves 2 and 4 in Fig. 2 demonstrates the accelerating of the specific growth rate with the lowering of critical for culture growth oxygen dissolved concentration (Fig. 3).

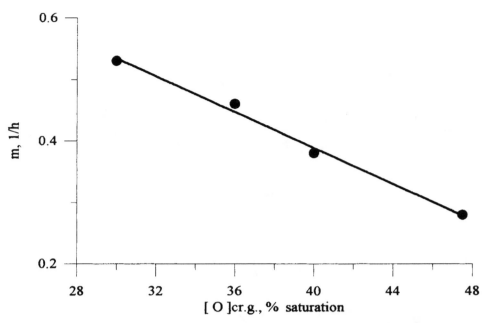

Figure 3 Dependence of the specific rate of *Pseudomonas putida-36* growth upon the critical concentration for culture growth of the dissolved oxygen $[O]_{cr.g.}$.

Thus, judging from the available data one can observe that the oxygen consumption and consequently, the provision of cells with oxygen, are reduced only after lowering the dissolved oxygen concentration to 6–18% of saturation. But at the same time oxygen becomes a factor that limits the specific growth rate and causes the retard of culture growth even at 30–47% from saturation.

Comparing figures 1 and 2 with Fig. 4 one can observe the dependence of the absolute value of the electrophoretical mobility (EM) in the periodic process of microbiological cultivation of *Pseudomonas putida-36* upon concentration of oxygen dissolved in the medium. An accelerating of EM takes place until the concentration of oxygen dissolved in the medium reaches the critical concentration for culture growth of the dissolved oxygen $[O]_{cr.g.}$. But the speed of EM lowering decreases after the critical concentration for oxygen consumption $[O]_{cr.c}$ achieves. From curve 1 in Fig. 1 and curves 1, 2, 5 in Fig. 2 in compare with the curves 1–4 in Fig. 4, one can notice that the more the maximal specific growth rate, the higher the maximal value of EM. This dependence is reflected in Fig. 5. Thus, EM depends on both specific growth rate and the rate of oxygen consumption.

C. Biodegradation of Oil in Soil

The oil biodegradation processes in the soil have not been studied thoroughly so far. Difficulties appear, particularly, due to the variability of the soil content.

The use of pure biopreparation *Pseudomonas putida-36* for soil remediation is possible at the oil concentration no more than 25 g/kg, which is the limiting value for oil oxidation by bacteria. Higher oil concentrations lead to inactivation of the oxidizing activity of biopreparation and then to its death owing to inhibiting influence of the hydrocarbon substrate. Using of oxygen transferors makes it possible to apply the

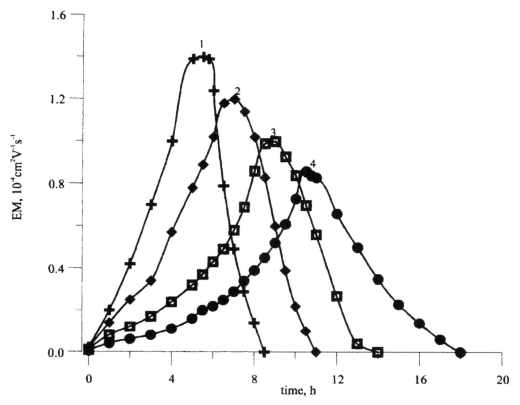

Figure 4 Variations in electrophoretical mobility of *Pseudomonas putida-36* under the periodic growth conditions of the culture. The initial oil concentration 0.8 g/L: (1), without oxygen transferors. The initial oil concentration 2 g/L: (2), in the presence of P168; (3), in the presence of P268; and (4), without oxygen transferors.

hydrocarbon-degrading microorganisms to biodegradation of oil with the concentrations exceeding the limiting one.

An influence of P168 and P268 transferors on oil spills biodegradation depends on the physical, chemical, and biological properties of the soil.

In chernozem with the initial oil concentration of 25 g/kg there is no oil degradation at all without injecting the Putidoil preparation, whereas using an active preparation results in the oxidizing of oil hydrocarbons to residual concentration of 0.85 g/kg over 67 days (curve 3 in Fig. 6). An adding of P168 and P268 proxanoles causes the 100% biodegradation by Putidoil biopreparation over 45 or 35 days, respectively. An increase in oil concentration to 62 g/L in chernozem inhibits hydrogencarbon-degrading microorganisms of Putidoil biopreparation. Injecting P168 and P268 in the same soil without Putidoil biopreparation does not cause oil degradation. An adding of an active Putidoil biopreparation in the presence of P168 and P268 assists oil degrading to the residual concentration 0.25 g/kg over 98 and 75 days, respectively.

Another situation is when using P168 and P268 proxanoles in the sand soil with the initial oil concentration 50 g/kg. Proxanole P268 does not cause biodegradation and oil concentration remains constant (curve 1 in Fig. 7), whereas P168 is responsible for

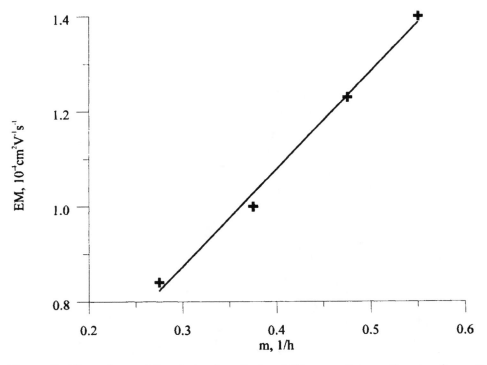

Figure 5 Dependence of the electrophoretical mobility upon the specific rate of growth.

oxidizing of oil hydrocarbons to the residual concentration of 0.75 g/kg over 78 days (curve 2 in Fig. 7).

Thus, the use of P168 and P268 oxygen transferors enables to exhibit activity of *Pseudomonas putida-36* at oil spills concentration of 50–60 g/kg in the soil within 70–90 days.

The results obtained were successfully applied to remove oil contamination from the territories of Vostochno-Suleevskii, Al'keevskii, and Romashkinskii deposits of the Tatoil production center.

III. CONCLUSIONS

1. A retarding of *Pseudomonas putida-36* culture begins when concentration of oxygen dissolved in medium falls below 30-50% of saturation of medium by oxygen. At the same time the concentration of oxygen dissolved in medium being critical for oxygen consumption by the same culture, lies in the range of 6–18% of saturation.
2. In the processes of microbiological cultivation of *Pseudomonas putida-36* proceeding at the extreme conditions under the excess of oil concentrations, Proxanole-168 and Proxanole-268 oxygen transferors lead to the lowering of critical concentration of dissolved oxygen for culture growth, and the increasing of the specific growth rate along with the critical concentration of oxygen consumption.

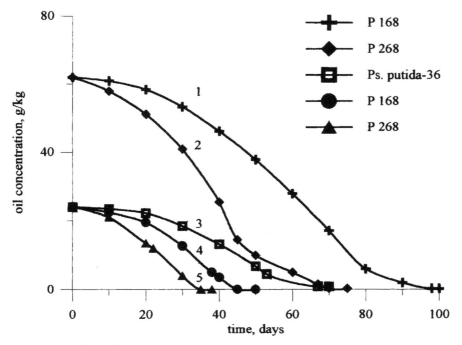

Figure 6 Oil biodegradation in chernozem by *Pseudomonas putida-36* microorganisms in the presence of oxygen transferors. The initial oil concentration 62 g/kg: (1), P168; (2), P268. The initial oil concentration 24 g/kg: (3), *Pseudomonas putida-36* without oxygen transferors; (4), P 168; and (5), P 268.

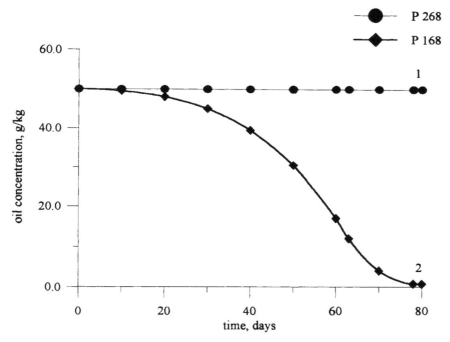

Figure 7 Oil biodegradation in the sand soil by *Pseudomonas putida-36* microorganisms using oxygen transferors: (1), P168; (2), P268.

3. An increasing of the maximal specific growth rate of *Pseudomonas putida-36* when using Proxanole-168 and Proxanole-268 in the extreme conditions assist the accelerating of the specific growth rate and the critical concentration of oxygen consumption.
4. The use of Proxanole-168 and Proxanole-268 oxygen transferors allows to shorten the biodegradation process and to realize the complete oil biodegradation at the emergency conditions when oil concentration excess the limiting value for oxidizing by microorganisms in 2.5 times.

REFERENCES

1. Dong Jin Kim and Ho Nam Chang. *Biotech. Let.*, *12*:289–294, 1990.
2. Sakellaris G. and Gikas P. *Biotech. Let.*, *13*:217–222, 1991.
3. Illarionov E. F., Maevskii E. I., Kirsh Yu. E. et al. Fluorocarbon gas-transferring media. *Pushchino*: 73–78, 1984.
4. Abramson A. A. SAS. Moscow: *Khimiya*, p. 303, 1981.
5. Zakupra V. P. An analysis of SAS. Kiev: *Tekhnika*, p. 186, 1972.
6. Afonin N. I., Rybolovlev Yu. R. and Utyatina T. K., Unionogen SAS as components of blood substituents-oxugen transferors on a basis of fluorocarbons. *Problemy Gematologii*, *10*:12–14, 1982.
7. Islamov B. I., Maevskii E. I., Vorob'ev S. I., et al., *Vestnik AMN USSR*, 2: 40–45, 1987.
8. Bentley B. K., Gates R. M. C., Rowe K. C. et al., *Biotech. Let.*, *11*:111–114, 1989.
9. King A. T., Lowe K. C. and Mulligan B. J. *Biotech Let.*, *10*:177–180. 1988.
10. Kumar V., Laonar L., Davey M. R., et al., *Biotech Let.*, *12*:937–940, 1990.
11. Lur'e Yu and Rybnikov A. I. Chemical analysis of industrial water and sewage. Moscow: *Khimiya*, p. 290, 1974.

14
Hydrocarbon Uptake and Utilization by *Streptomyces* Strains

György Barabás, András Penyige, István Szabó, György Vargha, Sándor Damjanovich, János Matkó, János Szöllősi, and Anita Mátyus
University Medical School, Debrecen, Hungary

Samir S. Radwan
Kuwait University, Safat, Kuwait

Tadashi Hirano
The Jikei University, Tokyo, Japan

I. INTRODUCTION

Streptomyces are typical aerobic soil bacteria. Although their micromorphology shows some similarities to certain microscopic fungi which form branching hyphae that develop into mycelium, they are true bacteria that have nucleoids instead of a real nucleus and cell walls containing peptidoglycan, characteristic of bacteria. *Streptomyces* are highly differentiated bacteria, with morphologically well-distinguished stages of development. They form vegetative and reproductive hyphae in submerged cultures, or vegetative, and aerial hyphae on solid media, completing the life cycle in both cases with spores (see Figs. 2–5). In the soil they have the important function of degrading plant and animal waste.

Since they produce a high proportion of known antibiotics, many enzymes, and other important substances, they are known in the fermentation industry, where they are cultivated in large quantities.

Streptomyces are present in different kinds of soils and can adapt to extreme environmental conditions (hot deserts, salt-marsh areas, and cold Alpine slopes). They represent a large part of the bacterial population in soil, where their proportion varies between 40% and 60%. Their versatility enables them to live and propagate under toxic circumstances, such as in oil-polluted soil. It is therefore surprising that their use for bioremediation of hydrocarbon-polluted areas has been neglected. There are few publications about their use in hydrocarbon transformation, most of them restricted to the application of screening experiments (1). In this chapter we describe our recently introduced research on elimination of oil pollution by *Streptomyces* strains isolated from Kuwaiti oil fields (2).

Although it is well established that individual alkanes are taken up as intact molecules, the uptake mechanism(s) is (are) still obscure. This chapter also gives data concerning the uptake of n-hexadecane and about the membrane anisotropy of oil utilizer and nonutilizer strains.

Micromorphology of oil-consuming strains shows special cytoplasmic membrane inclusions when cultivation is done in the presence of hydrocarbones. Throughout this chapter the designation KCC (Kuwait Culture Collection) is given to the oil utilizers, which were identified taxonomically as *Streptomyces* strains.

II. ISOLATION, CULTIVATION AND CHARACTERIZATION OF OIL UTILIZING *Streptomyces* STRAINS

The organisms were isolated from a highly polluted Kuwaiti desert oil field (2) on starch–casein medium (3) of the following composition (g/L): starch, 10; casein, 0.3; KNO_3, 2; NaCl, 2; K_2HPO_4, 2; $MgSO_4 \times 7H_2O$, 0.05; $CaCO_3$, 0.2; $FeSO_4 \times 7H_2O$, 0.01; Bacto agar (Difco), 18; Nystatin 50 µg/mL, pH adjusted to 7.5.

To test the oligocarbophilic nature of the isolates a loopful of a spore suspension was streaked on the surface of a solid inorganic medium free of any supplemented carbon source and examined for growth. The inorganic medium had the following composition (g/L): $NaNO_3$ 0.85; KH_2PO_4 0.56; Na_2HPO_4 0.86; K_2SO_4 0.17; $MgSO_4 \times 7H_2O$ 2.32; $MnSO_4 \times 4H_2O$ 1.78; H_3BO_3 0.56; $CuSO_4 \times 5H_2O$ 1.0; $Na_2MoO_4 \times 2H_2O$ 0.39; $CoCl_2 \times 6H_2O$ 0.42; KI 0.66; EDTA 1.0; $FeSO_4 \times 7H_2O$ 0.4; $NiCl_2 \times 6H_2O$ 0.004; Bacto agar 20; pH 7.0 (2).

n-Alkane utilization: The test strains were first grown on sterile cellophane sheets covering the surface of the starch–casein–agar medium. After incubation at 30°C, the medium-free biomass on cellophane was harvested and tested for its ability to utilize either n-hexadecane (C_{16}) or n-octadecane (C_{18}) (2,4,5). For this purpose 0.5 g of fresh biomass were inoculated in 25 mL of liquid inorganic medium containing 10 mg of n-alkane in sealed containers, and incubated at 30°C in an electric shaker. Control samples were prepared by following the same procedure after killing the biomass by autoclaving. Cultures were harvested after 30 min or 6 h, respectively, centrifuged to remove the biomass, and the residual n-alkanes were recovered by extracting three times with hexane. After volatilizing the solvent, the residual alkane was redissolved in exactly 1 mL of hexane and 2 µL of the solution was analyzed by gas–liquid chromatography (GLC) using a Chrompack CP-9000 instrument equipped with a flame ionization detector, a WCOT fused-silica CP-Sil-SCB capillary column, and a temperature program of 60–250°C, raising the temperature 16°C/min. The peak areas of the n-alkanes were measured, and the percent decrease based on the area of the control peak was calculated as a quantitative measure of alkane utilization (Fig. 1 and Ref. 2).

KCC strains were characterized as described previously (2,5), and they proved to be *Streptomyces* strains. The light microscopic morphology of them is given in Figs. 2–5. The interaction of n-hexadecane droplets with spores and germinating spores is shown in Fig. 3. The branching hyphae of the strains forming a network is characteristic of *Streptomyces* spp. The life cycle is completed by forming spores, sometimes embedded in clumps (Fig. 5).

Of the 40 isolates, some were characterized as follows: KCC 26, KCC 28, KCC 30, KCC 42 as *Streptomyces plicatus*, and KCC 25 as *Streptomyces griseoflavus*. The characteristics of the oil-nonutilizer *Streptomyces griseus* were published previously (6).

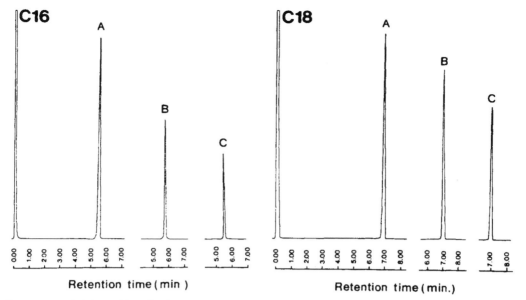

Figure 1 GLC profiles. Typical GLC profiles showing the consumption of *n*-hexadecane (C_{16}) and *n*-octadecane (C_{18}) by strains KCC 26 and KCC 42, respectively. (A) Control (original *n*-alkane supplemented). (B) *n*-Alkanes recovered after 30-min incubation. (C) *n*-Alkanes recovered after 6-h incubation (2).

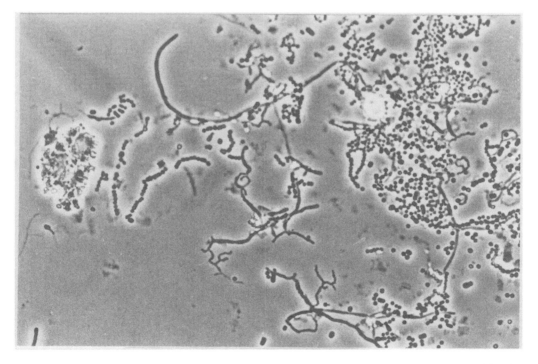

Figure 2 Spores and spore chains of KCC 42 strain. Phase-contrast microscopy, magnification 4500 ×. Cultivation was made on hexadecane-containing inorganic medium on agar plates.

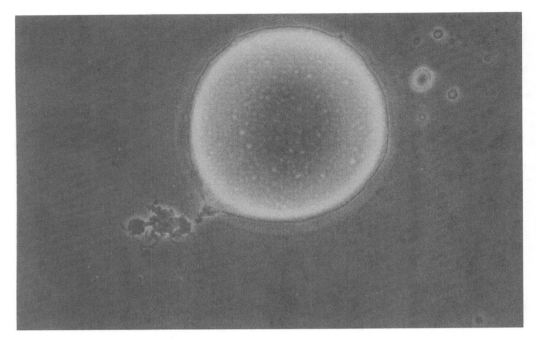

Figure 3 Interaction of germinating spores of KCC 42 with oil droplets. Phase-contrast microscopy magnification 4500 ×. The strain was cultivated as described in Fig. 2.

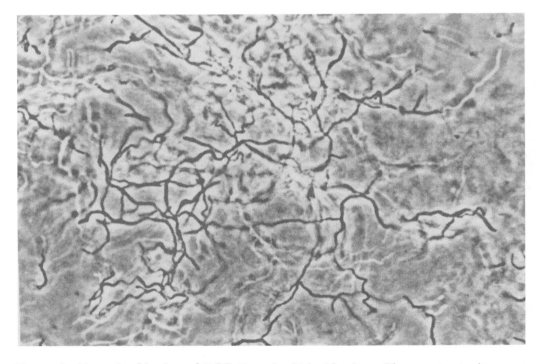

Figure 4 Network of hyphae of KCC 42 strain. 24-h-old culture. Phase-contrast microscopy, magnification 4500 ×. Cultivated as for Fig. 2.

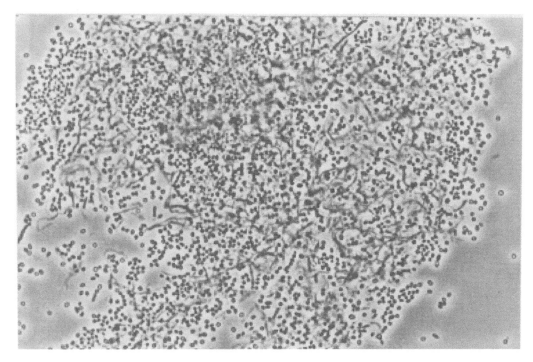

Figure 5 Spores in "clumps" of KCC 42 strain. 72-h-old culture. Phase-contrast microscopy, magnification 4500 ×. Cultivated as for Fig 2.

III. UPTAKE AND TRANSFORMATION OF n-ALKANES

The measurement of n-alkane (i.e., n-hexadecane, n-octadecane) uptake was carried out by gas–liquid chromatography (GLC) as described previously (2). Fresh biomass aliquots of 0.5 g were measured in 25 ml of inorganic medium (composition in Ref. 2) containing 10-mg aliquots of either n-hexadecane (C_{16}) or n-octadecane (C_{18}) in tightly sealed containers and incubated at 30°C in an electric shaker at 120 rpm. Control samples were prepared by following the same procedures after killing the biomass samples by autoclaving. Cultures were harvested at appropriate times, centrifuged to remove the biomass, and the residual alkanes were recovered from the medium by extracting three times with hexane. After volatilizing the solvent, the residual alkane fractions were dissolved in 1-mL hexane aliquots and 1-µL portions of the solution were analyzed by GLC. The peak areas of residual alkanes were measured and their decrease as compared to area of control peak was calculated (Fig. 1).

Data for some n-alkane utilizers are given in Table 1. It is clearly demonstrated that the four strains consumed both n-alkanes and that their utilization was higher in the case of 6-h incubations (Table 1). Comparing the data of *Streptomyces* strains with those of *Arthrobacter* and *Rhodococcus* spp., it is apparent that the KCC strains consume n-alkane much faster then *Arthrobacter* and *Rhodococcus* spp. since the latter two need four times longer incubation period (24 h) than our KCC *Streptomyces* strains (6 h) to consume approximately the same amount of alkane (Table 2).

Table 1 Consumption of n-Alkanes by *Streptomyces* Strains[a]

Streptomyces strain	n-Hexadecane (C_{16})		n-Octadecane (C_{18})	
	30 min	6 h	30 min	6 h
KCC 26	4.4	5.2	5.2	7.3
KCC 28	3.6	8.0	4.0	6.0
KCC 33	6.8	11.7	5.0	7.2
KCC 42	2.7	7.6	7.7	10.3

[a] Data are expressed in milligrams of n-alkane consumed by 1 g of fresh biomass after the given incubation period (2).

Table 2 Hexadecane Consumption by *Arthrobacter* sp., *Rhodococcus* sp., and *Streptomyces* strains[a]

Strain	KCC 26	KCC 28	KCC 30	KCC 42	*Arthrobacter* sp.	*Rhodococcus* sp.
Hexadecane consumed	5.2	8.0	9.2	7.6	10.0	7.2

[a] Data are expressed in milligrams of alkane consumed by 1 g of fresh biomass at 27°C. For *Arthorobacter* sp. and *Rhodococcus* sp. the incubation time was 24 h, for all the other strains (i.e., KCC) it was 6 h (5).

The results of analysis of constituent fatty acids of total lipids from strain KCC 33, a selected representative of the oil utilizers, are summarized in Table 3. The incubation of strain KCC 33 biomass with n-hexadecane (C_{16}) was associated with an increase in the proportion of n-hexadecanoic acid (16 : 1). On the other hand, the proportion of this fatty acid (16 : 1) in the lipid fraction from *S. griseus* biomass did not change after incubating the biomass with n-hexadecane. Similarly, the biomass of oil utilizers starts to accumulate C_{18}-fatty acids only on the n-octadecane (C_{18}) substrate.

The growth of four KCC isolates on n-hexadecane and n-octadecane is shown in Table 4. They grow quite well when n-alkane is the only carbon and energy source.

We studied the n-hexadecane elimination in the soil by strain KCC 42. Table 5 proves that our isolate actively eliminates n-hexadecane from the soil. The method previously described (1, 4) has been used in our experiment.

IV. INCORPORATION OF LABELED HEXADECANE INTO CO_2 AND THE MYCELIA

A quantity 2.5 µL hexadecane-1-^{14}C was added to 5 mL of 24-h-old culture of C_{16} utilizer strains. After 6 h of incubation the released radioactive CO_2 was measured. The quantity was less than 0.4% of the total radioactivity added to the culture (Table 6).

The mycelia were washed with hexane to remove the oil droplets attached to the surface or enclosed by the dense network of mycelia.

In order to determine the quantity of hexadecane incorporated or transformed by the cells, the washed mycelia were dissolved in scintillation liquid (Soluene 350, Packard Instrument Company, Meridon, CT) and its radioactivity was measured. In strain

Table 3 Constituent Fatty Acids of Total Lipids from Strain KCC 33 Grown With and Without n-Alkanes[a]

Fatty acids	Control, no n-alkanes	With n-hexadecane +C_{16}	With n-heptadecane +C_{18}
n 12 : 0	1.6	3.2	2.1
12 : 1	6.9	3.2	2.3
n 13 : 0	Trace	0.8	1.0
13 : 1	7.6	4.3	3.1
i 14 : 0	20.8	27.8	25.3
14 : 0	10.4	7.4	7.2
14 : 1	9.7	2.1	2.0
n 15 : 0	3.1	Trace	Trace
15 : 1	3.6	1.3	1.4
n 16 : 0	10.4	10.0	9.5
16 : 1	17.4	29.9	28.4
i 17 : 0	7.3	7.9	7.6
n 17 : 0	0.4	0.9	0.9
17 : 1	0.6	1.0	1.0
i 18 : 0	—	—	1.9
n 18 : 0	—	—	3.2
18 : 1	—	—	3.0

[a] Data are expressed in weight percent of the total fatty acids; n, normal acids; br, branched acids. Four-day-old biomass was incubated on the inorganic medium without and with n-alkanes for 6 h at 30°C, and subsequently analyzed (2).

Table 4 Growth of *Streptomyces* Strains on n-Hexadecane and n-Octadecane as Sole Sources of Carbon and Energy[a]

Strains	Biomass		
	No alkane	With n-hexadecane +C_{16}	With n-octadecane +C_{18}
KCC 26	Trace	112.6	101.4
KCC 28	Trace	111.3	132.1
KCC 33	Trace	100.5	98.9
KCC 42	Trace	79.7	82.3

[a] Data are expressed in milligrams of dry biomass per gram of alkane available in the medium. In the absence of alkanes definite albeit very slight growth was observed visually; the dry biomass was less than 1 mg/g alkane, hence is designated "Trace" (2).

Table 5 The change of residual n-Hexadecane Content in the Soil[a]

Incubation time [days]	0	14	42	68
Soil with n-hexadecane	100.0%	80.1%	18.3%	17.1%
Soil with n-hexadecane inoculated with KCC 42	100.0%	66.4%	11.5%	8.9%

[a] 5 g of sterile soil was supplied with 300 μL of n-hexadecane in sealed containers and kept at room temperature. Containers were loosely screwed for proper aeration. The percentage of residual n-hexadecane is expressed.

Table 6 Incorporation of ^{14}C-Hexadecane into $^{14}CO_2$ and Mycelia[a]

Strain[b]	$^{14}CO_2$ [DPM]	Radioactivity in Mycelia [DPM]
KCC 18	139	18,764
KCC 27	472	18,090
KCC 9/1	442	11,326
KCC 9/2	390	7,616
KCC 42	332	13,992
KCC 25	399	11,825

[a] Originally 130,0000 DPM was added to the culture. Thirty different *Streptomyces* strains were investigated; only six representatives are shown here.
[b] Different *Streptomyces* strains isolated in Kuwait.

KCC 18 the incorporated radioactivity proved to be 14.4% of the ^{14}C-hexadecane added to the culture medium. In other strains this value varies between 6% and 14% (Table 6).

Further investigations are needed to determine the ratio of incorporated but not transformed hydrocarbon to that of the transformed molecules. Nevertheless, the release of 0.4% radioactive CO_2 proves the transformation of *n*-hexadecane molecules, although the details of this transformation are unknown presently.

No correlation was found between the radioactive CO_2 formation and the quantity of incorporated ^{14}C-hexadecane into mycelia. In strain KCC 18 the mycelial radioactivity was the highest but the $^{14}CO_2$ was the lowest (Table 6).

At the same time, in strain KCC 27 the mycelial radioactivity was near to that of the former strain, and the quantity of $^{14}CO_2$ was also high. In strains KCC 9/1 and 9/2 the situation was the opposite of what was found with strain KCC 18 because high $^{14}CO_2$ release was accompanied by low mycelial incorporation of the radioactivity. These data prove that even closely related strains differ in their ability to use *n*-alkanes. Experiments are in progress to determine the chemical nature of the incorporated and transformed radioactive molecules originating from labeled hexadecane.

V. GTP-BINDING PROTEINS AND THEIR ROLE IN HYDROCARBON UPTAKE

GTP-binding proteins (GBP) were first recognized in eukaryotic cells. A common characteristic of these proteins is that they bind and also hydrolyze GTP due to their intrinsic GTPase activity (7,8). GBPs are involved in a variety of cellular processes, such as signal transduction, differentiation, cell division, vesicular fusion, and the regulation of the activity of a number of enzymes (7,9–13). Two classes of this superfamily of proteins could be distinguished, e.g., the classical membrane-associated heterotrimer G proteins and the "small" (21–30 kDa) GBPs (7,10).

It is known that G proteins play an essential role in mediating cellular responses to a wide variety of extracellular signals, such as hormones, growth factors, neurotransmitters, chemical signals, or light (7,12). These proteins transduce signals via a GTP-dependent mechanism to the effector systems (enzymes or ion channels) and thereby regulate these systems that often control production of intracellular second-messenger molecules. GBPs act as molecular switches, their activation is catalyzed by a ligand-activated receptor, and deactivation is established by the intrinsic GTPase activity of the GBP. According to this mechanism, the GTP-bound form is the active complex, which returns to inactive state by hydrolyzing its GTP to GDP. The exchange of GDP to GTP and the rate of GTP hydrolysis is regulated by specific regulatory proteins (7,12,14).

The study of their role in different cellular processes was greatly enhanced by using certain reagents, such as AlF_4^-, GTPγS, or the tetradecapeptide mastoparan. AlF_4^- mimics the effect of the γ-phosphate of GTP on the inactive GDP-bound form of the protein; GTPγS is a nonhydrolyzable GTP analog (15).

The presence of GTP-binding proteins in prokaryotes, although not so well documented as it is in eukaryotic systems, could turn out to be an important physiological factor. This notion is supported by the fact that the recent cloning of the complete genome of *Mycoplasma genitalium* revealed the presence of several genes coding for GTP-binding proteins different from the well-known elongation factors (16). Since this organism is thought to contain only the essential, minimal set of genes required to support the growth of a cellular organism, the presence of regulatory GTP-binding protein homologs emphasizes the importance of these proteins in prokaryotes, too.

Interestingly, in those bacteria which also undergo developmental cycles—*Bacilli* and *Myxobacteria*, for instance—GTP-binding proteins have already been recognized, such as the Obg protein encoded by the *spoOB* operon in *Bacillus subtilis*, the 54-kDa membrane-associated GTP-binding protein in *Stigmatella aurantiaca*, or the MglA protein in *Myxococcus xanthus*, which is essential for gliding motility (11,17–19). Beside the above-mentioned Obg protein, other, biochemically or genetically not characterized GTP-binding proteins were also detected in *B. subtilis* (20). The profile and abundance of these proteins varied in cells at different stages of the developmental cycle (20).

Several other functionally better-known GTP-binding proteins have also been described in prokaryotes: (a) elongation factor-like GTPases EF-G, EF-Tu, LepA, NodQ (11,28–30); (b) the signal-recognition particle (SRP) protein family FtsY, Ffh, Sso, and FlhF molecules (17,21–23); (c) the unique Ras-homolog Era protein in *Escherichia coli* (24); or the tubulin homolog FtsZ protein (25).

All these proteins are members of the GTPase superfamily since they all contain three groups of nearly invariant amino acid residues that define the guanine nucleotide-binding pocket of these proteins, designated as Region A [consensus sequence GXXXXGK(ST)], Region C [consensus sequence DXXG], and Region G [consensus sequence NKXD] (7,14).

GTP-binding proteins were reported to be present in *S. coelicolor A(3)2*, and we have shown the presence of these proteins in *S. griseus* and in several other *Streptomyces* strains (26,27). Our previous results suggested that in *S. griseus* A-factor—a γ-butyrolactone-type auto-regulator molecule produced by wild-type *S. griseus* cells and required for the normal differentiation process and antibiotic production in the producer strain (31)—could activate an intrinsic GTPase activity present in the cellular membrane of *S. griseus* NRRL B-2682 (27).

We have used the UV photocrosslinking technique (32,33) and [α^{32}P]GTP as substrate to detect GBPs in our hydrocarbon-utilizing strains KCC 25 and KCC 42. The cells were grown in surface cultures in starch–casein (S–C) medium or in inorganic medium in the presence or in the absence of the hydrocarbon compound *n*-hexadecane (C_{16}). The surfaces of agar plates were covered with cellophane sheets and cells were removed by scraping off the mycelium from these cultures at 44 and 120 h. Cellular crude

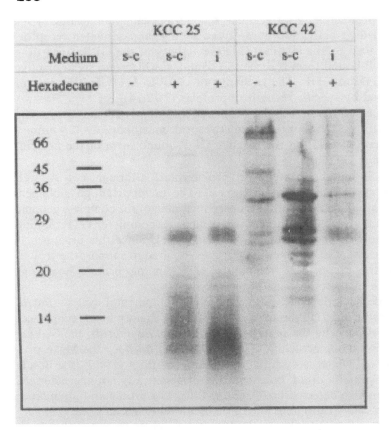

Figure 6 Identification of GBPs after denaturing polyacrylamide gel electrophoresis and Western blotting. GBPs were labeled with [α^{32}P]GTP in cellular crude extracts prepared from surface-grown KCC 25 and KCC 42 cultures. Strains were grown in starch–casein medium (lanes 1 and 4), in C_{16}-supplemented S–C medium (lanes 2 and 5), and in C_{16}-supplemented inorganic medium (lanes 3 and 6). Cells removed from these cultures were used to prepare cellular crude extracts for GTP detection.

extracts were prepared from these samples by sonication and differential centrifugation (27) and the supernatants of these samples were used in our labeling assays to identify GBPs. Labeled proteins were separated by SDS-PAGE, then electroblotted to Immobilon-P membranes using semidry Western-blotting technique. The labeled GBPs were visualized by autoradiography. Figure 6 shows the pattern of GBPs in our strains. It is noteworthy that these patterns are strain specific; the GBP complements of these strains are clearly different. The patterns also differ in the same strain when cells are cultivated under different conditions, as in S–C or inorganic medium. It might be important to emphasize that in both strains the addition of C_{16} to the culture medium resulted in qualitative and quantitative differences in the detectable GBP patterns (compare lanes 1 and 2 in the case of K25 or lanes 4 and 5 in the case of the K42 strain). These results suggest that GBPs are always present in *Streptomyces* and these proteins might play an important physiological role during the adaptation of these strains to different environmental conditions.

Table 7 The Effect of AlF_4^- on n-Hexadecane and n-Octadecane Uptake by Oil-Utilizing Microorganisms[a]

Microbial strain	n-Alkanes	$\mu M\ AlF_4^-$				
		0	50	75	100	125
KCC25	C16	7.2	10.0	12.5	12.0	11.6
	C18	6.3	ND	ND	ND	ND
KCC28	C16	8.0	9.4	9.8	9.7	9.9
	C18	5.2	4.9	5.3	7.8	10.2
KCC33	C16	9.2	10.5	10.9	11.8	13.6
	C18	7.2	ND	6.2	6.6	8.2
KCC42	C16	7.6	12.3	ND	13.7	15.7
	C18	10.3	ND	12.4	11.2	11.8
A. nicotiana	C16	5.9	4.9	6.7	12.9	10.8

[a] Data are expressed in milligrams of alkane consumed by 1 g of fresh biomass at 27°C in 6 h. ND: not determined. *Arthrobacter nicotiana* was used as control strain. In this case the incubation period was 24 h.

Table 8 The effect of GTPγS on n-Hexadecane Uptake by Oil-Utilizing Microorganisms[a]

	n-Hexadecane consumed	
GTPγS [μg/mL]	KCC 25	*Arthrobacter nicotiana*
0	3.2	1.8
10	12.3	4.2
20	12.8	6.9
50	ND	9.9
100	13.4	11.1

[a] Data are expressed in milligrams of alkane consumed by 1 g of fresh biomass at 27°C in 6 h. ND: not determined. *Arthrobacter nicotiana* was used as control strain. In this case the incubation period was 24 h.

The presence of these proteins in KCC 26 and KCC 42 cells prompted us to study the effect of GTPγS and AlF_4^-—stimulators of GBPs—on the uptake of the hydrocarbon molecules (C_{16} and C_{18}) in our strains. First, 0.5-g aliquots of mycelium precultivated in soybean medium for 36 h were resuspended in 25-ml aliquots of inorganic medium containing 10 mg of n-hexadecane or n-octadecane. The submerged cultures were supplemented with various amounts of GTPγS or AlF_4^- and incubated at 30°C with a shaking frequency of 250 rpm. Biomass was harvested after 6 h by centrifugation and the hydrocarbon uptake of the cells was determined from these samples as described above. The results in Table 7 and 8 show that the uptake of n-hexadecane (C_{16}) and n-octadecane (C_{18}) from the medium by hydrocarbon-utilizing microorganisms was enhanced by the addition of AlF_4^- and GTPγS, although the rate of uptake depends on strain specificity. Moreover, the magnitude of uptake was, in most cases, directly proportional to the concentration of these effectors in the medium. These results suggest that GBPs could fulfil important physiological functions in *Streptomyces* strains.

VI. ULTRASTRUCTURAL FEATURES

Because of the relatively recent discovery of oil-utilizer *Streptomyces* strains, their ultrastructural characteristics have not been investigated as yet. Direct electron-microscopic evidence for oil utilization is still missing, but there are some observations supporting the idea that oil droplets are taken up in the forms of inclusions in the cytoplasm. For some other hydrocarbon-utilizing bacteria, enrichment of the cytoplasm with intracellular membrane structures, microvesicles, hydrocarbon inclusions that make up approximately 40% of the total cell volume have been reported (34). We found similar structures in hydrocarbon-utilizing streptomycetes. Conventional transmission electron microscopic methods (fixation in 2% glutaraldehyde–0.1 M phosphate buffer, pH 7.4, followed by 6 times washing in 0.1 M phosphate buffer of the same pH containing 5% glucose, then postfixation with 1% osmium tetroxide–0.1 M phosphate buffer, pH 7.4, after which dehydration in ethanol and embedding in epoxy resin were made) using a JEOL 1010 TEM microscope revealed that hyphae of the hydrocarbon-utilizing strains (KCC 26, KCC 28, KCC 42, and *Arthorbacter nicotianae*, KCC B35) contained numerous electron-light areas making up to 50% of the cytoplasmic space when mycelia were previously incubated for 6 h in an inorganic medium containing *n*-hexadecane as sole carbon and energy source. In the control cells, i.e., on starch–casein complex medium and in the absence of *n*-hexadecane, these inclusions were missing or fewer in both number and size. Transmission electron micrographs of *Streptomyces* strain KCC 26 grown on inorganic medium supplied with *n*-hexadecane compared to cultures grown on starch–casein medium are shown as an example (Fig. 7).

The hydrocarbon-utilizing streptomycetes show ultrastructural characteristics different from the hydrocarbon-nonutilizing strains even when they are not cultivated on hydrocarbon-containing media. The cytoplasm of the former cells are richer in intracellular membrane structures, vesicles, than the hydrocarbon-nonutilizing ones. A transmission electron micrograph of a 24-h-old culture of *Streptomyces* strain KCC 25 is shown as an example compared to a conventional, hydrocarbon-nondegrading *Streptomyces griseus* 2682 strain (Fig. 8).

The findings do not provide direct evidence that the cytoplasmic inclusions are the real morphological equivalents of the amounts of hydrocarbon taken up, although the fact that they occur more frequently and are larger in size in the presence of hydrocarbons makes this suggestion very likely. It is to be investigated further whether the inclusions contain hydrocarbon pools for storage before breakdown or represent places where the breakdown process actually takes place and the hydrocarbon molecules are transformed.

VII. MONITORING THE KINETICS OF HYDROCARBON UPTAKE BY FLUORESCENT PROBES

Several oil-soluble fluorescent dyes, such as perylene or diphenyl-hexatriene (DPH), were designed for cell biological investigations in the 1970s, namely, to investigate fluidity (microviscosity) of various biological membranes (35,36). These dyes are practically insoluble in water and become highly fluorescent when dissolved in oils or in the phospholipid bilayer of biomembranes. The fluorescence properties of these dyes have been characterized in detail (35,36 and refs. therein). Based on these properties the diphenyl-hexatriene (DPH) was selected to follow uptake of hydrocarbons (the C_{16}

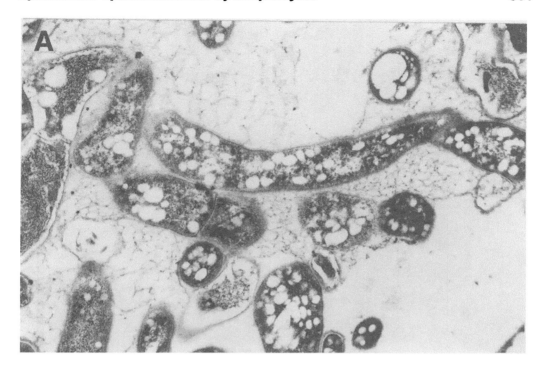

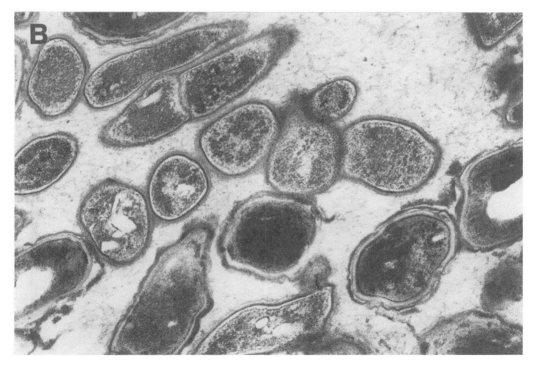

Figure 7 Transmission electromicrographs of *Streptomyces* strain KCC 26 grown on inorganic medium + C_{16} (A) and grown on starch–casein medium (B). Magnifications: 18,000 × (A); 36,000 × (B).

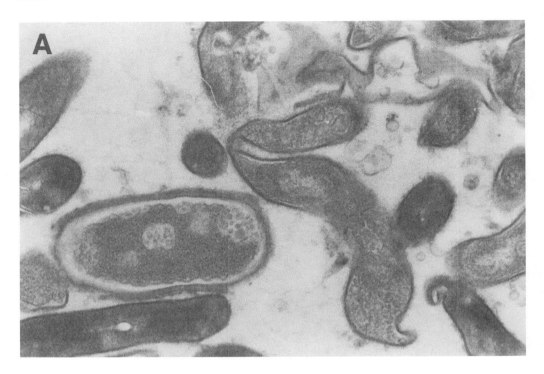

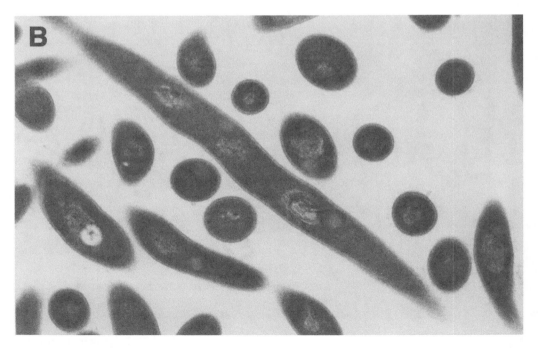

Figure 8 Transmission electromicrographs of 24-h-old cultures of *Streptomyces* strain KCC 25 (A) and *Streptomyces griseus* 2682 (B). Magnifications: 38,000 × (A); 20,000 × (B).

Hydrocarbon Uptake/Utilization by *Streptomyces*

hydrocarbon fraction) by *Streptomyces* strains using fluorescence spectroscopy. DPH is usually applied to membrane studies in form of a stock solution in organic solvent (most frequently in tetrahydrofuran, THF). This stock solution is mixed with aqueous physiological buffers and stirred until the THF evaporates. The cell suspension is added to this dispersion followed by an easy and relatively fast incorporation of DPH into the lipid bilayer (saturated in ca. 30–40 min) (35).

The different capacity of alkane-utilizing and alkane-nonutilizing *Streptomyces* strains to take up hydrocarbons was then studied by following the kinetics of DPH uptake in two ways:

1. DPH was dissolved in THF and injected into the inorganic medium and after evaporating THF the dispersion was added to cells.
2. DPH was dissolved in a C_{16} hydrocarbon fraction and then the mixture was injected and dispersed into the cell suspension.

In condition 1 the uptake of DPH by both 2682 (nonutilizer) and KCC 25 (utilizer) strains showed the same kinetics and the cells were saturable with DPH within 30–40 min (Fig. 9).

When the kinetics of DPH uptake were followed at condition 2, a remarkable difference could be observed between strains *S. griseus* 2682 and KCC 25. Although the initial part of the dye uptake showed very similar kinetics, after 40 min the *S. griseus* 2682 is saturated, while the uptake by KCC 25 continues further with a high rate and shows a plateau only after about 3 h (Fig. 10).

At both conditions the specific DPH fluorescence of cells was monitored using excitation and emission wavelengths (selected by monochromators) of 355 and 430 nm, respectively. The autofluorescence of cells was negligible at these wavelengths. The kinetics of dye uptake was obtained from the fluorescence intensities measured on cell samples taken from time to time and normalized to the cell density (optical absorbance at 355 nm).

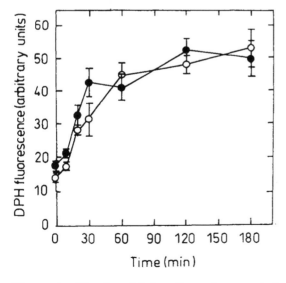

Figure 9 Kinetics of diphenylhexatriene uptake by *Streptomyces griseus* (filled circles) and strain KCC 26 (open circles) from tetrahydrofuran phase.

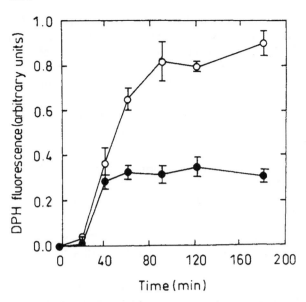

Figure 10 Kinetics of diphenylhexatriene uptake by *Streptomyces griseus* (filled circles) and strain KCC 26 (open circles) from *n*-hexadecane phase.

The localization of the internalized oil fraction can be followed either by the time dependence of DPH anisotropy or as an alternative by microscopic analysis of subcellular distribution of Nile Red, a selective fluorescent stain for intracellular lipid droplets (37). We used the anisotropy parameter of DPH fluorescence as well as the fluorescence anisotropy of the positively charged trimethyl-ammonium derivative of DPH (TMA-DPH) for this purpose. The fluorescence of the latter probe can be excited at 340 nm and its emission can be monitored at 425 nm (38). TMA-DPH was designed for cell-membrane studies. Because of the presence of the positively charged TMA group, the dye is localized in the plasma membranes even several (5–6) hours after its mixing with the cell suspension; meanwhile, the other dye, DPH, was already internalized into intracellular lipid droplets or organelle membranes (38) during this time interval.

The emission anisotropy of DPH and TMA-DPH fluorescence was measured with Polcoat polarizers (Perkin-Elmer) in a commercial spectrofluorimeter (Perkin-Elmer MPF 44B) and was calculated as follows:

$$r = \frac{I_{VV} - G * I_{VH}}{I_{VV} + 2G * I_{VH}}$$

where V or H stand for the vertical or horizontal position of the excitation and emission polarizers, respectively. G is an instrumental correction factor determined as I_{HV}/I_{HH} (39).

An anisotropy parameter proportional to the microviscosity of the medium embedding the dye, $(r_0/r_s - 1)^{-1}$, was derived from the measured steady-state anisotropy data. Here r_0 and r_s are the limiting anisotropy (characteristic of the completely immobile dye) and the emission anisotropy of the sample, respectively. The anisotropy parameter was used to characterize localization (environment) of the dyes (36,39) during and at the end of dye uptake.

Table 9 Anisotropy Parameters of DPH and TMA-DPH in C_{16} Utilizer and Nonutilizer Strains[a]

Bactrial strain	$(r_0/r_s - 1) - 1$[b]			
	DPH		TMA-DPH	
	60 min	180 min	15 min	180 min
Nonutilizer	0.95 ± 0.25	0.38 ± 0.09	1.57 ± 0.49	1.35 ± 0.33
Utilizer	0.18 ± 0.04	0.17 ± 0.04	3.11 ± 1.09	3.31 ± 0.9

[a] Data are given as mean $\pm$ S.D. (n: 3–5).
[b] $(r_0/r_s - 1) - 1$ values are proportional to the microviscosity of the medium where the probes are embedded. r_0 is limiting anisotropy for DPH equal to 0.362 (4), r_s is the measured anisotropy of the sample.

There was a considerable difference between the emission anisotropy parameter of DPH in the two strains (Table 9). The DPH anisotropy parameter in *S. griseus* 2682 (nonutilizer) is relatively high after 60 min, while in the KCC 25 (oil-utilizer) strain it has a low value after 60 min, and after 3 h as well. This finding strongly suggests that the DPH is mainly membrane-localized in *S. griseus* while in KCC 25 it is taken up quickly and the majority of the dye is localized in small intracellular lipid droplets/compartments, in agreement with the electron-microscopic observations (Fig. 7). This is further supported by the anisotropy parameters of plasma membrane-localized TMA-DPH, which was high in both strains and did not change significantly in time, at least in a 3-h interval (Table 9).

All these fluorescence data reflect that the KCC 25 (utilizer) strain transports the oil (C_{16}) droplets through the membrane at a much higher rate than *S. griseus* 2682, and the intracellular utilization of the hydrocarbon is also relatively fast.

VIII. CONCLUSION

The results of this study prove that several *Streptomyces* strains actively take up *n*-hexadecane and *n*-octadecane. The mechanism of the process is not known, although some details are clarified by these investigations. The hydrophobic spores and hyphae of *Streptomyces* interact directly with the oil droplets, as shown in Fig. 3. The membranes of the oil utilizers differ from those of the nonutilizers, as demonstrated in model experiments (Fig. 10, Table 9) proving that oil droplets can penetrate the membrane at a much higher rate than in the nonutilizers.

It has previously been published that in the presence of *n*-alkanes the oil utilizers tend to accumulate fatty acids with chains equivalent in length to that of the alkane substrate (40,41). This was true for our strains, too, as the data obtained with one representative isolate of ours proves (Table 3), using C_{16} *n*-hexadecane substrate.

The fatty acid patterns presented in Table 3 are similar to those reported for other *Streptomyces* strains (42,43).

Electron microscopy of the *n*-alkane utilizers showed significant accumulation of inclusions in the cytoplasm, and their proportion was reduced when the strains were cultivated on complex medium in the absence of *n*-alkanes (Fig. 7).

Our observation that GTP protein activators promoted *n*-alkane uptake is a new observation. As Tables 7 and 8 show, both AlF_4^- and GTPγS (both known as G-protein activators) increased the uptake although the rate of uptake differed in various strains.

Radioactively labeled n-hexadecane-^{14}C was converted to $^{14}CO_2$ and to some radioactive materials not yet identified (Table 6). This proves that the n-alkane was metabolized.

Soil experiments in laboratory also proved that *Streptomyces* actively eliminate hydrocarbons (Table 5).

All the above data prove that the application of *Streptomyces* for hydrocarbon elimination is promising and the possibility of their practical use is supported by the facts that these soil bacteria are indigenous in almost every kind of soils and grow under many different climatic conditions. They are useful in the soil by degrading plant and animal wastes and produce spores that resist hard environmental changes. Their future application in bioremediation is therefore of great promise.

ACKNOWLEDGMENT

The authors are grateful to Prof. I. M. Szabó (ELTE Microbiology Department, Budapest, Hungary) for the taxonomical identifications and to Anikó Ottenberger, Andrea Kiss, Katalin Kosztolányi, and Magdolna Szabó for their technical assistance. We wish to thank the editors of *Actinomycetologica* for permission to publish figures from Ref. 2 and *FEMS Microbiology Letters* for figures and tables from Ref. 5.

REFERENCES

1. Sorkhoh, N. A., Al-Hassan, R. H., Khanafer, M., and Radwan, S. S. 1995. Establishment of oil-degrading bacteria associated with cyanobacteria in oil-polluted soil. *J. Appl. Bacteriol.*, 78:194–199.
2. Barabás, Gy., Sorkhoh, N. A., Fardoon Fardia, and Radwan, S. S. 1995. n-Alkane utilization by oligocarbophilic actionomycete strains from oil-polluted Kuwaiti desert soil. *Actionomycetologica*, 9:13–18.
3. Williams S. T., and Davies, F. L. 1965. Use of antibiotics for selective isolation and enumeration of actinomycetes in soil. *J. Gen. Microbiol.*, 38:251–261.
4. Sorkhoh, N. A., Ghannoum, M. A., Ibrahim, A. S. Stretton, R. J., and Radwan, S. S. 1990. Crude oil and hydrocarbon-degrading strains of *Rhodococcus rhodochrous* isolated from soil and marine environments in Kuwait. *Environ. Pollution*, 65:1–17.
5. Radwan, S. S., Barabás, Gy., Sorkhoh, N. A., Damjanovich, S., Szabó, I., Szöllősi, J., Matkó, J., Penyige, A., Hirano, T., and Szabó, I. M. 1998. Hydrocarbon uptake by *Streptomyces*. *FEMS Microbiol. Lett.*, 169:87–94.
6. Barabás, Gy., and Szabó, G. 1968. Role of streptomycin in the life of *Streptomyces griseus*: Streptidine-containing fractions in the cell wall of *Streptomyces griseus* strains. *Can. J. Microbiol.*, 14:1325–1331.
7. Gilman, A. G. 1987. G proteins: Transducers of receptor-generated signals. *Annu. Rev. Biochem.*, 56:615–649.
8. Price, S. R., Barber, A., and Moss, J. 1991. Structure-function relationships of guanine nucleotide-binding proteins. In J. Moss and M. Vaughan (eds.), *ADP-Ribosylating Toxins and G-Proteins*. ASM Publications, Washington, D.C.
9. Firtel, R. A. 1991. Signal transduction pathways controlling multicellular development in *Dictyostelium*. *Trends Genet.*, 7:381–387.
10. Hall, A. 1992. Signal transduction through small GTPases—A tale of two GAPs. *Cell* 69:389–391.
11. March, P. E. 1992. Membrane associated GTPase in bacteria. *Mol. Microbiol.*, 6:1253–1257.

12. Taylor, C. V. 1990. The role of G Proteins in transmembrane signalling. *Biochem. J.*, 272:1–13.
13. Kumagai, A., Pupillo, M., Gunderson, K., Miake-lye, R., Devreotes, P. N., and Firtel, R. A. 1989. Regulation and function of Gα protein subunits in *Dictyostelium*. *Cell*, 57:265–275.
14. Itoh, H., Kozasa, T., Nagata, S., Nakamura, S., Katada, T., Ui, M., Iwai, S., Ohtsuka, E., Kawasaki, H., Suzuki, K., and Kaziro, Y. 1986. Molecular cloning and sequence determination of cDNAs for α subunits of the guanine nucleotide-binding proteins Gs, Gi and Go from rat brain. *Proc. Natl. Acad. Sci. USA*, 83:3776–3780.
15. Yatani, A., and Brown, A. M. 1991. Mechanism of fluoride activation of G protein-gated muscarinic atrial K^+ channels. *J. Biol. Chem.*, 266:22872–22877.
16. Fraser, C. M., Gocayne, J. D., White, O., Adams, M. D., Clayton, R. A., Fleischmann, R. D., Bult, C. J., Kerlavage, A. R., Sutton, G., Kelly, J. M., Fritchman, J. L., Weidman, J. F., Small, K. V., Sandusky, M., Fuhrmann, J., Nguyen, D., Utterback, T. R., Saudek, D. M., Phillips, C. A., Merrick, J. M., Tomb, J.-F., Dougherty, B. A., Bott, K. F., Hu, P.-C., and Lucier, T. S. 1995. The minimal gene complement of *Mycoplasma genitalium*. *Science*, 270:397–403.
17. Benaissa, M., Viegres-Lubochinsky, J., Odeide, R., and Lubochinsky, B. 1994. Stimulation of inositide degradation in clumping *Stigmatella aurantiaca J. Bacteriol*, 176:1391–1393.
18. Derijard, B., BenAissa, M., Lubochinski, B., and Cenatiempo, Y. 1989. Evidence for a membrane-associated GTP-binding protein in *Stigmatella aurautiaca*, a prokaryotic cell. *Biochem. Biophys. Res. Commun.*, 158:562–568.
19. Trach, K., and Hoch, J. A. 1989. The *Bacillus subtilis* spo0B stage 0 sporulation operon encodes an essential GTP-binding protein. *J. Bacteriol.*, 171:1362–1371.
20. Mitchell, C., and Vary, J. C. 1989. Proteins that interact with GTP during sporulation of *Bacillus subtilis*. *J. Bacteriol.*, 171:2915–2918.
21. Carpenter, P., Hanlon, B., and Ordal, G. W. 1992. flhF, a *Bacillus subtilis* flagellar gene that encodes a putative GTP-binding protein. *Mol. Microbiol.*, 6:2705–2713.
22. Ramirez, C., and Matheson, A. T. 1991. A gene in the archaebacterium *Sulfolobus solfataricus* that codes for a protein equivalent to the alpha subunits of the signal recognition particle receptor in eukaryotes. *Mol Microbiol.*, 5:1687–1693.
23. Samuelsson, T., and Olsson, M. 1993. GTPase activity of a bacterial SRP-like complex. *Nucleic Acids Res.*, 21:847–853.
24. Lin, Y-P., Sharer, D., and March, P. E. 1994. GTPase-dependent signalling in bacteria: Characterisation of a membrane-binding site for Era in *Escherichia coli*. *J. Bacteriol.*, 176:44–49.
25. Mukherjee, A., Dai, K., and Lutkenhaus, J. 1993. *Escherichia coli* cell division protein FtsZ is a guanine nucleotide binding protein. *Proc. Natl. Acad. Sci. USA*, 90:1053–1057.
26. Itoh, M., Penyige, A., Okamoto, S., and Ochi, K. 1996. Proteins that interact with GTP in *Streptomyces griseus* and its possible implication in morphogenesis. *FEMS Microbiol. Lett.*, 135:311–316.
27. Penyige, A., Vargha, Gy., Ensign, J. C., and Barabás, Gy. 1992. The possible role of ADP-ribosylation in physiological regulation in *Streptomyces griseus*. *Gene*, 115:181–185.
28. Laursen, R. A., L'Italien, J. J., Nagarkotti, S., and Miller, D. L. 1981. The amino acid sequence of Elongation Factor Tu of *Escherichia coli*. The complete sequence. *J. Biol. Chem.*, 256:8102–8109.
29. March, P. E., and Inouye, M. 1985. GTP-binding membrane protein of *Escherichia coli* with sequence homology to initiation factor 2 and elongation factor Tu and G. *Proc. Natl. Acad. Sci. USA*, 82:7500–7504.
30. Ovchinikov, Y. A., Alakhov, Y. B., Bundulis, Y. P., Bundulee, M. A., Dovga, N. V., Kozlov, V. P., Motuz, L. P., and Vinokurov, L. M. 1982. The primary structure of Elongation factor G from *Escherichia coli*. *FEBS-Lett.*, 138:130–134.
31. Khokhlov, A. S., Anisova, L. N., Tovarova, J. J., Kleiner, E. E., Kovalenko, O. S. Krasilnikova, O. S. Kornitskaya, E. Y., and Pliner, S. A. 1973. Effect of A-factor on growth of asporogeneous mutant of *Streptomyces griseus*, not producing this factor. *Z. Allg. Microbiol.*, 13:647–655.

32. Basu, A., and Modak, M. J. 1987. An affinity labelling of ras p21 protein and its use in the identification of ras p21 in cellular and tissue extracts. *J. Biol. Chem.*, 262:2369–2373.
33. Friedman, E., Butkerait, P., and Wang, H-Y. 1993. Analysis of receptor-stimulated and basal guanine nucleotide binding to membrane G proteins by sodium dodecyl sulphate-polyacrylamide gel electrophoresis. *Anal. Biochem.*, 214:171–178.
34. Radwan, S. S., and Sorkhoh, N. A. 1993. Lipids of n-alkane-utilizing microorganisms and their application potential.*Adv. Appl. Microbiol.*, 39:29–90.
35. Shinitzky, M., and Barenholz, Y., 1978. Fluidity parameters of lipid regions determined by fluorescence polarization. *Biochem. Biophys. Acta*, 515:367–394.
36. Szöllősi, J., 1994. Fluidity/viscosity of biological membranes. In: S. Damjanovich, M. Edidin, J. Szöllősi, and L. Trón (eds), *Mobility and Proximity in Biological Membranes*, pp. 137–209, CRC Press, Boca Raton, FL.
37. Greenspan, P., Mayer, E. P., and Fowler, S. D., 1985. Nile Red: A selective fluorescent stain for intracellular lipid droplets. *J. Cell. Biol.*, 100:965–973.
38. Kuhry, J.-G., Duportail, G., Bonner, C., and Laustriat, G., 1985. Plasma membrane fluidity measurements on whole living cells by fluorescence anisotropy of trimethylammonium-diphenylhexatriene. *Biochim. Biophys. Acta*, 845:60–67.
39. Matkó, J., Szöllősi, J., Trón, L., and Damjanovich, S., 1988. Luminescence spectroscopic approaches in studying cell surface dynamics. *Quart. Rev. Biophys.*, 21:479–544.
40. Klug, M. J., and Markovets, A. J. 1967. Degradation of hydrocarbons by members of the genus *Candida* II. Oxidation of n-alkanes and 1-alkanes by *Candida lipolytica*. *J. Bacteriol.*, 93:1847–1852.
41. Sorkhoh, N. A., Ghannoum, M. A., Ibrahim, A. S., Stretton, R. J., and Radwan, S. S. 1990. Sterols and diacylglycerophosphocholines in the lipids of the hydrocarbon-utilizing prokaryote *Rhodococcus rhodochrous*. *J. Appl. Bacteriol.*, 69:856–863.
42. Shim, M. S., and Kim, J. H. 1993. Fatty acid and lipid composition in mycelia from submerged or surface culture of *Streptomyces viridochromogenes*. *FEMS Microbiol. Lett.*, 108:11–14.
43. Kröppenstedt, R. M. 1985. Fatty acid and menaquinone analysis of actinomycetes and related organisms. In M. Goodfellow and D. E. Minnikin (eds.), *Chemical Methods in Bacterial Systematics*, pp. 173–199. Academic Press, London.

15
Enhanced Naphthalene Bioavailability in a Liquid–Liquid Biphasic System

J. A. G. F. Menaia and F. M. Gírio
National Institute for Engineering and Technology, Lisbon, Portugal

M. Rosário Freixo
University of Évora, Évora, Portugal

I. INTRODUCTION

Current knowledge supports that the removal of polycyclic aromatic hydrocarbons (PAHs) from soils and waters relies predominantly on biological mechanisms (1). Likewise, bioremediation strategies based on microbial degradation are presently perceived as the most advantageous for the treatment of soils, wastes, and waters with strong PAH contamination (2). Therefore, factors influencing the microbial utilization of PAHs similarly affect the rate and extent of the bioremedial treatments and the natural attenuation of these pollutants by the diverse PAH-degrading microorganisms. The ability to metabolize PAH aerobically occurs in a wide variety of bacteria (3) and Eukarya, including fungi and algae (4). The capability for anaerobic PAH degradation by bacteria was recently reported (5,6).

In addition to the environmental factors (e.g., water activity, pH, temperature, nutrient availability), and the types and distribution of microbial populations and predators, the physicochemical properties of PAHs are important parameters for the pace and intensity of their biodegradation (7), hence for their persistence in the environment.

PAHs are compounds with more than one benzene ring that are solids at standard environmental conditions. Because of their hydrophobicity, in the environment PAH occur mainly sorbed to soil and sediment particles (4,7). As they have very low solubility and transfer slowly to the aqueous phase, their actual concentrations in water depend on the dissolution rates and may limit bioavailability for active PAH-degrading microbial populations (8). Growth of bacteria on solid naphthalene was reported (9). Nevertheless, the general consensus is that microorganisms generally utilize hydrocarbon molecules that are dissolved in the aqueous phase, inasmuch as the degradation rates of PAHs correlate directly to their solubility in water (1).

Strategies used to enhance PAH bioavailability, by improving their transfer to aqueous media, include liposome encapsulation and the addition of surfactants or co-solvents. While the use of liposomes increases the bioavailability of alkanes, PAH encapsulation has no similar effect (30). Nonionic surfactants (e.g., Triton X, Tergitol NPX) are known by their ability to enhance the biodegradation of hydrocarbons by increasing their dispersion and solubility in water (10). By diminishing solid/liquid interfacial tension (11), surfactant concentrations below critical micelle concentration (CMC) are expected to improve particle and pore wettability, so that PAH dispersion and transfer to the aqueous phase is facilitated (12,13). However, published results on the effect of nonionic surfactants on PAH biodegradation differ greatly in their conclusions. While some authors reported enhancement of the microbial degradation of PAH by nonionic detergents at sub-CMC (10), others found no significant effects for such concentrations (14). The reports on the enhancement of PAH bioavailability with supra-CMC concentrations of nonionic surfactants are more consistent; at such concentrations however, inhibition of microbial activity may arise (14). Co-solvents (e.g., dimethyl formamide, dimethyl sulfoxide, acetone, ethanol) that are miscible in water and solubilize solid PAHs are used to improve PAH solubilization by dispersing the hydrocarbon particles, thus increasing the water–solid interface area (15). However, to prevent inhibitory effects due to co-solvent toxicity, these are ordinarily used at low concentrations.

Water-imiscible solvents, which have long been used in biotechnological transformations in biphasic biocatalysis (16), are particularly attractive for PAH degradation, since the efficiency of mass transfer from an organic solvent to water is generally much higher than that occurring between solid and water phases (17). In addition, such systems may allow the minimization of substrate and product toxicity by using a solvent whose partition properties enable the control of the aqueous concentrations of those compounds (18).

In order to investigate the effect of the presence of a biologically inert hydrophobic phase on the degradation of naphthalene by the PAH-degrading co-culture EXPO98, which was isolated from a soil contaminated with crude oil, studies were carried out in a batch system with dimethylpolysiloxane, a silicone oil, as the organic phase (20% v/v), with the microbial cells distributed predominantly within the bulk aqueous phase. Previously, cultures in mineral medium were done to study mechanisms of solid naphthalene bioavailability, effects and degradability of nonionic surfactants, and biodegradability and biocompatibility of dimethylpolysiloxane to EXPO98.

II. ISOLATION OF A PAH-DEGRADING CULTURE

The EXPO98 bacterial culture was isolated from the Lisbon-EXPO98 grounds, from a soil that had been heavily contaminated with crude oil for several decades. Modifications of the overlayer technique for the isolation of PAH-degrading bacteria (19) were used. Aseptically, 1 g of soil was suspended at 10% (w/v) in a 0.1 M sodium pyrophosphate solution. After being vigorously mixed with a vortex stirrer, the suspension was filtered to remove soil particles. The filtered suspension was then serially diluted up to 10^{-3}. Each decimal dilution was incorporated at 1:10 (v/v) in 7.5 mL of Vishniac mineral medium (MM) (20), supplemented (MMYE) with 0.01 g/L of yeast extract (Oxoid), with 1.5% (w/v) of purified agar (Difco). The medium was kept at 40°C after autoclave sterilization. The mixture was stirred and poured onto a naphthalene thin layer that was prepared by spreading an aerosol of ethyl ether saturated with naphthalene (Merck, purity >99%) over MMYE agar plates. By using this technique the exposure of cells to the solvent was prevented and the PAH layer homogeneity was improved so that clearer zones were readily distinguishable around a number of colonies, which were produced over 3 weeks of incubation at 30°C. The colonies that had apparently removed naphthalene from their surroundings were small in (diameter < 1 mm) and alike. For isolation, one of these colonies was streaked on MMYE agar plates with a few naphthalene crystals attached to the inner side of the cover. Colonies formed on naphthalene supplied in the form of vapor attained a larger diameter (~ 3 mm) and were similar in appearance. However, when colonies were streaked onto Triptone soy agar (TSA-Oxoid) or colony serial dilutions were spread on this solid medium, two types of colonies of different opacity were invariably formed. Under phase-contrast microscopy at 1000 × magnification, fresh mounts from both types of colonies always displayed small, nonmotile rods, as single cells or associated in short chains that appeared to belong to the same pure culture. However, systematic attempts to isolate colonies with the same opacity failed, even when serial dilutions of liquid cultures from both colonies in MM with solid naphthalene (0.01 mol/L) were spread onto TSA. Thereafter we assumed that the purified microbial cells constituted a co-culture. In addition to naphthalene, EXPO98 grew in MM on phenanthrene (Merck, purity >98%) or dibenzothiophene (Acrös, purity >99%), but not an anthracene (Merck, purity >98%).

III. GROWTH AND NAPHTHALENE DEGRADATION IN MINERAL MEDIUM

For studying EXPO98 growth and consumption of naphthalene added as solid-particles, twelve 250 mL Erlenmeyer flasks, with 75 mL of MM medium and 0.1 g of sterilized naphthalene particles (average diameter ~ 0.5 mm), were inoculated at 10% (v/v) with exponentially growing cultures that had been transferred more than five times in MM with naphthalene particles. The cultures were incubated at 30°C with orbital shaking (130 rpm). These cultivation procedures were used throughout this work.

At regular intervals, two flasks were outdrawn for determination of culture absorbance at 590 nm, protein, and aqueous naphthalene contents.

The Lowry method (21) was used to determine the protein in cell pellets obtained by centrifuging (6000g for 10 min) 4 mL of each culture, which were treated at 100°C with NaOH (1 M), for 5 min. For analysis, the naphthalene content of 50 mL samples of culture supernatant plus 250 µL of an internal standard solution of biphenyl in methanol (2 g dm^3), was concentrated in a Sep-Pak C18 column and eluted with 2 mL of ethanol. Naphthalene and biphenyl in 2 µL of the ethanol eluate were determined in a gas chromatograph Crompack-CP 9001 equipped with a Chromosorb W HP (100–120 mesh) column and a flame ionization detector. Nitrogen was used as the carrier gas at a flow rate of 15 mL/min. The injector was at 250°C and the detector at 335°C. For separation, a temperature gradient was set in the oven (100°C for 5 min, 5°C/min up to 180°C, and 10°C/min up to 300°C). Additionally, naphthalene concentrations were determined spectrophotometrically at 275 nm.

The evolution in the total cell protein contents (data not shown) and in absorbance (Figs. 1A and 1B) were correlated, and showed that growth of EXPO98 in MM medium with solid naphthalene was exponential during the first hours and became linear by the time the aqueous naphthalene concentrations started approaching zero (Fig. 1C).

These results indicate that EXPO98 grew on aqueous naphthalene, are in accordance with the data published by Weissenfels et al. (22), and are consistent with the model developed by Volkering et al. (8) stating that bacterial growth on solid PAH is linear and is rate-limited by the mass transfer from the solid phase to the liquid phase. Thus, the rates

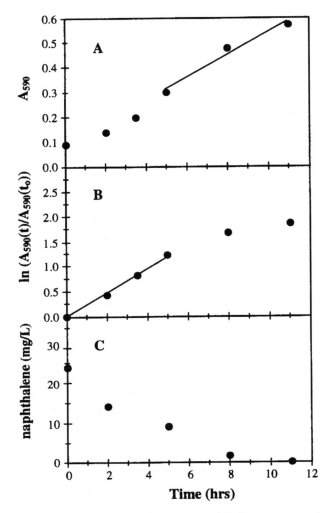

Figure 1 Growth and aqueous naphthalene consumption by EXPO98 in mineral medium.

of desorption/dissolution of PAH may control the bioavailability of this type of contaminant.

IV. EFFECT OF PARTICLE SURFACE AREA ON GROWTH

In order to further confirm the naphthalene dissolution rate as a limiting factor for PAH bioavailability, the growth of EXPO98 in 150 mL of MM medium with 0.2 g of naphthalene in the form of a single piece (diameter ~7 mm) or divided into small particles (average diameter ~0.5 mm) was investigated in 500 mL Erlenmeyer flask shaking cultures.

Since the rate of dissolution of a solid is proportional to the area of the solid–water interface (8), and naphthalene bioavailability depends on its rate of dissolution, faster growth and naphthalene consumption were expected to occur on solid

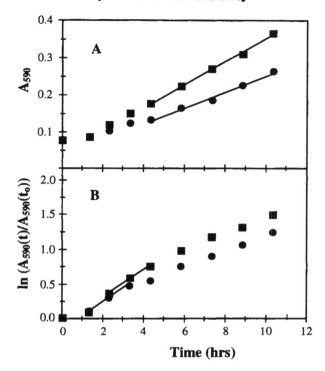

Figure 2 Growth of EXPO98 in 150 mL of mineral medium with 0.2 g of naphthalene added as a single piece (●) or divided into particles (■).

naphthalene with larger surface areas (smaller particles). By the same token, in a batch system with an initial equilibrium concentration of aqueous naphthalene, a lengthened exponential growth is expected for a larger naphthalene surface area, inasmuch as a higher rate of naphthalene dissolution will surpass the rate of substrate consumption for a longer period than a dissolution rate that is closer to the rate of naphthalene consumption.

As expected, growth of EXPO98 on the same mass of solid naphthalene (0.2 g) displayed a longer exponential period when the substrate had a larger area of solid–water interface (Fig. 2B). In addition, the subsequent linear growth was approximately 30% faster on the divided naphthalene (Fig. 2A).

In the experiment illustrated by Fig. 2, the same specific growth rate was observed both for divided ($\mu = 0.20 \pm 0.02$ h^{-1}) and undivided naphthalene ($\mu = 0.20 \pm 0.01$ h^{-1}). The same experiment was repeated twice, giving matching specific growth rates that were identical to the ones that were observed repeatedly for cultures on solid naphthalene ($\mu = 0.20 \pm 0.02$ h^{-1}). Therefore, it is likely that the initial concentrations of aqueous naphthalene were the same as well. Thus, most probably all the experiments started with equilibrium naphthalene concentrations as planned (culture media was allowed to equilibrate with solid naphthalene for at least 12 h prior to inoculation). On the other hand, while the rates of linear growth were invariably higher for smaller naphthalene particles, a variation of ca. 10% was observed among performed experiments, which was presumably due to the difficulty of preparing naphthalene particles of uniform surface area.

I. EFFECT OF SURFACTANTS ON NAPHTHALENE BIOAVAILABILITY

The effect of the presence of Tween 80 ($d = 1.080$, Merck) or Triton X-100 ($d = 1.080$, USB) on the bioavailability of 0.2 g of naphthalene divided into particles (average diameter ~ 0.5 mm) was studied with cultures of EXPO98 in 150 mL of MM medium in 500 mL Erlenmeyer flasks. Surfactants were tested at concentrations that were about 1, 10, and 100 times the CMC (14). Tween 80 was assayed at 0.003, 0.03, and 0.3% (v/v). Triton X-100 was tested at 0.031, 0.31, and 3.1% (v/v).

The utilization as growth substrates, or the possible toxic effects of the surfactant compounds, were tested with EXPO98 cultures on 0.1% (w/v) glucose, which were inoculated with cells grown on glucose or naphthalene. At the studied concentration ($\sim 10 \times$ CMC), none of the detergents produced any detectable effect on the growth of EXPO98 on glucose (Fig. 3). EXPO98 did not grow on Triton X-100. However, although in an irregular mode, cells, either pregrown on naphthalene or on glucose, were able to grow on Tween 80 (Fig. 3B).

None of the experiments with surfactants yielded any perceptible effect on the growth of EXPO98 on naphthalene (Fig. 4).

These results were unexpected, not only because of the ability of EXPO98 to grow on Tween 80, but also because either the enhancement of naphthalene bioavailabilty or the strong inhibition of microbial growth by nonionic surfactants at supra-CMC were reported previously (13,15).

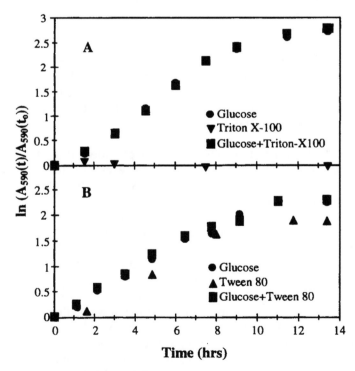

Figure 3 Biodegradability and biocompatibility of Triton X-100 (A) and Tween 80 (B) to EXPO98 (pregrown on naphthalene).

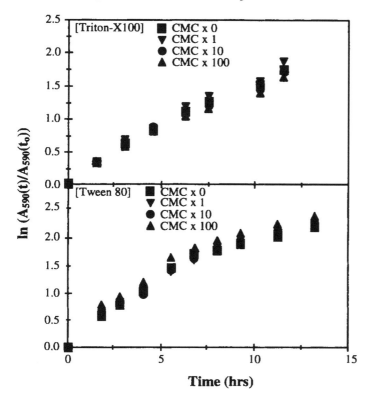

Figure 4 Growth of EXPO98 on naphthalene in mineral medium with Triton X-100 or Tween 80 at different concentrations.

Perhaps the expected effects on EXPO98 growth on solid naphthalene were too subtle to be detected and, therefore, have no practical significance, or were masked by the effect of emulsifying compounds that could possibly be produced by the co-culture. However, perceptible signs of biosurfactant excretion (e.g., foaming) by EXPO98 were never observable during growth on naphthalene.

VI. EFFECT OF AN ORGANIC PHASE ON NAPHTHALENE BIOAVAILABILITY

Many organic solvents, which are traditionally used as hydrophobic phases in liquid–liquid systems, have a toxicity to microorganisms that restricts their volumetric ratio to values far below those required for the full use of the potential performances of biphasic systems (23). Recently, alternative organic compounds of low toxicity have begun to be applied to such systems. These compounds include the polysiloxanes (PSs), which, because of their physicochemical properties, are suitable for use as the organic phase of biphasic liquid–liquid systems. PSs are physiologically inert and refractory to biodegradation, and they have good thermal stability and a high capacity for gas dissolution (24,25). The polydimethylsiloxanes are a group of PSs that comprises amphiphilic compounds with alternate polar and apolar groups in their polymeric chains (24). The viscosity and partition properties of the polydimethylsiloxanes can be controlled to a certain extent by selecting a polymeric chain with the hydrophobic groups of interest (25).

The effect of a polydimethylsiloxane (visc. = 20 centistoke, Sigma), hereafter designated as PDMS20, on the bioavailability of naphthalene was investigated in a liquid–liquid batch system. The experiments were done with 150-mL (total volume) cultures in MM medium, plus 20% (v/v) of PDMS20, inoculated with cells pregrown on naphthalene.

As PDMS are biodegraded in soils (26,27), and may have unknown effects on microorganisms, the ability of EXPO98 to grow on PDMS20, as well as possible effects of PDMS20 on the growth of EXPO98 on a soluble substrate, were investigated previously. Tests were done with 150-mL (total volume) cultures of EXPO98, with an inoculum pregrown on naphthalene, in MM medium with 0.1% (v/v) glucose, in MM medium with 0.1% (w/v) glucose plus 20% (v/v) of PDMS20, and in MM medium plus 20% (v/v) of PDMS20 and no glucose.

Growth on glucose alone matched growth on glucose in the presence of PDMS20. PDMS20 did not support growth of EXPO98 during the course of the experiment (14 h). Therefore, no inhibitory effects or co-consumption of PDMS were observed (results not shown). So, the presence of PDMS20 was also not expected to affect growth of EXPO98 on naphthalene.

In the studies on the effect of PDMS20 on naphthalene bioavailability, naphthalene was added in solution in the organic phase, at a concentration of 0.1% (w/v). These experiments included cultures with solid naphthalene added as a single particle or divided into particles, with no DPMS20. Solid naphthalene and PDMS were individually sterilized in an autoclave.

As in the cultures on solid naphthalene, growth on naphthalene transferred from DPMS20 was exponential during an initial period, which was followed by zero-order growth. The exponential growth rate in the liquid–liquid system ($\mu = 0.22 \pm 0.01$ h^{-1}) did not differ significantly from that on solid naphthalene ($\mu = 0.20 \pm 0.02$ h^{-1}). However, the exponential growth length, which was already significantly longer for apportioned than for undivided naphthalene, was extended substantially when naphthalene transfer occurred from the hydrophobic phase. In addition, in the biphasic system the rate of linear growth was considerably higher (Fig. 5).

Thus, the effect of incrementing the naphthalene area available for solid–water transfer, and the effect of providing an apolar phase supporting the liquid–liquid transfer of naphthalene in the form of a solute, are alike. However, while both the area increment and the presence of a hydrophobic phase improved naphthalene bioavailability by promoting its transfer to the aqueous phase, the bioavailability enhancement was greater in the biphasic system.

The occurrence of identical growth rates during the exponential periods of all cultures on solid naphthalene or naphthalene in DPMS20 suggests that the initial concentrations of bioavailable aqueous naphthalene were equilibrium concentrations, which probably approached the naphthalene solubility in water. Thus, no apparent increase in the solubility of naphthalene was produced by DPMS20.

The observed increase in the length of exponential growth may indicate that the liquid–liquid system had the ability to sustain the rate of naphthalene transfer, surpassing the rate of naphthalene consumption for a longer time, even if the last was then increasing exponentially over time.

The faster transition from exponential to linear growth in the liquid–liquid system suggests that the strength of the naphthalene concentration gradient fell rapidly to a level that was insufficient to support exponential growth. Still, over the course of the experiment, the subsequent rate of naphthalene transfer to water remained substantially higher in the liquid–liquid system, as was indicated by the significantly faster linear growth.

Because of the high O_2 solubility in PDMS, which available data (25) indicate to be roughly six times more than in water, PDMS20 may improve O_2 transfer in liquid–liquid systems (28). Hypothetically, therefore, the observed positive effect of DPMS20 on EXPO98 growth on naphthalene may have been due to an optimization of the O_2 transfer. However, the occurrence of matching specific growth rates of EXPO98 on solid naphthalene of different particle size and in the liquid–liquid system led us to discard this proposition.

On the other hand, the enhancement of substrate transfer rates in biphasic liquid–liquid systems is partially ascribed to the adhesion of the involved microbial cells to the interface between the water and the hydrophobic phase (17,29,30). However, microscopic observations of fresh mounts of EXPO98 cultures with DPMS20 exhibited no attachment of the microbial cells to the organic-phase droplets. To verify this observation, the cell distribution was investigated for all tested culture conditions by adding PDMS20 to EXPO98 cultures growing exponentially or linearly on solid naphthalene and measuring the absorbance of the aqueous phase after letting growth to proceed for 1 h. The addition of PDMS did not produce any associated decrease in the absorbance of the aqueous phase. Apparently the enhancement of naphthalene transfer in the liquid–liquid system did not depend on the attachment of cells to the organic phase. Therefore, observed results suggest that the absence of cell hydrophobicity does not preclude the advantageous utilization of PAH-degrading microorganisms in liquid–liquid biphasic systems.

VII. CONCLUSIONS

The suitability of bioremediation for the treatment of PAH-contaminated soils is well established (2). However, due to the very low solubility of PAH in water, the full use of this technology is still limited by the low rates of PAH biodegradation (1). So, bioremedial strategies to accelerate PAH removal from soils, sludge, and waters must involve the enhancement of PAH bioavailability. Reported strategies are not completely satisfactory (e.g., PAH encapsulation), do not consistently produce expected results (e.g., surfactant addition), or may be hindered by the toxicity of the solvents that are used to promote PAH bioavailability (30).

Jimenez and Bartha (30) produced encouraging data on the use of biologically inert solvents (e.g., squalene, paraffin oil) to improve the rate and extent of PAH microbial degradation by attached cells. As these authors achieved doubled pyrene mineralization rates by adding low volumetric concentrations of solvents (0.8%), they proposed the application of hydrophobic solvents in aqueous or slurry-type bioreactors.

In addition to a significant improvement (>30%) in the linear rate of naphthalene removal (Fig. 5A) by using a higher volumetric ratio (20%) of a silicone oil as the organic phase in a biphasic batch system with suspended cells, we observed that the hydrophobic solvent made up a PAH pool that could sustain the exponential degradation of naphthalene (Fig. 5B). Thus, while indicating that solvent-augmented biodegradation of PAH may be applied to suspended cell systems, our results suggest further that the use of hydrophobic solvents may support the biotechnological development of continuous

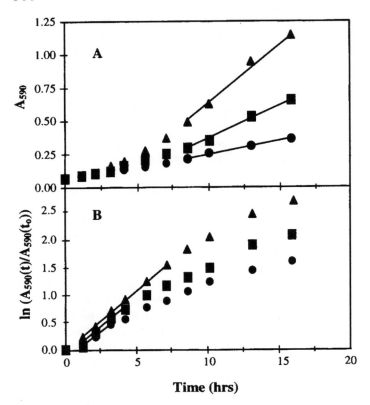

Figure 5 Growth of EXPO98 on naphthalene added in solution in PDMS20 (▲) or in the solid form, divided (■), or as a single particle (●).

biphasic reactors for the removal of PAH from waters or slurries. Since silicone oils, or other organic compounds of similar properties, separate easily from water, continuous systems with solvent recycling may be envisaged, provided that inhibitory metabolites are not accumulated or are scrubbed from the organic phase.

ACKNOWLEDGMENTS

This work was supported by EU research grant INCO-COPERNICUS Contract IC15 CT 96-0716.

REFERENCES

1. Shuttleworth, K. L., and Cerniglia, C. E. 1995. Environmental aspects of PAH degradation. *Appl. Biochem. Biotechnol.*, 54:291–302.
2. Cerniglia, C. E. 1993. Biodegradation of polycyclic aromatic hydrocarbons. *Curr. Opin. Biotechnol.*, 4:331–338.
3. Mueller, J. G., Devereux R., Santavy, D. L., Lantz, S. E., Willis, S. G., and Pritchard P. H. 1997. Phylogenetic and physiological comparisons of PAH-degrading bacteria from geographically diverse soils. *Antonie Van Leeuwenhoek*, 71:329–343.

4. Cerniglia, C. E. 1992. Biodegradation of polycyclic aromatic hydrocarbons. *Biodegradation*, 3:351–368.
5. Rueter, P., Rabus, R., Wilkes, H., Aeckersberg, F., Rainey, F. A., Jannasch, H. W., and Widdel, F. 1994. Anaerobic oxidation of hydrocarbons in crude oil by new types of sulphate-reducing bacteria. *Nature*, 372:455–458.
6. Coates, J. D., Woodward, J., Allen, J., Philp, P., and Lovley, D. R. 1997. Anaerobic degradation of polycyclic aromatic hydrocarbons and alkanes in petroleum contaminated marine harbor sediments. *Appl. Environ. Microbiol.*, 63:3589–3593.
7. Hughes, J. B., Beckles, D. M., Chandra, S. D., and Ward, C. H. 1997. Utilization of bioremediation processes for the treatment of PAH-contaminated sediments. *J. Ind. Microbiol. Biotechnol.*, 8:152–160.
8. Volkering, F., Breure, A. M., Sterkenburg, A., and van Andel, J. G. 1992. Microbial degradation of polycyclic aromatic hydrocarbons: Effect of substrate availability on bacterial growth kinetics. *Appl. Microbiol. Biotechnol.*, 36:548–552.
9. Guerin, W. F., and Boyd, S. 1992. Differential bioavailability of soil-sorbed naphthalene to two bacterial species. *Appl. Environ. Microbiol.*, 58:1142–1152.
10. Jahan, K., Ahmed, T., and Maier, W. J. 1997. Factors affecting the nonionic surfactant-enhanced biodegradation of phenanthrene. *Water Environ. Res.*, 69:317–325.
11. Pankow, J. 1991. *Aquatic Chemistry Concepts*, p. 307 Lewis Publishers, Chelsea, MI.
12. Grimberg, S., Nagel, J., and Aitken, M. 1995. Kinetics of phenanthrene dissolution into water in the presence of nonionic surfactants. *Environ. Sci. Technol*, 29:1480–1487.
13. Volkering, F., Breure, A., Andel, J., and Rulkens, W. 1995. Influence of nonionic surfactants on bioavailability and biodegradation of polycyclic aromatic hydrocarbons. *Appl. Environ. Microbiol.*, 61:1699–1705.
14. Laha, S., and Luthy, R. 1992. Effects of nonionic surfactants on the solubilization and mineralization of phenanthrene in soil-water systems. *Biotechnol. Bioeng.*, 40:1367–1380.
15. Sikkema, J., de Bont, J. A. M., and Poolman, B. 1995. Mechanisms of membrane toxicity of hydrocarbons. *Microbiol. Rev.*, 59:201–222.
16. Woodley, J., and Lilly, M. 1992. Process engineering of two-liquid phase biocatalysis. In: J. Tramper (ed.), *Biocatalysis in Non-Conventional Media*, pp. 147–154. Elsevier, Amsterdam.
17. Efroymson, R., and Alexander, M. 1991. Biodegradation by an *Arthrobacter* species of hydrocarbons partitioned into an organic solvent. *Appl. Environ. Microbiol.*, 57:1441–1447.
18. Kawakami, K., and Nakahara, T. 1993. Importance of solute partitioning in biphasic oxidation of benzyl alcohol by free and immobilized whole cells of *Pichia pastoris*. *Biotechnol. Bioeng.*, 43:918–924.
19. Bogart, A. H., and Hemmingsen, B. B. 1992. Enumeration of phenanthrene-degrading bacteria by an overlayer technique and its use in evaluation of petroleum-contaminated sites. *Appl. Environ. Microbiol.*, 58:2579–2582.
20. Vishniac, W., and Santer, M. 1995. The Thiobacilli. *Bacteriol. Rev.*, 21:95–113.
21. Lowry, O. H., Rosebrough, N. J., Farr, A. L., and Randall, R. J. 1951. Protein measurement with the Folin phenol reagent. *J. Biol. Chem.*, 193:265–275.
22. Weissenfels, W., Beyer, M., and Klein, J. 1990. Degradation of phenanthrene, fluorene and fluoranthene by pure bacterial cultures. *Appl. Microbiol. Biotechnol.*, 32:479–484.
23. Fernandes, P., and Pinheiro, H. M. 1996. Biocatálise em solventes orgânicos utilizando células inteiras. *Boletim de Biotecnologia*, 55:25–32.
24. Kosswig, K. 1994. Surfactants. In: B. Elvers, S. Hawkins, and W. Kussey (eds.), *Ullmann's Encyclopedia of Industrial Chemistry*, Vol. A25. Elsevier, Amsterdam.
25. Moretto, H. H., Schulze, M., and Wagner, G., 1994. Silicones. In: B. Elvers, S. Hawkins, and W. Kussey (eds.), *Ullmann's Encyclopedia of Industrial Chemistry*, Vol. A24. Elsevier, Amsterdam.
26. Carpenter, J. C., Cella, J. A., and Dorn, S. B. 1994. Study of the degradation of polydimethylsiloxanes in soil. *Environ. Sci. Technol.*, 29:864–868.

27. Sabourin, C. L., Carpenter, J. C., Leib, T. K., and Spivack, J. L. 1996. Biodegradation of dimethylsilanediol in soils. *Appl. Environ. Microbiol.*, 62:4352–4360.
28. Rols, J. L., Condoret, J. S., Fonade, C., and Goma, G. 1990. Mechanism of enhanced oxygen transfer in fermentations using emulsified oxygen-vectors. *Biotechnol. Bioeng.*, 35:427–435.
29. Ascon-Cabrera, M., and Lebeault, J. M. 1993. Selection of xenobiotic-degrading microorganisms in a biphasic aqueous-organic system. *Appl. Environ. Microbiol.*, 59:1717–1724.
30. Jimenez, I. Y., and Bartha, R. 1996. Solvent-augmented mineralization of pyrene by a *Mycobacterium* sp. *Appl. Environ. Microbiol.*, 62:2311–2316.

16
Intensive Treatment of Mineral Oil-Contaminated Drilling Cuttings

A. Noke, R. A. Müller, and Ulrich Stottmeister
UFZ Centre for Environmental Research Leipzig-Halle, Leipzig, Germany

I. INTRODUCTION

In oilfields, soil contamination and the production of oily drilling mud occurs regularly from oil recovery technology. Typical environmental problems result from broken pipelines, drilling holes, and the production of contaminated drilling cuttings. The most applied methods for soil and sludge treatment are a mechanical-chemical treatment (conditioning) followed by deposition in dumping sites.

Our biotechnological approach to manage these problems is ingrained in a bilateral technology transfer cooperation between Argentina and Germany. Whereas a biological treatment of sandy, loamy sand, or sandy loam soils can be described as a state-of-the-art technology, classical methods such as soil heaping or landfarming technologies fail with soils characterized as clay or loamy. Soil consistency prevents transportation of water and nutrients. For the specific environmental problem of mineral oil-contaminated drilling mud, a mobile and flexible bioreactor performance was tested (Fig. 1). The process consists of a mechanical pretreatment (mixing and sieving), biological treatment in the airlift reactor, and a dewatering step. Slurry-phase bioreactors are already applied for polycyclic aromatic hydrocarbon (PAH) degradation in contaminated soils (1,2).

Advantages of a reactor performance for soil treatment are as follows (3). Permanent mixing of the material ensures a high oxygen input and an improved bioavailability of the contaminant, leading to relatively short treatment times. Furthermore, important process parameters can easily be optimized by monitoring and control systems. At the end of biological remediation and after dewatering, the sludge can be refilled in the ground or might be reused for building projects.

II. EXPERIMENTAL DETAILS

A. Drilling Mud

Drilling mud, also called drilling cuttings, is a mixture of the drilling fluid which is used for perforation, and the drilled material. The tested drilling mud, deriving from a mud pit of

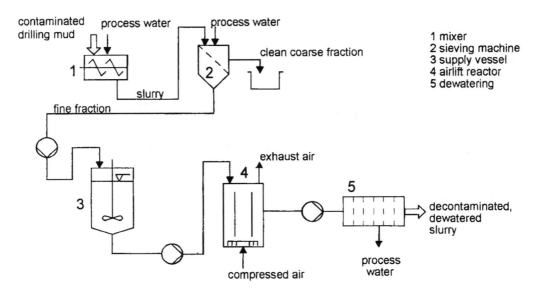

Figure 1 Schema of the bioreactor treatment of mineral oil-contaminated drilling mud.

an ancient oil recovery site in northern Germany, was characterized by its fine particle distribution (80% mass < 63 μm), a mineral oil contamination of approx. 20,000 ppm, and a high content of salt (approx. 20,000 ppm). The mud contained 60% dry matter.

High adsorption rates for nutrient compounds such as phosphate (adsorption of 3000 ppm) and ammonium (adsorption of 600 ppm) are caused by the high specific surface and ion-exchange capacity of the bentonite matrix of drilling mud.

Furthermore, the drilling mud was anoxic. At the beginning of the aeration a strong oxygen uptake was observed, which is probably due to the chemical oxidation of iron sulfide (FeS) to Fe^{3+} and SO_4^{2-}.

B. Experimentation

Microbial degradation of mineral oil-contaminated drilling mud was carried out on two different scales of airlift bioreactors, the first being a series of 4.5-L glass reactors and the second a 65-L stainless steel airlift bioreactor (Biolift®, Eimco, Salt Lake City). Airflow was adjusted via flowmeters to 600 L/h for the 4.5-L reactors and to 40 L/min for the 65-L reactor. The reactors were equipped with pH and redox potential devices for on-line measurement. Temperature was held at 30°C via a double coat. The nutrients ammonium chloride and dipotassium hydrogenphosphate were added at different times to guarantee sufficient nutrient supply.

The biological treatability of the drilling mud was tested at two different dry matter contents, 20% and 30%, in 4.5-L airlift reactors. The mud was previously sieved (< 1 mm); dry matter content was adjusted by dilution with tap water and controlled by an infrared dryer (Sartorius).

After a degradation period of 7 days, the first feeding (sequenced batch process) with untreated drilling mud in the ratio 1 : 2 was performed. On day 13 the second feeding in the ratio 1 : 4 was carried out (which means that three of four parts of the reactor content were replaced by untreated drilling mud of the same dry matter content). Sequenced batch process was applied to retain one part of the active degrading microbial biocenosis in the airlift reactor, which ensures a faster biodegradation in the following degradation cycle.

At the half-technical scale a short-term biodegradation test (6 days), without a second feeding, was carried out with 65 L of sludge (30% d.m., sieved < 1 mm).

The biodegradation of mineral oil hydrocarbons was determined daily according to the German norm DIN 38409 H18. One gram of lyophilized sludge was extracted by shaking with freone (1,1,2-trichlorotrifluoroethane) for 30 min. The extracts were filtered and passed through an aluminium oxide column prior to FTIR analysis (FTIR 8000, Shimadzu). Qualitative hydrocarbon analysis was performed on the 65-L run at days 0, 4, and 6 by GC-MS. Lyophilized sludge samples were extracted with hexane in an ultrasonic bath. The GC conditions permitted the detection of hydrocarbons in the range of C_{14} to C_{40} n-alkanes.

The volatile organic compounds (VOCs) of reactor outlet gas were analyzed in the 65-L degradation test by gas chromatography. Defined gas volumes were adsorbed on activated-carbon cartridges (Dräger). The adsorbed VOCs were extracted with carbon sulfide in an ultrasonic bath. The filtered extracts were analyzed by GC-FID.

Nutrient concentrations (nitrate, nitrite, ammonium phosphate) were determined frequently by Merck test kits. Bacterial growth was examined by counting of colony-forming units (CFUs) on R2A agar every 1–3 days.

III. RESULTS AND DISCUSSION

A. Experiments at 4.5-L Scale

The initial period of drilling mud treatment is characterized by enrichment of the anoxic mud with oxygen and subsequent chemical oxidation of reduced compounds. Redox potential increased from the negative to the positive range within the first 10 h. Following this period of oxidation, the autochthonous microorganisms were activated, biomass concentrations increased, and the biodegradation of contaminants was initiated.

The kinetics of biodegradation of mineral oil constituents in the drilling mud is shown in Fig. 2.

After a short period of microbial adaptation and chemical oxidation of the anoxic sludge, a fast diminution of the hydrocarbons was observed. The biodegradation process terminated at concentrations of 3500 and 4500 ppm, respectively. This accounts for degradation efficiencies of 82% and 77% within 5 days.

During slurry reactor treatment of diesel oil-contaminated soil, Croft et al. (4) reported 90% degradation within 5 days. In the case of aged soil contamination or polluted fine material, extents of degradation are generally lower and treatment times are much longer (5).

Static bioremediation procedures in nonsaturated systems exhibit generally much lower degradation rates. During a 9-month static microcosm experiment under optimal conditions, Chaîneaù et al. (6) obtained 75% of total petroleum hydrocarbon (TPH) degradation for drilling cuttings/soil mixtures. Despite the storage of the tested drilling mud for many years, the content of biodegradable compounds remained high, due to low permeability and anoxic conditions, which excluded aerobic microbial degradation.

Although the residual concentrations are still higher than the upper legislative limits, the importance of such strict levels was considered to be questionable (7). Biologically remediated soil material usually does not exhibit a dangerous potential, since it contains

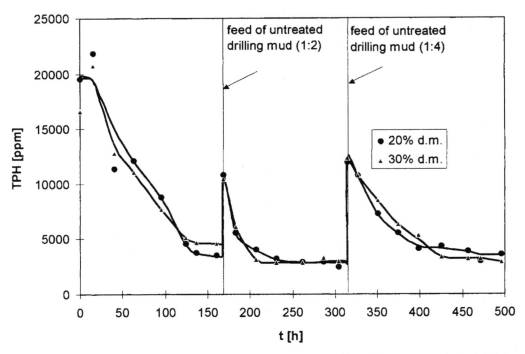

Figure 2 Kinetics of mineral oil degradation during biological treatment of contaminated drilling mud in the 4.5-L airlift reactor. Arrows indicate the partial exchange of reactor content against untreated material.

essentially hydrophobic, nonpolar compounds with high boiling points, which do not infiltrate groundwater aquifers and are not toxic to plants, animals, and microorganisms. Wang and Batha (8) state, that after a temporary increase in toxicity, hydrocarbon degradation leads to rapid and permanent detoxification.

After the first feeding, in the ratio 1 : 2 (168 h), the biodegradation process accelerated rapidly and the residual hydrocarbon concentration was reduced to 2800 ppm.

The second feeding (312 h), at a ratio of 1 : 4, resulted in lower degradation rates but comparable residual concentrations. The degradation rates for hydrocarbons were independent of the dry matter content. Thus, a higher dry matter content allows treatment of a larger quantity of sludge in the same time and is therefore economically preferable.

The degradation rates for the three batch cycles at the 4.5-L scale were calculated as 2700, 2800, and 1800 ppm/day. Only the first feeding led to increased degradation velocity. The second feeding was carried out after a 3-day plateau period (Fig. 2) and at a higher percentage of exchanged material. This caused lower biodegradative activity and emphasizes the importance of an optimized process regime. A fast process analytic is therefore a crucial prerequisite. In this context, the pH value was shown to be a practicable parameter for estimation of the hydrocarbon degradation.

The pH value (Fig. 3) was not stabilized and exhibited good temporal correlation with the degradation of mineral oil constituents. After an immediate increase in pH at the start, up to 8.5, the degradation process (Fig. 2) was accompanied by a decrease in pH to around 6.8–7.0. Stabilization and/or an increase of pH value indicated exhaustion of the biodegradable hydrocarbons. However, it has to be taken into account

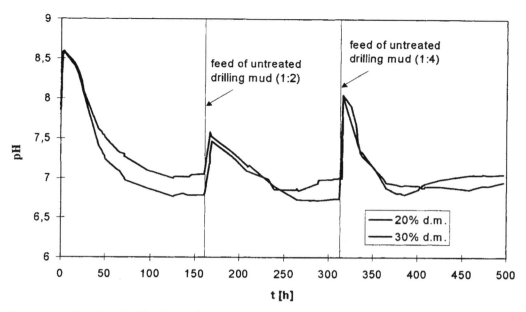

Figure 3 Kinetics of pH value during the biotreatment of drilling mud in the 4.5-L reactor at two different dry matter contents.

that additions of nutrients in high concentrations can change the pH value immediately, and that the oxidation of sulfide to sulfate causes acidification.

Figure 4 indicates the growth of the autochthonous microorganisms during sequenced batch treatment of contaminated sludge in the 4.5-L reactor. Autochthonous microorganism grew from initially 10^6 cells/g d.m. to 10^9 to 10^{10} cells/g d.m. during each of the three cycles.

B. Experiments at 65-L Scale

The scale-up to the 65-L airlift reactor confirmed the biodegradability of a high percentage of the drilling mud contaminants (Fig. 5).

The degradation rate during the 65-L test was determined to be 2200 ppm/day and is therefore similar to those obtained at the 4.5-L scale. The scaling-up procedure can be considered to be successful. Because of the progressive evolution of foam, the experiment had to be interrupted after 142 h. Analogous to the 4.5-L reactor, cell counts increased from 10^6 to 10^9 to 10^{10} cells/g d.m. and pH decreased from 8.2 initially to 6.6 at the end of the test.

In order to estimate the loss of contaminants via the gas phase during intensive aeration in the airlift reactor, exhaust analyses after different time intervals were performed and the quantity of VOCs integrated over time. The majority of the VOCs evolved within the first 5 h and account for less than 1% of the total initial hydrocarbon content in the reactor.

Process water investigations indicated that no significant quantities of hydrocarbons or metabolites accumulated in the liquid phase.

GC-MS analysis of hexane extracts of drilling mud samples demonstrated the high molecular nature ($>C_{40}$) of the residual contamination. In contrast to the start of the

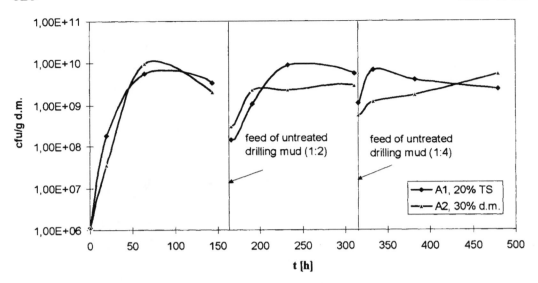

Figure 4 Evolution of CFU during sequenced batch treatment of drilling mud at 4.5-L scale at two different dry matter contents.

experiment, no hydrocarbons in the range C_{14} to C_{40} n-alkanes could be detected after 6 days of biological treatment. Limited bioavailability of compounds with boiling points higher than C_{40} n-alkane due to very long chain structures or due to sorption on inorganic material may be responsible for the low biodegradability of these residues.

The improvement of the sedimentation velocity of the solids is another positive effect of the biological drilling mud treatment. In comparison to untreated material, where no sedimentation could be observed within 48 h, the sludge volume of the treated mud (20% d.m.) was reduced to 62% within 6 h, which simplifies the following process step of dewatering.

IV. CONCLUSIONS

The biological treatment of mineral oil-contaminated drilling mud in 4.5-L airlift reactors led reproducibly to a diminution of ca. 80% of total petroleum hydrocarbons (TPHs) within 3–4 days. Final concentrations of 2800–4500 ppm were obtained. The scaling-up procedure to a 65-L airlift reactor was successful. Autochthonous microorganisms were activated by intensive aeration and sufficient nutrient supply, so that concentrations between 10^9 and 10^{10} cells/g d.m. and degradation rates between 1800 and 2800 ppm/day could be obtained. During the degradation tests, it was found that the pH was in temporal correlation to the TPH degradation, and could therefore be used as a simple on-line parameter for process description. That is of special importance for batch-feed processes, where the feeding with untreated sludge has to occur immediately as the biodegradation process reaches the plateau phase.

Less than 1% of the initial TPHs left the reactor system with the exhaust air. A total degradation of n-alkanes in the range of C_{14} to C_{40} could be detected via GC-MS analysis.

A major problem during the batch-feed treatment of drilling mud in the airlift bioreactors was the evolution of foam. Possible reasons are the limitation of nutrients

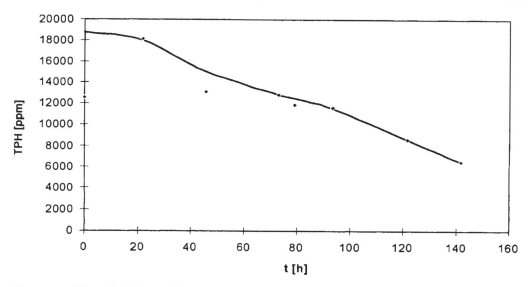

Figure 5 Mineral oil degradation during biological treatment of contaminated drilling mud in a 65-L airlift bioreactor.

and subsequent lysis of microbial cells and the formation of biotensides by bacteria to improve the bioavailability of nonpolar carbon sources. Further investigations should be directed to identification of the underlying mechanism and to means of reducing foam production by modifications in process design.

Microbial growth requires sufficient nutrient supply. Problems arise from the high adsorption of phosphate and ammonium on clay minerals.

Cooperation with industrial partners are currently in the planning phase for large-scale applications of airlift treatment techniques and treatment cost estimation.

REFERENCES

1. Lewis, R. F. 1993. Site demonstration of slurry-phase biodegradation of PAH contaminated soil. *Air & Waste*, 43:503–508.
2. Stinson, M. K., Skovronek, H. S., and Ellis, W. D. 1992. EPA site demonstration of the BioTrol soil washing process. *J. Air Waste Manage. Assoc.*, 42(1):96–103.
3. Koning, M., Hupe, K., Lüth, J.-C., Cohrs, I., Quandt, C., and Stegmann, R. 1998. Comparative investigations into the biological degradation of contaminants in fixed-bed and slurry reactors. In *Contaminated Soil '98*, 17–21 May 1998, Edinburgh, pp. 531–538.
4. Craft, B., Pittman, S., Davies, S. 1997. Combined ex-situ physical and bioreactor treatment of diesel in soil. *In Situ and On-Site Bioremediation*, Vol. 5, pp. 83–84. Batelle Press.
5. Elias, F., and Wiesmann, U. 1995. Biologische Behandlung von Reststoffen der Bodenwäsche in einer 4-stufigen Rührreaktorkaskade im Labormassstab, *Altlasten-Spektrum*, 3:148–157.
6. Chaîneaù, C.-H., Morel, J.-L., and Oudot, J. 1995. Microbial degradation in soil microcosms of fuel oil hydrocarbons from drilling cuttings. *Environ. Sci. Technol.*, 29(6):1615–1621.
7. Anghern, D., Gälli, R., Schluep, M., and Zeyer, J. 1997. Biologisch Saniertes Bodenmaterial aus Mineralölschadensfällen: Abfall oder Produkt? *Terra Tech*, 3:51–56.
8. Wang, X., and Bartha, R. 1994. Effects on toxicity, mutagenesis, and microbiota in hydrocarbon-polluted soils. In: D. L. Wise and D. J. Trantolo (eds.), *Remediation of Hazardous Waste Contaminated Soils*, vol. 10, pp. 175–197. Marcel Dekker, New York.

17
Ecological Consequences of Enhanced UV Radiation on the Phenolic Content of *Brassica oleracea*: A Review

Jeffrey M. Lynch and Alicja M. Zobel
Trent University, Peterborough, Ontario, Canada

I. INTRODUCTION

Many people underestimate the ability of plants to adapt to environmental change. For over 20 years, ozone depletion and the subsequent impact of increased levels of ultraviolet (UV) radiation, notably UV-B, on higher plants has been investigated (1). Many researchers fail to recognize that the plants they study are typically the product of millions of years of evolution and that increased levels of solar radiation will have little to no influence on the viability of most plant species. What needs to be investigated is the "chemical soup" that the plant cells produce and the mixture that they will produce in response to changing amounts of radiation. Will radiation change the chemical constituents of plant cells and tissues? Will nature's kitchen produce compounds yet unseen? Will the compounds in the plant tissues remain the same but the relative percentages change? Given that people are dependent on plants as a source of energy as well as for medical purposes, the above questions must be answered to assure our survival. In this investigation, we wanted to determine if a commercially important plant, *Brassica oleracea*, would exhibit internal and external qualitative and quantitative changes in secondary metabolites in response to short- and long-term enhanced low-intensity UV radiation.

A. *Brassica oleracea*

Among the oldest of vegetables grown for human food is the cabbage. Homer, almost 3000 years ago, wrote in the *Iliad* that Achilles washed cabbages (1). Cabbage and kale (2) have also been well documented for their medicinal qualities. In ancient Egypt, Pliny declared that cabbage would cure as many as 87 diseases (3). Cato the Censor (234–149 B.C.) wrote that cabbage leaves should be washed thoroughly with hot water and the crushed leaves applied to wounds or cancers to heal (4). Cabbage has been used since Dioscorides'* time

---

* Dioscorides, Pedanius (Greek physician). As a physician to Roman armies, Dioscorides collected information on plants in many countries.

(1st century A.D.), as a remedy for digestive ailments (e.g., ulcers), for skin problems (e.g., acne), for fevers, as a joint tonic, and even to prevent drunkenness (5). In folk medicine, cabbage leaves are used both externally and internally for numerous types of family illnesses. Ody (5) stated that the chemical constituents of cabbage treat antiinflammatory, antibacterial (6), and antirheumatic ailments as well as promote cell division (5). In modern times, *Brassica oleracea* has an alleged ability to protect against aging (7) and cancer (3,8–10).

In the modern era, scientific experiments suggest that dietary cabbage, particularly in the raw form (9), enhances the aromatic hydrocarbon hydroxylase (AHH) microsomal enzyme system and increases the rate of metabolism of certain drugs and carcinogens (4). Crude cabbage extracts have been found to stimulate several detoxification mechanisms, notably phase I and phase II enzymes, used to metabolize mutagens (10–12) Ho (13) and Salah et al (14) argued that the polyphenolic compounds within the vegetables are antioxidants and these compounds are responsible for the anticancer characteristic.

There are more than a hundred varieties, or cultivars, of *Brassica oleracea* (1). In this investigation, we decided to focus on flowering cabbage and flowering kale plants, as the leaves of these varieties are fully exposed to the environment as opposed to forming a compact head. These plants are of commercial importance not only because of their popularity as ornamental plants but also because human consumption of *Brassica* is increasing (8,15). Tookey et al. (15) stated that *Brassica oleracea* is the most important of the *Brassica* crops.

II. LOCALIZATION OF PHENOLIC COMPOUNDS IN PLANTS

Secondary metabolites are phenolic compounds located in the vacuole of a plant cell (16), in the intercellular spaces (17), and on the surfaces of tissues (18,19). The secondary metabolites are produced by the endoplasmic reticulum (20,21) and are typically more toxic than their precursors (22). These phenolic compounds, as glycosides, are transported via vesicles which can fuse with the central vacuole (23). Every plant cell deposits in its vacuole several different secondary metabolites. In the vacuole, the compounds (glycosides) are relatively inactive, but they are activated when released from this compartment (24). When exocytosis occurs, glucosidase located in the cell wall cleave the sugar constituent of the secondary metabolites, forming aglycones. The removal of the sugar constituent increases the toxicity of the phenolic compounds. Zobel et al. (17) proposed that phenolic compounds that are covering the external walls of cells surrounding intercellular spaces will form a defense barrier against microbial attack. Vickery and Vickery (25) found that plant phenolics will accumulate in cells adjacent to those infected by disease so as to protect against microorganisms. The aglycones can also be extruded to the surface of plant tissues.

External parts of plants are covered by a cuticle and epicuticular waxes. Esau (26) suggested that waxes form a water-resistant barrier, while Levitt (27) suggested that the cuticle and waxes protect against excess light. Baker (28) reported that phenolic compounds can be embedded in the epicuticular waxes. To date, two large groups of secondary metabolites, the flavonoids (18) and the coumarins (19), have been found to be embedded in the waxes. Wollenweber (29) stated that of the more than 400 species known to contain flavonoids, many species had these compounds on their surfaces. On the surface of the plant the phenolic compounds (a) will kill bacteria and fungal spores (22), (b) may render an unappealing taste to herbivores (30–32), (c) will protect the plant tissues from free radical damage, (d) may be responsible for some dispersion of UV radiation (33), (e) will form a barrier to protect the plant from UV radiation (34), and (f) may transduce UV radiation (absorbed energy) into lower-energy, longer-wavelength, usable (i.e., visible) light (32,35,36).

In *Brassica oleracea*, sinapic acid (37), anthocyanins, flavonoids, and glucosinolates have all been found in crude leaf extracts (9,12). Although the volatile compounds of cabbage have been investigated (38), the literature indicates that no research has been done on *Brassica* in which surface-deposited compounds were distinguished from compounds located within the interior of the leaves. Hence, it is not known if compounds exist on the surface of *Brassica* leaves. An objective of our investigation was to determine if phenolic compounds are present on the surface of *Brassica oleracea* leaves and, if present, what influence low-intensity UV radiation would have on the variety and quantity of these surface compounds.

A. Epidermal Cells

Strafford (39) stated that flavones and flavonols in the vacuoles of epidermal cells of leaves have a UV filter function and these compounds lessen the UV-A and UV-B irradiation reaching meosphyll cells. In plant mesophyll cells, all types of radiation have the potential to damage nucleic acids (i.e., DNA) and the photosynthetic machinery (40). Levitt (27) reported that tundra plants that were grown under UV-absorbing filters exhibited a rapid increase in UV filtration capacity when exposed to unfiltered solar radiation.

Levitt (27) and Bornman (41) reported that epidermal cells, which typically accumulate plant pigments, absorb approximately 90% of the incident UV. Levitt (27) attributed the high absorbance to the flavonoid compounds. Zobel and Lynch (34) found an increase in UV-absorbing phenolics in two *Acer* species following enhanced UV-A. Research by Flint et al. (42), Caldwell et al. (43), and Day et al. (40) found that UV-absorbing phenolics, specifically the flavonoids, increased in plant tissues in response to enhanced UV-B. Rhodes (22) stated that flavonoids increase in plant tissues following UV radiation treatment as UV increases the activity of phenylalanine ammonia lyase (PAL) as well as the other enzymes involved in flavonoid biosynthesis. Therefore, it would seem that most plant species respond to enhanced irradiation, regardless of whether the irradiation is UV-A or UV-B, by synthesizing flavonoids.

III. LEAF MORPHOLOGICAL STAGE AND PHENOLIC CONCENTRATION

Traditional medicine men have long known that leaves from medicinal plants can differ in potency from other leaves of the same plant (44) due to their location on the shoot apex (i.e., morphological stage) as well as due to the amount of sunlight (i.e., UV) the leaves receive (27). Rhodes (22) stated that during particular developmental stages of the plant or following UV radiation, plant cells experience a short-lived increase in PAL synthesis at a time when the rate of PAL degradation is constant. The increase in PAL results in a short-term increase in the biosynthesis of phenolic compounds. Zobel (45) and Zobel and Brown (46) found that during the vegetative period, leaves of *Angelica archangelica, Heracleum mantegazianum, Ruta chalepensis,* and *Ruta graveolens* showed changes with muturation, with the younger leaves having a greater concentration of furanocoumarins than the older leaves. Therefore, in any experiment, the morphological stage of the leaves must be taken into consideration.

Environmental changes can influence the toxicity of plant tissues; thus, the plants will have different proportions of phenolic compounds in the tissues at different times of the year. Harborne (47) and Levitt (27) states that anthocyanins and other flavonoids appear in young leaves and organs in response to environmental change. Zobel and Brown (48) reported that phenolic concentrations increased inside and on the leaves of *Helacleum lanatum* in response to warm (36°C) and cold ($\simeq 0$°C) environmental temperatures. Macleod (49) reported that glucosinolate content in cabbage leaves declined as the age of the plant increased. Therefore, for any experiment, the harvest time must be similar for all plants.

Growing conditions can influence the physiology of the plant and the phenolic content of the tissues (9,22,23,50). Flint et al. (42) and Caldwell et al. (43) stated that field-grown plants would exhibit little to no response to enhanced irradiation, whereas plants grown in greenhouses or growth chambers (i.e., low to no UV) are more sensitive to radiation because they have less developed protective mechanisms (i.e., for DNA and RNA repair and radical scavenging). Consequently, in our investigation both greenhouse-grown plants and field-grown flowering cabbage plants were used.

IV. SECONDARY METABOLITES

Plant cells are able to respond to many different environmental stresses because several isoenzymes exist for each enzyme involved in the biosynthesis (Fig. 1) of phenolic compounds (21). Chalcone synthase (CHS) is one enzyme which has several such isoenzymes (21,51). Zobel (36) stated that CHS is responsible for switching from the biosynthesis of aromatic amino acids, which are primary plant products, to the production of aromatic compounds, which are secondary products. Bell (52) stated that these compounds are secondary to the plant, as they are not present in all species and thus do not have an essential role in primary life processes. Phenolic compounds are one group of secondary metabolites typically present in higher plants (53).

Secondary metabolites are aromatic and most possess at least one hydroxyl group. It is the hydroxyl group(s) of phenolic compounds that enable them to absorb UV radiation (22). The presence of phenolic hydroxyl group(s) increases the hydrophilic character of the molecule (54), imparts an acidic nature to the compound, and enhances UV absorption (22).

Secondary metabolites are not waste products of metabolism (55). Harborne (31,56) stated that phenolic compounds are more complex than their precursors and typically more toxic. Barz and Koster (57) found that plants have a range of enzymes capable of degrading phenolics. Rhodes (22) suggested that some secondary metabolites undergo continued turnover, and Street and Cockburn (58) proposed an association between two different groups of secondary metabolites, the coumarins and the flavonoids. Given the energy required to synthesize and transport the secondary metabolites as well as the complexity of the compounds, it is probable that the secondary metabolites are of substantial value to the plant.

It has been suggested that phenolic compounds have been chosen by natural selection (59) and may play several important ecological roles including defense and protection from environmental stress. Harborne (47) suggested that phenolic compounds represent one method of defense utilized by plants against herbivores. Simple phenols are toxic and plants avoid the autotoxic effects of the phenols that they produce by coupling them with glycosides and storing them in the vacuole. However, the protective effect of plant

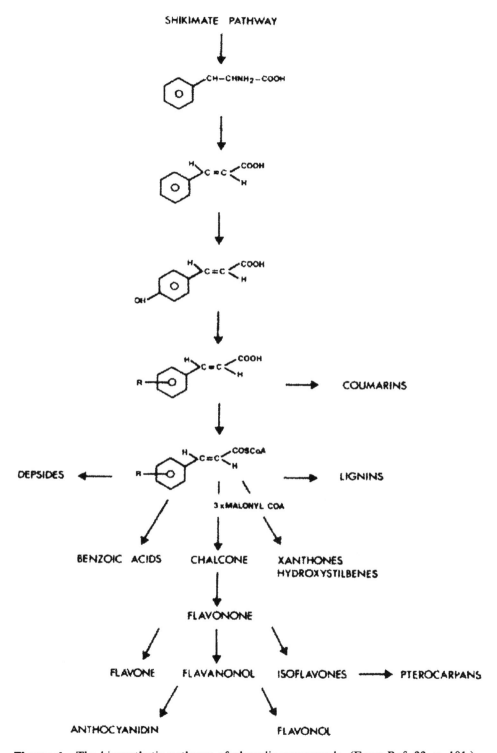

Figure 1 The biosynthetic pathway of phenolic compounds. (From Ref. 22, p. 101.)

Table 1 Some Factors that Influence the Biosynthesis of Phenolic Compounds

1.	Light/radiation	(a)	Ultraviolet
		(b)	Red/far red
		(c)	Blue light
2.	Stress	(a)	Wounding
		(b)	Infection
		(c)	Low temperature
		(d)	Waterlogging
		(e)	Toxic chemicals
3.	Plant growth regulators	(a)	Ethylene
		(b)	Auxins nutritional factors
4.	Nutritional factors	(a)	Nitrogen level
		(b)	Phosphorous level
		(c)	Boron deficiency

phenolics against predation depends on their localization within the plant. As such, the localization of the distasteful plant phenolics near to the surface or on the surface of tissues (i.e., either in hairs or embedded in the waxes) affects grazing as well as the ability of fungal spores to germinate. Therefore, an environmental change that results in a modification to the variety and quantity of secondary metabolites in or on plant tissues can significantly influence the health of the plant.

From Table 1, it can be seen that several factors can stimulate the biosynthesis of phenolic compounds. Rhodes (22) stated that the diversity of the secondary metabolites and their uneven distribution between and within plant families indicates that a single physiological role cannot be attributed to these compounds. Zobel (36) reported that a compound that is an allelochemical in one species does not have to be an allelochemical in another. Therefore, defining the physiological and biochemical role(s) of a secondary metabolite can be very complex; however, Harborne (60) suggested that compounds of the same group share a number of similar properties.

Given that secondary metabolites are not end products and can undergo degradation, we chose to consider several groups of secondary metabolites (coumarins, flavonoids, and glucosinolates) in our investigation. Flavonoids (61) and glucosinolates (2,50,62) are common in cabbage, whereas coumarins have not been reported. We decided to investigate coumarins based on the biosynthetic pathway of Street and Cockburn (58), which suggested a link between coumarins and flavonoids.

A. Coumarins

1. Chemistry, Distribution, and Function

The biosynthesis of coumarins is complex and involves several different enzymes (see Appendix). Hydroxycinnamic acids and coumarins are phenylpropanoids (58,60), as they contain at least one phenylpropane (C_6C_3) structure. All of these compounds are derived from the aromatic amino acid phenylalanine. Phenylalanine is synthesized by the shikimic acid pathway and is converted by PAL to cinnamic acid, a key intermediate in phenylpropanoid biosynthesis (63,64). From cinnamic acid, derivatives are synthesized by the substitution of hydroxyl and methoxy groups to the aromatic ring. Two such derivatives are *o*-coumaric acid and sinapic acid.

Lactonization of *o*-coumaric acid produces coumarin (58,65). The simplest coumarin (Fig. 2) has no hydroxyl groups (i.e., all R groups are hydrogen), thus coumarin itself does not absorb UV radiation nor does it possess an antioxidant potential. However, several hydroxylated derivatives of coumarin exist (Fig. 3) and these compounds possess can absorb UV radiation (32), scavenge free radicals (66), and transduce UV radiation into visible (usable) light (32).

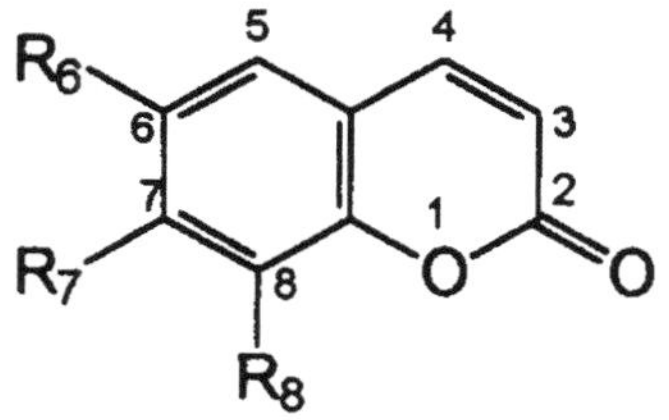

Figure 2 The typical coumarin ring. Coumarin ($C_9H_6O_2$) has a melting point of 70°C, a boiling point of 297°C, and contributes to the scent of many plants (65). It is a common volatile plant constituent and, depending on the specific coumarin, the R groups may be H, OH, OCH_3, or O-glucose. Coumarin is freely soluble in alcohol and ether (65).

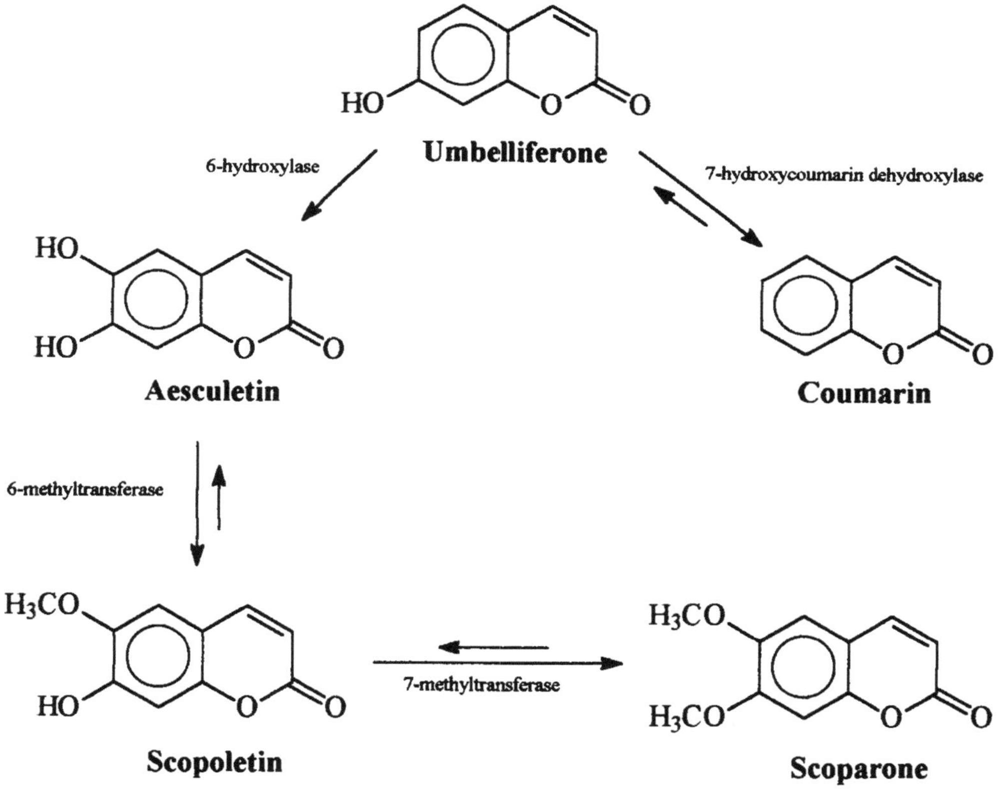

Figure 3 The biosynthetic pathway of coumarin and its derivatives from 7-hydroxycoumarin (24). All of the reactions are reversible.

Coumarins represent one phytochemical class that is widely distributed in plants, and Harborne (60) stated that the most widespread plant compound is the parent compound, coumarin, which occurs in over 27 plant families. Weinmann (67) reported that over 1000 coumarins have been discovered. Coumarin is synthesized and is metabolically active in shoots (68), in green tissue, and accumulates in woody tissue during dormancy. Zobel (32) stated that coumarins can occur in all parts of some plants (e.g., *Campanula* plant) yet be restricted to specific organs in other species. Thompson and Brown (69) stated that some species (e.g., *Rutaceae, Umbelliferae*) can have more than 20 different coumarins. On the surface of plant tissues, Zobel et al. (70) found that several different coumarins can exist. It is important to understand that the localization of the coumarins is correlated to their role in the plant (32). The wide distribution and different tissue locations of coumarins make these compounds an ideal class of toxins used by both primitive and higher plants for defensive purposes.

Towers and Yamamoto (71) stated that simple coumarins are phototoxic to viruses, bacteria, and mammalian cells. The coumarins serve to protect the plant against bacteria, fungi, and insects (32,72). Vickery and Vickery (25) and Zobel (32) suggested that coumarin also acts as a growth inhibitor in plants at high concentrations. Vickery and Vickery (25) stated that at high concentrations, coumarin interferes with plant growth compounds, particularly indolylacetic acid (IAA). The inhibition of growth by coumarin protects plants against fungal invasion and Vickery and Vickery (25) suggested that coumarins accumulate in cells adjacent to cells infected by disease.

Plants do not suffer from autotoxity because of compartmentalization of secondary metabolites. Produced at the endoplasmic reticulum, the phenolic acids, coumarin and coumarin derivatives can be transported via vesicles to the cell vacuole for storage as glucosides. Vacuolar storage of coumarin and coumarin derivatives as glucosides, as well as the deposition of coumarin outside of the cell (e.g., on the tissue surface), is advantageous to the plant as it minimizes the risk of autotoxicity. In the vacuole, o-coumaric-acid-β--glucoside remains until:

> Abrasion of the tissue (i.e., herbivores) occurs and the glucosidases located in the cell wall remove the sugar of o-coumaric-acid-β--glucoside and the compound spontaneously cyclizes to coumarin; and/or
> Coumarin is released via exocytosis into the intercellular spaces (i.e., to kill invading bacteria or viruses) or is extruded onto the surface of the root or leaf to kill bacteria, spores, and render an unappealing taste to animals.

B. Flavonoids

1. Chemistry, Distribution, and Function

Flavonoid is a general term that refers to a large group of secondary plant metabolites that exist as both glycosides and aglycones. The production of the enzyme CHS, which is responsible for initiating flavonoid and anthocyanin biosynthesis, is influenced by environmental factors, and different isoenzymes of CHS respond depending on the stress (21). Flavonoids are produced at the endoplasmic reticulum (Fig. 4) and, with glucose attached, transported via vesicles to the cell vacuole for storage. Flavonoids are stored in the vacuole of plant cells (16) or on the surface of tissues (18,22).

Harborne (56) stated that all flavonoids are structurally related to the parent compound, flavone (Fig. 5), and structural variation is due to (a) differing oxidation levels of the central ring, (b) the presence/absence of hydroxyl groups, and (c) the type of sugar

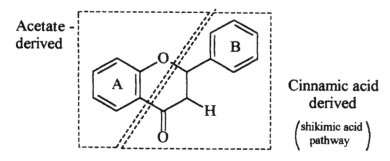

Figure 4 The general structure of a flavonoid. Both the acetate-malonate and the shikimic acid pathways are required for flavonoid biosynthesis. The cinnamic acid-derived compound is *meta*-hydroxyphenylacetic acid. (From Ref. 25, p. 183.)

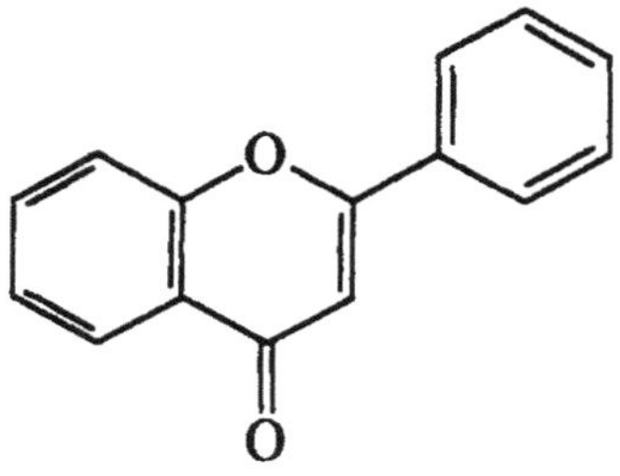

Flavone

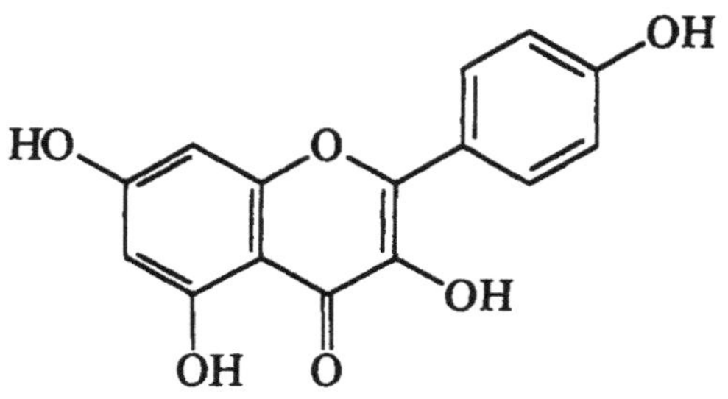

Kaempferol
(flavonol; pale yellow)

I   II

Figure 5 I. The parent compound of all flavonoids is flavone ($C_{15}H_{10}O_2$). II. Kaempferol ($C_{15}H_{10}O_6$) is a very common flavonoid commonly found in fruits. Budavari (65) stated that kaempferol (m.p. 278°C) is highly soluble in hot alcohol.

involved. Ten different classes of flavonoids are currently recognized; however, Harborne (60) suggested that only three subgroups are common: anthocyanins, flavones, and flavonols.

Structural transformations of anthocyanins and flavonoids, and at the same time the color of the pigment, are influenced by pH (73) in a manner such that a red color is associated with a low pH, a yellor or colorless appearance is indicative of a neutral pH, and a pH greater than 7 renders pigments that are blue in color and fade to ivory as the pH rises (22,47). Blom (74) suggested that the breakage of the β-glycosidic bond would cause the pigments to become unstable and susceptible to conversion into brown or colourless compounds. Due to the fact that the flavonoids are colored (i.e., visible) at low pH values, flavonoid extraction solutions are typically acidic.

Huang and Ferraro (75) stated that flavonoids are ubiquitous in plants and almost all plant tissues are able to synthesize flavonoids (58,76). In 1981, Wollenweber and Dietz (18) reported that over 2000 flavonoids had been identified. The flavonoids are typically stored in the plant cell vacuole. In the vacuole, flavonoids are generally found bound to sugar and one flavonoid aglycone may have several glycosidic combinations (76,60). For this reason, Harborne (60) suggested that when analyzing flavonoids, it is better to examine the aglycones before considering the complexity of the glycosides. Flavonoids are present in plants in mixtures and it is extremely rare to find only a single type of flavonoid in a plant tissue (60). As well, there is generally a mixture of flavonoid classes in the tissue. Harborne (60) suggested that the colored anthocyanins in flower petals are almost always accompanied by colorless flavones or flavonols. The flavones act as co-pigments, essential for the full expression of the anthocyanins. Mixtures of anthocyanins are also the rule, particularly in ornamental plants (60).

Anthocyanins are the most widely distributed subgroup of the flavonoids. Anthocyanins are known to occur in the seed, root, and stem tissues of many higher plants (47); thus, the energy expended by the plant in the production of anthocyanins is notable. Ollis (77) suggested that anthocyanins based on the structure of cyandin (Fig. 6) are the most numerous. In the red cabbage, Harborne (30) and Hrazdina et al. (61) attributed the red color to cyandin; however, this anthocyanin was found to be in 7-acylated and -glycosylated forms. Rhodes (22) noted that it is difficult to understand the advantage that the plant receives from having a mixture of derivatives in the tissues. Berenbaum et al. (78) suggested that having a mixture of compounds changes the toxicity of a plant and allows the plant to adapt to environmental conditions.

Composed of multiple phenolic rings, the primary role of flavonoids is in plant defense. The numerous hydroxyl groups of the flavonoids (e.g., kaempferol, shown in Fig. 5) enable them to absorb UV irradiation (22) as well as possess a radical-scavenging ability (79–82). Rhodes (22) suggested that the presence of the phenolic hydroxyl group(s) increases the hydrophilic character of the molecule, imparts an acidic nature to the molecule, and enhances the UV absorption.

In addition to a UV-filtering role, the flavonoid compounds serve other functions. Mulligan and Kevan (83) and Rieseberg and Schilling (84) suggested that the UV absorption by flavonoids in flower petals serves as a nectar guide, attracting birds and bees which are necessary for seed dispersion. In addition, Street and Cockburn (58) suggested that

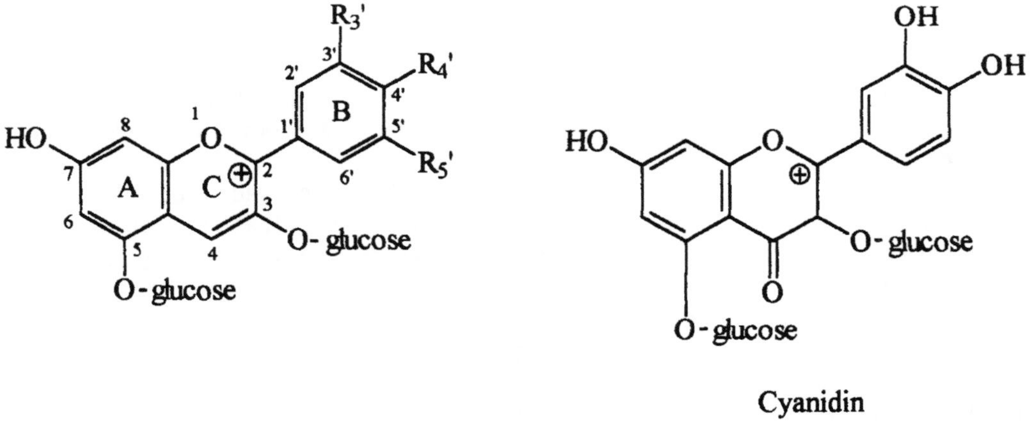

Figure 6 I. The general structure of the anthocyanins. II. A common anthocyanin is cyanidin ($R'_3 = OH$; $R'_4 = OH$; $R'_5 = H$). The glycoside components stabilize anthocyanins.

some flavonoids inhibit while others stimulate the enzyme IAA oxidase. This enzyme is responsible for the degradation of the plant growth compound IAA. The inhibition of growth by flavonoids may serve to protect plant tissues against fungal invasion (25), while the stimulation of IAA oxidase may be of benefit to young tissues.

C. Glucosinolates

1. Chemistry, Distribution, and Function

Glucosinolates occur in important crop plants and vegetables, particularly those from the family Cruciferae (85). Plants synthesize glucosinolates from amino acids through several reactions in which the carboxyl group is lost and the α carbon is transformed into the central carbon of the glucosinolate (86). The side chain R of the glucosinolate (Fig. 7) is identical to that on the α carbon of the amino acid. Larsen (86) stated that only seven glucosinolates are synthesized from protein amino acids (alanine, valine, leucine, isoleucine, phenylalanine, tyrosine, and tryptophan), while the remaining glucosinolates are derived from nonprotein amino acids or from chain-lengthened protein amino acids.

The glucose component and the ionic forms of the glucosinolates make these compounds hydrophilic. Majak (55) stated that glucosinolates are nonvolatile and stable in neutral pH. During isolation of the glucosinolates from plant material, myrosinases (e.g., thioglucosidase) can hydrolysize glucosinolates (49,50) and must be denatured. Consequently, Larsen (86) recommended extracting the glucosinolates using a boiling alcohol solution.

Isothiocyanates are produced when glucosinolates are degraded (Fig. 7). Glucosinolates are nonvolatile, whereas the isothiocyanates are volatile (2,88). Zhang and Talalay (10) stated that isothiocyanates possess a strong smell and taste and these compounds are chemically very reactive. Heaney and Fenwick (87) attributed the reactivity to the SH groups. Larsen (86) suggested that isothiocyanates with a small R side chain are more volatile than those with a large R group. Blaakmeer et al. (89)

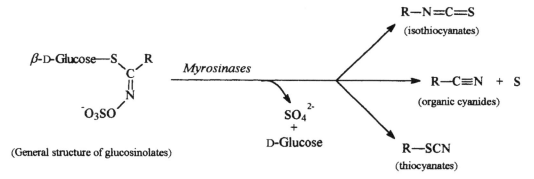

Figure 7 The general formula of the glucosinolates. Glucosinolates are enzymatically degraded by myrosinases to yield isothiocyanates, organic cyanides, and thiocyanates. Various glucosinolates degrade with different velocities, and Tookey et al. (15) and Preobrazhenskaya et al. (11) reported that ascorbic acid accelerates the reaction. Larsen (86) warned that there is substantial variability in the products of enzymatic decomposition. In the parent glucosinolate compound, R is typically the side group of the amino acid from which the glucosinolate was synthesized (62,87). (From Ref. 86, p. 502.)

found 21 volatiles in the headspace of intact control cabbage plants. Chin and Lindsay (38) found that the quantity of volatile compounds in the headspace of head cabbage increased following tissue disruption. Allyl isothiocyanate, methanethiol, methyl disulfide, and dimethyl trisulfide were identified in the headspace of cabbage, with the last three compounds being responsible for the objectionable sulfurous aromas (38). Macleod (49) suggested that the best method for qualitative identification of volatiles of the Cruciferae is to use gas chromatography/mass spectrometry (GC/MS).

More than 100 glucosinolates have been identified (50,90), and Steinmetz and Potter (9) reported that more than 20 glucosinolates occur in edible plants. Glucosinolates are a class of sulfur-containing glycosides. Kjaer (62) reported that plants possess a mixture of glucosinolates rather than one specific type. Glucosinolates occur in high concentration in seeds (91,92) and can be found in all parts of the mature plant (87,91). Larsen (86) reported that the glucosinolates are found primarily in the parenchymal tissue.

Given the medicinal and commercial importance of many glucosinolate-containing species, the influence of insects on these plants is well documented. Gross (93) found that glucosinolates and the decomposition products of glucosinolates (Fig. 7) have toxic effects on insects that do not normally feed on glucosinolate-containing plants. Microorganisms and animals that consume glucosinolates possess enzymes that degrade these compounds. Björkman (94) reported that bacteria, fungi, and the cabbage aphid (*Brevicoryne brassicae*) possess thioglucosidases which serve to hydrolyze glucosinolates. Larsen (86) reported that the glucosinolates (nonvolatile) are recognized by insects on contact, whereas the isothiocyanates (volatile) induce olfactory responses. Glucosinolates and isothiocyanates could be responsible for inhibition or stimulation of insect feeding (10,86) as well as oviposition (95). Finch and Skinner (96) found that some insect larvae were attracted by isothiocyanates from short distances.

Crop spacing can have an influence on headspace volatiles. Macleod (49) found that the closer together cabbage plants were grown, the greater was the relative abundance of glucosinolate degradation products, particularly the isothiocyanates. The same trend was confirmed for brussels sprouts (49). Macleod (49) attributed these findings to stressful growing conditions, which increased the biosynthesis of amino acids (i.e., glucosinolate precursors).

Isothiocyanates possess a variety of pharmacological and toxic activities. Zhang and Talalay (10) stated that some of these actions include goitrogenic activity, antibacterial and antifungal properties, the ability to attract or repel insects, cytotoxicity, induction of chromosome abnormalities and neoplasia, and the blocking of chemical carcinogenesis.

V. ULTRAVIOLET RADIATION

Ozone in the stratosphere absorbs solar radiation (UV-A, 321–399 nm, UV-B, 280–320 nm, UV-C, <280 nm) appreciably at wavelengths shorter than 300 nm. Therefore, UV-C is an artificial radiation to which plants are not exposed. It is also known that wavelengths as short as 315 nm reach the surface of the earth (43). With ozone depletion, an increase in UV-B radiation between 290 and 315 nm wavelengths is predicted (43,97,98). Caldwell et al. (43) stated that additional radiation from even sizable ozone depletion is trivial (Fig. 8) when compared to the total solar radiation.

Studies have found that 254-nm radiation, UV-C, stimulates the production of coumarins (99), while UV-A radiation of 366 nm stimulates the biosynthesis of coumarins

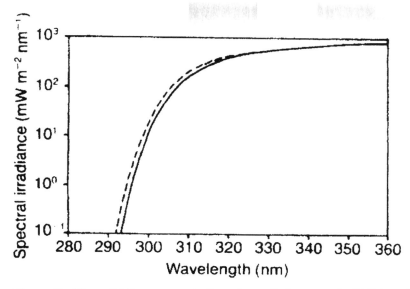

Figure 8 The solid line represents the solar radiation over the UV-A and UV-B portions of the spectrum at midday in the summer for temperate latitudes. The dashed line represents the same conditions, but with a high ozone depletion of 16%. Ozone in the stratosphere absorbs all wavelengths shorter than ~292 nm. (From Ref. 43, p. 363.)

(35) and phenolics (22). Levitt (27) reported that even the longest UV radiation may produce damage; however, destructive effects are produced by wavelengths less than 290 nm.

For the bulk of our experiments, UV-A radiation was used. In some instances, we considered the influence of UV-C on flowering cabbage leaves. UV-C radiation of 254 nm is near the peak absorption wavelength of proteins and nucleic acids (100), thus, UV-C can cause visible injury to plant tissues. *Brassica oleracea* is considered a hardy species, and it was not known if the plants would respond to UV-C as they do when irradiated with UV-A. Consequently, both types of irradiation were considered in our investigation.

A. Radiation Absorption by Phenolic Compounds

The presence of hydroxyl groups confers UV absorbance on secondary metabolites (22) as well as scavenged free radicals (79,80,82,101). From Murray et al. (24), it was known that different phenolic compounds have different spectral peaks. Zobel (36) suggested that a greater variety of phenolic compounds in any plant would increase the range of its UV absorption capacities.

Secondary metabolites are phenolic compounds that can possess one (phenolic acids), two (coumarins), three (flavonoids and anthocyanins), or more (tannins) benzene rings in the structure. Each carbon atom in a benzene ring has a 2p orbital, and these orbitals combine (overlap) to form six π orbitals that are spread uniformly over the ring (102). The bonding π orbital results when like charges overlap, but the antibonding π^* orbital is formed when opposite signs overlap (Fig. 9). Solomons (103) stated that the antibonding π^* orbital, in comparison to the bonding π orbital, is of higher energy and is not occupied by electrons when the molecule is in its ground state. However,

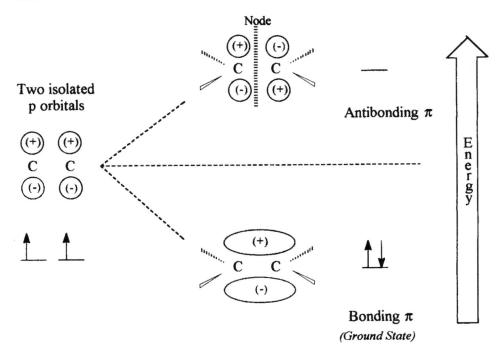

Figure 9 Two *p* orbitals combine to form two π orbits. The bonding π orbital is of a lower energy and contains both π electrons (with opposite spins) in the ground state. In comparison, the higher-energy antibonding π* orbital is not occupied by electrons in the ground state. Excitation of the molecule by light or radiation of a suitable frequency can promote an electron from the lower energy level to the higher energy level. (From Ref. 103, p. 59.)

the π* orbital can become occupied if the molecule absorbs light or radiation of the appropriate frequency and an electron is promoted from the lower-energy level to the higher one (i.e., an excited state). Upon absorption of UV radiation a compound acquires energy and, to revert to the ground state, this energy must be released (Fig. 10). The amount of energy reradiated equals the amount absorbed (1). This can be accomplished by emission of heat or longer-wavelength radiation, but if this is not done there is an increase in reactivity, and free radicals may be formed (32,101,104).

Coumarin is an example of a compound that has an abundance of π electrons and can be excited by UV-A radiation. Downum (105) suggested that the phototoxic mechanism of coumarins is via a type I mechanism (Fig. 11). In the excited state, coumarin can react with a suitable substrate to form radicals (Fig. 12). Hydroxylated derivatives of coumarin can be scavengers of free radicals in vitro (66).

B. Free Radicals

A free radical is a molecule containing one or more unpaired electrons. Free radicals are produced in the normal course of metabolism and are essential to many normal biological processes (101,103). For example, Rice-Evans et al. (101) and Stavric and Matula (81) suggested that free radicals can be intermediates and/or products of enzymatically catalyzed reactions. The highly reactive nature of free radicals makes them potentially

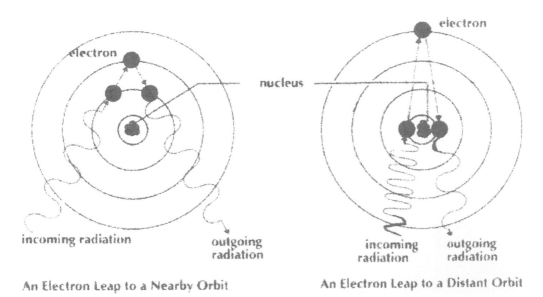

Figure 10 When radiation strikes an atom, an electron can absorb some of the energy and leap into a nearby orbit. Radiation is released when the electron returns to its original position. If the same situation occurs with high-energy radiation (< 300 nm), the electron may jump to a distant orbit. (Modified from Ref. 1.)

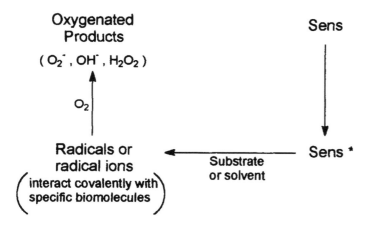

Figure 11 The type I phototoxic mechanism of action that occurs in coumarin. UV-A radiation renders the greatest toxicity of the excited coumarin compound. (From Ref. 105, p. 406.)

dangerous, particularly if they are not tightly controlled. Pine (104) suggested that free radicals can be formed in three general ways:

1. Homolytic cleavage of bonds:

$$A-B \longrightarrow A^{\cdot} + B^{\cdot}$$

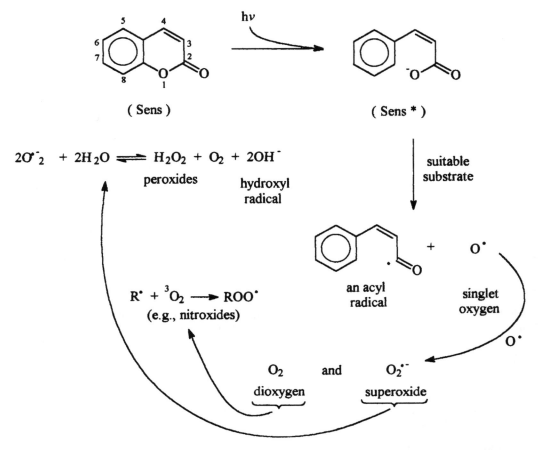

Figure 12 The radicals produced by the excitation of coumarin by UV-A radiation. Dioxygen (O_2) is dangerous in excess (i.e., toxic to the cell) and may combine with other radicals or broaden the effects of other radicals (101). Rice et al. (101) stated that singlet oxygen reacts as an electrophile, whereas superoxide is a weak base and has low reactivity as a radical. The hydroxyl radicals can damage DNA, whereas the peroxide radicals damage unsaturated membrane components.

2. Molecular reaction with other free radicals:

$$A-B + X^{\cdot} \longrightarrow A^{\cdot} + B-X$$

3. Addition of a free radical to an alkene:

$$\text{\Large$>$C=C$<$} + X^{\cdot} \longrightarrow -\overset{|}{\underset{\cdot}{C}}-\overset{|}{\underset{X}{C}}-$$

Reaction 1 may be induced thermally (104) or photochemically (101); Reaction 2 can be hydrogen ion transfer.

Free radicals are typically short-lived, but one radical can generate other free radicals until a termination reaction occurs (e.g., combination of two free radicals).

Targets of free radicals include proteins, DNA (105–107), carbohydrates, and the phospholipid component of cellular membranes (80,82,101) Depending on the cellular or extracellular target, free-radical damage can result in (a) modified ion transport, (b) modified enzyme activity, (c) mutations and translational errors, and (d) modification of the structural and functional integrity of the membrane. Most free radicals have a chronic effect on cells.

Antioxidant is a term given to a compound that scavenges free radicals. All radical scavengers are electron donors and possess at least one hydroxyl group. Zobel (36) stated that antioxidants were necessary for plants from the very beginning of land colonization. It is the hydroxyl group(s) of coumarin derivatives and flavonoids that enable these compounds to absorb UV radiation (22), emit longer-wavelength radiation following absorption (32), as well as possess a radical-scavenging ability (79,80,82). The large structure and numerous hydroxyl groups of the flavonoids make these compounds excellent antioxidants (14,108).

C. Short- and Long-Term Response to Ultraviolet Radiation

The first response of most plants to UV radiation is to increase the concentration of phenolic compounds in and on the tissues (17,34). Zobel et al. (99) reported that detached leaves of *Ruta graveolens* plants exposed to 254-nm radiation showed a dramatic decline in furanocoumarins after 48 h, while after 4 days the concentration of these compounds returned to control levels. Zobel (36) suggested that a recovery mechanism would be very beneficial to any plant because an accumulation of phenolic compounds inside the cell could damage the tonoplast of the vacuole and lead to cell death. In our investigation, we wanted to know if flowering cabbage and kale plants would exhibit a recovery response following UV treatment.

VI. CHANGES IN PLANT COMPOSITION AND THE POSSIBLE IMPACT ON ECOSYSTEMS

Small changes in the chemical composition of plants could have a dramatic impact on ecosystems around the world. For example, consider a change in the chemical composition of an important commercial crop plant that has medicinal applications. Fraenkel (109) and Waterman and Mole (110) suggested that food specificity of insects is dependent solely upon the presence or absence of particular secondary metabolites. It is possible that a reduction of a repellent compound or an increased concentration of attractants could result in decreased agricultural yields and/or additional usage of pesticides. Therefore, even small changes to the global climate could result in chemical changes in a plant that affect its toxicity and/or its medicinal attributes.

The above idea about altered food quality is not novel. Termura and Sullivan (111) found a large decline in soybean yield and quality following enhanced UV-B treatment. Caldwell et al. (43) constructed a diagram to suggest the possible consequences of enhanced radiation on the ecosystem (Fig. 13).

Weidenhamer et al. (112) and Pal (113) reported that phenolic compounds can become even more toxic due to abiotic and biotic (i.e., microbe) factors. Phenolic compounds extruded to the surface of plant tissues can be removed by rain and enter the soil, whereas the vacuolar compounds will enter the soil only after decay of the plant. White et al. (114) found that some phenolics become toxic only after the death of a plant.

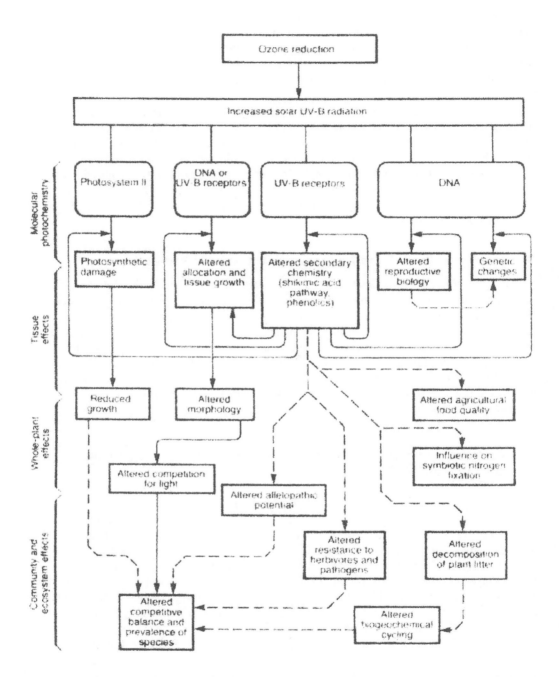

Figure 13 The potential effects of increased solar UV on higher plants are substantial and could dramatically influence the ecosystem. Solid lines indicate interactions for which there is experimental evidence, whereas the dashed lines indicate interactions for which there is as yet no scientific proof. Caldwell et al. (43) stated that the feedback loops indicate phenolics and flavonoids that are synthesized in response to the increased UV radiation and serve as UV-filtering agents. In this diagram we are not concerned with a specific stress, but rather we focus on the potential results of altered secondary products (highlighted).

Therefore, a change in the chemical composition of a plant can influence soil microbe activity synergistically or adversely. Zobel (36) suggested that more investigations are needed to determine the influence of these compounds on soil chemistry and on microbes. By any means, changes to the chemical composition of plants due to enhanced UV radiation could have a dramatic impact on the earth's ecosystem.

VII. CONCLUSION

A. Purpose of Our Investigation

We believe that although the suggested changes in UV wavelengths that are predicted to result from ozone depletion are trivial, the plants will detect the change. We predict that the variety and quantity of compounds within and on the surface of plant tissues will be modified. We believe that plants will adapt to the radiation stress and survive; however, the plants will differ chemically from the original species. The purpose of our investigation was to prove that short- and long-term low-intensity UV wavelength enhancement would change the secondary metabolite composition in the interior and on the surface of *Brassica oleracea* leaves. In our experiments, three groups of secondary metabolites were considered: coumarins, flavonoids, and glucosinolates.

For our investigation, we chose to use UV-A because the literature (43) suggested that excessive amounts of UV-A reach the surface of the earth and a change in UV-A radiation would have no impact on plants (i.e., no change to the chemical composition). Our aim was to prove that plants will detect trivial changes in UV wavelengths, even wavelengths to which they have been fully exposed in the field.

B. Findings

Our investigation proved very useful to demonstrate the influence of irradiation on the phenolic content of *Brassica oleracea* leaves. By segregating the interior and surface-deposited phenolic compounds, qualitative and quantitative differences between the interior and surface compartments, before and after treatment, were observed. Our investigation provided us with 18 findings, six of which have not been previously reported.

The ability of plants to detect and to respond to environmental changes, so as to adapt to stress and thereby assure the survival of the species, is truly a biochemical wonder that has been perfected through thousands of years of development. There is a need to determine the phenolic contents of important crop and medicinal plants before environmental changes (e.g., ozone depletion and the subsequent enhancement in UV-B; heavy-metal contamination) modify the phenolic contents of these plants. We are at a point in time when the secret ingredients in the "chemical soup" created in nature's kitchen by plant species may be lost forever. When we consider that people are dependent on plants as a source of energy as well as for medical purposes, humankind will no doubt suffer from such a loss.

REFERENCES

1. Compton's Interactive Encyclopedia. 1995. Compton's NewMedia, Carlsbad CA.
2. Fenwick, G., Heaney, R., and Mullin, W. 1983. Glucosinolates and their breakdown products in food and food plants. *CRC Crit. Rev. Food Sci. Nutr.*, 18(2):123–201.

3. Steinmetz, K., and Potter, J. 1991. Vegetables, fruit, and cancer. I. Epidemiology. *Cancer Causes Control*, 2:325–357.
4. Albert-Puleo, M. 1983. Physiological effects of cabbage with reference to its potential as a dietary cancer-inhibitor and its use in ancient medicine. *J. Ethnopharmacol.*, 9:261–272.
5. Ody, P. 1995. *The Complete Medicinal Herbal*. Key Porter Books, Toronto.
6. Kyung, K., and Fleming, H. 1994. S-methyl--cysteine sulfoxide as the precursor of methyl methanethioisulfinate, the principle antibacterial compound in cabbage. *J. Food Sci.*, 59:350–355.
7. Stohs, S., Lawson, T., Anderson, L., and Bueding, E. 1986. Effects of oltipraz, BHA, ADT, and cabbage on glutathione metabolism, DNA damage and lipid peroxidation in old mice. *Mechanisms Aging Develop.*, 37:137–145.
8. Ceponis, M., Cappellini, R., and Lightner, G. 1987. Disorders in Cabbage, bunched broccoli, and cauliflower shipments to the New York market, 1972–1985. *Plant Dis.*, 71:1151–1154.
9. Steinmetz, K., and Potter, J. 1991. Vegetables, fruit, and cancer. II. Mechanisms. *Cancer Causes Control*, 2:427–442.
10. Zhang, Y., and Talalay, P. 1994. Anticarcinogenic activities of organic isothiocyanates: Chemistry and mechanisms. *Cancer Res. Arch.*, 54:1976–1981.
11. Preobrazhenskaya, M., Bukhman, V., Korolev, A., and Efimov, S. 1993. Abcorbigen and other indole-derived compounds from *Brassica* vegetables and their analogs as anticarcinogenic and immunomodulating agents. *Pharmacol. Therap.*, 60:301–313.
12. Beecher, C. 1994. Cancer preventive properties of varieties of *Brassica oleracea*: A review. *Amer. J. Clin. Nutrition*, 59:1166–1170.
13. Ho, C. Phenolic Compounds in Food. In: M. Huang, C. Ho, and C. Lee (eds.), *Phenolic Compounds in Food and Their Effects on Health II: Antioxidants and Cancer Prevention*, pp. 2–7. American Chemical Society, Washington, DC.
14. Salah, N., Miller, N., Paganga, G., Tijburg, L., Bolwell, G., and Rice-Evans, C. 1995. Polyphenolic flavonols as scavengers of aqueous phase radicals and as chain-breaking antioxidants.
15. Tookey, H., Vanetten, C., and Daxenbichler, M. 1980. Glucosinolates. In: I. Liener (ed.), *Toxic Constituents of Plant Foodstuffs*, pp. 103–142, Academic Press, Toronto.
16. Stern, K. 1988. *Introductory Plant Biology*, 4th ed. Wm. C. Brown Publishers, Dubuque, IA.
17. Zobel, A. M., Crellin, J., Brown, S. A., and Glowniak, K. 1994. Concentrations of furanocoumains under stress conditions and their histological localization. *Acta Horticult.* 381:510–576.
18. Wollenweber E., and Dietz, U. 1981. Occurrence and distribution of free flavonoid aglycones in plants. *Phytochemistry*, 20:869–932.
19. Zobel, A. M., and Brown, S. A. 1988. Determination of furanocoumarins on the leaf surface of *Ruta graveolens* with an improved extraction technique. *J. Natural Prod.*, 51:8966–8970.
20. Hrazdina, G., and Wagner, G. 1985. Metabolic pathways as enzyme complex: Evidence for the synthesis of phenylpropanoids and flavonoids on membrane associated enzyme complexes. *Arch. Biochem. Biophys.*, 237:88–100.
21. Hrazdina, G., and Zobel, A. 1991. Cytochemical localization of enzymes in plant cells. *Bot. Rev.*, 129:269–322.
22. Rhodes, M. 1985. The physiological significance of plant phenolic compounds. *Annu. Proc. Phytochem. Soc. Europe*, 25:99–118.
23. Zobel, A. M. 1986. Sites of localization of phenolics in tannin coenocytes in *Sambucus racemosa* L. *Annu. Bot.*, 57:801–810.
24. Murray, R., Mendez, J., and Brown, S. A. 1982. *The Natural Coumarins: Occurrence, Chemistry and Biochemistry*. Wiley, Toronto.
25. Vickery, M., and Vickery, B. 1981. *Secondary Plant Metabolism*. Macmillan, London.
26. Esau, K. 1977. *Anatomy of Seed Plants*, 2nd ed. Wiley, Toronto.
27. Levitt, J. 1980. *Responses of Plants to Environmental Stresses*, 2nd ed. Academic Press, New York.

28. Baker, E. 1982. Chemistry and morphology of plant epicular waxes. In: D. Cutler, K. Alvin, and C. Priece (eds.), *The Plant Cuticle*, pp. 139–165. Academic Press, Toronto.
29. Wollenweber, E. 1986. Flavonoid aglycones in leaf exudate constituents in higher plants. In: L. Farkas, M. Gabor, and F. Kallay (eds.), *Studies in Organic Chemistry: Flavonoids, Bioflavonoids*, pp. 155–169. Elevier, Amsterdam.
30. Harborne, J. B. 1964. *Biochemistry of Phenolic Compounds*. Academic Press, New York.
31. Harborne, J. B. 1985. Phenolics and plant defence. *Annu. Proc. Phytochem. Soc. Europe*, 25:393–408.
32. Zobel, A. M. 1997. Coumarins in fruit and vegetables. In: F. Tomas-Barberan and R. Robins (eds.), *Phytochemistry of Fruit and Vegetables*, Clarendon Press, Oxford.
33. Cen, Y. P., and Bornman, J. F. 1993. The effect of exposure to enhanced UV-B radiation on the penetration of monochromatic and polychromatic UV-B radiation in leaves of *Brassica napus*. *Physiol. Plant.*, 87:249–255.
34. Zobel, A. M., and Lynch, J. M. 1997. Extrusion of UV-absorbing phenolics in *Acer* spp. in response to UV and freezing temperature. I. UV-A absorbing compounds on the surface of *Acer saccharum* and *Acer platanoides* autumn leaves. *Allelopathy J.*, 4:1–10.
35. Zobel, A. M., and Brown, S. A. 1993. Influence of low intensity ultraviolet radiation on extrusion of furanocoumarins to the leaf surface. *J. Chem. Ecol.*, 19:939–952.
36. Zobel, A. M. 1996. Phenolic compounds as bioindicators of air pollution. In: M. Yunas and M. Iqbal (eds.), *Plant Response to Air Pollution*, pp. 241–266. Wiley, Chichester, UK.
37. Towers, G. 1964. Metabolism of phenolics in higher plants and micro-organisms. In: J. Harborne (ed.), *Biochemistry of Phenolic Compounds*, pp. 249–294. Academic Press, New York.
38. Chin, H., and Lindsay, R. 1993. Volatile sulfur compounds formed in disrupted tissues of different cabbage cultivars. *J. Food Sci.*, 58:835–841.
39. Strafford, H. 1991. Flavonoid evolution: an enzymatic approach. *Plant Physiol.*, 96:680–685.
40. Day, T., Martin, G., and Vogelmann, T. 1993. Penetration of UV-B radiation in foliage: Evidence that the epidermis behaves as a non-uniform filter. *Plant, Cell, Environ.*, 16:735–741.
41. Bornman, J. 1989. Target sites of UV-B radiation in photosynthesis of higher plants. *J. Photochem. Photobiol., B: Biol.*, 4:145–158.
42. Flint, S., Jordan, P., and Caldwell, M. 1985. Plant protective response to enhanced UV-B radiation under field conditions. Leaf optical properties and photosynthesis. *Photochem. Photobiol.* 41:95–99.
43. Caldwell, M., Teramura, A., and Tevini, M. 1989. The changing solar climate and the ecological consequences for higher plants. *Trends Ecol. Evol.*, 4:363–367.
44. CBC International. 1995. *Nature of Things: Teaching New Doctors Old Tricks*.
45. Zobel, A. M. 1991. A comparison of furanocoumarin concentrations of greenhouse-grown *Ruta chalepensis* with outdoor plants later transferred to a greenhouse. *J. Chem. Ecol.*, 17:21–27.
46. Zobel, A. M., and Brown, S. A. 1991. Psoralens in senescing leaves of *Ruta graveolens*. *J. Chem. Ecol.*, 17:1901–1810.
47. Harborne, J. B. 1967. *Comparative Biochemistry of the Flavonoids*. Academic Press, New York.
48. Zobel, A. M., and Brown, S. A. 1990. Seasonal changes of furanocoumarin concentrations in leaves of *Heracleum lanatum*. *J. Chem. Ecol.*, 16:1623–1634.
49. Macleod, A. 1976. Volatile flavour compounds of the Cruciferae. In: J. Vaughan, A. Macleod, and B. Jones (eds.), *The Biology and Chemistry of the* Cruciferae, pp. 307–330. Academic Press, London.
50. Betz, J., and Fox, W. 1994. High-performance liquid chromatographic determination of glucosinolates in *Brassica* vegetables. In *Food Phytochemicals I: Fruits and Vegetables*, pp. 181–196. American Chemical Society, Washington, DC.

51. An, C., Ichinose, Y., Yamada, T., Tanaka, Y., Shiraiski, T., and Oku, H. 1993. Organization of the genes encoding chalcone synthase in *Pisum sativum*. *Plant Molec. Biol.*, 21:789–780.
52. Bell, E. 1981. The physiological role(s) of secondary (natural) products. In: E. E. Conn (ed.), *The Biochemistry of Plants: Secondary Plant Products*, pp. 1–20. Academic Press, Toronto.
53. Harborne, J. B. 1980. Plant phenolics. In: E. Bell and B. Charlwood (eds), *Encyclopedia of Plant Physiology, New Series*, pp. 329–402. Springer-Verlag, Berlin.
54. WINDOWCHEM^{RM} Software. 1995. *Molecular Modeling Pro^{RM} Version 2.14*. http://www.windowchem.com
55. Majak, W. 1992. Mammalian metabolism of toxic glycosides from plants. *J. Health Toxicol.*, 11:1–40.
56. Harborne, J. B. 1972. *Recent Advances in Phytochemistry*, Vol. 4. Appleton-Century-Crofts, New York.
57. Barz W., and Koster J. 1981. Turnover and degradation of secondary (natural) products. In: E. E. Conn (ed.), *The Biochemistry of Plants: Secondary Plant Products*, pp. 35–84. Academic Press, Toronto.
58. Street, H., and Cockburn, W. 1972. *Plant Metabolism*, 2nd ed. Pergamon Press, Toronto.
59. Brouillare, R., Dangles, O., Jay, M., Biolley, J., and Chirol, N. 1993. In: A. Scalbert (ed.), *Polyphenolic Phenomena*, pp. 41–48. Institut National De La Recherche Agronomique, Paris.
60. Harborne, J. B. 1984. *Phytochemical Methods: A Guide to Modern Techniques of Plant Analysis*, 2nd ed. Chapman & Hall, New York.
61. Hrazdina, G., Inedale, H., and Mattick, L. 1977. Light induced changes in anthocyanin concentrations, activity of PAL and flavonone synthase and some of their properties in *Brassica oleraceae*. *Phytochemistry*, 18:581–584.
62. Kjaer, A. 1976. Glucosinolates in the Cruciferae. In: J. Vaughan, A. Macleod, and B. Jones (eds.), *The Biology and Chemistry of the* Cruciferae, pp. 207–219. Academic Press, London.
63. Brown, S. A. 1981. Coumarins. In: E. E. Conn (ed.), *The Biochemistry of Plants: Secondary Plant Products*, pp. 269–300. Academic Press, Toronto.
64. Keating, G., and O'Kennedy, R. 1997. In: R. O'Kennedy and R. Thornes (eds.), *Coumarins: Biology, Applications and Mode of Action*, pp. 23–66. Wiley, Toronto.
65. Budavari, S. 1996. *The Merck Index*, 12th ed. Merck, Whitehouse Station, NJ.
66. Paya, M., Halliwell, B., and Hoult, J. 1992. Interactions of a series of coumarins with reactive oxygen species. *Biochem. Pharmacol.*, 44(2):205–214.
67. Weinmann, I. 1997. In: R. O'Kennedy and R. Thornes (eds.), *Coumarins: Biology, Applications and Mode of Action*, pp. 1–22. Wiley, Toronto.
68. Kosuge, T., and Conn, E. 1959. The metabolism of aromatic compounds in higher plants. *J. Biol. Chem.*, 234:2133–2137.
69. Thompson, H., and Brown, S. A. 1984. Separations of some coumarins of higher plants by liquid chromatography. *J. Chromatogr.*, 314:323–336.
70. Zobel, A. M. Brown, S. A., and Nighswander, J. E. 1991. Influence of salt sprays on furanocoumarin concentrations on the *Ruta graveolens* leaf surface. *J. Bot.*, 67:213–218.
71. Towers, G., and Yamamoto, E. 1985. Interactions of cinnamic acid and its derivatives with light. *Annu. Proc. Phytochem. Soc. Europe*, 25:271–288.
72. Ceska, O., Chaudhary, S., Warrington, P., Poulton, G., and Ashwood-Smith, M. 1986. Naturally occurring crystals of photocarcinogenic furanocoumarins on the surface of parsnip roots sold as food. *Experimentia*, 42:1302–1304.
73. Hrazdina, G. 1982. Anthocyanins. In: J. B. Harborne and T. J. Mabry (eds.), *The Flavonoids: Advances in Research*, pp. 135–188. Chapman & Hall, New York.
74. Blom, H. 1983. Partial characterization of a thermostable anthocyanin-β-glycosidase from *Aspergillus niger*. *Food Chem.*, 12:197–204.
75. Huang, M., and Ferraro, T. 1992. Phenolic compounds in food and cancer prevention. In: M. Huang, C. Ho, and C. Lee (eds.), *Phenolic Compounds in Food and Their Effects on Health II: Antioxidants and Cancer Prevention*, pp. 8–34. American Chemical Society, Washington, DC.

76. Hahlbrock, K. 1981. Flavonoids. In: E. E. Conn (ed.), *The Biochemistry of Plants: Secondary Plant Products*, pp. 425–456. Academic Press, Toronto.
77. Ollis, W. 1961. *Recent Development in the Chemistry if Natural Phenolic Compounds*. Pergamon Press, New York.
78. Berenbaum, M., Nitao, J., and Zangerl, A. 1991. Adaptive significance of furanocoumarins diversity in *Pastinaca sativa (Apiaceae). J. Chem. Ecol.*, 17:207–215.
79. Osawa, T., Namiki, M., and Kawakishi, S. 1990. Role of dietary antioxidants in protection against oxidative damage. In. Y. Kuroda, D. Shankel, and M. Waters (eds.), *Antimutagenesis and Anticarcinogenesis Mechanisms II*, pp. 139–154. Plenum Press, New York.
80. Das, N., and Ramanathan, L. 1992. Studies on flavonoids and related compounds as antioxidants in foods. In: A. Ong and L. Packer (eds.), *Lipid Soluble Antioxidants: Biochemistry and Clinical Applications*, pp. 295–306. Birkhauser Verlag, Boston.
81. Starvic, B., and Matula, T. 1992. Flavonoids in foods: Their significance for nutrition and health. In: A, Ong and L. Packer (eds.), *Lipid Soluble Antioxidants: Biochemistry and Clinical Applications*, pp. 274–294. Birkhauser Verlag, Boston.
82. Okuda, T. 1993. Natural polyphenols as antioxidants and their potential use in cancer prevention. In: A. Scalbert (ed.), *Polyphenolic Phenomena*, pp. 221–235. INRA Editions, Paris.
83. Mulligan, J., and Kevan, P. 1973. Colour, brightness, and other floral characteristics attracting insects to the blossoms of some Canadian weeds. *Can. J. Bot.*, 51:1939–1952.
84. Rieseberg, L., and Schilling, E. 1985. Floral flavonoids and ultraviolet patterns in *Viguiera* (Compositae) *Amer. J. Bot.*, 72:999–1004.
85. Heaney, R., and Fenwick, G. 1980. The glucosinolate content of *Brassica* vegetables. A chemotaxonomic approach to cultivar identification. *J. Sci. Food Agric.*, 31:794–801.
86. Larsen, P. 1981. Glusosinolates. In. E. E. Conn (ed.), *The Biochemistry of Plants: Secondary Plant Products*, pp. 502–525. Academic Press, Toronto.
87. Heaney, R., and Fenwick, G. 1987. Identifying toxins and their effects: Glucosinolates. In: D. Watson (ed.), *Natural Toxicants in Food: Progress and Prospects*, pp. 76–109. Ellis Horwood, London.
88. Heaney, R., and Fenwick, G. 1980. Glucosinolates in *Brassica* vegetables. Analysis of 22 varieties of brussel sprout (*Brassica oleracea* var. *gemmifera*). *J. Sci. Food Agric.*, 31:785–793.
89. Blaakmeer, A., Geervliet, J., Van Loon, J., Posthumus, M., Van Beek, T., and De Groot, A. 1994. Comparative headspace analysis of cabbage plants damaged by two species of *Pieris* caterpillars: Consequences for in-flight host location by *Cotesia* parasitoids. *Entomol. Exp. Appl.* 73:175–182.
90. Felix D'Mello, J. 1997. Toxic compounds from fruit and vegetables. In: F. Tomas-Barberan and R. Robins (eds.), *Phytochemistry of Fruit and Vegetables*, pp. 331–351. Clarendon Press, Oxford.
91. Heaney, R., and Fenwick, G. 1980. The analysis of glucosinolates in *Brassica* species using gas chromatography. Direct determination of the thiocyanate ion precursors, glucobrassin and neoglucobrassicin. *J. Sci. Food Agric.*, 31:593–599.
92. Macleod, A., Macleod, G., and Reader, G. 1989. Evidence for the occurrence of butyl- and isobutylglucosinolates in seeds of *Brassica oleracea*. *Phytochemistry*, 28:1405–1407.
93. Gross, D. 1993. Phytoalexins of the *Brassicaceae*. *J. Plant Dis. Protect.*, 100:433–442.
94. Björkman, R. 1976. Properties and functions of plant myrosinases. In: J. Vaughman, A. Macleod, and B. Jones (eds.), *The Biology and Chemistry of the* Cruciferae. pp. 191–205. Academic Press, New York.
95. Ma and Schoonhaven. 1973 .
96. French and Skinner. 1974 .
97. Bornman, J. 1991. UV radiation as an environmental stress to plants. *J. Photochem. Photobiol. B: Biol.*, 8:337–342.

98. Day, T. 1993. Relating UV-B radiation screening effectiveness of foliage to absorbing-compound concentration and anatomical characteristics in a diverse group of plants. *Oecologia*, 95:542–550.
99. Zobel, A. M., Chen, Y., and Brown, S. A. 1994. Influence of UV on furanocoumarins in *Ruta graveolens* leaves. *Acta Horticult.*, 381:355–360.
100. Becker, W., and Deamer, D. 1991. *The World of the Cell*, 2nd ed. Benjamin/Cummings, Don Mills, Ontario.
101. Rice-Evans, C., Diplock, A., and Symons, M. 1991. *Techniques in Free Radical Research*. Elsevier, New York.
102. McQuarrie, D., and Rock, P. 1987. *General Chemistry*, 2nd ed. Freeman, New York.
103. Solomons, T. 1992. *Organic Chemistry*, 5th ed. Wiley, Toronto.
104. Pine, S. 1987. *Organic Chemistry*, 5th ed. McGraw-Hill, Toronto.
105. Downum, K. 1992. Tansley Review 43: Light activated plant defence. *New Phytol.*, 122:401–420.
106. Mason, R., Stolze, K., and Flitter, W. 1990. Free radical reactions with DNA and its nucleotides. In: Y. Kuroda, D. Shankel, and M. Waters (eds.), *Antimutagenesis and Anticarcinogenesis Mechanisms II*, pp. 119–126. Plenum Press, New York.
107. Minnunni, M., Wolleb, U., Mueller, O., Pfeifer, A., and Aechbacher, H. 1992. Natural antioxidants as inhibitors of oxygen species induced mutagenicity. *Mutation Res.*, 269:193–200.
108. Cao, G., Sofic, E., and Prior, R. L. 1996. Antioxidant capacity of tea and common vegetables. *J. Agric. Food Chem.*, 44:3426–3431.
109. Fraenkel, G. 1959. The raison d'tre of secondary plant substances. *Science*, 129:1466–1470.
110. Waterman, P., and Mole, S. 1994. *Analysis of Phenolic Plant Metabolites*, 2nd ed. Blackwell Scientific, Boston.
111. Teramura, A., and Sullivan, J. 1998. Effects of ultraviolet-B radiation on soybean yield and seed quality: A six-year field study. *Environ. Pollution*, 53:466–468.
112. Weidenhamer, J., Hartnett, D., and Romeo, J. 1989. Density-dependent phytoxicity: Distinguishing resource competition and allelopathic interference in plants. *J. Appl. Ecol.*, 26:613–624.
113. Pal, S. 1994. Role of abiotic and biotic catalysts in the transformation of phenolic compounds through oxidative coupling reaction. *Soil Biol. Biochem.*, 26:813–820.
114. White et al. 1989.

APPENDIX: BIOSYNTHESIS OF COUMARIN

Enzymes:

1. Transaldolase
2. DAHP synthase
3. DHQ synthase
4. DHQase
5. Shikimate dehydrogenase
6. Shikimate kinase
7. EPS synthase
8. Chorismate synthase
9. Chorismate mutase
10. Prephenate dehydratase
11. Phenylalanine aminotransferase
12. Phenylalanine ammonia-lyase (PAL)
13. Cinnamic acid 2-hydroxylase

Consequences of Radiation on *Brassica oleracea*

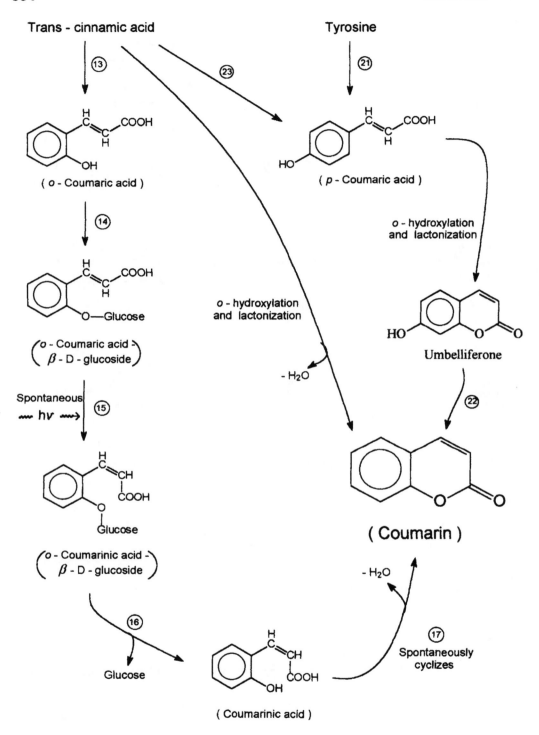

14. *o*-Coumaric acid glycosylase
15. *o*-Coumaric acid-β-isomerase
16. *o*-Coumarinic acid-β-glucosidase
17. Coumarinic acid dehydratase
18. Phenylalanine dehydrogenase
19. Tyrosine amino-transferase
20. Phenylalanine 4-hydroxylase
21. Tyrosine ammonia-lyase
22. 7-hydroxycoumarin dehydroxylase
23. Cinnamic acid 4-hydroxylase

Abbreviations:

DAHP: 3-Deoxy-D-arabino-heptulosonate 7-phosphate
DHQ: 3-dehydroquinate
DHQase: 3-dehydroquinate hydro-lyase
DHS: 3-dehydroshikimate
E4P: Erythrose-4-phosphate
EPSP: 5-Enolpyruvyishikimate-3-phosphate
EPS synthase: 5-Enolpyruvyishikimate-3-phosphate synthase
PEP: Phosphoenolpyruvate
SA: Shikimate
SA3P: Shikimate 3-phosphate

18
RC1 Consortium for Soil Decontamination: Its Preparation and Use

Dana M. V. Horáková and Miroslav Nemec
Masaryk University, Brno, Czech Republic

I. INTRODUCTION

Hydrocarbons are an important class of organic waste in soil. The interest in the focused use of special microorganisms for biological degradation of hydrocarbons has increased over the past few years. The literature on the subject of microbial degradation of petroleum and its products is quite voluminous. As early as 1903, the first microorganisms were described which used hydrocarbons as a source of carbons and energy. At present, more than 40 species of microorganisms are known which use aliphatic or aromatic hydrocarbons in the course of biological degradation. These microorganisms have been isolated from the oceans, from fresh waters, as well as from the soil.

The number of bacteria and fungi which are able to biologically degrade oil hydrocarbons increases rapidly and locally after oil spills. These substances occur mostly as selections gases. Natural degradation of oil by microorganisms is one of the most important long-term natural processes. Considerable basic knowledge about the factors that affect natural biodegradation, about the kinds of hydrocarbons capable of being degraded, and about the species and distribution of the microorganisms involved in biodegradation was developed in the early 1970s. The increased awareness of the ability of microorganisms to decompose hydrocarbons has stimulated the development of terrestrial bioremediation techniques (1,2). These techniques attempt to accelerate the remediation process by optimizing the environmental factors in control of the decomposition rate, which requires knowledge of biodegradation kinetics of organic contaminants (3). Alexander and Scow (4) have stressed that the application of most of the existing models to describe decomposition kinetics is often questionable, because many of the soil environment factors are omitted.

The 1980s were a period of rapid advances in the knowledge of genetics and molecular biology of bacterial degradation of different hydrocarbons, and of renewed interest in microbial ecology of pollution-stressed environments. From the genetic point of view it is clear that the ability of a major proportion of microorganisms to degrade oil hydrocarbons is partly controlled by plasmids. Many microorganisms, including several pseudo-monads, are able to use linear alkanes as their sole source of carbon and energy (5). The OCT-plasmid of *Pseudomonas oleovorans* contains two operons, alkBFGHJKL and alkST, which encode all proteins necessary for the degradation of *n*-octane and other 5- to 12-carbon linear alkanes. Branched isomers, such as isooctane, are less susceptible to biodegradation than *n*-octane (6). The conversion of linear alkanes to the corresponding alcohols is catalyzed by a group of proteins collectively referred to as the "alkane hydroxylase system." It has three main components: alkane 1-monooxygenase, and two soluble proteins, rubredoxin and rubredoxin reductase. Rubredoxin reductase transfers electrons from NADH to rubredoxin. This protein then passes electrons to alkane 1-monooxygenase, which is an enzyme localized in the cytoplasmatic membrane. The final product of this pathway, alkanoyl-CoA, enters the beta-oxidation cycle and is used as a carbon and energy source (5). The proof of the transfer of plasmid has made it possible to explain the rapid increase of the number of microorganisms with degrading ability.

As a substrate for microorganisms, raw oil is an extremely complex mixture of hydrocarbons. The saturated fraction of raw oil includes *n*-alkanes, branched alkanes, and cykloalkanes. Biodegradation of the *n*-alkanes up to C_{44} is in most cases realized by monoterminal oxidation. This process may lead to the accumulation of some more or less toxic fatty acids.

Branched isoprenoid alkanes are degraded by omega-oxidation under the creation of dicarbon acids. The terminal branched alkanes strongly resist biodegradation. They block beta-oxidation and tend to accumulate within the environment. After spontaneous degradation of hydrocarbon contamination, tripentacyclic compounds are often created, which are persistent in the environment.

II. ISOLATION OF OIL-DEGRADING MICROORGANISMS

In order to isolate oil and oil hydrocarbon-degrading microorganisms, samples of soil which were contaminated with crude oil over a long period of time were used. The samples were taken from around three abandoned oil wells, 1.2–3.0 m apart. After homogenization, chemical and microbiological analyses of the samples were carried out. The analyses proved that the concentration of nonpolar extractable compounds (NECs) decreased in proportion to the distance from the well. The concentration ranged from 153,000 to 25,000 ppm. Microbiological analysis was carried out for six homogenized soil samples taken at depths of 5–10 cm. The NEC concentration was 74,000–37,000 ppm. Microorganism species as well as the total number of aerobic heterotrophs varied (Table 1). All soil samples contained microorganisms of the *Pseudomonas* genus, particularly *Pseudomonas aeruginosa* and *Pseudomonas putida*, and the yeast-fungus *Geotrichum candidum*. With the single exception of one sample, there was also the lipolythic yeast *Yarrowia lipolytica*. The soil sample taken at a greater distance from well No. 2 showed the greatest variety of microorganisms.

A fresh soil extract was analyzed for *Pseudomonas* and *Acinetobacter* genus, as a large amount of oil degradators was expected. Six strains of *Pseudomonas aeruginosa*, 85 strains of *Pseudomonas putida*, one strain of *Pseudomonas stutzeri*, one strain of *Pseudomonas cepacia*, one strain of *Pseudomonas fluorescens*, one strain of *Acinetobacter haemolyticus* and one strain of *Acinetobacter hydrophylium* were isolated from the sample. Other tests were carried out using six isolated strains of *Geotrichum candidum* and five strains of *Yarrowia lipolytica*. The reason for the extended choice of micro-

Table 1 Chemical and Microbiological Analysis of Soil Samples

	Oil well 1		Oil well 2		Oil well 3	
Distance from oil well (m)	1.2	3	1.2	3	1.2	3
Concentration of NEC (ppm)	74,000	37,000	69,000	45,000	73,000	42,000
Microorganisms						
Pseudomonas aeruginosa	+	+	+	+	+	+
Pseudomonas putida	+	+	+	+	+	+
Pseudomonas sp.	−	+	−	+	−	+
Corynebacterium sp.	−	−	−	+	−	−
Acinetobacter sp.	−	+	−	−	−	−
Aspergillus niger	−	−	−	−	−	+
Aspergillus versicolor	+	+	−	−	−	+
Fusarium oxysporum	−	−	−	+	−	−
Penicillium chrysogenum	+	−	−	−	−	−
Penicillium expansum	−	−	+	−	−	−
Penicillium jensenii	−	−	−	+	−	−
Trichoderma koningii	−	−	−	+	−	−
Geotrichum candidum	+	+	+	+	+	+
Yarrowia lipolytica	+	+	+	+	+	−
Number of aerobic heterotrophs (CFU/mL)	2.1×10^3	5.3×10^6	1.8×10^5	4.6×10^7	1.2×10^3	5.1×10^6

organisms was the uniform occurrence of these microorganisms in all samples. The chosen races were selected by the modified Brown and Braddock method (7), which is based on the fact that emulsification of oil is a good indicator of hydrocarbon-degrading microorganisms. Microorganisms showing a positive emulgation reaction after 3 weeks of growth in 2 mL of a defined mineral medium (DMA) (8) with 0.05 mL of crude oil as the only source of carbon were subsequently tested for fast growth ability in a DMA with 1% (v/v) of hexadecane. The growth assessment was carried out after 7 and 14 days of submersive cultivation at 25–30°C. These additional selective tests were opted for *Pseudomonas putida* (strain PS4 and PS27), one strain of *Geotrichum candidum* (G2), and one strain of *Yarrowia lipolytica* (W1). Although all the isolated strains of *Pseudomonas aeruginosa* passed the selection, they were excluded from the subsequent experiments because of their possible pathogenous nature.

The selected microorganisms were tested both separately and in combinations for the ability to decompose crude oil from the locality from which the samples had been taken. Crude-oil degradation (18,000 ppm) was observed in DMA. The microorganisms were planted in the medium to obtain the resulting concentration of approx. 10^6–10^7 CFU/mL. The pH value was continuously adjusted to pH = 6.6. Submersive cultivation was carried out under the constant temperature of 25°C. The NEC concentration was determined after 4, 12, and 31 days of sample cultivation after extraction to tetrachloromethan and measured by infrared spectrometry according to DIN38409 H18.

This experiment showed that the crude oil used can be degraded most efficiently by a mixture of *Pseudomonas putida* PS27 and *Geotrichum candidum* G2. The presence of *Yarrowia lipolytica* W1 (i.e., combination PS27 + G + W1) did not improve the effect

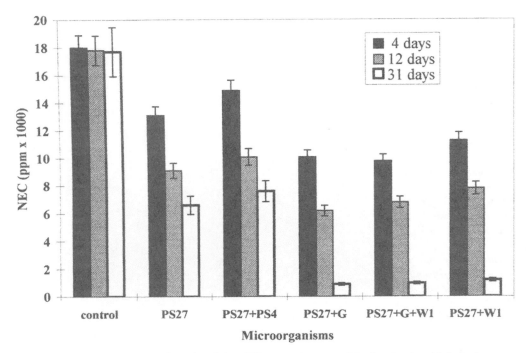

Figure 1 Biodegradation of crude oil by different mixtures of isolated microbial strains.

of PS27 + G. If *Yarrowia lipolytica* W1 was used with isolated pseudo-monads, its presence had a significant positive influence on the crude-oil biodegradation (Fig. 1). Based on previous experiments, an artificial microbial consortium, RC1, was produced CZ Patent 279 021, 1994; PCT/EP93/00897, 1994; U.S. Patent 5,579,998, 1996).

The rate of biodegradation of oily substances by RC1 can be increased insignificantly by a decrease in pH (Fig. 2), as the addition of oleic acid to the degradation mixture raises the rate remarkably. Enrichment of the liquid synthetic medium by 2% (v/v) oleic acid seems to be most favorable. A higher concentration of oleic acid has not shown any increase in the degradation rate. The increased degradation rate of oily substances in the presence of this fatty acid has proved not to result solely from co-oxidation (9–11), since it was significantly influenced by the fast buildup of the degrading biomass on this easily oxidated carbon substrate (Fig. 3).

The role of oleic acid in the induction was verified on *P. putida* PS27. Induction is one of the reasons for the frequently observed reduction of the acclimation period of microorganisms, despite the fact that the reduction is rather insignificant in the natural environment (12,13). The cells were treated by nutrient broth with oleic acid, washed, and eventually used for crude-oil degradation in a synthetic liquid environment. A comparison to control cells which were cultivated in the appropriate medium without oleic acid has proved that preinoculation induction plays a crucial role in the degradation activity of the microorganism used. In addition, Horáková and Pospíšilová (14) have proved that only a very low concentration of oleic acid, not exceeding 0.1% (v/v), is necessary for the actual cell induction, in the case where oleic acid is not the sole source of carbon.

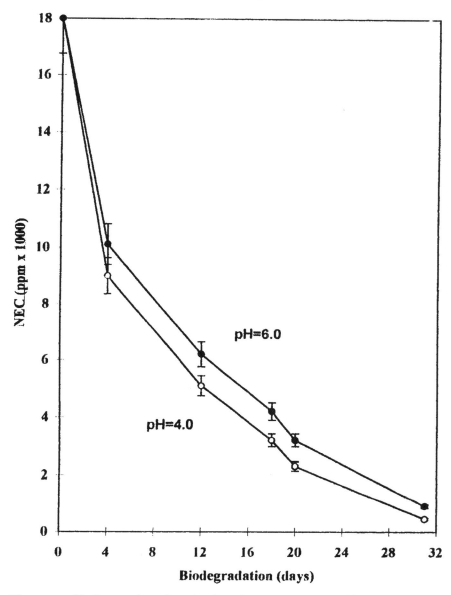

Figure 2 Biodegradation of crude oil in DMA by RC1 at different pHs.

All the experiments focusing on the rate of oily hydrocarbon degradation in a defined synthetic medium were carried out with the recommended C/N ratio within the range 10–60 (15).

The microorganisms PS27 and G2 were filed as patent cultures with registration numbers *Pseudomonas putida* CCM 4307 and *Geotrichum candidum* CCM 8170. *Yarrowia lipolytica* W1, which was filed as another patent (PCT/CH96/00205, 1996) and used in some of our studies, was the third component of RC1. *Yarrowia lipolytica* W1 was filed with registration number CCM 4510. The toxicity and pathogenity of RC1 were tested

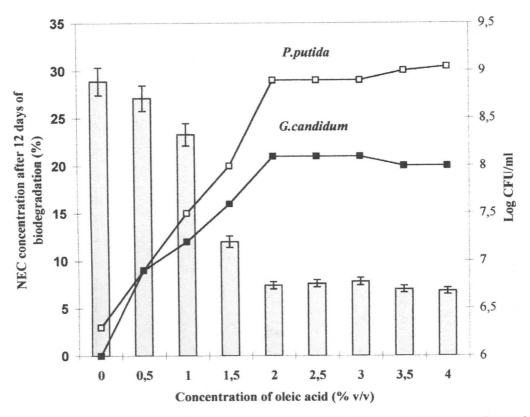

Figure 3 Effect of oleic acid on biodegradation of crude oil (18,000 ppm) by RC1 consortium and growth of *Pseudomonas putida* CCM 4307 and *Geotrichum candidum* CCM 8170.

by the state laboratories in the Czech Republic, with the following result: RC1 can be used as a biodegradation agent without any restrictions.

III. TECHNOLOGY OF RC1 PRODUCTION

The individual components of RC1 were stored separately on gel disks; cultivation was separate too. *P. putida* CCM 4307 inoculum was produced by the application of one gel disk to 100 mL of nutrient broth and subsequent 24-h cultivation at 30°C and vigorous aeration. For *G. candidum* CCM 8170, inoculum malt extract broth was used. The rest of the procedure was the same for all the components. The RC1 components (CFU/mL = approx. 10^6) were cultivated and prepared for soil application in an increasing volume of DMA enriched by 0.5% (v/v) of Maggi meat extract, 0.05% (w/v) of glucose, and 0.1% (v/v) of oleic acid. The Maggi meat extract was used as a cheap and easily available source of carbon and nutrients. Maggi has proved to have a positive influence on the growth of both the studied microorganisms even at very low concentration. In contrast to Maggi, the sugar component of the enriched DMA does not stimulate the growth of *P. putida* cells too much (Fig. 4), though playing an important role in reducing the surface tension of the medium during the process of cultivation

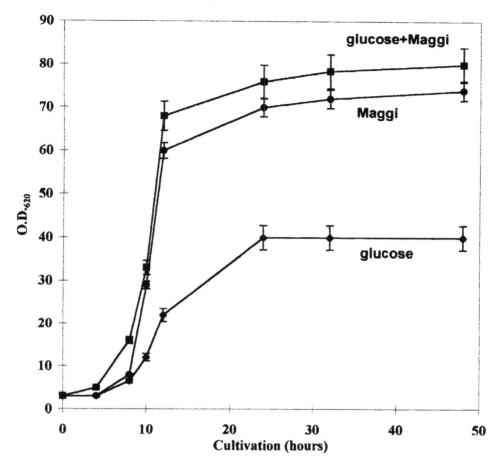

Figure 4 Growth of *Pseudomonas putida* CCM 4307 cells in enriched DMA.

(Fig. 5). Chemical analysis of DMA after the cultivation of *P. putida* CCM 4307 indicated the presence of glycolipidic biosurfactant. A similar reduction of surface tension was observed at DMA after the cultivation of *G. candidum* CCM 8170. Chemical analysis of DMA was not carried out. The importance of microbial production of biosurfactants in saving oil-contaminated soil has already been described, e.g., in (16–18). Thus the biosurfactants produced by RC1 components are likely to have a positive influence on pseudo-solubilization of hydrocarbons and hence on the rate of the saving process, particularly in soil. The results shown in Fig. 6 suggest that glucose-cultivated RC1 cells degrade crude oil in an intentionally contaminated sterile soil at a very high rate. The required limit concentration NEC = 200 ppm was reached after about 37 days of soil decontamination in the bed. RC1 cells cultivated in a glucose-free medium begin to degrade the hydrocarbons after an initial acclimation period; the biodegradation curve suggests that a certain part of the oily substances had become unavailable (and therefore undegradable) to the microorganisms. It is not possible to determine to what extent these results are affected solely by the presence of biosurfactants that enter the soil in low amounts together with the microorganisms and/or by biosurfactant produced by the microorganisms in the soil. The presence of residual glucose in the medium following

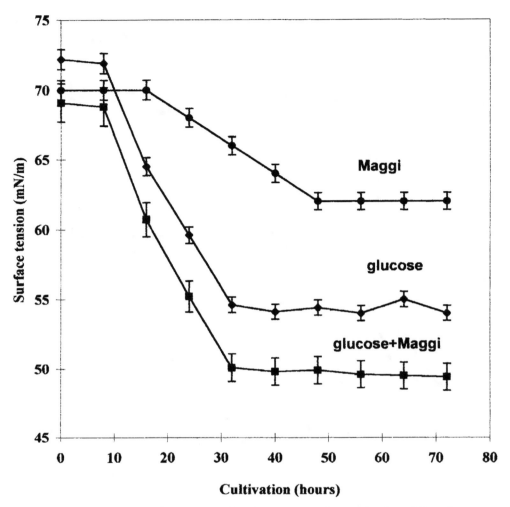

Figure 5 Reduction of surface tension of enriched DMA after cultivation of *Pseudomonas putida* CCM 4307.

the cultivation was excluded by a laboratory analysis before application to soil. The pH values of both RC1 samples were comparable due to good buffer characteristics of DMA; they ranged from 5.5 to 6.1. The RC1-cell concentration in the degraded soils did not differ too much; after 30 days of the degradation process the concentration of *P. putida* was 5.2–8.0 × 10^7 CFU/mL and of *G. candidum* was 2.2–5.1 × 10^6 CFU/mL.

The RC1 preparation can also be used in a dried form for field application. After both components were thickened in a flow centrifuge and mixed together in a 1 : 1 cell ratio, the resulting substance was sorbed onto corn chaff and dried in a fluid dryer. The drying process was supposed to keep both the microorganisms at the maximum vitality level. After 12 months of storing of the dried RC1 at 25°C, the viability of *G. candidum* decreased considerably compared to that of *P. putida* (Table 2). On the basis of another experiment, another two dried mixtures were produced, with cell ratios *P. putida* : *G. candidum* = 5 : 2 and 5 : 3. Both components were fairly active in terms of

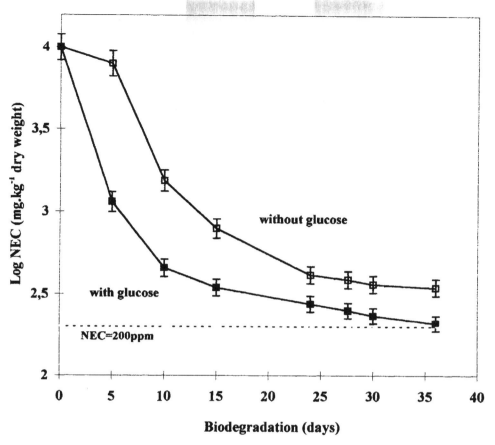

Figure 6 Degradation of crude oil in sterile soil by RC1. Before their application both components of RC1 were cultivated in DMA enriched with 0.05% (w/v) glucose.

Table 2 Survival of *Pseudomonas putida* CCM 4307 and *Geotrichum candidum* CCM 8170 Cells in Dry Preparation and Their Degradation Activity (starting concentration, NEC = 1000 ppm)

Time of storage (months)	Microorganisms (CFU/g d.w.)		Residual concentration of crude oil (ppm after 7 days, 25°C)
	P. putida	*G. candidum*	
1	1.4×10^8	1.2×10^8	530
2	1.3×10^8	8.2×10^7	520
3	1.2×10^8	5.2×10^7	520
4	9.5×10^7	2.3×10^7	490
5	9.3×10^7	9.2×10^6	490
6	9.0×10^7	6.5×10^6	550
7	5.0×10^7	4.1×10^6	600
8	4.2×10^7	2.1×10^6	680
9	1.2×10^7	5.7×10^5	750
10	9.5×10^6	9.3×10^4	760
11	8.6×10^6	3.0×10^4	780
12	5.2×10^6	2.5×10^3	780

Table 3 Biodegradation Activity of Dry RC1 Preparation with Different Ratios of Its Compounds and Time of Storage (starting concentration, NEC = 1000 ppm)

	Residual concentration of crude oil (ppm after 7 days of degradation at 25°C)		
	Time of storage, 25°C (months)		
P. putida : G. candidum	0	8	12
1 : 1	550	600	750
3 : 1	500	580	680
5 : 1	500	550	620
5 : 2	440	480	500
5 : 3	440	440	480

degradation ability even after 1 year of storage (Table 3). Although the actual outdoor application of the dried preparation is easier compared to that of the liquid form, the research on dried RC1 production was stopped because of high costs.

IV. RC1 POSSIBILITIES IN DEGRADATION OF VARIOUS OIL PRODUCTS

The RC1 substance with the cell ratio 2 : 1 was tested on oil products used predominantly in aeronautics (engine oil, aircraft-fuel, etc.). Their degradation took place in bottles sealed with a sorption-filter stopper which retained light volatile hydrocarbons during the biodegradation. At the end of the experiment the volatile hydrocarbons were extracted in order to correct the results. The process of hydrocarbon degradation took place in liquid DMA. The contamination was made fairly high (NEC = 10,000 ppm) in order to prove the convenience of RC1 for heavily polluted airport sites. The experiment under optimum laboratory conditions showed that all the selected carbohydrates were easily degradable by RC1 (Table 4). After 45 days the degradation process lowered the concentration of pollutants to NEC = 190–260 ppm. Oil products with large amounts of antioxidative compounds, e.g., aviation gasoline, were degraded at a somewhat lower rate.

A laboratory experiment with homogenized soil polluted by crude oil has proved that the time required to reach the limit concentration NEC = 200 ppm is dependent

Table 4 Biodegradation of "Aircraft Hydrocarbons" in DMA by Consortium RC1 (starting concentration, NEC = 10,000 ppm)

	Time of degradation (days)			
Oil substance	8	22	39	45
Cerosene RT	3580	340	210	210
Engine oil MS8P	6630	2090	350	200
Lubricating oil OLEMS-2	4230	1540	520	200
Brake fluid AMG-10	2120	800	240	190
Aviation gaseline BL-78	8130	3740	700	230
Aviation engine oil B-3V	3630	2070	800	260

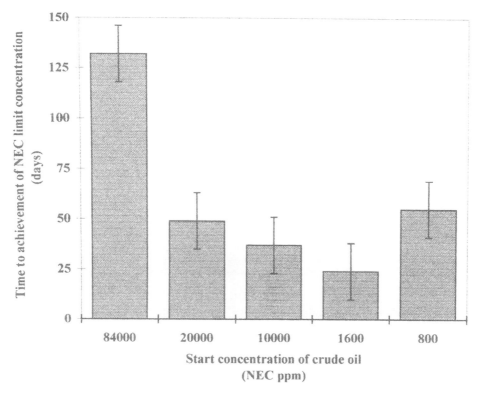

Figure 7 Effect of starting crude oil concentration on achievement of NEC limit concentration (NEC = 200 ppm).

on the pollutant concentration (Fig. 7). It has been found that soil contaminated with NEC < 1000 ppm is degraded at a lower rate than soil contaminated with higher concentrations of the tested pollutant. A parallel observation of microbial RC1 components proved that NEC < 1000 ppm concentration is insufficient for the RC1 cells to reproduce in the soil (Table 5). Hence it is possible that the actual process of oil hydrocarbon degradation took place only in a series of accidental collisions of a low number of cells with a pseudo-solubilized or sorbed carbon substrate. It has been proved that RC1 enriched with *Yarrowia lipolytica* W1 cells significantly lowers the concentration of PCBs in soils polluted with a variety of pollutants (NEC, PAHs, PCBs, metals). *Yarrowia lipolytica* W1 was observed to be highly dehalogenase-active under lab conditions (not published).

V. VERIFICATION OF BIODEGRADATION ACTIVITY OF RC1 IN PRACTICE

The biodegradation activity of RC1 was verified using soil obtained from a site heavily polluted over a long period of time. According to the record, the soil came from a heavy-oil loading platform which had been in use for over 20 years. The soil was also contaminated with PAHs and heavy metals. Pure oil product was available for comparison to the car-

Table 5 Effect of Starting Concentration of Crude Oil on RC1 Cell reproduction in Soil

Starting concentration of crude oil (ppm)	Components of mixture RC1	CFU/g of soil ($\times 10^5$)		
		Start	After 20 days	After achievement of limit concentration (NEC = 200 ppm)
84,000	P	20	52	1,000
	G	9.8	20	290
20,000	P	32	960	1,200
	G	10	520	670
10,000	P	29	1,200	2,000
	G	12	590	900
1,600	P	22	1,000	1,200
	G	9.7	600	620
800	P	20	60	68
	G	15	62	65

Note: P = *Pseudomonas putida* CCM 4307, G = *Geotrichum candidum* CCM 8170.

bohydrates contained in the soil sample. Figure 8 represents the amounts of aliphatic hydrocarbons C_9–C_{28} in the contaminated soil after RC1 biodegradation and a comparison to the amounts of these compounds in the actual pollutant. The biodegradation was carried out ex situ on two beds with 5 m³ of soil each. One moisturized and fertilized (Cererit N-P-K fertilizer, CZ) bed was treated with 10 L/m³ of fresh RC1 with the cell ratio *P. putida* : *G. candidum* = 2 : 1. The other bed was only moisturized and fertilized (N-P-K). The N-P-K concentration was chosen to keep the C/N ratio above 10 during the process. Both beds were moisturized and plowed on a regular basis. The purpose was to push the NEC concentration down, to values lower than 200 ppm.

After 2 weeks the sum of NEC was decreased from the initial value 5680 ± 80 ppm to 2820 ± 60 ppm. Chemical analyses for the determination of the total NEC concentration in the RC1-treated soil and in the control sample were carried out for 5 weeks; carbohydrate profile was determined only in the RC1-treated soil after 2 and 4 weeks of degradation. The results of this pilot experiment showed that 2 weeks after the RC1 application there were no aliphatic hydrocarbons C_{16}–C_{21} left; there was an increased percentage of C_{27} and C_{28} among the detected hydrocarbons (i.e., C_{22}–C_{28}) (Fig. 8). After 4 weeks of degradation there were no C_{24}–C_{28} and only C_{22} and C_{23} carbohydrates were detected. NEC concentration decreased to 620 ± 30 ppm. The objective of the experiment was fulfilled after 5 weeks of degradation with RC1 (NEC = 180 ± 30 ppm).

At the beginning and at the end of the experiment the amount and concentration of polyaromatic hydrocarbons (PAHs) were also determined. The control sample showed only a minor decrease in PAH concentration, which ranged from 0.2% to 1.85% of the initial concentration. Anthracene represented an exception, since its concentration decreased by 8.93%. Such a decrease is insignificant comparing to its concentration after the RC1 biodegradation process in the treated soil (Table 6). Chemical analysis showed that the total concentration of PAHs decreased considerably after the application of RC1. The RC1 consortium is relatively weaker in lowering the concentration of DB[a,h]anthracene and B[g,h,i]perylene with ind[1,2,3-c,d]pyrene. Even this relatively good result of PAH degradation suggests that it is going to be fairly hard to accomplish the limit concentration of max. 5 mg/kg using RC1. It is possible to expect that the result

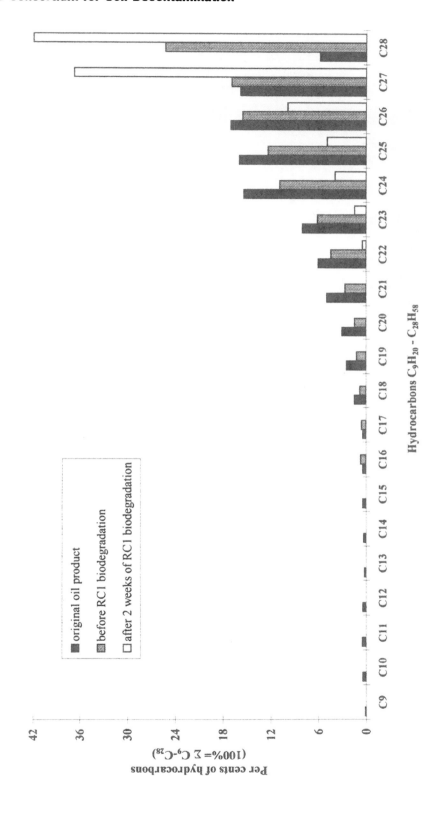

Figure 8 Biodegradation of mixture of hydrocarbons C_9-C_{28} by RC1 carried out ex situ.

Table 6 Biodegradation of PAHs in Soil by RC1

PAH	Sample 1 With mixture RC1	Sample 2 With mixture RC1	Control without RC1	Original PAH concentration
	mg/kg d.w.			
Naphthalene	<0.3	<0.3	<0.3	<0.3
Acenaphthene + fluorene	<0.5	<0.5	4.5	4.6
Phenanthrene	2.1	2.6	9.2	9.3
Anthracene	1.5	1.1	5.1	5.6
Fluoranthene	16	12.9	63.8	65
Pyrene	9.7	5.4	18.2	18.2
B[a]anthracene + pyrene	8.2	7.8	22.1	22.2
B[b]fluoranthene	19.4	14.2	30.7	31
B[k]fluoranthene	5.1	4.8	14.3	14.5
B[a]pyrene	22.6	23.8	49.4	49.5
DB[a,h]anthracene	8.8	10.9	11.5	11.5
B[g,h,i]perylene + indeno [1,2,3-c,d]pyrene	63.2	52.1	79.6	79.6
	157	136	308	311

of PAHs degradation will be influenced by concentration of the above-mentioned polyaromatic hydrocarbons and the presence and concentration of other hydrocarbons and heavy metals.

All decontaminations so far carried out using RC1 consortium have resulted in a significant decrease in toxicity, genotoxicity, and mutagenity of the contaminated soil. The soil was in most cases considered to be completely cleaned up even according to strict environmental standards. Table 7 shows some results of the practical use of RC1. The data suggest that nonoily pollutants influence the oil degradation rate in the presence of RC1.

VI. CONCLUSION

On the basis of results to date of experiments and decontaminations carried out in practice, it is possible to conclude that biodegradation of hydrocarbons using RC1 consortium is a highly dynamic process, with its rate depending not only on physiologically sufficient hydrocarbon concentration but also on conditions suitable for the reproduction of the inserted microbial population. RC1 has been used to decontaminate more than 20,000 tons of polluted soil. Decontamination technologies have always been designed for a specific polluted site, according to the results of chemical analyses and laboratory tests of biodegradability. In a number of studies RC1 has been compared to other commercial biodegradation preparations that are used for oily hydrocarbon degradation. In all cases the studies has proved the RC1 to have a very high degradation rate as well as a high level of tolerance to the presence of heavy metals.

Decontamination of soils polluted by oily hydrocarbons and PAHs has also shown a significant decrease in polyaromatic hydrocarbons concentration. In lab experiments RC1 enriched with *Yarrowia lipolytica W1* (approx. 10^9 cells/L of RC1, ratio 2 : 1)

Table 7 Practical Use of RC1 Consortium (selective examples)

Contamination	NEC concentration (ppm) Before sanation	NEC concentration (ppm) After sanation	Time of sanation (days)	Notes
Fuel oil	813	137	55	Clayey soil
Mixture of oils contaminated by PCBs	1,626	128	67	Sand soil (old contamination)
Oil from current transformer without PCBs	16,180	379	66	Clay-sandy soil
Mixture of oil compounds contaminated by heavy metals	7,166	430	113	Mud from washing bridges (military)
Mixture of oils	3,800	260	90	Clayey soil (storage of fuel)
Heavy oils contaminated by PAHs and PCBs	11,320	4,930	30	Pilot experiment
Soil sludge contaminated by heavy metals	25,682	1,250	152	Limit NEC = 1,500 ppm
Oil compounds contaminated by PAHs	5,680	180	35	Degradation in sludge
Crude oil	84,000	200	132	Degradation in clayey sludge from oil platform
Lubricating oils	3,314	126	42	Pilot experiment

has been used for decontamination of soils polluted by both oily hydrocarbons and PCBs. Decreased concentration has been proved.

ACKNOWLEDGMENTS

The research was supported by Biorem A.G., Zug, Switzerland, which is the owner of the RC1 consortium. RC1 is also used commercially by Ekohelp v.o.s, Brno, Czech Republic. Both companies have made the research and pilot experiments possible by their support and by putting the necessary chemical analysis data at our disposal during decontamination work at the polluted sites.

REFERENCES

1. Harmsen, J. 1991. Possibilities and limitations of landfarming for cleaning contaminated soils. In: R. E. Hinchee and R. F. Olfenbuttel (Eds.), *On-Site Bioreclamations*, pp. 255–272. Butterworth-Heinemann, Stoneham, MA.
2. Bower, E. J., 1993. Microbial remediation: Strategies, potentials, and limitations. In: H. J. P. Eijsacker and T. Hamers (eds.), *Integrated Soil and Sediment Research: A basis for proper protection*, pp. 533–544. Kluwer, Dordrecht.

3. Simkins, S., and Alexander, M. 1984. Models for mineralization kinetics with the variables of substrate concentration and population density. *Appl. Environ. Microbiol.*, 47:1299–1306.
4. Alexander, M., and Scow, K. M., 1989. Kinetics of biodegradation in soil. In: B. L. Sawhney and K. Brown (eds.) *Reactions and Movement of Organic Chemicals in Soil*, SSSA Spec. Publ. 22, pp. 243–269. ASA and SSSA, Madison, WI.
5. Beilen, J. B., Wubbolts, M. G., and Witholt, B. 1994. Genetics of alkane oxidation by *Pseudomonas oleovorans*. *Biodegradation*, 5:161–174.
6. Schaeffer, T. L., Cantwell, S. G., Brown, J. L., Watt, D. S., and Fall, R. R. 1979. Microbial growth on hydrocarbons: Terminal branching inhibits biodegradation. *Appl. Environ. Microbiol.*, 38:742–746.
7. Brown, E. J., and Braddock, J. F. 1990. Sheen screen, a miniaturized most-probable-number method for enumeration of oil-degrading microorganisms. *Appl. Environ. Microbiol.*, 56:3895–3896.
8. Pirth, S. J. 1967. A kinetic study of the mode growth of surface colonie of bacteria and fungi. *J. Gen. Microbiol.*, 47:181–197.
9. Atlas, R. M. 1981. Microbial degradation of petroleum hydrocarbons: An environmental perspective. *Microbiol. Rev.*, 45:180–209.
10. Herbes, S. E., and Schwall, L. R. 1978. Microbial transformation of polycyclic aromatic hydrocarbons in pristane and petroleum-contaminated sediments. *Appl. Environ. Microbiol.*, 35:306–316.
11. Horvath, R. S. 1972. Microbial co-metabolism and the degradation of organic compounds in nature. *Bacteriol. Rev.*, 36:146–155.
12. Barkay, T., and Pritchard, H. 1988. Adaptation of aquatic microbial communities to pollutant stress. *Microbiol. Sci.*, 5:165–169.
13. Alexander, M. 1994. *Biodegradation and Bioremediation*, pp. 16–38. Academic Press, San Diego, CA.
14. Horáková, D., and Pospíšilová, L. 1992. Estimation of possibilities for biodegradation of petroleum substances in locality Hodonín-Nesyt. *Agriculture*, 38:775–781.
15. Dibble, J. T., and Bartha, R. 1979. The effect environmental parameters on the biodegradation of oil sludge. *Appl. Environ. Microbiol.*, 37:729–739.
16. Goswami, P., and Singh, H. D. 1991. Different modes of hydrocarbon uptake by two *Pseudomonas* species. *Biotechnol. Bioeng.*, 37:1–11.
17. Falatko, D. M., and Novak, J. T. 1992. Effects of biologically produced surfactants on the mobility and biodegradation of petroleum hydrocarbons. *Water Environ. Res.*, 64:163–169.
18. Zhang, Y., and Miller, R. M. 1995. Effect of rhamnolipid (biosurfactant) structure on solubilization and biodegradation of *n*-alkanes. *Appl. Environ. Microbiol.*, 61:2247–2251.

# 19
Distribution and Fate of Organic and Inorganic Contaminants in a River Floodplain—Results of a Case Study on the River Elbe, Germany

Kurt Friese, Barbara Witter, Werner Brack, Olaf Buettner, Frank Krueger, Maritta Kunert, and Holger Rupp
UFZ—Centre for Environmental Research Leipzig-Halle, Magdeburg, Germany

Guenter Miehlich, Alexander Groengroeft, and René Schwartz
University of Hamburg, Hamburg, Germany

Andrea van der Veen and Dieter W. Zachmann
Technical University of Braunschweig, Braunschweig, Germany

I. INTRODUCTION AND GENERAL ASPECTS

In recent years awareness has grown concerning the significance of floodplains for a number of hydrological and ecological processes. As part of the river–floodplain system, wetlands have a valuable function in controlling surface and subsurface hydrological processes, such as by controlling flood water and groundwater recharge. Moreover, it is known that floodplains may play an important role within nutrient and pollutant dynamics as buffer zones (1,2).

During floods, an exchange of water, sediments, chemicals, and biota takes place between the main channel and the floodplain. Organic and inorganic toxicants in surface waters can be transferred to floodplains and the groundwater, and can lead to unfavorable changes in aquatic and terrestrial communities. Furthermore, as floodplains are often used in agriculture as pastures and for cultivation, pollutants can be introduced into the food web via the contamination of soil, water, or plants.

Although floodplains show similarities to terrestrial and aquatic ecosystems, there are several differences, particularly the importance of the continuous storage of polluted sediments (3). Chemicals are hydrologically transported to floodplains via precipitation, groundwater, and high floods. Floodplain soils can become highly reduced when submerged, but usually have a narrow oxidized surface zone that allows aerobic processes. The biotic and abiotic transformations that occur within the floodplain environment may change the effect and reaction of several compounds: some cause toxic conditions, while

Figure 1 Elbe floodplain during high flood.

others induce the loss of chemicals to the atmosphere or the underground. The unique and diverse hydrological conditions in floodplains result not only in changes in the chemical forms of materials but also in the spatial movement of material within the floodplains, such as in water–sediment exchange and plant uptake.

One of Europe's most polluted rivers is the Elbe, which is severely affected by human activities. The worst inputs came from the former GDR and Czechoslovakia, where environmental precautions were almost nonexistent. For many years industrial effluents and sewage were discharged into the river untreated, resulting in a strong reduction of the ecosystem. Owing to the construction of sewage treatment plants and especially the closure of several factories, the water quality of the River Elbe has greatly improved since German reunification. Nevertheless, the river sediments still contain pollution from the past, and will continue to be contaminated until the material is remobilized by high floods and transported downstream. This process is expected to take about 20 years.

In contrast to most other European rivers, the river Elbe still has extended floodplains that are valuable biotopes. These wetlands reduce the consequences of high flooding, for the floodplains are regularly flooded, mainly in winter and spring (see Fig. 1). River sediments are deposited on the floodplains during flooding, forming the substratum for soil formation. After being covered with plants, the sediments are resistent to erosion, so that contaminated materials are retained in the floodplain much longer than in the river itself. We consider the floodplain soils to be the river's long-term "memory" of sediment pollution.

Contaminants in a River Floodplain

Both inorganic and organic contaminants have so far been investigated in a number of studies, and some findings are presented here. The investigation area is shown in Fig. 2. Basic information on soil characterization and flooding and hydrological regimes is essential if contaminant distribution is to be understood. Our study is augmented by an ecotoxicological characterization of typical Elbe pollutants and by investigations into the transport and accumulation of contaminants in wetland plants.

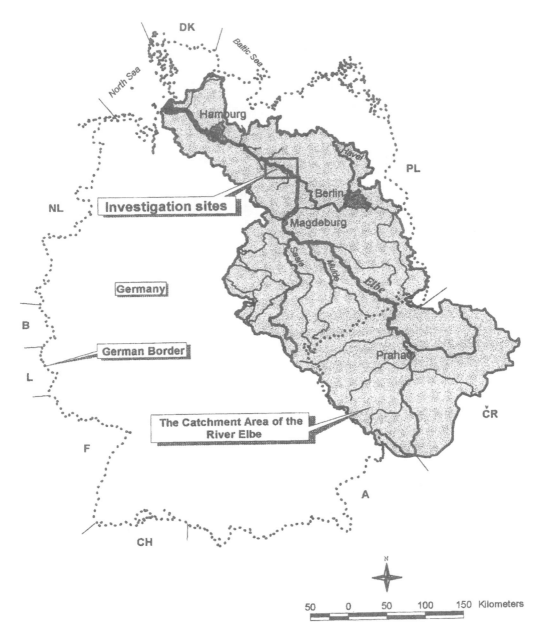

Figure 2 Investigation area at the Elbe near Wittenberge.

II. HYDROLOGICAL REGIME OF THE RIVER ELBE AND ITS FLOODPLAINS

The River Elbe is one of the most important rivers in central Europe. The total length of the River Elbe (from its source in the Czech Republic to its mouth in the North Sea) is 1091 km, and its catchment area is 148,000 km². The research area is situated in the region of stream km 435–485. It is characterized by long-term average annual precipitation of 541 mm and an average temperature of 8.5°C.

To describe the hydrological regime, we used the Elbe gauging station Wittenberge at stream km 454.6. The zero point of the gauging station (PN) is situated at a geodetic elevation of 16.59 m above sea level (NN, Amsterdam). The catchment area (AE) of the River Elbe in Wittenberge is about 123,532 km².

The hydrological regime of the River Elbe in Wittenberge is characterized by the following main hydrological parameters based on the period 1900–1995 (4):

Mean low water discharge: (MNQ) 272 m³/s
Mean discharge: (MQ) 681 m³/s
Mean high water discharge: (MHQ) 1920 m³/s

The long-term mean water level (MWL) of the River Elbe at the gauging station Wittenberge is 3.17 m. This corresponds to a water level around 22 m above sea level in the experimental area.

High flood events occur during winter, spring, and summer. They are fed predominantly by snow melting (winter and spring) and heavy rainfall (spring and summer), and they are closely related to the hydrometeorological conditions in the catchment area (i.e., snow melting in the whole catchment area or only in a subcatchment area, amount of precipitation, etc.). Figure 3 describes the monthly probability of flooding and MWL. The flooding probability is high in April and low in September.

The floodplain area shows a distinct microrelief, as ascertained by geodetic measurements. Several parts of the area (hollows, old arms) are located at geodetic elevations below the MWL. The geodetic relief elevation varies between 1.3 m below

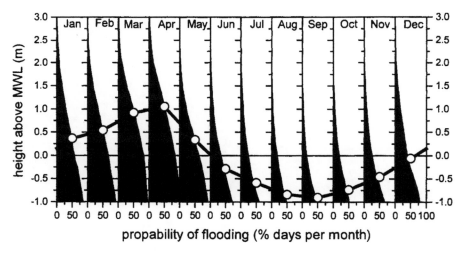

Figure 3 Monthly probability of flooding and mean water level (derived from daily water level measurement at Wittenberge 1899–1997).

and 3.5 m above mean water. Right next to the river bed are levees. These morphological elements have an important influence on the flooding regime. According to the microrelief of the floodplain area, the following different flooding phases can be distinguished for the left-hand (western) part of the experimental area.

1. Flooding starts if the water table in Wittenberge exceeds 4.0 m (0.83 m above MWL), which is equal to a discharge of 1000 m^3/s. During this first phase, water from the River Elbe enters the dead arms. The hydrostatic pressure in the groundwater is increased by the River Elbe, too. In the sandy subsoils, water conductivity is high. If the conductivity in the topsoils allows, the dead arms and hollows are filled by the rising groundwater.
2. If the high flood water level is in the range of 0.9–1.6 m above the MWL, flooding enters the area by passing the mouth of a dead arm. At this point the flood water flows in the opposite direction to the Elbe River into the floodplain. Because of the high levees, the areas near the riverbanks are not flooded.
3. 1.3 m above MWL, the River Elbe exceeds the levee in an upstream position. The high floods of the river stream along the dike. Any remaining hollows and dead arms are filled up. During this hydrological situation, the water enters the floodplain from two main directions.
4. The two main high flood streams are connected if the water level reaches 1.6 m. At this point the whole area is flooded. The rising water level unites both floodings in an upstream direction. The relief elements are flooded depending on their geodetic elevation.

III. SOIL CHARACTERIZATION

The Elbe landscape was molded by the last two ice ages (the Saalian glacial stage until 130,000 years ago and the Weichselian glacial stage until 10,000 years ago). The glaciers of the Saalian glacial stage completely covered today's Elbe Valley with till. The Weichselian glacial stage stopped farther in the north, and the melting water of the Weichselian glacial stage washed a Pleistocene watercourse into the Saalian moraines up to 20 km wide. About 15,000–20,000 years ago, in the middle of the last ice age, the amount of melting water was reduced, and gravel and sand were deposited in the Elbe Valley. At the end of the ice age, the stream of melting water increased again, and the Elbe cut into her own sediments without removing them completely. These old sediments from the Weichselian glacial stage nowadays still form the deeper underground on the left and right banks of the river. Several side arms and channels are also left over from the melting water.

Since the beginning of the Holocene (10,000 years ago), sedimentation in the Elbe floodplains has been determined by the changing water level of the river. Frequent high floods brought sand into the meadow, forming natural levees on the banks. In areas with a lower speed, fine-grained materials were deposited.

The population of the Elbe area grew, and with the beginning of the Middle Ages large forests were cleared. This led to dramatic erosion in the upper part of the river and to thick deposits of fine-grained sediments in the floodplains. This "meadow loam" forms the surface of large areas of today's Elbe meadows. New channels have been formed by the following high floods, so that nowadays we have a complex mixture of sand,

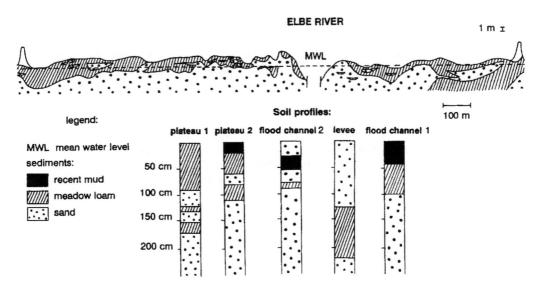

Figure 4 Cross section of the experimental area.

meadow loam, and recent mud deposits. Recent materials can be distinguished from older meadow loam by the higher levels of organic carbon and in particular by organic and inorganic pollution.

Figure 4 shows a cross section of the sediments of today's Elbe floodplain, revealing the great heterogeneity. Three main morphological types (levee, plateau, and flood channel) can be distinguished. On both sides of the river we have a sandy natural levee, which is often the highest natural elevation of the floodplain. The largest parts of the floodplains are flat plateaus formed from layers of meadow sand and meadow loam of variable thickness. Over the last 100 years mud has been deposited in both the flood channels and in small valleys, forming a layer up to 50 cm thick. Below the mud sands and meadow loams follow. As the grain size diagram (Fig. 5) shows, the soil texture within the uppermost 2 m of the soils belongs to the textural classes from sand to silty loam and silty clay. Three types of sediments occur: Holocene and recent sands with a typical mean grain-size diameter of 0.3 mm (medium sand, hereafter referred to as "sand"); Holocene loamy or clayey alluvial deposits with strong variation in grain-size distribution, and a small amount of primarily organic carbon, up to 2% (meadow loam), and recent silty sediments with large amounts of primarily organic matter (recent mud) (Fig. 6).

With the emerging industrial and municipal use of the Elbe as a river for wastewater discharge at the end of the nineteenth century, the sedimentation of the fine-grained material changed from meadow loam to recent mud. As the content of soil organic matter has partly changed since the sedimentation of the material due to humus enrichment or decomposition, the two types of fine-grained deposits cannot always be distinguished by field criteria (color). Nevertheless, as the measurements of heavy metal contents show (see section V), the analytical separation of meadow loam and recent mud is clearly possible.

Contaminants in a River Floodplain

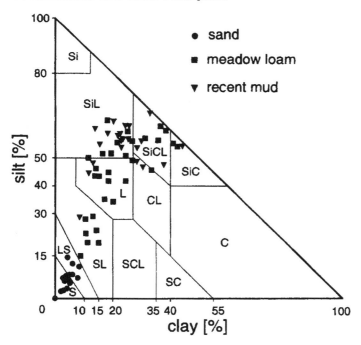

Figure 5 Grain-size diagram of soil samples from recent floodplains.

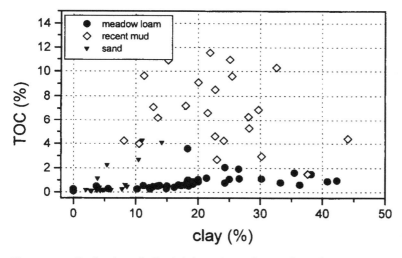

Figure 6 Grain size of alluvial deposits and organic carbon.

To demonstrate the influence of the morphological and the hydrological situation on soil composition and processes, we chose five typical soil profiles (plateau 1, plateau 2, flood channel 1, flood channel 2, levee).

Two profiles (plateaus 1 and 2) are typical of high plateau areas within the floodplain. With a height of 1.8 and 1.2 m above the MWL, the probability of flooding is small (on average 7% and 17% of the year). A flooding probability of more than 10% is restricted to the months January to May. The distance between river and site has caused the sedimentation of fine-grained, brown meadow loam over a long period. There are no hydromorphic features except in the deepest horizons (>1.5 and >0.8 m depth, respectively). In the case of profile plateau 2, a flat channel within the plateau has led to the sedimentation of a thin layer of recent mud.

Two profiles (flood channels 1 and 2) are situated in flood channels which are cut from the river by small anthropogenic walls or levees and lie 0.4 m below and 0.7 m above the MWL, respectively. These profiles are characterized by layers of recent mud, which are raised on meadow loam in case of low flood water velocities (profile flood channel 1) in 0.4-m thickness or which are integrated as small layers in a profile built up of recent sands (flood channel 2). These profiles represent the most flooded sites in the floodplain (on average, 30–55% of the year) and show strong gleyzation except in the topsoil horizon of flood channel 2.

One profile (levee) lies 2.5 m above the MWL and is rarely flooded (2% of the year). This profile is typical of natural levees and the upper part is composed of recent sands. The brown color of the sand indicates soil development (weathering at least between 20 and 60 cm) within the profile.

Figure 7 shows the soils' main properties. The topsoil which is most fine-grained (83% fines < 20 µm) and has the highest humus level (10.3% C of soil organic matter, SOM) occurs in profile flood channel 1, demonstrating the influence of recent mud sedimentation. When recent mud and sands are mixed in thin layers, the properties of the horizons depend on the mixing percentage of both materials.

For the example of profile flood channel 2, sands predominate, and the proportion of both fines and SOM are relatively low.

The most dominant profile type within the recent floodplain, profile plateau 1, consists of a homogenous layer of meadow loam with 35% fines (< 20 µm). The depth distribution of SOM shows that humus accumulation is restricted to the first horizon.

The properties of profile plateau 2 demonstrate a particular morphological situation. Although the probability of flooding is relatively low, the position in a small channel-shaped hollow without an outlet near the River Elbe has led to the recent sedimentation of sandy mud. Due to the mud input and the long duration of topsoil wetness, the percentage of SOM is high (7.1% C).

In the depth distribution of the fine-grain share of the profile levee, a value of 7% of the grain size fraction < 20 µm occurs in a layer of parent material sand (20–60 cm depth). Together with the brown color of this horizon, this is interpreted as in-situ weathering. High primary production of the riverine woodland and intensive bioturbation leads to a topsoil of medium humus enrichment and low bulk density.

The differences in pH values between the profiles are low. For profiles without recent sedimentation and with in-situ soil development (levee, plateau 1), pH values ($CaCl_2$) of 4.8–5.2 in the topsoil are typical. For the other profiles the pH value in the topsoil varies between 5.5 and 6.0. The cation exchange capacity (CEC) corresponds closely to the grain-size distribution and the percentage of SOM. After a multiple regression procedure with 39 samples ($r = 0.929$), the CEC of eight missing values was calculated using the formula:

$$\text{CEC (mmol eq/kg)} = -14.5 + 7.276 * \text{clay content (\%)} + 12.798 * \text{SOM(\%)} \quad (1)$$

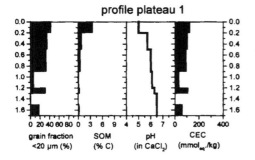
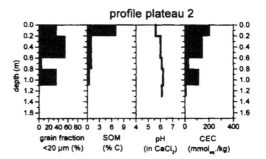
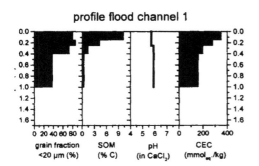
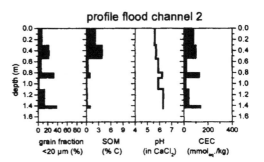
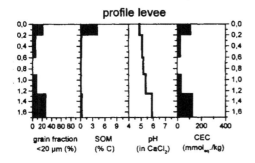

Figure 7 Depth distribution of soil properties in selected profiles (CEC partly calculated, see Eq. (1)).

The CEC is lowest within sandy horizons without SOM and highest within horizons of recent mud.

IV. SOIL HYDROLOGICAL REGIME

The soil hydrology in the recent floodplain is characterized by the influence of flooding. As an example, Fig. 8 shows the results of a summer flood (high water level 1.47 m above MWL) on water content and the redox potential in two topsoil layers at positions with different heights. Whereas the plateau (1.8 m above MWL) is not reached by the

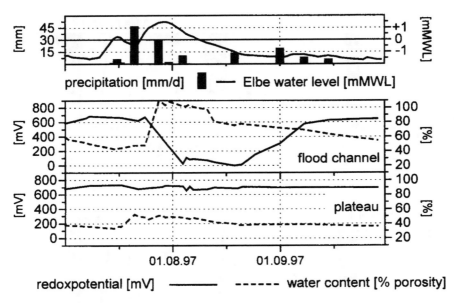

Figure 8 Precipitation, Elbe water level, water saturation, and redox potential for two topsoils during a summer high water (flood channel: +0.3 m MWL; plateau: +1.8 m MWL).

high-water wave, a flood channel profile situated just 0.3 m above MWL is inundated by the second peak of the high-water. Consequently, the water content reaches 100% of the pore volume and the redox potential drops sharply to about 0 mV. The topsoil remains waterlogged until the level of the Elbe water drops below MWL. Afterwards a slow decrease in the water content due to evapotranspiration occurs combined with an increase of the redox potential. The water content of the plateau site is affected only by precipitation. Saturation is always below 60% of pore volume, and thus the redox potential remains high.

Figure 9 shows the results of soil hydrological investigations throughout the depth of a typical plateau profile (plateau 1). During two periods, summer 1997 and winter to early summer 1998, the soil moisture levels depending on precipitation were recorded at three different soil depths by using TDR probes (time-domain reflectrometry).

The predominant soil textures of plateau 1 are sandy loam at a depth of 0–90 cm, and a mixture of loamy and sandy layers in the range 90–120 cm.

As expected, the depths of 30 cm and 60 cm show remarkable dynamics in terms of soil moisture content owing to precipitation events and the water demands of growing plants in spring and summer. At 30 cm in particular, the soil moisture content was decreased by the water consumption of the vegetation cover. However, as a consequence of heavy rainfalls, the soil moisture content at 60 cm was more strongly affected compared to the 30-cm level during the vegetation period. Possible reasons for this could be (a) preferential water flow taking place via macropores and (b) the exhausted water storage capacity in the top soil. Rainwater was required to refill the soil water storage capacity. However, when more water is available, plant water consumption increases, too.

In the period February–May 1998, the soil moisture content at the depths of 30 cm and 60 cm reached or exceeded field capacity (34.6 vol%). Seepage water in the lower soil layers (90–120 cm) can occur only in this period. A seepage water volume of 45 mm

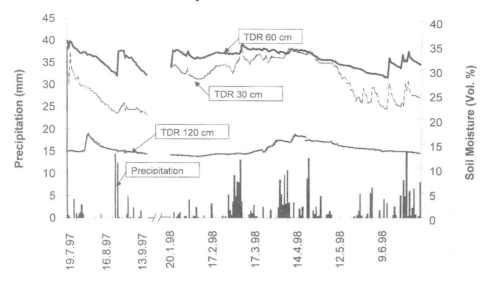

Figure 9 Course of soil moisture at various depths depending on precipitation in the experimental area.

was estimated based on the TDR measurements for this period. Therefore, the transport of harmful substances with the seepage water from the topsoil into the subsoil horizons cannot be ruled out.

The soil depth of 120 cm, which consists of sandy material, displays a different behavior. The wetness of the soil was not directly affected by rainfall events. A distinct interaction between the soil water content and the water level of the river can be observed during high flood periods. The summer high flood event of 1997 and the period of higher water levels in spring 1998 were accompanied by a significantly increased soil moisture content in this layer. It seems that the water of the River Elbe is able to migrate very rapidly into, or causes backwater in the sandy soil layer of 90–120 cm, which is characterized by relatively high Darcy permeability. The soil moisture content of the loamy soil layers above were not influenced by these processes.

V. SOIL POLLUTION

Floodplain soils are sinks for polluted sediments. The most harmful substances enter the meadows as suspended particulate matter (SPM). Elbe river soils are highly polluted by inorganic and organic contaminants (5–10). Miehlich (1983) (6) gave a description of heavy-metal contamination profiles of typical Elbe River floodplain soil. Meissner (8) and Krueger (5) investigated topsoils in the middle course of the River Elbe and worked out correlation coefficients between heavy metals and organic carbon. Neumeister and Villwock (11) described correlations between the MWL and soil contamination in the floodplain of an important Elbe tributary.

This part contains a survey of the range of heavy metal pollution of typical Elbe floodplain soils, describing the input via high flood, a square-grid geochemical characterization of the topsoils, and the depth distribution. It is augmented by investigations into the depth distributions of some organic pollutants.

When the floodplain is flooded, SPM can settle on the soil surface. This process causes the contamination of suspended solids to become part of the top layer of the soils. The pollutant input of individual soils depends on the contamination of the SPM and the intensity of sedimentation as a function of stream velocity, roughness of the vegetation, and other factors. To investigate the recent input of particulate matter and pollutants during flooding, two methods were applied: the investigation of SPM in the flooding water and the geochemical characterization of recent mud deposits.

A. Input of Suspended Particulate Matter

1. Materials and Methods

During the flood in Spring 1997, suspended particulate matter was investigated at five locations in the floodplain (sites *a–e*, Fig. 10). Three sites are situated directly on the riverbank (*a–c*) and two sites at the bottom of the dike (*d, e*). For one sample 2 L of flood water were pressure-filtered (3 bar). Cellulose nitrate filters with a pore size of 0.45 μm were used. The filters were dried for 2 h at 105°C and digested with HNO_3 and H_2O_2 in the microwave field (Paar, Austria).

Inductively coupled plasma mass spectrometry (ICP-MS, Elan 5000, Perkin-Elmer/Sciex) was used to analyze the elements As, Cd, Co, Cr, Cu, Li, Mn, Mo, Ni, Pb, Rb, Sb, Sn, Sr, Ti, U, and Zn.

2. Results and Discussion

The concentrations of SPM during the spring flood in 1997 (Fig. 11) show that the highest concentrations of SPM occur before the high flood vertex. These results are especially clear at the riverbank positions *a* and *b* and comparable with investigations by Wilken (12) and

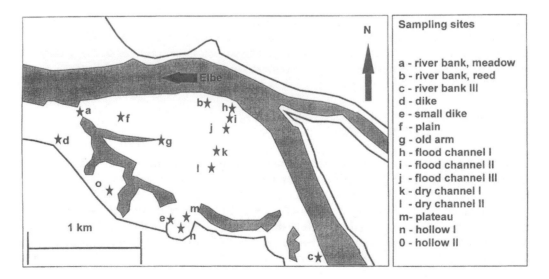

Figure 10 Sampling sites for suspended particulate matter determination and collection of recent mud.

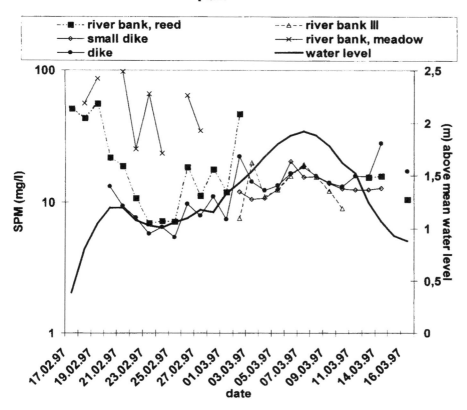

Figure 11 Concentration of suspended particulate matter during the spring flood in 1997.

Spott (13) in the River Elbe. The concentrations at the riverbank are higher than at the dike.

Figure 11 also shows the start of flooding at the different sampling positions. Positions *c* and *e* were flooded at water levels of more than 1.33 m above the MWL. Rising concentrations of SPM were determined at the dike during the high flood vertex owing to the remobilization of sediments in the dead arms of the River Elbe, which occurs only at higher drainage situations. This revealed that during a high flood period a dead arm could be a sink, and at higher discharges a source of SPM.

The metal concentrations transported of all elements studied depend with only few exceptions on the SPM concentration. By way of example, Fig. 12 describes uranium concentrations.

The change of metal contents of SPM during floods (or in a high-water wave) provides information about the source of suspended matter pollution. Figure 13 shows the uranium content (or specific loading) of SPM during the spring flood in 1997 as an example, for a first group of elements (As, Cd, Co, Li, Pb, Rb, Ti, and U).

During the spring flood the concentrations of the elements As, Cd, Co, Li, Pb, Rb, Ti, and U showed no significant influence of the various sampling sites. A plateau of element concentrations was found until a few days before the high flood vertex. Afterwards the specific element levels decreased regularly. Only at the dike (*d*) did some element levels increase at the end of the flood. This phenomenon cannot be explained at present.

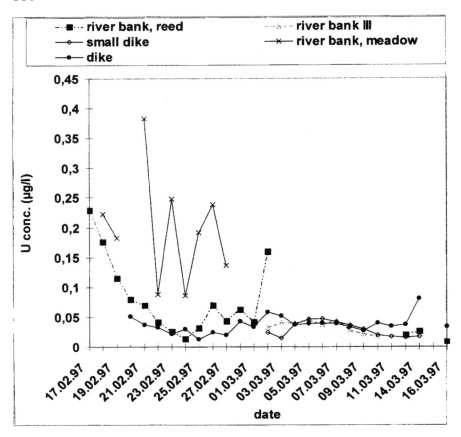

Figure 12 Uranium concentration of suspended particulate matter during the spring flood in 1997.

The second group of elements includes Cu, Mn, Mo, Sb, Sn, Sr, and Zn. They are characterized by sharply varying contents during the high flood period (Fig. 14).

Before the high flood vertex, the content of these elements in the SPM partly increased, especially at the dike (*d*). One reason for this could be grain-size fractionation by the flowing water. However, grain-size fractionation affects the elements of group 1 too. Because of the increased levels of Mn and Mo, which are often linked with organic material (14), a higher content of organic material in the SPM was assumed. This may be due to remobilization of old sediments with a high amount of organic material. Molybdenum in Fig. 14 shows as an example for the elements of the second group that a dead arm may be not only a sink for harmful substances but also a source.

During the summer flood in 1997, the metal content was lower than during the spring flood. The spring flood had already eroded a large proportion of the contaminated sediment. In addition, the summer flood had several small peaks before the investigation area was flooded. A great amount of the polluted SPM thus evaded the investigation sites. Consequently, a rapidly draining flood is a more important source of polluted SPM for the floodplains than a slow-draining flood with several lower high flood verticles.

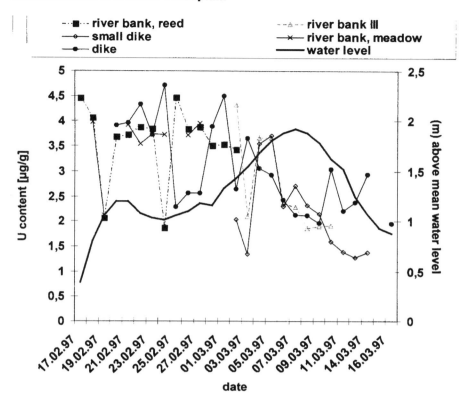

Figure 13 Uranium content of suspended particulate matter during the spring flood in 1997.

Although it is not possible to calculate the relation between the metal content of SPM and recent mud, the range of levels can still be compared. Table 1 gives a survey of the concentrations measured in Spring 1997.

B. Deposited Sediments—Recent Muds

1. Materials and Methods

The investigation of SPM provides information about the variability of heavy-metal inputs during flood periods. However, it is not possible to characterize the sedimentation processes from these results. Therefore sediment traps were installed to study the sedimentation process directly. Pieces of synthetic lawn (30 × 40 cm) were exposed in the recent floodplain. The synthetic lawn material has bristles nearly 3 cm long to simulate the roughness of wetlands surfaces (15) and trap the high flood sediments. At the end of the flooding period the synthetic lawns were collected. The sediments were rinsed out using tap water.

The water was collected in open bins, the sediment was allowed to settle for 24 h, and finally the water was decanted. The samples were oven-dried (24 h, 105°C), weighed, and used for elemental analysis with ICP-MS after aqua regia digestion in the microwave field.

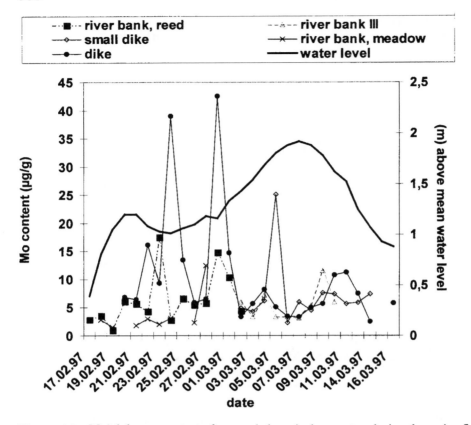

Figure 14 Molybdenum content of suspended particulate matter during the spring flood in 1997.

2. Results and Discussion

In the recent floodplain, 12 sites of varying height and vegetation (Fig. 10, Table 2) were each fitted with nine sediment traps. Five traps were collected after the flood; four traps were left in the field to collect information about the influence of the next high flood on the sedimentation or erosion processes.

The following data were collected in Spring 1997, when the Elbe water level achieved a maximum height of nearly 2 m above the MWL. Figure 15 shows the development of the water level and the times of sediment collection at the riverbank.

The synthetic lawns were first collected in mid-March (a/I; b/I; f/I). The other sites were still submerged. After saving the fresh muds, the pieces of lawn were replaced in their original position. In the second half of March they were flooded again. The sediments were collected for the last time in April (a/II; b/II; f/II). All other sediments were collected gradually as the water level sank.

The highest mass inputs occur with the highest discharge. Except at the site a, both flooding periods resulted in the same mass input. The mass input at sites b and f is much smaller during the second flooding period (Fig. 16). This is caused by the different roughness of landscape surfaces and by the larger distance between river and the sampling sites b and f. The transport capacities of river water decrease depending on the sinking water discharges.

Table 1 Range of Specific Element Concentrations of Suspended Particulate Matter in Floodplain Areas (Spring 1997)

	Min (µg/g)	Max (µg/g)
Ag	1.3	18.4
As	26	129
Cd	1.5	8.6
Co	7.2	42.1
Cr	61	522
Cu	62	1,184
Li	12	71
Mn	1,184	14,819
Mo	1.0	42.5
Ni	38	370
Pb	58	303
Rb	19	92
Sb	0.06	1.43
Sc	4.1	9.0
Sn	1.4	17.8
Sr	61	278
Ti	240	1,323
U	1.3	4.7
Zn	402	2,988

Table 2 Information About Sites of High Flood Sedimentation

Site		Distance between site and river (m)	Height (m) above mean water level	Vegetation
River bank, meadow	a	25	0.4	Meadow
River bank, reed	b	45	0.4	Reed
Plain	f	135	0.5	Meadow
Old arm	g	315	−0.3	Reed
Flood channel I	h	180	0.5	Reed
Flood channel II	i	173	0.1	Reed
Flood channel III	j	60	0.3	Reed
Dry channel I	k	334	0.8	Meadow
Dry channel II	l	518	0.9	Meadow
Plateau	m	905	1.1	Meadow
Hollow I	n	944	0.7	Reed
Hollow II	o	748	−0.3	Meadow

The input of dry matter varies over a wide range—from 40,000 kg/ha at the riverbank to 40 kg/ha in areas with high-velocity flows over a large stretch of the river. The average dry matter input (median) was about 2200 kg/ha (Table 3). The most important factor concerning the dry matter input is the distance between site and river (Fig. 18).

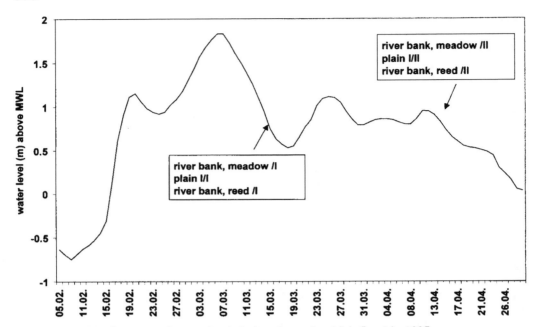

Figure 15 Development of water level during the spring high flood in 1997.

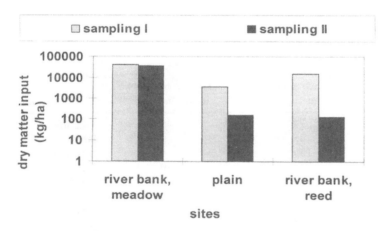

Figure 16 Mass inputs at different sampling locations from two flooding periods in the spring of 1997.

Because of the large range of mass input, the differences of heavy-metal concentrations in recent muds is less important concerning the heavy metal input. The heights above the MWL do not play such an important role (Fig. 19).

From the results of site a, it is apparent that element concentrations in fresh deposits from the first high flood period are much higher than in the following flooding periods (Fig. 17).

The concentrations of some elements in relation to the loss on ignition are not typical, according to investigations by Meissner et al. (8). The organic carbon contents

Table 3 Input and Concentrations of Recent High Flood Sediments in Spring 1997

	Input minimum (kg/ha)	Input maximum (kg/ha)	Input median (kg/ha)	Conc. minimum (mg/kg)	Conc. maximum (mg/kg)	Conc. median (mg/kg)
Dry matter	46.6	38,104	2167	—	—	—
Loss on ignition	—	—	—	124	600	243
Ag	0.0001	0.058	0.006	1.5	3.6	2.6
As	0.0015	1.006	0.103	26	107	41
Cd	0.0003	0.158	0.0123	4.1	10	6.6
Co	0.0004	0.561	0.0487	9.1	43	19
Cr	0.0035	2.8	0.2512	35	134	106
Cu	0.0052	2.9	0.2415	69	146	112
Li	0.0014	1.008	0.0782	7.2	50	29
Mn	0.0484	39	4.3097	534	5,542	1,189
Mo	0.0002	0.085	0.0071	2.2	5.7	3.1
Ni	0.0022	1.6	0.1336	42	75	53
Pb	0.0196	2.8	0.3002	61	188	127
Rb	0.0026	1.5	0.1429	14	83	56
Sc	0.0003	0.194	0.0149	1.9	9.4	5.9
Sn	0.0010	0.372	0.0305	3.8	21	14
Sr	0.0054	3.2	0.2624	81	264	111
Ti	0.0414	30	2.17	276	1,233	921
U	0.0001	0.086	0.006	1.3	4	2.8
Zn	0.2044	26	2.0129	651	1,981	1,150

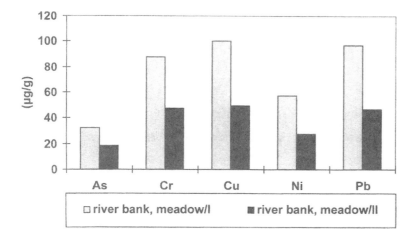

Figure 17 Element concentrations from two flooding periods in the spring of 1997.

(as loss on ignition) of the investigated deposits in this study are much higher. This is interpreted as the in-situ growth of algae on the sediment traps. If the input of mineral deposits on the traps is low, the algae might assume a high percentage of mass.

The comparison of element concentrations in recent muds with their concentrations in SPM show that the range is nearly the same for many elements: As, Cd, Co, Li,

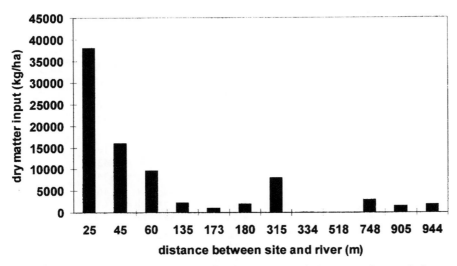

Figure 18 Dry matter input depending on the distance between sites and river.

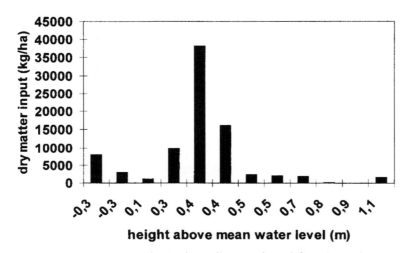

Figure 19 Dry matter input depending on the heights above the mean water level.

Rb, Sn, Sr, Ti, U, and Zn. Cr, Cu, Mn, Mo, Ni, and Pb concentrations sometimes reach much higher values in SPM. Only Sb concentrations are higher in recent deposits.

In any case, the high concentrations of cadmium, zinc, lead, and copper in fresh muds (Table 3) show that the enrichment of heavy metals in recent Elbe River floodplains will continue, despite continued improvement in the purification of sewage and industrial effluents.

C. Spatial Distribution of Heavy Metals

The description of the spatial distribution of heavy metals is based on the geochemical mapping of a 9-ha large section of the Elbe floodplain near Wittenberge (16). The nearly flat plain borders on the southeastern shore of the River Elbe. The remains of an eighteenth-century dam lie in the investigated area, rising about 1.5 m above the plain. Generally, the height differences in this certain area are in the range of 50 cm.

A regular square grid with intervals set at 30 m defines the position of the 100 sample sites. The uppermost 10 cm of the topsoil represent the maximum sample depth. For reasons of comparability, geochemical analysis is limited to the fine fraction of the samples (< 63 μm; silt and clay fraction). Soil profile description, groundwater level, and altitude specify the basic sample properties. The soil composition is dominated by loamy or silty materials in the topsoil. The grain-size distribution was determined for an exemplary number of specimens. According to the results, the soil composition varies over a wide range (sand: 11–55 wt%). The fine fraction in itself is rather homogeneous. Silt contributes about 80% and clay 20% to the fraction smaller than 63 μm. Recent mud builds up the topsoil in the lower regions of the floodplain, while the higher parts are dominated by meadow loam. More details on the general soil parameters are contained in Section III.

1. Materials and Methods

The trace elements considered in this study are cobalt, chromium, copper, nickel, lead, and zinc. The geochemical evaluation for the distribution of heavy metals in this study includes, furthermore, iron, manganese, aluminum, and potassium, together with total organic carbon (SOM) and sulfur as possible factors influencing the distribution of trace elements in the fine fraction. The contents of these components allow the influence of pedogene oxides (Fe, Mn, Al), clay minerals (K, Al), and organic substances (SOM) on the enrichment processes of heavy metals to be assessed. Except for carbon and sulfur, element concentrations were measured by ICP-OES after sample dissolution in aqua regia. O_2-pyrolysis of the samples and the IR detection of CO_2 and SO_2 (Leco, USA) allowed the measurement of carbon and sulfur. Table 4 lists the element concentrations of the analyzed alluvial soil fraction.

Table 4 Statistical Data and Local Background Values for pH, K, Al, Fe, Mn, C, S, and Heavy Metals in the Alluvial Soils of the Elbe Floodplain Near Wittenberge (< 63 μm)

		Mean	Range	Local background (< 63 μm; aqua regia)	Elbe background (< 20 μm; HF extract) (17)
pH		5.5	4.8– 6.1	—	—
C	(wt%)	7.4	2.6– 11.9	2.6	—
S	(wt%)	0.16	0.06– 0.27	0.06	—
K	(μg/g)	2,510	1,189– 3,885	3,229	—
Al	(μg/g)	19,319	13,341–25,440	19,572	88,600
Fe	(μg/g)	29,537	19,631–38,825	27,329	47,600
Mn	(μg/g)	667	261– 1,157	697	850
Co	(μg/g)	15	7– 23	11	19
Cr	(μg/g)	115	37– 205	37	117
Cu	(μg/g)	136	19– 260	19	32
Ni	(μg/g)	44	26– 64	27	53
Pb	(μg/g)	196	31– 360	31	29
Zn	(μg/g)	595	210– 945	112	150

2. Results and Discussion

To verify heavy-metal contamination by the flood muds, the higher parts of the old dam served as a source for element background concentrations. The surrounding flats in the preindustrialized area supplied the construction material for the dam. As the floods of the River Elbe rarely reach the old dam, its influence on the river's heavy-metal content is low. Comparison of the element levels determined with literature data (Table 4) (17) supports the hypothesis that they resemble background concentrations valid for the River Elbe, although the data presented refer to different sample fractions. Calculations based on the background concentrations from Wittenberge show that the silt and clay fraction of the floodplain chiefly accumulated copper, lead, and zinc. The mean enrichment factors for these elements range from 5 to 7 [Eq. (2)].

$$\text{Enrichment factor } (x) = \frac{\text{element concentration X of the plain}}{\text{element concentration X of the dam}} \quad (2)$$

where X = specific element. Lead has a maximum factor of 16 (495 µg/g Pb), which confirms (a) its character as an immobile element and (b) the long-term and/or intense input of lead. On the other hand, the silt and clay fraction also accumulated zinc, which is a relatively mobile trace element (max. enrichment factor 10; 945 µg/g Zn). Cobalt and nickel show the lowest enrichment factors (max. 2).

The geochemical maps (Figs. 20 and 21) show that the investigated area is usually characterized by three zones of high and two zones of low element concentration. Comparison of the individual geochemical maps and the topographic map provides initial clues on the enrichment processes and distribution of heavy metals in the soil fraction examined.

The correlation (Table 5) between topographic height and contents of heavy metals indicates altitude to be one of the primary factors affecting the heavy-metal content in the topsoil. Correlation analysis (Bravais-Pearson type; r = coefficient of correlation; Table 5) confirms this factor for some of the elements.

Table 5 Correlation Matrix of the Elbe Floodplain Topsoil Samples (< 63 µm) (Bravais-Pearson correlation)[a]

	Height	pH	C	S	Al	K	Fe	Mn	Co	Cr	Cu	Ni	Pb	Zn
Height		−.34	−.07	−.18	−.36	.03	−.46	−.52	−.64	−.40	−.43	−.55	−.58	−.45
pH	.27		−.05	.03	.18	.32	.33	.11	.18	.00	.04	.24	.27	.18
C	−.01	−.24		.91	.16	.34	.43	.45	.52	.73	.70	.57	.30	.58
S	−.44	−.42	.82		.34	.39	.52	.51	.59	.70	.70	.64	.54	.58
Al	−.14	−.03	.06	.05		.42	.64	.34	.42	.37	.41	.60	.60	.39
K	−.01	.01	.18	.19	.54		.32	.15	.20	.31	.32	.38	.26	.30
Fe	−.07	.05	.08	.03	.63	−.07		.59	.83	.72	.77	.82	.60	.80
Mn	.60	.14	−.11	−.41	.23	−.26	.51		.87	.57	.61	.66	.68	.55
Co	.49	.06	.12	−.17	.21	−.32	.60	.88		.81	.84	.84	.70	.80
Cr	−.23	−.31	.77	.76	.16	−.06	.30	−.10	.21		.98	.89	.40	.95
Cu	−32	−26	.77	.79	.15	.04	.25	−.17	.15	.97		.91	.49	.96
Ni	−.10	−.04	.70	.65	.12	−.05	.37	.05	.37	.88	.86		.65	.91
Pb	−.34	−.36	.73	.79	.17	.15	.22	−.22	.04	.78	.79	.68		.39
Zn	.13	.13	.67	.49	.22	.11	.32	.11	.43	.83	.82	.93	.56	

[a] The lower triangle represents the coefficients for the area lower than 0.7 m above MWL ($n = 52$). The upper triangle represents the coefficients for the area higher than 0.7 m above MWL ($n = 46$).

Contaminants in a River Floodplain

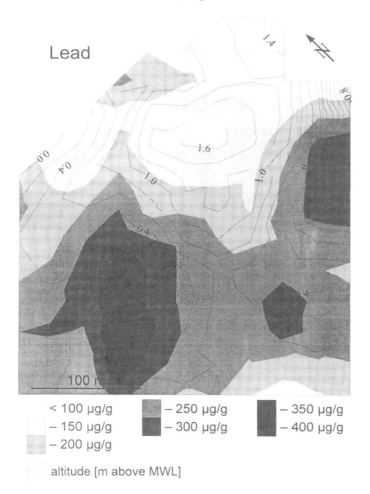

Figure 20 Spatial distribution of lead in the alluvial soils (< 63 μm) and topography of the investigation area (16).

In general, it can be stated that the lower the altitude, the higher is the concentration of a specific element; lead, for example shows a steady decline of contents with an increase in altitude (Fig. 22). On the other hand, for some elements the correlation between concentration and altitude applies only above a certain height (approximately 0.7 m above MWL). The scatter diagram for zinc allows this behavior to be verified (Fig. 23). The correlations between height and element concentrations of these samples are more scattered below the 0.7 m above MWL level than those above the limit. Flooding events and the fluctuation of the groundwater level expose the lower parts of the Elbe floodplain to more frequent changes in the soil properties, e.g., redox potential, than the higher regions. Therefore, chemical equilibria are more often disturbed in the soil profiles which are close to the groundwater.

As can be seen from the very low correlation coefficients for soil acidity (Table 5), the pH can be excluded from the potential controlling factors for the distribution of the trace elements in the Elbe floodplain.

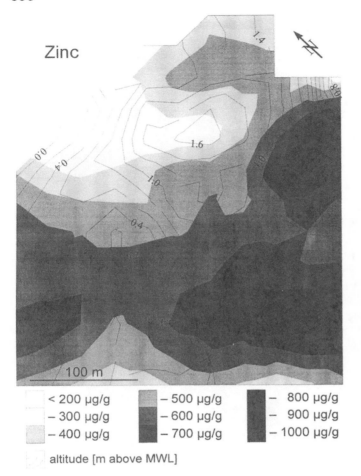

Figure 21 Spatial distribution of zinc in the alluvial soils (< 63 μm) and topography of the investigation area (16).

The control of clay-mineral proportions on the trace-element concentrations in the alluvial soils cannot be derived from the collected data. Apart from the correlation between potassium and aluminum, which can be regarded as an expression of the presence of clay minerals, there are no correlations between these elements and other elements.

Besides topography, correlation analysis indicates additional controlling factors by the trace-element levels in the examined soil fraction. However, the multitudes of possible additional enrichment controls overlap with others. Therefore, multivariate statistical methods such as factor analysis (19) and cluster analysis (18) were used to evaluate the data collected.

The altitude seems to have a significant influence on the trace-element content. As shown before, it is reasonable to split the investigation area by topographical height into areas below and above 0.7 m above MWL. The topographical height can better be described as the distance from the middle groundwater level. Splitting the data yielded more definite results. Each factor analysis retrieved two factors. The total variance explained reaches 76%.

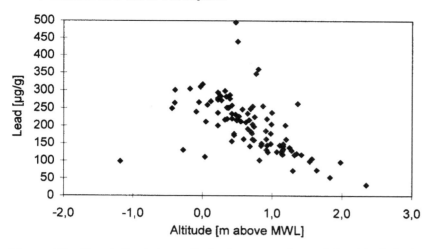

Figure 22 Correlation between the lead concentration and the altitude in the samples (< 63 μm) (16).

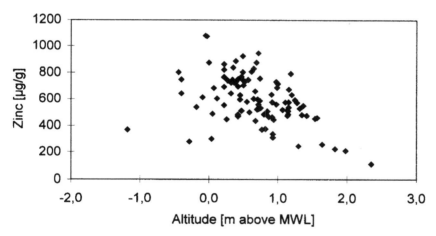

Figure 23 Correlation between the zinc concentration and the altitude in the samples (< 63 μm) (16).

Factor analysis for the area lower than 0.7 m above MWL produced a simply structured matrix with high loadings on the single factors (Table 6). The variables considered can be divided into two groups: (a) trace elements (Cu, Cr, Ni, Pb, Zn) together with carbon and sulfur which load high on factor 1; and (b) pedogene oxides (Fe, Mn) and Co which load relatively high on factor 2. The dendrogram of cluster analysis (Fig. 24) shows a comparable pattern. Again there is one group with the trace elements, C and S, and another group with the pedogene oxides. Both groups are related to a relatively low correlation coefficient ($r = 0.4$).

The results of the area higher than 0.7 m above MWL clearly differ from those of the lower area. In contrast to the lower parts, the differences between the loadings of each element on the two factors are much smaller (Table 7). In the case of nickel, they are almost identical. This means that nickel is indifferent to both extracted factors.

Table 6 Rotated Factor Matrix for the Area Lower than 0.7 m above MWL of the Elbe Floodplain Near Wittenberge ($n = 52$; R-mode factor analysis; Kaiser's varimax rotation; explained total variance: 75% [factor 1: 51%; factor 2: 24%])

	Factor	
	1	2
Cu	.973	−.034
Cr	.969	.048
Ni	.908	.230
S	.834	−.363
Pb	.826	−.139
Zn	.816	.329
C	.816	−.022
Co	.183	.970
Mn	−.145	.950
Height	−.238	.536
Fe	.265	.487

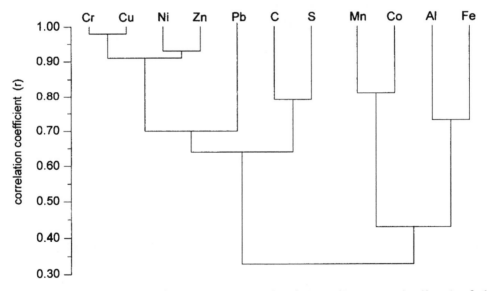

Figure 24 Dendrogram of the cluster analysis of the soil samples (< 63 μm) of the Elbe floodplain (16).

Factor 1 includes almost the same elements as the preceding analysis. This indicates the dominating control of the bonding forms of heavy metals with organic carbon and sulfur (adsorption, sulfides). Factor 2 obviously represents the pedogene oxides in both factor analysis with the difference of the inclusion of the height to factor 2 of the upper area. In both areas, factor analysis distinguishes the groups: (a) heavy metals and (b) pedogene oxides.

Table 7 Rotated Factor Matrix for the Area Higher than 0.7 m Above MWL of the Elbe Floodplain near Wittenberge ($n = 46$; R-mode factor analysis; Kaiser's varimax rotation; explained total variance: 76% [factor 1: 40%; factor 2: 36%])

	Factor	
	1	2
C	.887	.057
Cr	.849	.450
Cu	.814	.522
S	.789	.236
Zn	.727	.532
Co	.503	.841
Height	−.036	−.748
Pb	.225	.710
Mn	.360	.696
Ni	.646	.690
Fe	.500	.669

In the area higher than 23 m, lead shifted from factor 1 (trace elements) to factor 2. Factor 2 now represents lead together with aluminum, iron, manganese, and the altitude. Hence it may be concluded that lead exists in different speciations in the discriminated areas. Organic matter and reduced sulfur could bind lead in the lower parts of the floodplain, whereas it forms oxides or sulfate in the soils of the upper area.

The high loadings on factor 1 at both height levels and the dendrogram of cluster analysis indicate that the trace elements apart from cobalt and lead (≥ 0.7 m above MWL) are associated mainly with carbon and sulfur. A distinction between trace elements bound to either carbon or sulfur is not possible because of the high correlation between these elements in the soil (r C–S = 0.79). Lead in particular may exist as sulfates or sulfides depending on the redox potential, as well as bound to organic material. Similarly, other studies in the Elbe floodplain confirm the high correlation of organic carbon and heavy metals (5,8). The association of the trace elements and the soil organic matter is understood to be a result of the anthropogenic input of the contaminants into the soil (20).

In the Elbe floodplain near Wittenberge, heavy-element distribution can be summarized as being caused by repeated flooding together with the fluctuation of the groundwater level and the levels of organic matter (SOM) in the soils. Lead exists in different speciations depending on the topographical height. In the investigated soil fraction, the pedogene oxides are of little influence on the enrichment of trace elements.

D. Distribution of Heavy Metals by Depth

The heavy-metal contamination of River Elbe floodplain soils varies over a wide range, as do the other soil parameters. The heavy-metal concentrations in depth profiles depend on the composition of the deposited sediments. It is possible to identify the soil material analytically. While the native heavy-metal concentrations in sands and meadow loams correlate with the < 20-μm share of the grain-size fraction, the recent muds show correlations with the amount of primarily organic matter (Figs. 25 and 26).

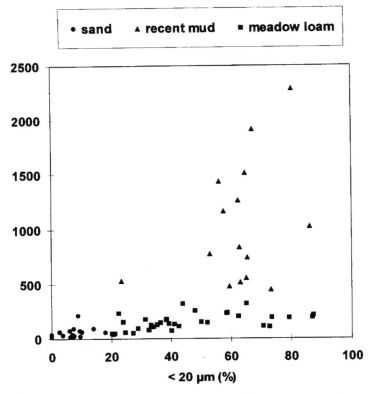

Figure 25 Zinc concentrations in sands, meadow loam, and recent muds.

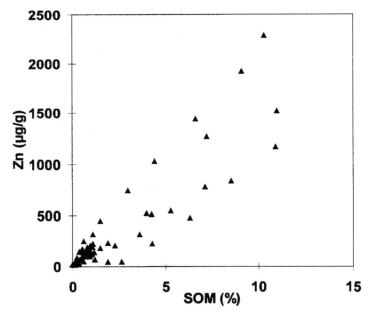

Figure 26 Zinc concentrations of topsoils and recent muds in correlation to the organic carbon content.

Table 8 Equations to Determine Background Concentrations According to the <20-μm Grain-Size Fraction (Y=background concentration; X=share of <20-μm fraction)

Cr	$Y(\mu g/g) = 0.9301 * X(\%) + 11.143$	$R^2 = 0.9516$
Cu	$Y(\mu g/g) = 0.282 * X(\%) + 6.7369$	$R^2 = 0.8078$
Ni	$Y(\mu g/g) = 0.4503 * X(\%) + 5.8836$	$R^2 = 0.9474$
Zn	$Y(\mu g/g) = 1.2817 * X(\%) + 24.256$	$R^2 = 0.7989$
Pb	$Y(\mu g/g) = 0.236 * X(\%) + 12.763$	$R^2 = 0.4052$
As	$Y(\mu g/g) = 0.162 * X(\%) + 5.1172$	$R^2 = 0.4038$
Cd	$Y(\mu g/g) = 0.0022 * X(\%) + 0.2257$	$R^2 = 0.1062$
Hg	$Y(\mu g/g) = 0.0018 * X(\%) + 0.0842$	$R^2 = 0.1791$

1. Materials and Methods

Because of the high correlation of heavy-metal concentrations to the <20-μm share of grain-size composition, it is possible to calculate background concentrations by linear regressions with the help of data from inner dike soil depth profiles. The equations in Table 8 can be used to calculate background concentrations in River Elbe floodplain soils for Cr, Cu, Ni, and Zn. Chromium, copper, nickel, zinc, and lead concentrations are totals (X-ray fluorescence), while arsenic, cadmium, and mercury concentrations were measured by atomic absorption spectrometry after aqua regia digestion.

2. Results and Discussion

To show the influence of recent and historical sedimentation on heavy-metal pollution, three typical Elbe floodplain soil depth profiles were chosen. Figures 27–32 show the depth profiles of some heavy metals in levee, plateau 1, and flood channel 1 and their anthropogenic enrichments.

Plateau 1 is a typical profile of meadow loam. Because of its height of 1.8 m above the MWL, the input of heavy metals with recent muds is small—for instance, the high flood in Spring 1997 did not flood this site. Higher contamination can just be found in the topsoil. The anthropogenic enrichment of heavy metals varies among the elements. Whereas chromium, nickel, copper, and cadmium enrichments are more or less limited to the topsoils, the enrichments of zinc, lead, arsenic, and mercury in particular reach depths of 50 cm or more.

Flood channel 1, the site at the lowest altitude (0.4 m below MWL), is mostly polluted because of the highest input of recent muds. Because of the different enrichment of heavy metals it is possible to categorize the muds into older and recent muds. Anthropogenic enrichment concerns the whole profile, especially for cadmium, zinc, lead, arsenic, mercury, and to some extent copper enrichments characterize the older mud up to a depth of 40 cm. Merely the topsoil, consisting of recent mud, is also characterized by anthropogenic enrichments of chromium and nickel.

The profile levee, which is 2.5 m above the MWL, has the lowest heavy-metal concentrations. The different concentrations of elements characterize the different soil parent materials. The sandy part of the profile is characterized by the lowest concentrations above the part of the profile consisting of meadow loam with higher natural contents. The heavy-metal concentrations show the differences of horizons with more and with less

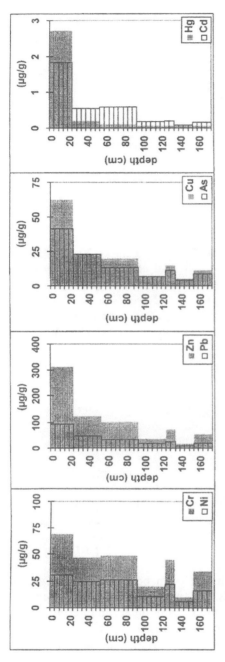

Figure 27 Heavy-metal concentrations of plateau 1.

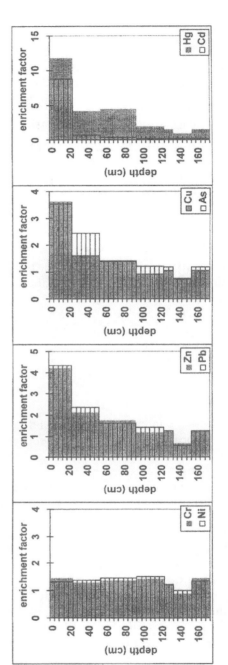

Figure 28 Heavy-metal enrichments of plateau 1.

Contaminants in a River Floodplain

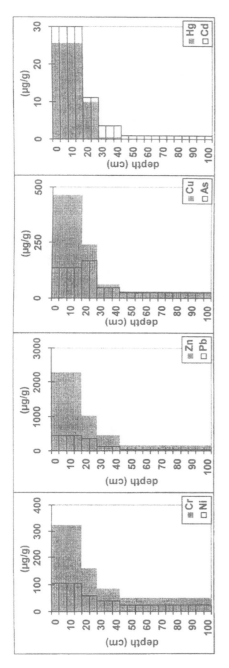

Figure 29 Heavy-metal concentrations of flood channel 1.

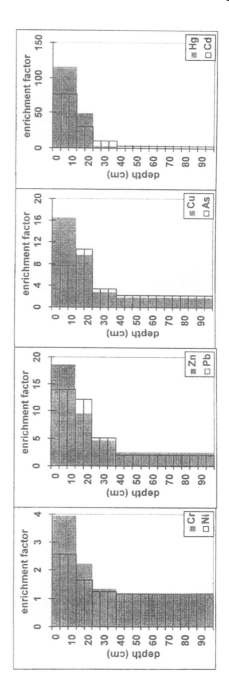

Figure 30 Heavy-metal enrichments of flood channel 1.

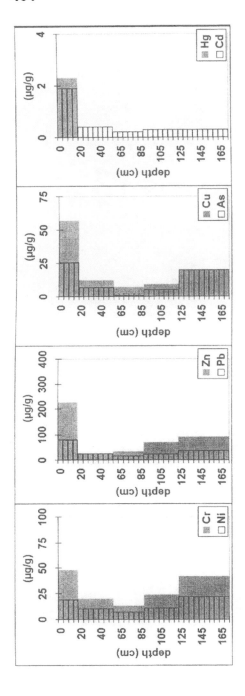

Figure 31 Heavy-metal concentrations of levee.

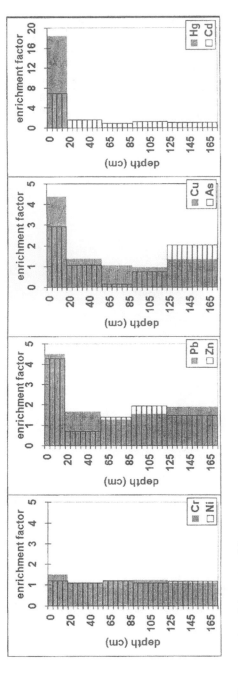

Figure 32 Heavy-metal enrichments of levee.

Table 9 Topsoil and Recent Mud Features ($n = 23$)

	Mean	Median	Min	Max
< 20 µm (%)	48	58	7.3	86
SOM (%)	5.0	4.3	0.25	11
CEC (mmol eq/kg)	194	202	26	352
Cr (µg/g)	124	108	17	326
Cu (µg/g)	165	127	10	466
Ni (µg/g)	46	44	9	104
Zn (µg/g)	714	529	46	2287
Pb (µg/g)	161	160	25	400
As (µg/g)	74	69	6	167
Cd (µg/g)	7.9	3.8	0.3	30
Hg (µg/g)	9.3	8	0.5	26

Table 10 Features of Subsoil Horizons ($n = 48$)

	Mean	Median	Min	Max
< 20 µm (%)	31	29	0	87
SOM (%)	0.5	0.5	0	1.5
CEC (mmol eq/kg)	116	106	12	360
Cr (µg/g)	41	44	6	90
Cu (µg/g)	20	16	2	64
Ni (µg/g)	21	22	5	48
Zn (µg/g)	107	72	15	451
Pb (µg/g)	36	25	5	131
As (µg/g)	20	13	0.8	79
Cd (µg/g)	0.6	0.3	0.08	3.6
Hg (µg/g)	0.2	0.14	0.05	0.7

weathering features (Bv and BvC, respectively, according to the German soil classification). The anthropogenic enrichment factors of the topsoil are in the same scale as at plateau 1.

As is shown in Figs. 27–32, the topsoils are mostly polluted because they consist predominantly of recent muds. Although there are sites which are not flooded every year, critical heavy-metal concentrations can still be found. Table 9 gives an overview of the heavy-metal pollution situation of the topsoils. The concentrations reach values that are typical of industrial regions (flood channel 1) in Germany (21). The average topsoil concentrations are not critical, with the exception of cadmium and mercury. In topsoils, high correlations (Pearson) exist between nearly all features; there are no significant correlations (0.1%) only between arsenic and the cation-exchange capacity (CEC), cadmium and < 20-µm fraction, and mercury and < 20-µm fraction. The correlation between soil organic matter (SOM) and CEC is nearly as high as between the < 20-µm fraction and the CEC. The link between heavy metals and soil organic material predominate in the correlation to the < 20-µm fraction with the exception of lead and arsenic.

In subsoil horizons, heavy-metal concentrations are not critical with respect to the mean and median values (Table 10), because they consist predominantly of meadow loam and sand. However, maximum concentrations show the anthropogenic enrichment of some heavy metals and the muds' influence. All subsoil features show significant correlations (Pearson). The lowest correlations exist between As, the <20-μm fraction, and SOM; Cd and the <20-μm fraction. The correlation between SOM and CEC is much lower than between <20-μm and CEC. The mineral compounds assume greater influence concerning bonds to heavy metals. There are still weak correlations between As and Cd with the <20-μm fraction. This may be due to the relatively high mobility of cadmium (22), as well as the high correlation of arsenic species to iron oxides (23). High mobilities of heavy metals are not expected because of the high pH values and the high buffer capacities of highly polluted soils. Problems for the groundwater and the enrichment in higher plants are merely predicted for cadmium with a threshold pH for mobility of 6.5 and ubiquitous distribution (24, 25).

E. Distribution of Organic Micropollutants by Depth

When talking about the pollution of floodplain soils, it is also important to consider organic compounds. All groups of contaminants that can be found in River Elbe sediment and seston are represented in these soils as well. Even some polar pesticides have been found that are well soluble in water and therefore must have been enriched in the soil by the infiltration of river water, not by sedimentation of polluted seston (10). More significant in terms of concentration are, for example, polycyclic aromatic hydrocarbons and chlorinated hydrocarbons (CHCs).

Here we will only focus on chlorinated unpolar compounds of low volatility, such as the pesticides DDT and lindane, and compounds such as hexachlorobenzene and polychlorinated biphenyls, which have industrial and technical sources. Hexachlorobenzene is one of the main contaminants of the Elbe, and is (along with pentachlorobenzene and octachlorostyrene) a by-product of chemical synthesis. Hexachlorobenzene can also be employed as a fungicide. DDT and lindane were used as relatively cheap and effective insecticides in the GDR for much longer than in most other countries, probably until the reunification of Germany in 1989. This explains the high levels of DDT, DDT metabolites (referred to here as DDX), lindane (g-HCH), and technical by-products of lindane (a-, b-, d-HCH) in the entire Elbe system. The results are shown for p,p'-DDT, p,p'-DDD, a-HCH, b-HCH, hexachlorobenzene (HCB), octachlorostyrene (OCS), pentachlorobenzene (QCB), and the sum of the PCBs 28, 52, 101, 138, 153, and 180.

The depth profiles of these compounds at two locations in our Elbe floodplain are presented. Each has a unique distribution of individual compounds, which may be explained by the respective conditions of flooding and sedimentation. The two profiles have been characterized above with respect to soil parameters and hydrology (plateau 2 and flood channel 2; see section III).

1. Materials and Methods

The samples were collected in the field in glass bottles. The samples were kept frozen until homogenization and lyophilization. The dry samples were sieved (2 mm) and 0.5 g was mixed with copper powder and sodium sulfate prior to extraction (26). All soil extractions were carried out on an SFE extractor (Isco, USA) equipped with a modifier pump and a fraction collector with variable restrictor and liquid trap. Our SFE methods comprised 20-min static extraction followed by dynamic extraction at a flow rate of 1.5 mL/min with 30 g of CO_2. The temperature was maintained at 100°C. The extraction pressure was 450 bar. Methanol (5%) was used as a modifier during static and dynamic extraction. All extracts were cleaned up with silica gel with good recovery (>90%) (10). Analysis was carried out on a gas chromatograph with an electron-capture detector (GC/ECD, Hewlett-Packard, USA; columns DB5 and DB1701, J&W, USA) with a detection limit of 10 µg/kg soil.

2. Results and Discussion

At the sampling location plateau 2 (Figs. 33 and 34) we have 20 cm of recent mud at the top of the profile. As was expected, we find an enrichment of CHCs in this top 20 cm, with a sharp decrease of concentrations below; between 40 and 65 cm only traces of p,p'-DDD can be detected. However, in the second lower meadow loam layer, once again certain amounts of DDT, DDD, and even PCBs are found. Hexachlorobenzene, which

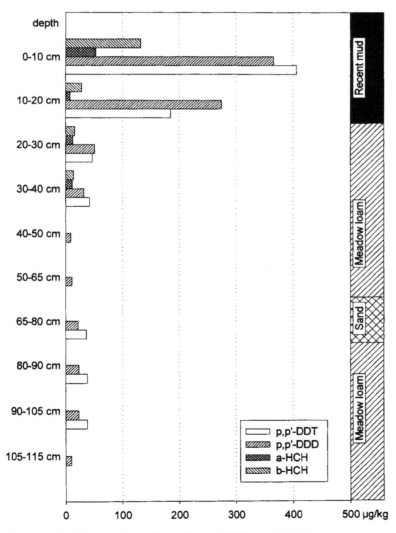

Figure 33 Plateau 2, depth profile of DDX and HCH.

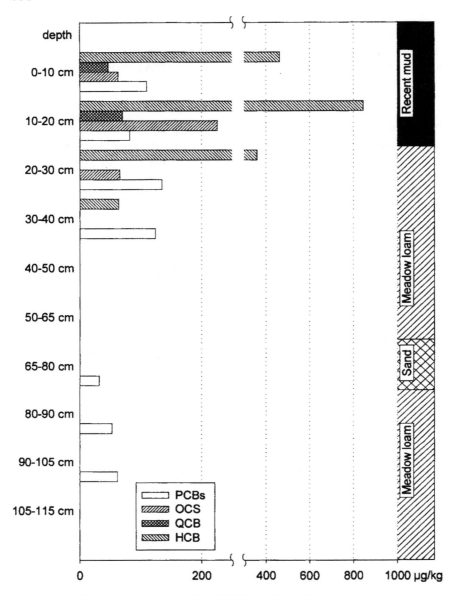

Figure 34 Plateau 2, depth profile of HCB, OCS, QCB, and PCBs.

is the main pollutant with up to 800 µg/kg in the second sample, cannot be detected in the deeper samples. The same is true of pentachlorobenzene and octachlorostyrene. The PCBs do not show a certain peak in one layer, but appear all over the profile in concentrations of 40–120 µg/kg. DDD and DDT always display similar concentrations, which is an important difference from Elbe River sediments, where normally only little DDT but high amounts of DDD can be found (27). The detection of DDX and PCBs in the old layer at a depth of 1 m can hardly be explained without taking the possibility of mobilization into account. This has already been found in another Elbe floodplain profile (10).

Contaminants in a River Floodplain

At the sampling location flood channel 2 (Figs. 35 and 36) we have 30 cm of recent mud covered by 30 cm of sand. Consequently, the highest concentrations of the compounds under investigation are found between 30 and 40 cm. The concentrations of most contaminants are low at the surface, with only hexachlorobenzene and PCBs showing a different pattern: they display similar concentrations at the top and at a depth of 30–40 cm, probably due to young inputs via flood sediments. Octachlorostyrene and pentachlorobenzene are hardly detectable, but the DDX and HCH compounds reach very high levels in the first 10 cm of the recent mud layer. This is especially obvious when comparing their concentrations with the concentration of HCB, the ratio of DDX to

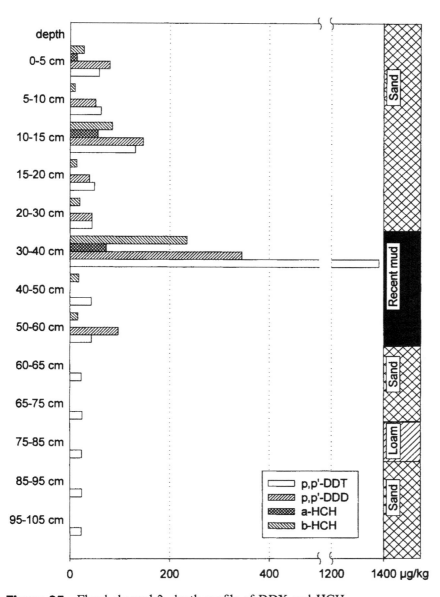

Figure 35 Flood channel 2, depth profile of DDX and HCH.

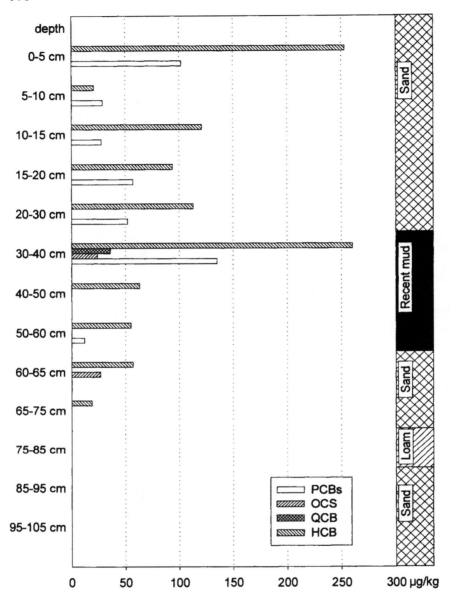

Figure 36 Flood channel 2, depth profile of HCB, OCS, and QCB, and PCBs.

HCB being < 1 in most samples. In the recent mud layer of flood channel 2 the ratio is up to 5 (DDT 1380 µg/kg/HCB 260 µg/kg). Again we have similar concentrations of DDD and DDT. Only in the unusual layer between 30 and 40 cm is the amount of DDT four times greater than that of DDD (DDT 1380 µg/kg/DDD 340 µg/kg). A layer containing fresh, unmodified DDT was probably deposited here, quickly covered with sand, and thus preserved from any degradation. Traces of DDT can be found in the whole profile, even in the sandy layers at the bottom. Again, this might indicate a certain mobility on the part of the pollutant.

These results leave a lot of questions unanswered. Why are DDT and b-HCH in the profile flood channel 2 found in such high concentrations? Why is the metabolism of DDT so slow? What will happen to the compounds in the future—will they stay in the soil, or might they be mobilized and transported to the groundwater?

In order to answer these questions, further investigations are to follow which tackle aspects of mobilization. As has been shown, quite unpolar compounds can in particular be mobilized under floodplain conditions (28). The next steps will include the analysis of seepage water for these compounds.

VI. THE TRANSFER OF HEAVY METALS TO PLANTS

The transfer and accumulation of heavy metals and other toxic elements from soil to plants can adversely affect the quality of food and fodder plants. The uptake of heavy metals by plants from contaminated soils is an important issue since the dietary intake of these metals could have a negative impact on the health of humans and animals (29,30). It is well known that plant concentrations of heavy metals are influenced by a number of soil and biological factors, including soil texture, organic matter content, oxidation–reduction potential, pH, and plant species. Several plant species (e.g., grasses and herbs) are able to evolve tolerances against increasing amounts of heavy metals.

Cd is generally known as one of the most toxic heavy metals for humans, animals, and plants, even at low concentrations (31). Compared to Cd, the phytotoxicity of Pb to plants is relatively low, due to the limited bioavailability of Pb from the soil (30,32). Although Zn is supposed to be less toxic than the other heavy metals studied, some sensitive plant species showed symptoms of toxicity at relatively low levels (33). The dominating factor determining the bioavailability of the heavy metals is believed to be soil pH affecting all adsorption/desorption mechanisms and the speciation of the metals in the soil solution (30,34). In this part we present data on the accumulation of Fe, Mn, Cd, Cr, Cu, Pb, Zn, and As by the meadow grass *Poa*.

1. Materials and Methods

All sample preparation and analytical techniques used for soil and plants in this study followed wherever possible the DIN German National Standards; they are briefly explained below. Soil samples (topsoil) are taken at 12 sites along a transect running from the dam to the river bank about 1 km long. The sampling locations were selected depending on geographic position in order to obtain a mean distribution of possible soil contamination. The plant samples were taken at the same position as the soil samples. We chose the meadow grass *Poa* as a representative of the fodder grass of the floodplain studied. Plants were taken from an area of 1 m^2 and mixed to form a single plant sample.

Soil samples were dried (105°C), homogenized, and ground by hand in an agate mortar, and then passed through a 2-mm plastic sieve prior to soil analysis. Plant samples were shock-frozen with liquid nitrogen, ground by hand in an agate mortar, and lyophilized prior to digestion. The soil pH was determined in a 0.01 M CaCl$_2$ solution. The organic carbon content (SOM) was measured as total carbon after treatment of a sample aliquot with HCl. To analyze metal contents, the soil samples were digested in aqua regia using a microwave oven. Various techniques of atomic absorption spectrometry were used to determine the heavy-metal and arsenic concentrations. To analyze metal content in the meadow grass, digestion was performed in a mixture of 2 : 1 HNO_3/H_2O_2 with a high-pressure asher (35). In order to evaluate the accuracy of the data obtained, standard reference materials (SRM, NIST) were analyzed. The results obtained corresponded satisfactorily with certificate or recommended values.

2. Results and Discussion

Soil characterization and soil pollution are discussed in Sections III–V and are not repeated here. The pH ($CaCl_2$) of the topsoil samples analyzed for this investigation varied between 3.9 and 5.7, while the span of SOM varied in the range 3–6% d.w. (dry weight). Table 11 shows the variation of the metals measured in the topsoil samples by maximum, minimum, mean and medium concentrations. In most cases these concentrations exceeded German threshold values for soils (36,38). No clear link was found between the element concentration of the soil and the topographic position of the sampling site. Moreover, no correlation was found between pH and the measured element concentration either. By contrast, positive correlations with SOM can be shown for all the elements. In addition, we found the well-known interelement correlations between the heavy metals (except Mn and Pb) and geochemical correlations between the heavy metals and arsenic except Mn, Zn, and Cd (Table 12).

Table 11 Variation of Heavy Metals and Arsenic in Topsoil Samples (dry weight) Given by Maximum, Minimum, Mean, and Median Values; For Comparison, German Threshold Values (TV) for Cd, Cr, Cu, Pb, and Zn [36] and As [38] are given

Topsoil	SOM	pH	Fe (g/kg)	Mn (mg/kg)	As (mg/kg)	Cd (mg/kg)	Cr (mg/kg)	Cu (mg/kg)	Pb (mg/kg)	Zn (mg/kg)
Max	6.0	5.7	35.5	839.0	56.6	4.2	103.0	114.0	201.0	651.0
Min	3.1	3.9	11.8	183.0	16.9	0.6	21.0	25.0	37.1	106.0
Mean	4.6	4.8	21.7	513.9	33.6	2.3	59.1	70.8	99.1	363.6
Median	4.6	4.9	22.2	517.5	33.0	2.3	57.5	67.5	98.5	343.5
TV			—	—	40	1.5	100	60	100	200

Table 12 Correlation Matrix for Topsoil Samples Showing Interelement Relationship and Dependence of Heavy Metals and Arsenic from the Organic Carbon Content

Topsoil	SOM	pH	Cr	Fe	Mn	Cu	Zn	As	Cd	Pb
SOM	1									
pH	−0.30	1								
Cr	0.75	0.18	1							
Fe	0.76	−0.03	0.94	1						
Mn	0.59	0.005	0.52	0.58	1					
Cu	0.75	0.24	0.98	0.89	0.58	1				
Zn	0.71	0.36	0.89	0.77	0.70	0.95	1			
As	0.57	−0.10	0.80	0.83	0.39	0.71	0.55	1		
Cd	0.58	0.38	0.79	0.64	0.70	0.86	0.94	0.46	1	
Pb	0.66	0.13	0.97	0.94	0.40	0.94	0.79	0.80	0.68	1

Table 13 Concentrations of Heavy Metals and Arsenic in the Meadow Grass *Poa* Given by Maximum, Minimum, Mean, and Median Values; All Values for Dry Weight Samples

Poa	Fe (µg/g)	Mn (µg/g)	As (µg/g)	Cd (µg/g)	Cr (µg/g)	Cu (µg/g)	Pb (µg/g)	Zn (µg/g)
Max	898.5	166.0	2.1	0.6	25.0	23.1	5.5	110.7
Min	50.9	20.1	0.5	0.1	0.8	4.3	0.5	46.0
Mean	293.0	71.2	0.8	0.3	7.5	11.7	1.9	78.0
Median	188.5	67.0	0.6	0.3	2.6	11.4	1.6	72.9

Table 14 Correlation Matrix for the Meadow Grass *Poa*, showing Interelement Relationships of the Heavy Metals and Arsenic

Poa	Cr	Fe	Mn	Cu	Zn	As	Cd	Pb	SOM	pH
Cr	1									
Fe	0.68	1								
Mn	0.50	0.68	1							
Cu	0.34	0.70	0.84	1						
Zn	0.36	0.53	0.24	0.55	1					
As	0.65	0.96	0.63	0.67	0.57	1				
Cd	0.43	0.66	0.43	0.69	0.68	0.62	1			
Pb	0.69	0.94	0.77	0.79	0.58	0.91	0.76	1		
SOM	−0.03	0.30	0.20	0.29	0.56	0.33	0.11	0.33	1	
pH	0.30	−0.14	−0.56	−0.51	0.03	−0.17	0.15	−0.13	−0.30	1

The plant concentrations of the heavy metals analyzed and arsenic vary by factors of up to 10 and for Fe by a factor of nearly 20 (Table 13). We found several interelement correlations (Table 14) in the plant concentrations. For example, Fe and Pb closely correlate positively with all the other elements measured, whereas Cr shows correlations (positive) only with Fe, As, and Pb. When combining the concentrations of the elements in the plant species *Poa* with the general parameters pH and SOM of the topsoil samples, we found a positive correlation only between the Zn concentration in the plant and the SOM content in the topsoil. A negative correlation was calculated between the Mn and Cu concentrations in the plant and the pH of the topsoil (Table 14). These results point to the assumption that the Zn concentration is directly dependent on the SOM content of the topsoil and the Mn and Cu concentrations in the plant are a direct response to the pH in the topsoil. To confirm these findings we first calculated the correlations between the element concentrations in *Poa* and in the topsoil (Table 15), and then the transfer factors for the elements measured by dividing the concentrations in the meadow grass by the concentration of the elements in the topsoil samples (Table 16).

The most interesting result for the relationship between the concentration in the plant and the concentration of an element in the topsoil is the high positive correlation of Zn in the plant and nearly all other elements in the topsoil (except Mn and As).

It is likely that the high positive correlation between Zn in the plant and Zn in the topsoil is the result of a direct linear dependency (Fig. 37). The same was observed for As and Fe, albeit with lower correlation coefficients. Linear correlations between

Table 15 Correlation Matrix for Heavy Metals and Arsenic in the Topsoil and Meadow Grass *Poa*[a]

Soil/*Poa*	Fe	Mn	As	Cd	Cr	Cu	Pb	Zn
Fe	0.55	0.16	0.59	0.48	0.24	0.33	0.49	0.78
Mn	0.02	−0.01	−0.06	0.31	−0.21	0.14	0.09	0.42
As	0.50	0.06	0.55	0.36	−0.02	0.23	0.35	0.43
Cd	0.001	−0.40	0.003	0.12	−0.08	−0.19	−0.05	0.53
Cr	0.46	−0.03	0.52	0.37	0.24	0.15	0.39	0.73
Cu	0.36	−0.10	0.39	0.30	0.21	0.09	0.30	0.71
Pb	0.51	0.04	0.58	0.35	0.34	0.18	0.40	0.72
Zn	0.18	−0.22	0.18	0.29	0.08	0.01	0.18	0.66

[a] The first row represents the elements from the meadow grass *Poa*, whereas the first column represents the elements from the topsoil.

Zn in the plant and other elements in the topsoil as well as between As or Fe in the plant and other elements in the topsoil is believed to be a statistical consequence of the interelement correlations in the topsoil. This interpretation is confirmed by the other elements, which show no correlation between their concentrations in the plant and in the topsoil.

When calculating transfer factors (37), we found low values for most elements (see Table 16); only for Zn did we generally count values between 10% and 50%. This finding also supports the thesis of a direct relationship between the Zn concentration in the plant and the topsoil. In individual cases we observed higher transfer factors for heavy metals (e.g., Cr 0.63, Mn 0.44, Cu 0.49, Cd 0.64), which may be due to specific site conditions as yet unidentified. The correlation of transfer factors with SOM and pH (CaCl$_2$) gave significant results only for Zn and Cu with the pH (Fig. 38). Again, this finding revealed the close relationship between Zn concentration in the meadow grass *Poa* and Zn content in the topsoil, depending on the soil pH.

Table 16 Transfer Factors for Heavy Metals and Arsenic for Different Sample Locations of the Meadow Grass *Poa*; The pHCaCl$_2$ of the Topsoils of the Different Sample Locations is Given for Comparison

Sample no.	pH	Fe	Mn	As	Cd	Cr	Cu	Pb	Zn
SD 7	4.97	0.01	0.10	0.02	0.09	0.18	0.11	0.01	0.18
SD 12	5.06	0.004	0.04	0.01	0.04	0.01	0.05	0.00	0.10
SD 17	5.66	0.04	0.43	0.06	0.25	0.63	0.19	0.04	0.25
SD 21	4.77	0.01	0.07	0.02	0.14	0.04	0.14	0.02	0.23
SD 25	4.78	0.01	0.11	0.01	0.15	0.02	0.15	0.01	0.18
SD 32	4.34	0.03	0.23	0.04	0.25	0.19	0.29	0.05	0.25
SD 34	4.67	0.02	0.26	0.04	0.18	0.17	0.13	0.02	0.25
SD 36	4.77	0.01	0.20	0.02	0.64	0.18	0.38	0.05	0.30
SD 37	5.08	0.01	0.09	0.02	0.13	0.04	0.16	0.01	0.23
SD 39	4.96	0.003	0.05	0.02	0.18	0.06	0.25	0.01	0.29
SD 44	5.2	0.01	0.04	0.02	0.09	0.03	0.10	0.01	0.17
SD 49	3.87	0.01	0.44	0.03	0.10	0.11	0.49	0.02	0.50

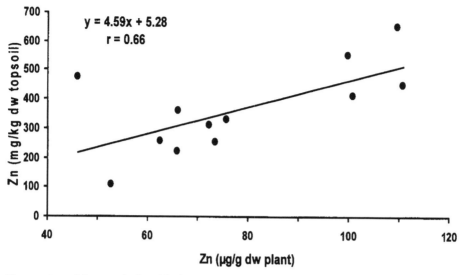

Figure 37 Linear relationship between the amount of Zn in the topsoil and the concentration of Zn in the meadow grass *Poa* (dw: dry weight).

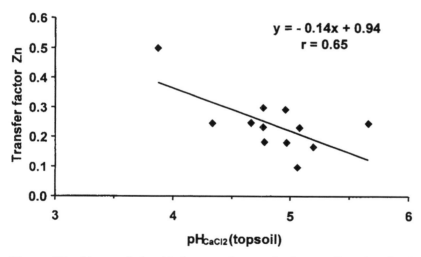

Figure 38 Linear relationship between the transfer factor soil to plant for the element Zn and the pH of the topsoil.

VII. IDENTIFICATION OF TOXIC ORGANIC COMPOUNDS IN SEDIMENTS—A CONTRIBUTION TO ENVIRONMENTAL HAZARD ASSESSMENT

In this section an example of toxicity identification in a contaminated sediment in the small river Spittelwasser in the catchment area of the River Elbe will be presented. This river received untreated effluents from chemical plants in the region of Bitterfeld (Saxony-Anhalt, Germany) for several decades, until production was largely discontinued

Table 17 Biotest Battery Applied to Acetonic Fractions of Spittelwasser Sediment

Organism	Endpoint	Incubation time	Effect type
Photobacterium phosphoreum	Bioluminescence	30 min	Acute inhibition of energy metabolism
Daphnia magna	Immobilization	24 h	Acute invertebrate lethality
Scenedesmus vacuolatus	Inhibition of cell multiplication	24 h	Chronic photo-autotrophic toxicity
Danio rerio	Lethality of embryos	48 h	Fish developmental toxicity
Rainbow trout gonad (RTG) cell line	Neutral red retention	2 h	Cytotoxicity
Rainbow trout liver (RTL) cell line	EROD, inhibition and induction	24 h	Xenobiotic biotransformation activity
Rainbow trout liver (RTL) cell line	DNA strand breaks, COMET assay	2 h	Genotoxicity

about 8 years ago. Since the industrial region of Bitterfeld was shown to be a significant source of toxic pollutants in the northern part of River Elbe its sediments, and its floodplains (50), the major contaminants in the Spittelwasser sediment are also expected to be relevant for the contamination of the floodplain in Wittenberge. Toxic effects to the fauna and flora in situ due to those compounds may not be excluded.

Contaminated sediments subsequently themselves act as a source of toxicants for the aquatic environment by the resolution and resuspension of contaminated particles. Aquatic organisms may be exposed to toxicants associated with sediments and suspended particles via different pathways such as interstitial or pore water, and the ingestion of and direct contact with contaminated particles (39–42). The bioaccumulation of toxicants from sediments and suspended particles in benthic organisms and plankton may play an important role for the introduction of sediment-associated toxicants into the aquatic food web. Higher trophic levels such as fish may be affected via biomagnification (43).

Possible effects of organic toxicants associated with sediments include acute toxicity on aquatic organisms of all trophic levels, effects on reproduction and sublethal effects such as genotoxicity, effects on xenobiotic biotransformation activity, endocrine effects, and immunotoxicity. Acute effects include nonspecific narcotic effects exhibited in every organism as well as specific effects, for example, on photosynthesis or the nerve system, which inhibit distinct groups of organisms. Therefore, the risk assessment of contaminated sediments must ideally consider all these different effects.

The hazard assessment of contaminated sediments has been performed using several approaches. Benthic community structure analyses provide highly ecologically relevant data when investigating biodiversity in situ and looking for bioindicators for a healthy or stressed environment (44). Bioavailability, synergistic, and antagonistic effects are considered. However, the high variance of benthic communities in natural sediments may make the recognition of toxicant effects very difficult and prevent the clear identification of cause–effect relationships. Therefore, in order to decide whether the differences observed are due to toxic effects or are caused by different nutrient supply, grain size of sediments, or competing organisms, this approach must be combined with analytical and biotesting methods such as in the TRIAD approach (44,45).

Another approach is to biotest bulk sediment, pore water, aqueous elutriates, and organic extracts in the laboratory using standard test organisms. As in benthic community structure analysis, effects are measured directly. Individual biotests may overlook important effects and toxicants because each biotest is able to detect only some of the possible ecotoxicological effects. Therefore a battery of biotests is necessary (46).

The most powerful tools for the hazard assessment of contaminated sediments are combinations of the principal approaches discussed above. The TRIAD approach (44) combines benthic community structure analysis with laboratory biotests, the chemical analysis of priority pollutants and multivariate statistical analysis providing plenty of environmental information about the sediment. However, analytical and toxicological data are combined solely by statistical correlation, which provides no real cause–effect–relationship. It remains unclear whether the pollutants analyzed cause the effects detected.

The toxicity identification evaluation (TIE) approach, according to the concept of the U.S. EPA, (47–49) combines the laboratory biotesting of liquid samples such as pore water, elutriates, and extracts and chemical analysis with fractionation procedures focusing on the identification of effective toxicants. In-situ studies and bulk sediment tests are dispensed with. The main advantage of TIE is the identification of real cause–effect relationships and effective toxicants. Samples are sequentially biotested and fractionated, separating a limited number of compounds with defined chemical properties according to the fractionation procedures applied. These chemicals can in a further step be identified analytically and confirmed as toxicants. This confirmation may be done, for example, by applying artificial mixtures to biotesting containing the expected toxicants in concentrations equivalent to those in the fraction, by removing the expected toxicant and testing again, by calculating the additive toxicity of the components using the toxic unit approach, by comparing symptoms caused by the fraction and the individual compounds or the mixture, or by comparing species sensitivity.

The ability to detect and identify ecotoxicologically relevant compounds and therefore the ecological strength of such an approach depends very much on the selection of biotests.

A. Identification of Toxicants in Spittelwasser Sediment

Toxicants in Spittelwasser sediment extracts were identified using the procedure summarized in Fig. 39 (46). The biotest battery applied to the fraction contains acute toxicity tests with organisms of every trophic level as well as sublethal tests (Table 17).

Relative toxic potencies of primary fractions to bacteria, daphnia, algae, and fish cells are presented in Figs. 40 and 41. Absolute values will be presented in another paper (46).

The toxicity of fractions to these test organisms representing different trophic levels and different effect types are evidently not caused by the same fractions and therefore the same compounds. Major toxicity to luminescent bacteria is detected in F1, major toxicity to daphnia and algae is due to F1 and F3, while F5 is the most toxic fraction to the RTG cell line. Major toxicity to fish embryos is detected in F5, while EROD induction is divided among fractions F1, F2, and F4. All of the primary fractions exhibit similar numbers of DNA strand breaks.

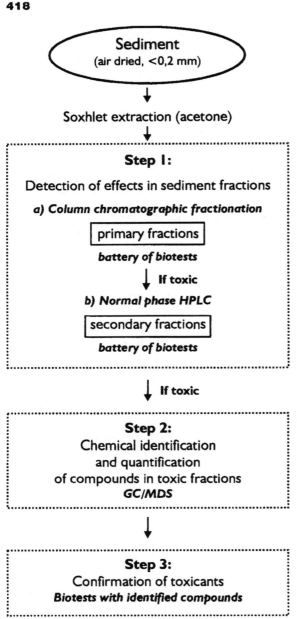

Figure 39 Principal steps of the TIE procedure (46).

Further fractionation shows that the toxicity of both F1 and F3 is caused by different subfractions and therefore toxicants for the different biotests (Figs. 42–44). Whereas the major toxicity of F1 to luminescent bacteria and algae is found in F1.2, F1.1 is very toxic to daphnia. Major EROD induction and major genotoxicity are detected in F1.2. Major toxic subfractions of F3 to daphnia and algae are F3.6 and F3.4, respectively. It is apparent that none of the tests can be abolished without a drastic loss of information. Each test is necessary because effects at any trophic level may affect the entire ecosystem.

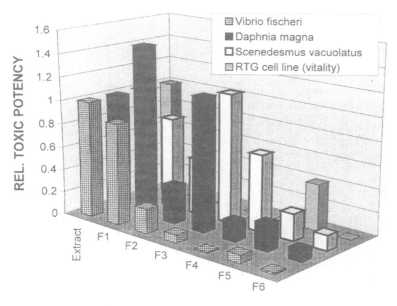

Figure 40 Relative acute toxic potencies of primary fractions of acetonic sediment extract.

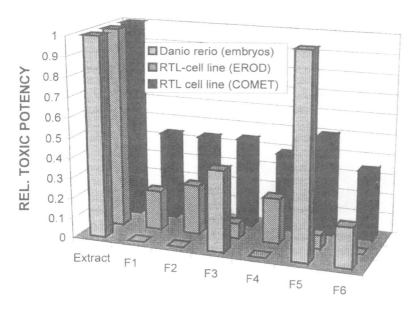

Figure 41 Relative sublethal and embryo toxic potencies of primary fractions of acetonic sediment extract.

For hazard assessment and appropriate remedial actions we have to know which compounds cause the effects in order to be able to consider their environmental properties concerning sorption, degradability, and volatility, and to identify sources of contamination (51).

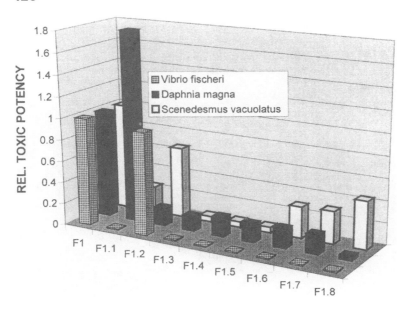

Figure 42 Relative acute toxic potencies of secondary fractions of F1.

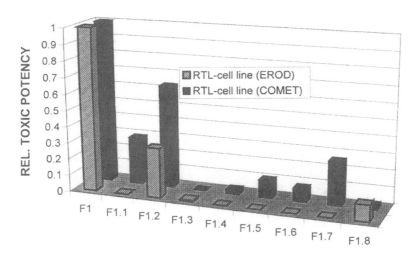

Figure 43 Relative sublethal toxic potencies of secondary fractions of F1.

In many cases, the most powerful tool to identify toxicants is gas chromatography equipped with mass specific detection (GC/MSD). For our example, the expected acute toxicants identified in the toxic fractions are shown in Table 18. In a further step these compounds in the fractions were confirmed as the cause of the effect. This was done by testing them in the various biotests as solutions of standard compounds in concentrations corresponding to those in the fractions (Table 18) or, as in the case of sulfur, by removing the toxicant and testing again (Table 19).

As is seen in Table 18, many of the major toxicants were identified and confirmed with this approach, including PAHs, an insecticide (methyl parathion), a herbicide

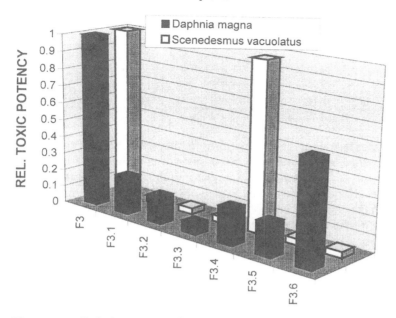

Figure 44 Relative acute toxic potencies of secondary fractions of F3.

(prometryn), and the biocide n-tributyltin. Additionally, N-phenyl-β-naphthalene amine, a compound which is very toxic to algae and which is normally not considered in hazard assessment studies, was confirmed as a major toxicant. Whereas sulfur was confirmed as the toxicant in fraction F1.2 inhibiting *Vibrio fischeri*, Table 19 clearly indicates that toxicity to algae is caused to only a minor extent by this compound. In addition to sulfur, F1.2 contains numerous diaromatic compounds such as PCBs, PCNs, 4,4′-dichlorodiphenylsulfide, methyl and ethyl biphenyls, etc. (46). The identification

Table 18 Identified and Confirmed Toxicants in Acetonic Fractions

Fraction	Biotest	Toxicants	Correspondence (%)
F1.1	*Daphnia*	n-alkanes (C13–C18), n-tetrabutyltin	28
F1.6	*Scenedes mus*	Phenanthrene, anthracene, 2-phenylnaphthalene	68
F1.7	*Scenedes mus*	Fluoranthene, pyrene	99
F1.8	*Scenedes mus*	Benzo[ghi]fluoranthene, benz[a]anthracene, triphenylene 2-Ethylhexyl phthalate	92
F3.4	*Scenedes mus*	N-Phenyl-β-naphthalene amine	246
F3.6	*Daphnia*	Methyl parathion	103
F4	*Scenedes mus*	Prometryn	80
F5	*Vibrio*	n-Tributiltin	157
	Daphnia	n-Tributiltin	960
	Scenedes mus	n-Tributiltin	66
	RTG	n-Tributiltin	82
	Danio rerio	n-Tributiltin	0.59–1.7

Table 19 Indirect Confirmation of Sulfur Toxicity by Removal; ED_{50} Values Given in Dilution Factors

Fraction	Biotest	ED_{50} (dilution)	ED_{50} after sulfur removal
F1.2	*Vibrio*	3500	< 100
F1.2	*Scenedesmus*	1370	1149

and confirmation of individual toxicants in F1.2 requires further fractionation and in some cases the synthesis of standard compounds which are not commercially available.

The hazard assessment of environmental samples requires a combination of the analysis of biological effects and chemical analysis. To avoid misinterpretations, one important prospective of hazard assessment studies must be to establish real cause–effect relationships. The most powerful tool for this is toxicity identification evaluation studies combining biotests, fractionation, and chemical analysis. As the example shows, in many cases this approach allows the identification and confirmation of individual compounds as toxicants. However, the example also illustrates that a TIE study with the prospective of detection and identification of the whole spread of effective toxicants requires the application of a broad array of biotests, including different trophic levels and different effect types taking into account acute as well as chronic and sublethal effects.

VIII. CONCLUSIONS

Floodplains are an integral part of natural rivers in lowland areas. They represent a unique ecosystem which is characterized by a permanent but irregular exchange between the aquatic and terrestrial milieus. Floodplains perform an important function for both the discharge regulation of the river and (via retention, sedimentation, and resuspension) for the system's material balance.

In order to ascertain the flooding frequency of certain points within a floodplain area and to produce models illustrating the transport of suspended particulate matter, detailed topographic land surveys are required. Flooding frequencies and the amounts of suspended particulate matter (including their degree of contamination) input recently must be determined so that focal points of pollution in the floodplain can be localized.

To estimate the risk of pollution penetrating floodplain areas via the river–groundwater pathway, detailed hydrologic knowledge of the area studied is necessary. In the case of rising floodwater levels, the input of harmful substances from the interstitial cannot be ruled out, especially in areas with sandy subsoil.

Assessing the retention capacity of floodplains with respect to various nutrients and pollutants must take into account transfers within a flood zone. The frequency and degree of flooding are decisive for the transport into the floodplain of suspended particulate matter and the pollutants adsorbed thereon. Consequently, rapidly rising floodwater is a more serious source of contaminated suspended particulate matter for the central Elbe region than slowly rising water levels with numerous small advance peaks.

As long as highly contaminated sediments continue to persist in the river system and its catchment area, the floodplain ecosystem will continue to be affected by pollutants. Adsorption on organic material such as algae can result in sustained fixing. Factor and cluster analysis of a geochemical topsoil survey showed that most trace metals are adsorbed on phases rich in carbon and/or sulfur, with only lead and cobalt occurring together with pedogenic oxides of iron and manganese. No links were found between the levels of heavy metals and the soil's pH or clay content. The topographic position in the flood area is decisive for the level of pollution.

High accumulations of heavy metals and organic pollutants are thus encountered in flood channels with recent deposits. The reduction of organic contaminants seems to be prevented under certain conditions, such as when contaminated mud deposits are covered by a sandy layer. The accumulations of certain organic pollutants at greater depths indicate the possibility of transfers and transport into the groundwater. The transfer of heavy metals appears to assume a critical dimension only in the presence of certain location factors not yet precisely identified. Only for the (essential) element zinc were direct links found between soil content and accumulation in the plant. Work will be continued in order in particular to better understand aspects of depth distribution and the mobilization of organic and inorganic substances.

The major toxicants affecting bacteria, algae, daphnids, and fish embryos in the sediment of the river Spittelwasser were identified and confirmed as methyl parathion, prometryn, N-phenyl-β-naphthalene amine, n-tributyltin, and various PAHs. In addition, diaromatic compounds such as PCBs and PCNs were shown to exhibit algal toxicity and EROD induction in fish cells, even though the individual toxicants could not be identified and verified. It can be assumed that the toxicants emitted by the industrial Bitterfeld region are transported down the Elbe. Depending on their physicochemical and biological properties, they may be transported in solution or adsorbed on suspended particulate matter, bio- or phototransformed, or volatilized into the atmosphere, or they may accumulate in sediments and floodplains. As the relative ecotoxicological relevancy of the toxicants may be altered, it must be determined which of the toxicants emitted in Bitterfeld are still relevant for the floodplain ecosystem in Wittenberge, and which additional toxicants may contribute to the toxic stress of the fauna and flora in this area.

Our extensive work has shed light on the diverse factors of influence affecting the material balance of a floodplain ecosystem. It is designed to highlight the critical aspects which should be taken into account while studying the pollution of floodplains when agricultural usage and floodplain management are at issue. The devastating floods all over the world over the past few years and the resulting demand for the designation of additional large flood zones highlight the need to fathom the biogeochemical processes in floodplain systems. We hope that our work will further our understanding of these processes.

ACKNOWLEDGMENT

The authors gratefully acknowledge the active help in field and laboratory work by the following persons: Dr. R. Altenburger, S. Berner, D. Bösel, I. Christmann, Dr. U. Ensenbach, C. Hoffmeister, H. Hagemann, U. Kayser, R. Luedtke, M. Mages, Dr. Möder, K. Muhs, S. Nehls, J. Patzer-Seifriedt, I. Ränker, J. Seegert, Dr. H. Segner, P. Schonert, A. Sperreuter, M. Voss, and Z. Zielinska.

During our case study several colleagues gave advices and help by thorough discussions on various aspects of our work. We thank warmly Prof. Dr. W. Geller, Dr. H. Guhr, Prof. Dr. Markert, Prof. Dr. R. Meißner, and Dr. M. Rode.

The current work would not have been possible without financial support by grants of the German ministry for education, science, research, and technology (BMBF, Fkz. 02 WT 9617/0 and Fkz. 03 395 571).

REFERENCES

1. Brunke, M., and Gonser, T. 1997. The ecological significance of exchange processes between rivers and groundwaters. *Freshwater Biol.*, 37:1–33.
2. Mitsch, W. J. and Gosselink, J. G. 1993. Wetlands, Van Nostrand Reinhold, New York.
3. Bubb, J. M., and Lester, J. N. 1994. Anthropogenic heavy metal inputs to lowland river systems, a case study. The River Stour, U.K. *Water, Air Soil Pollution*, 78:279–296.
4. *Deutsches Gewaesserkundliches Jahrbuch, Elbegebiet, Teil III, Untere Elbe ab der Havelmuendung, Hrsg.: Freie und Hansestadt Hamburg, Wirtschaftsbehoerde, Strom und Hafenbau Hamburg*, ISSN 0949 3654, 1997.
5. Krueger, F., Buettner, O., Friese, K., Meissner, R., Rupp, H., and Schwartz, R. 1997. Lokalisation der Schwermetallbelastung durch Simulation des Ueberflutungsregimes. *DBG-Mitteilung.*, 85:949–952.
6. Miehlich, G. 1983. Schwermetallanreicherung in Boeden und Pflanzen der Pevestorfer Elbaue, Abh. naturwiss. *Verein Hamburg*, 25:75–89.
7. Miehlich, G. 1994. Auen und Marschen als Senken fuer belastete Sedimente der Elbe. In: H. Guhr, A. Prange, P. Puncochar, R. D. Wilken, and B. Buettner (eds.), *Die Elbe im Spannungsfeld zwischen Oekologie und Oekonomie*, pp. 307–312. Teubner, Stuttgart.
8. Meissner, R., Guhr, H., Rupp, H., Seeger, J., and Spott, D. 1994. Schwermetallbelastung von Boeden und Elbsedimenten in ausgewaehlten Gebieten Ostdeutschlands, *Z. Kulturtech. Landentwick.*, 35:1–9.
9. Witter, B. 1995. Untersuchung organischer Schadstoffe in Auen der mittleren und unteren Elbe unter Anwendung der SFE. Ph.D. thesis, University of Hamburg.
10. Witter, B., Franke, W., Franke, S., Knauth, H.-D., and Miehlich, G. 1998. Distribution and mobility of organic micropollutants in River Elbe floodplains. *Chemosphere*, 37:63–78.
11. Neumeister, H., and Villwock, G. 1997. Stoffquellen und -dynamik in der Muldeaue. In: R. Feldmann, K. Henle, H. Auge, J. Flachowsky, S. Klotz, and R. Krönert (eds.), *Regeneration und nachhaltige Landnutzung*, pp. 53–57. Springer-Verlag, Berlin.
12. Wilken, R.-D., Fanger, H.-U., and Guhr, H. 1994. Ergebnisse der Hochwassermessungen 1993/1994. In: H. Guhr, A. Prange, P. Puncochar, R. D. Wilken, and B. Buettner (eds.), *Die Elbe im Spannungsfeld zwischen Oekologie und Oekonomie*, pp. 125–135. Teubner, Stuttgart.
13. Spott, D. 1994. Schwebstoff- und Schwermetallbelastung der Elbe bei Hochwasser—Untersuchungen am linken Ufer von Magdeburg im Zeitraum Dezember bis Mai 1994. In: H. Guhr, A. Prange, P. Puncochar, R. D. Wilkin, and B. Buettner (eds.), *Die Elbe in Spannungsfeld zwischen Oekologie und Oekonomie*, pp. 499–502. Teubner, Stuttgart.
14. Smith, K. A., and Paterson, J. E. 1992. Manganese and cobalt. In: B. J. Alloway (ed.), *Heavy Metals in Soils*, 2nd ed., pp. 224–244. Blackie, Glasgow.
15. Schwartz, R., Duwe, J., and Groengroeft, A. 1997. Einsatz von Kunstrasenmatten als Sedimentfallen zur Bestimmung des partikulaeren Stoffeintrags in Auen und Marschen. *DBG-Mitteilung.*, 85:353–356.
16. van der Veen, A. 1998. Geochemische Schwermetallkartierung in der Elbaue bei Wittenberge/Sachsen-Anhalt. Unpublished diploma thesis, Technical University of Braunschweig.
17. Prange, A., Boessow, E., Jablonski, R., Krause, P., Lenart, H., Meyercordt, J., Pepelnik, R., Erbsloeh, B., Jantzen, E., Krueger, F., Leonhardt, P., Niedergesaess, R., and von Tuempling, W., Jr. 1997. Geogene Hintergrundwerte und zeitliche Belastungsentwicklung. In: *Erfassung und Beurteilung der Belastung der Elbe mit Schadstoffen*, Vol. 3/3, pp. 152–156. GKSS.
18. Davis, J. G. 1973. *Statistics and Data Analysis*. Wiley, New York.
19. Backhaus, K., Erichson, B., Plinke, W., and Weiber, R. 1996. *Multivariate Analysemethoden*, p. 591. Springer-Verlag, Berlin.
20. Bruemmer, G. W., Zeien, H., Hiller, D. A., and Hornburg, V. 1994. Bindungsformen und Mobilitaet von Cadmium und Blei in Boeden, pp. 197–217. Dechema, Frankfurt/Main.

21. Lux, W. 1986. Schwermetallgehalte und -Isoplethen in Boeden, subhydrischen Ablagerungen und Pflanzen im Suedosten Hamburgs—Beur-teilung eines Immissionsgebietes. Diss. Hamburger Bodenkundliche Arbeiten Bd. 5.
22. Schachtschabel, P., Blume, H.-P., Bruemmer, G., Hartge, K.-H., Schwertmann, U., Fischer, W. R., Renger, M., and Strebel, O. 1992. *Lehrbuch der Bodenkunde*, 13th ed., p. 322. Ferdinand Enke Verlag, Stuttgart.
23. O'Neill, P. 1992. Arsenic in heavy metals in soils. In: B. J. Alloway (ed.), *Heavy Metals in Soils*, 2nd ed., pp. 105–121. Blackie, Glasgow.
24. Blume, H. P., and Bruemmer, G. 1987. Prognose des Verhaltens von Schwermetallen in Boeden mit einfachen Feldmethoden. *DBG-Mitteilung.*, 53:111–118.
25. Koester, W., and Merkel, D. 1985. Schwermetalluntersuchungen landwirtschaftlich genutzter Boeden und Pflanzen in Niedersachsen. Final report, Landwirtschaftskammer Hannover, pp. 96–98.
26. Boewadt, S., and Johansson, B. 1994. Analysis of PCBs in sulfur-containing sediments by off-line supercritical fluid extraction and HRGC-ECD. *Anal. Chem.*, 66:667–673.
27. Gandrass, J., and Zoll, M. 1996. Chlorinated hydrocarbons in sediments of the Elbe catchment area—Analytical methods and status of pollution. *Acta Hydrochim. Hydrobiol.*, 24:212–217.
28. Pardue, J. H., Masscheleyn, P. H., De Laune, R. D., and Patrick, W. H. 1993. Assimilation of hydrophobic chlorinated organics in freshwater wetlands: Sorption and sediment-water exchange. Environ. Sci. Technol., 27:875–882.
29. Dudka, S. 1994. Effect of concentrations of Cd and Zn in soil on spring wheat yield and the metal contents of the plants. *Water, Air Soil Pollution*, 76:333–341.
30. Adriano, D. C. 1986. *Trace Elements in the Terrestrial Environment*. Springer, New York.
31. Bergmann, W. 1992. Nutritional disorders of plants. Gustav Fischer Verlag, Jena, Germany.
32. Kabata-Pendias, A., and Pendias, H. 1986. *Trace Elements in Soils and Plants*. CRC Pres., Boca Raton, FL.
33. Bahlsberg-Pahlson, A. M. 1989. Toxicity of heavy metals (Zn, Cu, Cd, Pb) to vascular plants. *Water, Air Soil Pollution*, 47:287–319.
34. Alloway, B. J. 1995. *Heavy Metals in Soil*. Chapman & Hall, London.
35. Markert, B. 1996. *Instrumental Element and Multi-Element Analysis of Plant Samples—Methods and Applications*. Wiley, Chichester, U.K.
36. AbfKlär, V. 1992. *Klärschlammverordnung (idF v. 15.04.1992) Bundesgesetzblatt*, Teil I, Nr. 21, p. 912.
37. Nagel, R., and Loskill, R. 1991. *Bioaccumulation in aquatic systems*. VCH, Weinheim, Germany.
38. LÖLF. 1988. Landesanstalt für Ökologie, Landschaftsentwicklung und Forstplanung: Mindestuntersuchungsprogramm Kulturboden zur Gefährdungsabschätzung von Altablagerungen und Altstandorten im Hinblick auf eine landwirtschaftliche oder gärtnerische Nutzung, Recklinghausen.
39. Knezovich, J. P., Harrison, F. L., and Wilhelm, R. G. 1987. The bioavailability of sediment-sorbed organic chemicals: A review. *Water, Air Soil Pollution*, 32:233–245.
40. Ankley, G. T., and Mount, D. R. 1996. Retrospective analysis of the ecological risk of contaminant mixtures in aquatic sediments. *Human Ecol. Risk Assess.*, 2:434–440.
41. Harkey, G. A., Landrum, P. F., and Klaine, S. J. 1994. Comparison of whole-sediment, elutriate and pore-water exposures for use in assessing sediment-associated organic contaminants in bioassays. *Environ. Toxicol. Chem.*, 13:1315–1329.
42. Belfroid, A. C., Sijm, D. T. H. M., and Van Gestel, C. A. M. 1996. Bioavailability and toxicokinetics of hydrophobic aromatic compounds in benthic and terrestrial invertebrates. *Environ. Rev.*, 4:276–299.
43. Maguire, R. J., and Tkacz, R. J. 1985. Degradation of the tri-*n*-butyltin species in water and sediment from Toronto harbor. *J. Agric. Food Chem.*, 33:947–953.
44. Carr, R. S., Chapman, D. C., Howard, C. L., and Biedenbach, J. M. 1996. Sediment quality triad assessment survey of the Galveston Bay, Texas system. *Ecotoxicology*, 5:341–364.

45. Chapman, P. M. 1996. Presentation and interpretation of sediment quality triad data. *Ecotoxicology*, 5:327–339.
46. Brack, W., Altenburger, R., Ensenbach, U., Möder, M., Segner, H., and Schüürmann, G. 1999. Bioassay-directed identification of organic toxicants in river sediment in the industrial region of Bitterfeld (Germany)—A contribution to hazard assessment. *Arch. Environ. Contem. Toxicol.*, 37:164–174.
47. Norberg-King, T. J., Mount, D. I., Durhan, E. J., Ankley, G. T., Burkhard, L. P., Amato, J. R., Lukasewycz, M. T., Schubauer-Berigan, M. K., and Anderson-Carnahan, L. (eds.). 1991. Methods for aquatic toxicity identification evaluations. Phase I. Toxicity characterization procedures, 2nd ed., EPA/600/6-91/003. U.S. Environmental Protection Agency, Washington, DC.
48. Mount, D. I., and Anderson-Carnahan, L. 1989. Methods for aquatic toxicity identification evaluations. Phase II. Toxicity identification procedures, EPA/600/3-88/035. U.S. Environmental Protection Agency, Washington, DC.
49. Mount, D. I. 1989. Methods for aquatic toxicity identification evaluation. Phase III. Toxicity confirmation procedures, EPA/600/3-88/036. U.S. Environmental Protection Agency, Washington, DC.
50. Kuballa, J., Wilken, R.-D., Jantzen, E., Kwan, K. K., and Chau, Y. K. 1995. Speciation and genotoxicity of butyltin compounds. *Analyst*, 120:667–673.
51. Kosian, P. A., Makynen, E. A., Monson, P. D., Mount, D. R., Spacie, A., Mekenyan, O. G., and Ankley, G. T. 1998. Application of toxicity-based fractionation techniques and structure-activity relationship models for the identification of phototoxic polycyclic aromatic hydrocarbons in sediment pore water. *Environ. Toxicol. Chem.*, 17:1021–1033.

20
Formation, Long-Term Stability, and Fate of Nonextractable ^{14}C-PAH Residues in Contaminated Soils

Annette Eschenbach and Bernd Mahro
Hochschule Bremen, Institut für Technischen Umweltschutz, Bremen, Germany

Reinhard Wienberg
Umwelttechnisches Büro und Labor Dr. R. Wienberg, Hamburg, Germany

I. INTRODUCTION

Bioremediation is one of the most attractive clean-up methods for soils which are contaminated with polycyclic aromatic hydrocarbons (PAH), since it has been demonstrated that a wide variety of microorganisms is able to degrade PAH (1–3). To enhance the biodegradation of PAHs in soil, the effect of pure cultures of PAH-metabolizing microorganisms [bacteria (4–6), white rot fungi (7–9)] or organic supplements [e.g., compost (10–12)] has been studied intensively in recent years.

When soils from contaminated sites were investigated, it was often not possible to reduce the extractable PAH concentration completely. The degradation proceeded to a certain threshold and parts of the contaminants remained undegradable in soil (13,14). It was shown that the biodegradation of PAH in contaminated soils is often limited by a lack of bioavailability, which results from a lack of mass transfer (15,16).

On the other hand, one has to be aware that the apparent analytical depletion of the extractable amount of PAH in soil may be the result of a strong adsorption or binding of the xenobiotics to the soil matrix. These substances can be partly released by more rigid extraction procedures (17). However, it was shown by the use of ^{14}C-labeled PAH that a relevant part of the substances becomes unextractable (12,18). This formation of nonextractable residues (occasionally referred to as "bound residues") is a well-known phenomenon from investigations with radiolabeled pesticides (19,20). In this context nonextractable residues were defined according to the IUPAC as "chemical species (active ingredient, metabolites and fragments) originating from pesticides . . . that are unextracted by methods which do not significantly change the chemical nature of these residues, but which remain in the soil. These nonextractable residues are considered to exclude fragments recycled through metabolic pathways leading to naturally occurring products" (21).

The mechanisms of the binding or sorption processes of PAH in soil are under research and the chemical structure of these nonextractable PAH residues is almost unknown. A new and successful tool to evaluate interactions between organic pollutants and soil organic matter is the use of ^{13}C-labeled chemicals and ^{13}C-NMR spectrometry (22–24).

The chemical association of organic xenobiotics, such as PAH and their metabolites, with macromolecular organic matter in soils may be based on both sorptive processes (e.g., hydrophobic sorption, charge-transfer interactions, hydrogen bonds) and on covalent binding (ester, ether, carbon–carbon bonds) (25–27). Since adsorption is considered to be primarily a reversible process, the bulk of the substances remains available and solvent extractable (27,28). However, there is evidence that with longer residence time in soil, adsorbed substances tend to become more resistant to extraction and degradation "aging" of the chemicals in soil (29,30). The covalent binding of xenobiotics to organic matter should result in more persistent associations. It is presumed that covalently bound xenobiotics become integral components of the humic substances (31).

It was shown in several in-vitro experiments with different pesticides that free radicals were formed after an initial oxidation by exoenzymes. The highly reactive molecules can undergo coupling or polymerization (28,32,33). This kind of polymerization process is pointed out to be one of the most important mechanisms during the natural humification process of soil organic matter (34). For this reason the formation of nonextractable residues of xenobiotics is also described as (controlled or induced) humification.

The entrapment or sequestration of the pollutants in macromolecular humus substances is another mechanism which has received more attention recently (35). It is believed that the sequestration involves slow partitioning of the hydrophobic compounds into organic matter or slow diffusion into micropores (36).

Beside stimulation of the mineralization, the increased immobilization of the pollutants by the formation of nonextractable residues has become a major point of interest in bioremediation studies. A model of the behavior of PAH degradation in soil was deduced from previous studies (Fig. 1). PAH can be mineralized, which includes

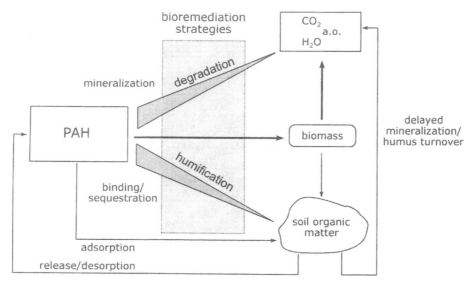

Figure 1 Model of different bioremediation strategies for the reduction of extractable PAH in contaminated soils. (Modified from Ref. 13.)

the formation of biomass, CO_2, and H_2O, or can introduced in the humification process. It is presumed that, depending on the binding mechanisms, the PAH or PAH metabolites can undergo either an adsorption–desorption cycle or a delayed mineralization as is to be expected for the natural humus turnover. It has therefore been suggested that intentional enhancement of the binding of xenobiotics to soil organic matter can be used as a method for detoxifying organic pollutants and therefore as an alternative bioremediation strategy (37,38). The application of this strategy however, requires studying factors which influence the formation of these nonextractable residues in more detail. The fate and stability of these nonextractable PAH residues should also be monitored carefully for long-term risk assessment. Points to be considered are whether the nonextractable residues can be remobilized, in which time periods significant amounts of previously "bound" residues can be released, or whether the residues remain stable in the soil at all and therefore long-term detoxification can be presumed.

II. FORMATION OF NONEXTRACTABLE PAH RESIDUES IN CONTAMINATED SOILS

In the first set of experiments the formation of nonextractable residues and some factors of influence were investigated. These experiments were conducted in the context of a bioremediation of a PAH-contaminated site near Hamburg, Germany. The bioremediation technique employed by the remediation company was based on the addition of the white rot fungus *Pleurotus ostreatus*. White rot fungi like *P. ostreatus* are known to metabolize PAH (39) and to promote radical reactions which may result in an enhanced formation of nonextractable residues. To our knowledge this was the first bioremediation of PAH-contaminated soil where *P. ostreatus* was used on a technical scale.

In our simultaneously conduced laboratory studies these processes were simulated by mass balance experiments with ^{14}C-labeled PAH. The soil samples of the contaminated site were spiked with either ^{14}C-naphthalene, ^{14}C-anthracene, ^{14}C-pyrene, or ^{14}C-benzo(a)pyrene and were incubated after the addition of the white rot fungus, precultivated on straw, for about half a year in bioreactor systems as described by Wienberg et al. (40) (Fig. 2).

During this incubation the fate of the ^{14}C-PAH was balanced in the different fractions, namely, mineralization as ^{14}C-CO_2, release of volatile compounds, the solvent-extractable fraction, and the fraction of the nonextractable residues. The extractable ^{14}C-activity was analyzed by the use of a two-step extraction procedure composed of a solvent extraction and an alkaline hydrolysis (17). The nonextractable residues were determined either by combustion of the remaining soil pellet or by combustion of an unextracted parallel probe. Since the analytical variation was less than 6%, the nonextractable residues were in most cases calculated as the difference between the total ^{14}C-activity of the parallel probe and the extractable ^{14}C-activity.

By determining the ^{14}C-activity in the different pathways we were able to get complete mass balances with a recovery rate of $100 \pm 7\%$ of the initial ^{14}C-activity. It was confirmed that the mineralization and the humification of the PAH were the two main processes of PAH depletion in soil (Fig. 3). As shown for ^{14}C-naphthalene, ^{14}C-anthracene, and ^{14}C-pyrene, a relevant part of the ^{14}C-activity was mineralized and another part was immobilized as nonextractable residues. If ^{14}C-naphthalene and ^{14}C-pyrene were applied, the mineralization was the more important process. About

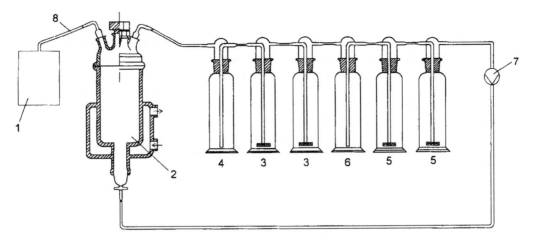

Figure 2 Bioreactor system to estimate the fate of ^{14}C-labeled xenobiotica in soil samples (1, oxygen reservoir; 2, bioreactor with soil; 3, gas-washing vessels to adsorb volatile compounds; 4 and 6, empty gas-washing vessels; 5, gas-washing vessels with NaOH to adsorb CO_2; 7, pump; 8, glass tubing).

65–70% of the ^{14}C-activity was mineralized to $^{14}CO_2$ at the end of the incubation time of about half a year. The formation of nonextractable residues was not so pronounced as in the case of ^{14}C-anthracene. At the end of this incubation more than 45% of the ^{14}C-activity from anthracene was not extractable, while this fraction ranged between 15% and 25% for ^{14}C-naphthalene and ^{14}C-pyrene. The evolution of volatile compounds was negligible in all experiments. The volatilization was less than 1% of the total ^{14}C-activity, also in the experiments with the most volatile ^{14}C-naphthalene. Even ^{14}C-benzo(a)pyrene was mineralized in this set of experiments, however to a smaller extent than the other ^{14}C-PAH, obviously due to the minor water solubility of the compound.

Control experiments with sterilized soil showed that the degradation was a biogenic process. In Fig. 4 the fate of, e.g., ^{14}C-anthracene and ^{14}C-pyrene is shown for soil samples that were sterilized by fumigation with chloroform. At the end of the incubation the release of $^{14}CO_2$ was less than 0.5% of the initial ^{14}C-activity. The second important conclusion of these sterile control experiments was that after an initial spontaneous formation of some amount of bound residues, the formation of nonextractable residues depended on biogenic activity as well. The further increase of the nonextractable residues fraction during the incubation of the sterilized soil samples was either not detectable or very small (Fig. 4). In some cases this fraction was even slightly reduced during the incubation.

However, since a small part of the total ^{14}C-activity was not extractable at the beginning of the incubation (day 0), it can also be assumed that different processes are involved in the formation of nonextractable residues: (a) a fast and presumably chemical process, e.g., a process which may result from sorption and/or reaction with radicals present in the soil humic substances; and (b) a slower, biogenic-mediated process which may result from a (covalent) binding of partially transformed PAH metabolites to the soil organic matter.

PAH themselves are not readily susceptible to oxidative coupling. A covalent binding of PAH can occur only after transformation by radical reactions or after incomplete degradation. The functional groups of such metabolic transformation products can then

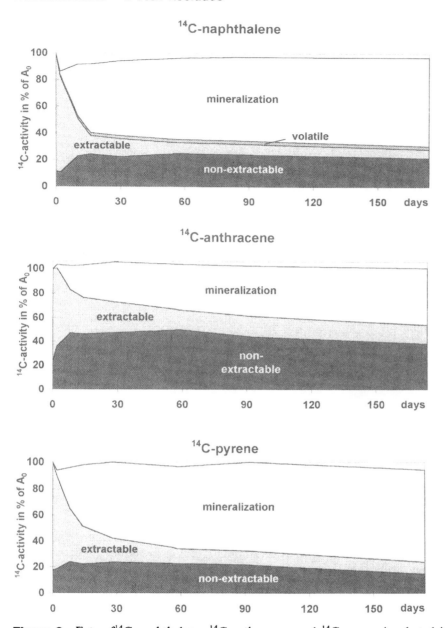

Figure 3 Fate of ^{14}C-naphthalene, ^{14}C-anthracene, and ^{14}C-pyrene incubated in contaminated soils after the addition of the white rot fungus *P. ostreatus*.

react with the surrounding organic soil matrix (27). The formation of covalent bondings (e.g., ester bonds) between partly degraded PAH or PAH metabolites and soil organic matter was also proven recently by selective chemical degradation experiments and by the use of ^{13}C-labeled PAH (23,27).

The assumed role of the necessity of a metabolic activation of PAH is further confirmed by the fact that the mineralization correlated well with the formation of

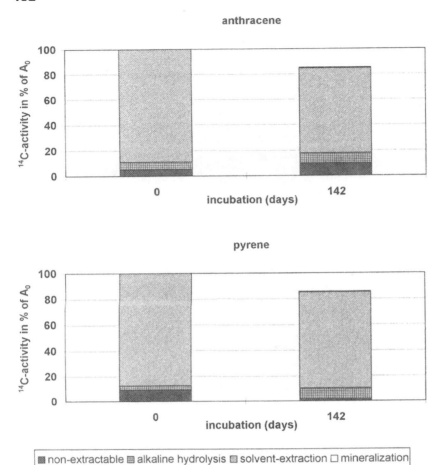

Figure 4 Fate of ^{14}C-anthracene and ^{14}C-pyrene in sterilized soil samples.

nonextractable residues. All experiments carried out so far have shown a clear relationship between the mineralization and the formation of nonextractable residues. The first observation was that the formation of residues did not occur if no mineralization took place. However, as soon as the degradation started, indicated by low mineralization rates, an enhanced formation of nonextractable residues began (Fig. 5). Therefore it can be concluded that the mineralization or partial degradation of the xenobiotics is a requirement to the further formation of nonextractable residues. On the other hand, high mineralization rates reduced the formation of nonextractable residues. The formation slowed down or stopped if the mineralization became dominant. The degradation of ^{14}C-naphthalene and ^{14}C-pyrene are examples for this case (Fig. 5). We conclude from this observation that an appreciable mineralization may also become a competitive process to the formation of nonextractable residues.

The original assumption that the white rot fungus *P. ostreatus* would be able to enhance the immobilization of PAH in soil due to its ligninolytic enzymes could not be confirmed in our experiments. Specific stimulation of the humification process was not observed (Fig. 6). The formation of nonextractable residues did not differ regardless

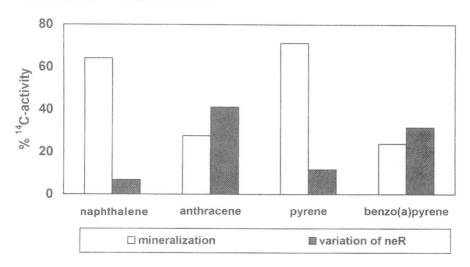

Figure 5 Comparison of mineralization rates and formation of nonextractable residues in soil samples with the PAH ^{14}C-naphthalene, ^{14}C-anthracene, ^{14}C-pyrene, and ^{14}C-benzo(a) pyrene.

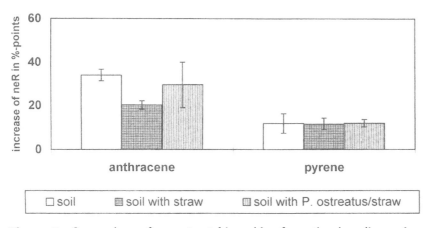

Figure 6 Comparison of nonextractable residue formation in soil samples supplemented with *Pleurotus ostreatus*, straw, and unsupplemented soil.

of the supplementation of the soil samples. It was as high in soil samples supplemented with *P. ostreatus* as in soil samples that had not been supplemented at all (controls).

Further experiments with mature compost showed that this organic supplement also had no influence on the extent of nonextractable residue formation. Even in this case the mineralization was the decisive factor, as documented in Fig. 7. In this experiment the concentration of ^{14}C-anthracene was stepwise enhanced to investigate the adaptation of the soil microflora to metabolize PAH. During the first 150 days the addition of ^{14}C-anthracene reacted a nearly constant level of ^{14}C-activity in the soil sample. At day 200 the ^{14}C-activity in the system was nearly doubled. The mineralization and the amount of nonextractable residues were compared for a compost-supplemented soil and for an unsupplemented control as well. The addition of mature compost led to

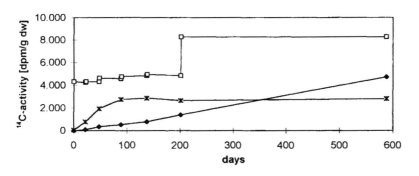

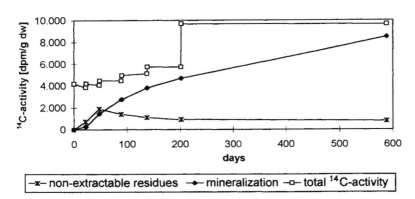

Figure 7 Fate of ^{14}C-anthracene added stepwise in unsupplemented soil and compost-supplemented soil.

enhanced mineralization of ^{14}C-anthracene. In the unsupplemented control experiment the mineralization was very low at the beginning of the incubation time and went on steadily. Within the first 140 days only minor parts of the ^{14}C-activity were transformed to ^{14}CO$_2$. However, the formation of a nonextractable residue fraction was well established at that time. The nonextractable radioactivity increased in the course of incubation up to a stable concentration after 90 days, which was not influenced by the further addition of anthracene. In the soil sample that was supplemented with compost, the transformation to ^{14}CO$_2$ started with high rates, after a small lag phase of about 20 days. Nearly 88% of the stepwise-added ^{14}C-anthracene was mineralized at the end of the incubation. The high mineralization rate resulted at the same time in a decrease of the further formation of bound residues. While after 49 days of incubation 46% of the added ^{14}C-activity was nonextractable, the mineralization became clearly predominant afterwards. This experiment confirmed again the presumed relationship between the mineralization and nonextractable residue formation.

To obtain more information about the character of the bound residues, a humus extraction of the nonextractable residue fraction was carried out. The humic acids and the fulvic acids were extracted by a 0.5 N sodium hydroxide solution. After

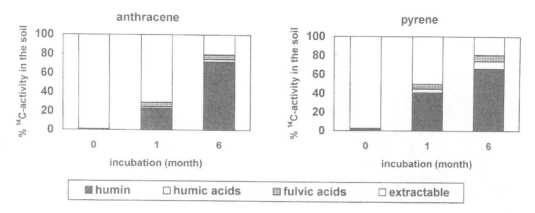

Figure 8 Differentiation of the ^{14}C-activity in the soil material in the fractions extractable, humic acids, fulvic acids, and humin.

centrifugation the supernant was acidified with diluted hydrochlorid acid to precipitate the humic acids. This fraction was redisolved in sodium hydroxide solution and the ^{14}C-activity in the humic acid and fulvic acid fraction was analyzed. The part of the organic substances which could not be extracted is called humin fraction. Because the soil originated from a contaminated site, the fraction of the humin is not well defined: it may include nonresolvable parts of the contamination such as coal and coke as well. The distribution of the 14C-activity in the different fractions (extractable part, humic acids, fulvic acids, and humin) is given in Fig. 8 for three incubation times (0, 1, and 6 months) for ^{14}C-anthracene and ^{14}C-pyrene. At the beginning only minor parts of the ^{14}C-activity were not extractable by solvents or were carried into the fraction of humic acids, fulvic acids, and humine. The nonextractable radioactivity increased in the course of incubation. The highest ^{14}C-activity of the nonextractable residue fraction was analyzed in the humin, while only minor parts of the radioactivity were located in the humic and fulvic acids fractions. At the end of the incubation (6 months), about 80% of the ^{14}C-activity of the soil remained in the humin fraction. It is assumed that just this fraction is the more stable part of the soil organic matter [e.g., (41)].

III. FATE AND STABILITY OF NONEXTRACTABLE ^{14}C-PAH RESIDUES

The use of pollutant immobilization in soil requires determining in advance the fate and stability of the nonextractable residues formed. One way to study the stability of nonextractable residues is to test the remaining residues under different ecological stress conditions such as biological, physical, and chemical worst-case conditions. The applied experimental concept of the remobilization studies is demonstrated in Fig. 9. The remobilization or the kind and amount of transformation determine the remaining risks. While strong mineralization to CO_2 would indicate a harmless detoxification of the residues, a slight mineralization rate could also indicate that the nonextractable residues became involved in natural turnover processes of the soil organic matter. A critical point would be the release as original PAH or PAH metabolites. Extractable pollutant residues are potentially bioavailable and could therefore also cause harm to the environment or to human health.

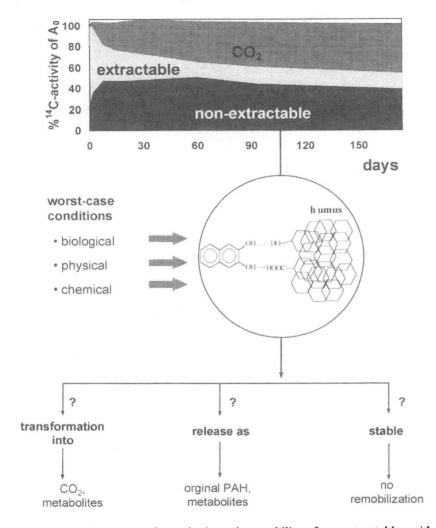

Figure 9 Scheme to evaluate the long-time stability of nonextractable residues.

Soil samples with stable fractions of nonextractable residues were used for these remobilization experiments. Soil samples being spiked with ^{14}C-PAH were preincubated after the addition of mature compost for about 200 days. The soil material originated (a) from a previous uncontaminated soil in agricultural use (reference soil "Muttergarten") and (b) from a PAH-contaminated site in Eastern Germany ("Wülknitz"). At the end of the preincubation period, 60–80% of the total ^{14}C-activity was no longer recoverable by extraction. The impact of a wide variety of different worst-case conditions was tested with these prepared soil materials, as given in Table 1. The effect of each treatment was analyzed in comparison to an untreated control by measuring the ^{14}C-activity in the different fractions (mineralization, water and solvent

Table 1 Worst-Case Conditions to Examine the Stability of Nonextractable Residues

Biological treatments
    Impact of organic supplements (compost and bark chips)
    Impact of isolated and enriched microorganisms (PAH- and humus-degrading)
    Impact of isolated enzymes
    Simulation of the "priming effect"

Physical Treatments
    Alternating temperatures
    Alternating soil water contents
    Mechanical disruption of soil aggregates

Chemical treatments
    Destabilization of metal-organic complexes
    Repeated extraction procedures
    Impact of a tenside

extractable part, nonextractable residues). The analytical procedures of the treatments have been described in more detail by Eschenbach et al. (42).

A. Impact of Biological Treatments

Organic supplements such as mature compost and bark chips were employed because it is known that they contain microorganisms which are able to degrade organic matter, such as litter and humic substances. Pregrown pure cultures of *Trametes versicolor* (DSM 3086) and *Rhodococcus erythroplois* (DSM 1069, Deutsche Sammlung für Mikroorganismen und Zellkulturen, Braunschweig, Germany) were used for the same reason [lignin-decomposing (43,44), PAH-degrading (45)]. The impact of an isolated peroxidase enzyme (HRP; Sigma, Deisenhofen, Germany) was determined in a further set of the experiment. Peroxidases are well known for their ability to catalyze the breakdown of organic polymers such as lignin and humic substances in soil. They are involved in humification processes as well.

After the addition of the diverse biological supplements, the soil samples ("Muttergarten" with 87% and "Wülknitz" with 82% nonextractable ^{14}C-residues of ^{14}C-anthracene) were incubated for about 200 days in a bioreactor system, shown in Fig. 2. The mineralization of the ^{14}C-activity was rather delayed, and no distinguishable differences in the mineralization rates were detectable among the different soil samples. The evolution of $^{14}CO_2$ after 206 days of incubation varied between 7.5% and 9.3% of the initial ^{14}C-activity, but in the soil sample with bark chips the mineralization was slightly higher (14.3% $^{14}CO_2$) (Fig. 10).

The amount of extractable ^{14}C-activity was also not influenced by the biological treatments. Initially and at the end of the incubation, about 13% of the ^{14}C-activity was extractable by the two-step extraction procedure in each of the treated soil samples as well as in the untreated control. This result reveals that no additional ^{14}C-activity was transformed into extractable fraction during the incubation.

However, the amount of nonextractable residues was slightly reduced. While 87% of the initial ^{14}C-activity was not extractable by the extraction with acetone and alkaline hydrolysis at the beginning, this fraction decreased to 82% up to 74% after the incubation,

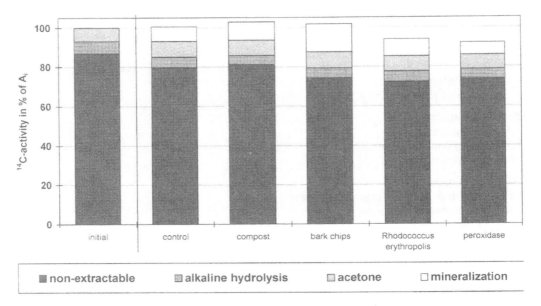

Figure 10 Impact of organic supplement (compost and bark chips), of the humus-degrading microorganism *R. erythropolis*, and of the enzyme peroxidase on the fate of nonextractable ^{14}C-anthracene residues (soil: Muttergarten). Mineralization to $^{14}CO_2$ and fractionating of the ^{14}C-activity in the soil (extractable by acetone alkaline hydrolysis, and nonextractable) at zero time (initial) and at the end of the incubation.

as shown in Fig. 10. These small differences were undistinguishable. The fraction of nonextractable residues was reduced regardless of the impact of all biological worst-case treatments. The same results were obtained after the inoculation of *T. versicolor* (data not shown).

These results agree very well with previous investigations which were conducted with other soil materials from a contaminated site near Hamburg, Germany, and which were preincubated with the white rot fungus *Pleurotus ostreatus* to form nonextractable residues from ^{14}C-anthracene (42). Both observations, i.e., the slight reduction of the nonextractable residue fraction and its correlation to a slow mineralization of ^{14}C-activity, indicate that the bound PAH residues may have become involved in slow humus turnover and degradation processes as shown in Fig. 1. However, the most important result for long-time risk assessment is that the extractable fraction of the ^{14}C-activity did not change in any way.

B. Impact of Physical Treatments

Soil samples with stable amounts of nonextractable residues were treated by alternating both, temperature and water content, to investigate the impact of physical stress on the release of nonextractable residues. The background of these experiments was that the aggregate structure of a soil is altered by these treatments, which simulate natural conditions such as frost damage.

To induce disruption of the spatial structure of soil aggregates, a soil sample with 66% nonextractable residues was exposed to daily temperature alternations of $-20°C$

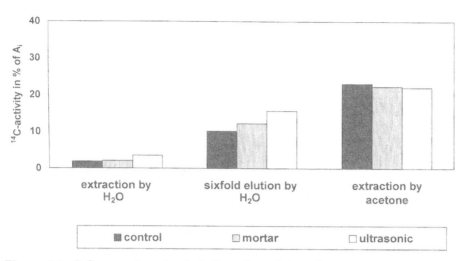

Figure 11 Influence of mechanical disruption of the soil aggregates on the extractability of ^{14}C-activity in a soil sample with nonextractable ^{14}C-anthracene residues.

to $+40°$C cycles. The distribution of ^{14}C-activity was measured at the beginning and after a total incubation time of 4 months. The nonextractable residue fraction was not influenced by this treatment, and 63% of the initial ^{14}C-activity was not extractable by the two-step procedure at the end of the incubation. The fraction of the extractable ^{14}C-activity decreased regardless of the treatment, by about 5 percentage points.

Another technique applied to study the effect of a physical disruption of soil aggregates on the fate of nonextractable residues was to crush soil aggregates by mechanical means. Soil samples with 70% nonextractable residues of ^{14}C-anthracene were treated for this purpose by either an ultrasonic desintegrator or grinding in a mortar. The treated soil samples and a nontreated aliquot of soil were subsequently extracted by water and acetone, respectively, and a sixfold elution with water was conducted.

The extractable amount of ^{14}C-activity was equal in all soil samples if one compares the ^{14}C-activity present in the acetone extract (Fig. 11). Release of parts of the nonextractable PAH residues could not be detected. However, the soil samples that had been treated by either of the mechanical means showed a slight tendency to release more ^{14}C-activity in the water-extractable fraction than in the untreated control. The sixfold elution with water confirmed this result: while only 10% of the initial ^{14}C-activity could be eluted from the control sample after the sixth elution cycle, it was possible to elute 16% of the ^{14}-activity in the water phase from the soil sample that had been treated by ultrasonics. The results could be confirmed by comparable remobilization experiments with other soil samples containing nonextractable residues. The results have therefore shown that efficient mechanical disruption of soil aggregates might actually contribute to a slight increase of the water-elutable ^{14}C-activity.

C. Impact of Chemical Treatments

The third type of worst-case treatments we tested was the destabilization of the humic material by chemical agents. The soil organic matter is stabilized by metal organic complexes which may form chelates or produce a spatial aggregate structure of humic substances (46). Therefore the effect of a complexing agent (EDTA), which should be able to destabilize these naturally occurring metal organic complexes, was studied. Different soil materials with nonextractable residues from ^{14}C-anthracene were extracted with EDTA solutions at different concentrations. The control soil was extracted by water rather than by EDTA solution. After this extraction step the treated soil samples were sequentially extracted by methanol/water, acetone, and alkaline hydrolysis.

The application of increasing concentrations of EDTA solutions led to increasing ^{14}C-activity in the EDTA soil extracts, while rather low ^{14}C-activity was detectable in the water-treated control samples. This fraction amounted in both water-treated soil samples to less than 6%. The following extraction step by methanol/water was not able to extract more than additional 2% in the control samples (Fig. 12). However, considerable amounts of ^{14}C-activity were found in the EDTA fractions of the treated soil samples. In the soil sample from the contaminated site (soil "Wülknitz," Fig. 12B), about 12% of the initial ^{14}C-activity was extractable by the 0.05 M EDTA solution; another 7% was detectable after the subsequent extraction by methanol/water.

The increasing extractability of ^{14}C-activity at increasing concentrations of EDTA solution correlated with a reduction of the nonextractable residue fraction. In the soil sample which was treated at the highest EDTA concentration, only 63% of the initial ^{14}C-activity remained nonextractable, whereas the fraction of nonextractable residues amounted to 75% in the untreated control sample. The fraction of nonextractable residues was reduced to a similar extent, by 13%, after treatment by the EDTA solution in the reference soil ("Muttergarten") as well (Fig. 12A). The chemical dissolution of metal organic complexes led to an appreciable reduction of the nonextractable residue fraction and to a transfer of ^{14}C-activity in the extractable fractions (extraction by EDTA solution and methanol/water).

The remobilization effect of EDTA on the nonextractable ^{14}C-activity raises the question of the chemical structure of the released ^{14}C-activity. One explanation could be that the ^{14}C-activity consists mainly of those ^{14}C-PAH molecules that had been physically entrapped within the soil matrix and that were subsequently set free due to the destruction of the spatial structure of the soil organic matter. It has been shown recently that humus may contain such cagelike structures in which hydrophobic substances such as PAH become sequestered or encapsulated (47–49). An alternative hypothesis is that EDTA may have cause the preferential release of soluble colloidal organic soil substances. In this case the ^{14}C-activity of the EDTA solutions would result from ^{14}C-atoms that become attached to or become part of these soluble humic substances.

Some further tests have shown clearly that the latter hypothesis is more plausible. A liquid/liquid extraction of the EDTA solutions by hexane yielded no ^{14}C-activity in the hexane fraction. This indicated that the released ^{14}C-activity did not result from the original PAH molecules or other hydrophobic PAH metabolites. A cross-check confirmed that spiked PAH were completely extractable by hexane from EDTA solutions.

Precipitation of the dissolved or colloidal organic carbon by acidifying the EDTA solution with concentrated HCl (pH < 1) demonstrated that the ^{14}C-activity was linked at least partly to the organic matter. The ^{14}C-activity in the supernatant was reduced by more than 50% by the precipitation. For the interpretation of this result it has to be considered that soil organic matter fractions cannot be precipitated completely by the precipitation methods applied (e.g., HCl precipitates only humic acids).

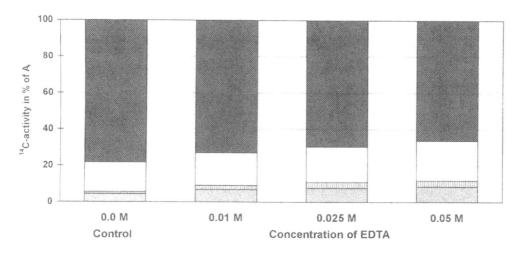

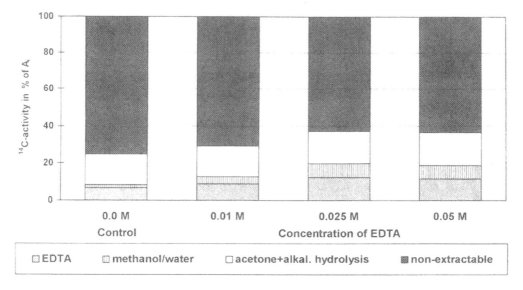

Figure 12 Effect of resolving metal-organic complexes by an extraction with EDTA at different concentrations on the fate of nonextractable residues of ^{14}C-anthracene (A) in the reference soil ("Muttergarten") and (B) in the soil from a contaminated site ("Wülknitz").

Though the measured ^{14}C-activity of ^{14}C-PAH is probably attached to the dissolved organic matter, this does not imply that the released ^{14}C-activity is harmless. It is well known that water-soluble humic substances can act as carriers for organic xenobiotics, transporting them over long distances (50,51). In addition, it cannot be excluded yet that the ^{14}C-PAH or their metabolites could be released from the water-soluble humic substances.

IV. CONCLUSIONS

The results presented have shown that humification is a major sink of PAH in soil. Both the formation of nonextractable residues and mineralization represent the major depletion pathways of previously available PAH. Both processes, degradation and humification, are also linked to microbial activities. No enhanced formation of residues was detectable in sterilized soil samples. However, a spontaneous nonbiological derived formation of residues was observed to a minor extent at the very beginning of the incubation.

It was demonstrated in this study that the metabolic potential of the soil microorganisms is one of the main factors controlling nonextractable residue formation. The formation of nonextractable residues is closely related to the mineralization of the substances. We assume that the partial transformation of PAH is a requirement of PAH-residue formation, because no further residue formation was observed if the PAH were not mineralized. On the other hand, mineralization at high rates was discerned as a competitive process. The amount of nonextractable residues was low if the mineralization was distinct.

Kästner et al. (12) presumed that compost might function as an additional binding substrate in soil, since they found greater residue formation after the addition of this organic supplement. However, we could not confirm this relationship. The experiments presented revealed that the supplementation of soil with compost, bark chips, or the white rot fungus *Pleurotus ostreatus* does not induce higher rates of nonextractable residue formation. However, in some cases the addition of supplements can enhance the biological potential to mineralize the PAH and therefore also the formation of bound residues. The mentioned experiments of Kästner et al. (12) support this assumption as well, if the CO_2 evolution and the nonextractable residue formation kinetics are regarded more closely. Experiments with stable isotope-labeled anthracene confirmed that the metabolic potential of soil microflora is the main factor controlling nonextractable residue formation (52).

Concerning the mechanism of nonextractable residue formation of PAH, it is assumed that the microbial degradation results in PAH metabolites which can subsequently be bound to the polymer soil organic matter. Bond-specific chemical degradation techniques allow us to identify covalent bonds between PAH degradation products and soil organic matter (23,27). The chemical processes which result in the spontaneous occurrence of nonextractable residues at the very beginning of the incubation are not yet elucidated. It seems possible, for example, that noncovalent types of association or nonbiological radical reactions are involved.

The covalent binding of xenobiotics to humic substances is considered to be relatively stable. However, one has also to take into account that covalent binding is proven so far only for minor parts of the total residues. Therefore it is essential to estimate by empirical investigations the remobilization potential and the long-term stability of the residues formed. We studied the long-term fate of these nonextractable PAH residues by treating them with a wide variety of worst-case conditions. Except for the special case of some release after treatment with metal-complexing agents such as EDTA, it can be said that the release of nonextractable PAH residues was generally very limited. Only very small amounts of nonextractable PAH-soil residues were transformed at all, and most were converted to $^{14}CO_2$ and thereby detoxified. The applied microbial and mechanical treatments were unable to convert nonextractable PAH residues into an extractable form. The stability of the nonextractable PAH residue fraction in soil indicates, therefore, that a very stable association of PAH molecules and humic substances must have been created during the preceding bioremediation period. This supports the assumption that PAH or rather their metabolites could be covalently bound to humic polymers and thereby become an indistinguishable part of the humic substances (33,38,53). Based on this hypothesis, the turnover of nonextractable PAH residues could be compared to humus turnover rates. Investigations with artificially produced complexes of covalently bound aniline and chlorinated phenols have shown that the mineralization of these compounds occurred at approximately the same rates as the mineralization of the humic substances themselves (54,55).

Therefore it can be concluded that nonextractable PAH residues will probably not cause harmful effects to human health or the environment. The given results support the concept of using PAH humification and the formation of bound residues as a decontamination strategy. However, for a more extended risk assessment one should also elucidate the chemical structure of the particular type of bonds between the ^{14}C-PAH-activity and the soil organic matter in more detail. The risk that may be involved with the attachment of ^{14}C-activity to dissolved organic matter, as assumed after the treatment with complexing agents (EDTA), must be studied further and in more detail as well.

ACKNOWLEDGMENTS

We thank Karen Johannsen, Volker Kleinschmitt, Momme Petersen, and Marco Silla for experimental support. We also gratefully acknowledge the funding of the research by the Bundesministerium für Bildung, Wissenschaft, Forschung und Technologie (BMBF) (Project No. 1480937 and No. 14810925). Parts of the investigation were sponsored by a grant of the Umweltbehörde Hamburg (WUP-Veringstrasse).

REFERENCES

1. Cerniglia, C. E. 1992. Biodegradation of polycyclic aromatic hydrocarbons *Biodegradation*, 3:351–368.
2. Smith, M. R. 1990. The biodegradation of aromatic hydrocarbons by bacteria. *Biodegradation*, 1:191–206.
3. Kästner, M., Wienberg, R., and Mahro, B. 1993. *Biologischer Schadstoffabbau in kontaminierten Böden.* Economic Verlag, Bonn.
4. Grosser, R. J., Warshawsky, D., and Vestal, J. R. 1991. Indigenous and enhanced mineralization of pyrene, benzo(a) pyrene and carbazole in soils. *Appl. Environ. Microbiol.*, 57:3462–3469.
5. Mueller, J. G., Chapman, P. J., Blattmann, B. O., and Pritchard, P. H. 1990. Isolation and characterization of fluoranthene-utilizing strain of *Pseudomonas paucimobilis*. *Appl. Environ. Microbiol.*, 56:1079–1086.
6. Mueller, J. G., Lantz, S. E., Ross, D., Colvin, R. J., Middaugh, D. P., and Pritchard, P. H. 1993. Strategy using bioreactors and specially selected microorganisms for bioremediation of groundwater contaminated with creosote and pentachlorphenol. *Environ. Sci. Technol.&d*, 27:691–698.

7. Brodkorb, T. S., and Legge, R. L. 1992. Enhanced biodegradation of phenanthrene in oil tar-contaminated soils supplemented with *Phanerochaete chrysosporium*. *Appl. Environ. Microbiol.*, 58:3117–3121.
8. Morgan, P., Lee, S. A., Lewis, S. T., Sheppard, A. N., and Watkinson, R. J. 1993. Growth and biodegradation by white-rot fungi inoculated into soil. *Soil Biol. Biochem.*, 25:279–287.
9. Bumpus, J. A. 1993. White rot fungi and their potential use in soil bioremediation processes. In: J.-M. Bollag and G. Stotzky (eds.), *Soil Biochemistry*, Vol. 8, pp. 65–100. Marcel Dekker, New York.
10. Kästner, M., and Mahro, B. 1996. Microbial degradation of polycyclic aromatic hydrocarbons in soils affected by the organic matrix of compost. *Appl. Microbiol. Biotechnol.*, 44:668–675.
11. Hupe, K., Lüth, J.-C., Heerenklage, J., and Stegmann, R. 1996. Enhancement of the biological degradation of soils contaminated with oil by the addition of compost. *Acta Biotechnol.*, 16:19–30.
12. Kästner, M., Lotter, S., Heerenklage, J., Breuer-Jammali, M., Stegmann, R., and Mahro, B. 1995. Fate of ^{14}C-labeled anthracene and hexadecane in compost-manured soil. *Appl. Microbiol. Biotechnol.*, 43:1128–1135.
13. Mahro, B., Schaefer, G., and Kästner, M. 1994. Pathways of microbial degradation of polycyclic aromatic hydrocarbons in soil. In: R. E. Hinchee, A. Leeson, L. Semprini, and S. K. Ong (eds.), *Bioremediation of Chlorinated and Polycyclic Aromatic Hydrocarbon Compounds*, pp. 203–217. Lewis Publishers, Boca Raton, FL.
14. Weissenfels, W. D., Klewer, H.-J., and Langhoff, J. 1992. Adsorption of polycyclic aromatic hydrocarbons (PAHs) by soil particles: Influence on biodegradability and biotoxicity. *Appl. Microbiol. Biotechnol.*, 36:689–696.
15. Bosma, T. N. P., Middleldrop, P. J. M., Schraa, G., and Zehnder, A. J. B. 1997. Mass transfer limitation of biotransformation: Quantifying bioavailability. *Environ. Sci. Technol.*, 31:248–252.
16. Luthy, R. G., Dzombak, D. A., Peters, C. A., Roy, S. B., Ramaswami, A., Nakles, D. V., and Nott, B. R. 1994. Remediating tar-contaminated soils at manufactured gas plant sites. *Environ. Sci. Technol.*, 28:266A–276A.
17. Eschenbach, A., Kästner, M., Bierl, R., Schaefer, G., and Mahro, B. 1994. Evaluation of a new, effective method to extract polycyclic aromatic hydrocarbons from soils samples. *Chemosphere*, 28:683–692.
18. Qiu, X., and McFarland, M. J. 1991. Bound residues formation in PAH contaminated soil composting using *Phanerochaete chrysosporium*. *Hazard. Waste Hazardous Mater.*, 8:115–126.
19. Calderbank, A. 1989. The occurrence and significance of bound pesticide residues in soil *Rev. Environ. Contam. Toxicol.*, 108:71–103.
20. Khan, S. U. 1982. Bound pesticide residues in soil and plants. *Residue Rev.* 84:1–24.
21. Roberts, T. R., Klein, W., Still, G. G., Kearney, P. C., Drescher, N., Desmoras, J, Esser, H. O., Aharonson, N., and Vonk, J. W. 1984. Non-extractable pesticide residues in soil and plants. *Pure Appl. Chem.*, 56:945–956.
22. Hatcher, P. G., Bortiatynski, J. M., Minard, R. D., Dec., J., and Bollag, J.-M. 1993. Use of high-resolution ^{13}C NMR to examine the enzymatic covalent binding of ^{13}C-labeled 2,4-dichlorophenol to humic substances. *Environ. Sci. Technol.*, 27:2998–2103.
23. Richnow, H. H., Eschenbach, A., Mahro, B., Seifert, R., Wehrung, P., Albrecht, P., and Michaelis, W. 1998. The use of ^{13}C-labeled polycyclic aromatic hydrocarbons for the analysis of their transformation in soil. *Chemosphere*, 36:221–2224.
24. Dec. J., Haider, K., Schäffer, A., Fernandes, E., and Bollag, J.-M. 1997. Use of a silylation procedure and 13C-NMR spectroscopy to characterize bound and sequestered residues of Cyprodinil in soil. *Environ. Sci. Technol.*, 31:2991–2997.
25. Hassett, J. J., and Banwart, W. L. 1989. The sorption of nonpolar organics by soils and sediments. In: B. L. Sawhney and K. Brown (eds.), *Reactions and Movement of Organic Chemical in Soils*, SSA Spec Publ. 22, pp. 31–44. Soil Science Society of America, Madison, WI.

26. Senesi, N. 1992. Binding mechanisms of pesticides to soil humic substances. *Sci. Total Environ.*, 123/124:63–76.
27. Richnow, H. H., Seifert, R., Hefter, J., Kästner, M., Mahro, B., and Michaelis, W. 1994. Metabolites of xenobiotica and mineral oil constituents linked to macromolekular organic matter in polluted environments. *Adv. Org. Geochem.*, 22:671–681.
28. Bollag, J. M., Myers, C. J., and Minard, R. D. 1992. Biological and chemical interactions of pesticides with soil organic matter. *Sci. Total Environ.*, 123/124:205–217.
29. Hatzinger, P. B., and Alexander, M. 1995. Effect of aging of chemical in soil on their biodegradability and extractability. *Environ. Sci. Technol.*, 29:537–545.
30. Pignatello, J. J. 1989. Sorption dynamics of carbonic compounds in soils and sediments. In: B. L. Sawhney and K. Brown (eds.), *Reactions and Movement of Organic Chemicals in Soils*, SSSA. Spec. Publ. 22, pp. 45–80. Soil Science Society of America, Madison, WI.
31. Wais, A. 1998. Non-extractable residues of organic xenobiotics in soils—A review. In: Deutsche Forschungsgemeinschaft (eds.), *Pesticide Bound Residues in soil, Senate Commission for the Assessment of Chemicals Used in Agriculture*, 2, pp. 5–33. Wiley-VCH Verlag, Weinheim, Germany.
32. Tatsumi, K., Freyer, A. Minard, R. D., and Bollag, J.-M. 1994. Enzyme-mediated coupling of 3,4-dichloroaniline and ferulic acid: A model for pollutant to humic materials. *Environ. Sci. Technol.*, 28:210–215.
33. Dec, J., and Bollag, J.-M. 1988. Microbial release and degradation of catechol and chlorophenols bound to synthetic humic acid. *Soil. Sci. Soc. Am.*, 52:1366–1371.
34. Bollag, J. M., and Loll, M. J. 1983. Incorporation of xenobiotics into soil humus. *Experientia*, 39:1221–1230.
35. Senesi, N. 1993. Organic pollutant migration in soils as effected by soil organic matter. Molecular and mechanistic aspects. In: D. Petruzelli and F. G. Helfferich (eds.), *Migration and Fate of Pollutants in Soils and Subsoils*, pp. 47–74. Springer-Verlag, Berlin.
36. Steinberg, S. M., Pignatello, J. J., and Sawhney, B. L. 1987. Persistence of 1,2-dibromoethane in soils: Entrapment in intraparticle micropores. *Environ. Sci. Technol.*, 21:201–1208.
37. Mahro, B., Kästner, M. 1993. PAK-Altlasten—Bewertung der mikrobiellen Sanierung. *Spektrum der Wissenschaft*, 10:97–100.
38. Bollag, J.-M. 1992. Decontaminating soil with enzymes. *Environ. Sci. Technol.*, 26:1876–1881.
39. Bezalel, L., Hadar, Y., and Cerniglia, C. E. 1996. Mineralization of polycyclic aromatic hydrocarbons by the white rot fungus *Pleurotus ostreatus*. *Appl. Environ. Microbiol.*, 62:292–295.
40. Wienberg, R., Eschenbach, A., Nordlohne, L., Kästner, M., and Mahro, B. 1995. Zur Erfordernis vollständiger stoffspezifischer Bilanzen bei der biologischen Bodensanierung. *Altasten Spektrum*, 5:238–243.
41. Stevenson, F. J. 1982. *Humus Chemistry. Genesis, Composition Reactions*. Wiley, New York.
42. Eschenbach, A., Wienberg, R., and Mahro, B. 1998. Fate and stability of nonextractable residues of [^{14}C]PAH in contaminated soils under environmental stress conditions. *Environ. Sci. Technol.*, 32:2585–2590.
43. Trojanowski, J., Haider, K., and Sundman, V. 1977. Decomposition of ^{14}C-labelled lignin and phenols by a *Nocardia* sp. *Arch. Microbiol.*, 114:149–153.
44. Eggeling, L., and Sahm, H. 1980. Degradation of coniferyl alcohol and other lignin-related aromatic compounds by *Nocardia* sp. DSM 1069. *Arch. Microbiol.*, 126:141–148.
45. Collins, P. J., Kottermann, M. J. J., Fields, J. A., and Dobson, A. D. 1996. Oxidation of anthracene and benzo (a) pyrene by laccases from *Trametes versicolor*. *Appl. Environ. Microbiol.*, 62:4563–4567.
46. Senesi, N. 1994. Spectroscopic studies of metal ion-humic substance complexation in soil. In: *International Society of Soil Science and Mexican Society of Soil Science 15th World Congress of Soil Science. 3A: Commission II: Symposia*, July 1994, Acapulco, Mexico, pp. 384–402.
47. Engebrestson, R. R., Amos, T., and von Wandruszka, R. 1996. Quantitative approach to humic acid associations. *Environ. Sci. Technol.*, 30:990–997.

48. Engebrestson, R. R., and von Wandruszka, R. 1994. Microorganization in dissolved humic acids. *Environ. Sci. Technol.*. 28:1934–1941.
49. Nanny, M. A. Bortiatynski, J. M., and Hatcher, P. G. 1997. Noncovalent interactions between acenaphtenone and dissolved fulvic acid as determined by ^{13}C NMR T_1 relaxation measurements. *Environ. Sci. Technol.*, 31:530–534.
50. Magee, B. R., Lion, L. W., and Lemley, A. T. 1991. Transport of dissolved organic macromolecules and their effect on the transport of phenanthrene in porous media. *Environ. Sci. Technol.*, 25:323–331.
51. Deschauer, H., and Kögel-Knabner, I. 1992. Binding of a herbicide to water-soluble soil humic substances. *Sci. Total Environ.*, 117/118:393–401.
52. Richnow, H. H., Eschenbach, A., et al. 1999. Formation of nonextractable soil residues: a stable isotope approach. *Environ. Sci. Technol.*, 33:3761–3767.
53. Mahro, B., Eschenbach, A., Schaefer, G., and Kästner, M. 1996. Possibilities and limitations for the microbial degradation of poly-cyclic aromatic hydrocarbons (PAH) in soil. In: *Wider Application and Diffusion of Bioremediation Technologies. The Amsterdam '95 Workshop*, OECD, Paris, pp. 297–307.
54. Arjmand, K., and Sandermann, H. J. 1985. Mineralization of chloroaniline/lignin conjugates and free chloroanilines by the white rot fungus *Phanerochaete chrysosporium*. *J. Agric. Food Chem.*, 33:1055–1060.
55. Haider, K., and Martin, J. P. 1988. Mineralization of C-labeled humic acids and of humic-acid bound C-xenobiotics by *Phanerochaete chrysosporium*. *Soil Biol. Biochem.*, 20:425–429.

21
Industrial Trickling Bed Biofilters for Abatement of Volatile Organic Compounds from Air Emissions

Vladimir O. Popov and Alexey M. Bezborodov
A. N. Bakh Institute of Biochemistry, Russian Academy of Sciences, Moscow, Russia

Phillip Cross
Waterlink Sutcliffe Croftshaw Ltd., Ashton-in-Makerfield, England

Adrian Murphy
BIP Limited, Oldbury, England

I. INTRODUCTION

Since the Industrial Revolution, enormous quantities of industrial chemicals have been released into the environment. These xenobiotics belong to various classes of organic compounds and comprise aliphatic and aromatic hydrocarbons, carbonyl, chlorinated compounds, different malodorous substances, etc. While some of these chemicals are readily biodegraded, many of them are quite persistent in the environment. The observed accumulation of hazardous recalcitrant compounds emphasizes that the natural diversity of microbes is not adequate to protect the environment from pollution. Selection and/or development of stable bacterial strains and consortia able to mineralize toxic organic pollutants under a variety of environmental conditions and proper application of these organisms in specialized hardware to produce effective and economic biodegradation purification processes is a high priority in the environment protection area.

One of the major sources of environment pollution are industrial air emissions containing volatile organic compounds (VOCs) (1 and references therein). Both European and U.S. legislation demands substantial reduction of VOC emissions from existing and newly established businesses and imposes deadlines for introducing either new technologies or end-of-pipe solutions. Many of these deadlines will come into effect before the end of the millennium. There is an urgent demand for environmental friendly, nonexpensive, affordable techniques for VOC control in industrially developed countries.

Biological methods of air purification are the methods of choice when large and diluted volumes of VOC-laden air are to be purified [Fig. 1 (2)]. Their main advantages over traditional methods (incineration, activated carbon adsorption, etc.) are as follows:

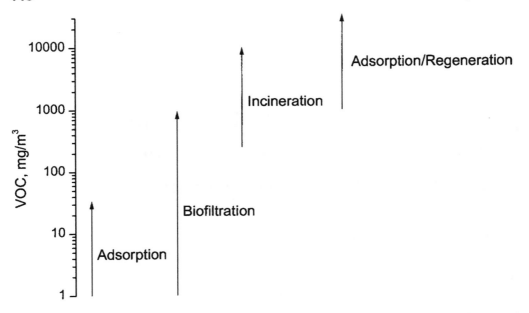

Figure 1 Application range of various VOC control technologies.

Moderate capital investment and low operational costs (operation at normal temperature and atmospheric pressure, low energy consumption, relatively long service life, no regeneration of the bed required)
Ecological safety, producing no secondary pollutants
Ready acceptance by regulating bodies and the general public

The efficiency of the biological methods of eliminating VOCs from gaseous industrial effluents depends primarily on the degradation ability of the microorganisms employed, as well as on the capacity of the specially engineered hardware to utilize the potential of microorganisms to degrade volatile pollutants. Several techniques of abatement of VOCs from industrial air emissions are known to date, including conventional biofiltration, bioscrubbers, and trickling bed biofilters (Table 1). A number of review articles are available covering recent advances in these areas (1–5 and references therein). Each of the techniques has its own advantages and limitations and occupies a well-defined market niche (Table 2).

The technology identified in the present chapter belongs to the subtype of so-called trickling bed biofilters (TBBs). TBBs operate by providing an efficient mechanism for contacting polluted air with a biocatalyst obtained by immobilization of viable, specially selected and adapted microbial consortia on the inert carrier. While passing through the biocatalyst bed, the VOCs in the contaminated air are continuously degraded to water and carbon dioxide. During the process of contaminant degradation, the biocatalyst is supplied with a nutrient solution which serves to irrigate the bed to provide the required nutrients to promote microbial activity and to maintain the system pH and other conditions at optimum levels for the microbial culture to function.

TBBs allow closer process control, and compared to conventional biofilters provide in some instances more efficient solutions to the problem of VOC abatement from air emissions. The contact times between the VOC-laden air and the biocatalyst in TBBs

Table 1 Types of Biological Air Purification Plants

	Biofilter	Bioscrubber	Trickling bed biofilter
Working medium	Filter bed—microorganisms immobilized on natural support (peat, bark, compost, soil, etc.)	Water, active sludge	Biocatalyst—microorganisms immobilized on synthetic carriers
Watering mode	No circulation	Circulation	Circulation
Basic stages of VOC abatement	Adsorption onto a filter bed followed by destruction by microorganisms	Absorption by a mobile phase (water) in absorber with subsequent destruction in aerotank by active sludge	Gas–liquid transfer followed by destruction in a biolayer
Source of nutrients	Filter bed material	Mineral salts added to water	Mineral salts added to water

Table 2 Comparison of Gas-Phase Bioreactors

Property/type	Biofilter	Bioscrubber	Trickling bed biofilter
Operation/maintenance	Simple	Complex	Simple/complex
Investment cost	Low	Moderate/high	Moderate
Operating costs	Low	Moderate	Low
Process control	Poor	Good	Good
Flow rates	Low	High	High/moderate
Susceptibility to channeling	High	—	Moderate
Susceptibility to fouling with excess biomass	Moderate	High/moderate	High
Capability to handle high VOC loads	Low/moderate	High	Low/moderate
Capability to handle irregular VOC loads	Moderate	High	Moderate
Capability to handle insoluble VOCs	Good	Low	Good
Capability to control odors	Good	Low/moderate	Moderate

may be reduced without a decrease in degradation efficiency, thus advantageously affecting the overall system dimensions and power requirements (6). However, these advantageous properties are counterbalanced by the TBB's restricted specificity, subject to fouling with excess biomass and reduced adaptability to the change of the inlet parameters.

Many theoretical models have been suggested to describe the performance of gas-phase bioreactors (7–14 and references therein). However, most of the models are of limited value because they rely on parameters that are not readily measurable and are based on a simplified assumptions. Recent insights into biofilm architecture (15), distribution of VOC degraders inside the biofilm (16), and population dynamics (15,17,18) reveal much more complex types of processes proceeding in the gas-phase bioreactors than was previously envisaged, thus requiring reevaluation of existing theories. Even the most sophisticated models, accounting for the biomass accumulation and disposal (19,20) require special calibration procedures for the given type of the plant. Thus devising a practical gas-phase bioreactor is still based mainly on an extensive prior pilot test program with subsequent scaling up.

While biofiltration and bioscrubbing have already established themselves as standard methods of VOC and odor control (21,22), with more than probably 1000 plants installed in the United States and Europe, TBB applications are confined mainly to pilot studies, with only a few full-scale systems reported (3), and they have yet to prove their technological potential and economic feasibility.

The above-mentioned considerations justify for further microbiological, macrokinetic, and technological research into VOC abatement in gas-phase bioreactors. The main goal of the present chapter is to summarize the activities in the field of the A. N. Bakh Institute of Biochemistry of the Russian Academy of Sciences and its British collaborator, Waterlink Sutcliffe Croftshaw Ltd., which resulted in the commercialization in the Russian Federation and the United Kingdom of the VOC-abatement technology based on the idea of TBBs, and which we call Bioreactor.

II. RESULTS AND DISCUSSION

Research in the field of microbiological VOC abatement was initiated in Moscow by the A. N. Bakh Institute of Biochemistry in the early 1980s. From the very beginning, it has been an application-oriented project, originally aimed at developing a miniature system capable of removing VOCs emitted by humans in enclosed compartments (spacecraft, submarines, etc.). The principle of TBBs has been used as a starting point for the R&D efforts, as it enabled most stringent process control and, if successful, could lead to the development of an intensive process with high throughput characteristics. Economic reforms undertaken in Russia in the early 1990s opened new market opportunities and were the driving force behind the evolution of the project into the VOC control field and commercialization of the technology. Since 1994 Waterlink Sutcliffe Croftshaw Ltd. of the United Kingdom has joined efforts with the Moscow team in developing the Bioreactor technology which culminated in a successful demonstration project of an industrial TBB for VOC control commissioned at BIP Ltd., U.K., in 1997.

To promote a Bioreactor project from the level of an idea to the point of commercially available technology, several problems were formulated and solved:

Isolation and selection of microorganisms (screening of existing collections of microorganisms and/or isolation and selection of efficient VOC degraders, maintaining collection of VOC degrading strains that proved suitable for practical applications, studying of VOC degradation pathways)

Biocatalyst development (choosing a proper carrier and development of cell immobilization techniques)

Lab-scale studies with individual VOCs and multicomponent mixtures (determining basic operating parameters and limiting factors)

Pilot plants (engineering and long-term monitoring)

Industrial plants

A. Isolation and Selection of Microorganisms

Efficient and robust microorganisms that degrade toxic VOCs to carbon dioxide and water form the foundation of the Bioreactor technology. It is well known that using specialized microbial strains in gas-phase bioreactors can provide significant advantages over activated sludge in terms of process efficiency (23,24). However, little is known about the fate of the initial inoculum and proliferation of the resulting consortia under real operational conditions existing in gas-phase bioreactors. Usually, after some time, a complex type of biocenosis is established, represented by the main strain, contaminant microflora, fungi, protozoa, and higher organisms. Recently, attention has been drawn to the importance of all, even minor members of the establishing ecosystem (3) in the stable performance of the process of biodegradation of the target pollutant(s). Thus the constraints applied to selection of the microorganisms to be used in TBBs are somewhat different from those for conventional biofilters and should be properly identified and understood.

The microorganisms to be used in industrial plants should satisfy very stringent requirements. They should allow efficient and complete mineralization of the target VOCs, should be characterized by high genetic stability, adaptivity, viability, and resistance to stress, contaminant microflora, and high levels of pollutants. High rates of xenobiotic utilization taken separately without considering all other important properties are not sufficient to justify the use of a microorganism in a particular application. Thus the strains

Table 3 Culture Collection of Industrial Microorganisms Maintained at A. N. Bakh Institute of Biochemistry, Moscow

Strain	Substrates	Code*
Pseudomonas fluorescens 16N2	Naphthalene, anthracene, salicyclic acid, benzene, toluene, *m*-, *p*-xylene, benzoic acid	B – 3343
Pseudomonas fluorescens ALK	Hexadecane, undecane, heptane, decane, naphthalene	B – 4097
Pseudomonas fluorescens STY	Styrene, ethyl benzene, 1-phenyl ethanol, 2-phenyl ethanol, phenyl acetic acid	B – 6477
Pseudomonas oleovorans TA/UN	Benzene, toluene, *m*-xylene, *p*-xylene, benzoic acid, naphthalene, anthracene, salicylic acid, ethyl acetate, butyl acetate, acetaldehyde, ethanol	B – 6655
Pseudomonas fluorescens PH	Phenol, *o*-xylene, cresol mixture	B – 6592
Pseudomonas sp. YZI	Acetone, 1-propanol	B – 6592
Pseudomonas oleovorans Sol	Solvent	—
Pseudomonas fluorescens WS	White spirit	—
Pseudomonas fluorescens sp. C1	Solvent	—
Pseudomonas sp. STY2	Styrene	—
Pseudomonas sp. XYL	Xylene	—
Rhodococcus sp. C6	Hexane	—
Rhodococcus sp.	Cyclohexanone, dimethyl formamide, kerosine, diethyl hexyl formate,	—
Methylobacterium organophilium	Methanol, formaldehyde	—
Methylobacterium extorquens	Methanol, formaldehyde	B – 2029
Nocardia rugosa	Methyl ethyl ketone, dimethyl formamide	—
Nocardia sp. EC2	Epichlorohydrine, methylene chloride, dichloroethane, chloroform	—
Nocardia sp. C7	Heptane	—

*National Collection of Industrial Microorganisms.

which have been finally selected for further use and maintained in an in-house strain collection (Table 3) are not the absolute record holders in terms of efficiency but rather are robust and versatile species preserving their high activity under hostile and competitive environment of the gas-phase bioreactor. These wild-type strains were isolated from contaminated sites and specially adapted to treat specific pollutants. The metabolic pathways of VOC degradation operating in these strains have been studied (25–28).

At first glance, selection of a panel of microorganisms for industrial application seems to be a rather labor-intensive task, because of a large variety of potential contaminants and thus the necessity to maintain vast collections of specific strains. However, this is not exactly the case. Simple statistical analysis shows that only a limited range of microorganisms is required (Fig. 2) to cover the majority of the potential applications suitable for TBB technology (excluding odor-control applications). Of course, these statistics are biased by the country of origin, targeted segments of the market, and local

Industrial Trickling Bed Biofilters for VOCs

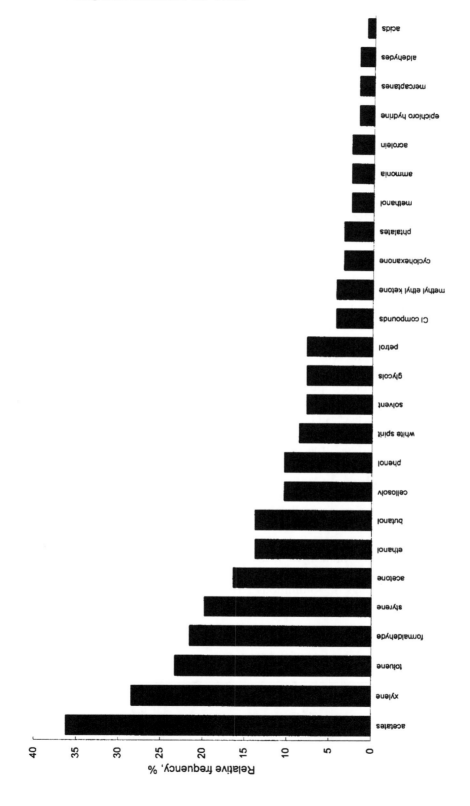

Figure 2 Frequency of inquiries received concerning particular VOCs (Russia, 1993–1996).

legislation. However, the trend probably will remain the same. Only a few types of VOCs—aromatic hydrocarbons, carbonyl compounds, alcohols, ketones, and esters—constitute the majority of the potential applications. Thus, to service such markets only a relatively limited stock of microorganisms will be sufficient.

B. Biocatalyst Development

One of the key factors determining the overall efficiency of microbial systems of VOC abatement in TBBs is an efficient biocatalyst, i.e., the combination of a microorganism and a carrier. The ideal biocatalyst should satisfy rather stringent and sometimes contradicting requirements: (a) maximal efficiency, i.e., high rates of xenobiotics utilization determined by the viable cell load per carrier unit volume and developed gas/liquid-biofilm contact area; (b) low back-pressure across the biocatalyst bed; (c) resistance to chemical, mechanical, and biological damage, or durability; (d) good rheological properties—resistance to compression over long-term operation; (e) availability and low cost.

Several methods of cell immobilization (physical adsorption, physical entrapment, chemical binding to derivatized carriers, etc.) and a number of commercially available carriers (sand, activated carbon, silica gel, various polymer structured packings and foams, various synthetic fibers) have been tested (29,30). As a result of extensive studies the following features were determined to be critical for efficient cell immobilization: (a) hydrophobicity of the cells; (b) surface charge of the carrier and the value of the free surface energy; and (c) accessible surface area and surface defects.

Comparison of some popular granular and fibrous carriers (Table 4) reveals that the latter are superior with respect to cell load. This is easily accounted for by much more developed surface area accessible to cell attachment in the case of fibrous carriers. Numerous surface defects in fibers are advantageous for tight cell adhesion, while the highly developed internal surface area of, e.g., activated carbon remains mainly inaccessible to cells because of rapid plugging of the narrow pores of the carrier. About 10% of the cells loaded on fibers are very tightly bound and could not be desorbed even after hours of intense rinsing of the carrier.

No direct correlation is usually observed between the cell load and accessible surface area in the case of granular carriers. When the granular size is reduced severalfold, the cell load increases only by several percent. However, such a reduction drastically increases the back-pressure of the bed, making it unsuitable for practical applications. On the contrary, very high surface areas per unit volume (3000–5000 m^2/m^3) can be achieved with special types of fibrous packings of voluminous structure employed in Bioreactor plants. Such

Table 4 Maximal Cell Load of Selected Granular and Fibrous Carriers

Carrier	Fiber (granule) diameter (mm)	Cell load (mg of dry biomass/g)
Silica gel	1.0 ± 0.1	2.5–3.5
Activated carbon	2.0 ± 0.5	4.0–6.5
Sand	1.0 ± 0.7	2.0–2.5
Polypropylene fiber	0.05–0.12	80–160
Polyester fiber	0.05–0.12	90–140
Triacetate fiber	0.09–0.11	65–70

packings ensure good retention of the microorganisms, are highly permeable to the air flow (differential pressures in the range of 50 to 150 mm of WG per meter of bed height), are light (packing density without biomass < 100 kg/m^3 of packing), durable, and readily available. These figures can be compared to 100–400 m^2/m^3 of specific surface area reported for conventional commercially available types of carriers currently employed in TBBs (6,13,24,31–33).

C. Laboratory Microreactors

Laboratory studies have been and remain a crucial step in developing the Bioreactor technology and investigating new applications. In the course of these studies, answers to the following questions are obtained: (a) which pollutants, in what concentrations, with what efficiency could be degraded; (b) which microorganisms could be used; (c) what are the critical factors (temperature, nutrients, maximal xenobiotic concentrations, etc.) limiting system performance; (d) what happens on transition from one-component to multicomponent VOC systems; (e) how the biocatalyst comprising several microbial strains will develop; etc.

A simple microreactor bench-top test system has been developed which enabled rapid screening of all the above-mentioned parameters and comprised all the major elements of the TBB. About 25 cm^3 of the biocatalyst (fibrous carrier with immobilized microorganisms) is placed in a thermostated glass vessel of 20 mm internal diameter. The biocatalyst is irrigated with a nutrient solution continuously circulated through the bed with a peristaltic pump. Controlled flow of VOC-laden air supplied by pollutants in a predetermined concentration from a special dosing system is provided by a microcompressor.

Of course, this simple system mimics only some of the features of the real TBB. It can be viewed as an "ideal" TBB operating under strictly controlled conditions in the absence of any complications inherent to real plants (Table 2). In spite of all these limitations, microreactors proved very instrumental in rapid screening of the various biocatalysts for particular applications and determining the maximal possible parameters of the process of VOC abatement (contact times, efficiencies, maximal loads, uptakes, competition between microbial strains, etc.).

D. Pilot Plants

The necessity to model real operational conditions existing in industrial TBBs, study long-term performance of the system, and optimize the technology led to design and construction of several pilot Bioreactor plants.

At the moment, two pilot Bioreactor plants are in operation. A stationary plant rated up to 200 m^3/h is located at the A. N. Bakh Institute of Biochemistry, Moscow, and is dedicated to R&D. A mobile, skid-mounted plant rated for 50–100 m^3/h (Fig. 3), is owned and utilized by Waterlink Sutcliffe Croftshaw (U.K.), and is moved from customer to customer for testing and demonstration purposes. Both pilot plants have a general layout as outlined in Fig. 4. They comprise an air pretreatment block including a heater, humidifier, carbon filter for VOC spike buffering (U.K. plant only), and VOC dosing device (if the plant is operated with the model artificial pollutant feed); the bioreactor block including the bioreactor itself, the nutrient dosing and spray systems, the device for reversing the air flow (optional), and demister; various control and measuring equipment.

Figure 3 Mobile, skid-mounted plant operated by Waterlink Sutcliffe Croftshaw Ltd.

The gas-phase reactor of the R&D pilot plant is made of stainless steel and has the form of a cylindrical vessel with dimensions 150×45 cm (cross section 0.159 m^2). It is equipped with a hatch (door) spanning the height of the vessel and made of a transparent material enabling visual monitoring of the process. The bioreactor has three shelves supporting special fibrous packing used for biomass immobilization. Each shelf is equipped with individual spray system consisting of an array of nozzles ensuring even

Industrial Trickling Bed Biofilters for VOCs

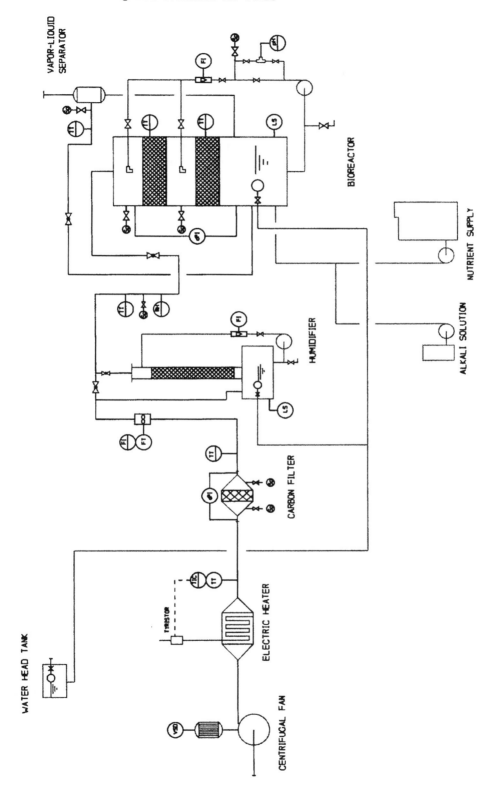

Figure 4 Flow diagram of bioreactor pilot plant.

moistening/rinsing of the whole surface of the carrier. The spray density over each level of the biocatalyst can be regulated manually. The irrigation of the biocatalyst bed with the nutrient solution is performed in a discontinuous manner. The duration of the "wet" and "dry" periods is programmed through a digital controller which also controls the operation of the heater and humidifier.

Concentrated stock mineral nutrient solution comprising a source of nitrogen, phosphorous, and sometimes—e.g., during the start-up periods—microelements and other additives (if not sufficient in the local tap water) is diluted continuously by a peristaltic pump into the storage tank (~ 40 L). The feeding rate of the stock solution determines the concentrations of the essential nutrients in the system. The resulting diluted solution is circulated via the bed by the irrigation pump, providing nutrients to the biocatalyst and ensuring its rinsing. Some minor part of the nutrient solution is drained continuously (up to 1–2% an hour) or at certain time intervals (daily or weekly). The level of the liquid in the system is replenished by tap water.

The overall volume of the packing (specific area 4900 m^2/m^3) in the R&D plant is about 0.07 m^3. At an air flow rate of 100 m^3/h the contact time between VOC-laden air and the biocatalyst is about 2.6 s. The following parameters of the process are collected routinely from the pilot plant on a daily basis: inlet and outlet concentrations of VOCs, obtained with a gas chromatograph (GC) equipped with a flame ionization detector (FID); temperature, back-pressure of the bed, air flow, pH, ammonium ion concentration, and water supply. Every other day, the value of COD in the drain is determined and the waste water is subjected to visual microscopy. Every second day, concentrations of the VOC between the shelves of the biocatalyst are measured. From time to time, samples of the carrier are withdrawn for microscopic investigation and quantitation of the biomass accumulated.

The results of several case studies are presented.

1. One-Component Emission: Xylene

Xylene is a rather abundant pollutant. Usually it is found in a mixture with other aromatic hydrocarbons, benzene, and toluene (so-called BTX). However, in some instances it could constitute a major source of contamination by itself. According to the reported degradation rates, BTX hydrocarbons can be ranked in order as benzene < xylene < toluene (34,35). Toluene is one of the most widely used model xenobiotics to study the processes of biological VOC abatement. A compilation of the reported toluene degradation rates is provided in Ref. 6.

The aim of the work that was accomplished using the Moscow pilot plant was to test the new microbial strain specially adapted to ensure high xylene degradation rates and to develop recommendations for further system scale-up. Technical-grade xylene, which is composed mainly of *m*- and *p*-isomers and contains only a minor part ($< 3\%$) of the *o*-isomer was used as a pollutant source. [It is well known that *o*-xylene is much less susceptible to biodegradation than the *m*- and *p*-isomers and requires specialized strains (34).]

Pseudomonas oleovorans var XYL specially adapted for xylene utilization has been used in the present study. The air flow rate was kept at 50–70 m^3/h (contact time ~ 3.6–5.0 s), while inlet xylene concentration was varied from 50 to 250 mg/m^3 (up to 200 $g/m^3 * h$ load). During the first half of the test period the system runs for 8–10 h/day, mimicking one shift operation of the real business. In the second half of the test period the system runs in continuous mode 24 h/day.

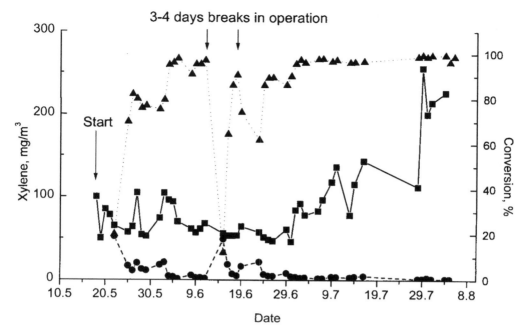

Figure 5 Pilot-plant operation data. *Ps. oleovorans* var. XYL immobilized on fibrous carrier. Flow rate 50 m^3/h, contact time ~5 s; ■, inlet xylene concentration, mg/m^3; ●, outlet xylene concentration, mg/m^3; ▲, efficiency, %; 18.05–26.06, discontinuous operation, 8 h/day, weekend breaks; 29.06–03.08, continuous operation, 24 h/day.

Figure 5 presents operation data obtained over a period of 2.5 months. After initial start-up the system reached steady state in about 2 weeks, and after fixing minor technical problems showed >95% efficiency with inlet xylene concentrations of 50–100 mg/m^3. Two times the system was halted, the first time for 4 days and the second time for 3 days, to imitate short-term breaks in business operation. Each time the system required from 4 to 5 days to regain its full efficiency. When run in continuous mode the system showed consistently high efficiency with respect to xylene removal (95–99%), while inlet concentrations were gradually increased from 50 to more than 200 mg/m^3.

The maximal values of xylene uptake observed at highest inlet concentrations were around 150 g/m^3 * h (contact time ~5 s). This compares very favorably with the data reported for a more readily degradable aromatic hydrocarbon, toluene (Table 5). On average, the rates of toluene utilization obtained with conventional biofilters are in the range 20–40 g/m^3 * h and residence times are usually around 30–90 s. Only recent, rates higher than 100 g/m^3 * h with contact times of 10 s or less were reported for a special type of stirred TBB with periodic removal of excess biomass (36).

2. Mixed Solvents

An example is presented by a U.K. printer, which manufactures a range of printed products that involve a gravure printing process during which inks (held in a solvent carrier) are applied to paper/acetate. The production process results in the generation of gas-phase emissions containing isopropyl acetate (IPAc), *n*-propanol, *n*-propyl acetate, butyl

Table 5 Bioreactor Performance Toward Selected VOCs

VOC	Contact time (s)	Uptake (g/m^3 * h)	Efficiency (%)	Site/reference	Strain
Toluene	3.3–3.0	130–220	80–90	Pilot, 100 m^3/h	*Pseudomonas oleovorans* TA
Xylene	5.0–3.6	100–150	>95	Pilot, 50–70 m^3/h	*Pseudomonas oleovorans* var XYL
Formaldehyde	2.4	750	>98	BIP Ltd. 15,000 m^3/h	*Methylobacterium extorquens*
Methanol	2.4	450	65	BIP Ltd. 15,000 m^3/h	*Methylobacterium extorquens*
Ethyl acetate	2.4–1.4	>130	80–90	"Zarya," Moscow 7,000–10,000 m^3/h	*Pseudomonas oleovorans* TA
	1.9	940–445	60	"Kauchuk," Moscow 5,600 m^3/h	
Mixed solvents (80% IPAc)	3.3	195	>90	Pilot, 75 m^3/h	*Pseudomonas oleovorans* TA
Toluene Various trickling filters and biofilters	1.2–480 (av. 30–90)	9–84 (av. 20–40)		(6)	
Stirred trickling filter	10	100	~100	(36)	Mixed culture
	7.7	160–210 275 max	25–60		

acetate, and periodically toluene. The first three VOCs comprise the major part of the solvent emissions (at around 90%), IPAc being the dominant pollutant. In accordance with U.K. environmental emission legislative restrictions, emissions of the above solvents should not exceed 100 mg/m^3 total VOC (expressed as carbon). The preliminary monitoring survey indicated a maximum VOC output of 250 mg/m^3 with a total air flow of 14,000 m^3/h. Bioreactor could be a suitable technology to treat the emission. To verify its applicability in terms of "best available technology not entailing excessive cost" pilot-plant studies were undertaken by Waterlink Sutcliffe Croftshaw Ltd. employing the skid-mounted SC Bioreactor mobile plant.

Commercially available packing (250 m^2/m^3) arranged in the form of a deep bed has been used for *Pseudomonas oleovorans* TA/UN immobilization. The pilot plant was operated continuously at 75 m^3/h (contact time 1–2 s) for 3 weeks, using a proportion of the real exhaust.

The data presented in Figs. 6 and 7 clearly indicate that at solvent loading < 200 mgC/m^3 the SC Bioreactor has shown an average of 84–95% overall removal of solvent from the air stream. The resulting outlet emission averaged < 20 mgC/m^3, which is well below the current consent limit. Based on the results of the pilot-plant studies, Bioreactor technology has been approved by the customer for scaling up. A full-scale Bioreactor plant rated for 14,000 m^3/h to treat mixed solvents manufactured by Waterlink Sutcliffe Croftshaw Ltd., was commissioned in 1999.

Industrial Trickling Bed Biofilters for VOCs

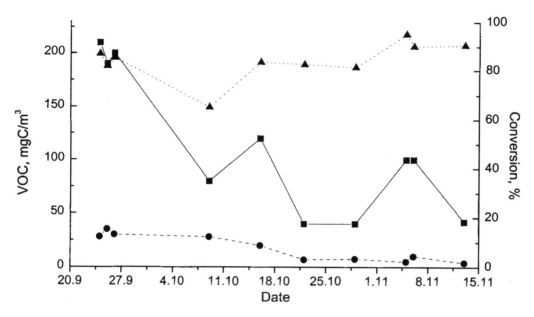

Figure 6 Pilot-plant operation data. *Ps. oleovorans* TA/UN immobilized on commercial packing. Mixed solvent (~80% isopropyl acetate). Flow rate 75 m^3/h, contact time 2 s; ■, inlet solvent concentration, mgC/m^3; ●, outlet solvent concentration, mgC/m^3; ▲, efficiency, %. Continuous operation, 24 h/day.

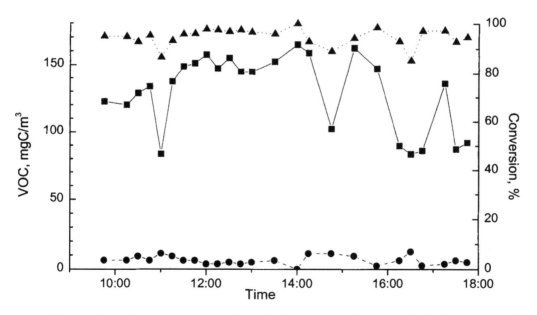

Figure 7 One-day pilot-plant operation data. *Ps. oleovorans* TA/UN immobilized on commercial packing. Mixed solvent (~80% isopropyl acetate). Flow rate 75 m^3/h, contact time 2 s; ■, inlet solvent concentration, mgC/m^3; ●, outlet solvent concentration, mgC/m^3; ▲, efficiency, %.

3. Odor Control

Odor control is one of the most mature areas of application of biofiltration methods. Conventional biofilters utilizing natural packings, e.g., peat, compost, etc., have proved their high efficiency in the field (22). Contrary to VOC abatement, odor control applications have a number of special features: a complex, multicomponent mixture of volatiles usually belonging to various classes of organic and sometimes inorganic (H_2S) compounds has to be neutralized; odor perception is rather individual and difficult to quantitate; the "multilayer" origin of odors—removal of a major part of the contaminants can unmask underlying odors, etc.

The applicability of Bioreactor technology for odor control of pet food plant emissions (Lujniki, Moscow region) has been tested. Initially the obnoxious smell originating during production of dry pet food was imitated in the laboratory. Experiments with laboratory microreactors allowed us to choose microbial strains of *Nocardia* sp. and *Rhodococcus* sp. that (a) considerably reduced the original odor intensity; (b) could operate under elevated temperatures (40–42°C), which was crucial for this particular application; and (c) had neutral background odor.

A pilot bioreactor plant rated for 300 m^3/h with overall dimensions of L : H : W = 3 : 2.3 : 0.6 m, comprising a fan, rotary humidifier/cooler, gas-phase reactor (rectangular, three shelves, fibrous packing), nutrient dozing and spray systems, and demister, has been erected at the company premises and connected to the source of malodorous emission. The plant was inoculated by the mixed culture of *Nocardia* sp. and *Rhodococcus* sp. and operated continuously for about 6 months. The biocatalyst temperature varied from 28 to 35°C, sometimes increasing to 45–50°C. The contact time was about 1.8 s. Organoleptic tests were performed daily by the local personnel and verified by obtaining GC/MS profiles (Fig. 8).

The organoleptic tests performed by a number of individuals showed that considerable decrease in odor intensity occurred. The specific intense obnoxious smell of pet food subjected to high temperatures has been completely neutralized, and the resulting emission after Bioreactor treatment was characterized in terms of a smell "of a pond," "of a cellar," etc. The instrumental data corroborate the organoleptic observations of the panelists. GC/MS profiles show that initial emission comprised at least 20 identifiable compounds. The major component was 2,6-ditretbuthyl-4-methyl phenol. A 10-fold overall reduction of volatiles has been observed, while some of the contaminants—hexanal, methyl propionate, valeric acid, benzaldehyde, 2-phenyl phurane—were not detected at the outlet.

E. Industrial Plants

Industrial VOC-abating Bioreactor plants should satisfy a number of criteria: (a) comply with local safety and environment legislation; (b) operate with minimal manual interference; (c) allow easy and rare maintenance; (d) have long service life; and (e) be cost competitive against existing alternative technologies (biofiltration, incineration, carbon adsorption, etc.)

The design of industrial Bioreactor plants for VOC abatement in general follows the concept implemented in pilot plants. However, as each application in VOC control is specific, the particular layout of the full-scale plant, especially auxiliary and control equipment, may vary from customer to customer.

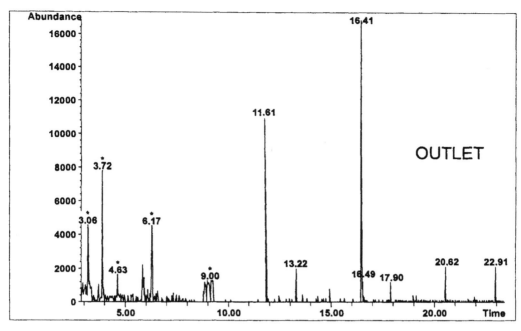

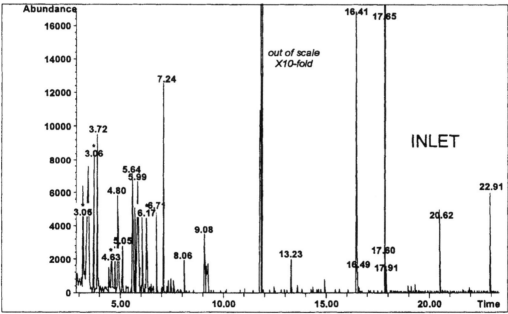

Figure 8 GC/MS profiles of the inlet and outlet of the desodoration pilot plant. Retention times of the identified components, s: hexanal 3.55; methyl propionate 4.30; 2-methyl propanol 4.63; valeric acid 4.80; heptanal 5.05; heptanol 5.64; benzaldehyde 5.81; 2-penthyl phurane 5.99; octanol 6.71; ether 7.16; nonanol 7.24; paraphine 9.10; 2,6-ditertbuthyl-4-methyl phenol 11.61; dibuthyl phtalate 16.41b; phtalate 16.49; methyl ester of linolenic acid 17.60; ether 20.62; di(2-ethyl hexyl)phtalate 22.91. [a]Compounds originating from hexane used for sample extraction are marked with an asterisk. [b]16.41 peak (dibuthyl phthalate, plastisizer) originates from the rubber parts of the pilot plant.

Figure 9 Bioreactor plant installed at BIP Ltd., Oldbury, U.K. 15,000 m^3/h, 2.4-s contact time; average load, 2 kg of formaldehyde and 5 kg of methanol per hour.

A typical Bioreactor unit, as shown in Fig. 9, rated for 5,000–15,000 m^3/h of VOC-laden air will have a footprint of 3.5 × 3.5 m or less excluding auxiliary and control equipment and a height of about 3.5–4.5 m. Usually it is manufactured in the form of a cylindrical vessel 2.0–2.5 m in diameter, made of stainless steel or plastic, with a sump to hold up to 1000 L of a nutrient solution. It contains about 10 m^3 of packing distributed over a number of levels. The dry weight of the plant is 1000–2000 kg. Normally, Bioreactor plants operate at contact times of 2–4 s (1000–2000 h^{-1}, m^3 of VOC-laden air per m^3 of biocatalyst per hour), consume 10–100 L of tap water an hour to compensate for evaporation and loss, and utilize ~10 kg of inexpensive inorganic salts per week. The electrical energy consumption is about 2–20 kWh (excluding fan energy required for air pretreatment). Typical operating back-pressure range across the bed is 100–200 mm WG. A bioreactor plant produces a waste water stream equal to about 1–2% of the sump volume a day, depending on the particular application.

More than 10 Bioreactor plants have been erected in Russia since the early 1990s to treat various VOCs. The first Bioreactor plant erected outside Russia was commissioned by Waterlink Sutcliffe Croftshaw Ltd. for their customer, BIP Ltd. (Oldbury, U.K.) in 1997 for abating emissions of C1 compounds—formaldehyde and methanol. In 1999 another SC Bioreactor plant was erected in the U.K. to control mixed solvent emission in the print industry.

The results of several case studies are presented.

1. Ethyl Acetate: "Zarya" (Moscow) Shoe-Making Factory

The shoe-making factory "Zarya" is located close to the city center, only 2 mil from the Kremlin. Because of its sensitive geographic location it is subject to close attention from environment protecting authorities. The main problem of the factory management in the area of gaseous emissions in 1992 was ethyl acetate. Ethyl acetate is emitted during various technological operations, being a major component of the glues used at the workshops. Each workshop of the factory is equipped with a network of local hoods collecting fumes from individual workplaces and/or automatic production lines, channeling them into a common collector, and finally disposing into the atmosphere.

Two options were considered by the factory management: erecting a conventional biofilter with a capacity of 180,000 m^3/h to process emissions from the whole factory, or equipping each workshop/production line with an individual Bioreactor. The biofilter designed for the factory was to be located in a separate building and had dimensions of $55 \times 15 \times 4.2$ m. The biofilter bed was to be made of a mixture of straw and hay, weighted 40 t, had a life time of 1 year, and required servicing (loosening) 5 times a year.

The cost analysis showed that while capital investment in the hardware was comparable for the biofilter and Bioreactor units, the cost of renovation of the premises and annual operating costs in the case of Bioreactors were much less (10–20% of that with conventional biofilter). The main reasons were the necessity to heat the building housing the biofilter in winter, extensive renovation of the duct lines, and complicated servicing of the biofilter bed.

The Bioreactor plant (2.0 m in diameter, 3.5 m in height without ducting, 1200 kg dry weight), rated for $\sim 10,000$ m^3/h, has been installed as the end-of-pipe solution indoors of one of the workshops (sixth floor of the industrial property building) to neutralize ethyl acetate emissions from an automatic shoe-manufacturing line. The plant was primed with *Pseudomonas oleovorans* TA/UN and operated with contact times of 1.4–2.4 s, while inlet ethyl acetate concentration varied from 10–20 to 100–150 mg/m^3 (average values ~ 40 mg/m^3).

Figure 10 presents the results obtained over the period of about 1 month in April–May 1995. On average, the plant performed with 70% efficiency and ensured compliance with local environmental legislation. With increase of the load the efficiency also increased, to 85–90%, probably because of overcoming stress due to biomass starvation.

2. Styrene: NPO "Mozaika," Balakovo, Russia

NPO "Mozaika" manufactures plastic sanitary equipment using polystyrene as one of the components. During the manufacturing process low-level (10–50 mg/m^3) styrene emission is generated. In regard to styrene release into the environment, Russian environmental legislation is much stricter than the European one, which allows up to 50 mg/m^3 at the end of the pipe. Thus, some protective measures had to be undertaken. To solve the problem three Bioreactor units with capacities of 12,000 m^3/h each have been installed and inoculated with *Pseudomonas fluorescens* Sty. These plants were among the first to be erected in Russia.

Plants normally operated continuously, three shifts a day. However, the business was often halted for periods lasting from several days to several weeks. The plants were monitored and operated manually by a technician, while general supervision was performed by a factory microbiologist.

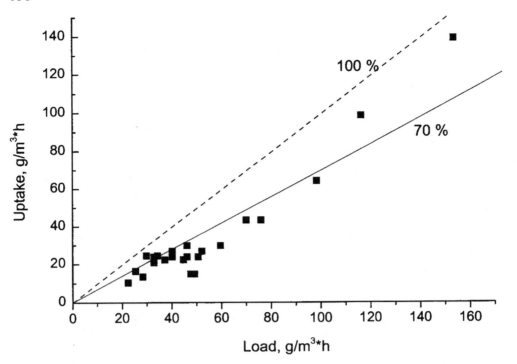

Figure 10 Bioreactor plant operation data. Shoe-making factory "Zarya," Moscow. 7,000 m³/h, 2.4-s contact time, ethyl acetate emission. Dashed line, 100% efficiency; solid line, least-squares fit of experimental data (70% efficiency).

Figure 11 presents the data obtained averaged over a period of 9 months in 1993. All three plants behaved similarly, showing average efficiencies of 69–73% and decreasing outlet styrene concentrations below the approved limit. Frequent stoppages of production did not affect the system performance significantly, as the biocatalyst during such periods was continuously supplied with styrene into the circulating nutrient solution and thus maintained in the active state. If the plants were run continuously, the efficiency gradually increased and reached 80–90%.

3. Formaldehyde and Methanol Removal at BIP Ltd., Oldbury, U.K.

BIP Ltd. is an international business operating globally, with manufacturing operations in the U.K., Mexico, the United States, Canada and Brazil. BIP produce a range of materials for use in the plastics, coatings, and manufactured goods industries. The processes which BIP carry out at their main site located in Oldbury, U.K., covers two main areas: amino molding powders and speciality resin production. The processes involved in the production of these materials create a number of chemical by-products and gaseous and liquid-phase emissions. While these emissions have traditionally been treated by liquid-phase biological treatment or released to the atmosphere, it is the objective of BIP that contaminants from their processes are removed at the source by an appropriate technology, which (in keeping with the HMIP/EA objectives and EPA 1990 legislation) is "best available technology not entailing excessive cost."

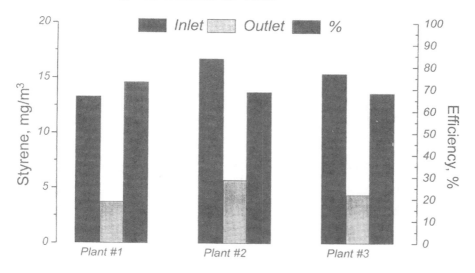

Figure 11 Bioreactor plant operation data. NPO "Mozaika," Balakovo, Russia. $3 \times 12{,}000$ m^3/h, 2-s contact time, styrene emission.

In order to fulfill these objectives BIP has been involved in a range of process rationalization and emission assessment operations which have identified potential emission control strategies. The Bioreactor technology offered by Waterlink Sutcliffe Croftshaw Ltd. was identified by BIP as both most suitable and the most cost-effective. The amino molding powder production-line emissions containing formaldehyde and methanol were chosen as a priority for a SC Bioreactor demonstration project. Emission limits on these compounds which are to be imposed in accordance with current EPA 1990 legislation require that atmospheric release is no greater than 5 mg/m^3 for formaldehyde and 80 mg/m^3 for methanol.

Initially, in 1996, a 2-month pilot-plant study was undertaken by Waterlink Sutcliffe Croftshaw to demonstrate the effectiveness of the SC Bioreactor in treating a proportion of dryer exhaust containing formaldehyde and methanol. The study provided essential performance data used in the subsequent design of a full-scale SC Bioreactor system to handle the total exhaust from the process.

Following successful completion of pilot studies, the full-scale system was designed and erected in 1997 by Waterlink Sutcliffe Croftshaw. Severe constraints were imposed on the intended VOC control system because of weight and space (roof installation). Superb characteristics of the Bioreactor in terms of footprint and weight/capacity allowed it to overcome these stringent limitations. Up to four systems, each capable of treating 15,000 m^3/h of contaminated air, can be positioned on the roof of the BIP industrial building without undertaking any additional reinforcements.

A general view of the plant is presented in Figure 9. The SC Bioreactor is preceded by a wet scrubber (not shown in the photo) that removes particulates present in the air emission. The purified air is discharged via a demister (to the left of the reactor body) and exhaust stack, preventing buildup of the plume (at the rear left). The system is computer controlled and equipped with necessary measuring instruments and sensors to enable functioning in automatic mode.

The SC Bioreactor plant is characterized by the following parameters:

Biocatalyst	*Methylobacterium extorquens* on fibrous packing
Air flow (nominal)	15,000 m^3/h
Dimensions	2.5 m diam × 4.5 m tall
Operating weight	~7 t
Volume of the carrier	9.6 m^3
Contact time	2.3 s
Bed temperature	30–35°C
Mineral salt consumption	2.5 kg/day
Water consumption	~1 m^3/day
Energy consumption	~8 kwh
Average formaldehyde load	1.6 ± 1.2 kg/h
Average methanol load	5.1 ± 1.4 kg/h

Close monitoring of the SC Bioreactor was performed over the whole period of operation since it was commissioned in August 1997. All the key operating parameters remained remarkably stable. The differential pressure of the bed did not change considerably over the period of 12 months and remained the same as at start-up (125 mm WG). Buildup of excess biomass resulting in reactor clogging and efficiency decline is the weak point of all TBBs (3,19,33,34,36). Careful optimization of the operating parameters allowed effective control over biomass development in the Bioreactor, ensuring its stable long-term performance. The microbial consortium also remained stable. Microscopic investigation of the samples of the carrier removed from the system after 7 months of continuous operation revealed that the microbial consortium was close to a monoculture.

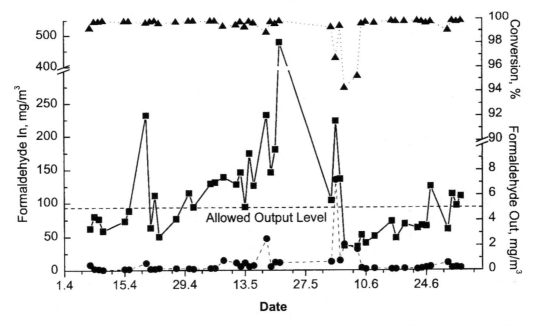

Figure 12 BIP Ltd. bioreactor plant operation data: ■, inlet formaldehyde concentration, mg/m^3; ●, outlet formaldehyde concentration, mg/m^3; ▲, efficiency, %.

Figure 12 shows the results obtained for formaldehyde over a period of nearly 3 months. The preliminary projections were that the inlet formaldehyde concentrations would not exceed 50–100 mg/m^3, and that methanol would not exceed 100–200 mg/m^3. In real life, average inlet formaldehyde concentration was twice as high, ~100 mg/m^3, and sometimes it reached even 250–500 mg/m^3. Average inlet methanol concentration was 340 mg/m^3 and varied from 200 to nearly 500 mg/m^3. However, in spite of all the oscillations in the load, the end-of-pipe concentration of formaldehyde remained within the approved 5 mg/m^3 range. An average 96% efficiency with respect to formaldehyde and 56% with respect to methanol has been achieved. At present, efforts to further improve system performance toward methanol are in progress.

The Bioreactor technology is approved by the Environment Agency in the U.K. and is compliant with the EC Council Directive on Limitation of Emissions of VOCs 97/C 99/02. The Groupe Europe des Plastiques Renforces Materiaux Composite estimates that around 1500 factories in Europe that could experience problems with formaldehyde emissions are potential users of the system. Particularly significant is that these are small to medium-size companies, which usually cannot afford much more expensive incinerators. We estimate that an incinerator of the same capacity as a Bioreactor in our particular case would be 2.5-fold more expensive with regard to capital investment and 10-fold more expensive to run.

III. CONCLUSIONS

Trickling bed gas-phase bioreactors are a rapidly emerging new technology in the field of biological methods of waste gas purification. TBBs could provide a number of advantages over existing types of biological VOC control techniques, the main ones being high removal efficiency per biocatalyst unit volume and efficient process control. High specific characteristics of TBBs significantly reduce space/energy requirements and thus overall costs. Sustained long-term efficient performance of TBBs can be achieved provided efficient control of the process parameters and adequate hardware design.

The concept of the TBB has been successfully realized in the bioreactor technology. A record of successful pilot-plant tests and industrial system applications position bioreactor as a versatile tool to solve environment protection problems.

ACKNOWLEDGMENTS

The invaluable contribution of Mr. J. H. Mole in promoting the technology transfer and implementing the project is highly appreciated.

The authors are grateful to W. Jackson and his project staff at Waterlink Sutcliffe Croftshaw Ltd. for efficient management of the project, to J. Lamond and M. Cavanagh of Waterlink Sutcliffe Croftshaw Ltd. for stimulating discussions, and to H. Whitcomb of BIP Ltd. for continued interest in the work.

The contributions of Dr. V. Jukov, A. Kurlovitch, Dr. I. Rogojin, Dr. I. Ulezlo, and Dr. N. Zagustina of the A. N. Bakh Institute of Biochemistry and Dr. N. Ushakova of the A. N. Severtzev Institute of Ecology and Evolution of the Russian Academy of Science are appreciated.

The engineering backup of METTEM-Technologies Ltd. of Balashiha, Russia (Dr. A. Orlov, V. Kopyilov, and A. Rastegaev) in designing the first bioreactor plants and the contribution of Innovational Biotechnologies Ltd. of Moscow (M. Piskunov) in commercialization of the project in Russia is acknowledged.

The authors thank Mrs. S. Kutzaeva of NPO "Mozaika," Balakovo, for providing long-term monitoring data.

V.O.P. and A.M.B. are grateful for financial support of the Russian part of the project provided by The State Fund for Promotion of Small Businesses in High Technological Areas (Chairman Mr. I. Bortnik) and International Business Technology Incubator (Chairman Prof. N. Fonstein).

The BIP demonstration project was supported in part by grant LIFE96ENV/UK/000403 from the European Union under the LIFE program.

REFERENCES

1. Edwards, F. G., and Nirmalakhadan, N. 1996. Biological treatment of airstreams contaminated with VOCs: An overview. *Water Sci. Technol.*, 34:565–571.
2. Leson, G., and Winer, A. M. 1991. Biofiltration: An innovative air pollution control technology for VOC emissions. *J. Air Waste Manage. Assoc.*, 41:1045–1054.
3. Cox, H. J., and Deshusses, M. A. 1998. Biological waste air treatment in biotrickling filters. *Curr. Opin. Biotechnol.*, 9:256–262.
4. Deshusses, M. A. 1997. Biological waste air treatment in biofilters. *Curr. Opin. Biotechnol.*, 8:335–339.
5. Dawson, D. S. 1993. Biological treatment of gaseous emissions. *Water Environ. Res.*, 65:368–371.
6. Pedersen, A. R., and Arvin, E. 1997. Toluene removal in a biofilm reactor for waste gas treatment. *Water Sci. Technol.*, 36:69–76.
7. Ottengraph, S. P. P., and Van den Oever, A. H. C. 1983. Kinetics of organic compound removal from waste gases with a biological filter. *Biotechnol. Bioeng.*, 25:3089–3102.
8. Andrews, G. F., and Noah, K. S. 1995. Design of gas-treatment bioreactors. *Biotechnol. Prog.*, 11:498–509.
9. Shareefdeen, Z., Baltzis, B. C., Oh, Y-S., and Bartha, R. 1992. Biofiltration of methanol vapor. *Biotechnol. Bioeng.*, 41:512–524.
10. Shareefdeen, Z., and Baltzis, B. C. 1994. Biofiltration of toluene vapor under steady-state and transient conditions: Theory and experimental results. *Chem. Eng. Sci.*, 49:4347–4360.
11. Hodge, D. S., and Devinny, J. S. 1995. Model removal of air contaminants by biofiltration. *J. Environ. Eng.*, 121:21–32.
12. Deshusses, M. A., Hamer, G., and Dunn, I. J. 1995. Behavior of biofilters for waste air biotreatment. 1. Dynamic model development. *Environ. Sci. Technol.*, 29:1048–1058.
13. Pedersen, A. R., and Arvin, E. 1995. Removal of toluene in waste gases using a biological trickling filter. *Biodegradation*, 6:109–118.
14. Diks, R. M. M., and Ottengraf, S. P. P. 1991. Verification studies of a simplified model for the removal of dichloromethane from waste gases using a biological trickling filter. *Bioprocess Eng.*, 6:93–99 (Part I); 131–140 (Part II).
15. Moller, S., Pedersen, A., Poulsen, L. K., Arvin, E., and Molin, S. 1996. Activity and three-dimensional distribution of toluene-degrading *Pseudomonas putida* in a multispecies biofilm assessed by quantitative *in situ* hybridization and scanning confocal laser microscopy. *Appl. Environ. Microbiol.*, 62:4632–4640.
16. Villaverde, S., Mirpuri, R. G., Lewandowski, Z., and Jones, W. L. 1997. Physiological and chemical gradients in a *Pseudomonas putida* 54G biofilm degrading toluene in a flat plate vapor phase bioreactor. *Biotechnol. Bioeng.*, 56:361–371.
17. Pedersen, A. R., Moller, S., Molin, S., and Arvin, E. 1997. Activity of toluene-degrading *Pseudomonas putida* in the early growth phase of a biofilm for waste gas treatment. *Biotechnol. Bioeng.*, 54:131–141.

18. Mirpuri, R., Jones, W., and Bryers, J. D. 1997. Toluene degradation kinetics for planktonic and biofilm-grown cells of *Pseudomonas putida* 54G. *Biotechnol. Bioeng.*, 53:535–546.
19. Alonso, C., Suidan, M. T., Sorial, G. A., Smith, F. L., Biswas, P., Smith, P. J., and Brenner, R. C. 1997. Gas treatment in trickle-bed biofilters: Biomass, how much is enough? *Biotechnol. Bioeng.*, 54:583–594.
20. Alonso, C., Suidan, M. T., Kim, B. R., and Kim, B. J. 1998. Dynamic mathematical model for the biodegradation of VOC's in a biofilter. Biomass accumulation study. *Environ. Sci. Technol.*, 32:135–142.
21. VDI Richtlinien 3478. 1985. *Biologische abluftreinigung Biowascher*. VDI, Dusseldorf.
22. VDI Richtlinien 3477. 1989. *Biologische abluftreinigung Biofilter*. VDI, Dusseldorf.
23. Kirchner, K., Schlachter, U., and Rehm, H.-J. 1989. Biological purification of exhaust air using fixed bacterial monocultures. *Appl. Microbiol. Biotechnol.*, 31:629–632.
24. Kirchner, K., Gossen, C. A., and Rehm, H.-J. 1991. Purification of exhaust air containing organic pollutants in a trickle-bed bioreactor. *Appl. Microbiol. Biotechnol.*, 35:396–400.
25. Utkin, I. B., Yakimov, M. M., Matveeva, L. N. Kozliak, E. I., Rogozhin, I. S., Solomon, Z. G., and Bezborodov, A. M. 1990. Catabolism of naphthalene and salicilate by *Pseudomonas fluorescens*. *Folia Microbiol.*, 35:557–560.
26. Utkin, I. B., Yakimov, M. M., Matveeva, L. N., Kozliak, E. I., Rogozhin, I. S., and Bezborodov, A. M. 1992. Degradation of benzene, toluene and o-xylene by *Pseudomonas* sp. Y13. *Appl. Biochem. Microbiol.* (Russia), 28:367–370.
27. Yakimov, M. M., Rogozhin, I. S., Matveeva, L. N., Kurlovitch, A. E., Zaikina, I. V., and Bezborodov, A. M. 1994. Direct dioxygenation of the aromatic moiety in the course of degradation of ethyl benzene and toluene by *Pseudomonas* sp. TA2. *Appl. Biochem. Microbiol.* (Russia), 30:224–249.
28. Utkin, I. B., Yakimov, M. M., Matveeva, L. N., Kozliak, E. I., Rogozhin, I. S., and Bezborodov, A. M. 1991. Degradation of styrene and ethyl benzene by *Pseudomonas* species Y2. *FEMS Microbiol. Lett.*, 77:237–241.
29. Kozliak, E. I., Solomon, Z. G., Yakimov, M. M., Fadushina, T. V., Germanskiy, G. I., Utkin, I. B., Rogozhin, I. S., and Bezborodov, A. M. 1991. Entrapment of the cells of *Pseudomonas fluorescens* 16N2 into triacetate cellulose fibers. *Appl. Biochem. Microbiol.* (Russia), 27:210–215.
30. Kozliak, E. I., Solomon, Z. G., Yakimov, M. M., Fadushina, T. V., Utkin, I. B., Rogozhin, I. S., and Bezborodov, A. M. 1993. Sorption of the cells of *Pseudomonas* sp. 16N2 on different carriers. *Appl. Biochem. Microbiol.* (Russia), 29:138–143.
31. Wittorf, F., Klein, J., Korner, K., Unterlohner, O., and Ziehr, H. 1993. Biocatalytic treatment of waste air. *Chem. Eng. Technol.*, 16:40–45.
32. Schindler, I., and Friedl, A. 1995. Degradation of toluene/heptane mixtures in a trickling-bed bioreactor. *Appl. Microbiol. Biotechnol.*, 44:230–233.
33. Weber, F. J., and Hartmans, S. 1996. Prevention of clogging in a biological trickle-bed reactor removing toluene from contaminated air. *Biotechnol. Bioeng.*, 50:91–97.
34. Sorial, G. A., Smith, F. L., Suidan, M. T., Pandit, A., Biswas, P., and Brenner, R. C. 1997. Evaluation of trickle bed air biofilter performance for BTEX removal. *J. Environ. Eng.*, 123:532–537.
35. Arcangeli, J.-P., and Arvin, E. 1995. Biodegradation rates of aromatic contaminants in biofilm reactors. *Water Sci. Technol.*, 31:117–128.
36. Laurenzis, A., Heits, H., Wubker, S.-M., Heinze, U., Friedrich, C., and Werner, U. 1998. Continuous biological waste gas treatment in stirred trickle-bed reactor with discontinuous removal of biomass. *Biotechnol. Bioeng.*, 57:497–503.

22
Air Stripping and Soil Vapor Extraction as Site Remediation Measures

Constantine J. Gregory and Frederic C. Blanc
Northeastern University, Boston, Massachusetts

I. INTRODUCTION

Air stripping, soil vapor extraction, and air sparging are basic stripping processes that may be used for the successful removal of volatile hazardous waste contaminants from soils and associated groundwater. Such stripping may be conducted in situ or in stripping towers as used in other gas–liquid-system industrial operations. Stripping tower systems may be constructed to provide a mobile technology for hazardous waste site remediation. The gas transfer is based on the gas–liquid equilibrium described by Henry's law. The process is capable of removing volatile organic and gaseous solutes present in low concentrations in the solvent liquid. This chapter describes air stripping, soil vapor extraction, and air sparging in detail.

II. PROCESS DESCRIPTIONS

A. Air Stripping

Air stripping is a commonly used process for removing volatile organic compounds (VOCs) from contaminated groundwater. Air stripping is not an in-situ remediation technology. This process is generally performed in a chemical tower-type reactor in which the tower is packed with a medium over which the contaminated water flows in a liquid thin-film fashion. The packing thus provides a large amount of air/liquid surface area across which the transfer of the gas or volatile organic substance can take place. Such packed towers are generally arranged with the liquid and air flows in a countercurrent fashion, as illustrated below in Fig. 1. The air stream usually enters the column at the bottom and leaves at the top, while the liquid stream enters at the top and leaves at the bottom. The removal or mass transfer of the VOC or gas may be described using mass balance relationships.

There are alternative air stripping methods to the packed tower system, such as diffused air aeration or liquid spray reactors, which will provide a large amount of liquid–air interface surface area.

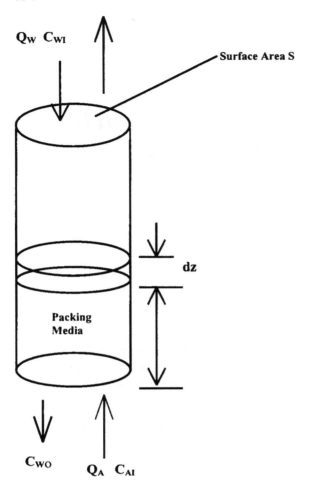

Figure 1 Basis for air stripping theory.

B. Soil Vapor Extraction

Soil vapor extraction (SVE), illustrated below, in Fig. 2, is a relatively new, in-situ remediation technique used to remove VOCs from contaminated subsoils in the vadose zone which lies above the groundwater table. The groundwater table may also be lowered by pumping, with or without the aid of subsurface slurry curtain wall containment barriers, to create an expanded vadose zone. The vapors, which are extracted in an air stream, are often passed through a treatment unit which removes or destroys the vapors before discharging the air. Included in the process train are containers in which moisture removed by the air stream can be condensed. In the typical case a number of subsurface extraction wells are constructed and connected by means of manifold piping which is connected to a vacuum system. As air flows through the soil to the extraction wells, volatile organics desorb from the soil particles as vapors which are carried by the air flow. Clean air enters the soil naturally or is introduced through a reverse venting system as the contaminated air is removed by the vapor extraction system. In some instances, impermeable barrier materials may be placed on the ground surface at site to limit short-circuiting of air flows through the system. For the soil vapor extraction system to be effective, the subsoils must have a high permeability. In such circumstances it offers a low-cost, easy-to-implement method of subsoil VOC removal which can be utilized at sites contaminated with gasoline, industrial solvents, and other volatile chemical substances.

C. Air Sparging

Air sparging is an in-situ technology used to remediate contaminated groundwater and soil. Sparging involves agitation of a liquid by blowing air through it. In the case of site remediation, the liquid is in the saturated subsoil below the groundwater table. If it remains in the gas bubble form this is induced air will accumulate and drive VOCs to the vadose zone, where SVE techniques can be used to collect the vapors if required. Air sparging is also used in some cases to provide oxygen required for in-situ bioremediation of soils. The process is applicable to high-permeability, large-grain-sized soils.

III. THEORY OF AIR STRIPPING

Air stripping for groundwater treatment is a mass transfer process that usually takes place in a countercurrent reactor such as a packed tower, tray tower, spray unit, or diffused aeration tank. A packed tower reactor will be used to illustrate the process theory.

Figure 1 depicts such a situation. The design of such a unit is based on mass transfer calculations:

Mass of contaminant out of liquid = mass of contaminant into air stream

For a specific volatile contaminant such as a volatile organic compound using the parameters in Fig. 1, the mass balance can be expressed as

$$Q_W(C_{WI} - C_{WO}) = Q_A(C_{AO} - C_{AI}) \tag{1}$$

Generally the inlet air stream does not contain any VOC, therefore $C_{AI} = 0$. The concentrations expressed by the C terms should relate to the number of VOC molecules and are therefore molar concentrations, such as moles/m^3.

The water effluent concentration of VOC is sometimes negligible compared to the water influent concentration. In such cases $C_{WO} = 0$. Alternatively, a degree of removal desired could be used, for example, $C_{WO} = 0.05 C_{WI}$ for 95% removal. If the effluent water and influent air VOC concentrations are taken as zero, Eq. (1) is reduced to

$$Q_W(C_{WI}) = Q_A(C_{AO}) \tag{2}$$

According to Henry's law, under conditions of equilibrium the partial pressure of the VOC gas phase in the air stream will be proportional to the concentration C_L of the VOC molecules in the liquid stream. This relationship is illustrated as

$$p_g = HC_L \tag{3}$$

where C_L is the concentration in the liquid stream in mol/m^3 and H is the Henry's law constants in units of atm-m^3/mol. The value of H may also be considered as the ratio of the concentration in the gas to the concentration in the liquid. In some cases, Henry's law is derived for transferring the gas from an air stream into water. Then a reciprocal constant is obtained which has the units of mol/atm-m^3. There is also a Henry's law constant H', which is termed the dimensionless Henry's law constant. This incorporates the effects of temperature as illustrated by the relationship

$$H' = \frac{H}{RT} \tag{4}$$

where

H = Henry's constant in atm-m^3/mol

R = universal gas constant in atm-m^3/mol K

T = absolute temperature in K

Applied to this case. Eq. (3) becomes

$$(p_g)_{\text{out}} = HC_{WI} \tag{5}$$

Assuming the conditions of Eq. (2) and equilibrium between the air out and liquid in and using the dimensionless Henry's law constant, Eq. (5) becomes

$$C_{AO} = H'C_{WI} \tag{6}$$

Then Eq. (2) becomes

$$Q_W(C_{WI}) = Q_A(H'C_{WI})$$

which is reduced to

$$Q_W = H'Q_A \tag{7}$$

When the terms are regrouped, one can write Eq. (7) as

$$1 = \frac{Q_A H'}{Q_W} \tag{8}$$

This arrangement of the expression is for the ideal transfer case, For real conditions more air flow is required and Eq. (8) is often converted to the expression

$$R = \frac{Q_A H'}{Q_W} \tag{9}$$

In Eq. (9), the term R is referred to as a *stripping factor*. Values of this factor, which always has values greater than 1 (quite often greater than 10), must be estimated or determined by pilot testing. This stripping factor is sometimes described by the equation

$$R = H'\left(\frac{G}{L}\right) \tag{10}$$

where G is the molar flow rate of air and L is the molar flow rate of water.

The design of an air stripping reactor is based on mass transfer principles. On a volumetric basis, the mass transfer rate per unit volume M for the packed tower shown in Fig. 1 may be related to the liquid flow rate Q_w, the surficial surface area S, the column depth Z, and the concentration gradient dC/dz by the equation

$$M = \frac{Q_W dC}{S \, dz} \tag{11}$$

Describing the same mass transfer in terms of the air–water interface, the equation

$$M = K_{La}(C_W - C_{WEQ}) \tag{12}$$

Air Stripping and Soil Vapor Extraction

may be written, where

$$C_{WEQ} = \frac{C_A}{H'} \tag{13}$$

The concentration gradient is the driving force. Setting the expressions for M in Eqs. (11) and (12) equal, one gets

$$\frac{Q_W}{S}\frac{dC}{dz} = K_{La}(C_W - C_{WEQ}) \tag{14}$$

Let the expression Q_w/S be expressed in molar units L/M_w, where

L = liquid water loading rate, kmol/s − m²

M = molar density of water, 55.6 kmol/m³

Now,

$$\frac{L}{M_W}(dC/dz) = K_{La}(C_W - C_{WEQ}) \tag{15}$$

Rearranging, this becomes

$$dz = \frac{L}{M_W K_{La}}\left(\frac{dC}{C_W - C_{WEQ}}\right) \tag{16}$$

The expression $L/M_w K_{La}$ is a separate term for a specific air stripping tower application for which the overall mass transfer coefficient K_{La} must be obtained by pilot-testing measurement or estimated based on experience. It is called the height of transfer unit or HTU. Thus,

$$\text{HTU} = \frac{L}{M_W K_{La}}$$

and

$$dz = (\text{HTU})\left(\frac{dC}{C_W - C_{WEQ}}\right) \tag{17}$$

From the bottom of the column up to section dz,

$$Q_A(C_A - C_{AI}) = Q_W(C_W - C_{WO}) \tag{18}$$

where $C_{AI} = 0$ and $C_A = C_{WEQ}H'$. Then Eq. (18) can be written to become

$$C_{WEQ} = \frac{Q_W}{H'Q_A}(C_W - C_{WO})$$

or

$$C_{WEQ} = \frac{1}{R}(C_W - C_{WO}) \tag{19}$$

Integrating to solve Eq. (17) with Eq. (19) and replacing the C_{WEQ} term results in

$$Z = (\text{HTU})\left(\frac{R}{R-1}\right)\ln\left\{\left[\left(\frac{C_{WI}}{C_{WO}}\right)(R-1)+\right]/R\right\} \tag{20}$$

where the expression in large parentheses is referred to as NTU or number of transfer units. Hence,

$Z = (HTU)(NTU)$

To use this equation one must obtain a set of values for K_{La} and the stripping factor R from pilot testing. The value of the absolute Henry's law constant is included in the stripping factor. Thus this absolute constant is evaluated along with the stripping factor. Care should be exercised when examining any published values of H'. In general, it is better to estimate based on published values of H than H'. Chemicals with higher values of H ($\gg 100$ atm) tend to be stripped more easily, while values between 100 and 1000 atm indicate a more resistant group with some possibility. Vinyl chloride, carbon tetrachloride, and 1,1-dichloroethylene are examples of chemicals in the easily strippable group, while tetrachloroethylene, toluene, and benzene are examples of compounds in the second category.

The preceding air stripping theory may be used for the design of air stripping towers or for subsoil formations containing contaminated groundwater. There is, however, one other important design consideration to be evaluated for each application. This is the flow hydraulics. If, for example, too great a water flow rate is attempted in an air stripping tower and air flow is increased correspondingly, the air flow will block the water flow, causing flooding of the tower. Pressure drops for air flowing through the system should also be estimated and evaluated during operation. The manner in which feasible air/water ratios on a g/L basis can be determined for a specific tower packing, temperature, and gas pressure drop can be found in various reference texts (1,2). For compounds which are readily stripped, 99% removal efficiencies are reported to be typical for the air stripping process (3). Air stripping is especially good for removing petroleum products such as gasoline from groundwater. Gasoline contains as much as 20% BTX (benzene, toluene, and xylene) by weight. BTX typifies hazardous VOCs. Removals of BTX from groundwater by air stripping have been reported to exceed 99.9% (4).

IV. SITE CONSIDERATION FOR SOIL VAPOR EXTRACTION

Consideration of soil vapor extraction for remediation of contaminated soils requires knowledge of site geology, soil, and contaminant characteristics.

The geology of a contaminated site will be a major factor in the determination of process suitability. The degree of soil stratification, soil type, and depth to groundwater all impact the effectiveness of soil vapor transport. Since soil vapor extraction is dependent on the attainment of a significant vapor flow rate through the vardose zone, the process is most effective for soils composed of sand, gravel, coarse silt, or fractured bedrock.

Sites with complex soil stratification are generally not suited for soil vapor extraction. The structural characteristics can promote vapor flow patterns that result in short-circuiting, thereby dispersing contaminants to regions outside the radius of influence being generated through the applied vacuum. The result may be fugitive losses of contaminants to the atmosphere or increased treatment times.

Soil characteristics are evaluated based on properties that influence transport of gases through the vardose zone Table 1 summarizes the suitable ranges of parameters representing transport in porous media.

Of particular importance are soil permeability, moisture content, and organic carbon content. Soil permeability affects the rate of vapor migration through the medium.

Air Stripping and Soil Vapor Extraction

Table 1 Soil Parameters Related to Selection of Soil Vapor Extraction

Parameter	Value
Soil permeability	$>10^{-6}$ CM^2
Hydraulic conductivity	$>10^{-6}$ cm/s
Soil moisture content	$<50\%$
Aquifer type	Unconfined
Depth to groundwater	2–100 m

High permeability values promote efficient air transport and contaminant phase transfer. Soil moisture content is related directly to permeability values. Increases in moisture content will retard vapor flow. Low-moisture soils or soils with high organic carbon content will enhance the sorption of VOCs, thereby reducing the phase transfer and ultimate removal efficiency.

Soils contaminated with volatile and semivolatile compounds (principally light petroleum products such as gasoline) are candidates for remediation by soil vapor extraction. The specific chemical parameters of molecules that relate to the suitability of soil extraction are vapor pressure and Henry's law constants. Henry's law coefficients represent the equilibrium distribution of compounds between gaseous and liquid phases. They are derived mathematically from solubility and vapor-pressure values for a compound or, alternatively, are estimated from experimental data.

Soils contaminants with Henry's coefficients in the range 10^{-5} to 10^{-3} atm-m^3/mol will demonstrate significant volatilization.

Table 2 provides examples of organic compounds that have been successfully remediated by soil vapor extraction.

Soil vapor extraction relies on vacuum pressure to promote movement of vapor-phase contaminants through a soil matrix. In order to assure the integrity of the negative soil pore pressures, soil vapor extraction is generally considered for minimum depths of 2 m. The performance of extraction wells is enhanced for deep wells; depths to 100 m have performed satisfactorily. Sites where the groundwater is less than 2 m often

Table 2 Representative Compounds Amenable to Soil Vapor Extraction

Compound	Henry's law coefficient (atm-m^3/mol)
Benzene	5.59×10^{-3}
Ethylbenzene	6.43×10^{-3}
Toluene	6.37×10^{-3}
Xylene isomers	7.04×10^{-3}
Tetrachloroethene	2.59×10^{-2}
Trichloroethene	9.10×10^{-3}
1,1-Dichloroethene	3.40×10^{-2}
Vinyl Chloride	8.19×10^{-2}
1,1,1-Trichloroethane	1.44×10^{-2}
1.2-Dichloroethane	1.00×10^{-3}
Freon-113	5.26×10^{-1}

create problems of water infiltration into the vardose zone due to upwelling created by the applied vacuum. Various procedures can be employed to manage water table levels. Two specific options are (a) use of pumping to draw down the water table and (b) installation of horizontal wells.

V. PRELIMINARY DESIGN REQUIREMENTS FOR SOIL VAPOR EXTRACTION

The design of soil vapor extraction systems is primarily an empirical process. Dispute the availability of a number of mathematical models developed to provide design guidance, the design of soil vapor extraction ultimately derives from data obtained from site-specific pilot studies. A properly conceived pilot study will provide data leading to optimum estimates of the radius of influence for each potential extraction well and, more significantly, allow detection of site-specific irregularities in soil vapor movement.

A typical pilot installation would utilize a limited combination of extraction and monitoring wells. The extraction wells provide data on the initial concentrations of VOCs that would be generated through the operation of various extraction rates. Monitoring wells allow for analysis of in-situ gas concentrations and measurement of vacuum efficiency. Pilot testing periods should allow opportunity to access performance through a range of extraction rates and wellhead vacuums. The actual duration of a pilot phase will vary from 10 to 180 days, depending on the size and complexity of a site. The outcome of pilot studies will be design parameters for full-scale systems. Specific outcomes will be well spacing and well screening intervals, and vacuum requirements (air flow rates). Additionally, a specific part of the pilot effort should be directed to the evaluation of methods suitable for treatment of soil vapor gas streams.

A. Extraction Wells

A typical vertical extraction system is usually a 4-in.-diameter well constructed with casings of schedule 80 PVC (polyvinyl chloride). The screening may be either stainless steel or PVC. The integrity of the well is enhanced through use of coarse sand or gravel filter packing encased with bentonite pellets and cement grout. These aspects of well construction serve to limit short-circuiting of soil vapor. The well is piped to a vacuum source. The exact nature of a system will vary according to the site characteristics, and may include items such as air/water separators, pressure sensors, vapor treatment systems, and discharge stacks.

The operation of an extraction well as a pilot system serves to provide opportunities to determine the radius of influence of the well and to investigate specifically the relationship of various air extraction rates and applied vacuum on this parameter. Concurrent analyses of contaminant levels in the extracted air stream serves to provide loading rates as a function of vacuum pressure and flow rate. Evaluation of this data assists in the determination of optimum process conditions and additional well placements.

B. Monitoring Wells

The inclusion of monitoring wells as part of a pilot study and ultimately in the system design for soil vapor extraction systems generally enhances the optimization of the design process. The monitoring wells may be employed for a variety of purposes, including:

Refined assessment of the radius of influence through inclusion of vacuum piezometers,

Determination of in-situ contaminate levels for tracking extraction performance or detecting preferential air flow pathways

Conversion to air injection wells to enhance contaminant transport in low permeability soils

Conversion for use with groundwater depression pumps in order to reduce either groundwater upwelling induced by the application of vacuum or expansion of the depth of the vadose zone.

C. Radius of Influence

The most important aspect of soil vapor extraction design is the determination of the effective radius of influence. The U.S. Environmental Protection Agency (EPA) characterizes this parameter as an indicator of the greatest distance from an extraction well at which a sufficient vacuum and vapor flow can be induced to adequately enhance volatilization and extraction of the contaminants in the soil. In the final configuration of a system, it is important that if multiple extraction wells are employed, there be overlap of the radii of influence over the entire area of contamination. The operating vacuum and air flow rates are the major parameters affecting both treatment performance and costs. Representative values of these parameters given in various cost and performance reports published by the U.S. EPA are given in Table 3.

D. Contaminant Treatment Options

The treatment of extracted soil vapor is determined by contaminate concentration and composition. Typical treatment options are activated carbon adsorption, thermal oxidation, catalytic oxidation, biofiltration, and no treatment.

Activated carbon adsorption is the most common choice for treatment of soil vapor extraction off-gas. The process is cost-effective and efficient for vapor streams with low levels of contaminants. Vapor streams with high either contaminant or moisture levels present a problem with premature exhaustion of adsorptive capacity.

Thermal oxidation is an effective treatment for vapor steams with high levels of combustible contaminants. The contaminant levels in the vapor stream generally decline with time, thereby requiring an increase in amounts of auxiliary fuel required for maintaining proper combustion. Incorporation of a thermal oxidizer for off-gas treatment may dictate the overall operating characteristics of a soil vapor extraction system. Effective operation of a thermal oxidizer, i.e., maximum efficiency of contaminant destruction, will be a function of flame temperature and gas flow rate. Hence, attainment of maximum efficiency determines the operating vacuum applied to extraction wells. When chlorinated

Table 3 Range of Operating Parameters for Soil Vapor Extraction Wells

Parameter	Range of values
Air flow rate	28–2500 scfm
Vacuum	3–15 in. Hg

hydrocarbons are part of the contaminant matrix, secondary production of acids will require consideration of additional treatment protocols.

Catalytic oxidation should generally be coupled with thermal oxidizers. Thermal systems for destruction of off-gas are most appropriate during the start-up phase of treatment. During the later stages of system operation, contaminant levels decline, making the use of catalytic oxidizers more economical. The savings are due to the reduced amounts of supplemental fuel required by the low-temperature catalytic systems as compared to thermal oxidizers.

Biofiltration utilizes microorganisms to degrade organic compounds.

No treatment of the vapor stream is an acceptable option when the contaminant levels are below regulatory requirements for emissions to the atmosphere.

E. Operational Modes

The operational protocols of soil vapor extraction systems may change as a function of accumulated time online. As removal rates decrease, conversion from continuous to intermittent operation can enhance total vapor removal. Provision for changes in operation mode should be part of the initial design, primarily by incorporation of a flexible operating system and an appropriate selection of a contaminant treatment system. A typical example would be the operational characteristics of thermal/catalytic oxidizers. The use of these systems is based on consistent process conditions; therefore, they are not tolerant of intermittently operated extraction protocols. In contrast, activated carbon systems are relatively insensitive to variations in extraction well operation.

VI. A REPRESENTATIVE SOIL VAPOR EXTRACTION SYSTEM

The following illustrative example of the application of a soil vapor extraction system is based on the remediation of soil beneath pits utilized for fire training at Luke Air Force Base in Arizona (5). The site was utilized for fire fighting exercises involving JP jet fuel from 1973 to 1990. Table 4 compares the analyses of soil boring samples from the contaminated soil at the site, taken at the beginning and end of a soil vapor extraction remediation.

The test borings indicated that the contamination extended to a depth of 55 ft. The depth to the water table exceeded 350 ft. The pilot study performed at this site produced mass removal rates greater than 40 lb/day at 20 acfm.

The remediation system at this site was operated for 6 months. At the end of the remediation, analysis of the VOC levels in the extracted soil vapor stream indicated sub-

Table 4 Soil Boring Sample Concentrations at a Depth of 10 ft

Contaminant	Pretreatment concentrations (mg/kg)	Posttreatment concentrations (mg/kg)
Benzene	15	.02
Toluene	470	100
Ethylbenzene	360	100
Xylenes	520	290

Air Stripping and Soil Vapor Extraction

Table 5 Reductions in Soil Gas Contaminant Concentrations

Contaminant	Percent reduction
Benzene	96
Toluene	81
Ethylbenzene	74
o-Xylene	72
p- and m-xylene	79

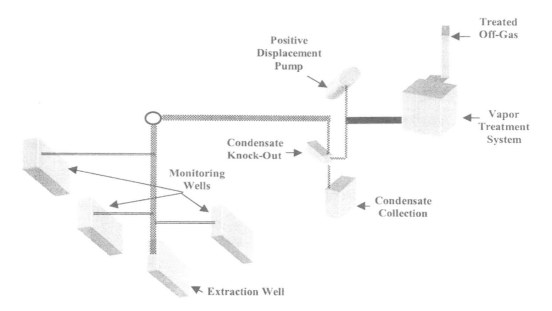

Figure 2 Schematic of the major components in soil vapor extraction (5).

stantial reduction in the contaminant concentrations. Representative values are given in Table 5.

Figure 2 illustrates the major components of the soil vapor extraction system utilized at this site. The overall performance of this system, as estimated from on-site gas chromatographic analyses and extraction well flow rates, indicate that 12,000 lb of contaminants were removed and destroyed.

The details of placement of monitoring and extraction wells in a soil vapor extraction system vary widely according to the radius of influence. Figure 3 illustrates a typical system of monitoring wells and a single extraction well. A summary of the overall advantages and disadvantage of the soil vapor extraction process is given in Table 6.

VII. SITE CONSIDERATIONS FOR AIR SPARGING

Air sparging is an in-situ process, which involves the direct injection of air into a contaminated saturated zone. The air will rise into the vadose zone and disperse through

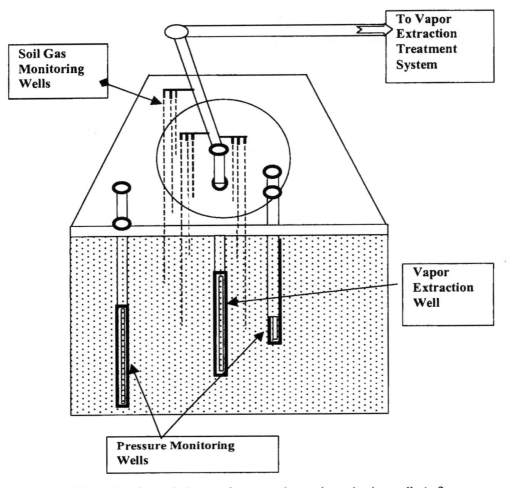

Figure 3 Schematic of a typical setup for extraction and monitoring wells (ref).

Table 6 Characteristics of Soil Vapor Extraction Systems (ref)

Advantages	Disadvantages
Established technology; standard equipment; simple construction	May require sophisticated and expensive off-gas treatment
Minimal disturbance of treatment site	Sensitivity to soil permeability and stratification
Optimum conditions give reasonable treatment times (6 months to 2 years)	Decreased effectiveness with reduced contaminant concentrations; overall efficiencies greater than 90% difficult
Attractive economics ($20–$40 per ton of contaminated soil)	Site-specific pilot studies generally required
Flexible applications for in-situ or ex-situ configurations	Restricted to unsaturated zone; sites with contaminated groundwater in saturated zone require other methods

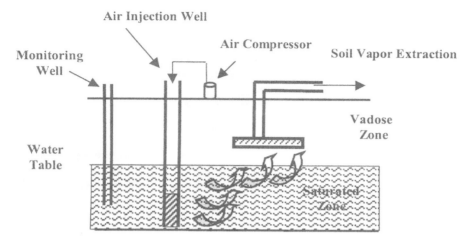

Figure 4 Schematic of a typical air sparging system.

the unsaturated region according to the specific soil permeability and structural channels present at the location. The migration of air produces the equivalent of an in-situ air stripper that volatilizes contaminants with high volatility and low solubility. Figure 4 illustrates a typical air sparging system.

When accessing remediation options at a specific site, sparging can be eliminated from consideration immediately in a number of situations.

> The site has free product contaminates. Air sparging often creates groundwater mounding, which in this situation would lead to the possibility of spreading the contamination zone. Proximity to subsurface confined spaces such as basements or sewers. The enhanced soil vapor migration produced by air sparging could lead to unacceptable contaminant levels in these spaces. An air sparging system coupled with an appropriately placed soil vapor extraction well can sometimes overcome the problem. The contaminated zone is located in a confined aquifer. Air injected into this type of aquifer would remain trapped.

The considerations for applicability of air sparging systems are similar to those utilized in accessing soil vapor extraction. Table 7 summarizes the major aquifer parameters required for utilization of air sparging. The heterogeneity of a site is of particular

Table 7 Aquifer Parameters Related to Selection of Air Sparging

Parameter	Value
Soil character	Coarse-grained
Heterogeneity	Absence of impervious layers above sparge zone
Permeability in saturated zone	1×10^{-4} cm^2
Hydraulic conductivity	1×10^{-3} cm/s
Depth to groundwater	>5 ft
Aquifer type	Unconfined
Saturated thickness	5–30 ft

Table 8 Representative Contaminate Properties Amenable to Air Sparging

Parameter	Required values
Volatility	High (Henry's coefficient $>10^5$ atm-m^3/mol)
Solubility	Low ($<20{,}000$ mg/L)
Presence of dense nonaqueous-phase liquids	None
Biodegradability	High (BOD$^5>0.01$ mg/L)

importance, as the major object of air sparging is to enhance volatilization and transport of contaminants. Air sparging is inappropriate for sites with highly stratified soils with low permeability, fractured rock, or clay composition. Soil heterogeneity may cause some zones to be unaffected in the remediation process. Another problem due to nonuniformity would be the uncontrolled movement of vapors through the saturated zone.

Air sparging is effective only for a specific class of contaminants. Table 8 summarizes the contaminant properties compatible with air sparging.

VIII. DESIGN REQUIREMENTS FOR AIR SPARGING

Many of the same considerations applied to the design of soil vapor extraction systems are applicable to air sparging design. Site-specific considerations coupled with field pilot studies are essential to a successful design. The remediation objectives are often tempered by concerns about the integrity of the contaminant plume. The introduction of pressurized air into the saturated zone can promote unwanted migration of either the dissolved or vapor-phase contaminant plumes. As a consequence, the placement and utilization of monitoring wells is especially important in air sparging systems.

The fundamental objective of air sparging is to promote transfer of dissolved-phase contaminants to the vapor phase. The mass transfer rate is a function of the rate of air injection into the saturated zone; soil permeability and structure are, therefore, major factors in the effectiveness of the process. The decision as to ultimate placement and quantity of sparging wells is defined by the radius of influence estimated from pilot data.

A. An Alternative Design

Unlike conventional air sparging systems, which inject pressurized air into soil pore space, density-driven systems provide for aeration of the groundwater inside the wellbore. A sparging well is configured with an air injection tube positioned to deliver pressurized air into the base of the well. This process will create a driving force for groundwater circulation throughout the well. Therefore the dissolved oxygen content of the water will be increased, thereby promoting biodegradation in the saturated zone. Additionally, it will increase both oxygen and contaminant transfer in the vadose zone.

A schematic of a density-driven groundwater sparging system utilized at Amcor Precast is shown in Fig. 5. The remediation process utilized at this site included soil vapor extraction and a groundwater pumping system in addition to groundwater sparging. Specific design considerations were utilized to prevent migration of contaminants beneath an office building adjacent to the contaminated area. Down-gradient pumping wells were utilized to intercept migrating contaminants in the saturated zone. The groundwater pumped from these wells was delivered to an up-gradient injection gallery, thereby elim-

Air Stripping and Soil Vapor Extraction

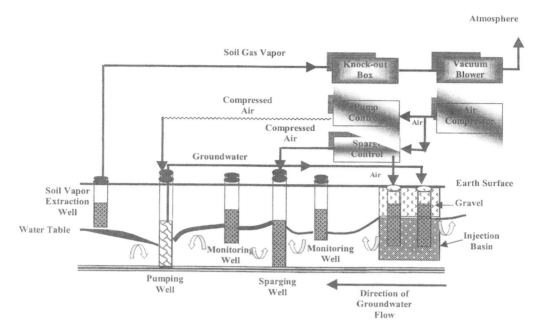

Figure 5 Schematic of density-driven groundwater sparging.

inating the need for surface treatment. The system also employed soil vapor extraction wells to further reduce the migration of contaminants through the vadose zone.

REFERENCES

1. LaGrega, M. D., et al. 1994. *Hazardous Waste Management*, pp. 457–460. McGraw-Hill, New York.
2. Watts, R. J. 1997. Hazardous Wastes: Sources—Pathways—Receptors, pp. 602–607. Wiley, New York.
3. U.S. Environmental Protection Agency. 1993. *Remediation Technologies Screening Matrix and Reference Guide*, EPA 543-B-93-005.
4. McFarland, W. 1989. Sir Stripping removes petroleum from groundwater. *Water/Eng. Manage.*, May, pp. 48–52.
5. U.S. Department of Defense. 1994. Technology Application Analysis: Soil Vapor Extraction At North Fire Training Area (NFTA) Luke AFB, Arizona, prepared by Stone & Webster Environmental Technology & Services.
6. U.S. Environmental Protection Agency. 1995. Cost and performance report: Soil vapor extraction at the Verona Well Field Superfund Site, Thomas Solvent Raymond Road (OU-1), Battle Creek, Michigan. U.S. Office of Solid Waste and Emergency Response Technology Innovation Office.
7. U.S. Environmental Protection Agency. 1996. Engineering Forum Issue Paper: Soil Vapor Extraction Implementation Experiences, EPA 540/F-95/030. Office of Emergency and Remedial Response.

8. U.S. Environmental Protection Agency. 1996. How to evaluate alternative cleanup technologies for underground storage tank sites: A guide for corrective action plan reviewers, EPA 510-B-94-003.
9. U.S. Department of Defense. 1994. Technology application analysis: Density-driven groundwater sparging at Amcor Precast, Ogden, Utah, prepared by Stone & Webster Environmental Technology & Services.

23
Structural and Functional Aspects of Artificial Soils Construction

Natalia Zaimenko, T. Cherevchenko, and G. Rusin
Central Botanical Garden of Ukraine NAS, Kyiv, Ukraine

I. INTRODUCTION

Anthropogenic effects cause global transformation of the environment. Ecological aftereffects of technogenesis pass all regional and state bounds and become a powerful factor in landscape formation and geopolitical change. Mankind has come close to the line on the other side of which environment changes can have catastrophic and irreversible character. In this connection the main problems for people are nature conservation and food. To solve them is possible on the basis of complete study of mechanisms of soil formation and function as a component of the biosphere. Therefore a modeling study of soil biosystems as objects of control needs detail research. Actual approaches to ecobiosystem simulation do not allow analysis of soil interaction with the environment as a direct, developing, and adaptive system in full measure because of soil adaptation to certain bioprocesses is hard to approximate. Moreover, the instability of soil biosystems makes parametric identification of processes of function and structure formation impossible using simulation theory.

Study of relations in nature is an urgent problem. Solution of this problem will make it possible to forecast soil changes under the effect of any factors. Here we found some difficulties caused by lack of data about the set of parameters that gives the most complete characteristics for the soil. Besides this, many features of soil are in dynamic interaction and soils are nonhomogeneous by nature. Therefore even minute transgression of relations between soil and factors of the outer environment make formulation difficult. Dynamic soil parameters (acidity, density, humus content metabolism bases, mobile forms of different compounds) also prevent establishing relations between the environment and soil features.

To study and simulate soil processes we use artificial soil substitutes with controlled physical and chemical parameters as model objects. The most perspective for this aim are inert fibrous substrates with selected contents of substrate and specific structure which determine efficient use of material for cultivation of plants with different ecotypes and monostructures.

Currently, for industrial plant growing in greenhouses, different artificial substrates are used as soil substitutes. As substrates can be used high-molecular weight organic compounds of the types of cellular polystyrene (1), polyurethane, polypropylene, thermoplastic polymer, and ion-exchange resin (2,3). In addition, such nonorganic materials as perlite, vermiculite, keramsite, basalt gravel, sand, granite detritus, volcanic slag, dolomite limestone (4–6), and zeolite (7,8) may be used.

In many countries the basic substrates for greenhouses are various types of peat (9). Along with peat in vegetables growing, waste from wood processing industries, brown coal, and sweepings are also used (10). Today, fibrous materials (glass, fiber, asbestos fiber, etc.) are widely used (11).

The main advantage of mineral-wadded substrates over traditional soils are small mass, sterility, economic water and fertilizer supply, and the possibility of using highly effective drop watering.

II. BODY OF MATERIALS PRESENTED

Among the variety of materials used as soil substitutes, fibrous substrates are the most suitable for plant cultivation. To determine components we examined the chemical content of different mineral fibers. It was found that the best are fibers from basalt, porphyrite, and diabase. These fibers have high indices of elasticity and unlimited sources of raw materials. Most slag fibers, including those from wastes of metal industries, contain a critical level of unhealthy admixtures (chrome, arsenic, and lead compounds), unstable by chemical content. It was found that among organic fibers the optimal for plants are polyamidic and polyacrylnitrilic ones, notable for absorption of biogenic elements, especially potassium, nitrogen, and iron. Using mineral and organic fibers in certain proportions and new technologies of substrate production allows development of a number of soil substitutes with a variety of physical and chemical features (Table 1). Joining basalt fibers to synthetic ones of different origin creates the preconditions for regulation of water–air balance in the root medium, which has some effect on plant productivity. Thus, in *Anthurium andreanum*, *Aglaonema commutatum*, and *Diffenbachia picta* growing on polyamidic and polyacrylnitrilic fibers, plant growth and root volume increased in comparison with other fibrous material by 1.4–1.5 and 1.2–3.4 times, (Fig. 1).

To increase water capacity of fibrous substrates and to form some structures we used different polymers: nontoxic latexes of natural and synthetic rubbers, polyacrylamide, polyvinyl alcohol, starch, dialkylphenylic ether of polyetherglycol, and phenolformaldehydic alcohol. Agrophysical analysis of fibrous materials showed that to produce formed substrates the most suitable are mixtures of latex–dialkylphenylic ether of polyetherglycol and latex–polyacrylamide (Table 2).

Simulated tests proved the possibility of controlling the water–air regime of the substrate by changing the relation between fibers and coherent. It was determined that the best for plants is a substrate consisting of a mixture of organic and mineral fibers in the ratio 75:25 with 5% latex. Germination of vegetable seeds after applying 5% latex increased by 2.5–7.4%, and the weight of plants increased by 2.7–9.4%. The same dependence was observed for ornamental plants. In 60 days *Anthurium andreanum* grew an average of 2.4–15.3%, *Aglaonema commutatum* an average of 5.0–13.4%, and *Dieffenbachia picta* an average of 1.8–10.1%.

The physical and chemical features of fibrous substrates determine the absorptive ability of the material relative to biogenic elements. The highest absorption among applied fertilizers is characteristic for potassium sulfate, phosphoracidic ammonium, and urea. It

Table 1 Agrophysical Character of Soil Substitutes Depend on Origin and Correlation of Fibrous Components[a]

Indices	Fibrous components								
	BSF+PAN	BSF+Pa$_b$	BSF+Pa$_n$	BSF+U	BSF+PF	BSF+CRT	BSF+PAN+Pa$_n$	BSF+PAN+CRT	
Density, kg/m^3	105	90	125	98	122	189	123	194	
Solid phase, %	7	6	10	7	9	15	9	19	
Common porosity (%)	93	94	90	93	91	85	91	81	
Porosity of aeration (%)	16	12	29	56	34	50	28	30	
Water capacity (%)	77	82	61	37	57	35	63	51	
Water absorption (%)									
30 min	8	28	5	1	1	1	25	2	
60 min	11	30	7	2	2	1	27	2	
120 min	23	32	8	2	2	1	28	2	
180 min	26	32	10	2	3	1	28	3	
240 min	57	33	11	2	4	1	30	3	

[a] BSF basalt super-thin fiber; PAN, polyakrylonitric fiber; PA$_b$, polyamide fiber (thin, 20–25 μm); Pa$_n$, polyamide fiber (thick, 140–170 μm); U, carbonic fiber; PF, polyformaldehydic fiber; CRT, cut rubber thread. Mineral and organic fibers in ratio 30:70.

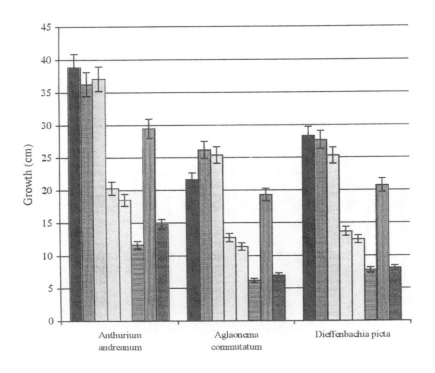
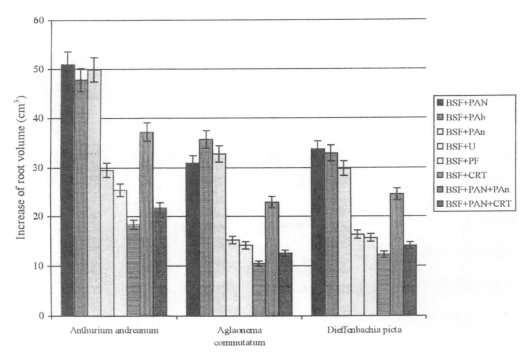

Figure 1 Effect of agrophysical parameters of fibrous substrates on growth and development of ornamental plants (calculation after 60 days).

Table 2 Effect of Polymeric Coherent Phase on Agrophysical Character of Fibrous Substrates[a]

	PNH + BSF						PA + BSF						
						Polymeric coherent phase							
Indices	L	Ph	S	PAA	PV	DPh+L	PAA+L	L	Ph	S	PAA	PV	DPh+L	PAA+L
Density (kg/m³)	97	94	138	136	99	97	99	87	85	121	119	92	90	91
Solid phase (%)	7	7	10	9	7	7	7	6	5	8	10	6	7	7
Common porosity (%)	93	93	90	89	93	93	93	94	95	92	90	94	94	95
Porosity of aeration (%)	22	11	7	6	7	21	20	13	11	17	8	10	18	20
Water capacity (%)	71	82	83	85	86	72	73	81	84	75	82	84	75	72
Water absorption (%)														
30 min	21	4	94	2	37	1	3	34	23	92	20	43	2	3
240 min	25	20	99	8	49	3	5	42	35	98	33	57	5	5

[a] L, latex; Ph, phenol alcohol; S, starch; PA, polyacrylamide; PV, polyvinyl alcohol; DPh, dialkylphenylic; ether of polyethylene glycol. Mineral and organic fibers in ratio 30:70.

was determined that certain organic fibers absorb the elements of mineral fertilization. Polyacrylnitrilic fibers are best to keep potassium, and polyamidic is best for nitrogen.

On the basis of results obtained, it was determined the main approaches to structural organization of fibrous soil substitutes were determined, and substrates with different agrophysical features were developed. Physical and chemical characteristics of fibrous substrates are, to a certain extent, under the influence of their structural peculiarities, taking into account the exchange capacity of soil substitutes and the effectiveness of biogenic ion sorption by donor atoms in functional groups of organic fibers (–COOH, –OH, –C, –CO, –H) block (with polymeric coherent) and granular substrates.

One way to stabilize agrophysical parameters of a soil substitute in the form of a briquette is to improve the fibrous substrate structure by crimping. This increases material solidity and decreases shrinkage (even with low substrate density, equal 70–110 kg/m^3) and secures high absorptive ability of capillaries. The last is attained by orienting the fibrous layers to the material surface at an angle of 60–85° and forming a large-pore, macroheterogenic structure. Our experiments show that crimping the substrate makes it most suitable for plants with pores filled by air and water and simultaneously increases the absorptive ability of the substrate up to 38–45% of volume and decreases material shrinkage to 2–4% by high resistance to pressure. Moreover, a crimped structure allows fibrous soil substitute to be used many times and for large assortments of plant species.

Mineral and organic fibers are characterized by sterility and lack of biologically active systems present in soil. To supply sterile and inert substrates with biological activity it is advantageous to treat them with amino acids, carbohydrates, and organic acids. Siliceous fibers—the result of treatment with organic acids—are characterized by chemically active centers of absorptive soil complex and supply inert material with features closed to that of real soil. Therefore we can consider fibrous materials as a soil biosystem with character absorptive complex created by microsporous structures of fiber and biologically active compounds (Fig. 2).

Organic matter in fibrous substrates are mainly residues and root exudates of plants. Transformation of plant material is a very complicated biological process, in which not only decomposition but synthesis of complex organic compounds takes place. The speed of this process and the main direction depend on chemical compounds of decomposing plants, the water–air regime of the substrate, and microorganisms. Study of microbial colonization of substitutes under long-term permanent plant growing shows absence of phytotoxic microflora, high fermentative activity, and an optimal agrophysical regime of the root medium (Table 3). When *Anthurium andreanum* was grown on fibrous material, activity of phosphatase, urease, and nitroreductase in the substrate at the third year of growing increased by 38%, 32%, and 33%, respectively.

We determined the effect of fiber substrate color on its biological activity. Optimal for plant growth and development are substrates with color in the 450–650 nm range of the visible part of the electromagnetic spectrum. For example, for winter wheat (*Triticum vulgare*) this is a green color of substrate, for leaves of beet (*Beta vulgaris*) and Chinese cabbage (*Brassica chinensis*) it is yellow, for algaonema (*Aglaonema commutatum*) and fennel (*Foenculum vulgare*) it is red, and for anthurium (*Anthurium andreanum*) and dieffenbachia (*Dieffenbachia picta*) it is blue. Agrophysical features of substrates with colors of this spectrum, such as high water-holding ability, quite stable temperature on the surface and into the substrate, have a positive effect on development of microflora useful for agronomy. It was found to decrease the level of phytotoxic microorganisms. Content of phytotoxic fungi bacteria and actinomyces in fibrous

Artificial Soils Construction

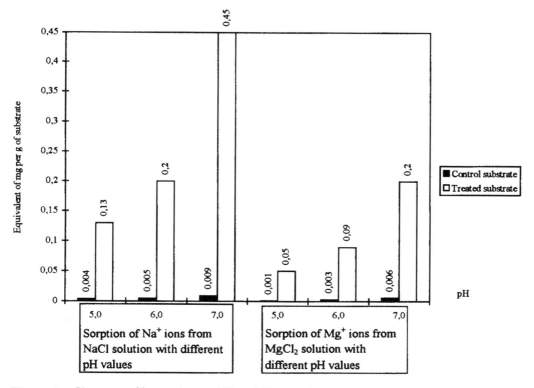

Figure 2 Character of ion-exchangeability of fibrous substrates treated by 1 N solution of oxalic acid.

Table 3 Substrate Toxicity and Content of Phytotoxic Microflora After 3 Years of *Anthurium andreanum* Growing

Variant	Period of cultivation (years)	Toxicity, (CCU)[a]	Phytotoxic microflora (%)		
			Bacteria	Fungi	Actinomices
Typical soil mixture (turf-soil, peat, humus, 1:1:1:1)	1	33.9	12.3	9.7	2.6
	2	47.5	21.5	18.3	3.9
	3	69.3	34.9	29.2	5.7
Fibrous substrate (polyacrylnitrilic and basalt fibers, 70:30)	1	4.5	1.3	2.8	0
	2	10.7	2.7	4.0	1.0
	3	15.2	3.5	5.1	1.5
ISD$_{0.95}$		0.87	0.36	0.59	0.11

[a] CCU, conventional cumari units.

substrate of green color after growing of winter wheat was 1.2–4.6 times lower than in other variants of the test. Substrate color had a significant effect on biosynthesis of photosynthetic pigments in plants. Maximal number of chlorophylls and carotinoids in leaves of winter wheat was characteristic for plants cultivated on soil substrate of green color; for fennel it was red and for chinese cabbage it was yellow. For example, chlorophyll content in leaves of test plants was 5.2–51.6% 5.5–66.0, and 10.6–58.9% higher, respectively (Fig. 3).

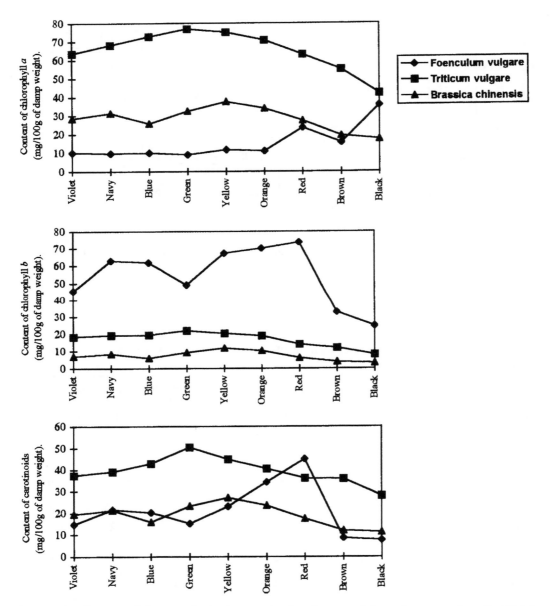

Figure 3 Content of photosynthetic pigments in plant leaves and dependence on color of artificial soil.

Artificial Soils Construction

Table 4 Toxicity of Substrates and Presence of Phytotoxic Microflora Treated by Gamma-radiation

Doze of radiation (kGy)	Time of treatment (min)	Toxicity (CCU)[a]	Phytotoxic microflora (%)		
			Bacteria	Fungi	Actinomyces
		One treatment			
3	15	49.2	4.0	15.8	1.4
5	15	35.4	3.4	13.1	1.2
8	15	38.7	3.9	14.5	1.3
		Two treatments			
0.5–2.5	25	13.7	2.3	5.3	1.3
1.0–4.0	20	13.0	2.1	4.6	1.2
1.5–3.5	15	12.5	2.4	4.2	1.1
2.0–6.0	30	12.8	2.7	4.4	1.2
0.3–2.3	10	32.2	4.0	12.6	1.4
2.2–6.8	35	28.6	5.2	5.2	2.7
		Control (without any treatment)			
		51.3	6.4	17.3	1.5
$ISD_{0.95}$		3.74	0.81	1.02	0.33

[a] CCC, conventional cumari units.

We also developed a method of fibrous substrate regeneration after long-term exploitation by treatment of this material with a gamma-radiation doze of 3–8 kGy (Table 4). The result of this treatment was stabilization of root medium by decreasing substrate toxicity and increase of mobility of mineral alimentation elements with positive effect on plant productivity.

New fertilizers were developed to supply plants cultivated on fibrous substrates with mineral elements of nutrition. These universal organomineral fertilizers of prolonged action consist of balanced nutrient matter needed for plant organisms, including poultry droppings as the organic component. Development of new fertilizer has been done by analysis of initial organic matter, selection of materials for prolonged-activity application of target-oriented additions, and study of physical and chemical characteristics.

Physical and chemical properties of granulated organomineral compositions depends on their origin. Polymer used in fertilizer has two functions: it quarantines necessary granule solidity, and it controls speed of dissolution. Nontoxic polymer compounds include starch, polyacrylamide, and polyvinyl alcohol, which were selected as polymers to use in producing organomineral fertilizers. Results of study of structural fertilizers show that using polyvinyl alcohol best prolongs the activity of the fertilizer. This is explained by the low ratio of fertilizer dissolution and uniform capsule formation, which causes gradual transition of biogenic elements into water solution (Table 5).

The decisive criterion in formation of formula parameters for fertilizers is a certain ratio of organic and mineral components, taking account of the physical and chemical features of fibrous substrates. Chemical analysis of poultry droppings showed a high level of biogenic element content. For example, the content of calcium phosphorous and potassium reached 9.2, 6.1, and 4.2 g/L, respectively, and of magnesium and

Table 5 Egress of Biogenic Elements into Water Medium from Compositions of Organomineral Fertilizers

	Coherent								
	Starch			Polyacrylamide			Polyvinyl alcohol		
	Time of exposition (days)								
Element	2	16	64	2	16	64	2	16	64
N	240	450	500	175	300	450	150	175	240
P	218	1016	1123	168	517	838	55	168	281
K	362	927	980	204	415	638	98	154	237
Ca	869	1533	1605	697	933	1175	563	718	792
Mg	377	891	914	208	556	696	112	139	166
Fe	100	500	550	75	250	400	50	100	127
Cu	5	23	25	4	12	20	2	4	5
Zn	39	60	63	30	43	50	25	31	34

sulfur, 3.9 and 1.0 g/L of waste. Also quite high was the content of nitrogen, iron, and manganese.

For normal growth and development, plants need a controlled, balanced composition of substrates with a certain correlation between macro- and microelements. To make qualitative compost on the base of poultry droppings with balanced elements of nutrition and to reduce nitrogen loss and increase biological activity we used phosphogypsum and superphosphate as application chalk. On the basis of analysis of data obtained, it was determined that poultry droppings with chalk and phosphogypsum in amounts of 10% and 20% of mass, respectively, stimulated plant growth and development.

According to synergistic and antagonistic interaction of ions, the chemical composition of organic raw material was by application of the following mineral solts: NH_4NO_3, 36.3; KNO_3, 10.0; K_2SO_4, 4.0; $CaCO_3$, 162.8; $MgCO_3$, 84.0; $FeSO_4$, 30.0; $MnSO_4$, 13.2; $CuSO_4$, 1.2; $ZnSO_4$, 1.8; $CoSO_4$, 1.2; and H_3BO_3, 0.6 g/L of wastes.

An optimal system of fertilizers increases plant productivity and optimizes the absorptive soil complex too. Effectiveness of this complex activity is related to the effectiveness of exchange processes, which depend on the presence in the soil of small-crystal, dispersible silicate and aluminosilicate minerals. These minerals include natural zeolites (clinoptilolites). We used zeolites modified by salts of metals as nonorganic sorbents–ion exchangers in composition of organomineral fertilizers. Comparative analysis of initial and modified zeolite compositions and the Bulgarian analog Balkanite showed high ion-exchange power of modified clinoptilolite. Content of copper in modified zeolite was 23.1 times higher than in initial cliniptilolite and 3.7 times higher than in Balkanite. Analysis of kinetic relations of biogenic elements from granules of modified zeolite into water medium proved efficient as one of the components to prolong fertilizer activity. In addition, there was a positive effect of modified zeolite on plant growth and development.

It is known that simple salts of metals are characterized by small biological activity; they are weakly assimilated by plants because of formation of insoluble forms of metals.

Lack of microelements in soil causes a decrease of productivity and plant diseases. To control plant activity it is advantageous to form and use biologically active complexes of microelements—metals with organic ligands stable under different values of pH, which are soluble and easily penetrate through cell membranes. Selection of compounds of d-changeable metals with N-2,3-diphenylantranilic acid depends on ions which are biogenic elements and lack the tendency to transmit in insoluble hydroxophorms under changes of pH in the medium. Study of electron transportation reactions in simulated photosynthetic membranes with involvement of d-metal complexes with HL acids and an analog of chlorophyll Mg-*tetra*-4-tributylphthalocyanine show that the ML complex and binary compound introduced into membranes decreased the level of photoresistance of electron acceptors by 37%, in comparison with control. Synthesized complexes have evident biostimulative effect. They increase biosyntheses of photosynthetic pigments and decrease nitrates accumulation in plants. The highest indices are characteristic of binary compounds, for example, a compound based on the Fe–Mn couple. So, content of photosynthetic pigments in leaves of test plants on a background binary compound is 1.6–2.6 times higher than in simple salts and 1.9–3.2 times higher in comparison with control. As to the level of decreasing nitrate accumulation, binary compounds compare favorably with all objects of comparison, as 1.5–13.5, 1.3–12.0, and 6.0–36.0 times, respectively. An antimicrobial effect of metal complexes was found. Organomineral fertilizers containing d-metals according to the index of bacterial flora development have an order conformed to an average level characteristic for aerated rooms.

Comparative study of physical and chemical properties of organomineral fertilizers (OMF) in the form of tablets and granules show that granules from all lots of fertilizers have high solidity, exceeding 6 kg/granule.

Optimal temperature regime for drying granules and tablets fertilizers is 150°N with low liming of organic matter and conservation of fertilizer biological value. Such a fertilizer has a positive effect on growth processes and microflora development, gives fibrous substrate buffer character, and creates active chemical centers.

A technological scheme for industrial manufacture of granulated organomineral was designed. Laboratory methods of granulated compositions of OMF were tried out.

Results obtained show that new organomineral fertilizers of prolonged activity with balanced chemical composition were developed. It is shown that use of desired admixtures on the basis of zeolite and d-complexes (changeable metals) is effective.

On the basis of research carried out, we developed the main approach to soil substitutes detailing. Optimal content of substrate was determined, physical, chemical, and biological transformations of fibrous material were studied, and interactions with plants and the environment were analyzed. The main qualitative and quantitative criteria for inert soil substitute optimization were found.

Biosystems at any level of hierarchy can be described in terms of a conceptual structural model, which shows general principles of life activity, adaptation, and evolution. Development of a model is needed to solve the problem of life activity and structural organization of plants according to the purposes and conditions for the system "model organism." This system is represented by the following structures:

GSRE: global source of solar energy resources
M[OELP]: model of the local environment within the bounds of the planet system
M[PRSD]: plant model within the bounds of purposeful biosystem structure

Each substructure is described by the model of structural organization, by cause–effect relationships, by operators of physical and power transformation of

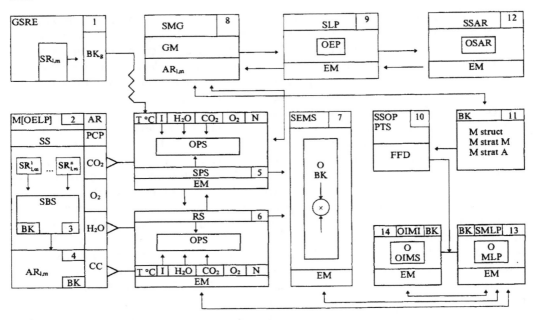

Figure 4 The plant structural model as a purposeful system: GSRE, global source of energy resources; $Sr_{e,m}$, source of power and material resources; BK, knowledge bank; M[OELP], A model of local environment; SS, soil system; PCP, physical and chemical properties; AR, resource accumulator; SBS, soil biosystem; CC, chemical compounds; BAM, biological and active matter; OPS, operator of power transformations; SPS, photosynthesis system; RS, root system, preliminary resources preparation; OPC, operator of physical and chemical transformations; SMG, growth control system resources accumulation system; AR_e, power resources accumulator; AR_m, material resources accumulator; SEMS, power and material synthesis system; SLP, vital capacity system (metabolic processes); SSOP, system of structural organism organization and imitation to adaptation with formation; PTS, purpose testing system; FFPS, dynamic situation formation system; OOIMS, observing informative and measuring system; SSAR, synthesis and accumulation resources system for providing with emergency regime; SMLP, system of organism vital capacity resources control; OEP, metabolic processes operator; EM, executive mechanism; SR, resources source; OMLD, operator of vital capacity resources; M struct, model of system structural organization; M strat M, model of system conduct strategy; M strat A, model of system control strategy.

resources streams, and also by information operators, which characterize transformations in the volume of substructure (Fig. 4).

Investigations conducted permitted us to develop a conceptual model of artificial soil and to determine a series of structural-functional formation, consisting of definition of purposes and criteria of identification, determination of a conceptual model structure of the object, where every subsystem has its own information model and adequate parameters of condition, determination of class of information matrix parameters of condition and parameters' orthodoxy, the process of initial information and its working according to current time, and testing of system effectiveness. In the general model are presented the main features of artificial soil (Figs. 5 and 6), determined parameters of the plant biosystem condition and their parameters for identification, and a conceptual notion about the system "Environment–Soil–Plant."

Artificial Soils Construction

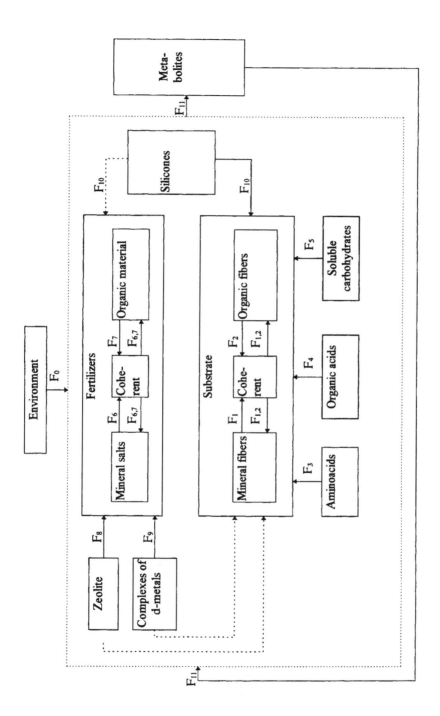

Figure 5 Compartmental model of soil substitute.

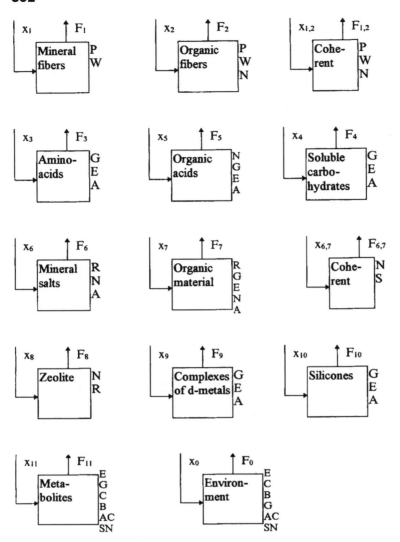

Figure 6 Functional model of soil substitute: P, porosity; W, water absorption; N, chemically active centers; E, photosynthetic pigments; A, microbiological activity; G, growth regulation; S, rate of fertilizes dissolutions; R, source of biogenic elements; B, breath; C, carbon-hydrated exchange; AC, protein exchange; SN, mobility of biogenic elements.

III. CONCLUSIONS

Science and engineering have great potential of knowledge and experience to estimate achievements in various studies of ecobiosystems. Available data from literature and the authors' study of this problem show that artificial soil substitutes as models for soil biosystems allow analysis of physical, chemical, and biological processes in the root zone in interaction with changes in the environment. Prognostication of soil biosystem behavior under the effect of outside factors, including ecological ones, made it possible to develop a mechanism of rapid reaction to remove negative after-effects to study the processes in soil providing a further hitch in researches on allelopathic plant interaction. Moreover, soil substitutes are perspective for plant growing on territories polluted by high metals and unsuitable for plant growing—droughts, mountains—and for studying closed biosystems and future assimilation in outer space.

REFERENCES

1. Agafonov, O. A., Aktasova, A. D., and Katicheva, I. A. 1984. Sterol effective structure former and hydrofuger. *Pochvovedenije*, 4:109–112.
2. Tishchenko, T. E. 1972. Polymer materials in plant growing. *Plasmassy*, 7:12–17.
3. Soldatov, V. S., Peryshkina, N. G., and Khoroshko, R. B. 1978. *Ionic Soils*. Nauka I Technica, Minsk.
4. Huitt, E. 1963. *Sand and Water Culture*. Mir, Moscow.
5. Bojko, L. A., and Levitsry, V. V. 1976. *Vermiculite in Hydroponics*. Nauka, Moscow.
6. Bobkova, L. P., and Makhanko, A. A. 1972. Using of large-fractional gravel in hydroponical culture. *Reports of Institutes of Agrochemical Problems and Hydroponics, AS of Armenian SSR*, 12:95–100.
7. Mitko, L. V., Khirsanova, I. F., and Khoroshko, R. B. 1991. Productive ability of substrates and using of clinoptiolites with anionides of different types. *Ispolzovanije Prirodnych Zeolitov v Narodnom Khozjaistve*, Part. 1: 178–181.
8. Lomakina, L. G. 1990. Flower culture growing in greenhouses on substrate with zeolite and biological additions. In: *Natural Zeolites in National Economic. Theses of All-Union Conf.*, 18–19 April 1990, Kemerovo, pp. 94–95.
9. Uuitis, V. V. 1973. Peat—Substrate for vegetable culture and flowers in greenhouses. *Izvestija AN Latv. SSR*, 1:143–146.
10. Abushajev, G. I. Straw packages as substrate for vegetables growing. *Proc. Sci. Papers of Karaganda Agr. Sta.*, 4:62–64.
11. Jorgnson, E. 1975. Grodan stone wool as medium for propagation and culture. *Acta Horticult.*, 54:137–141.

24
Soil Trace Metal Remediation Using Solid Additives: Thermodynamic and Physical-Chemical Aspects

Isabelle Yousfi
Institut de Protection et de Sûreté Nucléaire, Fontenay-aux-Roses, France

Alain Bermond
Institut National Agronomique—Laboratoire de Chimie Analytique, Paris, France

I. INTRODUCTION

Over the past decade, soil contamination has emerged worldwide as a key environmental issue, and it has become increasingly apparent that soil contamination is placing human and environmental health at risk. The reasons for soil contamination are diverse. On the one hand, coal tar pits, mine tailing wastes, landfill sites, wastewater discharge accidents and, due to agricultural activities, excess of fertilizers, sewage sludges, and pesticides all contribute to soil contamination. On the other hand, atmospheric deposition is also an important source of contamination. The array of contaminants to be addressed—trace metals and toxic elements, excess nutrients, volatile and nonvolatile organic compounds, radioactive isotopes—is today very large.

Regardless of their origin and the reasons for the increase in their concentration in soils, trace metals are liable to contaminate food chains by migrating toward groundwater or by accumulating in plants. Thus, increasing contamination of a part of agricultural soil by toxic compounds constitutes an important health and economic issue.

With this in mind, several clean-up methods have been investigated. The techniques currently used for this purpose can be divided into two groups: techniques which involve removal of contaminants and techniques which transform pollutants into harmless forms, by fixation, oxidation, etc. In the latter case, we talk about immobilization, as opposed to removal in the first case. These clean-up methods can be applied on or off-site.

Elimination and immobilization involve three kinds of treatment. Biological treatments are generally in-situ remediation and consist of biodegradation relying on microbiological activity in order to remove contaminants, particularly in the case of organic pollutants; this technique is used to remove HAP or PBC (1), or to transform pollutants into harmless forms. This is followed by an immobilization technique such as photolysis of organic compounds, for instance. Another biological clean-up approach is the cultivation of plants, known as hyperaccumulators, which are able to fix large amounts of one or several toxic elements. This method has already been used successfully for different toxic trace elements, for instance, Cd and Zn (2,3).

Physical techniques include treatments involving removal of contaminants by physical means. Among these approaches, thermal desorption or vapor stripping devoted to semivolatile or volatile compounds and electrokinetic migration, valid for anionic and cationic compound extraction, can also be mentioned.

Chemical treatment includes all techniques involving reagents or external compounds. We can mention soil washing with different reagents able to solubilize toxic elements, coupled with removal of leaching solution, and adding chelator which makes cations less labile and then makes it possible to immobilize toxic elements in less bioavailable form. For instance, EDTA is one of the proposed complexing reagents (4).

Lastly, the addition of solid phases should also be mentioned. Solid phases are chosen because of their ability to fix toxic elements (irreversibly); solid waste can be sometimes used to diminish remediation costs. Actually, adding solid phases seems to be an interesting and low-cost in-situ remediation method. In addition, it seems to be easier to carry out.

For this purpose, the effects of different kinds of amendments added to contaminated soils have been investigated: clay minerals, hydrous iron oxides (HFOs), hydrous manganese oxides (HMOs), fly ashes, etc. Two kinds of study have been carried out. The first type consists of plant studies in which some authors attempted to determine the effect of amendments by studying the uptake of trace elements by plants and by comparing the amounts of cadmium, lead, copper, or zinc in plants grown on contaminated soils with or without amendments (1–11).

For instance, Gworek (5) studied the effect of adding synthetic zeolites to the Cd in lettuce, oats, and rye grass. She noted a significant drop in the amount of Cd in plant tissue: up to 85% in lettuce leaves, and about 45% in lettuce roots. Chlopecka and Adriano (9) worked with four different types of soil and studied the effects of two kinds of zeolite and apatite on the amount of Cd, Zn, and Pb in maize tissue. The results obtained revealed a decrease of about 50% in the Pb in maize leaves grown on a zeolite-amended soil, and a 20% reduction in the Cd content in maize grown on apatite-amended soil.

Fly ash is also used as an agricultural amendment, particularly because it is a low-cost waste. Petruzzelli et al. (8) investigated the effect of the addition of fly ash between 0% and 5% (w/w) to different types of soil and studied the amount of Cu, Ni, Cr, and Zn in wheat seedlings (8). Amendments seem to have a useful effect and result, for instance, in a decrease of about 65% in the Zn in roots when 5% fly ash is added. In another study (7), fly ash and mixed coal fly ash at rates of 0, 3%, 6%, and 12% were added to two soil samples. While the amount of Zn in vegetables (cabbages) was divided by a factor of 2, an increase in soil pH, and a decrease of the vegetal yield were noticed in several cases. It was concluded that fly ash had a possible phytotoxic effect. However, the liming effect produced by adding fly ash, implying reduced bioavailability of oligo-elements, could be another assumption (12,13). In another study, the effect of lime, HFO, HMO, and zeolite on different types of soil was studied (15). Among the results presented, it can be seen that the amount of Cd decreased by up to 75% in rye grass roots grown on HMO-amended soils and up to 50% in tobacco. In addition, the lead content in plant tissue decreased considerably with HMO (−75%) or lime (−65%). Whatever the phase added, the trace metal plant content was wholly affected by adding phases.

Although the range of effects varied with the plants studied, in most cases the amendments brought about a reduction in trace-metal concentrations in plant tissue. However, in the case of fly ashes, an important variation in soil pH was noticed after the application of phases and a correlation between trace-element concentrations in the plant and the pH of the amended soil was found (7).

In another approach, some authors studied the amounts of trace elements in plants and also their fractionation in the soils. In other terms, the quantities of trace metals in the soil were determined using chemical fractionation procedures, before and after the amendments had been added (14–18). By means of these protocols, a redistribution of trace elements between the soil fractions studied was observed; the amendments used resulted in changes in the fractionation of trace elements.

Few examples can be found. Sappin and Gomez (19) studied the effect of adding 1% (w/w) of steel shot to two different soil samples on the amount of Ni, Cr, and Pb in rye grass and tobacco. Among the results presented, it can be noted that the decrease in Cd could reach up to 40% in tobacco. Simultaneously, the fractions of H_2O-extractable and $Ca(NO_3)_2$-extractable Cd decreased by about 75% and 85%, respectively.

Lastly, the effect of adding agricultural limes (13,18,20) has also been studied; these phases were contaminated with different amounts of trace metal. The EDTA-extractable fractions of Cd, Pb, and Zn were studied before and after amendments were added; in addition, the total content of trace elements in radishes was investigated. For a single application, a decrease of about 20% in the Zn and Pb in vegetative tissue was observed, while the EDTA-extractable fraction of zinc and lead decreased.

Amendments with solid phases such as fly ash, limestone, K_2PO_4 or hydrous manganese oxides, and hydrous iron oxides, and zeolites can influence plant uptake but can also, influence the physical-chemical parameters of the soil, soil fractionation, and also pH. These changes may be predominant when assessing the effectiveness of amendments, but they are rarely taken into account.

Several studies mention the effect on pH value of adding solid phases. For instance, it was shown that the pH value could increase from 1.5 to 2 units when fly ash was added (7,13,17), although it must be noted that adding phases does not result, in some cases, in an increase of pH value—the liming effect is not generalized. The effect of ammonium sulfate, calcium carbonate, acid peat, HFO, and steel shot on the total content of Cd, Ni, Zn, and Mn in rye grass was studied (21). It was observed, for example, that Zn in the first cut of rye grass was multiplied by 1.6 when ammonium sulfate was added, while the pH value decreased from 5.5 to 4.9, but decreased by about 25% when lime was added, with the pH increasing to 7.4. Close observations have been made for Cd and Ni. These results are summarized in Table 1.

Table 1 Effect of Different Amendments on Soil pH Value and Amount of Cadmium in Rye Grass (First Cut) in $\mu g \cdot kg^{-1}$*

Soil	pH	[Cd] (effect)
	5.5	18 (blank)
+ $(NH_4)_2SO_4$	4.9	27 (+50%)
+ $CaCO_3$	7.4	12 (−33%)
+ Acid peat	5.4	19 (+5%)
+ HFO	5.2	16 (−18%)
+ Steel shot	6.4	6 (−66%)

*Effect calculated as: (soil − soil + amendement)/(soil).
Source: From Ref. 21.

Finally, the possible changes in the physical-chemical parameters make it difficult to propose a validated interpretation of amendment action in terms of the immobilization of trace metals. It is not easy, for instance, to distinguish between the true fixation of trace elements onto added solid phases and reducing availability of elements due to the effect of an increasing pH which enhances the fixation of trace metals onto the soil compartments and reduces their bioavailability (22). Moreover, comparison of the effects of different amendments is quite difficult if, for the same sample, pH and added phase vary simultaneously. Lastly, it can be assumed that plant study alone does not give any information on effectiveness of immobilization in preventing migration toward waters (groundwater and surface waters).

According to these conclusions, our aim was to determine the main parameters involved in this remediation approach using immobilizing soil additives. An experimental study of the change of extracted cation quantities from polluted soils in $Ba(ClO_4)_2$ medium, due to different immobilizing additives, was compared with the results of a thermodynamic approach using a simple model to represent the soil sample and the additives.

II. MATERIALS AND METHODS

A. Soil Characteristics

Two soil samples contaminated by trace metals were used in this study. The first, known as Couhins, is a sewage sludge-treated soil; the other, known as Evin, is contaminated by industrial atmospheric deposits. Table 2 gives their main physical-chemical characteristics.

In a first step, several experiments were carried out in order to plot the amounts of trace elements released as a function of pH at equilibrium time, pHe, from soil samples without additive. The resulting graphs will be our references for estimating, at a given pH, the effect of immobilizing additives. For this purpose, the soil sample (1 g) and the reagent, $Ba(ClO_4)_2$ 0.5 M (50 mL), were shaken with a mechanical stirrer, for equilibration time, determined by a preliminary kinetic study, i.e., 140 min. After mixing, the pH of the solution, was measured; the amounts of extracted trace elements (Zn, Cd,

Table 2 Physical Chemical Characteristics of the Studied Soil Samples

	Evin	Couhins
pH	8.56	7.64
Organic matter (%)	1.8	2.2
Clay (%)	16.5	2.9
Zinc ($\mu g.g^{-1}$)	1415	151
Cadmium ($\mu g.g^{-1}$)	23	94.9
Lead ($\mu g.g^{-1}$)	1120	44.8
Copper ($\mu g.g^{-1}$)	43.5	45.3
Iron ($\mu g.g^{-1}$)	20 900	3526
Manganese ($\mu g.g^{-1}$)	411	40.4
Calcium ($\mu g.g^{-1}$)	7 600	1486
Magnesium ($\mu g.g^{-1}$)	33 000	212

Table 3 Total Trace Element Contents of the Studied Solid Additives ($\mu g \cdot g^{-1}$)

Element	HMO	HFO	Clay	O.M	EDF	Valenton
Zinc	10.7	0	57.7	49.1	849	2912
Cadmium	9	0	4.7	0	0.71	15.4
Lead	0	0	0	0	323	1270
Copper	0	4.8	10	0	233	1900
Iron	55.7	28120	710	10400	60150	20670
Manganese	622200	39.2	358	44.4	291	13441
Calcium	77	0	6150	10106	5233	77300
Magnesium	6.2	0	6500	1097	5175	4334

Cu, and Pb) or major elements (Ca, Mg, Fe, and Mn) were determined in order to establish "reference curves": $[C] = f(pHe)$.

B. Rapid Characterization of the Solid Phases Used as Amendments

Six solid phases were used:

Synthetic compounds: a hydrous manganese oxide (HMO), a goethite (HFO), a clay (a montmorillonite), and fulvic acids (OM)

Wastes: two kinds of fly ash; one is a coal fly ash (EDF), the other (Val.) is a product of sewage sludge combustion from the Valenton wastewater plant, referred to as Valenton in this study

The trace-metal composition (total concentration) of these phases is given in Table 3. Synthetic additives such as HMO, HFO, clay, or OM are only slightly polluted by trace metals, contrary to fly ashes, which are strongly contaminated. Nevertheless, these compounds could have a good remediation capacity if trace elements are strongly bound and it was checked that no one of the studied trace metals was released in these physical chemical conditions [$Ba(ClO_4)_2$ 0.5 M].

In order to assess the ability of these phases to rehabilitate polluted soils and to fix trace elements, a rapid study of their capacity to fix several trace elements was performed. An example of the global trend obtained is given in Fig. 1, where the percentage of cations adsorbed by HFO at different pH values is shown. The solid-to-solution ratio (1–2% w/w) and the equilibration times used were equal to those used for experiments carried out with soil samples. The initial concentrations of trace elements were 0.4 mg/L for zinc and 0.2 mg/L for cadmium.

As expected, in most cases the fixation of trace metals increased when pH increased. Moreover, some results do not show the fixation effect of the phase. For instance, clay minerals do not adsorb zinc or cadmium in $Ba(ClO_4)_2$ medium. These trace elements may be in competition with Ba^{2+} ions, which could explain the results. Zn and Cd display similar behavior. Lastly, from a general point of view, it can be seen that polluted phases are able to reduce the total content of trace metals in solution.

C. Experimental Study of the Effect of Additives

To assess the amendment effect, a second set of experiments was undertaken, in each medium, for a mixture of soil sample (1 g) plus solid phase in 50 mL of solution. In this

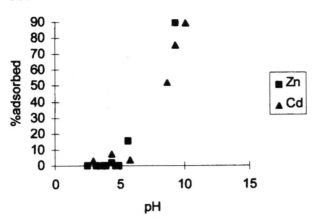

Figure 1 Amount of cations adsorbed (%) in $Ba(ClO_4)_2$ medium onto HFO.

case, the amount of phase added was 10 mg for EDF, Valenton, OM, and HMO, 12.5 mg for HFO, and 20 mg for clay. After equilibration (i.e., 140 min), the pHe was recorded and trace-element concentrations in solution were determined.

D. Analytical Methods

Solutions were analyzed with a flame atomic absorption spectrophotometer equipped with an air-acetylene flame. The following wavelengths were used: 213.9 nm for zinc, 228.8 nm for cadmium, 324.8 nm for copper, and 283.3 nm for lead. The external standards method was chosen and standards were made with $Ba(ClO_4)_2$ (0.5 mol/L) solutions obtained from analytical-grade salts, titrisol solutions, and pure milliQ water.

E. Thermodynamic Simulations

Thermodynamic simulations were performed with a commercially available software, called TOT, provided for calculations of equilibrium in solutions. The software can perform titration curves for a chosen group of equilibria, initial conditions, and titrating solution composition. It calculates the concentrations of all species in solution and therefore the final pH of the equilibrium. Titration curves can be plotted for varying parameters as a function of pH or of any other component.

An example of initial data required for a case involving one ligand, A^-, the conjugated base of the weak acid AH, able to complex a divalent cation M^{2+} with a stability constant K, is given in Table 2. In this example, the solution contains A^- and the metallic cation at concentrations (Co solution) equal to 10^{-4} M and 10^{-6} M, respectively; the reagent (titrant) is OH^- (1.5 M). The equilibrium constants are input, and TOT calculates step by step, for each increment of titrant volume, the amount of all species issued from the different equilibria; in addition, it provides the pH resulting from the selected conditions.

To avoid dilution effect in simulations, the total titrant volume (V_{max} titrant) is smaller than the solution volume (V_o solution). In this case, initial pH is defined by the choice of Co solution for H^+ (10^{-3} M); if the concentration of H^+ is equal to zero, TOT calculates the initial pH resulting from the entered equilibria set.

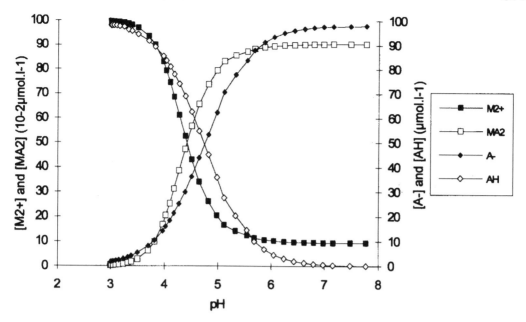

Figure 2 Example of titration curves obtained with the software TOT (data are provided in Table 4).

Figure 2 presents the different curves plotted with TOT for the chosen initial conditions. It should be noticed that, from a formal point of view, these results could have been obtained for a solid model compound (clay mineral, FeOOH, etc.) exhibiting the same constants and properties.

III. RESULTS AND DISCUSSION

A. Experimental Study

Finally, the possible changes in the physical-chemical parameters and particularly the change of final pH make it difficult to propose a validated interpretation of amendment action in terms of the immobilization of trace metals. It is not easy, for instance, to distinguish between the true fixation of trace elements onto added solid phases and reduced availability of elements due to the effect of an increase in pH which enhances the fixation of trace metals onto the soil compartments. In order to propose estimation of the true effect of additives, we compared the quantities of extracted cadmium and zinc in perchlorate medium, with and without additives, at the same pH.

According to this approach, the results obtained for zinc and cadmium from the experimental study performed with the Evin soil sample and different additives are presented in Figs. 3 and 4. The first result to consider is the evolution of extracted amount of these trace metals from soil before amendment (T) versus pH: the extracted amount increases as pH decreases. Regardless of the trace element, the importance of the solution pH is a well-known effect (22) for most of the divalent metallic cations, illustrating the competitive extracting effect of protons.

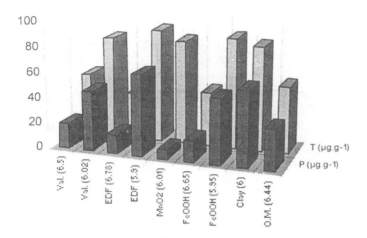

Figure 3 Comparison of zinc extracted from Evin soil sample with or without immobilizing soil additives, respectively, notice T and P at different pHe (noted in brackets).

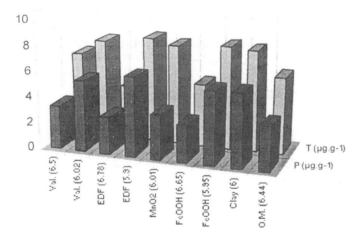

Figure 4 Comparison of cadmium extracted from Evin soil sample with or without immobilizing soil additives, respectively; notice T and P at different pHe (noted in brackets).

If we compare now, for each trace metal, the effectiveness of the decrease of metal amount in the solution, we obtain the following order (at a given pH of approximately 6) for zinc and cadmium, respectively:

$$MnO_2 \gg Valenton = FeOOH > clay = EDF$$

$$MnO_2 > (FeOOH = clay = Valenton)$$

Soil Trace Metal Remediation

MnO_2 is for this sample and for both studied cations the most effective additive to limit their concentration in the solution. For the other added solid compounds, cadmium behavior is different from the behavior of zinc.

The results obtained for the Couhins soil sample using the same additives are presented in Figs. 5 and 6. As discussed previously, the quantities of cadmium and zinc extracted in the $Ba(ClO_4)_2$ solution (T) are increasing as pH decreases

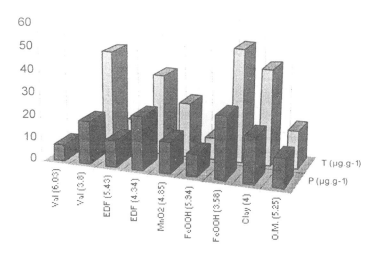

Figure 5 Comparison of zinc extracted from Couhins soil sample with or without immobilizing soil additives, respectively; notice T and P at different pHe (noted in brackets).

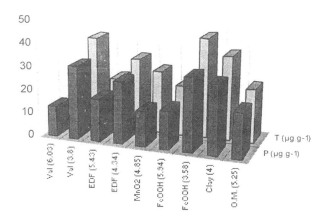

Figure 6 Comparison of zinc extracted from Couhins soil sample with or without immobilizing soil additives, respectively; notice T and P at different pHe (noted in brackets).

From a general point of view, the results obtained with this soil sample seem to indicate less effective fixation of cadmium or zinc onto the different additives. The acidity of this soil probably explains this effect, and we can notice in this case that the final pH ranged from 3.5 to 6.

If we compare now, for each trace metal, the effectiveness of the reduction of metal in the solution at different pH levels, we obtain the following order:

For zinc and cadmium, Valenton > FeOOH (at pH = 6.0)
For cadmium, FeOOH > Valenton (at pH = 3.7)
For zinc, Valenton > FeOOH (at pH = 3.7)

However, it should be noticed that the best results (decrease of cadmium concentration in the solution) were obtained with MnO_2 at a pH of approximately 4.8.

Another approach to the interpretation of these experimental results may be given by the effect of a given added solid phase (E), which was calculated as follows:

$$E(\%) = 100 * [(R - A)/R] \qquad (1)$$

where

A is the amount of trace metal released from mixing soil sample+additive at a given pHe; A was adjusted to the input of trace elements by the phase, but in fact this correction was not necessary because the amounts of trace elements brought by the additive was always (very much) smaller than the amount released by the soil.

R is the amount of trace metals released from the soil sample without additive, read on the reference curves at the same pH.

The variation of E calculated for zinc, according to different experimental conditions (pH, additives, and soil sample), is given in Fig. 7.

We can see that the different parameters involved in this experimental study may change the effectiveness of the remediation, but it is not possible in this case to classify

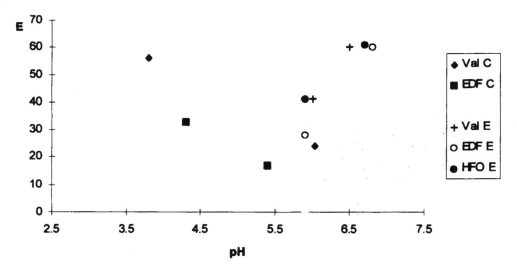

Figure 7 Effect E of immobilizing additives according to the different experimental conditions studied (case of zinc).

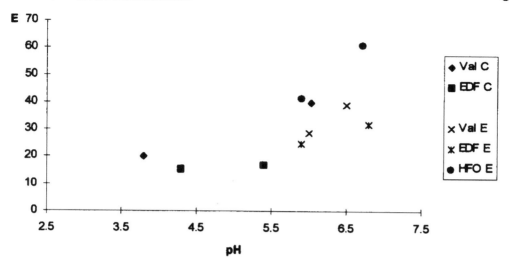

Figure 8 Effect E of immobilizing additives according to the different experimental conditions studied (case of cadmium).

the different parameters from the point of view of their importance in the remediation phenomenon.

For cadmium, the variation of E according to these same experimental conditions is given in Fig. 8. For this trace metal the results are different, and it can probably be said that the main parameter involved in this case is the acidity level: a general trend which corresponds to the decrease of the effect of additives when pH decreases seems to be involved. However, the effect of other parameters cannot be excluded: for instance, the influence of the additive or of the soil sample can be checked on this figure for a few experimental situations.

From a general point of view, it can be said from this experimental study that the remediation of trace-metals polluted soils using solid additives—when measured in terms of the decrease of extracted trace metals—involves numerous parameters such as, at least, pH, soil, cation, and additive. In order to give a best understanding of the role of these different parameters, we would like in the second part of this study to apply a thermodynamic approach, using a very simple model for describing the soil sample and the additives.

B. Thermodynamic Simulations

Excluding the occlusion reactions, the fixation of a metallic cation M^{2+} in a soil may be described by a set of equilibria (one equilibrium for each given soil component S_i) and/or one precipitation reaction. For a given set of equilibria and the corresponding apparent constants, calculation of metallic cation distribution between two forms—free cation in solution (M^{2+}), and bound cation [$M(S_i)_2$, $M(OH)_2$]—would be simple in a classical problem of speciation in solution.

Unfortunately, in the case of soil samples, the number of constituents S_i, and especially the constants of the mass action law, are unknown. However, it is always possible to use arbitrary constants to perform the calculations quoted earlier. Ligands (A, B, etc.) and the free cation (M^{2+}) involved would correspond, respectively, to soil compartments (S_i) and to the extracted metal in given physical chemical conditions. The information provided must, of course, be considered qualitatively, and therefore, we will use the word "simulation."

We have aimed to use this approach using a very simple model of soil including exchange. In this work simulations will involve two ligands, A^- and B^-. Simulations with a greater number of ligands would be possible, but the use of two ligands seems to be sufficient to simulate all cases of cation concentration evolution due to a change in the physical-chemical conditions.

The two ligands are supposed to form complexes with a metallic cation M^{2+} according to the following reactions:

$$2A^- + M^{2+} \Leftrightarrow MA_2 \tag{2}$$

$$2B^- + M^{2+} \Leftrightarrow MB_2 \tag{3}$$

Both ligands A^- and B^- have acido-basic properties according to the following equilibria:

$$A^- + H^+ \Leftrightarrow HA \tag{4}$$

$$B^- + H^+ \Leftrightarrow HB \tag{5}$$

In our simulations, we will take into account the competition of other cations in the complexation of M^{2+} with A^- and B^-, represented by Ca^{2+}:

$$2A^- + Ca^{2+} \Leftrightarrow CaA_2 \tag{6}$$

$$2B^- + Ca^{2-} \Leftrightarrow CaB_2 \tag{7}$$

The simulation needs the following constants:

$$K_{MA_2} = \frac{[MA_2]}{[A^-]^2}[M^{2+}] \tag{8}$$

$$K_{MB_2} = \frac{[MB_2]}{[B^-]^2}[M^{2+}] \tag{9}$$

$$K_{HA} = \frac{[HA]}{[A^-]}[H^+] \tag{10}$$

$$K_{HB} = \frac{[HB]}{[B^-]}[H^+] \tag{11}$$

In order to simulate barium extraction of trace metal, one must take into account the

Soil Trace Metal Remediation

complexation of Ba^{2+} with the ligands A^- and B^-, according to the following equilibria and thermodynamic apparent constants:

$$2A^- + Ba^{2+} \Leftrightarrow BaA_2 \tag{12}$$

$$K_{BaA_2} = \frac{[BaA_2]}{[A^-]^2}[Ba^{2+}] \tag{13}$$

$$2B^- + Ba^{2+} \Leftrightarrow BaB_2 \tag{14}$$

$$K_{BaB_2} = \frac{[BaB_2]}{[B^-]^2}[Ba^{2+}] \tag{15}$$

Therefore, the metallic cation will be considered to be distributed among two forms:

Free M^{2+} in solution (considered as representing the amount of trace metal extracted and determined experimentally in a soil study)

Complexed MA_2 and MB_2 by the two ligands A^- and B^-

From the simulation point of view, the presence of an additive may simply be treated by introducing a new ligand C, with its own constants for the different equilibria in which it is involved.

In all cases, the conclusions drawn from the simulations are based on the free-form cation M^{2+}. The first simulation (see Table 5) compares the barium-extracted amount of metal M in the presence or not of the additive C. The result is presented in Fig. 9. First, it can be seen that the simulation of the extracted metal *versus* pH (reference curve)

Table 4 Example of the Set of Data Required for TOT Calculations

Component name	Co solution	Concentration of reagent	pK (H)	pK (M)
H^+ (strong acid)	1.00E-3	0.00		
OH^- (strong base)	0.00	1.50		
M^{2+}	1.00E-6	0.00		
A^-	1.00E-4	0.00	4.75	9

The following data is also required: Temperature (°C): 25.00; pKe 13.97; Vo solution (ml): 10.00; Vmax titrant (ml): 1.00E-2.

Table 5 Set of Data Required for TOT Calculations Simulating a Two-Compartment Soil Sample and the Effect of an Immobilizing Additive C

Component name	Co solution	pK (H)	pK (M)	pK (Ca)	pK (Ba)
Ca^{2+}	5.00E-5				
Ba^{2+}	1.00E-1				
M^{2+}	1.00E-6				
A	1.00E-4	5	8	3	4
B	1.00E-4	6	8	4	4
C	2.00E-5	5	9	3	4

Following data is required: Vo solution (ml): 10.00; Vmax titrant (ml): 1.00E-2.

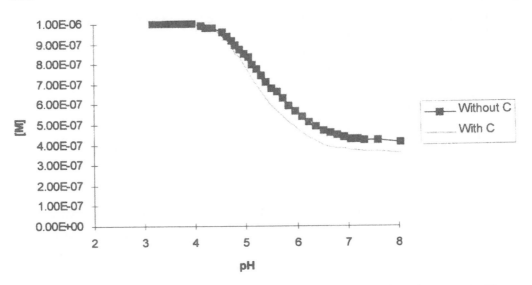

Figure 9 One example of the thermodynanmic simulation of the extraction of a trace metal from a soil sample with or without an immobilizing additive C (see also Table 5).

agrees with known results: the extracted trace metals increases as pH decreases; the competition of protons for adsorption sites are the main reactions involved to explain this change. When an additive, C, is added in the simulation, the concentration of the extracted metal (i.e., not bound to a ligand) is also increasing as pH decreases. However, its concentration is, as expected, lowered compared to the result obtained for the reference curves; it can also be seen that the effect of the additive is pH-dependent. The next simulations will try to put in evidence the roles of the additive and of the soil compartments.

In Fig. 10 are presented some results dealing with the study of the metal-binding effectiveness of the additive. It is obvious that the effectiveness of the additive is, for a given pH, increasing with the constant K_{MC2}, which measures in the simulation the extent of the metal binding by the additive. The effect E, calculated from the blank and the curves corresponding to the concentrations of M^{2+} for pK_{MC2} ranging from 8 to 10, ranges, at the optimum pH (pH $\geq$ 6), from 10% to 70%.

If we now consider the role of the soil compartments, i.e., one of the constants K_{MA2} or K_{MB2} in our simulations, the results obtained are given in Fig. 11. Similar evolution of the extracted metal is observed. The extracted metal M decreases when its affinity for soil compartment A is increased by the way of the constant pK_{MA2}. The effectiveness of the additive seems to correspond in fact to a less extent; this is due to the increasing affinity of the metal toward soil compartments.

IV. CONCLUSION

A very simple thermodynamic approach to the use of additives to immobilize soil trace metals and the results of an experimental study have shown that at last three parameters

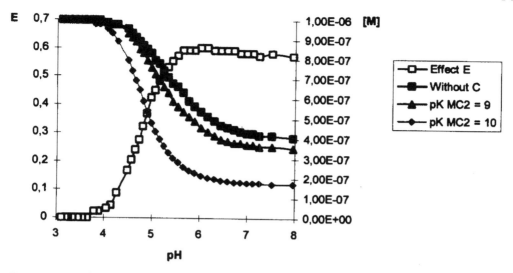

Figure 10 Effect of the constant K_{MC2} on the effectiveness of the immobilizing additive C.

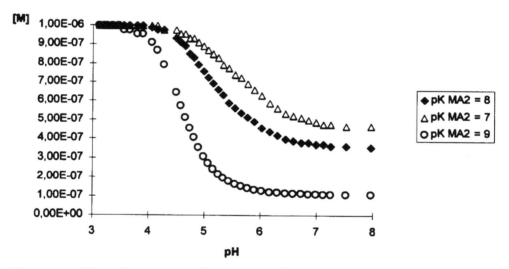

Figure 11 Effect of the constant K_{MA2} on the effectiveness of the immobilizing additive C.

are involved in in-situ remediation studies. From a chemical-physical point of view, the effectiveness of an additive to immobilize trace metals depends at the same time on the acidity level, the soil compartments, and obviously on the additive itself.

As a general conclusion from this study, it can be said that in-situ remediation of trace metal-contaminated soils should take into account not only the accumulation of metals in plants but also such a chemical-physical approach in order to optimize the conditions of this remediation and to choose the best additive.

REFERENCES

1. Hamby, D. M. 1996. Site remediation techniques supporting environmental restoration activities—A review. *Sci. Total Environ.*, 191:203–224.
2. Keller, C., Attinger, W., Furrer, G., Kayser, A., Keller, A., Lothenbach, B., Ludwig, C., Merki, M., Stenz, B., and Schulin, R. 1997. Extraction of metals from contaminated agricultural soils by crop plants. In S. E. Iskandar, A. C. Hardy, and G. M. Chang (eds.), *Proceedings of Fourth International Conference on the Biogeochemistry of Trace Elements*, Berkeley, CA, 1997, pp. 473–474.
3. Homer, F. A., Morrison, R. S., Brooks, R., Clemens, J., and Reeves, R. 1991. Comparative studies of nickel, cobalt and copper uptake by some nickel hyper accumulators of the genus *Alyssum*. *Plant Soil*, 138:195–205.
4. Li, Z., and Shuman, L. M. 1996. Extractability of zinc, cadmium, and nickel in soils amended with EDTA. *Soil Sci.*, 161:226–233.
5. Gworek, B. 1992. Inactivation of cadmium in contaminated soils using synthetic zeolites. *Environ Pollution*, 75:269–271.
6. Gworek, B. 1992. Lead inactivation in soils by zeolite. *Plant Soil*, 143:71–74.
7. Wong, J. W. C., and Wong, M. H. 1990. Effects of ash on yields and elemental composition of two vegetables, *Brassica parachinensis* and *B. chinensis*. *Agr. Ecosyst. Environ.*, 30:251–264.
8. Petruzzelli, G., Lubrano, L., and Cervelli, S. 1987. Heavy metal uptake by wheat seedlings grown in fly ash-amended soils. *Water Air Soil Pollution*, 32(4):389–397.
9. Chlopecka, A., and Adriano, D. C. 1997. Inactivation of metals in polluted soils using natural zeolite and apatite. In S. E. Iskandar, A. C. Hardy, and G. M. Chang (eds.), *Proceedings of Fourth International Conference on the Biogeochemistry of Trace Elements*, 1997, Berkeley, CA, pp. 415–416.
10. Adriano, D., and Berthelin, J. 1998. Evaluation et faisabilité de procédés biologiques, chimiques et physiques appliqués à la réhabilitation des sols. In: *Proceedings of Soil Science Congress*, Montpellier, France, 21–26 August, 1998.
11. McLaughlin, M., Maier, N., and Smart, M. 1998. Use of industrial by-products to remediate saline cadmium contaminated soils to protect the food chain. In: *Proceedings of Soil Science Congress*, Montpellier, France, 21–26 August 1998.
12. Waren, C. J., Evans, L. J., and Sheard, R. W. 1993. Release of some trace elements from sluiced fly ash in acidic soils with particular reference to boron. *Waste Manage. Res.*, 11:3–15.
13. Davies, B. E., Paveley, C. F., and Wixson, B. G. 1993. Use of limestone wastes from metal mining as agricultural lime: Potential heavy metal limitations. *Soil Use Mange.*, 9(2):47–52.
14. Alloway, B., Singh, B., and Bochereau, F. 1998. The use of adsorptive minerals to minimise the uptake of cadmium by food crops on contaminated lands. In: *Proceedings of Soils Science Congress*, Montpellier, France, 21–26 August 1998.
15. Mench, M. J., Didier, V. L., Löffer, M., Gomez, A., and Masson, P. 1994. A mimicked *in situ* remediation study of metal contaminated soils with emphasis on cadmium and lead. *J. Environ. Qual.*, 23:58–63.
16. Berti, W., and Ryan, J. 1998. Use of phosphorus to inactivate Pb in contaminated soils. In: *Proceedings of Soil Science Congress,* Montpellier, France, 21–26 August 1998.
17. Pierzynski, G. M., and Schwab, A. P. 1993. Bioavailibity of zinc, cadmium, and lead in a metal-contaminated alluvial soil. *J. Environ. Qual.*, 22:247–254.
18. Luz Mora, G., and Demanet, F. 1998. Gyspum and liming effect on surface reactivity in chilean acid soils. In: *Proceedings of Soil Science Congress*, Montpellier, France, 21–26 August 1998.
19. Sappin-Didier, V., and Gomez, A. 1994. Réhabilitation des sols pollués par des métaux toxiques; exemple de la grenaille d'acier. *Analusis*, 22:28–30.
20. Lehoczky, E., Marth, P., and Szabados, I. 1998. Effect of liming on the heavy metal uptake of lettuce. In: *Proceedings of Soil Science Congress*, Montpellier, France, 21–26 August 1998.

21. Juste, C., and Soldâ, P. 1998. Influence de l'addition de différentes matières fertilisantes sur la biodisponibilité du cadmium, du manganèse, du nickel et du zinc contenus dans un sol sableux amendé par des boues de station d'épuration. *Agronomie*, 8(10):897–904.
22. Sims, J. T. 1986. Soil pH effects on the distribution and availability of manganese, copper and zinc. *Soil Sci. Soc. Am. J.*, 50:367–373.

25
Surfactants-Enhanced Photolysis of Polychlorinated Biphenyls (PCBs): A Solution for Remediation of PCB Contaminated Soils

Zhou Shi, Mriganka M. Ghosh, and Kevin G. Robinson
The University of Tennessee, Knoxville, Tennessee

I. INTRODUCTION

Polychlorinated biphenyls (PCBs) are remarkable pollutants due to their ubiquitous environmental distribution and chronic toxicity (1–4). Though PCBs are generally recalcitrant to chemical and biological degradation, they are photodegradable (5–7). Most PCBs have weak absorption bands that partially overlap the solar spectrum in the ultraviolet (UV) region at wavelengths above 280 nm (5). Photodegradation is reported as an important process of PCB decomposition in the environment (7–8). The loss of PCBs in natural water due to solar radiation has been estimated to be of the order of 10 to 1000 g/km^2/year (9). Artificial lights with wavelengths less than 280 nm are absorbed strongly by PCBs and are therefore useful in decontaminating PCB-containing wastewaters. However, direct application of photolysis to remediation of PCB-contaminated soil is limited due to lack of light penetration.

Soil washing followed by UV irradiation of PCBs in surfactant solutions could overcome the limitation of photolysis in soil remediation. Surfactants, found to effectively desorb PCBs from contaminated soils (10–13), may potentially promote photolysis. Serving as H-atom sources, surfactants can enhance photodechlorination and minimize hydroxyl substitution reactions (14–19). Additionally, at dosages above the critical micelle concentration (CMC), surfactant micelles may impose a localized "cage effect" which can promote PCB photolysis. The hydrocarbon-like micellar core or the "cage" favors partitioning of the hydrophobic PCB molecules while the polar photoproducts such as chloride ion are expelled. Thus, the common-ion effect is eliminated and PCB molecules can react with the surfactant (H-atom source) longer at a higher local concentration within the "cage" (14–17). It is well documented (8,9,20–23) that PCB photochemistry primarily undergoes C–Cl cleavage from the lowest triplet state, forming an aryl radical and a chlorine radical. After abstracting H-atoms from H-donors, the aryl radical forms less chlorinated products and chlorine radical forms HCl. HCl dissociates to give H$^+$ and

Cl⁻. The *ortho*-chlorines are removed preferentially. Thus, from the standpoint of surfactant-assisted pump-and-treat technology, photodechlorination of PCBs in soil washings could detoxify or even completely destroy the compounds, eliminating the step of PCB separation from the washings. Further, the preferential removal of *ortho*-chlorines, which are among the most biologically persistent (24), may benefit the integrated biodegradation process (25).

Several considerations are very important in application of photolysis in soil washings: attenuation of light by suspended particles and possible reactions catalyzed at the particle surface and/or sensitized by humic materials from soils. Miller and Zepp (26) examined the effect of suspended sediment upon the photolysis rate of dissolved pollutants in several natural water bodies. They found photolysis rates to be generally more rapid in low-turbidity waters than in clear waters, by a factor of up to 1.8 in clay suspensions. Light scattering was suggested as being responsible for the observed rate enhancement. Light attenuation due to the presence of colloidal particles in natural waters was attributed to two independent mechanisms: scattering and absorption. Photochemistry at the surface of soil particles may mimic that at the surface of TiO_2, which involves electron transfer and free-radical reactions. Occhiucci and Patacchiola (27) demonstrated dechlorination of PCBs adsorbed on montmorillonite, a clay mineral. Draper et al. (28) reported that irradiation of a concrete surface contaminated with Aroclor 1260 (81 ± 31 g/100 cm²) resulted in 47% degradation in 21 h. In soil slurries, with TiO_2 added, 2-h irradiation with UV and sunlight resulted in up to 77% degradation of PCBs (29,30). Similarly, Hawarl et al. (31) reported that photolysis of soil-extracted Arcolor 1254 in alkaline 2-propanol led to effective dechlorination, especially when phenothiazine, a sensitizer, was added. Direct UV irradiation of PCB (Aroclor 1248)-contaminated soils at a surfactant dose up to 5% of the weight of the dry soil led to less than 15% PCB reduction (32). After adding Fenton's reagent, the PCB degradation increased up to 55% (32). The cost of soil treatment was estimated to be $120–$250/m³ for treatment of 1 m³ of soil per day using a 100-kW system (33).

The objectives of this study are to evaluate the mechanism of PCB photolysis in surfactant solutions, and to test the feasibility of PCB photolysis in surfactant soil washing solutions.

II. MATERIALS AND METHODS

A. Photolysis of 2,4,4'CB in Surfactant Solutions

Solutions of 2,4,4'-trichlorobiphenyl (2,4,4'CB) (97% pure, Ultra Scientific Co.) in a nonionic surfactant, polyoxyethylene (10) lauryl ether [POL(10), Sigma Chemical Co.], were used to study the mechanism of PCB photolysis in surfactant solution. PCB crystals were placed into 150-mL centrifuge bottles containing 120 mL POL(10) (4 g/L) solution. After shaking for 3 days on an end-over-end rotor (Wheaton), the solutions were centrifuged at 10,000 rpm ($7280 \times g$) for 30 min to separate undissolved PCB crystals. After suitable dilution with POL(10) (4 g/L), the supernatants of PCB-POL(10) solutions were photolyzed without pH adjustment, aeration, or degassing. PCB-POL(10) solutions in 5-mL quartz tubes (id = 10 mm) were irradiated in a Rayonet photoreactor (PRP-200, Southern New England Ultraviolet Company) fitted with a merry-go-round unit. The UV light intensity within the test tubes was determined from separate tubes containing the potassium ferrioxalate [$K_3Fe(C_2O_4)_3$] actinometer (0.006 M), following the suggested procedures (34). Thus, the light intensity reaching the sample was 1.09×10^{-5} einsteins/L-s^{-1}.

To determine the concentration of the remaining PCB and products, a 1-mL aliquot, carefully collected from mid-depth of the irradiated solutions, was transferred to a 15-mL glass tube containing 4 mL of hexane. To prevent emulsion formation during mixing, 0.3 g of a silica-based powder, tC_{18} (Waters, Inc.), was then added to solutions containing nonionic surfactants. For experiments involving anionic surfactants, 0.3 g of an anionic exchange resin, QMA (Waters, Inc.), was used instead of the tC_{18}. The samples were mixed for 30 min in a reciprocal shaker followed by centrifugation for 10 min at 1000 rpm ($168 \times g$). The hexane phase was carefully collected, suitably diluted, and analyzed for PCB and products by a GC (Shimadzu model GC-14A) with an electron-capture detector (ECD) and by a GC (HP5890) with a mass spectrum (MS) detector (HP5989B). The GC-ECD was equipped with a DB-1 capillary column (id × length × film thickness = 0.25 mm × 30 m × 0.25 m) (J & W Scientific, Inc.). The GC-ECD column temperature was programmed as follows: it was initially held at 60°C for 2 min, then raised to 180°C at a rate of 10°C/min and thereafter to 225°C at a rate of 6°C/min. After 14 min at 225°C, the temperature was further raised to 280°C at a rate of 5°C/min and held constant for 13.5 min. The GC-MS was equipped with a HB-5 column (id × length × film thickness = 0.45 mm × 30 m × 0.25 m). The column temperature of the GC was programmed as follows: it was initially held constant at 70°C for 1 min, then raised to 150°C at a rate of 10°C/min, and at a rate of 2°C/min thereafter to 230°C. Finally, the temperature was further raised at a rate of 10°C/min to its final value, 280°C, where it was held constant for 15 min. The PCB concentration in a sample was determined by comparing with PCB standards. 2,2′,3,3′,4,5,6,6′-Octachlorobiphenyl (Ultra Scientific) was used as the internal standard in all GC analyses. The products were identified by matching their retention times and mass spectra with those of standards.

To estimate PCB extraction efficiency, known amounts of PCB dissolved in ether were added to 15-mL test tubes. The amount of PCB added ranged from 90% to 110% of that needed to saturate the surfactant solution. After vaporizing the ether by gently purging with N_2, 2 mL of surfactant solution was added. The tubes were capped with Teflon-lined caps and shaken for 3 days. PCBs in each surfactant solution (2 mL) was then extracted with 8 mL of hexane following the procedures outlined above. Either 0.6 g of tC_{18} or QMA was used. The PCB extracted in the hexane phase was analyzed by the GC. The ratio of the total recovered PCB to the total added was taken as a measure of extraction efficiency. The extraction efficiency of all PCB-surfactant solutions tested was greater than 90%.

The concentration of chloride in photolyzed solutions was measured by using a chloride-selective electrode (Orion model 96-17): 3 mL of the photolyzed sample was placed in a 22-mL glass vial and measured immediately. The chloride concentration was determined by comparing with chloride standards prepared by dissolving KCl in POL(10) (4 g/L). In addition, the chloride concentration was occasionally verified by a colorimetric method (35). The detection limits for Cl^- by both methods were ~ 0.05 mg/L.

The POL(10) concentration was measured colorimetrically at 620 nm using the standard method (36) without any pretreatment of samples. The cobalt reagent used in the test was prepared by carefully dissolving 36 g of $Co(NO_3)_2 \cdot 6H_2O$ and 240 g of NH_4SCN in 1000 mL of deionized (DI) water. The test sample (1 mL) was mixed with 3 mL of cobalt thiocyanate reagent and 8 mL of methylene chloride (CH_2Cl_2) in a 15-mL glass tube and shaken for 10 min. After centrifuging at 1000 rpm for 5 min, 1.2 mL from the CH_2Cl_2

phase was placed in a quartz cuvette and the absorbance was measured at 620 nm using a UV-VIS spectrophotometer (Beckman DU-70).

B. Photolysis of PCBs in Surfactant Soil Washing Solutions

To evaluate the feasibility of photolysis of PCB in surfactant soil washings, two surface soils, collected from commercial PCB-contaminated sites, were used to produce surfactant soil washings. Soil A had a clayey texture, while soil B had a sandy texture. These soils were air-dried and storied in amber glass bottles at 4°C. In this study, the soils were further prescreened using a standard sieve (No. 35, opening 0.50 mm). The soil particles with size 0.50 mm and less were collected for soil washing experiments.

PCB contamination levels in the soils were determined according to EPA Method 3541. A small amount (2.5 g) of soil was extracted using 50 mL of a 1:1 mixture of acetone: hexane in a Soxhlet extraction unit (HT 1043, Perstop Analytical, Inc.). The time and temperature for boiling and rinsing were set at 1 h and 140°C, respectively. The solvent remaining in the extracting cup, at the end of extraction, was concentrated to 5–8 mL and transferred into a 15-mL glass tube. The hexane from cup rinsing was combined with the extracting solvent to bring the total solvent volume to 10 mL. After the addition of 0.5 g of anhydrous Na_2SO_4, the sample was shaken for 15 min and then centrifuged for 10 min at 1000 rpm. Aliquots of the solvent were transferred into GC vials with suitable dilution for PCB analysis. The profile of the extracted PCBs was close to that of Aroclor 1248, a commercial PCB mixture. Thus, Aroclor 1248 standard (hexane solution) was used to identify and quantify PCB congeners as discussed in Section III.B.1. PCB contamination levels for soil A and soil B were determined at 11,730.8 and 317.9 g/g soil, respectively.

The PCB-surfactant soil washing solutions were prepared as follows: 1 g of screened (US sieve no. 35) PCB-contaminated soil was shaken for 3 days with 20 mL of 4 g/L surfactant [POL (10) or sodium dodcyl sulfate (SDS)] solution. The resulting suspension was either settled by gravity for different time periods or centrifuged for 30 min at 10,000 rpm ($7280 \times g$). Thus, suspensions having various turbidities were obtained. Nephelometric measurements of turbidity (NTU) were made using a turbidimeter (HACH Chemical Co. model 2100A). The samples were then irradiated in the Rayonet photoreactor and the PCBs in the samples were analyzed in the same fashion as described for the photolysis of 2,4,4'CB in POL(10) solution.

III. RESULTS AND DISCUSSION

A. Photolysis of 2,4,4'CB in Nonionic Surfactant [POL(10)] Solutions

1. Products and Mechanism

To study the mechanism of PCB photolysis in surfactant solutions, 2,4,4'CB was chosen as a modal compound. The congener (23.5 mg/L) in a POL(10) (4 g/L) solution was irradiated at 254 nm. The UV-VIS absorbance spectra of the PCB congener [after 1 : 1 dilution with POL(10)] at various photolysis times are shown in Fig. 1. It was observed that the major absorbance band (K-band) of 2,4,4'CB in the POL(10) solution had a red shift from 252 nm to 259 nm after 1 min of photolysis. Further photolysis (5 min) resulted only in the K-band decreasing in intensity, without discernible shifting. Reportedly, the K-band of 2,4,4'CB, compared to that of 4,4'CB has a blue shift due to substitution

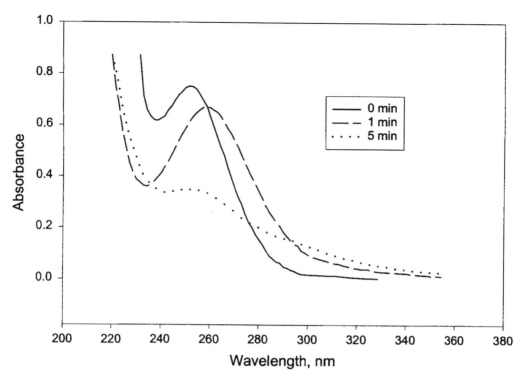

Figure 1 UV-VIS spectra of 2,4,4'CB in POL(10) (4 g/L) solutions at various photolysis times.

of the *ortho*-chlorine (37). Therefore, the red shifting of the K-band of 2,4,4'CB after photolysis suggests a possibly preferential removal of the *ortho*-chlorine.

Gas chromatograms (GC) of 2,4,4'CB and its photoproduct are shown in Fig. 2. 4,4'CB (peak 2) was identified as the major photoproduct of 2,4,4'CB by matching the retention time as well as the mass spectrum (GC-MS) with the 4,4 standard. Consequently, it was confirmed that the *ortho*-chlorine of 2,4,4'CB was preferentially removed by photolysis. In addition to the 4,4'CB, small amounts of 4CB (2.5% of the initial 4,4'CB) and biphenyl (5% of the initial 4,4'CB) were also identified by GC-MS analysis after 5 and 30 min of irradiation, respectively. Formation of 4,4'CB, 4CB, and biphenyl further suggested a stepwise dechlorination pathway of 2,4,4'CB. To further confirm the PCB dechlorination, free chloride produced was monitored using a chloride ion-selective electrode (Orion 96-17). The concentration of 2,4,4'CB and its photoproducts (4,4'CB and Cl^-) are shown as a function of time in Fig. 3. After 1 min of photolysis, there were 1.64 mg/L (6.37×10^{-6} M) of 2,4,4'CB remaining and 9.48 mg/L (4.25×10^{-5} M) of 4,4'CB produced. No 4CB or biphenyl was identified. In the same time period, the measured free chloride ion was 2.94 mg/L (8.28×10^{-5} M). Thus the total recovered chlorine (free chloride plus bonded chlorine on PCBs) was 1.87×10^{-4} M, accounting for 68% of the initial amount of chlorine from 2,4,4'CB. After 60 min, the PCBs were completely degraded, and 7.8 mg/L (2.2×10^{-4} M) of Cl^- were measured. The measured Cl^- accounted for 80% of the initial amount of chlorine of 2,4,4'CB. This result indicated that dechlorination was the major pathway of 2,4,4'CB photodegradation.

In summary of the experimental results discussed previously, the photodegradation of 2,4,4'CB mainly undergoes stepwise dechlorination with preferential removal of the

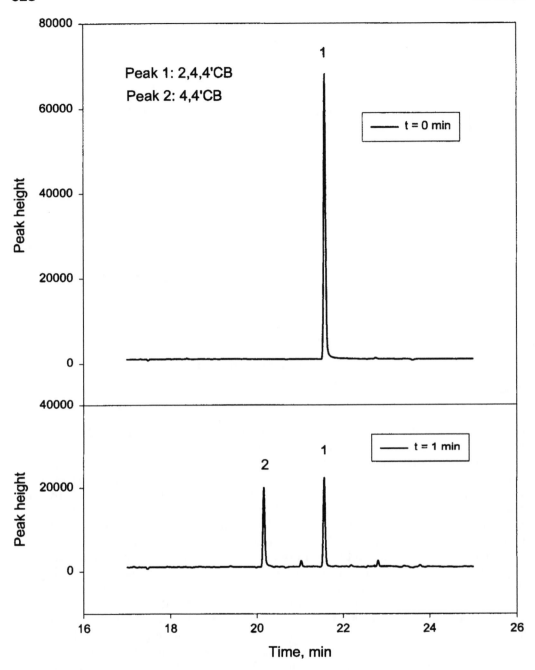

Figure 2 Gas chromatograms of 2,4,4'CB and its photoproducts.

ortho-chlorine. This result agrees well with the reported pathway of PCB photolysis in nonionic surfactant solutions (16,19) and in organic solvents (20–23). Consequently, the mechanism of 2,4,4'CB degradation in surfactant solution is proposed to be the same as in the nonpolar solvents (20,22,23) (Fig. 4): upon absorption of UV light, 2,4,4'CB molecules are promoted from ground state to singlet excited state, followed by

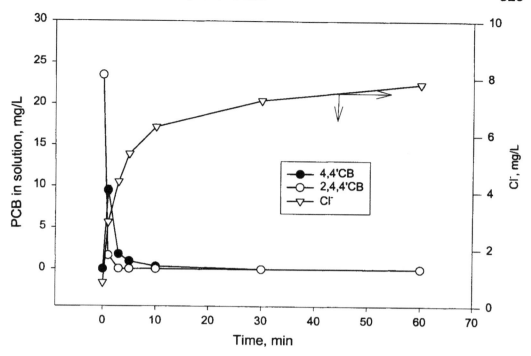

Figure 3 Concentrations of 2,4,4'CB and its photoproducts (4,4'CB and Cl$^-$) in POL(10) (4 g/L) solution at various photolysis times.

singlet-to-triplet conversion via intersystem crossing. In the triplet excited state, the C–Cl bond at the ortho position will preferentially cleave to produce an aryl radical and a chlorine radical. After abstracting H-atoms from H-donors which are likely the surfactant molecules (14–19), the aryl radical forms 4,4'CB and the chlorine radical forms HCl which dissociates to give H$^+$ and Cl$^-$. Similarly, prolonged photolysis will further transform the 4,4'CB to 4CB and biphenyl. In this study, since the chloride mass balance accounts for only 80% of the initial amount of chlorine based on the proposed pathway, other unknown minor pathways involving unidentified chlorinated intermediates may also be involved.

It was observed that 8% of the initial POL(10) was degraded after 1 h of photolysis. The consumption of the surfactant during photolysis is possibly attributed to the role of the surfactant as an H-atom donor. Multiple H-atoms per surfactant molecule and the high molar ratio of the surfactant to the PCB helped to minimize such kind of consumption. Direct UV degradation is unlikely to contribute to the surfactant consumption simply because the surfactant did not absorb the light [the molar extinction coefficient (ε) of POL(10) is 2 M^{-1} cm^{-1}]. These results indicate that the surfactant is reusable for the PCB photolysis.

2. Rates and Quantum Yields

Since the initial absorbance ($\lambda = 254$ mm) of 2,4,4'CB in POL(10) was greater than 1.5, almost all incident light was absorbed by the compound. Consequently, the initial photolysis reaction should obey zero-order kinetics (38,39):

$$C_0 = C_t + k_0 t \tag{1}$$

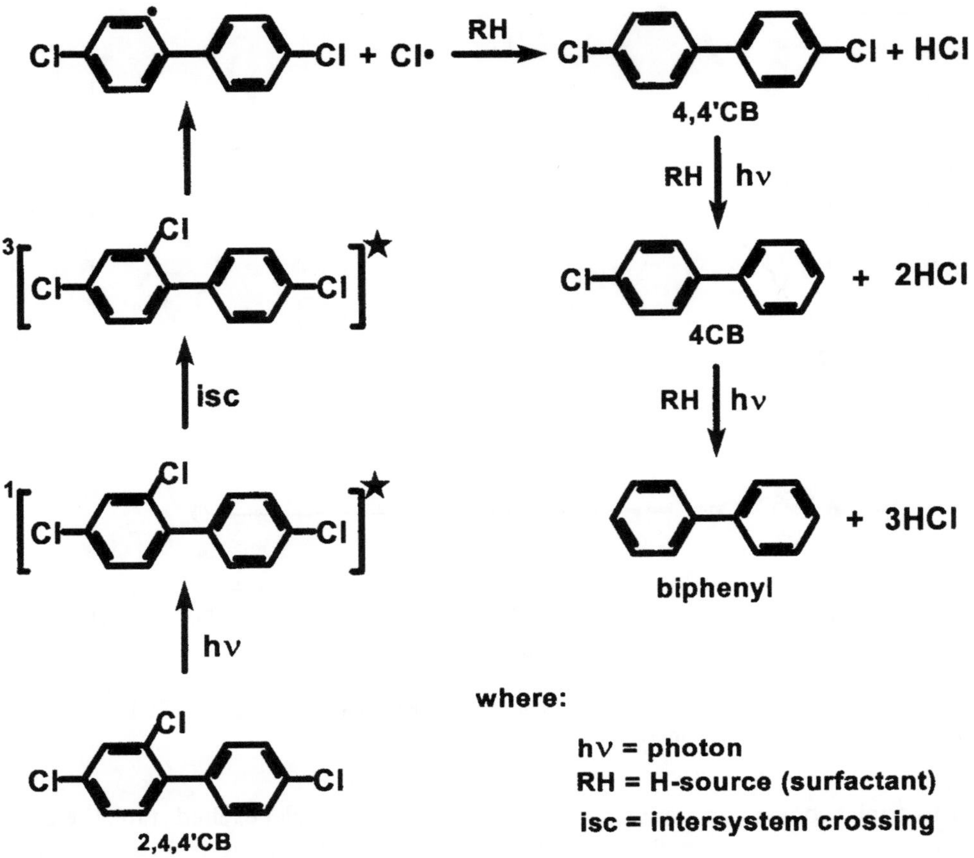

Figure 4 Mechanism of 2,4,4'CB photolysis in POL(10) solution.

where

C_0 and C_t = concentration of PCB at time 0 and t, respectively (M)
k_0 = zero-order rate constant (M/s)
t = time (s)

To minimize error due to competitive absorption of light by photoproducts, the initial photolysis rate constants (k_0) were estimated using only the data within 1 min of the initial irradiation period. k_0 was calculated as $(1.41 \pm 0.03) \times 10^{-6}$ M/s. The corresponding quantum yields (Φ) at $\lambda = 254$ nm is calculated as follows (38,39):

$$\Phi = \frac{k_0}{I_0} \qquad (2)$$

where

I_0 = average irradiance of incident light at 254 nm (einsteins/L-s) $\times 10^{-5}$ (einsteins/L-s)
k_0 = zero-order rate constant (M/s) = $(1.41 \pm 0.03) \times 10^{-6}$ (M/s)

The quantum yields of 2,4,4'CB in POL(10) solutions was calculated as 0.129 ± 0.003. The reported quantum yield (16) of 2,4,4'CB in CH_3CN/H_2O (v : v = 9 : 1) was 0.014. Thus, the quantum yield of 2,4,4'CB was enhanced by a factor of 9.2 in POL(10) solution relative to mixture of CH_3CN/H_2O. Surfactant-assisted PCB photolysis was also observed by other researchers. Epling et al. (16) reported that quantum yields of PCBs were enhanced in the present of surfactants and sodium borohydride in the mixture of CH_3CN/H_2O. Chu et al. (19) found that quantum yield of 2,3,4,5-tetrachlorobiphenyl (2,3,4,5CB) photolysis was sixfold greater in nonionic surfactant solution (Brij 58, 0.5 mM) than that in water alone.

B. Photolysis of PCB in Surfactant Soil Washing Solutions

1. Identification and Quantification of PCBs in Surfactant Soil Washing Solutions

The gas chromatogram (GC) of Aroclor 1248 (a commercial PCB mixture) shown in Fig. 5 compares with GCs of PCBs washed from soil A and B using POL(10) as shown in Figs. 6 ($t=0$) and 7 ($t=0$), respectively. The congener profiles (GC peaks) of soils A and B are similar to that of Aroclor 1248 standard. The congener profiles of PCBs washed from the same soils using SDS (not shown) are also similar to that of Aroclor 1248. Aroclor 1248 was chosen as the standard to identify and quantity the PCB congeners in all surfactant soil washings. The major peaks of the PCB soil washing samples were assigned numbers from 1 to 34. The sum of concentration of all individual congeners (peaks) was taken as the concentration of the PCBs in a given soil washing sample:

$$\sum [X]_s = \sum \frac{[X]_{std}}{[IS]_{std}} \frac{A_{ISstd}}{A_{Xstd}} \frac{A_{Xs}}{A_{ISs}} [IS]_s \qquad (3)$$

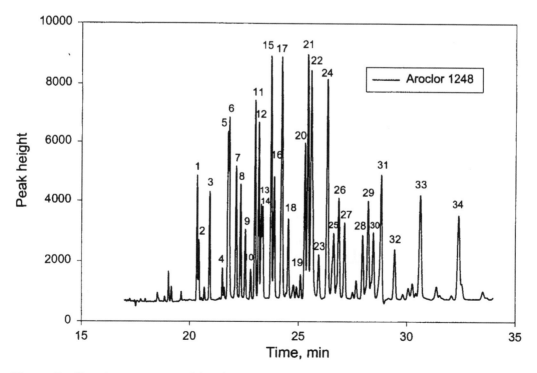

Figure 5 Gas chromatogram of Aroclor 1248 standard (0.99 mg/L) in hexane.

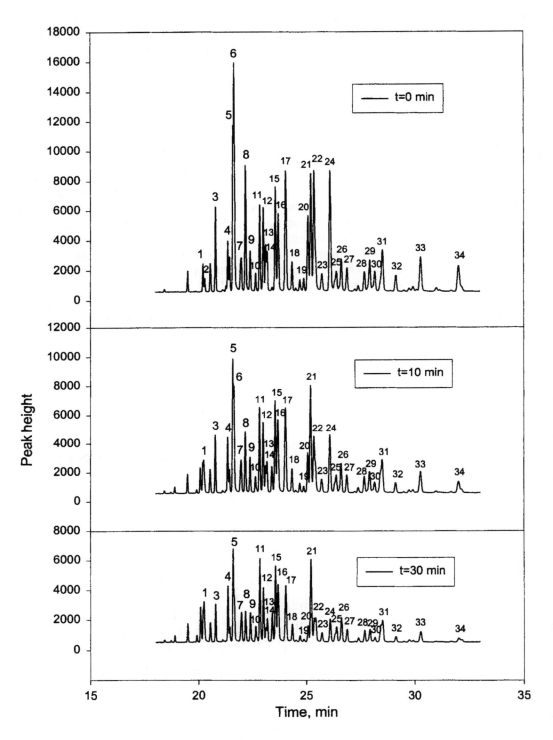

Figure 6 Comparison of gas chromatograms of PCBs in soil A washing solutions [4 g/L(4)] at various photolysis times.

Surfactants-Enhanced Photolysis of PCBs

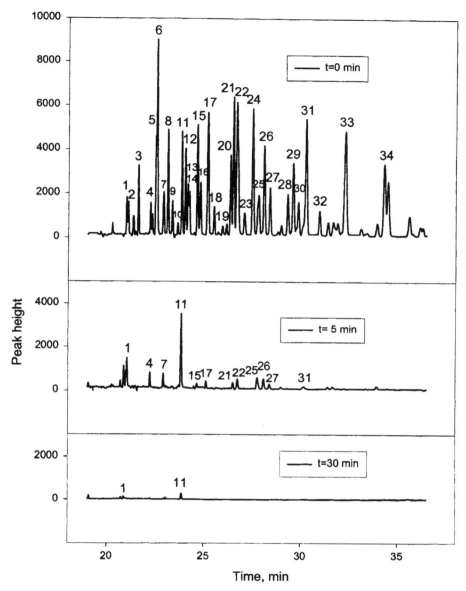

Figure 7 Comparison of gas chromatograms of PCBs in soil B washing solutions [4 g/L(4)] at various photolysis times.

In this study, $[IS]_{std} = [IS]_s$ and $[X]_{std} = [\text{total PCB}]_{std} \times \text{wt\%}$.

$$\sum [X]_s = [\text{totalPCB}]_{std} \sum \text{wt\%} \frac{A_{ISstd}}{A_{Xstd}} \frac{A_{Xs}}{A_{ISs}} \tag{4}$$

where

$[X]_s$ and $[X]_{std}$ are the concentrations of peak X in sample and standard, respectively (mg/L)

$[IS]_{std}$ and $[IS]_s$ are the concentrations of internal standard (2,2′,3,3′,4,5,6,6′CB) in sample and standard, respectively (mg/L)

A_{ISstd} and A_{ISs} are peak areas of internal standard in standard and sample, respectively

A_{Xstd} and A_{Xs} are peak areas of peak X for standard and sample, respectively

[total PCB]$_{std}$ is total concentration of commercial PCB standard (mg/L)

wt% is weight percent of peak X in the commercial PCB, using the data given by Frame et al. (40) (Table 1)

All the PCB concentrations in the soil washings were calculated using Eq. (4).

Table 1 Weight Distribution of GC Peaks in Aroclor 1248 (40)

Peak no.[a]	PCB congener	Weight (%)
1(14)	4,4′	2.95
2(15)	2,2′,4	1.22
3(17)	2,2′,3; 2,4′,6	1.62
4(21)	2,3′,5	0.77
5(23)	2,4′,5	4.76
6(24)	2,4,4′	6.09
7(25)	2,3′,4′; 2,2′,5,6′	2.87
8(26)	2,3,4′; 2,4,2′,6′	1.60
9(27)	2,2′,3,6	1.17
10(29)	2,2′,3,6′	0.84
11(31.1)	2,2′,5,5′	5.91
12(32)	2,2′,4,5′	4.59
13(33)	2,2′,4,4′	2.55
14(34)	2,2′,4,5	2.19
15(37.2)	2,2′3,5′	5.00
16(38)	2,2′3,4′	2.91
17(39)	2,3,4′,6	6.10
18(42)	2,2′3,3′	1.18
19(45)	2,3,4′,5	0.78
20(46.1)	2,4,4′,5	4.50
21(47)	2,3′,4′,5	7.55
22(48.1)	2,3′,4,4′	9.63
23(49)	2,2′,3,4′,6	0.83
24(50)	2,3,4,4′	5.34
25(51)	2,2′,3,3′,6	1.06
26(53)	2,2′,4,5,5′,2,2′,3,5,4′	2.07
27(54)	2,2′,4,4′,5	1.76
28(57.1)	2,2′,3′4,5	1.05
29(58.1)	2,2′3,4,5′	1.35
30(59)	2,2′,3,4,4′	1.25
31(61)	2,3,3′,4′,6	2.73
32(63)	2,2′,3,3′,4	0.85
33(69)	2,3′,4,4′,5	2.35
34(74)	2,3,3′,4,4′	1.82

[a] Peak numbers assigned by Frame et al. (40) are given in parentheses.

2. Photolysis of PCBs in Soil Washing Solutions

The GCs of PCBs washed from soil A and B using POL(10), before and after photolysis, are shown in Figs. 6 and 7. Highly chlorinated congeners with larger retention times tend to be degraded faster than less chlorinated congeners with small retention times. This conclusion also applied for the PCBs washed from soil A and B using SDS. Additional results of PCB photolysis in soil A washings of POL(10) and SDS are summarized in Table 2. After 30 min of UV irradiation in the POL(10) washing, all PCB congeners, except 4,4'CB (peak 1), were 19.4% to 100% photolyzed. Peak 1 showed a net increase of 99%. Presumably, 4,4'CB is formed as a major by-product of incomplete dechlorination of many highly chlorinated congeners such as 2,2',4,4'CB (peak 13). Two peaks, peak 2 (2,2',4CB) and peak 6 (2,4,4'CB), were absent in the photolyzed washing solution. Presumably, these congeners were completely photolyzed. PCBs with *ortho*-chlorines such as 2,2',4,4'CB (peak 13) and 2,4,4'CB) (peak 6) were degraded faster than PCBs without *ortho*-chlorine such as 4,4'CB; this corresponds well with the results of photolysis of pure 2,4,4'CB in POL(10) solutions discussed previously. The literature (16,19,20,23) is replete with such observations. Further, a significant decrease in both the concentration and mass (% weight) of congeners in peaks 27 to 34 indicated that highly chlorinated congeners (pentachlorobiphenyls) represented by these peaks were more photodegradable than the less chlorinated ones (peaks 1, 4, etc.). Results obtained in SDS washings were similar (Table 2). Thus, photolysis of highly chlorinated congeners may benefit subsequent aerobic biodegradation of PCBs in an integrated two-step photolysis/biodegradation remediation scheme.

The total concentration of the PCB congeners in different soil washing solutions with different surfactants is shown as a function of photolysis time in Figs. 8 and 9. Since the absorbance (at 254 nm) of the PCB-surfactant solution was greater than 2, the solution was optically dense. Thus, initially (≤ 5 min), photodegradation should follow zero-order kinetics (38,39) as shown in Eq. (1). Based on these data, "apparent" initial zero-order rate constant (k_0) and relative quantum yields (Φ) of the initial photolysis ($t \leq 5$ min) were estimated using Eqs. (1) and (2), respectively (Table 3). In application of Eq. (2), the unit of k_0 was converted from mg/L-min to M/s assuming molecular weight (MW) of the PCBs in soil washings equal to that of Aroclor 1248 (MW = 288.1, calculated from the nominal formula $C_{12}H_{6.11}Cl\{3.89\}$ with 48% Cl content). For soil A washings with turbidity of 6.5 (NTU), 5-min photolysis resulted in 96 and 31 mg/L of total PCBs being degraded in POL(10) and SDS solutions, respectively (Fig. 8). The "apparent" initial rate constants and quantum yields were 19.3 mg/L-min and 2.07×10^{-2} for POL(10) solution, and 6.2 mg/L-min and 6.65×10^{-3} for SDS solution, respectively. For soil B, almost all of the PCBs were degraded after 5 min of photolysis in both POL(10) and SDS solutions (Fig. 9). The photolysis rate and quantum yield in POL(10) washing solution were 1.43 mg/L-min and 1.53×10^{-3}, while for SDS washing solution they were 0.435 mg/L-min and 4.68×10^{-4}, respectively. Nonionic surfactant, POL(10), is better than the anionic surfactant, SDS, in the enhancement of PCB photolysis in soil washings.

To investigate the effect of soil particles on photolysis, POL(10)-soil A washing solutions containing different turbidities were irradiated (Fig. 8). The initial ($t \leq 5$ min) photolysis rate constants and quantum yields are presented in Table 3. After 5 and 30 min of irradiation, nearly 18% and 62% of the PCB initially present in the soil washing solution containing a turbidity of 6.5 NTU was photolyzed, respectively. The photolysis of PCBs in the same soil washing solution decreased to 3% and 21% when the turbidity was 3488

Table 2 Photolysis of PCB Congeners in Soil A Washing Solution (6.5 NTU)

Peak no.	POL(10) solution				SDS solution			
	0 min (mg/L)	(wt%)	30 min (mg/L)	Photolyzed (wt%)	0 min (mg/L)	(wt%)	10 min (mg/L)	Photolyzed (wt%)
1	7.24	1.36	14.41	−99.0	1.00	1.60	0.53	−46.96
2	3.07	0.58	0.00	100.0	0.42	0.66	0.00	100.00
3	13.48	2.53	4.46	66.9	1.87	2.98	0.04	97.73
4	14.51	2.73	12.25	15.6	1.85	2.95	1.26	32.05
5	41.89	7.90	28.78	31.3	5.23	8.35	4.30	92.12
6	83.50	15.74	0.00	100.0	9.93	15.83	0.00	100.00
7	10.36	1.95	5.66	45.4	1.33	2.12	3.57	74.28
8	18.81	3.55	3.51	81.3	2.50	3.99	0.79	96.96
9	6.79	1.28	3.78	44.3	0.93	1.49	0.31	66.46
10	5.22	0.98	4.21	19.4	0.64	1.02	0.31	51.79
11	24.67	4.65	17.02	31.0	3.14	5.00	1.62	48.32
12	22.00	4.15	10.43	52.6	2.63	4.20	0.36	86.15
13	13.00	2.45	2.83	78.2	1.53	2.45	0.00	100.00
14	9.55	1.80	4.54	52.5	1.13	1.80	0.17	84.77
15	21.20	4.00	11.45	45.9	2.55	4.07	0.71	72.23
16	19.82	3.74	12.74	35.8	2.39	3.82	0.58	95.94
17	30.17	5.69	10.82	64.1	3.63	5.78	0.15	95.94
18	4.05	0.76	1.97	51.5	0.48	0.77	0.05	88.69
19	4.14	0.78	0.78	81.1	0.40	0.64	0.00	100.00
20	21.31	4.02	3.34	84.3	2.23	3.56	0.00	100.00
21	33.01	6.22	18.12	45.1	3.66	5.83	1.01	72.36
22	42.37	7.99	9.43	77.8	4.72	7.53	0.36	92.38
23	3.36	0.63	1.35	59.7	0.36	0.58	0.04	88.63
24	27.96	5.27	3.91	86.0	3.13	4.99	0.00	100.00
25	3.30	0.62	1.82	44.7	0.35	0.57	0.12	65.97
26	6.24	1.18	3.45	44.7	0.66	1.06	0.21	68.89
27	5.24	0.99	2.21	57.8	0.53	0.84	0.02	96.83
28	3.14	0.59	1.47	53.4	0.32	0.51	0.04	88.44
29	4.10	0.77	1.33	67.6	0.42	0.66	0.00	100.00
30	3.03	0.57	0.58	80.7	0.34	0.54	0.00	100.00
31	8.88	1.67	2.74	69.1	0.89	1.42	0.18	80.12
32	2.64	0.50	0.76	71.1	0.28	0.44	0.00	100.00
33	6.55	1.24	1.64	75.0	0.59	0.94	0.00	100.00
34	5.83	1.10	0.73	87.5	0.53	0.84	0.00	100.00

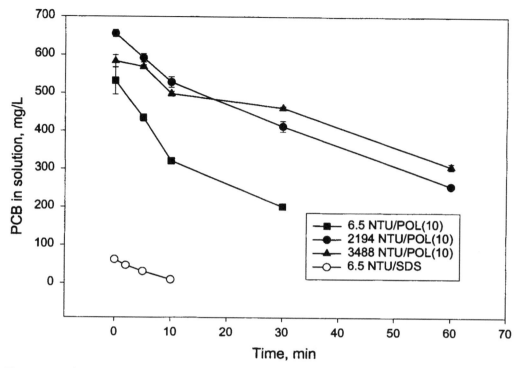

Figure 8 Photolysis of PCBs in soil A washing solutions (4 g/L surfactant) at various turbidities.

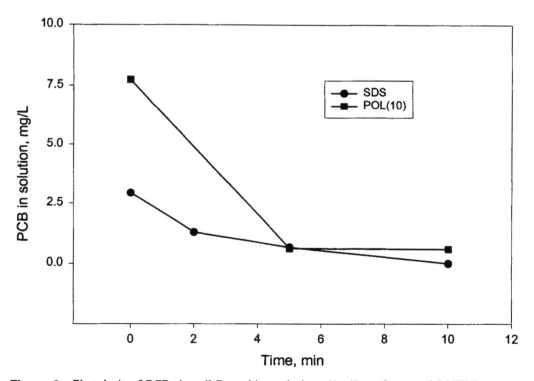

Figure 9 Photolysis of PCBs in soil B washing solutions (4 g/L surfactant, 6.5 NTU).

Table 3 Apparent Initial Photolysis Rate Constants (k_0) and Quantum Yields (Φ) of PCBs in Surfactant Soil Washing Solutions

Soil	Surfactant	Turbidity (NTU)[a]	(mg/L)	PCB Conc. (mg/L-min)	k_0
A	POL(10)	6.5	531.7	19.28 ± 7.25	2.07×10^{-2}
		2194	655.7	12.76 ± 2.93	1.37×10^{-2}
		3488	583.6	3.08 ± 3.36	3.31×10^{-3}
	SDS	6.5	62.7	6.17 ± 0.52	6.65×10^{-3}
B	POL(10)	6.5	7.7	1.43 ± 0.01	1.53×10^{-3}
	SDS	6.5	2.9	0.44 ± 0.85	4.68×10^{-4}

[a] Nephelometric turbidity unit (36).

NTU. The "apparent" initial rate constant and quantum yield were 19.3 mg/L-min and 2.07×10^{-2} in the solution with a turbidity of 6.5 NTU. As the turbidity of solution increased from 6.5 to 3488 NTU, both the rate constant and quantum yield decreased by 84%. However, when the turbidity of the solution increased to 2194 NTU, the rate and quantum yield decreased by only 34%. Since the solutions with 6.5 NTU were prepared by centrifugation ($7280 \times g$) for 30 min, particles with radius >0.08 µm were removed (41). The solutions with 2194 and 3488 NTU were prepared by settling for 3 and 0.5 h, respectively. Therefore, the increased turbidity from 2194 to 3488 NTU was contributed by relatively large particles. Thus, it seemed that smaller soil particles have less adverse effect on PCB photolysis than the larger particles. Therefore, satisfactory photolysis of PCBs may be achieved in moderately turbid waters.

IV. CONCLUSIONS

Photolysis of 2,4,4'CB in nonionic surfactant (POL(10)) solution follows a stepwise dechlorination pathway with preferential removal of *ortho*-chlorine. Surfactant molecules serving as H-atom donors promote PCB photodechlorination, and enhanced both the rate and quantum yield of PCB degradation. In the soil washings, PCBs can be effectively degraded, even in the presence of moderately high turbidities. The highly chlorinated congeners that are usually more biorecalcitrant tend to be more photodegradable. Nonionic surfactant, POL(10), is better than the anionic surfactant, SDS, in promoting quantum yield of PCB photolysis in soil washings. Less than 8% of the initial POL(10) was degraded after 1 h of photolysis. Conceivably, surfactant, following small augmentation, can be recycled in field applications. Surfactant washing followed by photolysis appears to be a viable solution for remediation of PCB contaminated soils.

ACKNOWLEDGMENTS

The authors gratefully acknowledge Dr. Michael Sigman of Oak Ridge National Laboratory for his insight and helpful suggestions. This study was supported by the Waste Management Research and Educational Institute of the University of Tennessee at Knoxville and the U.S. Department of Energy, Grant DE-FG02-97ER62350.

REFERENCES

1. DHEW Subcommittee. 1978. Final report of the Subcommittee on the Health Effects of Polychlorinated Biphenyls and Polybrominated Biphenyls of the DHEW Committee to Coordinate Toxicology and Related Programs. *Envir. Health Perspectives*, 24:173–184.
2. Callahan, M. A., Slimak, M. W., Gabel, N. W., May, I. P., Fowler, C. F., Freed, J. R., Jennings, P., Durfee, R. L., Whitmore, F. C., Maestri, B., Mabey, W. R., Holt, B. R., and Gould, C. 1979. Water-related environmental fate of 129 priority pollutants. U.S. EPA-440/4-79-029b, Vol. II, pp 84–1 to 84–8.
3. Safe, S. 1990. Polychlorinated biphenyls (PCBs), dibenzo-p-dioxins (PCDDs), dibenzofurans (PCDFs), and related compounds: Environmental and mechanistic considerations which support the development of toxic equivalency factors (TEFs). *Crit. Rev. Toxicol.*, 21:51–88.
4. Silberhorn, E. M., Glauert, H. P., and Robertson, L. W. 1990. Carcinogenicity of polyhalogenated biphenyls: PCBs and PBBs. *Crit. Rev. Toxicol.*, 20:439–496.
5. Hutzinger, O., Safe, S., and Zitko, V. 1974. *The Chemistry of PCB's*. CRC Press, Cleveland, OH.
6. Hutzinger, O., Sundstrom, G., Karasek, F. W., and Safe, S. 1976. Photodecomposition of halogenated aromatic compounds. In: L. H. Keith (ed.), *Identification and Analysis of Organic Pollutants in Water*, pp. 35–47. Ann Arbor Science Publishers, Ann Arbor, MI.
7. Sawhney, B. L. 1986. Chemistry and properties of PCBs in relation to environmental effects. In J. S. Waid (ed.), *PCBs and the Environment*, Vol. I, pp. 47–64. CRC Press, Boca Raton, FL.
8. Safe, S., Bunce, N. J., Chittim, B., Hutzinger, O., and Ruzo, L. O., 1976. Photodecomposition of halogenated aromatic compounds. In: L. H. Keith (ed.), *Identification and Analysis of Organic Pollutants in Water*, pp. 35–47. Ann Arbor Science Publishers, Ann Arbor, MI.
9. Bunce, N. J., Kumar, Y., and Brownlee, B. G. 1978. An assessment of the impact of solar degradation of polychlorinated biphenyls in the aquatic environment. *Chemosphere*, 7:155–164.
10. Kile, D. E., and Chiou, C. T. 1989. Water-solubility enhancement of DDT and trichlorobenzene by some surfactants below and above the critical micelle concentration. *Environ. Sci. Technol.*, 23:832–838.
11. McDermott, J. B., Unterman, R., Michael, J. B., Ronald, E. B., David, P. M., Charles, C. S., and Dietrich, D. K. 1989. Two strategies for PCB soil remediation: Biodegradation and surfactant extraction. *Environ. Prog.*, 8:46–51.
12. Abdul, A. S., and Gibson, T. L. 1991. Laboratory studies of surfactants-enhanced washing of polychlorinated biphenyl from sandy material. *Environ. Sci. Technol.*, 25:665–670.
13. Abdul, A. S., Gibson, T. L., Ang, C. C., Smith, J. C., and Sobczynski, R. E. 1992. *In situ* surfactant washing of polychlorinated biphenyls and oils from a contaminated site. *Groundwater*, 30:219–231.
14. Ramamurthy, V. 1986. Organic photochemistry in organized media. *Tetrahedron*, 42:5753–5839.
15. Kalyanasundaram, K. 1987. *Photochemistry in Microheterogeneous Systems*, pp. 2–141. Academic Press.
16. Epling, G. A., Florio, E. M., Bourque, A. J., Qian, X., and Stuart, J. D. 1988. Borohydride, micellar, and exciplex-enhanced dechlorination of chlorobiphenyls. *Environ. Sci. Technol.*, 22:952–956.
17. Shi, Z., Sigman, M. E., Ghosh, M. M., and Dabestani, R. 1997. Photolysis of 2-chlorophenol dissolved in surfactant solutions. *Environ. Sci. Technol.*, 31:3581–3587.
18. Chu, W., and Jafvert, C. T. 1994. Photodechlorination of polychlorobenzene congeners in surfactant micelle solutions. *Environ. Sci. Technol.*, 28:2415–2422.
19. Chu, W., Jafvert, C. T. Diehl, C. A., Marley, K., and Larson, R. A. 1998. Phototransformations of polychlorobiphenyls in Brij 58 micellar solutions. *Environ. Sci. Technol.*, 32:1989–1993.

20. Ruzo, L. O., Zabik, M. J., and Schuetz, R. D. 1974. Photochemistry of bioactive compounds. Photochemical processes of polychlorinated biphenyls. *J. Am. Chem. Soc.*, 96:3809–3812.
21. Bunce, N. J. 1978. Photolysis of 2-chlorobiphenyl in aqueous acetonitrile. *Chemosphere*, 7:653–656.
22. Bunce, N. J., Kumar, Y., Ravanal, L., and Safe, S. 1978. Photochemistry of chlorinated biphenyls in iso-octane solution. *J. Chem. Soc., Perkin Trans.*, No. 9:880–884.
23. Bunce, N. J. 1982. Photodechlorination of PCB's: Current status. *Chemosphere*, 11:701–714.
24. Bedard, D. L., Wagner, R. E., Brennan, M. J., Haberl, M. L., and Brown, J. F., Jr. 1987. Extensive degradation of Aroclors and environmentally transformed polychlorinated biphenyls by *Alcaligenes eutrophus* H850. *Appl. Environ. Microbiol.*, 53:1094–1102.
25. Shi, Z., LaTorre, K. A., Ghosh, M. M., Layton, A. C., Luna, S. H., Bowles, L., and Sayler, G. S. In print. Biodegradation of UV-irradiated polychlorinated biphenyls in surfactant micelles. *Water Sci. Technol.*, in press.
26. Miller, G. C., and Zepp, R. G. 1979. Effects of suspended sediments on photolysis rates of dissolved pollutants. *Water Res.*, 13:453–459.
27. Occhiucci, G., and Patacchiola, A. 1982. Sensitized photodegradation of absorbed polychlorobiphenyls (PCB's). *Chemosphere*, 11:255–262.
28. Draper, W. M., Stephens, R. D., and Ruzo, L. Q. 1987. Photochemical surface decontamination: Application to a polychlorinated biphenyl spill site. In: M. J. Comstock (ed.), *Solving Hazardous Waste Problems*, pp. 350–366. American Chemical Society, Washington, DC.
29. Chiarenzelli, J., Scrudato, R., Wunderlich, M., Rafferty, D., Jensen, K., Oenga, G., Roberts, R., and Pagano, J. 1995. Photodecomposition of PCBs absorbed on sediment and industrial waste: Implications for photocatalytic treatment of contaminated solids. *Chemosphere*, 31:3259–3272.
30. Zhang, P.-C., Scrudato, R. J., Pagano, J. J., and Roberts, R. N. 1993. Photodecomposition of PCBs in aqueous systems using TiO_2 as catalyst. *Chemosphere*, 26:1213–1223.
31. Hawarl, J., Demeter, A., and Samson, R. 1992. Sensitized photolysis of polychlorobiphenyls in alkaline 2-propanol: Dechlorination of Aroclor 1254 in soil samples by solar radiation. *Environ. Sci. Technol.*, 26:2022–2027.
32. IT Corporation. 1994. Photolysis/biodegradation of PCB and PCDD/PCDF contaminated soils. *Emerging Technology Bulletin*, EPA/540/F-94/502.
33. Wekhof, A. 1991. Treatment of contaminated water, air and soil with UV flashlamps. *Environ. Prog.*, 10:241–246.
34. Calvert, J. G., and Pitts, J. N., Jr. 1966. *Photochemistry*, pp. 783–786. Wiley, New York.
35. Iwasaki, I., Utsumi, S., and Ozawa, T. 1952. New colorimetric determination of chloride using mercuric thiocyanate and ferric ion. *Bull. Chem. Soc. Jpn.*, 25:226.
36. American Public Health Assoc. 1995. *Standard Methods for the Examination of Water and Wastewater*, 19th ed., pp. 5-45 to 5-47.
37. MacNeil, J. D., Safe, S., and Hutzinger, O. 1976. The ultraviolet absorption spectra of some chlorinated biphenyls. *Bull. Environ. Contam. Toxicol.*, 15:66–77.
38. Zepp, R. G. 1978. Quantum yields for reaction of pollutant in dilute aqueous solution. *Environ. Sci. Technol.*, 12:327–329.
39. Leifer, A. 1988. *The Kinetics of Environmental Aquatic Photochemistry: Theory and Practice*. American Chemical Society, Washington, DC.
40. Frame, G. M., Wagner, R. E., Carnaham, J. C., Brown, J. F., Jr., May, R. J., Smullen, L. A., and Bedard, D. L. 1996. Comprehensive, quantitative, congener-specific analyses of eight aroclors and complete PCB congener assignments on DB-1 capillary GC columns. *Chemosphere*, 33:603–623.
41. Shi, Z. 1998. Surfactant-enhanced dissolution and photolysis of chlorinated aromatic compounds (CACs). Ph.D. dissertation. The University of Tennessee, Knoxville.

26
Decontamination of Heavy-Metal-Laden Sludges and Soils Using a New Ion-Exchange Process

Sukalyan Sengupta
University of Massachusetts Dartmouth, North Dartmouth, Massachusetts

Arup K. SenGupta
Lehigh University, Bethlehem, Pennsylvania

I. INTRODUCTION

The presence in the United States of large soil masses contaminated with "heavy metals" poses a serious threat to life and the environment. The mitigation of this hazard nationwide has become a matter of high priority for state and federal regulatory agencies. Besides being toxic to human beings and other living organisms, the heavy metals affect ecological cycles under various redox and chemical environments (1). The most common heavy metals of concern are Pb, Zn, Cd, Cu, Hg, and Ni.

In a typical heavy-metal-contaminated site, the mass or volume of heavy-metal compounds is a minuscule fraction of the total mass/volume of the soil. However, because of the possibility of generation of a leachate that would violate toxicity characteristic leaching procedure (TCLP) standards (2), the entire soil is considered hazardous. Thus, if only the heavy metal is removed from the soil, this will constitute an effective treatment process because it would render the soil nonhazardous. Moreover, if the heavy metal can be recovered and reused in an industrial process, this would add economic justification to the solution of an environmental problem. Therefore, the primary challenge is selective heavy-metal removal. This challenge is compounded by the fact that the heavy metals can be present in the soil in a number of scenarios, each with its own ramifications in terms of physical-chemical interaction of the heavy metals with the background nontoxic species (of the soil). Figure 1 shows the different cases of heavy-metal-contaminated solid phase that can exist, along with the contribution of the background environment in each case.

In removing heavy metals from a soil, the most common ex-situ methods employed include soil washing (3,4), ligand extraction (5,6), and vitrification (7). These methods suffer from the disadvantages of (a) being unable to concentrate the heavy metal for recycle/reuse, (b) requiring input of chemicals or high energy, (c) generating a liquid (in the case of soil washing or ligand extraction) or solid waste (for incineration) disposal

Case	Scenario	Schematic	Remarks
1.	Chemically non-interacting solid phase species. HM ppt. In sand, no BC	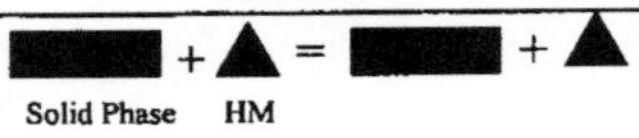 Solid Phase + HM = ▇ + ▲	HM dissolution independent of accompanying soil constituents
2.	Chemically interacting solid phase species. HM ppt. with CaCO$_3$, high BC	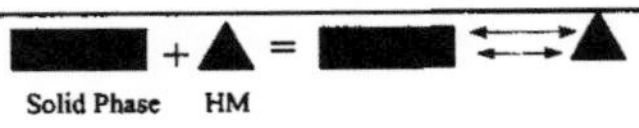 Solid Phase + HM = ▇ ←→ ▲	HM dissolution dependent on accompanying solid phase in the soil
3.	Ion-exchanging solid phase. HM bound to I-X sites of soil	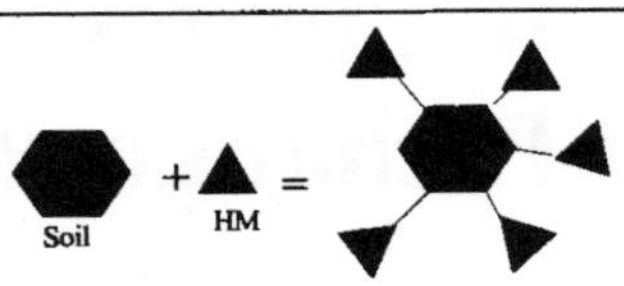 Soil + HM = (hexagon with HM attached)	HM dissolution dependent on sorption/desorption phenomenon
4.	Interaction with H/F materials; dissolved HM chelated by HA or FA	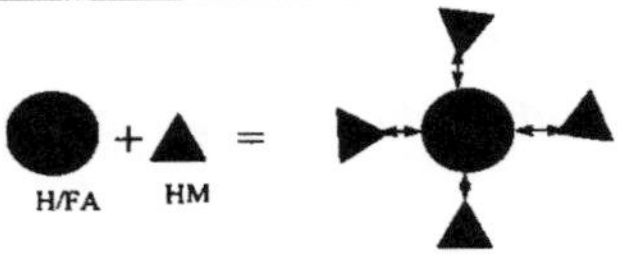 H/FA + HM = (complex)	HM decontamination dependent on the intensity of HM complexation with H/FA

HM = Heavy Metal   BC = Buffering Capacity   I-X = Ion-Exchanging
HA = Humic Acid   FA = Fulvic Acid

Figure 1 Heavy-metal-contaminated soil: various scenarios.

problem, and (d) being not versatile enough to be applied to all situations referred to in Fig. 1. Some in-situ methods include vacuum extraction. (8), thermal desorption (9), hydraulic fracturing (10), electrokinetic decontamination (including the "Lasagna" process) (11,12), biotreatment (13–15), immobilization by encapsulation, and placement of barrier systems (16,17). However, most of the methods enumerated above are employed for removal of organics present in soil, none of them is heavy-metal selective, the possibility of reintroduction of the contaminants (e.g., by the breaching of the barrier) cannot be ruled out in some cases, and all of them would be highly inefficient for cases 2 and 3 in Fig. 1. Thus, there is a need for a technology that is heavy-metal selective and is versatile enough for all the cases described in Fig. 1. Our work with composite ion-exchange material (CIM) has demonstrated its ability for use in heavy-metal decontamination of soils, as either an ex-situ or an in-situ process.

II. CIM CHARACTERISTICS

It is well known that chelating ion exchangers have high selectivity for heavy metals over alkali and alkaline earth metal cations, the competing but nonregulated (and innocuous) species in cases of heavy-metal-contaminated soils. However, the physical configuration of conventional ion exchangers (spherical or granular) mandates that they be used in fixed-bed systems with clear solutions, and this limitation makes their use inappropriate for remediating soil. We have identified a material which shows potential as being appropriate under such conditions. It is a new class of sorptive/desorptive composite ion-exchange material (CIM) available commercially as thin sheets (approximately 0.5 mm thick) and is suitable for heavy-metal decontamination from soil. The morphology of the material—along with its physical texture and tensile strength—makes it compatible for use with soil. This property of the material—combined with the adaptation of reactor chemistry—makes it possible to selectively remove heavy metals from a soil against a background composed primarily of nonregulated elements.

The CIM is a thin sheet prepared by comminuting a cross-linked polymeric ion exchanger to a fine powder, and fabricating mechanically into a microporous composite sheet consisting of ion-exchange particular matter enmeshed in polytetrafluoroethylene (PTFE). During this mechanical process, the PTFE microspheres are converted into microfibers that separate and enmesh the particles (18). When dry, these composite sheets consist of >80% particles (polymeric ion exchanger) and <20% PTFE by weight. They are porous (usually >40% voids, with pore size distributions that are uniformly below 0.5 µm). The ion-exchange microspheres are usually <100 µm in diameter with total thickness ≈0.5 mm. As such, they are effective filters that remove suspensoids >0.5 µm from permeating fluids. Because of this sheetlike configuration, this material can be easily introduced into or withdrawn from an external reactor containing contaminated soil, with the target solutes being adsorbed onto or desorbed from the microadsorbents. The chelating functionality of the microspheres chosen was iminodiacetate (IDA), a functional group universally acknowledged to be very selective toward heavy-metal cations vis-à-vis cations such as Ca^{2+}, Mg^{2+}, and Na^+. Figure 2 shows an electron microphotograph of the composite IDA membrane along with a schematic depicting how the microbeads are trapped within the fibrous network of PTFE. Table 1 provides the salient properties

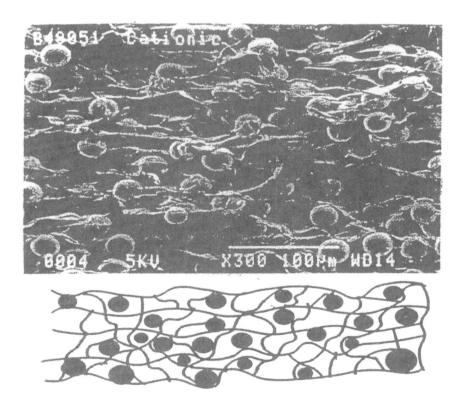

Figure 2 Scanning electron microphotograph (300×) of the composite membrane (top) and schematic representation of the ion-exchanger beads present in a network of interlaced PTFE (bottom).

Table 1 Properties of the CIM

Composition	90% Chelex chelating resin, 10% Teflon
Pore size (nominal)	0.4 μm
Nominal capacity	3.2 meq/g dry membrane
Membrane thickness	0.4–0.6 mm
Ionic form (as supplied)	Sodium
Resin matrix	Styrene–divinylbenzene
Functional group	Iminodiacetate
pH stability	1–4
Temperature operating range	0–75°C
Chemical stability	Methanol; 1N NaOH, 1N H_2SO_4
Commercial availability	3M Corp. (MN); Bio-Rad (CA)

of the composite IDA membrane used in the study. Note that the chelating microbeads constitute 90% of the composite membrane by mass. This feature allows the membrane to achieve the same level of performance as the parent chelating beads used in a fixed-bed operation. More details about characterization of the membrane are available in Ref. 19 and are not repeated here. It may be noted that this material differs fundamentally from traditional ion-exchange membranes used in industrial process such as Donnan dialysis (DD) and electrodialysis (ED) because of its high porosity. DD and ED membranes have very low porosity and are strongly influenced by the Donnan coion exclusion principle (20), which does not allow anions to pass through cation-exchange membranes and vice versa. However, in the case of the CIM, large gaps between ion exchangers allow anions to pass through freely even though it is a cation-exchange membrane. The suspended solids that are >0.5 μm are not able to penetrate across the skin of the membrane because of the pore size of the material, as explained above. However, water molecules and ions can easily move in and out of the thickness of the sheet, thus allowing for unimpeded ion-exchange reactions between heavy metals in the slurry and the counterions of the CIM, as shown schematically in Fig. 3 and

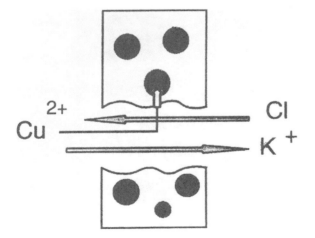

Figure 3 Schematic diagram showing the porosity of the CIM to cations and anions and selective entrapment of heavy-metal cations.

explained in detail elsewhere (19,21,22). After a design time interval, the membrane can be taken out and chemically regenerated with strong (3–5%) mineral acid solution.

III. PROCESS CONFIGURATION

Figure 4 shows a conceptualized process schematic where a CIM strip is continuously run through the contaminated soil (sorption step) and an acid bath (desorption step). Such a cyclic process configuration is relatively simple and can be implemented by using the composite membrane as a slow-moving belt. For a sludge containing heavy-metal hydroxide, say $Me(OH)_2$, the process works in two steps:

1. *Sorption.* The CIM, when in contact with the sludge, selectively removes dissolved heavy metals from the aqueous phase in preference to other nontoxic alkali and alkaline earth metal cations. Consequently, fresh heavy-metal hydroxide dissolves to maintain equilibrium, and the following reactions occur in series:

    Dissolution: $\quad Me(OH)_2(s) \leftrightarrow Me^{2+} + 2OH^-$ (1)

    CIM uptake: $\quad \overline{RN(CH_2COOH)_2} + Me^{2+} \leftrightarrow \overline{RN(CH_2COO^-)_2 Me^{2+}} + 2H^+$ (2)

2. *Desorption.* When the CIM is immersed in the acid chamber, the exchanger microbeads are efficiently regenerated according to the following reaction:

    $\overline{RN(CH_2COO^-)_2 Me^{2+}} + 2H^+ \leftrightarrow \overline{RN(CH_2COOH)_2} + Me^{2+}$ (3)

The regenerated CIM is then ready for sorption again and the cycle is repeated. For such an arrangement to be practically feasible, the CIM sheet needs to be physically tough

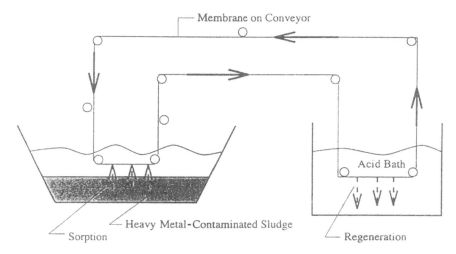

Figure 4 Conceptualized continuous decontamination process for heavy-metal removal from a sludge reactor using CIM.

enough to withstand the tension of the conveyor belt and resilient to the chemical forces that are created during sorption and regeneration. Previous studies conducted in this area (23,24) have confirmed that the CIM sheet is durable enough to withstand cyclical forces (physical and chemical) for higher than 200 cycles.

The CIM can also be used in an in-situ decontamination process. This process, termed electromembrane partitioning (EMP), is a physicochemical process that involves placing the electrodes radially in a confined volume of the contaminated soil. The anode is placed in the contaminated soil and the cathode is placed at a predetermined distance from the anode. The CIM is wrapped around the cathode. Successful use of this process requires synchronization of the following steps:

1. Inducement of ion movement by the application of DC potential; the heavy-metal cations move toward the cathode and the anions move toward the anode.
2. Entrapment of only the heavy-metal cation by the heavy-metal-selective, but at the same time highly porous, CIM during its migration toward the cathode.

In a practical application situation, the flux of heavy-metal cation would be due to the combination of Fickian diffusion gradient, electric potential gradient, and the rate of heavy-metal cation generated. Since there is no hydraulic gradient, the advective flux is absent in this case. When the potential gradient is applied, highly oxidizing conditions develop at the anode and this solubilizes the heavy-metal precipitates, giving rise to free heavy-metal cations, which move toward the cathode. However, to maintain electroneutrality, reducing conditions develop at the cathode, giving rise to OH^-. The CIM wrapped around the cathode is present originally in hydrogen form; therefore, it will neutralize the OH^- generated. The electric potential applied induces all cations to move toward the cathode, thus Me^{2+}, Na^+, and Ca^{2+} will travel toward the cathode. OH^- generated at the cathode is neutralized immediately by the protonated functional group of the CIM. The neutralization step is important since it keeps the heavy-metal cations in solution due to the lowering of pH and prevents their reprecipitation near the cathode. The CIM placed at the cathode will selectively trap only Me^{2+} ions since the chelating functionality present in the CIM is highly selective for Me^{2+} over Ca^{2+} or Na^+, as will be shown later. Thus Me^{2+} is removed from the solid phase and transferred to the CIM phase. Note that the overall process involves selective transfer of heavy metal, Me(II), from one solid phase (soil precipitate) to another (CIM), which is attained primarily by the migration of H^+ ions generated at the anode. Note also that all other cations except Me^{2+} will be in solution near the cathode. After the end of one run, the CIM is chemically regenerated, resulting in the composite membrane being converted to hydrogen form. Thus, the CIM can be used again for another cycle.

IV. EFFECT OF BACKGROUND COMPOSITION

The background nontoxic materials that constitute the bulk of the solid phase are unimportant from a regulatory viewpoint, but their physical-chemical properties strongly influence the selective separation of heavy metals. On the basis of physical-chemical interaction, the scenarios provided in Fig. 1 are discussed below.

1. *Heavy-metal precipitate present with chemically inert solid phase.* This case is the simplest in terms of treatment or removal of the heavy-metal contaminant. For example, if $Me(OH)_2(s)$ is present in a background of sand with no buffer capacity, addition of a

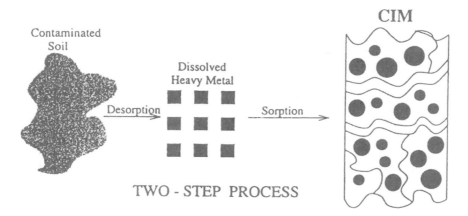

Figure 5 Conceptual diagram showing the sequential process of desorption of heavy metal from contaminated soil followed by uptake of the dissolved heavy metal by the composite membrane.

mineral acid will dissolve the precipitate, bringing Me^{2+} into solution. After this stage, introduction of the CIM sheet into the soil reactor will result in movement of the heavy-metal cations from the liquid phase (reactor solution) to the solid phase (CIM surface). The use of the CIM allows the entire treatment operation to be conducted in a single reactor as opposed to other, alternative processes.

2. *Heavy metals present amidst a background of high buffer capacity*. The background soil can interact chemically with the dissolved heavy metals. For example, if a heavy-metal precipitate, $Me(OH)_2$, is present as a minor contaminant in a background of calcite soil, the mass of $CaCO_3(s)$ will be orders of magnitude higher than that of $Me(OH)_2$. Because of the high buffer capacity of $CaCO_3$, the pH of the system will be alkaline. This will result in the concentration of heavy metal in the aqueous phase being orders of magnitude lower than that of Ca^{2+}. Any physicochemical treatment process for such a scenario that depends primarily on the aqueous-phase Me^{2+} concentration for separation from Ca^{2+} would be inefficient in this case. As described later, the CIM can be used in a novel scheme to achieve efficient heavy-metal separation in such cases.

3. *Heavy-metal cations bound to the ion-exchange sites of soil*. An example is a clay mass where the cation-exchange sites are loaded with heavy-metal cations. Removal of the heavy metal in this case entails a two-step process:

(a) Desorption of the heavy metal from adsorption/ion-exchange sites
(b) Selective removal of heavy-metal cations from a background of innocuous cations, e.g., Na^+, Ca^{2+}, Mg^{2+}

The underlying steps are shown schematically in Figure 5.

V. EXPERIMENTAL SECTION

A. Equilibrium Isotherms of Free Metal(II) Cations at Alkaline pH with Addition of Aqueous-Phase Ligand

For case 2 in Fig. 1, an increase in aqueous-phase heavy-metal concentration is accomplished by introducing an aqueous-phase ligand. The coordination energy of the complex between the heavy metal and the aqueous-phase ligand is less than that of the complex between the heavy metal and the functional group of the CIM. For our experiments simulating such a situation, copper was chosen as the heavy-metal contaminant and oxalate was the representative aqueous-phase ligand. The equilibrium isotherms at $23 \pm 2°C$ were obtained by contacting a piece of the membrane weighing ~ 0.1 g in basic (Na^+) form with 200 mL of aqueous solution of different concentrations of $Cu(NO_3)_2 \cdot 2H_2O$ and the required amount of $Na_2C_2O_4$, at a pH of 9.0 maintained by a computer-aided titrimeter (Fisher Scientific model CAT), for 48 h. Following equilibration and short rinsing with deionized (DI) water, the CIM was regenerated with 3% v/v concentrated HNO_3 solution, and the metal-ion Cu(II) concentration in the spent regenerant and in the reactor were analyzed using an atomic absorption spectrophotometer (Perkin-Elmer model 2380). Membrane uptake for each case was determined by mass balance.

B. Cyclic Process to Buffered Sludge at Alkaline pH with Aqueous-Phase Ligands

In order to simulate a cyclic process for the above-referred situation, a laboratory soil sample was prepared by mixing $CaCO_3(s)$ and CuO in the ratio 10:1. The mass of the two compounds was such that $CaCO_3(s)$ and $Cu(OH)_2(s)$ were always the governing solid phases in the sludge. The solution volume was 200 mL and the pH was maintained at 9.0 throughout the experiment. Six such solutions were prepared and a ligand (acetate, citrate, or oxalate at concentration of 0.05 M) was added (as its sodium salt) to each of them. Sorption/desorption steps were carried out for 10 cycles, each cycle comprising of a 4-h exhaustion period followed by a 1-h regeneration period. The choice of the aqueous-phase ligand is critical, and the effect of the ligand is discussed later under Results and Discussion. The three ligands were chosen because:

1. They encompass a wide range of stability constant values with the metals of interest, i.e., Cu(II) and Ca^{2+}
2. They are innocuous to the environment and also biodegradable and thus can be efficiently removed from the reactor after the last cycle
3. Most organic matter contains carboxylate functional groups, so they can be used as surrogate representatives of complexing organic matter in the contaminated water

Samples were taken from the sludge and the regenerant solution after the end of each cycle and analyzed for Cu(II), Ca(II), and the ligand. Ca(II) was analyzed by AAS (Perkin-Elmer model 2380), oxalate and citrate were analyzed by ion chromatography (Dionex model 4500i) with a conductivity detector, and radiolabeled citrate was used for analysis on a scintillation counter.

Another experiment was conducted in a higher-suspended-solids-content reactor. The solid phase in this experiment was prepared by mixing 5 g of $CaCO_3(s)$, 45 g of fine sand (-200 mesh), 13.4 g of $Na_2C_2O_4$, and 0.38 g of $CuO(s)$. One liter of deionized water was added to this mixture, and the pH was adjusted to 9.0. Aqueous-phase oxalate concentration was 4000 mg/L, and $< 1\%$ $CuO(s)$ was present in the solid phase of the sludge. Sorption/desorption steps were subsequently carried out for 10 cycles; with Cu(II), Ca(II), and oxalate being analyzed after each cycle. Oxalate was analyzed using an ion chromatograph (Dionex model 4500i) with conductivity detector and standard carbonate/bicarbonate eluent.

C. Cyclic Process for Heavy Metals Extraction from Ion-Exchange Sites of Soil

In order to determine the viability of the CIM to extract heavy metals bound to the cation-exchanging sites of soil, analytical-grade bentonite was chosen as the soil type because of its high ion-exchange capacity and was loaded with Cu(II) by equilibrating it with an aqueous phase containing 400 mg/L Cu(II) concentration added as $Cu(NO_3)_2 \cdot 2.5\ H_2O$ at a pH of 5.5. The total exchange capacity of bentonite was found by mass balance. Cu(II)-loaded bentonite was then introduced into a plastic container containing 200 mL of 500 mg/L Na^+ (added as NaCl) solution. Cu(II)-loaded bentonite was the only solid phase present in the sludge, and the suspended solids loading was 2.5% (m/v). A strip of conditioned CIM weighing 0.112 g was introduced into the plastic container and run for 30 cycles. In each cycle, the sorption step involved rigorous shaking of the plastic container for 3 h at a constant pH of 5.0, while the desorption step used 5% H_2SO_4 (v/v) for 1 h for regeneration. Cu(II) concentration in the regenerant was used to compute the percentage Cu(II) recovery from the bentonite phase.

D. Ion Selectivity of Bentonite

Since bentonite was chosen as the representative clay sample, the ion selectivities of bentonite for Cu^{2+}, Ca^{2+}, and Na^+ were determined. Binary separation factors Cu/Na and Ca/Na were obtained experimentally by the following process. For the Cu^{2+}/Ca^{+2} system:

1. Dry analytical-grade bentonite was rinsed 2–3 times with DI water to purge the bentonite slurry of any cations that might be generated from the congruent/incongruent dissolution of any solid phase present in the original powder.
2. Next, 12 g of conditioned and dried bentonite was loaded in Ca^{2+} form by equilibrating it with 600-mg/L Ca^{2+} [added as $Ca(NO_3)_2 \cdot 4H_2O$] at a pH ≈ 5.5. After an equilibration period of 48 h, filtered solutions of initial and final slurry were analyzed for concentrations of Ca^{2+}, Mg^{2+}, and Al^{3+}. Calcium loading on the bentonite was determined by mass balance.
3. The calcium-loaded bentonite slurry was dewatered and dried. Different weights of the dried sample (1.75–2.5 g) were then equilibrated with 200 mL of 160-mg/L Cu^{2+} solution [present as $Cu(NO_3)_2 \cdot 2.5H_2O$] in plastic bottles where the pH in each bottle was ≈ 4.7.

E. Sorption/Desorption Batch Kinetic Studies

Sorption kinetics of Cu^{2+}, Pb^{2+}, and Ni^{2+} onto the CIM and desorption kinetics of Cu^{2+} from Cu(II)-loaded bentonite were studied at a pH of 3.0 or 5.0. Each experiment was performed in a 1.0-L baffled cylinder Plexiglas vessel and a computer-aided titrimeter to control the pH. For the sorption run, initial concentration of heavy metal in the reactor was 200–230 mg/L added as its nitrate salt. About 0.5 g of previously conditioned CIM in basic form was added at the start, following which approx. 2-mL volumes of samples were collected from the reactor at predetermined time intervals. For the desorption run, the stripping solution was 400 mg Ca(II) added as $Ca(NO_3)_2 \cdot 4H_2O$. Approximately 6.3 mg of Cu-loaded bentonite was introduced into the baffled cylinder and 2-mL samples were taken from the cylinder at regular time intervals for Ca(II) and Cu(II) analysis.

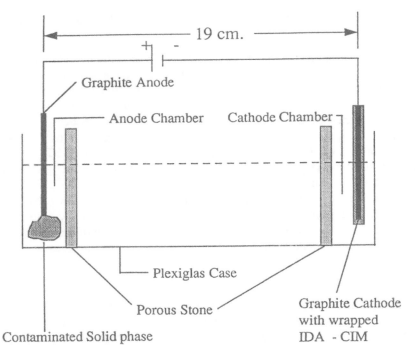

Figure 6 Schematic of laboratory-scale EMP cell.

F. EMP Cell Experiments

All EMP experiments were conducted in a Plexiglas reactor, the diagram of which is provided in Fig. 6. Both the anode and cathode were composed of graphite strips and were placed in the appropriate chambers. The simulated contaminated soil was placed in the anode chamber, and the CIM was either placed midway between the electrodes or wrapped around the graphite cathode. In a typical run, a potential difference of 10 V was applied over an electrode-to-electrode distance of 19.0 cm for a period greater than 4 h. Samples were taken at definite time intervals form the anode, cathode, and middle chambers during the course of an EMP run and analyzed for various cations and anions using a Dionex 4500i ion chromatograph or a Perkin-Elmer 2380 AAS. The pH was measured periodically in each chamber.

VI. RESULTS AND DISCUSSION

A. Heavy Metals Uptake at Alkaline pH by Addition of an Aqueous-Phase Ligand

Figure 7 shows the Cu(II) uptake isotherms at two different oxalate concentrations. In both cases, there was no Cu(II) in the solid phase. To maintain this condition, the following procedure was adopted. The desired oxalate solution (0.005 M and 0.1 M) was prepared by adding $Na_2C_2O_4$ to DI water. The pH was adjusted to 9.0 and the maximum $\{Cu_t\}$ that can be supported for each oxalate solution was determined

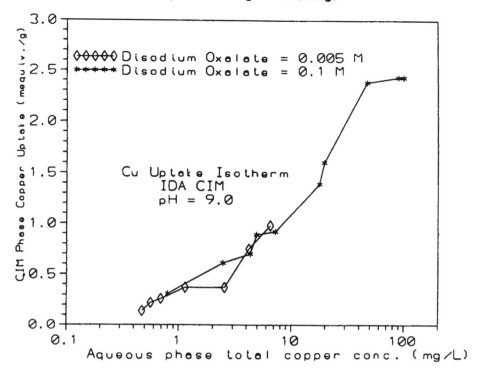

Figure 7 Copper(II) uptake isotherms at two oxalate concentrations.

experimentally. This was done by adding CuO(s) in mass much higher than that obtained by theoretical calculations based on Cu complexes with OH^- and $C_2O_4^{2-}$. The system was allowed to attain equilibrium, after which a sample was taken from the reactor, filtered, and analyzed for Cu(II). After determining the maximum $\{Cu_t\}$ for each case (8 mg/L for 0.005 M and 100 mg/L for 0.1 M), different amounts of CuO were added for each run after making sure that all the Cu added would remain in the aqueous phase.

It can be observed that the uptake of Cu(II) by the CIM depends only on the total aqueous-phase Cu(II) concentration; for $\{Cu_t\} < 8$ mg/L, concentration of Na^+ or $C_2O_4^{2-}$ in the system makes practically no difference in Cu uptake even though they are 20 times higher in one case than the other. This is due to the extremely high affinity of the IDA functional group toward Cu as compared to that for Na^+. The high affinity is caused by strong Lewis acid–base interaction of the IDA functionality toward Cu(II) (25–29), whereas with Na^+ the interaction is only electrostatic.

Figures 8, 9, and 10 compare the performance of the system with three ligands at the same concentration (0.05 M), i.e., acetate, citrate, and oxalate. It can be seen that citrate provided the highest sludge-phase concentrations of calcium, while the acetate-added system showed the lowest reactor concentration of calcium. Because of this, the uptake of calcium by the CIM in descending order is citrate>oxalate>acetate. This is understandable from Table 2, which provides the stability constants of the ligands used in this study. Note that N-B-IDA is the closest monomeric analog to the polystyrene-divinylbenzene IDA functionality of the CIM. However, the case of copper uptake does not strictly follow the hierarchy as may be expected from the stability constants in Table 2. Acetate provided the lowest recovery of copper from the solid phase, but an anomalous situation exists for

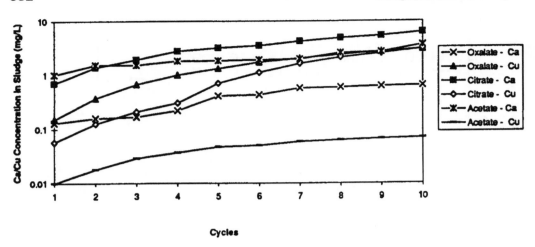

Figure 8 Aqueous–phase metals concentration over number of cycles with different ligands.

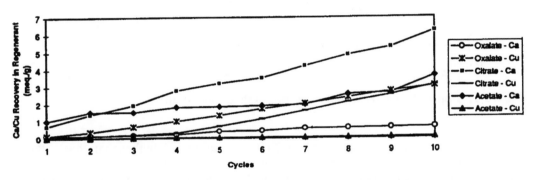

Figure 9 Metal uptake by IDA CIM over number of cycles with different ligands.

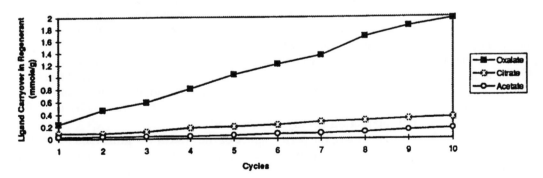

Figure 10 Ligand carryover in regenerant.

the case of oxalate and citrate. Although aqueous-phase copper concentration is lower for the oxalate system as compared to citrate, uptake by CIM is comparable, especially in the latter cycles. Also, it may be observed that oxalate tended to be taken up consistently by the CIM, a phenomenon not observed for citrate or acetate. It may be pertinent to describe

Table 2 Stability Constants for Cu^{2+}–Ligand and Ca^{2+}–Ligand Complexes[a]

Ligand	Metal : ligand	Log K for Cu–ligand	Log K for Ca–ligand
Oxalate	1:1	6.23	1.66
Oxalate	1:2	10.27	2.69
Citrate	1:1	5.90	3.50
Acetate	1:1	2.16	0.6
N-B-IDA	1:1	10.62	3.17
N-B-IDA	1:2	15.64	—

[a] All values from Ref. 30.

the experimental framework for the oxalate system in more detail. In this case, to a 5% (w/v) sludge, fine sand (70%), $Na_2C_2O_4$ (21%), calcite (7.8%), and CuO (0.6%) were mixed. The sludge pH was maintained at 9.0, and CaC_2O_4 and $Cu(OH)_2$ were the controlling solid phases under these conditions. Thus, the amount of oxalate added was much more than 0.05 M. The oxalate added reacted with calcium to form calcium oxalate, a compound with low solubility. Precipitation of calcium oxalate reduced the aqueous-phase oxalate concentration to 0.05 M. After the formation of $CaC_2O_4(s)$, the cyclic process was started.

Figure 11 shows the recovery of Cu, Ca, and C_2O_4 in the regenerant solution for the oxalate system. Although Cu is present primarily as Cu–C_2O_4 complex in the sludge phase, Cu recovery was significant and increased steadily with every cycle. Ca recovery was much lower and tended to approach an asymptotic concentration in the regenerant with an increase in the number of cycles. Dissolved sludge-phase concentrations of Cu and Ca remained fairly constant with the number of cycles (Fig. 12), suggesting that they are

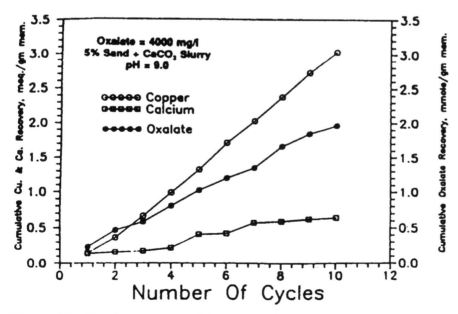

Figure 11 Cumulative copper, calcium, and oxalate recovery with increase in number of cycles at alkaline pH.

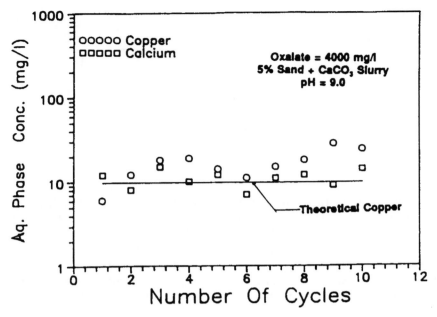

Figure 12 Dissolved copper and calcium concentrations during the recovery process at pH = 9.0 for oxalate concentration = 4000 mg/L.

controlled by the solubility products of the solid phases. Heavy-metal uptake by the CIM is substantial even when free Cu ion, Cu^{2+}, is practically absent and most of the dissolved Cu exists primarily as anionic or neutral $Cu\text{-}C_2O_4$ complexes. Table 3 presents data about the computed percentage distribution of important Cu(II) species at three different oxalate concentrations based on stability constants data in the open literature (30). Significant Cu uptake by the CIM under such conditions seems counterintuitive.

Also, Fig. 11 shows a consistently increasing oxalate recovery in the regenerant. Oxalate exists primarily as the divalent anion $C_2O_4^{2-}$ at a pH of 9.0 and should be rejected by the chelating cation exchanger according to the Donnan exclusion principle. One possible explanation for this counterintuitive behavior could be that the suspended particles of the insoluble calcium oxalate and copper oxide were probably trapped in the large pores of the CIM and subsequently carried over to the regenerant phase, thus exhibiting significant copper and oxalate recovery. In order to eliminate such a possibility, another sorption isotherm was carried out in the absence of suspended solids at pH = 9.0 and total

Table 3 Percentage Distribution of Various Copper(II) Species

Species	Oxalate = 0.005 M	Oxalate = 0.045 M	Oxalate = 0.1 M
Cu^{2+}	2.25E − 04	4.37E − 06	1.26E − 06
$Cu(OH)^+$	2.25E − 02	4.37E − 04	1.26E − 04
$Cu(OH)_2^0$	1.08E + 00	0.02E + 00	6.04E − 03
$Cu(OH)_3^-$	2.25E − 02	4.37E − 04	1.26E − 04
$Cu(C_2O_4)^0$	1.84E + 00	0.26E + 00	0.14E + 00
$Cu(C_2O_4)_2^{2-}$	97.04E + 00	99.72E + 00	99.85E + 00

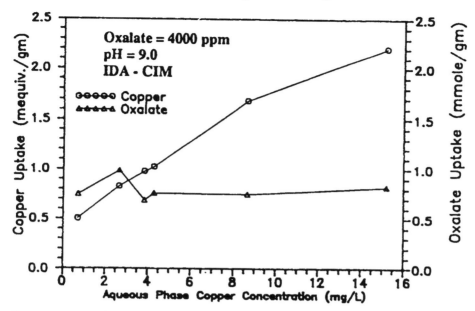

Figure 13 Equilibrium uptake of copper(II) and oxalate by the CIM at pH = 9.0.

aqueous-phase oxalate concentration of 4000 mg/L but at varying dissolved copper concentrations. Figure 13 shows Cu and C_2O_4 uptakes under these experimental conditions free of suspended solids. Note that the C_2O_4 uptake by the CIM is significant, and Cu uptake increases with an increased Cu concentration—a pattern very similar to those obtained in Figs. 8, 9, and 10. These observations suggest strongly that Cu and C_2O_4 are removed from the solution (or sludge) phase primarily through sorption processes. The following are identified as plausible binding mechanisms for C_2O_4 and Cu(II) onto the composite IDA membrane.

As already indicated, the neutral copper–oxalate complex $[Cu(C_2O_4)]^0$ was significantly present in the aqueous phase under experimental conditions, and only two of the four primary coordination numbers of Cu(II) are satisfied in this complex. Since these complexes are electrically neutral, they can permeate readily to the sorption sites containing nitrogen donor atoms, which can favorably satisfy the remaining coordination requirements of Cu(II). Figure 14 shows how electrically neutral 1:1 copper–oxalate complexes can be bound to the neighboring nitrogen donor atoms of the iminodiacetate moieties through formation of ternary complexes. This mode of sorption (i.e., formation of ternary complex through metal–ligand interaction) is believed to be the primary pathway for sorption of oxalate onto the CIM and subsequent carryover into the regenerant as exhibited in Fig. 8.

Although free Cu ions (Cu^{2+}) were practically absent under the experimental conditions of Figs. 8, 9, 11, and 13, Cu uptake was quite significant. It is very likely that ligand substitution was a major mechanism by which Cu(II) was sorbed onto the chelating microbeads of the CIM for oxalate as aqueous-phase ligands. Table 2 shows 1:1 and 1:2 metal–ligand stability constants of Cu(II) with oxalate, citrate, acetate, and N-benzyl iminodiacetate (N-B-IDA), and the much higher ligand strength of IDA can be easily noted. Also, Cu(II) complexes are known to be labile. Therefore, the ligand substitution

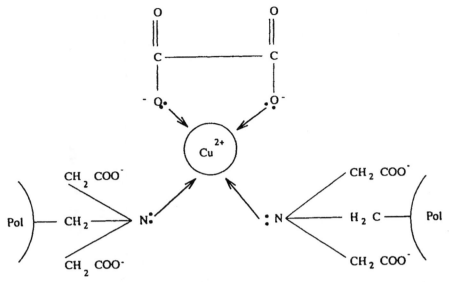

Figure 14 Sorption of neutral copper–oxalate complex onto chelating microbeads through formation of ternary complex.

reaction, where oxalate in the aqueous phase is replaced by IDA in the exchanger phase, is favorable and results in an increased Cu(II) uptake as shown by the following reaction:

$$[Cu(C_2O_4)_2]^{2-} + \overline{R-N-(CH_2COO^-Na^+)_2} \leftrightarrow \overline{R-N-(CH_2COO^-)_2Cu^{2+}} + 2Na^+ + 2C_2O_4^{2-} \quad (4)$$

Thus, ligand substitution and ternary-complex formation are the major mechanisms for heavy-metal uptake in the case of oxalate as the aqueous-phase ligand. However, in the case of citrate or acetate, ligand substitution is the only mechanism for transfer of copper from the soil to the regenerant. This is verified by the almost total absence of citrate or acetate from the regenerant. The trace amounts of acetate and citrate noticed in the regenerant is attributed to their entrapment in the pores of the CIM. Therefore, Cu(II) uptake from an environment where citrate is the aqueous-phase ligand can be represented solely by the following equation:

$$[Cu(C_6O_7H_5)]^- + \overline{R-N-(CH_2COO^-Na^+)_2} \leftrightarrow \overline{R-N-(CH_2COO^-)_2Cu^{2+}} + 2Na^+ + C_6O_7H_5^{3-} \quad (5)$$

A comparative study of the three ligands reveals that

> Acetate is not an effective ligand for selective heavy-metal removal in a situation like the one simulated, because it is too weak a ligand to affect aqueous-phase heavy-metal concentration, and thus is ineffective for subsequent CIM uptake.
> The heavy-metal uptake profile is similar in the cases of citrate and oxalate, even though the aqueous-phase heavy-metal concentration is higher for the citrate system. This is because heavy metal is taken up by the CIM functionality by two parallel mechanisms in the case of oxalate addition, ligand substitution and ternary complex formation, whereas ligand substitution is the only mechanism for heavy-metal uptake for the case of citrate addition.

Calcium uptake by the CIM is much higher in the case of citrate as the aqueous-phase ligand compared to oxalate. This is because (a) the ligand citrate has a much higher affinity for calcium than oxalate; and (b) calcium oxalate is sparingly soluble and thus most of the calcium remains in the solid phase. Since free calcium concentration, $\{Ca^{2+}\}$, is lower because of $CaC_2O_4(s)$, the total calcium aqueous-phase concentration, $\{Ca_t\}$, is also lower for oxalate relative to citrate, as can be noted from Fig. 8.

The major repercussion of the choice of ligand is in the suitability of the decontamination process. If one of the goals of soil decontamination is recycle/reuse of the heavy metal, purity of the heavy metal in the acid regenerant stream is important, and in such cases, oxalate would be a better candidate because of lower calcium carryover. However, if removal of heavy metal from the soil is the only goal, both oxalate and citrate would be suitable.

Oxalate sorption by the CIM is much higher than citrate for the reasons discussed earlier. Thus, if the ligand is undesirable in the regenerant, citrate would be a better choice than oxalate. Also, since citrate remains in the aqueous phase, stoichiometric amounts of it are needed in the reactor. In contrast, oxalate is removed from the liquid phase as a precipitate and thus the amount needed is much higher than stoichiometric.

From an application viewpoint, the concept is very important because it proves scientifically that heavy-metal-laden soil with high buffer capacity can be decontaminated even at alkaline pH in the presence of organic ligands without addition of acid to reduce pH; in fact, acid addition will not lower the pH, because of the high buffer capacity. The main concept employed here is to increase the aqueous-phase concentration of the heavy metal with minimum increase of the concentration of the competing cation in the solid phase. The stability constant values of the two ligands (aqueous ligand added and solid-phase ligand of CIM functionality) with the metals of interest play an important role in the suitability of the process.

Another methodology that can be employed is the use of an ion exchanger which has very low affinity for Ca(II) as compared to that for heavy metal, say Cu(II). If one can identify a chelating functionality with affinity for heavy metal that is orders of magnitude higher than that for calcium, the need to add an aqueous-phase ligand will be obviated. In this regard, the work done with chelating polymers containing multiple nitrogen donor atoms (31,32) is significant. The authors have reported the presence of two chelating ion exchangers with pyridine base functionality that have very high affinity for strong Lewis acids and very low affinity for "hard" acids such as Ca(II). If such a functionality could be impregnated in the CIM, the process of cyclic extraction would be efficient.

B. Sorption Kinetics

Figure 15 shows almost identical uptake rates for Cu(II), Ni(II), and Pb(II) by the CIM, although they are not equally labile from a chemical reaction viewpoint. It is therefore likely that chemical reaction kinetics is not the rate-limiting step. Previous investigations in this regard with spherical chelating exchangers have demonstrated intraparticle diffusion to be the most probable rate-controlling step (33,34). Also, since the PTFE fibers do not sorb any metal ions, solute transport by surface diffusion is practically absent. However, a significantly different physical configuration of the CIM may introduce additional diffusional resistance within the CIM. As may be observed from the schematic diagram in Fig. 16, fairly stagnant pore liquid is present in the channels of the CIM

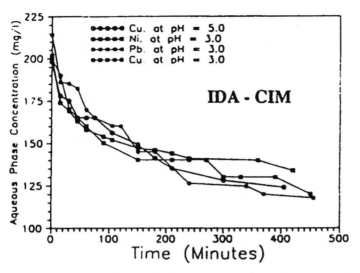

Figure 15 Plot of aqueous-phase metal concentration versus time during the batch kinetic study of the IDA-CIM.

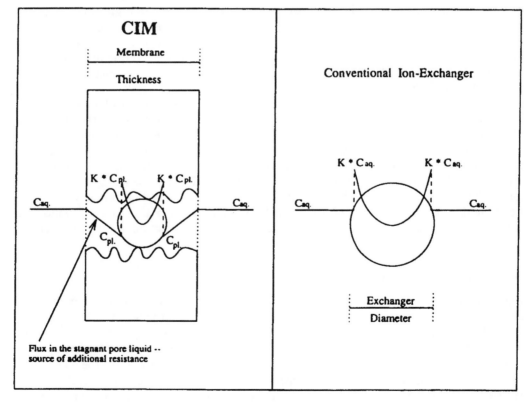

Figure 16 Schematic diagram showing the difference in the flux curves for a conventional (bead-shaped) ion exchanger and CIM.

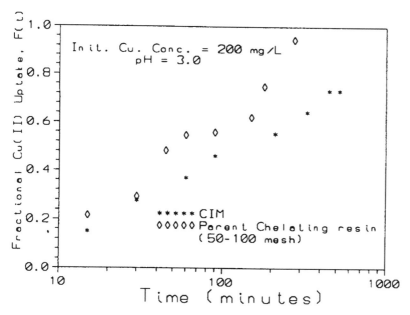

Figure 17 Copper uptake rates for the composite membrane versus the parent chelating exchanger under otherwise identical conditions.

between individual microbeads, and the solutes need to be transported through this pore liquid for sorption or desorption. This additional resistance to sorption/desorption is likely to retard kinetic rate.

Figure 17 shows the results of a batch kinetic study (fractional metal uptake versus time) comparing the parent chelating exchanger with the CIM under otherwise identical conditions. Fractional metal uptake, $F(t)$, is dimensionless and is defined as the ratio of the metal uptake $q(t)$ after time t and the metal uptake at equilibrium, q^0, that is, $F(t) = q(t)/q^0$. Although the parent chelating microbeads were bigger in average diameter (50–100 mesh) than the microbeads within the CIM, the copper uptake rate, as speculated, was slower with the CIM. In order to overcome the complexity arising due to the heterogeneity of the CIM (chelating microbeads randomly distributed in nonadsorbing teflon fibers), a model was proposed (19,21) in which the thin-sheet-like CIM may be reviewed as a flat plate containing a pseudo-homogeneous sorbent phase as shown in Fig. 18. Under the experimental conditions, it may be assumed that:

1. The surface of the CIM is in equilibrium with the bulk of the liquid phase.
2. The total amount of solute in the solution and in the CIM sheet remains constant as the sorption process is carried out.
3. The solute (heavy metal) has high affinity toward the thin-sheet sorbent material.

This is a case of diffusion from a stirred solution with limited volume. The CIM is considered as a sheet of uniform material of thickness $2w$ placed in the solution containing the solute, which is allowed to diffuse into the sheet. The sheet occupies the space $-w \leq x \leq +w$, while the solution is of limited extent and occupies the space $-w-a \leq x \leq -w$, $w \leq x \leq +w+a$. The concentration of the solute in the solution is always uniform and is initially C_0, while initially the sheet is free from solute. Considering

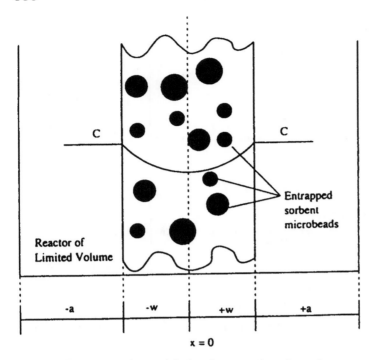

Figure 18 Schematic model showing sorption through an assumed flat plate with constant diffusivity from a reactor of limited volume.

an apparent metal-ion diffusivity within the CIM phase, the partial differential equation (PDE) that must be solved in this case of metal uptake through a plane sheet (the thickness of the CIM) from a solution of limited volume is given as

$$\frac{\partial C}{\partial t} = \overline{D}\frac{\partial^2 C}{\partial x^2} \tag{6}$$

where
  x = axial space coordinate in the direction of CIM thickness
  C = concentration of solute in the solution
The initial conditions are

$$C = 0 \quad -w < x < +w \quad t = 0 \tag{7}$$

and the boundary conditions express the fact that the rate at which the solute leaves the solution is always equal to that at which it enters the sheet over the surfaces, $x = \pm w$. These conditions are expressed mathematically as

$$\left(\frac{a}{K}\right)\frac{\partial C}{\partial t} = \pm \overline{D}\frac{\partial C}{\partial x} \quad x = \pm w \quad t > 0 \tag{8}$$

where K is the partition factor between the CIM and the solution, i.e., the concentration just within the sheet is K times that in the solution.

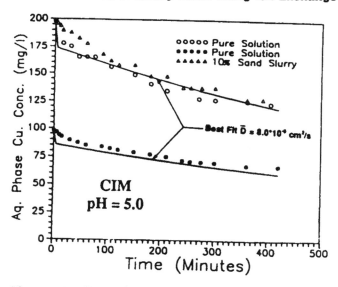

Figure 19 Comparison of kinetic model predictions (solid lines) with three independent experimental data sets.

An analytical solution of this problem is given as (35)

$$F(t) = \frac{q_t}{q^0} = 1 - \sum_{n=1}^{\infty} \frac{2\alpha(1+\alpha)}{1+\alpha+\alpha^2 q_n^2} \exp\left(\frac{-\overline{D} q_n^2 t}{w^2}\right) \quad (9)$$

where q_n values are the nonzero positive roots of

$$\tan q_n = -\alpha q_n \quad (10)$$

and $\alpha = a/(K \times w)$.

Figure 19 shows the results of three independent kinetic studies in which aqueous-phase copper concentrations are plotted against time; two sets had different initial Cu concentration (200 mg/L versus 100 mg/L), while the third one was a 10% fine-sand slurry. Solid lines Fig. 19 indicate model predictions for an apparent copper ion diffusivity of 8.0×10^{-9} cm^2/s. Note that although the exact morphology, i.e., distribution of voids and particles within the PTFE fiber, is not known, the suggested pseudo-homogeneous plane sheet model showed good agreement for the three sets of independent kinetic data produced with different initial concentration and level of suspended solids.

C. Cyclic Desorption of Heavy Metals from Ion-Exchange Sites of Soil

Figure 20 shows the plot of percentage recovery of Cu(II) and the aqueous-phase Cu(II) concentration versus number of cycles for the case of Cu(II)-loaded bentonite. Note that >60% recovery of Cu(II) was achieved in less than 30 cycles.

Bentonite clays (alumino silicates) derive their cation-exchange capacity primarily from the isomorphous substitution within the lattice of aluminum and possibly phosphorus for silicon in tetrahedral coordination and/or magnesium, zinc, lithium, etc., for aluminum in the octahedral sheet (36). Divalent metal ions are strongly held onto

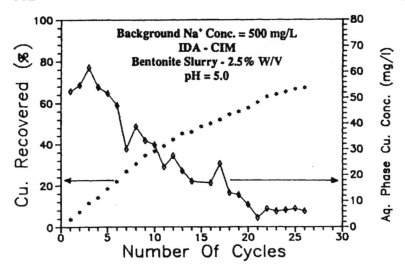

Figure 20 Copper(II) recovery from the ion-exchange sites of bentonite clay during the cyclic process.

these ion-exchange sites primarily thorough electrostatic interactions. Removal of heavy metals from such contaminated soils during the sorption step essentially involves (a) desorption from the ion-exchange sites of the soil into the aqueous phase by a counterion, and (b) selective sorption from the liquid phase onto the composite membrane. As in any sequential process with varying kinetic rates, the slower step will govern the rate of the overall process. Figure 21 compares the results of fractional desorption of Cu(II) from the ion-exchange sites of bentonite at a pH of 5.0 using a 400-mg/L Ca(II) solution with that of fractional uptake of Cu(II) from a pure solution with an initial Cu(II) concentration of 200 mg/L carried out at the same pH value. From this figure, it can safely be concluded that the desorption rate of heavy metal from the ion-exchange sites of clay is much faster than the sorption rate of the same by the composite membrane. For example, for the experimental conditions of Fig. 21, the time taken for 50% desorption from bentonite is 6 min, while that for 50% sorption by the membrane (both expressed as a ratio of the equilibrium capacity) is $\simeq 240$ min. Thus,

$$\frac{t_{1/2}\,\text{sorption}}{t_{1/2}\,\text{desorption}} \approx \frac{240}{6} = 40 \qquad (11)$$

Therefore, in any system where two solid phases coexist, the sorption of heavy metal from solution is most likely to be the rate-limiting step. Also, any practical application would involve cyclic use of the CIM. Since the heavy metal is removed from the CIM into an acid solution with each cycle, the CIM is totally free of heavy metal at the beginning of each cycle. This creates a maximum possible concentration gradient between the aqueous solution phase and the CIM phase, thus facilitating faster kinetics. Therefore, for the case of heavy-metal decontamination from ion-exchange sites of clay, the kinetic limitations encountered would be due to the composition of the CIM and not from the desorption of heavy metal from the functional groups of the clay material.

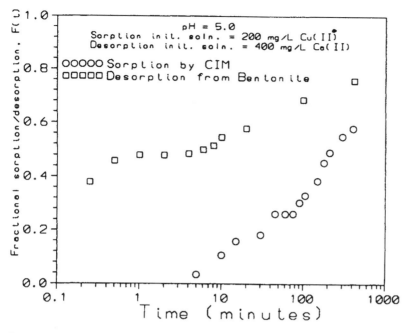

Figure 21 Comparison of fractional desorption rate of Cu(II) from the ion-exchange sites of bentonite with the fractional sorption rate of dissolved Cu(II) from pure solution by the CIM.

D. Ion Selectivity of Bentonite

Figure 22 shows the results of Cu/Ca selectivity of bentonite at a pH of 4.5, and Table 4 details coefficient and separation factor values for different composition of the aqueous and CIM phases. In Fig. 22, the aqueous-phase total cation concentrations $\{Ca^{2+}\} + \{Cu^{2+}\}$, remained almost the same. This proves that cation exchange was the major physicochemical process occurring.

For the Cu^{2+}/Ca^{2+} system, separation factor of copper over calcium is given by

$$\alpha_{Cu/Ca} = \frac{y_{Cu} x_{Cu}}{x_{Cu} y_{Ca}} \tag{12}$$

where y represents the faction of the element in the solid phase and x represents the fraction in the aqueous phase.

From Fig. 22 it may be noted that

1. $\alpha_{Cu/Ca} < 1.0$, meaning that the bentonite has more affinity for Ca^{2+}. This matches the observation reported in the literature (20,37) for strong-acid cation-exchange resin. The primary reason attributed for calcium selectivity over copper is that Ca^{2+} ion causes less swelling of the resin than Cu^{2+}, or water uptake by the resin/clay is smaller for the Ca^{2+} form than for the Cu^{2+} form. This value of separation factor is in sharp contrast with $\alpha_{Cu/Ca} \approx 1000$ for iminodiacetate functionality chelating resin (38).
2. $\alpha_{Cu/Ca}$ decreases with increasing x_{Cu}. This phenomenon has been observed in almost all strong acid and strong base ion exchangers and is attributed to the nonuniformity of ion-exchange sites (39).

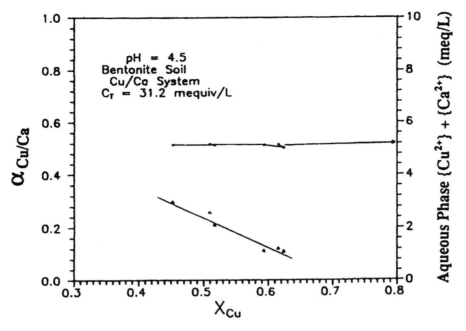

Figure 22 Copper–calcium selectivity of bentonite at pH = 4.5.

3. The aqueous-phase total cation concentration, $\{Ca^{2+}\} + \{Cu^{2+}\}$, remained almost the same in each case, proving that ion exchange between Cu^{2+} and Ca^{2+} is the major physicochemical process occurring. A similar procedure was carried out to determine the selectivity of bentonite for the Cu/Na system. Since this is a heterovalent ion-exchange system, the "selectivity coefficient" is thermodynamically a more rigorous parameter, and this value is reported in Table 4. The selectivity coefficient $K_{Cu/Na}$ can be expressed mathematically as

$$K_{Cu/Na} = \frac{y_{Cu} \cdot x_{Na}^2 \cdot C_T}{y_{Na}^2 \cdot x_{Cu} \cdot Q} \tag{13}$$

where

Table 4 Cu/Na Binary System[a]

y_{Cu} (dimensionless)	x_{Cu} (dimensionless)	$K_{Cu/Na}$ (g/L)	$\alpha_{Cu/Na}$ (dimensionless)
0.550	0.028	6,389	42.43
0.600	0.031	7,309	46.89
0.647	0.039	7,283	45.16
0.739	0.044	11,722	61.52
0.780	0.064	10,009	51.85
0.794	0.087	7,812	40.45
0.864	0.123	10,532	45.30

[a] Average $K_{Cu/Na} = 8,722$ (g/L); average $\alpha_{Cu/Na} = 47.66$ (dimensionless).

C_T = total aqueous-phase cation concentration, $\{Cu^{2+}\} + \{Na^+\}$, which ranged from 19 to 36 mequiv/L

Q = bentonite cation-exchange capacity = 0.5 meq/g

The average separation-factor values calculated for the Cu/Ca and Cu/Na systems are

$$\text{Avg}_{\alpha_{Cu/Ca}} = 0.325$$

$$\text{Avg}_{\alpha_{Cu/Na}} = 47.66$$

From the above observations, it can be concluded that the cation-exchange behavior of smectite clay is similar to that of strong-acid cation exchanger. Thus for any system containing bentonite loaded with heavy metals in its ion-exchange sites, the CIM can be represented mechanistically by the interaction of two cation exchangers, one with selectivity preference the same as that of strong-acid cation exchanger ($\alpha_{Cu/Na} \approx 50$) and the other with chelating functionality ($\alpha_{Cu/Na} \approx 10{,}000$) (20,38). For the case of Cu(II) desorption from ion-exchange sites of bentonite by addition of Na^+ in the aqueous phase, the exchange reactions involved can be summarized as follows:

$$\overline{(Z^-)_2 Cu^{2+}} + 2Na^+(aq) \leftrightarrow 2\overline{Z^- Na^+} + Cu^{2+}(aq) \tag{14}$$

$$\overline{R-N-(CH_2COO^- Na^+)_2} + Cu^{2+}(aq) \leftrightarrow \overline{R-N-(CH_2COO^-)_2 Cu^{2+}} + 2Na^+(aq) \tag{15}$$

Overall,

$$\overline{(Z^-)_2 Cu^{2+}} + \overline{R-N-(CH_2COO^- Na^+)_2} \leftrightarrow \overline{R-N-(CH_2COO^-)_2 Cu^{2+}} + \overline{2Z^- - Na^+} \tag{16}$$

where Z and R represent the bentonite clay and the CIM, respectively.

In order for the proposed process to succeed, the overall reaction [Eq. (15)] has to be thermodynamically favorable. Considering ideality, the equilibrium constant for such a reaction in terms of soil-phase and membrane-phase considerations is given as

$$K_{\text{overall}} = \frac{q_{Na}^Z q_{Cu}^R}{q_{Cu}^Z q_{Na}^R} \tag{17}$$

where superscripts Z and R denote the soil phase and the CIM phase, respectively, while q_{Na} and q_{Cu} represent the sodium and copper concentrations in the corresponding solid phase. Multiplying both the numerator and denominator of Eq. (17) by C_{Na}/C_{Cu} (C_i denotes the aqueous-phase concentrations of species i), we obtain

$$K_{\text{overall}} = \left(\frac{q_{Cu}^R / C_{Cu}}{q_{Na}^R / C_{Na}}\right) \left(\frac{q_{Na}^Z / C_{Na}}{q_{Cu}^Z / C_{Cu}}\right)$$

$$= \frac{\alpha_{Cu/Na}^R}{\alpha_{Cu/Na}^Z} \tag{18}$$

Thus, K_{overall} is the ratio of the Cu/Na separation factor between the CIM and the bentonite clay. As discussed earlier, due to the presence of the chelating functional group (iminodiacetate moiety), the dimensionless Cu/Na separation for the CIM is about two to three orders of magnitude greater compared to bentonite. As a result, the overall process is quite selective for decontamination of clays with high ion-exchange capacity.

Thus, advantage is being taken of the fact that the bentonite clay ion-exchanging functionality has much less selectivity for Cu(II) over Na (by introducing a high concentration of Na in the aqueous phase), whereas the chelating CIM has very high affinity for Cu(II) over Na [due to which only Cu(II) is taken up by the membrane and concentrated in the regenerant solution].

E. EMP Process

In order for the EMP process to succeed, the flow of ions in the aqueous solution inside the EMP cell toward the respective electrodes is mandatory, since this phenomenon completes the electrical circuit. Therefore, it is extremely important to ascertain if the placement of the CIM will create resistance to the flow of current in the aqueous circuit. This was verified by filling the EMP cell with 10 mequiv/L of KCl solution. Two runs were carried out, the first without any membrane and the second with ~ 35 cm^2 of IDA functionality CIM around the cathode. Figure 23 shows the plot of current carried by the reactor as a function of time for both cases. It may be noted that the current flowing through the system showed a difference of only a few milliamperes. Thus, it can be stated that the presence of CIM will not significantly reduce the current flowing through the system.

The EMP process will be successful only if the CIM is porous to let nonselective cations and anions pass through, but at the same time can trap heavy-metal cations. To verify this, an experiment was conducted with the same EMP cell as described above. A base solution of 200-mg/L K$^+$ was added (as KCl) to the background solution, but at the anode chamber was added 3.0 mequiv of Cu^{2+} and 3.0 mequiv of Li$^+$ (as their nitrate salts). A CIM sheet of IDA functionality with area ~ 40 cm^2 was placed midway between the two electrodes and 10 V of potential difference was passed through the system. Samples were taken from different points along the length of the reactor every

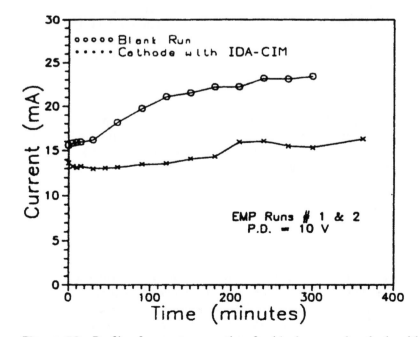

Figure 23 Profile of current versus time for blank run and cathode with IDA-CIM.

30 min and analyzed for Cu^{2+}, Li^+, and K^+. Figure 24 shows the Cu^{2+} and Li^+ profiles in the different chambers as a function of time. It can be seen that Cu^{2+} concentration decreased in the anode chamber with concomitant increase in the chamber adjacent to the anode chamber, but the concentration of Cu^{2+} beyond the CIM always remained very low (<30 mg/L, compared to an initial anode concentration of ~600 mg/L). However, Li^+ shows a different behavior, with its concentration at the cathode chamber showing a steady increase with time. Analysis of the membrane regenerant confirmed the earlier observation. The CIM was initially in Na^+ form. The regenerant solution contained only Na^+ and Cu^{2+} but no Li^+ or K^+. This proves that the process is able to let nonselective cations and anions pass though the pores of the CIM but sorbs or holds back the heavy-metal cations.

Finally, the ability of the CIM membrane wrapped around the cathode to neutralize OH^- generated at the cathode is a major advantage to the process, since it is this feature that prevents heavy metals from reprecipitating in the reactor. An experiment was conducted to verify the buffering capacity of the CIM and the ability of the process to remove heavy metal from the system. In the laboratory EMP cell, a 500-ppm NaCl solution was introduced as the background electrolyte. Then 1.0 g of amorphous $CuO(s)$ was added at time 0 and a potential difference of 10 V was applied. Two runs were made, one with 113 cm^2 of CIM in hydrogen form wrapped around the cathode (case A) and the other without the membrane (case B). The pH at the anode and cathode chambers were noted and is shown in Figs. 25 and 26. It may be noted that the pH at the cathode chamber was <4.0 even after applying the potential difference for 300 min for case A, whereas for case B the pH at the cathode continued to increase with time, reaching a value >11.5 after 300 min. However, the pH at the anode was very similar in both cases. After the end of the run, the CIM from case A was regenerated with 5% H_2SO_4 and analyzed for Cu^{2+} and K^+. The mass of copper sorbed on the CIM at the cathode was 3.5% of the CIM capacity. While this number seems to be very low, it may be pointed out that the electric potential was passed only for 390 min. Thus, only the leading edge of the front reached the CIM. In order to run this conceived process in a practical scenario, the time of run has to be balanced by the capacity of CIM to neutralize hydroxide generated at the cathode. The best way this balance can be maintained is by running the process in cycles.

To confirm this, a simulated contaminated soil sample was placed in the anode chamber by mixing 150 g of sand and 208 mg of $CuCO_3(s)$. The background electrolyte solution was 2900-mg/L NaCl. The distance between electrodes was 8 cm, and 72 cm^2 of IDA CIM was wrapped around the cathode. The potential difference applied was 5 V; five cycles were run, with each cycle consisting of 5 h of electric gradient followed by 45 min of chemical regeneration with 1 N H_2SO_4. To maintain continuity with the electric potential applied, two cathodes with CIM wrapped were synthesized. During chemical regeneration of one, the other was placed in the reactor. It took only 2–3 min to replace electrodes in the experimental cell. After the end of the fifth run, the regenerating solution was analyzed for copper and it was calculated that it contained about 67% of the copper that was originally added as $CuCO_3(s)$. Aqueous samples near the anode and cathode showed copper concentrations of 93 and 57 mg/L respectively. This observation proves that copper dissolved from the solid phase existed in the aqueous phase and the process needed to be run for more cycles; nevertheless, the experiment shows that it is possible to separate heavy metals from a precipitate solid phase by the use of this method. The pH values near the anode and the cathode were recorded every 30 min (after temporarily disconnecting the power supply, since the electrical circuit was interfering

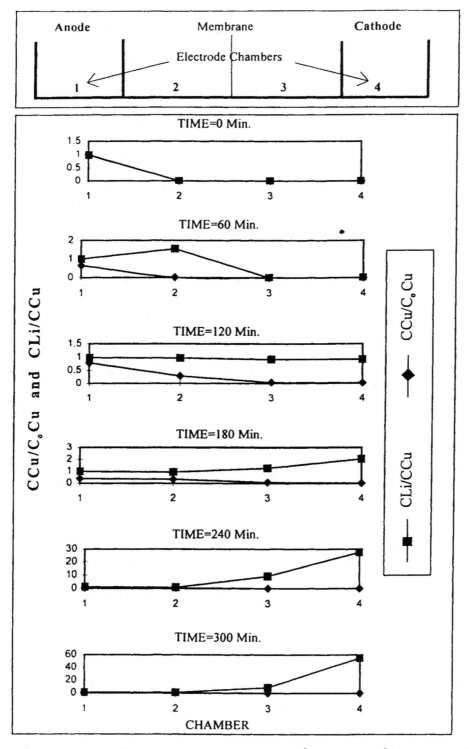

Figure 24 Normalized concentration profiles for Cu^{2+} and Li^+/Cu^{2+} ratio (in equivalents) at different chambers of EMP cell versus time.

Decontamination of Heavy Metal Using Ion Exchange

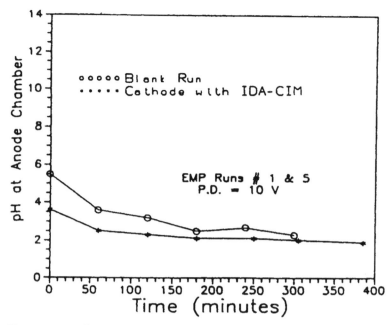

Figure 25 pH at the anode chamber versus time for blank run and cathode containing IDA-CIM.

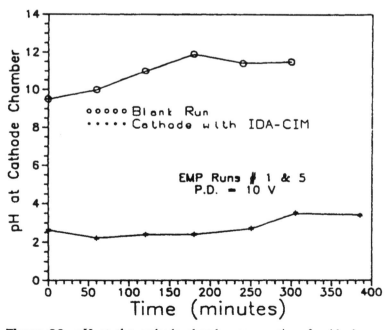

Figure 26 pH at the cathode chamber versus time for blank run and cathode containing IDA-CIM.

with the millivolt measurement for pH determination). The pH at the cathode never increased beyond 7.5, proving once again that the hydrogen-form CIM is able to act as a buffer, neutralizing the OH^- generated at the cathode. Calculations with solubility-product values of $Cu(OH)_2(s)$ and $CuCO_3(s)$ were done to ascertain that copper did not reprecipitate anywhere along the length of the Plexiglas reactor.

VII. CONCLUSIONS

The presence of toxic and hazardous waste sites in which the soil contains heavy metal(s) as the target contaminant is a widespread problem. In cases where the soil contains heavy-metal precipitates along with other bulk nontoxic but chemically interacting solids, the selective removal of the target heavy-metal(s) would be an ideal solution but also is a challenging separation problem. Chelating ion exchangers can selectively remove dissolved heavy metals, but their morphology (spherical beads of diameter < 2 mm) does not permit their use in a soil decontamination scenario.

We have identified and studied extensively the suitability of a process which uses chelating ion exchangers but in a form (CIM) such that they can be used in a reactor containing high concentrations of suspended solids. Moreover, by tailoring the chemical constituents in the reactor with the functionality of the ion-exchanger group on the CIM, an effective process can be devised which has the potential of selective and reversible separation of heavy metals from a soil with an unfavorable background. The material can also be applied in a scheme using an electric gradient—termed EMP—which has shown promise in being used as an in-situ process for heavy-metal decontamination from soil or sludge. An extensive number of laboratory experiments were carried out to investigate some key aspects of the proposed process. Primary conclusions of the study can be summarized as follows.

> The CIM was able to remove Cu(II) from a synthesized soil with high buffer capacity by the introduction of an aqueous-phase ligand. Comparative studies showed that of the three ligands (citrate, acetate, and oxalate) studied, citrate was most effective in heavy-metal removal but also demonstrated more competing ion uptake in the regenerant solution. Oxalate was effective in avoiding competing ion uptake while still maintaining heavy-metal uptake nearly equal to citrate, but had a tendency to be sorbed by the ion-exchange membrane. Acetate, the weakest ligand (in terms stability-constant values versus calcium and copper) was ineffective in removing the heavy metal from the solid phase present in the reactor.
>
> The CIM was capable of removing Cu(II) from Cu(II)-loaded bentonite clay. Copper ions were selectively transferred from the ion-exchange sites of bentonite clay to the ion-exchange sites of the CIM.
>
> The physical composition of the CIM has a significant effect on the kinetic of the process. A kinetic model, in which the CIM is conceived as a pseudo-homogeneous plate of thickness of the CIM sheet, is capable of predicting the kinetics of heavy-metal uptake from a sludge. The CIM is composed of ion-exchanger beads, but stagnant pores inside the membrane present an extra diffusional resistance to the path of dissolved heavy-metal ions involved in the exchange reaction. This phenomenon is primarily responsible for our observation that the kinetics of sorption of heavy metal by the membrane is much slower than that of desorption (of the heavy metal) from

bentonite clay; therefore, sorption by the composite membrane is the rate-limiting kinetic step. This observation suggests that the process efficiency can be improved by eliminating or reducing the intramembrane stagnant pores in the composite membrane.

When the CIM is used in an EMP application scenario, the following salient features are observed.

The CIM is able to selectively sorb heavy metals when used in the EPCIM process.
The presence of CIM in the EPCIM reactor results in a very moderate reduction in the current passing through the system.
The CIM is porous enough to allow free transport of nonselective cations and anions, i.e., the Donnan coion exclusion phenomenon is practically absent in this process.
The CIM-wrapped cathode is able to buffer the hydroxide generated at the cathode.
The CIM-wrapped cathode is able to sorb heavy metal cations selectively, and if it is used in a cyclic operation it is able to remove heavy metals from the solid phase of the soil and concentrate them in an acid solution.

REFERENCES

1. Campbell, P. G., and Tessier, A. 1987. Metal speciation in natural waters: Influence of environmental acidification. In: *Sources and Fates of Aquatic Pollutants*. Advances in Chemistry Series 216, R. A. Hites and S. J. Eisenreich (eds.) American Chemical Society, Washington, DC.
2. U.S. Environmental Protection Agency. 1986. *Test Methods for Evaluating Solid Waste, Physical/Chemical Methods*. Method 1310, 3rd ed., SW 486 and additions thereto.
3. U.S. Environmental Protection Agency. 1992. *Soil Sediment Washing System*, EPA/540/MR-921075.
4. Bardos, P., et al. 1990. Contaminated land treatment concepts under development at Warren Spring Laboratory. Euroenviron Contaminated Land Workshop, September.
5. Raghavan, R., Coles, E., and Dietz, D. 1990. *Cleaning Excavated Soil Using Extraction Agency: A state-of-the-Art Review*, EPA/600/S@-89/034.
6. Shah, D. B., Phadke, A., and Kocher, W. M. 1995. Lead removal from foundry wastes by solvent extraction. *J. Air Waste Manage. Assoc.* 45.
7. U.S. Environmental Protection Agency. 1992. *The Babcock and Wilcox Cyclone Furnace Soil Vitrification Technology*, EPA/540/F-92/010.
8. Hiller, D., and Gudeman H. 1989. Analysis of vapor extraction data from applications in Europe. *Proc. Third Int. Conf. on New Frontiers in Hazardous Waste Management*, U.S. EPA, September.
9. Iben, I., et al. 1995. Field test of in-situ thermal desorption of PCBs using a thermal blanket. In: *Extended Abstracts of American Institute of Chemical Engineers 1995 Summer National Meeting*, July.
10. Leach, B. 1995. Hydraulic fracturing for enhanced in-situ remediation. *Extended Abstracts of American Institute of Chemical Engineers 1995 Summer National Meeting*, July.
11. Hamid, J., Acar, Y. B., and Gale, R. J. 1980. Pb(II) removal from kaolinite by electrokinetics *J. Geotechn. Eng. Div. ASCE*, 106 (GT10).
12. Ho, A. V., et al. 1995. Integrated in-situ soil remediation technology: The Lasagna process. *Environ. Sci. Technol.*, 29(10).
13. Baldi, F., Vanghan, A. M., and Olson, G. 1990. Chromium(VI)-resistant yeast isolated from sewage treatment plant receiving tannery wastes. *Appl. Environ. Microbiol.*, 56(4).
14. Neufeld, R. D., and Hermann, F. G. 1975. Heavy metal removal by acclimated activated sludge. *J. Water Pollution Control Fed.* 47.

15. Volesky, B. 1990. *Biosorption of Heavy Metals*. CRC Press, Boca Raton, FL.
16. Peters, R. W., and Shem, L. 1992. Use of chelating agents for remediation of heavy metal contaminated soil. In: *Environmental Remediation—Removing Organic and Metal Ion Pollutants*, G. F. Vandergrift, D. T. Reed, and I. R. Tasker (eds.), ACS Symposium Series 509, American Chemical Society, Washington, DC.
17. Mott, H. V., and W. J. Weber. 1989. Solute migration in soil-bentonite containment barriers. *Proc. 10th Natl. Conf. on the Management of Uncontrolled Hazardous Waste Sites*, pp. 526–533. HMCRI, Washington, DC.
18. Errede, L. A., et al. 1986. Swelling of particulate polymers enmeshed in poly(tetrafluoroethylene). *J. Appl. Polymer Sci.*, 31:2721–2737.
19. Sengupta, S. 1993. A new separation and decontamination technique for heavy-metal-laden sludges using sorptive/desorptive ion-exchange membranes. Ph.D. dissertation, Lehigh University, Bethlehem, PA.
20. Helfferich, F. 1961. *Ion-Exchange*, Xerox University Microfilms, Ann Arbor, MI.
21. Sengupta, S., and Sengupta, A. K. 1993. Characterizing a new class of sorptive/desorptive ion-exchange membranes for decontamination of heavy-metal-laden sludges. *Environ. Sci. Technol.*, 27.
22. Sengupta, A. K., et al. 1995. Selective recovery of alum from clarifier sludge using composite ion-exchange membranes. Am Water Works Assoc. Final Report.
23. Li, P., and Sengupta, A. K. 1995. Selective recovery of alum from clarifier sludge using composite ion-exchange membranes. *Proc. 27th Mid-Atlantic Industrial Waste Conf.*, Lehigh University, Bethlehem, PA.
24. Sengupta, A. K., and Shi. B. 1992. Selective alum recovery from clarifier sludge. *J. Am. Water Works Assoc.*, 84.
25. Dorfner, K. 1961. *Ion Exchange, Properties and Applications*. Ann Arbor Science, Ann Arbor, MI.
26. Hudson, M. J. 1986. *Coordination Chemistry of Selective Ion-Exchange Resins*, Ion-Exchange Science and Technology (NATO ASI Series).
27. Schmukler, G. 1965. Chelating resins—Their analytical properties and applications. *Talanta*, 12.
28. Loewenschuss, H., and Schmuckler, G. 1964. Chelating properties of the chelating Ion-exchanger Dowex A-1. *Talanta*, 11.
29. Leyden, D. E., and Underwood, A. L. 1964. Equilibrium studies with the chelating ion-exchanger resin Dowex A-1. *J. Phys. Chem.*, 68.
30. Smith, R. M., and Martell, A. E. 1974. *Critical Stability Constants*, Vol. 2. Plenum Press, New York.
31. Zhu, Y., et al. 1990. Toward separation of toxic metal(II) cations by chelating polymers: Some noteworthy observations. *Reactive Polymers*, 13.
32. Sengupta, A. K., et al. 1991. Metal(II)-ion binding onto chelating exchangers with nitrogen donor atoms. *Environ Sci. Technol.*, 25.
33. Hoell, W. 1984. *Reaction Polymers*, 2:93–101.
34. Helfferich, F. 1990. *Reaction Polymers*, 13:191–194.
35. Crank, J. 1975. *The Mathematics of Diffusion*. Oxford University Press, London.
36. Grim, R. E. 1968. *Clay Mineralogy*, 2nd ed. McGraw-Hill, New York.
37. Bonner, O. D., and Livingston, F. L. 1956. Cation exchange equilibria involving some divalent ions. *J. Phys. Chem.*, 60.
38. Roy, T. K. 1989. Chelating polymers: Their properties and applications in relation to relation to removal, recovery and separation of toxic metal cations. M.S. thesis, Lehigh University, Bethlehem, PA.
39. Reichenberg, D., and McCauley, D. J. 1955. Properties of ion-exchange resins to their structure. Part VI: Cation-Exchange equilibria on sulfonated polystyrene resins of varying degrees of crosslinking. *J. Chem. Soc.*, III.

27
Cost-Effective Method of Determining Soil Respiration in Contaminated and Uncontaminated Soils for Scientific and Routine Analysis

Markus Robertz
Aqualytic® GmbH & Co., Langen, Germany

Thomas Muckenheim, Susanne Eckl, and Leslie Webb
Research Centre of Jülich GmbH (FZJ), Jülich, Germany

I. INTRODUCTION

The increasing awareness of the importance of soil for the maintenance of our environment has been emphasized by the number of national and international seminars held to discuss the problems of contaminated sites, bioremediation, and soil protection. This was confirmed in Germany in February 1998 with the passing of legislation covering soil protection (1). Since soil cannot be created and self-regeneration is a very slow process, the declared aim is to protect soil functions as an essential basis for human existence. Appropriate analysis processes are essential, to determine soil characteristics and to provide accurate monitoring.

Unlike the fields of air and water analysis, there are few standard processes available for soil analysis or processes which are suitable for routine investigation work. Existing processes are often extremely complicated and do not provide reproducible results. It is therefore not possible to draw firm conclusions from them and they are often too expensive for routine investigations in compliance with regulations. Physicochemical methods of analysis allow specific groups of compounds and their concentrations in soil to be determined. However, it is not possible to investigate the biological properties and the various functions in soil accurately by these methods. Individual chemical analysis cannot encompass the wide variety of substances in contaminated soil nor detect all metabolites produced when these substances are degraded. As a result, hazardous soil conditions may not be recognized.

This analytical gap can be closed in a cost-effective way by the use of simple physiological investigation methods. With the methodology described in this chapter, the influences on biological soil functions can be detected and biologically effective contaminants can be monitored without the need for expensive analytical procedures. By a combination of physical, chemical, and biochemical investigation methods, therefore, it is possible to detect potential danger to soil functions or to determine the effective exploitability of soils in a more dependable manner. This also enables those soil characteristics which are significant for mankind to be evaluated (2).

It was thus our aim to develop a biological monitoring system for soil analysis which would allow conclusions regarding biological functions and possible danger potential of soils (contamination), while at the same time providing a low-cost, user-friendly investigation tool suitable for the use with legislative regulations. This objective was achieved by the cooperative efforts of the Research Center of Jülich GmbH (FZJ) and Aqualytic® GmbH.

II. SOIL RESPIRATION

Soil respiration is defined as the absorption of oxygen (O_2) or the discharge of carbon dioxide (CO_2) by microbial activity and includes gas exchange from aerobic and anaerobic metabolism. Soil respiration results from the degradation of biologically accessible organic substances and their mineralization. This biological soil reaction is the summery of many individual activities which result from the condition of the particular soil and its constituents.

Under normal conditions, an ecological balance is achieved in the soil between the organisms and their activities. In these conditions, basic respiration is obtained (3). If the balance is disturbed (for example, by the addition of degradable organic substances), a change in soil respiration will be observed as a result of more intense microbiological growth and mineralization activity. The presence of poisonous substances, such as may be found in wastes, can lead to limitation of soil respiration. The ability of a soil to discharge CO_2 is thus an indication of the biological function of the soil in its entirety, taking into consideration all the constituents of the soil (4,5).

With appropriate modifications, soil respiration can be employed to carry out further biological determinations, including the biological degradability of substances (such as pesticides, plastics, and fertilizers), toxicity tests (3) to assess contaminated soils, substrate-induced respiration rate (SIR), microbial biomass, etc. (see below).

III. PRINCIPLE OF MEASUREMENT

The principle of operation is based on the measurement of a difference in pressure occurring within a closed system (see Fig. 1), as a result of microbial respiration activity (SR), biodegradable organic substances (C_{org}) are converted to carbon dioxide (CO_2) in the course of oxygen consumption (O_2) after Eq. (1):

$$C_{org} + O_2 \xrightarrow{\text{microorganisms}} CO_2 + \text{Salts} \tag{1}$$

The evolving CO_2 is linked to an absorber (KOH solution or soda lime with saturation indicator) in the unit, so that a pressure drop, proportional to the level of soil respiration, is generated in the reaction vessel. This is determined and stored periodically by a digital pressure sensor. The oxygen consumption can be calculated by reference to the selected volume for the reactor and soil sample.

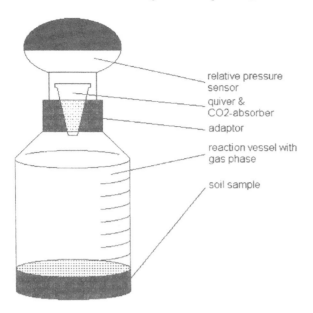

Figure 1 Components of the manometric measurement system. The mobile unit is incubated at a constant temperature. The sensor records measured values on-line; an integrated infrared interface provides a link to a central monitoring unit.

IV. THE MEASUREMENT SYSTEM

The system described here has been developed from the established BOD measurement technique for water analysis, which has been brought up to the latest modern standards and adapted for the determination of soil respiration. The suitability of the new measurement system has been demonstrated by comparing measured data with figures obtained using the electrochemical method (ECM; see Figs 2, 3, and 5). The measurement system (see Fig. 1) comprises a relative pressure sensor (Aqualytic® IR Sensor), a quiver filled with CO_2 absorber, an adaptor, and a reaction vessel (Schott Duran flask, GL-45). Because the vessel is sealed, measurements are not affected by air pressure (manometric principle). To ensure stable temperature conditions, the samples are incubated in an incubator.

V. MEASUREMENT RANGE

Due to the test design, the measurement range is dependent on the quantity of oxygen in the gas phase, the activity of the sample, the incubation temperature, and the quantity of CO_2 absorber. The quantity of O_2 determines the maximum possible respiration, while the activity of the sample determines the duration of the incubation period. Because of this, the final design of the test will depend on the type of test to be performed (see below). Because soil respiration is based on the linear absorption of oxygen, a measurement can be carried out within 24 h (2). However, a test for complete biological breakdown will generally require several days or weeks. The measurement range is thus

Table 1 Measurement Ranges, Depending on Soil Activity and Test Design

Measurement range[a] (mg O_2/kg ds)	V_{flask}[b] (mL)	Soil sample (g)	45% KOH solution[c] (drops)
0–200	500	200	10
0–400	500	100	10
0–800	500	50	10
0–1600	1000	50	20
0–3200	2000	50	20
0–∞[d]	500	50	10 (refresh)

[a] Calculated with 2 g/mL soil density, 20°C incubation temperature, and maximum 30% consumption of the initial oxygen amount (100%); ds, dry substance.
[b] Nominal volume (the actual volume is greater).
[c] Present in excess.
[d] Open the system to renew the KOH solution and to re-create the initial oxygen quantity (100%).

determined by the duration of measurement and the activity of the sample (see Table 1). The CO_2 absorber is present in surplus and therefore has no influence on the measurement range. The constant incubation temperature also does not influence the measurement.

The test period can be extended to a considerable degree for long-term investigations, by opening the sample vessel before the upper measurement range is reached and renewing the CO_2 absorber. In this way, the atmosphere in the measurement system regains the initial oxygen concentration (see Fig. 5), so that automatic recording of results can continue on-line over a lengthy period. If investigations are made over a period of weeks, it may be advisable to regulate a loss of soil humidity.

VI. SAMPLING AND PREPARATION

A representative number of soil samples is taken from the site to be investigated and these are blended into a single homogeneous sample. Various factors come into play here: the time at which the samples are taken; the choice of representative areas; removal and transportation of the samples, etc. (4):

Pretreatment of the sample material could differ due to the type of determination to be performed, or to comply with regulations and guidelines. As a general rule, biological activity is determined by tests on untreated, naturally moist soils. The water-holding capacity (4) can be set to $50 \pm 10\%$ in order to achieve optimum respiration activity (2). Soils which are too moist can be dried at room temperature until they are in a condition to pass through a sieve (4).

For greater standardization, samples can be mixed and sieved (sieve pits diameter 2 mm) until a homogenous sample is obtained. Only a portion of the homogenized sample is used for chemical analyses (volume, ion and grain size determination). If it is not possible to carry out the analysis within one month, the sample can be deep-frozen for subsequent investigation (4). For investigations related to soil respiration, such as SIR, toxicological tests, or degradation studies (see below), various substances may be added to a portion of the samples.

VII. SOIL VOLUME DETERMINATION

In determining soil respiration by manometric means, the proportion of the soil related to the total volume of the test flask is of prime importance. The volume of soil (V) can be calculated from its weight (m) and its density (ρ) by Eq. (2):

$$V = \frac{m}{\rho} \tag{2}$$

For simplicity's sake, the density of the soil may generally be taken as 2. The density may, in fact, be checked using Archimedes' principle: place 100 mL of distilled water into a measurement cylinder and then add 100 g of soil. The amount of water displaced by the soil represents the volume of the soil. The density is also required in order to determine soil respiration in relation to the weight of the sample (see below).

VIII. CARRYING OUT MEASUREMENTS

The prepared soil samples (see above) are weighed into the sample flasks and leveled off so that they are all at the same height. They are then acclimatized at $20 \pm 1°C$ for 24 h in an incubator (4). The adapters are then screwed into the sample flasks. The quivers are filled with CO_2 absorber (see Table 1) and placed in the adapter openings. The flasks are then sealed with the sensors.

It is then merely a matter of pushing a button to start the measurements. This automatically initiates the on-line recording of all the data measured during the predetermined duration of the test (up to 99 days). A graph of the entire course of all soil respiration activities can be displayed and printed out at any time via the infrared interface, at the touch of a button. The incubation temperature of $20 \pm 1°C$ is held constant throughout the investigation.

IX. DOCUMENTING AND ASSESSING THE RESULTS

All measured data are recorded on-line and can be transferred to a PC (Windows 3.1 or higher) for further evaluation.

X. CALCULATING SOIL RESPIRATION

The measurements made via the pressure sensor can be calculated by Eq. (3), depending on the individual test conditions:

$$SR = \frac{M(O_2)}{R \cdot T_{inc}} \cdot \left(\frac{V_{ges} - V_{sample}}{V_{sample}} \right) \cdot \frac{1}{\rho} \cdot 3.25 \cdot \text{digit} \tag{3}$$

[mg O_2/kg ds soil]

(where SR = soil respiration, $M(O_2)$ = molecular weight of oxygen, R = gas constant, T_{inc} = incubation temperature, V_{ges} = volume of reaction vessel, V_{sample} = volume of inserted soil, ρ = soil density, digit = value measured by the BSB/BOD sensor (using the range 0–40 mg/L), and ds = dry substance.)

Table 2 Simplified Calculation of Soil Respiration, Using Factors for the Standard Measurement Ranges

Measurement range[a] (mg O_2/kg ds)	V_{flask}[b] (ml)	Soil sample (g)	Factor[c]
0–200	500	200	11
0–400	500	100	24
0–800	500	50	50
0–1600	1000	50	96
0–3200	2000	50	186
0–∞[d]	500	50	50 (refresh)

[a] Calculated with 2 g/mL soil density, 20°C incubation temperature, and maximum 30% consumption of the initial oxygen amount (100%); ds, dry substance.
[b] Nominal volume (the actual volume is greater).
[c] The quantity in mg O_2/kg ds represented by one scale division (digit) on the sensor.
[d] Open the system to renew the KOH solution and to re-create the initial oxygen quantity (100%).

Calculation example: $M(O_2) = 32,000$ mg/mol, $R = 83,144$ L-mbar/mol-K, $T_{inc} = 293.15$ K, $V_{ges} = 0.609$ L, $V_{sample} = 0.0875$ L, $\rho = 2.285$ kg/L, digit = 2. Then SR = 22.3 mg O_2/kg ds.

XI. SIMPLIFIED CALCULATION

Based on the test designs given in Table 1, soil respiration can be calculated simply with the aid of factors. The volumes of soil and air in the test flask are fixed so that the pressure drop which occurs as a result of the respiration can be calculated directly into mg O_2/kg ds (see Table 2). Here, an incubation temperature of 20°C and a soil density of 2 are presupposed. Differing soil densities effect only light changes of the factors: per each 0.5 difference in soil density, the factor needs to be changed by 1. For example, if the sensor is operating in the 0–800 range and gives a reading of 10.0 digit, then the soil respiration will be (10 digit × factor 50) = 500 mg O_2/kg ds. With a soil density of 1.5 (10 digit × factor 49), 490 mg O_2/kg is obtained.

XII. CALIBRATION, FUNCTION CHECK, AND GLP

The sensor calibration can be checked using a quick-test device. The effectiveness of the sealing to the measurement equipment can be checked over a period of several days by a simulated consumption measurement. At the same time, all the various components affecting measurement can also be calibrated. The measurement system will support regular checks on the calibration. Measured data can be documented, either by a printout via the controller or using the PC.

XIII. COMPARATIVE MEASUREMENTS

Measurements with the method presented in this chapter were compared with those obtained in parallel investigations using a much more complicated and expensive analysis process. This involved a respirometer operating on the electrochemical method (ECM), whereby the carbon dioxide generated in the course of soil respiration is absorbed and the oxygen consumed is replaced periodically by oxygen, created by chemical electrolysis.

Naturally moist soil (water-holding capacity = $50 \pm 5\%$) was used for the tests. By way of pretreatment (see above), the samples were sieved through 2-mm mesh. Because a sieving process mobilizes substrates from the aggregates, initiating a temporary rise in soil activity, the samples were stored for 7 days at 4°C before measurements began (4). Before the tests started, the samples were weighed into the measurement flasks, then leveled off to the same uniform height and acclimatized for 24 h at an incubation temperature of 20 ± 1°C. The following results are shown as being typical of the data gained from the series of tests.

XIV. ORTHIC LUVISOL

The soil respiration of an orthic luvisol (brown soil) was tracked for a period of 20 days with both measurement systems, using the measurement range of 0–200 mg O_2/kg ds (see Table 1). Both the gradient of oxygen consumption and the absolute values measured by the two systems were the same (see Fig. 2). In neither case did the reduction in oxygen concentration or the partial pressure in the flask to 70% of the initial concentration (air) have any effect on the measurements. The respiration rate was 9 mg/kg ds per day. The measured data were recorded on-line by the system described in this chapter, over a period of 20 days, with no human intervention. The determination made with this system were carried out quickly and easily.

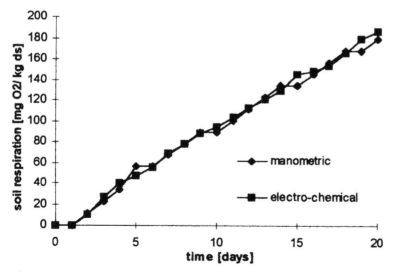

Figure 2 Measuring cumulative soil respiration in an ortic luvisol using the new Aqualytic® measurement system (–◆–) compared with the electrochemical system (ECM, –■–). The mean figures from parallel measurements with the new system presented in this chapter ($n = 12$) are shown, together with those obtained using the ECM ($n = 6$). The data were recorded on-line by the new measurement system over a period of 20 days without any human intervention. The measured data and the gradients were identical with both systems.

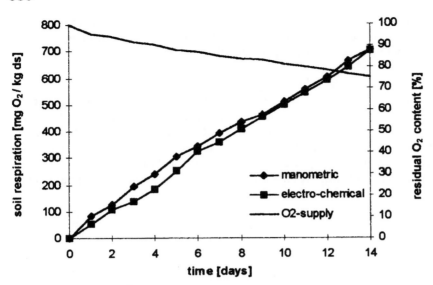

Figure 3 Cumulative soil respiration in an intensely active woodland soil using the new Aqualytic® measurement system (-◆-) compared with the electrochemical system (ECM, -■-). The mean figures from parallel measurements with the new system presented in this chapter ($n = 5$) are shown, together with those obtained using the ECM ($n = 6$). The reduction in oxygen concentration in the system (———) is shown as a percent of the initial value (air = 100%). Over the selected measurement range of 0–800 mg O_2/kg ds, both the measured data and the gradients were identical with both systems.

XV. WOODLAND SOIL

The organic horizon of a woodland soil usually displays a high level of basal respiration. Comparative checks were made on a sample of this type of soil, using a measurement of range of 0–800 mg O_2/kg ds (see Table 1). Figure 3 shows that the results obtained by the two methods corresponded with each other. The respiration rate was ca. 54 mg/kg ds per day. The measured data were recorded on-line by the new system over a period of 14 days, with no human intervention. At the upper end of the measurement range, the oxygen concentration in the air within the sample flask was 70% of the initial figure (100%).

XVI. REPRODUCIBILITY

In parallel with the measurement of soil respiration in a woodland soil over a measurement range 0–800 mg O_2/kg ds (see Fig. 3), measurements were also taken in the range of 0–400. These raw data in are displayed in Fig. 4 and demonstrate a high level of reproducibility of the parallel sets, both within the one measurement range and between the two different ranges. The scatter of the parallel sets within the measurement range is extremely low, with an average of ±0.9 scale unit (digit). On the 14th day of the test, the final figure using the 0–800 range was (15.0 digit × factor 50) = 750 mg O_2/kg ds, which is very close to the figure obtained in the 0–400 range: (30.3 digit × factor 24) = 727 mg O_2/kg ds.

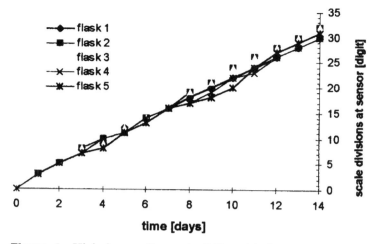

Figure 4 High degree of reproducibility with the new measurement system. All measured data from parallel sets ($n = 5$) in the measurement range of 0–400 are shown. The mean deviation is extremely low, at ±0.9 scale divisions (digit). The mean final figure for the measurements in the 0–400 range is (30.3 digit × factor 24) = 727 mg O_2/kg ds, which corresponds closely with the (15.0 digit × factor 50) = 750 mg O_2/kg ds obtained in the 0–800 range.

XVII. POTENTIAL AND LIMITS

Within the specified measurement ranges it has been shown that soil respiration can be determined quickly and reliably using the new system. However, at very high consumption rates, such as may be expected where easily consumable substrates are added (i.e., SIR), or where a determination is made of complete degradation of substances, the measurement ranges could be exceeded and special adjustments were required. The option of periodic ventilation of the system permits a significant extension to the measurement range (see Table 1): here, before the end of a measurement range is reached, the system is opened, both in order to renew the carbon dioxide absorber and to reestablish the initial partial oxygen pressure. In this way, the data collected before the system is ventilated can be accumulated, together with those obtained after the re-start.

In long-term investigations, the water content of the soil sample should be balanced, in order to ensure its biological functions. This technique is suitable for increasing the measurement range or the test duration to a significant degree—or to react quickly to unexpectedly high respiration activity. This is shown in Fig. 5.

XVIII. APPLICATIONS: VARIANTS ON SOIL RESPIRATION

In addition to soil respiration, it is possible to carry out biomass determination by substrate-induced respiration. SIR measurements are carried out in order to draw conclusions regarding the potential activity of a particular soil and usually are long-term tests. To do so, a substrate which is easily converted by microorganisms (such as glucose) is added to the samples (4). Since unexpectedly high respiration levels can occur here, a higher measurement range should be selected.

A further modification to the soil respiration measurement will permit testing of the biological degradability of substances, such as pesticides or fertilizers, organic substances,

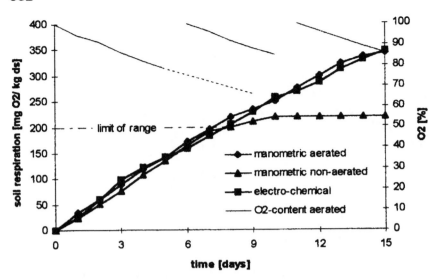

Figure 5 Ventilating the samples in order to increase the measurement range or to extend the test duration. The mean figures for cumulative soil respiration are shown, with a comparison of data obtained with the new system ($n = 10$) and the electrochemical system (ECM, $n = 6$). The oxygen decrease in the system is given in percent from the initial value (————/———————). Left: in the range 0–200, the data obtained with the new system correspond in all cases with those obtained using the ECM (–■–). Right: Ventilating the flasks enabled the same data to be obtained even outside a measurement range (–◆–). Without ventilation, only a limited amount of respiration activity is displayed, because the measurement range was exceeded (–◇–).

or chemicals. This enables the degree and speed of mineralization of samples in long-term respiration tests to be compared with figures for untreated, control samples (4). In such circumstances it is recommended that the system be ventilated periodically and the test results accumulated.

Equally, the toxicity of a test substance on the soil can be compared with that of an untreated control sample, by determining the inhibition of soil respiration.

REFERENCES

1. BBodSchG. 1998. German Federal Ministry for the Environment, Nature Conservancy and Reactor Safety, Bonn, Germany.
2. G. Kreysa and J. Wiesner. 1995. *Biological Test Methods for Soils (Biologische Testmethoden für Böden)*, Dechema, Frankfurt/Main, Germany.
3. VAAM. 1994. Regeneration of microbial activity in soils in natural stress situations: Evaluation criteria for soil quality (Regeneration mikrobieller Aktivität in Böden nach natürlichen Stresssituationen—Bewertungskriterium für die Bodenqualität). The Technical Group "Environmental Microbiology" of the VAAM Group "Eco-Systems, Soils", *BioEngineering*, 10:38–41.
4. F. Schinner, R. Öhlinger, E. Kandeler, and R. Margesin. 1993. *Biological Soil Operations Methods (Bodenbiologische Arbeitsmethoden)*. Springer-Verlag, Berlin, Germany.
5. F. Schinner and R. Sonnleitner. 1996. *Soil Ecology: Soil Biology and Soil Enzymatics (Bodenökologie: Bodenbiologie und Bodenenzymatik)*, Volume 1. Springer-Verlag, Berlin, Germany.

28
Remediation of Soils Contaminated with Chromium Due to Tannery Wastes Disposal

K. Ramasamy and S. Mahimairaja
Tamil Nadu Agricultural University, Coimbatore, India

R. Naidu
CSIRO Division of Land and Water, Adelaide Laboratory, Glen Osmond, South Australia, Australia

I. INTRODUCTION

Tanning is the process through which animal skins and hides are converted into the nonbiodegradable and tough material known as leather. The tannery industry is one of the major export industries in India, and the export income from leather products is approximately 7000 crore rupees. At present in Tamilnadu alone more than 1008 small-scale tanneries are functioning, with an employment potential of about 1 million and accounting for 6% of world leather production (1).

Once India was a net exporter of raw hides and skins. After World War II, necessity compelled the government to start a tannery for its requirements. This was the beginning of the growth of the leather industry in India. Available techniques and labor-oriented skills paved the way for setting up many tanneries. In the beginning, the tanneries processed vegetable-tanned or East India (EI)-tanned leathers. These leathers attracted good demand and price at that time. It was conceived that E.I. leathers would be very conducive as a basic raw material for importers to reprocess any type of finished leather in which they were interested.

During the tanning process, approximately 30–40 L of water are used for every 1 kg of leather processed (2). It is estimated that about 17,000 tons of hides and skins are converted into leather, which results in generation of approximately 680 million liters of waste water (effluent) and 50,000 tons of solid waste (sludge) per year.

II. TANNING PROCESSES

There are different types of tanning processes, namely, chrome tanning, vegetable tanning, aluminum tanning, and zirconium tanning. In India mostly vegetable tanning and chrome tanning are practiced, which are briefly discussed here.

A. Vegetable Tanning

Vegetable tanning was the original tanning method and remained the only method until the development of commercial tanning at the beginning of the twentieth century. The leathers made by full vegetable tanning are used for shoe soles, belting, saddlery, upholstery, lining, and luggage. Vegetable tannins are the water-soluble extracts of various parts of plant materials including wood, bark, leaves, fruits, and roots.

The mechanism for vegetable tanning is through hydrogen bonding between the CO–NH linkage of the protein and the phenolic hydroxyl group of vegetable tannin. The availability of hydrogen bonds on the protein and on the vegetable tanning material is of prime importance. The pH of the vegetable tanning ranges from 5 to 7 (3). Vegetable tanning has been widely replaced by chrome tanning due to the material availability and the cost of labor in harvesting the vegetable tanning materials.

B. Chrome Tanning

Chrome tanning is used to prepare light and more resistant leather. The chemical reaction involved is the formation of a stable compound between the hide protein and chromium [Cr(III)] ions. Salts containing Cr(III) ions are soluble in the pH range below 4–5. Chromium has variable affinity for hydroxyl ions over the entire pH range, and at pH 2–3 the fixation of Cr is moderate. During tanning the hide is brought to an acid condition (pH 4), chrome tanning salts are added, and then tanning starts. As the solution is strongly acidic, Cr salts penetrate the hide without excessive surface fixation. After a period of time, when the Cr has penetrated the hide sufficiently, the pH is raised to promote the reaction of the Cr with the hide. The reaction is very strong and the resulting leather is nonbiodegradable and resistant to mechanical damage. The tanning is done in a salt solution such as sodium chloride (3–5%) to prevent osmotic swelling of the untanned hide. The solution contains sodium sulfate at about 2%, from the sulfuric acid and chromium sulfate used.

The tanning action of Cr salts depends on chemical variations which take place during alkalization by sodium hydroxide with substitution of one or more water molecules by hydroxyl ions and with formation of basic Cr salts. The hide protein contains free carboxyl groups and other reactive sites which form coordinate complexes with Cr salts. The importance of Cr salts as mineral tanning agents derives from the fact that Cr(III) in aqueous solution may bind to a water molecule coordinately (4).

In chrome tanning, 276 chemicals with 14 heavy metals are used. It is estimated that approximately 32,000 tons of basic chromium sulfate salts are used annually in Indian tanneries. This amounts to an annual loss of nearly 2000–3200 tons of chromium (when expressed as element). The concentration of chromium in the effluents from chrome tanning yards is in the range of 2000–5000 mg/L. International regulations stipulate that the chromium concentration in industrial waste shall not exceed 2 mg/L.

Chromium is a potential soil, surface water, ground water, sediment, and air contaminant. High soil chromium is usually associated with anthropogenic contamination, mainly from industrial operations such as tanning, and direct disposal of sludge and waste water for land application. Chromium toxicity and mobility depend on oxidation state. The trivalent forms [Cr(III)] are relatively immobile, more stable, and much less toxic than the hexavalent [Cr(VI)] forms. Chromium forms a large number of relatively kinetically inert complexes which can be isolated as solids. Amines are well-known complexes. Chromium is not a significant contaminant of plant tissues, except at site-specific discharge points. Oxidation of Cr(III) to Cr(VI) by abiotic means, mainly by Mn oxides and dissolved oxygen, are known (5,6). Biological reduction of Cr(VI) to Cr(III) is common in soils, but biological oxidation of Cr(III) to Cr(VI) is unknown. High concentrations of Cr(VI) are the result of man-made pollution. Cr(VI) is highly soluble and toxic, and it is a carcinogen. A more positive feature is that the midpoint potential of the Cr(VI)/Cr(III) couple (ca. 1.3 V) indicates that the known physiological electron donors for microbial metabolism can potentially serve as electron donors for the reduction of Cr(VI) to stable Cr(III). Because of the toxicity of Cr(VI), on-site remediation aims at reducing it to Cr(III).

Residues from freshwater industrial zone sources generally range up to 50 mg/kg. Unlike some metals, Cr does not concentrate in specific tissues and it does not normally accumulate in fish; as a result, the burden is low in marine and freshwater species. Concentrations in fish tissues are below 0.25 mg/kg. Under most conditions, Hg, Cd, Cu, Pb, Ni, and Zn are more toxic than Cr. Growth inhibition in aquatic plants generally occurs at 0.5–5 mg/L of Cr(VI), whereas potassium dichromate may stimulate the growth of some species. However, little is known about the factors influencing the uptake of Cr in aquatic plants. Toxicity to plants depends on the pH of the medium and hence the availability of free and chelated ions. Other factors such as the presence of organic chelators, cations, nutrients, and other heavy metals in solution influence the toxicity to plants.

Chromium is not acutely toxic to humans. This is due to the high stability of natural Cr complexes in abiotic matrices; in addition, the acid nature of chromium imparts strong affinity for oxygen donors rather than sulfur donors present in biomolecules. However, the hexavalent form Cr(VI) is more toxic because of its high rate of absorption through the intestinal tract. In contrast to the biomagnification of methyl-Hg and Cd, Cr has not been shown to accumulate and biomagnify, but biomagnification through the food chain is known. However, accumulation of Cr far above the average levels are known in both the plant and animal kingdoms. In normal soils with total Cr concentrations between 20 and 100 mg/kg, dry weight of plants grown in it was less than 1 µg/g and seldom exceeded 5 µg/g. Only exception is serpentine soil which contains 1000–50,000 mg/kg Cr, and plants that are endemic to this soil will accumulate toxic concentrations of Cr in tissues. Like plants, the majority of invertebrates die before accumulating amounts of Cr that might prove toxic to predators. Among animals, gastropods are known accumulators of Cr to the concentration of 140–440 mg Cr/kg dry weight including shells.

III. POLLUTION DUE TO DISPOSAL OF TANNERY WASTES

The leather industry is among the major sources of pollution in Tamil Nadu. The effluent and sludge disposed from these industries into rivers and onto land has led to extensive degradation of productive land. The tannery wastes typically consist of high concentrations of salts (sodium, chloride, sulfates etc.,) and chromium, both of which threaten surface and subsurface land and water resources. The fate of Cr in tannery waste contaminated soil and its impact on the environment is shown in Fig. 1. High concentrations of heavy metals in soil due to effluent or sludge is of concern because of possible phytotoxicity or increased movement of metals into the food chain (7) and the potential for surface and ground water contamination. Even in cases where there is low or no metal contamination, pollution of surface and ground waters may also be of concern due to

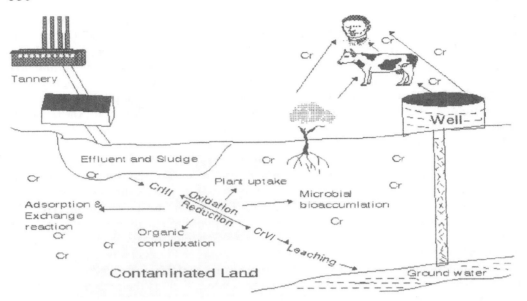

Figure 1 Fate of chromium in tannery waste-contaminated soil and its impact on the environment.

excessive nitrogen (N) and phosphorus (P) loading of potable water supplies. Before we discuss the various technologies available for the remediation of Cr-contaminated soil, it is imperative to describe the characteristics of the tannery wastes and Cr-contaminated soil.

IV. CHARACTERIZATION OF TANNERY WASTES

The effluent collected from selected tanneries in Tamil Nadu varied considerably in pH and electrical conductivity (EC) (Table 1). The pH ranged from 6.17 to 8.34 and the raw effluent had acidic pH. The EC was generally high and ranged between 10.4 and 23.0 dSm^{-1} (90). The effluent derived during the washing of preserved skin is highly saline, with the concentration of Na exceeding 2000 mg/L at all sites. The high salt content originated from the sodium chloride and sodium sulfide used in the washing process. A wide variation was also observed in the Cr concentration of the effluent collected at different locations. The total Cr ranged from 620 to 26,200 µg/L (Table 1). The difference in the chemical constituents may possibly be due to various processing methods employed by different tanneries (90).

The pH of sludge measured both in water and 0.01 M CaCl$_2$ were relatively higher (7.41–8.46) than effluent, but EC values were found to be lower in most sludge samples (Table 2). This may be due to the nature of the salts present in sludge. The concentration of Na varied from 9917 to 62,781 mg/kg, and the total Cr content ranged from 1179 to 16,158 mg/kg. Other heavy metals such as copper (Cu) and zinc (Zn) were also observed in the sludge. Disposal of this effluent and sludge onto land is of concern since high salinity is linked to both sodicity as well as salt-associated land degradation and

Table 1 Mean pH, EC, Total Na, and Cr Content of Tannery Effluent

Sample	pH	EC (dSm^{-1})	Na (mg/L)	Cl (mg/L)	Cr (µg/L)
Ambur					
Raw effluent	6.17	11.4	2,280	1,805	26,200
Treated effluent	8.17	10.4	2,040	1,644	8,800
Vaniambadi					
Treated effluent	7.02	23.0	9,000	3,674	900
Walajapet					
Treated effluent	8.17	13.5	2,600	3,143	2,300
Pernampet					
Raw effluent	7.26	20.0	3,150	3,191	7,200
Treated effluent	7.43	20.4	3,250		620
Vaduganthangal					
Treated effluent	7.56	13.0	2,900		760

Table 2 Some Important Characteristics of Tannery Sludge

Sample	pH H$_2$O	pH 0.01 M CaCl$_2$	EC (dSm^{-1})	Na (mg/kg)	Cr (mg/kg)	Cu (mg/kg)	Zn (mg/kg)
Ambur							
Site 1	7.81	7.65	16.0	62,781	5,728	58	106
Site 2	8.28	8.01	6.0	44,875	1,179	13	38
Site 3	7.76	7.65	15.5	40,698	9,312	54	218
Vaniambadi							
Site 1	8.16	7.77	2.3	10,125	1,314	21	60
Site 2	8.05	7.85	5.7	9,917	1,895	16	48
Vaduganthangal	8.46	—	8.3	12,231	8,241	42	67
Pernampet	7.70	7.41	20.8	29,239	16,158	32	51

related constraints on crop productivity. According to one estimate, such disposal in the past has already degraded over 50,000 ha of agricultural land in Tamil Nadu (8).

V. CHROMIUM CONTAMINATION IN THE ENVIRONMENT

A. Chromium in Soil

The mean values of pH, EC, Na, and Cr in surface and subsurface soils around some selected tanneries at Ambur and Vaniambadi are presented in Table 3. The pH (H$_2$O) of surface and subsurface soil at Ambur ranged from 8.11 to 8.57 and from 8.29 to 9.67, respectively, and the EC from 1.28 to 12.32 and from 0.35 to 3.60 dSm^{-1}, respectively (90). At Vaniambadi, the pH and EC at different soil depths varied from 7.68 to 8.26 and from 4.99 to 18.86 dSm^{-1}, respectively. The surface and subsurface soils both at Ambur and Vaniambadi had very high Na, ranging from 8838 to 77,711 mg/kg, and tended to decrease with depth. The salt-rich effluent and sludge disposed on the sampling sites is responsible for the accumulation of Na in soils.

Table 3 Range in pH, EC, Na, Cl, and Cr of Surface Soil

Location	pH (1:5 H$_2$O)	EC (dSm^{-1})	Na (mg/kg)	Cl (mg/kg)	Cr (mg/kg)
1. Ambur	7.96–8.57 (8.27)	0.15–12.3 (3.36)	8,838–77,711 (39,707)	21.3–4,587 (3,240)	924–16,731 (3,791)
2. Vaniambadi	7.68–8.87 (8.28)	0.43–20.6 (9.94)	2,405–74,398 (18,952)	60.9–8,175 (5,165)	569–79,865 (20,164)
3. Reference soil	7.96–8.23 (8.14)	0.32–0.59 (0.48)	1,022–2,697 (1,648)	45.4–84.0 (58.0)	5.2–8.6 (6.7)

Figures in parentheses refer to mean values.

At both locations the Cr content did not show a definite pattern with soil depth but accumulated in soil (Fig. 2). This may be due partly to the varying texture of the soil and hydrological features of the sampling sites, which may have influenced the mobility of Cr and resulted in differential accumulation in the soil profile. Nevertheless, at Vaniambadi the Cr contamination exceeded 70,000 mg/kg at depths exceeding 70 cm. The difference in the chemical constituent of effluent/sludge also contributed to the variation in Cr accumulation.

The data clearly showed that the soils around tannery industries are severely contaminated with Cr. In nature Cr can occur in both hexavalent [Cr(VI)] and trivalent [Cr(III)] forms, depending on the nature of the soil environment. Cr(VI), which exists primarily as CrO_4^{2-} and $HCrO_4^-$ in natural systems (9), is of particular concern because even at low concentrations it is toxic to both plants and animals (10), and it is more soluble and more mobile in the environment than Cr(III). The Cr accumulation in soils of Ambur and Vaniambadi exceeded the maximum threshold limit prescribed in different developed countries (Table 4).

1. Speciation of Cr

The speciation of Cr as determined by a sequential fractionation procedure indicated that over 85–99% of Cr was extractable by HNO$_3$. The Cr that was extractable by NaOH (organic) and EDTA (remaining organics plus iron oxide bound) constituted only 0.5–15% (91). Though no Cr (soluble) was detected in the water extracts, a concentration ranging from 5.5 to 128 µg/kg obtained in KNO$_3$ extract is of concern, since this fraction represents the exchangeable and sorbed Cr and is likely to be transformed into soluble fraction.

B. Chromium in Groundwater

The concentration of total Cr in groundwater from selected sites ranged from 51 to 996 µg/L, with a pH variation from 6.78 to 8.42 (Table 5). Wide variation (1.0–11.2 dSm^{-1}) was also observed in EC. Chromium concentration was below the detectable limit in water samples from uncontaminated sites in Vinnamuthi village (92). A low concentration of only 50 µg/L was recorded for the samples from Pernampet where tanneries have been recently established and function with a common effluent treatment plant (CETP). The well waters at Vaduganthangal, where only one large-scale chrome tannery without any treatment plant is located, contained a relatively high concentration

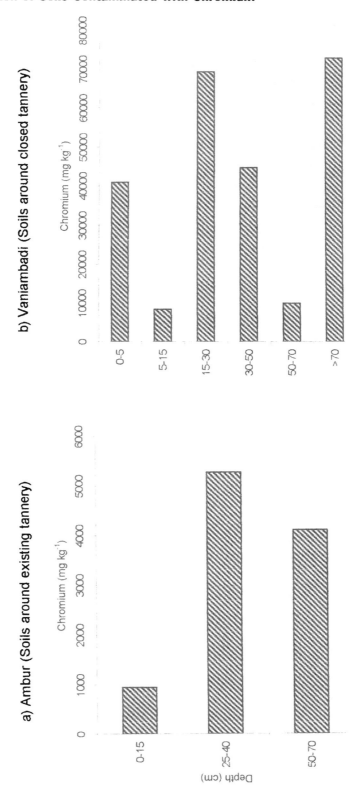

Figure 2 Chromium contamination in soil due to disposal of tannery waste.

Table 4 Threshold Levels of Cr in Soils of Some Developed Countries (mg/kg)

	Australia (18)	New Zealand (90)	U.K. (91)	Canada (92)	U.S. (29,93)	Germany (94)
Cr(III)		600	600	250	1500	200
	50 (Total)					
Cr(VI)			10	25	8	—	—

Table 5 Mean pH, EC, and Total Na and Cr Concentrations of Borehole Water Samples

Site	pH	EC (dSm^{-1})	Na (mg/L)	Cl (mg/L)	Chromium (µg/L) Total	Chromium (µg/L) Cr(VI)
Ambur						
Site 1	7.40	6.5	1200	946	178	159
Site 2	6.78	2.7	4850	395	511	480
Walajapet						
Site 1	8.42	1.0	50	105	996	922
Site 2	8.40	1.1	100	147	923	895
Vaniambadi						
Site 1	7.71	2.5	765	464	720	640
Vaduganthangal						
Site 1	7.23	11.2	1600	1628	256	231
Pernampet						
Site 1	7.13	6.5	780	947	54	52
Site 2	8.10	6.9	960	1006	51	48
Vinnamuthi (uncontaminated site)	7.90	1.5	198	221	ND	ND

of Cr (256 µg/L). At Ambur, where a large number of both small-scale and large-scale tanneries are located, the concentration varied from 178 to 511 µg/L. The result showed that Walajapet and Vaniambadi are the worst-affected areas, where the Cr concentration in the groundwaters ranged from 720 to 996 µg/L. Exceptionally high Cr content was observed in water samples from Walajapet.

A concentration of >500 µg Cr per liter in the groundwater sampled at 10-m depth inside the Palar River (Ambur, site 2) and >950 µg Cr per liter in groundwater sampled from a bore well located approximately 2 km away from the effluent outlet of a closed tannery at Walajapet (site 1), clearly indicated the extent of Cr pollution in the groundwater. It was surprising to note that 33–97% of the Cr in groundwaters was toxic Cr(VI), which ranged between 48 and 922 µg/L (Fig. 3). Except at site 1 in Ambur, all other water samples had a higher proportion of Cr(VI). Similar to the total Cr, the Cr(VI) was not detected in borewell waters from an uncontaminated site (Vinnamuthi). This indicates that the oxidation of Cr(III) to Cr(VI) had occurred in soil, therefore the Cr(VI) should have dissolved in the soil water, moved down into the soil profile, and contaminated the groundwater (92).

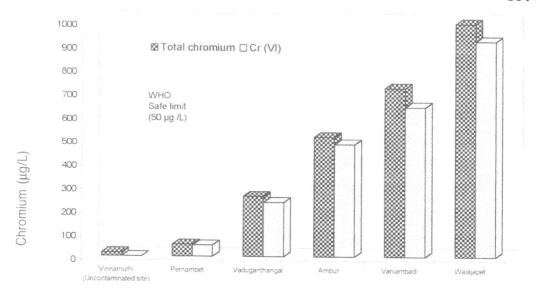

Figure 3 Chromium contamination in groundwater around selected tanneries.

The Cr concentration in groundwaters was much higher than the normal average background value of 4–7 µg Cr per liter reported in different parts of India (11). A high Cr content in groundwater has been reported by several workers. Davids and Lieber (12) reported an average of 450 µg/L in groundwater, whereas Forstner and Wittman (13) reported a mean value of >10,000 µg/L of Cr in groundwater of Tokyo. They suggested 0.5 µg/L as the background content of Cr in surface water. Drinking-water standards have been set at 50 µg/L total Cr because of the toxic effect of Cr(VI) and the possibility of oxidation of Cr(III) to toxic Cr(VI) (14). The Cr content in groundwaters examined in our study also exceeded the maximum permissible limit of the Australian National Health and Medical Research Council (ANHMRC) (83).

1. Mobility of Cr

To examine the transport of Cr in soil and its impact on groundwater pollution, a nest of piezometers was installed at different locations around Vaniambadi (92). During first sampling after moderate rains, the piezowaters (150–210 cm depth) at the contaminated site had a concentration of only 67 µg Cr per liter, which increased during subsequent sampling but decreased slightly at third sampling. A similar trend was observed with Na and Cl^-.

The soils at different sampling sites are mostly sandy loam to sandy clay loam with saline/alkaline conditions. The presence of Cr in groundwater depends also on its solubility and its tendency to be adsorbed by soil or aquifers. Our results showed evidence that Cr, which is often regarded as immobile, can leach under these soil conditions. The Cr species most frequently found in groundwater studies elsewhere include chromate (CrO_4^{2-}) and cationic hydroxo complexes such as $Cr(OH)^{2+}$, depending on the pH and Eh (15). In the tannery industries, however, Cr(III) is the predominant species used during the tanning process. Given the nature of the soil environment, which consists of minerals and organic matter, many previous studies indicate rapid transformation of Cr(VI) to Cr(III). Therefore the presence of Cr(VI) in borehole waters (as determined by the colorimetric diphenylthiocarbazide indicator method) is surprising and raises questions about the potential mechanisms that can enhance transition of Cr(III) added to soils in tannery wastes to Cr(VI) in groundwaters. Since the water sampled in such locations had microbial activity, there is positive indication that the oxidation–reduction system might have favored the conversion. However, it remains to ascertain the roles of particular types of organism.

VI. REMEDIATION OF CHROMIUM-CONTAMINATED SOIL

Several technologies involving physical, chemical, or biological approaches are available for the remediation of Cr-contaminated soils due to waste disposal from tannery industries, as depicted in Fig. 4. A large variety of methods has been developed due to the diverse chemical properties of the contaminants, particularly heavy metals, which can also be used for the remediation of Cr-contaminated soil. The selection and adoption of these technologies depend on

Extent and nature of Cr pollution
Type of soil
Characteristics of the polluted site
Cost of operation
Availability of materials
Regulations which apply in the country

The remediation options vary from the minimum of reducing the toxic levels (bioavailability) of Cr to the maximum of either complete clean-up of the soil or its removal from the site.

A. Physical In-Situ Treatment

The major physical in-situ treatment technologies in practice are soil mixing, soil washing, and solidification.

1. Soil Mixing

The simplest technique for reducing the hazard associated with Cr in soils is mixing the contaminated soil with uncontaminated soil. This results in the dilution of Cr to levels below which exposure is not considered a risk. This can be achieved by importing clean soil and mixing it with Cr-contaminated soil, or redistributing clean materials already on site (16). Another dilution technique relies on deep plowing, during which the vertical mixing of surface contaminated soils with less contaminated subsoils reduces the surface contamination. However, in this method the total concentration of Cr loading of the soil will remain the same (17).

2. Soil Washing

Soil washing or extraction is applied widely for the remediation of heavy metal-contaminated soils in Europe (18), and this procedure is also applicable for

Remediation of Soils Contaminated with Chromium

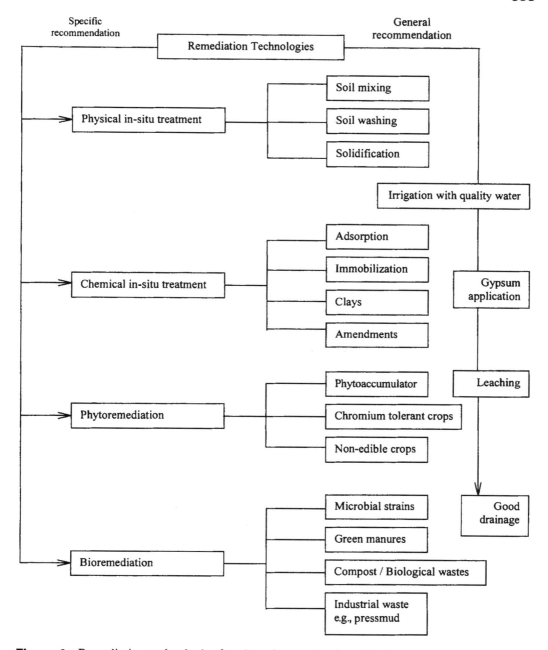

Figure 4 Remediation technologies for chromium-contaminated soil.

Cr-contaminated soils. It is based on the desorption or dissolution of Cr from the soil inorganic and organic matrix during washing with acids and chelating agents. Although soil washing is very suitable for off-site treatment of soil, it can also be used on-site using mobile equipment (19). However, the high cost of chelating agents and choice of extractants makes this technology not feasible.

3. Solidification

Solidification involves the stabilization of the contaminated soil into a solid mass using cement, fly ash, gypsum, asphalt, or vitrification, and the technique has been applied in field-scale remediation. Blast furnace slag-modified grouts were used for in-situ stabilization of Cr-contaminated soil (20). However, there is a major question of long-term resistance to breakdown of the solidified soil mass (21).

B. Chemical In-Situ Remediation

The bioavailable fraction of Cr poses toxicological or environmental hazards, as this chemical form will affect the life cycles of plant and other organisms. The Cr in contaminated soils can be adsorbed to ion-exchange sites and can be precipitated with other minerals or as insoluble chromite, or it may form complexes with organic matter, or be present in the soil solution phase as free Cr ion or coupled with organic or inorganic ligands.

1. Adsorption

Adsorption of Cr on soil mineral particles and organic matter is one of the main processes determining the mobility of Cr and bioavailability. Organic and inorganic amendments and chemicals, because of their unique characteristics, are known to immobilize a range of heavy metals. Chelate ion-exchange resin, cation-exchange resin, ferrous sulfate and silica gel, lime, gypsum, and naturally occurring clay minerals such as bentonite, kaolin, zeolite, and green sand have been found to immobilize heavy metals. A few chemicals, such as ferrous sulfate, are used to reduce Cr(VI) to Cr(III) and then precipitate Cr as chromium hydroxide (22). Commercial surfactants (surface-active agents) such as Dowfax 8390 were found to be effective in remediating subsurface Cr contamination (23). Chemical in-situ immobilization is achieved by complexing contaminants using chemical additives and thereby reducing the solution-phase concentration of Cr. If bioavailability is a key point for remediation technologies, then immobilization may be a preferred option (24).

The most common chemical remediation technique is by liming the soil to pH 7 or higher, thereby rendering the Cr less mobile and bioavailable (25). The anionic species of Cr(VI), viz., $HCrO_4^-$ and CrO_4^{2-}, are generally mobile in most neutral to alkaline soils. In acid soil, Cr(VI) species can be removed from solution by adsorption onto positively charged sorption sites (26,27). The species of Cr(VI) are also strong oxidants that are reduced to the trivalent state by aqueous Fe(II), organic compounds, and H_2S (10,28). Cr(VI) reduction by Fe(II) minerals and organic materials is most rapid under acidic conditions (10,28,29). Eary and Rai (30) reported that the rates of Cr(VI) reduction in subsoils increased with a decrease in pH below 4.0. Their data suggested that the reduction of Cr(VI) by Fe(II) species is of equivalent importance as reduction by organic material under acidic conditions.

Arnfalk et al. (31), after comparing several clay minerals and soils for their effectiveness in metal uptake, concluded that the relative order of metal uptake was the same for Cr(III), Cd, and Pb:

Na-type bentonite > smectite > Ca-type bentonite > illite > kaolinite

The most commonly recommended practice for minimizing Cr uptake by crops is raising soil pH by liming (25,32). The influence of pH on minimizing metal uptake can be attributed to the roles that ionic speciation in soil solution and surface charge characteristics of soil particles play in metal adsorption (33). The effectiveness of raising soil pH to minimize Cr uptake by crops has been variable.

A major problem associated with chemical immobilization techniques is that although Cr becomes less bioavailable, the concentration of Cr in soil remains unchanged and the immobilized Cr may become plant available with time through natural weathering processes.

2. Clay and Minerals

Clays are naturally occurring, nontoxic materials with large specific surface areas and a significant electric charge. They also may be modified, particularly in order to render them hydrophobic and organophilic. Attempts have been made to use both natural and modified clays for the treatment of waste waters. Churchman et al. (34) reported that many Australian clays have an appreciable sorption capacity for heavy metals. A zeolite mineral synthesized from an Australian kaolin sorbed Cr strongly.

Wang (35) opined that among all the soil amendments, zeolite, an aluminosilicate mineral, was very promising for the remediation of heavy metals in China. The advantages of zeolite application are not only its high efficiency for retention of heavy metals in soils, but also its low cost and convenience in application.

C. Phytoremediation

Phytoremediation technology is an emerging technology for the remediation of contaminated soils. It involves the use of metal-accumulating plants for remediating contaminated soil. Growing certain crops in contaminated soils is the simplest form of remediation. Some plants have inherent capacity to absorb and hyperaccumulate heavy metals in their tissues (36,37). Such capacity can be harnessed to remove toxic heavy metals from contaminated soils in a process referred to as phytoremediation. Simply, phytoremediation is the use of plants to remove pollutants from soil. Several plants of terrestrial and aquatic ecosystems are known to bioconcentrate toxic heavy metals. For example, rice, wheat, barley, oats, corn, chick pea, pigeon pea, beetroot, radish, carrot, tomato, spinach, peanut, soybean, tobacco, alpha-alpha, and oil seeds are reported to accumulate metals when grown in metal-contaminated agro-ecosystems (38–41). Based on data from published literature, Siedlecka (40) divided the metal accumulation in plant parts into three categories:

1. Accumulate more in roots/rhizomes than in shoots (stems/leaves): Cd, Co, Cu, Fe, and Mo in beetroot, carrot, radish, jerusalem artichoke, potato
2. Accumulate more in shoots (stems/leaves) compared to roots/rhizomes: Ag, Cr, Pb, Sn, and V in cabbage, cauliflower, tomato, rice, barley, oats, wheat, corn, pigeon pea, chick pea, soybean, peanut, broccoli, lettuce, spinach, amaranthus, etc.
3. More or less uniform distribution in roots/shoots of the plant: Mn, Ni and Zn in bush bean, broad bean, mung bean and cucumbers, etc.

Wetland plants may also be used in phytoremediation. Both natural and artificial wetlands have been shown to substantially reduce the levels of biological oxygen demand (BOD), nutrients, metals, and organic pollutants in water passing through them. There is evidence that wetland plants can hyperaccumulate trace elements in their tissues when grown in polluted waters.

Table 6 Hyperaccumulators of Heavy Metals

Heavy metal	Hyperaccumulating plants	Maximum metal concentration in leaves (mg/kg)	Reference
Zn	*Thlaspi calaminare*	39,600	Reeves (89)
Cu	*Aeolanthus biformifollus*	13,700	Brooks (90)
	Lemna minor L.	3,360	Zayed et al. (91)
		15,000	Bassi and Sharma (92)
		5,230	Jain et al. (42)
	Vigna radiata	9,415	Macnair (93)
Ni	*Phyllanthus serpentinus*	38,100	Zaranyika and Ndapwadza (98)
	Lemna minor L.	1,790	Zayed et al. (91)
Co	*Haumaniastrum robertil*	10,200	Brooks (95)
Se	*Astragalus* sp.	>10,000	Rosenfield (96)
Mn	*Alyxia rubricaulis*	11,500	Brooks (97)
Cd	*Thlaspi caerulescens*	1,020	Brown et al. (98)
	Cardaminopsis halleri	60,000	McGrath et al. (99)
Se	*Lemna minor* L.	13,300	Zayed et al. (91)
	Lemna minor L.	4,270	Zayed et al. (91)
Cr	*Lemna minor* L.	2,870	Zayed et al. (91)
	Ceratophyllum demersum L.	870	Garg and Chandra (100)
	Hydrilla sp.	590	Rai et al. (101)
	Cyperus esculentus	730	Lee et al. (99)
Hg	*Lemna minor* L.	25,800	Zayed et al. (97)

Floating wetland plants [e.g., duckweed (*Lemna minor* L.) and water velvet (*Azolla pinnata* R. Br)] have been found to bioconcentrate heavy metals such as Fe and Cu up to 78 times their concentration in the wastewater (42). Pinto et al. (43) and Singaram (44) demonstrated that water hyacinth (*Eichhornia crassipes*) can remove silver and Cr, respectively, from industrial wastewater.

Metal accumulation by vascular plants can range from a slight elevation relative to background concentrations to a degree where the accumulated metal constitutes a significant portion of the plant dry matter. Some plant species endemic to metalliferous soils have been demonstrated to accumulate exceptionally high metal concentrations. Brooks et al. (45) coined the term *hyperaccumulator* for serpentine plants capable of concentrating Ni to more than 1000 µg/g (0.1%) in their leaf dry matter. Critical lower limits for the hyperaccumulation of Cr are yet to be established.

Certain plants tolerate about 10–100 times higher metal concentrations in their shoots than agronomic species. For example, the definition of a Ni hyperaccumulator is the occurrence of >1000 mg/kg of Ni in dry leaves collected at the native habitat of the plant, while normal plants start to suffer Ni phytotoxicity at 50–100 mg/kg (46). Some of the hyperaccumulators are given in Table 6. Three characteristics have been found in these species: (a) hypertolerance of soil metals and shoot metals, (b) extreme uptake of the metal from soils, and (c) hypertranslocation of metal from roots to shoots.

Although most metal-tolerant plant species exclude metals from roots to shoots, hyperaccumulator plants apparently evolved this ability to limit predation by chewing insects and to gain resistance to plant diseases caused by microbial pathogens (47). The maximum limit of accumulation of heavy metals that must be achieved by a plant species to be classified as a hyperaccumulator and hence a good phytoremediator of that metal is not yet well defined. The following concentrations in the dry matter of the aboveground tissues have been suggested by Reeves et al. (48) as threshold to define hyperaccumulation in plants:

Zn and Mn: 10,000 mg/kg (1.0%)
Ni, Co, Cu, Cr, and Pb: 1000 mg/kg (0.1%)
Cd and Se: 100 mg/kg (0.01%)

However, the use of metal concentration in dry plant tissues as a criterion for identifying plant species that can be good phytoremediators does not take into account the metal concentration in the roots. Unfortunately, little information is available on the maximum uptake and accumulation of Cr in crops.

The existence of these phytoaccumulators was known during the 1970s (49). These plants are considered to be biogeoprospecting plants, or botanical prospecting plants, since the plants occur on soils containing high concentrations of the metal of concern (45). Most hyperaccumulators have been found to be endemic metallophytes or plants which occur only on soils with very high metal concentrations (46).

According to Chaney and Oliver (46), no plant has been found to hyperaccumulate Cr under conditions which would produce meaningful biomass yields. Many of the metal-tolerant species occur in low-fertile soils and have slow growth. It is necessary to breed improved hyperaccumulator crops specifically for Cr. Cultivars or species which do not accumulate high concentrations of Cr should be identified. This could also be possible by breeding low-Cr-accumulating cultivars or species.

The uptake of Cr occurs in either chromate or chromic form by the plant. Chromate uptake is an active process and follows Michelis-Menton kinetics via a sulfate carrier and is mostly accumulated by the shoot. In contrast, the uptake of chromic ion is a passive process occurring mostly in roots (50). The chemical form of Cr inside the plant varies among plant species. The predominant form of soluble Cr in serpentine plant extracts was found as trioxalato-chromate(III) ion, which was not present in the xylem sap. Hence, the transport of Cr as chromate in xylem sap might be analogous to that of SO_4^{2-} and PO_4^{3-}, principally as inorganic anions. The uptake of Cr by plants is influenced by the soil pH, organic matter and cation exchange capacity (CEC).

Crop species differ in their ability to accumulate Cr in either shoot or root. Beetroot and maize plants were found to be good accumulators of Cr (51). The Cr content observed in some crops is presented in Table 7. According to Keller et al. (52), the most efficient plants in taking up heavy metals were spinach, salad, potato and beet shoots, which accumulated more heavy metals in shoots than in storage or fruit parts (53).

It has been recognized for more than two decades that plant uptake could be exploited as a biological clean-up technique for various polluted rooting media including soils, composted materials, effluents, and drainage waters.

Metal tolerance is the ability of plants to survive in concentrations of metals that are toxic to nontolerant individuals or species. Accumulation is the concentration of metals in the shoot as a net result of root uptake and translocation. These two characters are of importance when considering the potential to use plants in remediating metal-contaminated soil. Different technologies have been proposed (54).

Table 7 Chromium Content in Some Crops

Crops	Cr content (mg/kg)	References
Serpentine plants	20,000	Lyon et al. (58)
Shrubs and trees	0.2–0.6	Hanna and Grant (100)
Orchard leaves	2.6	Bowen (101)
Pine needles	2.6	Peterson and Girling (50)
Tomato	4.5	Peterson and Girling (50)
Maize	4.6	Renan et al. (102)
Spinach	10	Peterson and Girling (50) Misra and Jaiswal (103)
Fruits	20*	Mertz (104)
Cereals	40*	Peterson and Girling (50)
Peas, beans, sweet corn	10–40*	Cary et al. (105)

*PPb.

1. *Phytostabilization.* Plants can be planted to revegetate the land, stabilize the soil, and reduce bioavailability of metals. Plants need to be tolerant of metals in this case, but accumulation may be disadvantageous.
2. *Rhizofiltration.* The roots of plants can be used to adsorb or absorb metals from a contaminated solution and the metals removed by harvesting the whole plant. In this case tolerance and translocation are largely irrelevant.
3. *Phytoextraction.* Plants can be grown on contaminated soil and the aerial parts (and the metals they contain) harvested. In this case plants need to be tolerant only if the soil metal content is very high, but they need to accumulate very high concentrations (more than 1000 mg/kg) in their aerial parts, and in practice need to be hyperaccumulators.

1. Flower Crops

To examine the level of Cr accumulation in flower plants, viz., *Jasminum sambac* (Gundumalli), *Jasminum grandiflorum* (Jathimalli), *Polianthus tuberosa* (Tuberose), and *Nerium oleander* (Nerium), Anandhkumar (89) conducted a field experiment on a sandy loam soil (Typic Ustropept). The results have shown that a considerable amount of Cr was accumulated in flower crops due to irrigation with tannery effluent. The Cr content in plants varied from 0.74 to 4.83 mg/kg in flowers, from 1.69 to 7.85 mg/kg in leaves, and from 2.83 to 14.02 mg/kg in roots. Invariably, in all crops, roots accumulated higher levels of Cr than leaves and flowers (Fig. 5). Irrigation with 100% effluent resulted in the highest accumulation of Cr (14.0 mg/kg) in roots of *P. tuberosa*, followed by *J. grandiflorum* (7.89 mg/kg), *J. sambac* (7.82 mg/kg) and *N. oleander* (7.32 mg/kg). Leaves of Jathimalli appeared to have accumulated significantly higher Cr than leaves of other crops. A concentration of Cr up to 14.02 mg/kg did not exhibit any toxicity symptoms in flower crops.

Large amounts of Cr were found to be accumulated in different parts of crops. Roots of all crop accumulated higher levels of Cr, followed by leaves and flowers. The translocation of Cr from root to flower was found to be low in flower crops. This suggests that Cr was relatively less mobile, due mostly to its being in chromic form.

Remediation of Soils Contaminated with Chromium

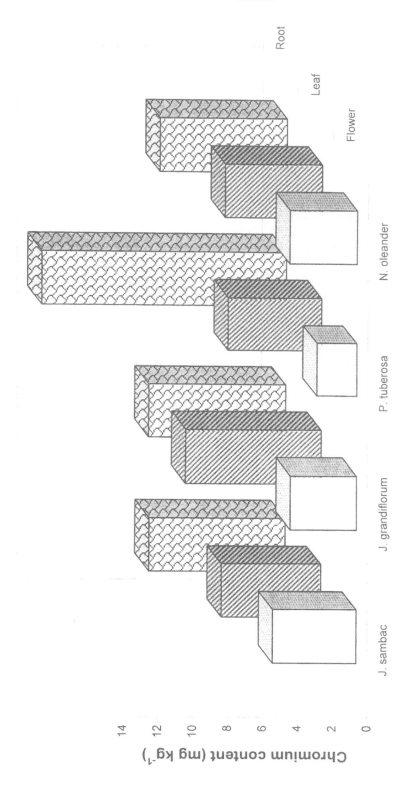

Figure 5 Chromium content in selected flower crops due to tannery effluent irrigation.

In another glasshouse experiment with three flower crops, Ramasamy (59) observed that *Jasminum auriculatum* was relatively tolerant up to 1000 µg/g Cr in soil than *Crossandra infundibuliformis* and *Jasminum sambac*, which were found to be very sensitive at this concentration.

2. Plantation Trees

Plantation trees offer another avenue for the remediation of Cr-contaminated soils. Similar to flower crops, certain tree species could withstand high concentrations of Cr in soil, besides showing tolerance to salts. Observation from a long-term field experiment showed that *Eucalyptus* spp. were found to be highly sensitive to tannery effluent irrigation, whereas *Acacia* and *Casuarina* were found to withstand effluent irrigation moderately. This was clearly seen from the reduced plant growth with effluent irrigation. The stem girth was reduced with increasing effluent concentration. Effluent with high concentrations of salts (mostly sodium chloride) and Cr affected plant growth. In general, the crop tolerance toward Cr-rich effluent irrigation was in the order: *Casuarina*>*Acacia*>*Eucalyptus* (95).

3. Vegetable Crops

Large accumulation of Cr in roots of beans, peas, tomato, maize and barley were reported by Ramachandran et al. (55). In aerial parts of the plants a higher uptake of Cr was observed when Cr(VI) was added to soil (56). According to Shewry and Peterson (57), barley roots accumulated more Cr and did not show any preferential areas of location of radioactive Cr (^{51}Cr) when present mostly in soluble nonparticulate forms in roots. Most of the Cr present in root was in the form of trioxalatechromate(III) ions, which was also present in stem and leaf (58). Similarly, Hernandez et al. (51) reported higher accumulation of Cr in the seedlings of cauliflower and beetroot than in barley and beans. There is a risk for Cr to accumulate in the edible parts of plants and enter into the food chain. Therefore, it is preferable to cultivate nonedible crops in Cr-contaminated soils.

Glasshouse studies were conducted to examine the Cr accumulation in selected vegetable crops such as Bhendi (*Abelmoschos esculentus*), Brinjal (*Solanum melongena*), and chillies (*Capsicum annum*). The results revealed that there was no deleterious effect of sludge-Cr on yield of vegetables up to a concentration of 750 mg/kg, above which a sharp decline was observed in all Bhendi and Brinjal (96). Increase in Cr levels in soil increased the Cr uptake by the crops (40–130 mg/kg), and the critical concentrations of Cr in plants was found to be 30–40 mg/kg. The Cr accumulated more in the shoots than in fruits. The water-extractable Cr in soil was found to correlate with the Cr translocation to the fruits.

4. Field Crops

A number of crops and varieties were evaluated at Tamil Nadu Agricultural University for their tolerance to tannery effluent-polluted soil. The results showed that the following crops and varieties were tolerant to some extent and recorded reasonable yields in contaminated soils (59):

Rice (*Oryza sativa*), varieties TRY-1, CO43, Paiyur-1, ASD-16
Maize (*Zea mays*), variety CO1
Finger millet (*Eleusine coracana*), varieties CO12, CO13

Remediation of Soils Contaminated with Chromium

Fodder grass, variety BN-2
Desmanthus (*Desmanthus vergatus*)
Sugarcane (*Saccharam officinalis*) varieties COG-94076, COG-88132, and COC-771
Korai grass (*Cyperus corymbosus*)

D. Bioremediation

Bioremediation of contaminated soils is a widely accepted technology in which native or introduced microorganisms and/or biological materials such as compost, animal manures, and plant residues are used to detoxify or transform Cr. Though it has several limitations, this technology holds continuing interest because of its potential for cost-effectiveness. The unique aspect in the bioremediation is that it is a natural approach and does not require the addition of chemical amendments other than microbial cultures and biological wastes. The different mechanisms involved in bioremediation are depicted in Fig. 6.

1. Use of Microorganisms

As Cr undergoes biological transformation in soil, the use of appropriate microorganisms offers bright scope for the remediation of Cr-contaminated soils. Existing and developing

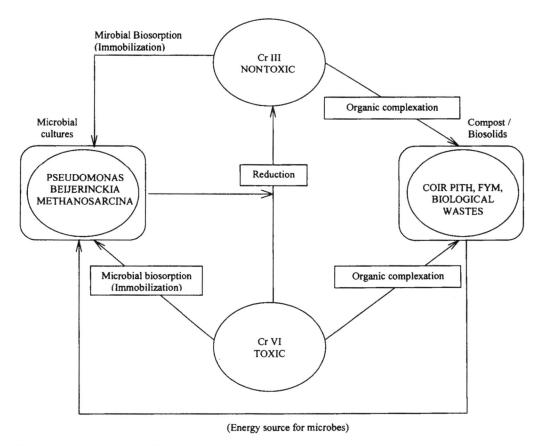

Figure 6 Mechanism of bioremediation using microbial strains and biological wastes/compost.

in-situ bioremediation technologies may be grouped into the following two broad categories (60):

1. *Intrinsic bioremediation*—where the essential materials required to sustain microbial activity exist in sufficient concentrations and amounts that naturally occurring microbial communities are able to degrade the target contaminants without the need for human intervention. This technique is the more likely choice for remediation of soils with low levels of contaminants over an extensive area.
2. *Engineered bioremediation*—which relies on various approaches to accelerate in-situ microbial degradation rates. This is accomplished by optimizing the environmental conditions by adding nutrients and/or electron acceptors, thus promoting the proliferation and activity of existing microbial consortia. It is favored for highly contaminated soils.

Microbial reduction of Cr(VI) is attractive for several reasons. Microbes reduce Cr under either aerobic or anaerobic conditions. The reason that some microbes have developed a capacity for Cr(VI) reduction has not yet been adequately explained. It has been suggested that:

The reduction may be a mechanism for chromate resistance.
Cr(VI) reduction may just be a fortuitous reaction carried out by enzymes that have other physiological substrates.
Cr(VI) reduction may provide energy for a few microbes.

Chromium reductase is widespread in microbes, but it is present primarily in the soluble fractions of the microbial cell. It reduces the Cr(VI) to Cr(III) with the oxidation of 3 moles of NADH per mole of Cr reduced. It has been reported that NADH can nonenzymatically reduce Cr(VI) to Cr(V) in the absence of enzyme. A membrane protein also appears to be important for Cr(VI) reduction by bacteria. Direct and indirect microbial-mediated bioreduction of hexavalent chromium has been observed. Anaerobic bacterial strains with accelerated Cr(VI)-reducing capabilities have been isolated from chromate-contaminated water and sludge. It is evident that anaerobic bacteria can effect preferential bioreduction of chromates under reducing conditions.

Toxic and mutagenic effects of chromium on microbes are well documented. Soil bacteria were inhibited with 10–12 mg/L of Cr(VI) (30,89). Chrome-electroplating waste was toxic to saprophytic and nitrifying bacteria. Nature is also kind to bacteria that are supposed to live in various environments where heavy metals occur naturally at high concentrations. Microbes in such environments evolve with resistance mechanisms including transformation of toxic species through processes such as

Methylation, demethylation, oxidation, reduction, hydrogenation, hydroxylation, dehydroxylation
Changes in uptake or transport of metals
Anaerobic growth with acetate, ethanol, malate, succinate, glycerol favor reduction of Cr

There is a need to search for innovative methods for minimizing the pollution. What is needed is a practical and early-implementable methodology to curtail the much-feared chromium, the foul-smelling sulfides, and the level of common salt in tannery effluents. Many of our tanning industries are small-scale industries and hence constructing wastewater treatment plants is not economical for them. Treating the wastewaters chemically is a costly affair for them. It is necessary at this stage to work out an economical treatment process which will alleviate the problem effectively. One alternative could be biological wastewater treatment processes. It is known that all biological processes work more quickly in tropical conditions. Employing microorganisms is possible to degrade the pollutants, and their fate in the natural environment could be worked out.

Alexander (61) lists the following criteria in judging in-situ bioremediation as a practical alternative (a) existence of microbial consortia possessing the requisite catabolic activity; (b) ability to sustain adequate microbial transformation rates for decreasing metal concentrations to regulated levels within an acceptable period of time; (c) absence of metabolites that are either toxic to the microbial communities, or absence of by-products that are themselves of environmental concern; (d) target contaminants are bioavailable; (e) presence of environmental conditions that are conducive to microbial activity, including adequate concentrations of nutrients, electron acceptors, and energy sources; and (f) cost-effective when compared to other technologies.

No actual degradation of Cr occurs during bioremediation, but the oxidation state Cr(VI) or Cr(III) is changed. This can lead to an increase in water solubility and subsequently removal by leaching or precipitation and reduction in bioavailability.

Cifuentes et al. (62) studied the microbial sorption and reduction of Cr in soil with implications to bioremediation. They reported that *Aspergillus niger* sorbed more Cr(VI) than Cr(III). Binding of Cr(III) by *A. niger* depended on pH and was greatest at pH 9, but binding of Cr(VI) was not pH-dependent. Soil amendment with freeze-dried *A. niger* increased sorption of Cr(III) significantly. Organic amendment of Cr(VI)-contaminated soil resulted in both direct sorption of Cr(VI) and reduction to Cr(III). Reduction was primarily by indigenous soil microorganisms, and inoculation with Cr-tolerant *Pseudomonas maltiphilia* was of little benefit. Yeast extract and Bermuda grass were more effective than cow manure in removing Cr. Stimulating the soil microflora with an energy source should result in reduction of Cr(VI), thus immobilizing Cr and preventing leaching or plant uptake. Microbial amendment becomes necessary only if the indigenous microflora are killed by high levels of Cr contamination. According to these authors, the application of an organic amendment to reduce and bind Cr appears feasible as a technique to immobilize Cr(VI) in contaminated soils, and the inoculation of Cr-tolerant microorganisms does not appear feasible or necessary to immobilize Cr(VI).

Through a laboratory solution culture experiment, Ramasamy (59) studied the microbial reduction of Cr and observed that Cr reduction was 36.8% in *Pseudomonas* and 68.5% in *Beijerinckia* during exponential growth phase. The toxic Cr(VI) was completely transformed to Cr(III) after 72 h. Further, the results showed that the sorption of Cr(VI) by *Pseudomonas* sp. was only 23–27%, whereas by *Beijerinckia* sp it was 18.7–23.7%. Bopp and Ehrlich (63) and Horitsu et al. (64) also reported the reduction of Cr(VI) by *Pseudomonas* isolates while growing on a variety of substrates. Since psuedomonads constitute a substantial portion of bacteria in soil, Cr(VI)-contaminated soils may enrich for strains that reduce Cr(VI) as a detoxification mechanism.

Sudha Nair and Krishnamurthy (65) reported that *Pseudomonas aeruginosa* not only tolerated but also accumulated Cr. Their result showed that this organism sorbed roughly 10 times more Cr(III) than Cr(VI), and the sorption was linked to electrostatic attractions between oppositely charged Cr complexes and the side-chain sites.

One important factor to remember in bioremediation is that microbial species do not exist in isolation but in interactive communities. Therefore the metabolites produced by one species will often serve as growth and energy substrates for another (61).

The sorption of Cr and the rate of desorption are rate-limiting factors for bioavailability of Cr in soil. It is possible that some microorganisms are better than others at accessing adsorbed substrate—for example, those microbes with extracellular enzymes that penetrate clay lattices to bind with and release products for microbial metabolism. Other microbes, by virtue of their metabolic activities, will release protons, change the microenvironment pH, and stimulate the desorption of a previously adsorbed potential substrate.

It has been suggested that the introduction of genetically modified microorganisms to soil will revolutionize bioremediation due to the expression of novel metabolic pathways. Such manipulations may also help in the development of microorganisms that are capable of tolerating compounds that would otherwise be toxic.

As has already been mentioned, Cr toxicity and mobility depend on its oxidation state. Cr(III) is relatively immobile, more stable, and much less toxic than Cr(VI). Abiotic oxidation of Cr(III) to the more mobile and toxic Cr(VI) by Mn oxides has been demonstrated (5,67), but it appears to be fairly specific, and few cases of Cr(III) oxidation are reported in the literature (68). Oxidation of Cr(III) to Cr(VI) by dissolved oxygen is also reported in an aquatic environment (6).

Because of the adverse effects of Cr(VI) in the environment, its removal or transformation to immobile and less toxic Cr(III) is necessary. This can be achieved through microbial remediation. Some of the potential forms of on-site remediation of Cr(VI)-contaminated soils may include phytoremediation with Cr accumulators, gaseous bioreduction (68), organic matter simulation of Cr(VI) reduction by soil microorganisms (10), inoculation with Cr(VI)-tolerant bacteria for reduction of Cr(VI) to Cr(III) (64,69). Reduction and immobilization of Cr(VI) in soils may negate expensive remedial measures at certain sites (70).

Although reduction of Cr(VI) to Cr(III) would not remove Cr from soil, it would limit the mobility and toxicity of Cr in contaminated soils greatly.

A number of bacterial strains have demonstrated high Cr(VI) tolerance, and some exhibit the ability to facilitate reduction to Cr(III) (63,64). Some of these strains carry out the reduction of Cr(VI) only under anaerobic conditions, while others are capable of reducing Cr(VI) in both anaerobic and aerobic environments. *Enterobacter cloacae* strain HO1, which reduces Cr(VI) anaerobically, has shown promise in detoxifying Cr-contaminated water (71,72). This could be a practical approach for bioremediation, whereas chemical or electrochemical methods, which require high energy or extensive reagent application, may not be appropriate. The studies of Losi and Frankenberger (73) showed that at the higher concentrations (500 and 1000 mg/L), the isolates obtained from the contaminated sediments were more tolerant to Cr(VI) in their environment. Chromate tolerance mechanisms include methylation, reduction, and precipitation at the cell surface (74), blocking cellular uptake by altering the uptake pathway, and removal from cytoplasm by efflux pumps (75).

Reduction of Cr(VI) to less toxic Cr(III) can be considered as an additional chromosome-determined mechanism of resistance to CrO_4^{2-}. Chromate reduction by bacteria occurs under aerobic or anaerobic conditions and has been associated with soluble or membrane-bound enzyme activities. This bacterial chromate reduction has the potential to become a useful bioremediation system. Campos et al. (76) found that cell suspensions of *Bacillus* strain QC1-2, which is a soluble NADH-dependent enzyme, rapidly reduced

Cr(VI), in both aerobic and anaerobic conditions, to Cr(III). The Cr(VI) reduction was dependent on the addition of glucose, but sulfate, an inhibitor of chromate transport, had no effect.

2. Compost and Other Organic Amendments

The application of farm yard manure (FYM), compost, or biological waste materials is a common practice and traditionally followed in Indian agriculture. This technology can also be exploited for the remediation of Cr-contaminated soils. The in-situ immobilization of Cr using these composts or biological waste materials to remediate contaminated soil is a viable and cheap option. Large quantities of compost or organic matter can be added to contaminated soil with the aim of immobilizing Cr as stable complexes with organic colloids (77). During the decomposition of organic matter, compounds such as citric acid or gallic acid are formed which have the potential for chelating Cr(III) or reducing Cr(VI), and thereby reducing the toxicity of Cr (78,79).

Application of organic amendments such as cow dung, Bermuda grass, and yeast extract have been found to be effective in the reduction and immobilization of Cr(VI)-contaminated soils (62,73). According to Cifuentes et al. (62), yeast extract-amended soil removed more Cr(VI) than grass and manure amendments. The addition of an easily degradable substrate with a low C:N ratio stimulated more Cr(VI) reducers compared with the less-easy-to-degrade manure. Their study also suggested that although manure encourages reduction of Cr(VI), more readily decomposable substrates are better for remediation.

In a study of chromate reduction by various soils Eary and Rai (30) concluded that Cr(VI) reduction by organic matter and Fe(II) were roughly equivalent in terms of total reduction capacity. Soil humic substances, which constitute the majority of the organic fraction in most soils, thus represent a significant reservoir of electron donors for Cr(VI) reduction. Humic substances are complex mixtures of biogenic molecules consisting of aromatic components extensively substituted with oxygen-containing functional groups (80). Humic acids (HA) are soluble in basic solutions and have higher molecular weights and carbon contents than fulvic acids (FA). Reduction reactions have been observed between several inorganic ions and humic acid and therefore it will serve as a reductant of Cr(VI). Wittbrodt and Palmer (81) reported that soil humic acid can reduce Cr(VI) in aqueous solution in the pH range 2–7, with the Cr(VI) reduction rate increasing with decreasing pH, decreasing concentration of Cr(VI), and increasing amount of soil humic acid.

Application of composts rich in the chemical factors can inactivate toxic Cr(VI) and reduce its potential for phytotoxicity and mammalian toxicity and therefore could offer great potential for solving soil remediation needs in many countries. Compost would also correct the severe infertility of metal contaminated sites, correct phytotoxicity caused by co-contaminants, and provide improved soil physical properties and organic-N which would facilitate development of a remediated ecosystem at a contaminated site (46).

Singaram et al. (82) evaluated the potential of different amendments for remediating the Cr-contaminated soil at different locations in Tamil Nadu. Their field experiment with maize showed that the application of composted coir pith at a rate of 10 t/ha performed better than other amendments (Table 8) and attributed that coir pith kept the soil moist,

Table 8 Effect of Soil Amendments on Maize Yield in Cr-Contaminated Soil

	Yield (kg/ha)					
	Vaniambadi[a]		Pernampet[a]		Madanapally[a]	
Amendments	Grain	Straw	Grain	Straw	Grain	Straw
1. Composted coir pith	2615	5825	3580	7190	3308	6275
2. Pressmud	2477	5400	3411	7173	3023	6400
3. FYM	2431	5350	3359	7100	3053	6575
4. Gypsum	2349	5375	3264	6863	2778	6363
5. Control	2125	5025	3065	6175	2585	5975
CD (P + 0.05)	112	317	92	361	119	NS

[a] Location.

prevented the capillary rise of salts from subsurface layers, and aided greater leaching of salts from the root zone.

VII. TECHNOLOGIES TO MINIMIZE CHROMIUM CONTAMINATION IN SOIL

A. Biological Treatment of Effluent

1. Anaerobic Treatment

a. Anaerobic Digestion. Anaerobic digestion of vegetable tanning liquor indicated a BOD reduction of 95%, but this was reported at a low loading of 0.8 kg/m^3 of volatile solids. Here the retention period was also high. Our group has isolated and characterized an anaerobic bacterium, *Bacteroides* sp., capable of degrading tannin from the combined effluent (59). We have also characterized the intermediary compounds formed by this bacterium and developed a treatment strategy to remove residual tannin from composite effluent from tanneries. It was observed that the reduction of Cr by anaerobes and specific sorption by aerobes were a possibility for treatment through bioremediation.

b. Anaerobic Lagoon. An anaerobic lagoon employs the principles of anaerobic digestion with no gas collection. The retention period will be about 15–20 days. These lagoons can be operated at high loading, varying from 100 kg/ha/day to 250 kg/ha/day. BOD reduction of tannery waste is 70–80%. Once the lagoon stabilizes and methane fermentation sets in, it becomes self-sustaining. Studies have indicated that the BOD of the settled tannery waste came down from 1500 mg/L to 190 mg/L. This process involves low capital and running cost (59).

2. Aerobic Treatment

a. Oxidation Pond. The depth of pond varies from 1.0 to 1.5 m and BOD loading is 250 kg/ha/day while the retention period is 40 days. The BOD of the settled and diluted waste mixed with sewage in the ratio of 3 : 1 came down to 218 mg/L from 802 mg/L (59).

b. Activated Sludge Process. Studies have indicated that at a mixed liquor suspended solid (MLSS) concentration of 3000–4000 mg/L, the BOD of the pretreated effluent came down to 34 mg/L from 988 mg/L within 24 h of aeration. Generally, a BOD removal of 85–95% can be expected from this system. We have isolated and characterized several floc-forming bacteria associated with the sewage sludge digestion. Same cultures adopted to agro-industrial wastes have performed well in association with other heterotrophic bacteria to reduce the protein concentration and metals. Most of the metal removal is by sorption and sedimentation (59).

c. Trickling Filtration. The BOD of pretreated vegetable tannery effluent came down from 900 mg/L to 158 mg/L. Settling the filtered effluent brought the BOD down to 56 mg/L. In the case of chrome tanning waste, the BOD was brought down to 88 mg/L from 129 mg/L and further reduction in BOD was obtained at a value of 56 mg/L, after setting the effluent from trickling filtration (59).

d. Aerated Lagoon. A BOD removal of 70–90% can be achieved with this system.

e. Oxidation Ditch. An oxidation ditch is an extended aeration system with low organic loading and high mixed-liquor suspended solids concentration of the order of 400 mg/L. A BOD removal of 90–95% can be expected from this system (59).

B. Reed Bed Technology

Macrophytes grown in reed beds are known to accumulate large amounts of salt and organic pollutants and therefore could be used to reduce the pollutant load in the effluent. We examined the effectiveness of some macrophytes for the treatment of tannery effluent. The macrophyte plants [*Canna indica*, *Colacacia* sp., *Cyperus* sp., *Typha* sp., and Cumbu Napier hybrid grass (CO 3)] were collected from the Bhavani River basin and transplanted into reed beds of size 6 m × 1 m × 0.5 m. The plants were irrigated with good-quality water until their establishment, for about 2 months. To adopt these plants for tannery effluent irrigation, the plants were irrigated with 50% water and 50% treated effluent mixture at a rate of 60 L per bed. It was found that only the *Typha* sp. and Cumbu Napier hybrid (CO 3) could withstand the tannery effluent. The *Canna indica*, *Colacacia* sp., and *Cyperus* sp. failed to establish with tannery effluent irrigation. With the standing crop of *Typha* sp. and Cumbu Napier hybrid, the treated effluent (100%) was allowed to pass through the reed beds at a rate of 60 L per bed. Cumbu Napier hybrid (CO 3) and *Typha* sp. were found to withstand high concentration of effluent irrigation. The results suggest that these two macrophyte plants could be used for treating the tannery effluent with a reed bed system. The root system absorbs and accumulates the Cr 2 in both epidermal and cortex layers (59).

Laboratory experiments were also conducted to quantify the salts removal reduction of TDS with floating macrophytes, namely *Eichornia* sp., *Pistia* sp., *Salvenia* sp., and *Azolla* sp. The study revealed that these floating macrophytes removed between 40–65% of the total dissolved solids from the treated effluent (*Eichornia*, 48%; *Pistia*, 56%; *Salvenia*, 52%; and *Azolla*, 20%). Analysis of the plant parts indicated that these plants accumulated Na up to 1% of its dry weight in the shoots and 1.25% of its dry weight in the roots. The TDS removal was related directly to biomass production and inversely related to Cr concentrations (59).

C. Microbial Reduction of Chromium

Traditional techniques for remediating chromate-contaminated tannery effluent involve reduction of Cr(VI) to Cr(III) by chemical or electrochemical means at pH > 5, followed by precipitation and finally filtration or sedimentation. These processes are intensive, as suggested by the fact that present costs for disposal are very high. The discovery of microorganisms that preferentially reduce Cr(VI) has led to applications in the bioremediation field, which are potentially more cost-effective than traditional methods.

1. Biosorption/Reduction

Russian researchers first proposed the use of Cr(VI)-reducing bacterial isolates in the removal of chromates from industrial effluents. Since then, various reduction parameters have been evaluated for a diverse group of microorganisms which accelerated Cr(VI)-reducing capabilities with the prospect of developing commercially viable bioremediation techniques exploiting these organisms. Bioreactors are used which basically consist of a reduction phase with Cr(VI)-reducing bacteria immobilized on inert matrices within the reactor, followed by a settling or filtration phase to remove Cr(III) precipitates. Chromate-contaminated effluent is pumped into the reactor and supplemented with various carbon sources and nutrient additives; the Cr(VI) is then reduced, precipitated, and removed. A disadvantage here is that the lowest achievable effluent Cr concentrations are probable around 1 mg/L, which is considerably higher than the U.S. EPA drinking-water standard of 0.05 mg/L. Also, the reaction rate is slow (59).

Land application method for remediating Cr-containing effluent was investigated which consists of passing the effluent through an organic matter-enriched soil where reduction, precipitation, and immobilization would take place (59).

Another application of direct reduction was demonstrated using anaerobic Cr-reducing bacteria. The cultures were contained within dialysis tubes and submerged in contaminated water. Chromate diffusing into the tubes was reduced and precipitated and thus was unable to diffuse out. Laboratory studies using this system showed that 90% of the Cr was removed from the wastewater (59).

2. Gaseous Bioreduction

Gaseous bioreduction involves microbial reduction of Cr(VI) by a metabolic by-product, which is produced in an anaerobic environment by sulfate-reducing bacteria. By this method effluent Cr levels as low as 0.01 mg/L are attainable. An advantage of this system over biosorption/bioreduction is that the CrO_4^{2-} need not come into contact with the cells for reduction to occur. H_2S diffuses out into the medium, which can increase the reduction rate as well as protect the cells from the toxic effects of Cr(VI). This could promote faster reduction rates and may explain the lower effluent concentration achieved (59).

D. Composting of Tannery Sludge

The possibility of composting tannery sludge for its use as an organic manure has been examined. The role of earthworms and microorganisms on sludge decomposition under field conditions has also been studied. The sludge beds for the experiment were prepared to a size of $1 \times 1 \times 0.3$ m. Earthworms and a mixed consortia of microorganisms were introduced into the sludge compost beds and watered to maintain a moisture level of 50–60% in the beds for 120 days.

The worms *Eudrilus euginae* and *Lambito mourtii* were found suitable for vermicomposting of tannery sludge. Worms appear to accumulate Cr in their tissues. A 50–80% reduction in C/N ratio was achieved during 120 days of composting. Vermicomposting enriched the N, P, and K contents of sludge composts. A 40% reduction in Cr content was observed in tannery composts due to vermicomposting.

There was an increase in the number of bacteria and fungi, whereas the population of actinomycetes was reduced. *Pseudomonas*, *Bacillus*, *Penicillium*, *Aspergillus*, *Beijerinckia*, and *Streptomyces* were the predominant organisms observed in the sludge composts. The results of this study clearly indicated the possibility of composting tannery sludge using earthworms and microorganisms, which could play an important role in the mitigation of the Cr-rich sludge disposal problem.

E. Constructed Wetlands

The use of constructed wetlands using native water plants and floating macrophytes is catching the imagination and eye of the pollution abatement system. Some of the macrophytes accumulate heavy metals at higher rates and are suitable candidates for tertiary treatments. Since these provide an aerobic and microaerophilic environment, it is a positive system to translate so as to reduce the heavy-metal contamination and organic loading in wastewater.

VIII. CONCLUSIONS

In India, where there is population increase, land and water scarcity are major issues. Any change or degradation to these ecosystems will definitely hamper progress. Our fragmented knowledge is to be pooled and test-verified so as to formulate combined treatment initiating with anaerobic processing of effluent and aerobic polishing. Tertiary treatment with macrophytes should follow the previous treatment systems. Demonstration of this combined system is in progress with cooperation from the tanneries, planners, and researchers. While we attempt to treat wastewater from tanneries, sincere attempts are being made to reduce the release of chromium into the environment. However, these attempts can only reduce or control future damage to the ecosystem. What about the damaged ecosystems wherein chromium and salt concentrations are above the permissible limits? Remediative measures may be chemical, physical, or biological. Plant and microbial remediation is intensively investigated. Selective sorption and removal from contaminated sites through microbial products and macrophytes attempted by the scientific community will pave the way for sustained growth of tanneries in India. What is more encouraging is the reduction of toxic forms of chromium to nontoxic forms by anaerobes in underground water and consequent precipitation preventing village populations from the drastic toxicity of Cr(VI). Another intriguing finding is the presence of an oxidized form of chromium in underground soil. Is it due to microbial transformation or chemical speciation? A complete understanding of this process will help us to design better strategies to cope with the chromium pollution.

ACKNOWLEDGMENT

The authors acknowledge help from the CSIRO Division of Land and Water, Australia, Australian Centre for International Agricultural Research, and Tanners Association, Tamil Nadu, India.

REFERENCES

1. Ramasamy, K. 1997. Tannery effluent related pollution on land and water ecosystems in India. In: I. K. Iskandar et al. (eds.), *Proc. Extd. Abstrc. from the Fourth Int. Conf. on the Biogeochemistry of Trace Elements*, University of California, Berkeley, June 23–26, 1997, pp. 771–772.
2. Manivasakam, N. 1987. *Industrial Effluents: Origin, characteristics, effects, analysis, and treatment*. Sakthi Publications, Coimbatore, India.
3. Thomas, C. T. 1993. *Practical Leather Technology*. 4th Edition. Krieger Publishing Company, Florida.
4. Sandro, S. 1989. Chrome tanning. In: *Agricultural Use of Leather Working Residue*. Italian Tanners Association. La Concerla Srl, Via Brisa, Milano.
5. Fendorf, S. E., and Zasoski, R. J. 1992. Chromium (III) oxidation by δ-MnO_2. 1. Characterization. *Environ. Sci. Technol.*, 26:79–85.
6. Francoise, C. R., and Bourg, A. C. M. 1991. Aqueous geochemistry of chromium: A review. *Water Res.*, 25:807–816.
7. Chaney, R. L. 1990. Public health and sludge utilization. Part 2. *Biocycle*, 31:68–73.
8. Naidu, R., and McLaughlin, M. J. 1993. Heavy metal contamination of agricultural land and ground water in Tamil Nadu, India and in Australia. Report to ACIAR.
9. Richard, F. C., and Bourg, A. C. M. 1991. Aqueous geochemistry of chromium: A review. *Water Res.*, 25:807–816.
10. Bartlett, R. J., and Kimble, J. M. 1976. Behaviour of chromium in soils. 1. Trivalent forms. *J. Environ. Qual.*, 5:379–383.
11. Handa, B. K. 1978. Occurrence of heavy metals and cyanides in groundwater from shallow aquifers in Ludhiana. *IAWPC Tech. Annu.*, 5:109–115.
12. Davids, H. W., and Lieber, M. 1951. Underground water contamination by chromium wastes. *Water Sewage Works*, 98:525–530.
13. Forstner, U., and Wittmann, G. T. 1983. *Metal Pollution in the Aquatic Environment*. Springer-Verlag, Berlin.
14. U.S. Environmental Protection Agency. 1976. National interim primary drinking water regulations, EPA/570/9-76/003. U.S. EPA, Washington, DC.
15. Robertson, F. N. 1975. Hexavalent chromium in ground water in Paradise Valley, Arizona. *Ground Water*, 13:516–527.
16. Musgrove, S. 1991. An assessment of the efficiency of remedial treatment for metal polluted soil. In: M. C. R. Davies (ed.), *Land Reclamation: An End to Dereliction*, Int. Conf. on Land Reclamation, April 1991, University of Wales, pp. 248–255. Elsevier, Dordrecht.
17. Burns, R. G., Rogers, S., and McGhee, I. 1996. Remediation of inorganics and organics in industrial and urban contaminated soils. In: R. Naidu. (Ed.), *Proc. 1st Int. Conf. on Contaminants and the Soil Environment in the Australasia-Pacific Region*, Adelaide, February 18–23, 1996, pp. 411–449. Kluwer, London.
18. Tuin, B. J. W., and Tels, M. 1991. Continuous treatment of heavy metal contaminated clay soils by extraction in stirred tanks and counter current column. *Environ. Technol.*, 12:178–190.
19. Bromhead, C. J., and Beckwith, P. 1994. Environmental dredging of the Birmingham canals: Water quality and sediment treatment. *J. Inst. Water Eng. Manage*, 8:350–359.

20. Allan, M. L., and Kukacka, L. E. 1995. Blast furnace slag-modified grouts for in-situ stabilization of Cr contaminated soil. *Waste Manage.*, 15:193–202.
21. Peters, R. W., and Shem, L. 1992. Use of chelating agents for remediation of heavy metal contaminated soil. In: G. F. Vandegrift, D. T. Reed, and I. R. Tasker (eds.), *Environmental Remediation Removing Organic and Metal Ion Pollutants*, pp. 70–84. American Chemical Society, Washington, DC.
22. Czupyrna, G., Levy, R. D., MacLean, A. L., and Gol, H. 1989. In-situ immobilization of heavy metal contaminated soils. *Pollution Technology Review No. 173*, pp. 39–53. Noyes Data Corporation, New Jersey.
23. Thirumalai, N., Sabatine, D. A., Shiau, B. J., and Harwell, J. H. 1996. Surfactant enhanced remediation of subsurface chromium contamination. *Water Res.*, 30:511–520.
24. Mench, M. J., Didier, V. L., Loffer, M., Gomez, A., and Masson, P. 1994. A mimicked in-situ remediation study of contaminated soils. *J. Environ. Qual.*, 23:785–792.
25. Alloway, B. J. 1995. *Heavy Metals in Soils*. Blackie and Wiley, New York.
26. Bartlett, R. J., and James, B. 1988. Mobility and bioavailability of chromium in soils. In: J. O. Nraigu and E. Nieboer (eds.), *Chromium in the Natural and Human Environments*, pp. 267–303. Wiley, New York.
27. Zachara, J. M., Ainsworth, C. C., Lowan, C. E., and Resch, C. T. 1989. Adsorption of chromate by surface soil horizons. *Soil Sci. Soc. Am. J.*, 53:418–428.
28. Eary, L. E., and Rai, D. 1989. Chromate reduction by subsurface soils under acidic conditions. *Soil Sci. Soc. Am. J.*, 55:676–683.
29. Amacher, M. C., and Baker, D. E. 1982. Redox reactions involving chromium, plutonium and manganese in soils. *Report of the U.S. Dept. of Energy*, DOE/DP/04515-2, Work performed in the contract DE-AS08-77DPO4515, Pennsylvania State University, University Park, PA.
30. Eary, L. E., and Rai, D. 1991. Chromate reduction by subsurface soils under acidic conditions. *Soil Sci. Soc. Am. J.*, 55:676–683.
31. Arnfalk, P., Wasay, S. A., and Tokunaga, S. 1996. A comparative study of Cd, Cr(III), Cr(VI), Hg and Pb uptake by minerals and soil materials. *Water, Air, Soil Pollution*, 87:131–148.
32. Page, A. L., Bingham, F. T., and Chang, A. C. 1981. Cadmium. In: N. W. Lepp (ed.), *Effects of Heavy Metal Pollution on Plants*. Vol, 1: *Effects of Trace Metals on Plant Function*. Applied Science Publisher: pp. 77–103.
33. Naidu, R., Bolan, N. S., Kookana, R. S., and Tiller, K. G. 1994. Ionic strength and pH effects on the sorption of cadmium and the surface charge of soils. *Eur. J. Soil Sci.*, 45:419–429.
34. Churchman, G. J., Slade, P. G., Self, P. G., Keeling, J. L., Raven, M. D. and Kimber, R. W. L. 1996. Australian clays for the uptake of pollutants. *Extended Abstracts, 1st Int. Conf. on Contaminants and the Soil Environment*, Adelaide, February 18–26, 1996, pp. 319–320.
35. Wang, H. K. 1996. Heavy metal pollution in soils and its remedial measures and restoration in mainland China. *Extended Abstract, 1st Int. Conf. Contaminants and the Soil Environment*, Adelaide, February 18–23, 1996, pp. 313–314.
36. Baker, A. 1981. Accumulators and excluders—Strategies in the response of plants to heavy metals. *J. Plant Nutr.*, 3:643–654.
37. Rosenfeld, I., and Beath, O. A. 1964. *Selenium Geobotany, Biochemistry, Toxicity and Nutrition*. Academic Press, New York.
38. Rivai, I. F., Koyama, H., and Suzuki, S. 1990. Cadmium content in rice and its daily intake in various countries. *Bull. Environ. Contam. Toxicol.*, 44:910–916.
39. Rivai, I. F., Koyama, H., and Suzuki, S. 1990. Cadmium content in rice and rice field soils in China, Indonesia and Japan, with special reference to soil type and daily intake from rice. *Jpn. J. Health Human Ecol.*, 56:168–177.
40. Siedlecka, A. 1995. Some aspects of interactions between heavy metals and plant mineral nutrients. *Acta Soc. Bot. Pol.*, 64:265–272.
41. Wagner, G. J. 1993. Accumulation of cadmium in crop plants and its consequences to human health. *Adv. Agron.*, 51:173–212.

42. Jain, S. K., Vasudevan, P., and Jha, N. K. 1989. Removal of some heavy metals from polluted water by aquatic plants: Studies on duck weed and water velvet. *Biol. Wastes*, 28:115–126.
43. Pinto, C. L., Caconia, A., and Souza, M. 1987. Utilization of water hyacinth for removal and recovery of silver from industrial waste water. In: D. Athie and C. C. Cerri (eds.), The use of macrophytes in water pollution control. *Water Sci. Technol.*, 19(10):89–102.
44. Singaram, P., 1996. Mitigation of chromium pollution by water weeds. In: *Extended Abstract, 1st Int. Conf. Contaminants and the Soil Environment*, February 18–23, 1996, Adelaide, pp. 309–310.
45. Brooks, R. R., Lee, J., Reeves, R. D., and Jaffr, T. 1977. Detection of nickeliferous rocks by analysis of herbarium specimen of indicator plants. *J. Geochem. Explor.*, 7:49–57.
46. Chaney, R. L., and Oliver, D. P. 1996. Sources, potential adverse effects and remediation of agricultural soil contaminants. In: R. Naidu (Ed.), *Proc. 1st Int. Conf. on Contaminants and the Soil Environment in the Australasia-Pacific Region*, Adelaide, February 18–23, 1996, pp. 323–359. Kluwer, London.
47. Boyd, R. S., and Martens, S. N. 1994. Nickel hyperaccumulated by *Thlaspi montanum* var montanum is acutely toxic to an insect herbivore. *Oikos*, 70:21–25.
48. Reeves, R. D., Baker, A. J. M., Borhidi, A., and Berazain, R. 1996. Nickel-accumulating plants from the ancient serpentine soils of Cuba. *New Phytol.*, 133:217–224.
49. Baker, A. J. M., and Brooks, R. R. 1989. Terrestrial higher plants whch hyperaccumulate metallic elements—A review of their distribution, ecology and phytochemistry. *Biorecovery*, 1:81–126.
50. Peterson, P. J., and Girling, C. A. 1981. Other Trace Metals. In: N. W. Lepp (ed.), *Effect of Heavy Metal Pollution on Plants*, Vol. 1, pp. 13–279. Applied Science Publishers, London.
51. Hernandez, T., Moreno, J. J., and Costa, F. 1991. Influence of sewage sludge application on crop yields and heavy metal availability. *Soil Sci. Plant Nutr.*, 37:201–210.
52. Keller, C., et al., 1997. Extraction of heavy metals from contaminated agricultural soils by crop plants. In: I. K. Iskandar et al. (eds.), *Proc. Extd. Abstrc. from the Fourth Int. Conf. on the Biogeochemistry of Trace Elements*, University of California, Berkeley, June 23–26, 1997, pp. 473–474.
53. Sara Parwin Banu, K. S., and Ramasamy, K. 1998. Chromium chemistry and uptake by plants. *Madras Agric. J.*, 84:522–528.
54. Cunningham, S. D., Berti, W. R., and Huang, J. M. 1995. Phytoremediation of contaminated soils. *TIBTECH*, 13:393–397.
55. Ramachandran, V., D'Souza, T. J., and Mistry, K. B. 1980. Uptake and transport of chromium in plants. *J. Nucl. Agrl. Biol.*, 9:126–128.
56. Mishra, S., Shankar, K., Srivastava, M. M., Srivastava, S., Srivastava, R., Sahah, D., and Prakash, S. 1997. A study on the uptake of trivalent and hexavalent chromium by paddy (*Oryza sativa*): Possible chemical modifications in rhizosphere. *Agric., Ecosyst. Environ.*, 62:53–58.
57. Shewry, P. R., and Peterson, P. J. 1974. The uptake and transport of chromium by barley seedlings (*Hordeum vulgare* L.). *J. Exp. Bot.*, 25:785–797.
58. Lyon, G., Peterson, P. J., and Brooks, R. P. 1969. 51Cr distribution in tissues and extracts of *Leptospermum scoparium. Planta*, 88:282–287.
59. Ramasamy, K. 1997. Effluent from agro-industries: Problems and prospects. *Sixth Natl. Conf. on Environment*, Coimbatore, India, pp. 1–7.
60. National Research Council. 1993. *In Situ Bioremediation, When Does It Work?* National Academy Press, Washington, D.C.
61. Alexander, M. 1994. *Biodegradation and Bioremediation*. Academic Press, New York.
62. Cifuentes, F. R., Lindemann, W. C., and Barton, L. L. 1996. Chromium sorption and reduction in soil with implications to bioremediation. *Soil Sci.*, 161:233–241.
63. Bopp, L. H., and Ehrlich, H. L. 1988. Chromate resistance and reduction in *Pseudomonas fluorescens* strain LB300. *Arch Microbiol.*, 150:426–431.

64. Horitsu, H., Futo, S., Miyazawa, Y., Ogai, S., and Kawai, ?. 1987. Enzymatic reduction of hexavalent chromium by hexavalent chromium tolerant *Pseudomonas ambigua* G-1. *Agric. Biol. Chem.*, 51:2417–2420.
65. Sudha Nair and Krishnamurthi, V. S. 1991. Accumulation of chromium by *Pseudomonas aeruginosa*. *Indian J. Environ. Health*, 33:230–236.
67. Eary, L. E., and Rai, D. 1987. Kinetics of chromium (III) oxidation to chromium (VI) by reaction with manganese dioxide. *Environ. Sci. Technol.*, 21:1187–1193.
68. Losi, M. E., Amrhein, C., and Frankenberger, W. T., Jr. 1994. Environmental biochemistry of chromium. *Rev. Environ. Contam. Toxicol.*, 136:91–121.
69. Luli, G. W., Talnagi, J., Strohl, W. R., and Pfister, R. M. 1983. Hexavalent chromium-resistant bacteria isolated from river sediments. *Appl. Environ. Microbiol.*, 46:846–854.
70. Palmer, C. D., and Puls, R. W. 1994. Natural attenuation of hexavalent chromium in ground water and soils, EPA/540/S-94/505.
71. Komori, K., Rivas, A., Toda, K., and Ohtake, H. 1990. A method for removal of toxic chromium using dialysis-sac cultures of a chromate-reducing strain of *Enterobacter cloacae*. *Appl. Microbiol. Biotechnol.*, 33:117–121.
72. Fujii, E., Toda, K., and Ohtake, H. 1990. Bacterial reduction of toxic hexavalent chromium using a fed-batch culture of *Enterobacter cloacae* strain HO1. *J. Ferment. Bioeng.*, 69:365–369.
73. Losi, M. E., and Frankenberger, W. T., Jr. 1994. Chromium-resistant microorganisms isolated from evaporation ponds of a metal processing plant. *Water, Air, Soil Pollution*, 74:405–413.
74. Wood, J. M., and Wang, H. K., 1983. Microbial resistance to heavy metals. *Environ. Sci. Technol.*, 17:582A.
75. Silver, S., Laddaqa, R. A., and Misra, T. K. 1989. Plasmid-determined resistance to metal ions. In: R. K. Poole and G. M. Gadd (eds.), *Metal-Microbe Interactions*, pp. 8, 16–17. IRL Press, Oxford.
76. Campos, J., Pacheco, M. M., and Cervantes, C. 1995. Hexavalent-chromium reduction by a chromate-resistant *Bacillus* sp. strain. *Antonie van Leeuwenhoek*, 68:203–208.
77. Harrison, R. M. 1992. Land contamination and reclamation. In: *Understanding our Environment: An Introduction to Environmental Chemistry and Pollution*. Royal Society of Chemistry, Cambridge.
78. Hale, M. G., Moore, L. D. and Griffen, G. J. 1978. Root exudate and exudation. In: Y. R. Dommergues and S. V. Krup (eds.), *Interactions Between Non-pathogenic Soil Microorganisms and Plants*. Elsevier, Amsterdam.
79. James, B. R., and Bartlett, R. J. 1983. Behavior of chromium in soils. VII. Adsorption and reduction of hexavalent forms. *J. Environ. Qual.*, 12:177–181.
80. Stevenson, F. J. 1985. Geochemistry of soil humic substances. In: G. R. Aiken, D. M. McKnight, R. L. Wershaw, and P. MacCarthy (eds.), *Humic Substances in Soil, Sediment, and Water*, pp. 13–52. Wiley, New York.
81. Wittbrodt, P., and Palmer, C. D. 1996. Reduction of Cr(VI) by soil humic acids. *Eur. J. Soil Sci.*, 47:151–162.
82. Singaram, P., Avudinayagam, S., and Chinnasamy, K. N. 1992. Tannery pollution studies in North Arcot district of Tamil Nadu. In: National Agricultural Research Project Report, pp. 118–185.
83. ANZECC/NHMRC. 1992. *Australia and New Zealand Guidelines for the Assessment and Management of Contaminated Sites*, p. 57.
84. Ministry of Environment/Ministry of Health. 1995. Draft health and environmental guidelines for selected timber treatment chemicals. Public Health Regulation Services, Ministry of Health, Wellington, New Zealand.
85. U.K. Department of Environment. 1987. ICRCL-Guidance on the assessment and redevelopment of contaminated land. Guidance note 59/83, 2nd ed.

86. Canadian Council of Ministry of Environment. 1990. Interim remediation criteria for contaminated sites.
87. U.S. Environmental Protection Agency. 1993. Standards for the use or disposal of sewage sludge. *Fed. Reg.*, 58:210–238.
88. Marshall, S., 1994. World-wide limit for toxic and hazardous chemical in air, water and soil. Parch Ridge, New Jersey.
89. Anandhkumar, S. P. 1998. Studies of treated tannery effluent on flower crops and its impact on soil and water quality. M.Sc. thesis, Tamil Nadu Agricultural University, Coimbatore, India.
90. Mahimairaja, S., Sakthivel., S., Ramasamy, K., Thangavel, P., and Naidu, R. 1997. Chromium contamination in soil and ground water due to disposal of tannery waste. *Proc. Sixth Natl. Symp. on Environment*, Coimbatore, India, pp. 250–255.
91. Mahimairaja, S., Sakthivel, S., Divakaran, J., Naidu, R., and Ramasamy, K. 1998. Extent and severity of contamination around tanning industries in Vellore district. In: *Abstr. of the Proc. ACIAR Workshop on Towards Better Management of Chromium-Rich Tanning Wastes*, Tamil Nadu Agricultural University, Coimbatore, 31 Jan–5 Feb 1998, p. 3.
92. Mahimairaja, S., Divakaran, J., Sakthivel, S., Naidu, R., and Ramasamy, K., 1998. Chromium contamination of ground water in Vellore district: Evidence of chromium mobility at contaminated sites. In: *Abstr of the Proc. ACIAR Workshop on Towards Better Management of Chromium-Rich Tanning Wastes*, Tamil Nadu Agricultural University, Coimbatore, 31 Jan– 5 Feb 1998, p. 4.
93. Mertz, E., Angino, E. E., Cannon, H. L., Hambidge, K. M., and Voors, A. W. 1974. Chromium. In: *Geochemistry and the Environment. V. 1 The Relation of Selected Trace Elements to Health and Disease*, pp. 29–35. National Academy of Sciences, Washington, DC.
94. Ramesh, P. T., Ramasamy, K., Mahimairaja, S., Gunathilagaraj, K., and Naidu, R. 1998. Tannery sludge disposal using earthworms and microbes: Preliminary investigation. In: *Abstr. of the Proc. ACIAR Workshop on Towards Better Management of Chromium-Rich Tanning Wastes*, Tamil Nadu Agricultural University, Coimbatore, India, 31 Jan–5 Feb 1998, p. 8.
95. Sakthivel, S., Mahimairaja, S., Divakaran, J., Saravanan, K., Ramasamy, K., and Naidu, R. 1998. Tannery effluent irrigation for tree plantations—Preliminary observations from field experiment. In: *Abstr. of the Proc. ACIAR Workshop on Towards Better Management of Chromium-Rich Tanning Wastes*, Tamil Nadu, Agricultural University, Coimbatore, India, 31 Jan–5 Feb 1998, p. 7.
96. Sara Parwin Banu, K., Ramesh, P. T., Ramasamy, K., Mahimairaja, S., and Naidu, R. 1998. Is it safe to use tannery waste sludge for the growth of vegetables? Glasshouse study. In: *Abstr. of the Proc. ACIAR Workshop on Towards Better Management of Chromium-Rich Tanning Wastes*, Tamil Nadu Agricultural University, Coimbatore, India, 31 Jan–5 Feb, 1998, p. 6.
97. Zayed, A., Gowthaman, S., and Terry, N. 1998. Phytoaccumulation of trace elements by wetland plants. 1. Duckweed. *J. Environ. Qual.*, 27:715–721.
98. Zaranyika, M. F., and Ndapwadza, T. 1995. Uptake of Ni, Zn, Fe, Co, Cr, Pb, Cu and Cd by water hyacinth (*Eichhornia crassipes*) in Mukuvisi and many rivers. *Zimbabwe. J. Environ. Sci. Health*, A30 (1):157–169.
99. Lee, C. R., Sturgis, T. C., and Landin, M. C. 1981. Heavy metal uptake by marsh plants in hydroponic solution culture. *J. Plant Nutr*, 3 (1–4):139–151.
100. Hanna, W. J., and Grant, C. L. 1962. Spectrochemical analysis of the foliage of certain trees and ornamentals for 23 elements. *Bull. Torrey Bot. Club*, 89:293–302.
101. Bowen, H. J. M. 1979. *Environmental Chemistry of the Elements*. Academic Press, New York.
102. Renan, M. J., Drennan, B. D., Keddy, R. J., and Sellschop, J. P. F. 1979. Oesophagal cancer in the Transkei: Determination of trace elements concentration in selected plant material by instrumental neutron activation analysis. In: *Nuclear Activation Techniques in Life Sciences*, International Atomic Energy Agency, Vienna: pp. 479–495.

103. Misra, S. G., and Jaisal, P. C. 1982. Extraction of added Cr vi and its uptake by spinach plant from soil. *Ind. J. Plant. Nutr.*, 1:93–99.
104. Mertz, W. 1969. Chromium occurrence and function in biological systems. *Physio. Rev.*, 49:163–239.
105. Cary, E. E., Allaway, W. H., and Olson, O. E. 1977. Control of chromium concentration in food plants. Ii. Chemistry of Cr in soil and its availability to plants. *J. Agri. Food Chem.*, 25:305–309.

29
Aqueous Solvent Removal of Contaminants from Soils

James C. O'Shaughnessy
Worcester Polytechnic Institute, Worcester, Massachusetts

Frederic C. Blanc
Northeastern University, Boston, Massachusetts

I. INTRODUCTION

Cleansing of contaminated soils associated with hazardous waste sites has been a major area of interest in environmental engineering practice for the last two decades. The use of aqueous solutions to accomplish this is the subject of this chapter, which reviews the technology that has evolved in the areas of soil flushing and soil washing.

II. SOIL FLUSHING

Soil flushing is primarily an in-situ treatment technology for the removal of sorbed-phase contaminants from the vadose soil zone. Soil flushing can also be applied to treatment of contaminated soils located in the saturated zone. The process involves injection of a "solvent" or "chemical countermeasure" to enhance contaminant solubility and transport the contaminant via the aqueous phase. The mobilized contaminant is then collected and put through a treatment process which removes and or treats the contaminants. The treated flushing water is either recycled or reinjected to the groundwater. In essence this phase of the treatment resembles classical "pump-and-treat" technology. Treatment of the collected groundwater is required to remove the dissolved contaminants and the solvent. The treated groundwater can be reinjected and used to mobilize the remaining contamination. If the collected groundwater cannot be reused, other disposal methods will be required and remediation cost will likely increase.

The "solvent" is a chemical addition that solubilizes, emulsifies, or chemically modifies the contaminants. To date the focus has been on using surfactants for organic contaminants and chealating agents and/or acids for heavy-metal contaminants. The term "countermeasures" has often been used when describing these chemical additions. Soil flushing is not a standalone technology, and it results in phase transfer of a given contaminant. Soil flushing has been used on both organic and inorganic contaminants. It has been selected as a treatment technology at a number of Superfund sites and for mining spoil locations. Applications have been evaluated for contaminated soils both above and below the groundwater table.

A. Theory

Soil flushing is the mobilization of contaminants from the soil into the aqueous phase. The mobilized contaminant and groundwater must then be collected and further treated. Treatment can include further phase transfer such as vaporization, precipitation, adsorption, or contaminant destruction via bioremediation or thermal destruction. A successful soil flushing remediation requires a number variables be addressed and as a technology is site specific.

B. Background

In-situ soil treatment offers many advantages, including long-term effectiveness, easy applicability to soils with high permeabilities, moderate costs (depending on the flushing agent), and addresses a wide variety of contaminants. Like most technologies, in-situ soil flushing is not applicable for cleaning all sites, but it has been the best remedial technology selected for a number of sites. In-situ treatment of soil is generally divided into three categories: biological, chemical, and physical. The biological aspect of in-situ treatment is used mainly to biodegrade organic compounds. The physical category of in-situ treatment encompasses the solidification and stabilization technologies. Although the physical category can be successful in containing the contaminant, this technology leaves the contaminant on site. Over time the stabilized contaminant complex may degrade and the remobilized contaminant may leach into the surrounding soil and groundwater. As a result, there are usually long-term environmental monitoring and related long-term costs, and the treatment is not successful in permanently treating the soils. The chemical category of in-situ treatment is the most appropriate for remediating organic and heavy metal-contaminated soil. The process includes applying a flushing agent to the soil, which is allowed to percolate downward through the contaminated soil and interact with the contaminants. During this interaction, the flushing agent, or chemical countermeasure, removes the contaminant from the soil and combines with it to form a mobile compound by means of the desorption, chelation, solubilization, and complexation properties of the flushing agent. The resulting contaminant solution is collected by a down-gradient well or other recovery method. This solution is stored before treatment, then treated, and possibly reapplied to the contaminated soil. Figure 1 presents a description of the in-situ soil flushing technology.

For contamination located in the vadose zone, the mobilizing solution or countermeasure is applied to the surface of the soil volume to be treated using spray or trickle irrigation. The downward percolation of he flushing solution through the soil leaches sorbed contaminants from the soil matrix and flushes dissolved contaminants to a collection system. The collection system is usually a relief or interceptor drain for collection above the groundwater table, or to the water table for removal via a recovery (pumping) well.

C. Suitable Applications

The crucial factor governing the effectiveness, implementation, and cost of in-situ soil flushing is the permeability of the soil. Permeability is not a fundamental property of

Aqueous Solvent Removal of Contaminants

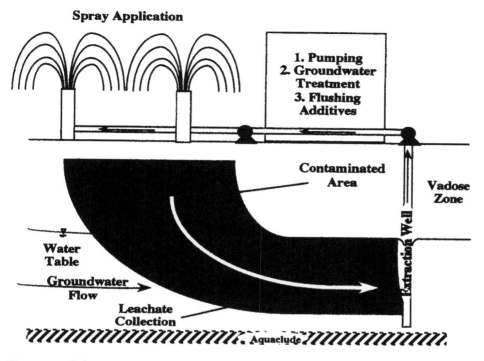

Figure 1 Schematic of a soil flushing system.

a soil, but depends on a number of important factors, including particle size distribution, particle shape and texture, mineralogical composition, voids ratio, degree of saturation, soil fabric, nature of fluid, type of flow, and temperature. The first three items are consistent for a certain type of soil. The void ratio and degree of saturation are dependent on the placement of the soil, and the last three relate to the fluid flowing through the soil (1).

The particle size distribution greatly influences the permeability of a granular soil. As smaller particles increase in the voids of a granular soil, the permeability of the soil correspondingly decreases because the flow of water is inhibited by the smaller particles. Not only does the size of a soil particle affect permeability, the shape and texture of the soil particle can affect the permeability. Elongated or irregular particles create flow paths that are not laminar and therefore inhibit the flow of water. The texture of a particle, smooth or rough, can impart an added frictional resistance to the water's flow and therefore decrease the permeability of the soil. Since a site's soil characteristics and properties will differ, evaluations of these criteria are required before the practicality of a pump-and-treat technology can be instituted.

D. Soil Types

Soil flushing is best suited for porous and well-drained soils. Soils with hydraulic conductivity greater than 10^{-3} cm/s are recommended as possible candidates for soil flushing technology. Soil flushing is best adapted to sandy and gravel type soils. Soil flushing can be applied to any remediation site where a pump-and-treat system has been determined to be applicable.

E. Contaminant Types

The contaminant found in contaminated soils can be classified as either organic or inorganic. Various contaminants found at Superfund sites are listed in Table 2.

Soil flushing is best suited for heavy metals, halogenated volatile organic compounds and nonhalogenated organic compounds. In the case of organic contaminants, compounds with low octonal/water partitioning coefficients able best suited to soil flushing. Phenols,

Table 1 Critical Soil Flushing Parameters

Factors influencing technology selection	Optimum in-situ treatment conditions	Basis	Data needs
Equilibrium partitioning of contaminant between soil & extractant fluid	—	High contaminant partitioning into extractant is desirable. High partitioning concentrations decrease flushing fluid volumes	Equilibrium partitioning coefficient
Complex waste mixture	—	Complex mixtures increase the complexity of a successful flushing fluid formulation	Contaminant composition
Soil-specific surface area	$< 0.1 \, m^2/g$	Higher surface areas increase contaminant soil sorption	Specific surface area of the soil
Contaminant solubility in water	$> 1000 \, mg/L$	Soluble compounds can be removed with water flushing	Contaminant solubility
Octonal/water (K_{ow}) partitioning coefficient	10–1000	Highly soluble contaminants are mobilized by natural processes	Octonal/water partitioning coefficient
Spatial variation in contaminant composition	—	Contaminant composition changes may require re-formulation of the extraction fluid	Statistical sampling of contaminant volume
Hydraulic conductivity	$> 10^{-3} \, cm/s$	Higher hydraulic conductivities produce better distribution of the flushing fluid	Hydrogeologic flow regime
Clay content	—	Low clay content is desirable. Clay increases contaminant sorption and reduces flow	Soil composition, color, and texture. Grain size distribution
Cation-exchange capacity (CEC)	—	Low CEC preferred. CEC increases sorption and reduces desorption	Cation-exchange capacity
Flushing fluid characteristics	Low toxicity, low cost, treatability and reuse	Toxicity increases health risks. Non-reuse increase costs	Fluid characterization. Bench- and pilot-scale testing
Soil total organic carbon (TOC) content	< 1% by weight	Soil flushing is more effective in soils that have low organic content	Soil total organic carbon content
Contaminant vapor pressure	< 10 mm Hg	NAPL volatile compounds will partition to the vapor phase	Contaminant vapor pressure at operating temperatures
Fluid viscosity	< 2 centipoise	Low fluid viscosity enhances flow through the soil	Contaminant viscosity at operating temperatures
Organic contaminant density	$> 2 \, g/cm^3$	Prediction of contaminant migration	Contaminant density at operating temperatures

Source: Adapted from Ref. 1.

Table 2 Hazardous Soil Contaminants at Superfund Sites

Heavy metals contamination						47 sites	
Chromium	9	Cadmium	4	Mercury	2	Vanadium	1
Arsenic	8	Iron	3	Selenium	2	Fly ash	1
Lead	7	Copper	2	Nickel	1	Plating wastes	1
Zinc	5						
Other inorganics						26 sites	
Cyanides	6	Acids	7			Alkalis	6
Radioactives	3						
Hydrophobic organics						38 sites	
PCBs	15	Oil, Grease	11	VOCs	6	Pesticides[a]	5
PNAs[b]	1						
Slightly soluble organics						64 sites	
Aromatics							
Benzene	9	Toluene	8	Xylene	5	Other aromatics	3
Halogenated hydrocarbons							
Trichloroethylene	11	Ethylene dichloride	6	Vinyl chloride	4	Methylene chloride	3
Other	15						
Hydrophilic organics						20	
Alcohols	4	Phenols	12	Other hydrophilics			4
Unspecified organics solvents and other compounds						30	

[a] Chlorinated pesticides.
[b] Polynuclear aromatics.
Source: Ref. 2.

lower-molecular-weight alcohols, and carboxylic acids can be removed from soils using soil flushing.

The soil contaminant determines the type of flushing solution to be applied. Three types of flushing solutions are used:

1. Water only
2. Water plus additives
3. Organic solvents

While some research has evaluated in-situ solvent soil flushing, most of the current literature suggests that organic solvents, should be used in a contained ex-situ application. This position is supported by the authors, who also conclude that the release and mobilization of a solvent contaminant complex in any unpredictable groundwater environment is not an acceptable risk. Therefore, in-situ solvent flushing will not be covered as an acceptable technology.

Soil flushing is one of the few technologies that are effective in the remediation of heavy-metal-contaminated soils, when the clean-up objective is the removal of the heavy metals from the soil. Soil flushing, soil washing, and phytoremediation are

Table 3 Contaminants Considered for Treatment by in-situ Soil Flushing

Contaminants	Industries
Heavy metals (cadamium. chromium, lead, copper, zinc)	Metal plating, battery recycling
Aromatics (benzene, toluene, cresol, phenol)	Wood treating
Gasoline and fuel oils	Automobile, petroleum
Halogenated solvents (TCE, trichloroethane)	Drycleaning, electronics assembly, metals finishing
PCBs and chlorinated phenol	Pesticide, herbicide, electric power

the most effective heavy-metal removal technologies for nonvolatile heavy-metal contamination. Soil flushing is also the only currently cost-effective, field-ready technology that can be applied to heavy-metal contamination at greater depths below the groundwater table.

Soil flushing is not well suited to nonaqueous-phase liquids (NAPLs) either light (LNAPLs) or dense (DNALPLs), polychlorinated biphenols (PCBs), dioxins/furans, highly volatile organics (VOCs), and asbestos. Research in remediating these contaminants has identified more effective current technologies for on-site remediation. These technologies include thermal destruction, vapor extraction, and stabilization. If these technologies are not financially feasible for a given site, then soil flushing may prove to be a cost-effective treatment method. Some of the contaminant types and industries associated with these wastes are summarized in Table 3.

F. System Equipment Requirements

The equipment required in a soil flushing operation consists of three components. The first component is the application of the flushing solution to the soil. The second component is the subsurface flush-water collection system. The third component is the flush-water treatment system. There is also a need for the contaminated area to be enclosed, usually with a physical barrier although hydraulic containments have also been used. Figure 2 shows the different components of the soil flushing system.

The soil flushing system can be applied either to the ground surface or via subsurface injection. Surface application methods include flooding, ponding, ditches, and sprinkler systems. These methods are used for contaminants at depths up to 15 ft. The surface contour may require regrading to assure a uniform application rate and movement down through the vadose zone.

When using surface application, ground topography should have slopes of less than 3%, and should have a uniform contour without ridges and gullies. Sandy soils are best suited for surface application methods, and soils with larger hydraulic conductivities (10^{-3} cm/s) are preferred.

The ditch method is used where topography limits other methods of application, or where the entire surface area does not require wetting. Most ditch systems employ flat-bottomed, shallow ditches to transport and disperse all the flushing solution. Sprinklers can be used to cover either an entire area or several subareas. Sprinkler systems have been reported to deliver moisture to depths of 50 ft.

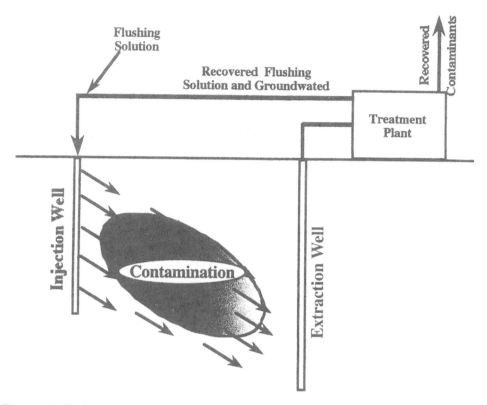

Figure 2 Various components of an in-situ soil flushing system.

Subsurface gravity delivery systems use infiltration galleons (trenches) and infiltration beds. Subsurface systems consist of excavated areas filled with porous media (coarse sands or gravel) that will distribute solutions to the contaminated areas. Filtration galleries consist of a pit through which the solution is distributed to the surrounding soils in both horizontal and vertical directions. Forced subsurface delivery systems can also be used for flushing solution distribution. These pressurized systems deliver the flushing solution into the contaminated area using either open-end or slotted vertical and horizontal pipes. Pressurized systems are applicable to soils with hydraulic conductivities $>10^{-4}$ cm/s and high porosities $>25\%$. Some examples of surface and subsurface flushing solution distribution systems are presented in Fig. 3.

The collection system for the flushing-solution contaminant mixture may use barriers, subsurface collection trenches, or recovery wells. In many instances a combination of these techniques is used. Each collection system is site specific and is a function of both the geohydraulic issues and the contaminant characteristics.

In all instances the contaminant is collected down-gradient of the site of mobilization and precautions must be taken to ensure that the mobilized contaminant is not allowed to leave the remediation site. The more complex the subsurface environment, the more complex the collection system design. The collection systems are similar to the "pump-and-treat" systems used at many remediation sites. Examples of horizontal injection and collection wells are shown in Fig. 4.

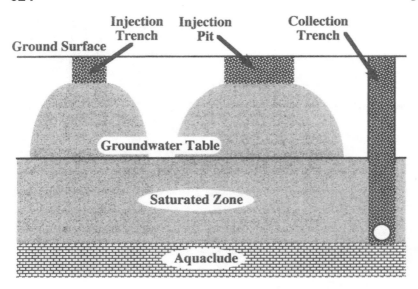

Figure 3 Trench and pit applications in soil flushing systems.

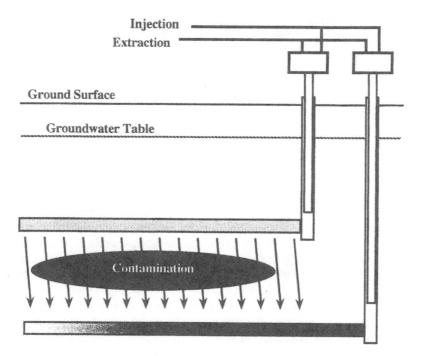

Figure 4 Horizontal injection and collection wells.

Equipment and controls common to injection and collection wells include:

1. Injection pumps
2. Submersible
3. Pipes

Aqueous Solvent Removal of Contaminants

Table 4 Recommended Wastewater Treatment Processes

Type of contaminant	Intermediate treatment	Final treatment requirement
Volatile organics	Air stripping	Activated-carbon columns
Nonvolatile organics	Filtration	Activated-carbon columns
Nonvolatile organics	Filtration	Biological treatment
Heavy metals	PH adjustment	Chemical precipitation
Heavy metals	Filtration	Ion exchange
Complexed heavy metals	Hydroxide precipitation	Sulfide precipitation

4. Valves
5. Fillers
6. Flow meters and water-level sensors
7. Control panels
8. Filters
9. Tanks
10. Safety equipment

The treatment system is a function of the contaminant. If the contaminant is volatile, a vapor system is usually the best removal. Vapor stripping and capture in activated-carbon columns are commonly used in pump-and-treat and in flushing systems. If the organic contaminants are not volatile, activated-carbon columns are also used to remove the contaminants. However, removal from water is less efficient than removal from the gas phase.

Fuel oils and lighter distillates can be recovered using dissolved-air floatation. Phenols can be treated using biological wastewater treatments, as can other organics if the concentration is high enough. Often a supplemental carbon source is required if biological treatment is being considered. Heavy metals are removed using chemical precipitation or ion exchange. Table 4 list some of the various treatment technologies used in soil flushing applications.

G. Additional Treatment Requirements

As soil flushing is an in-situ process, extensive knowledge of the geohydraulic profile is required. Site characterizations including test wells, soil coring, and proper soil logs are part of any soil flushing investigation. Background water quality conditions must be collected both upgradient and downgradient of the contaminated area. Often both laboratory and field pilot tests are required to properly evaluate the feasibility of soil flushing at each specific site. Following these studies, site modeling and prediction of results obtained from the pilot studies should be performed. If the technology appears to be feasible, then well field design and/or design for application and collection of the flushing solution can be completed.

Additional steps include well drilling, design and construction of the flushing solution treatment facility, and determination of final management of the collected contaminates. Following site remediation, a long-term monitoring program should be initiated.

For vadose-zone contamination, the decision to allow the mobilized contaminant to reach the saturated zone or capture prior to that event happening must be resolved. Depth to water table and both vertical and horizontal permeability are key factors. If vadose-zone capture is determined, capture zones (trenches, pits, etc.) must be designed. If the contamination will be recovered in the saturated zone, "horizontal" flow using sets of pumping/injection wells are often used.

Posttreatment processing will involve various unit operations, based on the nature of the mobilized contaminants. For organic contaminants volatilization, activated-carbon adsorption, biological treatment, and thermal destruction are options. For heavy metals, ion exchange and chemical precipitation are the two most common technologies employed.

In the United States, groundwater monitoring following remediation is covered by a number of laws and regulations. RCRA, CERCLA, SARA, the Clean Water Act, the Safe Drinking Water Act, and the Underground Storage Tanks Technical Standards and Requirement Act all address various aspects of remediation. The provisions under RCRA require owners and operators to monitor groundwater quality for at least 30 years following site remediation.

The minimum monitoring network requires one up-gradient and three down-gradient wells. The monitoring network usually requires sampling wells within and around the regulated area, as a way to track any further residual contamination movement. The design of the monitoring network is site specific and depends on the geohydraulic conditions and the contaminants to be monitored.

In addition to federal requirement, each state has established its own soil and groundwater standards. These standards can be obtained from the appropriate state regulatory agency. Judge, Kostecki, and Calabrese (3,4) have reviewed both the soil and groundwater standards for most states.

H. Design Considerations

Soil flushing techniques are considered to be either innovative or conventional. Conventional soil flushing is defined as involving the following activities:

- Vadose-zone well and capture methods
- Pump-and-treat systems in the saturated zone
- A combination of pump-and-treat and vadose-zone flushing
- Natural attenuation

Innovative soil flushing is defined as either secondary recovery or tertiary recovery. Innovative soil flushing employs methods developed by the U.S. Environmental Protection Agency, the petroleum industry, and the mining industry. These methods usually can remove over 90% of the contamination. Secondary methods use water flooding and pressure techniques. Tertiary methods use surfactants and other solvents which are injected with the soil flushing solution.

The design of a soil flushing system is the process of putting together a number of subsystems in order to facilitate site remediation. The components of the soil flushing system that have to be designed are

1. The contaminant containment barriers
2. The flushing solution application system

Aqueous Solvent Removal of Contaminants

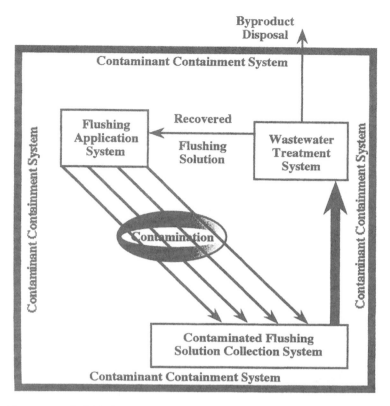

Figure 5 Soil flushing subsystems.

3. The contaminant-flushing solution collection system
4. The flushing solution recovery and recycle system
5. The contaminant treatment-disposed system

The interrelation of subsystems is illustrated in Fig. 5.

A soil flushing system cannot be designed until a complete site characterization study has been completed. As the technologies used in soil flushing systems will mobilize the contaminants, improper application could increase environment risk rather than decrease the risk.

1. Contaminant Containment

The contaminants at a remediation site are usually contained using cutoff walls or hydrodynamic controls. Sheet walls and slurry walls are common cutoff methods used to control groundwater movement. Injection wells and pumping wells are the two most popular methods of hydraulic containment.

2. Flushing Solution Application

The flushing solution application system must optimize the contact of the flushing solution with the contaminated soils. Groundwater models and VOC concentration maps are helpful in defining in the extent of the problem. A complete site history and site

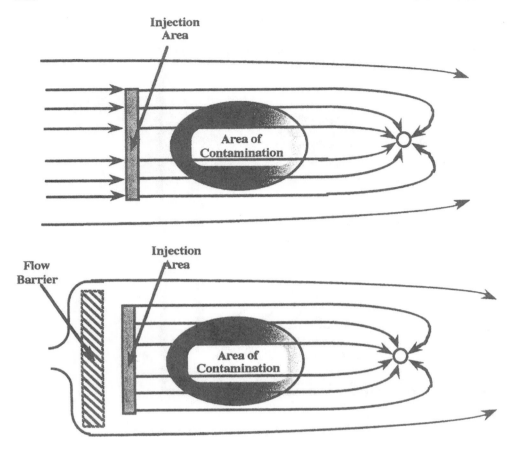

Figure 6 Example of a flow barrier to reduce flushing solution dilution.

characterization are essential if optimization of the flushing solution application system is to be achieved.

The best solution will be site specific, and governed by the type of contaminant, soil type, and geohydrology. Either surface or subsurface systems are effective, and as a rule, the more flexibility in the application system, the more efficient the remediation.

In the case of NAPLS, a pulsing system often has proven to be more efficient than a steady-state system. As a result, alternating application systems may be designed where NAPLS are the major contaminant.

3. Contaminant-Flushing Solution Recovery

If the proper steps are taken in designing the containment and solution application systems, the only criteria for the recovery system is that it be down-gradient of the contaminated soil. Trenches, pits, and collection wells are the most common technologies. Again, site characterization is the critical issue. The collection system must be designed to handle the maximum hydraulic conditions within the remediation zone.

When the total flow that must be collected is excessive, barriers up-gradient of the contaminated areas are often employed. Up-gradient barriers can help improve overall

Aqueous Solvent Removal of Contaminants

system efficiencies. An example of an up-gradient barrier is shown in Fig. 6. With the installation of the barrier up-gradient of the contamination, a number of advantages are obtained. These advantages include:

1. Reduced volume in the collection system
2. Longer contact between the flushing solution and the contaminated soil
3. Higher contaminant concentrations in the collected groundwaters
4. Better control of the overall system

4. Recovery and Recycle of the Flushing System

Design information for the recovery and recycle of the flushing solution can only be obtained by either laboratory or field studies. The effectiveness of each "solvent" or "chemical countermeasure" will be reduced with each use. As the flushing solution passes through the soil, it will react with anything in the soil–groundwater complex. In the case of acids and bases, the chance of recycle and reuse is minimal.

Column and pilot studies are the best methods to obtain data necessary to design the recovery system. It is suggested that a number of flushing solutions be tried at the laboratory scale and the relative recovery be evaluated along with their contaminant desorption effectiveness.

5. Contaminant Treatment Systems

The contaminant treatment system is a function of the contaminant, its concentration, and its interaction with the flushing solution. The major factors for organic contaminants are

1. Volatilization
2. Biodegradation
3. Concentrations

As a rule, the higher the propensity for any of these characteristics, the easier it is to design a successful treatment system.

For heavy metals, the major design issue is the ionic form of the metal. If the metal ion is not complexed, then most treatment technologies will work. If the metal ion is complexed, then treatment methods are limited to selective precipitation or ion-exchange technologies.

The overall soil flushing system can be evaluated based on the overall contaminant concentration following the implementation of the remedial effort. The optimal system will have high contaminant concentration in the collected flushing solution for a relatively short period of time. A less effective system will have lower contaminant concentration over a longer time period.

I. Case Studies

1. United Chrome Products

The United Chrome Products site located in Corvallis, Oregon, uses soil flushing of chromium from contaminated soil and groundwater (5). Soil contamination exceeded 60,000 mg/kg chromium and groundwater concentrations were observed. The upper contaminated soil was 18 ft in depth and comprised of coarse to fine silt. The clean-up operations included the following actions:

Table 5 Chromium Removal from the United Chrome Superfund Site

Parameter	Total	Average daily
Groundwater extraction	6,700,000 gal	11,400 gal
Influent Cr(VI) concentration range	146–19,233 mg/L	
Mass of Cr(VI) removed	24,300 lb	41 lb
Infiltration recharge	4,700,000 gal	8,000 gal
Average effluent Cr(VI) concentration		1.7 mg/L
Sludge produced (@ 25% solids)	6,070 ft^3	10 ft^3

Source: Ref. 5.

1. Excavation and off-site disposal of 1100 tons of heavily chromium-contaminated soils
2. Installation of 23 groundwater extraction wells and 12 monitoring wells
3. Installation of two infiltration basins over the highest contamination areas
4. Construction of an infiltration trench through the contaminant plume area
5. Construction of a wastewater treatment facility for chromium removal
6. Changing the surface drainage to bypass the site

The infiltration basins had rates of 7600 gal/day and 3000 gal/day during dry weather periods. The infiltration rates were decreased by more than 50% during wet periods. The infiltration trench operated at a rate of 2500 gal/day.

Chromium was removed by using chromium reduction, chemical precipitation and sedimentation. The results of these actions are summarized in Table 5. This data are for project results conducted from August 1988 through December 1990.

2. Borden Canadian Forces Base

At Borden Canadian Forces Base near Alliston, Canada, a field test of a surfactant-enhanced pump-and-treat system was conducted from June 1990 to August 1991. The test was conducted in a 27-m^3 cell which was self-contained and contaminated with 231 L of tetrachloroethylene (PCE). Five injection and five withdrawal wells were installed in the cell.

First, free-phase PCE was removed with direct pumping, followed by water flooding for free-phase and dissolved PCE removal, and then surfactant flushing. Following the first two phases, about 50% of the PCE remained in the soil. Fifty percent of the remaining PCE was removed with surfactant-enhanced flushing fluid within 15 pore volumes (6,7).

3. Hill Air Force Base

At Hill Air Force Base, Utah, surfactant flushing was used to removed an estimated 99% of residual TCE at a pilot test plot. The surfactant used was sodium dihexyl sulfosuccinate, which was injected into three injection wells on one side of a 20 ft by 20 ft test plot. The soil is classified as sandy. Three extraction wells were located on the opposite site of the test plot.

Table 6 Examples of Superfund Sites Using Soil Flushing (as of August 1995)

Name of site	Status	Type of facility	Contaminants
Lipari Landfill, Piman, NJ	Operational	Landfill	Volatile organic compounds (VOCs), phenol, semivolatile organic compounds (SVOCs), metals (chromium, lead, nickel, mercury)
Vineland Chemical, Vineland, NJ	Design (on hold)	Pesticide manufacturing	VOCs (dichloromethane), arsenic
Ninth Avenue Dump, Gary, IN	Completed	Industrial landfill	VOCs (BTEX, TCE), phenols, total metals, lead, polyaromatic hydrocarbons (PAHs)
Lee Chemical, Liberty, MO	Operational	Solvent recovery	VOCs (TCE)
Idaho Pole Company, Butte, MT	Operational	Wood preserving	SVOCs, PAHs
United Chrome Products, Corvallis, OR	Operational	Chrome plating	Metals (chromium)
Umatilla Army Depot, Hermistan, OR	Design complete	Explosives storage	Explosives, propellants

Source: Adapted from Ref. 8.

Researchers in this study included personnel from the University of Texas, Radian International, Intera, Inc., and the U.S. Air Force. Laboratory testing and modeling using the UTCHEM chemical transport model preceded the tests. Pre- and posttest residual saturations were measured using tracer techniques. The retardation of the tracer that was used is dependent on the amount of residual DNAPL present, as it moves from the injection to extraction wells. Based on tracer transport, 99% of the residual DNAPL was estimated to have been removed by surfactant flushing.

Other applications of soil flushing remediation efforts are summarized in Table 6.

III. SOIL WASHING

Soil washing is an ex-situ process in which the contaminated soil is excavated and laundered with water and water-soluble cleansing agents. The washwater and rinsewater which now contain the contaminants are then treated. The clean soil then may be used as backfill at the excavation location or off site at another location. Often the soil is physically separated into various fractions. Once separated, the various fractions are cleansed to different degrees based on reuse function or ultimate disposal requirement. Size-fractional separation occurs early in the treatment operations sequence, and the various size fractions may be washed separately.

In some applications not all of the separated size fractions are washed. If a separated soil fraction represents a small percentage of the total soil volume and most of the contaminants are sorbed to this fraction, then direct disposal of that fraction may be the most economical alternative. Soil washing always produces a concentrate or sludge which requires some form of ultimate disposal.

Soil washing is performed in containment vessels, which allows for process control and performance evaluation. In many cases, the contamination is concentrated largely in the fines fraction of the soil mixture, which represents a small fraction of the total volume to be treated.

A. Theory

Soil washing is accomplished by a series of physical unit operations and chemical processes similar to the technologies that have long been employed in the mining industry. Water is used to separate and wash various soil fractions. This is accomplished by a series of ex-situ operations and processes in secure, controlled-volume reactors. All contaminants removed from the soil are transferred to the aqueous phase. This contaminated water is treated, resulting in concentrated, residual streams similar to the concentrates resulting from industrial wastewater treatment.

B. Background

Soil washing technology was developed beginning around 1980 in the United States by the U.S. Environmental Protection Agency and by governmental agencies in other countries. The Netherlands, where soil is a valued resource, was one of the countries active in soil washing development. Initially, attempts at treating all the soil as excavated (rocks, gravel, sands, silt, and clay) used one soil washing treatment tank. This approach proved difficult, and screening and separating the soil into various size fractions was attempted. The soil size-fraction approach prior to washing resulted in better feasibility and performance.

The particle-size-fraction separation process can provide an advantage by establishing a more uniform feed for subsequent treatment operations and eliminate difficulties due to the fact that excavated contaminated subsoils are often not homogeneous from a soil characterization standpoint as one moves from point to point in the excavation. Water is the main solvent in the washing process. The water may be modified as with acid addition to drive metals attached to soil particles into the water solution. Chelating agents may also be used to promote metal solubility. Various detergents may be used to bring organic contaminants into solution.

C. Suitable Applications

To categorize what may be termed "suitable applications", one must generalize. It should be remembered that generalizations are subject to exceptions and one should always determine if the generalization applies to a specific case before applying it as an ironclad rule.

Soil washing may be used as a pretreatment or treatment for soils contaminated with radioactive materials, organic substances or mixtures of organic substances, metals, or other inorganics. The technology may be used successfully to clean soils simultaneously contaminated with many types of inorganics and organics.

D. Soil Types

In general, soil washing seems to be more successful when the soil particles in the contaminated soil are larger. Sand, gravel, silty sand, and similar soil groupings are more likely to be successfully treated. Clays and silts are difficult to treat. In general, if a soil contains more than 25–30% silt and clay, it can be considered difficult to treat. This is because fine-grained soils have more surface area per kilogram of soil and relatively greater amounts of contaminants on the soil surfaces. When subjected to agitation in a slurry form, larger soil particles can be subjected to greater hydraulic shear forces due to their relative velocities than the smaller colloidal particles. Larger sand particles will collide with one another in a kind of scrubbing action, while colloidal particles will not collide. This makes it difficult to wash clay and silt suspensions. Fine suspensions are also more difficult to separate by means of gravitational force. Economics defines the processes used on each soil fraction. Small-particle-size fractions are often treated in a more cost-effective manner by biodegradation treatment processes when organic contaminants are the problem.

In an attempt to create a soil with uniform-sized particles prior to treatment, various screening and separation techniques are utilized as a pretreatment operation. With a uniform soil one can better predict the performance when a soil is subjected to a certain washing operation for a given contact time. Clay and silt particles do not always exist as discrete particles in the soil. Such colloidal particles may be bound together to form large lumps of soil which will have to be broken or may be bound to sand and larger pieces of gravel. Thus the removal of colloidal soils will occur in the treatment operations which deal with preseparated large-grain-sized batches.

Debris such as plastics, wood, metal, and other materials in the bulk soil will also be removed by prescreening operations to prevent interference with the downstream treatment processes. Vegetation collected in the bulk contaminated soil should also be removed in the same manner as the debris.

Many of the soil washing operations involve hydraulically suspending the particles or aggressively agitating the particles mechanically. The practical particle size limit for this agitation is about 3/8 in. Cobbles and particles larger than this size are usually easily rinsed for contaminant removal. When suspension and continuous agitation is not cost effective for large particles, the large-particle fraction may be placed in a pile which is then placed in contact with the water. This soil leaching process may or may not fall under the category of soil washing.

In addition to contaminated surface and subsurface soils from industrial and spill sites, soil washing techniques may be applied to clean contaminated sediments found in waterways, industrial slags, and mine tailings.

E. Contaminant Types

Soil washing is applicable to various types of contaminants, such as heavy metals, radioactive elements, and an assortment of organic compounds including petroleum hydrocarbons, volatile organic compounds, polychlorinated biphenyls (PCBs), and polynuclear aromatics (PNAs). The soil washing operation must be tailored to the specific contaminated soil or sediment based on treatability studies, and the washing and rinse solutions may also require study and treatment or concentration. Contaminants in a soil may exist in a separate particulate form. Metal particles, slag, scrapings, and dusts deposited as a result of sandblasting operations, particulate deposition from industrial exhaust systems and process stacks, or dumping are examples of such free soil particles. Lead from bullets embedded in a military firing range berm are another example. Such particles may be physically separated based on specific gravity after initial size separation.

F. Washing Solutions

Washing solutions consist of water with other agents added. Acids, alkali detergents, chelating agents, and other chemicals may be added to the water. Different solutions may be used for different soil fractions and stages of the cleaning operation. In each case, when choosing the additive or agent the effect on the residual washwater waste treatment should be taken into account.

Acids are commonly added to produce weak acid solutions in cases where metals in ionic form are attached to soil particle surfaces. This increases metal solubility. In some cases, weak acid solutions can also be used to remove chemical precipitate coatings from particles. Hot acid solutions are sometimes utilized to increase washing effectiveness.

Alkali detergents can be added to remove a variety of organic chemicals, such as petroleum compounds, which may be coating soil particles. In some cases increasing the washwater solution temperature is used to improve cleansing performance in the same fashion as domestic laundering with hot water.

Brine solutions may be utilized for ion-exchange purposes in washing operations. This operation is similar to the regeneration of an ion-exchange unit with brine. Caustic solutions can be used to effect hydroxide ion exchange and remove various anions. In some cases carbonates and bicarbonates have been used for such regenerative anion removal.

Surfactants in water solutions have also been used successfully in soil washing. Surfactants are useful in bringing dense nonaqueous-phase liquids (DNAPLs) and light nonaqueous-phase liquids (LNAPLs) from contaminated soils into the washing liquid. Many commonly used detergents contain surfactants. Sodium dihexyl sulfosuccinate and sodium lauryl sulfate are examples of surfactants. The surfactants are long-chain organic compounds which are polar at one end of the chain and nonpolar at the other end. One end of the surfactant which is a hydrocarbon attaches to the hydrophobic organic contaminant molecule, while the other end—which may be composed of a sulfonic group—is attracted to the water molecules bringing the organic contaminants into solution. In this solubilization process, the surfactant molecules with the attached organics form globule-like clusters, termed micelles, with the ionic water-soluble ends on the outer surface of the globule and the hydrophobic ends with attached organics on the interior. For each application and surfactant there is a minimum concentration of surfactant required for such micelle to form. This is termed the critical micelle concentration (CMC). Required CMC values are reported to range from 500 to 1500 mg/L (9).

Chelating agents are helpful in removing metals from soil particles. A chelating agent is an organic that contains an anionic portion termed a ligand, which forms a bond with a metal cation. This structure is termed a metal complex and allows more metal to enter into the washwater solution. A ligand that can form two bonds to a lone metal ion is called a bidentate ligand, while ligands that can form more than two bonds to a metal ion are called polydentate ligands. A polydentate ligand which can form six bonds to a metal ion is termed a hexadentate ligand. Ethylenediaminetetraacetate (EDTA) is a hexadentate ligand which has been used as a scavenger to remove heavy metals such as lead from the human body. EDTA is relatively safe and is used in consumer food products, soaps, and cleaners. The authors have used EDTA to remove metals from contaminated soils (10–12). Naturally occurring organic ligands form organometal complexes which impede metal removal in water and wastewater treatment. Similarly, once the chelating agent metal complex has been created in the soil washing operation, treating the resulting washwaters may become more difficult, requiring extensive pH alteration to break the complex.

G. System Equipment Requirements

Equipment is available for soil washing which can be transported and set up on-site. Such equipment utilizes unit operations such as screening, sedimentation, centrifugal separation, elutriation, filtration, acid/base extraction, mechanical shearing, flotation,

Table 7 Soil Washing Equipment

Bar screens
Trommel screens
Vibrating screens
Conveyor systems
Slurry pumps
Grit washers
Aerated grit chambers
Cyclonic and spiral separators
Attrition reactors
Clarifiers
Flotation units
Drum filters
Centrifuges
Filter presses
Fluidized-Bed washers
Bioslurry reactors

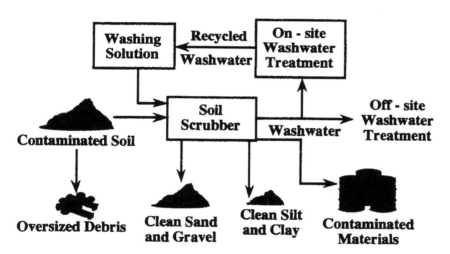

Figure 7 Schematic diagram of a soil washing process.

and suspended biological slurry reactors. Mechanical shearing, for example, may be accomplished in mixing reactors which are sometimes termed attrition reactors or high-intensity scrubbers. Some of the equipment utilized is listed in Table 7.

In addition, the washwater treatment operations feature all of the unit operations used to treat industrial wastewater. Air is also a fluid, and in that sense one might categorize air stripping of soils as a washing process. A schematic drawing of the soil washing process is shown in Fig. 7.

H. Design Considerations

The design and operation of soil washing systems depends on the remediation goals and standards which the "cleaned soil" must meet. This involves debate on the issue of what

Table 8 Achievable Treatment Levels for Contaminants Commonly Treated in Soil Washing Applications

Contaminant	Dutch 'B' level (mg/kg)
Metals	
Chromium	250
Nickel	100
Zinc	500
Arsenic	30
Cadmium	5
Mercury	2
Lead	150
Organics	
Total polynuclear aromatics (PNAs)	20
Carcinogenic PNAs	2
PCBs	1
Pesticides	<1
Total petroleum hydrocarbons (TPH)	100

Source: Ref. 8.

is considered a clean soil. The U.S. Environmental Protection Agency (EPA) and state regulatory agencies usually set standards on a site-by-site basis. In the Netherlands, standards were developed which have been termed the Dutch A/B/C levels. The A/B/C standards are expressed in terms of concentrations of some 50 common organic compounds and inorganic elements. The A level is for unrestricted reuse, while the B level reflects residual concentrations associated with limited reuse of a site. Table 8 illustrates some concentrations associated with the Dutch B-level standards.

I. Case Studies

There have been many successful applications of soil washing technology at sites in Europe and North America (Table 9). Many of these have been full scale and resulted in site remediation.

1. King of Prussia, Pennsylvania

One of the more famous applications was at the King of Prussia Superfund Site located in Winslow Township, New Jersey. This was the EPA's first full-scale application of soil washing at a Superfund site (13). The 10-acre site was unsuccessfully used by the KOP Technical Corporation as an industrial waste recycling center before being declared a Superfund site. The soil around and sludge in the abandoned industrial waste lagoons were contaminated with arsenic, beryllium, cadmium, chromium, copper, lead, nickel, and zinc. Some 19,000 tons of soil and sludge were cleaned by soil washing. Chromium, copper, and nickel were the metals that posed the greatest environmental problem. The peak concentrations of the three metals in the sludge exceeded 10,000 mg/kg for each metal. After treatability studies and pilot demonstration operations, a full-scale operation

Table 9 Examples of Superfund Sites Using Soil Washing (as of August 1995)

Name of site	Status	Type of facility	Contaminants
Myers Property, NJ	In design	Soil, sediment	Metals
Vineland Chemical, NJ	In design	Soil	Metals
GE Wiring Devices, PR	In design	Soil, sludge	Metals
Cabot Carbon/Koppers, FL	In design	Soil	Semivolatile organic compounds (SVOCs), polyaromatic hydrocarbons (PAHs), metals
Whitehouse Waste Oil Pits	Predesign	Soil, sludge	Volatile organic compounds (VOCs), PCBs, PAHs, metals
Caper Fear Wood Preserving	Design completed	Soil	
Moss American, WI	Predesign	Soil	PAHs, metals
Arkwood, AR	In design Complete	Soil, sludge	

Source: Adapted from Ref. 8.

consisting of screening, hydrocyclonic separation, air flotation, and a series of sludge-thickening and dewatering operations for the fines. Average concentrations in the cleaned soil were 25 mg/kg for nickel, 73 mg/kg for chromium, and 110 mg/kg for copper.

2. Monsanto Site

The Monsanto Site in Everett, Massachusetts, used soil washing on an 84-acre brownfield location. The contaminants at this site included naphthalene, bis(2-ethlhexyl)phthalate (BEHP), arsenic, lead, and zinc. Remediation began in 1996, and a 15-ton/h soil washing plant was used. Oversized materials were set at larger than 2 in. Treated soils were split into >2 mm which contained the PAPR contaminant, and the <2 mm slurry. The wet slurry was put through a hydrocyclone separation unit which produced coarse sand and fine fractions. The coarse sand was tested and was used on-site as clean backfill. The fines fraction was treated in a bio slurry reactor. The overall volume reduction of materials that had to be removed off-site was 93%.

The total volume of soil treated was 9600 tons and the remediation took 6 months. The cost of soil washing and bioremediation feed preparation was $900,000 (8).

REFERENCES

1. U.S. Environmental Protection Agency. 1993. Remediation Technologies Screening Matrix and Reference Guide, EPA 542-B-93-005.
2. Ellis, W. D., and Payne, J. R. 1983. Chemical countermeasures for in-situ treatment of hazardous materials releases, U.S. EPA Contract 68-01-3113, NJ Oil and Hazardous Materials Spills Branch, Edison, NJ.
3. Judge, C., Kostecki, P., and Calabrese, E. 1997. State summaries of soil cleanup standards. Soil Groundwater Cleanup, Nov., pp. 10–34.
4. Judge, C., Kostecki, P., and Calabrese, E. 1998. State by state groundwater standards. Soil Groundwater Cleanup, May, pp. 11–31.

5. McPhillips, L. C., Pratt, R. C., and McKinley, W. S. 1991. Effective Groundwater Remediation at the United Chrome Superfund Site. Air and Waste Management Association, U.S. EPA, Pittsburgh, PA.
6. Fountain, J. C., Waddell-Sheets, C., Lagowswki, A., Taylor, C., Frazier, D., and Byrne, M. 1995. Enhanced removal of dense nonaqueous-phase liquids using surfactants, capabilities and limitations from field trials. In: D. A. Sabatini, R. C. Knox and J. H. Harwell. (eds.), Surfactant-Enhanced Subsurface Remediation-Emerging Technologies, American Chemical Society, Washington, DC.
7. Freeze, G. A., Fountain, J. C., Pope, G. A., and Jackson, R. E. 1995. Modeling the surfactant-enhanced remediation of perchloroethylene at the Borden test site using the UTCHEM compositional simulator. In: D. A. Sabatini, R. C. Knox, and J. H. Harwell (eds.), Surfactant-Enhanced Subsurface Remediation-Emerging Technologies. American Chemical Society, Washington, DC.
8. Mann, M. J., et al. 1998. Liquid Extraction Technologies, Vol. 3. WASTECH and the American Academy of Environmental Engineers.
9. Watts, R. J. 1997. Hazardous Wastes: Sources, Pathways, Receptors. Wiley, New York.
10. Connick, C., Blanc, F. C., and O'Shaughnessy, J. C. 1985. Adsorption and release of heavy metals in contaminated soils. Proc. ASCE Environmental Engineering Division Specialty Conference, Boston, MA, pp. 1045–1052.
11. Bailey, S. J., O'Shaughnessy, J. C., and Blanc, F. C. 1986. Effects of organic compounds on the adsorption and desorption of metals in soils. Proc. ASCE Environmental Engineering Specialty Conference, Cincinnati, OH.
12. Blanc, F. C., and O'Shaughnessy, J. C. 1991. Leaching of heavy metals from contaminated soils. Proc. Aquifer Reclamation and Source Control Conf., USEPA/N.J. Inst. of Tech./ASCE/NEWPCA, Boxboro, MA.
13. US Environmental Protection Agency. 1995. Remediation Case Studies: Thermal Desorption, Soil Washing, and In-Situ Vitrification, EPA-542-R-95-005. Office of Solid Waste and Emergency Response, Washington, DC.

30
Relationships Between Heavy-Metal Mobility in Soils and Sediments and the Presence of Organic Wastes

Luis Madrid, Encarnación Díaz-Barrientos, and María Bejarano
Instituto de Recursos Naturales y Agrobiología (CSIC), Seville, Spain

I. INTRODUCTION

The recycling of wastes has become a major concern during the last decades of this century. It is well established that some wastes, especially those rich in organic compounds, have beneficial effects on several soil characteristics, and they often contain nutrients in plant-available forms which give them a definite agronomic value. These circumstances have induced farmers to use such wastes as soil amendments, either in their original form or after some composting process. Since the plant nutrient content is often low, large amounts of these wastes or composts are frequently added to soils, which may cause a buildup of any undesirable component present if high doses are used. A well-known example of this is the case of sewage sludge, which frequently contains significant quantities of heavy metals, so that some restrictions to its use in soils exist in most countries. Such restrictions refer to maximum permissible concentrations of potentially toxic elements in soil after application of sewage sludge (which depend on the soil pH), maximum annual rates of addition of metals, and maximum concentrations in the sludges themselves.

Another possible consequence is that when such wastes are added to soils they may adversely influence, in one sense or another, some physicochemical equilibria of environmental or nutritional significance to the soil or the relationships between the soil and other environmental reservoirs. An important example of such equilibria is the immobilization of potentially toxic metals, even though their total contents in the soils might not be significantly altered. Traditionally, soils have been considered highly efficient sinks for heavy metals as a final result of several reactions, including precipitation, adsorption, complexation, and inclusion in various minerals as solid solutions (1). Through such reactions, metals from various sources tend to become insoluble and accumulate in soils. How the addition of several organic wastes to soils may change the status of the metals, either by increasing their solubility or contributing to their retention, is described below.

II. ORGANIC SUBSTANCES AND METALS: TWOFOLD INTERACTION

Even though soils may act as sinks for heavy metals, the accumulation of the latter implies a long-term risk which demands that several soil properties (especially pH, redox potential, and organic matter content) should be maintained within certain limits to prevent metals from being remobilized (2). It has been known since the end of the last century that compounds with functional groups able to form strong bonds with transition metals may increase the solution concentration of heavy metals. Formation of these soluble complexes, often of high stability, may favor metal mobilization. Some selectivity between different metals is frequently reported, as shown by Xue et al. (3) in the case of a eutrophic lake, where the ratio of free to total Cu^{2+} was about 10^{-6}, while a significant proportion of Zn was present in free form or as weak complexes. According to Salomons and Förstner (4), the action of strong complexing agents, especially synthetic compounds including EDTA, DTPA, NTA, etc., may increase dissolved metal concentrations by two different but concurrent causes, namely, by desorbing heavy metals from the solid surfaces and by hindering the naturally occurring adsorption processes. The case of NTA is of particular interest, as it was considered a possible alternative to the polyphosphate in detergents, and significant concentrations may occur in sewage treatment plants.

The metal-mobilizing power of these complexing agents and other weak organic acids identifies them as possible candidates for remediation of metal contamination. Davis et al. (5,6) have shown that, in soil columns artificially contaminated with Zn or Cd oxides, EDTA and DTPA remove a high proportion of the metal added, and Wasay et al. (7) concluded that these two complexants are much more efficient for removing Pb than for some other metals, e.g., Hg or Cr. Other complexants, e.g., citrate, oxalate, or tartrate (7,8), or polyethylenimine (9), were very effective in removing extremely toxic metals such as Cd, Hg, or Pb. Toxicity of Cu^{2+} to algae occurs at concentrations as low as 10^{-10} M, and higher Cu concentrations induce the release of strong ligands from green algae and consequently an increase in complexing capacity of the solution (3).

The action of natural organic polymers present in soils on metal mobilization is more complex than that observed for simple complexing agents. Humic substances and other natural organics can act on metal solubility in two opposite ways. Their soluble components are often responsible for an increase in solubility (10–12), although this action is not uniform with respect to either the metals or the organics. For example, it has been shown that major cations such as Ca^{2+} compete with some metals such as Cu^{2+} or Pb^{2+} for the ligands, but not with others such as Cd^{2+} (11). Moreover, Spark et al. (12) observed that a humic acid contributed to metal insolubilization at slightly acid pH, while at higher pH it reduced precipitation of metals, suggesting that the interaction occurs through carboxyl groups. On the other hand, large molecules, solid particles, or organic matter covering mineral surfaces frequently contribute to the incorporation of metals in the solid phase (13–16). Some metals, e.g., Cu (13), compete for sorption sites on solid humic substances much more efficiently with protons than Ca or Cd, and the sorption preference between different metals can generally be predicted using the hard/soft acid/base principles (14).

Whether the net result of the metal/organics interaction favors increasing solubility or immobilization appears to be determined mainly by the size and solubility of the organic substances. According to Förstner and Wittman (17), fulvic acids play an important role in the transport of heavy metals in water, due to their lower molecular masses, larger number of functional groups, and much greater solubility of fulvic than humic fraction, while the humic fraction plays a role in trace-element transport and retention by sediments. In general, a definite relationship is found between natural dissolved organic matter and the mobilization of metals (18–20). Lu and Johnson (18) identified three distinct Cu complexes with humic substances in swamp water for different humic/Cu ratios using electron spin resonance. Kalbitz and Wennrich (19) observed that Cr, Hg, Cu, and As concentrations in soil percolates were correlated with dissolved organic carbon, and Witter et al. (20), using anodic stripping voltammetry, determined that 99% of the dissolved copper in rice field, river, and catchment basin waters was complexed by natural organic matter.

Metals bound to solid organic surfaces are unlikely to be immediately available to plants, unless the solid matter is degraded or some change in the chemical conditions breaks the bonds between the metal and the surface. On the contrary, uncomplexed, free metal ions in solution or weakly bound to small ligands will be the most easily taken by plants. When the solution contains soluble ligands forming strong metal complexes, the complexed metals do not usually enter the plant roots, although there are some lipid-soluble complexes which easily penetrate cells (21).

Due to their large number of active groups, differing in their nature, surroundings, etc., natural organic polymers are heterogeneous from the viewpoint of metal complexation. Thus a single value for the stability constant of the complex formed with a given metal ion cannot be defined, and although several methods can be used [e.g., anodic stripping voltammetry (20), ion-selective electrodes (22), solvent extraction (23), etc.], only apparent or average values can be estimated. The complexation capacity, defined as the maximum amount of a given metal that can be bound per gram of complexant (e.g., a humic acid) seems to be a more straightforward description of an organic polymer and can also be estimated by various techniques. However, it is unlikely that all the available binding sites will be saturated with metal under a given set of experimental conditions due to the relatively weak strength of the complexes formed by many of the functional groups (24).

III. ORGANIC RESIDUES AS METAL-MOBILIZING AGENTS

Many organic residues have been found to contain polymers with chemical characteristics similar to those of natural humic substances. Therefore, the common practice of adding them to soils may have a similar influence to that of humic substances on metal mobility. In soils receiving liquid animal manure, a correlation was found between dissolved organic carbon and the concentration of metal in aqueous extracts (25,26), as well as in soils flooded with sewage (27). In all these examples, complexation was one of the proposed causes for metal mobilization. Indeed, Japenga et al. (25) and Del Castilho et al. (28) concluded that Cu and Zn in the liquid fractions of pig slurry added to soils were associated mainly with the higher-molecular-weight fraction. Moreover, Barrado et al. (29) also found that metals formed complexes with extracts from plant litter, and could estimate their complexing constants (vide infra), while Wenzel et al. (27) concluded that soils contaminated with heavy metals could be purified by percolation with sewage of low metal content.

The practice of addition of sewage sludge to soils is of particular interest with respect to metal mobility as related to wastes, due to the frequently high metal contents of this waste. Several authors have shown that this practice increases metal mobility, and some have provided evidence of association between metals and organics in soils amended with sewage sludge. Cheshire et al. (30) concluded that Cu was in coordination with at least two N atoms in the fulvic and humic acid fractions. Keefer et al. (31) observed the association of amino acids with Zn, Cr, Ni, and Pb, carbohydrates with Zn, Cd, Cr, Ni, and Pb, and phenols especially with Cu and Pb. Previously, Keefer and Singh (32) had found that Zn was associated with the hydrophobic base fraction, whereas Ni was associated with the hydrophilic base fraction. They also found some evidence of clay/organometallic association. Munier-Lamy et al. (33) related the biodegradation of organic matter to metal availability. However, the relatively high metal contents of sewage sludge have caused many authors to attribute the apparent higher metal mobility to the metals added with the residue. Consequently, no legal restriction to the use of sewage sludge exists as long as the amounts of metals added do not add up to the permitted maximum contents. In fact, some contradictory results are found in the literature, insofar as some authors have reported increased solubility of metals present in amounts below the legal limits (34,35), whereas others have found a decrease in metal mobility (36,37).

IV. METAL-MOBILIZING ACTION OF OLIVE MILL WASTEWATER

Millions of tons of olive mill wastewater (OMW) are generated every year in Mediterranean countries from olive oil manufacturing, although in the last few years new production techniques using much smaller volumes of water has reduced the quantity. OMW is a very dark aqueous phase with a fetid smell, containing about 15% organic matter, and showing high chemical and biological oxygen demands. Its disposal into rivers has been prohibited since 1981, and the oil manufacturing plants store it in ponds, leaving it to concentrate by evaporation, before the resulting sludge is composted to generate an agronomically utilizable product. From these ponds, the soluble components of OMW may migrate slowly toward surface or ground waters, and also accidental discharges to water courses do occur. Moreover, in countries with low OMW production, its discharge into rivers is not uncommon. The average composition of this residue (38) suggests that metal complexation by some of its components can occur, and thus making it a good candidate as an efficient metal-mobilizing agent. Actually, Cabrera et al. (39) showed that OMW had a complexing capacity of 0.66 mmol of Cu per gram of freeze-dried residue.

Gonsior et al. (40) showed that complexants such as EDTA caused a definite mobilization of metals present in river sediments, although the concentrations of this chelate commonly found in soils were lower than those necessary for a significant metal solubilization. However, in occasional discharges of OMW in rivers, the concentration of chelating agents can be high enough to cause metal solubilization at a particular point and favor metal transport downstream. Bejarano and Madrid (41–44) studied the solubilization of heavy metals from contaminated river sediments by dilute solutions of OMW at several slightly acid pH values, and found that Cu, Fe, and Pb from the sediments were solubilized, while other metals, e.g., Mn or Zn, were not. In the case of Mn, the sediment even retained part of the metal originally present in the OMW. Pb was released in the presence of OMW at all the pH values studied, while Cu or Fe were released only at pH 5 and, to a lesser extent, at pH 4, while at pH 3 the OMW retained these two metals from the amounts originally present in the residue. Given that the dissolution of these metals is easier at lower pH, it is likely that removal of Cu or Fe from the sediment is related to the formation of soluble complexes with OMW rather than to the influence of pH. Figure 1, from Bejarano and Madrid's data (41), illustrates their results for Pb and Cu.

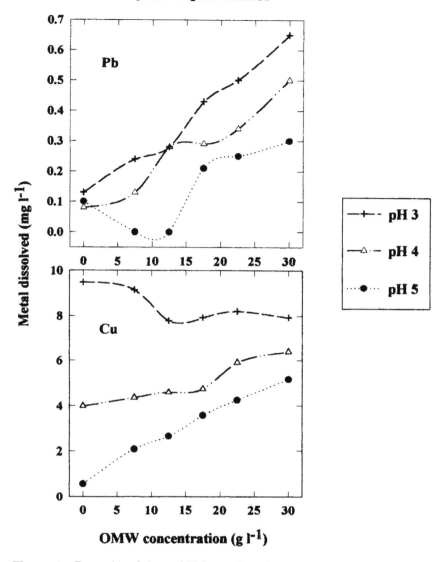

Figure 1 Examples of the mobilizing action of solutions containing OMW on metals of a river sediment at several pH values.

In order to test whether complexation was responsible for metal mobilization from the sediments, Bejarano and Madrid (43) studied the release of metals by several more diluted solutions of OMW. The resulting solutions were passed through C-18 reverse-phase cartridges, which retained most of the organic matter from the solution, and the metals complexed to it were also assumed to be retained. Previous experiments had proven that free metals were not immobilized by the cartridges. A high proportion of several of the metals present in the OMW solutions after contact with the sediment, especially Cu and Zn (Fig. 2), was retained by the cartridges, confirming that most of the metal in the OMW solution was in the complexed form. Other metals were retained by the cartridges to a considerably lower extent, showing that either complexation

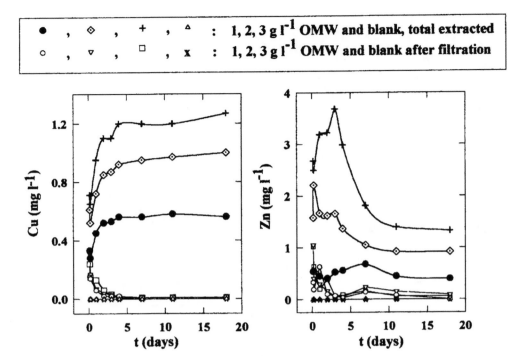

Figure 2 Time dependence of the total amounts of Cu and Zn extracted by several OMW dilute solutions from a river sediment and comparison with the amounts in solution after passing through C-18 reverse-phase cartridges.

was less significant or the complexes formed were more labile. The behavior of Zn suggests a fast dissolution and a slow readsorption of the complexes, while the Cu complexes remain in the aqueous phase.

V. METAL COMPLEXES WITH ORGANIC MATTER IN RESIDUES

Several authors have tried to estimate the stability parameters of the complexes which are postulated to form between metals and components of the residues, utilizing techniques or models previously applied to complexes with humic substances. In most cases, the basic principles described by Buffle (45) have been followed. Frequently the reaction is considered to occur between a metal ion M and a ligand L, which is considered as a single component even though it may be part of larger molecules, and a 1 : 1 stoichiometry is assumed for the complex. The known heterogeneity of the ligands implies that L is not a unique species, and it is often expressed as L_i, which represents a variable species corresponding to many different degrees of complex stability. With these ideas in mind, many authors represent the ith stability constant as

$$K_{M_i} = \frac{ML_i}{M \cdot L_i^f}$$

where ML_i is the activity of the metal–ligand complex, M is the activity of the free metal ion, and L_i^f is the activity of the free ligand of the kind i. In most cases, the latter activity

Table 1 Maximum Complexation Capacity, Conditional Stability Coefficients and Parameter n for Complexes of Several Metals with OMW, Determined by the Techniques of Zunino et al. (46,48)

Metal	MCC (mmol metal)(g OMW)$^{-1}$	Log K	n
Cu	0.470 ± 0.004	4.59 ± 0.54	1.04 ± 0.09
Zn	0.314 ± 0.015	3.36 ± 0.21	0.94 ± 0.04
Pb	0.068 ± 0.002	4.23 ± 0.15	0.88 ± 0.02
Mn	0.300 ± 0.011	2.43 ± 0.12	0.85 ± 0.02

and that of ML_i cannot be found, therefore only stability constants for an "average" complexing site are estimated, and the calculated parameter K_M will be a conditional stability coefficient, which will depend on the experimental conditions:

$$K_M = \frac{(ML)_{\text{total}}}{M \cdot L^f_{\text{total}}}$$

where $(ML)_{\text{total}}$ and L^f_{total} are the total amounts of complexed and uncomplexed ligand, and correspond to (ML_i) and L^f_i for all i (24,45).

Despite the large number of nonthermodynamic assumptions implicit in most papers which try to estimate the stability "constants" of metal complexes with components of organic residues (and, for that matter, with humic substances), a remarkable agreement is found among the results of many authors for different wastes. Bejarano and Madrid (44) used a simple model for complexation of Cu and Zn by dilute OMW solutions, assuming that the interaction occurred through formation of a mononuclear, bidentate complex and that each ligand L was a divalent anion. The total amount of L corresponded to the value of maximum complexation capacity (MCC) estimated by a cation-exchange resin method (46). The amount of metal in free form, M, was assumed to be the amount of metal in solution after passing through C-18 reverse-phase cartridges, and the amounts of complexed metal, $(ML)_{\text{total}}$, and free ligand, L^f_{total}, were estimated by difference. The calculated log K_M for Cu for various experimental conditions was about 3.5, and for Zn varied between 3.5 and 4.8. The same authors (47) estimated the stability constants of several metals with the same residue by the method of Zunino et al. (48), which assumed that the complexes were of the form M_nL, and determined the MCC for each metal. Table 1 lists these values, together with those of the parameter n, which shows that the mononuclear model fits their data very well. It was shown (47) that the stability constants were related directly to the electronegativity of the corresponding metal ion, while the MCCs were inversely related to the ion size. Although the technique and the conditions were different from those previously used (44), reasonably congruent values for the stability coefficients were obtained and they were close to those previously obtained for Cu^{2+} and Pb^{2+} by voltammetric techniques and in much lower concentrations (49). Becker and Peiffer (50) studied heavy-metal complexation by leachates from solid waste using several techniques, and distinguished at least four different groups of ligands. They obtained values for log K_M, defined as above, for a number of metals. Several of the values for Pb^{2+} were generally higher than that obtained by Bejarano et al. (49), but one of the ligand groups distinguished by Becker and Peiffer (50) gave log $K_M = 4.3$, very close to the value for Pb^{2+} in Table 1. For Zn^{2+}, Becker and Peiffer obtained values of 3.0 and 4.7 for two of their groups of ligands, in good agreement with those given by Bejarano and Madrid (44) for that metal and that shown in Table 1. By voltammetric titration, Barrado et al. (29) obtained values of 4.26 and 5.32 for Cu complexation with eucalyptus and oak extracts, respectively, again very similar to those estimated in the literature. Finally, Esteves da Silva et al. (51,52) observed Cu complexation by fulvic acids extracted from raw and composted sewage sludge, and from urban and livestock wastes, and estimated log K_M values for the 1:1 complexes of 4.18, 4.22, 4.21, and 4.51, respectively, which were also close to the estimates of other authors.

These values for log K_M are considerably lower than those reported for Cu and natural organic matter by Witter et al. (20), but these were obtained at much lower Cu concentrations than those used in the preceding examples. Using other conditions, values between 2 and 10 have been reported for Cu and aquatic humic substances over a wide range of metal concentrations (18). Values around 5 for log K_M have also been found for Fe/fulvic acid complexes (53), and between 3 and 5 for Cu/fulvic acid complexes at pH varying between 3 and 7 (54).

VI. EFFECTS OF RESIDUES ON THE METAL SORBING PROPERTIES OF SOILS

In the preceding sections, the mobilizing action of organic residues on metals has been preferentially considered on the basis of a direct interaction of organic components with metal ions, especially through formation of metal complexes. However, metal mobilization is also indirectly affected by changes in the metal-sorbing properties of soils by the presence of organic amendments obtained from various residues. As previously mentioned for natural organic matter, these changes can also occur in two different ways, either by decreasing the metal-sorbing power of the soil, e.g., through blocking sorbing sites, or by favoring metal retention, e.g., through addition of new sorbing sites or causing the precipitation of insoluble compounds. Good examples of this twofold action of residues added to soils are the contrasting data obtained for metal availability in two soils receiving OMW. Pérez and Gallardo-Lara (55) reported a decrease in Zn availability and no effect on Cu availability, with a significant residual increase in the latter, whereas Martín-Olmedo et al. (56) observed significant increases in available Cu, Zn, and other metals determined by DTPA extraction. However, the latter authors used OMW composted with other plant refuse, so that the differences may be related to modifications of the residue when being composted.

Madrid and Díaz-Barrientos (57) studied the effect of the presence of raw or composted OMW on metal adsorption by soils. One of their soils had been manured with the same OMW compost used by Martín-Olmedo et al. (56), while another had received raw OMW, as in (55). Samples of the untreated soils were aged in vitro with freeze-dried OMW in a proportion corresponding to twice the dose received in the field, and the interaction of all the samples with solutions of different metals was studied. The adsorption of Cu and Zn was significantly depressed in the aged samples, while only Zn adsorption was decreased in the manured soils, and Cu adsorption even increased slightly in the soil manured with composted OMW (Fig. 3). The authors suggested that large doses of freeze-dried OMW could coat the sorbing surfaces, while manuring with lower rates of the waste and no effect or, in the case of OMW compost, could add relatively insoluble

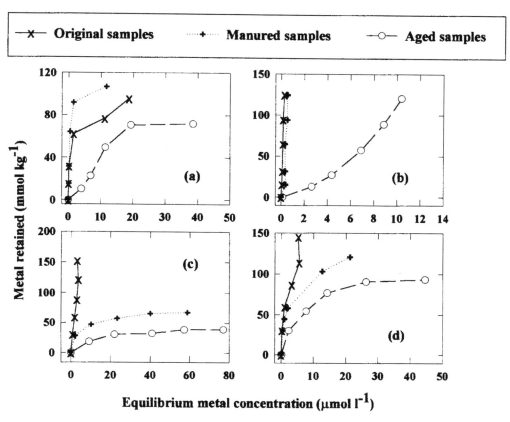

Figure 3 Adsorption isotherms of Cu (a, b) and Zn (c, d) by soils receiving amendments with composted OMW (a, c) or raw OMW (b, d), and comparison with data for the untreated soils and for the soils aged in vitro with a larger dose of freeze-dried OMW.

organic matter with sites able to bind some metals, e.g., Cu. The effect of composted (58) or raw OMW (59) on the soil sorbing surfaces can be visualized by studying the kinetics of Cu interaction with the soils using a multireaction model, which considers the existence of several kinds of sites able to retain metals (60). The model assumes that some sorbing surfaces react "instantaneously" with the metals in solution, a second group of sites reacts reversibly and kinetically, and other components of the sorbing sites retain metal by a time-dependent, irreversible reaction. Table 2 shows the values of the relevant parameters of the model estimated by Madrid and Díaz-Barrientos (58) for soils which have received composted OMW and the corresponding untreated soils: K_d is the "instantaneous" distribution coefficient, k_1 and k_2 are the forward and reverse rate constants of the time-dependent, reversible component of the sorbing surfaces, and k_s is the rate constant of the irreversible component. The instantaneous component, represented by its distribution coefficient K_d, is spectacularly increased by soil amendment, whereas the irreversible component becomes irrelevant when the soil receives the compost (Table 2). The rate constants of the reversible component, especially the forward constant k_1, were also significantly reduced upon addition of the compost. Other components proposed in the original model (60) are shown to be nonsignificant for the data or are not affected

Table 2 Values of the Kinetic Parameters of Cu Retention by Untreated Soils and by Soils Amended with Composted OMW, According to the MRM Model (60)

Kinetic component	Parameters of each fraction of the model			
	Instantaneous	Reversible		Irreversible
Soil	K_d	k_1	k_2	k_s
A orig.	9	24.5	2.79	0.04
A comp.	440	1.82	0.79	0
C orig.	18	28.4	1.15	0.15
C comp.	925	9.7	0.45	0

by the soil amendment. When a similar approach was applied to adsorption data of soils treated with raw OMW (59), K_d was not affected, while the decrease in k_1 and the disappearance of k_s were also evident. The irreversible fraction of the model can be related with the "sink" action of the soil for added metals, and is clearly affected by either amendment. On the other hand, the "instantaneous" component, which responds readily to changes in the solution concentration of the metal, can represent those sites which bind metals less strongly, and may correspond to sites on the solid components of the composted residue, which are likely to be much less relevant in the raw OMW (which is considerably richer in soluble components). The binding of metals by the solid components of composted OMW has also been reported by Gharaibeh et al. (61), who tested the processed solid residue of olive mills for the removal of heavy metals present in aqueous solutions in trace concentrations. They concluded that the residue was highly efficient in adsorbing Pb and Zn, but much less efficient for other metals, e.g., Cr, Ni, and Cd.

Various authors have provided evidence for the influence of sewage sludge on metal retention by soils. Soon (62) concluded that Cd sorption either decreased or increased after sludge application, but the changes were attributed mainly to pH changes caused by the sludge. Neal and Sposito (63) also found a decrease in Cd adsorption when soils were treated with sewage sludge, and attributed the effect to the formation of soluble, nonsorbing complexes. However, the same authors found a strong increase in Cd sorption by a soil which had been treated with sludge for a long time. They concluded that, although sewage sludge decreased the ability of soils to retain Cd, an excessive dose of the residue can have the opposite effect. Hooda and Alloway (37) found an increased Cd adsorption in soils amended with sewage sludge, while they found an increased Pb retention for some of their soils, and a decline in Pb retention for other soils. They also found that metal retention decreased as the soil/sludge mixtures were aged for up to 450 days. The same authors (64) concluded that the increase in metal retention by sludge-amended soils was the result of increases in the organic matter content and pH following the application of the residue. Long-term irrigation of soils with sewage effluent increases the contents of total and dissolved organic carbon (TOC and DOC) and decreases the contents of manganese oxides (65). Thus the final result will depend on the balance between these three parameters, as DOC enhances metal solubility (due to formation of soluble complexes), while TOC and oxides favor metal adsorption. The influence of DOC is likely to occur at low metal solution concentrations, while the influence of TOC and oxides will be more noticeable at high loading rates.

VII. EFFECTS OF ORGANIC RESIDUES ON METAL MOBILITY IN SOIL COLUMNS

Although knowledge of metal mobility in soils in situ implies the use of undisturbed soil columns, it is a well-known fact that metals or other solutes can be transported through cracks and macropores to lower depths than those reached if only reactions with soil components were involved, by what has been called nonmatrix water flow (66), or more frequently, preferential flow (67). Preferential flow is known to favor the movement of chemicals in soils, so that it may bias any study intending to characterize the chemical characteristics of metal mobilization, and causes experiments with undisturbed soil columns to be not easily reproducible. Consequently, this section will concentrate on experiments where homogenized soil columns were used. However, it must be kept in mind that homogeneous soil columns are likely to underestimate the metal mobility expected in soil profiles (67).

As in other examples in the preceding sections, experiments with soil columns are not coincident in their conclusions. For example, columns of a silty loam were shown (67) to retain all the metals added in solution to the columns, regardless of whether the metal solution contained soluble organics or not. On the contrary, Bourg and Darmendrail (68) found that a column of river sediment strongly retained Zn from a solution leached through the column, but if the Zn solution contained fulvic acid, most of the Zn passed through the column and was not retained. Moreover, Madrid and Díaz-Barrientos (69) found significantly higher quantities of Cu or Zn released from homogeneous columns of a sandy soil which had received a solution containing metal and dilute OMW compared with solutions which did not contain the organic residue (Table 3). Nevertheless, in the latter case (69), metal leaching was significantly lower if the soil had been previously treated with composted OMW in field experiments, suggesting that the compost contributed to the retention of the metals via solid surfaces with metal-binding sites. In contrast, previous treatment in the field with concentrated sugarbeet vinasse, which contains practically no solid components, generally caused a greater release of metal than in the soil with composted OMW. The authors concluded that solid organic amendments improve the metal-retaining power of the soil, while the excessive abundance of soluble organic components should be either avoided or decreased by previous treatments. Similar observations were found with Cd (70), in contradiction with the conclusions of Neal and Sposito (63), who believed that Cd was preferentially bound to complexants in solution rather than to solids.

Table 3 Percentage of Metal Added to Soil Columns Which Is Released by Leaching with Indifferent Electrolyte (69)

		Soil		
Metal	Experiment[a]	Untreated	Composted OMW	Concentrated vinasse
Cu	A	0.24	0.07	0.05
	B	11.2	4.90	7.20
Zn	A	11.3	2.93	14.9
	B	26.2	8.2	17.2

[a] A, addition of Zn and leaching with 0.01 M KNO_3; B, addition of Zn in dilute OMW solution and leaching with 0.01 M KNO_3.

Release of the metals previously retained by soil columns, when the latter are leached with soluble components of residues, has also been shown to occur. Bourg and Darmendrail (68) leached a column of Zn-treated sediment with a solution of fulvic acid and found that the metal concentration in the leachates was significantly higher than in the absence of organics. The formation of soluble complexes is probably responsible for this mobilizing action, as with synthetic chelators such as EDTA or DTPA for Zn (5), Cd (6,71), or Pb (71). Mobilization of previously retained metals by diluted solutions of OMW also occurs (69,70), and it was shown that a large proportion of the metals released were associated with the dissolved organic matter (69,72). Mobilization of metals by poultry litter extract (PLE) has also been studied from columns of both uncontaminated and metal-contaminated soils (73). No mobilization of metal from the uncontaminated soil was observed, and some Zn initially present in the PLE was retained by the soil, while the PLE leachates for the metal-contaminated soil released significant amounts of Zn and Cd. Release of Pb from the contaminated soil by PLE was not significant, and a strong complexant such as EDTA was necessary to mobilize this metal. Giusquiani et al. (74) found that urban waste compost enhanced the level of several metals in soil columns leachates. The authors concluded that the soluble organic fraction of the compost was relevant only for Cu and Zn leaching, while the compost did not affect Pb or Cd mobility.

VIII. EFFECT OF WASTES ON SEQUENTIALLY EXTRACTED FRACTIONS OF SOIL METALS

As described above, the presence of organics in soils and sediments causes changes in metal "speciation," using this term in the meaning of "describing the distribution and transformation of metal species in various media" (75). Quantitative estimation of the various chemical species in these solids is often very difficult or even impossible due to the complexity and heterogeneity of the systems involved, and therefore in this context speciation is used in a more limited meaning, as "operational procedures for determining typical metal species in environmental samples." Typical solid speciation schemes are the various sequential extraction procedures, which subject a soil or sediment to a series of increasingly strong reagents. Frequently, some particular chemical nature is attributed to the metals extracted in each fraction, e.g., exchangeable, bound to carbonate, bound to organic matter, or some others. Sequential extraction procedures have received a lot of well-based criticisms, especially on the grounds that they have not been sufficiently validated with well-defined samples, and that the fractions obtained are "operationally defined" (76). However, even though fractions with a real, definite chemical nature cannot be truly defined, sequential extraction often can serve for estimating differences in metal solubility, especially between samples of similar nature which differ only in a given property or between samples before and after a given treatment. Since many of the observations described above can be related to metal/organic associations and imply changes in solubility of the soil metals, sequential extraction can be a tool to estimate such changes in solubility or in metal bonding due to the presence of wastes. In the examples commented on below, different procedures of sequential extraction were used, so comparison between different authors is difficult. However, it is to be expected that application of organic wastes will affect mainly those fractions assumed to be related to organic matter. As shown below, this is often the case, regardless of the specific technique used.

Sposito et al. (77) found that application of anaerobically digested sewage sludge to soils over several years tended to reduce the "sulfide" fraction (extraction with 4 M HNO_3) and increased the "organic" and "carbonate" fractions (extractions with 0.5 M NaOH and 0.05 M Na_2EDTA, respectively), and they concluded that sludge applications provided metals in labile chemical forms. In sewage sludge–soil incubated mixtures, Obrador et al. (78) observed that the potentially bioavailable fractions of all the metals studied (extracted with $MgCl_2$ or NaOAc) increased at the beginning of the incubation period, whereas the organically bound fraction (extracted with an oxalic/oxalate mixture) decreased during the incubation, especially in the case of Cu, probably by decomposition of the organic matter. In a soil amended with sewage sludge and Zn or Cu carbonates, Nyamangara (79) found that most of the Cu added to the soil became complexed with organic matter, as shown by the organically bound fraction (digestion with H_2O_2), while the added Zn was incorporated to the exchangeable fraction (diluted HOAc). In mangrove soils irrigated with synthetic wastewater (80), the metals present in the wastewater were retained by the soils, and while Cu was incorporated only to the organic fraction (extracted with $Na_4P_2O_7$ solution), other metals were also bound to the soils in water-soluble and exchangeable fraction. Addition of poultry litter extract to an uncontaminated soil was found to increase Zn in the organic fraction (extracted with NaOCl), whereas if the soil was metal-contaminated, this residue solubilized Zn and Cd from the exchangeable fraction (81). On the contrary, in metal-contaminated sediments, Bejarano and Madrid (42,43) found that diluted solutions of OMW released metals mainly from the carbonate- and oxide-bound fractions.

IX. CONCLUDING REMARKS

Thus it can be concluded that the addition of organic wastes to soils can significantly influence metal mobility. The numerous examples in the literature show that metal solubility, and hence metal mobility, is considerably enhanced by the presence of organic residues, especially the water-soluble fractions, although the effect is not identical for the various toxic metals. This effect seems to be frequently related with formation of soluble complexes between metals and organic components of the wastes. However, the influence of such wastes on metal mobility does not always manifest itself as an increase in metal solubility. Often the capacity of the soils as sinks for metals is enhanced, especially when the residues contain a significant proportion of solid components. This twofold action of organic residues on metal solubility means that the effectiveness of a particular soil amendment must be evaluated in each specific case. For example, a specific waste may be a good tool for remediation of metal pollution in a particular soil, but the possible pollution of groundwater must be considered. The higher the content of soluble organic components of the waste, the greater the risk of groundwater pollution. In other cases, an increase of the fixation power of a soil for soluble forms of metals may be convenient in order to minimize the transfer of metals to the food chain through plant crops. This latter action of organic wastes appears to be improved when the residues undergo some composting treatment, which often decreases the soluble organic contents and enhances the polymerization of the organic matter.

REFERENCES

1. Alloway, B. J. 1990. Soil processes and the behaviour of metals. In: B. J. Alloway (ed.), *Heavy Metals in Soils*, pp. 261–279, Blackie, Glasgow.

2. Hesterberg, D. 1998. Biogeochemical cycles and processes leading to changes in mobility of chemicals in soils. *Agric. Ecosys. Environ.*, 67:121–133.
3. Xue, H. B., Kistler, D., and Sigg, L. 1995. Competition of copper and zinc for strong ligands in a eutrophic lake. *Limnol. Oceanogr.*, 40:1142–1152.
4. Salomons, W., and Förstner, U. 1984. *Metals in the Hydrocycle.* pp. 176–177, Springer-Verlag, Berlin.
5. Davis, A. P., and Singh, I. 1995. Washing of zinc(II) from contaminated soil column. *J. Environ. Eng.*, 121:174–185.
6. Davis, A. P., Matange, D., and Shokouhian, M. 1998. Washing of cadmium(II) from a contaminated soil column. *J. Soil Contam.*, 7:371–393.
7. Wasay, S. A., Barrington, S. F., and Tokunaga, S. 1998. Organic acids to remediate a clay loam polluted by heavy metals. *Can. Agric. Eng.*, 40:9–15.
8. Krishnamurti, G. S. R., Cieslinski, G., Huang, P. M., and Van Rees, K. C. J. 1997. Kinetics of cadmium release from soils as influenced by organic acids: Implication in cadmium availability. *J. Environ. Qual.*, 26:271–277.
9. Rampley, C. G., and Ogden, K. L. 1998. Preliminary studies for removal of lead from surrogate and real soils using a water soluble chelator. Adsorption and batch extraction. *Environ. Sci. Technol.*, 32:987–993.
10. Van den Berg, C. M. G. 1995. Evidence for organic complexation of iron in seawater. *Marine Chem.*, 50:139–157.
11. Cao, Y., Conklin, M., and Betterton, E. 1995. Competitive complexation of trace metals with dissolved humic acid. *Environ. Health Perspec.*, 103:29–32.
12. Spark, K. M., Wells, J. D., and Johnson, B. B. 1997. The interaction of a humic acid with heavy metals. *Aust. J. Soil Res.*, 35:89–101.
13. Benedetti, M. F., Milne, C. J., Kinniburgh, D. G., Van Riemsdijk, W. H., and Koopal, L. K. 1995. Metal ion binding to humic substances: Application of the non-ideal competitive adsorption model. *Environ. Sci. Technol.*, 29:446–457.
14. Jin, X., Bailey, G. W., Yu, Y. S., and Lynch, A. T. 1996. Kinetics of single and multiple metal ion sorption processes on humic substances. *Soil Sci.*, 161:509–520.
15. Muller, F. L. L. 1998. Colloid/solution partitioning of metal-selective organic ligands, and its relevance to Cu, Pb and Cd cycling in the firth of Clyde. *Estuar. Coastal Shelf Sci.*, 46:419–437.
16. Schroth, B. K., and Sposito, G. 1998. Effect of landfill leachate organic acids on trace metal adsorption by kaolinite. *Environ. Sci. Technol.*, 32:1404–1408.
17. Förstner, U., and Wittman, G. T. W. 1983. *Metal Pollution in the Aquatic Environment*, 2nd ed., p. 223. Springer-Verlag, Berlin.
18. Lu, X. Q., and Johnson, W. D. 1997. The reaction of aquatic humic substances with copper(II) ions: An ESR study of complexation. *Sci. Total Environ.*, 203:199–207.
19. Kalbitz, K., and Wennrich, R. 1998. Mobilization of heavy metals and arsenic in polluted wetland soils and its dependence on dissolved organic matter. *Sci. Total Environ.*, 209:27–39.
20. Witter, A. E., Mabury, S. A., and Jones, A. D. 1998. Copper(II) complexation in northern California rice field waters: An investigation using differential pulse anodic and cathodic stripping voltammetry. *Sci. Total Environ.*, 212:21–37.
21. Carlson-Ekvall, C. E. A., and Morrison, G. M. 1995. Toxicity of copper in the presence of organic substances in sewage sludge. *Environ. Technol.*, 16:243–251.
22. Fitch, A., and Stevenson, F. J. 1984. Comparison of models for determining stability constants of metal complexes with humic substances. *Soil Sci. Soc. Am. J.*, 48:1044–1050.
23. Takahashi, Y., Minai, Y., Ambe, S., Makide, Y., Ambe, F., and Tominaga, T. 1997. Simultaneous determination of stability constants of humate complexes with various metal ions using multitracer technique. *Sci. Total Environ.*, 198:61–71.
24. Perdue, E. M. 1988. Measurements of binding site concentrations in humic substances. In: J. R. Kramer and H. E. Allen (eds.), *Metal Speciation: Theory, Analysis and Application*, pp. 135–154, Lewis Publishers, Chelsea, MI.

25. Japenga, J., Dalenberg, J. W., Wiersma, D., Sheltens, S. D., Herterberg, D., and Salomons, W. 1992. Effect of liquid animal manure application on the solubilization of heavy metals from soil. *Int. J. Environ. Anal. Chem.*, 46:25–39.
26. Del Castilho, P. Chardon, W. J., and Salomons, W. 1993. Influence of cattle-manure slurry application on the solubility of cadmium, copper, and zinc in a manured acidic, loamy-and soil. *J. Environ. Qual.*, 22:689–697.
27. Wenzel, W. W., Pollak, M. A., and Blum, W. E. H. 1992. Dynamics of heavy metals in soils of a reed bed system. *Int. J. Environ. Anal. Chem.*, 46:41–52.
28. Del Castilho, P., Dalenberg, J. W., Brunt, K., and Bruins, A. P. 1993. Dissolved organic matter, cadmium, copper and zinc in pig slurry- and soil solution- size exclusion chromatography fractions. *Int. J. Environ. Anal. Chem.*, 50:91–107.
29. Barrado, E, Vela, M. H., and Pardo, R. 1995. Determination of acidity and metal ion complexing constants for eucalyptus and oak extracts. *Commun. Soil Sci. Plant Anal.*, 26:2067–2078.
30. Cheshire, M. V., McPhail, D. B., and Berrow, M. L. 1994. Organic matter complexes in soils treated with sewage sludge. *Sci. Total Environ.*, 152:63–72.
31. Keefer, R. F. Mushiri, S. M., and Singh, R. N. 1994. Metal-organic associations in two extracts from nine oils amended with three sewage sludges. *Agric Ecosys. Environ.*, 50:151–163.
32. Keefer, R. F., and Singh, R. N. 1986. Correlation of metal-organic fractions with soil properties in sewage-sludge-amended soils. *Soil Sci.*, 142:20–26.
33. Munier-Lamy, C., Adrian, Ph., and Berthelin, J. 1991. Fate of organo-heavy metal complexes of sludges from domestic wastes in soils: a simplified modelization. *Toxicol. Environ. Chem.*, 31–32:527–538.
34. McBride, M. B. 1994. Toxic metal accumulation from agricultural use of sludge: Are USEPA regulations protective? *J. Environ. Qual.*, 24:5–18.
35. Evans, L. J., Spiers, G. A., and Zhao, G. 1995. Chemical aspects of heavy metal solubility with reference to sewage sludge amended soils. *Int. J. Environ. Anal. Chem.*, 59:291–302.
36. Emmerich, W. E., Lund, L. J., Page, A. L., and Chang, A. C. 1982. Solid phase forms of heavy metals in sewage sludge-treated soils. *J. Environ. Qual.*, 11:178–181.
37. Hooda, P. S., and Alloway, B. J. 1994. Sorption of Cd and Pb by selected temperate and semi-arid soils: Effects of sludge application and ageing of sludged soils. *Water, Air Soil Pollution*, 74:235–250.
38. González-Vila, F. J., Verdejo, T., and Martín, F. 1992. Characterization of wastes from olive and sugarbeet processing industries and effects of their application upon the organic fraction of agricultural soils. *Int. J. Environ. Anal. Chem.*, 46:213–222.
39. Cabrera, F., Soldevilla, M., Osta, F., and Arambarri, P. 1986. Interacción de cobre y alpechines. *Limnética*, 2:311–316.
40. Gonsior, S. J., Sorci, J. J., Zoellner, M. J., and Landenberger, B. D. 1997. The effects of EDTA on metal solubilization in river sediment/water systems. *J. Environ. Qual.*, 26:957–966.
41. Bejarano, M., and Madrid, L. 1992. Solubilization of heavy metals from a river sediment by a residue from olive oil industry. *Environ. Technol.*, 13:979–985.
42. Bejarano, M., and Madrid, L. 1996. Solubilization of heavy metals from a river sediment by an olive mill effluent at different pH values. *Environ. Technol.*, 17:427–432.
43. Bejarano, M., and Madrid, L., 1996. Release of heavy metals from a river sediment by a synthetic polymer and an agricultural residue: Variation with time of contact. *Toxicol. Environ. Chem.*, 55: 95–102.
44. Bejarano, M., and Madrid, L. 1996. Solubilization of Zn, Cu, Mn and Fe from a river sediment by olive mill wastewater: Influence of Cu or Zn. *Toxicol. Environ. Chem.*, 55:83–93.
45. Buffle, J. 1988. *Complexation Reactions in Aquatic Systems: An Analytical Approach*. Ellis Horwood, Chichester, U.K.
46. Zunino, H., Peirano, P., Aguilera, M., and Escobar, I. 1972. Determination of maximum complexing ability of water-soluble complexants. *Soil Sci.*, 114:414–416.

47. Bejarano, M., and Madrid, L. 1996. Complexation parameters of heavy metals by olive mill wastewater determined by a cation exchange resin. *J. Environ. Sci. Health*, B31:1085–1101.
48. Zunino, H., Galindo, G., Peirano, P., and Aguilera, M. 1972. Use of the resin exchange method for the determination of stability constants of metal-soil organic matter complexes. *Soil Sci.*, 114:229–233.
49. Bejarano, M., Mota, A. M., Gonçalves, M. L. S., and Madrid, L. 1994. Complexation of Pb(II) and Cu(II) with a residue from the olive-oil industry and a synthetic polymer by DPASV. *Sci. Total Environ.*, 158:9–19.
50. Becker, U., and Peiffer, S. 1997. Heavy-metal ion complexation by particulate matter in the leachate of solid waste: A multi-method approach. *J. Contam. Hydrol.*, 24:313–344.
51. Esteves da Silva, J. C. G., Machado, A. A. S. C., and Pinto, M. S. S. D. S. 1997. Study of the complexation of Cu(II) by fulvic acids extracted from a sewage sludge and its compost. *Fresenius J. Anal. Chem.*, 357:950–957.
52. Esteves da Silva, J. C. G., Machado, A. A. S. C., and Pinto, M. S. S. D. S. 1997. The complexation of Cu(II) by anthropogenic fulvic acids extracted from composted urban and livestock wastes. *J. Environ. Sci. Health*, B32:469–482.
53. Esteves da Silva, J. C. G., Machado, A. A. S. C., and Oliveira, C. J. S. 1997. Quantitative study of the interaction of Fe(III) with soil and anthropogenic fulvic acids. In: M. Kaemmerer, J. R. Bailly, M. Guiresse, and J. C. Revel (eds.), *Humic Substances in the Environment*, Toulouse. IHSS, pp. 85–90.
54. Lamy, I., Cromer, M., and Scharff, J. P. 1988. Comparative study of copper(II) interactions with monomeric ligands and synthetic or natural organic materials from potentiometric data. *Anal. Chim. Acta*, 212:105–122.
55. Pérez, J. D., and Gallardo-Lara, F. 1993. Direct, delayed and residual effects of applied wastewater from olive processing on zinc and copper availability in the soil-plant system. *J. Environ. Sci. Health*, B28:305–324.
56. Martín-Olmedo, P., Cabrera, F., López, R., and Murillo, J. M. 1995. Successive applications of a composted olive oil mill sludge: Effect of some selected soil characteristics. *Fresenius Environ. Bull.*, 4:221–226.
57. Madrid, L., and Díaz-Barrientos, E. 1994. Retention of heavy metals by soils in the presence of a residue from the olive-oil industry. *Eur. J. Soil Sci.*, 45:71–77.
58. Madrid, L., and Díaz-Barrientos, E. 1996. Nature of the action of a compost from olive mill wastewater on Cu sorption by soils. *Toxicol. Environ. Chem.*, 54:93–98.
59. Madrid, L., and Díaz-Barrientos, E. 1997. Modelling the mobilizing effect of olive mill wastewater on heavy metals adsorbed by a soil. In: D. Rosen, E. Tel-Or, Y. Hadar, and Y. Chen (eds.), *Modern Agriculture and the Environment*, pp. 443–447, Kluwer, Dordrecht.
60. Selim, H. M. 1992. Modeling the transport and retention of inorganics in soils. *Adv. Agron.*, 47:331–384.
61. Gharaibeh, S. H., Abu-El-Sha'r, W. Y., and Al-Kofah, M. M. 1998. Removal of selected heavy metals from aqueous solutions using processed solid residue of olive mill products *Water Res.*, 32:498–502.
62. Soon, Y. K. 1981. Solubility and sorption of cadmium in soils amended with sewage sludge. *J. Soil Sci.*, 32:85–95.
63. Neal, R. H., and Sposito, G. 1986. Effects of soluble organic matter and sewage sludge amendments on cadmium sorption by soils at low cadmium concentrations. *Soil Sci.*, 142:164–172.
64. Hooda, P. S., and Alloway, B. J. 1998. Cadmium and lead sorption behaviour of selected English and Indian soils. *Geoderma*, 84:121–134.
65. Siebe, C., and Fischer, W. R. 1996. Effect of long-term irrigation with untreated sewage effluents on soil properties and heavy metal adsorption of Leptosols and Vertisols in Central Mexico. *Z. Pflanzenernähr. Bodenk.*, 159:357–364.
66. Dowdy, R. H., Latterell, J. J., Hinesly, T. D., Grussman, R. B., and Sullivan, D. L. 1991. Trace metal movement in an Aeric Ochraqualf following 14 years of annual sludge applications. *J. Environ. Qual.*, 20:119–123.

67. Camobreco, V. J., Richards, B. K., Steenhuis, T. S., Peverly, J. H., and McBride, M. B. 1996. Movement of heavy metals through undisturbed and homogenized soil columns. *Soil Sci.* 161:740–750.
68. Bourg, A. C. M., and Darmendrail, D. 1992. Effect of dissolved organic matter and pH on the migration of zinc through river bank sediments. *Environ. Technol.*, 13:695–700.
69. Madrid, L., and Díaz-Barrientos, E. 1998. Release of metals from homogeneous soil columns by wastewater from an agricultural industry. *Environ. Pollution*, 101:43–48.
70. Madrid, L., and Díaz-Barrientos, E. 1998. Mobility of cadmium added to a soil treated with agricultural residues. *Fresenius Environ. Bull.*, 7:849–858.
71. Kedziorek, M. A. M., Dupuy, A., Bourg, A. C. M., and Compère, F. 1998. Leaching of Cd and Pb from a polluted soil during the percolation of EDTA: Laboratory column experiments modeled with a non-equilibrium solubilization step. *Environ. Sci. Technol.*, 32:1609–1614.
72. Díaz-Barrientos, E., Bejarano, M., and Madrid, L. 1997. Use of C-18 reverse-phase cartridges for estimation of metal complexation by a residue from olive oil industry. *Fresenius J. Anal. Chem.*, 357:1164–1167.
73. Li, Z., and Shuman, L. M. 1997. Mobility of Zn, Cd and Pb in soils as affected by poultry litter extract. I. Leaching in soil columns. *Environ. Pollution*, 95:219–226.
74. Giusquiani, P. L., Gigliotti, G., and Businelli, D. 1992. Mobility of heavy metals in urban waste-amended soils. *J. Environ. Qual.*, 21:330–335.
75. Förstner, U. 1993. Metal speciation—General concepts and applications. *Int. J. Environ. Anal. Chem.*, 51:5–23.
76. Nirel, P. M. V., and Morel, F. M. M. 1990. Pitfalls of sequential extractions. *Water Res.*, 24:1055–1056.
77. Sposito, G., Lund, L. J., and Chang, A. C. 1982. Trace metal chemistry in arid-zone field soils amended with sewage sludge: I. Fractionation of Ni, Cu, Zn, Cd, and Pb in solid phases. *Soil Sci. Soc. Am. J.*, 46:260–264.
78. Obrador, A., Rico, M. I., Alvarez, J. M., and Mingot, J. 1998. Mobility and extractability of heavy metals in contaminated sewage sludge-soil incubated mixtures. *Environ. Technol.*, 19:307–314.
79. Nyamangara, J. 1998. Use of sequential extraction to evaluate zinc and copper in a soil amended with sewage sludge and inorganic metal salts. *Agric. Ecosys. Environ.*, 69:135–141.
80. Tam, N. F. Y., and Wong, Y. S. 1997. Retention and distribution of heavy metals in mangrove soils receiving wastewater. *Environ. Pollution*, 94:283–291.
81. Li, Z., and Shuman, L. M. 1997. Mobility of Zn, Cd and Pb in soils as affected by poultry litter extract. II. Redistribution among soil fractions. *Environ. Pollution*, 95:227–234.

31
Field Techniques for Sampling and Measurement of Soil Gas Constituents at Contaminated Sites

Alan J. Lutenegger and Don J. DeGroot
University of Massachusetts Amherst, Amherst, Massachusetts

I. INTRODUCTION

In the last 15 years the nature of site investigation work has expanded significantly in areas involving environmental-related problems. Many of these problems have been related to the underground storage and accidental release of petroleum products and other contaminants into the subsurface. An attractive site investigation technique that may be used in situations involving light nonaqueous-phase liquid (LNAPL) petroleum contaminants is the measurement of soil gas constituents in the unsaturated (vadose) zone above the water table.

A number of different methods for sampling and measuring soil gas have evolved, largely from trial and error, to where the level of confidence from field measurements is now very high and the results may be used for (a) delineating a zone of contamination, (b) evaluating the degree of contamination, and (c) monitoring changes in the contaminant concentration as remediation progresses. The measurement of soil gas constituents is not a standalone site investigation technique but should be used in conjunction with groundwater monitoring and soil sampling to provide a more complete view of subsurface conditions.

In addition to determining total hydrocarbon concentrations, most measurements of soil gas constituents at contaminated sites involving petroleum will also require determination of oxygen and carbon dioxide. The combined measurement of these three constituents will assist in the design of remediation strategies and will also help to evaluate the progress and success of the remediation. The measurements also provide site-specific data for use in modeling the effects of remediation and in making accurate and reliable estimates of the necessary life of the remediation.

This chapter presents a brief background summary of the use of soil gas sampling for the delineation of contaminants in the subsurface environment. A description of various tools that have been used successfully in the field is given along with a discussion of potential problems that may be encountered. Results of soil gas measurements performed at four field sites are presented to illustrate the use of this technology.

II. BACKGROUND

Soil gas sampling has proved to be a useful means of delineating the areal extent of gasoline (1), diesel fuel (2), heating oil (3), chlorinated solvents (4,5), and other light petroleum distillate spills beneath the ground surface. LNAPLs are immiscible with and lighter than water. Subsurface contamination from surface spills, leaking underground storage tanks, or failed distribution systems initially spread out over the water table. Some of the contaminants may remain in the unsaturated zone above the water table and be trapped. Because of their volatility, evaporation is a preferential transport mechanism, which tends to produce elevated concentrations of hydrocarbon vapors in the soil gas above the separate-phase hydrocarbons. This means that soil gas surveys can be a useful approach to delineating and mapping the separate-phase source and movement of petroleum contamination.

Soil gas sampling is best suited to sites that are composed of coarse-grained (granular) soils with low organic matter content (6). Infiltrating surface water, perched water tables, or fluctuations in the water table, low-permeability lenses, surface cover, fluctuations in barometric pressure, fluctuations in soil temperature, and biodegradation may all significantly change hydrocarbon vapor concentrations with increasing distance (both lateral and vertical) from a separate-phase source. Many of these factors are climatic and often seasonal, so frequent soil gas sampling is needed for an accurate quantitative measure of the state of the contamination at any given time. Aerobic biodegradation of hydrocarbon vapors consumes oxygen (7) and generates carbon dioxide (8), whereas anaerobic degradation generates methane gas (9).

Ostendorf and Kampbell (7) suggested that the occurrence of high hydrocarbon and low oxygen partial pressures is an indication of aerobic degradation from active microbes residing in soil moisture in the vadose zone. If the site is unpaved, atmospheric oxygen will diffuse downward from the ground surface. Active biodegradation reduces total hydrocarbon vapor pressures near the ground surface, especially in the presence of a surficial organic soil layer, making it difficult to detect LNAPLs by soil gas observations. At paved sites there may also be active biodegradation, but oxygen will diffuse toward the center of the source of separate-phase contaminants, where it is deleted by reactions; whereas hydrocarbon vapors diffuse away from the center, since exchange with the atmosphere is limited by the restricted surface. This means that the most favorable conditions for detecting LNAPL spills using total hydrocarbon vapors are when:

1. The soils are granular.
2. The LNAPL is volatile.
3. The water table is shallow.
4. The soil gas sampling point is located immediately adjacent to the spill.
5. The ground surface is paved.

III. STATE-OF-PRACTICE

In the past, traditional site investigation practice at petroleum-contaminated sites has relied largely on conventional test drilling and soil sampling using truck-mounted drill rigs. Much of the early site work related to soil gas originated during the test drilling for the installation of groundwater monitoring wells. Soil samples were obtained at selected intervals or locations during routine drilling using a split spoon sampler as the drilling progressed. Standard penetration test equipment was typically used to drive the spoon into

Field Sampling of Soil Gas Constituents

the ground. Not only was a soil sample retrieved for visual field classification and laboratory testing, but a check of the soil gas could be made by simply passing a portable hydrocarbon detector over the sample. This procedure could only be used to give a "yes–no" approach to qualitatively identifying the presence of contamination. Since vapors can escape, volatilize, or mix with surface atmospheric air, quantitative measurement of contamination was not possible. Additionally, this method could not be used to measure oxygen or carbon dioxide levels.

A. Headspace Measurements

True headspace analysis of soil samples involves placing the material from the split spoon inside a clean glass jar, typically a 1-qt Mason jar, breaking the soil up lightly, and placing a seal on the top. The jar is filled about halfway, and the gap between the soil and the top of the jar is called the "head space." The seal is often aluminum foil. After the jars come to ambient surface temperature the seal is pierced with the sampling end of a portable hydrocarbon meter and a sample of the headspace gas is analyzed. This procedure is considered only slightly better than passing a meter over an open sampler, since the exact procedure is not standardized and the results may be affected by the volume of the jar, the amount of soil in the jar, the degree to which the soil is broken up, the seal on the jar, temperature, and waiting time. The results should be considered qualitative and again may be best used to give a "yes–no" response. Both of these techniques require that a soil sample be obtained and therefore the nature of the soil gas in situ is compromised as a result of the sampling and handling of the sample.

B. Gas Sampling with Probes

To avoid some of the problems of dealing with soil samples, and because of the increased desire to obtain soil gas samples directly from specific locations in the subsurface, a number of sampling probes have been suggested and used in the field. Probes as simple as a hand-driven, 1.3-cm-diameter, stainless steel, thick-walled tube tipped with either a loose-fitting carriage bolt (10) or an expendable conical drive point (1,11) have been successfully deployed in the field, as have more elaborate systems with side ports (12,13) or retractable shields (14). One problem encountered with probes that are designed with an expendable tip is that only one depth may be sampled at each location. Other depths must be sampled by redriving the probe with a new point at an adjacent location. Additionally, in some cases the point may become jammed into the end of the drive rods as a result of the driving.

Deployment of soil gas sampling probes has included a number of techniques. Probes may be either driven or pushed into the ground using a manual drop hammer or hand-held electric impact hammer (10,13), using the hydraulic thrust from a drill rig or cone penetration rig (15), or using a pneumatic or hydraulic impact hammer mounted on the back of a rig or the Standard Penetration Test equipment available on most commercial drill rigs.

On most current field probes, a filter element is used to prevent soil from clogging the ports on the probe and allowing soil gas to enter the probe body. Soil gas is drawn into the probe by a hand-operated manual vacuum pump (13) or by electric sampling pumps, which may also be incorporated directly into some portable meters. For samples pulled to the surface, a small-diameter stainless steel or flexible Teflon tube is connected from the probe to the pump. At the surface, samples may be trapped for subsequent laboratory analysis (16), subsampled by syringes (13), fed directly into portable field meters (10), or fed directly into field gas chromatographs.

Permanent sampling points installed at specific locations are also used for long-term monitoring of soil gas. In most cases, permanent points are installed in drilled boreholes and are constructed of either stainless steel tubing (1) or plastic tubing (e.g., 17).

IV. SOIL GAS SAMPLING AND MEASUREMENT SYSTEMS

Use of soil gas information at contaminated sites requires that a representative sample of soil gas be obtained at a discrete position in the subsurface and that a reliable measurement of pertinent constituents be obtained for that sample. This involves methodologies related to both the sampling protocol and the measurement protocol.

A. Sampling Methods

The American Society for Testing and Materials (ASTM) *Standard Guide for Soil Gas Monitoring in the Vadose Zone* (18) divides soil gas sampling into six basic systems based on the method of collection. These are: (a) whole air—active, (b) sorbed contaminants—active, (c) whole air—passive, (d) sorbed contaminants—passive, (e) soil sampling for subsequent headspace atmosphere extraction sampling, and (f) soil pore liquid headspace gas. Active methods involve the forced movement of soil gas from the sampling horizon to a collection device, whereas passive methods rely on the passive movement of soil gas to a collection device. The headspace methods measure the soil gas present in the headspace of a contained soil or soil pore liquid sample.

The whole air—active method is typically performed with a drive probe or similar device. Once drawn to the surface and contained, soil gas samples can either be analyzed on site or transported to a laboratory for subsequent analysis. Alternatively, the sampling device can be coupled directly to a portable field detection meter. A whole air—active sampler can be designed for permanent installation (usually in a predrilled borehole) or can be driven/pushed to the required sampling depth (rapid deployment drive/push probe). This method is most suitable for granular soils that have relatively high air permeability and low moisture content. The whole air—passive method typically involves the use of a flux chamber for collection of vapor emissions at the ground surface. This method is used mostly for public health monitoring and typically cannot be used accurately to indicate contaminant concentrations at depth.

The sorbed contaminants active and passive approaches use an adsorbent material that traps soil gas contaminants. Soil gases come in contact with the adsorbent material either through an active sample stream or by passive movement. Typical applications of the passive method involve installation of samplers in the ground surface, where they are left in place for a period of time before recovery and subsequent analysis. The function of these sorbed methods is to concentrate the contaminant components of interest, which is necessary in cases where in-situ concentrations are lower than the detectable limit for the whole-air methods. However, the sorbed methods can be used only to determine the presence or absence of certain contaminants and cannot be used as a measure of in-situ concentrations.

The soil sample headspace method requires collection of a soil sample in a containment device for measurement of contaminant concentrations present in the headspace above the soil sample. Since the collection procedure is intrusive, a large percentage of soil gas and even some percentage of the solute and sorbed-phase contaminants may be lost. Furthermore, the headspace gas is not a true representation of in-situ conditions, since it is derived from the soil sample itself. Soil pore liquid headspace samples are typically collected using a device such as a suction lysimeter. This method, however, has limited use because of the relatively high expense and complexity of installing suction lysimeters (18).

The whole air—active method is the most suitable approach for soil gas sampling when measurements of volatile organic compounds (VOCs), oxygen, and carbon dioxide are required at contaminated sites containing relatively low moisture content—high air permeability soils. In such cases, soil gas samples are easily conveyed to the ground surface through probes or permanently installed sample points, either directly by soil gas meters or into a collection container. Additionally, oxygen and carbon dioxide do not lend themselves to sampling by sorptive methods, and passive sampling techniques cannot produce the necessary spatial and temporal concentrations.

B. Sampling Equipment

A variety of soil gas samplers consisting of both permanent installations and rapid-deployment probes may be used for whole-air active sampling. Rapid-deployment probes typically consist of a variety of different designs including flush filter probes, recessed filter probes, retractable-tip probes, and downhole sensor probes. Permanent installations points usually consist of clusters of stainless steel tubes with a sintered stainless steel filter element. Permanent sampling points typically require a borehole for installation, but allow temporal observations without mobilizing a drilling rig for each sampling event. They are usually placed at strategic locations so that sufficient data may be obtained to evaluate the effects of remedial action.

C. Rapid-Deployment Samplers

Rapid-deployment methods involve direct-push or direct-drive probes that may include either downhole sensors or a sampling port for bringing a soil gas sample to the surface for measurement. The advantage of this method is that during the initial phase of a site investigation, a large area can be investigated quickly, increasing the investigation without increasing the cost. In most cases, probes are deployed without drilling a borehole, so there is minimal intrusion, no drill cuttings to dispose of, and minimal surface restoration required. A probe designed with a filter element allows a vertical profile of soil gas sampling to be performed at a single location. Between profile locations, the probe is dismantled, checked for damaged parts, and all components are thoroughly cleaned and decontaminated prior to reassembly. Prior to use, a field blank of atmospheric gas is taken and analyzed to ensure that the probe is clean.

1. Flush Filter Probe with Surface Sampling

The flush filter drive probe with surface sampling consists of a tip and an intake section, with a tube attached to the filter section for sampling (Fig. 1). The design of the probe is intentionally simple with no moving parts to allow rapid field disassembly, cleaning, and repair. Probes are constructed entirely of stainless steel, with a replaceable conical tip machined with a 60° apex. The cone tip threads onto the probe body and also holds the filter element in the proper location. The filter intake section is machined to accept

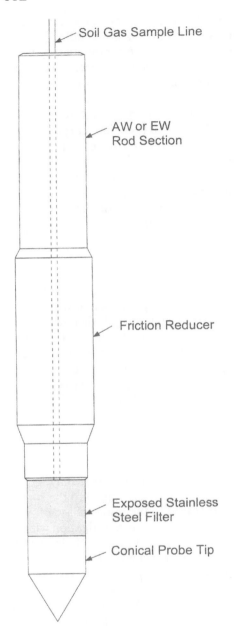

Figure 1 Schematic of flush filter probe.

a stainless steel filter element and is also equipped with an end thread to accept a compression fitting. A 3-mm-diameter flexible Teflon or polyproplene line is attached to the probe and is used to draw soil gas samples to the surface. A sampling tube is used to eliminate contamination of the gas sample from gas leakage at joints between drive rods.

Field Sampling of Soil Gas Constituents

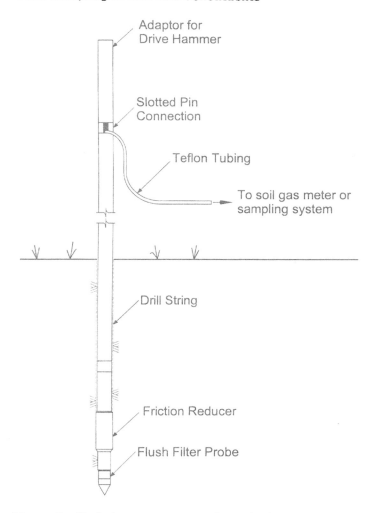

Figure 2 Typical test arrangement for rapid-deployment probe.

In order to drive the probe into the ground, a drill rod adapter with appropriate threads is first attached to the probe. In most cases, the probe is either designed with an enlarged section behind the filter element or a friction reducer is provided on the drill rod adapter. This provides a seal around the filter element and also reduces friction along the drive rods which allows for easier advance of the probe. Typically, the probe is advanced using the dynamic penetration energy from a drill rig using conventional Standard Penetration Test driving equipment available on most rigs. Standard size drill rods (AW or EW) are used to drive the probe. The flexible sampling line is threaded through the center of the drill rods and is brought out at the ground surface through a slotted drill rod sub. A typical arrangement of the test is shown in Fig. 2.

For very shallow or remote testing where it may be difficult or not cost effective to mobilize a drill rig, the size of the probe is scaled down. Typically, smaller-sized standard drill rods (EW) are used, or even smaller rods may be fabricated. In these cases, the probe may be advanced by hand using either a manual drive hammer or a portable electric impact hammer as shown in Fig. 3. Recently, a number of commercial rigs outfitted with either hydraulic or pneumatic impact hammers have become available. In each of these cases, the arrangement of the test is essentially the same as shown in Fig. 3.

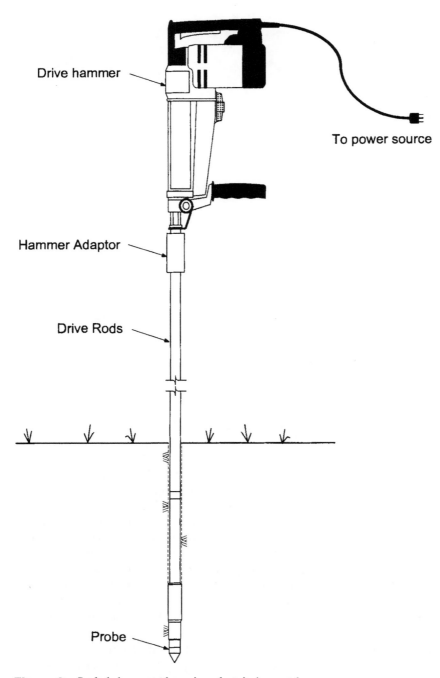

Figure 3 Scaled-down probe using electric impact hammer.

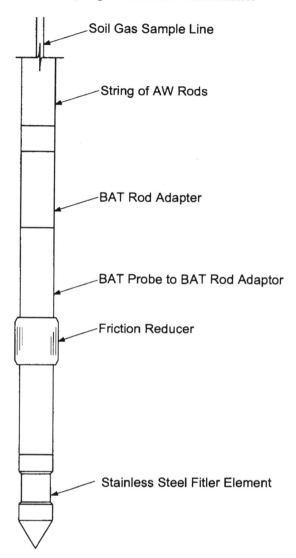

Figure 4 Schematic of recessed filter probe.

2. Recessed Filter Probe with Surface Sampling

An alternative design to the flush-mounted filter is a probe equipped with a recessed filter element. An example of this design is shown in Fig. 4. The probe is actually a commercially available BAT groundwater sampling probe (19) that was retrofitted by the authors for soil gas sampling. A stainless steel threaded fitting equipped with a stainless steel compression fitting is substituted for the conventional BAT rubber septum and fitting that is supplied with the BAT probe. A 3.2-mm-diameter flexible Teflon sampling line is then attached to the compression fitting and threaded inside the drive rods as described before. Difficulties have been encountered using a recessed filter probe, especially if the probe is driven through organic or fine-grained lenses. These materials tend to adhere to the filter element and stay attached as the probe is advanced. This does not appear to be the case with a flush-mounted filter, as soil is removed from the face of the probe during driving to the next test depth.

3. Retractable-Tip Probe with Surface Sampling

Some investigators have used a probe with an enclosed filter element that is protected during driving and then extended to allow sampling. One thought in using this type of probe is to reduce cross-contamination between sample depths. The retractable-tip probe with surface sampling used by the authors consists of a sliding tip, a filter housing containing a sintered stainless steel filter, and a tip housing (Fig. 5). The tip is retained

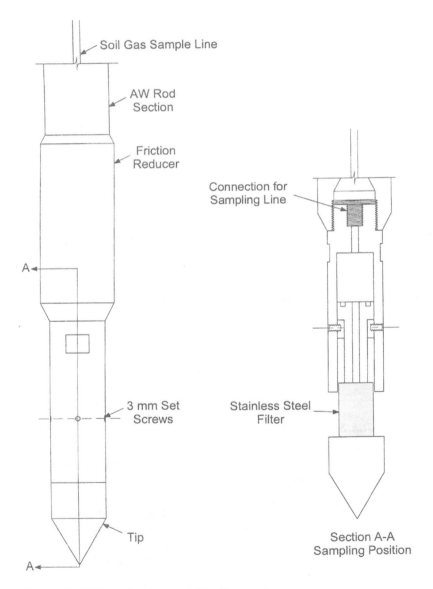

Figure 5 Schematic of retractable filter probe.

Field Sampling of Soil Gas Constituents

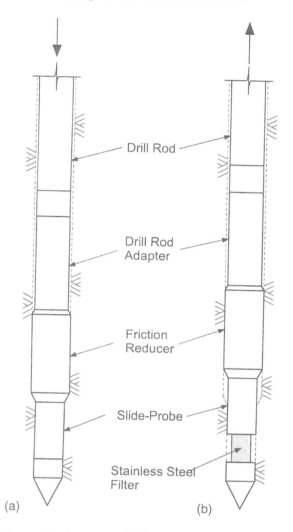

Figure 6 Sequence of using a retractable-tip probe.

in the housing by a ring held in place by four set screws. The ring limits the travel of the tip when the probe is retracted to expose the filter. A series of 3-mm-diameter holes are located around the circumference of the filter housing to allow a sample of soil gas to be drawn into the probe. The back end of the filter housing is tapped to accept a compression fitting. Soil gas samples are drawn to the surface through a flexible line as described before.

During driving, the probe is in the closed position with the filter element inside the probe body. After driving to the desired sampling depth, the drill string is pulled back approximately 5 cm to expose the sampling ports. After sampling, the probe is then driven to the next sampling depth. This sequence is shown in Fig. 6. There are several difficulties that may be encountered using retractable probes. In order to expose the filter element and obtain a sample, the drill string must be pulled backwards. If there is any soil material inside the probe body, the probe may not open. During the retraction of the drive rods, the driller may inadvertently pull too far, creating an open pathway for short-circuiting of atmospheric gas. Once the probe is extended, soil material may enter in between the body and tip so that the probe may not close for driving to the next depth. There are some mechanical ways to reduce some of these problems, but in general, there is really no advantage to using a retractable-tip probe.

4. *Flush Filter Probe with Downhole Oxygen Sensor*

A simple method to eliminate the need to bring a soil gas sample to the ground surface is to equip the probe with a downhole sensor. An example of this type of probe, as used by the authors, involves a flush filter drive probe with downhole sensor consisting of a tip, an intake assembly with a filter element, and a sample chamber containing a remote oxygen sensor (Fig. 7). The probe is constructed of stainless steel with a 60° apex conical tip on one end and a 13-mm fine thread on the other end. The tip is threaded onto a filter intake body which accepts a stainless steel filter element. This is the same design

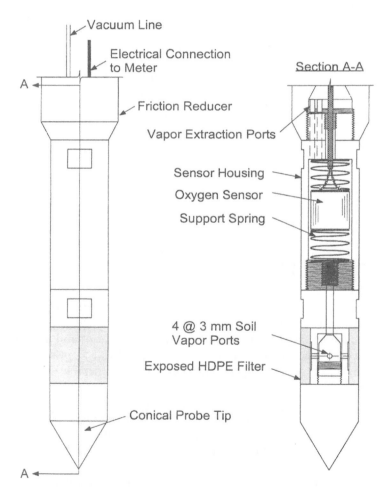

Figure 7 Schematic of flush filter probe with downhole oxygen sensor.

described previously. The intake body is attached to the main body, which contains a sample chamber that houses an electrochemical oxygen sensor. A 2-mm-diameter stainless steel tube is soldered to the top of the chamber to create an extraction port, and a 2-mm-diameter flexible tube is connected to this port and to a hand vacuum pump at the surface. The oxygen cell is connected to a digital oxygen meter with a no. 2 conductor wire soldered to the terminals on the cell and fed through a compression fitting at the top of the chamber. The soil gas oxygen concentration at each sampling depth is determined by drawing a sample into the downhole chamber with the vacuum pump and then reading the digital meter. Ambient air is used as the control for calibration of the sensor. To avoid damage to the oxygen sensor during driving, the sensor is cushioned using inert padding material and/or isolation springs.

D. Permanent Sampling Points

The design of permanent monitoring points for use at petroleum-contaminated sites requires that the points be constructed of inert materials that will not degrade over time and compromise the quality of samples. For most of the field studies performed by the authors during the past 5 years a simple design consisting of rigid stainless steel tubing with an attached sintered stainless steel filter element has proven to be very useful. Figure 8 illustrates the construction. In most cases, 6.4-mm-diameter tubing with an internal diameter of 4.0 mm is used. This size of tubing provides sufficient rigidity for installation and gives a small volume (13 mL per meter of length) for dead gas volume. The lower end of the tube is equipped with a compression fitting with a 12.7-mm-diameter, 40 μm sintered steel filter element. The upper end of the tube is equipped with a compression union connector with a removable cap. During sampling, the cap is removed and a flexible sampling tube is attached.

Installation of permanent points is accomplished using a drill rig and hollow-stem augers. After drilling to the desired depth, a bed of clean sand (2 mm diameter) is placed at the bottom of the borehole. The lowest point is inserted, additional sand pack is placed, and the hollow-stem augers are withdrawn to the next level. During withdrawal of the augers, care is taken not to disturb the point just installed. This procedure is repeated until all points are installed and a cluster of points has been created. To complete the construction, a well cover is installed. Figure 9 illustrates the completed construction of a permanent sampling point cluster. Typically, the points are installed at intervals of 0.5 m.

An alternative installation procedure is to assemble all of the points into a single cluster unit and place the entire cluster inside the hollow-stem augers. Using this technique, the boring is advanced to the required depth and the assembled sampling cluster is placed inside the augers. The augers are withdrawn, taking care to ensure that the sampling ports remain at the correct elevation, and the borehole is backfilled with clean native soil cuttings. During backfilling, the cuttings are continuously tamped around the cluster using a small-diameter capped polyvinylchloride (PVC) pipe to prevent short-circuiting. Ideally, the backfill should consist of native soil as close as possible to the same porosity as the surrounding in-situ soil. Consideration can be given to using a bentonite seal in between each point, but if the native soil is a low-moisture-content granular material, significant desiccation cracking of the bentonite would likely result after initial installation, making the seal ineffective.

An alternative design for the installation of permanent sampling points is to use a sacrificial tip and filter element on the end of a drive rod. As shown in Fig. 10, the

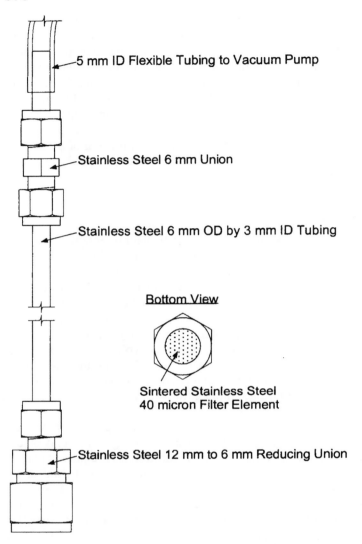

Figure 8 Schematic of stainless steel permanent sampling point.

tip has a flexible tube attached to the back and the point essentially resembles a very small flush-mount filter probe. After driving to the desired depth, the rod is withdrawn, leaving the tip, filter element, and tubing in the ground. Clean backfill is then placed in the hole and some type of surface protection is provided. This method eliminates the need for drilling a borehole for placement of the sampling point, but allows only one point to be installed at each location.

In some situations, where the contamination does not involve volatile organic compounds, an alternative to using stainless steel as a permanent sampling point is to use sampling points constructed of inexpensive materials. DeGroot et al. (20) described the use of oxygen sampling points constructed of polypropelene or nylon sampling tubes attached to the outside of a schedule 40 PVC pipe using nylon cable ties as shown in Fig. 11. The ends of the individual tubes are located at increments along the length

Field Sampling of Soil Gas Constituents

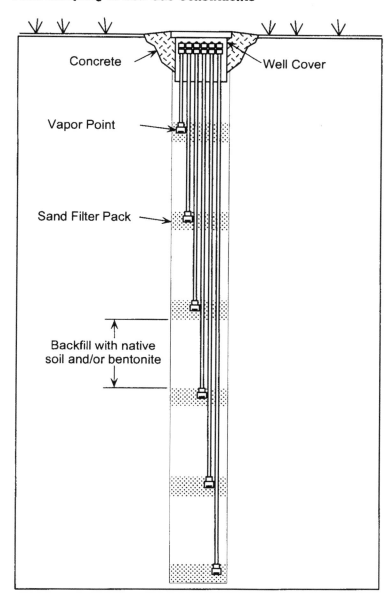

Figure 9 Completed permanent sampling point cluster.

of the PVC pipe. The open end of the sampling tubes are covered with 20-μm nylon mesh to prevent clogging. The other end of the tubes are capped with compression fittings to provide a means of attaching the surface sampling equipment. Alternatively, rigid PVC sampling tubes may be used and can be constructed from 13-mm-ID schedule 40 PVC pipe. In addition to using the tube for measuring oxygen, in some cases these tubes have been equipped with a moisture cell and a thermocouple (Fig. 12). A compression union is provided at the top of each tube. At the base of each tube, a series of 3-mm-diameter holes is drilled to create sampling ports. The sampling ports are protected from clogging

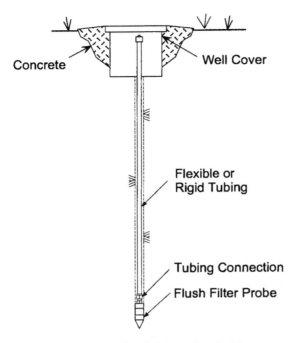

Figure 10 Schematic of driven detachable permanent sampling point.

by wrapping 20-μm nylon mesh around the tube and securing it with nylon cable ties. These sampling tubes are installed inside hollow-stem augers in the same manner used for stainless steel sampling points.

V. MEASUREMENT OF SOIL GAS CONSTITUENTS

A. Instruments

Analysis of soil gas samples can be performed using a variety of instruments ranging from portable hand-held meters to sophisticated laboratory-based analytical instruments. Hand-held meters are relatively inexpensive, easy to use, and portable. However, they suffer from limited sensitivity and accuracy, and they cannot give specific compound identification. Therefore, they are best suited for use as a screening tool and for detection of relative changes (spatial and temporal) in soil gas concentrations. Portable and laboratory gas chromatographs (GC), and laboratory gas chromatograph-mass spectrometry (GS-MS) devices are the most versatile and accurate for soil gas analysis and can provide detailed chemical composition information. However, these devices are expensive, require a high skill level to operate, and are somewhat sensitive to operation in the field, even in an enclosed vehicle. Portable hand-held meters and GCs have the advantage of providing immediate measurements, eliminating the need for storage and transportation of samples.

Portable hand-held meters are available for measurement of VOCs, O_2, and CO_2. There are many types of portable total VOC meters available, but the most common ones used for soil gas analysis are the flame ionization detector (FID), photoionization detector

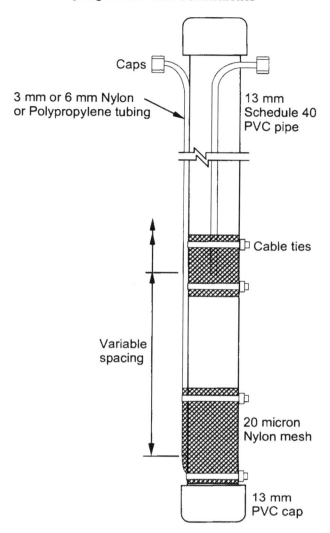

Figure 11 Schematic of flexible polypropylene sampling tubes.

(PID), and the catalytical combustion detector. The meters are usually battery powered and contain a pump that allows for whole-air active sampling of soil gas. Typical pump flow rates are 1–2 L/min, and full-scale range of most meters is usually from 0 to 10,000 ppm. Most meters are calibrated using a reference gas such as methane or hexane. When testing gas with a high contaminant concentration and low oxygen concentration, the detectors in these meters do not operate properly, and can even shut down. Typical oxygen meters use an electrochemical sensor (potentiometric or galvanic cell) to determine oxygen concentrations. Most meters have a built-in pump like VOC meters and display oxygen concentration in the range from 0 to 20.9%. Calibration is usually performed using ambient air with an assumed oxygen concentration of 20.9%. Carbon dioxide meters typically use an infrared absorption and are calibrated using a cylinder of gas with a known percentage of CO_2 in a carrier gas such as nitrogen.

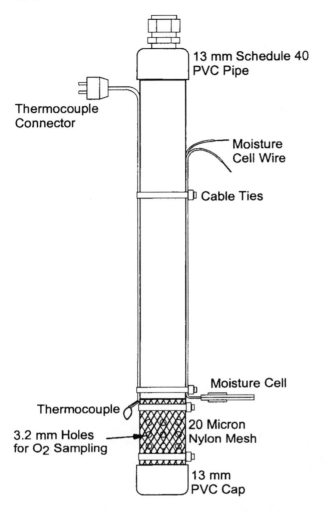

Figure 12 Schematic of rigid PVC sampling tubes.

B. Direct Read Versus Sample Collection

Measurement of soil gas constituents with whole-air active samplers can be made using either the direct-read technique or sample collection with subsequent measurement technique (hereafter called the Tedlar bag protocol). The direct-read method consists of connecting a portable meter directly to the sampler tubing and using the meter's internal pump to draw the soil gas sample into the meter. The Tedlar bag protocol (21) uses a separate vacuum pump, regulator, and flow meter system that is connected to the sampler tubing as shown in Fig. 13. This system is used to draw a soil gas sample to the surface at a controlled rate into an evacuated 3-L nitrogen-cleaned Tedlar bag. Once filled with a sample, the Tedlar bag is disconnected from the pump assembly and connected to a gas meter (or subsampled for gas chromatographic analysis) to make the soil gas reading. A minimum of 1.5 L is required when making a set of VOC, O_2, and CO_2 readings using portable meters.

Field Sampling of Soil Gas Constituents

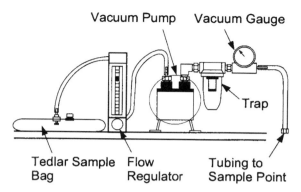

Figure 13 Tedlar bag sampling system.

The direct-read method is quick, simple to perform, and involves a minimum number of connections. However, the method suffers from potential meter failure and/or erratic readings under low-flow conditions due to low soil air permeability or the sampler clogging with in-situ moisture. This problem arises from the fact that the vacuum pumps built into portable meters are relatively weak and are designed for open-air, ambient conditions. The Tedlar bag sampling system can largely overcome this problem, since the sample is first drawn into the Tedlar bag using a separate vacuum pump. In addition, it allows for conducting serial dilution (described in the following section) and subsampling for gas chromatographic analysis. This system requires more line connections (Fig. 13) than the direct-read method, which can produce leaks to ambient air especially when the vacuum pressure is high. It is therefore important that a vacuum gauge and flow meter be used together to monitor and adjust the sample stream rate. The flow meter allows the flow rate to be decreased when the vacuum pressure becomes too high so as to minimize the potential for a short-circuit to occur. It is also important that all line connections be checked frequently for leaks. This method is more time consuming than the direct-read method, and the Tedlar bags need to be cleaned with inert gas between each reading to ensure sample quality; however, these problems are considered to be a minor trade-off considering its advantages as noted above.

C. Serial Dilution

Soil gas VOC concentrations can vary significantly from low-level (undetectable) background readings to very high concentrations directly above a contaminant plume. Background readings can often be below the lower detection limit of some meters, whereas high concentrations can saturate some sensors and adversely effect instrument performance. Deyo et al. (1) and Robbins et al. (21) investigated a serial dilution method for high-concentration samples to overcome problems such as off-scale readings and/or to avoid problems with low oxygen samples for flame-based sensors. Figure 14 shows a schematic of the serial dilution setup used at UMass Amherst for dilution of high-concentration VOC and CO_2 samples. For VOC readings, a 200-mL soil gas sample is first extracted from its Tedlar bag using a Teflon magnum syringe controlled with a pneumatic piston and pumped into a second, clean, Tedlar bag. The sample is diluted by adding 1.8 L of zero-grade compressed air. For CO_2 readings, the same procedure

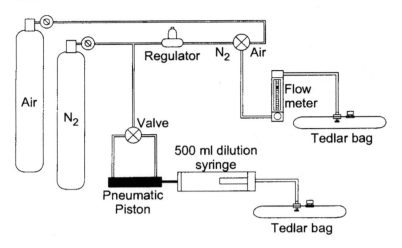

Figure 14 Schematic of serial dilution procedure.

is followed except that nitrogen is used as the dilution gas. In both cases, the actual VOC or CO_2 reading is 10 times higher than the diluted value read by the meter.

VI. QA/QC

Quality control/quality assurance (QA/QC) procedures must be developed as part of any soil gas sampling program to ensure accuracy and reliability of the results. During sampling, false negatives (i.e., falsely low values) can be caused by dilution due to leaks or short-circuits, improper purging of the samplers, and biological and chemical degradation. False positives (i.e., falsely high values) can be caused by cross-contamination and improper decontamination of sampling equipment, containers, and instruments. Many natural in-situ conditions such as perched water tables, clay layers, etc., can lead to false negatives or positives. Collection and analysis of QA/QC samples, such as field blanks, travel blanks, sample container blanks, and sample replicates are necessary for validation of the sample program (18). Field blanks are used to monitor the sampling system for contamination, whereas travel blanks are used to monitor handling and transportation procedures when off-site measurements are made. Replicate samples are used to evaluate the precision of the sampling and analysis and should consist of no less than 10% of the total number of samples collected (18). Instruments should be regularly calibrated and checked with sample blanks to check for contamination.

VII. TEMPERATURE

Soil temperature influences volatilization of hydrocarbons, soil gas diffusivity, and the rate of biodegradation. Accurate interpretation of soil gas measurements is therefore affected by temperature and seasonal variations in subsurface temperature should be accounted for. Measurement of ground temperatures using a nested thermocouple cluster is a relatively inexpensive and simple technology that has proven to be reliable for long-term measurements. Thermocouples are constructed using copper-constantan wire fitted with a miniature connector at the top. A typical installation uses thermocouples

Field Sampling of Soil Gas Constituents

installed at 0.5-m intervals in the upper 3 m and at intervals of 1 m thereafter. The thermocouple cluster is created by inserting individual thermocouples beginning with the deepest one in a drilled hole and then backfilling with clean native soil that is lightly tamped. A short section of lead wire is used at the top to allow connection to a digital thermocouple readout. The readout is calibrated using an external thermocouple and mercury thermometer to obtain ground surface ambient air temperature when measurements are made. The installation is completed by placing a locking well cover at the ground surface. A typical installation is shown in Fig. 15. An alternative installation,

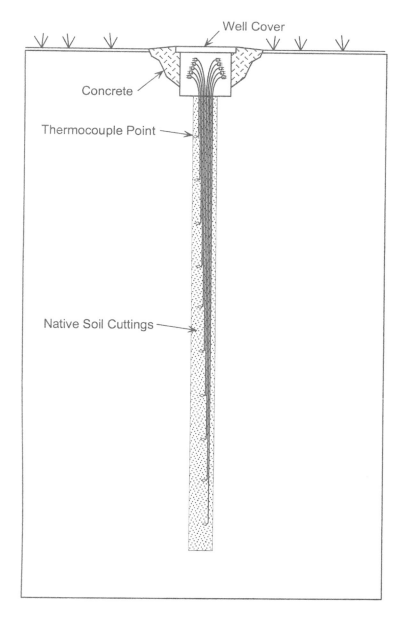

Figure 15 Typical installation of thermocouples.

used in conjunction with a permanent rigid PVC soil gas sampling cluster, is shown in Fig. 12.

VIII. FIELD CASES

The reliability of soil gas sampling based on measurements from whole-air active soil gas samplers was tested over several years at four contaminated sites located in New England. Two of the sites were locations of service stations with hydrocarbon contamination from leaking underground storage tanks and waste oil, one was a location of jet fuel contamination, and the fourth was the location of a project investigating the effect of the highway deicer calcium magnesium acetate on groundwater quality. To illustrate some of these problems, results of measurements made at the four field sites are used. The sites are research locations investigated over the past several years by UMass Amherst researchers.

A. Site Descriptions

Site A is located in eastern Massachusetts along Route 128 (22,23). In 1987, separate-phase petroleum was identified in several monitoring wells, apparently the result of leaking underground tanks located on the site. The site is predominantly covered by an asphalt parking lot. Subsurface conditions are highly variable and consist of heterogeneous nonuniform sands with gravel and variable amounts of fines. The water table varies seasonally from 2 to 3 m below the ground surface.

Site B is located on a decommissioned Air Force Base in northern New York. By contrast with Site A, this site consists of very uniform fine to medium sand with a very small amount of fines. The water table is located at a depth of about 11.5 m below ground surface and fluctuates only about 0.3 m seasonally. Contamination at the site consists of residual aviation jet fuel that has infiltrated from the ground surface periodically over the past 25 years. The surface of the site is unpaved.

Site C is located in eastern Massachusetts on a decommissioned Army base. The site was previously used as a service station, with the location of test drilling used as a dump for waste motor oil. The soil conditions at the site typically include mixed coarse sand and gravel with a water table located at a depth of about 6 m.

Site D is located in southeastern Massachusetts along the shoulder of state Route 25 (20). Winter deicing along this section of highway has been accomplished using calcium magnesium acetate (CMA) as an alternative to conventional road salt. Because of potentially sensitive vegetation adjacent to the roadway, the fate of the CMA in the groundwater is of concern. The site consists of very dense, compacted granular road material overlying native coarse-grained sand and gravel. The water table is located at a depth of about 3–4 m, depending on the shoulder topography.

B. Typical Soil Gas Results

Soil gas data collected at all four sites were typically obtained using drive probes and permanent sample points. For Sites A, B, and C, which had hydrocarbon contamination, varying combinations of stainless steel permanent sample point clusters (Figs. 8 and 9), and flush filter (Fig. 1), recessed filter (Fig. 4), and retractable-tip drive probes (Fig. 5) were used. Readings of VOC, O_2, and CO_2 were typically made using the Tedlar

Field Sampling of Soil Gas Constituents

bag sampling protocol (Fig. 13) and, in cases of high VOC/low O_2 concentrations, the serial dilution procedure (Fig. 14) was used. For Site D, which did not involve VOC contamination, O_2 and CO_2 data were collected from clusters of permanent 3-mm flexible sampling tubes (Fig. 11) and rigid PVC tubes (Fig. 12) and a flush filter probe (Fig. 1; using an HDPE filter element instead of stainless steel). Readings at this site were taken using both direct-read and the Tedlar bag system.

Figures 16 and 17 show soil gas data for Site A taken from permanent stainless steel sample clusters, flush and recessed filter tip drive probes, and soil headspace analysis of SPT samples. The direct-read method was used for the headspace samples. The data in Fig. 16 show high VOCs, very low O_2, and high CO_2 readings. These readings were taken directly above the footprint to the separate-phase petroleum plume. The data do not show any trend with depth due to the surface asphalt layer, which allows little to no vertical movement of VOC and CO_2 gas from the subsurface or atmospheric O_2 into the subsurface. The headspace data in Fig. 16 indicate the probable influence of atmospheric dilution of the soil gas during collection and transfer of the soil samples to Mason jars for subsequent headspace analysis.

Figure 17 plots data showing the temporal variation of soil gas data for two locations at Site A: one directly above the separate phase plume and one 60 m away in a region outside its footprint. Both sets of data were collected using stainless steel permanent sampling points set at 2 m below ground surface. The data clearly show the influence of the separate-phase plume location on soil gas VOC, O_2, and CO_2 readings. This type of soil gas data collected at the site was successfully used to estimate the spatial distribution of the separate-phase plume that was subsequently confirmed through a groundwater sampling program.

Figure 18 shows VOC, O_2, and CO_2 data versus depth at Site B obtained with a flush filter drive probe. These soil gas profiles show typical trends of increasing VOC, decreasing

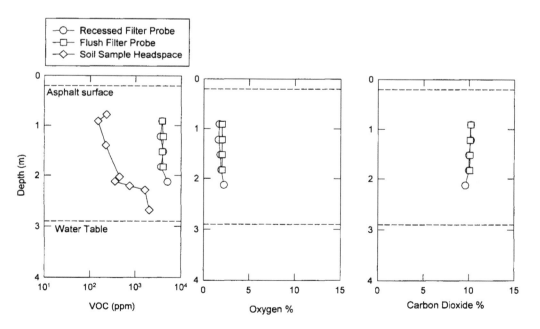

Figure 16 Vertical distribution of soil gas at Site A.

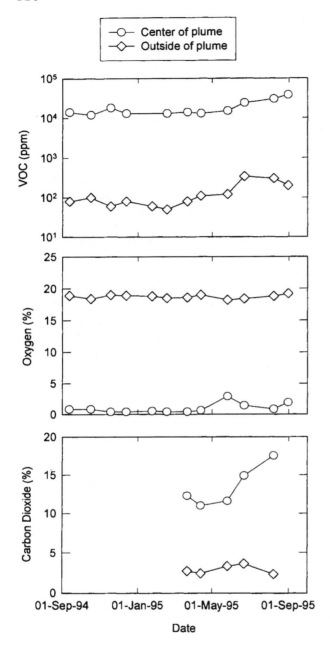

Figure 17 Variation in soil gas constituents with time at Site A.

O_2, and increasing CO_2 as the sample depth increases toward the water table. In contrast to the data collected at Site A, the soil gas measurements at this site are near ambient values at and near the ground surface, i.e., near 0 ppm for VOC, 20.9% for O_2, and 0% for CO_2. The unpaved condition of the site allowed for free movement of soil gas at the ground surface. Similar data trends were measured at Site C using stainless steel permanent sampling points and the flush filter, recessed filter, and retractable-tip drive

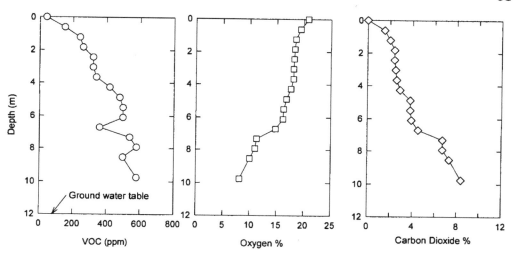

Figure 18 Vertical distribution of soil gas at Site B.

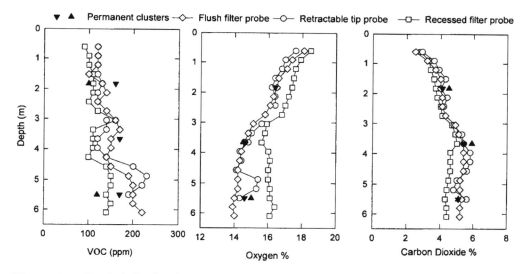

Figure 19 Vertical distribution of soil gas at Site C.

probes (Fig. 19). Although Site C was paved, there were open slopes around the perimeter of the site that allowed for connection between subsurface soil gas and atmospheric air. The measurements for Sites B and C were taken directly over their separate-phase contamination plumes, but the VOC readings are lower at Site C due in part to the contaminant at that site consisting of waste motor oil, which is less volatile than the JP-4 jet fuel at Site B.

Figure 20 plots oxygen concentrations measured for a cluster of 3-mm flexible polypropylene sampling tubes installed at Site D. The data show the presence of relatively high oxygen concentrations near the ground surface, followed by a gradual decrease in the concentrations with depth. The data also indicate a temporal variation, with the lowest

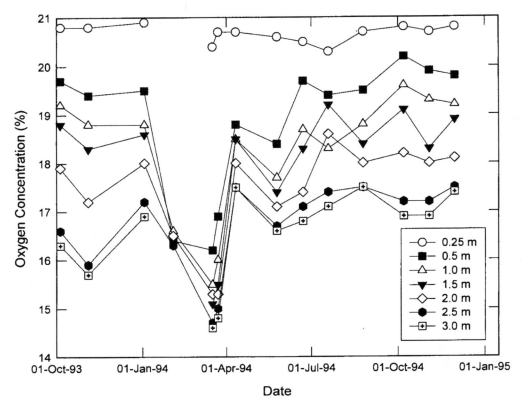

Figure 20 Variation in oxygen with time at Site D.

values being measured during the month of March 1994. These results indicate the probable influence of calcium magnesium acetate (CMA) being used on the highway during January–March 1994. During these months there were numerous precipitation events that required the use of CMA on the highway. However, the ground was frozen during January and February, thus preventing much infiltration of snow melt. As temperatures increased during March, the near-surface soil thawed, allowing infiltration of CMA-laden snow melt and rain. This can result in a decrease in the oxygen concentration within the soil, which was found to be the case for the two sampling dates in March. As the CMA is aerobically degraded and/or passes through the unsaturated zone to the water table, the oxygen concentrations should eventually return to the prewinter values. This was also found to be the case, with the 1994 summer and fall readings being nearly identical to the prewinter values. Similar temporal and spatial trends were also observed in a number of flush filter probe (Fig. 1; using a HDPE filter element) profiles as shown in Fig. 21.

Temperature measurements were made at three of the sites using the thermocouple system described previously and shown in Fig. 15. Temperature data are important when analyzing temporal trends in soil gas concentrations and are a necessary input to contaminant transport modeling. Figure 22 plots data measured at Site B, which are typical for northern climates. Temperatures vary seasonally at shallow depths but show little variation at greater depth. The data in Fig. 23 display a sinusoidal temporal variation that is dampened with depth due to the insulating effect of overlying soil.

Field Sampling of Soil Gas Constituents

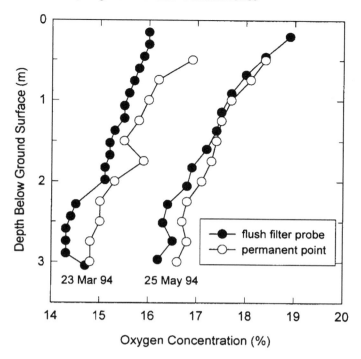

Figure 21 Vertical distribution of oxygen at Site D.

IX. SAMPLING ISSUES

The previous sections of this chapter describe a variety of meters and samplers that can be used for soil gas sampling. While this variety provides users with a broad selection of options, variations in equipment and procedures used can result in significant differences in measured values. Consistency in equipment and procedures at a site is important for obtaining meaningful data. DeGroot and Lutenegger (24) discussed several issues related to reliability of obtaining accurate soil gas measurements. Some important issues that influence sample quality, including purge volume, meter limitations, short-circuiting, and direct-read versus sample collection with subsequent measurement, need to be considered. Variations in these procedures can sometimes result in significantly misleading results and therefore it is important to consider potential sources of error so that sampling errors can be minimized, thereby increasing the reliability of soil gas measurements. This section presents examples that highlight some of these important issues. All data were obtained from Site A, previously described. UMass Amherst researchers obtained some of the data as part of field work that was explicitly investigating soil gas sampling issues. However, other data were obtained during the early stages of the work as experience in soil gas sampling was being developed. Hindsight allowed for inconsistencies in the measured data to be recognized and the sampling issues that can adversely affect data to be identified.

A. Purge Volume Prior to Sampling

Recommendations in the literature vary as to how soil gas sample systems should be purged prior to taking readings. Christy and Sprandlin (14) suggest that readings be taken

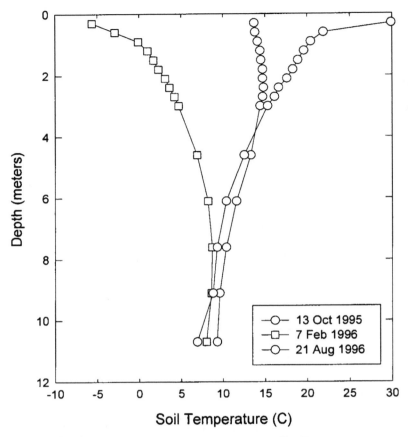

Figure 22 Variations in ground temperature at Site B.

on the gas stream immediately upon sampling, whereas Burris et al. (25) suggest that several purge volumes first be drawn before taking a reading in order to get a more standard sample concentration. Figures 24 and 25 present results showing the influence of purge volume on readings from clusters of the stainless steel permanent samplers installed at Site A. Figure 24 shows data from samples that involved taking a reading on the first 1.5 L drawn from the sampler, followed by a reading on a second 1.5-L sample taken immediately after the first. The results show slightly higher oxygen readings and lower carbon dioxide readings for the first sample and mixed results for the VOC data. However, the differences between the first and second sample readings are not significant and suggest that purging large volumes of soil gas prior to taking readings is unnecessary, especially if the sampler volume is small.

Further evidence that purging large volumes is unnecessary is shown in Fig. 25a, which plots readings for eight 1.5-L Tedler bag samples taken in rapid succession. Other than a slight change in readings from the first to the second sample, the readings remain approximately constant with continued purging. This was not the case for another sampler, as shown in Fig. 25b. However, the effect shown here, of increasing oxygen and decreasing carbon dioxide and VOC, was determined to be due to a short-circuit leaking ambient air into the samples. The short-circuit was found to be along the outside

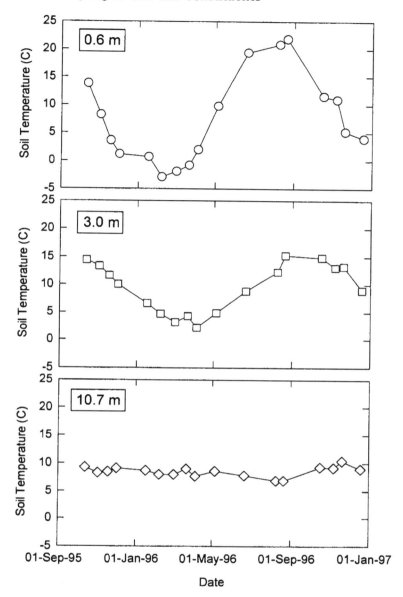

Figure 23 Seasonal variations in ground temperature at Site B.

wall of the sample riser tube from the ground surface to the sampler port. The change in values from the second to the third reading was due to a 4-h time delay between these readings, allowing in-situ conditions to reestablish—further evidence of the presence of the short-circuit. All other readings were taken in rapid succession.

The data in Figs. 24 and 25 indicate that only a small volume of soil gas needs to be purged prior to taking readings. In fact, purging large volumes can cause a number of problems, including increasing the likelihood for a short-circuit to occur, dilution by creating a large zone of influence, and overlapping zones of influence when making depth-varying measurements. Good practice should involve some purging of system dead

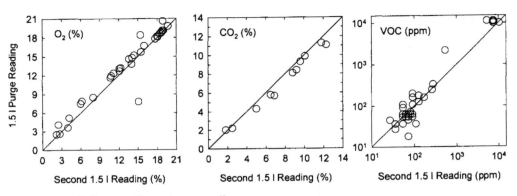

Figure 24 Influence of purging on soil gas measurements.

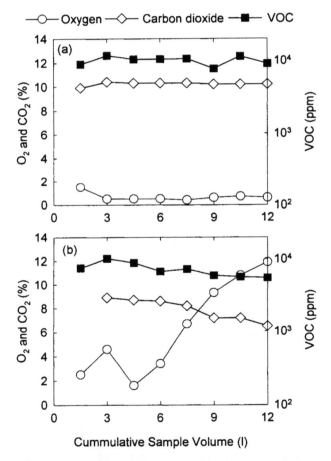

Figure 25 Influence of continuous sampling on soil gas measurements.

gas prior to sampling, especially if small samples are being collected (e.g., for gas chromatograph analysis). Another important issue, particularly when making temporal comparisons of soil gas data, is to be consistent in the procedure used for each sampling trip. With these factors in mind and the type of results shown in Figs. 24 and 25, the

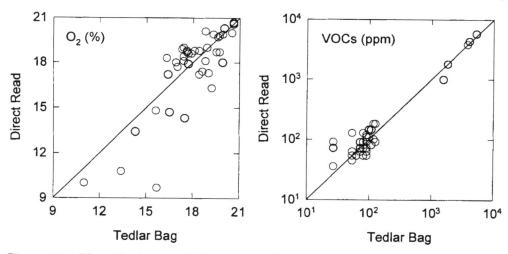

Figure 26 Direct-Read versus Tedlar bag sampling.

standard practice adopted at UMass Amherst for permanent samplers is to purge the first 1.5-L sample and use the second 1.5-L sample for readings. The 1.5-L sample is approximately the minimum amount of gas needed for taking a set of three portable meter readings, i.e., O_2, CO_2, and VOC. The flow rate during purging and sampling is adjusted depending on the vacuum pressure being developed; i.e., the flow rate is decreased if the vacuum pressure increases significantly. This is done because higher vacuum pressures increase the likelihood of a short-circuit developing, as described previously. Typical flow rates range from 0.5 to 1.5 L/min. Between each set of readings the Tedlar bags are cleaned with nitrogen and the meters are purged with ambient air.

B. Direct-Read Versus Tedlar Bag Sampling

In the early stages of developing soil gas sampling procedures at UMass Amherst, all soil gas readings were taken using the direct-read method. This was gradually changed to the Tedlar bag procedure because the method was believed to be more accurate. The Tedlar bag method allows three gas constituents to be measured from the same 1.5-L sample. It also allows for serial dilution as described previously for high-VOC, low-oxygen samples. However, the change to the Tedlar bag system did create some problems affecting the reliability of the measured data. Figures 26a and 26b show a comparison of direct-read versus Tedlar bag measurements of oxygen and VOC from a stainless steel sampling point at Site A. While there is some scatter in the results, they indicate a trend in that lower oxygen concentrations in the Tedlar bag method gave higher oxygen and lower VOC values. This result was caused by leakage of ambient air into the Tedlar bag through the sample-drawing system; the leakage increased the oxygen concentration and diluted the VOC concentration. Subsequent installation of a vacuum gauge to monitor the sampling stream and periodic checks of all connections were found to minimize the leakage problems in the Tedlar bag system.

C. Meter Limitations

Catalytic combustion meters rely on sufficient oxygen for complete combustion and therefore meter accuracy. As a result, VOC readings in zones of high hydrocarbon/low oxygen

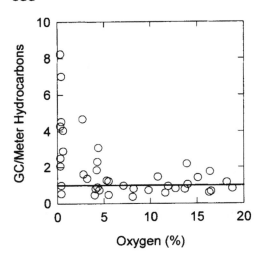

Figure 27 Comparison of meter and gas chromatograph readings.

can often be unreliable. This effect is shown in Fig. 27 for data from stainless steel permanent clusters at Site A. Measurements of total hydrocarbons were made on each Tedlar bag sample using both the TLV meter and an HNU portable gas chromatograph (GC). The data in Fig. 27 suggest that the ratio of GC to meter hydrocarbon reading increase with decreasing oxygen when the oxygen concentration is below approximately 5%. These results clearly show that portable catalytic combustion meters must be calibrated when used in high-hydrocarbon/low-oxygen environments; otherwise, hydrocarbon concentrations can be significantly underpredicted.

D. Short-Circuiting

The potential for a short-circuit to develop is a problem for both permanent and rapid-deployment samplers. Figure 25b shows clear evidence of a short-circuit problem with permanent samplers. This type of problem can be minimized with careful installation of the samplers, including the use of bentonite seals to isolate the sampling zone and concrete surface plugs to keep the samplers fixed in place. However, even with these precautions, problems can arise. Bentonite plugs must remain hydrated, or significant desiccation cracking can occur. Concrete surface plugs tend to break down, especially in areas that experience freeze–thaw cycles. Excessive handling of the sampler during each sampling trip can exacerbate these problems. Using quick-connect-type connections rather than threaded or push-on connections allows for easy connection of the sampling riser pipe to the instrument or vacuum pump system with a minimal amount of handling.

Short circuiting can also be a problem for rapid-deployment profiling probes. Evidence of this is shown in Fig. 28, which presents VOC and oxygen data taken using stainless steel permanent clusters and a retractable-tip drive probe at Site A. This drive probe has a retractable tip that protects the filter element during driving, which is subsequently retracted a few centimeters at each sampling depth to expose the filter element. The probe readings for VOC plotted in Fig. 28a are very low near the surface and again at a depth of approximately 2 m, clearly showing the diluting effect of ambient air being drawn from the surface along the drill string. The oxygen data in Fig. 28b for the retractable-tip probe are very high compared to the permanent samplers. Furthermore,

Field Sampling of Soil Gas Constituents

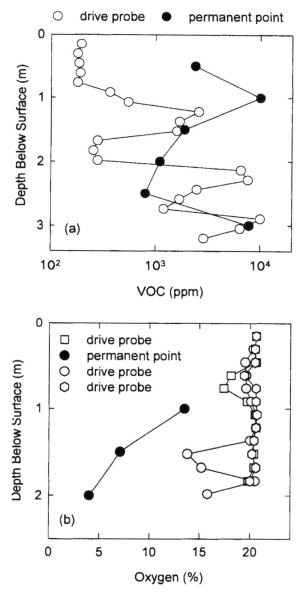

Figure 28 Examples of short-circuiting.

most of these readings are almost equal to ambient conditions, which is highly unlikely in this area, which is known, from monitoring well observations, to have a significant amount of free-floating petroleum product at a depth of about 2.5–3 m. It was determined that these short-circuits were produced because the drill rods were pulled at an angle during retraction of the drill string at each sampling depth. This broke the contact between the drill string and native soil, allowing ambient air to travel down the annulus and be drawn into the sampler. Similar problems can occur with short friction reducers located just above the sample port, because this often creates an open annulus between the native soil and drill string.

Being careful to ensure that the drill string always remains vertical during driving and retraction can minimize short-circuit problems with drive probes. In fact, because of this difficulty, the retractable-tip probe is not recommended for soil gas sampling. Flush filter-type probes such as that shown in Fig. 1 are preferred because the probe does not need to be pulled back at each sampling depth, thus decreasing the likelihood of a short-circuit developing. The flush filter probes can clog periodically in fine-grained soils but rarely do so in coarse-grained soils like those encountered at the sites studied in this research. Friction reducers should be relatively long and located some distance away from the filter element or, if possible, eliminated completely.

X. SUMMARY

Sampling and measurement of soil gas samples at contaminated sites can be an extremely useful site investigation technique at sites involving volatile organic compounds. In order to obtain reliable data, strict sampling and measurement protocols should be adhered to. Many of the systematic errors associated with soil gas sampling can be minimized if careful attention is given to the equipment and procedures used. The following recommendations are suggested for improving the reliability of soils gas measurements.

1. Careful attention to sampler construction, installation, and handling should help to minimize dilution problems arising from short-circuits. Probes with a flush-mounted filter has consistently been the most successful probe for repeated site investigations.
2. Permanent sampling points at petroleum-contaminated sites should be constructed from stainless steel tubing and should be equipped with a porous filter element.
3. The Tedlar bag sampling system is preferred over direct-read measurements provided that a flow meter and vacuum gauge are used to monitor the flow stream.
4. Standard sampling protocol should include a small dead gas purge volume prior to sampling, but purging large volumes prior to readings does not appear to be necessary and may give misleading results.
5. Soil gas sampling at petroleum-contaminated sites should include measurements of the three primary indicator gasses: VOCs, oxygen, and carbon dioxide.
6. Combustion-type hydrocarbon meters should be calibrated using gas chromatograph data for high-hydrocarbon/low-oxygen samples.
7. Measurement of ground temperatures should be included in a site investigation to allow for variations in temperature to be included in transport modeling.

REFERENCES

1. Deyo, B. G., Robbins, G. A., and Binkhorst, G. K. 1993. Use of portable oxygen and carbon dioxide detectors to screen soil gas for subsurface gasoline contamination. *Groundwater*, 31:598–604.
2. Downey, D. C., and Guest, P. H. 1991. Physical and biological treatment of deep diesel contaminated soils. *Proc. Conf. on Petroleum Hydrocarbons and Organic Chemicals in Groundwater*, NGWA/API, pp. 361–376.

3. Folkes, D. J., Bergman, M. S., and Herst, W. E. 1987. Detection and delineation of a fuel oil plume in a layered bedrock deposit. *Proc. Conf. on Petroleum Hydrocarbons and Organic Chemicals in Groundwater*, NGWA/API, pp. 279–304.
4. Bishop, P. K., Burston, M. W., Lerner, D. N., and Eastwood, P. R. 1990. Soil gas surveying of chlorinated solvents in relation to groundwater pollution studies. *Quart. J. Eng. Geol.*, 23:255–265.
5. Joyner, S., and Thomsen, K. O. 1990. Areal and vertical soil gas sampling techniques used to located and characterize a VOC source. *Proc. 4th Outdoor Action Conf. on Aquifer Restoration, Ground Water Monitoring and Geophysical Methods*, pp. 235–246.
6. Marrin, D. L. 1988. Soil gas sampling and misinterpretation. *Groundwater Monit. Rev.*, 8:51–54.
7. Ostendorf, D. W., and Kampbell, D. H. 1991. Biodegradation of hydrocarbon vapors in the unsaturated zone. *Water Resources Res.*, 27:453–462.
8. Kerfoot, H. B., Mayer, C. L., Durgin, P. B., and D'Lugosz, J. J. 1988. Measurement of carbon dioxide in soil gases for indication of subsurface hydrocarbon contamination. *Groundwater Monitor. Rev.*, 8:67–71.
9. Marrin, D. L. 1991. Subsurface biogenic gas ratios associated with hydrocarbon contamination. *Proc. In Situ Bioreclamation*, pp. 546–560.
10. Kampbell, D. H., Wilson, J. T., and Ostendorf, D. W. 1990. Simplified soil gas sampling techniques for plume mapping and remediation monitoring. *Proc. 3rd Natl. Conf. on Petroleum Contaminated Soils*, pp. 125–139.
11. Weinig, W. T. 1992. Monitoring the vadose zone during pneumatic pumping tests. *Proc. 6th Natl. Outdoor Action Conf. on Aquifer Restoration, Ground Water Monitoring and Geophysical Methods*, pp. 133–146.
12. Swallow, J. A., and Gschwend, P. M. 1983. Volatilization of organic compounds from unconfined aquifers. *Proc. 3rd Natl. Symp. on Aquifer Restoration and Ground Water Monitoring*, pp. 327–333.
13. Kerfoot, H. B. 1987. Soil gas measurement for detection of groundwater contamination by volatile organic compounds. *Environ. Sci. Technol.*, 21:1022–1024.
14. Christy, T. M., and Spradlin, S. C. 1992. The use of small diameter probing equipment for contaminated site investigation. *Proc. 6th Natl. Outdoor Action Conf.*, pp. 87–101.
15. Tillman, N., and Leonard, L. 1993. Vehicle mounted direct push systems, sampling tools, and case histories: An overview of an emerging technology. *Proc. Conf. on Petroleum Hydrocarbons and Organic Chemicals in Groundwater*, NGWA/API, pp. 177–188.
16. Moyer, E. E., Ostendorf, D. W., Kampbell, D. H., and Xie, Y. F. 1994. Field trapping of subsurface vapor phase petroleum hydrocarbons. *Groundwater Monitor. Remed.* 14:110–119.
17. Kittel, J. A., Hinchee, R. E., Miller, R., Vogel, C., and Hoeppel, R. 1993. In situ respiration testing: A field treatability test for bioventing. *Proc. 1993 Petroleum Hydrocarbons and Organic Chemicals in Ground Water: Prevention, Detection, and Restoration*, pp. 351–366.
18. American Society for Testing and Materials. 1994. *Annual Book of Standards*, Vol. 4.08. ASTM, Philadelphia.
19. Torstensson, B. A., and Petsonk, A. M. 1988. A hermetically isolated sampling method for ground-water investigations. In: *Ground Water Contamination: Field Methods*, ASTM STP 963, pp. 274–289. ASTM, Philadelphia.
20. DeGroot, D. J., Lutenegger, A. J., Panton, J. G., Ostendorf, D. W., and Pollock, S. J. 1996. Methods of determining in situ oxygen profiles in the vadose zone of granular soils. In: *Sampling Environmental Media*, ASTM Spec. Tech. Publ. 1282, pp. 271–288.
21. Robbins, G. A., Deyo, B. G., Temple, M. R., Stuart, J. D., and Lacy, M. J. 1990. Soil gas surveying for subsurface gasoline contamination using total organic vapor detection instruments 2. Field experimentation. *Groundwater Monitor. Rev.*, 10:110–117.
22. Ostendorf, D. W., Lutenegger, A. J., and Pollock, S. J. 1995. Soil gas sampling and analysis in petroleum-contaminated transportation department right of way. *Transport. Res. Rec. No. 1475*, pp. 110–120.

23. Ostendorf, D. W., Lutenegger, A. J., Suchana, R. J., Cheever, P. J., and Pollock, S. J. 1996. LNAPL detection, measurement, and distribution in the subsurface environment. In: *NAPLS in Subsurface Environment*, pp. 91–101. American Society of Civil Engineers.
24. DeGroot, D. J., and Lutenegger, A. J. 1998. Reliability of soil gas sampling techniques. *Proc. Int. Symp. on Site Characterization*, 1:629–634.
25. Burris, D. R., Wolf, D. M., and Reisinger, J. 1988. Examination of soil vapor sampling techniques. *Proc. 2nd Natl. Outdoor Action Conf.*, Las Vegas, NV, p. 2.

32
Speciation of Heavy Metals: An Approach for Remediation of Contaminated Soils

G. S. R. Krishnamurti
C.S.I.R.O. Land and Water, Glen Osmond, Australia

I. INTRODUCTION

The industrial revolution of recent times has resulted in the pollution of the biosphere. The primary sources of this pollution are mining and smelting industries, municipal wastes, fertilizers and sewage sludges (1,2). Toxic metal contamination of soil poses a major environmental and human health concern which is still in need of an effective and affordable technological solution for remediation strategy.

Several approaches are currently used for remediation of soils contaminated with toxic metals: (a) land filling—excavation, transport, and deposition of contaminated soil in a permitted landfill site; (b) fixation—the chemical processing of soils to immobilize the metals; (c) leaching—using acid solutions or complexing leachants to desorb and leach metals from soil from the contaminated soil followed by the return of the soil residue to the site; (d) bioremediation—use of microorganisms to degrade pollutants in-situ (since the heavy metals cannot be chemically degraded, application of microbial remediation to the in-situ removal of heavy metals from contaminated substrates is limited mainly to their immobilization by precipitation or reduction) (3); (e) phytoremediation—use of specially selected and engineered metal-accumulating plants for environment clean-up either by phytoextraction or by phytostabilization (4).

The ecotoxicological significance of the environmental impact of heavy metals in soils is determined by the specific form and reactivity of their association with particulate forms rather than by their accumulation rate. Estimation of plant transfer and prediction of long-term effects on soil solution and groundwater quality, therefore, should be based on the proportion of the potentially "active species" of the heavy metals. Qualitative assessment of these fractions involves speciation of both the soil solution and the active particulate fraction of the soils.

The ecological consequences of heavy-metal pollution of soils relate largely to the heavy-metal solubility and mobility within the soil profile. These interrelated factors determine the leaching, availability to microbes and plants and ultimately to humans. However, leaching of heavy metals through the soil profile onto the groundwater, even in highly polluted soils treated with sewage sludge, does not occur to any appreciable extent (5,6). Movement of heavy metals within the soil profile will be mainly in the solution phase. Soil chemical reactions controlling mobility of heavy metals and their subsequent uptake by plants can be broadly classified as adsorption/desorption or solubility/precipitation, which are in turn controlled by the concentration and ionic species in solution and the nature of surface phases present. Adsorption/desorption has provided a useful means of describing heavy-metal reactions in soils, but the boundary between adsorption and solubility is sensitive to small changes in soil components and soil conditions (7). The important role of soluble metal–organic complexes on ionic speciation and it s application to soil chemistry (8) through the use of advanced computer models such as GEOCHEM (9), SOILCHEM (10), and MINTEQL (11) have improved our understanding of heavy-metal reactions in soils.

Soil biological processes which can be considered sensitive to heavy metals are mineralization of N and P, cellulose degradation, and possibly N_2 fixation. However, there is little evidence to date to suggest that soil biological processes are affected in most heavily polluted soils by heavy metals (12).

For classification of soil types or identification of nutrient-deficient zones, determination of total content of selected elements often suffices. However, the identification of particular species present tends to be far more informative than mere total elemental percentages. For agricultural and environmental purposes, the amount of nutrient "available" to plants or the amount of pollutant easily "mobile" becomes more important since this fraction, even though it represents only a small fraction of the total content greatly influences plant growth and the quality of groundwater. The generalized schematic geochemical cycle of heavy metals in agrosystems can be represented as shown in Fig. 1.

Many studies (e.g., (13)) have indicated the existence of different binding forms and a strongly pH-dependent solubility effects of the trace elements. Plant uptake of trace elements occasionally correlates well with the amount of metal extracted by a specific chemical reagent. Unfortunately, the nature of the most appropriate reagent seems to vary with the type of plant being chosen, the element being considered, and the soil type. The diverse responses arise, in part, from the fact that the trace elements of soil are bound to individual components in different ways, so that chemicals differ in their ability to effect their release. In addition, the fraction present in the associated soil solution can also possess chemical forms which are not readily transformed into the root systems and translocated into other parts of the plant (14). It can therefore be argued that elemental analysis should be supplemented by a "speciation scheme" which facilitates (a) assessment of the chemical form in the aqueous phase (solution speciation) and (b) identification of bonding modes or component associations (soil particulate-bound speciation).

It should be possible to summarize the principal forms of mobile and mobilizable toxic elements in soil (15) as

1. Soil solution: ionic, molecular, and in chelated forms
2. Exchange interface: readily exchangeable ions in inorganic and organic fractions
3. Adsorption complex: more firmly bound ions
4. Fe, Mn oxides: incorporated in precipitated Fe, Mn, and Al oxides
5. Detrital (residual): fixed in crystal lattices of secondary minerals

and as illustrated in Fig. 2. In order to quantify these fractions and to assess the mobility and bioavailability of different forms, scientists have attempted to extract the different fractions using a range of chemical extractants.

Speciation of Heavy Metals

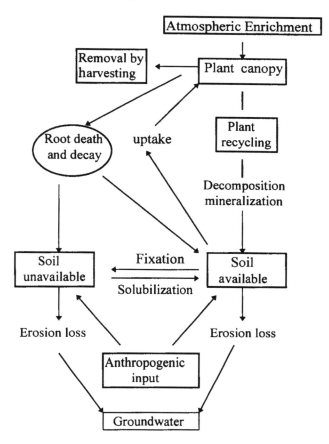

Figure 1 Generalized biogeochemical cycle for trace elements in crop ecosystems.

II. SOLUTION SPECIATION

Metal-ion speciation in the soil solution may well play a role in its bioavailability. The element bioavailability is reported to be a function of at least three parameters (16):

1. The total amount of potentially available elements (the quantity factor)
2. The concentration or activity and ionic ratios of elements in the soil solution (the intensity factor)
3. The rate at which elements transfer from solid to liquid phases and to plant roots (reaction kinetics)

The fate of toxic metals in soils depends mainly on the initial chemical form of the metal even through the environmental and edaphic conditions such as pH, redox status, and soil organic matter content have significant influence. The metal speciation, metal complexation, and different metal fractions in soils can be broadly discussed based on the following approaches:

1. Computer-based modeling, based on geochemical principles, using either the equilibrium constants or using Gibbs free-energy values. Both approaches are subject to the conditions of equilibrium and mass balance.

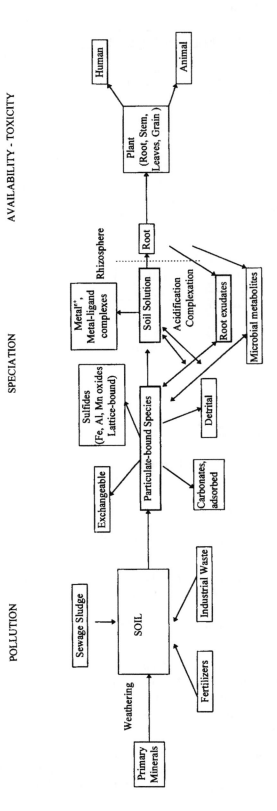

Figure 2 Nutrient–heavy-metal pollution–speciation–availability/toxicity cycle in terrestrial agro-ecosystems.

2. Soil environmental constraints, based on the understanding of the soil processes and conditions which control the formation and transformation of metal species
3. Chemical analysis using extractants, based on the reactivity of the extractants with the metal under study

These different approaches, individually, provide an idea of the processes involved in the retention and mobility of toxic metals. There are only a few examples of soil studies that have utilized a combination of these approaches to understand the fate of metals in soil ecosystems. Numerous studies have used chemical extractants to quantify different metal fractions, but only a limited few have attempted to characterize metal phytoavailability by correlating soil-extractable metal fractions with plant metal uptake. The ones that are available relate to soils with high total metal contents (e.g., sewage sludge-treated soils, soils spiked with high concentrations of heavy metals).

Using basic chemical theory, Sposito (17) and Morgan (18) have outlined the principles for elucidating the metal speciation in natural environmental systems; this was the basis for the later development of the computer-based geochemical modeling approach of Sposito and Coves (10) specifically for soils.

Two different thermodynamic approaches can be used to calculate metal-ion speciation, either by using equilibrium constants or by using Gibbs free-energy values, which are subject to conditions of equilibrium and mass balance. Even though the same thermodynamic calculations can be applied to both inorganic and organic metal complexes, the structure and binding mechanism of the organic ligands is not clearly understood at the present time and hence we do not have reliable thermodynamic data for realistic organic species.

Using the equilibrium constant approach, the dominant species of the metal which can form in a given soil environmental conditions can be assessed using activity graphs. These graphs plot the log [(solid phase)/(free metal cation)] against any soil characteristic such as pH or pCO_2. They methods to obtain such information were dealt in detail by Sposito (19).

The change in the Gibbs free energy for selected processes is also used to assess the stability of compounds and the dissolution sequence. Lindsay (20) used this method to calculate the log K_0 values of an extremely large number of trace-metal reactions in soils.

Computer-based chemical equilibrium models of natural systems, such as soils, have undergone a great deal of development in recent years and have become useful tools for studying water quality criteria (21). The widely used programs GEOCHEM (22), used exclusively for soil systems, and MINEQL (23,24), applied for the study of the water bodies, are progeny of the program REDEQL2 (25,26). Details of the REDEQL2 program were discussed by Legatt (27) and by Ingle et al. (28). Some typical applications of GEOCHEM include (19):

1. Prediction of the concentration of inorganic and organic complexes of a metal cation in a soil solution
2. Calculation of the concentration of a particular chemical form of a nutrient element in a solution bathing plant roots so as to correlate that form to nutrient uptake
3. Prediction of the chemical fate of pollutant metal added to a soil solution of known characteristics

4. Estimation of the effect of changing pH, ionic strength, redox potential, water content, or the concentration of some constituent on the solubility of a chemical element of interest in a soil solution.

GEOCHEM differs from REDEQL2 principally in containing more than twice as much thermodynamic data; in using thermodynamic data that have been selected critically, especially for soil systems; in containing a method for describing the cation exchange; and in employing different subroutines for correcting thermodynamic equilibrium constants for the effect of nonzero ionic strength. MINEQL differs from REDEQL2 in the subroutines that describes solid phases and the adsorption phenomena.

In recent years several computer models have become widely available and are used successfully, e.g., SOILCHEM [the updated version of GEOCHEM (10)], HYDRAQL (29), ECOSAT (30), and MINTEQA2 (11).

An increasing number of studies have focused on modeling the impact of anthropogenic metal inputs, such as sewage sludge (31), geothermal brines (32), acid rain (33), and coal gasification (34), into soil systems. Elemental speciation was predicted by GEOCHEM. These studies provided valuable insights into the chemistry of speciation of metals in soils systems and a basis for predicting the behavior of the elements in contaminated soils.

It is also generally recognized that metal-ion availability and uptake by plants is controlled by several factors such as pH, ionic strength, redox potential, composition of solution, ionic size, and valence (35). Therefore elemental speciation is a major factor in controlling the availability and uptake of various nonessential elements by plants (36). Speciation of 10 metals and 13 ligands in saturation extracts of soils was computed by GEOCHEM, and the uptake of Cd by sweet corn grown on these soils subjected to known additions of Cd was studied (36). The results showed that Cd uptake by sweet corn was highly correlated with the concentration of $CdCl^+$ in soil solution and not with Cd^{2+} concentration in soil solution (Fig. 3). The authors concluded that reduced charge on Cd through complex formation appeared to enhance Cd uptake.

In recent years the understanding of colloid surfaces and soil constituents has increased tremendously. Surface coordinating functional groups on particulate inorganic and humic materials are viewed as complexant ligands (37). The surface complex models (SMC) are now finding increased application in the fields of pollutant retention behavior (38), the chemistry of plant nutrient retention (39), and transport of pollutants by colloids (40). The elegance of the surface complexation approach lies in the fact that it can be incorporated into the thermodynamic speciation models used for soluble complexes. The commonly used surface complex models are the diffuse double-layer model, DDLM (41,42), the constant-capacitance model, CCM (43–46), the triple-layer model, TLM (47–51), and the 1pK basic Stern model (52–54). The application of many of the commonly used computer models in the determination of the speciation in solution phase was dealt with exhaustively by Lumsdon and Evans (55).

III. PARTICULATE-BOUND METAL SPECIATION

Metals in soil solutions are generally undersaturated with respect to the least soluble mineral species evaluated (5,6). However, the solid phase that is controlling the activity of a metal in solution is unknown. Thus, the identification of solid phases in soils that control the activity of metals in soil solutions needs to be ascertained.

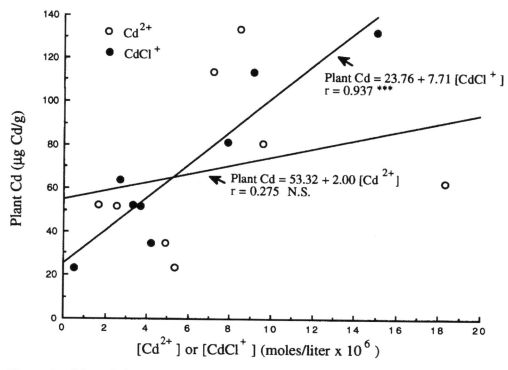

Figure 3 Cd uptake by corn shoots against molar concentrations of $CdCl^+$ and Cd^{2+} in saturation extracts of the soils. (Redrawn from Ref. 36.)

The distribution of trace elements among soil components such as organic matter or hydrous metal oxides is important of assessing the potential of soil to supply sufficient micronutrients to plant growth or to contain toxic quantities of trace metals, and for determination of amelioration procedures for soils at risk of causing trace-metal contamination. Metal cations may be soluble, readily exchangeable, complexed with organic matter, or hydrous oxides, substituted in stoichiometric compounds, or occluded in mineral structures. Delineating the speciation of metals in soils is considered essential for understanding the mobility, bioavailability, and toxicity of the metals, and to develop useful environmental guidelines for potential toxic hazards (56,57). Trace elements in soils are normally found in many different physicochemical forms (58,59) and may be associated with the various soil components, such as clays, metal hydrous oxides, carbonates, and soil organic matter. The nature of this association is referred to here as "speciation." Speciation may also consider the type of bonding between an element and other solid components. For example, an element in ionic form may bind to colloidal-size fractions of soils such as clay minerals or organic matter by coulombic forces, whereas covalent bonds may be formed with surface ligands on hydrous oxide surfaces (60,61). Ligands can form either inner- or outer-sphere complexes with cations on an absorbent (62). Factors affecting distribution of an element among different forms include, pH, ionic strength of the solution, the solid and solution components and their relative concentration and affinities for an element, and time (62–68).

A. Single Chemical Extractants

Single chemical extractants are generally used to determine "available" amounts of soil metals and usually aim to extract one and/or another species, such as the water-soluble, easily exchangeable, bound to organic matter or hydrous oxides, and occluded in mineral structures.

1. Water-Soluble and Exchangeable

Use of electrolytes, such as $CaCl_2$, $MgCl_2$, NH_4Cl, KNO_3, $Mg(NO_3)_2$, and $Ca(NO_3)_2$ as extractants promotes displacement of metal ions held by electrostatic attraction to negative sites on particle surfaces (exchangeable). Usually, 1 M solutions are employed as extractants 1 M $MgCl_2$, (69–75); 1 M KNO_3, (76–80); 1 M $Mg(NO_3)_2$, (81,82)], though more dilute solutions have been preferred [0.5 M KNO_3, (6,83); 0.05 M $CaCl_2$, (81,84,85); 0.5 M $CaCl_2$, (86); 0.25 M $Ca(NO_3)_2$, (87); 0.5 M $Ca(NO_3)_2$, (88,89); 0.5 M $MgCl_2$, (90,91)], since they more closely resemble the electrolyte concentrations that can occur in natural systems. Use of nitrate salts were preferred to chloride salts on the grounds that chloride ion has a specific complexing effect (82,88).

Use of 1 M NH_4OAc was also advocated as an effective reagent for the extraction of exchangeable phase (92–98). However, acetate ion has a special complexing effect, particularly with heavy metals.

2. Specifically Sorbed Carbonate-Bound

Significant concentration of trace elements can be associated with sediment carbonates. A mixture of 1 M NaOAc with HOAc at pH 5 has been shown to extract >99% of total carbonate present in the soils (92) and was used as a specific extractant for the determination of carbonate-bound heavy metals (69,73,82,99–101).

Other reagents which were also used include 2.5% HOAc (84,93,102) and 0.05 M Na_2-EDTA (83,103,104).

3. Bound to Hydrous Oxides of Fe and Mn

Fe and Mn oxides, present in soils as nodules, concretions, matrix components, cement between particles, or as coatings on particles, are excellent scavengers of trace metals (105). A mixture of sodium dithionite, sodium citrate, and sodium bicarbonate, buffered at pH 7.3, was suggested as a suitable reagent for determining the total free iron contents (106) and has been used widely in soil/sediment studies (69). This reagent dissolves both the crystalline and amorphous oxyhyroxides. However, the dithionite salt can be contaminated with metal impurities such as Zn (107).

Hydroxylamine hydrochloride, dissolved in acetic or nitric acid, was shown to extract selectivity Mn oxyhyroxides and Mn oxides as well as amorphous Fe oxides (108–110), and was used for extraction of trace metals associated with hydrous oxides of Fe and Mn (69,76,80,82,87,89,93,98,100).

For total recovery of amorphous Fe oxides, treatment of samples with acidified ammonium oxalate (pH 3), in the absence of the catalyzing effect of light, has been proposed (111–113). This reagent has been used extensively for selectively extracting trace elements associated with amorphous Fe oxides (71,82,94,95,97).

4. Organically Bound Metals

Hydrogen peroxide is the commonly used extractant (69,73,82,93,98,114), even though it dissolves Mn oxides (71,81,115). However, Orsini and Bermond (116) found the kinetics

Speciation of Heavy Metals

of the destruction of organic matter was slower and took almost 24 h. Other reagents used include NaOCl buffered at pH 9.5 (81,90,117) and 0.1 M sodium or potassium pyrophosphate (79,80,82,86). Even the C removal with NaOCl is said to be higher than that achieved by H_2O_2 (118), but the associated precipitation of released metal ions in the alkaline medium and the possible alteration of mineral constituents are major disadvantages.

Beckett (119) has an excellent exhaustive review on the use of various extractants used in the selective extraction of trace metals associated with different fractions of soils.

B. Sequential Selective Extraction

Sequential chemical extraction techniques have been used to determine the chemical forms of trace elements in soils, sediments, and dissolved solids in natural waters and are based theoretically and experimentally on more than 100 years of research (120). A basic requirement of any extraction procedure should be the ability of the extractant to dissolve specific component of a soil or sediment (121). Many different methods have been employed to fractionate trace elements, and these have proved useful for metal speciation. Reviews of the fractionation methods used to determine the chemical forms of trace elements in soils and sediments (e.g., 14,122–124), geochemical exploration (121), and in natural waters (125) are available.

Gupta and Chen (93) were the first to use a sequential extraction procedure for the speciation of Cd in sediments. The scheme delineated Cd as exchangeable (1 M NH_4OAc), carbonate-bound (1 M HOAc), easily reducible metal oxide-bound (0.1 M $NH_2OH \cdot HCl$ in 0.01 M HNO_3), organics-bound (30% H_2O_2, NH_4OAc), iron oxide-bound (0.04 M $NH_2OH \cdot HCl$ in 25% HOAc), and residual (HF, $HClO_4$). However, the procedure used extensively, with minor modifications, by various environmental researchers for the speciation of particulate-bound heavy metals in soils and sediments is that of Tessier et al. (69), which delineates the metal species sequentially as exchangeable, carbonate-bound, Fe and Mn oxide-bound, organically bound, and residual.

Shuman (107,126), working on the speciation of Cu, Mn, Fe, and Zn, modified the sequence of the extraction as: exchangeable, acid-soluble, easily reducible (Mn oxides-bound), moderately reducible (amorphous Fe oxides-bound), strongly reducible (crystalline Fe oxides-bound), oxidizable (organically-bound), and residual. The low percentages of the oxidizable fractions obtained following the sequential extraction scheme of Tessier et al. (69) as modified by Shuman (126) were attributed to the lack of selectivity of the method used (127).

The differentiation of the metal-organic-complex-bound Cd species as distinct from the other organically bound species was the innovation in the selective sequential extraction scheme suggested by Krishnamurti et al. (82), which proportionates the particulate-bound Cd species in soils as exchangeable, carbonate-bound, metal-organic complex-bound, easily reducible metal oxide-bound, organic-bound, amorphous mineral colloid-bound, crystalline Fe oxide-bound, and residual. The metal-organic complex-bound Cd species were selectivity extracted using 0.1 M pyrophosphate as the extractant in the sequential extraction scheme detailed below:

Fraction 1. Exchangeable M $Mg(NO_3)_2$ at pH 7 (1:10), 4 h
Fraction 2. Carbonate-bound M NaOAc at pH 5 (1:25), 6 h
Fraction 3. Metal-organic complex-bound 0.1 M $Na_4P_2O_7$ (pH 10) (1:30), 20 h
Fraction 4. Easily reducible metal oxide-bound 0.1 M $NH_2OH \cdot HCl$ in 0.01 M HNO_3 (1:20), 30 min

Fraction 5. Organically bound 5 ml 30% H_2O_2 and 3 ml 0.02 M HNO_3, 2 h at 85°C (two times); cool and add 10 ml M $Mg(NO_3)_2$ in 20% HNO_3, 30 min

Fraction 6. Amorphous mineral colloid-bound 0.2 M $(NH_4)_2C_2O_4$ (pH 3) (1 : 10), 4 h (dark)

Fraction 7. Crystalline Fe oxide-bound 0.2 M $(NH_4)_2C_2O_4$ in 0.1 M ascorbic acid (1 : 25), 30 min at 95°C

Fraction 8. Residual $HF/HClO_4$ acid digestion

Krishnamurti et al. (82) determined the distribution of particulate-bound Cd species in a few typical soils of southern Saskatchewan, Canada, following the schemes of Tessier et al. (69) and the modified scheme. Cadmium in these soils was found to be predominantly in the form metal-organic complex-bound, accounting for on an average of 40% of the total Cd present in the soils, whereas on average 42.6% of Cd in the soils was observed to be in the form Fe, Mn oxide-bound following the fractionation scheme of Tessier et al. (69) (Fig. 4).

Ma and Uren (104) argued that the high concentrations of Cd reported by Hickey and Kittrick (73) and Ramos et al. (128) in contaminated soils which were extracted with 1 M NaOAc at pH 5.0 (carbonate-bound) following extraction with 1 M $MgCl_2$ at pH 7.0 (exchangeable) probably came from not only the carbonate-bound fraction but also from the specifically adsorbed and easily reducible Mn oxide-bound fraction. The specifically adsorbed metals were extracted using 1% NaCaHEDTA in MNH_4OAc at pH 8.3; followed by the easily reducible Mn oxide-bound fraction, which was extracted by 0.1 M $NH_2OH \cdot HCl$ in acidic medium as was used earlier (82,115). The specifically adsorbed heavy metals could also be extracted by pyrophosphate, which was used for the selective extraction of organic matter-bound fraction (129) and metal-organic complex-bound fraction (82). Ma and Uren (104) compared the distribution of metals (Cu, Pb, and Cd) following the Tessier et al. (69) method and a method using 1% NaCaHEDTA in 1 M NH_4Oac at pH 8.3 for extracting the specifically adsorbed metal fractions before extraction for carbonate-bound metals (103). They argued that the high proportion of metals associated with Fe-Al oxides (36.3 mg/kg) following Tessier et al. (69), as compared to the low proportion (2.1 mg/kg) obtained following the modified method, could possibly be associated not only with Fe-Al oxides but also with organic matter and layer silicates, since the specifically adsorbed fraction was not removed earlier. The specifically adsorbed fraction as obtained using 1% NaCaHEDTA in 1 M NH_4Oac at pH 8.3 was observed to be 31.5 mg/kg.

The possible adverse effects of trace-element accumulation by soils amended with sewage sludge has recently been the focus of research. Sequential extraction procedures have been applied to fractionate trace metals in sewage sludges, and the most commonly used procedure is that of Stover et al. (76) with minor variations. The procedure was developed to fractionate the metals as "exchangeable," "organic-bound," "carbonate-bound," and "sulfide mineral-bound or residual" through sequential extraction with 0.5 M KNO_3, 0.5 M NaOH, 0.05 M Na_2EDTA, and 4 M HNO_3 (83,130). The "organic-bound" fraction was also fractionated using $Na_4P_2O_7$ (76–79). The low percentage of "exchangeable" fraction observed could signify low availability of the element to the plant, since this fraction is often regarded as the most bioavailable. However, it is possible that the "exchangeable" fraction represented the potential mobility more than potential bioavailability in sludge-amended soils, as suggested by many reports (77,131), and other extracted fractions such as $Na_4P_2O_7$ (82), DTPA (132), or EDTA (133) may be better correlated.

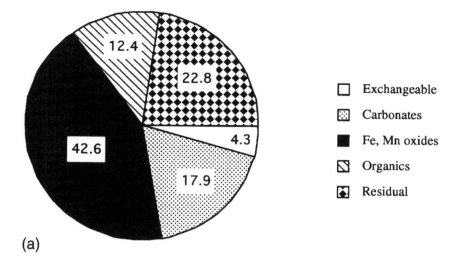

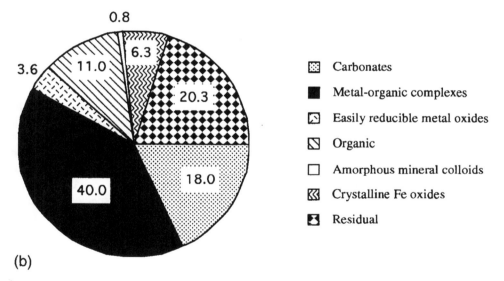

Figure 4. Average percent distribution of particulate-bound Cd species in selected soils of Saskatchewan, Canada, following the method of (a) Tessier et al. (69) and (b) Krishnamurti et al. (82). (Redrawn from Ref. 82)

Application of sewage sludge to soils generally causes a shift of less available to easily available forms that might be potentially more bioavailable than those in unamended soils (58). Sludge application decreased the sulfide/residual forms from Cu (65–82%); Ni (80–94%), and Zn (92–96%) to Cu (19–24%), Ni (61–69%), and Zn (28–35%) of the soils studied (6). Marked increase in the amounts of the amounts of Cd, Cu, Pb, and Zn extractable by DTPA (believed to estimate "available" metal species) were observed after sewage sludge was applied to the soils (77,134). The apparent increase in the "available" forms of sludge-amended soils was ascribed to changes in pH, oxidation of metal sulfides to sulfates, and release of metals complexed with sludge organic matter resulting from microbial activity. On the other hand, percolate water, groundwater quality, and foliar metal concentrations appeared to be unaffected by sewage sludge used for reclaiming soils drastically disturbed by surface mining for coal, provided the sludge contained low to median metal concentrations (135).

IV. SPECIATION, AVAILABILITY AND IMMOBILIZATION

Elaborate sequential chemical extraction schemes, designed to release a particular geochemical fraction, have frequently been used to identify the distribution of the different species of the element among the various forms of solid species identified. The specific particulate solid species that is immediately available to the plant is dependent on the microenvironment conditions of the soil rhizosphere, which modifies the distribution of the heavy metal. Broadly, the availability can be summarized as shown in Fig. 5.

However, very few attempts have been made to identify the particular species of the element which may contribute to the bioavailability. An attempt in this direction was made by Krishnamurti et al. (82), who carried out multiple regression analysis between cadmium available index (CAI), as measured by ABDTPA-extractable Cd (136), and different forms of particulate-bound Cd. The importance of metal-organic complex-bound Cd species in the bioavailability of Cd and the nature of bonding of Cd sites was also worked out in detail using multiple regression analysis and differential FTIR analysis (82,137).

The beta coefficients [standard regression coefficients (138)] of the different species obtained from the multiple regression analysis of the data were in the order: Fe, Mn oxide-bound–organic-bound>carbonate-bound [following the Tessier et al. (69) scheme], indicating the importance of both Fe, Mn oxide-bound and organic-bound Cd species in estimating CAI of the soils. In the scheme of Krishnamurti et al. (82), the Fe, Mn oxide-bound and organic-bound Cd species were subfractionated into five distinct species, viz., metal-organic complex-bound, easily reducible metal oxide-bound, organic-bound, amorphous mineral colloid-bound, and crystalline Fe oxide-bound Cd species. The beta coefficient of metal-organic complex-bound Cd species was at least 4–5 times higher than any of the other species, clearly bringing out the importance of metal-organic complex-bound Cd species in estimating the CAI.

Even though Cd in the metal-organic complexes was associated strongly with Mn, the importance of the Al-organic complex-bound Cd in influencing the bioavailability

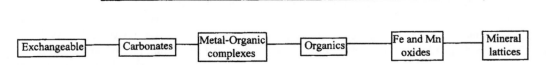

Figure 5 Availability of particulate-bound metal species in soil–plant system.

of Cd in the soils studied was clearly brought out by their beta coefficients in the multiple regression analysis. Based on the differential FTIR spectra of the metal-organic complexes extracted by the 0.1 M sodium pyrophosphate extractant used in the speciation scheme, Krishnamurti et al. (137) showed that Cd in the soils was apparently bonded at the COO- or carboxyl and the OH of the phenolic groups and the OH of the Fe, Al, and Mn in the metal-organic complexes.

A logical solution to minimize plant uptake and also to protect the quality of the food chain is to render the trace metals in the soil immobilie (139). Immobilization of heavy metals in soils was achieved by adding additives that may not produce any detrimental by-product nor alter the physicochemical environment of the soils to affect the plant growth, but could fix the heavy metals into their lattice structure. A number of zeolites were used to reduce Pb, Zn, and Cd contamination. The 1A- and 13X-type zeolites was shown to reduce the Pb content of plant tissues of several species (140). The zeolite Valfor G100 in combination with ferrous sulfate had been shown to be quite effective in immobilizing Cd (141). Beringite, a modified aluminosilicate, also indicated to be a useful additive for reduction or elimination of phytotoxicity of Zn and Cd in bean (*Phaseolus vulgaris* L.) (142). Hydrous oxides of Fe, Al, and Mn are known to specifically adsorb heavy metals, thus enhancing the metal immobility in soils (143,144). Clay-Al-hydroxy polymers were also shown to specifically adsorb Zn and Cd, rendering the metals immobile (145–147).

V. SUMMARY AND OUTLOOK

Polutant transfer to plant and organisms takes place predominantly via the solution phase. However, with regard to remedial options on in-situ sediment contamination, the study of speciation and availability of particulate-associated contaminant should also be undertaken together with the solution phase. The surface-bound pollutants could become partly mobilized in the aquatic mileau by changes of the pH and redox conditions, increased salinities, or concentrations of organic chelators (148).

In connection with the problems arising from the remediation of contaminated material, sequential extraction procedures have been developed which include successive leaching of metals from interstitial waters and from different forms of particulate-bound sediment fractions (69,82). Despite clear advantage of the sequential extraction scheme over that of total analysis, it should be borne in mind that the various extraction steps are not as selective as stated and one has to realize the many pitfalls that could be encountered in the application of these methods (69,149,150).

The present state of knowledge on solid-species speciation is still in its infancy with respect to the assessment of bioavailable metal concentrations. The exchangeable and leachable fractions do not necessarily correspond to the amount available. Information is also lacking about the specific mechanism by which the plant roots participate in the removal of metals from the solid substrate. The redox changes, pH alterations, and organic complexing that are encountered in the root–soil interface modify the availability. Thus, for assessment of mobility of heavy metals in sediments under field conditions, solid speciation experiments have to be complemented by the respective pore solution chemistry.

Various methods are advocated to immobilize heavy metals in highly polluted soils, such as soil excavation followed by soil washing and subsequent disposal of treated soils

(151). However, soil removal is prohibitively expensive, so more cost-effective in-situ technologies are highly desirable. The knowledge of the speciation of heavy metals, both solution and solid state, should assist in finding ways to develop in-situ remediation technologies of highly contaminated soils.

REFERENCES

1. Nriagu, J. O. 1979. Global inventory of natural and anthropogenic emissions of trace metals to the atmosphere. *Nature*, 279:409–411.
2. Alloway, B. J. 1995. Cadmium. In: B. J. Alloway (ed.), *Heavy Metals in Soils*, pp. 122–151. Blackie, Glasgow.
3. Summers, A. O. 1992. The hard stuff: Metals in bioremediation. *Curr. Opin. Biotechnol.*, 3:271–276.
4. Salt, D. E., Blaylock, M., Kumar, P. B. A. N., Dushenkov, V., Ensley, B. D., Chet, I., and Raskin, I. 1995. Phytoremediation: A novel strategy for the removal of toxic metals from the environment using plants. *Biotechnology*, 13:468–474.
5. Emmerich, W. E., Lund, L. J., Page, A. L., and Chang, A. C. 1982. Movement of heavy metals in sewage-sludge treated soils. *J. Environ. Qual.*, 11:174–178.
6. Emmerich, W. E., Lund, L. J., Page, A. L., and Chang, A. C. 1982. Solid phase forms of heavy metals in sewage sludge-treated soils. *J. Environ. Qual.*, 11:178–181.
7. Brummer, G., Tiller, K. G., Herms, U., and Clayton, P. M. 1983. Adsorption/desorption and/or precipitation dissolution processes of zinc in soils. *Geoderma*, 31:337–354.
8. Hodgson, J. F., Geering, H. R., and Norvell, W. A. 1965. Micronutrient cation complexes in soil solution: Partition between complexed and uncomplexed forms by solvent extraction. *Soil Sci. Soc. Am. Proc.*, 29:665–669.
9. Parker, D. R., Zelazny, L. W., and Kinraide, T. B. 1987. Improvements to the program GEOCHEM. *Soil Sci. Sci. AM. J.*, 51:488–491.
10. Sposito, G., and Coves, J. 1988. *SOILCHEM: A Computer Program for the Calculation of Chemical Speciation in Soils*. Kerney Foundation of Soil Science, University of California, Berkeley, CA.
11. Allison, J. D., Broen, D. S., and Nova-Gradac, K. J. 1990. *MINTEQA2/PRODEFA2: A Geochemical Assessment Model for Environmental Systems*, EPA/600/3-91/021. U.S. Environmental Protection Agency, Athens, GA.
12. Domsch, K. H. 1984. Effects of pesticides and heavy metals on biological processes in soil. *Plant Soil*, 76:367–378.
13. Sauerbeck, D. R., and Rietz, E. 1983. Soil chemical evaluation of different extractants for heavy metals in soils. In: Comm. Europe Communities Report EUR 8022. Environ. Eff. Org. Inorg. Contam. Sewage Sludge, CA 99: 193726, pp. 147–160.
14. Pickering, W. F. 1986. Metal ion speciation—Soils and sediments (A review). *Ore Geol. Rev.*, 1:83–146.
15. Berrow, M. J., and Burridge, J. C. 1980. Trace element levels in soil: Effects of sewage sludge. In: *Inorganic Pollution and Agriculture*, MAFF Reference Book No. 326, pp. 159–190. HMSO, London.
16. Brummer, G. W. 1986. Heavy metal species, mobility and availability. In: M. Bernhard, F. E. Brinkman, and P. J. Sadler (eds.), *The Importance of Chemical Speciation in Environmental Processes*, pp. 169–192. Springer-Verlag, Berlin.
17. Sposito, G. 1986. Distribution of potentially hazardous trace metals. In: H. Siegel (ed.), *Metal Ions in Biological Systems, Vol. 20*, Marcel Dekker, New York.
18. Morgan, J. J. 1987. General affinity concepts, equilibria and kinetics in aqueous metal chemistry. In: J. W. Patterson and R. Passino (eds.), *Metal Speciation, Separation and Recovery*, pp. 27–61. Lewis Publishers, Michigan, MI.

19. Sposito, G. 1983. The chemical form of trace metals in soils. In: I. Thornton (ed.), *Applied Environmental Geochemistry*, pp. 123–170. Academic Press, London.
20. Lindsay, W. L. 1979. *Chemical Equilibria in Soils*. Wiley, New York.
21. Jenne, E. A. 1979. *Chemical Modelling in Aqueous Systems*, ACS Symposium Series No. 93. American Chemical Society, Washington, DC.
22. Sposito, G., and Mattigod, S. V. 1980. GEOCHEM: *A Computer Program for the Calculation of Chemical Equilibria in Soil Solutions and other Natural Water Systems*. Kearney Foundation of Soil Science, University of California, Berkeley, CA.
23. Westall, J. C., Zachary, J. L., and Morel, F. M. M. 1976. MINEQL: *A Computer Program for the Calculation of Chemical Equilibrium Composition of Aqueous Systems*. Tech. Note 18, Ralph M. Parsons Laboratory, Massachussets Institute of Technology, Cambridge, MA.
24. James, R. O., and Parks, G. A. 1978. Application of the computer program MINEQL to solution of problems in surface chemistry. Technical Note, Standord University, Stanford, CA.
25. Morel, F., and Morgan, J. 1972. A numerical method for computing equilibria in aqueous chemical systems. *Environ. Sci. Technol.*, 6:58–67.
26. McDuff, R. E., and Morel, F. M. M. 1973. *Description and Use of the Chemical Equilibrium Program REDEQL2*, Tech. Rep. EQ-73-02, California Institute of Technology, Pasadena, CA.
27. Legatt, D. J. 1977. Machine computation of equilibrium concentrations—Some practical considerations. *Talanta*, 24:535–542.
28. Ingle, S. E., Schuldt, M. D., and Schults, D. W. 1978. *A User's Guide for REDEQL-EPA*, EPA-600/3-78-024. U.S. Environmental Protection Agency, Corvallis, OR.
29. Papelis, C., Hayes, K. F., and Leckie, J. O. 1988. HYDRAQL: *A Program for the Computation of Chemical Equilibrium Composition of Aqueous Batch Systems including Surface Complexation Modelling of Ion Adsorption at the Oxide/Solution Interface*, Tech. Rep. 306, Department of Civil Engineering, Stanford University, Stanford, CA.
30. Keizer, M. G. 1991. *ECOSAT: A Computer Program for the Calculation of Speciation in Soil-water Systems*. Department of Soil Science and Plant Nutrition, Agricultural University, Wageningen, The Netherlands.
31. Mattigod, S. V., and Sposito, G. 1979. Chemical modelling of trace metal equilibria in contaminated soil solutions using the computer program GEOCHEM. In: E. A. Jenne (ed.), *Chemical Modelling in Aqueous Systems, Speciation, Sorption, Solubility and Kinetics*, ACS Symp. Ser. No. 93, pp. 837–850. American Chemical Society, Washington, DC.
32. Sposito, G., Page, A. L., and Mattigod, S. V. 1979. *Trace Metal Speciation in Saline Waters Affected by Geothermal Brines*. Kearney Foundation of Soil Science, University of California, Berkley, CA.
33. Sposito, G., Page, A. L., and Frink, M. E. 1980. *Effects of Acid Precipitation on Soil Leachable Quality—Computer Calculations*. Environ. Res. Lab., Corvallis, U.S. Environmental Protection Agency, Ore.
34. Ireland, R. R., McConachie, W. A., Stuermer, D. H., Wang, F. T., and Koszykowski, R. F. 1982. The analysis of inorganic constituents in the groundwater at an underground coal gassification site. In: R. H. Filby (ed.), *Atomic and Nuclear Methods in Fossil Energy Research*, Proc. Topical Conf. American Nuclear Society, American Chemical Society, Dec. 1980, Mayaguez, Puerto Rico. pp. 175–188. Plenum Press, New York.
35. Frausto da Silva, J. J. R., and Williams, R. J. P. 1976. The uptake of elements by biological systems. *Structure and Bonding*, 29:67–121.
36. Sposito, G., and Bingham, F. T. 1981. Computer modelling of trace metal speciation in soil solutions: Correlation, with trace metal uptake by higher plants. *J. Plant Nutr.*, 3:81–108.
37. Stumm, W. 1992. *Chemistry of the Soil-Water Interface. Processes at the Mineral-Water and Particle-Water Interface in Natural Systems*. Wiley, New York.

38. Zachara, J. M., Smith, S. C., Resch, C. T., and Cowan, C. E. 1992. Cadmium sorption to soil separates containing layer silicates and iron and aluminum oxides. *Soil Sci. Sci. Am. J.*, 56:1074–1084.
39. Goldberg, S., and Traina, S. J. 1987. Chemical modelling of anion competition on oxides using the constance capacitance model-mixed-ligand approach. *Soil Sci. Sci. Am. J.*, 51:929–932.
40. Goldberg, S. 1992. Use of surface complexation models in soil chemical systems. *Adv. Agron.*, 47:223–329.
41. Huang, C. P., and Stumm, W. 1973. Specific adsorption of cations on $\gamma - Al_2O_3$. *Sci. Total Environ.*, 57:129–150.
42. Dzombek, D. A., and Morel, F. M. M. 1990. *Surface Complexation Modelling: Hydrous Ferric Oxide*, Wiley, New York.
43. Stumm, W., Huang, C. P., and Jenkins, S. R. 1970. Specific chemical interactions affecting the stability of dispersed systems. *Croatica Chimica Acta*, 42:223–231.
44. Stumm, W., Hohl, H., and Dalang, F. 1976. Interaction of metal ions with hydrous oxide surfaces. *Croatica Chimica Acta*, 48:491–499.
45. Stumm, W. Kummertt, R., and Sigg, L. 1980. A ligand exchange model for adsorption of inorganic and organic ligands at hydrous oxide interfaces. *Croatica Chimica Acta*, 53:291–312.
46. Schindler, P. W., Furst, R. D., and Wolf, P. U. 1976. Ligand properties of surface silanol groups. 1. Surface complex formation with Fe^{3+}, Cu^{2+}, Cd^{2+} and Pb^{2+}. *J. Colloid Interface Sci.*, 55:469–475.
47. Davis, J. A., James, R. O., and Leckie, J. O. 1978. Surface ionization and complexation at the oxide/water interface. I. Computation of electrical double layer properties in simple electrolytes. *J. Colloid Interface Sci.*, 63:480–499.
48. Davis, J. A., and Leckie, J. O. 1978. Surface ionization and complexation at the oxide/water interface. II. Surface properties of amorphous iron oxyhydroxide and adsorption of metal ions. *J. Colloid Interface Sci.*, 67:90–107.
49. Davis, J. A., and Leckie, J. O. 1980. Surface ionization and complexation at the oxide/water interface. III. Adsorption of anions. *J. Colloid Interface Sci.*, 74:32–43.
50. Hayes, K. F., and Leckie, J. O. 1987. Modelling ionic strength effects on cation adsorption at hydrous oxide/solution interfaces. *J. Colloid Interface Sci.*, 115:564–572.
51. Hayes, K. F., Papelis, C., and Leckie, J. O. 1988. Modelling ionic strength effects on anion adsorption at hydrous oxide/solution interfaces. *J. Colloid Interface Sci.*, 125:717–726.
52. Bolt, G. H., and Van Riemsdijk, W. H. 1982. Ion adsorption on inorganic variable charge constituents. In: G. H. Bolt (ed.), *Soil Chemistry. B. Physico-Chemical Models*, 2nd ed., pp. 495–502. Elsevier, Amsterdam.
53. Van Riemsdijk, W. H., Bolt, G. H., Koopal, L. K., and Blackmeer, J. 1986. Electrolyte adsorption on hetergeneous surfaces: adsorption models. *J. Colloid Interface Sci.*, 109:219–228.
54. Van Riemsdijk, W. H., De Witt, J. C. M., Koopal, L. K., and Bolt, G. H. 1987. Metal ion adsorption on heteregeneous surfaces: adsorption models. *J. Colloid Interface Sci.*, 116:511–522.
55. Lumsdon, D. G., and Evans, L. J. 1995. Predicting chemical speciation and computer simulation. In: A. M. Ure and C. M. Davidson (eds.), *Chemical Speciation in the Environment*, pp. 86–134. Blackie, London.
56. Davies, B. E. 1980. Trace element pollution. In: B. E. Davies (ed.), *Applied Soil Trace Elements*, pp. 287–351. John Wiley, Chichester, U.K.
57. Davies, B. E. 1992. Trace elements in the environment: Retrospect and prospect. In: D. C. Adriano (ed.), *Biogeochemistry of Trace Metals*, pp. 1–17. Lewis Publishers, Boca Raton, FL.
58. Lake, D. L., Kirk, P. W. W., and Lester, J. N. 1984. Fractionation, characterization, and speciation of heavy metals in sewage sludge and sludge-amended soils: A review. *J. Environ. Qual.*, 13:175–183.

59. Tessier, A., and Campbell, P. G. C. 1988. Partitioning of trace metals in sediments. In: J. R. Kramer and H. E. Allen (eds.), *Metal Speciation: Theory, Analysis and Application*. pp. 183–199. Lewis Publishers, Chelsea, MI.
60. Fendorff, S. E., Sparks, D. L., Lamble, G. M., and Kelley, M. J. 1994. Application of X-ray absorption fine structure spectroscopy to soils. *Soil Sci. Sci. Am. J.*, 58:1583–1595.
61. Manceau, A. M., Harge, J. C., Sarret, G., Hazelamnn, J. L., Boisset, M. C., Mench, M., Cambier, P. H., and Prost, R. 1995. Direct determination of heavy metal speciation in soils by EXAFS spectroscopy. In: R. Prost (ed.), *Contaminated Soils*, Proc. 3rd Int. Conf. Biogeochemistry of Trace Elements, May 15–19, 1993, Paris, France, pp. 99–120. INRA, Paris.
62. Ritchie, G. S. P., and Sposito, G. 1995. Speciation in soils. In: A. M. Ure and C. M. Davidson (eds.), *Chemical Speciation in the Environment*, pp. 201–233. Blackie, London.
63. Jones, L. H. P., and Jarvis, S. C. 1981. The fate of heavy metals. In: D. J. Greenland and M. H. B. Hayes (eds.), *The Chemistry of Soil Processes*, pp. 593–620. Wiley, New York.
64. Tiller, K. 1983. Micronutrients. In: CSIRO (ed.) *Soils: An Australian Viewpoint*, pp. 365–387. CSIRO/Academic Press, New York.
65. McBride, M. B. 1989. Reactions controlling heavy metal solubility in soils. *Adv. Soil Sci.*, 10:1–56.
66. McBride, M. B. 1991. *Processes of Heavy and Transition Metal Sorption by Soil Minerals*. Kluwer, Dordrecht.
67. Alloway, B. J. 1990. Soil processes and the behavior of metals. In: B. J. Alloway (ed.), *Heavy Metals in Soils*, pp. 7–28. Blackie, Glasgow.
68. Foerstner, U. 1991. Soil pollution phenomena-mobility of heavy metals in contaminated soil. In: G. H. Bolt, M. H. B. Hayes, and M. B. McBride (eds.), *Interactions at the Soil Colloid–Soil Solution Interface*, pp. 543–582. Kluwer, Dordrecht.
69. Tessier, A., Campbell, P. G. C., and Bissom, M. 1979. Sequential extraction procedure for the speciation of particulate trace metals. *Anal. Chem.*, 51:844–850.
70. Hoffman, S. J., and Fletcher, W. K. 1979. Extraction of Cu, Zn, Mo, Fe and Mn from soils and sediments using a sequential procedure. In: J. R. Watterson and P. K. Theobald (eds.), *Geochemical Exploration*, pp. 289–299. Association of Exploration Geochemists, Rexdale, Ontario, Canada.
71. Shuman, L. M. 1979. Zinc, manganese and copper in soil fractions. *Soil. Sci.*, 127:10–17.
72. Maher, W. A. 1984. Evaluation of a sequential extraction scheme to study associations of trace elements in estuarine and oceanic sediments. *Bull. Environ. Contam. Toxicol.*, 32:339–344.
73. Hickey, M. G., and Kittrick, J. A. 1984. Chemical partitioning of Cd, Cu, Ni and Zn in soils and sediments containing high levels of heavy metals. *J. Environ. Qual.*, 13:372–376.
74. Nielsen, D., Hoyt, P. B., and MacKenzie, A. F. 1986. Distribution of soil Zn fractions in British Colombia interior orchard soils. *Can. J. Soil Sci.*, 66:445–454.
75. Elliot, H. A., Dempsey, B. A., and Maille, M. J. 1990. Content and fractionation of heavy metals in water treatment sludges. *J. Environ. Qual.*, 19:330–334.
76. Stover, R. C., Sommers, L. E., and Silviera, D. J. 1976. Evaluations of metals in waste water sludge. *J. Water Pollution Control Fed.*, 48:2165–2175.
77. Silviera, D. J., and Sommers, L. E. 1977. Extractability of Cu, Zn, Cd and Pb in soils incubated with sewage sludge. *J. Environ. Qual.*, 6:47–52.
78. Schalscha, E. B., Morales, M., Ahumada, I., Schirado, T., and Pratt, P. F. 1980. Fractionation of Zn, Cu, Cr and Ni in waste water solids and in soil. *Agrochimica*, 24:361–368.
79. Schalscha, E. G., Marlaes, M., Vergara, I., and Chang, A. C. 1982. Chemical fractionation of heavy metals in waste-water effected soils. *J. Water Pollution Control Fed.*, 54:175–180.
80. Miller, W. P., and McFee, W. W. 1983. Distribution of Cd, Zn, Cu and Pd in soils of industrial north western Indiana. *J. Environ. Qual.*, 12:29–33.
81. Shuman, L. M. 1983. Sodium hyochlorite methods for extracting microelements associated with soil organic matter. *Soil Sci. Sci. Am. J.*, 47:656–660.

82. Krishnamurti, G. S. R., Huang, P. M., Van Rees, K. C. J., Kozak, L. M., and Rostad, H. P. W. 1995. Speciation of particulate-bound cadmium of soils and its bioavailability. *Analyst*, 120:659–665.
83. Sposito, G., Lund, L. J., and Chang, A. C. 1982. Trace metal chemistry in arid-zone field soils amended with sewage sludge I: Fractionation of Ni, Cu, Zn, Cd and Pb in solid phases. *Soil Sci. Sci. Am. J.*, 46:260–264.
84. McLaren, R. G., and Crawford, D. V. 1973. Studies on soil copper. I. The fractionation of Cu in soils. *J. Soil Sci.*, 24:172–181.
85. Iyengar, S. S., Martens, D. C., and Miller, W. P. 1981. Distribution and plant availability of soil zinc fractions. *Soil Sci. Sci. Am. J.*, 45:735–739.
86. McLaren, R. G., Lawson, D. M., and Swift, R. S. 1986. The forms of cobalt in some Scottish soils as determined by extraction and isotope exchange. *J. Soil Sci.*, 37:223–234.
87. Miller, W. P., Martens, D. C., Zelazny, L. W., and Kornegay, E. T. 1986. Forms of solid phase copper in copper-enriched swine manure. *J. Environ. Qual.*, 15:69–72.
88. Tiller, K. G., Honeysett, J. L., and de Bries, M. P. C. 1972. Soil zinc and its uptake by plants. II. Soil chemistry in relation to prediction of availablity. *Austral. J. Soil Res.*, 10:165–182.
89. Miller, W. P., Martens, D. C., and Zelazny, L. W. 1986. Effect of sequence in extraction of trace metals from soils. *Soil Sci. Sci. Am. J.*, 50:558–560.
90. Gibbs, R. J. 1973. Mechanisms of trace metals transport in rivers. *Science*, 180:171–173.
91. Gibbs, R. J. 1977. Transport phases of transition metals in Amajon and Yukon rivers. *Geol. Soc. Am. Bull.*, 88:829–843.
92. Jackson, M. L. 1958. *Soil Chemical Analysis*. Elsevier, Englewood Cliffs, NJ.
93. Gupta, S. K., and Chen, K. Y. 1975. Partitioning of trace elements in selective chemical fractions of nearshore sediments. *Environ. Lett.*, 10:129–158.
94. Salomans, W., and Foerstner, U. 1980. Trace metal analysis of polluted sediments. II. Evaluation of environmental impact. *Environ. Technol. Lett.*, 1:506–517.
95. Salomans, W., and Foerstner, U. 1984. *Metals in the Hydrocycle*. Springer-Verlag, New York.
96. Schoer, J., and Eggersgluess, D. 1982. Chemical forms of heavy metals in sediments and suspended matter of Weser, Elbe and Ems Rivers. *Mitt. Geol. Paleontol. Inst., Univ. Hamburg*, 52:667–685.
97. Kersten, M., and Foerstner, U. 1986. Chemical fractionation of heavy metals in anoxic estuarine castal sediments. *Water Sci. Technol.*, 18:121–130.
98. Rule, J. H., and Alden, R. W. 1992. Partitioning of Cd in geochemical fractions of anaerobic estuarine sediments. *Estuarine Coastal Shelf Sci.*, 34:487–499.
99. Foerstner, U., Calmano, K., Conrad, H., Jaksch, H., Schimkus, C., and Schoer, J. 1981. Chemical speciation of heavy metals in solid waste materials (sewage sludge, mining wastes, dredged materials, polluted sediments) by sequential extraction. *Proc. 3rd Int. Conf. Heavy Metals in the Environment*, pp. 698–704. CEP Consultants Ltd., Edinburgh, UK.
100. Harrison, R. M., Laxe, D. P. H., and Wilson, S. J. 1981. Chemical associations of lead, cadmium, copper and zinc in street dusts and road side soils. *Environ. Sci. Technol.*, 15:1378–1383.
101. Robbins, L. M., Lyle, M., and Heath, G. R. 1984. A sequential extraction procedure for partitioning elements among coexisting phases in marine sediments. College of Oceanography Rep. 84-3, Oregon State Univ., Corvallis, Ore.
102. Garcia-Mirgaya, J., Castro, S., and Paolini, J. 1981. Lead and zinc levels and chemical fractionation in roadside soils of Caracas, Venezuela. *Water Air Soil Pollution*, 15:285–297.
103. Ma, Y. B., and Uren, N. C. 1995. Application of new fractionation scheme for heavy metals in soils. *Commun. Soil Sci. Plant Anal.*, 26:3291–3303.
104. Ma, Y. B., and Uren, N. C. 1998. Transformation of heavy metals added to soil—Application of new sequential extraction procedure. *Geoderma*, 84:157–168.

105. Jenne, E. A. 1968. Controls on Mn, Fe, Co, Ni, Cu and Zn concentrations in soils and waters: The dominant role of hydrous Mn and Fe oxides. In: *Trace Inorganics in Water*, Advances in Chemistry Series 73, pp. 337–387. American Chemical Society, Washington, DC.
106. Mehra, O. P., and Jackson, M. L. 1960. Iron oxide removal from soils and clays by a dithionite-citrate system buffered with sodium bicarbonate. *Clays and Clay Miner.*, 7:317–327.
107. Shuman, L. M. 1982. Separating soil iron- and manganese oxide fractions for microelement analysis. *Soil Sci. Sci. Am. J.*, 46:1099–1102.
108. Chester, R., and Hughes, M. J. 1967. A chemical technique for separation of ferro-manganese minerals, carbonate minerals and adsorbed trace elements from pelagic sediments. *Chem. Geol.*, 2:249–262.
109. Chao, T. T. 1972. Selective dissolution of manganese oxides from soils and sediments with acidified hydroxylamine. *Soil Sci. Soc. Am. Proc.*, 36:764–768.
110. Chao, T. T., and Zhou, L. 1983. Extraction techniques for selective dissolution of amorphous iron oxides from soils and sediments. *Soil Sci. Sci. Am. J.*, 47:225–232.
111. LeRiche, H. H., and Wier, A. H. 1963. A method for studying trace elements in soil fractions. *J. Soil Sci.*, 14:225–235.
112. Schwertmann, U. 1964. Differenzierung der Eisenoxide des Bodens durch Extraktion mit Ammoniumoxalat-losung. *Z. Pflanzenernahr. Dung Bodenk.*, 105:194–202.
113. McKeague, J. A., and Day, J. H. 1966. Dithionite and oxalate extractable Fe and Al as aids in differentiating various classes of soils. *Can. J. Soil Sci.*, 46:13–22.
114. Gibson, J. J., and Farmer, J. G. 1986. Multistep sequential chemical extraction of heavy metals from urban soils. *Environ. Pollution Ser. B*, 11:117–135.
115. Keller, C., and Vedy, J.-C. 1994. Heavy metals in the environment: Distribution of copper and cadmium fractions in two forest soils. *J. Environ. Qual.*, 23:987–999.
116. Orsini, L., and Bermond, A. P. 1994. Copper biodisponibility in calcareous soil samples. Part 1. Chemical fractionation of copper. *Environ. Technol.*, 15:695–700.
117. Hoffman, S. J., and Fletcher, W. K. 1981. Detailed lake sediment geochemistry of anomalous lakes on the Nechako Plateau, central British Colombia—Comparison of trace metal distributions in Capoose and Fish Lakes. *J. Geochem. Explor.*, 14:221–244.
118. Lavkulich, L. M., and Wiens, J. H. 1970. Comparison of organic matter destruction by hydrogen peroxide and sodium hypochlorite and its effect on selected mineral constituents. *Soil Sci. Soc. Am. Proc.*, 34:755–758.
119. Beckett, P. H. T. 1989. The use of extractants in studies on trace metals in soils, sewage sludges, and sludge-treated soils. *Adv. Soil Sci.*, 9:143–176.
120. Jackson, M. L. 1985. *Soil Chemical Analysis—An Advanced Course*, 2nd ed. Published by the author, Department of Soil Science, University of Wisconsin, Madison, WI.
121. Chao, T. T. 1984. Use of partial dissolution techniques in geochemical exploration. *J. Geochem. Explor.*, 20:101–135.
122. Pickering, W. F. 1981. Selective chemical extraction of soil components and bound metal species. *CRC Crit. Rev. Anal. Chem.*, 12:233–266.
123. Ross, S. M., 1994. Sources and forms of potentially toxic metals in soil-plant systems. In: S. M. Ross (ed.), *Toxic Metals in Soil-Plant Systems*, pp. 3–26. Wiley, New York.
124. Sheppard, M. I., and Stephenson, M. 1997. Critical evaluation of selective extraction methods for soils and sediments. In: R. Prost (ed.), *Contaminated Soils*, Proc. 3rd Int. Conf. Biogeochemistry of Trace Elements, Paris, France, May 15–19, 1995, pp. 69–97. INRA, Paris.
125. Florence, T. M., and Batley, G. E. 1977. Determination of chemical forms of trace metals in natural waters with special reference to copper, lead, cadmium and zinc. *Talanta*, 24:151–158.
126. Shuman, L. M. 1985. Fractionation method for soil micoelements. *Soil Sci.*, 140:11–22.

127. Charlatchka, R., Cambier, P., and Bourgeois, S. 1997. Mobilization of trace metals in contaminated soils under anaerobic conditions. In: R. Prost (ed.), *Contaminated Soils*, Proc. 3rd Int. Conf. Biogeochemistry Trace Elements, Paris (France), May 15–19, 1995, pp. 159–174. INRA, Paris.
128. Ramos, L., Hernandez, L. M., and Gonzalez, M. J. 1994. Sequential fractionation of copper, lead, cadmium and zinc in soils from or near Donana National Park. *J. Environ. Qual.*, 23:50–57.
129. Han, F. X., Hu, A. T., and Qi, Y. H. 1995. Transformation and distribution of forms of zinc in acid, neutral and calcareous soils of China. *Geoderma*, 66:121–135.
130. Lund, L. J., Betty, E. E., Page, A. L., and Elliott, R. A. 1981. Occurrence of naturally high cadmium levels in soils and its accumulation by vegetation. *J. Environ. Qual.*, 10:551–556.
131. Latterel, J. J., Dowdy, R. H., and Larson, W. E. 1978. Correlations of extractable metals and metal uptake of snap beans grown on soil amended with sewage sludge. *J. Environ. Qual.*, 7:435–440.
132. Soltanpour, P. N. 1991. Determination of nutrient availability and elemental toxicity by AB-DTPA soil test and ICPS. *Adv. Soil Sci.*, 16:165–190.
133. Lakanen, E., and Ervio, R. 1971. A comparison of eight extractants for the determination of plant available micronutrients in soils. *Acta Agral. Fenn.*, 123:223–232.
134. Petruzzelli, G., Lubrano, L., and Guidi, G. 1981. The effect of sewage sludge and composts on the extractability of heavy metals from soil. *Environ. Technol. Lett.*, 2:449–456.
135. Seaker, E. M. 1991. Zinc, copper, cadmium and lead in minespoil, water, and plants from reclaimed mine land amended with sewage sludge. *Water Air Soil Pollution*, 57–58:849–859.
136. Soltanpour, P. N., and Schwab, A. P. 1977. A new soil test for the simultaneous extraction of macro- and micro-nutrients in alkaline soils. *Commun. Soil Sci. Plant Anal.*, 8:195–207.
137. Krishnamurti, G. S. R., Huang, P. M., Van Rees, K. C. J., Kozak, L. M., and Rostad, H. P. W. 1997. Differential FTIR study of pyrophosphate extractable material of soils: implication in Cd-bonding sites and availability. In: R. Prost (ed.), *Contaminated Soils*, Proc. 3rd Int. Conf. Biogeochemistry of Trace Elements, May 15–19, 1995, Paris, France. INRA, Paris.
138. Snedecor, G. W., and Cochran, W. G. 1980. *Statistical Methods*. The Iowa State University Press, Ames, IA.
139. Kabata-Pendias, A., and Pendias, H. 1992. *Trace Elements in Soils and Plants*. CRC Press, Boca Raton, FL.
140. Gworek, B. 1992. Lead inactivation in soils by zeolites. *Plant Soil* 143:71–74.
141. Czupyrna, G., Levy, R. D., MacLean, A. I., and Gold. H. 1989. *In situ Immobilization of Heavy-Metal-Contaminated Soils*. Noyes Data Corp., Park Ridge, NJ.
142. Vangronsveld, J., Van Assche, F., and Clijsters, H. 1990. Immobilization of heavy metals in polluted soils by application of a modified alumino-silicate: Biological evaluation. In: J. Barcelo (ed.), *Environmental Contamination*, Proc. 4th Int. Conf., Barcelona, pp. 238–285. CEP-Consultants, Edinburgh, U.K.
143. McKenzie, R. M. 1980. The adsorption of lead and other heavy metals on oxides of manganese and iron. *Austral. J. Soil Res.*, 18:61–73.
144. Fu, G., Allen, H. E., and Cowan, C. E. 1991. Adsorption of cadmium and copper by manganese oxide. *Soil Sci.*, 152:72–81.
145. Janssen, R., Bruggenwert, M., and Van Riemsdijk, W. 1993. Adsorption of zinc by complexes of clay/Al-hydroxide polymers. In: H. J. P. Eijsackers and T. Hammers (eds.), *Integrated Soil and Sediment Research: A Basis for Proper Protection*, pp. 261–262. Kluwer, Dordrecht.
146. Mench, M. J., Didier, V. L., Loeffler, M., Gomez, A., and Masson, P. 1994. A mimicked in-situ remediation study of metal contaminated soils with emphasis on cadmium and lead. *J. Environ. Qual.*, 23:58–63.

147. Sakurai, K., and Huang, P. M. 1995. Cadmium adsorption on the hydroxyaluminum-montmorillonite complex as influenced by oxalate. In: P. M. Huang, J. Berthelin, J.-M. Bollag, W. G. McGill, and A. L. Page (eds.), *Environmental Impact of Soil Component Interactions. Volume II. Metals, Other Inorganics, and Microbial Activities*, pp. 39–46. CRC Lewis Publishers, Boca Raton, FL.
148. Foerstner, U. 1984. Metal pollution of terrestrial waters. In: J. O. Nriagu (ed.), *Changing Metal Cycles and Human Health*, pp. 71–94. Springer-Verlag, Berlin.
149. Kheboian, C., and Bauer, C. F. 1987. Accuracy of selective extraction procedures for metal speciation in model aquatic sediments. *Anal. Chem.*, 59:1417–1423.
150. Foerstner, U. 1987. Sediment-associated contaminants—An overview of scientific basis for developing remedial options. *Hydrobiologia*, 149:221–246.
151. U.S. Environmental Protection Agency. 1991. Rod annual report FY 1990, USEPA Rep. 540/8-91/067. U.S. EPA, Washington, DC.

33
Arsenic-Contaminated Soils: I. Risk Assessment

A. Brandstetter
University of Agricultural Sciences, Vienna, Austria

E. Lombi
IACR-Rothamsted, Harpenden, Hertfordshire, England

W. W. Wenzel
University of Agricultural Sciences, Vienna, Austria

D. C. Adriano
Savannah River Ecology Laboratory, University of Georgia, Aiken, South Carolina

I. INTRODUCTION

Recent awareness of the toxicity of arsenic to humans at much lower concentrations than was previously deemed to be dangerous has led to increased interest in the environmental chemistry of As (1). In addition, the anionic nature of As, and therefore its relatively higher mobility in alkaline soils compared with metals in cationic form, has compelled scientists and regulators to look at its mobility from the soil–plant system into groundwater.

Background levels of total As in soils are typically below 10 mg/kg (2). Larger concentrations of As are found in soils that have received large doses of As-based pesticides, or have been affected by industrial emissions, such as from mining and smelting (3). Localized areas can also be contaminated, some at high levels, in connection with the use of As pesticides (bathing solutions) to control certain infestation in the cattle industry. This is rather common in large cattle-producing countries (e.g., New Zealand and Australia) (4).

Anomalously large concentrations of As in soils and sediments due to geogenic origin or mining and smelting operations are common in some regions of the Austrian Alps (5–7). The As contamination at many sites is several centuries old. Levels of As in soil and freshwater (river/streams/groundwater) at these sites may exceed permissible limits, even though atmospheric deposition of As in Central Europe has decreased significantly during the past decade (8). In view of the occurrence, in some countries, of arsenic poisoning, and the rather high natural occurrence of As in some areas, including the Austrian Alps, it is essential to establish a risk assessment protocol for As that integrates its mobility to groundwater and bioavailability to plants.

II. POTENTIAL RISKS ASSOCIATED WITH As-CONTAMINATED ENVIRONMENTS

The primary exposure pathways of As to the population are ingestion by drinking water and food (Fig. 1). In some cases, the use of contaminated groundwater for cooking can exacerbate the exposure of humans to this element, as will be discussed later.

The occurrence of inorganic As in drinking water has been identified as the main risk for human health (9). Most drinking water standards for As range between 10 and 50 µg As L^{-1} (10–12). Although the U.S. Environmental Protection Agency (EPA) permissible concentration is currently 50 µg As L^{-1}, WHO (11) suggests only 10 µg As L^{-1}. Because of recent observations on As toxicity at much lower concentrations than were previously reported to be dangerous, the German standard in drinking water has been lowered to 10 µg As L^{-1} (12), which is also the target standard for the EU. Consequently, the EPA is also considering lowering its limit to 10 µg As L^{-1} (10).

Arsenic is ubiquitous is nature and is found in detectable concentrations in all soils and nearly all other environmental media. Toxicity of As to biota varies considerably and is largely influenced by the form and speciation of As (9). The most predominant As species in soils are not the most toxic. In general, organo-As compounds are less toxic than inorganic As compounds. As(III) is considered to be more mobile and more toxic than As(V) (13).

The solubility and bioavailability of As depends on several key soil properties. Arsenic is less soluble and bioavailable in soils that contain high amounts of clay and sesquioxides. In these soils strong adsorption maintains low concentrations of As in the soil solution. Arsenic solubility can be enhanced by the addition of competing anions, e.g., phosphate or molybdate. Arsenic solubility and speciation is also related to soil pH and Eh. Reduced As toxicity has been reported for soils treated with Al, Fe, and Zn compounds, S, lime, and organic matter (14).

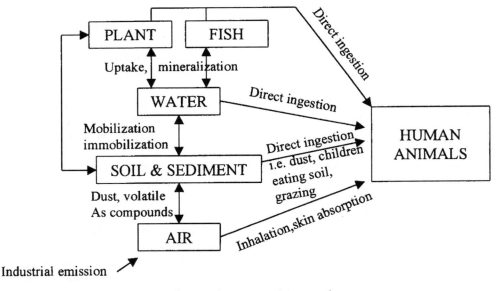

Figure 1 Exposure pathways of As to humans and mammals.

Because of the potential mobility and exposure of organisms to naturally occurring As, guideline values need to be developed to safeguard the general population. In Germany and other Central European countries, guideline values for total As in soils range between 7 and 40 mg As kg^{-1} for plant production and between 7 and 30 mg As kg^{-1} in playgrounds (15–19). Eikmann and Kloke (15) defined As guideline values according to land use on total As concentration in soils. Multifunctional land use is possible with background concentrations below 20 mg As kg^{-1}. Maximum tolerable concentrations are 20 mg As kg^{-1} for children's playgrounds, 35 mg As kg^{-1} for sport grounds, 40 mg As kg^{-1} for backyards, gardens, parks, recreational areas, agricultural land, and nonagricultural ecosystems, and 50 mg As kg^{-1} for industrial sites. A need for remediation is defined at 50 mg As kg^{-1} in playgrounds and agricultural land, 60 mg As kg^{-1} in nonagricultural ecosystems, 80 mg As kg^{-1} in backyards, gardens, parks, and recreational areas, 90 mg As kg^{-1} in sport grounds, and 150 mg As kg^{-1} at industrial sites (15).

Arsenic is naturally present in plants, but concentrations in plant tissues rarely exceed 1 mg kg^{-1}. Arsenic uptake in terrestrial plants is typically low, with transfer coefficients ranging from 0.01 to 0.1 (7,20). The Austrian threshold value for As in fodder is 2 mg kg^{-1} DW. However, certain soil properties, such as high pH and low redox potential, may promote As solubility and thus uptake by plants.

III. MAJOR REGIONS AFFECTED BY UNUSUAL ENVIRONMENTAL LEVELS OF As

Large concentrations of As in groundwater have been identified in some regions of Europe, the Americas, Asia, and Oceania (Fig. 2).

Long-term effects on humans exposed to high concentrations of As compounds have been reported from West Bengal, India, where more than 1,000,000 people are exposed to As by drinking As-contaminated well waters. An estimated 200,000 people already have As-induced skin lesions, many of them having hyperkeratoses and hardened patches of skin that may develop into cancer. Indian scientists have also found symptoms not usually associated with As poisoning, such as liver diseases and respiratory problems. Part of the problem, however, is believed to be confounded by the poor diet of some West Bengalis (21–23).

In Bangladesh a survey report of 26 districts indicates that groundwater of 21 districts contains more than 50 μg L^{-1} As, and in 18 districts (covering 40,000 km^2, 35 million population) people suffering from arsenical skin lesions including skin cancer have been identified. Chakraborti et al. (23) collected about 2550 water samples from Bangladesh and 35,000 from West Bengal, India, and some hair and nail samples of the As victims of the affected districts. The As concentration in hair in 49% of the affected West Bengal population was above the toxic level of 0.08–0.2 mg kg^{-1}, and in Bangladesh 92% were above the toxic level. The nail concentrations of 74% of the tested persons in West Bengal and 98% in Bangladesh were larger than the normal range of 0.4–1.1 mg g^{-1} (23).

In India as well as in Bangladesh the source of As is primarily geological. The As-affected area is in the region where sediment deposition took place during the Quaternary Period, i.e., 25,000 to 80,000 years ago, commonly known as the Younger Deltaic Deposition. This sediment contains As-rich pyrite and covers almost the entire alluvial region of the river Ganges (24,25). In this region, millions of cubic meters of underground water are extracted for irrigation by shallow and deep tube wells. Due

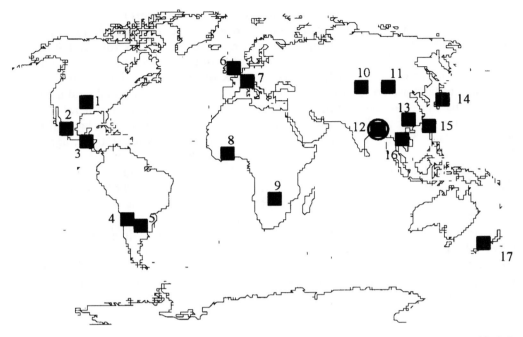

Figure 2 Areas around the world where environmental problems with As have been identified (1, several areas of the United States; 2, Mexico; 3, Nicaragua; 4, Chile; 5, Argentina; 6, England, U.K.; 7, Alps, Central Europe; 8, Ghana; 9, Zimbabwe; 10, Xinjiang, China; 11, Inner Mongolia, China; 12, West Bengal, India & Bangladesh; 13, Yunnan and Guizhu, China; 14, Japan; 15, Taiwan; 16, Thailand; 17, New Zealand).

to such high groundwater withdrawal, air enters the aquifer, and due to available oxygen, As-rich pyrite is oxidized and As is mobilized from the vadoze zone (26).

While the As poisoning in West Bengal (India) and Bangladesh is the most extensive and catastrophic in the history of environmental toxicology, less extensive and less severe As poisoning has also been observed elsewhere. In a 1000-km^2 large basin area in Inner Mongolia, drinking water from more than 50% of deep wells and more than 20% of shallow wells contained As exceeding 50 mg L^{-1}, with maximum concentrations up to 1860 μg L^{-1}. Since no arsenical pesticides have been used, and no factories, mines, or other industries discharge As into the environment, it is concluded that sediments release As into the groundwater. The subterranean waters occur in a Quaternary earth stratum and have a high concentration of As from the local rock. People using this water show chronic arsenicism such as hyperkeratosis, hyperpigmentation, and other severe skin lesions. The prevalence of these symptoms increased with exposure to As concentrations in water of 350 μg L^{-1} or more (27). However, surface water is not contaminated.

Elevated As concentrations in drinking water of some regions in Taiwan have also occurred (28,29). Drinking As-enriched water accounts for about 70% of As in human urine in northern Taiwan (30).

Arsenic poisoning caused by domestic burning of coal containing 90–2100 mg As kg^{-1} was reported from southern China (31). Since coal is burned in open pits in households for daily cooking, As is accumulated and circulated in indoor air which affects people by inhalation. In the southern part of the Chinese province of Yunan, the daily As intake of local residents was calculated to be in the range of 0.1–1.1 mg, while WHO suggested that a daily intake of 2 µg of inorganic As per kilogram of body weight should not be exceeded (32). Emission of As from a nonferrous smelter in Yunan is contaminating the air and food (31).

In central Slovakia, power plants burned locally mined coal containing 800–1500 mg As kg^{-1}. Since around 500 kg As have been emitted daily, elevated As concentrations have been found in soil, plants, animals, and honey. Consequently, health problems such as carcinogenocity, immunotoxicity, and neurotoxicity have occurred. The benefits of reduced emission of As through air pollution control devices on long-term health effects in the population is now being evaluated (33).

Farago et al. (34) reported that there is some cause of concern over large As concentrations in agricultural soils, garden soils, and house dust in southwest England. However, no evidence of health-related symptoms caused by this high As levels has been reported.

In Japan, As pollution of soils has been caused by mining activities and agricultural use of pesticides (35,36). Phytotoxic effects on rice growing on soils with elevated As concentrations are causing yield reductions. Further, areas with high As concentration in soils and/or groundwater are known in some regions of North America (37–39), Mexico (40), Chile (41), Argentina (42), Ghana (41), and New Zealand (4,43).

IV. THE BANGLADESH CASE STUDY

The As calamity in the Indian subcontinent received wider attention after Indian scientists reported data on As levels in groundwaters (concentrations up to 3700 µg L^{-1}) along with As-related health problems from seven districts in West Bengal, India, close to the Bangladesh border (23–25). Shortly after this announcement our team initiated an effort, in conjunction with Bangladesh colleagues, to evaluate the distribution of arsenic in soil–groundwater systems in some districts of Bangladesh where As poisoning of the population due to the utilization of arsenic-contaminated groundwaters is evident. Soil samples from five highly suspect areas were collected in 1996. In addition, water samples utilized by the inhabitants for drinking and for irrigation were collected. Soil and water samples were analyzed for total As concentrations. Soil samples were additionally analyzed to obtain information on the biogeochemical behavior of As.

Deep groundwaters contaminated with As have been used increasingly as drinking and irrigation water in a number of regions in Bangladesh. There is confirmed evidence that the utilization of these As-contaminated groundwaters caused As poisoning and attendant diseases in some of the regions concerned. Initially, As contamination of groundwaters was attributed to geogenic and/or anthropogenic sources. Recently, however, geogenic processes have been considered, from data obtained from West Bengal, India, and Bangladesh, to be the primary cause. In contrast, elevated As levels in the vicinity of Dhaka have been suspected to be from both anthropogenic and geogenic sources. For example, an important but highly localized anthropogenic As source is related to the use of As-treated wooden poles for electric power lines.

Highly suspect groundwater samples were collected and measured in Narayanganj near Dhaka and in 11 other districts of Bangladesh next to the Indian province of West Bengal (44). More than two out of three of the water samples exceeded the WHO drinking water standard for As (10 µg L^{-1}) (Fig. 3). People exposed to these elevated As concentrations in drinking water are at risk of As poisoning.

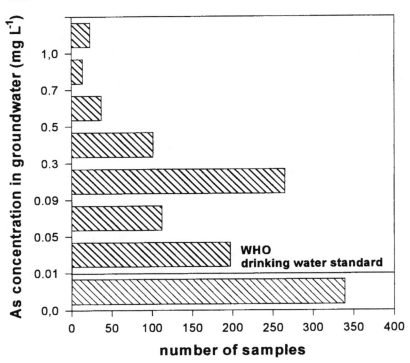

Figure 3 Frequency of As in 1088 groundwater samples from 16 districts in Bangladesh where As poisoning is evident (*Source*: School of Environmental Studies, Jadavpur University Calcutta, India.)

Arsenic poisoning primarily affects rural villagers in Bangladesh who use untreated water and consume a protein-poor diet. Moreover, their water uptake is greater than that of the rest of the population, because they are subsistence farmers who toil under tropical conditions.

Annual irrigation of a paddy soil with 500 mm water containing 35 µg As L^{-1} (well 1) is equivalent to an input of 150 g As ha^{-1} y^{-1} and would increase the As concentration in the top 15 cm of the soil by 0.1 mg kg^{-1} y^{-1}. Utilizing the same amount of water containing 1100 µg L^{-1} (well 2) would increase the As levels, through an input of 4.5 kg ha^{-1} y^{-1}, by about 3 mg kg^1 y^{-1} (45). On a specific site, the As concentration in the top soil under rice cultivation is larger than in soil under production of vegetables (Fig. 4), because rice is more intensively irrigated.

Enhanced accumulation of As in top soils could promote its transfer into the food chain [soil–crop–(animal)–human] and decreases rice yield in paddy soils (36). The rate of transfer is not only a function of the As concentration in soil, but is also largely influenced by its geochemical behavior. In short, the fate and effect of As in the soil–plant system is affected by arsenic speciation and soil characteristics, such as pH and redox potential; at a given pH, As(V) is reduced to the more soluble and toxic As(III) under low redox potentials. Such conditions are common in paddy soils used for rice cultivation (46), and there is a positive correlation between soil As and As in the tissues of rice plants (47). Rice yields as affected by As under flooded conditions were decreased more substantially than under intermittently flooded systems in Japan (36). This is probably

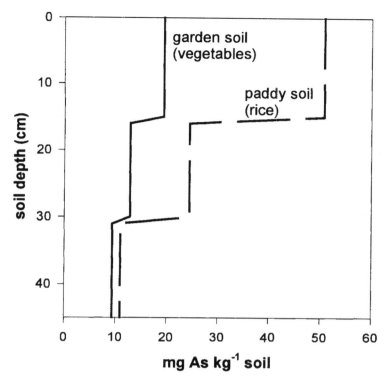

Figure 4 Total As concentration in soils that were collected from an irrigated garden and periodically flooded (paddy) site in Faridpur, Bangladesh (45).

due to the maintainance of an oxidative state in the flooded soil. It has been reported that in river sediments (48) under oxidizing conditions (200–500 mV), the major As species is As(V) and the solubility is low. Reducing conditions lead to mobilization of As. At redox < 100 mV, As mobilization is controlled by the dissolution of hydrous iron oxides. The amount of As released increase 10- to 13-fold compared to aerobic conditions, and As(III) is the dominant species released. Because the net redox potential of soils depends on the redox potentials of all the reducing and oxidizing systems occurring in the soils, these relationships are complex, and the redox value for soils is not directly proportional to the As(III):As(V) ratio (49). Water management in Bangladesh must address these problems. However, the more serious problem in Bangladesh arises from the drinking of water from wells with large levels of As (Fig. 5).

In addition to the problem of As-contaminated groundwaters, the Bangladesh Rural Electrification Board (REB), established in 1997 to supply industry and households with electricity, laid out power lines using wooden poles treated with chromated copper arsenate, ammoniacal copper arsenate, or ammoniacal copper-zinc arsenic as preservative to protect the poles from biodegradation (50). This prompted us to also collect soil samples adjacent to the utility poles for As assay. The Bangladesh Arsenic Investigation Committee (50) found that soils around installed poles, especially at 0–30 cm depth, show higher As concentrations than garden or agricultural soils from the same area. However, no leaching of As to the groundwater from utility poles was evident. In contrast, the storage yards for wooden poles show increased As concentration in the top soil by leaching

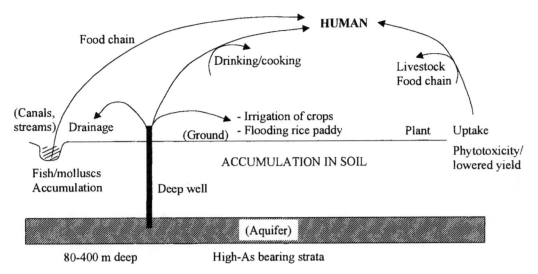

Figure 5 Pathways of As in well waters pumped from deep aquifers depicting the scenario in the sub-Asian continent of West Bengal and Bangladesh.

of As from the poles. In this case, direct ingestion of this As-contaminated soil and dust by children as well as livestock grazing in the yard is considered the main pathway of As to human (Fig. 6).

V. THE AUSTRIAN CASE STUDY

Various processes lead to accumulation of As in soil and freshwater to levels exceeding permissible limits (7), even though atmospheric deposition of As has decreased significantly during the past decade (8). Anomalously high levels of As in soils and sediments due to geogenic origin or mining and smelting operations are common in some regions of the Austrian Alps. Thalmann et al. (5) found, by analyzing stream sediments, that As is a characteristic element of the ore-bearing parts of the Central Alps in Austria. The As distribution in soil in this area follows mainly the distribution of Fe and its associated elements.

The results of the Austrian soil inventory program show that soils in the Central Alps such as around Salzburg have more elevated As concentrations than soils from other Austrian regions (Table 1) (51). For risk assessment and monitoring of As distribution and mobility in contaminated Austrian soils, soil samples were collected according to genetic horizons at 38 As-contaminated sites in Carinthia, Salzburg, Tirol, and Styria (Fig. 7)

In addition to characterizing soil parameters such as pH, total element concentration, carbonate, exchangeable cations, and oxides of Fe, Al, and Mn, we fractionated As into various forms: total As by microwave digest using HNO_3/H_2O_2, exchangeable form by NH_4NO_3 (DIN 19730) extraction and 0.05 M $(NH_4)_2SO_4$ extraction, and the specifically sorbed fraction using 0.05 M $(NH_4)_2HPO_4$ extraction.

At selected contaminated sites, As concentrations in plants and As fluxes were measured. Additionally, methods for chemical fractionation and speciation of As in con-

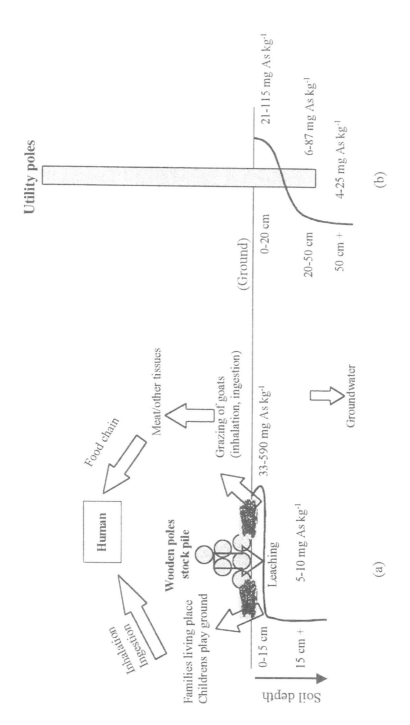

Figure 6 Pathways of As contamination arising from the use of As-treated wooden (electric utility) poles (data from Bangladesh Arsenic Investigation Committee, 1996): (a) As leaching profile under a wooden poles stock pile; (b) As leaching profile beneath an installed utility pole.

Table 1 Relative Distribution (in percent) of As Concentration in Some Austrian Regions (Danneberg et al., 1997)

	N	Range of total As[a] concentration in soils (mg kg^{-1})				
		< 5	5–10	10–15	15–20	> 20
Lower Austria	1449	20.9	48.6	13.3	5.1	2.1
Upper Austria	880	55.6	27.2	12.4	1.8	3.0
Burgenland	174	4.0	20.1	46.0	22.4	7.5
Styria	84	23.8	26.2	22.6	13.1	14.3
Salzburg	462	3.6	22.3	20.8	14.2	39.1

[a] Aqua regia.
The soils were sampled in a 4 × 4 km basis grid.

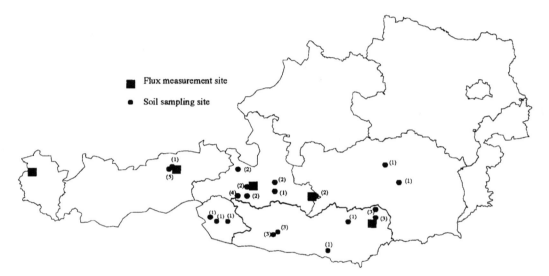

Figure 7 Overview of the soil sampling sites (schematic) in the Austrian Alps. Numbers in parentheses indicate the number of soil profiles at a site.

taminated soils were developed and applied to selected samples. These methods integrate chemical fractionation and speciation of As, As distribution and fractionation in particle-size fractions, and As localization in the solid phase (Fig. 8).

A. Arsenic Distribution and Mobility in Contaminated Soils of Austria

The pH values of the investigated soils cover a range from strongly acidic forest soils to soils with free carbonate. Base saturation in these soils varied between 5.5% and 99.2%. The total As contents of 191 soil samples from contaminated sites ranged from 1 to 2995 mg As kg^{-1} (Table 2). The mean value is greater than the median value, indicating deviation of the data from normal distribution due to anomalously high concentrations of As. A comparison of the ammonium oxalate-extractable As fraction to the total As content suggests that, on average, 50% of As is bound to amorphous hydroxide. Even

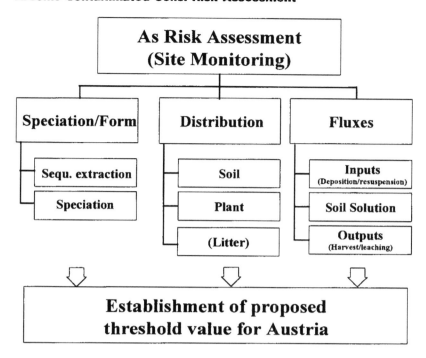

Figure 8 Flow diagram assessing the form, inventory, and fluxes of As as a prelude to risk assessment.

Table 2 Summary of As Concentrations in Different Fractions of Soils Collected from the Austrian Alps

As fractions	Unit	N	Mean	Median	Minimum	Maximum
Total As	(mg kg^{-1})	191	235	77	1.00	2995
NH$_4$ oxalate-extractable As	(mg kg^{-1})	171	76	19	0.00	726
(NH$_4$)$_2$HPO$_4$-extractable As	(mg kg^{-1})	170	14	2.4	0.00	191
NH$_4$NO$_3$-extractable As	(µg kg^{-1})	13	180	147	27.3	566
(NH$_4$)$_2$SO$_4$-extractable As	(mg kg^{-1})	191	0.7	0.13	0.00	11

though the soils contain large amounts of total As, the extractability by ammonium nitrate is typically low. (NH$_4$)$_2$SO$_4$, and (NH$_4$)$_2$HPO$_4$-extractable As are significantly correlated with total As. (NH$_4$)$_2$SO$_4$ is considered to extract mainly the exchangeable As pool and is more efficient than NH$_4$NO$_3$, which is used primarily for extraction of exchangeable metal cations.

B. Sequential Extraction of Arsenic

Sequential extraction of As based on successive extraction steps is a useful tool to predict As mobility and may enhance the understanding of the ecotoxicological significance of anomalously large As concentrations in soil.

Table 3 Summary of the Results of the Sequential Extraction Procedure (SEP) Performed for 24 Soils of the Austrian Alps (mg As kg^{-1})

	$(NH_4)_2SO_4$[a]	$NH_4H_2PO_4$[b]	Oxalate[c]	Oxalate/Asc.[d]	Residual[e]	Sum
Mean	1.81	57.5	190.2	193.0	84.9	527.4
Median	0.71	28.7	118.8	63.5	30.2	222.3
Max.	11.13	179.9	695.6	763.4	343.8	—
Min.	0.02	3.1	17.7	8.7	2.3	—
Percent of total	0.34	10.9	36.1	36.6	16.1	100.0

[a] $(NH_4)_2SO_4$ extraction: 1:25, 0.05 M $(NH_4)_2SO_4$, 4 h shaking at 20°C.
[b] $(NH_4)H_2PO_4$ extraction: 1:25, 0.05 M $(NH_4)_2HPO_4$, 16 h shaking at 20°C.
[c] NH_4 oxalate: 1:25, 0.2 M NH_4 oxalate pH 3.25 [0.2 M $(NH_4)_2$ oxalate H_2O + 0.2 M oxalic acid 2 H_2O, pH adjusted with diluted NH_3 solution], 4 h shaking in the dark. Wash step: 1:12.5, 0.2 M NH_4 oxalate pH 3.25, 10 min shaking in the dark.
[d] NH_4 oxalate/ascorbic acid: 1:25, 0.2 M NH_4 oxalate + 0.1 M ascorbic acid pH 3.25 (0.1 M ascorbic acid + 0.2 M $(NH_4)_2$ oxalate H_2O + 0.2 M oxalic acid 2 H_2O, pH adjusted with diluted NH_3 solution), 30 min in the water bath at 96°C in the light. Wash step: 1:12.5, 0.2 M NH_4 oxalate pH 3.25, 10 min shaking in the dark.
[e] Acid digestion: $HNO_3 + H_2O_2$, closed-vessel microwave digestion.

A procedure described by Sletten et al. (52) was adopted to fractionate As in 24 Austrian soils that vary considerably in chemical composition and extent of contamination. The extraction steps include (a) $(NH_4)_2SO_4$-extractable As for the exchangeable fraction, (b) $(NH_4)H_2PO_4$-extractable As for the specifically sorbed fraction, (c) NH_4 oxalate-extractable As for amorphous Fe-bound fraction, (d) NH_4 oxalate/ascorbic acid-extractable As for the crystalline Fe-bound fraction, and (e) acid digestion (HNO_3/H_2O_2)-extractable As for the residual fraction. A summary of the results is reported in Table 5, below.

The results show that As was most abundant in the two oxalate fractions, indicating that As is associated primarily with amorphous and crystalline Fe oxides. The fraction of As extracted by $(NH_4)_2HPO_4$ represents only about 10% of total As. This fraction may provide a relative measure of specifically sorbed As in soils. The amount of readily mobile As extracted by $(NH_4)_2SO_4$ is generally small, but represents the most important fraction related to risk assessment. The residual fraction is typically small, since most arsenic is bound to amorphous and crystalline Fe oxides (Table 3).

C. Arsenic Distribution and Fractionation in Particle-Size Fractions

Positive correlations between clay and As content in the soil have been reported by several investigators (53,54). In contrast, soils with sandy texture, or enriched in silicon and alumina, generally show low contents of As (54,55). Therefore, in regions with soils derived from similar parent material, soil texture may affect the background level of As in soil (56). Arsenic toxicity was found to be less pronounced in fine-textured soils (57,58), inferring low bioavailability of As in such soils.

Differences in sesquioxide and clay mineral contents among textural fractions suggest that As distribution and mobility may be controlled by the particle-size distribution. Moreover, many soil properties are affected by surface phenomena at the interface between the liquid and solid phases. The surface area increases in the order sand < silt < clay, indicating the predominant role of clay-sized minerals in As biogeochemistry.

Six bulk soil samples containing from 214 to 1900 mg kg^{-1} total As were used to investigate the As distribution among the particle size fractions. Applying sequential extraction analysis on the particle fractions, most As was found in fractions targeting amorphous and crystalline Fe oxides. The most effective extractant for As was NH_4 oxalate, since it reacts primarily with Fe oxides. Thus, it appears that As distribution and mobility in soil is controlled by Fe oxides.

The largest As concentration was found in clay fractions, but due to the large proportion of silt present in the study soils, most As was typically associated with the silt fraction.

Since As is least strongly bound in the sand fractions, this fraction could be important in the short-term mobilization of As in soils.

D. Arsenic Localization in the Solid Phase of Contaminated Soils

Bulk soil samples (< 2 mm) and sand, silt, and clay fractions of different soils were investigated by electron microprobe analysis (EDAX). The samples were first surveyed using a backscatter electron (BSE) detector. Spectrochemical X-ray analyses were performed using energy dispersive spectrometer (EDS) and wavelength dispersive systems (WDS).

It is apparent from the spectroscopic characterization of the soil samples that As is generally associated with iron oxides (see Fig. 10, below). Association of As with weathered minerals has been also observed. It is not always clear whether the As in the weathering rinds may be due to its presence in the original mineral or to adsorbed phases. In some samples, we identified weathering products of arsenopyrite, such as scorodite and sarmientite, and in this case their content of As is imputable to the composition of the primary mineral. In other samples we found weathered grains of pyrite. The absence of As in the pyrite, but its presence in the weathering rinds, indicates that As had been adsorbed on the surface of the weathered grains. Only a few primary minerals were detected in the samples (59).

The results of these analyses conducted on the particle-size fractions reveal that Fe oxides represent the major sink for As in all soil fractions investigated (Fig. 9). The SEM

 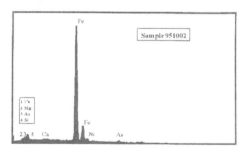

Figure 9 Backscatter image of an As-enriched soil from the B horizon of a eutric cambisol derived from schist and electron diffraction spectra confirming the association of As with Fe oxides.

images show marked differences between silt and clay-size fractions. The images of the clay-size fractions indicate that this fraction is fairly homogeneous. In this fraction, As is diffusely scattered throughout the whole material. Sand and silt fractions appear to have been constituted of separate grains of various minerals. In these size fractions, the only detectable As was associated with Fe, apparently in secondary minerals including oxide deposits and weathering rinds. There was no evidence of As being associated with primary minerals, with the exception of arsenosiderite which was found in one soil sample.

Microprobe analyses corroborate the results of the chemical fractionation of As; i.e., As was associated mainly with Fe oxy/hydroxides. Both particles of Fe oxides and weathering rinds of Fe oxides surrounding primary minerals seem to provide specific accumulation sites for As. Primary minerals containing As are generally scarce and limited to arsenopyrite and arsenosiderite. A different As distribution pattern was observed for sand/silt and clay fraction. Arsenic is distributed homogeneously in the clay fraction, whereas in silt and sand it is associated primarily with discrete particles or coatings of Fe oxides.

E. Arsenic Concentrations in Plants

Arsenic is toxic to nontolerant plants, but toxicity can be alleviated by addition of high concentrations of phosphate to the soil (60). Some plants are tolerant to large concentrations of arsenate but not arsenite (61). Arsenic is naturally present in most plants, but the concentration in plant tissues rarely exceeds $1\,mg\,kg^{-1}$ (62). Few higher plants are known to accumulate As in their tissues. Its distribution, in general, decreases from root to stem and leaf to edible parts (63).

In order to assess whether the transfer of As from soil to plants increases with increasing concentration of As in the soil, we collected plant and soil materials at selected sites for As analysis. The fodder quality at grassland/pasture sites as well as single species were evaluated in order to study the variation of As uptake of different plants.

A total of 128 plant samples were collected at 10 contaminated sites. Sampling included 31 different plant species (see Fig. 11), forest litter, and 14 bulk samples of grass and herbaceous species at grassland sites. All sites had total As concentrations in soil greater than $90\,mg\,kg^{-1}$.

Arsenic concentrations in plant tissues ranged from 0.03 to $34.5\,mg\,kg^{-1}$ (DW), with an average of $1.9\,mg\,kg^{-1}$ ($N=128$). The median was $0.8\,mg\,kg^{-1}$. About 30% of the plant samples exceeded the regulation limit of $2\,mg\,As\,kg^{-1}$ according to the Austrian legislation on fodder quality (64). We found no correlation between total or $(NH_4)_2SO_4$-extractable As in soils and As in plants [$R^2=0.20$ for total As; $R^2=0.24$ for $(NH_4)_2SO_4$-extractable As; $N=88$]. Some species, as well as the mixed fodder, displayed large variation of As concentrations among specimens collected at the same time from the same site (Figs. 10 and 11). Soil–plant transfer coefficients were typically less than 0.01, the maximum being 0.05.

Figures 11b and 11c show the seasonal variation of As uptake by fodder plants. The large As concentrations in the fodder of site c is associated with soil total As of more than $2000\,mg\,kg^{-1}$. The increased concentrations in fodder tissues toward the end of the growing season may be due to resuspension of contaminated soil particles (Fig. 11c). Resuspension and direct ingestion of soil particles by livestock (grazing, hay) are expected to contribute more to As exposure than As ingested with fodder, because plant tissue concentrations are usually less than 1% of the soil concentration.

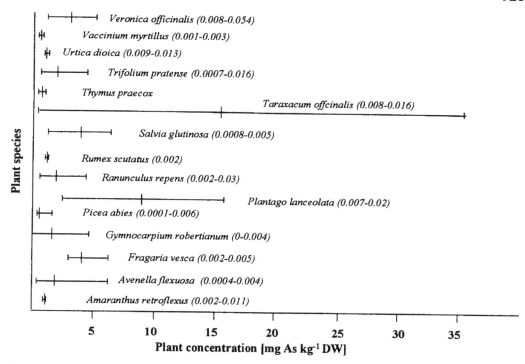

Figure 10 Arsenic concentration (minimum, mean, maximum) in different plant species ($n = 128$) from 10 contaminated sites in the Austrian Alps. The ranges in parentheses show the soil–plant transfer coefficient.

F. Arsenic Fluxes in Contaminated Ecosystems of Austria

Flux measurements represent a powerful tool to identify key processes involved in trace element mobility and to estimate future trends of leaching. With this approach, changes in soil internal and external pools can be estimated and linked to soil impacts (e.g., atmospheric deposition) and related ecosystem compartments such as groundwater, vegetation, and the food chain. Thus, we integrated the measurement of the current input of As to contaminated soils, redistribution of As within the plant–soil system, and output via leaching to groundwater and harvest of fodder grass and trees. Based on this inventory, changes in As pools in the study soils were calculated.

Five experimental plots at four sites in the Austrian Alps were selected (Fig. 8). These sites varied with respect to the source and extent of contamination, land use and soil characteristics, such as soil texture and pH, and were chosen to represent a wide range relevant to risk assessment. Each experimental plot was fitted with a bulk collector for wet deposition, a litter collector, and tension lysimeters at three different soil depths (65).

Water samples were collected from streams and springs in the vicinity of the experimental plots. In addition, wood and fodder grass samples were collected in order to estimate plant uptake of As and removal by harvest. Arsenic fluxes in the soil solution were calculated using a simple water balance approach for the individual sites (66).

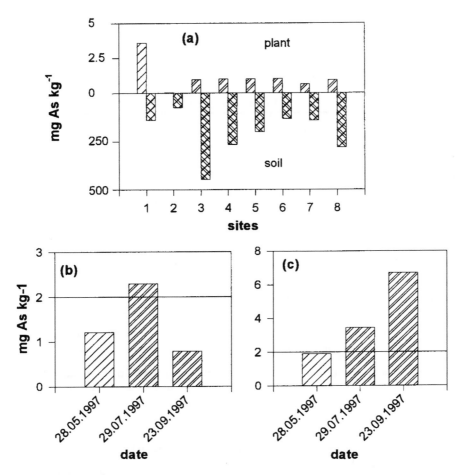

Figure 11 Arsenic concentration in green fodder and the corresponding soil on (a) eight noncalcareous sites and the seasonal variation of As in green fodder on a (b) noncalcareous (110 mg As kg^{-1}) and (c) highly contaminated (2000 mg As kg^{-1}) calcareous soil. The Austrian threshold value for As concentration in fodder of 2 mg kg^{-1} is indicated by a line.

1. Input of Arsenic via Wet Deposition

Wet deposition of As ranged between 0.5 and 5.7 mg m^{-2} y^{-1}. Compared to data reported for six NW German forest sites (8), atmospheric input of As to some of the experimental plots was exceptionally large. Deposition of 57 g ha^{-1} y^{-1} at a geochemically enriched pasture site recorded in 1997 was about 20 times greater than the average As input to forest sites in NW Germany in 1993 (about 3 g ha^{-1} y^{-1}). The latter have been impacted primarily by anthropogenic sources of As that have been decreasing since 1983. Because the studied geochemically enriched site is located in a relatively pristine Alpine area, one might suspect even lower deposition of As from anthropogenic sources than in NW Germany. Therefore, we conclude that the high deposition rates of As at the Austrian sites are due primarily to resuspension of As-laden soil material. Soil resuspension, in particular at nonvegetated sites, apparently represents an important process of As redistribution in contaminated environments.

2. Cycling of Arsenic via Litter Fall

Arsenic fluxes via litter fall at forested sites ranged between 0.2 and 0.8 mg m^{-2} y^{-1}. This corresponds to 25–40% of the As deposited from the atmosphere, indicating that litter fall contributes substantially to As cycling in the forest ecosystems investigated.

3. Arsenic Removal via Harvest

Arsenic removal via harvest of wood is typically small (0.02–0.09 mg m^{-2} y^{-1}) and is not expected to contribute significantly to the decontamination of the study soils, because it represents only 0.7–18% of the atmospheric input. On the other hand, wood grown on As-contaminated sites displays As concentrations in the normal range, indicating that no restriction in the utilization of this wood is required.

The maximum removal rate of As expected at grassland sites is about one magnitude larger than at forested sites (about 1–2 mg m^{-2} y^{-1}), thus being in the same range as As input via wet deposition. If no recycling due to utilization of manure is assumed, the removal of As via harvest of fodder grass may at best balance the atmospheric input.

4. Arsenic Mobilization/Immobilization and Transport in the Soil Solution

We calculated As fluxes between soil horizons (i.e., soil and parent material), via the transport by soil solution, ranging from 0.2 to about 61 mg m^{-2} y^{-1}. The flux patterns are indicative of the slight mobilization of As in the topsoil of noncalcareous soils and immobilization in deeper mineral horizons (Fig. 12a). The forest soils are characterized by mobilization of As from the organic surface layer, but As is readily immobilized in the uppermost mineral horizons. In highly contaminated calcareous soils, an unusually large mobilization of As in the mineral topsoil was observed, with virtually no immobilization in the lower mineral horizons (Fig. 12b).

A pronounced mobility of As in the organic layers is in accord with possible competition between arsenate and negatively charged organic ligands for adsorption sites. However, As appears to be relatively immobile in the acidic mineral soil layers. The fate of As in such environments can basically be related to adsorption on hydrous iron oxides (67). Investigations by Xu et al. (68,69) revealed that the adsorption of As(III) and As(V) were strongly influenced by pH. They explained decreasing adsorption on solid surfaces at pH above 7 due to their negatively charged surfaces. Thus, as well as ions competing with OH$^-$ ions at the geochemically contaminated calcareous sites, will induce enhanced mobility of As.

We found substantial leaching of As only from calcareous soils in two experimental plots, which were geochemically enriched in As. Arsenic mobility in these soils is due to large total concentrations of As and the preserving chemical conditions (in particular, high pH) limiting As adsorption onto hydrous iron oxides. Arsenic is substantially removed from these soils because leaching rates exceed As input via wet deposition by 6–10 times (Fig. 12b). In contrast, at the investigated acidic sites, leaching rates of As are relatively small, even though the soils contain up to several hundred milligrams of As per kilogram. Arsenic deposition is balanced or slightly exceeded As leaching at these sites. Accordingly, total As concentrations in these soils are expected to be stable or to slightly increase with time (Fig. 12a).

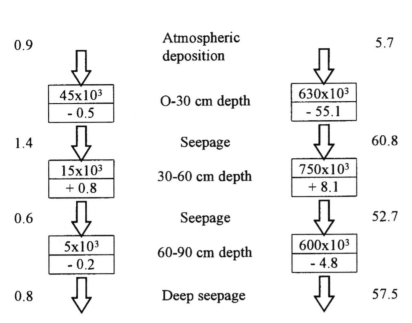

Figure 12 Annual As fluxes at a (a) noncalcareous and (b) a calcareous site in the Austrian Alps. Output by vegetation is not taken into consideration, since both sites are extensive pastures, where manure is recycled. The fluxes by atmospheric deposition and seepage (indicated by arrows) have the unit mg As m^{-2} y^{-1}. The numbers in the upper part of the boxes indicate the As pool (mg As m^{-2}); its annual changes (mg As m^{-2} y^{-1}) are indicated in the lower part.

G. Guidelines for Risk Assessment and Evaluation of As-Contaminated Sites in Austria

Relationships between As in soil solutions as the dependent variable and soil total (As$_t$), or (NH$_4$)$_2$SO$_4$-soluble As(As$_s$) as independent variables, were calculated. The coefficient of simple determination (r^2) is 0.938 for As$_s$, compared to $r^2 = 0.833$ for As$_t$. From this relation, models (1) and (2) were derived to predict the As concentration in the soil solution, based on simple extraction with 0.1 M (NH$_4$)$_2$SO$_4$, or total arsenic concentrations in the soil.

$$As_{soil\ solution}(\mu g\ L^{-1}) = 0.279 + 15.6\ As_{(NH4)2SO4}(mg\ kg^{-1}) \qquad \text{Model 1}$$

$$As_{soil\ solution}(\mu g\ L^{-1}) = 1.272 + 0.043\ As_{total}(mg\ kg^{-1}) \qquad \text{Model 2}$$

Linking the regression models (1) and (2) to drinking water standards (DWS), soil standards that can have bearing on freshwater protection can be derived. Currently, DWS in Austria are set at 30 µg L^{-1} and 50 µg L−1. Corresponding standards in other countries are 50 µg L^{-1} in the United States and 10 µg L^{-1} in Germany and WHO. Dependent on the DWS under consideration, derived standards for As$_s$(mg kg^{-1}) range between 0.62 (for DWS = 10 µg L^{-1}, WHO, Germany) and 3.19 (for DWS = 50 µg L^{-1}, United States and uppermost standard for Austria). Corresponding

Table 4 Maximum Allowable Total As Concentration for Groundwater Protection, Derived from Three Different Drinking Water Standards and Percentage of Soils from Contaminated Sites of the Austrian Alps Exceeding These Levels

Max. allowable water concentration (μg As L^{-1})	Derived max. total soil concentration (mg As kg^{-1})	Percent of total soil samples exceeding max. concentration	Percent of subsoil samples exceeding max. concentration
50	1130	5	1
30	668	10	2
10	203	29	4

standards for As$_t$ (mg kg^{-1}) are 203 (DWS = 10 μg L^{-1}) and 1133 (DWS = 50 μg L^{-1}). Assuming a DWS of 30 μg L^{-1}, a total concentration of 668 As mg kg^{-1} soil is considered safe according to model (2). These standards are valid for the soils in deeper strata, since they are closest to the groundwater table. The considerations demonstrate that changes in legislation on DWS may have dramatic impact on acceptable concentrations of As in soil.

Based on these standards, derived from models (1) and (2), 55 of 191 investigated soil samples in the Austrian Alps display As$_t$ somewhere in the soil profile that may violate the DWS of 10 μg L^{-1} (Table 4). Applying higher standards decreases the number drastically. Many soils in the Austrian Alps are affected by anthropogenic activities, and hence As has accumulated in top soils. Therefore, only 7 of 38 contaminated soils are at risk to release As into the groundwater (Table 4).

In general, soil solution concentrations calculated by models (1) and (2) were in the same range as As concentrations measured in associated surface waters. Larger concentrations in spring waters at some sites were likely due to further mobilization in deep soil horizons. At some sites no evidence for a relation between soil and surface water contamination was found. Data support the hypothesis that As is mobilized from ancient mine dumps and/or deposits or mining tunnels through As-bearing layers. In this case, the soil pathway does not play a role.

VI. CONCLUSIONS

Numerous areas throughout the world display unusually large concentrations of As in aquifers associated with geogenic processes. None, however, has the extent and catastrophic nature of the one now happening in the Asian subcontinent of Bangladesh and West Bengal (India), where millions of people are at risk of arsenicosis. The main exposure pathway in the population in this area is via drinking of As-enriched well water. Minor contributions arise from the use of the well water for cooking and irrigation, but may become relevant in the long term.

Although small in comparison to the sub-Asian continental calamity, several areas in the Austrian Alps have As-enriched soils that can potentially endanger organisms locally with As transfer in the food chain or leaching to deeper strata. Modeling of As in the soil–fodder (plant) and soil–water systems, using chemical extraction of As, provided some basis on which to redefine soil standards for As in relation to drinking water standards.

REFERENCES

1. Smith, A., Hopenhayn-Rich, C., Bates, M. N., and Goeder, H. M. 1992. Cancer risks from arsenic in drinking water. *Environ. Health Perspect.*, 97:259–267.
2. Adriano, D. C. 1986. *Trace Elements in the Terrestrial Environment*. Springer-Verlag, New York.
3. Buat-Menard, P. 1987. Group report: Arsenic. In: T. C. Hatchinson and K. M. Meen (eds.), *Lead, Mercury, Cadmium and Arsenic in the Environment*, pp. 43–48. Wiley ???.
4. McLaren, R. G., Naidu, R., Smith, J., and Tiller, K. G. 1998. Fractionation and distribution of arsenic soils contaminated by cattle dip. *J. Environ. Qual.*, 27:3448–3454.
5. Thalmann, F., Schermann, O., Schroll, E., and Hausberger, G. 1989. *Geochemical Atlas of the Republic of Austria 1:1,000.000*. Geologische Bundesanstalt, Vienna.
6. Sattler, H., and Wenzel, W. W. 1993. Ursachen and Wirkungen grenzüberschreitender Arsen-u. Vanadiumgehalte in Böden. Research report for A. d. Salzburger Landesregierung.
7. Brandstetter, A., Jockwer, F., and Wenzel, W. W. 1995. Arsenanomalie Feistritz am Wechsel. Research report for A. d. NÖ Landesregierung.
8. Schulte, A., Gehrmann, J., and Wenzel, W. W. 1995. Entwicklung der Niederschlags-Deposition von Arsen in Waldökosystemen. *Forstarchiv*, 66:86–90.
9. Borum, D. R., and Abernathy, C. O. 1994. Human oral exposure to inorganic Arsenic. In: W. R. Chappel, C. O. Abernathy, and C. R. Cothern (eds.), *Arsenic Exposure and Health*. Science and Technology Letters, Northwood, IL.
10. Abernathy, C. O., and Ohanian, E. V. 1922. Non-carcinogenic effects of inorganic arsenic. *Environ. Geochem. Health*, 14:35–41.
11. WHO. 1994. *Guidelines for Drinking Water Quality. Vol. 1*. World Health Organization, Geneva.
12. Jekel, M. R. 1994. Removal of arsenic in drinking water treatment. *Arsenic in the Environment*, Part I:119–153.
13. Fowler, B. A. 1977. Toxicology of environmental arsenic. In: R. A. Goyer and M. A. Mehlman (eds.), *Toxicology of Trace Elements*, pp. 79–122. Wiley, New York.
14. Brannon, J. M., and Patrick, W. H. 1987. Fixation, transformation, and mobilization of arsenic in sediments. *Environ. Sci. Technol.*, 21(5):450–459.
15. Eikmann, Th., and Kloke, A. 1991. Nutzungs-und schutzgutbezogene Orientierungswerte für Schadstoffe in Böden. In: Rosenkranz et al. (eds.), *Handbuch des Bodenschutzes*. Erich Schmidt Verlag, Berlin.
16. Berliner Liste. 1990. *Bewertungskriterien für die Beurteilung kontaminierter Standorte in Berlin*. 40. Jg., Nr. 65.
17. Minister für Arbeit, Gesundheit und Soziales des Landes Nordrhein-Westfahlen. 1990. Metalle auf Kinderspielplätzen. Erlass des Ministers, VB-4-0292.5.3, Düsseldorf.
18. Niederländische Liste. 1988. Niederländischer Leitfaden zur Bodenbewertung und Bodensanierung. In: Rosenkranz et al. (eds.), *Handbuch des Bodenschutzes*. Erich Schmidt Verlag, Berlin.
19. König, W. 1990. Untersuchung und Beurteilung von Kulturböden bei der Gefährdungsabschätzung von Altlasten. In: Rosenkranz et al. (eds.). *Handbuch des Bodenschutzes*. Erich Schmidt Verlag, Berlin.
20. Blume, H. P. 1990. *Handbuch des Bodenschutzes*. Ecomed, Landsberg/Lech.
21. Bagla, P., and Kaiser, J. 1996. India's spreading health crisis draws global arsenic experts. *Science*, 274:174–175.
22. Chowdhury, T. R., Mandal, B.Kr., Samanta, G., Basu, G.Kr., Chowdhury, P. P., Chanada, C. R., Karan, N. Kr., Lodh, D., Dhar, R.Kr., Das, D., Saha, K. C., and Chakraborti, D. 1997. Arsenic in groundwater in six districts of West Bengal, India: The biggest arsenic calamity in the world: The status report up to August 1995. In: R. L. Abernathy, W. R. Calderon, and W. R. Chappel (eds.), *Arsenic Exposure and Health Effects*, pp. 93–111. Chapman & Hall, London.

23. Chakraborti, D., Mandal, B. K., Dhar, R. K., Biswas, B., Samanta, G., and Saha, K. C. 1997. Groundwater arsenic calamity in West Bengal, India and Bangladesh. *Proc. of Extended Abstracts from the Fourth Int. Conf. on the Biogeochemistry of Trace Elements*, Berkeley, CA, pp. 769–770.
24. Das, D., Chatterjee, A., Samantha, G., et al. 1994. Arsenic contamination in groundwater in six districts of West Bengal, India: The biggest arsenic calamity in the world. *Analyst*, 119:168–170.
25. Das, D., Samanta, G., Mandal, B. K., Chowdhury, T. R., Chanda, C. R., Chowdhury, P. P., Basu, G. K., and Chakraborti, D. 1996. Arsenic in groundwater in six districts of West Bengal, India. *Environ. Geochem. Health*, 18:5–15.
26. Chakraborti, D. 1997. Bangladesh's arsenic calamity. A report from the School of Environmental Studies, Jadavpur University, Calcutta, India, *Holiday*, Jan. 17.
27. Luo, Z. D., Zhang, Y. M., Ma, L., Zhang, G. Y., He, X., He, L., Grumski, H., and Lamm, S. H. 1997. Chrionic arsenicism and cancer in Inner Mongolia—Consequences of well-water arsenic levels greater than 50 µg/l. In: R. L. Abernathy, W. R. Calderon, and W. R. Chappel (eds.), *Arsenic Exposure and Health Effects*, pp. 55–68. Chapman & Hall, London.
28. Lu, F. 1990. Blackfoot disease: Arsenic or humic acid? *Lancet*, 336:115–116.
29. Choprapawon, C., and Dodcline, A. 1997. Chronic arsenic poisoning in Ronpibol Nakhon Sri Thammarat, the southern province of Thailand. In: R. L. Abernathy, W. R. Calderon, and W. R. Chappel (eds.), *Arsenic Exposure and Health Effects*, pp. 69–77. Chapman & Hall, London.
30. Hsu, K.-H., Froines, J. R., and Chen, C.-J. 1997. Studies of arsenic ingestion from drinking water in northeastern Taiwan: Chemical speciation and urinary metabolites. In: R. L. Abernathy, W. R. Calderon, and W. R. Chappel (eds.), *Arsenic Exposure and Health Effects*, pp. 190–209. Chapman & Hall, London.
31. Niu, S., Cao, S., and Shen, E. 1997. The status of arsenic poisoning in China. In: R. L. Abernathy, W. R. Calderon, and W. R. Chappel (eds.), *Arsenic Exposure and Health Effects*, pp. 78–83. Chapman & Hall, London.
32. WHO. 1994. *Guidelines for Drinking Water Quality, Vol. 2: Health Criteria and Other Supporting Information*. World Health Organization, Geneva.
33. Bencko, V. 1997. Health aspects of burning coal with high arsenic content: The central Slovakia experience. In: R. L. Abernathy, W. R. Calderon, and W. R. Chappel (eds.), *Arsenic Exposure and Health Effects*, pp. 78–92. Chapman & Hall, London.
34. Farago, M. E., Thornton, I., Kavanagh, P., Elliott, P., and Leonardi, G. S. 1997. Health aspects of human exposure to high arsenic concentrations in soil in south west England. In: R. L. Abernathy, W. R. Calderon, and W. R. Chappel (eds.), *Arsenic Exposure and Health Effects*, pp. 210–226. Chapman & Hall, London.
35. Terade, H., Katsuta, K., Sasagawa, T., Saito, H., Shirata, H., Fukuchi, K., Sekiya, Yokoyama, Y., Hirokawa, S., Watanabe, Y., Hasegawa, K., Oshina, T., and Sekiguchi, T. 1960. Clinical observation of chronic toxicosis by arsenic. *Nihon Rinsho*, 118:2394–2403.
36. Tsutsumi, M. 1991. Arsenic pollution in arable land. In: K. Kitagishi and I, Yamane (eds.), *Heavy Metal Pollution in Soils of Japan* pp. 181–192. Japan Scientific Societies Press, Tokyo.
37. Feinglass, E. J. 1973. Arsenic intoxication from well water in the United States *N. Engl. J. Med.*, 288:828–830.
38. Grantham, D. A., and Jones, F. J. 1977. Arsenic contamination of water wells in Nova Scotia. *J. Am. Water Works Assoc.*, 69:653–657.
39. Pereya, F. J., and Creger, T. L. 1994. Vertical distribution of lead and arsenate in soils contaminated with lead arsenic pesticide residues. *Water Air Soil Pollution*, 78:297–306.
40. Cebrian, M. E., Albores, A., Garcia-Vargas, G., and Del Razo, L. M. 1994. Chronic arsenic poisoning in humans: The case of Mexico. In: J. O. Nriagu (ed.), *Arsenic in the Environment, Part II: Human Health and Ecosystem Effects*, Wiley.
41. Thornton, I., and Farago, M. 1997. The geochemistry of arsenic. In: R. L. Abernathy, W. R. Calderon, and W. R. Chappel (eds.), *Arsenic Exposure and Health Effects*, pp. 1–16. Chapman & Hall, London.

42. Astolfi, E., Maccagno, A., Fernandez, J. C. G., Vaccara R., and Stimola, R. 1981. Relation between arsenic in drinking water and skin cancer. *Biol. Trace Element Res.*, 3:133–143.
43. Ritchie, J. A. 1961. Arsenic and antimony in New Zealand thermal waters. *New Zealand J. Sci.*, 4:218–229.
44. Bangladesh Department of Public Health Engineering. 1993 and 1995. Unpublished data.
45. Ahmed, M., Brandsetter, A., Wenzel, W. W., and Blum, W. E. H. 1997. The arsenic calamity in Bangladesh. *Proc. 4th Int. Conf. on the Biogeochemistry of Trace Elements*, Berkeley, CA, pp. 263–264.
46. Tensho, K. 1973. Studies on behaviour of trace elements in flooded soil-rice systems by radio-isotope technique. *Stud. Soils Fert. Mod. Agric.*, 4:65–71.
47. Koyama, T., Awano, H., and Shibuya, M. 1976. Soil-plant relation studies on arsenic: I. Correlations between the forms of arsenates in soil and the response of rice plants. *Nippon Dojo Hiryogaku Zasshi*, 47:85–92.
48. Mok, W. M., and Wai, C. M. 1994. Mobilization of arsenic in contaminated river waters. In: J. O. Nriagu (ed.), *Arsenic in the Environment, Part I: Cycling and Characterization*. Wiley.
49. Vinogradov, A. P. 1959. *The Geochemistry of Rare and Dispersed Chemical Elements in Soils*, 2nd ed. New York.
50. Arsenic Investigation Committee. 1996. Report on possibility of arsenic pollution in soils and water from REB wooden poles in Bangladesh. Ministry of Energy and Mineral Resources, Bangladesh.
51. Danneberg, O. H., Aichberger, K., Puchwein, G., and Wandl, M. 1997. Bodenchemismus. In: W. E. H. Blum, E. Klaghofer, A. Köchl, and P. Ruckenbauer (eds.), *Bodenschutz in Österreich*, pp. 55–110. Ferdinand Berger & Söhne.
52. Sletten, R. S., Jockwer, F., Wenzel, W. W., Prohaska, T., and Stingeder, G. 1997. Sequential extraction optimized for arsenic fractionation in soils. *Proc. 4th Int. Conf. on the Biogeochemistry of Trace Elements*, Berkeley, CA, pp. 243–244.
53. Yang, G. Z. 1983. Cluster analysis of some elements in soils of Tainijin area. *Acta Sci. Circumstantiae*, 3:207–212.
54. Shen, B. Z., Chen, L. G., and Zhao, Z. D. 1983. Correlation between the content of some elements and the mechanical composition of the soils in Tianjin region. *Turang Xuebao*, 20:440–444.
55. Koyama, T. 1975. Arsenic in soil-plant system. *Nippon Dojo Hiryogaku Zasshi*, 46:491–502.
56. Huang, Y.-C. 1994. Arsenic distribution in soils. In: J. O. Nriagu (ed.), *Arsenic in the Environment. Part 1: Cycling and Characterization*, pp. 17–49. Wiley.
57. Jacobs, L. W., and Keeney, D. R. 1970. Arsenate-phosphorus interation on corn. *Commun. Soil Sci. Plant. Anal.*, 1:85–94.
58. Woolson, E. A., Axley, J. H., and Kearney, P. C. 1973. The chemistry and phytotoxicity of arsenic in soil: 2. Effects of time and phosphorus. *Soil Sci. Soc. Am. Proc.*, 37:254–259.
59. Lombi, E., Sletten, R. S., and Wenzel, W. W. 2000. Arsenic sequentially extracted in sand, silt, and clay of contaminated soils. *Water, Air & Soil Pollution* (in press).
60. Macnair, M. R., and Cumbes, Q. 1987. Evidence of arsenic tolerance in *Holcus lanatus* L. is caused by an altered phosphate uptake system. *New Phytol.*, 107:387–394.
61. Bhumbla, D. K., and Keefer, R. F. 1994. Arsenic mobilization and bioavailability in soils. In: J. O. Nriagu (ed.), *Arsenic in the Environment, Part II: Human Health and Ecosystem Effects*, pp. 51–82. Wiley ???.
62. Kiss, A. M., Oncsik, M., Dombovari, J., Veres, S., and Acs, G., 1992. Dangers of arsenic drinking and irrigation water to plants and humans. Antagonism of arsenic and magnesium. *Acts Agron. Hung.*, 41:3–9.
63. Liu, G.-L., Cheng, F.-X., Gao, S.-D., and Li, M.-Q. 1985. Effect of arsenic in soil on plants. *Zhongguo Nongye Kexue* (Beijing), 4:9–16.
64. Bundesgesetzblatt für die Republik Österreich. 1994. 84. Stück, Futtermittelverordnung. Österreichische Staatsdruckerei.

65. Brandstetter, A., Wenzel, W. W., Wutte, H., Lombi, E., Prohaska, T., Stingeder, G., and Adriano, D. C. Arsenic in contaminated soils: Predicting solute concentrations. Submitted for publication.
66. Baumgartner, A., Reichel, E., and Weber, G. 1983. *Der Wasserhaushalt der Alpen*. Oldenburg-Verlag, Müchen-Wien.
67. Misra, S. G., and Tiwari, R. C. 1963. Studies of arsenite-arsenate system adsorption of arsenate. *Soil Sci. Plant Nutr.*, 9:216–219.
68. Xu, H., Allard, B., and Grimvall, A. 1988. Influence of pH and organic substance on the adsorption of As (V) on geologic material. *Water Air Soil Pollution*, 40:293–305.
69. Xu, H., Allard, B., and Grimvall, A. 1991. Effects of acidification and natural organic materials on the mobility of arsenic in the environment. *Water Air Soil Pollution*, 57–58:269–278.

34
Arsenic-Contaminated Soils: II. Remedial Action

E. Lombi
IACR-Rothamsted, Harpenden, Hertfordshire, England

W. W. Wenzel
University of Agricultural Sciences, Vienna, Austria

D. C. Adriano
Savannah River Ecology Laboratory, University of Georgia, Aiken, South Carolina

I. INTRODUCTION

Arsenic (As) is a ubiquitous element, and its presence in soils is due to both natural and anthropogenic inputs. The properties of some arsenic compounds have been known and used for thousands of years. Arsenic sulfide was mentioned by Aristotle in the fourth century B.C., and Discorides first named it *arsenicum* in the first century A.D. (1). Arsenic compounds have been used for extremely varied purposes: as medicine, poison, insecticide, herbicide, fungicide, and recently, to produce semiconductors. The ancient Greeks used As as a medicine (2), and for many years Flower's solution (containing 1% As trioxide) was one of the most frequently dispensed medicine in Western countries (3). Calcium and lead arsenates were the most commonly used insecticides at the beginning of this century, until the advent of organic pesticides in the 1940s (4). Arsenical herbicides such as monosodium methylarsine (MSMA), disodium methylarsine (DMSA), and cacodylic acid were largely produced in the mid-1970s (5). The organic arsenicals are still registered in the United States as herbicides (6). While the use of arsenical insecticides and herbicides has substantially decreased during the last decades, the utilization of As-containing wood preservatives is increasing (2).

Various studies have shown that As in drinking water may be toxic at smaller concentrations than was previously thought (e.g., 7,8). At present, a lowering of the limit of As in drinking water is under consideration in various countries: Germany has reduced the limit to $10\,\mu g$ As l^{-1}, and the European Union will follow soon; in the United States decrease to as little as $2\,\mu g$ As l^{-1} is under active consideration (9). Frey and Edwards (10) have reported that approximately 25% of all community water supplies in the United States would violate an As standard of $2\,\mu g$ As l^{-1}, and between 6 and 17% were predicted to violate a standard of $5\,\mu g$ As l^{-1}. In the previous chapter it is demonstrated that a decrease of the current drinking water standard of $30\,\mu g\,L^{-1}$ in Austria to $10\,\mu g\,L^{-1}$

as defined by the European Union will have implications on the threshold values of As in the soils. As a consequence, more soils would require remedial actions. Remediation of As-polluted soils is already required at numerous sites worldwide. In the United States, As is the second most common inorganic contaminant after Pb at Superfund sites, being present at 41% of the sites (11).

In this chapter, soil pollution with As is discussed in relation to its different sources. Moreover, the environmental chemistry of As is briefly reviewed in order to provide a basis for the discussion of potential remedial options relevant to As clean-up.

II. SOURCES OF As IN THE ENVIRONMENT

Arsenic is generally present in soils at concentrations below 50 mg kg^{-1}, with a mean value of about 5–6 mg kg^{-1} (2,5). Anomalously large concentrations of As in soils are due to both natural and anthropogenic processes. Nriagu (12) has calculated that the global flux of As resulting from human activity is approximately 50% larger than that from natural fluxes. Analysis of As distribution in the profiles of ombrotrophic peat bogs in Canada and Switzerland show that the anthropogenic fluxes have exceeded the natural fluxes for over 2000 years, yielding enrichment factors (calculated using the crustal abundance of As) ranging from 10 to 60 (13,14). The biogeochemical cycle of As is presented in Fig. 1, in which the relative distribution of As among environmental compartments (expressed as relative amount to soil) according to Mackenzie et al. (15) is also indicated.

The largest reservoir of As occurs in rocks in which this element can substitute Si, Al, or Fe in the crystalline structure of phyllosilicates (16). Arsenic is commonly associated

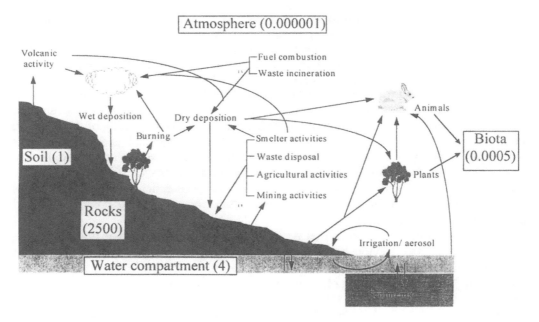

Figure 1 Biogeochemical cycle and relative distribution of As among environmental compartments.

with sulfides, and soils overlaying sulfidic ore deposits can contain as much as 8000 mg As kg^{-1}. The process of weathering is largely responsible for anomalous elevated concentrations of As in soil. Soils, sediments, and associated freshwater resources containing naturally large concentrations of As have been identified in various regions of the world, including Argentina, Bangladesh, Chile, India, Mexico, Taiwan, Thailand, and the United States (17). The biggest calamity associated with the utilization of As-contaminated groundwater is evident in some districts of West Bengal, India, and adjacent areas in Bangladesh, where more than 1 million people are exposed to drinking water contaminated with As that derives from sediments containing arsenopyrite (18). Other natural sources of As derive from volcanic activities, wind-borne soil particles, sea salt sprays, and biological volatilization of As compounds at low temperatures (12). Among the anthropogenic sources of As, perhaps the most important is associated with mining and smelter activities. Arsenic is contained in Cu, Pb, Co, and Au ores in amounts ranging from 20 to 30 g kg^{-1} in Cu ores to up to 110 g kg^{-1} in some Au ores (1). Large amounts of As in mining deposits can lead to elevated concentrations of As in surface soils developed on these materials. Concentrations of As up to 160 g kg^{-1} in soils on or near mine wastes have been reported for southwest England (19–21). During smelting processes As is readily vaporized (sublimation temperature $= 615°C$) and can be deposited from stack emission or generate As-rich by-products (22). The combustion of fossil fuels is another important source of As. The content of As in coal is variable, ranging from less than 1 $\mu g\, g^{-1}$ to more than 10 g kg^{-1} (23). The content in petroleum fuels tends to be lower than in coal (< 0.2 mg kg^{-1}), but the large consumption of these fuels make them a significant source of As (24). The utilization of As compounds in agriculture has generally decreased in the last years. However, arsenic-based herbicides are still in use as desiccants and defoliants in the cotton industry (2), and ammoniacal and chromated copper arsenates are used in wood preservatives (25). Arsenic-containing pesticides have been used to treat animal hides, with the consequence of large As concentrations in tannery wastes (26). The agricultural use of sewage sludge may represent another source of As. In the United States the Standards for the Use or Disposal of Sewage Sludge (27) set the concentration limit for Class-A sludge at 41 mg kg^{-1}. According to this disposition, the maximum permitted As concentration in soil treated with sewage sludge was calculated to be 20 mg kg^{-1} (28). Irrigation of paddy fields with As-contaminated water is another agricultural practice that can increase the concentration of As in the soil and therefore decrease its fertility (29).

Since the mid-1980s, the emission of As to the atmosphere and its subsequent deposition rate has been diminishing in developed countries. For instance, As deposition onto forest sites in NW Germany, an area that previously had been heavily affected by emissions from the Ruhr industrial area, has decreased considerably since 1984 (Fig. 2) (30).

The identification of As sources and pathways causing soil contamination at a specific site is a prerequisite for a successful remedial action. In fact, natural contamination may differ from anthropogenic contamination in terms of species of As involved, concentrations of the different species in the soil, presence of other pollutants, distribution of the pollutants along the soil profile, and other features. Figure 3 displays the distribution of As in three different soil profiles of the Austrian Alps (31). The geogenic As-rich soil shows an even As distribution with depth, whereas in the anthropogenically contaminated soil As concentrations substantially decrease with depth. It is obvious that in these cases different remedial action will be required.

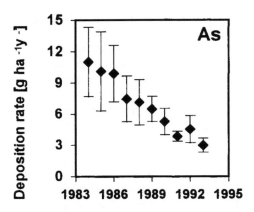

Figure 2 As deposition on forest in NW Germany.

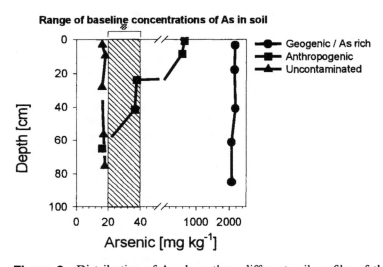

Figure 3 Distribution of As along three different soil profiles of the Austrian Alps.

III. ENVIRONMENTAL BIOCHEMISTRY OF As

Arsenic (atomic no. 33) has an atomic weight of 74.922, an outer electronic configuration of $4s^2 4p^3$, and is a member of the Group V-A, along with N, P, Sb, and Bi. Its chemistry is similar to that of P because both, in the oxidation state V, form oxyanions in soils. Unlike P, As is present in nature also in the oxidation state III; other oxidation states are 0 and −III.

Reduction/oxidation, sorption/desorption, precipitation/dissolution, and methylation/demethylation reactions lead to a complex biochemical cycle of As in soils and associated environments. Changes in oxidation state are due to both chemical and biological processes, whereas transformation between organic species is typically associated with biological activities (32). Biological methylation can occur in anaerobic and aerobic conditions (2). This process was first described by Gmelin in 1815 and provided an explanation of As poisoning of people. At that time, As was an essential constituent in many organic and inorganic pigments, and it was therefore released from wallpaper in the form of toxic As-methylated compounds.

In soil, As is present in organic and inorganic species that differ in their toxicities and mobilities:

- Arsenic III: This oxidation state is stable in reduced soils. It is more toxic, soluble, and mobile (33) than As V. This oxidation state may be the dominant form in moderately reducing conditions (33). As_2O_3 is one of the most common As compounds in anthropogenically contaminated soils.
- Arsenic V: This is the most common oxidation state of As in aerobic condition. Its geochemistry is similar to that of P in soils.
- Methylarsenic acids: The most common in the soil is generally the dimethylarsinic acid that seems to be naturally present in the soil (34). The Na salts of these compounds (MSMA, DSMA) were used extensively in agriculture. In the soil these substances can be transformed to inorganic As or to arsines (6). In reduced condition, methanogenic bacteria reduce and methylate arsenate to methylarsenic acids. These compounds are less toxic than As III (35).
- Arsines: Arsines are inorganic (AsH_3) and organic (mono-, di-, and trimethylarsine) volatile compounds. Arsenic trihydride is an extremely toxic gas.
- Arsenobetaine and arsenocholine: These organic compounds are virtually nontoxic to many species (36).

In the previous chapter we concluded that the risk associated with As-polluted soils is more related to easily extractable fractions of As than to its total content. Arsenic mobility and bioavailability in the soil are controlled by a number of chemical, physical, and biological parameters, such as pH, redox potential, texture, content in organic matter, oxides, carbonates, and interaction with other elements, e.g., phosphorus.

In most soils, arsenite and arsenate are the dominant species of As. Therefore only the fate of these species in soils will be considered in detail in this report. The presence of As III or V in soils is controlled primarily by pH and redox potential: As V is the most common oxidation state of As at $pe + pH > 9$, As III is the dominant species in relatively anoxic conditions at $pe + pH < 7$ (37). A simplified stability diagram for As is displayed in Fig. 4.

The pH and redox potential are key factors controlling As speciation and mobility in soil. The charge of As oxyanions is pH dependent (Fig. 4). In the normal range of soil pH (4–8), the dominant As species in solution are $H_3AsO_3^0$ in the case of As III, and $H_2AsO_4^-$ and $HAsO_4^{2-}$ in the case of As V (38). Because the surfaces of Mn, Al, and Fe oxides/hydroxides and clay minerals are positively charged only below soil pH 3, 5, 8, and 4, respectively (37), As adsorption on soil colloids is more pronounced at low pH. At high pH, most of the colloids show a net negative charge, thus favoring As solubility. In oxic soils, $FeAsO_4$ and $AlAsO_4$ may form in As-contaminated soils under highly acidic conditions (37); in alkaline conditions $Ca_3(AsO_4)_2$ may form, but its solubility product is larger than those of As-Fe and As-Al minerals occurring in soils (38). Iron arsenate has a solubility of only 10^{-11} M, whereas Ca and Mg arsenates have solubilities of 10^{-5} M (39). The instability of Ca arsenate is due to the absorption of CO_2 from the air, resulting in the formation of Ca carbonate with a consequent release of As. Accordingly, $Ca_3(AsO_4)_2$ was identified as a primary mineral present in coarse fractions of As-contaminated, calcareous soils, but was not detected in the weathered fine earth fraction (< 2 mm) of the same soils (40).

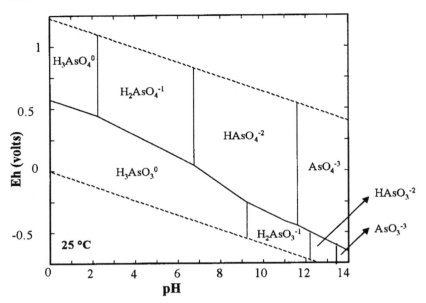

Figure 4 Stability diagram for As.

Arsenic speciation in soils is also influenced by the redox conditions. In aerobic environments As V is the dominant species, while under mildly reducing conditions arsenious acid species are more abundant. Arsenic can react with sulfide to form As_2S_3, which is a slightly soluble, stable solid under more reducing conditions. In extremely reduced environments elemental As and volatile arsine (AsH_3) can occur (11). An increased As mobility at low redox potentials is associated with the reduction of Fe III to Fe II and subsequent dissolution of Fe arsenates (33), or to the reduction of As V as found in paddy soils by Xie and Huang (41). Masscheleyn et al. (42) studied the effect of pH and redox potential on the solubility of As in lake sediments. A decrease in redox potential from $+500$ mV (pH 4.0) to -200 mV (pH 6.9) increased As solubility by about 25 times.

Because the chemical and physical characteristics of soils, e.g., surface area, are strongly related to texture, the distribution and fate of arsenic can be affected by the particle-size distribution. Lombi et al. (43) found that the As concentration in the particle-size fractions of five Austrian soils decreased in the order clay $\gg$ silt $>$ sand. This is consistent with a positive correlation between clay and arsenic content in the soil (44). In contrast, As concentrations are typically small in sandy soils (45,46).

The interactions between As and organic matter are not well understood. Humic substances have been shown to preferentially adsorb As V compared to As III, with an adsorption maximum at about pH 5.5. The adsorption mechanism was probably related to the capacity of humic acids to act as anion exchangers (47). Sadiq (37) suggested that a limited interaction between As oxyanions and organic matter is expected because of the similar type of electrical charge. Because simple organic acids and phenolics can be adsorbed by Fe oxides (48,49), low-molecular-weight organics may compete with As oxyanions for adsorption sites on Fe oxides.

Arsenic, like P, interacts strongly with Fe oxy/hydroxides, but shows less affinity for Al oxides than phosphate (5). Arsenic mobility in soil is regulated primarily by adsorption on oxy/hydroxides of Fe and Al (43,50–52). Arsenate tends to be more preferentially adsorbed than arsenite (53,54). Mn oxides are not considered very effective in adsorbing As (54,55). However, Mn oxides can play an important role in oxidising As III to As V. Arsenic oxidation is enhanced also on the surfaces of kaolinite and illite (56).

As illustrated in the previous chapter, a survey of 38 As-contaminated sites in Austria has demonstrated that As in soil is associated primarily with Fe oxy/hydroxides even in calcareous soils. Because of the importance of Fe oxy/hydroxides in controlling As mobility in soil, a number of studies have been carried out to understand the mechanisms of interaction between these minerals and As. Recent studies conducted using extended X-ray absorption fine structure (EXAFS) have demonstrated the formation of bidentate–binuclear complex of As V on the surface of goethite (57–59) and ferryhydrite (58). This complex appears to be the dominant form at large surface coverage while at lower surface coverage a monodentate complex was favored (57,58). A third type of complex, a bidentate–mononuclear complex, was also identified. In contrast, using wide-angle X-ray scattering (WAXS), Waychunas et al. (60) obtained data supporting the presence of a bidentate–binuclear complex but showing no indication of a bidentate–mononuclear complex. Sun and Doner (61) investigated arsenate and arsenite bonding structures on goethite by FTIR-based techniques. They found that most of As V and III replaced two singly coordinated surface OH groups (A-type) to form a binuclear–bidentate complex. On the other hand, As III was found to react preferentially with doubly coordinated surface OH groups (C-type), while As V reacted mainly with triply coordinated OH groups (B-type).

The importance of carbonates to influence As mobility in soil is also not so clear. Arsenic adsorption on carbonates may occur via ligand exchange or chemisorption with a maximum around pH 10 for calcite (62,63). On the other hand, Polemio et al. (64) found no correlation between lime content and As adsorption.

Fertilization of soil with P can directly affect the mobility of As (65). The application to soil of P fertilizers can lead to an increased mobility of As due to displacement of As from adsorption sites and to antagonistic effect between P and As for plant absorption (5). The competition between As and P for adsorption sites and plant uptake is due to their biogeochemical similarity: these elements have similar electron configuration and form triprotic acids with similar dissociation constants (66). Since the mechanism of As and P adsorption in soil is related mainly to ligand exchange on the surfaces of Al and Fe oxides, a competitive interaction between these elements is excepted (67,68). Phosphate in soil should be a better competitor than arsenate for adsorption sites because of the smaller size and the higher charge of phosphate species (38).

IV. REMEDIATION OF As-CONTAMINATED SOILS

Remedial decision-making process is discussed in the previous chapter relative to the assessment of the risk pertinent to soil contamination by As. Once a soil has to be remedied, the best available technology (BAT) has to be selected according to the nature, toxicity, and origin of the contaminant, the present and potential hazard related to the degree of contamination, the chemical and physical characteristics of the soil, the land use, the community acceptance, and cost–benefit analysis. The clean-up goal selected can be expressed in terms of total or leachable concentration of the pollutant. If the goal is expressed as total concentration, technology based on the removal of the pollutant will be preferred. In case of clean-up goals based on the leachable fraction, remediation can be achieved through technologies based on either the removal or the reduction of mobility

Table 1 Soil Clean-up Goals Adopted in the United States for Total and Leachable As

Total arsenic (mg kg^{-1})		Leachable arsenic (μg l^{-1})	
Background (range) (69)	1 to 50	TCLP threshold for RCRA waste (method 1311)^b	5000
Superfund site goals (69)	5 to 65	Extraction procedure toxicity test (EP Tox, method 1310)	5000
California total threshold limit (69)	500	California soluble threshold leachate concentration	5000
Theoretical minimum total As to ensure TCLP leachate < threshold (i.e., TCLP × 20)^a	100	Superfund site goal (69)	50

^{a} TCLP, Toxicity Characteristic Leaching Procedure.
^{b} RCRA, Resource Conservation and Recovery Act.
Source: Modified from Ref. 11.

and bioavailability of the pollutant which, in this chapter, will be called containment technologies. The clean-up goals adopted in the United States based on total and leachable As are listed in Table 1.

Since the primary pathway of As ingestion by humans is via drinking water, greater attention has been directed to the removal of As from water. Although the occurrence of As in water can be related to the presence of this element in the soil, in the present chapter only the technologies applicable for the reclamation of As-contaminated soils will be highlighted. Several reviews dealing with the technologies and principles for the removal of As from water are available (i.e., 70, 71). The technologies presented will be grouped according to the underlying processes. Our emphasis is on potentially viable technologies based on the biogeochemical behaviour of As in soils.

A. Engineering Techniques

1. Physical Barriers

The construction of physical barriers represents an in-situ containment technology to prevent or limit contaminant migration due to water flow. The physical barriers range from surface cap to vertical or horizontal barriers to avoid infiltration of uncontaminated water, and restrict lateral or vertical migration (72). Different materials and technologies are available for construction of the barriers; their selection has to take into account the nature of the pollutant. For instance, large concentrations of water-soluble As salts may significantly decrease the impermeability of cement-bentonite barriers (11), thus limiting the utilization of this material in the case of As.

Physical barriers do not clean up the soil or decrease the solubility of As, but only reduce its movement into the surrounding environments or the groundwater. Therefore, the multifunctionality of the polluted site is not recovered and, in the case of subsurface barriers, the revegetation of the site with tolerant plants can be a useful complementary action to prevent soil erosion.

2. Separation/Concentration Processes

This group of technologies aims at concentrating As in a smaller amount of material that can be eventually treated with further processes in order to reduce its hazard. These

Table 2 Arsenic Distribution Among the Sand, Silt, and Clay Fractions of Six Contaminated Austrian Soils (data in mg kg^{-1})

Soil type	pH	As in the fraction < 2000 μm	As in soil textural fractions[a]			
			Sand	Silt	Clay	
Eutric cambisol	5.62	1520	559	2002	4935	Pasture located near an abandoned smelter
Eutric cambisol	6.67	655	317	800	2431	Pasture located near an abandoned smelter
Dystric cambisol	3.39	214	70	170	1288	Forest/geogenic
Dystric cambisol	3.98	581	814	495	1774	Forest/active smelter
Dystric cambisol	4.2	335	250	529	1196	Pasture/old mining deposit
Calcaric cambisol	7.3	1900	818	2367	4400	Grassland/geogenic

[a] Sand = 2000–63 μm; silt 63–2 μm; clay < 2 μm.

Table 3 Physical Separation Techniques Based on Particle Characteristics

Particle characteristic	Technique	Basic principle
Particle size	Screening	Sieving of the material
Particle density	Sedimentation/thickening	Gravity separation
Magnetic properties	Magnetic separation	Magnetic susceptibility

Source: Modified from Ref. 72.

processes lead to the removal of As from the soil and may be conducted in situ or ex situ.

a. Ex-situ Technologies. These technologies are based on specific physical properties of the soil constituents. After excavation, the soil is treated to separate the particles that contain the pollutant in larger concentrations. Generally, the distribution of total As among various particle-size fractions of the soil is: clay $\gg$ silt > sand, which is consistent with a higher sorption capacity of fine-textured fractions due to greater surface area and greater content of Fe oxides. In Table 2 the distribution of As among the sand, silt, and clay fractions in six contaminated soils is reported. Energy-dispersive X-ray microanalysis (EDAX) performed on the various fractions of these soils has demonstrated a close association of As with Fe oxy/hydroxide.

Depending on the particle distribution of As in polluted soil, and the fact that As is associated with Fe oxy/hydroxides, different techniques can be used to clean up polluted soils (Table 3). The physically separated pollutant-rich fraction can be processed to recover As using pyrometallurgical or hydrometallurgical (soil washing) separation techniques. Pyrometallurgical technologies use thermal means to cause volatile compounds or elements, such as As, to separate from soil and accumulate in the fly ash. According to the U.S. EPA (11), As is not a good candidate for this technique due to the low commercial value of As trioxide. Soil washing processes involve the use of leaching solution to recover As; this technique is commonly associated with physical separation but is sometimes used as a stand-alone treatment technology (73). Legiec et al. (74) tested a soil washing technology to treat an acidic sandy soil contaminated with 149 mg kg^{-1} inorganic As. The fines fraction (clay-sized) contained large amounts of As in tightly bound forms and for this reason disposal of this fraction was suggested. Arsenic leaching was targeted for the sand fraction that appeared moderately contaminated. Based on tests of different leaching solutions, 0.02 M NaOH (pH 11.5) was identified as the most effective, resulting in the removal of 57% of the As. Up to 25% As recovery was obtained when phosphate was added in the presence of 0.015 M NaOH. Phosphate addition had the advantage of buffering the solution to a near neutral pH.

b. In-situ Technologies. These technologies recover As from the soil using soil flushing or electrokinetic treatment. Soil flushing is based on leaching with water or chemical solutions able to mobilize As in the soil and remove it with the percolate. The choice of the washing solution has to take into account the efficiency of As recovery, the economic outcome, and the possible side effects. Arsenic mobility can be enhanced by raising the pH or using solutions containing competitive anions such as phosphate. Woolson et al. (75) studied the effect of phosphate on As leaching from a sandy soil contaminated with 625 mg kg^{-1} of As. They found that a 0.05 M KH_2PO_4 solution could remove up to 77% of the total As. After percolation the contaminated fluid can be collected and pumped to the surface, where it can be recirculated, removed, or treated and reinjected. At the moment this technology has not yet been applied at Superfund sites in the United States, but it is considered promising in case of permeable, neutral to alkaline soils containing only small amounts of Fe oxides and clay minerals [11].

Electrokinetic treatment is suitable to remove charged contaminants by application of a low-intensity direct current between electrodes positioned in the soil (76). The electric field generated by anodes and cathodes placed in the soil causes the migration of anions and cations. The cathode and anode can be equipped with different circulation solution systems in order to maximize the recovery of specific ions (77). Similar to the case of soil flushing, treatments that increase the mobility of the contaminant in the soil will improve the efficiency of this technique. The evolution of oxygen and production of hydrogen ions at the anode can cause the formation of an acid front. Similarly, the formation of hydrogen gas (which escapes) and hydroxide ions may increase the pH in the cathode zone to above 13 (78).

This technique has been used in the former Soviet Union since the 1970s to concentrate metals and to explore for minerals in deep soils. Pilot field studies to assess the feasibility of electrokinetic treatments have been conducted in the Netherlands on soils contaminated with Pb, As, Ni, Hg, Cu, and Zn. Geokinetics International, Inc., has applied this technology to clean up a clay soil polluted with As by a former timber plant. The concentration of As in this soil decreased from more than 250 to less than 30 mg kg^{-1} as a result of the electrokinetic treatment—reported by the U.S. EPA (76).

3. Solidification/Stabilization (S/S) Technologies

S/S technologies are usually applied by mixing the contaminated soil with a physical binding agent to form a crystalline, glassy, or polymeric framework surrounding the waste particles (76). These technologies are more commonly applied ex situ, but may be alternatively employed in situ. S/S technologies include polymer microencapsulation, cement-based stabilization, and vitrification.

a. Polymer Microencapsulation. This technology is based on the physical encapsulation of polluted soil in water-insoluble organic resins, or more commonly, bitumen. It involves mixing the waste material with the resin at elevated temperatures. Excessive presence of organic pollutants at some sites may limit the applicability of this method (72). Polymer microencapsulation is the most suitable S/S technology for As, which is difficult to immobilize using cement-based stabilization (79).

b. Cement-Based Stabilization.
These methods are based on the treatment of contaminated soils with inorganic materials such as cements and siliceous pozzolans. As a result, the mobility of the contaminant is reduced by physical and chemical processes. Solidification of the polluted substrate with cement limits its contact with groundwater and air. Because of pH in the alkaline range and excessive heat of hydration, this treatment is not efficient for anions, and therefore not suitable for As (11,80).

4. Vitrification

In these technologies the soil is melted, either in situ or ex situ, and the pollutants are incorporated into a stable vitreous mass. The large amount of glass-forming materials, such as silica, in soil renders this substrate ideal for this technology without pretreatment. The in-situ processes are achieved by electric melter technology. Resistance heaters are placed into the soil to initiate the melting process. In ex-situ treatments the soil is first excavated and then the process of vitrification takes place in a glass melter, heated for example by fossil fuels, or in a plasma centrifugal furnace. Arsenic, lead, selenium, and chlorides are incorporated less efficiently than other inorganic pollutants (72). The sublimation of As at 615°C may cause As to volatilize, consequently any volatile emission during the treatment of As-polluted soils should be controlled and treated.

B. Biogeochemically Based Techniques

1. Soil Amendments

Soils contaminated with heavy metals have been traditionally ameliorated using amendments such as lime, phosphate, and organic matter (81). Although providing some positive results in treating heavy metals, these amendments may not be effective in the case of As-contaminated soils. This is due to the anionic nature of As in soil, where application of such amendments may even result in increased mobility and bioavailability of this metalloid because of competition between the As oxyanion and hydroxyl ions, phosphate, and organic ligands for adsorption sites. Oxy/hydroxides of Fe and Al have been identified as highly effective sorbents for As in soils and reduce its mobility (58,64,82–85).

Fixation of As in the soil may be achieved, enhancing this natural process and furnishing additional quantities of these As sinks to the soil. In the case of soil amendments, the stability and long-term efficiency of the treatment is of fundamental importance. To this end, a Fe oxide-coated sand (IOCS) was evaluated for its capacity of binding As and to assess the reversibility of sorbed As over time (86). More than 99% of the As III and all of the As V were removed from the solution by IOCS within 30 days. Moreover, the conversion of As to less mobile forms was demonstrated by the observation that the extraction with weaker extractants decreased with time, whereas the amount of As in more recalcitrant fractions increased (Fig. 5). These findings are in agreement with those of Onken and Adriano (87), who indicated that amending soils with Fe oxides may be a suitable technique to immobilize and contain As in polluted soils. In soils polluted with a mix of metals and metalloids, both heavy metals and As can be immobilized due to their high affinity for Fe oxides.

Ferrous sulfate has been mixed into As-polluted soils to produce sparingly soluble ferric arsenate ($FeAsO_4$) (88,89). This system has been known for a long time: in 1942 Kardos et al. (90) applied ferrous sulfate to reduce the mobility of As in orchard soils. Voigt et al. (91) used a similar technique followed by the addition of Portland cement to immobilize As in soils of a heavily polluted old dump site. Artiola et al. (92) tested

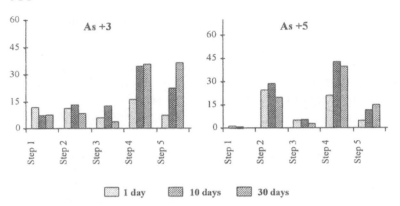

Figure 5 Time dependency of As extractability from IOCS.

several amendments for effectiveness in reducing leachable As in soils. The TCLP procedure indicated that only ferrous sulfate was able to fix As: the mobility of As decreased from 50 to 2 times the regulatory limit of 5 mg L^{-1} after only six 24-h wet–dry cycles. Hematite, gypsum, and siderite were ineffective in reducing As mobility.

Other possible amendments may be Mn oxides. These compounds are less efficient than Fe oxy/hydroxides in fixing As (37), but are effective in oxidizing As III to the less toxic and less mobile As V according to the following reactions (54):

$$HAsO_2 + MnO_2 \rightarrow (MnO_2) \cdot HAsO_2$$
$$(MnO_2) \cdot HAsO_2 + H_2O \rightarrow H_3AsO_4 + MnO$$
$$H_3AsO_4 \rightarrow H_2AsO_4^- + H^+$$
$$H_2AsO_4^- \rightarrow HAsO_4^{2-} + H^+$$
$$(MnO_2) \cdot HAsO_2 + 2H^+ \rightarrow H_3AsO_4 + Mn^{2+}$$

As pointed out by Oscarson et al. (54), Mn oxides can catalyze and thus enhance the adsorption of arsenic acid on Al and Fe oxides. Accordingly, a combined application of Mn oxides and Fe oxy/hydroxides may be efficient in immobilizing As in slightly reduced soils.

2. Biological Processes

Biological technologies for remediation of contaminated soils are mediated by specific processes of microorganisms (bioremediation) and plants (phytoremediation). For some processes, such as volatilization of metalloids from the soil, it is difficult to separate the microbial and the plant component. In these circumstances we advocate the term plant-assisted microbial remediation (93).

a. Bioremediation. Microbes are capable of mediating a variety of reactions (i.e., methylation, oxidation/reduction) to protect themselves from toxic pollutants or to use the contaminants as substrates to obtain energy. Biotechnologies that exploit these mechanisms are in a more advanced state of development for the remediation of organic compounds. Processes such as bioleaching, biosorption, biovolatilization, biological oxidation and reduction may provide in-situ treatments without the use of hazardous chemicals. Microbial methylation of As has been known for a long time and is common to both bacteria and fungi. Bacterial methylation seems to be favored by anaerobic conditions and may be employed only in ex-situ bioreactor systems (94). Fungal methylation seems to be important in the volatilization of As compounds used in agriculture (95,96). The microbial production of arsines from arsenicals in soil seems to contribute only marginally to the loss of As. Gao and Burau (6) reported a loss of volatile arsine in the range of only 0.001–0.4% from As applied to the soil as Na cacodylate or methanearsonic acid.

Ahmann et al. (97) found a bacterium (MIT-13) that reduced As V to As III in a drainage basin near a Superfund site in Massachusetts. This biological process may be applied in ex-situ treatment where the low-soluble As V can be reduced to the much more mobile As III and then removed from the soil by flushing. Another interesting bacterium was discovered in an acidic stream of a Pb-(Zn) mine at Gard, France (98). Arsenic was accumulated in rapidly growing bacteria-made structures similar to those of stromatolites. Living bacteria were recognized in the precipitates that contained between 90 and 200 g kg^{-1} As in the form of Fe arsenate and arsenate-sulfate compounds. This remarkable capacity to biologically bind As may lead to the development of new clean-up technologies based on microbiological processes.

b. Phytoremediation. Based on traditional terminology (99–101), we refer to phytoremediation as technologies that use green or higher terrestrial plants for treating chemically or radioactively polluted soils. Phytoremediation technologies involve containment (phytostabilization/immobilization) and clean-up processes such as phytoextraction, phytodegradation, and phytovolatilization (93). Phytostabilization refers to the use of pollutant-tolerant plants that mechanically stabilize polluted soils to prevent bulk erosion and airborne transport to other environments. This approach can be efficiently combined with technologies that decrease the bioavailability of As through the application of soil amendments, in particular Fe oxy/hydroxides. Phytovolatilization and phytoextraction have not yet been developed for As. This is surprising because of the similarities between Hg, Se, and As (93–102). Phytovolatilization implies the ability of plants to volatilize As, yet this has not been shown to be a plant-internal process. Plants may only assist and stimulate microbial volatilization via rhizospheric interactions with microorganisms.

Two basic strategies of phytoextraction have been developed, which, according to Salt et al. (103), can be defined as induced and continuous phytoextraction. Continuous phytoextraction refers to the use of hyperaccumulator plants to clean up the soil. At present the As concentration in plants that qualifies for hyperaccumulation has not been yet defined. Plants growing on uncontaminated soil typically contain As in the range of few milligrams per kilogram of dry matter (5). Indigenous plant species collected from a number of As-contaminated sited in Austria typically contained less than 10 mg As kg^{-1} dry matter, with soil plant transfer ratios of < 0.01. Only few plant species accumulating anomalously large concentrations of As in their tissues have been identified and may represent the raw material for the implementation of phytoextraction technologies. Arsenic accumulation has been found in plants grown on mine dumps in eastern Zimbabwe (104). Up to 3800 mg kg^{-1} As were found in the leaves of *Amaranthus hybridus*, a high-biomass crop plant, and 11,000 mg kg^{-1} in the roots of the ubiquitous grass species *Cynodon dactylon*. However, these findings need to be confirmed by further investigations. In Florida, Komar et al. (105) have identified the fern *Pteris vittata* as a potential As hyperaccumulator. This species, growing at a wood treatment site contaminated with Cu, Cr, and As, accumulates 3280–4980 mg As kg^{-1} dry matter, while As concentrations in other plant species collected from the same site were < 9 mg kg^{-1}.

A first attempt to employ phytoextraction in the remediation of As-polluted soils has been reported by Feller (106). The experimental area, located in Germany, was contaminated with As-based chemical warfare agents during World Wars I and II. The high-biomass plant *Polygonum sachalinense* was grown on this soil to test its ability to accumulate As; additional trials were conducted using soils spiked with arsenite and arsenate. One-year-old plants were able to accumulate up to 1900 mg As kg^{-1} in leaves and shoots and up to 430 mg kg^{-1} in roots when cultivated in these soils for about 1.5 months. In a 2-year field experiment the As concentration decreased from 30–600 mg kg^{-1} to 6.5–9.3 mg kg^{-1} when *Polygunum sachalinense* was cultivated. However, the results are preliminary, and further work needs to be done to confirm these findings.

Induced phytoextraction refers to the application of chemicals to the soil to increase the mobility of the pollutant and its uptake in high-biomass plants. This has been implemented for heavy metals using chelating agents such as EDTA (107,108) but has not been tested for anionic contaminants. Based on the low mobility of As in most soils, and the typically limited uptake by plants, induced phytoextraction may be the method of choice. Studies have to be conducted to evaluate the appropriate agents able to mobilize As and to identify high-biomass plants that can take up As in large amounts.

V. CONCLUSIONS

Along with Cd, Hg, and Pb, As is considered as one of the "big four" trace elements of environmental concern. The increasing concern about its ecotoxicity and several examples of regional significance of its disastrous effects on human health makes it mandatory to develop environmentally sound and economically viable technologies to clean up As-polluted soils. Available engineering technologies are effective in specific cases, but generally expensive and produce secondary wastes.

Contamination due to anthropogenic activities is often confined to the top horizon. When the top soil is heavily contaminated, the most suitable technologies are probably engineering-based clean-up processes or immobilization with soil amendments. In the case of moderate contamination, biologically based technologies hold promise, but have not been developed yet. A better understanding of the processes involved in the biological transformation of As is required to enhance these technologies for the remediation of As-contaminated soils. Biologically-based in situ technologies are considered time consuming and may therefore be less suitable as immediate-risk scenarios. However, ex-situ treatment of excavated soils and/or residuals of physical or chemical remediation may be a viable option.

Since the soil is a nonrenewable resource, technologies that restore its multifunctionality at low energy and material input, and avoid/minimize the production of secondary waste, should be preferred. The selection of the most appropriate technology is case specific and has to be based on several factors, including the level of contamination, soil conditions, the time available, and economical aspects. A decision tree for a quick lookup of the most appropriate alternatives to treat As-polluted soils is displayed in Fig. 6.

REFERENCES

1. Azcue, J. M., and Nriagu, J. O. 1994. Arsenic: historical perspectives. In: J. O. Nriagu (ed.), *Arsenic in the Environment. Part I: Cycling and Characterization*, pp. 1–15. Wiley, New York.

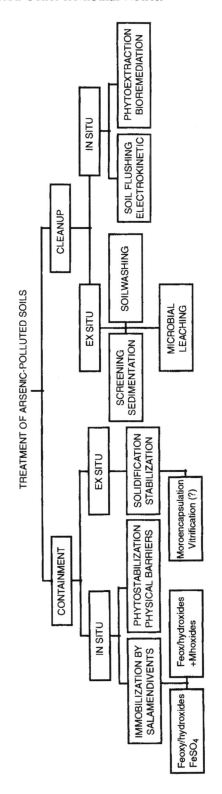

Figure 6 Decision tree of the alternatives to treat As-polluted soils.

2. Peters, G. R., McCurdy, R. F., and Hindmarsh, J. T. 1996. Environmental aspects of arsenic toxicity. *CRC Crit. Rev. Clin. Lab. Sci.* 33:457–493.
3. Nriagu, J. O. 1994. Preface. In: J. O. Nriagu (ed.), *Arsenic in the Environment. Part II: Human Health and Ecosystem Effects*, p. ix. Wiley, New York.
4. Alden, J. C. 1983. Arsenic industrial, biomedical, environmental perspectives. W. H. Lederer and R. J. Fensterheim (eds.), pp. 63–71. Reinhold, New York.
5. Adriano, D. C. 1986. *Trace Elements in the Terrestrial Environment*, pp. 47–72. Springer-Verlag, New York.
6. Gao, S., and Burau, R. G. 1997. Environmental factors affecting rates of arsine evolution from and mineralization of arsenicals in soil. *J. Environ. Qual.*, 26:753–763.
7. Chiou, H. Y., Hueh, Y. M., Liaw, K. F., Horng, S. F., Chiang, M. H., Pu, Y. S., Lin, J. S. N., Huang, C. H., and Chen, C. J. 1995. Incidence of internal cancers and ingested inorganic arsenic: A seven years follow up study in Taiwan. *Cancer Res.*, 55:1296–1300.
8. Smith, A. H., Hopenhayn-Rich, C., Bates, M. N., Goeden, H. M., Hertz-Picciotto, I., Duggan, H. M., Wood, R., Kosnett, M. J., and Smith, M. T. 1992. Cancer risk from arsenic in drinking water. *Environ. Health Perspect.*, 103:13–15.
9. Cantor, P. C. 1996. Arsenic in drinking water: How much is too much? *Epidemiology*, 7:113–114.
10. Frey, M. M., and Edwards, M. A. 1997. Surveying arsenic occurrence. *J. Am. Water Works Assoc.*, 89:105–117.
11. U.S. Environmental Protection Agency. 1997. Technology alternatives for remediation of soils contaminated with arsenic, cadmium, chromium, mercury, and lead. Engineering Bulletin EPA/540/s-97/500.
12. Nriagu, J. O. 1990. Global metal pollution. *Environment*, 32:28–33.
13. Shotyk, W., Cheburkin, A. K., Appleby, P. G., Fankhauser, A., and Kramers, D. 1996. Two thousand years of atmospheric arsenic, antimony and lead deposition recorded in an ombrotrophic peat bog profile, Jura Mountains, Switzerland. *Earth Planet. Sci. Lett.*, 145:E1–E7.
14. Shotyk, W. 1994. Natural and anthropogenic enrichments of arsenic through three Canadian ombrotrophic shagnum bog profiles. In: J. O. Nriagu (ed.), *Arsenic in the Environment. Part I: Cycling and Characterization*, pp. 281–402. Wiley, New York.
15. Mackenzie, F. T., Lantzy, R. J., and Paterson, V. 1979. Global trace metal cycle and predictions. *J. Int. Assoc. Math. Geol.*, 11:99–142.
16. Onishi, H., and Sandell, E. B. 1955. Geochemistry of arsenic. *Geochim. Cosmochim. Acta*, 7:1–10.
17. Ganapati, M. 1996. Arsenic poisons 220,000 in India. *Br. Med. J.*, 313:9.
18. Mandal, B. M., Chowdhury, T. R., Samanta, G., Basu, G. K., Chowdhury, P. P., Chanda, C. R., Lodh, D., Karan, N. K., Dhar, R. K., Tamili, D. K., Das, D., Saha, K. C., and Chakrabori, D. 1996. Arsenic in groundwater in seven district of West Bengal, India—The biggest arsenic calamity in the world. *Curr. Sci.*, 70:976–986.
19. Frizzell, P. M. 1993. Verification of a preliminary site inspection guidance manual for the identification of potential contaminated land. M.Sc. thesis, University of London.
20. Farago, M. E., Thornton, I., Kavanagh, P., Elliott, P., and Leonardi, G. S. 1997. Health aspects of human exposure to high arsenic concentrations in soil in south-west England. In: C. O. Abernathy, R. L. Calderon, and W. R. Chappell (eds.), *Arsenic—Exposure and Health Effects*, pp. 211–225. Chapman & Hall, London.
21. Kavanagh, P. J., Fernandes, S., and Freire, I. 1995. Arsenic, antimony and bismuth concentrations in soils of the Tamar Valley, S.W. England. Poster Book, Second Int. Conf. on Arsenic Exposure and Health Effects, San Diego CA, June 1995.
22. Davis, A., Ruby, M. V., Bloom, M., Schoof, R., Freeman, G., and Bergstrom P. D. 1996. Mineralogical constrains on the bioavailability of arsenic in smelter-impacted soils. *Environ. Sci. Technol.*, 30:392–399.
23. Bouska, V. 1981. *The Geochemistry of Coal*. Elsevier, Amsterdam.

24. Sato, K., and Sada, K. 1992. Effects of emissions from coal fired power plant on soil trace element concentrations. *Atmos. Environ.*, 26:325–337.
25. Jensen, G. E., and Olsen, I. L. B. 1995. Occupational exposure to inorganic arsenic by workers and taxidermists—Air sampling and biological monitoring. *Environ. Sci. Health*, A30:921–938.
26. Sadler, R., Olszowy, H., Shaw, G., Biltoft, R., and Connell, D. 1994. Soil and water contamination by arsenic from a tannery waste. *Water Air Soil Pollution*, 78:189–198.
27. U.S. Federal Regulation. 1993. Standard for the use of sewage sludge. *Fed. Reg.*, 58:9248–9415.
28. McGrath, S. P., Chang, A. C., Page, A. L., and Witter, E. 1994. Land application of sewage sludge: Scientific perspectives of heavy metal loading limits in Europe and United States. *Environ. Rev.*, 2:108–118.
29. Takamatsu, T. H., Aoki, H., and Yoshida T. 1982. Determination of arsenate, arsenite, monomethylarsonate, and dimethylarsinate in soil polluted with arsenic. *Soil Sci.*, 133:239–246.
30. Schulte, A., and Gehrmann, J. 1996. Development of heavy-metal deposition by precipitation in West Germany. 1. Arsenic, chromium, cobalt, and nickel. *Z. Pflanzen. Bodenk.*, 159:385–390.
31. Lombi, E., Wenzel, W. W., and Adriano, D. C. 1998. Soil contamination, risk reduction and remediation. *Land Contam. Reclam.*, 6:1–15.
32. Lemmo, N. V., Faust, S. D., and Belton, T. 1983. Assessment of the chemicla and biological significance of arsenical compounds in a heavily contaminated watershed. I. The fate and speciation of arsenical compounds in aquatic environments, a literature review. *J. Environ. Sci. Health*, A18:335–387.
33. Deuel, L. E., and Swoboda, A. R. 1972. Arsenic solubility in a reduced environment. *Soil Sci. Soc. Am. Proc.*, 36:276–278.
34. Braman, R. S. 1975. Arsenic in the environment. In: E. A. Woolson (ed.), *Arsenical pesticides*, pp. 108–123. American Chemistry Society, Washington, DC.
35. Hindmarsh, J. T., and McCurdy, R. F. 1986. Environmental aspects of arsenic toxicity. *CRC Crit. Rev. Clin. Lab. Sci.*, 33:315–347.
36. Vela, N. P., and Caruso, J. A. 1993. Potential of liquind cromatography inductively coupled mass spectrometry for trace metal speciation. *J. Anal. Atomic Spectrom.*, 8:787–794.
37. Sadiq, M. 1997. Arsenic chemistry in soils: An overview of thermodynamic predictions and field observations. *Water Air Soil Pollution*, 93:117–136.
38. O'Neill, P. 1995. Arsenic. In: B. J. Alloway (ed.), *Heavy Metals in Soils*, 2nd ed., pp. 105–121. Blackie, London.
38. Woolson, E. A., Axley, J. H., and Kearney, P. C. 1971. The chemistry and phytotoxicity of arsenic in soil. *Soil Sci. Soc. Am. Proc.*, 35:938–943.
39. Walsh, L. M., and Keeney, D. R. 1975. Behaviour and phytotoxicity of inorganic arsenicals in soils. *ACS Symp. Ser.*, 7:35–52.
40. Valersi, T. 1998. Kinetics and bonding structures of arsenic sorption in soils and model substances. Ph.D. thesis, University of Agricultural Sciences, Vienna.
41. Xie, Z. M., and Huang, C. Y. 1998. Control of arsenic toxicity in rice plants grown on an arsenic-polluted soil. *Commun. Soil Sci. Plant Anal.*, 29:2471–2477.
42. Masscheleyn, P. J., Delaune, R. D., and Patrick, W. H. 1991. Arsenic and selenium chemistry as affected by sediment redox potential and pH. *J. Environ. Qual.*, 20:522–527.
43. Lombi, E., Sletten, R. S., and Wenzel, W. W. 2000. Arsenic sequentially extracted in sand, silt, and clay of contaminated soils. *Water, Air and Soil Poll.* (in press).
44. Yang, G. Z. 1983. Cluster analysis of some elements in soils of Tainijin area. *Acta Sci. Circumstantiae*, 3:207–212.
45. Koyama, T. 1975. Arsenic in soil-plant system. *Nippon Dojo Hiryogaku Zasshi*, 46:491–502.

46. Shen, B. Z., Chen, L. G., and Zhao, Z. D. 1983. Correlation between the content of some elements and the mechanical composition of the soils in Tianjin region. *Turang Xuebao*, 20:440–444.
47. Thanabalasingam, P., and Pickering, W. F. 1986. Arsenic sorption by humic acids. *Environ. Pollution, Ser. B.*, 12:233–246.
48. Schwertmann, U., and Taylor, R. M. 1989. Iron oxides in the laboratory. In: J. B. Dixon and S. B. Weed (eds.), *Minerals in Soil Environments*, 2nd ed., pp. 379–438. Soil Science Society of America, Madison, WI.
49. McBride, M. B. 1987. Adsorption and oxidation of phenolic compounds by iron and manganese oxides. *Soil Sci. Soc. Am. J.*, 51:1466–1472.
50. Anderson, M. A., Ferguson, J. F., and Gavis, J. 1976. Arsenate adsorption on amorphous aluminium hydroxides. *J. Colloid Interface Sci.*, 54:391–399.
51. Huang, P. M. 1975. *Soil Sci. Soc. Am. Proc.* 39:271.
52. Norimoto, W., and Osamu, H. 1982. Adsorption of arsenite and arsenate by soils and charcoal. *Kankyo Gijutsu*, 11:565–571.
53. Pierce, M. L., and Moore, C. B. 1980. Adsorption of arsenite on amorphous iron hydroxide from dilute aqueous solution. *Environ. Sci. Technol.*, 11:89–99.
54. Oscarson, D. W., Huang, P. M., Liaw, W. K., and Hammer, U. T. 1983. Kinetics of oxidation of arsenite by various manganese dioxides. *Soil Sci. Soc. Am. J.*, 47:644–648.
55. Oscarson, D. W., Huang, P. M., Defosse, C., and Herbillon, A. 1981. Oxidative power of Mn (IV) and Fe (III) oxides with respect to As (III) in terrestrial and aquatic environments. *Nature* (Lond.), 191:50–51.
56. Manning, B. A., and Goldberg, S. 1997. Adsorption and stability of arsenic (III) at the clay mineral-water interface. *Environ. Sci. Technol.*, 31:2005–2011.
57. Fendorf, S., Eick, M. J., Grossl, P., and Sparks, D. L. 1997. Arsenate and chromate retention mechanisms on goethite. 1. Surface structure. *Environ. Sci. Technol.*, 31:315–320.
58. Waychunas, G. A., Rea, B. A., Fuller, C. C., and Davis, J. A. 1993. Surface chemistry of ferrihydrite: Part 1. EXAFS studies of the geometry of coprecipitated and adsorbed arsenate. *Geochim. Cosmochim. Acta*, 57:2251–2269.
59. Manceau, A. 1995. The mechanism of anion adsorption on iron oxides: Evidence for the bonding of arsenate tetrahedra on free $Fe(O, OH)_6$ edges. *Geochim. Cosmochim. Acta*, 59:3647–3653.
60. Waychunas, G. A., Rea, B. A., Davis, J. A., and Fuller, C. C. 1995. Geometry of sorbed arsenate on ferrihydrite crystalline FEOOH: Reevaluation of EXAFS results and topological factors in predicting geometry and evidence for monodentate complexes. *Geochim. Cosmochim. Acta*, 59:3655–3661.
61. Sun, X., and Doner, H. E. 1996. An investigation of arsenate and arsenite bonding structures on goethite by FTIR. *Soil Sci.*, 161:865–872.
62. Branson, J. M., and Patrick, W. J. 1987. Fixation, transformation and mobilization of arsenic in sediments. *Environ. Sci. Technol.*, 20:450–459.
63. Goldberg, S., and Glaubig, R. A. 1988. Anion sorption on a calcareous montmorillonite soil—Arsenic. *Soil Sci. Soc. Am. J.*, 52:1297–1300.
64. Polemio, M., Bufo, S. A., and Senesi, S. 1982. Minor elements in south east Italy soils—A survey. *Plant Soil*, 69:57–66.
65. Peryea, F. 1991. Phosphate-induced release of arsenic from soils contaminated with lead arsenate. *Soil Sci. Soc. Am. J.*, 55:1301–1306.
66. Melamed, R., Jurinak, J. J., and Dudley, L. M. 1995. Effect of adsorbed phosphate on transport of arsenate through an oxisol. *Soil Sci. Soc. Am. J.*, 59:1289–1294.
67. Hingston, F. J. 1981. A review of anion adsorption. In: M. A. Anderson and A. J. Rubin (eds.), Adsorption of inorganics at solid–liquid interfaces, pp. 51–90. Ann Arbor Science Publ., Ann Arbor, MI.
68. Sposito, G. 1984. *The Surface Chemistry of Soil*. Oxford University Press. New York.

69. U.S. Environmental Protection Agency. 1995. Contaminants and remedial options at selected metal contaminated sites, EPA/540/R-95/512.
70. Jekel, M. R. 1994. Removal of arsenic in drinking water treatment. In: J. O. Nriagu (ed.), *Arsenic in the Environment. Part I: Cycling and Characterization*, pp. 119–132. Wiley, New York.
71. Hlavay, J., and Polyak, K. 1997. Removal of arsenic ions from drinking water by novel type adsorbents. In: C. O. Abernathy, R. L. Calderon, and W. R. Chappell (eds.), *Arsenic—Exposure and Health Effects*, pp. 383–405. Chapman & Hall, London.
72. Smith, L. A., et al. 1995. Remedial options for metals-contaminated sites. L. A. Smith et al. (eds.). CRC Lewis Publishers, Boca Raton, FL.
73. U.S. Environmental Protection Agency. 1996. Soil washing treatment. Engineering Bulletin EPA/540/2-90/017.
74. Legiec, I. A., Griffin, L. P., Walling, P. D., Breske, T. C., Angelo, M. S., Isaacson, R. S., and Lanza, M. B. 1997. DuPont soil washing technology program and treatment of arsenic contaminated soils. *Environ. Prog.*, 16:29–34.
75. Woolson, E. A., Axley, J. H., and Kearney, P. C. 1973. The chemistry and toxicity of arsenic in soils: II. Effects of time and phosphorus. *Soil Sci. Soc. Am. Proc.*, 37:254–259.
76. U.S. Environmental Protection Agency. 1997. Recent development for in situ treatment of metal contaminated soils, EPA/542/R-97/004.
77. Acar, Y. B., and Alshawabken, A. N. 1993. Principles of electrokinetic remediation. *Environ. Sci. Technol.*, 27:2638–2647.
78. U.S. Environmental Protection Agency. 1990. *Handbook on In Situ Treatment of Hazardous-Waste Containing Soils*, EPA/540/2-90/002. Risk Reduction Engineering Laboratory, Cincinnati, OH.
79. Kalb, P. D., Burns, H. H., and Meyer, M. 1993. Thermoplastic encapsulation treatability study for a mixed waste incinerator off-gas scrubbing solution. In: T. M. Gilliam (ed.), *Third Int. Symp. Stabilization/Solidification of Hazardous, Radioactive and Mixed Wastes*, ASTM STP 1240. America Society for Testing and Materials, Philadelphia.
80. U.S. Environmental Protection Agency. 1990. Final best demonstrated available technology (BDAT) background document for mercury-containing wastes D009, K106, P092, and U151, EPA/530-SW-90-059Q; PB90-234170. Report prepared for the U.S. EPA Office of Solid Waste.
81. Kabata-Pendias, A., and Pendias, H. 1992. *Trace Elements in Soils and Plants*, 2nd ed. CRC Press, Boca Raton, FL.
82. Elkatib, E. A., Bennet, O. L., and Wright, R. J. 1984. Arsenite sorption and desorption in soils. *Soil Sci. Soc. Am. J.*, 48:1025–1030.
83. Elkatib, E. A., Bennet, O. L., and Wright, R. J. 1984. Kinetics of arsenite sorption in soils. *Soil Sci. Soc. Am. J.*, 48:758–762.
84. Fuller, C. C., Davis, J. A., Wachunas, G. A., and Rea, B. A. 1993. Surface chemistry of ferrihydrite: Part 2: Kinetics of arsenate adsorption and coprecipitation. *Geochim. Cosmochim. Acta*, 57:2271–2282.
85. Polemio, M., Senesi, S., and Bufo, S. A. 1982. Soil contamination by metals—A survey in industrial and rural areas of southern Italy. *Sci. Total Environ.*, 25:71–79.
86. Lombi, E., Wenzel, W. W., and Sletten, R. S. 1999. Arsenic adsorption on soil and iron oxides coated sand. *J. Plant Nutr. Soil Sci.*, 162:451–456.
87. Onken, B. M., and Adriano, D. C. 1997. Arsenic availability in soil with time under saturated and subsaturated conditions. *Soil Sci. Soc. Am. J.*, 61:746–752.
88. Sims, R., Sorenson, D., Sims, J., McLean, J., Mahmood, R., Dupont, R., Jurinak, J., and Wagner, K. 1986. *Contaminated Surface Soils in-Place Treatment Techniques*. Noyes Publications, Park Ridge, NJ.
89. U.S. Environmental Protection Agency. 1984. *Review of in-Place Treatment Techniques for Contaminated Surface Soils, Vol. 1, Technical Evaluation*, EPA/540/2-84/003a.

90. Kardos, L. T., Vandecaveye, S. C., and Benson, N. 1941. Causes and remedies of the unproductiveness of certain soils following the removal of mature (fruit) trees. *Wash. Agr. Exp. Sta. Bull.*, 410:25.
91. Voigt, D. E., Brantley, S. L., and Hennet, R. J. C. 1996. Chemical fixation of arsenic in contaminated soils. *Appl. Geochem.*, 11:633–643.
92. Artiola, J. F., Zabcik, D., and Johnson, S. H. 1990. In situ treatment of arsenic contaminated soil from hazardous industrial site: Laboratory studies. *Waste Manage.*, 10:73–78.
93. Wenzel, W. W., Salt, D., Adriano, D. C., and Smith, R. 1999. Phytoremediation: A plant-microbe-based remediation system. In: D. C. Adriano et al. (eds.), *Soil Remediation*. SSSA Monographs, Madison, WI, in press.
94. McBride, B. C., and Wolfe, R. S. 1971. Biosynthesis of dimethylarsine by Methanobacterium. *Biochemistry*, 10:4312–4317.
95. Cullen, W. R., McBride, B. C., Pickett, A. W., and Regalinski, J. 1984. The wood preservative chromated copper arsenate is a substance for trimethylarsine biosynthesis. *Appl. Environ. Microbiol.*, 47:443–444.
96. Baker, M. D., Inniss, W. E., Mayfield, C. I., Wong, P. T. S., and Chau, Y. K. 1983. Effect of pH on the methylation of mercury and arsenic by sediment microorganisms. *Environ. Technol. Lett.*, 4:89–100.
97. Ahmann, D., Roberts, A. L., Krumholz, L. R., and Morel, F. M. M. 1994. Microbe grows by reducing arsenic. *Nature*, 371:750.
98. Leblanc, M., Achard, B., Othman, D. B., Luck, J. M., Bertrand, L., Sarfati, J., and Personne, J. C. 1996. Accumulation of arsenic from acidic mine waste waters by ferruginous bacterial accretions (stromatolites). *Appl. Geochem.*, 11:541.
99. Baker, A. J. M., Reeves, R. D., and McGrath, S. P. 1991. In situ decontamination of heavy metal polluted soils. Using crops of metal-accumulating plants—A feasibility study. In: R. E. Hinchee and R. F. Olfenbuttel (eds.), *In situ Bioreclamation*, p. 539. Butterworth-Heinemann, Stoneham, MA.
100. McGrath, S. P., Sidoli, C. M. D., Baker, A. J. M., and Reeves, R. D. 1993. The potential for the use of metal-accumulating plants for the in-situ decontamination of metal-polluted soils. In: H. J. P. Eijsackers and T. Hamers (eds.), *Integrated Soil and Sediment Research: A Basis for Proper Prediction*, pp. 673–676. Kluwer, Dordrecht.
101. Raskin, I., Kumar, N., Dushenkow, S., and Salt, D. E. 1994. Bioconcentration of heavy metals by plants. *Curr. Opin. Biotechnol.*, 5:285–290.
102. McGrath, S. P. 1998. Phytoextraction for soil remediation. In: R. R. Brooks (ed.), *Plants That Hyperaccumulate Heavy Metals*, pp. 261–288. CAB International, Wallingford, U.K.
103. Salt, D. E., Smith, R. D., and Raskin, I. 1998. Phytoremediation. *Annu. Rev. Plant Physiol. Plant Mol. Biol.*, 49:643–668.
104. Jonnalagadda, S. B., and Nenzou, G. 1997. Studies on arsenic rich mine dumps. 2. The heavy element uptake by vegetation. *J. Environ. Sci. Health Part A*, 32:455–464.
105. Komar, K. M., Ma, L. Q., Rockwood, D., and Syed, A. 1998. Identification of arsenic tolerant and hyperaccumulating plants from arsenic contaminated soils in Florida. *Agron. Abstr.*
106. Feller, K. A. 1998. Future of phytoremediation. *Proc. Int. Workshop "Plant Impact at Contaminated Sites,"* Schmallenberg (D), Dec. 1–2, 1997, in press.
107. Huang, J. W., Chen, J., Berti, W. B., and Cunningham, S. D. 1997. Phytoremediation of lead-contaminated soils: Role of synthetic chelates in lead phytoextraction. *Environ. Sci. Technol.*, 31:800–805.
108. Blaylock, M. J., Salt, D. E., Dushenkov, S., Zakharova, O., Gussman, C., Kapulnik, Y., Ensley, B. D., and Raskin, I. 1997. Enhanced accumulation of Pb in indian mustard by soil-applied chelating agents. *Environ. Sci. Technol.*, 31:860–865.

35
Advanced Risk-Based Biodegradation Study Using Environmental Information System and the Holistic Macroengineering Approach

Stergios Dendrou and Basile Dendrou
ZEi Engineering, Inc., Annandale, Virginia

Mehmet Tumay
Louisiana State University, Baton Rouge, Louisiana

I. INTRODUCTION

The modern approach to remediation is the risk-based approach. The idea is to provide rationale to the environmental problem by linking health hazard risk at receptors to acceptable treatment levels at the contaminant source. The added advantage is that this allows accounting for the capacity of the medium for natural attenuation (assimilative capacity). Often, extracting the contaminated water for treatment can do more damage than letting it biodegrade in situ. The regulatory position is that natural attenuation can be acceptable, provided that short- and long-term monitoring proves the truthfulness of the theory in the field, and guarantees compliance at imposed checkpoints. Therein lies the difficulty, because whatever can be saved by reduced remediation can be squandered by a burdensome monitoring plan. Furthermore, sometimes the collection of data—drilling monitoring wells through various formations—actually creates artificial pathways which cause further contamination.

This situation can be made tractable only with the use of advanced simulation/prediction capability. Data are used to elucidate how the prototype behaves (model calibration), on the basis of which future evolution of the contamination episode can be predicted and, as the case may be, to compare various remediation alternatives and establish acceptable monitoring plans. However, credibility in the model predictions can be founded only on a comprehensive simulator. In addition to three-dimensional flow and contaminant migration via advection/dispersion, the model must also account for chemical interactions (reactions) among numerous dissolved and adsorbed species, contaminants, electron acceptors, nutrients, and their by-products.

Few models with such capability exist. Biorem3D (1) is the model presented here. Other models, such as RT3D (2), are experimental, depend on as many as three independent models for groundwater flow, transport, and even ordinary differential equation solver, and run into difficulties when applied to large real case studies (3). In contrast, Biorem3D is an industrial-strength code whose high level of integration makes it ideal for use in large-scale real case studies. It has been in continuous use since 1994, has been presented in short courses, and has been reviewed and applied by the EPA Kerr Lab in Oklahoma.

Second, such models make extensive use of data and therefore are ineffective when used in "batch" mode (i.e., run from an input file, not interactively). That is, they need some graphical user interface (GUI). Biorem3D is embedded in the Environmental Information System EIS Platform (1). Graphical support usually comes in the form of pre- and postprocessors. GMS is an example (5). The common feature in this type of software is that it is the user's responsibility to track the multitude of files—graphical, tabular, and most important, input files to simulators.

Unlike these graphical processors, EIS is the first entirely object-based, integrated-automated platform, in continuous use since 1992. The user deals with physical objects, geologic features, chemical substances, topographic distributions, and scenario definitions, and everything else is calculated, coordinated, and managed by the graphical platform. In this chapter, we present the theory, the implementation, and application of the EIS/biorem3D Platform from an industrial case study. We begin by introducing in detail the risk-based approach.

II. RISK-BASED APPROACH

The risk assessment method attempts to link required treatment or remediation levels to the health hazard that the contamination has caused. Early application of the method resulted in controversial situations, for example, where a remediation scheme was "chasing" the last exponent of risk, e.g., a pump-and-treat scheme resulting in health hazard risk reduction from groundwater contamination from 10^{-6} to 10^{-7}, all the while causing air pollution from VOCs emission of much larger proportions.

More recently, the practice of risk assessment has evolved to a more broadly based approach. This is characterized by consideration of multiple endpoints, sources, pathways, and routes of exposure, but also community-based decision making, flexibility in achieving goals, case-specific responses, a focus on all the environmental media, and, significantly, holistic reduction of risk. This more complex assessment involves cumulative risk assessment and is defined in each case according to who (or what) is at risk of adverse effects from identifiable sources (stressors), through several paths of exposure over varied time frames.

The EIS platform with its integrated framework is developed to fill the needs of a cumulative risk assessment. Indeed, with the help of EIS, risk assessors and risk managers can make judgments early in planning major risk assessments. EIS helps the assessor regarding the purpose, scope, and technical approach (that is, the conceptual model) by evaluating the full range of discernible human health and ecological dimensions of risk, that is, stressors, sources, effects, exposed populations, pathways of exposure, and time frames of risks). Special emphasis is put on cumulative risk, that is, the potential risks presented by multiple stressors in aggregate. The specific elements of risk evaluated under the EIS platform are an integral part of the planning and scoping (PS) stage of each risk assessment. With EIS during planning and scoping, risk assessors, technical expert ecologists, toxicologists, economists and engineers, and risk managers work together as a team, in coordination with stakeholders, to perform the following tasks:

EIS Macroengineering Approach

1. Determine the overall purpose and general scope of the risk assessment.
2. Delineate the information needed by management for risk decision making.
3. Review the risk dimensions and technical elements that may be evaluated in the assessment.
4. Determine the relationships among alternative end points and corresponding risk management options.

The risk assessment evaluation involves establishing the following elements:

1. Analysis plan and conceptual model
2. Resources—data, models, required or available
3. The identity of involved parties—technical, technical, legal, or stakeholder advisors
4. The schedule to be followed, including timely, and independent, external peer review

The overall EIS framework for integrated risk assessment and management is shown in Fig. 1. For implementation of the different stages shown in Fig. 1, there is a need to frame risk assessment in terms of the sources, stressors, pathways, populations, endpoints, and time frames. All these elements are identified and automatically processed under the EIS umbrella and can be automatically available to plan the risk assessment with the risk manager and explain the scope of the risk assessment to the interested and affected

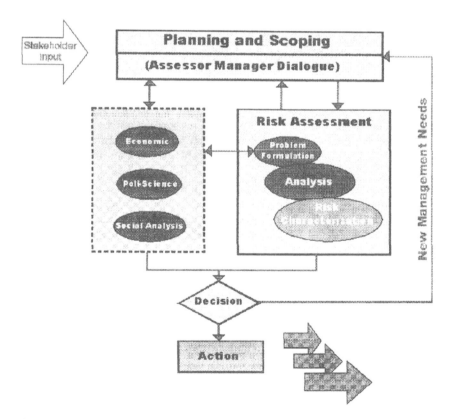

Figure 1 Stages in the EIS integrated risk assessment process.

parties. But EIS also supports the next step, which is the technical approach (also referred to as problem formulation), a process in which the analysis plan and preliminary hypotheses about the relationship between stressors and effects on populations are developed in a conceptual model. The various EIS protocols that are defined in conjunction with these risk components ("Dimensions" in the EPA's nomenclature) are summarized in Table 1.

A. Uncertainties of Cumulative Risk Assessment

Uncertainty is a pervasive element in all decision-making processes such as cumulative risk assessment. Every aspect of the decision-making process of cumulative risk assessment, whether in environmental engineering, government, or policy making, is uncertain: there is uncertainty in selecting all the relevant risk components and variables, in assigning them proper weight, in assessing the importance of these factors in selecting pieces of evidence indicative of the qualities we wish to express, and whether we are ready to accept the liability resulting from our decisions.

The engineering profession has well recognized the role of measurement error on the decision-making process, and the EIS platform has been a pioneer in providing a robust set of tools for the assessment of its effects. Quantifying these uncertainties with ease allows the assessor to focus attention on the total uncertainty that surrounds the cumulative risk assessment.

Substantial uncertainties reside in the very human process of establishing the relationship between that evidence and the decision that will need to be made. The language of uncertainty risk management is thus not only that of probability, but also of possibility, plausibility, ignorance, and belief. Methodologies exist, some esoteric—fuzzy sets, Bayesian probability theory—but the logic is drawn from common human experience.

This is a pertinent time to address issues of uncertainty risk management and decision strategy analysis. This is because in the early stages of the development of EIS, the focus was on complexity risk management. With complexity risk management, we sought to develop the means of coping with the enormous volume of data that typically need to be considered in a cumulative risk assessment study, and to extract useful information from the data.

With regard to cumulative risk assessment, the tools available in the early stages of EIS made only single solutions possible; they could not easily account for competing risk objectives, trade-off options, or levels of risk. Nor did they offer multiple allocation techniques or criteria weighting schemes. Presently the need is to incorporate these issues of uncertainty risk management and decision strategy analysis to build tools that allow decision makers to explore all possible solutions, not just to land allocation problems but to a wide variety of environmental issues.

All the above elements of the risk-based approach and their uncertainties were carefully examined in the case study presented below. The strategy for the risk-based approach is delineated in the flowchart of Fig. 2. Early in the process, models are used to identify (screen) environmental concerns. Screening is often based on a structured-value approach, and is designed for use with regional representative information. The purpose of the exercise is to determine if a situation is a problem, but not to provide a risk-based relative ranking of problems.

Detailed analyses require a highly specialized assessment of potential impacts. These methodologies require numerically based models. Most existing models are not physically

EIS Macroengineering Approach

Table 1 Elements of Cumulative Risk Assessment and EIS Protocols

Receptors (Population, Environment)
(Who/what/where is at risk?)

Humans [Pr-EIS-100-Hum]	Ecological entities [Pr-EIS-101-Eco]	Landscape or geographic concern [Pr-EIS-103-Land]

(a) Individual
(b) General population distribution or estimation of central tendency and high end exposure
(c) Population subgroups
   (1) Highly exposed subgroup (for example, due to geographic area, age group, gender, racial or ethnic group, or economic status)
   (2) Highly sensitive subgroups (for example, asthmatics or other preexisting conditions, age, gender)
	(a) Groups of individuals
	(b) Populations
	(c) Multiple species
	(d) Habitats or ecosystems
		(a) Groundwater aquifers
		(b) Watersheds (that is, surface water bodies and their associated terrestrial ecosystems)
		(c) Airsheds
		(d) Regional ecosystems
		(e) Recreational lands

Contamination Sources
(What are the relevant sources of stressors?)

Single source [Pr-EIS-200-SSour]	Multiple sources [Pr-EIS-201-MSour]
(a) Point sources (for example, industrial or commercial discharge, superfund sites) (b) Non-point sources (for example, automobiles, agriculture, consumer use releases) (c) Natural sources (for example, flooding, hurricanes, earthquakes, forest fires)	(a) Point sources (for example, industrial or commercial discharge, superfund sites) (b) Non-point sources (for example, automobiles, agriculture, consumer use releases) (c) Natural sources (for example, flooding, hurricanes, earthquakes, forest fires)

Contaminants, Stressors
(What are the stressors of concern?)

1. Chemicals [Pr-EIS-300-Chem]
   (a) Single chemical
   (b) Structurally related class of substances
   (c) Individual substances (that is, only one is present at a time)
   (d) Existing in a mixture
   (e) Structurally unrelated substances with similar mechanism of impact and/or same target organ
   (f) Individual substances
   (g) Existing in a mixture
   (h) Mixtures (that is, dissimilar structures or dissimilar mechanisms)

Table 1 (*continued*)

2. Radiation [Pr-EIS-301-Rad]
3. Microbiological or biological (range from morbidity to ecosystem disruption) [Pr-EIS-303-Micrb]
4. Nutritional (for example, diet, fitness, or metabolic state) [Pr-EIS-304-Nutr]
5. Economic (for example, access to health care) [Pr-EIS-304-Econo]
6. Psychological (for example, knowledge of living near uncertain risks) [Pr-EIS-305-Psyc]
7. Habitat alteration (urbanization, hydrological modification, timber harvest) [Pr-EIS-306-Hab]
8. Land-use changes (for example, agriculture to residential, public to private recreational uses) [Pr-EIS-307-Land]
9. Global climate change [Pr-EIS-308-Clim]
10. Natural disasters (for example, floods, hurricanes, earthquakes, disease, pest invasions) [Pr-EIS-309-Disast]

Pathways
(Environmental pathways and routes of exposure. What are the relevant exposures?)

Pathways (recognizing that one or more may be involved) [Pr-EIS-400-Pthways]	Routes of human and single species exposures [Pr-EIS-401-RtHum]	Routes of exposure within communities and ecosystems [Pr-EIS-402-RtExp]
(a) Air (b) Surface water (c) Groundwater (d) Soil (e) Solid waste (f) Food (g) Non-food consumer products, pharmaceuticals	(a) Ingestion (both food and water) (b) Dermal (includes absorption and uptake by plants) (c) Inhalation (includes gaseous exchange) (d) Nondietary ingestion (for example, "hand-to-mouth" behavior)	(a) Direct contact or ingestion (without accumulation) (b) Bioaccumulation (c) Biomagnification (d) Vector transfers (for example, parasites, mosquitoes)

Endpoints
(What are the assessment endpoints?)

Human health effects (for examples, as based on animal studies, morbidity and disease registries, laboratory and clinical studies, or epidemiological studies or data) [Pr-EIS-500-HlthHum]	Ecological effects [Pr-EIS-501-EcolEff]

(a) Carcinogenic
(b) Neurotoxicologic
(c) Reproductive dysfunction
(d) Developmental
(e) Cardio-vascular
(f) Immunologic
(g) Renal
(h) Hepatic
(i) Others
	(a) Population or species
	  (1) Loss of fecundity
	  (2) Reduced rate of growth
	  (3) Acute or chronic toxicity
	  (4) Change in biomass
	(b) Community
	  (1) Loss of species diversity
	  (2) Introduction of an exotic species
	  (3) Loss of keystone species
	(c) Ecosystem
	  (1) Loss of a function (for example, photosynthesis, mineral metabolism)

Table 1 (*continued*)

    (2) Loss of habitat structure
(3) Loss of a functional group of organisms (for example, grazers, detritivores)
(4) Climate change (for example, sunlight, temperature change)
(5) Loss of landscape features (for example, migration corridors, home ranges)

Time Frames
(What are the relevant time frames: frequency duration, intensity, and overlap of exposure intervals for a stressor or mixtures of stressors?)

1. Acute [Pr-EIS-600-TAcut]
2. Subchronic [Pr-EIS-601-Schron]
3. Chronic, or effects with a long latency period [Pr-EIS-602-ChrEffect]
4. Intermittent [Pr-EIS-603-Inter]

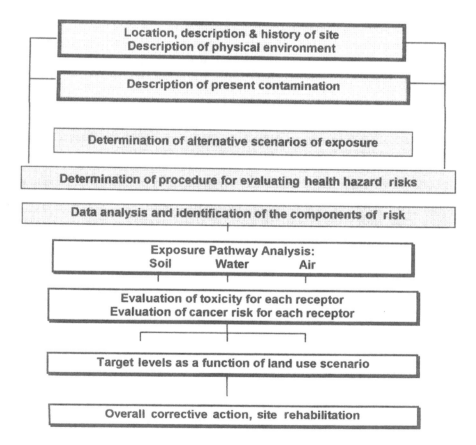

Figure 2 Strategy for risk-based approach.

linked and represent single-medium models, implemented independently in series. This approach usually is reserved for complex cases, and is data-intensive. Often such tools are appropriate for their intended application, but extension beyond site-specific applications is either difficult or cost-prohibitive. An alternative to these analytical/semianalytical/empirical-based multimedia models (designated as analytical models) is offered by the EIS platform, which includes a large pool of numerical models (scientific engines) that can be used for prioritization, preliminary assessments, and exhaustive risk assessment studies.

The EPA risk assessment methodology for exposure assessment suggests a series of standard default exposure routes and exposure assumptions/parameters for use in conjunction with discrete current and future land use scenarios. While the exposure routes themselves may be more or less applicable to a specific site, the majority of the standard exposure assumptions advocated for use in estimating chemical intakes are not site-specific, nor are they necessarily the most current, relevant numerical values. The use of alternative standard assumptions for the development of site-specific assumptions has been met with varying degrees of acceptance by regulatory agencies. Existing EPA guidelines advocate the use of site-specific information wherever possible. Site-specific information and viable exposure routes vary with the location, magnitude, and nature of the spill or leak, as well as the local human populations, regional topology and hydrogeology, and land use. Practical, site-specific considerations as implemented in the EIS platform are discussed below.

Rather than using point estimates in exposure assessment, the EIS simulation tools can be used to estimate distributions for exposure assumptions. Use of this methodology does not alter the basic structure of the exposure estimate. However, it does refine the way chemical intakes are calculated in the exposure assessment.

Figure 3 illustrates the overall approach as implemented in the EIS platform. The starting point is to establish the statistical characteristics of all pertinent input parameters characterizing a site. Such parameters include soil properties (soil layers, porosity, hydraulic conductivities, dispersion, etc.), chemical properties (adsorption, stoichiometry, etc.), as well as loading/source site-specific conditions. Then, their mean values and standard deviations automatically feed the scientific engines (simulation algorithms) available in the EIS platform. These algorithms simulate different natural processes that include:

Multiphase contaminant migration in the unsaturated zone

Groundwater flow with special features such as slurry walls, geosynthetics, and geologic faults

Multispecies contaminant migration processes such as advection (computed from fluxes produced by the flow module), dispersion, chemical reaction (sorption, ion exchange, chemical decay), and sink/source mixing

Migration and degradation of hydrocarbons and chlorinated compounds accounting for aerobic and anaerobic biodegradation and cometabolism occurring at the site

The outcome of the simulation is the spatial distribution of the mean values and standard deviations of the concentrations of the contaminant (BTEX, chlorinated compounds, etc.) throughout the site. The mean values are the conventional point estimates as produced by the corresponding algorithms activated on a given site. A first-order approximation is used to compute the standard deviation of the contaminant concentrations (stressor) C.

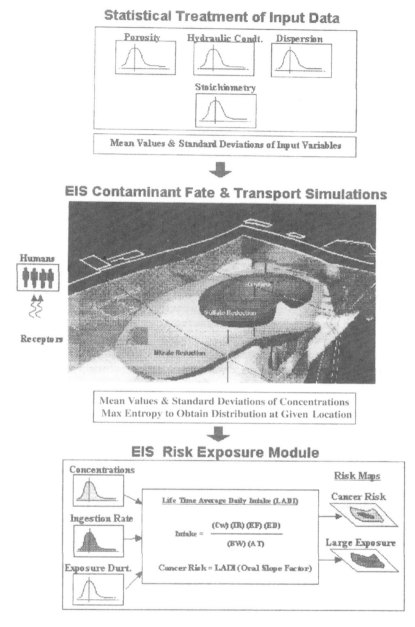

Figure 3 Uncertainty risk assessment implementation in EIS.

At this stage we know only the mean and variance of the contaminant concentration probability distribution. The entire probability distribution is derived by invoking the principle of maximum entropy. The concentration probability distribution is that which maximizes the information entropy subject to the additional constraints imposed by the given information, i.e., the mean and variance (6,7).

In the EIS simulation model, the analyst determines a continuous or discrete distribution to describe each random variable. This distribution is defined in terms of the probability density function (PDF) or the cumulative distribution function (CDF). Several distributions are defined with one, two, or more parameters. When running the EIS simulation model, the computer proceeds automatically to determine the distribution of daily intakes. From this distribution, a specific intake can be selected (e.g., the average or mean intake, median intake, or 95th percentile upper confidence limit on the intake) that, in combination with the appropriate toxicity benchmark concentration, is used to calculate risk.

EIS simulations can also include correlation between variables. For example, there is a correlation between body weight and ingestion rate. Using strongly correlated variables in deriving an estimate of exposure serves to strengthen the estimate by preventing nonsensical combinations of variables in its derivation.

In most cases, the daily human intake calculated using the EIS simulation is less than that calculated using point estimates. This is not to suggest the use of the platform because it produces lower estimates, but rather because its estimates can be associated with probabilities. This, results in increased confidence in the estimate of intake, thus ensuring increased confidence in public health protection.

The deterministic values of the contaminant concentrations typically result in a point estimate scenario for contaminant ingestion in water. However, to perform an uncertainty analysis a distribution is needed for all exposure parameters (such as the ingestion rate, the exposure frequency, the exposure duration, the body weight, and the cancer potency factor) and the contaminant concentration at "end points" (receptors). The distribution of the exposure parameters is obtained from databases of laboratory experiments. The distribution of the contaminant concentrations from the migration simulation is a more elaborate task, and several options are available as shown in Table 2, including point estimates, the Monte Carlo method, and the adopted uncertainty model.

A quick observation of the resources needed to estimate the contaminant distribution at "end points" favors the adopted uncertainty model. Invoking the maximum entropy principle allows a cost-effective evaluation of the contaminant concentration dis-

Table 2 Characteristics of Uncertainty Risk Analyses

Method	Fundamental principle	Analysis type and "end" results	Size [degrees of freedom]	Number of runs
Deterministic (point estimate)	One-value estimate	Nonlinear mean	6,912 DOF	1
Monte Carlo	Multiple estimates for range of probable values	Nonlinear distribution	6,912 DOF	5,000–10,000
Generalized point estimate	Equivalent discrete probabilities at reaction points	Nonlinear distribution	6,912 DOF	2,048
Stochastic finite element	Karhunen-Loeve expansion	Nonlinear distribution	37,600 DOF	1
Uncertainty model (adopted model)	Max. entropy principle	Nonlinear distribution	6,912 DOF	1

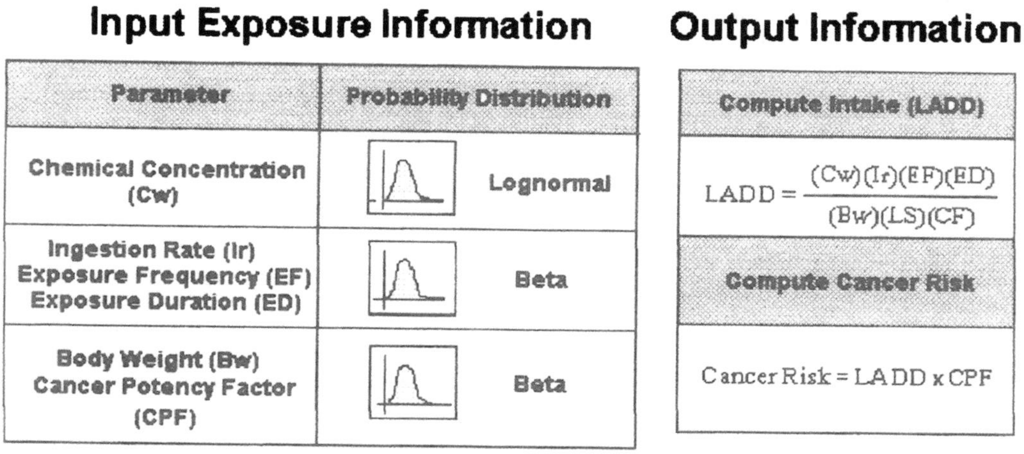

Input Exposure Information

Parameter	Probability Distribution	
Chemical Concentration (Cw)		Lognormal
Ingestion Rate (Ir) Exposure Frequency (EF) Exposure Duration (ED)		Beta
Body Weight (Bw) Cancer Potency Factor (CPF)		Beta

Output Information

Compute Intake (LADD)

$$LADD = \frac{(Cw)(Ir)(EF)(ED)}{(Bw)(LS)(CF)}$$

Compute Cancer Risk

$$Cancer\ Risk = LADD \times CPF$$

Figure 4 Estimating distribution of human daily intakes.

tribution without any computational penalty. At this stage we combine all the various distributions to arrive at the distribution of the "cancer risk" due to contamination of this particular site. This is shown in Fig. 4.

First we compute the lifetime average daily dose (LADD) based on the input distribution of the following parameters: the BTEX concentration obtained from the simulation, the ingestion rate, the exposure frequency, the exposure duration, the body weight, the life span, and the conversion factor. Then the cancer risk is estimated based on the following equation:

$$\text{Cancer risk} = (LADD) * (CPF) \tag{1}$$

where

$LADD$ = lifetime average daily dose (mg/kg/day)

CPF = cancer potency factor (mg/kg/day)

Typical input exposure values for a risk assessment are shown in Table 3. For the exposure parameters we assume log-normal and beta distributions as the most representative of the laboratory tests. These parameters are combined to estimate the cancer risk according to Eq. (1).

From the above it is evident that the simulation engines of the contaminant fate and transport are the key factors in evaluating the site risk exposures and their uncertainties. Therefore it is important to address the theoretical framework of the modeling effort with all the details of the chemical and biochemical interactions of the site contaminants. Because of their toxicity and/or carcinogenicity, emphasis is placed here in examining the behavior of BTEX and chlorinated solvents.

III. BIODEGRADATION OF BTEX COMPOUNDS

Numerous laboratory and field studies have shown that microorganisms indigenous to the subsurface environment can degrade a variety of hydrocarbons, including components of gasoline, kerosene, diesel, and jet fuel. Most petroleum hydrocarbons are biodegradable,

Table 3 Characteristics of Uncertainty Risk Analyses

Parameter	Distribution	Mean	Variance	Min-max range
Chemical concentration (Cw)	Log-normal	From simulation engine (mg/L)		
Ingestion rate (Ir)	Log-normal	2.0 L/day	0.25	
Exposure frequency (EF)	Beta	350 days/year		Min = 250, max = 365
Exposure duration (ED)	Beta	70 years		Min = 9, max = 70
Body weight (Bw)	Log-normal	70 (kg)		
Cancer potency factor (CPF)	Log-normal	0.029 (mg/kg)	0.67	

Table 4 Microorganisms Capable of Degrading Various Hydrocarbons

Contaminant	Microorganisms	Biodegradability
Benzene	*Pseudomonas putida, P. rhodochrous, P. aeruginosa Acinetobacter* sp., *Methylosinus trichosporium* OB3b, *Nocardia* sp., methanogens, anaerobes	Moderate to high
Toluene	*Methylosinus tricliosporium* OB3b, *Bacillus* sp., *Pseudomonas putida, Cunninghamella clegans, P. aeruginosa, P. mildenberger, Pseudomonas aeruginosa, Pseudomonas* sp., *Archromobacter* sp., methanogens, anaerobes	High
Ethylbenzene	*Pseudomonas putida*	High
Xylenes	*Pseudomonas putida*, methanogens, anaerobes	High
Jet fuels	*Cladosporium, Hormodendrum*	High
Kerosene	*Torulopsis, Caiididatropicalis, Corynebacterium hydrocarboclastus, Candida parapsilosis, C. guilliermondii, C. lipolytica, Trichosporon* sp., *Rhohosporidium toruloides, Cladosporium resinae*	High

Source: Modified from Ref. 12.

and under favorable conditions the biodegradation rates of the low- to moderate-weight aliphatic, alicyclic, and aromatic compounds can be very high. As the molecular weight of the compound increases, so does the resistance to biodegradation. Table 4 shows typical microorganisms that are known to degrade petroleum hydrocarbons.

During biodegradation, microorganisms transform nutrients and fuel hydrocarbons into forms useful for energy and cell reproduction. They obtain energy by facilitating the transfer of electrons from electron donors to electron acceptors. This results in oxidation of the electron donor and reduction of the electron acceptor. Electron donors include natural organic material and fuel hydrocarbon compounds. Electron acceptors are elements or compounds that occur in relatively oxidized states. The more important electron acceptors in groundwater include dissolved oxygen, nitrate, Fe(III), sulfate, and

Table 5 Trends in Concentrations During Biodegradation

Analyte	Terminal electron-accepting process	Trend in analyte concentration during biodegradation
BTEX	(contaminant–nutrient)	Decreases
Dissolved oxygen	Aerobic respiration	Decreases
Nitrate	Denitrification	Decreases
Iron(II)	Iron(III) reduction	Increases
Sulfate	Sulfate reduction	Decreases
Methane	Methanogenesis	Increases

carbon dioxide. In addition, Mn(IV) can act as an electron acceptor in some groundwater environments. Biodegradation of organic pollutants occurs when microbially mediated redox reactions cause a pollutant to be oxidized.

Biodegradation causes measurable changes in groundwater geochemistry. During aerobic respiration, oxygen is reduced to water and dissolved oxygen concentrations decrease. In anaerobic systems where nitrate is the electron acceptor, the nitrate is reduced to NO_2^-, N_2O, NO, NH^{4+}, or N_2 and nitrate concentrations decrease. In anaerobic systems where iron(III) is the electron acceptor, it is reduced to iron(II) and iron(II) concentrations increase. In anaerobic systems where sulfate is the electron acceptor, it is reduced to H_2S and sulfate concentrations decrease. In anaerobic systems where CO_2 is used as an electron acceptor, it is reduced by methanogenic bacteria and CH_4 is produced, as summarized in Table 5. Changes in dissolved oxygen, nitrate, iron(II), sulfate, and methane concentrations can be used to ascertain the dominant terminal electron-accepting process.

Fuel hydrocarbons biodegrade naturally when an indigenous population of hydrocarbon-degrading microorganisms is present in the aquifer and sufficient concentrations of electron acceptors and nutrients are available to these organisms. In most subsurface environments, both aerobic and anaerobic degradation of fuel hydrocarbons can occur, often simultaneously in different parts of the plume. For thermodynamic reasons, microorganisms preferentially utilize those electron acceptors that provide the greatest amount of free energy during respiration. The rate of natural biodegradation is generally limited by a lack of electron acceptors rather than by a lack of nutrients such as ammonia, nitrate, or, phosphate. Microorganisms often efficiently recycle nitrogen, phosphorous, and other trace nutrients.

When fuel hydrocarbons are utilized as the primary electron donor for microbial metabolism, they typically are completely degraded or detoxified. When fuel hydrocarbons are not present in sufficient quantities to act as the primary metabolic substrate, the fuel hydrocarbon can still be degraded (secondary utilization) but the microorganisms will obtain the majority of their energy from alternative substrates in the aqueous environment (organic matter).

Subsurface bacteria typically are smaller than their terrestrial counterparts, giving rise to a larger surface area-to-volume ratio. This allows them to utilize nutrients efficiently from dilute solutions. Bacteria that degrade fuel hydrocarbons have been known to withstand fluid pressures of hundreds of bars, pH conditions ranging from 1 to 10 standard units, temperatures from 0 to 75°C, and salinity greater than that of normal sea water.

The biodegradation processes release energy, which is quantified by Gibbs' free energy of the reaction (ΔG_r). The Gibbs free energy defines the maximum useful energy change for a chemical reaction at constant temperature and pressure. The sign of (ΔG_r) defines the state of a redox reaction relative to equilibrium. Negative values indicate that the reaction is exothermic (energy producing) and will proceed from left to right (i.e., reactants will be transformed into products and energy will be produced). Positive values indicate that the reaction is endothermic and, in order for the reaction to proceed from left to right, energy must be put into the system.

Like all living organisms, microorganisms are constrained by the laws of thermodynamics. They can only facilitate those redox reactions that are thermodynamically possible (8). Microorganisms will facilitate only those redox reactions that will yield some energy (i.e., $\Delta G_r < 0$). Microorganisms will not invest more energy into the system than can be released. In order to derive energy for cell maintenance and production from the BTEX compounds, the microorganisms must couple an endothermic reaction (electron donor oxidation) with an exothermic reaction (electron acceptor reduction). Most of the reactions involved in BTEX oxidation cannot proceed abiotically, even though they are thermodynamically favorable. These reactions require microorganisms to proceed. The microorganisms facilitate these redox reactions by providing the necessary activation energy. Without this input, these redox reactions would not occur spontaneously in groundwater.

Microorganisms are able to utilize electron transport systems and chemiosmosis to combine energetically favorable and unfavorable reactions with the net result of producing energy for life processes. Table 6 shows how microorganisms combine endothermic and exothermic reactions to release useful energy for cell maintenance and reproduction. By coupling the oxidation of the BTEX compounds to the reduction of electron acceptors, the overall reaction becomes energy-yielding, as indicated by the negative ΔG_r values. For example, the oxidation of 1 mole of benzene via oxygen reduction process 765.34 kcal of free energy that is available to the microorganisms. Thus, microorganisms derive a significant source of energy by facilitating this coupled reaction.

Table 6 Coupled Benzene Oxidation–Reduction Reactions

Coupled benzene oxidation–reduction reactions	ΔG_r (kcal/mole)	ΔG_r (kJ/mole)	Stoichiometric mass ratio of El. Acc./BTEX	Stoichiometric ratio of metabolic by-products/BTEX
$7.5O_2 + C_6H_6 \rightarrow 6CO_2(g) + 3H_2O$ Aerobic respiration	−765.34	−3202	3.07:1	—
$6NO_3^- + 6H^+ + C_6H_6 \rightarrow 6CO_2(g) + 6H_2O + 3N_2(g)$ Denitrification	−775.75	−3245	4.77:1	—
$30H^+ + 15MnO_2 + C_6H_6 \rightarrow 6CO_2(g) + 15Mn^{2+} + 18H_2O$ Manganese reduction	−765.45	−3202	16.7:1	10.56:1
$60H^+ + 30Fe(OH)_3(g) + C_6H_6 \rightarrow 6CO_2 + 30Fe^{2+} + 78H_2O$ Iron reduction	−560.10	−2343	41:1	21.5:1
$7.5H^+ + 3.75SO_4^{2-} + C_6H_6 \rightarrow 6CO_2(g) + 3.75H_2S + 3H_2O$ Sulfate reduction	−122.93	−514.3	4.61:1	—
$4.5H_2O + C_6H_6 \rightarrow 2.25CO_2(g) + 3.75CH_4$ Methanogenesis	−32.40	−135.6	—	0.77:1

Process	$\Delta G°_r$, Gibbs Free Energy, Benzene Oxidation, kJ/mole
Aerobic Respiration	-3202
Denitrification	-3245
Iron Reduction	-2343
Sulfate Reduction	-514
Methanogenesis	-136

Figure 5 Sequence of microbially mediated redox reactions and Gibbs free energy of the reaction.

Similar coupled reactions can be written for the other hydrocarbon components, toluene, ethylbenzene, and xylene.

Coupled redox reactions would be expected to occur in order of their thermodynamic energy yield, assuming that there are organisms capable of facilitating each reaction and that there is an adequate supply of BTEX and electron acceptors. Figure 5 illustrates the expected sequence of microbially mediated redox reactions. In general, reactions that yield the most energy tend to take precedence over those reactions that yield less energy. Variances occur; e.g., although denitrification yields slightly more energy than aerobic respiration, free dissolved oxygen is toxic to anaerobic bacteria in concentrations in excess of about 0.5 mg/L. Therefore, dissolved oxygen must be removed from the groundwater before denitrification can occur. Once the available dissolved oxygen is depleted and anaerobic conditions dominate the interior regions of the BTEX plume, anaerobic microorganisms can utilize other electron acceptors in the following order of preference: nitrate, manganese, iron(III), sulfate, and carbon dioxide. Stated simplistically, as each electron acceptor being utilized for biodegradation becomes depleted, the next most preferable electron acceptor is utilized. In reality, iron, manganese, and sulfate may alternate cyclically, as the prevailing acidity and toxicity of the medium favors one set of microbes over another.

The expected sequence of redox processes is also a function of the oxidizing potential of the groundwater, which measures the relative tendency of a solution or chemical reaction to accept or transfer electrons. As each subsequent electron acceptor is utilized, the groundwater becomes more reducing and the oxidation–reduction potential of the water decreases. Thus, measurement of the redox potential of the groundwater is important.

Redox potential Eh can be used as a crude indicator of which redox reactions may be operating at a site. Eh can be expressed as pE, which is the hypothetical measure of the electron activity associated with a specific Eh. High pE means that the solution or redox couple has a relatively high oxidizing potential. Figure 5 shows the relationship between the dominant terminal electron acceptor process and the redox potential of the groundwater. The sequence of reactions depicted in Figure 6 would take place if the following conditions were met:

The system is moving toward equilibrium.
Redox reactions occur in order of their energy-yielding potential, provided microorganisms are available to mediate the specific reactions. Reduction of a highly oxidized species decreases the pE of the system.

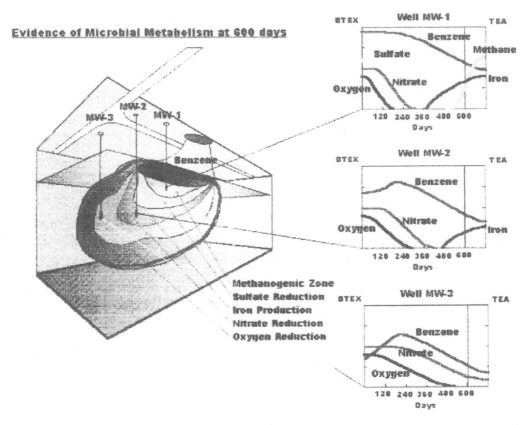

Figure 6 Typical microbial metabolism and redox conditions for LNAPL contaminants.

The sequence is paralleled by an ecological succession of biological mediators.

The reduction of highly oxidized species results in an overall decrease in the oxidizing potential of the groundwater. As shown in Figures 5 and 6, the reduction of oxygen and nitrate will reduce the oxidizing potential to levels where manganese and iron(III) reduction can occur. As each chemical species that can be used to oxidize the BTEX compounds is exhausted, the microorganisms are forced to use electron acceptors with lower oxidizing capacity. When sufficiently negative pE levels have been developed by these redox reactions, sulfate reduction and methane fermentation can occur.

During anaerobic biodegradation, aerobic degradation can still be occurring near the margins of the plume, where dispersion helps spread the plume out into more oxygenated regions of the aquifer. There is unequivocal evidence that microbial processes degrade fuel hydrocarbons. The most important thing to ascertain during the natural degradation demonstration is which mechanisms of biodegradation are most important and the rate at which biodegradation is occurring, because biodegradation rates can be limited by substrate availability as well as by the availability of electron acceptors.

The EIS platform offers two modeling options for the BTEX biodegradation, based on

A simple stoichiometry of the basic chemical constituents

Table 7 Mass Ratio of Electron Acceptors or Metabolic By-products to Total BTEX Degraded

Terminal electron-accepting process	Average mass ratio of electron acceptor to total BTEX	Average mass ratio of metabolic by-product to total BTEX	Mass of BTEX degraded per unit mass of electron acceptor used (mg)	Mass of BTEX degraded per unit mass of metabolic by-product produced (mg)
Aerobic respiration	3.14:1	—	0.32	—
Denitrification	4.9:1	—	0.21	—
Iron(III) reduction	—	21.8:1	—	0.05
Sulfate reduction	4.7:1	—	0.21	—
Methanogenesis	—	0.78:1	—	1.28

Monod kinetics algorithms that also model the mass production of the microorganisms

Details of the Monod kinetics algorithms are given in a subsequent modeling section of EIS. For the first option of a simple stoichiometric model, Table 7 summarizes the mass of total BTEX degraded per milligram of dissolved oxygen utilized. This table shows average values for BTEX according to Table 6 (simple average of all BTEX compounds based on stoichiometry). Based on the stoichiometry presented in Table 6, the ultimate metabolic by-products of aerobic respiration are carbon dioxide and water. However, this stoichiometry presents the final products of aerobic respiration only. In reality, several intermediate products are produced from the BTEX before the end result is achieved.

Actual oxygen requirements may vary from those predicted by the above stoichiometric relationships because they are dependent on the bacterial yield coefficient (Y_m) that describes the amount of biomass produced per unit mass of substrate biodegraded. Yields of microbial biomass vary depending on the thermodynamics of substrate biodegradation and on the availability of oxygen, nutrients, and substrate concentration. The Monod kinetics algorithms allow a better simulation of the above considerations.

IV. DEGRADATION OF CHLORINATED COMPOUNDS

Chlorinated solvents can be degraded both biologically by microorganisms indigenous to the subsurface environment and abiotically (i.e., degradation can also occur without the mediation of microorganisms). During biodegradation, dissolved chlorinated solvents are ultimately transformed into carbon dioxide, methane, and water. The path to complete biodegradation, however, is substantially more complex than that for BTEX compounds. Various biotic and abiotic processes associated with organic compounds are summarized in Table 8. Additional products that chlorinated solvents are degraded to include chloride, methane, and water. More often, intermediate products of these transformations may be more hazardous than the original compound. However, they in turn also tend to be more easily degradable. Ultimately, degradation of dissolved substances in groundwater results in a reduction in contaminant mass (and concentration) and slowing of the contaminant relative to the average advective groundwater velocity.

Table 8 Biologic and Abiotic Degradation Processes for Various Synthetic Organic Compounds

Compound	Degradation processes
PCE	Reductive dechlorination
TCE	Reductive dechlorination, cometabolism
DCE	Reductive dechlorination, direct biological oxidation
Vinyl chloride	Reductive dechlorination, direct biological oxidation
TCA	Reductive dechlorination, hydrolysis, dehydrohalogenation
1,2-DCA	Reductive dechlorination, direct biological oxidation
Chloroethane	Hydrolysis
Carbon tetrachloride	Reductive dechlorination, cometabolism, abiotic
Chloroform	Reductive dechlorination, cometabolism
Methylene chloride	Direct biological oxidation
Chlorobenzenes	Direct biological oxidation, reductive dechlorination, cometabolism
Benzene	Direct biological oxidation
Toluene	Direct biological oxidation
Ethylbenzene	Direct biological oxidation
Xylenes	Direct biological oxidation
1,2-Dibromoethane	Reductive dehalogenation, hydrolysis, direct biological oxidation, and photolysis

A. Abiotic Attenuation (Degradation) Mechanisms

Abiotic reactions are typically not complete and often result in the formation of intermediate by-products that may be at least as toxic as the original contaminant. The most common reactions are hydrolysis (a substitution reaction) and dehydrohalogenation (an elimination reaction). Other possible reactions include oxidation and reduction. However, no abiotic oxidation reactions involving typical halogenated solvents have been reported. Also, reduction reactions (which include hydrogenolysis and dihaloelimination) are commonly mediated microbially, although some abiotic reduction reactions have been observed. In general, attributing changes in either the presence or absence of halogenated solvents or the concentrations of halogenated solvents to abiotic processes is usually difficult. Often, chlorinated solvents may be undergoing both biotic and abiotic degradation, and discerning the relative contribution of each mechanism on the field scale is very difficult, if possible at all. As another example, to substantiate that hydrolysis is occurring, the presence of nonhalogenated breakdown products such as acids and alcohols should be established. In general, these products are more easily biodegraded than their parent compounds and can be difficult to detect.

Given the difficulties of demonstrating abiotic degradation on the field scale, it may not be practical to demonstrate that such processes are occurring and to evaluate the contributions of those reactions quantitatively (i.e., separately from biotic processes). However, while the rates of abiotic degradation may be slow relative to biotic mechanisms, the contribution of these mechanisms may still play a significant role in natural attenuation. Moreover, because some of the by-products of these reactions are chlorinated compounds that may be more easily or less easily degraded than the parent, the contributions of abiotic mechanisms may need to be considered when evaluating analytical data from a site.

Among abiotic attenuation mechanisms, hydrolysis and dehydrohalogenation reactions are the most important. The rates of these reactions are often quite slow within the range of normal groundwater temperatures, with half-lives of days to centuries.

1. Hydrolysis

Hydrolysis is a substitution reaction in which an organic molecule reacts with water or a component ion of water. A halogen substituent is replaced with a hydroxyl (OH^-) group, which leads initially to the production of alcohols. If the alcohols are halogenated, additional hydrolysis to acids or diols may occur. Also, the addition of a hydroxyl group to a parent molecule may make the daughter product more susceptible to biodegradation, as well as more soluble. The likelihood that a halogenated solvent will undergo hydrolysis depends in part on the number of halogen substituents. More halogen substituents on a compound will decrease the chance for hydrolysis reactions to occur. Locations of the halogen substituent on the carbon chain may also have some effect on the rate of reaction. The rate may also increase with increasing pH above a pH of 11.

Hydrolysis rates can generally be described using first-order kinetics, particularly in solutions in which water is the dominant nucleophile. This is a simplification of what is typically a much more complicated relationship. Chlorinated methanes and ethanes hydrolize readily. Monohalogenated alkanes have half-lives on the order of days to months, while polychlorinated methanes and ethanes have half-lives that may range up to thousands of years for carbon tetrachloride (as the number of chlorine atoms increases, dehydrohalogenation may become more important). Chlorinated ethenes and vinyl chloride do not undergo significant hydrolysis reactions. A listing of half-lives

Table 9 Approximate Half-Lives of Abiotic Hydrolysis and Dehydrohalogenation Reactions Involving Chlorinated Solvents (from literature)

Compound	Half-life (years)	Products
Chloromethane	No data	
Methylene chloride (dichloromethane)	704	
Trichloromethane (chloroform)	3500, 1800	
Carbon tetrachloride	41	
Chloroethane	0.12	Ethanol
1,1-Dichloroethane	61	
1,2-Dichloroethane	72	
1,1,1-Trichloroethane	1.7, 1.1	Acetic acid
	2.5	1,1-DCE
1,1,2-Trichloroethane	170, 140	1,1-DCE
1,1,1,2-Tetrachloroethane	380, 47	TCE
1,1,2,2-Tetrachloroethane	0.8, 0.4	1,1,2-TCA
	0.3	TCE
Tetrachloroethene	0.7, 1.3×10^6	
Trichloroethene	0.7, 1.3×10^6	
1,1-Dichloroethene	1.2×10^8	
1,2-Dichloroethene	2.1×10^{10}	

for abiotic hydrolysis and dehydrohalogenation of some chlorinated solvents is presented in Table 9.

One common chlorinated solvent for which abiotic transformations have been well studied is 1,1,1-TCA, which may be abiotically transformed to acetic acid through a series of substitution reactions, including hydrolysis. In addition, 1,1,1-TCA may be reductively dehalogenated to form 1,1-DCA, and then chloroethane (CA), which is then hydrolyzed to ethanol or dehydrohalogenated to vinyl chloride. Rates of these reactions as reported in the literature are summarized in Table 9.

2. Dehydrohalogenation

Dehydrohalogenation is an elimination reaction in which a halogen is removed from one carbon atom, followed by the subsequent removal of a hydrogen atom from an adjacent carbon atom. In this two-step reaction, an alkene is produced. Removal of a halogen decreases the oxidation state of the compound, but the loss of a hydrogen atom increases it. This results in no external electron transfer, and there is no net change in the oxidation state of the reacting molecule. Contrary to the patterns observed for hydrolysis, the likelihood that dehydrohalogenation will occur increases with the number of halogen substituents. Specifically, monohalogenated aliphatics apparently do not undergo dehydrohalogenation, unlike CA and polychlorinated alkanes, which have been observed to undergo dehydrohalogenation under normal conditions.

Dehydrohalogenation rates may also be approximated using pseudo-first-order kinetics as used to quantify the reaction rates. The rates do not depend only upon the number and types of halogen substituent, but also on the hydroxide ion concentration. Under normal pH conditions (near 7), interaction with water (acting as a weak base) may become more important. 1,1,1,-TCA is also known to undergo dehydrohalogenation, in which TCA is transformed to 1,1-DCE, which is then reductively dehalogenated to VC. The VC is then either reductively dehalogenated to ethene or consumed as a substrate in an aerobic reaction and converted to CO_2. In addition to 1,1,1-TCA and 1,1,2-TCA both degrading to 1,1-DCE, the tetrachloroethanes and pentachloroethanes degrade to TCE and PCE, respectively.

3. Reduction Reactions

Two abiotic reductive dechlorination reactions that may operate in the subsurface are hydrogenolysis and dihaloelimination. Hydrogenolysis is the simple replacement of a chlorine (or another halogen) by a hydrogen, while dihaloelimination is the removal of two chlorines (or other halogens) accompanied by the formation of a double carbon–carbon bond. These reactions are thermodynamically possible under reducing conditions, but they often do not take place in the absence of biological activity. Unlike cometabolism, such activity is only indirectly responsible for the reaction, i.e., microbes produce reductants that facilitate such reactions in conjunction with minerals in the aquifer matrix. Also, the reducing conditions necessary to produce such reactions are most often created as a result of microbial activity. So the vision becomes blurred in that, because of their reliance on microbial activity to produce reducing conditions or reactants, they should be considered to be a form of cometabolism.

Of interest is the reductive dehalogenation of chlorinated aliphatics using zero-valent iron, in which the iron serves as an electron donor in an electrochemical reaction. For example, TCE is reduced to ethene and carbon tetrachloride to methane in platinum-catalyzed reactions between elemental magnesium and water. Although

B. Biodegradation of Chlorinated Aliphatic Hydrocarbons

Biodegradation is the most important destructive process acting to reduce chlorinated contaminant concentrations in groundwater. Most important, estimation of the potential for natural biodegradation is useful also for selecting cost-effective remedial alternatives in eliminating or abating contamination in cases where natural attenuation alone is not sufficient. Chlorinated aliphatic hydrocarbons undergo biodegradation through three different pathways: use as an electron acceptor, use as an electron donor, or through co-metabolism, where degradation of the chlorinated organic takes place fortuitously as there is no benefit to the microorganism. One or all of these processes may be operating simultaneously. In most sites chlorinated aliphatic hydrocarbons are used as electron acceptors under natural conditions. In this case, biodegradation of chlorinated aliphatic hydrocarbons will be an electron *donor*-limited process. By comparison, biodegradation of fuel hydrocarbons (BTEX) is an electron *acceptor*-limited process.

Microorganisms are at work whether there is contamination or not. In a pristine aquifer, they use native organic carbon as an electron donor and dissolved oxygen (DO) as the primary electron acceptor. In contaminated areas, anthropogenic carbon (e.g., fuel hydrocarbons) is also used as an electron donor. After the DO is consumed, anaerobic microorganisms typically use additional electron acceptors (as available) in the following order of preference: nitrate, ferric iron oxyhydroxide, sulfate, and finally carbon dioxide. Evaluation of the distribution of these electron acceptors can provide evidence of where and how chlorinated aliphatic hydrocarbon biodegradation is occurring. In addition, because chlorinated aliphatic hydrocarbons may be used as electron acceptors or electron donors (in competition with other acceptors or donors), isopleth maps showing the distribution of these compounds and their daughter products can provide evidence of the mechanisms of biodegradation working at a site. The driving force behind oxidation–reduction reactions resulting in chlorinated aliphatic hydrocarbon degradation is electron transfer. Although thermodynamically favorable, most of the reactions involved in chlorinated aliphatic hydrocarbon reduction and oxidation do not proceed abiotically. Microorganisms are capable of carrying out the reactions, but they will facilitate only those oxidation–reduction reactions that have a net yield of energy.

C. Mechanisms of Chlorinated Aliphatic Hydrocarbon Biodegradation

1. Electron Acceptor Reactions (Reductive Dehalogenation)

The most important process for the natural biodegradation of the more highly chlorinated solvents is reductive dechlorination. The chlorinated hydrocarbon is used as an electron acceptor, not as a source of carbon, and a chlorine atom is removed and replaced with a hydrogen atom. In general, reductive dechlorination occurs by sequential dechlorination from perchloroethene to trichloroethene to dichloroethene to vinyl chloride to ethene. Depending on environmental conditions, this sequence may be interrupted, with other processes then acting upon the by-products. During reductive dechlorination, all three isomers of dichloroethene can theoretically be produced, but *cis*-1,2-dichloroethene is a more common intermediate than *trans*-1,2-dichloroethene, and 1,1-dichloroethene is the least prevalent of the three dichloroethene isomers when they are present as daughter products. Reductive dechlorination of chlorinated solvent compounds is associated with the accumulation of daughter products and an increase in the concentration of chloride ions.

Reductive dechlorination affects each of the chlorinated ethenes differently. Of these compounds, perchloroethene is the most susceptible to reductive dechlorination because it is the most oxidized. Conversely, vinyl chloride is the least susceptible to reductive dechlorination because it is the least oxidized of these compounds. The rate of reductive dechlorination also has been observed to decrease as the degree of chlorination decreases. This rate decrease may explain the accumulation of vinyl chloride in perchloroethene and trichloroethene plumes that are undergoing reductive dechlorination. Reductive dechlorination has been demonstrated under nitrate- and iron-reducing conditions, but the most rapid biodegradation rates, affecting the widest range of chlorinated aliphatic hydrocarbons, occur under sulfate-reducing and methanogenic conditions. Because chlorinated aliphatic hydrocarbon compounds are used as electron acceptors during reductive dechlorination, there must be an appropriate source of carbon for microbial growth in order for this process to occur. Potential carbon sources include natural organic matter, fuel hydrocarbons, or other organic compounds such as those found in landfill leachate.

2. Electron Donor Reactions

Trichloroethene and perchloroethene cannot be used as a primary substrate (electron donor) by microorganisms. Less oxidized chlorinated aliphatic hydrocarbons, on the other hand (e.g., vinyl chloride), can be used as the primary substrate in biologically mediated oxidation–reduction under both aerobic and anaerobic conditions. In this type of reaction, the facilitating microorganism obtains energy and organic carbon from the degraded chlorinated aliphatic hydrocarbon, as is the case for fuel hydrocarbon biodegradation. Dichloromethane also has the potential to function as a primary substrate under either aerobic or anaerobic environments. In addition, there has been evidence of mineralization of vinyl chloride under iron-reducing conditions so long as there is sufficient bioavailable iron(III). Aerobic metabolism of vinyl chloride may be characterized by a loss of vinyl chloride mass and a decreasing molar ratio of vinyl chloride to other chlorinated aliphatic hydrocarbon compounds.

a. Thermodynamic Considerations. Electron transfer results in oxidation of the electron donor and reduction of the electron acceptor and the production of usable energy. The energy produced by these reactions is quantified by the Gibbs free energy of the reaction (ΔG_r), which is given by

$$\Delta G_r^0 = \sum \Delta G_{f,\text{products}}^0 - \sum \Delta G_{f,\text{reactants}}^0 \tag{2}$$

where

ΔG_r = Gibbs free energy of the reaction at standard state

$\Delta G_{f,\text{products}}$ = Gibbs free energy of formation for products at standard state

$\Delta G_{f,\text{reactants}}$ = Gibbs free energy of formation for the reactants at standard state

The (ΔG_r) defines the maximum useful energy change for a chemical reaction at a constant temperature and pressure. Table 10 presents coupled oxidation–reduction reactions. In general, those reactions that yield the most energy tend to take precedence over less energy-yielding reaction.

3. Co-metabolism

When a chlorinated aliphatic hydrocarbon is biodegraded via co-metabolism, an enzyme or co-factor that is fortuitously produced by the organisms for other purposes catalyzes the degradation. The organism receives no known benefit from the degradation of the chlorinated aliphatic hydrocarbon; in fact, the co-metabolic degradation of the chlorinated aliphatic hydrocarbon may even be harmful to the microorganism responsible for the production of the enzyme or co-factor. Co-metabolism is best documented in aerobic environments, although it potentially could occur under anaerobic conditions. It has been reported that under aerobic conditions chlorinated ethenes, with the exception of perchloroethene, are susceptible to co-metabolic degradation. Furthermore, the co-metabolism rate increases as the degree of dechlorination decreases. During co-metabolism, bacteria indirectly transform trichloroethene as they use BTEX or another substrate to meet their energy requirements. Therefore, trichloroethene does not enhance BTEX or other carbon source degradation, nor does co-metabolism interfere with the use of electron acceptors involved in the oxidation of those carbon sources. Similar sets of coupled reactions can be written for toluene, ethylbenzene, xylene, naphthalene, and other chlorinated compounds.

D. Overall Evolution of Chlorinated Solvent Plumes

Chlorinated solvent plumes, depending on the amount of solvent, the amount of biologically available organic carbon in the aquifer, the distribution and concentration of natural electron acceptors, and the types of electron acceptors, can exhibit three types of evolution. Often, plumes may exhibit all three types of evolution simultaneously in different portions of the plume.

1. Evolution in the Presence of Anthropogenic Carbon (BTEX)

This situation arises where the primary substrate is anthropogenic carbon (BTEX, landfill leachate). Microbial degradation of this anthropogenic carbon drives reductive dechlorination. This evolution can be detected in the following situations:

1. When the electron donor supply is adequate to allow microbial reduction of the chlorinated organic compounds. Then, what are the microorganisms going to run out of first, the chlorinated aliphatic hydrocarbons (electron acceptors) or the primary substrate (anthropogenic carbon)?
2. Is there a series of competing electron acceptors [e.g., dissolved oxygen, nitrate, iron(III) and sulfate]?
3. What happens to vinyl chloride (a by-product), is it oxidized, reduced, or does it accumulate?

The presence of anthropogenic carbon leads to the rapid and extensive degradation of the more highly chlorinated solvents such as perchloroethene, trichloroethene, and dichloroethene.

Table 10 Example Coupled Oxidation–Reduction Reactions

Coupled benzene oxidation–reduction reactions	ΔG_r^0 (kcal/mole)	ΔG_r^0 (kJ/mole)	Stoichiometric mass ratio of El. Acc./primary substrate	Mass substrate/mass El. Acc. or metabolic by-product
$7.5O_2 + C_6H_6 \rightarrow 6CO_2(g) + 3H_2O$ Aerobic respiration	−765.34	−3202	3.07:1	0.326:1
$6NO_3^- + 6H^+ + C_6H_6 \rightarrow 6CO_2(g) + 6H_2O + 3N_2(g)$ Denitrification	−775.75	−3245	4.77:1	0.210:1
$30H^+ + 15MnO_2 + C_8H_6 \rightarrow 6CO_2(g) + 15Mn^{2+} + 18H_2O$ Manganese reduction	−765.45	−3202	10.56:1	0.095:1
$7.5H^+ + 0.75H_2O + 3.75NO_3 + C_6H_6 \rightarrow 6CO_2(g) + 15Mn^{2+} + 18H_2O$ Nitrate reduction	−524.1	−2193	2.98:1	0.336:1
$60H^+ + 30Fe(OH)_3(a) + C_6H_6 \rightarrow 6CO_2 + 30Fe^{2+} + 78H_2O$ Iron reduction	−560.10	−2343	21.5:1	0.047:1
$7.5H^+ + 3.75SO_4^{2-} + C_6H_6 \rightarrow 6CO_2(g) + 3.75H_2S + 3H_2O$ Sulfate reduction	−122.93	−514.8	4.61:1	0.217:1
$4.5H_2O + C_6H_6 \rightarrow 2.25CO_2(g) + 3.75CH_4$ Methanogenesis	−32.40	−135.6	0.77:1	1.30:1
$15C_2H_2Cl_4 + 12H_2O + C_6H_6 \rightarrow 6CO_2(g) + 15C_2H_3Cl_3 + 15H^+ + 15Cl^-$ PCA reduction	−374.56	−1570	31.9:1	0.03:1
$15C_2H_3Cl_3 + 12H_2O + C_6H_6 \rightarrow 6CO_2(g) + 15C_2H_4Cl_2 + 15H^+ + 15Cl^-$ TCA reduction	−377.86	−1580	25.4:1	0.04:1
$15C_2H_4Cl_2 + 12H_2O + C_6H_6 \rightarrow 6CO_2(g) + 15C_2H_5Cl + 15H^+ + 15Cl^-$ DCA reduction	−337.96	−1410	18.8:1	0.05:1

EIS Macroengineering Approach

Reaction			
$15C_2Cl_4 + 12H_2O + C_6H_6 \rightarrow 6CO_2(g) + 15C_2HCl_3 + 15H^+ + 15Cl^-$ Tetrachloroethylene reductive dehalogenation	−358.64	−1500	31.8:1	0.03:1
$15C_3HCl_3 + 12H_2O + C_6H_6 \rightarrow 6CO_2(g) + 15C_3H_2Cl_2 + 15H^+ + 15Cl^-$ Trichloroethylene reductive dehalogenation	−350.04	−1465	25.2:1	0.04:1
$15C_2H_2Cl_2 + 12H_2O + C_6H_6 \rightarrow 6CO_2(g) + 15C_2H_3Cl + 15H^+ + 15Cl^-$ cis-Dichloroethylene reductive dehalogenation	−278.64	−1166	18.6:1	0.05:1
$15C_2H_3Cl + 12H_2O + C_6H_6 \rightarrow 6CO_2(g) + 15C_2H_4 + 15H^+ + 15Cl^-$ Vinyl chloride reductive dehalogenation	−327.37	−1370	11.9:1	0.08:1
Coupled vinyl chloride oxidation-reduction reactions				
$2.5O_2 + C_3H_3Cl \rightarrow 2CO_2(g) + H_2O + H^+ + Cl^-$ aerobic respiration	−288.98	−1209	1.29:1	0.32:1
$2NO_3^- + H^+ + C_3H_3Cl \rightarrow 2CO_2(g) + 2H_2O + N_2(g) + Cl^-$ Denitrification	−292.44	−1224	2.00:1	0.21:1
$9H^+ + 5MnO_2 + C_3H_3Cl \rightarrow 2CO_2(g) + 5Mn^{2+} + 6H_2O + Cl^-$ Manganese reduction	−289.01	−1209	7.02:1	0.09:1
$19H^+ + 10Fe(OH)_3(a) + C_3H_3Cl \rightarrow 2CO_2 + 10Fe^{2+} + 26H_2O + Cl^-$ Iron reduction	−229.65	−960.9	17.3:1	0.05:1
$1.5H^+ + 1.25SO_4^{2-} + C_3H_3Cl \rightarrow 2CO_2(g) + 1.25H_2S + H_2O + Cl^-$ Sulfate reduction	−76.40	−319.7	1.94:1	0.21:1
$1.5H_2O + C_3H_3Cl \rightarrow 0.75CO_2(g) + 1.25CH_4 + H^+ + Cl^-$ Methanogenesis	−44.62	−186.7	0.44:1	1.28:1
$5C_2Cl_4 + 4H_2O + C_3H_3Cl \rightarrow 2CO_2(g) + 5C_2HCl_3 + 6H^+ + 6Cl^-$ Tetrachloroethylene reductive dehalogenation	−153.39	−641.8	13.4:1	0.03:1
$5C_3HCl_3 + 4H_2O + C_6H_5Cl \rightarrow 2CO_2(g) + 5C_3H_2Cl_2 + 6H^+ + 6Cl^-$ Trichloroethylene reductive dehalogenation	−150.54	−629.9	10.6:1	0.04:1
$5C_2H_2Cl_2 + 4H_2O + C_6H_5Cl \rightarrow 2CO_2(g) + 5C_2H_3Cl + 6H^+ + 6Cl^-$ cis-Dichloroethylene reductive dehalogenation	−126.74	−530.3	7.82:1	0.05:1

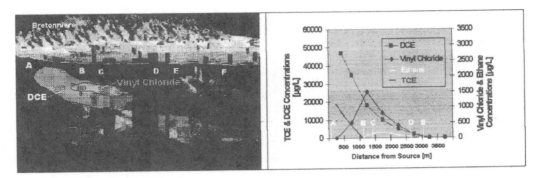

Figure 7 Typical microbial metabolism and redox conditions for chlorinated solvents.

2. Evolution in the Presence of Native Organic Carbon

This situation arises when there is a relatively high concentration of biologically available native organic carbon. Microbial utilization of this natural carbon source drives reductive dechlorination (i.e., it is the primary substrate for microorganism growth). This situation generally results in slower biodegradation of the highly chlorinated solvents than in the case of anthropogenic carbon, but under the right conditions (e.g., areas with high natural organic carbon contents), it also can result in efficient degradation of these compounds.

3. Evolution in the Absence of Anthropogenic or Native Carbon

In the absence of native and/or anthropogenic carbon, even with concentrations of dissolved oxygen greater than 1.0 mg/L, aerobic reductive dechlorination will still not occur (there is no removal of perchloroethene, trichloroethene, and dichloroethene). Only advection, dispersion, and sorption will cause natural attenuation. However, vinyl chloride can be rapidly oxidized under these aerobic conditions which favor cometabolism. This situation is depicted in Fig. 7.

E. Mixed Evolution

Often, a single chlorinated solvent plume can exhibit all three types of evolution in different portions of the plume. This can be beneficial for natural biodegradation of chlorinated aliphatic hydrocarbon plumes. Typically, a plume can co-metabolize with BTEX in the source area, then as BTEX is degraded allow vinyl chloride to degrade down-gradient from the source. A fortuitous scenario would involve a plume in which perchloroethene, trichloroethene, and dichloroethene are reductively dechlorinated in the presence of BTEX near the source, then vinyl chloride is oxidized, either aerobically or via iron reduction to carbon dioxide away from the source, and does not accumulate. Diagrammatically we have:

Perchloroethene → trichloroethene → dichloroethene → vinyl chloride →

carbon dioxide

Often, trichloroethene, dichloroethene, and vinyl chloride may attenuate at approximately the same rate, and thus these reactions may be confused with simple dilution. Note that no

Table 11 Sequence of Microbially Mediated Redox Reactions and Gibbs Free Energy of the Reaction

Process		ΔG_r^0, Gibbs free energy, benzene oxidation (kJ/mole)
Natural degradation	Anthropogenic	
Aerobic respiration		−3202
Denitrification		−3245
Iron reduction		−2343
	PCE reduction	−1500
	TCE reduction	−1465

ethene is produced during this reaction and vinyl chloride is removed from the system much faster under these conditions than it is under vinyl chloride-reducing conditions.

A less desirable scenario, but one in which all contaminants may be entirely biodegraded, involves a plume in which all chlorinated aliphatic hydrocarbons are reductively dechlorinated in the presence of organic carbon. Vinyl chloride is reduced to ethene, which may be further reduced to ethane or methane. The following sequence of reactions occurs in this type of plume:

Perchloroethene → trichloroethene → dichloroethene → vinyl chloride → ethene → ethane

In this case, however, vinyl chloride degrades more slowly than trichloroethene, and thus tends to accumulate.

Table 11 illustrates the expected sequence of microbially mediated redox reactions based on the Gibbs energy (ΔG_r). As can be seen, there is sufficient energy in the reaction of fuel hydrocarbons with chlorinated solvents to allow their use by microorganisms as physiological electron acceptors.

For spills of chlorinated aliphatic hydrocarbons, natural attenuation alone can be protective of human health and the environment for only a small fraction of contaminated sites. Most important, the data required to make the preliminary assessment of natural attenuation can also be used to aid the design of engineered remedial solutions if natural attenuation is not sufficient.

Of course, nature seldom recognizes man's classifications and nomenclature, and reactions happen not discretely, but rather in a continuum in 3D space. That is, part of a simulation model's prediction must include the development of the above-described zones of evolution. The EIS platform offers the Biorem3D groundwater simulator, which tracks, in addition to advection, dispersion, and sorption, the reactions and interactions of dozens of dissolved species contaminants, electron acceptors, nutrients, and daughter products. It is a numerical model in which time stepping is scaled to properly pace the advective, dispersive, and reactive processes. An operator-split procedure then allows isolating the reactive (time-dependent) processes from the spatial, advective, and dispersive processes. A system of ordinary differential equations is then solved to "sort out" the various reactive-kinetic rates and account for contaminant degradation and creation of by-products. Details of these kinetics algorithms are given in the next section on modeling.

V. THEORETICAL FRAMEWORK OF EIS MODELING

Soil constituents such as minerals, amorphous materials (e.g., metal oxides), soil organics, and carbonates, interact with each other in the presence of water with its dissolved solutes. The interactive processes involve either physical or chemical reactions or a combination of both. Figure 8 shows a possible configuration of interacting soil constituents.

The physical reactions are generally electrostatic. Attractive and repulsive forces develop between the soil constituents and also between the dissolved solutes in the pore water. Clay particles, for example, interact through the layers of adsorbed water, through the diffuse layers of exchangeable cations, and in some cases through direct particle contact. In the processes that result in chemical reactions, we are more concerned with chemical reactions between the constituents and the dissolved solutes. The surface reactivity of the soil constituents derives from the surface functional groups, which are the chemically reactive molecular units (organic or inorganic) attached to the surfaces of the constituents.

As can be observed, even for a small, representative volume ($1000\,\mu m^3$) corresponds a large number of interactions (~ 100) derived by the existing natural processes. A typical volume of contaminated soil in the unsaturated zone exceeds $100,000\,m^3$. Then the number of physical and chemical interactions exceeds 10^{17} interactions, a really huge number. Even with the latest computer advances, tracking down all these interactions is a huge, intractable task. An approach that is more appealing to the engineer is to lump together all soil constituents inside a variable representative volume according to the site needs.

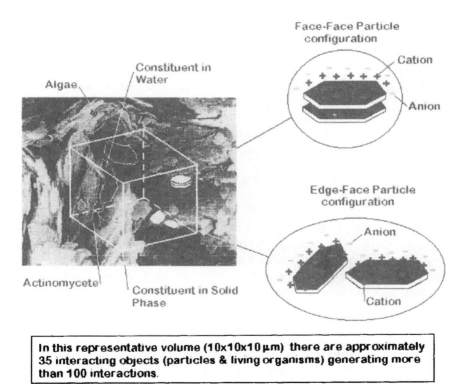

In this representative volume ($10\times10\times10\,\mu m$) there are approximately 35 interacting objects (particles & living organisms) generating more than 100 interactions.

Figure 8 Interacting objects in a representative volume.

EIS Macroengineering Approach

Now, instead of microinteractions we deal with macrointeractions following the definitions introduced below.

A. Notation and Definitions

The system in general is multicomponent and multiphase in nature and evolves in a variably saturated zone. Chemical transport within this zone means movement of the particular chemical species with respect to the other species. If the movement takes place within a single phase it is termed *intraphase transport*, and if it takes place between phases it is termed *interphase transport*.

1. Referential System

Chemical movement implies a species change of position with respect to a fixed point in space. A fixed 3D referential system is usually used, anchored at the ground surface, with the vertical dimension pointing downwards (Fig. 9).

2. Concentration and Phase Density

The existence of a chemical species within any phase of the environment is measured by its concentration. In multiphase systems, the concentrations of chemicals are "naturally" expressed in molar units. Mole fraction of chemical or microbial species (j) represents the moles of (j) with respect to the total molar quantity of the phase. The mole fractions represent the particular phase; specific definitions and notation are the following:

x_j = mole fraction of chemical species j in water (mol/mol)
y_j = mole fraction of chemical species j in air (mol/mol)

The mol in the denominator is of the mixture. Other mole fractions, such as species j in oil, are defined when needed. Note that mole fractions are subscripted, not to be confused with

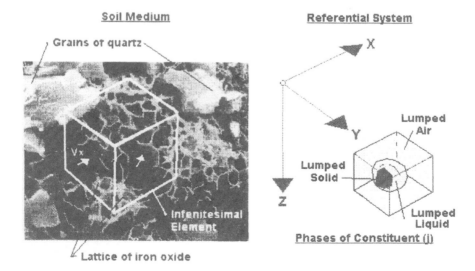

Figure 9 Adopted referential system.

space variables. The simplest multicomponent system is a binary system consisting of species i and j. For this system *the law of the whole* (i.e., the sum of the parts) requires that

$$x_i + x_j = 1 \qquad y_i + y_j = 1$$

In general, for an N-component system we have

$$\sum_{i=1}^{N} x_i = 1 \qquad \sum_{i=1}^{N} y_i = 1$$

Molar concentration is also a convenient quantitative measure of chemical species intensity. Using the symbol c for concentration,

$c_{1i} =$ molar concentration of species i in air (mol/L^3)
$c_{2i} =$ molar concentration of species i in water (mol/L^3)

The volume (L^3) in the denominator is that of the mixture. The second subscript, when it is an integer, denotes phase j with 1 for air, 2 for water, and 3 for soil. Species concentrations in other phases are defined by employing additional integers in a similar fashion when they are needed. The sum of the molar concentrations in a phase is the molar phase density. For the case of air,

$$c_1 = c_{11} + c_{12} + \cdots + c_{1N}$$

Similar molar densities exist for water.

Mass units are also useful and convenient for expressing concentration of environmentally important chemical species. The mass fraction of species i represents the mass of i with respect to the total mass quantity of the phase. The mass fraction should also represent a particular phase. Specific definitions and notations are

$\phi_i =$ mass fraction of species i in water (M/M)
$\psi_i =$ mass fraction of species i in air (M/M)
$\omega_i =$ mass fraction of species i in soil (M/M)

The mass (M) in the denominator is that of the mixture. Other mass fractions can be defined if needed. Mass concentration is denoted by ρ; specifically,

$\rho_{i1} =$ mass concentration of species i in air (M/L^3)
$\rho_{i2} =$ mass concentration of species i in water (M/L^3)
$\rho_{i3} =$ mass concentration of species i in soil (M/L^3)

The volume (L^3) in the denominator is that of the mixture.

Mass and molar concentrations are related by $c_i = \rho_i / M_i$. Just as for mole fraction and molar concentrations, the law of the whole demands that

$$\sum_{i=1}^{N} \varphi_i = \sum_{i=1}^{N} \psi_i = \sum_{i=1}^{N} \omega_i = 1$$

The sum of the mass concentration in a phase is the mass-phase density. In the case of air,

$$\rho_1 = \rho_{11} + \rho_{12} + \cdots + \rho_{1N}$$

3. Parts per Million, Parts per Billion in the Unsaturated Zone

Parts per million, parts per billion, and so on, in weight, volume, and mole ratios, are used for chemical concentrations. Sometimes the ratios are unclear. The following are commonly used. For gases, ppm is used on a volume basis (vol/vol). For liquids, both weight (wt/wt), and volume is used. For solids, a weight basis is normally used.

It is best for chemical concentrations in fluids to use a mass (or mole) per unit volume basis. For gases (including air), temperature and pressure must be specified precisely, so that the volume represents a known quantity of substance. The familiar standard conditions of chemistry and physics (standard temperature and pressure, STP) are 0°C and 760 mmHg and the molar volume of any gas is 22.4 L/mol. All gas concentrations given either as c_{il} or ρ_{il} should be referenced to STP or some other standard. Although most liquid volumes, including water, change only slightly with temperature and pressure, a precise definition is needed. Water at 1.0 atm pressure and 4°C has a mass volume of 1.0 L/kg, and is a natural point of reference.

The use of ppm, ppb, on the other hand, is ideal for specifying chemicals associated with soil or other solids, as long as weight ratios are used (g A/10e6 g soil). This measure is imprecise unless the soil is on a water-free (unbound) basis. Reporting and measuring concentrations with the soil dried in air for a specified time at 1 atm and 100°C will establish a precise standard for this case. Simple conversions exist between ppm, and ppb, and concentrations measured in SI units, provided that ppm and ppb are defined precisely for each phase.

Therefore, at this stage we have a precise mechanism to lump and quantify all constituents that exist in a representative volume. What remains to be done is to identify the natural processes that influence the various interactions.

B. Natural Processes Influencing a Representative Volume

Dissolved contaminants create a plume that expands by advection, dispersion, sorption, volatilization, and biodegradation. Complexation and chemical precipitation are also considered in the saturated zone. Integration of all these interacting natural processes into a concise model is a complex undertaking. The guiding light is to develop a model of the contaminant fate and transport that stresses those aspects of the problem that are important and omits all properties considered nonessential. Following these lines of thought, the general theory of the fate and transport of constituents is presented based on the law of mass conservation.

C. Mass Balance, the Way to Tune Rates of Interacting Processes

The driving goal is to analyze all the relevant processes simultaneously. To this end, the concept of mass balance serves as the means to link all interacting processes together. Mass balance is applied to "control volumes" (macroelements) which are linked with each other and with the rest of the world by fluxes. Next, for each macroelement, each chemical and microorganism, a mass balance equation is written:

$$\begin{array}{c}\text{Change of mass}\\\text{in macroelement}\\\text{with time}\end{array} = \begin{array}{c}\text{Sum of all}\\\text{inputs}\end{array} + \begin{array}{c}\text{Sum of all}\\\text{internal}\\\text{sources}\end{array} - \begin{array}{c}\text{Sum of all}\\\text{outputs}\end{array} - \begin{array}{c}\text{Sum of all}\\\text{internal sinks}\end{array}$$

(3)

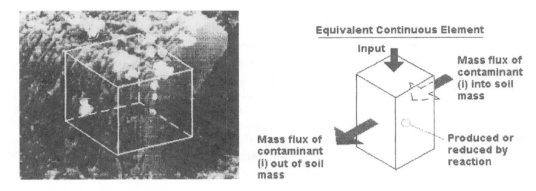

Figure 10 General configuration of macroelement and mass fluxes.

In mathematical terms for the air phase we have:

$$\frac{dM_{1i}}{dt} = I_{1i} + P_{1i} - O_{1i} - R_{1i} \tag{4a}$$

In mathematical terms for the liquid phase we have:

$$\frac{dM_{2i}}{dt} = I_{2i} + P_{2i} + O_{2i} - R_{2i} \tag{4b}$$

where M_{2i} is the mass of the chemical (i) in the macroelement for the liquid phase, and I_{2i}, P_{2i}, O_{2i}, R_{2i} are the sums of the rates of all inputs, internal production processes (sources), outputs, and internal removal mechanisms (sinks) of the chemical (i) within the macroelement (see Fig. 10). If the groundwater system is chosen to have only one macroelement, all inputs and outputs are fluxes across the system boundary. The right-hand side of Eq. (3) includes the following interacting processes (for the liquid phase; the air phase is similar):

$(I_{2i} - O_{2i})$ is the net mass balance of transport processes affecting constituent (i), which are the following:

$$(I_{2i} - O_{2i}) = (I_i - O_i)_{\text{ADV}} + (I_i - O_i)_{\text{DIS}} + (I_i - O_i)_{\text{SOR}} + (I_i - O_i)_{\text{INF}} + (I_i - O_i)_{\text{VOL}}$$
$$+ (I_i - O_i)_{\text{PAR}}$$

where

- $(I_i - O_i)_{\text{ADV}} = dM_{\text{ADV}}/dt$: The net mass balance due to advection of constituent (i) entering and exiting the macroelement during time dt. Expresses the liquid interaction of the constituent with the soil matrix. [Advection process].
- $(I_i - O_i)_{\text{DIS}} = dM_{\text{DIS}}/dt$: The net mass balance due to dispersion of constituent (i) entering and exiting the macroelement during time dt. Expresses the flow interaction of the constituent with the soil matrix. [Diffusion and Dispersion processes]
- $(I_i - O_i)_{\text{SOR}} = dM_{\text{SOR}}/dt$: The net mass balance due to sorption of constituent (i) entering and exiting the macroelement during time dt. Expresses the chemical interaction of the constituent with the soil matrix. [Sorption process]

$(I_i - O_i)_{\text{INF}} = dM_{\text{INF}}/dt$: The net mass balance due to infiltration of constituent (i) entering and exiting the macroelement in the vertical direction during time dt. [Infiltration process]

$(I_i - O_i)_{\text{VOL}} = dM_{\text{VOL}}/dt$: The net mass balance due to volatilization of constituent (i) entering and exiting the macroelement during time dt. Expresses the aqueous-vapor chemical interaction of the constituent. [Volatilization process]

$(I_i - O_i)_{\text{PAR}} = dM_{\text{POR}}/dt$: The net mass balance due to partitioning of constituent (i) entering and exiting the macroelement during time dt. Expresses the dissolution of free product through a chemical interaction producing constituent (i). [Partitioning process]

P_{2i} is the generation of by-products due to reactions with terminal electron acceptors. Expresses the biochemical interaction between constituent (i), electron acceptors, and by-product. [Biodegrading processes]

R_{2i} is the reduction due to reactions with terminal electron acceptors. Expresses the biochemical interaction between constituent (i), electron acceptors, and by-products. [Biodegrading processes]

Then,

$$P_{2i} - R_{2i} = dM_{\text{BIO}}/dt$$

If the system is subdivided into several macroelements, some of the input and output terms are external forces and others are fluxes connecting different macroelements and thus belong to the internal structure of the system.

Assuming that we have included all natural processes of the contaminant fate and migration, Eq. (3) can be rewritten as

$$\frac{dM_{2i}}{dt} = \frac{(dM_{2i})_{\text{ADV}}}{dt} + \frac{(dM_{2i})_{\text{DIS}}}{dt} + \frac{(dM_{2i})_{\text{SOR}}}{dt} + \frac{(dM_{2i})_{\text{BIO}}}{dt} + \frac{(dM_{2i})_{\text{INF}}}{dt} + \frac{(dM_{2i})_{\text{VOL}}}{dt} + \frac{(dM_{2i})_{\text{PAR}}}{dt} \quad (5)$$

We are looking specifically at changes in the concentration of solutes in groundwater as it moves through the soil medium. The water fluid moves through fractures, discontinuous micropores, and spaces along grain boundaries, Fig. 10. To express the change in concentration of a constituent (i) in the moving fluid, we introduce the *partial mass density*, C_{2i}, defined as the mass of constituent (i) per volume of water flow. At the scale of the differential element the time rate of change of the concentration of (i) in the fluid is written in terms of a partial differential equation as $(\partial \theta C_{2i}/\partial t)_{\text{fluid}}$. Note that this approach allows the contribution of all the various natural processes that take place at different geometric scales. Indeed dividing both sides of Eq. (5) by [kg] of groundwater, we obtain an expression that is independent of the particular geometry of a macroelement:

$$\frac{\partial \theta C_{2i}}{\partial t} = \frac{\partial \theta C_{\text{ADV}}}{\partial t} + \frac{\partial \theta C_{\text{DIS}}}{\partial t} + \frac{\partial \theta C_{\text{SOR}}}{\partial t} + \frac{\partial \theta C_{\text{BIO}}}{\partial t} + \frac{\partial \theta C_{\text{INF}}}{\partial t} + \frac{\partial \theta C_{\text{VOL}}}{\partial t} + \frac{\partial \theta C_{\text{PAR}}}{\partial t} \quad (6)$$

where

C_{2i} = solute transport concentration of constituent (i)
θ = volumetric water content
t = time
C_{ADV} = concentration due to advection
C_{DIS} = concentration due to dispersion

C_{SOR} = concentration due to sorption
C_{BIO} = concentration due to biodegradation
C_{INF} = concentration due to infiltration
C_{VOL} = concentration due to volatilization
C_{PAR} = concentration due to partitioning

Combinations of these processes are implemented in the EIS platform and reflect the contaminant migration process through an equivalent continuous medium. Similar equations are derived by Bear (9) and Domenico (10).

All these processes are measured in situ with devices that often do not capture the scale in which they evolve. These raw site data need to be further sorted and reduced to derive the characteristic modeling parameters of these processes. It is therefore important to know how to use and manage these raw data for modeling. Details can be found in the "EIS Software Protocol."

D. Overall Governing Equation for the Liquid Phase

Adding the above relations describing the natural processes in Eq. (6) leads to the following governing equation for the liquid phase:

$$\frac{\partial \theta C_{2i}}{\partial t} = -\frac{v_{2h}}{R_2}\frac{\partial \theta C_{2i}}{\partial h} - \frac{v_{2z}}{R_2}\frac{\partial \theta C_{2i}}{\partial z} + \frac{D_{2x}}{R_2}\frac{\partial^2 \theta C_{2i}}{\partial x^2} + \frac{D_{2y}}{R_2}\frac{\partial^2 \theta C_{2i}}{\partial y^2} + \frac{D_{2z}}{R_2}\frac{\partial^2 \theta C_{2i}}{\partial z^2} - \frac{\partial \theta C_{2i(BIO)}}{\partial t} + F_{2INF} + F_{2VOL} + F_{2PAR}$$

(where the first two terms are Advection, the next three Dispersion with Sorption via R_2, and the biodegradation term is labeled Biodegradation)

(7)

where
  C_{2i} = solute transport concentration
  θ = volumetric water content
  t = time
  v_{2h} = horizontal water seepage velocity
  v_{2z} = vertical water seepage velocity
  D_{2x} = hydrodynamic dispersion along flow path
  D_{2y} = hydrodynamic dispersion transverse to flow path
  R_2 = coefficient of retardation in the water medium

$\dfrac{\partial \theta C_{BIO}}{\partial t}$ = biodegradation term in the water medium

F_{2INF} = explicitly defined time function of the concentrations due to water "infiltration"

F_{2VOL} = explicitly defined time function of the concentrations due to "volatilization"

F_{2PAR} = explicitly defined time function of the concentrations due to "partitioning"

Similar equations are written for the air phase.

E. Solution of the Governing Equations

The numerical solution of each process is addressed separately in a Eulerian/Lagrangian framework, where by operator splitting, some processes are solved on a fixed (Eulerian) grid, whereas others use a moving framework. These procedures are general; that is, they are independent of the geometric discretization. A variable discrete element approach is adopted to discretize the unsaturated medium between the surface and the groundwater. Figure 11 illustrates the plume geometry and the overall configuration of the multiple vertical sources. Note that in the EIS platform the user needs only to determine the "contaminant" source, and everything else is computed automatically by the program.

The 3D solution of Eq. (7) leads to the time integration, Eq. (8). The first and second terms on the right-hand side depend on the contaminant fluxes through the source area, Fig. 11. The third and fourth terms depend on conditions applied throughout the entire semiinfinite domain of the aquifer and include the effects of biodegradation, infiltration, and volatilization. The concentration at any point in the semiinfinite domain defined by the referential system $(OXYZ)$ is computed considering the following steps.

$$dC_i = \int_{t_0}^{t} \left\{ -\frac{v_{2z}}{R_2}\frac{\partial \theta C_{2i}}{\partial z} + \frac{D_{2x}}{R_2}\frac{\partial^2 \theta C_{2i}}{\partial x^2} + \frac{D_{2y}}{R_2}\frac{\partial^2 \theta C_{2i}}{\partial y^2} + \frac{D_{2z}}{R_2}\frac{\partial^2 \theta C_{2i}}{\partial z^2} \right\} dt$$
$$+ \int_{t_0}^{t} \left\{ -\frac{v_{1z}}{R_1}\frac{\partial \theta_a C_{1i}}{\partial z} + \frac{D_{1x}}{R_1}\frac{\partial^2 \theta_a C_{1i}}{\partial x^2} + \frac{D_{1y}}{R_1}\frac{\partial^2 \theta_a C_{1i}}{\partial y^2} + \frac{D_{1z}}{R_1}\frac{\partial^2 \theta_a C_{1i}}{\partial z^2} \right\} dt$$
$$- \int_{t_0}^{t} \left\{ \frac{\partial \theta C_{2i(BIO)}}{\partial t} + F_{2INF} + F_{2VOL} \right\} dt + \int_{t_0}^{t} F_{2PAR} dt$$
$$- \int_{t_0}^{t} \left\{ \frac{\partial \theta_a C_{1i(BIO)}}{\partial t} + F_{1INF} \right\} dt + \int_{t_0}^{t} F_{1PAR} dt \qquad (8)$$

Solution Step 1

Compute the concentration due to advection, dispersion, and sorption using a time-marching scheme. For each of the discrete volumes (discrete elements, macroelements) characterizing the unsaturated zone as shown in Fig. 11, the first part of Eq. (8) becomes

$$C_i^A(t+1) = C_i^A(t)$$
$$+ \left[-\frac{\bar{v}_{2z}}{R_2}\overline{\theta C}_{2i}(t+1) + \frac{D_{2x}}{R_2}\overline{\theta C}_{2i}(t+1) + \frac{D_{2y}}{R_2}\overline{\theta C}_{2i}(t+1) + \frac{D_{2z}}{R_2}\overline{\theta C}_{2i}(t+1) \right] \Delta t$$
$$+ \left[-\frac{\bar{v}_{1z}}{R_1}\overline{\theta C}_{1i}(t+1) + \frac{D_{1x}}{R_1}\overline{\theta C}_{1i}(t+1) + \frac{D_{1y}}{R_1}\overline{\theta C}_{1i}(t+1) + \frac{D_{1z}}{R_1}\overline{\theta C}_{1i}(t+1) \right] \Delta t$$
(9)

where

$\bar{v}_{1z}$ = average velocity in the air medium between time $(t+1)$ and (t)
$\bar{v}_{2z}$ = average velocity in the water medium between time $(t+1)$ and (t)
$\overline{\theta C}_{1i}(t+1)$ = average concentration of constituent (i) in the air medium
$\overline{\theta C}_{2i}(t+1)$ = average concentration of constituent (i) in the water medium

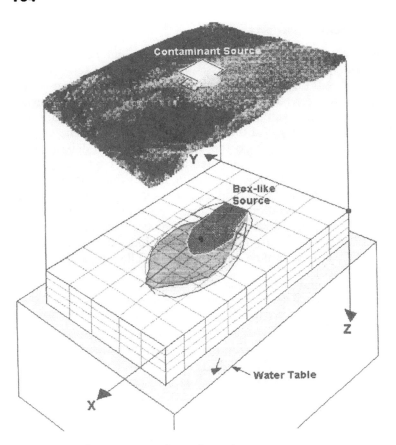

Figure 11 Source geometric configuration.

Lumping of all pertinent interacting components within each discrete element is automatically handled by the EIS platform computing for convenience the above averaged parameters at the center of each element.

Solution Step 2

Add the effects of biodegradation acting on the contaminant concentrations evaluated in step 1. Three modeling options are available: a first-order decay model, an instantaneous reaction model, and a modified Monod kinetics model.

a. First-Order Decay Model Solution (λ Term). The first-order decay of constituent (i) is the following expression:

$$\frac{\partial C_{\text{BIO}}}{\partial t} = \lambda_1 C_{1i} + \lambda_2 C_{2i}$$

This term may be incorporated to the general solution provided in Eq. (9) as shown below.

$$C_{\text{BIO}}(t+1) = C_{\text{BIO}}(t) + [\lambda_1 C_{1i}(t+1) + \lambda_2 C_{2i}(t+1)]\Delta t \tag{10}$$

EIS Macroengineering Approach

This equation implies a summation of the contribution of phases (from $j=1$ to $j=2$) with λ being the first-order decay term due to biodegradation in the air and water media, respectively.

b. Instantaneous Reaction Model Solution:

$$\frac{\partial C_{BIO}}{\partial t} = -R_{aer} - R_{anaer} + R_{product}$$

The model takes into consideration the stoichiometry of the equation of each chemical reaction resulting to the integration of the following expression (to better follow the development of these expressions we have considered an LNAPL contamination; similar expressions can be derived for other biodegrading processes):

$$dC_{BIO} = -R_{aer}\,dt - R_{anaer}\,dt + R_{product}\,dt \tag{11}$$

This expression starts with the reaction with the highest redox potential, the aerobic reaction,

$$R_{aer} = \frac{O_1}{S_{1iO}} + \frac{O_2}{S_{2iO}}$$

where O is the oxygen and S_{iO} the stoichiometry between contaminant (i) and oxygen. Note that the stoichiometry in the air and water media may be different for different chemical constituents.

At this point, the next most preferred reaction begins and continues until that electron acceptor is consumed. This leads to a hierarchical order where preferred electron acceptors are consumed one at a time, in a sequence of anaerobic reactions shown below (interaction of contaminant with nitrate N, manganese M, iron Fe, sulfate S only in the liquid phase).

$$\frac{N_2}{S_{2iN}} + \frac{M_2}{S_{2iM}} + \frac{Fe_2}{S_{2iI}} + \frac{S_2}{S_{2iS}} + \frac{Mt_2}{S_{2iMt}}$$

For LNAPLs we have:

$$C_{BIO}(t+1) = C_{BIO}(t)$$

$$-\left[\frac{O_1(t+1)}{S_{1iO}} + \frac{O_2(t+1)}{S_{2iO}} + \frac{N_2(t+1)}{S_{2iN}} + \frac{M_2(t+1)}{S_{2iM}} + \frac{Fe_2(t+1)}{S_{2iI}} + \frac{S_2(t+1)}{S_{2iS}} + \frac{Mt_2(t+1)}{S_{2iMt}}\right]\Delta t \tag{12}$$

We simplify by showing the computations for the liquid phase only, omitting the subscript (2) representing the liquid phase. Note that the expression inside the brackets is constant with time (assumption of instantaneous reaction). Therefore Δt can be neglected from the expressions that follow. S_{ij} is the stoichiometry between contaminant (i) and electron acceptor (j). The electron acceptors are updated at each time step according to the following relations in a sequence defined by the Gibbs free energy of the microbially mediated redox reactions:

$$O(t+1) = O(t) - C_{BIO}(t)S_{iO}$$

$$N(t+1) = N(t) - \left[C_{BIO}(t) - \frac{O(t+1)}{S_{iO}}\right]S_{iN} \quad \leftarrow A$$

$$Fe(t+1) = Fe(t) + \left[C_{BIO}(t) - \frac{O(t+1)}{S_{iO}} - \frac{N(t+1)}{S_{iN}}\right]S_{iF} \quad \leftarrow B \quad (13)$$

$$S(t+1) = S(t) - \left[C_{BIO}(t) - \frac{O(t+1)}{S_{iO}} - \frac{N(t+1)}{S_{iN}} - \frac{Fe(t+1)}{S_{iFe}}\right]S_{iS} \quad \leftarrow C$$

$$Mt(t+1) = Mt(t) + \left[C_{BIO}(t) - \frac{O(t+1)}{S_{iO}} - \frac{N(t+1)}{S_{iN}} - \frac{Fe(t+1)}{S_{iFe}} - \frac{Mt(t+1)}{S_{iMt}}\right]S_{iMt} \quad \leftarrow D$$

These reaction terms are subject to constraints that are dependent on the threshold concentrations O_{min}, N_{min}, Fe_{min}, S_{min}, Mt_{min}, below which no biodegradation takes place. Then the above expressions depend on the values inside the brackets as follows:

$O(t+1) = 0$ where $C_{BIO}(t) < O(t)/S_{iO}$
$N(t+1) = 0$ where $A < N(t)/S_{iN}$
$Fe(t+1) = 0$ where $B < Fe(t)/S_{iFe}$
$S(t+1) = 0$ where $C < S(t)/S_{iS}$
$Mt(t+1) = 0$ where $D < Mt(t)/S_{iMt}$

A numerical procedure is now adopted to implement these computations. The simulation domain of the unsaturated zone is discretized automatically according to the geologic strata along the vertical direction. Finally, the contaminant concentration is computed at the center of each discrete element as follows:

$$C_i(t+1) = C_i^A(t+1) + C_{BIO}(t+1) \quad (14)$$

where $C_i^A(t+1)$ is the contribution of advection, dispersion, and sorption [see Eq. (9)]. This concentration is computed in the global referential system $OXYZ$. $C_{BIO}(t+1)$ is the contribution of biodegradation obtained from Eq. (12) and is applied directly in the global referential system.

c. Monod Kinetics Model Solution:

$$\frac{\partial C_{BIO}}{\partial t} = -M_t k\left[\left(\frac{C_i}{K_h + C_i}\right)\left(\frac{O}{K_O + O}\right)\right]$$

Biodegradation of hydrocarbons is enhanced by the presence of microorganisms. Organisms enable such reactions to proceed via two important mechanisms illustrated in Fig. 12. These mechanisms are the following.

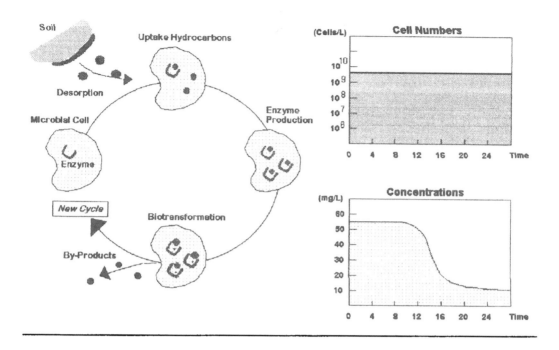
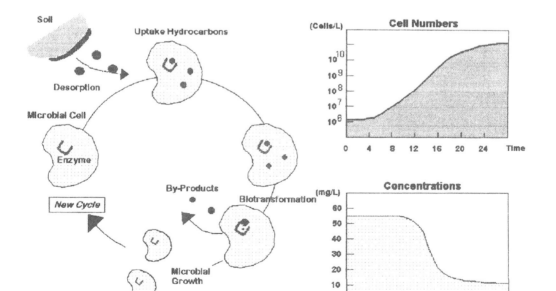

Figure 12 Sequence of events in the overall process of biodegradation.

Special proteins called enzymes serve as catalysts to lower the activation energy of chemical reactions by several tens of kilojoules per mole, thereby speeding the transformations by factors of 10^9 or more. The presence of the chemical may cause the organism to produce more enzyme units suited to degrading this substance. In this case, the biodegradation rate depends on the specific interactions of the compound with the enzyme. These interactions are best quantified using Michaelis-Menten kinetics.

If the metabolism of the chemical results in substantial energy yield and/or cell-building materials, then, in response, the microorganism will increase in cell numbers in just a few hours (more than 10^9 cells per liter). It is evident that these additional organisms will increase the biodegradation rate. To describe the time course of the contaminant concentration, the microbial population dynamics in the system has to be quantified. This is done using the Monod kinetics model.

In the EIS platform these microbial transformations are modeled by modifying the Monod kinetics model to lead to the following expression for LNAPLs and a unit time step dt:

$$C_{BIO}(t+1) = C_{BIO}(t) - C_{BIO}^A(t+1) - C_{BIO}^B(t+1) - C_{BIO}^C(t+1) - C_{BIO}^D(t+1)$$
$$- C_{BIO}^E(t+1) \tag{15}$$

where

$$C_{BIO}^A(t+1) = C_i(t+1) - M_t k \left[\left(\frac{C_i}{K_h + C_i} \right) \left(\frac{O}{K_O + O} \right) \right]$$

$$C_{BIO}^B(t+1) = C_{BIO}^A(t+1) - M_t k \left[\left(\frac{C_{BIO}^A}{K_h + C_{iBIO}^A} \right) \left(\frac{N}{K_N + N} \right) \right]$$

$$C_{BIO}^C(t+1) = C_{BIO}^B(t+1) - M_t k \left[\left(\frac{C_{BIO}^B}{K_h + C_{iBIO}^B} \right) \left(\frac{Fe}{K_{Fe} + Fe} \right) \right] \tag{16}$$

$$C_{BIO}^D(t+1) = C_{BIO}^C(t+1) - M_t k \left[\left(\frac{C_{BIO}^C}{K_h + C_{iBIO}^C} \right) \left(\frac{S}{K_s + S} \right) \right]$$

$$C_{BIO}^E(t+1) = C_{BIO}^D(t+1) - M_t k \left[\left(\frac{C_{BIO}^D}{K_h + C_{iBIO}^D} \right) \left(\frac{Mt}{K_{Mt} + Mt} \right) \right]$$

and where

O, N, Fe, S, Mt = concentration of oxygen, nitrate, iron, sulfate, and methane, respectively

k = maximum hydrocarbon utilization rate per unit mass microorganisms

K_h = hydrocarbon half-saturation constant

$K_O, K_N, K_{Fe}, K_S, K_{Mt}$ = oxygen, nitrate, iron, sulfate, and methane half-saturation constants, respectively

M_t = total microbial population attached to solid and in groundwater

EIS Macroengineering Approach

Also the following constraints apply:

$$C_{BIO}^A(t+1) = 0 \quad \text{if} \quad C_i(t) < M_t k\left[\left(\frac{C_i}{K_h + C_i}\right)\left(\frac{O}{K_O + O}\right)\right]$$

$$C_{BIO}^B(t+1) = 0 \quad \text{if} \quad C_{BIO}^A(t+1) < M_t k\left[\left(\frac{C_{BIO}^A}{K_h + C_{iBIO}^A}\right)\left(\frac{N}{K_N + N}\right)\right]$$

$$C_{BIO}^C(t+1) = 0 \quad \text{if} \quad C_{BIO}^B(t+1) < M_t k\left[\left(\frac{C_{BIO}^B}{K_h + C_{iBIO}^B}\right)\left(\frac{Fe}{K_{Fe} + Fe}\right)\right] \quad (17)$$

$$C_{BIO}^D(t+1) = 0 \quad \text{if} \quad C_{BIO}^C(t+1) < M_t k\left[\left(\frac{C_{BIO}^C}{K_h + C_{iBIO}^C}\right)\left(\frac{S}{K_S + S}\right)\right]$$

$$C_{BIO}^E(t+1) = 0 \quad \text{if} \quad C_{BIO}^D(t+1) < M_t k\left[\left(\frac{C_{BIO}^D}{K_h + C_{iBIO}^D}\right)\left(\frac{Mt}{K_{Mt} + Mt}\right)\right]$$

The estimation of these parameters is difficult. Some data are available for the basic aromatic hydrocarbons, but there are few data for other compounds and anaerobic conditions. As with the "instantaneous reaction" model, a numerical procedure is adopted to implement these computations. The simulation domain is discretized automatically according to the geologic strata along the vertical direction. The contaminant concentration is computed at the center of each discrete element as follows:

$$C_i(t+1) = C_i^A(t+1) + C_{BIO}^{Monod}(t+1) \quad (18)$$

where $C_i^A(t+1)$ is the contribution of advection, dispersion, and sorption [see Eq. (9)]. $C_{BIO}^{Monod}(t+1)$ is the contribution of biodegradation obtained from Eq. (17) and is computed directly in the global referential system. Note that in the EIS platform the microbial population can also be evaluated explicitly.

F. Uncertainty Analysis

In a risk assessment, descriptions of uncertainties may indicate the quality of information, range of knowledge, and level of confidence in the corresponding decisions. Several factors contribute to these uncertainties. They include limitations in the data that characterize sites and source terms; uncertainties in scenarios and choices of parameters to fit different scenarios; uncertainties in formulating the fate and transport model, and in physical parameters used as input to the models (e.g., dispersion coefficient); exposure parameters; and dose–response relationships. Therefore uncertainty arises from combinations of heterogeneity (variability), errors in measurement, and lack of knowledge.

> Heterogeneity is the variability within a parameter, such as the variability in the characteristics of a population. For example, it is relatively easy to determine the amount of water that individuals drink daily, but the amount will vary from day to day, and among individuals in a population.
>
> Error in measurement arises from inadequacy of sampling, sampling biases, errors in the measurements, and imprecision.
>
> Lack of knowledge can involve parameters that are expressed quantitatively, and components of a risk assessment that do not have numerical values per se. Major sources of uncertainty include inadequate knowledge of physical processes, such as environmental transport mechanisms, and gaps in qualitative

knowledge, such as future land-use scenarios. Parameters and their ranges of values can be profoundly affected by choices among these components of a risk assessment, in turn affecting the overall uncertainties of the risk estimates.

As part of a risk assessment, uncertainty analyses should be performed to determine which parameters exert a significant influence on the overall risk estimates, and the extent of influence of the range of uncertainty of each parameter on the risk.

Sensitivity analyses are used to assess which parameters are the most important contributors to the magnitude of an overall risk estimate. These analyses compare all parameters in an assessment for the overall effect of a specific degree of change (e.g., a 20% variation) in each parameter. Uncertainty analyses estimate the contribution of uncertainty associated with each variable to the overall uncertainty of a risk estimate. A sensitivity analysis can be performed as part of an uncertainty analysis to identify the parameters that contribute the most to the variance of the final risk estimates. In other words, the analysis quantifies the sensitivity of uncertainty of a risk estimate to a changed range or assumed type of distribution of a single variable.

Existing multimedia models presently use only deterministic (single) values for parameters. It is difficult in such assessments to sort out the contributions of individual parameters to the overall uncertainty of the risk estimates because the calculations can combine high (90th or 95th percentile) parameter values with lower (50th percentile or average) values. The EIS platform uses innovative concepts to allow the use of input statistical distributions on all natural parameters, leading to the determination of the statistical distribution of the concentrations of a particular contaminant.

The accuracy of the values of the parameters is difficult to verify, but in EIS such difficulties are mitigated by the user's ability to input alternative values. The platform also allows input of site-specific and region-specific data to reduce uncertainty for food chain and exposure parameters, and provide alternative choices for performing more accurate calculations based on the predicted concentrations of contaminants at the specific location of the site.

Most exposure estimates and risk calculations are a multiplicative combination of exposure assumptions resulting in a point estimate for the intake of a chemical. A "traditional" practice is to choose a combined variety of average, conservative, and worst standard assumptions. There are several disadvantages to this approach.

- There is no way of knowing the actual degree of conservatism in an assessment (i.e., no realistic depiction of the variation can be presented for the exposure estimate, and hence for the ultimate risk level).
- By selecting upper limits on many exposure variables, the assessment generally considers scenarios that will rarely occur (e.g., what is the likelihood of an individual ingesting the maximum soil amount per day, for every day of the maximum number of exposure days, for the maximum number of years the individual could live near a site?)
- Sensitivity analyses are of limited value since many of the variables are at or near their maxima.

An alternative approach to the use of conservative assumptions is an exposure distributional analysis as implemented in the EIS platform, in which ranges or distributions of individual exposure parameters (e.g., the distribution of adult body weights) are considered to produce an overall distribution of potential exposure.

Therefore, rather than using point estimates in exposure assessment, the EIS simulation tools can be used to estimate distributions for exposure assumptions. Use of this methodology does not alter the basic structure of the exposure estimate. However, it does refine the way chemical intakes are calculated in the exposure assessment.

The starting point is to establish the statistical characteristics of all pertinent input parameters characterizing a site. Such parameters include soil properties (soil layers, porosity, hydraulic conductivities, dispersion, etc.), chemical properties (adsorption, stoichiometry, etc.), as well as loading/source site-specific conditions. Then, their mean values and standard deviations automatically feed the scientific engines (simulation algorithms) available in the platform. These algorithms were described in details in the previous sections and include:

> Groundwater flow in the unsaturated and saturated soils with special features such as slurry walls, geosynthetics, and geologic faults
> Multiple species contaminant migration processes such as advection (Computed from fluxes produced by the flow module), dispersion, chemical reaction (sorption, ion exchange, chemical decay), and sink/source mixing
> Migration and degradation of hydrocarbons accounting for oxygen-limited biodegradation occurring at the site.

The outcome of the simulation is the spatial distribution of the mean values and standard deviations of the concentrations throughout the site. The mean values are the conventional point estimates as produced by the corresponding algorithms activated on a given site. A first-order approximation is used to compute the standard deviation of the concentrations C, assuming that all input are statistically independent.

The standard deviation for the concentration is written as follows:

$$S_c = \sqrt{S_h^2\left(\frac{\partial C}{\partial h}\right)^2 + S_n^2\left(\frac{\partial C}{\partial n}\right)^2 + S_p^2\left(\frac{\partial C}{\partial p}\right)^2 + \cdots + \cdots} \qquad (19)$$

where

> S_c = computed standard deviation for the concentration
> S_h = given standard deviation of the hydraulic conductivity
> S_n = given standard deviation of the porosity
> S_p = given standard deviation of any other parameter

The first-order derivative terms with respect to the hydraulic conductivity, porosity, and other parameters are computed explicitly in the simulation algorithms, thus considerably reducing the computational cost of evaluating S_c.

At this stage, through the use of various algorithms of the simulation engines, we have computed only the mean and variance of the multispecies concentrations. However, it is possible to identify an optimum probability distribution for the concentrations by invoking the principle of maximum entropy. This is done by identifying the concentration probability distribution that maximizes the information entropy subject to the additional constraints imposed by the given information (i.e., the computed by the simulation engine's mean and standard deviation of the multispecies concentrations). A summary of the EIS approach is given in Fig. 13.

As shown in Fig. 13, a beta distribution is the most unbiased distribution satisfying our constraints for the multispecies concentrations. In both the EIS/risk-"traditional" and EIS/risk-uncertainty approaches to risk exposure assessment, the analyst constructs

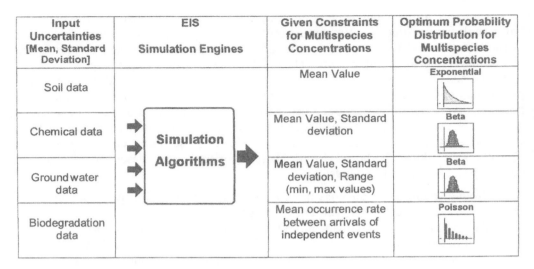

Figure 13 Approach to identifying probability distributions of multispecies concentrations.

a model consisting of relationships between random variables [e.g., the model for intake of a compound via ingestion of water (Ref. 11, Exhibit 6-11, p. 6-35)]:

$$\text{Intake(LADI)} = \frac{C_w \bullet IR \bullet EF \bullet ED}{BW \bullet AT} \tag{20}$$

where

intake (LADI) = intake (mg/kg/day)
C_w = chemical concentration in water (mg/L)
IR = ingestion rate (L/day)
EF = exposure frequency (days/year)
ED = exposure duration (years)
BW = body weight (kg)
AT = averaging time (days)

In the "traditional" approach to risk assessment, point estimates for each of the variables in the intake Eq. (20) are chosen (e.g., the 95th percentile upper confidence limit on the mean of the groundwater sample concentrations; the 90th percentile of the distribution of adult water consumption). This results in a point estimate for intake which, having used conservative estimates for each variable, suggests an intake quantity that is doubly indemnified by using 90th and 95th percentiles in its derivation. Because different percentiles are used, it is not possible to know what combined percentile to assign to the overall expression of intake.

In the EIS/risk-uncertainty simulation model, the analyst determines a continuous or discrete distribution to describe each of the random variables in Eq. (20). This distribution is defined in terms of the probability density function (PDF) or the cumulative distribution function (CDF). Several distributions are defined by one, two, or more parameters. When running the EIS simulation algorithms, the computer automatically proceeds to determine the distribution of daily intakes according to Eq. (20). The distribution usually falls in the category of beta distributions (often the log-normal distribution) or is asymmetric over the positive axis. From this distribution, a specific intake can be selected (e.g., the average or mean intake, median intake, or 95th percentile upper confidence limit on the intake) that, in combination with the appropriate toxicity benchmark concentration, is used to calculate risk.

VI. CASE STUDY

A. Location Description and History of the Site—Physical Environment

The site is an old chemical-industrial complex located in the floodplain of an ephemeral stream. Just 600 m down-gradient is a local community, which draws its water from the deep aquifer. Contamination has been observed in the wells, and the concern is that the pollution may have emanated from the industrial site. In addition, several houses are located in the immediate vicinity of the site.

Figure 14 shows the site located on the west side and the valley toward the east-northeast. Several monitoring wells were placed along the valley to establish potential paths of contamination. Figure 15 shows the layout of the site and plume configuration.

The site is located on top of an important deep aquifer formed of weathered calcium bedrock, overlain by alluvial deposits. The alluvial deposits rest on top of a thin, silty-clayee semipermeable layer, which separates it from the deep aquifer. The thin aquitard is nonhomogeneous and in many instances discontinuous. Therefore, a marked hydraulic connection exists with the deep aquifer, and this is corroborated by the presence of contamination in the deep aquifer. Furthermore, the regional groundwater flow is preferential in the general direction of the stream valley, with velocities of the order of meters per week. The location of municipal wells just 600 m down-gradient along the stream valley was all the more alarming.

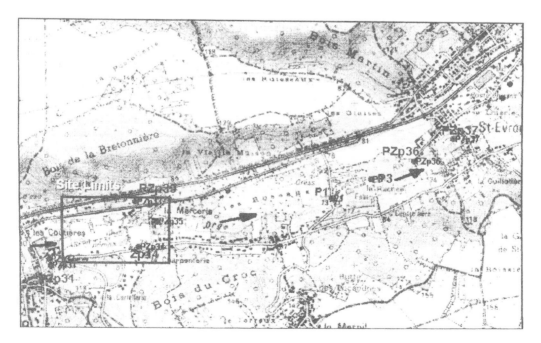

Figure 14 Case study layout and site limits.

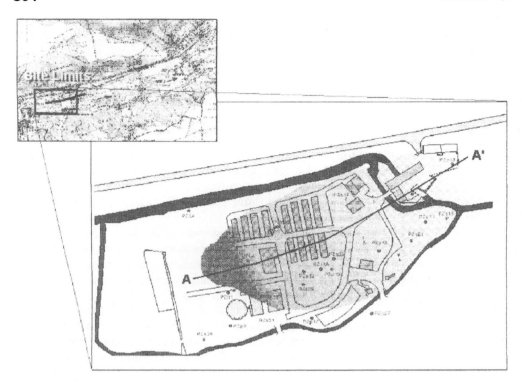

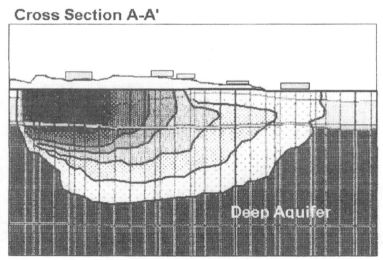

Figure 15 Site limits of case study.

B. Description of Present Contamination

The site is surrounded by the north and south branches of the stream. In Figure 15, the plume of benzene is shown in the alluvial (surficial) aquifer. The cross section is along the stream valley axis and clearly indicates contamination seeping through the thin-layer aquitard and migrating into the deep aquifer. Interestingly, the maximum concentration in the deep aquifer reaches approximately half the peak concentration in the alluvial aquifer. Most important, however, concentrations drop dramatically off site. This is due to the increased biodegradation potential off site caused by the influx of pristine waters on either side of the valley.

1. Groundwater Contamination

The most significant contaminants found in the groundwater include:

> BTEX: Benzene, toluene and xylenes
> Chlorinated solvents: ethyl-benzene, dichlomethane, trichloromethane, trichloroethane, trichloroethylene
> PCB
> Lead

Other contaminants such as phenols, creosols, chlorophenols, chlorobenzenes, phthalates, and heavy metals were found in very small, insignificant concentrations that are of no concern.

2. Air Contamination

Volatile organic compounds (VOCs) were found in most drillings. They were important (300–700 ppm) in approximately 75% of the site. Higher concentrations were observed near old landfill cells. Volatile readings persisted down to a depth of approximately 4 m.

3. Contamination of Soils

There is a strong correlation between VOCs and soil contaminants. These are BTEX, PCB, chlorinated solvents, pesticides, and some metals (Zn, Pb), summarized in Table 12. Additional chemicals found on the site include pesticides (lindane and hexachlorobenzene). In general:

> High concentrations were found near old landfills.
> High concentrations cover approximately 70% of the site.
> Highly contaminated soils were found to a depth of approx. 4–5 m.
> The entire volume of contaminated soil is 50,000–75,000 m^3.

Finally, an ecotoxicological study confirmed the degree and extension of contamination, namely, mutagenic effect in the soils of highest concentration in VOCs and BTEX, but no toxic or mutagenic effects from in living organisms in the soils in the southwest quadrant of the site, confirming that these soils are little contaminated.

4. Contamination of Surface Waters (Stream)

The following observations were made:

> TCA showed a peak value of 115 µg/L.
> TCE had a peak value of 5.4 µg/L.
> Chloroform had a peak value of 2.8 µg/L.
> PCE had a maximum value of 1.9 µg/L.

It is important to note that these values are many orders of magnitude lower than the values in the alluvial aquifer and that PCBs and BTEX are altogether absent. Very likely this contamination emanates from upstream.

Table 12 Contamination of Soils

	Frequency of detection	Minimum concentration (mg/kg)	Maximum concentration (mg/kg)
BTEX			
Benzene	20/28	1.3	242.8
Ethylene	24/28	0.2	1,769.4
Toluene	26/28	0.5	3,976.9
Xylenes	28/28	0.6	8,219.2
VOCs			
Chloroform	28/28	0.5	3,520.0
1,2-DCA	8/28	13.7	1,396.0
1,1-DCA	1/28	9.3	9.3
1,1-DCE	10/28	0.4	23.2
Halogenated volatiles			
Dichloromethane	18/28	3.2	2,415.0
PCE	17/28	0.3	285.7
1,1,1-TCE	20/28	0.3	2,355.0
TCE	21/28	1.8	4,484.0
Tetrachloroethylene	21/28	0.3	3,548.0
Chloroethylene	0/25	—	—
Phenols			
Total cresols—dimethylphenols, m-, p- and o-cresol	5/5	0.27	370.0
Phenol	5/5	0.18	1,400.0
Aromatic chlorinated compounds			
Chlorophenols	4/5	0.16	22.0
Chlorobenzenes	4/5	0.05	1,146.0
Phthalates	2/2	5,274	8,093.0
PAHs			
Total PAH	5/5	0.16	167.0
Benzo-a-pyrene	0/5	—	—
Total PCB	17/18	0.2	409.0
Heavy metals			
Lead	10/10	17.8	9,680.0
Zinc	10/10	55.6	20,100.0

C. Soil Profile

Twenty-four log-wells and 100 cone penetrometer locations allowed a good characterization of the soil conditions at the site. Five soil layers were identified, covering the extent of the site as shown in the cross section of the site (Fig. 16).

D. Groundwater Site Conditions

The site investigation revealed a recharging mechanism from the stream to the lower aquifer. The prevailing groundwater flow direction is from southwest to northwest. The site investigation also allowed the determination of the piezometric head contours, from which the hydraulic gradient was estimated to range from 4.4×10^{-3} in the upper aquifer (sand and gravel) to 3.7×10^{-3} in the lower aquifer. Pumping tests confirmed

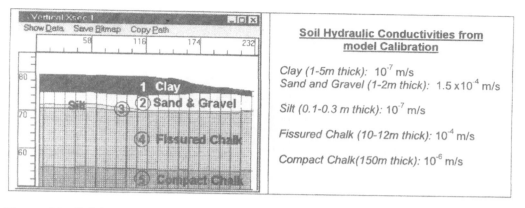

Figure 16 Soil layers at the site.

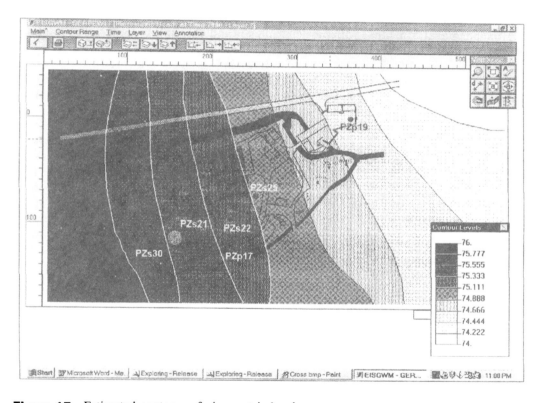

Figure 17 Estimated contours of piezometric heads.

the hydraulic conductivities of the model calibration shown in Fig. 16. They ranged from 1.1×10^{-4} m/s to 1.1×10^{-5} m/s.

Results of the groundwater flow model calibration are also shown in Fig. 17 and Table 13. The discrepancies between the estimated by EIS and the measured piezometric heads are seen to be of the order of 1 cm. Interestingly, the uncertainties of the estimated piezometric heads gave a standard deviation of 0.1, which gives an estimated discrepancy

Table 13 Estimated Discrepancies of the Piezometric Heads

Wells	Observed heads (m)	Estimated by EIS heads (m)	Discrepancy (m)
Pzs30	75.50	75.58	−0.08
Pzs21	75.22	75.30	−0.08
Pzs22	74.13	75.26	−0.13
Pzs25	74.21	75.09	−0.11
Pzs23	74.98	75.10	+0.12
Pzp28	74.87	74.81	+0.06
Pzp33	74.60	74.56	+0.04
Pzp19	74.62	74.54	+0.08
Pzp17	75.10	74.92	−0.18

of 10 cm, demonstrating that both the deterministic and uncertainty analysis flow modules of EIS are properly calibrated with respect to the flow module.

The site limits of the simulation model shown in Fig. 14 cover an area of 12.5 ha. The computational grid contains 55 columns, 23 rows, and 5 layers, for a total of 6325 macroelements. Each macroelement is 10 m wide, which gives adequate computational accuracy.

The objective of the study is to develop site-specific target concentration levels (SSTLs) for all critical site contaminants, in relation to critical exposure pathways, and according to projected land-use patterns in and around the site. Target concentration levels can be based on prevailing regulations, regardless of exposure pathways, target land use, and risk of health hazard. However, the starting point of any risk exposure study is an in-depth evaluation and sorting of the available site data. This diagnostic phase of the study is supported and complemented by the EIS simulation modules to help identify:

> The historical background of the fate and transport of the major groundwater contaminants such as BTEX and chlorinated compounds
> The impact of the prevailing contamination to future migration
> Possible remedial action with a pump/injection groundwater treatment

The numerical grid of the simulation model is shown in Fig. 18. The grid has 6325 macroelements. Each macroelement has a size of 10 × 10 m, which gives adequate computational accuracy.

The EIS modules simulate the contaminant source covering an area of 1800 m² located south of the industrial sector. The source area represents a recharging mechanism spanning a 20-year period with a maximum concentration of 74.4 mg/L. The 20-year period corresponds roughly to the start of the industrial operations and the leaking conditions. Table 14 lists all the simulations that were considered to identify the most probable scenario of the major contaminant fate and transport.

E. Simulation Results

The simulation results are summarized in Figs. 19–26. They include cases for BTEX and for chlorinated solvent plumes, calibration runs as well as remediation scenario analyses. Note the similarity in the depleted oxygen and nitrate zones due to the assumed uniform

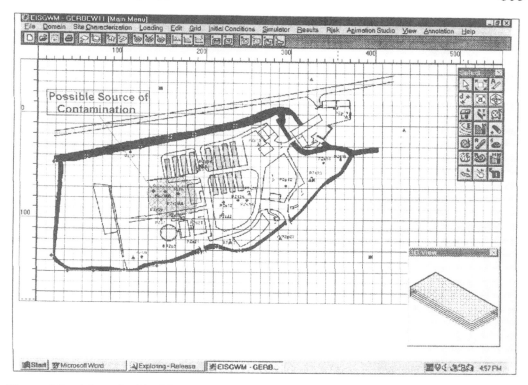

Figure 18 Adopted grid of fate and transport simulation models.

Table 14 Simulations of Fate and Transport

ID case	Description
BTEX	
Sim_BTEX_01	Continuous source without dispersion and biodegradation
Sim_BTEX_02	Continuous source with dispersion and without biodegradation
Sim_BTEX_03	Continuous source with dispersion and biodegradation
Sim_BTEX_05	Removal of the source mechanism
Sim_BTEX_06	Removal of the source mechanism and groundwater treatment
Chlor. Solv.	
Sim_OHV_01	Biodegradation of the tetrachloroethylene (PCE)
Sim_OHV_02	Biodegradation of the tetrachloroethylene (PCE), trichloroethylene (TCE), dichloroethylene (DCE), and vinyl chloride (VC)
Sim_OHV_03	Continued source with dispersion and without biodegradation (1977–1997)
Sim_OHV_04	Predictive migration of OHV (1998–2003)
Sim_OHV_05	Removal of polluted soils and groundwater treatment

initial distribution throughout the study area. The accuracy of the results of this simulation is essential to explain the importance of the biodegradation processes of the BTEX contaminant plume as it evolves with time, and to validate the assumptions on which the simulation model is based. A prognosis now can be made on the plume migration from year 1997 to 2002 (next 5 years). The simulation indicates that contaminant plumes

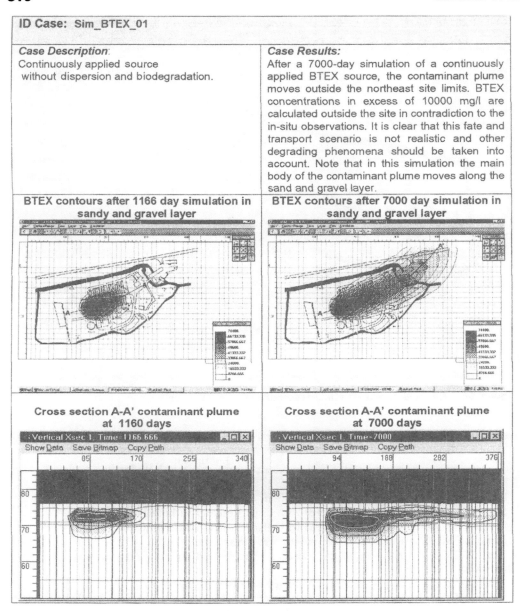

Figure 19 Simulation of BTEX plumes, Case 01.

will reach the site limits, with benzene concentrations exceeding 50 to 500 µg/L. Without groundwater remediation over the next 5 years, chances are that the contamination will migrate beyond the site limits. Therefore, the following two remedial alternatives are considered:

Removal of the contaminated soil
Removal of the contaminated soil and treatment of the groundwater (pump and treat)

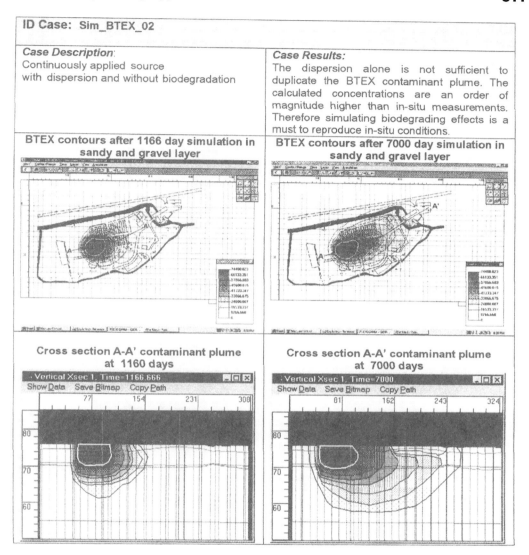

Figure 20 Simulation of BTEX plumes, Case 02.

Table 15 provides a summary of the main results of this pump-and-treat simulation, indicating the great reduction in BTEX concentrations.

1. Fate and Transport of Halogenated Compounds

Starting from the in-situ distribution of the halogenated compounds, several simulations were conducted taking into consideration all biodegrading processes and the transformation of tetrachloroethylene (PCE) to trichloroethylene (TCE), dichloroethylene (DCE), and vinyl chloride (VC) according to the following reactions:

$d(PCE)/dt = -K_1 \cdot PCE$ with degradation coefficient $K_1 = 0.0027/\text{day}$
$d(TCE)/dt = (Y_1 \cdot K_1 \cdot PCE) - (K_2 \cdot TCE)$ with a stoichiometry
  $Y_1(TCE/PCE) = 0.79$

$$d(\text{DCE})/dt = (Y_2 \cdot K_2 \cdot \text{TCE}) - (K_3 \cdot \text{DCE})$$
$$d(\text{VC})/dt = (Y_3 \cdot K_3 \cdot \text{DCE}) - (K_4 \cdot \text{VC})$$

The major results of these simulations are illustrated in Figs. 24–26.

At this stage all simulations for the diagnostic phase are successfully implemented and the most probable scenarios of the contaminants fate and transport are identified. Therefore we are ready to proceed with an in-depth risk exposure analysis of the site.

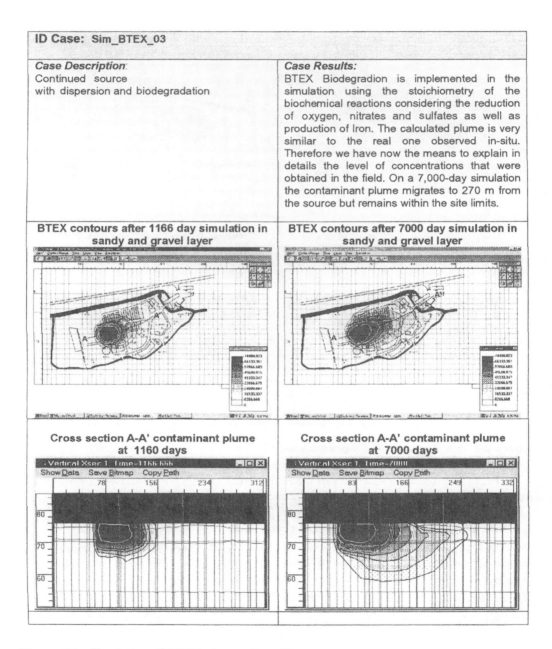

Figure 21 Simulation of BTEX plumes, Case 03.

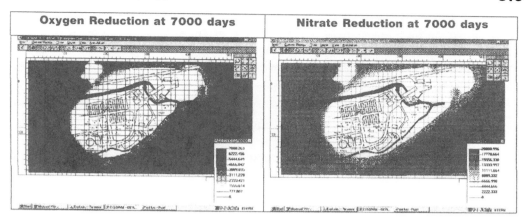

Figure 21 (*continued*)

Table 15 Results of BTEX Reduction with Time

Product	Maximum original concentration		Maximum concentration after 1 year of pumping		Maximum concentration after 2 years of pumping		Maximum concentration after 3 years of pumping	
	Sand layer	Chalk layer	Sand layer	Chalk layer	Sand layer	Chalk layer	Sand layer	Chalk layer
Benzene (µg/L)	74,000	27,000	30,000	12,900	23,500	8,732	<100	<100
Ethylbenzene (µg/L)	21,000		9,500	<100		<100	<100	<100
Toluene (µg/L)	584,000	226,000	292,000	114.00	192,000	80,000	<100	<100
Xylene (µg/L)	20,530	24	16,500	<100	8,265	<100	<100	<100

F. Summary of Risk Study Results—Target Corrective Action Concentrations as a Function of Ultimate Site Use Scenarios

The objective of the study is to develop site-specific target concentration levels (SSTLs) for all critical site contaminants, in reaction with critical exposure pathways, and according to projected land-use patterns in and around the site. Target concentration levels can be based on prevailing regulations, regardless of exposure pathways, target land use, and risk of health hazard, or they can be based on a risk study. The risk approach is summarized hereafter.

SSTLs were established for those constituents whose cancer risk associated with a particular medium (air, water, soil) exceeded 1×10^{-6}, or who's toxic risk index exceeded 1. These are trigger levels indicating that corrective action is necessary.

SSTLs are established on the basis of the ratio between trigger values and site specific concentration values. In the first iteration we assume:

> Linearity between concentration at the source and risk of health at receptors
> The use of one, average, or maximum concentration value at the source, as opposed to the actual concentration distribution at the site

Both limitations may result in overconservatism, which lacks the flexibility to propose imaginative land-use alternatives, but by relying on the simulation modules of EIS we can in a second iteration overcome these limitations and be more precise. In the present study, three alternative land-use scenarios were investigated, as listed in Table 16. These

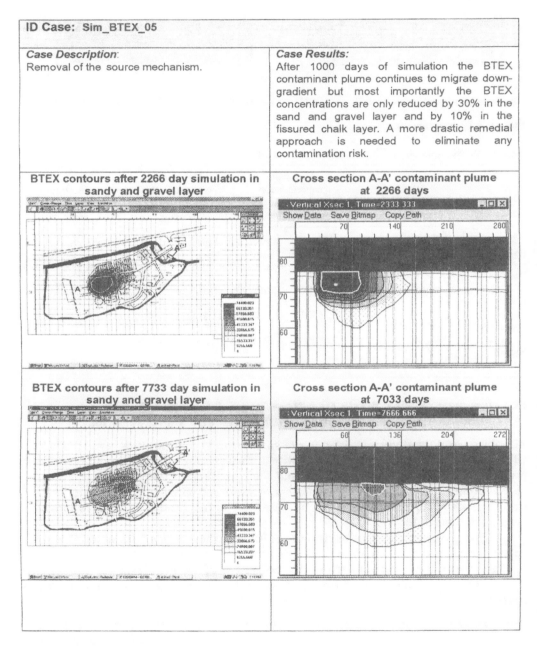

Figure 22 Simulation of BTEX plumes, Case 05.

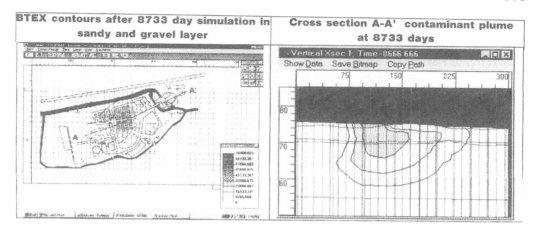

Figure 22 (*continued*)

Table 16 Land Use Scenarios

Scenario no. 1: Present land use	Residents on site + workers + residents in neighborhood of site + recreational use
Scenario no. 2: Present land use with restrictions	As in 1, with restrictions on vegetable ingestion and vapor intrusion to houses
Scenario no. 3: No residents on site	No residents on site; just maintenance workers; residents off site, plus recreational uses

scenarios represent realistic options available for a meaningful decision-making exercise. The three media of interest are the top soil layer (surface to depth of 2 m) and deep soil layer (2–6 m deep); air; and groundwater. The receptors and exposure pathways associated with the above three scenarios are listed in Table 17.

Table 18 gives the SSTLs calculated for the three media: surficial soil layer (0–2 m), deep soil layer (2–6 m), and groundwater, for the three scenarios above. Four exposure modes are investigated, substance ingestion, contact with substance, inhalation, and ingestion of vegetables. All calculations are made for all constituents whose toxic factor or cancer risk exceeded trigger values. The final corrective action is based on the most critical (toxic or cancer risk) among the constituents.

G. Conclusions Based on Risk Study Results

In general, regardless of the scenario retained, the risk study indicates that there is ground for concern with regard to both toxic risk and cancer risk for populations that can be potentially exposed to contaminants emanating from the site. However, the risks are mostly confined to the site. This is due in large part to the site location in the floodplain of a stream. As a consequence of lateral influx of pristine waters, contaminant plumes that migrate down-gradient of the site undergo sustained natural attenuation action, resulting in concentrations being in a dynamic equilibrium at levels below standards. This is true both for BTEX compounds and for chlorinated solvents (co-metabolism). To control the toxic and cancer risks at the site, corrective action must be undertaken at the site, paced in time as follows.

1. Immediate action:
   1.1. Restrict land use on site (residential, agricultural use)
   1.2. Secure premises of site with fence
   1.3. Inform population of potential risk associated with contamination by ingestion or contact with groundwater
   1.4. Cap zones where soil contamination is the highest
   1.5. Monitor water quality in stream (no pollution detected presently).
2. Short-term measures:
   2.1. Confine zones where free product exists (restricted areas)
   2.2. Skim-pump free-product floating in several wells
   2.3. Remove remaining drums from landfill areas
   2.4. Monitor groundwater quality down-gradient, offsite
3. Longer-term measures:
   3.1. Soil treatment in situ to reduce levels of benzene, PCE, chloroform, DCA, creosol, naphthalene, phthalates, phenol, and PCBs, according to recommendations from risk study.

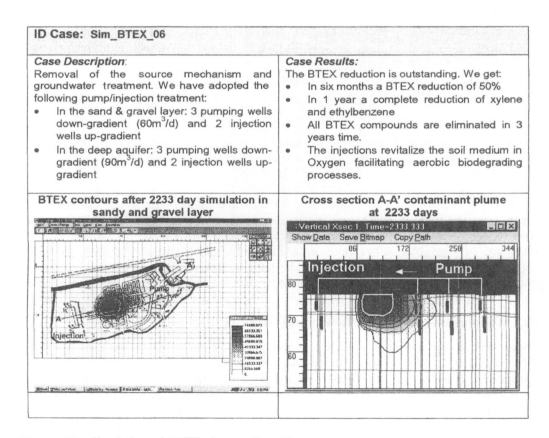

Figure 23 Simulation of BTEX plumes, Case 06.

EIS Macroengineering Approach

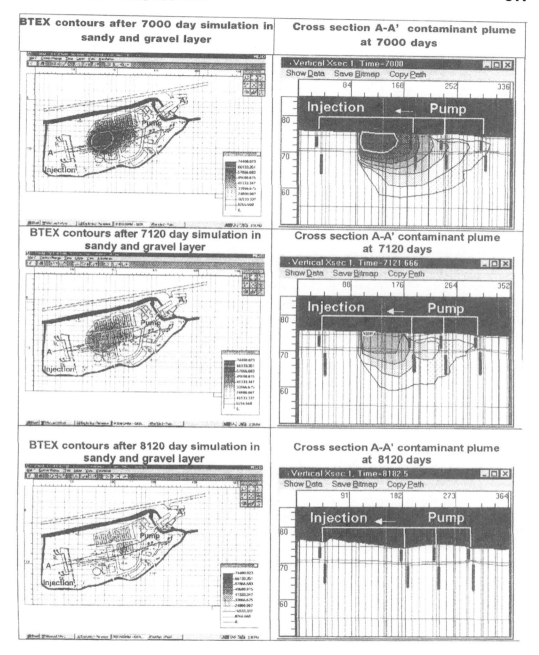

Figure 23 (*continued*)

3.2. Pump-and-treat both surficial (alluvial) and deep aquifers to reduce concentrations of benzene, PCE, chloroform, TCE, creosol, naphthalene, phthalates and PCBs, according to recommendations from risk study (taking into account effect of natural attenuation). Figure 27 illustrates the effect of the pump-and-treat system as simulated with EIS/ Biorem3D.

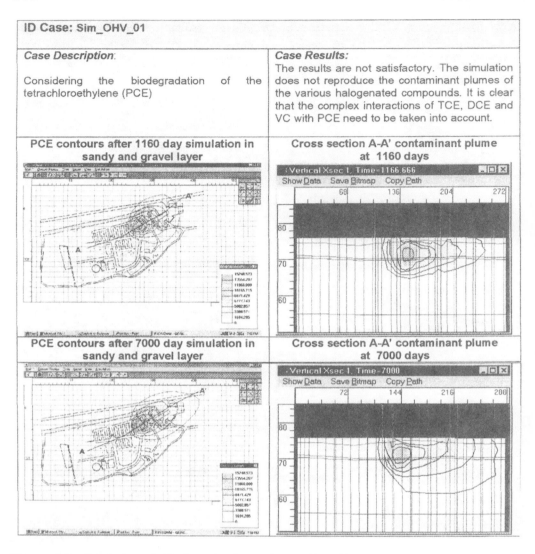

Figure 24 Simulation of chlorinated solvent plumes, Case 01.

VII. CONCLUSIONS

The risk-based approach provides rationale to the environmental problem by linking health hazard risk at receptors, to acceptable treatment levels at the contaminant source. The added advantage is that this allows accounting for the capacity of the medium for natural attenuation (assimilative capacity). Extracting the contaminated water for treatment often causes more damage than by letting it biodegrade in situ. Nevertheless, natural attenuation can only be accepted if short- and long-term monitoring proves the truthfulness of the theory in the field and guarantees compliance at imposed checkpoints. This is costly and can become tractable only with

EIS Macroengineering Approach

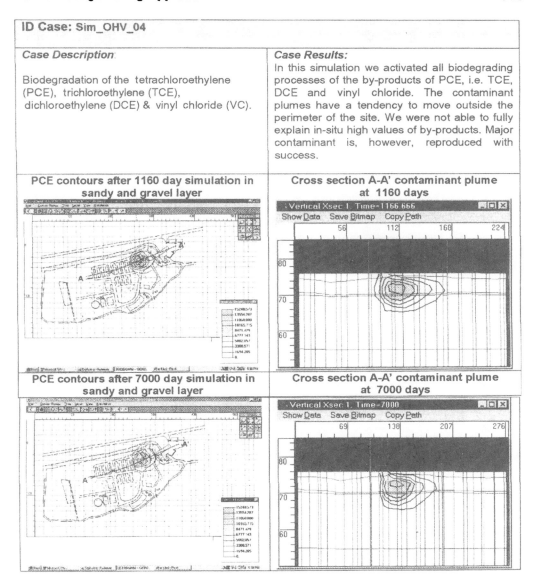

Figure 25 Simulation of chlorinated solvent plumes, Case 04.

the use of advanced simulation/prediction capability. Data are used to elucidate how the prototype behaves (model calibration), on the basis of which to predict future evolution of the contamination episode, and to compare various remediation alternatives and establish acceptable monitoring plans. However, credibility in the model predictions can only be founded on a comprehensive simulator. Biorem3D is the model presented here.

In addition to 3D flow and contaminant migration via advection/dispersion, Biorem3D also accounts for the chemical interaction (reactions) among 12 dissolved and adsorbed species, contaminants, electron acceptors, nutrients, and their by-products. It is an industrial-strength code whose high level of integration makes it ideal for use in large-scale real case studies. Furthermore, Biorem3D is embedded in the Environmental Information System—EIS platform. EIS is the first entirely object-based, integrated-automated platform, in continuous use since 1992. The user deals with physical objects, geologic features, chemical substances, topographic distributions, and scenario definitions, and everything else is calculated, coordinated, and managed by the graphical platform. In this chapter, we presented the theory, the implementation, and a successful application of the EIS/Biorem3D platform from an industrial case study. Natural attenuation embedded in a risk-based approach was used effectively to arrive at efficient and reliable remediation alternatives.

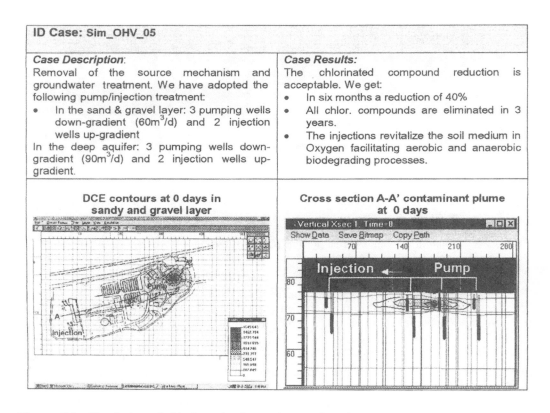

Figure 26 Simulation of chlorinated solvent plumes, Case 05.

EIS Macroengineering Approach

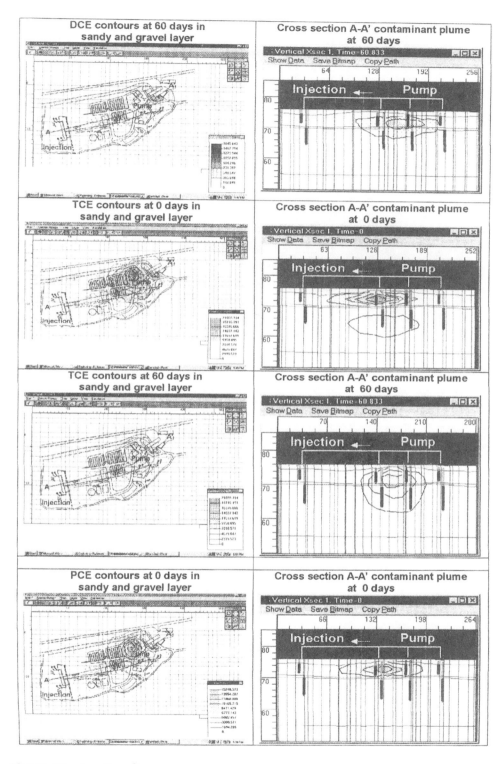

Figure 26 (*continued*)

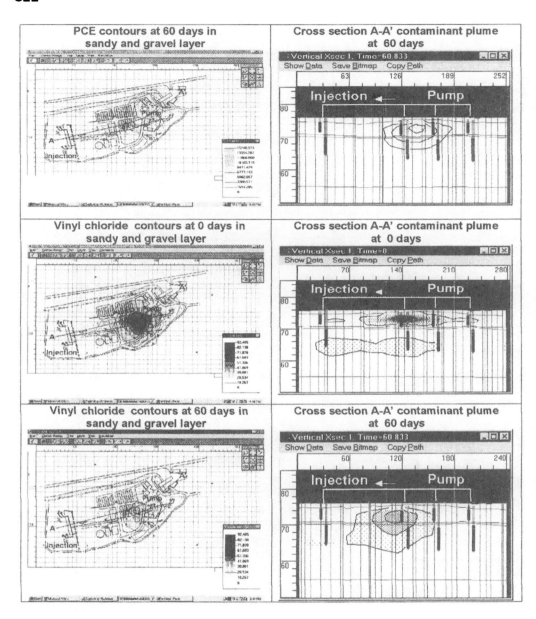

Figure 26 (*continued*)

Figure 27 Reduction of pump-and-treat procedure at 360 and 540 days.

Table 17 Scenarios, Pathways, Receptors, and Exposure Modes

	Residents on site	Workers on site	Residents off site, recreational uses
Scenario no. 1	Ambient vapor inhalation Indoor vapor inhalation Dermal exposure to groundwater and surficial soils Ingestion of fugitive dusts Accidental ingestion of groundwater Consumption of vegetables grown on site	Dermal exposure or ingestion of small quantities of groundwater Dermal exposure and/or ingestion of soils	Vapor inhalation Accidental ingestion of small quantities of groundwater Accidental dermal exposure to groundwater
Scenario no. 2	Ambient vapor inhalation Dermal exposure to groundwater and surficial soils Ingestion of fugitive dusts Accidental ingestion of groundwater	Ambient vapor inhalation Dermal exposure to or ingestion of small quantities of groundwater Dermal exposure and/or ingestion of soils	Vapor inhalation Accidental ingestion of small quantities of groundwater Accidental dermal exposure to groundwater
Scenario no. 3	Not applicable	Ambient vapor inhalation Dermal exposure to or ingestion of small quantities of groundwater Dermal exposure and/or ingestion of soils	Ambient vapor inhalation at moderate distances Ingestion of small quantities of groundwater Dermal exposure to small quantities of groundwater

Table 18 Summary of SSTLs by Polluted Medium and by Type of Scenario Considered

Scenario No.	UCL observed (mg/kg)	SSTL by scenario (mg/kg)			
		Scenario 1 (mg/kg)	Scenario 2 (mg/kg)	Scenario 3 Inhalation at 50 m (mg/kg)	Scenario 3 Inhalation at 250 m (mg/kg)
Surficial soil layer chemical species					
Benzene	72	0.025	1.2	2.9	19.0
Toluene	660	63.5	523.0	558.0	558.0
Ethylene	420	65.0			
Xylenes	1800	1465.0			
PCE	36	0.015	0.34	1.6	10.0
Chloroform	150	0.114	0.93	1.0	6.7
1,1-DCA	0.03				
1,2-DCA	21	0.0038	0.37	0.91	6.0
Dichloromethane	140	0.037	6.7	47.0	
Tetrachloroethylene	390	0.028	1.1	9.39	270.0
1,1,1-TCE	43	21.7			
TCE	200	0.171	0.6	0.6	90.0
PCB	290	0.052	0.052	0.61	4.0
Zn	6.2×10^4				
Deep soil layer chemical species					
Benzene	6.3	0.023	1.1	1.1	
Toluene	150.0	150.0			
Ethylene	88.0				
Xylenes	0.480				
PCE	1.2	0.032	0.58	0.6	
Chloroform	3.6	0.025	0.38	0.36	2.4
1,1-DCA	0.0025				
1,2-DCA	0.085	0.042			
Dichloromethane	7.7	0.29			
Tetrachloroethylene	79.0	0.52	16.0	16.0	
1,1,1-TCE	3.4				
TCE	19.0	0.067	4.7	4.7	
PCB	230.0	5.6	46.0	46.0	
Zn	2500.0				
Groundwater chemical species					
Benzene	20.0	0.021	0.7	3.06[a]	3.06[a]
Toluene	47.0	23.0	47.6		
Ethylene	6.9				
Xylenes	13.0				
PCE	0.027	0.0027			
Chloroform	2.7	0.01	0.39		
1,1-DCA	0.23				
1,2-DCA	0.95	0.022	0.095		

Table 18 (*continued*)

Scenario no.	UCL observed (mg/L)	Scenario 1 (mg/L)	Scenario 2 (mg/L)	Scenario 3 inhalation at 50 m (mg/L)	Scenario 3 inhalation at 250 m (mg/kg)
Dichloromethane	5.9	0.54			
Tetrachloroethylene	0.73	0.072	0.141	0.42[a]	0.42[a]
1,1,1-TCE	0.62				
TCE	3.7	0.047	1.162	3.51[a]	3.51[a,b]
PCB	6.6	2.37E-04	2.37E-04	7.2 E-4[a] 0.049[b]	7.2 E-4[a] 0.049[b]
Zn	43.0				

[a] For scenario 3, SSTLs for groundwater are controlled by exposure of workers on site.
[b] When concern for on-site workers is eliminated (e.g., permanent protective gear), the controlling population becomes residents off site. The critical chemicals then are PCB and TCE.
UCL, upper confidence limit.
Notes: (1) Empty cells signify that the observed concentration for that compound do not lead to toxic risk or cancer risk above the threshold levels for the concerned populations for the scenario investigated.
(2) Target concentration levels are more lenient for scenario 3. Possibly, that scenario could be further restricted to portions of the site only. The corresponding soil treatment targets could be achieved with one pass of in-situ treatment according to the Mobile Injection Treatment Unit (MITU) technology. Normally, several passes are necessary, depending on the level of treatment required. A single pass could result in significant savings.

REFERENCES

AFCEE. Air Force Center for Environmental Excellence. 1998. Brooks Air Force Base, Texas. Personal communication.

Bear, J. 1979. *Hydraulics of Groundwater*. McGraw-Hill, New York.

Borden, R. C., Gomez, C. A., and Becker, M. T. 1995. Geochemical indicators of intrinsic bioremediation. *Ground Water*, 33(2):180–189.

Chapelle, F. H., McMahon, P. B., Dubrovsky, N. M., Fujii, R. F., Oaksford, E. T., and Vroblesky, D. A. 1995. Deducing the distribution of terminal electron-accepting processes in hydrologically diverse groundwater systems. *Water Resources Res.*, 31(2):359–371.

Clement, T. P., Sun, Y., Hooker, B. S., and Petersen, J. N. 1998. Modeling multi-species reactive transport in groundwater aquifers.

Dendrou, S., and Dendrou, B. 1998. Contaminant transport in the unsaturated zone. In: J. W. Delleur (ed.), *Groundwater Engineering Handbook*, chap. 16. CRC Press, Boca Raton, FL.

Dendrou, S., and Dendrou, B. 1998. Holistic macroengineering approach for environmental information systems. In: J. W. Delleur (ed.), *Groundwater Engineering Handbook*, chap. 19. CRC Press, Boca Raton, FL.

Dendrou, B., and Dendrou, S. 1997. EIS/GWM-SESOIL: An integrated automated computer platform for risk-based remediation of hazardous waste contamination. In: M. Bonazountas, D. Hetrick, P. Kostecki, and E. Calabrese (eds.), *SESOIL in Environmental Fate and Risk Modeling*, chap. 21. Amherst Scientific Publishers.

Dendrou, S., Dendrou, B., and Tumay, M. T. 1995. Theoretical framework for simulating sustainable geo-environmental technologies, In: ASCE Specialty Report No. 47, Geo-Environmental Issues Facing the Americas, San Juan, Puerto Rico, October 1995, pp. 69–78.

Dendrou, B., and Dendrou, S. 1994. Environmental risk assessment under the EIS/GWM platform, hazard identification, exposure assessment of risk characterization. ZEi Report Mi-95-M030.

Domenico, P. A., and Shwartz, F. W. 1990. *Physical and Chemical Hydrogeology*, J. Wiley, N.Y.

GMS. 1997. Department of Defense Ground Water Modeling System. Brigham Young University.

Harr, M. E. 1987. *Reliability-Based Design in Civil Engineering*. McGraw-Hill, New York.

Riser-Roberts, E. 1992. *Bioremediation of Petroleum Contaminated Sites*. C. K. Smoley, Boca Raton, FL.

Tumay, M. T. 1993. Current research emphasis in geo-environmental engineering. *Geotechnical News*, 11(1):36–40.

USEPA, 1989. Risk Assessment Guidance for Superfund, Volume 1. *Human Health Evaluation Manual* (Part A). 540, 1–89, 002. EPA.

Valocchi, A. J., and Malmstead, M. 1992. Accuracy of operator splitting for advection-dispersion-reaction problems. *Water Resources Res.*, 28:1471–1476.

Wiedemeier, T. H., Wilson, J. T., Kampbell, D. H., Miller, R. N., and Hansen, J. 1997. Technical protocol for implementing intrinsic remediation with long-term monitoring for natural attenuation of fuel contamination dissolved in groundwater, Volumes 1 & 2, Air Force Center for Environmental Excellence, Technology Transfer Division, Brooks AFB, San Antonio, TX.

Wiedemeier, T. H., Swanson, M. A., Moutoux, D. E., Gordon, E. K., Wilson, J. T., Wilson, B. H., Kampbel, D. H., Hansen, J. and Haas, P. 1997. Technical protocol for evaluating natural attenuation of chlorinated solvents in groundwater, Air Force Center for Environmental Excellence, Technology Transfer Division, Brooks AFB, San Antonio, TX.

Wilson, J. T., Sewell, G., Caron, D., Doyle, G., and Miller, R. N. 1995. Intrinsic bioremediation of jet fuel contamination at George Air Force Base. *Bioremediation*, 3(1):91–100.

ZEi Engineering. 1994–1998. The EIS/GWM Integrated Computer Platform: "Environmental Information System for Contaminant Migration Simulations and Risk Assessment," Theoretical Manual, User's Guide, Biorem3D Modeling, Reports No. Mi-96-G001, G002, Mi-96-M010, M012.

Zhang, Z., and Tumay, M. T. 1999. The non-traditional approaches to soil classification derived from the cone penetration test. *ASCE J. Geotech. GeoEnviron. Eng.*, 125(3).

36
Application of Carbon Dioxide in Remediation of Contaminants: A New Approach

Katta J. Reddy
University of Wyoming, Laramie, Wyoming

I. INTRODUCTION

Pollution of natural resources (air, soil, and water) is a concern for the public because of potential health effects. In the mid-twentieth century, Rachel Carson warned of the consequences of pollution of natural resources, which enlightened public awareness of the need for environmental protection (1). There were a number of pollution incidents in which human health was endangered. As an example, the Love Canal in New York was a site for disposal of chemical wastes for a period of over 30 years. Subsequent inhabitance of the site resulted in serious health problems in residents. Similar incidents were reported in other places in the United States, Europe, and Asia (2).

When pollution effects persist and natural processes are unable to keep pace, serious ecological problems can arise. A unique example of such phenomena is the rise in the concentration of Earth's atmospheric CO_2. Before the Industrial Revolution, the CO_2 concentration in the atmosphere was around 280 ppm. Current atmospheric CO_2 concentration is approximately 365 ppm, and it is increasing at the rate of 0.5% per year (3). Coal constitutes a large portion of the fossil fuel consumed in the world and is therefore responsible for a large share of CO_2 emissions (4–6). Carbon dioxide in the atmosphere plays an important role. It regulates global climate by retaining some of the heat radiation. CO_2 from the atmosphere is also converted to organic matter by plants through photosynthesis, which control many chemical and biological processes in water and soil. Thus, an increase in atmospheric CO_2 could interfere with physical, chemical, and biological processes which occur in the atmosphere, hydrosphere, and geosphere.

Another potential issue associated with the coal combustion process is the generation of solid wastes such as fly ash and bottom ash. Currently, a small portion of the total solid wastes produced from coal combustion processes is recycled in cement products and road construction. The remaining is stored in surface impoundments and/or landfills, which could potentially contaminate soil, surface water, and groundwater. To minimize such effects, it is necessary to evaluate techniques for solving pollution problems in situ. In-situ control of contaminants in solid wastes, before they migrate from disposal environments to soils and groundwater, is preferred in any remediation actions, because it is more effective and less costly. Thus, the specific objectives of this chapter are to (a) review remediation methods for different contaminants, (b) evaluate use of CO_2 in remediation of contaminants in coal combustion solid wastes, and (c) identify future research needs for CO_2-based remediation methods.

II. REMEDIATION METHODS

The National Research Council (7) reported that approximately 300,000 to 400,000 contaminated sites in the United States require remediation, at an estimated cost of $500 to $1 trillion. Remediation approaches can be divided into three groups: (a) intrinsic remediation; (b) engineered remediation; and (c) in-situ immobilization of contaminants. In-situ remediation techniques include both natural bioremediation (intrinsic bioremediation) and engineered bioremediation. It is widely recognized that the most significant factor in the time-dependent decrease of volatile BTEX (benzene, toluene, xylene) compounds in subsoils is degradation by soil microorganisms. Evidence suggests that microorganisms of subsoil inherently biodegrade organic contaminants at varying rates (5–50% per day) (8). Pioneering work by Raymond et al. (9) led to increased use of in-situ engineered bioremediation to clean-up of soils and aquifers contaminated with petroleum products. Chambers (10) presented an excellent overview of various chemical and biological remediation techniques, including soil flushing, pump and treat, electrokinetics, and soil vapor extraction. This publication also provided information regarding the cost economics of different remediation approaches.

In-situ immobilization of contaminants in solid wastes will likely be an attractive alternative in many places, due to a lack of available clean-up technologies and/or high costs. One of the most commonly used techniques to immobilize pollutants is stabilization. Cullinane et al. (11) described this technique as follows. Stabilization refers to the process of limiting the solubility or mobility of contaminants with or without change or improvement in the physical characteristic of the waste. Sorbents such as alkaline fly ash, limestone, kiln dust, and bentonite are commonly used to immobilize contaminants, because sorbents are inexpensive and leaching of contaminants is greatly reduced. However, certain sorbent materials can be toxic to plants and microorganisms.

III. REMEDIATION OF CONTAMINANTS WITH CO_2

In earlier sections, importance of natural resources and consequences of environmental pollution, as well as different remediation approaches, were discussed. In this section, a new process for remediation of contaminants in coal combustion solid wastes is discussed. In this process, calcite will be precipitated in coal combustion solid wastes with CO_2. Calcite is a very reactive mineral and can influence organic and inorganic contaminant availability and transport in porous media through its effects on pH, adsorption, co-precipitation, and biomineralization processes. Studies have shown that calcite can adsorb both inorganic as well as organic contaminants from aqueous solutions. These contaminants include cadmium (Cd), copper (Cu), lead (Pb), manganese (Mn), zinc (Zn), arsenic (As), selenium (Se), carboxylic acids, alcohols, amines, amino acids, and carboxylated polymers. In addition, unpublished data from our laboratory suggest that in-situ precipitation of calcite in solid wastes enhances biological activity, which in turn helps biodegradation of organic contaminants.

A number of studies have suggested that calcite can immobilize contaminants in different environments. Zavarin and Doner (12) examined Se, Ni, and Mn interaction with calcite. The results suggested that calcite can immobilize these contaminants in soils through adsorption and coprecipitation processes. van Proosdij and Reddy (13) studied immobilization of contaminants with in-situ calcite precipitation. In this study, calcite was precipitated in aqueous solutions by injecting 5% CO_2 in the presence of Cd, Cu, and Pb. Results suggested a 98% reduction in the concentration of Pb within 70 min of reaction. With the same reaction time, a 62% reduction in the concentrations of Cd and Cu was observed. These results suggest that in-situ precipitation of calcite is effective in removing contaminants from aqueous solutions. In another study, van Proosdij and Reddy (14) observed significant adsorption of As and Se from aqueous solutions by the calcite precipitated by bubbling CO_2 through a 1 M calcium chloride ($CaCl_2$) plus 0.5 g of calcium oxide (CaO) solution. Studies by Cowan et al. (15) and Reeder et al. (16) also suggest that calcite effectively immobilizes selenite (SeO_3^{2-}) and selenate (SeO_4^{2-}) in soils through adsorption processes. Thomas et al. (17) and Frye and Thomas (18) studied adsorption by calcite of a wide variety of organic compounds. These studies used carboxylic acids, alcohols, amines, amino acids, and carboxylated polymers. Among these, calcite strongly adsorbed carboxylated polymers and fatty acids, followed by aromatic carboxylates and alcohols.

Coal-burning power plants in the United States account for almost 55% of electricity production. The increased use of coal-burning power plants in production of electricity is predicted to continue well into the twenty-first century. Thus, coal-burning power plants play a key role in the economic future of the United States (5). To comply with the Clean Air Act, coal-burning power plants are developing a variety of clean coal technologies (CCTs) to reduce SO_x emissions. Such technologies include furnace/duct sorbent injection, atmospheric fluidized-bed combustion, spray drying, and wet flue gas desulfurization (FGD) scrubbing. These technologies are currently in various stages of commercial or pilot scale and will likely be used in the future. In a CCT process, an alkaline sorbent is used to scrub SO_x from flue gas. Additionally, during the combustion process, high temperature drives off CO_2 from carbonate phases. Subsequently, the pH of the CCT solid waste aqueous extracts increases; this affects the solubility and mobility of contaminants.

Like any other process, the coal combustion process also generates by-products. For example, in 1990, coal-burning power plants produced approximately 100 million tons of solid wastes. About 62% was fly ash, 23% was bottom ash and boiler slag, and 15% was flue-gas desulfurization ash. Currently, a small portion (20–30%) of the total solid wastes produced is used in cement products, and the remaining (70–80%) is placed in landfills (19). Adriano et al. (20) reported that accumulation of contaminants and soluble salts in fly ash-amended soils are the most significant problems that affect the land application of fly ash to soils. Carlson and Adriano (21) reviewed environmental impacts of coal-combustion solid wastes and reported that land application of fly ash may create trace element imbalances in soils due to high pH. In another study, Pichtel (22) reported that application of alkaline fly ash to soils could inhibit normal nutrient cycling processes because few microorganisms survive under high-alkaline conditions. Even though coal-combustion solid wastes are classified as nonhazardous under the Resource Conservation Recovery Act of 1976, increased environmental awareness and subsequent natural resource policy changes may challenge these regulations (23).

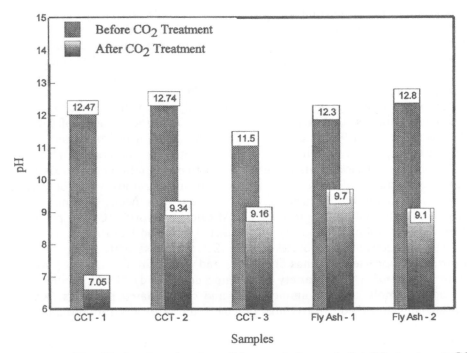

Figure 1 The pH of coal combustion solid wastes before and after CO_2 treatment. CCT-1 = lime injection process; CCT-2 = atmospheric fluidized-bed combustion process; CCT-3 = sodium carbonate injection process; fly ash-1 and fly ash 2 = conventional coal combustion process.

Calcium (Ca^{2+}) is one of the most abundant elements in coal-combustion solid wastes because it is not volatilized during the combustion process (24). Therefore, reactions involving Ca^{2+} and CO_2 are expected to control the pH, solubility, and mobility of inorganic contaminants in coal combustion solid wastes (25,26). Tawfic et al. (27) and Reddy et al. (28) examined the pH of CCT and fly-ash solid waste samples. In these studies CCT samples were collected from lime injection, atmospheric fluidized-bed combustion, and sodium carbonate injection processes, whereas fly ash samples were collected from a conventional coal combustion process. The pH was measured in a saturated paste with an Orion combination electrode. The pH of these samples is presented in Fig. 1. The pH before CO_2 treatment ranged between 11 and 13, depending on the type of process used. Commonly, lime injection, atmospheric fluidized-bed combustion, and conventional coal combustion processes result in alkaline solid waste when compared with sodium carbonate injection process. The high alkalinity of coal-combustion solid wastes is attributed to the hydrolysis of calcium and magnesium oxides, which were produced during the high-temperature combustion process (20,29). These phases react rapidly with water, and as a result, the pH of aqueous extracts of coal-combustion solid wastes approach 12.0.

The pH of alkaline solid wastes decreases slowly as these CO_2-deficient materials absorb CO_2 from the atmosphere (natural recarbonation). As CO_2 is absorbed by the alkaline solid wastes, calcite precipitates, which releases H^+ ions and lowers the alkalinity of the solid wastes to around 8.0 (30). Several studies examined the effects of CO_2 treatment on the pH of coal-combustion solid wastes. Results from some of these studies are shown in Fig. 1. The pH of CO_2-treated samples is ranged between 7 and 9.7.

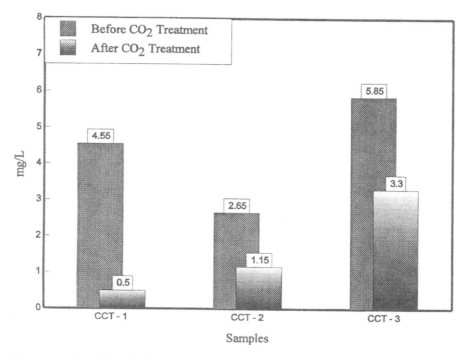

Figure 2 The effect of CO_2 treatment on soluble concentration of Pb in CCT solid wastes samples.

The reduction in pH was attributed to the precipitation of calcite. However, natural recarbonation of alkaline solid wastes occurs slowly. Such slow reactions may not prevent the transport of contaminants from waste disposal environments into soils, surface water, and groundwater.

To enhance the recarbonation process of coal-combustion solid wastes, Tawfic et al. (27) subjected CCT solid waste samples to different CO_2 treatment conditions. Treated and untreated samples were subjected to contaminant mobility studies. The effect of CO_2 treatment on the soluble concentrations of Pb and As are shown in Figs. 2 and 3. These results show that CO_2 treatment effectively reduced soluble concentrations of Pb and As. Furthermore, Tawfic et al. (27) found that CO_2 treatment was more effective for lime injection and atmospheric fluidized-bed combustion-processed samples than sodium carbonate injection-processed samples. Other studies (25,28) reported that CO_2 treatment effectively reduces the mobility of contaminants such as Cd, Pb, Mn, Zn, As, Se, F, and Mo in alkaline solid wastes. As an example, the effect of CO_2 treatment on fly-ash solid wastes for Mn is presented in Fig. 4. These results show that CO_2 treatment lowered the concentration of Mn from 2.2 to 0.8 mg/L for fly ash 1. For fly ash 2, the Mn concentrations was reduced from 15.5 to 2.7 mg/L. These results represent a 64–89% reduction in the concentration of Mn in CO_2-treated samples.

Several studies examined calcite solubility in CO_2-treated samples under different conditions (25,27,30). In these studies, the pH and total elemental concentrations of aqueous extracts from CO_2-treated samples were used as input to either the GEOCHEM (31) or MINTEQA2 (32) speciation model to calculate ion-activity product (IAP) to determine the solubility of calcite. Results suggest that IAPs in CO_2-treated samples were near saturation with respect to calcite (Fig. 5). In addition, X-ray diffraction analysis in these

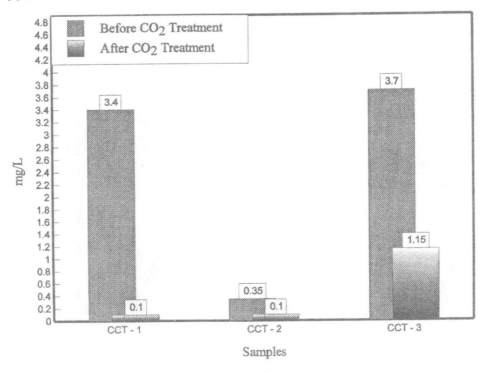

Figure 3 The effect of CO_2 treatment on soluble concentration of As in CCT solid wastes samples.

studies confirmed the formation of calcite in CO_2-treated samples. These results suggest that reacting alkaline solid wastes under CO_2 causes the precipitation of calcite. The Ca^{2+} ions from the dissolution of CaO combine with CO_2 and precipitate as calcite. Precipitation of calcite releases H^+ ions, which neutralize alkalinity. Rai et al. (24), on the basis of thermodynamic evaluations, suggested that dissolved Ca^{2+} concentrations in untreated fly ash samples are controlled by CaO and/or $Ca(OH)_2$ (calcium hydroxide). However, in CO_2-reacted samples, the dissolved Ca^{2+} concentrations would be controlled by calcite. Long-term leaching experiments (33) also have confirmed these predictions. Several other studies also reported similar findings (25,28,30,34).

A possible explanation for the decrease in the solubility of cationic contaminants (e.g., Pb, Mn) in CO_2-treated samples is the precipitation of metal carbonates. In alkaline solid wastes, these metals present as oxides and/or hydroxides [e.g., PbO, $Mn(OH)_2$] due to the coal combustion process. When these samples were reacted under CO_2, oxides and/or hydroxides were converted to carbonates (e.g., $PbCO_3$, $MnCO_3$). This causes a reduction in the solubility because carbonate phases are generally less soluble than oxides or hydroxides between pH 8.0 and 9.0 (35). A reduction in the concentration of solubility of anionic contaminants (e.g., As) is attributed to an increased sorption of these contaminants by the iron oxides associated with the coal combustion solid wastes.

Theis and Wirth (36) conducted experiments to examine the potential contamination of surface water and groundwater associated with the land disposal of coal-combustion solid wastes. Their data suggest that various contaminants including As, Cd, Cr, Cu, Pb, Ni, and Zn become more soluble in alkaline conditions (pH 12). As pH decreases from 12 to 9 these contaminants become less soluble, due to both adsorption and pre-

Application of Carbon Dioxide in Remediation

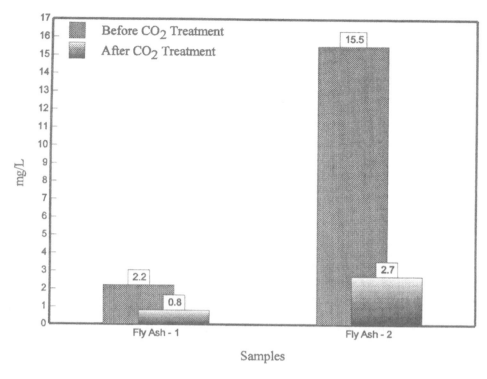

Figure 4 The effect of CO_2 treatment on soluble concentration of Mn in fly ash solid wastes samples.

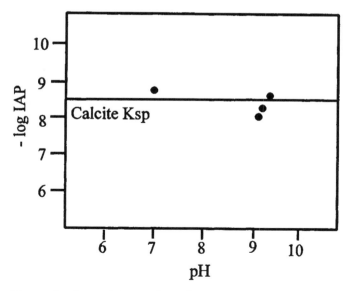

Figure 5 Ion activity products (IAPs) of calcite in CO_2-treated solid waste samples. The solubility product of calcite (K_{sp}) is $10^{-8.48}$ (37).

cipitation processes. Thus, observed reduction in the soluble concentrations of contaminants in CO_2-treated coal-combustion solid wastes is due to the mineral phases formed as pH is reduced through the uptake of CO_2, which in turn enhances both sorption and precipitation processes of contaminants. The proposed in-situ calcite precipitation for remediation of contaminants in solid wastes is an ecologically beneficial process. Calcite can adsorb and coprecipitate different contaminants and reduce their transport from disposal environments into soils, surface water, and groundwater. Additionally, calcite precipitation in alkaline solid wastes buffers pH around 8.5 and permits the growth of microorganisms. Enhancing the growth of microorganisms can help biodegrade organic contaminants.

IV. CONCLUSIONS

There is no question that all segments of society now face and will continue to face environmental problems, particularly those which may effect human health. The need to protect our environment while developing and using of our nation's energy resources has become obvious in recent years. Generation of electricity from coal combustion results in various by-products, including flue gases (SO_x, CO_2) and solid wastes. The new Clean Air Act enacted by the U.S. Congress mandated the reduction of SO_x emissions from coal-burning power plants. As a result, a variety of CCT processes will likely be implemented. In addition, there has been much discussion recently of proposals to reduce atmospheric CO_2 emissions, possibly by enacting carbon taxes. Increase in the concentration of atmospheric CO_2 has advantages and disadvantages. It may increase plant growth rate and food production in moderately weathered soils, but the opposite is true for soils that are in an advanced stage of weathering, particularly acidic soils which comprise 30% of global soils. Furthermore, it could affect biogeochemical processes that take place in atmosphere, soils, and waters.

Currently, a small portion of the total solid wastes produced from coal-fired power plants is recycled. The remaining is stored in surface impoundments and/or landfills, which could potentially contaminate soil, surface water, and groundwater. Several approaches can be developed to immobilize contaminants in coal-combustion solid wastes. However, the application of CO_2 for in-situ precipitation of $CaCO_3$ in coal-combustion solid wastes appears to be the most promising approach. Such CO_2 treatment processes not only reduce the transport of contaminants in coal-combustion solid wastes but also help reduce CO_2 emissions to the atmosphere. Additionally, these processes can enhance potential reutilization of CO_2 treated solid wastes in remediation of environmental problems in other areas. These areas include, but are not limited to, acidic mine spoils and soils, acid mine drainage, contaminated soils, and hazardous waste sites associated with the U.S. Department of Energy and U.S. Department of Defense. Often sorbents are used in these environments to reduce the transport of contaminants into soils and groundwater. Thus, calcite-precipitated solid wastes may be used as a sorbent material in these areas in remediation of contaminants.

V. FUTURE RESEARCH NEEDS

The CO_2 treatment process for remediation of contaminants has demonstrated its potential to reduce the soluble concentrations of contaminants (As, Cd, Pb, and Se) in various solid wastes, which should prevent their migration from disposal environments into soils and groundwater. Most of the CO_2 treatment studies reported in the literature have been conducted under laboratory conditions. Further research to utilize flue gas CO_2 in remediation of contaminants under field conditions will be valuable.

REFERENCES

1. Carson, R. 1962. *Silent Spring*. Library of Congress Catalog Card Number, 60-5148.
2. Gibbs, L. M. 1982. *Love Canal: My Story*. State University of New York Press, Albany, NY.
3. Brown, L. R., Renner, M., and Flavin, C. 1998. *Vital Signs 1998. The Environmental Trends That Are Shaping Our Future*. Worldwatch Institute.
4. Kane, R. L. 1991. Greenhouse gas emissions from coal combustion: A global perspective. In: *Proc. Int. Conf. on Coal, the Environment and Development: Technologies to Reduce Greenhouse Gas Emissions*. International Energy Agency, France.
5. National Research Council. 1995. *Coal Energy for the Future*. National Academy Press, Washington, DC.
6. Hileman, B. 1997. Fossil fuels in a greenhouse world. *Chem. Eng. News*, 75:34–37.
7. National Research Council. 1997. *Innovations in Ground Water and Soil Cleanup*. National Academy Press, Washington, DC.
8. Chiang, C. Y., Salanitro, J. P., Chai, E. Y., Colthart, J. D., and Klein, C. L. 1989. Aerobic biodegradation of benzene, toluene and xylene in a sandy aquifer-data analysis and computer modeling. *Groundwater*, 27:823–834.
9. Raymond, R. L., Jamison, V. W., and Hudson, J. O. 1977. Beneficial stimulation of bacterial activity in groundwater containing petroleum hydrocarbons. *AIChE Symp. Ser. 73*, 166:390–404.
10. Chambers, C. D. 1991. *In-Situ Treatment of Hazardous Waste Contaminated Soils*, 2nd ed., Pollution Technology Review No. 199. Noyes Data Corporation, Park Ridge, NJ.
11. Cullinane, M. J., Jr., Jones, L. W., and Malone, P. G. 1986. *Handbook for Stabilization and Solidification of Hazardous Waste*. Environmental Laboratory, USAE Waterways Experiment Station, Vicksburg, MS.
12. Zavarin, M., and Doner, H. E. 1997. Selenium, nickel, and manganese interactions with calcite. Sorption and desorption of trace elements. In: I. K. Iskandar, S. E. Hardy, A. C. Change, and G. M. Pierzynski (eds.), *Proc. 4th Int. Conf. on the Biogeochemistry of Trace Elements*, pp. 19–20. U.S. Army Cold Regions Research and Engineering Laboratory, Hanover, NH.
13. van Proosdij, E. M. H., and Reddy, K. J. 1997. Immobilization of contaminants with in-situ calcite precipitation: A preliminary evaluation. In: R. Prost (ed.), *Contaminated Soils, 3rd Int. Conf. on Biogeochemistry of Trace Elements*. INRA, Paris.
14. van Proosdij, E. M. H., and Reddy, K. J. 1995. In-situ immobilization of contaminants with calcite: A preliminary evaluation. Agronomy Abstr., 1995: 340.
15. Cowan, C. E., Zachara, J. M., and Resch, C. T. 1990. Solution ion effect on the surface exchange of selenite on calcite. *Geochim. Cosmochim. Acta*, 54:2223–2234.
16. Reeder, R. J., Lamble, G. M., Lee, J. F., and Staudt, W. J. 1994. Mechanism of SeO_4^{2-}-substitution in calcite: An XAFS study. *Geochim. Cosmochim. Acta*, 24:5639–5646.
17. Thomas, M. M., Clouse, J. A., and Lougo, J. M. 1993. Adsorption of organic compounds on carbonate minerals 1. Model compounds and their influences on mineral wettability. *Chem. Geol.*, 109:201–213.
18. Frye, G. C., and Thomas, M. M. 1993. Adsorption of organic compounds on carbonate minerals 2: Extraction of carboxylic acids from recent and ancient carbonates. *Chem. Geol.*, 109:215–226.
19. U.S. Environmental Protection Agency. 1988. Waste from the Combustion of Coal, EPA Rep. 530-SW-88-002.

20. Adriano, D. C., Page, A. L., Elseewi, A. A., Chang, A. C., and Straughan, R. I. 1980. Utilization and disposal of fly ash and other coal residues in terrestrial ecosystems: A review. *J. Environ. Qual.*, 9:333–334.
21. Carlson, C. L., and Adriano, D. C. 1993. Environmental impacts of coal combustion residues. *J. Environ. Qual.*, 22:227–247.
22. Pichtel, J. R., 1990. Microbial respiration in fly ash and sewage sludge amended soils. *Environ. Pollution*, 63:225–237.
23. Tyson, R. 1997. Scientists link coal fly ash disposal to amphibian abnormalities. *Environ. Sci. Technol. News*, 31:408A.
24. Rai, D., Ainsworth, C. C., Eary, L. E., Mattigod, S. V., and Jackson, D. R. 1987. Inorganic and organic constituents in fossil fuel combustion residues. Electric Power Research Institute EA-5176.
25. Reddy, K. J., Drever, J. I. and Hasfurther, V. R. 1991. Effects of a CO_2 pressure process on the solubilities of major and trace elements in oil shale solid wastes. *Environ. Sci. Technol.*, 25:1466–1469.
26. Schramke, A. J. 1992. Neutralization of alkaline coal fly ash leachates by CO_2. *Appl. Geochem.*, 7:481–492.
27. Tawfic, T. A., Reddy, K. J., Gloss, S. P., and Drever, J. I. 1995. Reaction of CO_2 with clean coal technology solid wastes to reduce trace element mobility. *Water, Air, and Soil Pollution*, 84:385–398.
28. Reddy, K. J., Gloss, S. P., and Wang, L. 1994. Reaction of CO_2 with alkaline solid wastes to reduce contaminants mobility. *Water Res.*, 28:1377–1382.
29. Waren, C. J., and Dudas, M. J. 1984. Formation of secondary minerals in weathered fly ash. *J. Environ. Qual.*, 13:530–538.
30. Reddy, K. J., Lindsay, W. L., Boyle, F. W., and Redente, E. F. 1986. Solubility relationships and mineral transformations associated with recarbonation of retorted oil shales. *J. Environ. Qual.*, 15:129–133.
31. Sposito, G., and Mattigod, S. V., 1980. GEOCHEM: A computer program for the calculations of chemical equilibria in soil solutions and other natural water systems. The Kearney Foundation of Soil Science, University of California, Riverside, CA.
32. Brown, D. S., and Allison, J. D. 1992. MINTEQA2: An equilibrium metal speciation model, EPA/600/3-87/D12. U.S. Environmental Protection Agency, Athens, GA.
33. Dudas, M. J. 1981. Long-term leachability of selected elements from fly ash. *Environ. Sci. Technol.*, 15:840–843.
34. Essington, M. E. 1989. Trace element mineral transformations associated with hydration and recarbonation of retorted oil shale. *Environ. Geol. Water Sci.*, 13:59–66.
35. Lindsay, W. L. 1979. *Chemical Equilibria in Soils.* Wiley, New York.
36. Theis, T. L., and Wirth, J. L. 1977. Sorptive behavior of trace metals on fly ash in aqueous systems. *Environ. Sci. Technol.*, 11:1096–1100.
37. Plummer, L. N., and Busenberg, E. 1982. The solubilities of calcite, aragonite, vaterite in CO_2-H_2O solutions between 0 and 90°C, and an evaluation of the aqueous model for the system $CaCO_3$-CO_2-H_2O. *Geochim. Cosmochim. Acta*, 46:1011–1040.

37
TORBED® Process Reactor Technologies: A Novel Treatment Approach for Contaminated Soils

C. E. Dodson and R. G. W. Laughlin
Torftech Canada, Oakville, Ontario, Canada

I. INTRODUCTION

The first TORBED process reactor was invented and patented in 1981. (TORBED® is a registered trademark of Torftech Ltd.). The technology has since been extensively developed to provide process techniques in gas/solid contacting. Over 80 plants have now been installed for applications as diverse as gas scrubbing, fat-free snack production, clinical waste pasteurization, mineral processing, combustion, and other industrial uses.

During more recent years this process reactor technology has been developed for the thermal treatment of highly hydrocarbon-contaminated solids, wood wastes, coal, and spent pot lining. One major advantage of these reactors is their generic ability to flash-process fine particles, often down to a few micrometers, in diameter, allowing high specific throughputs and precision in thermal treatment of hitherto difficult-to-process materials.

II. TORBED PROCESS REACTOR TECHNOLOGIES

The principal characteristics, advantages, and limitations of these reactors are set out below.

A. Compact TORBED Reactor

The Compact TORBED reactor retains a compact shallow packed bed of particles suspended above an annular ring of stationary blades or vanes (somewhat similar to a static set of turbine blades), through which a process gas stream is passed at high velocity (see Fig. 1).

Compact TORBED® Reactor

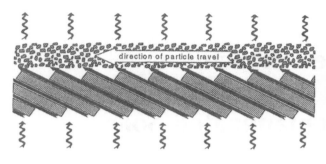

shallow packed bed of a few centimeters in depth

high velocity jets impinge on the underside of the shallow bed providing both lift and horizontal motion

fixed blades generate high velocity jets

Figure 1 Compact TORBED reactor principles.

The high-velocity gas jets (generated in the restriction between the blades) exchange energy on impact with the particles on the underside or base layer of the bed, providing both vertical lift and horizontal motion. This high-velocity impingement enhances the heat and mass transfer to that base layer. The blades and bed are arranged in such a way that the bed mixes rapidly in a controlled fashion, thus continually presenting material into the base layer and thus to the process gas stream (see Fig. 2).

Unlike fluidized beds, where a particle's diameter, density, and geometry dictate the minimum process gas velocity, the process gas mass flow through a Compact TORBED reactor can be set to suit the process—a smaller process gas mass flow can be used but at a higher velocity at exit from the blades to keep the bed in proper motion.

Compact TORBED reactors have similar superficial freeboard velocity restrictions (somewhat increased due to the cyclonic effect of the gas in the freeboard) as bubbling fluidized beds, to minimize elutriation. However, they achieve higher specific throughputs (due to enhanced heat and mass transfer rates) without the inherent high pressure drop, long retention time, and large solids inventory issues associated with fluidized beds. Unlike bubbling fluidized beds, Compact TORBED reactors are not lim-

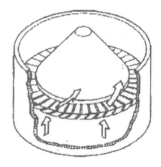 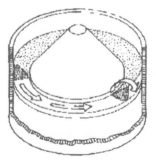

the passage of gas through the fixed blades .... produces a toroidal movement of the particles

Figure 2 Compact TORBED reactor principles.

ited to near-spherical closely sized particles. Indeed, these reactors accept widely graded feedstocks and irregularly shaped feedstocks including shredded, flaked, and complex-shaped extruded materials. Packed beds of widely graded feedstocks can be suspended by a process gas mass flow lower than that required for fluidization of the largest particles.

The small solids holds-up in the Compact reactors is both an advantage and a disadvantage. For processes that can be undertaken in milliseconds, seconds, or at most a few minutes, these reactors can provide real-time process control, allowing the process limits to be explored. The advantages that these reactors bring are:

- Heat/mass transfer rates higher per unit volume, allowing smaller reactor size with rapid start-up and program change.
- Particles are processed faster and with more precision, giving consistent product or process.
- Low process gas stream pressure losses facilitate process gas recirculation and operation with neutral, reducing, or other special atmospheres at high temperatures.
- Ability to process widely graded and irregularly shaped feedstocks.
- Real-time control that allows simplicity in operation and precise and simple automation.

When process retention time (for example, where phase changes are involved) is by necessity more than a few minutes, the small bed mass of the TORBED reactors is unlikely to be economically viable and conventional fluidized bed reactors or rotary kilns will be more applicable. It is worth noting, however, that perceived residence-time requirements derived from other gas–solid contactors are often many times those needed in a TORBED reactor, because of its enhanced heat and mass transfer characteristics.

The Compact reactors surprisingly produce minimal particle degradation due to reduced interparticulate motion (all particles are traveling in the same general direction and are not in collision) and short retention times.

Some applications require an inert resident bed of particles to be held in the reactor into which materials to be processed can be introduced. Liquids, slurries, and sludges can be pumped directly into such a bed for evaporation, combustion, or similar processes (where the bed remains predominantly dry since if the bed becomes fully saturated with liquid, it will cease to operate and will slump). Catalysts have also been used to enhance combustion and dissociation reactions within the reactors.

B. Expanded TORBED Reactor

The Expanded TORBED reactor was developed to retain an expanded diffuse bed of particles. The particles follow a toroidal circulation pattern. Initially they are entrained in a high-velocity central vortex (the process gas stream) whose cyclonic motion creates forces that cause the particles to separate radially outward. The particles are then transferred in an outer downward direction, back to the base of the reactor to be reentrained in the process gas stream again (see Fig. 3).

An Expanded TORBED reactor provides as fast and efficient gas/solid contacting as existing circulating fluidized beds (CFB) and provides the advantages outlined in the comparison shown in Table 1. These reactors can, if required, be configured for high gas and particle velocities that inevitably cause relatively high attrition rates, as also happens

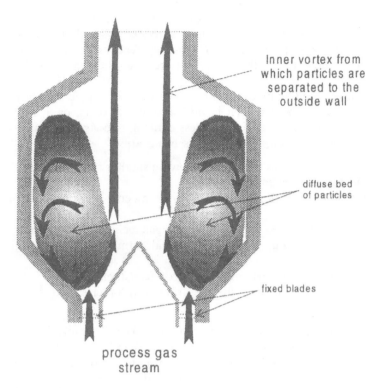

Figure 3 Principles of the expanded TORBED reactor.

Table 1 Comparison of the Characteristics of Expanded TORBED Reactors with Circulating Fluid Beds

Aspect	Circulating fluid bed	Expanded TORBED reactor
Size	Tall vertical vessel to give required retention time of particle in gas in a single pass.	Shorter reactor vessel since gas flow is helical in pattern, giving equivalent contact time.
Recirculation	Normally external to the reactor by cyclone, with solids returned to the base of bed for recirculation.	Solids are separated by centrifugal force directly within the reactor and recirculated internally, with no requirement for a cyclone.
Pressure drop	Reactors have a relatively high pressure drop.	The pressure drop can be as low as 100 mm (4 in.) water gauge.
Solids	It is not possible to selectively discharge differing particle size ranges.	Differing particle size ranges can be selectively removed from the reactor.
Fuel injection	To present particles in a predictable path through a zone of closely controlled combusting gases is difficult.	Particles can be continually circulated through a combustion path in the inner vortex at up to stoichiometric temperatures.

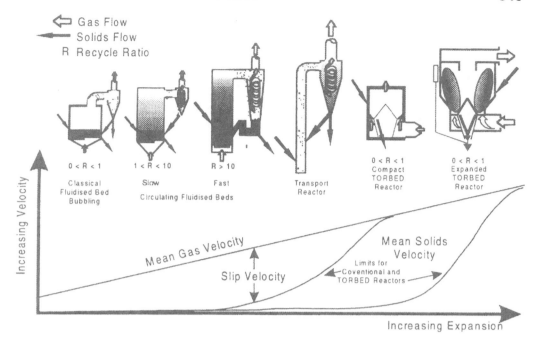

Figure 4 Illustrative comparison of a range of reactor types.

within a CFB. Such attrition can be an advantage in applications where the renewal of a particle surface is beneficial, as in dry gas scrubbing or combustion.

III. GENERAL CHARACTERISTICS

Both the Compact and Expanded TORBED reactors exhibit co-current or modified cross-flow heat transfer characteristics (i.e., the processed material leaves the reactor at the same temperature as the off-gases). A general comparison between a range of reactors and the TORBED reactors is shown in Fig. 4.

IV. PARTICLE RESIDENCE TIME DISTRIBUTION

Particle residence time distribution curves usually approximate to fully stirred reactors. The exception is when there is a physical characteristic of the processed material that can be used to differentiate it and allow separation when processing is complete (e.g., change in density, vaporization, or particle size reduction).

V. FUEL INJECTION

Both the above reactor types can have gaseous fuel injected at blade level, which generates a combustion reaction directly within the bed of material in the reactor.

VI. FINE POWDER PROCESSING

It has been found that fine powders (typically less than 50 μm in diameter) can be injected into the reactor between the blades and the bed such that the powder is transported through a resident packed bed of inert particles (typically 1–2 mm in diameter) held in the reactor. By this means the fine powder can be "flash processed" with retention times within 10–50 ms. Such a processing technique allows temperature lifts of the fine powder of 1000°C or more within this short residence time.

Frequently the retention time is so short that high-surface-area materials are produced with particle morphology that has not been achieved before. This is due to either exfoliation, formation of fine fissures (caused by rapid expansion of gases), or increased porosity. This characteristic of the TORBED process has distinct advantages in situations such as the flash roasting of sulfide ores, where the increased surface area allows for much more effective leaching of the oxidized metal species.

VII. PILOT-SCALE EVALUATION OF TORBED TECHNOLOGY FOR THE REMEDIATION OF CONTAMINATED SEDIMENTS AND SOILS

The TORBED reactor was evaluated in a pilot-scale study to determine its effectiveness in remediating soils contaminated with hazardous organic compounds, such as poly-aromatic hydrocarbons (PAHs). Many of the design features of the TORBED technology discussed earlier support pilot-scale evaluation of this technology for soil remediation; these include

> The ability to handle a wide range of particle sizes, including the ability to handle extremely fine particles of less than 50 μm
> The very precise control of process conditions
> The compact size of the reactor, which allows it to be transportable
> The closed nature of the reactor, which allows total control over emissions

A. Pilot-Scale Study 1: Harbor Sediment

In late 1997 a demonstration of the Compact TORBED reactor technology was undertaken on harbor sediments from Lake Superior adjacent to a wood preserving plant. This work was undertaken in the TORBED pilot plant at ORTECH, and was sponsored by GL & V Process Equipment Group Ltd., Torftech Ltd., and ORTECH Corporation (1).

Two different samples of waste sediment were obtained from the site; both had very similar physical characteristics, typical of sediments. They were about 40% solids content, and the solids had an average particle diameter of 15–16 μm. One sample contained 32,400 ppm total hydrocarbon and 5,500 ppm PAH; the second material contained 8,800 ppm total hydrocarbon and 1,268 ppm PAH.

The samples were dried at 40°C (which resulted in minimal changes to the PAH concentration in the material) prior to testing in the pilot-scale TORBED reactor located at ORTECH, a schematic of which is shown in Fig. 5.

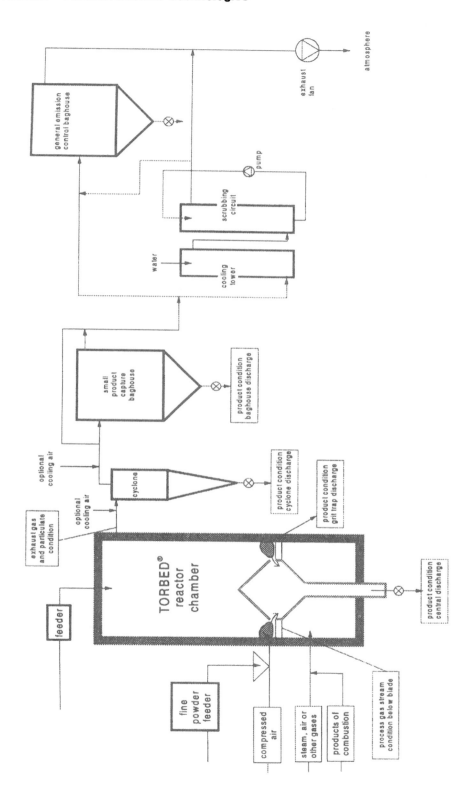

Figure 5 Illustration of the 400 mm TORBED reactor pilot at ORTECH.

Table 2 Summary of Gas Analysis Data, Monitoring, and Audit Tests, TORBED Reactor Test on Harbor Sediments, November 05/97

Test	CO (ppm)	THC (ppm)	NO (ppm)	NO_x (ppm)	SO_2 (ppm)
1. Low PAH feed					
Average	3	2	17	18	9
Minimum	1	1	7	8	3
Maximum	7	12	19	20	21
2. High PAH feed					
Average	6	2	17	17	7
Minimum	2	1	8	7	3
Maximum	21	6	19	19	10

Table 3 Summary of Particulate Emission Data

Test	Solid feed rate (g/s)	Particulate emission rate (g/s)	Percent of particulates captured in cyclone	Emission concentration (mg/Rm^3)
1. Low-PAH material	1.4	0.28	80	1221
2. High-PAH material	1.4	0.28	80	1276

The dried harbor sediments were introduced directly into a resident bed of fused alumina. The powder was fed using a compressed air-operated pneumatic feed system. After treatment the powders exited the reactor with the exhaust gas stream, and were captured in the cyclone and scrubber used to treat this exhaust gas.

The results showed promise. All tests run at 900 to 1050°C resulted in undetectable amounts (< 0.4 ppm) of residual PAH associated with the treated solids.

Two extended test runs were undertaken in which the off-gases were monitored and samples were taken for particulate and trace organic analysis immediately downstream of the cyclone. Table 2 shows a summary of the gas monitoring results on the two runs. The total hydrocarbon levels are reported as ppm methane equivalent.

Table 3 reports the particulate emissions measured after the cyclone. These show an apparent emission rate of 20% of the feed. Obviously, for a commercial application of this technology, a scrubber, hot bag house, or ESP would be used downstream of the cyclone. Our interest in monitoring was directed toward a good understanding of PAH destruction, rather than compliance monitoring. Hence, monitoring at this point, in conjunction with analysis of the product captured in the cyclone, gave us a good estimation of total PAH destruction.

Table 4 summarizes emission data for PAH. Since no PAH was detected on the product from the cyclone, the emission of PAH at this point compared to the feed rate of PAH shows that a destruction efficiency of about 99.99% was accomplished. Monitoring for chlorophenols was also undertaken, with none detected in either the feed or the effluent. Analysis for PAH and chlorophenols was undertaken using a Hewlett Packard 7693 A/I5890 II GC 5971A MSD combination. The detection limit for individual

Table 4 Summary of PAH Emission Data

Test	PAH feed rate (µg/s)	PAH emission rate (µg/s)	Percent destruction	Emission concentration (µg/Rm³)
1. Low-PAH material	1715	0.328	99.981	1.44
2. High-PAH material	7700	0.626	99.992	3.09

PAHs was 0.4 ppm for solids and 0.02 ppm for liquids. The detection for chlorophenols was 0.2 ppm for solids and 0.1 ppm for liquids.

In a test designed to determine how sensitive these high destruction levels were to the temperature of operation in the reactor, the temperature was lowered to 800 and then 750°C. The measured gaseous total hydrocarbon level increased to 24 and 36 ppm, respectively. The residual PAH associated with the treated solid was still undetectable in the product from the run at 800°C and only increased to 8.2 ppm at 750°C (this test having been run with the starting material containing 5500 ppm of PAH). The increase in gaseous hydrocarbon levels at lower temperature without significant increases in residual PAH is probably due to a reduction in the gas-phase oxidation rates of volatilized organic species.

From the results of these tests we concluded that the TORBED reactor technology was excellent for the treatment and remediation of very fine harbor sediments, and that its performance would not be adversely effected by quite large decreases in operating temperatures down to 750°C. However, the pilot runs on this material were quite short and therefore would not predict operational reliability of a commercial plant.

B. Pilot-Scale Study 2: Processing Manufactured Gas Plant (MGP) Wastes

A second program to test the applicability of this technology to processing wastes from old MGP disposal sites is just in the very preliminary stages. This material contains, PAHs, oils, wood chips, cyanides, and other soil materials. Unlike the harbor sediments, this waste contains a very mixed range of sizes, from very fine silt to large chunks of wood. The sample we tested contained about 100 ppm total PAH, 580 ppm cyanide, and 140,000 ppm solvent-extractable hydrocarbons. One preliminary test was run at 1000°C. Table 5 summarizes the level of destruction of PAH, total extractable hydrocarbon, and cyanide achieved in this study.

We believe that the TORBED technology shows promise in this application. The pilot work undertaken on the controlled combustion of wood waste by N. Bolt et al. (2) has shown that the processing of wood waste is readily achieved in a TORBED reactor, which gives confidence in the eventual commercial success of this application.

VIII. CONCLUSIONS

Pilot-scale studies with contaminated soil and harbor sediment wastes have demonstrated that a pilot-scale Compact TORBED Reactor technology can

Destroy PAHs to undetectable levels (<0.4 mg/kg) in solids
Significantly reduce cyanide levels in soils

Table 5 Results of Processing of MGP Purifier Box Waste in a Compact TORBED Reactor Pilot Plant

Parameter	Feed (mg/kg)	Product streams from TORBED reactor	
		Bottom discharge (mg/kg)	Cyclone discharge (mg/kg)
Naphthalene	< 20	< 0.21	< 1
Acenaphthylene	< 20	< 0.2	< 1
Acenaphthene	< 20	< 0.2	< 1
Fluorene	< 20	< 0.2	< 1
Phenanthrene	30	< 0.2	< 1
Anthracene	< 20	< 0.2	< 1
Fluoranthene	23	< 0.2	< 1
Pyrene	23	< 0.2	< 1
Benzo (a) anthracene	< 20	< 0.2	< 1
Chrysene	< 20	< 0.2	< 1
Benzo (b) floranthene	< 20	< 0.2	< 1
Benzo (k) floranthene	< 20	< 0.2	< 1
Benzo (a) pyrene	< 20	< 0.2	< 1
Dibenzo (a,h) anthracene	< 20	< 0.2	< 1
Benzo (ghi) perylene	< 20	< 0.2	< 1
Indeno (1,2,30cd) pyrene	< 20	< 0.2	< 1
Cyanide	580	< 0.5	21
Total extractable hydrocarbons	140,000	< 50	< 50

Destroy total extractable hydrocarbons to below detection levels (< 50 mg/kg) in soils

all without generating unacceptable organic contaminants in the exhaust gases from the process.

The characteristics of the TORBED reactors seem ideal to handle difficult-to-process materials such as the harbor sediments and MGP wastes described above. Their ability to handle both very small particle sized material, and materials with a broad range of particle size and shape, are key in these two applications.

REFERENCES

1. Laughlin et al. 1997. TORBED reactor tests on harbour sediments. ORTECH Rep. 97-11311.
2. Bolt et al. 1996. New reactor provides effective means of processing alternative fuels for electrical power generation. In: *Power Gen. '96 Conf. Proc. Book 3C-Vol. 7*: 291–300.

38
Dispersing by Chemical Reactions Remediation Technology

F. Bölsing
University of Hannover, Hannover, Germany

I. INTRODUCTION: HAZARDOUS WASTE PROBLEMS

This chapter provides a brief overview of the strategy for the application of dispersing by chemical reaction (DCR) technology in the field of environmental remediation and deals, in particular, with the clean-up of toxic waste sites and contaminated land, comprising the immobilization and fixation of heavy metals, the immobilization and detoxification of hazardous organic material, and the DCR-supported biological degradation of nonpersistent organics.

Toxic waste sites means in this context any anthropogenic accumulation of harmful materials in contact with a natural environment. The harmful materials referred to here are rarely uniform in a chemical or physical sense. Usually there are complex toxic mixtures that are often interspersed with discarded junk, such as metal, rubble, timber, and other trash. They may have been deposited at specific waste disposal sites, or accumulated in dumps or lagoons generally convenient to their point of manufacture. The location of such sites often encourages secondary dumping unmarked drums of other potentially hazardous materials (Fig. 1).

The toxic materials may be loosely categorized in one of the three following groups:

> Predominantly inorganic, such as heavy metal compounds in residues from ore processing, flue dust and ash from incineration plants, etc.
> Predominantly organic, such as those derived from the petroleum and petrochemical industries
> Mixtures of inorganic and organic materials, such as contaminated river and harbor dredgings, disused gasworks sites, and acid tars

Their physical form may be as aqueous solutions, immiscible liquid oily phases, sludges, solids, or mixtures of these. Their acknowledged hazard is due to their potential toxicity to living cells, whether in plants or animals, either directly or by accumulation in the food chain.

Figure 1 Some examples of typical waste sites: a pit close to a railway station with some 10,000 tons of bituminous waste (top, left); pastelike mix of soil and oily waste from a petrochemical firm (top, right); stream of oily waste from a refinery site (bottom left) and a landfill charged with more than 100,000 m^3 of an indefinable mixture of organic and inorganic residues.

In many cases it is the fluid oily phase which serves as a carrier for toxic compounds such as PCAH, phenols, polychlorinated aromatics (PCB, dioxins,[1] etc.). An example is given in Fig. 2. The oil lagoon shown contains a mix of water and oily residues with a considerable concentration of the defoliants 2,4,5-T and 2,4-D (Agent Orange). After having passed the earth wall, the oily phase migrates in the form of a thin layer on water into the environment, defoliating the trees around through the toxic compounds inside.

A. Hazard Potential and Endangering Value

A hazardous waste site containing the quantity Q of hazardous compounds, which, if homogeneously distributed into the surrounding soil, would be able to enter n space units of said soil to an extent that a complete intoxication of any local living system will take place, represents a hazard potential HP expressed through N_{pot}, the total number of space units that could *potentially* be affected:

$$Q \cdot N_s = N_{\text{pot}} \qquad (1)$$

where $Q = am$ means the number a of mass units m representing the hazardous waste quantity Q, which, by definition, shall comprise the mass of only liquid and/or leachable hazardous compounds; $N_s = n/m$ means the specific intoxication number N_S representing

---

[1] Any mix of chlorinated dibenzo-*p*-dioxins and the related dibenzofurans will be abbreviated as "dioxins."

Figure 2 Oily waste lagoon surrounded by an embankment as a very simple protection against spreading liquid hazardous contents (top, left). Since the content of this oil lagoon is a low-viscous mix of oily waste and water (top, right), the liquid phases nevertheless penetrate the earth wall (bottom left) and migrate into the environment (bottom right).

the number n of soil space units which can be charged with 1 mass unit m of hazardous waste to an extent that a complete local intoxication takes place.

Since N_S is linked with the toxicity T of a hazardous compound, for instance, via the LD_{50} value, and N_{pot} is the equivalent of the hazard potential HP, Eq. (1) can be written as

$$HP = Q \cdot T \tag{2}$$

Equation (2) states neither more nor less than the simple fact that a quantity Q of a toxic material would just *potentially* be able to intoxicate a certain number of living system units. However, as long as no interaction can take place between the content of the hazardous waste site and the environment, there is no chance for said content to really damage a biological system.

Figures 3 and 4 illustrate what is meant. A hazardous waste site with liquid and leachable components spreads toxic substances into the environment. This complicated three-dimensional process is depending on a lot of variables. In order to simply define the hazard potential it is not necessary to be extremely precise and it is, therefore, permitted to reduce the three-dimensional view of a motion to a one-dimensional one. Thus, a toxic waste accumulation in front of a row of space units of soil represents just a *potential* danger as long as the container will be kept shut (Fig. 3, bottom).

An *actual* hazardous impact can occur only if the toxic content is able to move off-site, for instance, in the form of a gaseous, dustlike, or liquid phase. In the case of liquid phases, oily and/or aqueous ones, the number of soil space units on the migration

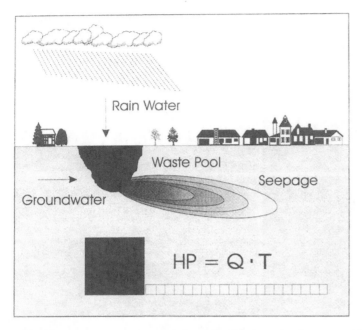

Figure 3 Schematic representation of soil contamination caused by a migrating liquid phase. In the one-dimensional graphic description it is stated that no real intoxication takes place if the waste container remains locked.

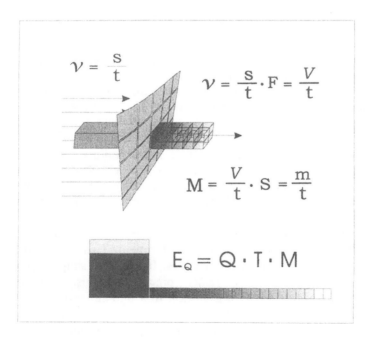

Figure 4 Definition of the mobility M of hazardous compounds by means of the migration velocity v of a contaminated phase and the solubility S of leachable materials. An actual local intoxication occurs if the waste container can interact with the row of soil space units.

pathway effectively intoxicated in a period of time will depend on the migration velocity v, the concentration c of hazardous compounds in the liquid phase, and again on N_S, representing the toxicity of said compounds. A liquid phase migrating with speed v transports volume V per area unit F and per time unit t (Fig. 4).

According to the concentration[2] c or, what comes to the same thing, the solubility S of the toxic compounds in the liquid phase, the mass m passing per area unit F and per time unit t thus represents the mobility M of migrating toxic materials. Hence, the actual endangering value, E_Q, can be defined as

$$E_Q = Q \cdot T \cdot M \tag{3}$$

Equation (3) reads: the higher the quantity Q, the toxicity T, and the mobility M, the higher is the real damaging impact on the environment. But if one of the factors is zero, E_Q will be zero. From this follows that if either the toxicity or the mobility of a given quantity of hazardous compounds has been cut to zero, the endangering impact of a source of danger will have been eliminated.

The quantity of hazardous material leached out in a period of time can be found by multiplying the mobility M by the flow time t_{fl}. The number of actually intoxicated soil space units is then $N_{act} = T \cdot M \cdot t_{fl}$. If t_{com} is the flow time in which the quantity Q has been completely leached out, then, of course, $E_Q = 0$, because the content of the hazardous waste accumulation has left its site.[3] However, in this final state $N_{act} = N_{pot} = HP$, i.e., the potentially highest possible number of intoxicated soil space units has been reached[4]:

$$N_{act} = T \cdot M \cdot t_{com} = TM \frac{Q}{M} = an = N_{pot} = HP$$

B. Appropriate Remediation Strategy

The way to solve environmental problems, as far as hazardous waste sites and contaminated soil is concerned, emerges clearly from the preceding. There are just three possibilities to do remediation work: to remove the quantity Q entirely, or to completely detoxify or to irreversibly immobilize hazardous waste compounds and toxic contaminants respectively (Fig. 5).

1. Removal of Hazardous Waste ($Q \to 0$)

The removal of hazardous substances from a waste site and from contaminated soil is restricted to some special cases. There are mainly two ways to follow this line of

---

[2] Here c refers to the concentration of inorganic and organic toxic compounds in an aqueous phase as well as to the concentration of organic toxic compounds in a relatively harmless organic phase, for instance, heavy metal compounds in rainwater and PCB, dioxins, etc., in mineral oil, respectively. If the organic phase itself must be regarded as the hazardous material, c has to be replaced by the density $d = g \cdot v^{-1}$. The concentration of hazardous compounds in a migrating phase depends mainly on their solubility. Consequently, the solubility S can be substituted for the concentration c.

[3] The same would be true if the waste site were completely dug out, for instance, for the production of a DCR solid fuel.

[4] The description of the motion of a contaminated liquid phase migration in soil completely detached from its original site is very complicated and beyond the scope of this contribution.

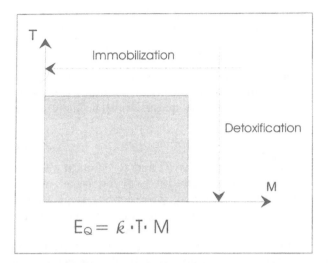

Figure 5 Graphic representation of any sound remediation concept. The actual endangering value, E_Q, of a toxic waste site under given environmental conditions can be cut down to zero by annulling either Q, T, or M. With a given quantity Q of waste, the originally three-dimensional graph becomes two-dimensional, indicating that remediation problems can be solved either through the irreversible immobilization of mobile contaminants or through complete detoxification of the toxic compounds.

remediation work, i.e., the incineration of pure organic waste material such as contaminated waste oil, acid tar, plastics, etc., and the separation of organic and inorganic contaminants in soil and sludge through so-called washing techniques. Incineration, which aims at a utilization of the energy content, is worthwhile only if there are large quantities of organic waste to be burned and if, in addition, the material can be worked up in a way that the resulting products can be reused as an industrial fuel.

Washing techniques[5] are, for physical-chemical reasons, strictly limited by the physical structure of the contaminated matrix and the chemical nature of the contaminants. Since the soil washing process is based on the distribution of a contaminant between at least two phases, in most applications an aqueous washing phase and a solid matrix, the effective distribution coefficient becomes more unfavorable the smaller the particle size and the higher the adsorption capacity of the solid matrix particles and, on the other hand, the lower the solubility of the contaminants will be. This trivial interrelationship allows us to predict the well-known result that, for instance, heavy metal compounds, even of relatively high solubility, can, by no economic means, be separated from a binding soil to an appreciable extent.

2. Toxicity and Detoxification ($T \to 0$)

Paracelsus (1493–1541) seems to have been the first to recognize a dose–response relationship when he made the following statement on toxicity:

---

[5] "Washing" with organic solvents is an extraction process governed mainly by the solubility of the contaminant in the solvent applied and its specific affinity to the solid matrix particles.

> All thynges are poisoun, and no thyng is wythout poyson;
> onely the dose makes a thyng unpoysonous.[6]

It is true that it is first of all of the dose that makes a compound a poison. But since those late-medieval times the description of toxicity has become much more sophisticated. Toxicity is not a question of extremes, toxic or nontoxic, but is a complex quantity comprising not only the relationship between dose and the corresponding observable toxic effects but including a lot of other toxicodynamic and toxicokinetic characteristics which, for instance, refer to molecular biologic interaction quality, exposition conditions, resorption and elimination rates, biotransformation mechanisms, and so on. However, as far as mere detoxification processes are concerned, there is no need to include considerations on details of intoxication mechanisms.

The toxicity of organics can be attributed to characteristic structure elements. A simple example is dimethyl ether, which is a relatively harmless substance, whereas its halogenated counterpart, bis(chloromethyl) ether, is extremely toxic because of its exceptional carcinogenic properties. It is perfectly obvious that the carcinogenicity of the chlorinated ether is generated through the presence of chlorine in the molecule. As a consequence, it is possible to detoxify the toxic chlorinated ether through the removal of the chlorine atoms from the molecule, or, in broader terms, through the destruction of the structural feature responsible for toxicity.

$$H_3C-O-CH_3 \underset{\text{decreased toxicity}}{\overset{\text{extremely increased}}{\rightleftarrows}} ClH_2C-O-CH_2Cl$$
Dimethyl ether   bis(chloromethyl) ether

Unlike the organics, the toxicity of inorganic compounds is in general an invariable property of the ions involved. Therefore, inorganic poisonous materials cannot, as a rule, be detoxified. Well-known exceptions are the cyanide and chromate ions. The first can be oxidized to yield the less toxic cyanate, and the latter might be reduced to the less toxic Cr^{3+}. In most cases, and this mainly includes the large group of heavy metals, any detoxification possibility is completely excluded.

3. Mobility and Immobilization ($M \rightarrow 0$)

In order for damage to be caused to biological systems, there has to be a physiological interaction between a harmful substance and a living organism. This, in turn, not only means that a potentially dangerous material has to be introduced to the biological system in a biologically available form but, first of all, that it has to be mobile. Only in case a mobile toxic material has found its way to the living organism will the above-quoted toxicological parameters be able to play a role, for instance, resorption by, distributing in and biological interaction with a biological system.

An illustration of this basic statement can be provided by considering the way heavy metals, such as lead, cadmium, mercury, etc., can cause environmental problems. This is only possible if they are present as water-soluble, bioavailable compounds. As has been shown in Fig. 4, the mobility M will be zero if the solubility S of heavy metal compounds is zero. As long as the same heavy metals originally were present as natural deposits in their ore forms, i.e., with $S=0$, there was generally no danger to the environment.

---

[6] Kindly translated from medieval German into Middle English by Manfred Markus, Institute of English Language and Literature, University of Innsbruck, Austria.

It is only with the processing of those ores and the subsequent technical application of the heavy metals that they become mobilized and are thus capable of physiological contact with biological systems.

Similar reasoning applies to the potential danger to the environment from accumulated toxic organic compounds. It is not only the specific toxicity N_S or the total amount Q present but much more the mobility and bioavailability as a *conditio sine qua non* which cause their negative impact on living organisms. For example, an oil lagoon composed of resinified material without mobile phases, and with components which cannot be dissolved neither through an oily phase nor through water, does not necessarily represent a danger to the environment.[7]

On the other hand, the hazard potential (HP) of a landfill which contains liquid and/or soluble toxic material is always given through the product QT, independent of whether or not the site of the toxic waste accumulation is sealed, lined, or "stabilized." If just one small part of the protective embedding material (a foil, a compacted layer of clay, concrete, drums, etc.) breaks down, and this may happen at any time, said toxic waste accumulation will constantly release its harmful substances into the subsoil. Even if this occurs at concentrations frequently regarded as permissible, its hazardous content will finally drain away completely. The remobilized substances initially contained in the waste will once again become biologically available and will create an intoxication of the environment to the full amount of N_{pot} living system units.

Stockpiling hazardous waste in a liquid or potentially soluble form is therefore really nothing more than a reduction in the rate at which harmful substances can be released into the environment to create pollution in an ever-widening area. This is expressed through the E_Q value in which the mobility factor M plays the dominating role.

Rainwater, surface water, and groundwater are the most common leaching liquids considered, but liquid organic phases can present problems of the same degree. For example, oil seeping through a waste disposal site can pick up organic contaminants such as polycyclic aromatic compounds, dioxins, and polychlorinated biphenyls inside a waste disposal site. Similarly, seepage water may dissolve toxic metal compounds or organics such as phenols, or may entrain particles of water-insoluble organic substances, or float such materials out from the dumping site as thin surface films and suspensions, respectively.

If surface erosion and the formation of gaseous reaction products from chemical and biochemical interactions are disregarded, the sole mechanism removing contaminants from waste disposal sites and causing pollution of the natural environment is that of mobilization via fluid phases. Remediation concepts that fail to recognize this simple fact will not provide a true "cure" for the problem, as they are aimed only at dealing with the symptoms, not preventing damage at its source.

a. Immobilizing Hazardous Waste by Technical Means. Processes that physically change liquid or semiliquid waste have been defined by the U.S. Environmental Protection Agency (EPA) as waste stabilization and waste solidification. In a stabilization process the hazard potential of waste is reduced by converting the contaminants into their least soluble, mobile, or toxic form[8]; the physical nature and handling characteristics of

---

[7] The hazard potential of an accumulation of toxic material which itself is not liquid and which cannot be dissolved is zero. Cf. the definition of Q in Eqn. (1).

[8] As has been shown, the toxicity can on no account be altered through immobilizing and fixation techniques, but only through the destruction of toxicity-causing structural molecular features.

the waste are not necessarily changed by stabilization. Solidification refers to techniques that encapsulate the waste in a monolithic solid of high structural integrity. Solidification does not necessarily involve a chemical interaction between the wastes and the solidifying reagents, but may bind the waste into the monolith mechanically (1).

A variety of techniques has been tried with varying success to convert toxic liquids and sludges into solid materials. In most cases these stabilization and solidification processes include the application of inorganic materials with high adsorption capacities and/or pozzolanic characteristics. Accordingly, the long-term properties of stabilized and solidified waste depend on the characteristics of the corresponding adsorption/desorption equilibrium and on the mechanical stability of the monolithic systems involved. Typical methods have been based on producing a hardened matrix to incorporate the contaminants using, for example, cement/silicate, lime/fly ash or other pozzolanic material, and gypsum, sometimes along with organic additives.

Problems have occurred as a result of misapplication or when the presence of certain materials, such as certain inorganic compounds or high concentrations of organic matter, have retarded or even prevented the solidification of the mixture, for instance, when cement was employed to solidify PCB-contaminated oil or oily soil. Mixes like these may form "monolithic" solids at the moment but, as experience shows, will disintegrate spontaneously in a few years.

Additionally, some of these techniques have failed under test conditions after repeated wet/dry and freeze/thaw cycles, such as deposited materials would have to withstand in practice.

Similar mechanical breakdowns must be taken into consideration with respect to attempts at sealing mobile phases as a whole at a landfill. Cosmetic measures taken at the peripheral areas of a waste disposal site do not annul the basic laws of physics, hydrogeology, and soil mechanics that apply within it.

One of the sealing measures most frequently used consists of lining the waste disposal site with one or more layers of impermeable plastic foils. Although this procedure may seem to be advantageous, there are nevertheless indications that there are considerable risks that place doubts on the long-term success of this type of remediation measure. A plastic liner can unexpectedly lose its initially optimum chemical and mechanical properties after long-term contact with solid or fluid substances, chiefly with dry solids, and spontaneously disintegrate. Effects of this kind cannot be foreseen, as illustrated by the example of polyethylene in contact with solids, and therefore cannot be accounted for during the planning stage. A further serious problem can arise with mechanical sealing using lining when cracks occur as a result of inhomogeneous settling properties of different landfill materials![9]

Finally, all the peripheral sealing measures require drainage systems and the treatment of seepage water. This produces indefinite long-term costs, especially considering that the processed hazardous substances from a slowly seeping landfill will eventually have to be moved to another deposit or be disposed of in some other way.

b. Immobilizing Through Dispersing Chemical Reaction (DCR). The disadvantages of the techniques described above could be avoided if it were possible to transfer any liquid hazardous waste phase into a solid preparation. Figure 6 illustrates what is

---

[9] In practice, it has become clear that containers, even ones made with high-grade polyenes, unexpectedly become brittle within a few years and finally disintegrate spontaneously if they are used for storing fine-grained dry solids.

Figure 6 Transformation of liquid phases of organic waste (top) of different viscosity (middle) into solid materials (bottom) by chemical means. The reaction products do not contain any traces of the original liquid phases. Accordingly, the mobility M of any former liquid waste compound is zero.

meant. The picture at the top shows a typical hazardous waste site, a former clay pit, with about 100,000 m^3 of a fluid mix of everything that can be regarded as hazardous organic and inorganic waste. Through careful trial dredging, various layers of different viscosity can be isolated. The first one is a two-phase mix of water with a low-viscous bituminous liquid charged with any sort of oleophilic organic toxic compounds (middle/left). The deepest layer consists of resinified acid tars and the like and is hard enough to be cut by a hatchet (middle/right).

All these fractions can, indeed, be reacted to form solids in which the organic phase has disappeared (Fig. 6, bottom). This does mean that, in accordance with the basic requirements of a sound remediation strategy, a complete and irreversible immobilization of nonaqueous fluid phases can be achieved by transforming them into dry pulverulent solid preparations. Hazardous compounds dissolved in an aqueous phase can be, due to the increased reactivity of finely dispersed products, also immobilized through an irreversible chemical and physical fixation of water-leachable harmful substances by removing their solubility, and finally, toxic organic ingredients can be detoxified by chemical or biological means. Thus the DCR technology proves useful to solve a great variety of environmental problems.[10]

This counts not only for the conversion of oily waste materials into finely dispersed solid preparations in order to cancel out the mobility M of a liquid phase, it is also a means to remove the quantity Q, for instance, through the conversion of oily waste into a granulated solid fuel. In addition, DCR technology opens up very effective and at the same time very economic ways for detoxification processes, so that the E_Q value can be nullified through the cancellation of Q as well as of T and M. What DCR variant in particular will be applied depends only on the situation in any given case, mainly with reference to the technical applicability and the remediation target, but never to the costs. This will become clear as soon as the principle of dispersing by chemical means has been explained in detail.

II. DISPERSING BY CHEMICAL REACTION: THE CHEMICAL BACKGROUND

There are a number of chemical reactions in which solid reaction products are formed from solid or fluid starting materials with the aid of a reaction partner, said solid reaction products having a significantly larger specific surface area in comparison with the educts. Chemical reactions of this kind can be applied to homogeneously disperse other substances or substance mixes by chemical means. In order to do this, it is only necessary, as a first step, to charge the educt of a dispersing chemical reaction with the substance to be dispersed (predistribution step) and then, in a second step, to allow the actual chemical reaction to take place (dispersing step). In this way fluid phases, for instance, oily phases as well as aqueous or nonaqueous solutions, are converted into extremely finely dispersed, pulverulent solid preparations[11] (2).

Among the numerous dispersing chemical agents that produce finely dispersed solid reaction products with a large specific surface, and which thus fulfill the requirements for dispersing by chemical means, calcium oxide is by far the most important, especially in the form of a commercially available, highly reactive, pulverized quicklime. This lime is available almost everywhere in unlimited quantities as a raw material that is used on a large scale in many technical fields. After reaction with water, with calcium hydroxide as the corresponding reaction product, it provides a carrier material that is completely problem-free from an ecological point of view. This is why calcium oxide, or quicklime,

---

[10] In order to avoid any misunderstanding, it should be emphasized that the term DCR technology is referring to a simple but very effective method for the homogeneous distribution of coherent phases in the course of a chemical reaction and that it does not mean any sort of "new" chemistry.
[11] In certain cases the activation of the surface area of a suspension of inert fine-grained solids, for instance, red mud, through the reaction of a water-consuming reaction partner is already sufficient to obtain analogous results.

Table 1 Solubility Products L, Calculated Solubility in g/L According to L, and the Total Solubility (including the molecular solubility) in g/L of Some Selected Heavy Metal and Calcium Compounds in Water[a]

Compound	Solubility product L		Solubility Calc. from L		Total solubility[a]	
$PbCO_3$	3.3	E-14	4.9	E-5	11	E-4
$Pb(OH)_2$	4.2	E-15	2.5	E-3	1,550	E-4
PbO					230	E-4
PbS	3.4	E-28	4.4	E-12	8	E-4
$Pb_3(PO_4)_2$	3.0	E-44	7.8	E-7	1	E-4
$Hg_3(PO_4)_2$	1.4	E-26	2.6	E-3		
HgS	1.6	E-54	3.0	E-25	Insoluble	
HgO					530	E-4
$CaSO_4$	2.4	E-5	6.7	E-1		
$Ca(OH)_2$	3.9	E-6	7.3	E-1	16,500	E-4
$CaCO_3$	4.7	E-9	6.9	E-3	140	E-4

[a] For the purpose of comparison, the solubility in the last column is given in units of 10^{-4}.
Sources: Refs. 3 and 4.

is used in many different ways under the heading of DCR technology and related processes in the clean-up of toxic waste sites.

In contrast to the alkali hydroxides, calcium hydroxide, which is to a certain extent water-soluble, forms, on reaction with carbon dioxide in an aqueous solution, calcium carbonate, which is de-facto water-insoluble (see Table 1). This means that when calcium oxide is used in an aqueous system, one will be observing a temporary steep rise in the pH value, but this effect will quickly be compensated for as the hydroxide is neutralized by carbon dioxide from the surrounding medium, for instance, from the atmosphere or from a biologically active surrounding soil. This also means, on the other hand, that any application of calcium oxide as the starting material of a dispersing chemical reaction finally ends up with the formation of a water-insoluble product, which, in the form of limestone, is a natural and harmless basic component of our environment.

The sequence calcium oxide/calcium hydroxide/calcium carbonate occurred in a geological time scale during the birth of the continents, the main product of the carbonating step being dolomite [$CaMg(CO_3)_2$]. Accordingly, the DCR process makes use of the sequence natural limestone → calcium oxide → calcium hydroxide → calcium carbonate (Fig. 7).

The formation of low-soluble calcium carbonate occurs on all exposed calcium hydroxide surfaces, which are being covered with a coherent carbonate crust. The crust acts as a protective and isolating inert layer, which in the course of time increases in thickness due to the continuing influence of the carbon dioxide. Thus, the very last fate of any compacted body of calcium hydroxide will be to become fully carbonated and to form an inert and water-insoluble body of limestone.

The in-situ carbonate formation described here is essential to the long-term behavior of compacts consisting of DCR-treated toxic waste as well as to the mechanisms of the DCR-supported microbiological degradation of certain harmful substances, as will be shown later. The overall DCR immobilizing process is summarized in a laboratory experiment as shown in Fig. 8. This greatly simplified but instructive representation requires some additional remarks. For instance, it is of importance for the practical application

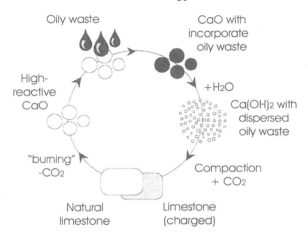

Figure 7 Graphic representation of the separate processing steps involved in the DCR immobilization mechanism utilizing the hydration reaction of CaO. The formation of limestone in the form of calcite from CaO resembles the corresponding natural process during the birth of the Earth.

of DCR processes that not only mixes of an oily phase with an excess of water can be reacted but also dry and nearly dry materials. In case of the latter, water has to be added in at least a stoichiometric quantity.

In order to disperse an almost anhydrous fluid substance such as used mineral oil, bituminous waste, or asphaltlike materials by chemical means, the waste, as can be learned from Fig. 8, must in any case be predistributed in the calcium oxide. In technical-scale applications this can be achieved by means of stirrers, mixers, or kneading machines, depending on the viscosity of the phase to be dispersed. The subsequent addition of the stoichiometric amount of water will then set the exothermic reaction in motion, and within a few minutes the initially fluid or pasty suspension will turn into a dry powder.

If organic waste is to be processed in the form of an aqueous two-phase system, it must be ensured that the organic phase to be dispersed is predistributed in the calcium oxide first, i.e., selectively and in competition with the water present. This can be achieved by first pretreating the calcium oxide with a reaction-retarding substance or, mostly, with a hydrophobizing additive. As a first step, the treated calcium oxide, which is now temporarily inactive or hydrophobic or, in other words, oleophilic, selectively adsorbs the oily organic phase. Only after the organic phase has completely been taken up, and only then, does the charged calcium oxide react in a second step with the water present, forming the dust-dry solid preparation already mentioned. This can also be learned from the sequences in Fig. 8.

If untreated, that is, hydrophilic and therefore automatically oleophobic, calcium oxide were used, it would preferentially and immediately react with the water present in the mixture because of its high reactivity. Therefore, a uniform predistribution of the organic phase is then impossible. The reason for this is that, under the circumstances described, the simple but absolutely essential condition has not been fulfilled, according to which a substance can only be dispersed during the course of a chemical reaction if it has previously been homogeneously incorporated by the starting material of the corresponding reaction. In the example described here, the oil would not be affected, and with the calcium hydroxide that will have been produced, it would form an

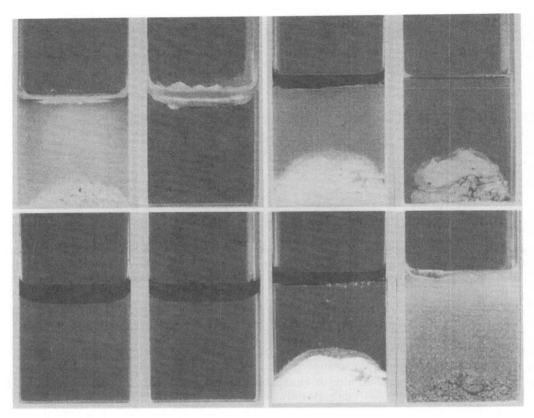

Figure 8 Laboratory demonstration of the basic principle of the DCR process with an oily phase in the presence of water serving as an example. In order to homogeneously disperse the oily phase into the solid carrier $Ca(OH)_2$, it is essential that the oil is incorporated by the CaO before the hydrating reaction commences. In the two cuvettes top/left, CaO has been poured on water—a normal, i.e., nonhydrophobized sample into the left one and a hydrophobized sample into the right one. It is self-evident that the normal CaO will fall to the bottom and immediately react to form $Ca(OH)_2$. The hydrophobized sample, however, floats on the water surface without interacting with the water. The question is, then, what will happen if the same samples are poured on an oil layer floating on water as shown in the couple of cuvettes bottom/left. The result can be seen in the pair of cuvettes at the top/right. Since the normal CaO is hydrophilic, or, what is the same, oleophobic, it will pass the oily layer without any interaction, will fall down to the bottom, and will immediately react with the water to form the hydroxide again. It is important to notice that the CaO has reacted to $Ca(OH)_2$ with its increased surface area but that, nevertheless, the oil layer is still present without any alteration (left cuvette). On the other hand, the hydrophobized, i.e., oleophilic CaO does not pass the oil layer but immediately incorporates it and sinks to the bottom as a uniformly charged pasty mix (right cuvette). After about 20 min, the CaO in the pasty mix reacts with water and forms $Ca(OH)_2$ in the form of a dry, fine-grained powder uniformly charged with the constituents of the oily phase forming a suspension in the surplus of the water. The oily phase has completely disappeared, whereas the oil reacted in the cuvette with the hydrophilic CaO is still present (cuvettes bottom/right).

inhomogeneous, pastelike mass. This result can be derived from the situation in the corresponding cuvette (Fig. 8, bottom, second from right-hand side) if one imagines what would have happened if the water phase between the oil layer and $Ca(OH)_2$ had been removed, i.e., if the excess of water had been decreased.

DCR Remediation Technology

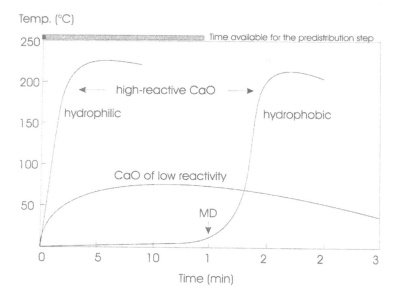

Figure 9 Reaction profiles of the hydration reaction of various types of calcium oxide: high-reactive hydrophilic (left curve), high-reactive hydrophobic (right curve), and low reactive (flat curve). Highly reactive but hydrophobized calcium oxide can be left on water for many hours without any recognizable reaction. In accordance with a DCR specification, appropriately hydrophobized calcium oxide is not permitted to react with water until the hydrophobizing structure has been mechanically destroyed. In the qualification test, the result of which is shown here, the hydrophobic CaO was softly blended with quartz sand and the mix saturated with water. With a test period of 15 min at room temperature there is only a negligible increase of temperature. However, as soon as the mix of CaO, water, and quartz sand has, within a few seconds, been stirred vigorously, a steep rise in temperature takes place, indicating the full reactivity of the liberated CaO. The bar at the top represents the time available for effective incorporation of oily waste. Under the same conditions this time shrinks to a few seconds (dark bar) if the CaO is hydrophilic. CaO of low reactivity cannot be used as a DCR reactant at all.

The dispersing of an aqueous two-phase system, such as an oily sludge containing water in the range of stoichiometric quantities, by means of untreated calcium oxide, might also be successful on condition that a suspension of calcium oxide is produced and homogenized so quickly in both fluid phases of the sludge that the synchronous adsorption of both the oily and the aqueous phases will have been completed before the hydration reaction can start to any extent worth mentioning. The period of time necessary for the homogenizing step can be increased through the addition of hydrophilic organic or inorganic reaction retarders, for instance, through the addition of alcohols or amines or alkali and earth alkali salts, respectively. If, however, for reasons that will be explained later, a hydrophobic reaction product is desired while using untreated calcium oxide, it is necessary to add a hydrophobizing agent to the reaction mixture during homogenization. The principle of the individual procedural variations can be found in the relevant reaction profiles in Fig. 9.

In all three examples, namely, dispersing an anhydrous oil with hydrophilic calcium oxide, an aqueous mix of oil with hydrophobized calcium oxide, and an aqueous mix of oil with hydrophilic calcium in a high-speed mixing step, the DCR process furnishes an equally homogeneous solid preparation, with the original fluid phase vanishing

Figure 10 Microphotograph of a sample of Ca(OH)$_2$ homogeneously chemically charged with the same quantity of a black colored paraffin oil. (Courtesy of David Brown, Neville Pulverized Lime, Inc., Pittsburgh, PA.)

completely. An examination under the microscope proves the absence of a free organic phase (Fig. 10).[12]

As already outlined, it should be expected that this immobilization step will reverse again during the course of time to the extent that the carrier substance calcium hydroxide is dissolved in accordance with its water solubility, even if this is small. In order to exclude this unwanted mechanism completely, solid DCR preparations always have to be made hydrophobic for any application of DCR technology in the field of environmental remediation. The calcium hydroxide rendered water repellent effectively prevents any dissolution, which would lead to at least a partial release of harmful substances while the first thin layers of in-situ-grown insoluble calcium carbonate crystals are being formed. This is the reason why a hydrophobizing agent must be applied even if the dispersing process variant utilizing hydrophilic CaO does not require it.

Dispersing by chemical reaction not only changes the physical properties of the dispersed phase in the sense that coherent phases are disintegrated and liquid phases are converted into pulverulent dry preparations; the chemical properties of all the components are also changed. The reactivity of chemical compounds increases with the degree of their dispersion. Since substances that have been dispersed by chemical means are in an extremely finely dispersed form, they show significantly increased reactivity. Chemical reactions can now be performed that in other cases cannot take place at all or can occur only under drastic conditions; this is of outstanding importance for DCR-supported detoxification processes.

---

[12] The results of a series of systematic microscopic examinations on DCR reacted materials of different types and various concentrations have kindly been provided by David Brown, Neville Pulverized Lime, Inc., Pittsburgh, PA.

This means, for instance, that polychlorinated biphenyls as contaminants in waste oil, seepage oil, oily sludges, and soil can be completely dehalogenated and thus easily made harmless under the conditions of a dispersing reaction (see Section VII.C).

Harmful substances, like any other chemical compounds, have their own characteristic chemical properties that can be used for chemical fixation. Among these are in particular precipitation reactions, complexing, condensation, and addition reactions, but also fixation through adsorption, especially on the basis of donor–acceptor interactions.

It is of practical significance that not only toxic oily, oil-like, and oil-containing phases can be dispersed by chemical means, but also the specific reactive reaction partners, which are then available in a finely dispersed, highly reactive state for chemical fixation of toxic contaminants.

It should be pointed out again that the chemical steps involved here are not at all new and that the fixation measures mentioned here are based on well-known and proven reactions, in which each harmful substance, with its characteristics as a chemical compound, is transformed into a resultant product that can be regarded as harmless from an ecological point of view because its mobility or biological availability has been eliminated.

What is new is that through the application of DCR technology not only the chemical reactivity of the reaction partners is significantly increased, but that, because of the hydrophobic properties of the carrier, the finely dispersed reaction partners are capable of effectively approaching each of the harmful substances, even in heterogeneous and aqueous multiphase systems. A particular advantage is provided by a process in which the precipitation reagents are homogeneously ground into the calcium oxide; in this way they will be transformed into an again optimized highly reactive state during the dispersing reaction in situ, thus increasing their chemical effectiveness. This will be illustrated in the next section by way of example.

III. IMMOBILIZATION THROUGH DCR TECHNOLOGY: HEAVY METALS

A. Heavy Metal Immobilization Through Precipitation

As has been discussed in detail, the mobility M, as the most important factor responsible for the endangering value E_Q, can be abolished in two ways: through the transformation of a mobile liquid phase into an immobile solid pulverulent material or through the cancellation of the solubility S of contaminants soluble in a migrating aqueous or oily phase. As far as heavy metals are concerned, it is only the cancellation of S that can be effected.

The potential toxic impact from physiological interactions with heavy metals can basically be eliminated by converting water-soluble heavy metal compounds into insoluble derivatives, thus immobilizing them, either by precipitation reactions with the formation of insoluble compounds or, if necessary, by reduction to free metal. Heavy metals as such are insoluble in water; but it is known that in a finely dispersed form and under suitable conditions, mainly in the presence of oxygen, they can be reoxidized spontaneously, forming water-soluble ions again.

It is not a simple matter to solve the problems mentioned here, even by means of precipitation reactions with the formation of water-insoluble derivatives. One reason is that, strictly speaking, there are no water-insoluble heavy metal compounds, but only ones of low solubility. Even the heavy metal sulfides, which are so often mentioned in this context, are, with the exception of the black mercuric sulfide, from an environmental viewpoint at best only of low solubility.

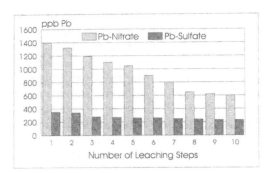 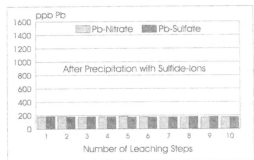

Figure 11 A clay/sand matrix contaminated with 100 ppm Pb in the form of PbNO$_3$ and the same mix contaminated with PbSO$_4$ show the expected leachability with water in a multiple standard leaching test in accord with the different solubilities of the salts. After precipitation with sulfide ions, PbS is formed, which is released continuously in a stream of 180 ppb Pb in the form of PbS.

Here, inappropriate arguments using the relevant solubility products fakes relationships that are completely irrelevant to the question under discussion. The solubility product L for PbS, for example, is given as 3.4×10^{-28} mol^2/L^2. This means no more and no less than that its ionogenic solubility, i.e., the solubility based on the formation of ions in a diluted solution according to the law of mass action, is not more than 4.4×10^{-12} g/L. The fact that lead sulfide is also molecularly soluble, or in other words that PbS is being dissolved in its undissociated form, is frequently overlooked. The information in the literature on the molecular solubility of heavy metal sulfides varies greatly, but there is agreement with regard to the order of magnitude. PbS values range between 3×10^{-4} g/L and 8.4×10^{-4} g/L. This means that the actual solubility of lead sulfide is not only dependent on ion formation and that it is, therefore, eight orders of magnitude higher than the solubility calculated from the solubility product; for numerical comparison in units of parts per million, the solubility product indicates a ionogenic solubility of 0.000000004 ppm, whereas the total solubility is measured to be in the range of 0.8 ppm.

The real situation can easily be demonstrated in a simple leaching test. A mix of clay and sand was contaminated with 100 ppm Pb in the form of readily soluble PbNO$_3$ and relatively low-soluble PbSO$_4$, respectively. The first diagram in Fig. 11 shows the leaching properties of both compounds as expected. After addition of a stoichiometric (+10%) quantity of sulfide ions, PbS is being formed and, with respect to the ionogenic solubility derived from the solubility product, one could expect the release of Pb to vanish completely. This is not true; the second diagram in Fig. 11 shows a continuous release of Pb of about 180 μg/L leachate caused through the molecular solubility of PbS in water.

Most heavy metals also form phosphates, carbonates, and hydroxides or, if the latter are not stable, oxides of low solubility, as can be seen from the example of the corresponding lead and mercury compounds. On the other hand, the numerical values for molecular solubility can still far exceed the extent that is justifiable from a toxicity point of view. Additionally, these solubility values determined for pure aqueous solutions decline in significance if one is dealing with the solubility of certain heavy metal compounds in such heterogeneous systems as natural soils, ore processing residues, harbor and river mud, etc. (see Table 1).

Among the natural processes in microbiologically active soils there are so many effects, and partly contradictory ones, that it is practically impossible to deal with the problem from a theoretical point of view. Here we find mobilizing mechanisms competing with immobilizing ones, such as adsorption into different soil components, the formation of complexes of low solubility with organic soil components, and precipitation with organic sulfur compounds, e.g., in the form of cysteine, and on the other hand, an increase in the solubility through acidic soil components, the formation of water-soluble complexes via hydrogen sulfide linking, and desorption by ion exchange as examples of reversing interactions.[13]

Thus, heavy metal compounds of low solubility obtained through precipitation, especially hydroxides, carbonates, and sulfides, can be mobilized again under ambient in-place conditions that are relevant to practical application. Here, in connection with the fate of heavy metal sulfides in particular, redox processes play the main role, under abiotic conditions as well as under aerobic conditions with the participation of microorganisms.

For instance, the biological oxidation of heavy metal sulfides of low solubility with *Thiobacillus ferrooxidans* into easily soluble sulfates is so effective under suitable acidic conditions, but only then, that it is used technically for obtaining heavy metals, such as copper, from poor sulfide ores (6).

Finally, heavy metal sulfides can also form hydrosols. This causes a mobility of the relevant metals that once again goes far beyond that attributed to molecular solubility.

The discussion so far has shown that when all the circumstances are considered, including the economic ones, the permanent and complete immobilization and removal of the biological availability of heavy metals is not possible solely by simple chemical means. The example of the biologically inert ore deposits, however, shows us that heavy metals can be permanently and safely removed from every kind of interaction with biological systems if the processing includes a combination of chemical and soil mechanical methods.

Summarizing, one can state that precipitation, adsorption, complexing, etc., can be used to convert water-soluble heavy metal compounds into products having no bioavailability. Under unfavorable circumstances, however, the metal contaminant immobilization can be canceled again by secondary chemical processes.

The remobilizing secondary processes require an aqueous, solubilizing medium and, insofar as they are microbiologically induced, require the necessary aerobic conditions. In addition, water must be present as the mobile phase in order to remove the water-soluble products that might have been formed again.

Remobilizing mechanisms can be excluded provided the heavy metal-contaminated waste is treated in such a way that, in a first step, homogeneous mixtures are produced in which all the heavy metals can be reached to be converted into water-insoluble compounds, and if, in a second step, these mixtures are compacted to form an inert body of soil according to the maximum criteria of soil mechanics. The criteria of soil mechanics mentioned here refer mainly to producing the minimum permeability and the optimum soil mechanical properties such as bearing capacity and density of the soil compact.

Cohesive soils, such as clay-like ones, can be effectively stabilized with calcium ions. The cohesive soil components coagulate as a result of ion exchange, so good compress-

---

[13] A detailed discussion of the state of techniques concerned with the immobilization of heavy metals can be found in Ref. 5.

ibility can be achieved even with wet conditions. Noncohesive soils, such as sandy ones, can be stabilized with bituminous bonding agents, which through adhesive forces produce the desired soil mechanical properties. Both cases produce solid matrices showing low water adsorption and permanent compaction under natural conditions.

Calcium oxide, calcium hydroxide, and calcium carbonate serve as the source for ions, which, as shown, can also be considered as precipitation reagents for heavy metals. For the bituminous fixation agent one uses DCR-dispersed bituminous auxiliary substances in the form of dust-dry solid preparations (7).[14]

Independent of whether the soils are cohesive or noncohesive or the stabilizing medium contains finely dispersed bitumen, the reagents are used in amounts above the values necessary according to the criteria of soil mechanics. In this way about 10–20% CaO is always employed. These concentrations stop all microbiological remobilizing activities. This still applies when the complete carbonating of the hydroxide finally has taken place after periods of time comparable with a geological time scale.

The homogeneity of the mixture consisting of precipitation reagent, soil stabilization agent, and soil necessary for effective chemical immobilization can only be achieved by simple means in exceptional cases, and that is with a dry and noncohesive soil. In order to reach all the metals chemically in a substrate of this kind, it is sufficient to mix in a pulverulent DCR preparation that has been produced by dispersing bituminous fixation agents in the presence of sulfides such as calcium sulfide. Compaction leads to an elastic solid phase of high strength with irreversibly chemically converted toxic substances physically fixed through dissolution in an asphaltlike material.

The relations undergo a decisive change if cohesive soils are introduced. In accordance with the EPA report already cited (5), every attempt at chemical immobilization with aqueous precipitation reagents or with the hydrophilic mixes of stabilizing agent and precipitation reagent previously described is destined to fail, because under these circumstances heavy metals can, at best, be reached chemically in the boundary layers of cohesive soil structures, but not, for instance, within a compact clay phase. In fact, chemical fixation does at first occur in a superficial precipitation step; but after the precipitation reagent has passed through, the heavy metals will be freely released as before.

This applies even more to contaminated soil-like materials with a relatively high water content, e.g., harbor mud and river sediment or simply wet clay. There can be no comprehensive chemical fixation of harmful substances with the usual precipitation reagents, because as a result of the premature interaction between additive and aqueous soil the components form lumps and clods, and the required homogeneity of the whole system cannot be achieved. Neither the complete capture of heavy metal ions nor the optimum compaction step is possible by these means.[15]

When all this is considered, it is not surprising that under practical conditions heavy metal immobilization through mere precipitation reactions is thought to be impossible.

---

[14] A quick and excellent review containing numerous literature references is given in Ref. 8; see also Ref. 9.

[15] Since insufficient attention has apparently been paid to these correlations so far, one can find the following remarks concerning immobilization techniques with sulfides in Ref. 5, p. 63: Status of technology: The treatment of soils by precipitation of sulfides is purely conceptual at the present time. Ease of application: Theoretically, one mole of sulfide reacts with one mole of divalent metal .... Excess sulfide must be added to ensure that the precipitation is as complete as possible. Calcium sulfide may be added as a slurry [sic] and incorporated.

However, as soon as one grinds the precipitation reagent, e.g., potassium polysulfide, together with calcium oxide in the presence of a hydrophobizing agent and mixes this hydrophobic solid preparation into a contaminated heavy clay, the clay disintegrates within a short time into a homogeneous, friable structure, in which all the heavy metals can be comprehensively reached and precipitated.

Through the hydrophobization of all the additives, the precipitation reagents and the soil stabilization agents, there is no premature interaction of these reagents possible with the water contained in the soil. Here the hydrophobic solid preparation serves to a certain extent as the vehicle to allow the homogeneous incorporation of all the components needed for chemical immobilization and fixation.

The amount of dispersing agent required is not particularly large; even the addition of only 10% hydrophobic calcium oxide to a heavy clay renders the clay into a homogeneously disintegrated matter. This is because the CaO does not react prematurely with the water, but only after the completion of the mixing procedure, at the same time forming a finely dispersed system.

Due to the friable structure of the treated clay, the mix can easily be compacted to a dense, solid block that is almost impermeable to water. The carbonating process described earlier commences on its surfaces, forming a particularly effective insulating carbonate layer, which protects the compacted body of soil permanently.

The use of precipitation reagents together with calcium oxide and/or other finely dispersed carrier substances in the form of hydrophobized solid preparations is of essential importance to the success of the comprehensive, irreversible chemical immobilization and fixation of heavy metals. It is only through the DCR step, which takes place inside the soil, that soil stabilization agents and precipitation reagents can be homogeneously worked into cohesive aqueous systems and the precipitation reagent can be converted into a finely dispersed, reactive state (Fig. 12).

The solubility and leachability of Pb as a typical heavy metal in dependence on the ingredients of a soil-like matrix have been carefully investigated.[16] In a first experimental run, continuous leaching tests on a lead salt deposit in a sand matrix were carried out (Fig. 13). In order to provide a comparison, in each case 1000 mg of lead acetate, corresponding to 546 mg of Pb, was dissolved in 100 mL of water and this solution introduced into a deposit of 100 g of dry quartz sand: without the addition of any further component (13.1), after blending the sand with 10 g and 20 g of bentonite (13.2 and 13.3, respectively), with 10 g calcium carbonate (13.4), with 2 g potassium sulfide (13.5), and with 2 g potassium sulfide dispersed in 10 g calcium oxide (13.6) as precipitating agents. The leaching rate was 150 mL/h (see Section VIII). The blank test yields a total quantity of 0.06 mg Pb in 4 h of leachate.

The total quantity of Pb released by the indicated volume is given in milligrams at the end of the bars, the actual Pb concentration in the course of the leaching process is represented schematically through the intensity of color. In order to make the distinction clearer, the graphical representation of the measurement results is a black-based color in the ppm range, and a light-gray-based color in the ppb range. In order to obtain a clear understanding of the immobilization test results, it is useful to imagine that the two scales in the original diagram differ by a factor of 1000. Thus, to give an example,

---

[16] The research work has been executed by a number of graduate students in state examination programs at the Hannover University. The exact quantities of released heavy metals per volume unit as well as a exact description of the experimental procedures can be taken from these papers.

Figure 12 The heavy metals in a smeary, clay-type soil from a mining site (top left) could on no account be immobilized through a mere addition of precipitation agents. The spreading car with the DCR agents had to been hauled by a heavy tractor to get through the pasty mud (top right). As soon as the CaO-based additives had been merged, the clay disintegrated within a few minutes, now allowing easy interaction of the precipitating compounds with the heavy-metal ions (bottom left). After compacting the strongly hydrophobic reaction product (bottom right), heavy-metal release becomes impossible.

the first leaching values after treatment with potassium sulfide, expressed in the original figure by columns a few millimeters high, are contrasted with leaching values from an untreated sample, which, if the same scale were applied, would have to be reproduced with a height of about 200 cm. These relationships correspond to a 99.9975% immobilization of the heavy metal by chemical immobilization.

Both the release of Pb as a function of time and the eluted total amount reveal that even in quartz sand a considerable proportion of lead is initially fixed by adsorption. The major part of the lead ions is leached out of the deposit at the very beginning, i.e., with the first low volume of the passing aqueous phase. The remaining concentrations are of only a few parts per million.

If an increasing amount of cohesive material is added, in this case bentonite, then, as expected, under otherwise identical conditions, the extent of the adsorptive capability of the solid matrix increases. But the measurements also confirm the fact that purely adsorptive immobilization only slows down the rate of release of heavy metal ions (Figs. 13.2 and 13.3 as examples with, respectively, 10% and 20% bentonite).

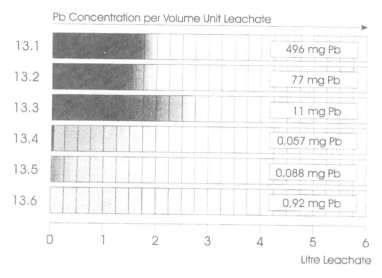

Figure 13 Schematic representation of the results of continuous leaching tests on a lead salt deposit in a sand matrix. The bars show the leaching properties with clean sand (13.1), with a blend of sand and 10% bentonite (13.2), with a blend of sand and 20% bentonite (13.3), with a sand endowed with 10 g of calcium carbonate (13.4), with 2 g of potassium sulfide (13.5), and with a mix of 2 g of potassium sulfide in 10 g of calcium oxide (13.6). The total quantity of leachable Pb is 546 mg.

After the lead salt solution has been treated with the precipitation reagents "insoluble" calcium carbonate (Fig. 13.4) and potassium sulfide (Fig. 13.5), only concentrations in the lower ppb range are obtained. These quantities are far below the saturation concentrations of lead carbonate (1100 ppb) and lead sulfide (800 ppb). The conditions remain practically unchanged even if the elution contact time is considerably increased.

The precipitation reagent calcium hydroxide occupies a special position if one produces it from calcium oxide in the lead salt solution. Depending on the concentration and reaction temperature, one obtains greatly varying leaching curves. In fact this is to be expected, because under these conditions it is possible for several processes to take place at the same time, in particular, the formation of lead oxide and plumbite together with a precipitation as hydroxide.

The result that can be obtained with the precipitation reagent CaO/K_2S also appears, at first sight, to be somewhat unusual (Fig. 13.6); it nevertheless seems to be plausible and this is due to the complex relationships between the individual ions in the system as well as the possibility that the mobilizing effect of the aqueous alkaline medium on lead sulfide can play a part here.

The relationships between the precipitation reagents within a given experimental setup with sand as the model soil, as discussed here, remain basically unchanged if one performs the same experiment on a cohesive soil, as would normally be found in practice.

In order to do this, lead acetate was worked homogeneously into a dry cohesive soil in the form of an aqueous solution, producing 50 g of damp soil containing 1 g of lead acetate. Figure 14 lists the leaching results produced under different conditions, in particular, before any further treatment (14.1), after treatment with calcium carbonate (14.2), and with potassium sulfide/calcium oxide without compaction (14.3) and after additional

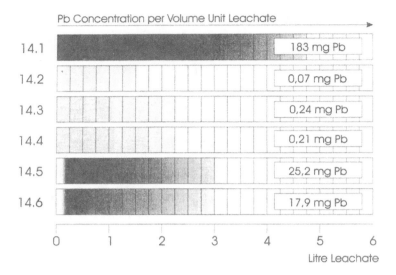

Figure 14 Schematic representation of the results of continuous leaching tests on a lead salt deposit in a clay-type soil matrix. The bars show the leaching properties of the original soil without further additions (14.1), after treatment with 10 g of calcium carbonate (14.2) and with 2 g of potassium sulfide in 10 g of calcium oxide without compaction (14.3), and compacted (14.4). Also shown are the leaching properties of the same contaminated soil in the form of clods in a sand matrix before (14.5) and after (14.6) treatment with an aqueous solution of potassium sulfide.

compaction (14.4). The results show that although the release of heavy metals from a cohesive soil is slowed down with respect to the rate of release of a sandy soil (Fig. 13.1); it nevertheless occurs relatively quickly (Fig. 14.1). The addition of $CaCO_3$ (Fig. 14.2) and CaO/K_2S (Fig. 14.3) leads to the expected heavy metal immobilization; the results correspond closely to those from the tests with sand. On compaction, however, the rate of release is significantly lower, due to the decrease of the active surface area, as can be seen by comparing Figs. 14.3 and 14.4 as an example.

If one kneads the contaminated cohesive soil and forms clods from it (Fig. 14.5), then the reduced permeability will in fact mean that less lead appears than is the case with the nonkneaded test sample; yet heavy metal is still being leached in a continuous flow (compare Fig. 14.1 with Fig. 14.5).

If potassium sulfide solution is added to a bed of sand in which contaminated clods of soil has been placed, an easily recognizable black sulfide layer is immediately formed on the surface. In accordance with the natural conditions, however, during further leaching larger proportions of the sulfide solution are carried through the bed of sand, passing the area with the clods of soil without penetrating the clayey structures. Thus, comprehensive immobilization of the heavy metal does not take place. It is therefore not surprising that the original concentration of heavy metal, as it is released to the value of several thousand parts per billion, immediately recurs after sulfide treatment has ceased (see Figs. 14.5 and 14.6).

If hydrophobic solid preparations with immobilizing reagents are worked in, then it is possible to process a soil that was compact, damp, and also cohesive. Figure 14.4 proves that the heavy metal is effectively immobilized by chemical means through both comprehensive precipitation and compacting applying soil mechanical techniques.

In practical applications, precipitation reagents predominate, which are milled in the presence of a hydrophobizing agent and which use calcium oxide as the carrier. The calcium carbonate that seems to be particularly appropriate in the examples given here is, in fact, far less important than sulfides, which have the advantage of being able to immobilize far more heavy metal compounds than carbonates can.

In general, one can also do without the special addition of carbonate, since it is formed by the hydroxide in the presence of CO_2 anyway. The carbonate developed in situ then not only prevents the calcium hydroxide from starting to dissolve, it also acts as a "trace catcher" for heavy metals in the boundary layers. The "trace catcher" results shown here are provided by experiments in which the boundary layers of contaminated soils were in direct contact with water. Even in optimized samples it is still possible to prove the presence of heavy metals, although the concentrations lie in the lower ppb range. As soon as soils or soil-like materials that, in a first step, have been dispersed together with precipitation reagents are, in a second step, once again compacted into a body of soil, however, then no traces of heavy metals whatsoever are detectable emanating from the contaminated soils, provided the body of soil is surrounded by an inert layer, which is produced from noncontaminated soils with the addition of hydrophobic calcium oxide and, if required, with the relevant precipitation reagents.

Hydrophobic precipitation reagents can also take the place of hydrophobized solid preparations. For example, long-chain organic compounds with an SH group are automatically hydrophobic. These include, for instance, substituted dithiocarbamic acids and thiuramdisulfides; they form heavy metal salts that are particularly difficult to dissolve (10). These compounds, however, do have the disadvantage that they can be easily hydrolyzed, which considerably restricts their application. But there are certain dialkylaryldithiophosphoric acid esters that, depending on the length of the hydrocarbon chain, are relatively stable and have already proved useful in the extraction of heavy metals from aqueous solutions (11). Compounds of this kind are practically water-insoluble; in order to make them suitable for immobilizing heavy metals they have to be transformed into a finely dispersed solid preparation along the lines of the overall concept introduced here. If their own hydrophobic character is not sufficient to keep the whole mix hydrophobic in the concentrations that are to be used, then hydrophobizing has to be undertaken as already described.

If heavy metals are present in addition to organic impurities or are even a part of organic phases of this kind, then the precipitation reagent is bedded directly in a hydrophobic inert material such as high-melting paraffin or bitumen. In this way, both the heavy metals and harmful organic substances are fixed (12).

B. Heavy Metal Immobilization by Means of a Trapping Matrix

There are some aspects concerning the immobilization of heavy metals through precipitation reactions as discussed so far that might be regarded as disadvantageous, in particular, regarding the employment of water-soluble sulfides as precipitating agents. Sulfides on their own are toxic substances, and this is why the application of this very effective agent might be restricted to immobilization techniques, which can be carried out only in cases where no hazard potential to the groundwater can arise.

Another point is that sulfides generate a toxic, foul smell, so that, for instance, thiophosphoric acid esters, which are endowed with an extremely low molecular solubility, cannot be used at all for the immobilization of heavy metals in soil.

Table 2 The Solubility Products of Some Toxic Heavy Metal Sulfides and Iron(II) Sulfide, Indicating the Possibility for Reprecipitation Processes

Heavy metal sulfide	Solubility product
Iron sulfide	4×10^{-17}
Zinc sulfide	1×10^{-20}
Nickel sulfide	1×10^{-22}
Lead sulfide	4×10^{-26}
Cadmium sulfide	6×10^{-27}
Copper sulfide	4×10^{-36}
Mercury sulfide	1×10^{-50}

In the foregoing paragraphs we have learned that water-insoluble precipitation reagents, for instance, calcium carbonate, can be utilized for the immobilization of heavy metals, since the solubility product of most of the toxic heavy metal carbonates is significantly lower than that of calcium carbonate. Thus, on contact with calcium carbonate particles, heavy metal ions will form the corresponding heavy metal carbonates through reprecipitation, provided the calcium carbonate is in a finely dispersed, highly reactive state.

It is also well known that freshly prepared "insoluble" iron sulfide is able to precipitate most of the toxic heavy metals, in correspondence with the mechanisms regarding calcium carbonate as a precipitating agent (see Table 2), but that it will lose its chemical reactivity within a short period of time through a spontaneous aging process. Since this aging step is based on recrystallization processes in which small crystals are being transferred into inactive coarse agglomerates, it should be possible to maintain the original reactivity through storing the reactive form inside a solid matrix.

By means of DCR technology it is possible to achieve both the preparation of extremely finely dispersed, highly reactive iron sulfide as well as the conservation of its original reactivity. In order to do so, stoichiometric quantities of a water-soluble iron salt, for instance, iron(II) sulfate, and a water-soluble sulfide, for instance, potassium sulfide, are dissolved in two different proportions of natural clay by means of a kneader, the ions being adsorptively fixed to the active sites of the layer lattice silicates. Upon kneading both the charged clay proportions, iron(II) sulfide is formed in situ closely incorporated in the structure of the clay constituents in an extremely finely dispersed form. Since there is no mobility of the solid FeS in the pasty matrix, aging processes, i.e., recrystallization, can be excluded. The final mix of FeS-charged clay is then reacted with hydrophobized calcium oxide to form a finely dispersed, pulverulent DCR preparation homogeneously charged with highly reactive iron sulfide.

The heavy metals in contaminated soil, soil-like materials, industrial residues, and so on, can be immobilized through simply mixing in this so-called trapping matrix, for instance, by means of a rotary mixer. When the uniform mix is compacted, reprecipitation becomes possible, and after a certain curing time the toxic heavy metal sulfides will be formed. Figure 15 shows the two steps necessary for a complete and irreversible immobilization and fixation of heavy metals. The left picture represents a sample of heavy metal-contaminated mud and the next one in the center a pulverulent DCR trapping matrix containing finely dispersed high-reactive FeS. On the right is the compacted test body which was formed after mixing the mud with the appropriate quantity of trapping

Figure 15 Schematic representation of a DCR heavy-metal immobilization process through the application of a finely dispersed trapping matrix charged with high-reactive FeS (center) on a contaminated mud (left). After compaction a body has been formed which is practically impermeable to water and which does not release even traces of heavy metals (right).

agent. It would be a misconception to believe that the precipitating compounds would have to surround each particle of the contaminated material. It is only necessary that the precipitating capacity is available in the compacted mix in order to trap each heavy metal ion which would have the idea to start migrating. Sooner or later it will run into an FeS molecule and will then be stopped and trapped through reprecipitation.

Likewise, it is not necessary to crush heavy metal-contaminated coarse materials in order to catch the heavy metal ions inside. Upon embedding the contaminated bricks, concrete parts, chimney breakaway, foundation brickwork etc., in any fine-grained solid enriched with at least the quantity of trapping matrix required stoichiometrically, the hazard potential will immediately be abolished. Again, as soon as heavy metal ions leave their cage, they are immediately immobilized in the form of sulfides. This can easily be demonstrated through a simple experiment. In Fig. 16 (left), parts of a brick lining charged with Hg, Cd, and other heavy metal compounds have been embedded in a

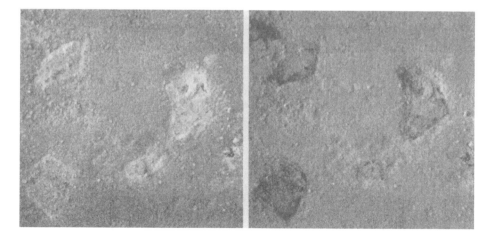

Figure 16 Mobile heavy-metal ions inside coarse pieces (left) are, when appearing on the surface, immediately captured through FeS finely dispersed in the embedding matrix (right).

mix of sand and trapping matrix. The pieces have been saturated with water to give the heavy metal ions a chance for motion. Some days later a black layer covers the pieces, indicating that the ions appearing outside their housing have been trapped.

This procedure works reliably even if said solids are themselves contaminated, such as waste soil, sand, ashes, slags, red mud, etc. The only difference is that the quantity of precipitating compounds has to be adjusted to the heavy metal concentration in the embedding material.

Heavy metal-contaminated matter frequently contains toxic organic compounds, for instance, ashes from incinerators which contain, in addition to the metals, dioxins, PCAHs, and the like. Copper slag is another well-known example. Soil from industrial areas may contain heavy metals as well as toxic organics. As long as the concentration of organics is not as high as to form a liquid phase, these toxic compounds may be immobilized in the same processing step. All one has to do is to disperse about 10% of bitumen in the course of manufacturing the trapping matrix. When the resulting pulverulent DCR product, which looks like the powder in Fig. 15 (center), is mixed with the to some extent double-contaminated material, the heavy metals, after compaction, will be immobilized through precipitation and the organics through incorporation in the resulting asphaltlike body.

As the heavy metal immobilization mechanism includes something like secondary chemical bonding of the sulfides to the layer lattices, it might be possible to talk to a certain extent in terms of chemical fixation. As a matter of fact, continuous leaching tests show that the release of heavy metal sulfides from an accordingly treated contaminated material is much lower than the molecular solubility would indicate (Fig. 17). The trapping matrix can be used not only in its pulverulent form, but also in the form of an aqueous suspension. The question is whether or not the low aging susceptibility of the immobilized FeS in the pulverulent DCR solid would survive in the form of an aqueous suspension.

For this purpose aqueous solutions of heavy metal salts, for instance, 100,000 ppb Cu in the form of copper acetate, were treated on one hand with a solid FeS-trapping

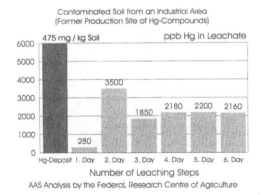 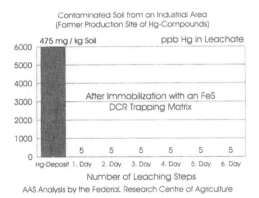

Figure 17 Irreversible immobilization and chemical fixation of mercury in a heavily contaminated soil (475 mg Hg/kg) from a former industrial site where Hg compounds had been produced. A continuous leaching test with the untreated material shows a permanent flow of mercury with a concentration in the range of 2000 ppb per day. After treatment with a DCR trapping matrix with FeS as the precipitating agent, the Hg concentration in the daily leachate is below the detection limit, being about 5 ppb.

Table 3 Aging Test with an FeS Trapping Matrix Applied as a Pulverulent Preparation and as an Aqueous Suspension

Storage time	Trapping matrix solid	Trapping matrix suspension
1 day	72 ppb Cu	105 ppb Cu
2 days	33 ppb Cu	51 ppb Cu
3 days	51 ppb Cu	89 ppb Cu
4 days	33 ppb Cu	72 ppb Cu
5 days	36 ppb Cu	53 ppb Cu

matrix and on the other hand with an aqueous suspension of the same material. Both the precipitation agents had been applied after a storage period of 1 to 5 days. Table 3 shows that the precipitating capacity in both cases does not change over time.

The ability of DCR trapping matrices in the form of suspensions to immobilize heavy metals through injection techniques in subsoil or in contaminated soil that, for technical reasons, cannot be excavated, for instance, soil beneath buildings, is a major advantage. The trapping suspension can also be tailored to solve specific immobilization problems with heaps of inorganic wastes and residues that are too big to be moved at all.

Needless to say, the solid trapping matrix as well as the suspension do not only work with FeS but can be adjusted to any heavy metal problem through the addition of carbonates, phosphates, sulfates, complexing agents, etc., using the same procedure described for the in-situ formation of iron sulfide. Additionally, other agents can be added to improve the physical and chemical properties of the final mix, for instance, to stabilize the suspension.

Besides the injection technique for heavy metal immobilization with water-insoluble agents, the application of water-soluble agents becomes important again as far as clay matrices are involved. For example, a preparation of a pulverulent DCR clay matrix homogeneously charged with potassium sulfide along with a hydrophobizing agent is a versatile tool for the remediation of huge industrial heaps. Through the strong adsorption as well as through the hydrophobic properties of the carrier, the preparation is nearly free from any unpleasant smell.

Figure 18 illustrates the principle of the different application techniques in a bench-scale test. A gradually heavy metal-contaminated soil with hot spots inside might be treated with a trapping matrix in the form of a fluid suspension by means of injection technique (A) or, alternatively, with an appropriately hydrophobized DCR pulverulent preparation of water-soluble precipitation agents in the form of a layer at the top of the contaminated material (B).

The interactions are as follows. In case A, the hot spots are saturated by injecting the precipitating agents, mainly FeS, with a precipitation capacity that had roughly been determined beforehand through drilling tests. In addition, a trapping layer is set up through systematic injections below the contaminated body of soil (not indicated). Since the precipitating agents are water-insoluble, it does not matter, from an ecological viewpoint, how large the injected absolute quantity is. Most of the heavy metal quantity in the hot spots is immobilized directly; the rest of the mobile ions from the hot spots as well as the ions from the contaminated soil around will be captured in the trapping matrix zone when trying to pass this barrier, with rain water as the mobile carrier phase.

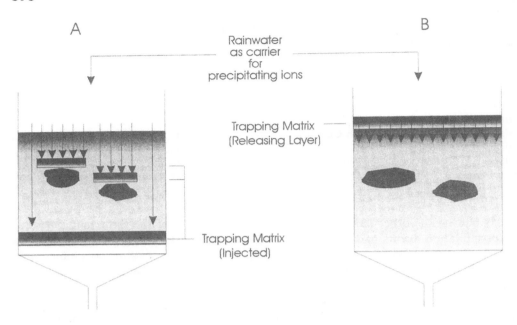

Figure 18 The two types of application of DCR trapping matrices: (a) through injection of water-insoluble precipitating agents in the form of a fluid suspension directly into the hot spots and, in particular, into the subsoil to form a trapping zone below (and/or around) the contaminated space; (b) through covering the contaminated area with a hydrophobic pulverulent DCR trapping matrix charged with water-soluble precipitating agents for continuous release of the precipitating agents.

In case B, the contaminated soil is covered with a pulverulent DCR preparation based on clay, which is charged with water-soluble precipitating agents, for instance, potassium sulfide. If possible, the pulverulent clay matrix might be mixed into the upper soil, e.g., by means of a rotary mixer.

Imagine a huge heavy metal-contaminated heap for decades exposed to the atmospheric conditions. During this time, pathways have been formed through which a continuous release of heavy metals into the environment takes place in accordance with the molecular solubility of the compounds present in the deposits. The rate of release can easily be determined through bench-scale tests employing the basic laws of multiplicative distribution. In most cases the seepage water is led in rivulets; this provides a chance for direct measurements, too.

All one has to do in order to stop this flow of heavy metals is to charge the mobile carrier phase, i.e., rain water, at the beginnings of the naturally formed pathways through which the heavy metals are released. For reasons that have been discussed in the context of Figs. 14.5 and 14.6, it is not at all permitted to add the full amount of the precipitation agents in the form of an aqueous solution, but only in such a way that a continuous but slow release of very small concentrations of the precipitating agents takes place, correlating with the rate of migration of the heavy metal ions in the pathways inside the heap. This can in fact be achieved through the addition of a special slow-release DCR trapping matrix containing the full precipitation capacity necessary for the total immobilization of all heavy metals inside the deposits in the heap.

The rate of release of the precipitating agents from the DCR matrix has to be adjusted carefully and this, actually, is possible not only through adjusting the adsorptive properties of the clay matrix but, much more effectively, through the adjustability of its hydrophobicity. Consequently, what we are dealing with is something like a "steady-state precipitation process," which does need, for a full remediation, longer periods of time, depending chiefly on the size of the heap (pathway distances), weather conditions (rainfall), quantity and solubility of heavy metals in the deposits (rate of migration), and, last but not least, the permeability of the solid waste matrix.

Once the immobilization process has been completed, the heap must be sealed off with a DCR isolating layer in order to avoid any further release of heavy metals, now in the form of the corresponding precipitates, which, as a matter of fact, do have an extremely low solubility, but are not, as we have learned, insoluble in a real sense.

It should be mentioned that the remediation concept for any environmental problem with heavy metals can be refined to any degree of sophistication. In the foregoing example, the foot of the waste heap will be surrounded with a trapping matrix charged with iron(II) compounds for trapping migrating sulfide ions if there should be some that are able to leave the heap.

C. Fixation of Heavy Metals by Means of Functionalized Macromolecules

As has been described in the preceding paragraphs, the main problem regarding heavy metal immobilization arises from two sorts of mobility, i.e., the ionogenic and the molecular solubility, the latter being the most critical. Through the fixation of immobilized heavy metals in a DCR clay matrix, the molecular solubility can be reduced significantly or, in the case of compaction being possible, even be removed. Nevertheless, it is also possible to completely abolish molecular solubility simply by introducing macromolecules with precipitating functional groups, the solubility now being determined solely by the solubility product based on the bonding in the functional group. This principle is well known and has found application especially in wastewater treatment.

Bifunctional organics, for instance, mercapto compounds, can undergo a condensation reaction with OH- groups of active silicates, forming giant molecules that are completely insoluble in aqueous systems (Fig. 19). DCR preparations based on this principle can easily be manufactured as described previously. It goes without saying that since these compounds are really insoluble, there always has to be a mobile aqueous phase that will lead the heavy metal ions to the precipitating functional groups.

IV. IMMOBILIZATION THROUGH DCR TECHNOLOGY: OILY PHASES

A. Converting Oily Phases into Dry Pulverulent Products

The first stage in the usual procedure for immobilizing organic phases consists of a dispersing step with the aid of strongly hydrophobized calcium oxide. Since it is in practice common for a temperature rise in a field process of more than 60°C to occur in the exothermic hydration reaction and inside a plant more than 100°C, caution must be taken if low-boiling or steam-volatile compounds are present. At this point a decision must be made as to whether the DCR treatment should be carried out in a field process on site or in a semiclosed on-site facility. The on-site process has the advantage of being more economical and having practically no limits to the throughput quantity. Treatment in

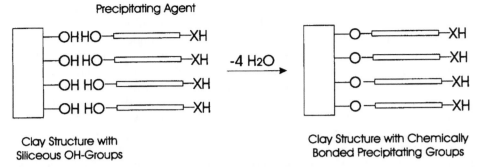

Figure 19 Schematic representation of the formation of a functionalized siliceous giant molecule as an insoluble precipitating agent.

a stationary or even mobile plant (at the present time, facilities of this kind can be supplied with a throughput of up to 20 tons/h), however, has the advantage that wastes containing volatile organics such as benzene and low-boiling chlorinated aliphatics can be processed without problem. These components are forced out of the reactor head space by air or, if necessary, by nitrogen flow for condensing.

From the standpoint of the process it does not matter what the constituents of the organic mixes are: since they are present in the form of a solution, any hazardous compound will be incorporated in the adsorbed layers of chemically dispersed organic phases, provided the quantity of inert material such as aliphatic and naphthenic compounds is relatively high.

It is also irrelevant whether the organic phase is interspersed with inorganic material such as sand, gravel, etc. If mixes like these will be reacted in a DCR process, the reaction product will consist of the "clean" inorganic material and the pulverulent DCR product.

For landfill, the hydrophobic, dry reaction product from the dispersing chemical reaction is built up layer by layer like an earth construction material and compacted, thus forming an inert DCR body of soil (Fig. 20).

As has been stressed previously, the outer surfaces of a non-hydrophobized calcium hydroxide carrier would be susceptible to slow dissolution by rain water or groundwater, thus possibly releasing harmful substances in accordance with the carrier's water solubility. This dissolution step, however, is prevented by the hydrophobizing agent, which

Figure 20 Schematic representation of the transformation of the hazardous oily content of a landfill (oil lagoon) into an inert DCR body of soil.

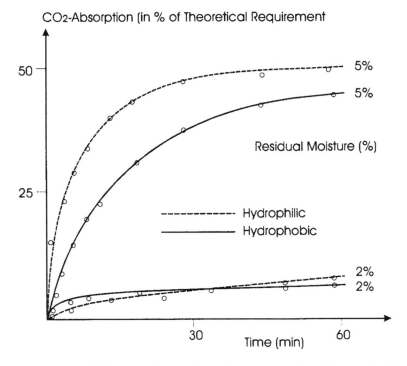

Figure 21 Calcium carbonate formation as a function of hydrophobicity and moisture, the percentage of which is given. Note, for example, the shape of the upper curve: after an initial period characterized through a ready interaction between free calcium hydroxide and CO_2, the rate of CO_2 uptake in the subsequent period slows down more as the protecting carbonate crust is formed. The reaction turns into a diffusion-controlled process, the rate of which is determined exclusively by the thickness of the growing carbonate layer.

has been simultaneously dispersed. It turns out that the hydrophobizing agent plays a double role in this context. Not only does it prevent the reaction of the calcium oxide from taking place with the water of an aqueous toxic waste until the organic phase has been completely incorporated for the DCR step, but it is also designed, in analogy with the chemical fixation system of heavy metals, to protect the calcium hydroxide charged with toxic substances from being attacked from outside until the first protecting insulating layer of water-insoluble calcium carbonate has been formed.

The formation of calcium carbonate, which is based on the interaction between ions in an aqueous medium, occurs surprisingly quickly even on highly hydrophobic calcium hydroxide surfaces. The rate of reaction is determined more by the extent to which moisture is present in the surface area than by its hydrophobic nature (Fig. 21) (13). For instance, hydrophobic calcium hydroxide with 5% residual moisture will plainly react with CO_2 faster than hydrophilic calcium hydroxide with 2% residual moisture inside the pulverulent preparation.

The shape of the curves is of special relevance. It is clearly indicated that under natural conditions, any DCR preparation ends up with the formation of a fully carbonated, i.e., insoluble material, the time necessary for this process being in the range of a few days for small-particle preparations and in the range of a geological time scale for compacted samples (Fig. 22).

Figure 22 A compacted test sample consisting of a finely dispersed water-soluble dyestuff incorporated in a $Ca(OH)_2$ matrix was weathered in the open air for 1 year and then opened. The first result was that no bleeding had taken place, indicating that water could not penetrate the test piece and that, as a consequence, no migration or leaching could take place. The second and, at first glance, more surprising result was that in this long period of time a $CaCO_3$ crust formed that was not thicker than about 1 mm. It was this thin layer that protected the material inside against any natural environmental impact.

The hydrophobic character of the DCR reaction product is retained for a long time; only with the mechanical destruction of the hydrophobic agglomerates does the system become wettable with water. This means that sufficient time for the formation of a carbonate layer is available even with very low CO_2 concentrations in the surrounding medium.

B. Immobilization of Organic Contaminants in Soil

This preceding statement is important with respect to immobilization of toxic organics as contaminants which are present in relatively low concentrations. If a contaminated soil is not, for example, polluted with harmful organics in the shape of a free phase, but contains them in low concentrations adsorbed on soil particles, e.g., in a sandy soil, it is recommended that irreversible fixation be undertaken with the aid of a water-insoluble organic component. Nearly all organic contaminants can most easily be permanently fixed by hard bitumen. Here the same principles apply as for the immobilization of heavy metals. (cf. Section III.B).

High-melting bitumen serves here as a solid solvent. In order to reach all the organic contaminants in a soil without leaving any residuals, it must be homogeneously transformed into a finely dispersed form. It is therefore converted in a DCR plant into a dry pulverulent solid preparation with the aid of hydrophobic calcium oxide. This solid preparation can then be easily incorporated into the contaminated soil.

If the soil contains water or is predominantly cohesive, then it has to be largely dried and dispersed beforehand with hydrophobic CaO. An asphaltlike material is produced after compaction, in which the organic auxiliary phase, the bitumen, has irreversibly incorporated the toxic compounds in the form of a solid solution.

If economic reasons so demand, finely dispersed hard bitumen may be replaced by bituminous waste substances without any detrimental effect, provided the latter do not contain any persistent harmful substances. As can be easily recognized, this whole process is of course merely the slightly altered DCR treatment of an organic phase with toxic components, since here the organic phase is not present a priori with the harmful substances, but is introduced from the outside in a finely dispersed form for the purpose of immobilization and fixation.

If the waste liquids contain toxic substances with active organic functional groups, it is possible for additional chemical fixation to be performed through the simultaneous dispersing of suitable reaction partners. Phenols, halogenated aromatic hydrocarbons, amines, and similar substances are examples among the classes of organics that can frequently create problems because of their good water solubility, especially in alkaline systems.[17]

Polycyclic aromatics and high concentrations of phenols are found in wastes at coking plant sites as well as with so-called ligneous tar. In experiments with ligneous tar, for example, one obtains DCR reaction products that in standard leaching tests still release phenol in concentrations of around 500 ppm.

Donor-substituted aromatics, such as phenols, not only easily dissolve in water but also possess a high level of chemical reactivity that can be used for addition, substitution, and condensation reactions. Through the addition of, for instance, formalin or, for the purpose of study, diisocyanates, finely dispersed in a dry clay-based trapping matrix, it is possible to achieve extensive chemical fixation; reactions of this type are known as clay-supported reactions (14).[18]

With regard to these phenolic compounds, there are considerable differences in reactivity depending on the substitution pattern. Accordingly, the condensation reaction of resorcinol or some alkylated monophenols with formalin in a clay matrix takes place much more readily than the reaction of phenolic derivatives of lower reactivity; this means in practice that the reaction temperature has to be in the range of 60 and 80°C at any rate. On the other hand, the reactivity of phenols in general is so high that this property can be utilized for complete oxidation, which takes place, due to the distinct difference in reactivity, even in surroundings with other reducing compounds such as humic soil (Table 4) (16).

Through the employment of DCR clay-supported chemical reactions, applying agents that cannot only easily be prepared but that are also in a highly reactive state, a great deal of "normal chemistry" can be carried out for detoxification and fixation of organic and inorganic contaminants under rough environmental circumstances, which otherwise would be possible only under clean conditions in the laboratory. Examples are the already-mentioned addition reaction of extremely moisture-sensitive isocyanates with low-concentration phenols and the condensation reaction of homogeneously adsorbed formalin with phenols and amines, the fixation of cyanide through the formation of iron(III) hexacyanoferrate(II) (Berlin bleu), and so on.

---

[17] At this point it should be mentioned that DCR technology is not, of course, restricted merely to the transformation of liquid phases into pulverulent solids, but that in the majority of cases it is only the DCR application of special (i.e., relevant to the particular problem at hand) inorganic reactions and methods as well as those of organic chemistry that permit problems to be solved in an ecologically and economically meaningful way.

[18] Regarding special reactions of phenols on clay, see, for instance, Ref. 15.

Table 4 Chemical Treatment of a Ligneous Tar Releasing Approximately 550 ppm Phenol in a Single Standard Leaching Test with Various Reaction Partners in a Standard DCR Reaction and in a Clay-Supported DCR Reaction (CS DCR)

Compound	DCR power	DCR compact
Formalin basic	258[a]	4.2
Formalin acidic	28	0.6
CS DCR: formalin	n.d.[b]	n.d.
Long-chain amine	13	0.5
CS DCR: amine	n.d.	n.d.
CS DCR: KMnO$_4$	n.d.	n.d.
Diisocyanate	9	0.4
CS DCR: diisocyanate	n.d.	n.d.
Blank value	546	9.8

[a] All figures are concentrations in ppm (4-aminoantipyrin analysis).
[b] n.d. = not detectable.

Since harmful substances with special functional groups relatively seldom occur in such high concentrations as in the examples given here, measures of this kind are not always necessary. In the majority of cases it is sufficient for compounds of this type to be incorporated into a dispersed organic inert phase such as heavy mineral oil or bituminous material and, if necessary, together with a trapping matrix, to achieve permanent immobilization (see blank values in Table 4).

C. Working up Oily Phases in Soil

With the aid of DCR technology, it is possible under certain circumstances to transform organic and inorganic toxic wastes into secondary raw materials and thus to make them available for reuse. This method is preferred to procedures that deposit the material.

There is already so little space available in waste dumps, and in the long term this space will become so scarce and so expensive, that it should be reserved exclusively for the storage of waste materials rendered inert for which there is absolutely no possibility of reuse.

The reuse of secondary raw materials produced with the aid of DCR technology is particularly economical when both the organic and the inorganic components contained in them can be reused within a single technical process. With regard to organic phases, this means that those technical processes can be considered for use in which the organic component either remains intact and, because of its physical properties, can be used together with the carrier or in which their energy content, and thus their use as a carrier-modified fuel, is of great importance. However, the term reuse means not only application of DCR reaction products is a technical sense, but also the reclamation of contaminated land for agricultural purposes.

1. Recultivation of Oil-Contaminated Land

Waste sites containing mobile or mobilizable toxic substances are a problem not only because they continually release harmful substances into the surrounding area, but because they also prevent valuable land from being used for industrial or agricultural purposes. This involves not only the area that is directly blocked by the waste site, but in general the much larger surrounding area as well. Appropriate processing thus means that not only is the actual environmental hazard potential removed, but that the remedied land can be made fully available as an economically usable area again. The actual method employed depends on the future land use. Industrial land use naturally demands soil mechanics of a completely different kind than for the preparation of a field or pasture for agricultural use.

In the first case, the method used will be determined by the same parameters of soil mechanics as those in soil compaction, i.e., high density of the body of soil with optimum strength and bearing capacity, a low level of water permeability and thus its associated resistance to frost, etc.

In the second case, if the remedied area is to be returned to agricultural use, then almost the reverse is required. In this case a loose soil structure is necessary, together with good water permeability and a sufficient capacity for ion exchange. This goal, however, would contradict the actual intent of the remediation process, which is the immobilization and fixation of harmful substances within an especially dense and especially water-impermeable body of soil.

In fact, every kind of immobilization of problematic harmful substances always follows the same principle, which is the compaction of the relevant DCR reaction products into an impermeable DCR body of soil. Reprocessing for agricultural and industrial use thus differ merely with regard to the way they place the compacted body of soil in its natural surroundings. In a recultivation measure the hydrophobic body of soil is incorporated into the prepared ground at a depth that is sufficient to isolate it from cultivated soil on the surface. Here some of the covering soil can come from the periphery of the waste site, as long as it contains no persistent harmful substances, thus forming the basis for its agricultural use at a later date. For this purpose the peripheral soil is previously treated as described for DCR-supported biological degradation in Section VII.D in order to decontaminate it. After a growing season has elapsed it may be necessary to till it with a rotary hoe in order to mechanically destroy any remaining hydrophobic structures of the calcium carbonate that have formed in the meantime. The noncompacted soil that now reacts neutral is subsequently prepared for sowing.

In some cases a suitable filling soil was produced by mixing sand with bentonite in the form of used fuller's earth, as is produced in the processing of vegetable fats. This filling soil is deposited to a depth of between 1 and 1.5 m on top of the compacted DCR body of soil and covered with topsoil. Even this covering layer can consist of the mixture of sand and bentonite described here, provided that compost, sewage sludge, and/or specially processed bark is added to it in order to improve the soil's structure.

Due to water impermeability of the body of soil containing the harmful substances, its surface must have a slope in the direction of the groundwater flow (cf. Fig. 20). In addition, in the case of waste sites with particularly large volumes, instead of a single block it is necessary to plan for several separate bodies of soil of a size that can be justified according to the local conditions. This should avoid any disturbance of the hydrogeological system.

All the procedural variants listed here were tested for the first time in a recultivation measure on the site of a refinery that had been closed. They were also investigated with regard to their effects from the long-term point of view. Figure 23 shows this site (ARAL) before and after DCR recultivation.

During World War II the former gasoline refinery in Dollbergen (Germany) produced, among other things, lubricating oils for airplane engines. The used fuller's earth

Figure 23 Former ARAL refinery site before and after DCR recultivation. Both pictures have been taken from almost the same point, the right one 2 years after completing the operations.

and acid tars resulting from the refining with sulfuric acid were stored in an area that was protected on its sides by raised embankments and insulated from the groundwater by a natural layer of marl.

When the refinery resumed production after the war, its manufacture of lubricating agents provided waste materials that were continuously incinerated in a cylindrical rotary kiln, and the usable heat produced was recycled into low-pressure steam. The refinery closed in 1956, and approximately 48,000 tons of waste materials from the war years were never incinerated, thus remaining on the refinery premises.

In the following period, during which tanks were used for storage of mineral oil products, all attempts to remove or transform these waste materials failed as a result of technical difficulties and enormous costs. Tests were made to determine whether they could, for example, be used in the cement industry or in the production of steel, or whether they could be burned, reprocessed, stored in special waste dumps, or used for sludge farming. The DCR process was also examined, and after thorough testing with the aid of the authorities responsible, it was finally selected and subsequently employed for clean-up.

For the total treatment of approximately 48,000 tons of oil-bearing harmful substances, approximately 8,000 tons of reagents were transported in tank trucks as needed each day over a distance of roughly 270 km from the lime producer RWK Dornap and used the day they were delivered. In order to improve the feasibility of working up asphaltlike toxic waste, approximately 3,500 tons of sand were mixed in with the tars found in an oil lagoon, so that the total amount dealt with in the recultivation process was approximately 90,000 tons when the sand previously removed from the transformation pits is included.

The cost of this was approximately DM 2.8 million, thus producing a unit price per ton of approximately DM 60 for the harmful substances themselves. When compared with other waste disposal processes, this price, in the case of Dollbergen, amounts to less than half the next more expensive form of waste disposal, which would be to deposit the waste in a special waste dump, a process that, moreover, would not at all deserve the name remediation.

The remediation was carried out on an area next to the waste site. After the embankment had been opened, it was possible to remove the harmful substances and place them on the site where they were to be treated. Bituminous acid tars were present. Because of the release of SO_2 when these acid tars were lifted, the machine operators had to wear

Figure 24 The ARAL waste site consisted of high-viscous waste oil, resinified acid tar, charged bleaching earth, and a mix of oil sludge and water (top, left, right, center, left). The photographs at center right and bottom left show a DCR-reacted layer of sludge before and after treatment. The soil mechanical properties of the compacted body of soil were measured, for instance, through the plate loading test (bottom right).

face masks. Metal containers and other bulky objects had to be removed with bulldozers. Liquid components that could not be pumped were transported by truck.

The finished DCR reaction product, which can be easily compacted, was pushed back into the site that contained the waste, the sides and bottom of which have now been isolated against the subsoil. The reaction product is then compacted to a body of soil. The testing results in Table 5 were obtained in plate load-bearing tests on each of the soil layers (Fig. 24).

The body of soil thus produced was finally covered with filling soil, which was produced from sand and spent fuller's earth, and prepared for biological degradation. In

Table 5 Soil Mechanical Data of a Stabilized DCR-Treated Highly Contaminated Soil (ARAL Site)

	Plate loading test (bearing values kp/cm^2)			Plate loading test soil density (Mp/m^3)			
Compaction	E_1	E_2	E_3	Sample	Section	Depth	Density
Immediate value	28	145	152	1 and 2	1	0.15 m	1.59/1.71
24 h later	83	256	275	3 and 4	2	0.15 m	1.33/1.35
5 days later	136	422	482	5	3	0.60 m	1.42

Compression test							
		Modulus of volume change of load stages					
Sample	Color	0–0.13	0.13–0.26	0.26–0.52	0.52–1.04	1.04–2.08	2.08–3.12
1a	Dark	18.2	18.4	20.1	42.8	62.9	107.9
1b	Light	20.2	33.4	34.2	61.2	63.3	104.0
5a	Dark	15.4	19.6	21.3	43.3	53.9	91.0
5b	Light	24.3	28.9	31.9	50.6	58.7	104.0

Density of DCR compact (acid tar): 1.62 t/m^3 Water permeability $k = 1.45 \times 10^{-7}$ m/s

addition, sewage sludge was milled into the upper layers. The remedied land is now used as pasture, after thorough experiments on vegetable material consisting of grasses, clover, vetch, potatoes, and tomatoes had produced no negative results.

Since the example described here was the first large recultivation measure in connection with the remediation of a waste site with an extraordinarily heterogeneous composition, for safety reasons and in order to provide an appliance for monitoring at both levels, i.e., of the hydrophobic body of soil as well as of the covering layer of filling soil, separate drainage systems were incorporated at the outset. These drainage systems were provided with observation wells in the flow direction of the groundwater. Regular control examinations, at first monthly and later annually, were carried out by the authorities responsible in order to determine the pH value and the hydrocarbon concentration, and confirmed that the measure was a complete success (see Fig. 25).[19]

It should be noted that the pH was completely unchanged over the time, indicating fast formation of an insulating, in-situ-grown carbonate layer. Well 6 seems to have a considerably high hydrocarbon (HC) concentration in the first 5 years after treatment. This well is in the center of the flow area of that part of groundwater, however, which has exactly passed the area of DCR-supported biological degradation, and contains the sum of all intermediates possible in the biological degradation pathway. Since the infrared-hydrocarbon (IR-HC) analysis yields the HC concentration of any organic compound, the figures do indicate a particularly high biological activity in the first years for fast biological degradation inside the upper layer containing a mix of DCR-treated,

---

[19] The preceding account is largely quoted from Ref. 26.

DCR Remediation Technology

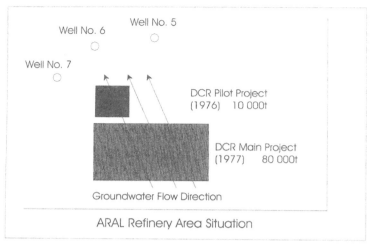

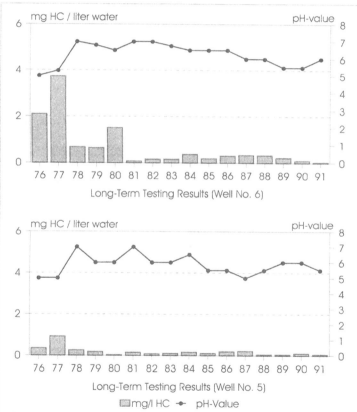

Figure 25 Schematic representation of the local situation of the ARAL refinery site (top) and monitoring results in the form of IR-HC concentrations and pH values of the two most relevant monitoring wells (with respect to the groundwater flow direction): No. 6 (center) and No. 5 (bottom). Since the analytical results referring to well No. 7 are nearly identical with those of well No. 5, that diagram has been omitted here. The average HC concentration in the whole refinery area is 7–8 mg/L.

heavily contaminated fuller's earth, sand, and sewage sludge. The figures do not resemble the oil-HC concentration, as was proved through mass spectrometry/gas chromatography (MS-GC) analysis, which was, with respect to any monitoring well, significantly lower than the oil-HC concentration in the whole surrounding area of the refinery site, which was measured to be between 7 and 8 mg/L.

2. Production of Earth Construction Materials

a. General Considerations. Cohesive soils are usually stabilized by the addition of calcium hydroxide or calcium oxide in concentrations that generally lie between 5% and 10%. Noncohesive soils are preferably compacted with bituminous binders. It is therefore to be expected that a DCR secondary raw material that consists of a highly viscous organic phase and calcium hydroxide will provide a usable soil stabilization agent in both cases (Table 5).

These results have now led to a series of earth construction measures being undertaken with DCR-treated wastes used for building roads and paths, as subsoil in building construction, and for the preparation of industrial land at the site of the waste impoundment. In this last case, especially if there is a large volume of waste, the question of increasing the volume as a result of the treatment gains importance. The extent of volume increase might be expected to correlate with the addition of an average 20% DCR reagent with a mean bulk weight of approximately 1.1 L/kg. Figure 26 shows the results produced by the treatment of the three soil types, sand, coarse clay, and clay, with hydrophobic calcium oxide. One hundred parts by volume of soil (this initial volume is marked by the horizontal line through all three parts of the figure) were in each case contaminated by 30 parts per volume of a bituminous oil. Here the waste volumes are given as dark-dotted areas.

In order to be able to represent the influence of DCR treatment on the total volume in a clear way, more than double the necessary amount of hydrophobic CaO was added, that is, 55 parts per volume in each case. This is indicated by the light-dotted areas. After compaction one obtains the final volumes given in the right-hand columns. According to this, the volume of a contaminated soil with 23% organic components increases by 8% with sandy soils and by 20% with coarse clay; with clay it is reduced by 8%. The figures are in complete agreement with practical experience gained with waste problems. This

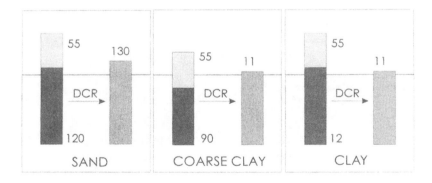

Figure 26 From bench-scale tests as well as of numerous large-scale applications, there is no alteration by volume of highly contaminated soil after DCR treatment and compaction.

demonstrates that after correct application according to data provided by soil mechanics, no volume increase is usually observed after DCR treatment.

The treatment methods that have been presented here for contaminated soils can be applied to other systems contaminated by heavy metals, such as contaminated red mud, dusts and ashes from various sources, and contaminated harbor and river mud or other excavated material. Materials of this kind are produced in large quantities. Because of their heavy metal content they cannot be profitably reused without further processing, although with regard to their other characteristics they can be classified as excellent earth construction materials for building roads and paths, as well as for landscaping.

As has been explained in Section III.B, it is usually only by compacting the body of contaminated soil following chemical fixation that heavy metals are immobilized permanently. In particular, this is the case when sulfide precipitations are performed in biologically active systems such as harbor and river mud. It is frequently found that a greater proportion of the heavy metals is already fixed to sulfur compounds that have developed biochemically. Here it is also possible for biologically induced remobilizing redox processes to occur. It is therefore important to remove the free mobility of all the reaction partners inside the system itself, through the immediate compaction of the earth construction material after treatment.

On the other hand, if precipitations are involved that are not performed with sulfides, then it is not necessary to compact the treated material immediately in order to prevent biological leaching processes. The solubility of most of the heavy metal hydroxides, oxides, or carbonates in water is so slight that remobilization from the hydrophobic mixes can be excluded with the required degree of certainty. One can therefore also treat biologically inactive materials and hold them in reserve for later use in construction.

This method of producing materials for earth construction can be optimized through the addition of small amounts of a specially tailored trapping matrix. In the case of a mix of inorganic and organic contaminants, the DCR preparations also contain finely dispersed bituminous substances. In this latter form of application, the DCR material has to be produced with the aid of cationic hydrophobizing agents. These additives provide not only a strong hydrophobizing effect, so that the earth construction materials obtained this way can be stored in the open air, but also excellent binding forces between inorganic and organic phases. Thus, they have the characteristics of highly effective stabilizing agents, which is revealed by the improvement of certain characteristic soil mechanical data when they are used for earth construction.

Since hydrophobic CaO is in the system in every case, in-situ calcium carbonate formation also occurs. This is particularly useful if a higher concentration of cadmium is to be fixed: the isomorphism of the two carbonates makes it especially easy to achieve complete fixation of this heavy metal (25). If metals such as chromium, which may be present in a higher state of oxidation, form compounds of low solubility only from ions with a lower oxidation number, then one can add ecologically safe reducing agents, such as aliphatic alcohols. Complete reduction and precipitation as chromium(III) hydroxide takes place under the conditions of a dispersing chemical reaction. After compaction, if necessary with the application of an excess of reducing agent, no reoxidation can take place.

b. Examples from Practice. In the case of remediation measures that are combined with tasks of soil mechanics, apart from the removal of the original problem and the reclamation of utilizable land it is the production of a soil construction material that stands in the foreground. In many cases it is only the further use of this soil that makes remediation measures of this kind possible at acceptable costs. Under certain circumstances just the gain in utilizable land and the saving of customary building materials can offset a large part of the costs of remediation.

Nor must the principle of the appropriateness of the means be forgotten when the remediation of waste sites is involved. For example, suggestions aimed at pushing heterogeneous waste sites through a high-temperature incinerator are unrealistic because of the extremely high costs involved, and they cannot seriously be justified.

The following contains three characteristic examples that demonstrate how flexibly concepts for remediation can be developed to fit the individual circumstances, and with the help of which not only is the actual problem of environmental protection solved, but at the same time so is the question of how the measure can be financed.

The remediation of a waste site with inorganic and organic harmful substances on the premises of the Vulcan shipyard at Bremen in 1975 is of particular significance when one considers that at the same time, in the course of a necessary extension to the shipyard, it was essential to create additional space for the temporary storage of segments for the construction of large tankers. This land had to withstand extremely heavy mechanical loads. Here it was possible to show that the regular road-building materials commonly used in such cases can be completely replaced to considerable technical advantage by the compacted DCR reaction product resulting from the remediation measure (Fig. 27).

The oil-contaminated soil was spread out for treatment. Bulky objects present in the waste site, such as steel cables, drums, machine units, building rubble, parts from railway

Figure 27 Construction of a heavy-duty industrial area in a shipyard. The inorganically contaminated organic phases of a waste site inside the shipyard area were excavated (top, left), shaped (top, right), DCR-reacted, and rebuilt as an earth construction material (bottom). Finally, the area was covered with a layer of asphalt.

Figure 28 Construction of an industrial area inside the harbor of Brest, making use of the mix of sand, water, and crude oil collected at the beach after the *Amoco Cadiz* tanker accident.

tracks, etc., were removed with the aid of construction machinery before the soil was moved; after the measure had been completed they were embedded well inside the reprocessed waste site with additional amounts of hydrophobic calcium oxide.

The reaction product was rebuilt under compaction. It was worked into the edge of the existing industrial premises. Although the hydrophobic body of soil lies in the groundwater of the Weser River, which is only a little more than 100 m distant, the subsoil properties have not changed over many years, as control measurements prove.

The second example shows that even crude oils in a mix with sand and water provide a valuable soil material after DCR reprocessing.

After the Amoco Cadiz tanker accident off the French coast in 1978, the authorities responsible were confronted with the question of what was to be done with the material consisting of oil sludge, sand, and water that had been collected in plastic bags by military and civilian helpers at the coast. It was finally decided that the DCR-treated material should be used to fill up a depression in the ground within the Brest harbor area. This piece of land had not been used so far, and filling up the depression thus provided a valuable industrial site (Fig. 28).

It was also possible to use the field process here, although the loads, which contained large amounts of water, consisted of numerous filled plastic sacks. It was decided at the time that the DCR body of soil should not be covered with asphalt, and this is the reason why, even today, on the site it is still possible to recognize remains of the milled plastic sacks inside the protective, still-growing carbonate layer.

The third example refers to a waste site found by the Federal Austrian Railways when the railway area was to be extended. Instead of excavating and transporting the oily waste to a landfill, the whole content was DCR-reacted and rebuilt according to the soil mechanical requirements. Since there as no space available for a DCR field process, the reaction was carried out on site in a mobile DCR plant transportable on two heavy motor trucks (Fig. 29).

The foregoing treatment technique is of general importance. The size of numerous waste sites is such that it is not worth erecting a stationary DCR treatment plant. The same applies when smaller deposits of harmful substances are spread over a wide area, so that it is more economical to reprocess them in situ and collect the more easily manageable reaction products for a particular use.

Special remediation plants have been developed for these remediation tasks. They are erected on site following the modular system and in accordance with the particular

Figure 29 Working up a waste site close to a rail track by means of a mobile DCR plant. The excavated mix of soil and oily waste passes a riddle sifter (top, left) and is then homogenized with hydrophobic CaO in the mixing device of the plant (top, right). The resulting homogenous, still dark mixture leaves the plant on a belt conveyer for final outside reaction (bottom, left). The light-brown reaction product is stockpiled and rebuilt as soon as the ground below is found to be clean.

requirements of the site. Since the plant has its own power supply, it is also possible to carry out remediation measures at remote sites.

3. DCR Products for Road Construction

A waste site was remedied in a narrow space inside a refinery in Rotterdam while it was working at full capacity. Previously an attempt had been made to consolidate liquid organics of high viscosity by mixing in finely dispersed, inorganic, inert pozzolanic material. Particular with summer temperatures, however, it became clear that the pozzolanic materials were not capable of a so-called solidification of an organic phase and thus settled and the organics continued to appear at the surface. This was why the mix of pozzolanic compounds, soil, and organic waste was to be reprocessed into road construction material (Fig. 30).

The material was excavated, deposited on a narrow strip of land made available for the treatment process, and spread so that the layer was 30 cm thick. A spreader with a capacity of 10 tons covered the spread layer of waste with CaO, and a heavy rotary mixer guaranteed that it was homogeneously mixed. The dispersing reaction started after approximately 20 min, yielding the pulverulent DCR product. Because of the bad weather conditions, most of the work had to be carried out during heavy rain, so that a higher degree of hydrophobicity of the calcium oxide was necessary; the reaction product therefore has particularly marked hydrophobic characteristics.

DCR Remediation Technology 895

Figure 30 DCR processing of a heavily contaminated soil in a field process comprising excavation of the polluted soil (top, left) formation of a layer ca. 30 cm thick for treatment (top, right), and covering the waste layer with a thin layer of CaO by means of a spreading car, followed by a rotary tiller (center, left). After abour 1 h, the DCR-reacted mix is ready for sampling (center right). It shows water repellent properties (bottom, left) and can be used for road construction as a high-value water-resistant subsoil (bottom right).

The total content of the waste site was to be utilized for the construction of a stable works site as well as for subsoil for road construction. For this reason samples were taken regularly for the necessary tests. Since it was possible for the planned construction measures to be carried out only during the following year, it was necessary to store the soil construction material temporarily.

The behavior of the temporarily stored DCR reaction product under extraordinarily unfavorable climatic conditions is very revealing. During continuous autumn rain, a protective carbonate crust developed so rapidly on the surface of the temporarily deposited, noncompacted but hydrophobic material that even in pouring rain none of the calcium hydroxide dissolved at all.

This confirmed what can be deduced from laboratory experiments on hydrophobic calcium hydroxide regarding carbonating processes. These experiments show that even under more difficult conditions no harmful substances, and not even any traces, are released at any time, and the system becomes increasingly safe as time goes on.

4. DCR Treatment of Soil-Like Materials

There is some sort of industrial waste which can be regraded as soil-like, for instance drilling mud, contaminated, as a rule, with oil and heavy metals, as well as used fuller's earth and red mud, a by-product of aluminum production. The advantage is that these materials are available in large or rather in huge quantities, so that it is worthwhile to establish a technical-scale manufacture of appropriate secondary products based on the application of DCR technology. Needless to say, all these materials can be worked to yield earth construction materials in a broad sense of highest quality, especially for road construction. But it is much more economical to produce additives for cement production or to manufacture building material, for instance, concrete parts and the like.

Figure 31 shows two stationary DCR plants in which oily drilling mud can be processed as well as any other soil-like waste to yield products on a large scale for road con-

Figure 31 DCR plants to produce road construction materials from industrial waste (left, top) and from oily drilling mud (left, bottom). The picture on the right-hand side provides a view at the high-speed mixing device as the central part.

Figure 32 Mixing device for blending DCR products with additives to manufacture secondary products for technical and soil mechanical use.

struction and for soil stabilization. The plant (left, top) is part of a remediation center processing oily industrial by-products exclusively for road construction. The waste is collected in a storage area as a buffer (left is that picture) and then transported to the compulsory riddle sifter by small spade lugs. The plant (left, bottom) reacts oily drilling mud only. By the addition of cement, concrete parts can be formed in an attached pugstream machine. Figure 32 allows a view into the opened kneading machine necessary for the addition of cement or other pozzolanic additives. Because of the contaminants present, i.e., oily phases and heavy metals, it is necessary not only to disperse the organic phase chemically but also to irreversibly immobilize the heavy metals. This can easily be achieved through the addition of appropriate precipitating reaction partners.

V. DCR PRODUCTION OF SOLID FUELS

Three industrial processes exploit combinations of organic fuels and inorganic, alkaline earth components. These are cement production, certain processes for desulfuration, and special processes in steel production. Presently, only the first two are economically significant.

Attention must be drawn in particular to possible ways of making further use of organic waste materials that have previously been deposited in waste dumps and that can be reconditioned together with acid tars to provide a valuable economic commodity. In this case acid tars, waste oils, and charged fuller's earth are processed into dry, hardened granulated secondary raw material in a quantitative composition that fulfills the demands of cement production. These granules have calorific values that correspond to those of brown coal. The use of $Ca(OH)_2$ as a basic carrier means that no harmful emissions occur.

Reprocessed waste mixes of this kind can also be advantageously burned in a fluidized-bed reactor. This application is based on the idea that in the first stage all the organic components burn as fuels, utilizing their full energy content, and in the second stage the mix of calcium hydroxide/calcium oxide formed by the prevailing temperatures becomes effective as a highly reactive desulfurizing agent. This is a result of the larger specific surface which is produced in the DCR process. If one remembers that in the dispersing chemical reaction it is also possible to simultaneously disperse inorganic substances with catalytic properties, then for this particular application special preparations

Figure 33 DCT plant inside a re-refinery producing a high-quality solid fuel from organic and inorganic waste, for use as a cement additive.

with specifically effective catalysts can be produced in a single-stage process. Of course, in this case sulfuric acids and sulfonic acid tars cannot be used.

Figure 33 shows a remediation center inside a re-refinery (top, left). Any oily waste, including acid tar, contaminated waste oil (contaminated with anything), bituminous materials, heavily charged used fuller's earth, and so on, are collected and stored in large tanks connected to a pump station (top, right), from where the mix is blended with the bleaching earth (center, left) and then processed in a system comprising a homogenizer and a granulator (center, right). The reaction product, almost dust-free granules (bottom, right) is loaded by a belt conveyer onto pickup trucks and shipped directly to a cement works (bottom, left).

The utilization of the DCR free-flowing granules for cement production is the reason the ingredients have to be mixed very carefully. It is the sulfate concentration which determines the rate of addition to the standard material flow. The valuable silicates of the charged fuller's earth increase some cement quality parameters significantly. This kind of reuse of waste is an excellent example for an optimum exploitation of all the constituents of the DCR product: the hydrocarbons as an energy carrier, the calcium hydroxide as the basic compound and the silicates as its sour counterpart, both necessary for cement production anyway. If the organic waste should be contaminated with chlorinated aromatics, PCB or the like, this will be of no consequence. The aromatics finely dispersed in $Ca(OH)_2$ are easily and completely dehalogenated under the reaction conditions in the rotary kiln.

VI. REMOVAL OF OILY PHASES

A. Oily Phases on Water

As has been outlined in Section II, an oily phase can be chemically dispersed only in case the oily phase was incorporated in the active CaO beforehand. This was shown in Fig. 8, in which the oily layer on water in the sequence of cuvettes on the left did not react with CaO which is hydrophilic. Only the hydrophobic variant incorporates the oil and leads, after the chemical dispersing reaction has commenced, to a finely dispersed suspension. As a result, the oily phase has vanished and the water surface is clean (sequence of cuvettes on the right). This result cannot be transferred, just like that on oil spills with an oil the volatile compounds of which have already evaporated. However, as could be shown in a research project of the German Federal Ministry of Science and Technology, even this sort of oil can be DCR-reacted provided the CaO is applied in its granulated form.

The sequence in Fig. 34 shows the separate steps of a process which results in a finely dispersed suspension in which any oily phase has disappeared. The original individual compounds of the oil, now adsorbed in the form of molecular layers by the calcium hydroxide and, later, by the calcium carbonate formed with the CO_2 of the environment, are highly reactive and biodegrade rapidly. In case charged carrier particles in deep areas sink down to a depth where there is no biological activity, the hydrophobic calcium hydroxide will nevertheless finally be reacted to carbonate, in particular through precipitation processes with carbonate ions being in solution from natural chalk in the surrounding environment, but will not, due to the hydrophobic properties, be redissolved in a way that an oily phase could be formed again.

As soon as the closed sealing oily phase has vanished, the oxygen of the air has access to the free water phase again, thus increasing the biological activity. In other words, the disastrous impact of the coherent oil phase to the water has completely and irreversibly been eliminated.

Figure 35 shows an application of this process on a technical scale. The water of a lagoon was covered with a layer of oil, 8–10 cm thick. In order to clean the water surface, the mix of skimmed oil and water was pumped into a chamber equipped with a baffle plate and a dosage facility for the CaO. Through the turbulence in the mix of oil, water,

Figure 34 Oil on water, polluting everything and, in the form of a coherent layer, preventing the oxygen of the air from migrating into the water, thus stopping any biological activity (top, left). Due to the oleophilic properties of their plumage, diving seabirds passing the oil layer take the immediately adsorbed oil down into the water (top, center). Through the addition of granulated CaO the oil is immediately adsorbed (top, right), thus stopping any further oil pollution (bottom, left). The story of normally applied adsorbents would end here; they would either sink to the ground to form an impermeable oily sludge that would damage plant and fauna growth, or, if the oil-charged adsorbent floats, would spread, for instance on a rough sea, so that it could not be collected. Not only would the problem as a whole not be solved, but the damage would increase at the end. The DCR process, however, works in quite another way. After the oil has been adsorbed, the dispersing chemical reaction takes place, forming a light suspension (bottom, center). Not only is there no further possibility for pollution (bottom, right), but also no oily sediment can be formed. The fate of the carrier is to become inert calcium carbonate and the fate of the hydrocarbons is to be biologically degraded. As a result the water regains its original clean quality.

and CaO when pumped against the baffle plate, the oily phase was first selectively incorporated and then chemically dispersed. After sedimentation the water was completely clean and clear.

Under normal conditions and if possible, the main proportion of the oil layer are separated mechanically from the water by special skimming devices for recovery, and only the remaining thin layer of oil is DCR-reacted through the addition of granulated CaO as shown in Fig. 34.

B. Oily Phases on Oil-Polluted Solids

The DCR process can easily be applied to mixes of oils and oil-like materials with nearly any sort of solid material, for example, to oil-contaminated sand and gravel, oily grinding

Figure 35 Oil on water was pumped into a reaction chamber (center of the left picture) and mixed with CaO (stored under a plastic foil) by means of a simple baffle plate. A mix of the finely dispersed DCR reaction product and water, free of oil, was conveyed into a second pond for sedimentation (right picture).

Figure 36 A heavily oil polluted sand (left) was DCR-reacted to yield a mix of pulverized oil and sand (center). Through separation, a clean sand can be recovered.

dust, and oily roll scale, in which the oily constituents pollute the solid materials. Through the dispersing reaction the oily phase is pulverized and can be removed in this form by known solid–solid separation techniques such as sieving, air classification, or flotation. The clean solid can be reused as usual.

Figure 36 shows the removal of oil from sand of a beach heavily contaminated with aged and therefore highly viscous crude oil. In this case the target of the operation is merely purification of polluted sand.

Oily roll scale contains iron and iron oxides as the basic material, which must be regarded as a valuable secondary raw material in connection with the production of steel. However, direct feed into the steel production process via the sintering plant runs into difficulties because of the hydrocarbons present.

Many attempts have therefore been made in the past in order to separate the oily parts from the iron compounds, mainly through thermal treatment, stripping, extraction with organic solvents, high-pressure washing, and flotation. These methods either do not lead to the expected results or they are endowed with secondary disadvantages which

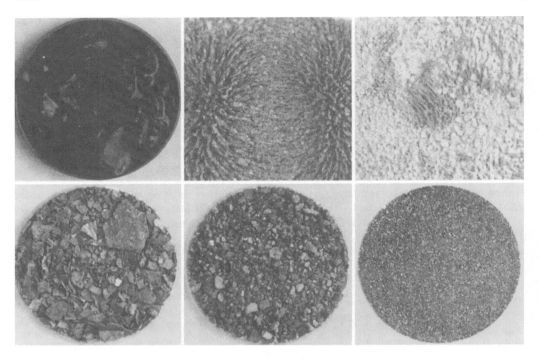

Figure 37 Heavily oil-contaminated roll scale before and after DCR treatment. The subsieve powder (top, right) contains still ferromagnetic parts which can be used as an additive for cement production, in particular if there is a lack of iron in the available raw materials.

have prevented large-scale application. An example of a drawback of this sort is the cost-effective water treatment associated with washing techniques.

The continuous addition of small proportions of oily roll scale directly into the agglomerating device has also been reported. Since the oily compounds vaporize from the sinter belt when approaching the combustion zone, this procedure may create serious problems in the filter system. In other words, the economic value of utilizing oily roll scale will be outweighed by the additional expenditure for the maintenance of the filter system.

Figure 37 shows a typical oil roll scale before DRC treatment (top, left), along with macroscopic photographs of three clean fractions from the sieve plant (bottom). After raising the magnetic constituents in a magnetic field, it becomes visible even with the naked eye that the DCR reaction product comprises two sorts of solid constituents which are easily separable (top, center). An additional proportion of iron/iron oxides can be recovered from the remaining dust fraction with particle size < 100 µm through magnetic extraction (top, right).

It is of great practical importance that the physical properties of the pulverized oil fraction can be exactly adjusted to the particular separating method through the addition of small quantities of reaction aids. For example, the efficiency of the separation of oil from roll scale through a combination of sieving and air classification can be increased significantly through the production of a particularly light, finely dispersed, and hydrophobic DCR reaction product which separates very easily

DCR Remediation Technology

Figure 38 DCR plant for the treatment of oil-polluted solids with a throughput of 20 tons/h. The photograph (right) shows an extra large reactor with its discharge unit. This plant is being run continuously and fully automatically.

from the iron and iron oxide particles. To what extent the separation is to be carried out depends exclusively on the clean-up target, particularly the requirements for reuse of both the purified roll scale and the pulverized oil. In this context the limit value for the residual hydrocarbon concentration on the roll scale determines both the separator plant layout and the acceptable quantity of iron oxide in the residual dust with respect to its intended reuse.

The above-reported DCR clean-up of oily roll scale is of particular importance in terms of soil remediation, inasmuch as it is known that huge quantities of this oily waste have been deposited in the form of mono-landfills over many decades in the past. It is possible now and economically reasonable to excavate the material, not only in order to recover the valuable iron and iron oxide components but also to stop the oil pollution caused by this source.

Various types of plants are available for the DCR treatment of oily roll scale, with throughputs of up to 100 tons an hour. The plant shown in Fig. 38, in which the spatial arrangement of the homogenizer and reactor is easy to recognize, is designed for the processing of oily bleaching earth for reuse in the cement industry but can be used for the treatment of oily roll scale as well.

Another example of the clean-up of oil polluted solids through DCR technique is polluted rocks. This sort of pollution may occur, for instance, after a tanker accident in a fjord. The rocks in this model, of course, may be replaced by the ballast of the track or any other similar solid (Fig. 39).

To continue the example of oil-polluted rocks in a fjord, the DCR process produces a suspension of an inert solid in clear water. There is no chance to form an oily phase again. On the other hand, when so-called oil dispersants are used, an emulsion may be formed from which oil will separate again on contact with solids or on dilution, so that renewed oil contamination occurs. Even an unbroken emulsion can cause oil pollution to the feathers of seabirds diving through it (cf. Fig. 34). In contrast, this is not the case with the DCR reaction product.

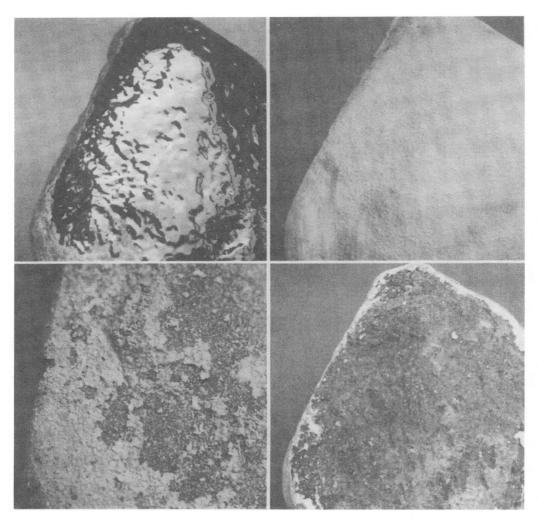

Figure 39 In an oil layer on a rock surface (top, left), granulated, strongly hydrophobized CaO is being blasted by means of a grit-blasting device. The granules incorporate the oil (top, right) and react with the moisture of the air to form a dry solid preparation, which separates spontaneously (bottom).

In order to incorporate oil on stones, rocks, and other solid surfaces into the pulverizing agent, a step that is here more essential than elsewhere, hydrophobized coarse particles of granulated CaO must be blasted into the oil layer by means of a sand-blasting device. Thus the polluting oil layer penetrates the mechanically applied granules and within about 24 h the DCR reaction commences, resulting in the disintegration of the coherent pasty cover. As time progresses, more and more scales are formed and at the end of the process the rock is cleaner than ever and, of course, free of any traces of oil. The dry scales drop to the ground or into the water, where they form a powdery solid in the shape of an inert suspension ready for fast biodegradation.

VII. DETOXIFICATION OF HAZARDOUS COMPOUNDS

A. Organochlorines

The organochlorine compounds which have been studied within the scope of a special remediation research program at the Hannover University[20] comprise PCBs, PCPs, dioxins, and a range of persistent organochlorine pesticides, mainly the biologically inactive hexachlorocyclohexanes, which have been found in monolandfills in the former GDR to the amount of some hundred thousand tons each. The objectives of the research and development activities were to provide methods capable for effective and economic removal and/or detoxification of said compounds from and in, respectively, any form of waste material including off-gases. In view of the enormous quantities to be treated worldwide and because of their exceptional environmental hazard potential, the main attention was turned to PCB and dioxin treatment.[21]

Polychlorinated biphenyls (PCB) have been employed for several decades in a variety of industrial applications. They are nonflammable and have low volatility and good viscosity characteristics at operating temperatures. Thus, they have been used primarily as additives in dielectric, hydraulic, and heat transfer fluids. It has been reported that over a half a million tons of PCB have been manufactured in the United States and it is estimated that 300 hundred million pounds of these are in chemical landfills and about 750 million pounds may be still in use. In the 1960s, the first signs of potential environmental problems appeared and subsequently, because of toxicity, environmental persistence, and suspected carcinogenicity, the EPA banned the continued manufacture and importation of PCB and issued regulations governing their storage and disposal.

PCBs can pose environmental problems in a number of different forms, for example, in the form of contaminated waste oil, contaminated soil from direct contamination with pure PCB solutions, PCB-contaminated waste oil in soil, and PCB-contaminated materials, such as equipment.

Dioxins are often found to be emitted from waste-burning facilities. It is known that, due to the ubiquity of inorganic chlorine compounds in the environment, any incineration process generates dioxins. Accordingly, particularly high concentrations of dioxins are always found in cases where the incineration takes place in the presence of higher concentrations of chlorides, for instance, in sintering plants.

PCB destruction technologies which can be used for the remediation of PCB-contaminated materials fall into four treatment classes. These classes are thermal treatment, chemical dechlorination, biological degradation, and physical separation (27–29).

Internationally, the most widely employed means to destroy PCB as a contaminant is high-temperature incineration (HTI). However, it is well recognized that by incomplete combustion of PCBs or by other thermal processes, ultratoxic dioxins and furans can be formed in significant quantities. This is the reason why public opposition to HTI has increased in recent years. Even if this thermal treatment,

---

[20] The research program was government-sponsored [Federal Ministry of Research and Development (BMFT) and Ministry of Economics, Technology and Traffic (Lower Saxony)].
[21] The research work on dioxins in the university has been carried out with appropriate model substances selected on the basis of MO calculations.

due to more and more effective off-gas cleaning technology, could be considered technically acceptable, it would still be the most expensive solution, especially for sites with large quantities of contaminated soil and with PCB concentrations from some parts per million up to 5000 ppm, where not only are the costs prohibitive but also applicability is not possible in practice: how to "incinerate," for instance, a landfill with some million cubic meters of contaminated aqueous soil, sludge, and water (30–32).

B. Chemical Detoxification Processes

It is well known that the toxicity of many organics increases tremendously when the original compound is substituted by halogens. A simple example is dimethyl ether, a relatively harmless substance compared to its extremely toxic halogenated counterpart, bis(chloromethyl) ether, which is regarded as one of the worst carcinogens. The same holds true for aromatic compounds such as dibenzodioxin, dibenzofuran, biphenyl, etc., which are nontoxic (biphenyl is used as a preservative for oranges), whereas the halogenated derivatives polychlorodibenzodioxins (PCDDs), polychlorodibenzofurans (PCDFs), and polychlorinated biphenyls (PCBs), are poisonous or even ultrapoisonous. Thus, detoxification of highly toxic halogenated organic compounds means in this context dehalogenation so as to reproduce the harmless parent compound.

Chemical dechlorination comprises catalytic dehydrochlorination, chlorolysis, catalyzed oxidation, and sodium- or potassium-catalyzed decomposition of the PCB molecules. Most of these processes are based on common organic chemical reactions elaborated for the treatment of clean halogen-containing hydrocarbons. When applied to the decontamination of an inorganic matrix such as sediment, soil, or sludge, the PCBs must be separated beforehand by thermal desorption processes, distillation, vacuum evaporation, or extraction.

Contaminated soils, without separation, can be treated with heated solvents consisting of polyglycols, capped polyglycols, potassium hydroxide, and dimethyl sulfoxide in a mixing device. This so-called KPEG process was selected as the highest-ranking alternative in an EPA study of PCB destruction technologies, despite the narrow applicability, high treatment costs, and negative environmental impact.

Halogenated aromatics may undergo dehalogenation reactions either in the form of direct displacement of the halogen by a nucleophilic ipso-substitution mechanism or by an elimination/addition reaction mechanism, in which arynes appear as discrete intermediates. The first mechanism demands an increased chemical reactivity of the substrate through electron-withdrawing substituents and, consequently, occurs only with polysubstituted arenes; the latter applies to monohaloarenes only and needs drastic reaction conditions because of the very low CH-acidity with which this type of compound is endowed. Nevertheless, complete dehalogenation reactions of pure halogenated aromatics such as PCDDs/PCDFs and PCBs have been, on several occasions, reported to take place as well with KOH in the presence of PEG/PEGM or PEG/DMSO as a solvent or catalyst, respectively (17–19).

As a matter of fact, an ipso-substitution of a halogen occurs only in case of activation by electron-attracting substituents such as NO_2 or additional halogens, exclusively giving products in which one and only one halogen has been replaced, even under drastic conditions. For instance, *o/m*-dichlorobenzene and monochlorobenzene give, on reaction with sodium methylate in the presence of HMPA, the corresponding monosubstitution products (20), e.g.,

$$\text{1,3-dichlorobenzene} \xrightarrow[\text{HMPA}]{\text{NaOR}} \text{3-chloroanisole} \tag{R 1}$$

Testaferri et al. (21) elucidated the dechlorination reaction mechanisms with respect to, for instance, tetrachlorobenzene and trichlorobenzene as model compounds. With sodium methoxide in the presence of HMPA at 120°C for 1 h the monosubstitution products formed are:

$$\text{1,3,5-trichlorobenzene} \xrightarrow[\text{HMPA}]{\text{NaOR}} \text{3,5-dichloroanisole} \tag{R 2}$$

$$\text{1,2,4,5-tetrachlorobenzene} \xrightarrow[\text{HMPA}]{\text{NaOR}} \text{2,4,5-trichloroanisole} \tag{R 3}$$

With an excess of sodium methoxide and with increased reaction times (2–7 h), the monosubstituted reaction products of the reactions (R2) and (R3) are transformed into the corresponding dimethoxy derivatives and into phenols, which, however, still contain chlorine (22):

$$\begin{array}{c} \text{3,5-dichloroanisole} \longrightarrow \text{3,5-dichlorophenol} \\ \downarrow \\ \text{3,5-dichloro-1,3-dimethoxybenzene} \longrightarrow \text{3-chloro-5-methoxyphenol} \end{array} \tag{R 4}$$

Since the chemical results provided in the papers of Brunelle and desRosiers (17–19) agree in the end with the basic chemical principles, one must conclude that the expressions "complete dehalogenation" and "successfully destroyed" as employed by the authors have been used in the meaning of "complete *monosubstitution* reaction."

Brunelle and Singleton (17), using tetrachlorobenzene as a model substance, have shown that in a reaction with 80 equivalents KOH and 6 equivalents PEG at 100°C for 16 h, trichlorophenol was formed and, under milder conditions, the corresponding phenol ether:

$$\text{(R 5)}$$

Accordingly, the reaction of any sort of PCB under related conditions gives the monosubstitution products as well, for instance,

$$\text{(R 6)}$$

Substitution for more than one chlorine has been reported to take place with, for instance, thiolates as the nucleophile, but this reaction might follow another pathway, possibly an $S_{RN}1$ mechanism.

Many attempts have been made since then to accelerate the dehalogenating process, chiefly through the employment of DMSO as a co-solvent or even as a catalyst. As a result, mixes have been applied comprising KOH/PEG/PEGM/DMSO or something like that, which have been reported to increase the dehalogenation rate, as found by Peterson and desRosiers, but it was not reported why (19,22).

In order to find out the role DMSO plays in this context, dehalogenating reactions have been investigated with some dichloro- and trichlorobenzenes and $NaOCH_3$/DMSO as the reagent. The results indicate that complete dehalogenation, in the real sense, can take place through the direct interaction of DMSO as a reaction partner at fairly increased temperatures and rather long reaction times (23). *m*-Dichlorobenzene, for instance, on reaction with 10 equivalents NaOMe and DMSO, at 80°C for 19 h, gives not more than about 18% completely dehalogenated products, whereas the same reaction carried out at the same temperature but for 8 days furnishes, surprisingly enough, about 82% completely dehalogenated compounds (yields for the second type of reaction in brackets):

$$\text{(R 7)}$$

However, much more important are some results achieved with some individual PCB congeners. For instance, 2,4-dichlorobiphenyl yields, on reaction with KOMe in the presence of DMSO at 80°C for 15 h, products that still contain chlorine besides OH, e.g.,

$$\text{(R 8)}$$

There is no doubt that any PCB congener substituted in the 2,2'-positions will undergo a condensation reaction to produce chlorinated dibenzofurans; from this point of view, methods applying KOH/PEG and/or DMSO will not only produce, under appropriate reaction conditions, ultrapoisonous dibenzofurans, but will additionally furnish still-toxic PCB derivatives with increased mobility (24).

All these preceding results can easily be understood. The first HO- or RO- group that enters the molecule of oligohaloarenes leads to a complete inversion of reactivity toward nucleophiles because of the profound electron-releasing effect these groups have, the consequence of which is that any further removal of halogen will be stopped immediately.

Aromatic halides without activating substituents do not easily give the corresponding substitution products. The replacement of chlorine in chlorobenzene, for instance, by the HO- group, carried out at 316°C with cuprous oxide as a catalyst, yields 92% phenol after a reaction time of 1 h. Even bromobenzene reacts incompletely with methanolic sodium methoxide at 220°C, forming anisol, phenol, and some benzene.

The complete dehalogenation of oligolialoarenes such as dichlorobenzene through nucleophilic substitution reaction partners in the presence of DMSO does not conflict with elementary chemical laws, but can be put down to the fact that mechanisms are involved that comprise the addition of DMSO to arynes in a pericyclic reaction (23).

The BCD technology[22] is a recent innovation developed and patented by the U.S. Environmental Protection Agency (UEPA) Risk Reduction Engineering Laboratory. BCD involves a reaction which sequentially strips chlorine atoms from organochlorine molecules and substitutes them with hydrogen atoms derived from an oil. The reaction, which requires the addition of a proprietary catalyst, takes place over several hours and at elevated temperatures to yield a completely dechlorinated organic molecule and common salt.

ADI Services, a BCD licensee in Australia, has developed a variant of the BCD reaction (called the ADOX reaction) in which the patented BCD catalyst is replaced by an "accelerator". In the ADOX reaction the nature of the reaction changes dramatically in that PCP molecules, for example, are decomposed completely to carbon. The reaction, which takes place rapidly, can be applied to much higher concentrations of organochlorines than the conventional BCD process can be used for and without the requirement of the addition of oil.

---

[22] The description of the BCD and EcoLogic processes has been taken from Bulletin No. 2. The Organochlorines Programme, Ministry for the Environment, Wellington, NZ (1995).

In developing technologies of this type, a significant challenge is to destroy the organochlorine contaminants bound up within soil and other solid materials. One approach is to pretreat the contaminated materials within a thermal desorption unit (TDU). The vaporized contaminants are then condensed and destroyed in the BCD reactor.

It should not be concealed that the chemistry the BCD process is claimed to be based upon is unknown. The postulated reaction mechanisms as well as the fate of the organochlorines have not been proven analytically. Since the expression *elevated temperatures* does mean temperatures in the range around 350°C, it becomes clear that this process is extremely costly, not to mention the question regarding applicability in practice (cf. contaminated oily waste as shown in Figs. 1 and 2).

Another technology has been developed by EcoLogic International, Inc. (ELI).[23] The technology has been successfully demonstrated to completely degrade PCP, dioxins, and other organochlorine substances as well as polycyclic aromatic hydrocarbons (PAHs). When applied to the treatment of soil, the contaminants must first be vaporized by a thermal reduction mill (TRM).

The unit adopted by EcoLogic comprises a ball mill (to achieve a finely divided soil) operating at high temperatures on a molten tin bath. The vaporized organics are swept into the EcoLogic reactor with a stream of gas (mainly hydrogen and steam). Chemical decomposition reactions take place in the reactor at 900°C in a hydrogen atmosphere to yield principally methane, carbon dioxide, and hydrogen chloride (recovered as hydrochloric acid). The technology has been successfully audited by both the U.S. EPA during a trial treatment of dioxin-contaminated soil and by Environment Canada during the treatment of PCB- and PAH-contaminated harbor sediments. Environmental Solutions International (ESI) and ELI have recently commissioned the first commercial-scale EcoLogic plant at Kwinana, south of Perth. This plant is destroying stockpiles of organochlorine pesticides (predominantly DDT) and PCBs. Independent indicates a greater than 99.9999% destruction removal efficiency. The success of both the BCD/ADOX and the EcoLogic technologies relies on effective materials handling and an efficient TRM pretreatment phase. With respect to the given data, i.e., the necessity to vaporize the PCB in a molten tin bath and to react the insulated material at 900°C with hydrogen, any further comment can be regarded as superfluous.

From a sound chemical point of view there is only one group of chemical reactions which can indeed completely remove every chlorine from any sort of aromatic compounds and thus effectively solve the problem of detoxification of hazardous materials; this is the group of reductive dehalogenation reactions. There are a number of methods in organic chemistry by which this type of dehalogenation can, in principle, be carried out on every kind of compound. However, most of the experimental instructions refer to the conversion of pure substances in a homogeneous medium, so that they are not immediately suitable for the solution of environmental problems. The economy of most chemical processes also usually plays only a subordinate part, for example, in the methods according to which halogenated aromatics can be dehalogenated with hypophosphorous acid in the presence of noble-metal catalysts. In contrast, substances and substance mixtures that come from the sphere of environmental pollution must be regarded from a completely different point of view. This is because of their usually extremely complex and inhomogeneous com-

---

[23] cf. footnote 22.

position. They demand methods which provide good yields with a high level of economy, and which are effective even for the smallest concentrations of harmful substances. In addition, the processes must be technically sound, thus guaranteeing that no secondary pollution of the environment is possible.

Many other reductively working methods have, therefore, been developed for application in the field of remediation chemistry, most of which are based on the use of alkali metals, especially dispersions of sodium metal. The difference between them is that the methods described make use of different hydrocarbon sources, mainly different alcohols, and that the particle size and thus the chemical reactivity of the metal varies. There are also processes known in which the sodium or any other alkali metal or mix of metals is applied in the form of finely dispersed preparations on inert carriers.

In most cases the known methods suffer from severe disadvantages, including the use of highly reactive sodium, which, if it is really extremely reactive, can only be handled in an atmosphere of protecting gases. Because of its high sensitivity to even traces of moisture, any application in a water-containing matrix is strictly prohibited. In case that no so-called accelerators, i.e., low concentrations of alcohols such as butanol or isopropanol, are involved, increased temperatures up to 300°C and extensive reaction times are indispensable. In addition, the use of hazardous sodium preparations requires complicated equipment and particular expensive safety precautions.

Finally, it should be mentioned that a number of studies have been performed to evaluate the biological degradation of PCB compounds. The results of most of these studies have been basically the same, that is, lower chlorinated PCB compounds, up to three chlorine atoms, can be biologically degraded, but higher chlorinated PCBs are very resistant to degradation. It should be noted that there is no evidence of any biological degradation process being commercially demonstrated for highly chlorinated PCBs beyond the bench or laboratory scale at the present time.

PCBs are strongly adsorbed onto soil matrices following spills. A number of solvents have been examined for separation of PCB by extraction from soils. While some of these solvents have proven to be successful on the laboratory scale, their full-scale implementation has not always been successful, for a variety of reasons. The only separation technology which can be called commercial at this time is a process called Basic Extractive Sludge Treatment. This BEST process does not degrade PCB compounds, or make them nontoxic; in the end, an ultimate treatment of the PCB materials is required.

C. DCR Detoxification

Not only can any sort of contaminated sludge, waste oil, mixes of any other solid waste with oil, water, contaminants, etc., be reacted to a dry solid material, but also any chemical agent, solvent, and additive applicable to detoxification reactions. According to the finely dispersed state, the reactivity of catalysts, reactants, etc., produced in this way is increased to the extent that reactions can be carried out which under standard conditions never would occur (2). These special features of DCR technology can be utilized for a variety of applications in the field of environmental detoxification chemistry, as will be shown in the following sections by examples.

1. DCR Treatment of PCB-Contaminated Waste Oil

In order to find out the most useful DCR variant it is necessary, as a first step, to define the treatment target. For instance, in the event of low concentrations of PCBs in waste oil, i.e.,

2–50 ppm PCBs, this material is not permitted by law to be re-refined in the usual way, for instance, for the production of fuel and lubricants. In this case the DCR method of choice is to *remove* the PCB from the waste oil and to feed back the PCB-free oil into the standard re-refining process.

This is one of the simplest processes and does not need any special industrial instrumentation. The contaminated waste oil is vigorously mixed with a small proportion of DCR reagent to functionalize the PCB and to remove the chemical derivative through absorption. The quantity of agent necessary depends on the PCB quantity present and is, in general, less than 1 kg per ton of waste oil. After a residence time of about 15 min at room temperature, a decolorizing additive should be added. The absorbed PCB derivative on the solid carrier separates spontaneously. After complete separation the clean oil can be isolated for reuse.

This reaction sequence has been given for better understanding of the simplicity of the process, which chemically comprises the following main step:

(R 9)

The PCB is removed in the form of a halogenated phenol and must, therefore, be dehalogenated in a second treatment step. The devices for a technical-scale application, which, in its most economical form, runs continuously, comprise a predrying reactor, a mixing device for the monosubstitution process, and a separator. The collected sediments from the separator are reacted in a dehalogenation unit as described in the following section.

If there are higher concentrations of PCB in waste oil, for instance, up to 20–40%, then it is still worthwhile to remove the PCB instead of changing the process to full dehalogenation. The advantage of the DCR PCB removal is that the clean oil is free from any trace of dehalogenated derivatives and can thus be reused as usual.

However, if one approaches concentrations of PCB which render the process uneconomical, then a direct and complete dehalogenation is preferred:

(R 10)

The PCB-free oil contains the full quantity of (harmless) biphenyl, and it might be uneconomical to try a separation. However, the mix can, of course, still be used as fuel.

If something like pure PCB, PCP, etc., has to be disposed of, it is recommended for economic reasons first to DCR "pulverize" the oil as shown in Fig. 40 (left) and then dehalogenate the pulverulent PCB directly in a reactor.

Direct dehalogenation of PCB in its liquid form or as a solid DCR preparation can be carried out either with DCR sodium as a precursor of the actual dehalogenating agent or with metals such as Mg at room temperature for from 15 min to 1 h, depending on the metal. It should be emphasized that coated DCR sodium on a carrier can be handled even in the presence of humid air.

DCR Remediation Technology

Table 6 DCR Removal of PCBs from Two Samples of Contaminated Oil[a]

	B(2500) before (mg/L)	B(2500) after (µg/L)	B(25000) before (mg/L)	B(25000) after (µg/L)
Q PCB 28	2,900	80	13,000	80
Q PCB 52	1,400	< 10	6,300	< 10
Q PCB 101	330	< 10	1,700	< 10
Q PCB 118	230	< 10	1,400	< 10
Q PCB 138	70	15	750	< 10
Q PCB 153	< 50	< 10	740	< 10
Q PCB 180	< 50	< 10	790	< 10
Arochlor 1242	38,000	600	190,000	380

[a] The reaction was carried out in a planetary ball mill at room temperature over 15 min.

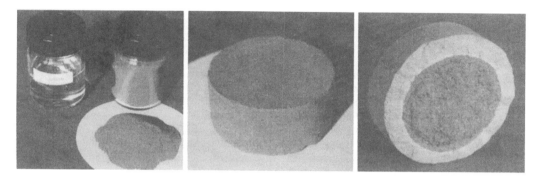

Figure 40 A solution of PCB in trichlorobenzene (Chlophen) before and after DCR treatment (left). The pulverulent reaction product can be dehalogenated in a reactor and then disposed of in a compacted form (center). Tests have proved that PCBs and PCB-containing waste can be taken in a landfill even without dehalogenation, provided the material has been DCR reacted in the presence of red mud (or any other finely dispersed waste) and if it will be built in under compaction (right: test specimen with insulating wrapping).

The first pilot-scale application of direct dehalogenation was carried out in a vibrating mill as the reactor to completely dehalogenate Arochlor 1242. On request of the authority in charge, the analysis before and after treatment had to be made through a specialized laboratory in the Netherlands.[24] The results are given in Table 6.

The extent of dehalogenation has also been tested with a crude waste oil of a re-refinery. The PCB concentration before and after treatment was 2.11 ppm and below the detection limit, respectively (33).

2. DCR Treatment of PCB-Contaminated Soil

It goes without saying that there is no conventional method available for the treatment of PCB-contaminated soil and sludge of the sort shown in Figs. 1 and 2. All proposals made so far are far away from any technological and economical applicability. However, with

---

[24] Witteveen & Bos, Deventer, The Netherlands.

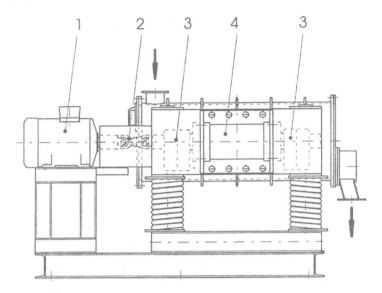

Figure 41 Siebtechnik Vibrating Mill driven by a three-phase motor (1) via a cardan shaft (2). The inhomogeneous vibrations are produced by an exciter unit (4) in the form of a bearing pedestal with eccentric weights (3). The arrows mark input and output devices.

regard to DCR technology, there is neither any problem with water nor a problem regarding accessibility, for instance, for the dehalogenating reagents to PCB in oil adsorbed in a wet heavy clay-type soil.

Sludge or, in other words, mixes of water and soil contaminated with an oily PCB-containing phase are first de-watered through the addition of small quantities of a dewatering agent and then predried in a continuously working centrifuge of the type used for dewatering harbor sediments. The pasty residue is then reacted in a first DCR step to furnish a dry pulverulent solid which is now ready for complete dehalogenation in a reactor, for instance, a vibrating mill (Fig. 41).

A first pilot-scale treatment of PCB-contaminated wet soil from a washing plant has been carried out as a NATO CCMS Pilot Study. The ECD-GC results are given in Fig. 42.

3. DCR Solutions for Problems with Dioxins

There are three main areas of application of DCR technology to solve dioxin problems, the clean-up of off-gases from waste incinerators, the treatment of filter dust with adsorbed dioxins, and the remediation of contaminated soil, ash, slag, etc. Since the parallels to the processes described above for PCB problems are obvious, the comments on this matter can be given in a very brief form.

Dioxins in off-gases must be chemically removed through absorption and not through adsorption. In order to remove dioxins from the gas phase, the same strategy has to be applied as set out above for PCB removal from a liquid phase:

(R 11)

DCR Remediation Technology

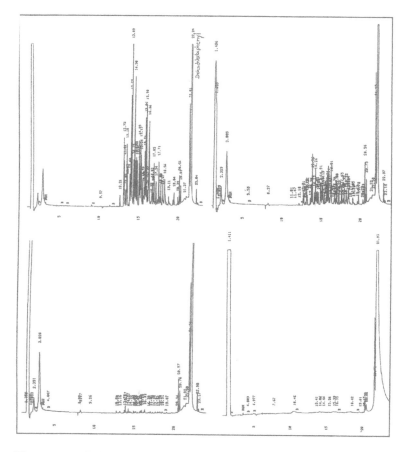

Figure 42 ECD-GC of PCB-contaminated soil before DCR treatment (top, left), and after treatment with stepwise increased quantities of dehalogenating agent, but with equal residence times and the same temperature (close to room temperature). The PCB concentration corresponding to the final diagram (bottom, right) is below detectability.

Monosubstitution of dioxins means that the halogenated compounds are removed in the form of halogenated phenols, thus avoiding any residual concentration through adsorption equilibrium. The absorbed dioxins must afterwards be reacted in a dehalogenation unit to yield a harmless product. It is economically favorable to simultaneously disperse precipitating compounds for trapping and adsorbing heavy metals present in the off-gases, mainly Hg and Cd.

(R 12)

Dioxins in sludge and soil and, of course, any other organochlorines have to be dehalogenated, as pointed out above for the same problem with PCB, i.e., dewatering in a centrifuge, disintegrating in a DCR reactor, and dehalogenation in a dehalogenation unit. Dioxins in ashes and slags can be dehalogenated directly, i.e., without pretreatment.

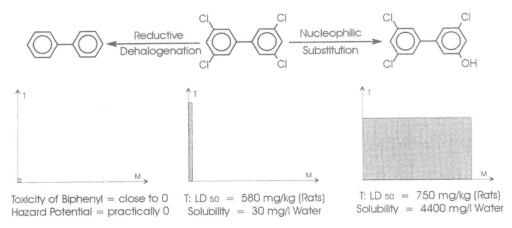

Figure 43 The endangering value, E_Q, as a function of toxicity T and mobility M with respect to tetrachlorobiphenyl as an example. The mobility is given through the solubility S. Accordingly, E_Q will be increased significantly in nucleophilic dehalogenation reactions in which just one chlorine has been replaced by OH. Unlike substitution reactions, the reductive dehalogenation furnishes a product for which the E_Q is close to zero.

Reductive dehalogenation reactions, as has been pointed out above, are well known in common organic chemistry and are described in detail in elementary textbooks. These reactions, however, can be carried out only under very special reaction conditions, for instance, in dry solution and with noble-metal catalysts on charcoal in the presence of hydrogen, which are not applicable in technical environmental dehalogenation processes.

If one makes use of the same reactions in the form of what is called DCR-supported reductive dehalogenation, a complete dehalogenation of any sort of chlorinated aromatic compound is achieved within a few minutes.

The complete reductive dehalogenation of PCB, halogenated dioxins/dibenzofurans, etc., carried out without addition of any problematic solvent such as DMSO or HMPA and yielding nothing else than the corresponding harmless hydrocarbons, can be regarded as the only process that deserves the name detoxification.

Methods designed to "completely" monosubstitute, e.g., hexachlorobiphenyl, giving the corresponding halogenated phenols, must undoubtedly be regarded as a means of tremendously increase the hazard potential. It is not important whether or not the toxicity of the halogenated phenols is as high as the toxicity of the parent compounds; what counts is the fact that the solubility of halogenated phenols is considerably higher than the solubility of halogenated aromatics without -OH group. In other words, hexachlorobiphenyl as such, to continue the example, with its very low solubility, is endowed with a much lower endangering value, E_Q, than in its solubilized and mobilized form, i.e., as the corresponding phenolic derivative (Fig. 43).

D. DCR-Supported Biological Degradation of Organics

Crude oil and the corresponding pure technical products manufactured of it such as lubricants, paraffin oil, diesel fuel, etc., are not genuinely toxic in the literal sense of the word. Nevertheless, it is no contradiction in terms to maintain the expression for E_Q as a function of Q, T and M. The reason is that the physiological toxicity can be replaced by a physical toxicity against the background that an oily layer on water or an oily phase in soil can secondarily cause the same symptoms of intoxication as a genuinely toxic substance can, for instance, anoxemia of fishes in water or microbes and superior living beings in soil, simply through the agglutinating properties of viscous oil phases.

As far as this sort of organic material is concerned, problem solving in the field of environmental protection must be considered optimal when it ends up with the formation of carbon dioxide and water and, possibly, inorganically fixed heteroatoms—i.e., when every organic structure has been removed by biological oxidation.

Since biological degradation of hydrocarbons takes place in the areas bordering aqueous systems, one can expect that a block of high-melting paraffin with its small specific surface and its water-repellent properties will be degraded into CO_2 and water only extremely slowly, even under favorable conditions. If the finely dispersed reaction product is left uncompacted and mechanically mixed into a biologically active soil instead, the high chemical reactivity of a finely dispersed organic phase predominates, causing rapid biological degradation.

As has been shown so far, the bioavailability of harmful substances can be abolished with the aid of dispersing chemical reactions. The harmful substances are immobilized or fixed in a compacted form and thus become inert with respect to their chemical reactivity. On the other hand, according to the same principle, the chemical reactivity of harmful substances can be significantly raised by forming a fine dispersion, so that among other things they can be subjected to rapid biological degradation, provided they are degradable in principle.

Let us look at the basic features of the two process variants, which at first glance seem to be contradictory. On the one hand, there is the formation of nonreactive bodies of soil after compaction of the DCR reaction product, i.e., ones that are immune from being chemically attacked under natural conditions and are thus permanent. On the other hand, there is the production of a highly reactive dispersion from the same reaction product, which, in order to increase the reactivity, including the reactivity in biological degradation processes, is not compacted. To clarify the matter, we will carry out an experiment with a piece of paraffin wax in humus. In a DCR reaction with calcium oxide, the hard paraffin is transferred into a finely dispersed solid preparation. There are now two options: either to compact the pulverulent DCR product so as to form a chemically inert block of filled paraffin wax, or making use of the significantly increased chemical reactivity of homogeneously finely dispersed paraffin hydrocarbons (Fig. 44).

In the case of compacted material, the physical and chemical properties are nearly identical with those of the original sample of paraffin wax: the inert block, like paraffin wax, floats on water, is insoluble in water, and, since the inorganic filler is uniformly surrounded by the paraffin wax, there is no reactivity of both the paraffin hydrocarbons and the calcium hydroxide with the exception that the calcium hydroxide, though it is embedded in a solid organic phase, forms a calcium carbonate layer, which grows increasingly slowly in the course of time, this carbonating step being possible only because, as is known, the "organic compound carbon dioxide" is able to migrate readily into organic phases.

The noncompacted material exactly resembles the properties of a finely dispersed organic phase. Some of the remarkable features of this preparation are: the fast and complete formation of calcium carbonate, an easy reaction of the paraffin hydrocarbons with oxygen with the possibility for spontaneous incineration commencing at ambient temperatures, and the outstanding reactivity of finely dispersed hydrocarbons in biological systems.

Figure 44 The question is how long a piece of paraffin wax (top, left) embedded in humus will survive. The estimated time varies between 100,000 years and infinity. The half-value for the biological degradation of DCR-dispersed wax (top, right) merged into the same humus (bottom, left) has been measured to be 28 days. If the pulverized product, however, was compacted, no degradation was detectable (bottom, right). The compacted specimen resembles the original paraffin sample (top, left). This is easy to understand: the physical and chemical properties are those of the original piece of paraffin wax merely filled with the inorganic filler $Ca(OH)_2$.

Of course, these remarks apply not only to hydrocarbons that are solid at room temperature, but also to liquid organics. Thus, it is known that crude oils that are present as an oil spill on a water surface can be biologically degraded rapidly and without secondary hazard potential if one proceeds as has been described in Section VI.

The mechanisms that are valid for the clean-up of oil spills on water are also effective in the case of oil spills on land. In a first immobilizing step, the organic phase is immediately converted into a highly hydrophobic solid preparation, so that the oil can neither seep away nor spread as a thin layer on groundwater and rain water. At this stage there are again two options: either to compact the pulverulent DCR product so as to form a chemically inert block like the filled paraffin wax in Fig. 44, or to make use of the significantly increased chemical reactivity of homogeneously finely dispersed hydrocarbons (Fig. 45).

If a decision is made to degrade the hydrocarbons microbiologically, then one has to wait for the formation of calcium carbonate and thus the neutralization of the calcium hydroxide to occur. This takes only a few days even with a highly hydrophobic reaction product; the carbonating step can be accelerated through mechanical mixing, e.g., with

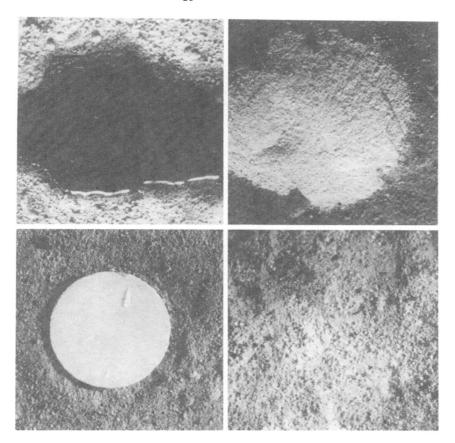

Figure 45 The two options for DCR treatment of oil on/in soil (top, left). After dispersing the oil chemically on site (top, right), the resulting product can either be compacted and thus irreversibly immobilized and rendered inert (bottom, left), or it can be blended with the biologically active soil of the surroundings for fast biodegradation.

a rotary tiller. The charged carbonate is finally mixed into the surrounding biologically active soil.

The DCR process is the only way to effectively and economically degrade compact phases of nonpersistent organics. All attempts made to biodegrade compact oily waste like the one shown in Fig. 46 (left) by what is called landfarming with educated bugs are far away from reality.

As is the case with all biological processes, the DCR-supported biological degradation method described here for the remediation of contaminated soils in the area of waste sites cannot be applied if hydrocarbon mixes with persistent compounds are present. This is of considerable importance in connection with the remediation of highly problematic waste sites, for instance, former coking plant sites and the like.

Persistent toxic organics, which are incorporated in, from an environmental viewpoint, inert bituminous hydrocarbons, are de-facto effectively immobilized. Through the biological degradation of the harmless embedding component these toxic substances become releasable and thus bioavailable, which is the opposite of what was originally intended. Instead, it is recommended that if mixtures containing persistent compounds

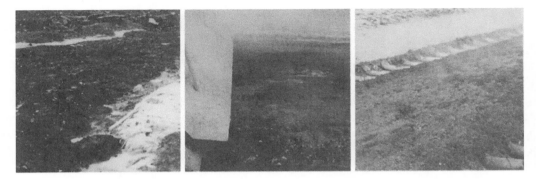

Figure 46 Landfarming on a former coking plant site. Not only is a degradation of a compact phase of organics impossible, nor is the subsoil being protected from continuous pollution by a plastic foil, nor is it altogether reasonable to do something like that (left). The nondegradable hazardous compounds (PCAH, phenols, etc.), which are really toxic, are, before treatment, immobilized in a harmless hydrocarbon matrix, which, however, has the property to be degradable. Landfarming of this sort does mean that the mobility M is significantly increased through the continuous release of persistent toxic compounds from the incorporating matrix of harmless organics being degraded, and it would have been much better if things had been left just as they were. The DCR treatment of (nonpersistent) oily waste comprises the immobilizing step (center) and blending of the reaction product with biologically active natural soil follows (right).

are involved, the harmful substance phases should be dispersed by chemical means and the reaction products should be compacted, as described in Section IV.B.

VIII. EVALUATION METHODS

Preliminary practice-relevant investigations carried out on the toxic waste to be treated will reveal which variant of the DCR process can be applied for the appropriate solution of a particular problem.

In connection with the reprocessing of organic harmful substances, the proportion of volatile components, and in particular ones that are volatile with water vapor, is extremely important in every case. The results obtained from the analysis performed in this connection form the basis for the decision in principle of whether or not transformation in a field process is justifiable or permissible, or whether a semiclosed plant is required. Considerations of this kind are also of importance when it is less a matter of the volatility of certain problem substances than of the possibility of their spreading in dust form after the production of the DCR solid preparations.

Seen from this point of view, of the organic phases it is only the noncontaminated oils and bituminous substances, preferably in the form of contaminated soils or soil-like materials, that can be treated in a field process. This also applies to the treatment of solids contaminated by heavy metals, but not quite so strictly, because the emission of harmful substances can be more easily avoided or restricted as a result of the better technical possibilities in this field.

Since in special cases, for instance, with acidic materials, temperatures of around 80°C can easily occur in the exothermic dispersing chemical reaction even during the reprocessing of contaminated soils in the field process, a steam distillation test with bath

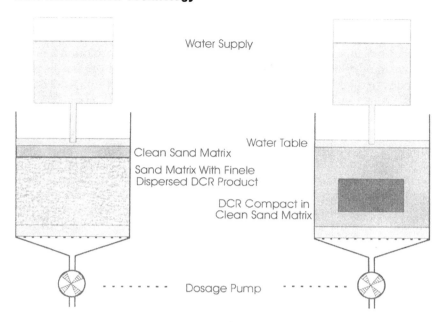

Figure 47 Apparatus for continuous leaching experiments with storage container for water supply, aqueous compensating phase, clean quartz sand matrix (left as a compensating layer), finely distributed DCR product in quartz sand (left), or, alternatively, in the form of a DCR compact (right), and multiple-channel dosage pump.

temperatures of around 120°C is always carried out in the laboratory in addition to a normal distillation test for the sake of orientation.

The analytical composition of the harmful substances is otherwise not of any special importance, since with incineration using up the inherent energy either all the organic substances, independent of their structure, are oxidized to the same extent to CO_2 and water, at the same time possibly forming inorganic secondary products containing the organic heteroatoms in an ionic form such as organically bonded chlorine and sulfur, which are transferred into calcium chloride and calcium sulfate, respectively; or, in the case of immobilization, all the organic components behave identically anyway.

In accordance with this concept, the absolute concentration of certain toxic compounds, e.g., heavy metals, has no significance at all. This is because it is only those parts that are soluble under natural conditions that are bioavailable and thus problematic.

Thus, a comprehensive chemical analysis for the determination of all the components, which destroys the original structure, is replaced by leaching tests under as natural conditions as possible, in order to determine the hazard potential of mobile toxic compounds in their current environment. The leaching apparatus shown in Fig. 47 is especially suitable for this, since in this way it is possible to compare numerical values for mobility before and after immobilizing treatment or treatment by chemical fixation under various conditions that are relevant to practical application.

The concept of the remediation of toxic waste sites introduced here is based on the idea that the bioavailability of harmful substances of an inorganic and organic nature can be removed by chemical means through immobilization or fixation. In this case, liquid organic harmful substance phases are converted to solid preparations in which calcium hydroxide functions as the carrier; in a similar way, inorganic harmful substances are transformed into water-insoluble derivatives.

It was shown that the superficial conversion of the calcium hydroxide into calcium carbonate is of decisive importance for the permanent immobilization of organic harmful substances. This applies to a remediation measure that includes microbiological degradation as well as to measures in which the harmful substance remains intact and the corresponding solid preparation is placed in a natural environment in the form of a compacted body of soil, either as an earth construction material or as a component of a solid landfill.

Thus, it had to be proved that in the interplay between the two reactions, i.e., the dispersing reaction of the calcium oxide with the formation of hydroxide, bearing harmful substances, and the formation of calcium carbonate on the surface of the hydroxide, not only the actual mobility of the liquid harmful substance phases is removed, but that this immobilization will still continue to exist in the foreseeable future, even when affected by environmental imponderables.

For this purpose, experiments were made on bodies of soil on a laboratory scale, in which the extent of the immobilization of test substances was examined and numerically determined with regard to the individual reaction steps and under changing strain.

In the case of the described procedure for the chemical fixation of toxic organics that do not occur as liquid phases, and also with the analogous fixation of inorganic harmful substances, especially heavy metals, it is a precondition that the chemical or even physicochemical step of fixation is irreversible under natural conditions. In this connection, experiments were also performed to examine the validity of these assumptions from the long-term point of view.

All the experiments were performed in an apparatus as shown in Fig. 47. It consists of a glass cylinder, a supply vessel for the leaching liquid, a dosage pump, and graduated cylinders to collect the eluate. As it is the actual leaching vessel, the glass cylinder has a flat bottom and a narrow outlet in order to keep the dead volume as small as possible. A bed of quartz sand lies on a removable filter plate. Depending on the experimental task, a deposit of original toxic material and the corresponding treated samples, both mixed up with additional quartz sand for uniform contact, is placed on the bed of sand (Fig. 47, left) or, if necessary, as a compacted test body in the middle of it (Fig. 47, right). If a layerlike deposit of a contaminated sample is used, then this has to be covered with an additional thin layer of sand. These arrangements guarantee a homogeneous throughflow, if necessary in both directions, with the original toxic waste samples or the relevant solid preparations being uniformly contacted from all sides.

In this experimental setup the inflow from the storage vessel takes place automatically. The inflow pipe is adjusted at such a height that a layer of leaching liquid 1 cm deep is formed above the uppermost layer of sand in the leaching vessel. Since the throughput is determined solely by the amount taken from the outlet pipe with the aid of the dosage pump, the material to be examined inside the leaching vessel is continually under water, so that artifacts caused by the formation of channels are excluded.

All the natural leaching processes can be simulated via the rate of removal. At the same time, however, the results obtained here can also be compared with the values from standard procedures such as the Swiss Leaching Test, American Standard Test (AST), and German Standard Specification (DEV), if, for example, one sets a throughflow rate at which 10 times the amount of leaching liquid, referred to the quantity of the sample, will have continuously passed within 24 h.

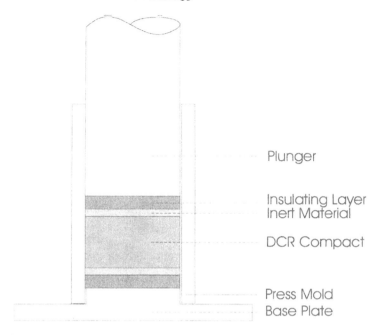

Figure 48 Compacting unit with plunger, DCR isolating layer, inert material, DCR compact, press mold, and base plate (from top to bottom) for the preparation of DCR compacts of soil on a laboratory scale. This diagram shows the preparation of a body of soil with an isolating layer and with an additional inert material as a buffer above and below it.

The procedure for immobilizing and fixing toxic waste material within the framework of measures using DCR technology demands that the DCR reaction products should be included in the form of compacted bodies of soil. They are built in according to the demands of soil mechanics, and here particular attention has to be paid to the construction of the isolating layers of DCR-treated but noncontaminated inert material. This refers mainly to the setting up of permeability gradients in a vertical direction along the slope at the surface and at the sides of the body of soil.

The technical measures described here can easily be simulated on a laboratory scale for test purposes. Seen from the other point of view, this means that model tests can be used for the deduction of very precise technical instructions and information concerning the special features of individual remediation projects.

In order to do this, one compacts the reaction products from DCR test approaches in a compacting tool, as shown in Fig. 48, under conditions relevant to those found in practice, i.e., with a pressure of $20-25 \text{ kg/cm}^2$ and a compaction time of 1 min. This compaction produces a test body, the weight of which is 50–100 g, depending on the task in hand.

For example, consider the preparation of a test compact manufactured in a DCR reaction with a solution of PCB in trichlorobenzene. Conversion into a solid was carried out with the addition of red mud for the simultaneous fixation of dioxins. The pulverulent product was compacted without further addition and with the addition of an isolating layer.

There are two ways to examine isolated bodies of soil, i.e., test samples that correspond as closely as possible to the procedure actually used in practice. One places a tube internally on top of the isolating layer within the compacting apparatus. The actual sample is deposited inside the tube, and the inert material is deposited outside, between the two tubes. After the filling height has been reached, one removes the inner tube, which has gradually been pulled upward during the filling process. Then a final covering layer is applied. Compacting then produces a test body.

One can also compact the test material in glass cylinders and provide it with an isolating layer only at the bottom and top. These test bodies are like a drilling core taken from an extended body of soil. In this variant the leaching test is performed at intervals in a beaker.

In order to make the processes during leaching tests easier to understand, it is of advantage to introduce extremely high concentrations of toxic compounds into bodies of soil with an isolating layer in the form of separate, highly concentrated deposits. The result of this is that the test results become considerably more meaningful. At the same time, it is possible in this test setup to examine the suitability of the inert materials that might be appropriate for a real remediation measure.

The problem of evaluating leaching tests suitable for practical application is that all the tests on bodies of soil that have been produced according to this procedure have the same result insofar as no harmful substance of any kind can be detected in the eluate, independent of the concentration in which it was presented.

It is natural that this fact does not alter when test bodies of this kind are examined in long-term experiments with extremely low throughput rates that last for a number of months. This is because the increasing formation of calcium carbonate in the interface of the isolating layer gradually isolates the system more and more.

Model tests were carried out in order to clarify this statement. These tests start with the most unfavorable case, which is the uncompacted, finely dispersed, and noncarbonated solid preparation. Here it is possible to demonstrate the extent of the first stage of immobilization by comparing the mobility of a liquid toxic phase with that of the appropriate solid preparation.

In an example dealing with the immobilization of a crude oil, the release of toxic compounds in the leaching apparatus, as shown in Fig. 47, amounts to 78% after a throughput of only 1500 mL of noncarbonated water. As soon as one uses the same amount of the same crude oil in the form of a dispersed solid preparation and leaches it under conditions that are otherwise the same, i.e., noncompacted sample, noncarbonated water, one obtains a mobility reduced to 0.0015%. These results must be qualified by the fact that the first leaching fractions still pick up all the components, which are contained on the test body in the form of traces of dust and burr, so that the measurement data are by no means characteristic of a compacted body, as is shown by a simple rough estimate that takes the specific surfaces into account. If one applies the much more realistic Swiss Leaching Test using carbonated water, however, there are, even with the noncompacted sample, only traces of hydrocarbons detectable, which have been released via the gas phase according to the vapor pressure of some volatile compounds. On the other hand, if one compacts the same quantity of the solid preparation and again applies the Swiss Leaching Test, no hydrocarbons can be detected at all.

Thus, some of the numerical values are seen to be the results of artifacts, and the immobilization results are far more favorable than shown here. Finally, if one surrounds the solid with an hydrophobic inert layer, then it is also natural that, from the outset, no hydrocarbons can be detected any more.

In accordance with this, one ought to compare the diagrams in Fig. 49, which describes the crude oil release already mentioned and thus the mobility of this harmful

DCR Remediation Technology

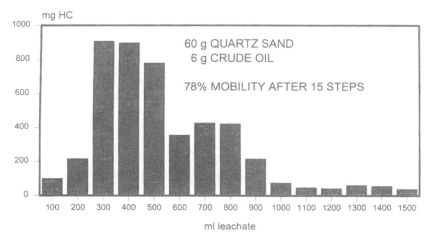

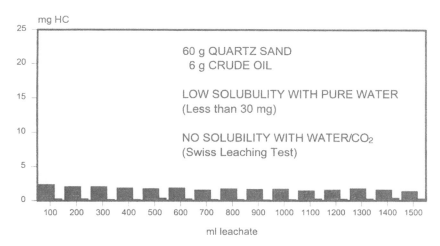

Figure 49 For a realistic representation of the immobilization of liquid toxic substance phases that can be achieved with the DCR process, the de-facto immobility (bottom) ought to be compared with the original mobility (top).

substance phase (top), with the diagram that contains de-facto no entry along the x axis in a leaching test with a uniformly distributed DCR preparation employing carbonated water according to the Swiss Leaching Test (bottom). If the leaching test is carried out with water free of carbon dioxide, the partial solubility of finely distributed hydrophobized calcium hydroxide and, moreover, the partial pressure of some volatile hydrocarbon compounds in the finely distributed preparation determine the hydrocarbon concentration in the leachate, which is, as experience shows, in the range of some parts per million.

In no case is the release of incorporated harmful substances to be observed in correctly produced model bodies with an isolating layer, independent of the type and concentration of the harmful substances inside them and independent of their contact time in the liquid phase of the leaching apparatus. For this reason, it is understandable that no negative long-term effects can be found due to the increasing effectiveness of the isolating layer as time progresses.

However, in long-term leaching tests, even bodies of soil without isolating layers provide results that, in the end, correspond to those obtained with test bodies with isolating layers. This is definitely predictable, since the release of toxic compounds must be reduced to almost zero within a short time with increasing formation of calcium carbonate and increasing impoverishment, i.e., with decreasing diffusion of harmful substances from the interface-active areas.

In order to prove this numerically, long-term tests were carried out with organic model substances in test bodies with isolating layers, inside which a dye deposit of Rhodamine B had additionally been placed. The dye is highly water-soluble and, because its color is so intense, it permits a very sensitive spectroscopic detection possibility in the parts per trillion (ppt) range.

Test bodies corresponding to the number of planned leaching cycles were distributed among beakers in sets and then mixed with 10 times their weight of water. The test bodies were held in the middle of the liquid phase by stainless steel wire attached at three points. The liquid phases were moved synchronously by means of magnetic stirrers in parallel connection.

After residence times each lasting 24 h the test bodies were raised from the leaching water simultaneously and exposed to the air with its CO_2 content for 6 days. After being returned to the beakers, which had been filled with fresh water, the first leaching cycle was completed. After each cycle a test body was removed and examined for bleeding; the concentration of hydrocarbons and the pH value were determined by infrared spectroscopy in the appropriate fraction (Fig. 50).

The results are in complete agreement with the model concepts that had been developed. In the first fraction, hydrocarbon concentrations were measured that correlate with the water solubility of the organic substances and that amount to approximately one-fiftieth of this solubility. This is not surprising, since the hydrophobic surface of the test body is able to adsorb a not inconsiderable amount of hydrocarbon from the initial release caused by the $Ca(OH)_2$ starting to be dissolved.

The proportion of hydrocarbons is significantly reduced in the following fractions; the jumps in the numerical values that are initially observed are due partly to unavoidable inhomogeneities in the production of the compacts. They are due mainly, however, to the hairline cracks occurring at irregular intervals in the surface of the compacts during the test period. This process must also be considered to be an artifact, because the test bodies are not firmly incorporated in a solid matrix as laid down for practical application, but they are free. Tests on firmly incorporated test bodies, in fact, prove that this is the reason for the observed diffusion.

After 20 test cycles, corresponding to a test period of 20 weeks, a test body was opened with a vertical and a horizontal cut. The sample showed no bleeding and the die lay unchanged in the deposit. This explains why no coloring was detectable in any of the leaching fractions. It also proves that no aqueous phase was able to move inside

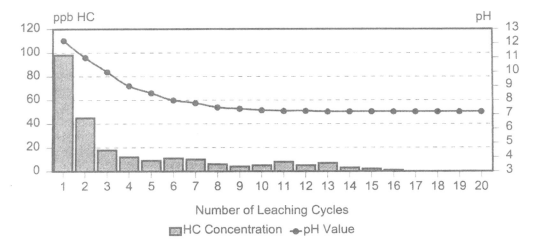

Figure 50 Leaching properties of 50% tetrachloronaphthalene as the model substance in liquid paraffin after DCR treatment and compaction. The cylindrical test piece contains a dye deposit (explanations in the text).

the test body. If one placed one-half of the opened test body in water, it was quickly stained red.

IX. SUMMARY

Materials can cause damage to the living environment and thus assume the quality of genuine toxic compounds only when they are brought into contact with living organisms in a bioavailable form and are taken up by them. As a result, difficulties with toxic waste sites occur only when the toxic compounds deposited in them leave the deposit and are able to spread out in the natural environment.

The mechanism for the release of toxic material is based on the mobility of liquid organic phases in the shape of seepage oils or aqueous phases in the form of seepage water containing harmful substances; because of the continuous flow of rain water and groundwater, this means that there are practically no limits to the spread of toxic contaminants.

Measures for the remediation of waste sites can thus only be meaningful and justifiable from an economics point of view if the damage is removed at the source through the immobilization of liquid toxic substance phases or through the fixation of mobile toxic contaminants within liquid phases.

With the aid of DCR technology, it is possible to convert liquid organic toxic material into solids and to permanently immobilize organic and inorganic harmful substances, e.g., heavy metals and phenolic compounds, respectively, which migrate via an aqueous phase, by chemical means.

Waste sites with toxic substances that are fluid or fluidizable as such, or that can be carried away with the assistance of water, after clean-up with the aid of the DCR technology described here contain, without exception, nothing else but solids. When compacted into a body of soil that cannot be attacked under natural conditions, no release of toxic compounds of any kind is possible.

With regard to the long-term behavior of DCR bodies of soil of this kind, it was possible to prove that they become increasingly safe during the course of time as a result of a natural carbonating process.

Environmental protection measures must be meaningful, appropriate, and, last but not least, financially acceptable. It is known that heavy metals can be incorporated into the crystal lattices of hydraulic systems; however, it is not very sensible to transform soils contaminated with heavy metals into lean concrete and to store it at great expense, at the same time taking up valuable waste deposit space, if, on the other hand, permanent and economically viable chemical fixation is possible in situ in accordance with the natural conditions, while maintaining the soil structure and without increasing the volume.

In addition, it is claimed that soils contaminated with organic harmful substances, such as polychlorinated biphenyls, can be "incinerated" in high-temperature installations. It is unnecessary, however, to accept the unrealistically high financial costs required for this if the same harmful substances can be permanently and safely immobilized with the same degree of success at a fraction of the cost, and if their bioavailability can thus be removed or, alternatively, if any sort of halogenated aromatics including PCBs, halogenated dioxins, and dibenzofurans can be detoxified, in the real sense of the word, in a DCR-supported dehalogenation reaction at ambient temperatures within, if necessary, a few minutes.

If soil and water are polluted with organic biologically degradable compounds, for instance, in the form of oil spills, DCR-supported biological degradation is a versatile tool for a fast and, moreover, "safe" clean-up under optimum reaction conditions—"safe" meaning in this context that, through the immediate transformation of any spilled liquid phase into a solid preparation, any secondary negative impact to the environment is stopped without delay. As a consequence thereof, it does not matter what period of time might be required for total degradation of all organic constituents.

Against this background one must pose the following questions: at what cost, with which technical means, and with which concrete result can entire toxic waste sites with their complex composition be processed by an incineration plant? Just a few examples illustrate how complex their composition is: metal drums, wooden beams, pieces of concrete, foundations, building rubble, rails. On the other hand, as has been shown, with the aid of DCR technology it is possible to solve tasks of this kind in a short time, without difficulty, under realistic conditions and with relatively small financial effort by using technical means that are currently available.

Environmental protection is generally unproductive. Even productive environmental protection becomes possible, however, with the application of DCR technology. The example described here of a remediation plant for organic toxic waste consisting both of waste sites and continuously produced residues must be regarded as an optimum solution, since an economic asset is produced here exclusively through the meaningful application of waste materials from different industrial processes. The same is also valid with regard to the use of economic assets from reprocessed materials containing harmful substances for applications connected with soil mechanics.

Finally, it must also be mentioned that measures can already be undertaken today in order to avoid future problems with toxic waste sites by establishing special so-called DCR waste deposits with stationary DCR plants, in which disposal of toxic waste is permitted solely in the form of DCR solid preparations.

REFERENCES

1. Conner, J. R. 1990. *Chemical Fixation and Solidification of Hazardous Wastes.* Van Nostrand Reinhold, New York.
2. Boelsing, F., and Hakim, A. 1978. *Z. Naturforsch.*, 33b:632.
3. *Handbook of Chemistry and Physics*, 1978/1979. CRC Press, Boca Raton, FL.
4. Lax, D'Ans. 1964. *Taschenbuch fuer Chemiker und Physiker*, 3 Aufl. Springer-Verlag, Berlin.
5. U.S. Environmental Protection Agency. 1984. *Review of in-place Techniques for Contaminated Surface Soils*, Vol. 1, EPA-540/2-84-003a.
6. Boseker, K. 1980. *Forum Mikrobiol.*, 2:98.
7. Linemann, K. 1966. *Erdstabilisierung in Theorie und Praxis.* VEB Verlag fuer Bauwesen, Berlin.
8. National Lime Association. 1976. *Lime Stabilization Construction Manual*, Bulletin 326. NLA, Washington, DC.
9. Bundesverband der Deutschen Kalkindustrie e. V., *Bodenverbesserung Bodenverfestigung mit Kalk.*
10. Buechel, K. H. 1977. *Pflanzenschutz und Schaedlingsbekaempfung.* Georg Thieme Verlag, Stuttgart.
11. Hoechst A. G. 1982. European Patent Application 0070415.
12. Edda Boelsing. 1988. Immobilization of heavy metals through long-chain alkylaryldithiophosphoric acid esters. State Examination Paper, Hannover University.
13. Doern, M. 1986. On the biological degradation of dispersed hydrocarbons. State Examination Paper, Hannover University.
14. Laszlo, P. 1987. *Preparative Chemistry Using Supported Reagents.* Academic Press, San Diego, CA.
15. Isaacson, P. J., and Sawhney, B. L. 1983. *Clay Minerals*, 18:253.
16. Brandt, K. 1992. Clay-supported immobilization of organic compounds. State Examination Paper, Hannover University.
17. Brunelle, D. J., and Singleton, D. A. 1983. *Chemosphere*, 12:183.
18. Brunelle, D. J., and Singleton, D. A. 1985. *Chemosphere*, 14:173.
19. desRosiers, P. E. 1989. *Chemosphere*, 18:343.
20. Shaw, J. E., Kunerth, D. C., and Swanson, S. B. 1976. *J. Org. Chem.*, 41:732.
21. Testaferri, L., Tiecco, M., Tingoli, M., Chianelli, D., and Montanucci, M. 1983. *Tetrahedron*, 39:193.
22. Peterson, R. L. 1986. U.S. Patent 4,574,013.
23. Boelsing, F., and Birke, V. 1991. On the dehalogenation of chlorinated aromatics, Dissertation of V. Birke, Hannover University.
24. Boelsing, F., and Birke, V. 1992. On the dehalogenation of chlorinated aromatics. Doctoral thesis of V. Birke, Hannover University.
25. Mueller, G., and Riethmayer, S. 1982. *Chem.-Ztg.*, 106:289.
26. Ising, U. (Aral AG). 1978. Remediation at Dollbergen: Recultivation of oil contaminated soil utilizing DCR technology, *OEL. Z. Mineraloelwirtschaft*, 78–80.
27. J. Britena. Report of the Gas Research Institute (unpublished).
28. U.S. Environmental Protection Agency. 1986. Draft guidelines for permit applications and demonstration test plans.
29. D. T. Kaschani et al. 1989. Analysis of polychlorinated biphenyls (PCB) in oil samples. BP AG, Institute of Research and Development, Wedel, Germany.
30. Rappe, Ch., et al. 1987. Sources and relative importance of PCDD and PCDF emissions. *Waste Manage. Res.*, 5:225.
31. Ballschmiter, K. et al. 1986. Experiments in high-temperature chemistry of organohalogens. *Chemosphere*, 15:1369.
32. Erickson, M. D. et al. 1986. Products of thermal degradation of dielectric fluids. *Chemosphere*, 15:1261.
33. Analytical reports of Mineralöl-Raffinerie Dollbergen GmbH, May 9, 1997, and June 20, 1997.

39
Processing of Vegetable Raw Material and Its Waste by Energy-Saving, Nature-Preserving Technology to Obtain Carbohydrate-Protein Fodder

I. Shakir, V. Panfilov, D. Kulinenkov, and M. Manakov
D. Mendeleyev University of Chemical Technology of Russia, Moscow, Russia

I. INTRODUCTION

One of the main problems of biotechnology is to work out effective and ecologically safe technologies for microbe protein production in order to lower the existing imbalance in fodder for agricultural animals and poultry.

Proteins of microorganisms are equal to proteins of animal origin in their biological value (1), and considerably surpass vegetable proteins in their quality, vitamin, and microelement content. Proteins of actinomycetes and bacteria form the main part of cells' dry weight (55–75% of dry weight). Cells of fungi contain on average 25–55% protein, and yeast cells contain 45–55%. Different kinds of raw material of vegetable origin, except soy, contain protein in dry weight in considerably smaller quantities.

Such peculiarities of microorganisms as their ability to utilize various carbon compounds, high speed of biomass synthesis, high protein content, possibility of strain selection with definite properties, and also the fact that microbiological protein production is less labor-consuming in comparison to agriculture production have determinative importance for wide inculcation of microbe protein in agriculture (2).

The volume of world production of one-celled protein for fodder purposes is prognosticated to be 106.3 million tons by 2000 (3). Considerable experience of large-scale microbe protein production on different substrates and use of protein for fodder purposes has accumulated in the world (4,5).

At the present time, renewable vegetable raw material and in particular carbohydrates obtained from its base is of special interest as a substrate for microbe protein production. Insofar as vegetable hydrocarbons are extraordinarily widespread and reproducible, they take a special place in nature and may serve as a capital raw material in particular for microbiological production.

In spite of the proved harmlessness of all industrially used types of one-celled proteins cultivated on carbohydrates of vegetable origin, yeasts raise fewer objections from consumers because food products of vegetable origin is a habit of ages. This is very important in terms of the use of vegetable raw material as a substrate for industrial microorganism growing.

Not only may specially grown crops be used as substrates for microbe protein production, but in some cases vegetable waste and by-products of agricultural raw material may also be used. In this case it is possible not only to solve the protein-deficit problem in stock breeding, but at the same time it is possible to utilize environment-polluting waste. It is known that only about 10% of vegetable raw material waste is applied in Russia, and unused waste considerably changes the ecological situation for the worse in some regions.

There is accumulation of great amounts of different sorts of unused or ineffectively used vegetable raw material in many regions of the world. As a rule, the most valuable vegetable wastes on small farms in many countries are used directly as animal feed. However, such waste utilization lowers the efficiency of stock breeding because most wastes have low raw protein content. This fact intensifies the problem of protein deficit and amino acid imbalance in fodder. At the same time, a considerable part of vegetable waste remains unused, polluting the environment.

Drawing on this waste for the production of microbe protein is of great importance from the point of view of the problem of protein deficit in stock breeding, removal of wastes, environmental pollution, and therefore, raising agricultural production productivity.

At the same time, we decided that it would be interesting to evaluate the expediency of vegetable growing—the sources of carbohydrates—for their further utilization by yeast to produce biomass of high quality which contains the necessary quantity of full-value protein.

We examined some perspective sources of raw material for fodder products processing enriched by microbe proteins. Calculation of possible raw protein increase was based on figures of middle crop capacity of raw material and rates of industrial waste formation (6,7). We examined 27 sources of vegetable carbohydrates, taking into consideration the peculiarities of the suggested technology for fodder products protein enrichment. The results of comparing a few kinds of practically interesting raw material are presented in Table 1. As can be seen in Table 1, the highest value of protein enrichment can be obtained by using such kinds of fodder cultures as root crops of beet, Jerusalem artichoke, and corn green mass. It should be mentioned that in the case of beet root crops processing, the efficiency of a field hectare as a basis for fodder protein production can be compared with the results of direct soy growing.

Thus, we consider that for regions where climate conditions do not permit culturing and vegetables with high protein content, in particular, soy, the production of microbe protein on vegetable carbohydrates is very perspective.

Combination of known technologies for obtaining food products from agricultural cultivation and microbe utilization of wastes, and enrichment of those wastes by protein to obtain valuable fodder, is especially perspective.

II. BODY OF MATERIAL

Practically all existing variants of industrial vegetable raw material processing are based on hydrolysis by diluted sulfuric acid and have a series of drawbacks, such as poor technology and formation of large amounts (in some cases up to 30% of the processing

Table 1 Quantity of Protein from 1 ha of Agricultural Land at growth and at Biotransformation of Agricultural Plants and Processing Waste

Raw material	Quantity of product or waste (tons/ha)	Raw material content (kg/ton)		Raw protein (kg/ha)		
		Carbohydrates	Protein	Yield	Increase[a]	Total yield
1. Soy (beans)	3.5	40.0	320.0	1120.0	—	1120.0
2. Pea (grain)	2.0	55.0	220.0	440.0	—	440.0
3. Corn green mass	38.8	27.0	13.0	450.0	530.0	980.0
4. Corn (cobs)	6.1	60.0	29.0	400.0	190.0	590.0
5. Beet (tubers):						
Sugar	30.2	120.0	16.0	480.0	1810.0	2290.0
Semisugar	38.2	80.0	16.0	610.0	1530.0	2140.0
Fodder	40.8	40.0	13.0	530.0	820.0	1350.0
6. Beet pulp	27.6	50.0	12.0	330.0	140.0	470.0
7. Beet broken material	3.0	2.8	16.0	48.0	95.0	143.0
8. Jerusalem artichoke	10.9	63.0	22.0	230.0	340.0	570.0
9. Corn pulp	4.8	65.0	25.0	120.0	70.0	190.0
10. Sugar cane	70.0	100.0	16.0	1120.0	3500.0	4620.0
11. Sugar cane bagasse	21.0	23.0	12.0	252.0	48.0	300.0

[a] Protein increment at yeast utilization of carbohydrates.

raw material) of lignin—the main waste of hydrolysis production—which mostly cannot be utilized. Also, progress in this sphere has been delayed by the absence of sufficiently profound technological elaboration of soft controlled hydrolysis processes or other ways of depolymerizing lignin–cellulose polysaccharide bonds, which could provide monosaccharide mixtures of required composition, and also by the high content of undesirable admixtures of compounds of the furan type in hydrolyzates, which inhibit the process of fermentation and lower the quality of the final product (4,5,8,9). Moreover, the low efficiency of microbe protein production processes on hydrolyzed media is considerably caused by the complexity and high energy expenditure of yeast cell concentration, which traditionally includes flotation, separation, and vacuum evaporation of microbe suspensions.

The most perspective technologies of vegetable raw material bioconversion are those which provide no waste or small waste production, yet raise the yield and quality of products while using simple apparatus with the possibility of automation. For the first turn there are technologies of solid-phase and heterophase submerged cultivation. We examined heterophase submerged cultivation of fodder yeast. In this case, production of carbohydrate-protein fodder from annually renewable vegetable raw material is one of the perspective ways to solve the protein deficit and the problems of human effects on nature.

One of the methods to intensify the process in total is to involve solid products of hydrolysis in fodder mixtures by providing submerged cultivation of yeast on liquid hydrolyzed media in the presence of nonutilizable solid phase and obtaining well-filtering suspensions. The product obtained by such technology is a carbohydrate-protein fodder (VCPF). In this case energy expenditure becomes lower because of the filtration process for the final product, which is evaporation instead of energy-capacious processes of separation (4,9). Depending on the type of raw material (carbohydrate-containing waste, vegetable raw material), the final product might contain 12–43% raw protein, 10–15% carbohydrates, and 8–12% ashes with humidity 10–12%.

Thus, use of renewable vegetable agricultural raw material which is widespread in a given region and the products of its processing is a perspective way to lower fodder protein deficit while feeding agricultural animals and poultry, to solve a series of ecological problems. This variant of VCPF production is economically expedient with good filtering properties of culture medium with cells and solid phase.

The variants of complex one-called protein enrichment of agricultural plants and the products of their processing, which we take into consideration below, are highly effective from both technical and ecological points of view. In particular, solution of these problems becomes easier because there is no need to create large-scale production with apparatus of high unit power and high energy expenditure, because agricultural raw material is widespread, and the merging of production reduces complex transport problems. Apparatus or plants to be created should be widespread in the region and solve local tasks of protein production, taking into account peculiarities of a specific raw material, the ecological situation, consumers, seasons, etc. Low capacity of such production demands maximum reduction of energy consumption (using a filtration stage) and realization of a closed water cycle (return of the greater part of the filtrate in the technological cycle), because such regions usually lack both energy resources and possibilities to create purifying constructions.

Investigations conducted by us (9–11) showed that filtration of yeast suspension is more effective in the presence of high dispersed particles of vegetable origin. Optimal regimes of solid-phase preparation and submerged heterophase cultivation conduction should be defined more precisely for each specific substrate and producer for more effective filtration of suspension obtained.

More attention has been paid in the past to the use of secondary water in the technological process of fodder protein production. Recycling of waste no-cell-containing culture liquid (filtrate, centrifugate, etc.) at the fermentation stage of microbe protein production promotes reduction of fresh water consumption, reduces expenditure of feeding components, and provides utilization of some metabolic products. The possibility of complete culture liquid recycling is usually impeded because of soluble metabolites which accumulate in it, inhibit culture growth, and lower the efficiency of the fermentation stage (12–14). That is why the quantity of recycled water on the plants is limited to 50–70% of the initial fermentation medium volume.

Some researchers (12) suggest using membrane ultrafiltration plants to remove inhibiting factors. Recycling of purified culture liquid promotes intensification of the process and increases the speed of yeast culture growth. Productivity of the fermentation process increases 7–9% (relative); biomass yield is increased 15–20% compared to analogous recycling without purification.

To realize a closed water cycle we suggest returning the filtrate to the feed medium preparation stage or to the fermentation stage with preliminary removal of inhibiting admixtures. In this case the inhibitors are glycoproteins with molecule masses of dozens of thousands of daltons, and their concentration could be easily reduced by not less then 10 times by one stage of ultrafiltration, for example, by use of membranes of type YAM-450 (produced in Russia). Therefore, it is possible to return to the cycle up to 90–95% of the technological water, and to send only 5–10% of it to biopurification, by using ultrafiltration of the culture liquid after the filtration stage. In some cases it is possible to use the rest of the water for watering of plantations on nearby farms.

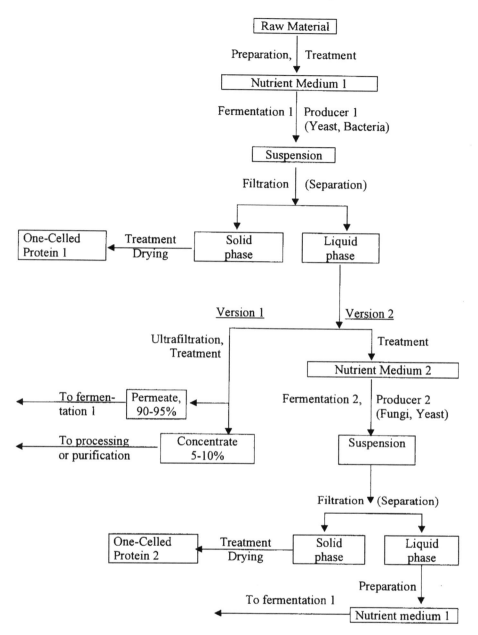

Figure 1 Scheme of microbe protein production on renewable vegetable raw material by energy-saving technology with culture liquid (filtrate) recycle.

Another possible way to use liquid technological waste is successive fermentation of various microorganisms, where the metabolic products of one of them do not inhibit the other (Fig. 1). In this case cultivation of the second producer on substrate based on the liquid waste of the first fermentation leads to utilization of metabolites (inhibitors) in it. This yields fodder microbe biomass of the second producer with no-cell-containing culture liquid with metabolites that do not inhibit the first producer. Corresponding thermal treatment of the waste formed ensures aseptic conditions for the process. Correct selection of producers at each stage (for example, the yeast genus *Candida* or the fungus genus *Fusarium*) permits realization of a "metabolically closed" technological process (15).

In a series of concrete cases when carbohydrate-containing raw material (waste of sugar and starch-syrup production, brewery mash) is used in the main technological process, extracting of the final product is conducted by filtration, and the part of filtrate in the complete water current is not more then 80%, variant 2 of the scheme (Fig. 1) could be essentially simplified. In these cases, filtrate recycle (90–95%) could be realized by use of only one producer with obligatory thermal treatment of the culture liquid (CL). Using different variants of the suggested scheme permits carrying out the return of not less than 90% of industrial liquid waste of fodder protein production to the stage of medium preparation.

Using a few kinds of vegetable raw material and their wastes as substrate, it is possible to produce fodder that is enriched by microbe protein, as we showed as an example.

Molasses and sugar squeezes (pulp) are the waste of food sugar production. As known, it is very hard to reuse the pulp. It has low fodder value, but hydrolytic treatment permits it to yield valuable carbohydrate media, and biomass yield could be increased by molasses or diffusion sap additions to the fermentation medium as a source of sugar and other nutritious substances (10,11). We investigated the process of protein production with use of both sugar and fodder beet root crops, and beet pulp as a raw material.

In extracting food starch from potato tubers, technological waste could be 10–24% of raw material mass. Starch production and simultaneous processing of potato pulp in fodder product enriched by microbe protein appears to be a real and practically significant complex solution of the problem (16).

Corn pulp is one of the most perspective wastes of cornstarch production, containing a considerable quantity of carbohydrate components. Special microbiological treatment of this waste could also increase its protein content.

Jerusalem artichoke tubers are a valuable inulin source. Waste of inulin extraction production and Jerusalem artichoke green mass could be comparatively easily enriched by microbe protein and used in stock breeding. It is known that Jerusalem artichoke haulm silage surpasses sunflower silage in its feed value. A protein–vitamin–mineral complex based on Jerusalem artichoke foliage and stems hydrolyzate was obtained (17,18). Our experiments show the possibility of complex processing all plant mass (tubers and green mass) to obtain VCPF of good quality.

Systematic investigations on VCPF production based on sugar cane bagasse carried out in our department seem to be very useful for the countries of southeast Asia, because great amounts of this waste accumulate in this region (19).

About 60 million tons of brewery mash remains as a waste of alcohol-producing plants of CIS countries at the present time. Brewery mash is of considerable fodder value, but it is not fully used. At some plants it is often poured out into the rivers or ejected to the fields, polluting the atmosphere, water, and topsoil. Direct biochemical purification of brewery mash is very complicated because it has a high content of organic substances. That is why much attention is being paid to the problem of its utilization. One of the methods to process it is to use it as a medium for growing fodder yeast (20,21).

We employed a unique algorithm of work to investigate the possibility of using these kinds of waste to produce fodder protein product by energy-saving technology with use of a filtration process at the stage of final product extracting and water recycle.

In the first stage of the work we investigated the influence of different regimes of fermentation medium preparation on the yield of reducing substances (RS) and filtration properties of the substrates obtained. An optimal regime of raw material preparation was selected considering receipt of well-filtering suspensions with the most complete RS yield.

Raw material preparation to conduct submerged heterophase yeast cultivation usually included such stages as washing, grinding, suspending of ground raw material in water, and also thermal suspension treatment.

At the next stage we selected a producer for heterophase cultivation among yeast strains widely used for industrial fodder protein production on hydrolyzates of vegetable raw material. As a criterion for comparison we used specific speed of growth (m, h^{-1}), completeness of RS consumption (%), period of log phase (h), and accumulation of protein in biomass (% of absolutely dry substance, ADS). We simultaneously determined filtration speed (G, L/m^2 h) and filter retentivity for microorganisms cells (penetration probability of filter, X, %) in experiments.

The process of yeast cultivation in the laboratory was conducted in rocker flasks and in an Ankum-2 laboratory fermenter supplied by a bubbler with air pumping up to 35 L/h, a turbine-type stirrer providing mixing intensity of 800–1200 turns/h, and a thermostat, in a working volume of 1.5 L. The pH was held at the desired standard by 5% water ammonia or sulfuric acid solution additives to the fermentation medium. In industrial conditions cultivation was conducted in apparatus of 5 and 320 m^3 volume under filling ratio 0.5–0.6.

Conduction of submerged heterophase cultivation also involved evaluation of the influence of solid-phase content in the medium on microorganisms' cultivation parameters, protein accumulation in biomass, and filtration efficiency of the yeast suspensions obtained.

Yeast cultivation was conducted in both periodic and removal–addition regimes.

Special attention was paid to filtration characteristics of obtained yeast suspensions (both filtration speed and filter retentivity) while carrying out the experiments. We studied the influence of the microorganisms' stage of growth, pH, and the temperature of the culture liquid on its filtration properties. We also conducted selection of filter material on the basis of the most widespread and evaluated them on such technological parameters as specific filtration speed, filter retentivity, and degree of CL purification ($N = 1/X$).

To elaborate the closed water cycle, we investigated the influence of filtrate formed after biomass extraction and returned to the feed medium preparation stage on the technological value of filtrate and parameters of the following fermentations.

When fodder and sugar beet were used as a substrate, their root crops contained 19% and 25% ADS, respectively, including 10% and 18% carbohydrates by mass. After washing and grinding they served as a solid phase in water suspension (ADS concentration, 1.3–5.2%). The suspension was standardized to pH 0.75–7.10 by dilute sulfuric acid or alkali and thermally treated in an autoclave under different degrees of pressure and temperature over 10–40 min. There was selected the regime of preliminary preparation of fodder and sugar beet root crops suspensions at temperature 121–128°C in the autoclave over 30 min, pH 4.0–4.5, hydromodulus 6–10. This produced suspensions containing 10–13 g/L RS, and their filtration speed was 390–450 L/m^2 h).

The process of dry beet pulp soft hydrolysis conducted at temperature 121–28°C provided high RS yield and high filtration speed of suspension, and also did not cause furfural accumulation. With increase of acid concentration from 0.4 to 5.0 g/L, RS yield increases, and pentose content increases, too, under well filtration speed. However, acid concentration of more than 3.0 g/L intensifies the process of pentose dehydration and causes accumulation of furfural, the main inhibiting admixture. Thus, hydromodulus 18–20, temperature 121–128°C, time of hydrolysis 30–60 min, and sulfuric acid concentration not more than 3.0 g/L are necessary conditions to achieve hydrolyzates on the base of dry beet pulp with RS yield not less than 40–45% and filterability 300–400 L/m^2 h) (22).

We also used potato pulp from a starch-syrup plant (Belgorod region) in the work. It was found that hydrolysis of potato pulp at pH = 1.0, $t = 128°C$ for 30 min provides high RS yield (up to 96% from maximum), doesn't cause formation of the products inhibited microorganism growth and provided cultivation parameters with sufficiently good filtration properties of the fermentation medium (filtration productivity 200–250 L/m^2 h) (23).

Samples of pressed corn pulp containing about 30% ADS and gluten water with pH = 4.2 were used to investigate the process of microbe protein enrichment of corn pulp by the method of submerged heterophase yeast cultivation. It was shown (10,11) that conduction of the thermohydrolysis process at a pressure of 0.25 MPa and pH = 3.5–4.5 for 30 min could yield well-filtered suspensions of corn pulp containing from 9.5 to 11.0 g/L of RS. This forms up to 11% ADS of corn pulp. At the same time, the pH of the medium permits using gluten water for thermohydrolysis, and this could increase protein content and utilize this industrial waste.

It was found during experiments that optimal conditions for thermal treatment of media on a base of Jerusalem artichoke tubers are pH = 2.0–3.5, $t = 121°C$, time = 30 min, considering PS yield (65–75%) and filtration speed of suspensions obtained (Table 2).

Taking into consideration PS yield and filtration speed of obtained microorganism suspensions, optimal conditions of bagasse preparation for fermentation were chosen as follow: pH = 2.1–3.0, $t = 121–128°C$, time 30–50 min, hydromodulus 7–8. This provided an RS yield of 60–70% (24).

Strain producers screening conducted on all mentioned substrates led us to recommend yeast strains of genus *Candida* and *Endomycopsis*, which provide high protein content and good filterability of culture liquid for further investigations.

Table 2 Influence of Thermal Treatment (30 min) and pH Value of Jerusalem Artichoke Tubers Suspensions (hydromodulus 6.25) on RS Yield and Productivity of the Process of Suspension Filtration Obtained Before and After Hydrolysis

pH value of suspension before hydrolysis	RS yield (g/L)			Filtration productivity (L/m^2 h)[a]		
	Initial	After thermal treatment		Initial	After thermal treatment	
	20°C	121°C	128°C	20°C	121°C	128°C
1.0	4.9	27.8	28.5	490	80	60
2.0	3.9	21.7	22.9	430	100	80
2.5	1.2	20.5	21.7	370	150	130
3.0	0.9	17.0	18.6	350	190	170
3.5	0.9	6.0	7.5	390	210	200
4.5	0.8	2.0	2.5	420	240	210
6.5	0.8	1.6	1.6	430	290	230

[a] Filtration was conducted through filter paper.

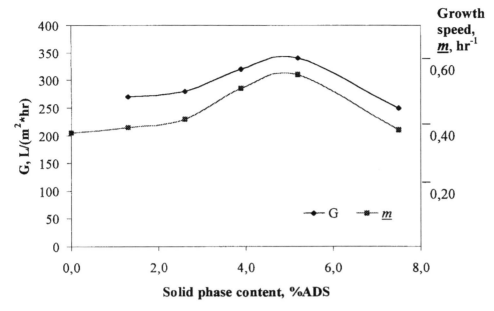

Figure 2 Influence of solid-phase content in the medium on growth speed and filterability of yeast *Candida tropicalis* CK-4 suspensions obtained by cultivation on beet tuber hydrolyzates.

So, strain *Endomycopsis fibuligera* C-2 was selected from the series of strains during cultivation on hydrolyzates of potato pulp, because there were registered the best parameters of filtration of culture liquid under sufficiently high percentage of RS consumption, high percentage of protein content, and specific speed of growth. As for cultivation on media based on corn pulp, the fastest-growing and most highly effective were strain producers *C. tropicalis* CK-4 and *C. scottii* KC-2, whereas for feed media based on Jerusalem artichoke tubers, the best characteristics were detected in the case of strains *C. tropicalis* CK-4 and *C. scottii* KCB. Strains *C. tropicalis* CK-4 and *C. scottii* KCB appeared to be the most productive on media based on bagasse of sugar cane.

By conducting submerged heterophase cultivation of the yeast *C. tropicalis* CK-4 on feed media based on fodder of sugar beet and dry beet pulp, it was detected that 3.9–5.2% of the solid-phase content is of the most practical interest. In this case high productivity is provided at the fermentation stage and also by good filterability of the culture liquid: 320–340 L/m² h. The results are shown in Fig. 2.

We also investigated the influence of the solid phase of potato content on the parameters of yeast cultivation and filterability of culture liquid. A regime of submerged heterophase cultivation with solid phase content 4.5–7.5% is of the most practical interest from the point of view of fermentation productivity and filtration properties of suspensions (23).

The highest values of specific cell growth speed and yeast suspension filtration, obtained on Jerusalem artichoke, were detected at solid-phase content 4.4% ADS. Such solid-phase content is provided by hydromodulus 5.25. That is why, considering the earlier data on substrate preparation, hydrolysis of ground tubers of Jerusalem artichoke in further experiments was conducted at pH 3.0–3.2, temperature 121–128°C, and hydromodulus 5.25 for 30 min (24).

Table 3 Yeast *Candida tropicalis* CK-4 Cultivation Conditions on Beet Tuber Hydrolyzates: Influence on Filterability of Yeast Suspensions Through Filter Paper

Time of growth (h)	RS content (g/L)	Titer of cells (mln/mL)	Filtration productivity (L/m² h)	Penetration probability of filter (%)
0	22.0	2.0	—	—
4	12.0	48.0	120	10.8
6	6.4	108.0	200	5.6
8	1.2	198.0	300	2.1
10	0.2	264.0	340	1.1

The maximum value of specific yeast growth speed and yeast suspension filterability obtained on hydrolyzates of sugar cane bagasse of were obtained at solid-phase concentration 8% ADS (24).

Thus, evaluation of solid-phase content in fermentation media influence on the basic characteristics of the fermentation process and medium filterability for different substrates showed that the optimal content of the solid phase is 4–8% and depends on the type of raw material used.

While conducting the experiments much attention was paid to investigation of filterability of microorganism suspensions. It was mentioned that, as a rule, the best filtration parameters (both for filtration speed and filter retentivity) during fodder yeast growing on hydrolyzates of vegetable raw material are shown by suspensions taken at the end of the logarithmic and at the beginning of the stationary growth stages, at the beginning of the growth limit by carbohydrate substrate (Table 3).

Investigation of pH and temperature of culture liquid influence on filtration properties showed that optimal values of pH for filtration are 4.0–5.5, practically conforming to cultivation pH (Fig. 3). With increase of CL temperature the viscosity decreases, which leads to considerable filtration speed increase. It should be mentioned that increase of filtered yeast suspension temperature does not lead to any significant decrease of culture liquid purification.

It was detected in the series of experiments that use of ground dry beet pulp as a filter bed at yeast suspension filtration increases productivity of the process up to 70%.

Among investigated filters for filtration of *C. scottii* KCB, yeast suspension obtained on hydrolyzates of Jerusalem artichoke nonwoven textile material "BTILP" (Russia) demonstrated high technological characteristics: high filtration speed and low filter penetration probability for yeast cells. Thus, in this case effective filtration of cultural suspensions could be conducted without any additional preparations.

If filter material is selected correctly, productivity of filtration during yeast culturing on hydrolyzates of corn pulp could rise up to 550 L/m² h.

Well-filterable yeast suspensions were obtained (the productivity of the process being up to 800 L/m² h), which formed a thick deposit with ADS content not less then 14% after filtration through paper filter in almost all fermentations on fodder beet hydrolyzates.

It was mentioned that yeast suspensions of *C. tropicalis* CK-4 obtained on a base of bagasse both before and after the fermentation were well filterable and showed a high degree of purification (24). This fact could be explained by the hard structure of bagasse, which forms a good filter bed.

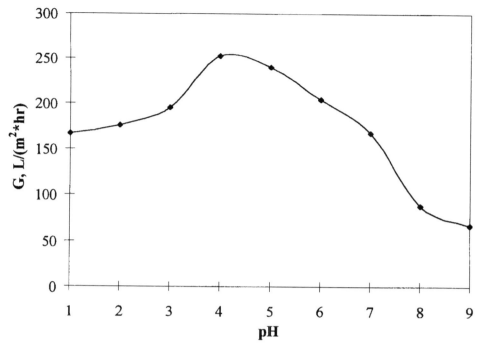

Figure 3 Dependence of filtration productivity of yeast suspension obtained from yeast *Endomycopsis fibuligera* C-2 on potato tuber hydrolyzate cultivation from pH value of the medium.

If a filtration process for preliminary thickening of fermentation medium obtained during yeast culturing on brewery mash is used, simple filtration is extremely ineffective (25). That is why it was suggested to investigate the influence of beet pulp additives on the filterability of culture liquid. This could increase productivity of the process and permit obtaining the final product according to a more economical scheme. Simultaneously, it could partially decrease the load on purifying constructions of alcohol-producing plants, because in this case extracted solid phase of brewery mash, enriched by yeast protein, is utilized as a valuable fodder, and at the same time, the content of organic and inorganic substances in filtrate is decreased because of their yeast consumption.

As experiments showed, beet pulp additives in quantities up to 3% in the fermentation medium practically do not inhibit yeast growth. Specific growth speed was 0.20–0.23 h^{-1}; RS consumption was up to 67%. At the same time, pulp additives led to increase of filtration productivity, from 95 L/m^2 h in the case of filtration without pulp to 240 L/m^2 h with 3% pulp addition. The filtration process was also characterized by a low percentage of cell penetrability—not more than 2%.

Thus introduction of additional solid phase in the fermentation medium increases productivity of CL filtration. It permits using this process for preliminary thickening of the product. Besides, it was shown that the filtrate obtained has a chemical oxygen demand (COD) 64–65% less a COD of initial brewery mash.

One of the main ways to raise the efficiency of fermentation stage, and the VCPF process in total, is to transfer from periodic to removal–addition and continuous regimes of microbe cell cultivation. Use of removal–addition regimes of cultivation could permit

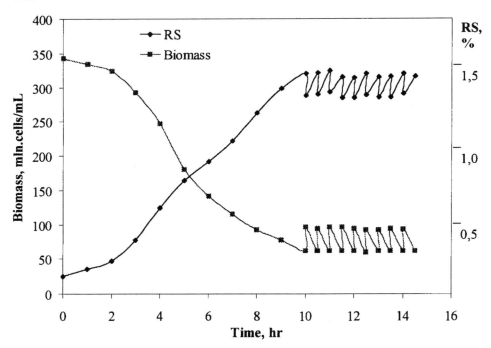

Figure 4 Yeast *Endomycopsis fibuligera* C-2 on potato tubers hydrolyzates cultivation in removal–addition regime.

(on some parallel-action apparatus) practically uninterrupted steam of yeast suspension on the stage of extraction (filtration, plasmolysis, drying).

In our experiments we also showed the possibility of submerged heterophase cultivation of yeast on hydrolyzates of vegetable raw material in removal–addition regime without decrease of process efficiency and final product quality.

For example, during cultivating on medium based on potato pulp in removal–addition regime (Fig. 4), raw protein content in the final product remained at a level of 42–44%. It should be mentioned that productivity of the filtration process in this case remained practically unchangeable and was 200–220 L/m^2 h.

We also conducted VCPF production based on beet root crops under industrial conditions (Table 4) The period of the process in removal–addition regime was 240 h. The content of raw protein achieved was 44%, and of real protein was 29–33% under optimal conditions. There was processed 300 tons of root crops and 42 tons of final product during all the tests.

At the next stage of the work we conducted experiments to realize closed-cycle water consumption by return of the greater part of the filtrate formed after biomass extraction to the main stages of the process.

While using fodder beet hydrolyzates as a substrate it was shown that at the fourth stage of filtrate recycle, without addition of technological water, the quantity of ADS in the filtrate increases by 8%, of sulfates by 9%, of soluble protein by 15%, and the quantity of volatile organic acids increases nearly twice. The results are presented in Table 5. We can see from the table that 3–4 rate recycle of filtrate under insignificant lowering of raw protein content (from 41% to 39% ADS) in the final product is the technologically optimal variant.

Table 4 Cultivation of Yeast *Candida tropicalis* CK-4 on Beet Tuber Hydrolyzates on Industrial Plant ($V = 320$ m^3) in Removal–Addition Regime

Time (h)	RS content (g/L)	Titer of cells (mln/mL)	Medium volume (m^3)	Raw protein (% ADS)
0	30.0	20	60	—
7[a]	0.7	190	65	—
14[b]	1.3	310	120	38.6
40	1.6	330	115	39.5
60	1.8	470	120	42.9
120	1.2	490	150	44.3
170	0.2	470	180	43.8
220	0.8	450	140	43.2
240	0.3	450	120	42.2

[a] Continuation of plant filling.
[b] Start of removal–addition regime.

Table 5 Results of Yeast *Candida tropicalis* CK-4 on Fodder Beet Hydrolyzate Cultivation with Filtrate Recycle

Characteristics	Stage of the process			
	1	2	3	4
1. RS concentration, initial/final (g/L)	23.0/1.4	24.5/2.9	25.6/3.6	29.0/4.8
2. RS consumption (%)	94.0	88.1	86.0	83.4
3. Titer of cells, initial/final (mln/mL)	8/281	7/270	8/254	8/247
4. Time of biomass doubling (h)	1.23	1.38	1.44	1.50
5. Growth speed (h^{-1})	0.56	0.50	0.48	0.46

While culturing yeast *C. tropicalis* CK-4 on a medium containing filtrate and technological water additions up to the initial volume in 6 cycles, RS consumption speed was on average of 5.5–6.0 g/L h (initial RS concentration was 20.0 g/L), raw protein concentration, 40% ADS and productivity of fermentation suspension filtration through paper filter, 320 L/m^2 h.

Thus, the principal possibility of creating a process of carbohydrate-protein fodder production on fodder beet hydrolyzates with a closed water cycle under multiple return of filtrate to the feed medium preparation stage was demonstrated as a result of these experiments.

We also investigated filtrate recycle influence on yeast *Endomycopsis fibuligera* C-2 fermentation parameters on potato pulp hydrolyzates. Seven cycles were carried out with complete return of CL filtrate, which is 80% of all water demand to feed medium preparation for the following stages. Results are presented in Table 6.

A certain increase of initial RS content appeared. This could be explained by the presence of some quantities of waste carbohydrates in the filtrate. Growth speed slightly decreased and was 0.65–0.68 h^{-1} after the third cycle. Degree of RS consumption (83–85%) and raw protein content (44.8–52.9%) remained sufficiently high during the whole experiment. The value of the economic coefficient Y (cells RS) decreased slightly, but after cycle 5 it stabilized at 25–26 × 10^9 cells/g RS. We also determined the concen-

Table 6 Results of Yeast *Endomycopsis fibuligera* C-2 on Potato Pulp Hydrolyzate Cultivation with Full Filtrate Recycle

Characteristics	Stage of the process						
	1	2	3	4	5	6	7
1. RS concentration, initial/final (g/L)	15.0/1.8	15.2/2.3	16.0/2.5	15.8/2.4	16.2/2.7	16.5/2.4	16.6/2.7
2. RS consumption (%)	88	85	84	85	83	84	84
3. Titer of cells, initial/final (mln/mL)	15/483	15/427	10/380	15/390	16/370	13/365	14/383
4. Growth speed (h^{-1})	0.82	0.72	0.68	0.68	0.68	0.65	0.66
5. Raw protein (% ADS)	52.9	50.2	49.9	46.0	44.8	45.1	45.4
6. Economic coefficient (10^9 cells/g RS)	35.5	31.9	27.4	28.0	26.2	25.0	26.5

tration of NH_4^+ and SO_4^{2-} ions in the medium, because we used ammonia water and sulfuric acid solution for pH correction. It was detected that concentration of these ions increased up to cycle 5, and practically did not change after that.

Thus the possibility of using *Endomycopsis fibuligera* C-2 yeast for carbohydrate-protein fodder production on potato pulp hydrolyzates by energy-saving technology is shown. This strain demonstrated the best parameters of filtration, and permits using a filtration process at the stage of final product extraction. This makes possible the following filtrate return to the stage of feed medium preparation, and, by that means, essentially decreases water consumption while utilizing food industry waste.

A closed cycle of water consumption while working with Jerusalem artichoke hydrolyzates permits using all that filtrate from the previous stage to prepare substrate for the next stage. Suspension volume was corrected by technological water. We examined six cycles of cultivation, and no significant decrease of process efficiency was observed. The results of these experiments are presented in Table 7.

There was a certain increase of initial RS consumption under recycle. Time of biomass doubling gradually increased from 1.5 to 3.0 h, and specific growth speed decreased from 0.56 to 0.45 h^{-1}. At the same time, RS consumption and raw protein content remained high during the whole experiment.

It is important to mention that throughout the experiment, the productivity of yeast suspension filtration decreased slightly at the beginning, and after that remained practically constant at all stages of the process.

Taking into consideration that the maximum volume of filtrate which could be used repeatedly did not exceed 70% of the water needed to prepare suspensions, we conclude that there is a possibility of using complete filtrate recycle to the stage of fermentation medium preparation without decrease of technical and economic characteristics of the process.

The results of closed-cycle water use on submerged heterophase fermentation of yeast genus *Candida* on bagasse of sugar cane are presented in Table 8 and in Figs. 5 and 6.

A small increase of the initial RS content and decrease of consumption level occurred during recycle. This could be caused by partial return of hard utilized carbohydrates

Table 7 Results of Yeast *Candida scottii* KCB on Jerusalem Artichoke Hydrolyzate Cultivation with Full Filtrate Recycle

Characteristics	Stage of the process					
	1	2	3	4	5	6
1. RS concentration, initial/final (g/L)	18.0/2.8	21.0/3.1	23.4/3.2	23.4/3.3	24.7/3.4	25.4/3.6
2. RS consumption (%)	84.4	85.2	86.3	86.0	86.2	85.5
3. Titer of cells, initial/final (mln/mL)	12/330	12/490	13/560	13/490	9/490	9/390
4. Time of biomass doubling (h)	1.5	1.5–2.0	2.0–2.5	2.0–2.5	2.0–2.5	2.5–3.0
5. Growth speed (h^{-1})	0.56	0.50	0.50	0.50	0.48	0.45
6. Raw protein (% ADS)	32	33	33	35	35	35
7. Economic coefficient (10^9 cells/g RS)	20.7	28.2	27.1	23.8	22.70	20.0
8. Filtration productivity (L/m^2 h)	270	270	250	260	250	240

Table 8 Results of Yeast *Candida scottii* KCB on Sugar Cane Bagasse Cultivation with Full Filtrate Recycle

Characteristics	Stage of the process					
	1	2	3	4	5	6
1. RS concentration, initial/final (g/L)	21.0/2.8	21.7/3.0	22.5/3.6	23.8/4.2	24.0/4.4	25.0/4.8
2. RS consumption (%)	86.7	86.2	84.0	82.0	82.0	81.0
3. Titer of cells, initial/final (mln/mL)	5/250	5/240	5/200	6/180	5/170	5/170
4. Time of biomass doubling (h)	1.5–2.0	2.0–2.5	3.0–3.5	3.0–3.5	3.0–3.5	3.0–3.5
5. Growth speed (h^{-1})	0.44	0.43	0.38	0.38	0.37	0.37
6. Raw protein (% ADS)	18	9	18	16	16	16
7. Economic coefficient (10^9 cells/g RS)	13.5	12.6	10.3	9.0	8.4	8.4
8. Filtration productivity (L/m^2 h)	300	300	280	280	280	270

during filtrate recycle. Phosphorus consumption was 24–36%, and it fell from stage to stage in the same way as the quantity of ammonia, which supported the pH of the medium. Time of biomass doubling increased from 2.5 to 3.5 h. Also, specific growth speed and economic coefficient value decreased slightly. All these changes happened only during the first three cycles; after cycle 4, all basic fermentation parameters became stabilized. This confirms the possibility of using full filtrate on this substrate recycle.

Thus, on the basis of experiments conducted, producer strains of protein were selected, and then were determined the main technological parameters of the process

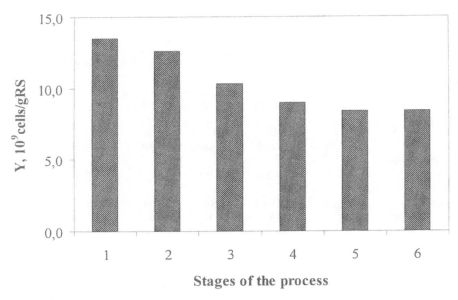

Figure 5 Variation of economic coefficient in the process of full filtrate recycle during cultivation of yeast *Candida tropicalis* CK-4 on sugar cane bagasse hydrolyzate.

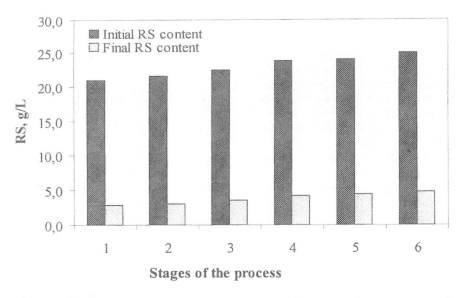

Figure 6 Variation of RS content in the process of full filtrate recycle during cultivation of yeast *Candida tropicalis* CK-4 on sugar cane bagasse hydrolyzate.

stages for vegetable carbohydrate-protein fodder production on the basis of the above-mentioned substrates. The results of investigation permit calculation of a technological scheme for VCPF production, and also the ability to select the main plant and equipment in order to realize the process on a model plant by low-waste technology.

III. CONCLUSIONS

Thus it is obvious that biotechnological utilization of carbohydrates allows not only biodestruction of vegetable raw material processing waste which pollutes the environment, but also the ability to obtain valuable protein fodder product, which could be used to help balance the fodder of agricultural animals.

It should be mentioned that protein fodder product obtained on media from vegetable raw material and the waste of its processing could replace, agricultural animal fodder using plants of high protein content (for example, soy) in regions where cultivation of such cultures is difficult or impossible.

Application of technology with filtration instead of flotation, separation, and evaporation with the later return of culture liquid filtrate to the starting production stages of the final product for fodder purposes is, undoubtedly, a perspective procedure.

On the basis of the experiments conducted, we suggest the energy-saving low-waste technology of VCPF production using submerged heterophase cultivation followed by filtration of microbe suspensions and filtrate recycling, which avoids using power-consuming methods of concentrating fermentation products from fodder production and essentially reduces the quantity of liquid waste. This technology could be realized on various capacity model plants, both as part of large-scale industrial production and on a small scale directly in regions with vegetable raw material or its processing waste.

REFERENCES

1. Garkavenko, A. I. and Kovaltchuk, A. P. 1973. Biologically active actinomycetes substances used in stock-breeding, *Shtiintsa*, Kishinev, pp. 92.
2. Spicer, A. 1971 (1972). *Trop. Sci.*, 13:4.
3. Dvoichenkova, E. U., and Kantere, V. M. 1987. Biotechnological utilization of keeping and processing fruit and vegetable products waste. *Minmedmicrobioprom SSSR*, Moscow, pp. 39.
4. Bykov, V. A. 1987. Problems and perspectives of industrial biotechnology. *Biotekhnologiya*, 6:692–700.
5. Ogarkov, B. I., Kiselev, O. I., and Bykov, V. A. 1985. Biotechnological directions of vegetable raw material use. *Biotekhnologiya*, 3:1–15.
6. Reference Book on Planning and Economics of Agricultural Production, Part 1, 2nd ed. 1987. Rosselkhozisdat, Moscow, pp. 512.
7. Ponomarev, A. P. 1988. Intensification of fodder production. *Rosagropromisdat*, Moscow, pp. 110.
8. Kholkin, U. I. 1989. The technology of hydrolytic productions. *Lesnaya prom.*, Moscow, pp. 496.
9. Bykov, V. A., Prichepov, F. A., and Manakov, M. N. 1985. Yeast genus *Candida* cultivation on wood hydrolyzates in the presence of unutilized solid phase. *Prikladnaya biokhimiya i microbiologiya*, 31(2):253.
10. Belenkyi, M. G. 1977. Recommendations on stock-breeding products and fodder biological evaluation using *Tetrahymena piryformis*. Moscow, pp. 20.
11. Bykov, V. A., Vinarov, A. U., and Sherstobitov, V. V. 1985. Microbiological production processes calculation. *Tekhnika*, Kiev, pp. 245.
12. Kurakov, V. V., Svitsov, A. A., and Manakov, M. N. 1987. Paraffin-assimilating yeast genus *Candida* cultivation with ultrafiltration use for cultural liquid recycle. *D. Mendeleyev University Chem. Technol.*, 149:108–112.

13. Zanko, N. G., Tanitcheva, R. V., and Makarov, V. L. 1989. Nutrient salts content influence on fodder yeast yield at no-flow water consumption scheme. *Gidroliznaya i lesotekchnicheskaya promishlennost*, 1:18.
14. Patent USA, 1977. #4048013.
15. Inventor's certificate USSR, 22 January 1990, #1566720.
16. Podgorskaya, V. S., and Ivanova, V. N. 1990. Plant growing waste biotechnological utilization. *Naykova Dymka*, Kiev, pp. 43.
17. Golubev, V. N., Volkova, I. V., and Kushalakov, H. M. 1995. Jerusalem artichoke: Content, properties, ways of processing, spheres of application. Moscow, pp. 82.
18. Solovieva, T. U. 1994. Reserve untraditional raw material resources investigation and baking yeast production technology on inulin-containing raw material elaboration. Author's abstract for Master of Science degree rivalry, Moscow, pp. 18.
19. Devendra, C. 1979. Ruminant nutrition and productivity in the ASEAN region. *Seminar on Animal Health and Nutrition in the Tropics*, ADAB, James Cook University, Queensland, Australia, pp. 169–189.
20. Yarovenko, V. L., and Belov, N. I. 1993. No-waste alcohol technology scheme. *Pishevaya promishlennost*, 2:35–39.
21. Gracheva, I. M., Ivanova, L. A., and Kantere, V. M. 1992. Microbe protein compounds, amino acids technology and bioenergy. Moscow, pp. 224.
22. Shakir, I. V., Panfilov, V. I., Markima, H. S., and Manakov, M. N. 1992. Use of renewable vegetable raw material for one-celled protein production. *Biotekhnologiya*, 2:47–53.
23. Kulinenkov, D. O., Mantsurova, I. V. Shakir, I. V. Panfilov, V. I., and Manakov, M. N. 1997. The obtaining of a carbohydrate-protein fodder using potato hydrolysates. *Biotekhnologiya*, 5:22–27.
24. Pham Ahn Kuong. 1998. One-celled protein production from vegetable raw material by low-waste and energy-save technology. Author's abstract for Master of Science degree rivalry Moscow, pp. 18.
25. Kulinenkov, D. O., Shakir, I. V., Panfilov, V. I., and Manakov, M. N. 1997. The primary purification of brewery mash with the use of heterophase submerged cultivation. *Biotekhnologiya*, 6:43–46.

40
Natural Attenuation of Explosives

Maurice V. Cattaneo*
Biotechnology Research Institute, National Research Council of Canada, Ottawa, Ontario, Canada

Judith C. Pennington, James M. Brannon, Douglas Gunnison, Danny W. Harrelson, and Mansour Zakikhani
U.S. Army Engineer Waterways Experiment Station, Vicksburg, Mississippi

I. INTRODUCTION

Much of the explosives contamination in the environment in the United States has resulted from manufacturing and load-assemble-package (LAP) processes conducted during and before World War II and the Korean conflict. Principal explosives waste products were 2,4,6-trinitrotoluene (TNT), 1,3,5-trinitro-1,3,5-hexahydrotriazine (RDX), octahydro-1,3,5,7-tetranitro-,1,3,5,7-tetrazocine (HMX), and N,2,4,6,-tetranitro-N-methylaniline (tetryl). Waste disposal practices were governed more by convenience and explosives safety concerns than by environmental awareness. Waste waters from manufacture and from LAP operations were often discharged into sumps, runoff and/or percolation ditches, and lagoons. These practices have resulted in contamination of groundwater with explosives and their degradation products.

To address long-term protection needs and cost-effectiveness, clean-up alternatives are actively being pursued by U.S. Department of Defense (DoD) components and others involved in the field of environmental protection. That the environment has some capacity to alter and assimilate natural and anthropogenic contaminants without unacceptable impacts has long been recognized. Monitored natural attenuation (MNA) is the formal incorporation of this capacity into protocols for long-term protection of the public health and environment while reducing the cost of site remediation. In fact, the U.S. National Contingency Plan (NCP) now requires that MNA attenuation be considered as a potentially acceptable remediation alternative (1).

In response to the recognition that natural attenuation may provide a realistic, protective, and cost-effective solution for the clean-up of some contaminated sites, extensive efforts have focused on evaluating the potential for MNA and developing various degrees of technical guidance to incorporate MNA into the evaluation, selection, and implementation of remedial alternatives. Thus far, guidance has been developed or proposed for application to sites contaminated with petroleum hydrocarbons (2), chlorinated

---
**Current affiliation*: Cambridge Scientific, Inc., Cambridge, Massachusetts

organics (3,4), and more recently, explosives (5). The U.S. Environmental Protection Agency (EPA) has also developed a directive, *Use of Monitored Natural Attenuation at Superfund, RCRA Corrective Action, and Underground Storate Tank Sites* (6).

The decision to support natural attenuation as a remedial option is site-specific. The burden of proof is on the proponent and not on the regulator. Monitored natural attenuation can be scientifically supported and does not imply "no action," but places its focus on monitoring as opposed to engineered treatments. In the United States, at least considering natural attenuation at all sites requiring remediation is mandated. Monitored natural attenuation may be an attractive alternative to available remediation technologies at sites that meet well-defined selection criteria and acceptable risk levels, and that satisfy specific regulatory concerns.

The objectives of this chapter are to review attenuation processes and considerations for implementation of MNA at explosives-contaminated sites. Examples are provided from laboratory and field demonstrations to illustrate (a) determination of transport properties, and (b) implementation of site characterization, monitoring, and modeling.

II. ATTENUATION PROCESSES

Explosives are subject to several environmental fate processes that many result in attenuation and reduction in the mass of the contaminants in the vadose zone or in the subsurface. These processes include transformation, immobilization, and biodegradation. Transformation occurs either biotically (7–10) or abiotically, (11–13) by reduction of the nitro moieties of TNT and RDX to amino groups. Some of these transformation products are of environmental concern and must be removed along with the parent compound to effect adequate site clean-up. Typical products of TNT transformation were first identified in a simulated compost system (Fig. 1) (9). The hydroxylamino products are rarely observed because of their instability. The azoxy dimers are rarely observed in natural systems.

Transformations are especially significant for TNT because the amino transformation products are capable of reacting with functional groups on organic matter in the soil to produce complex products that are no longer mobile. Evidence for these immobilization products has been observed in the high organic carbon matrix of compost (9,14,15) and in plant uptake studies (16). Recent studies using nuclear magnetic resonance techniques and TNT transformation products labeled with stable isotopes of nitrogen illustrate these reactions (Fig. 2) (17).

Many investigators have examined microbial degradation of nitroaromatic compounds using selected microbial communities (9,18–25). However, surprisingly little is known about the capability of microbial communities to degrade nitroaromatics in situ. In particular, transformation of nitroaromatics by microorganisms indigenous to aquifer systems has not been demonstrated conclusively (26,27).

A mineralization pathway for TNT has been proposed (28,29); however, TNT is only slowly mineralized in soils and groundwater (5,30). RDX is relatively unaffected by transformation compared to TNT and is more readily mineralized, especially under anaerobic conditions (31). An anaerobic mineralization pathway for RDX has been proposed by Kaplan (7) (Fig. 3).

Monitored natural attenuation relies heavily on modeling to answer two basic questions (a) What is the size of the groundwater contaminant plume? (requires a conceptual model), and (b) When will the contaminant plume reach potential receptors? (requires a numerical model to generate predictions of future plume behavior). An understanding

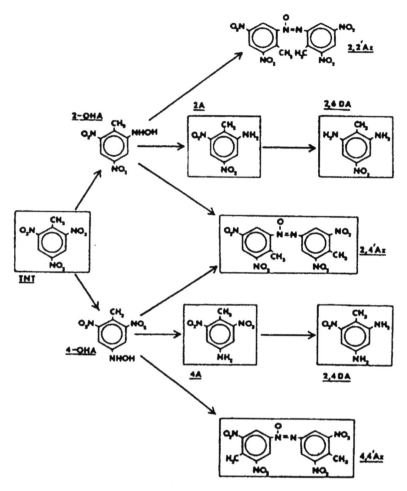

Figure 1 Biotransformation scheme for TNT in compost. Compounds boxed were identified in solvent extracts; other compounds are postulated (9).

of the following environmental processes is required to formulate the models: (a) advection or migration velocity of the groundwater; (b) dispersion rate of the contaminants; (c) retardation or sorption of contaminants to the aquifer matrix which includes both the (i) sorptive processes toward the mineral matrix, e.g., the clay fraction of the soil, and (ii) attachment to the organic fraction, e.g., humic substances present in the aquifer material, and (d) degradation rate, i.e., microbial degradation. The following laboratory case study illustrates procedures and findings concerning environmental transport at a site in Canada.

III. EXPLOSIVES TRANSPORT CASE STUDY

Column experiments were undertaken to better understand and quantify the relative importance of the processes characterizing the transport and fate of TNT and its metab-

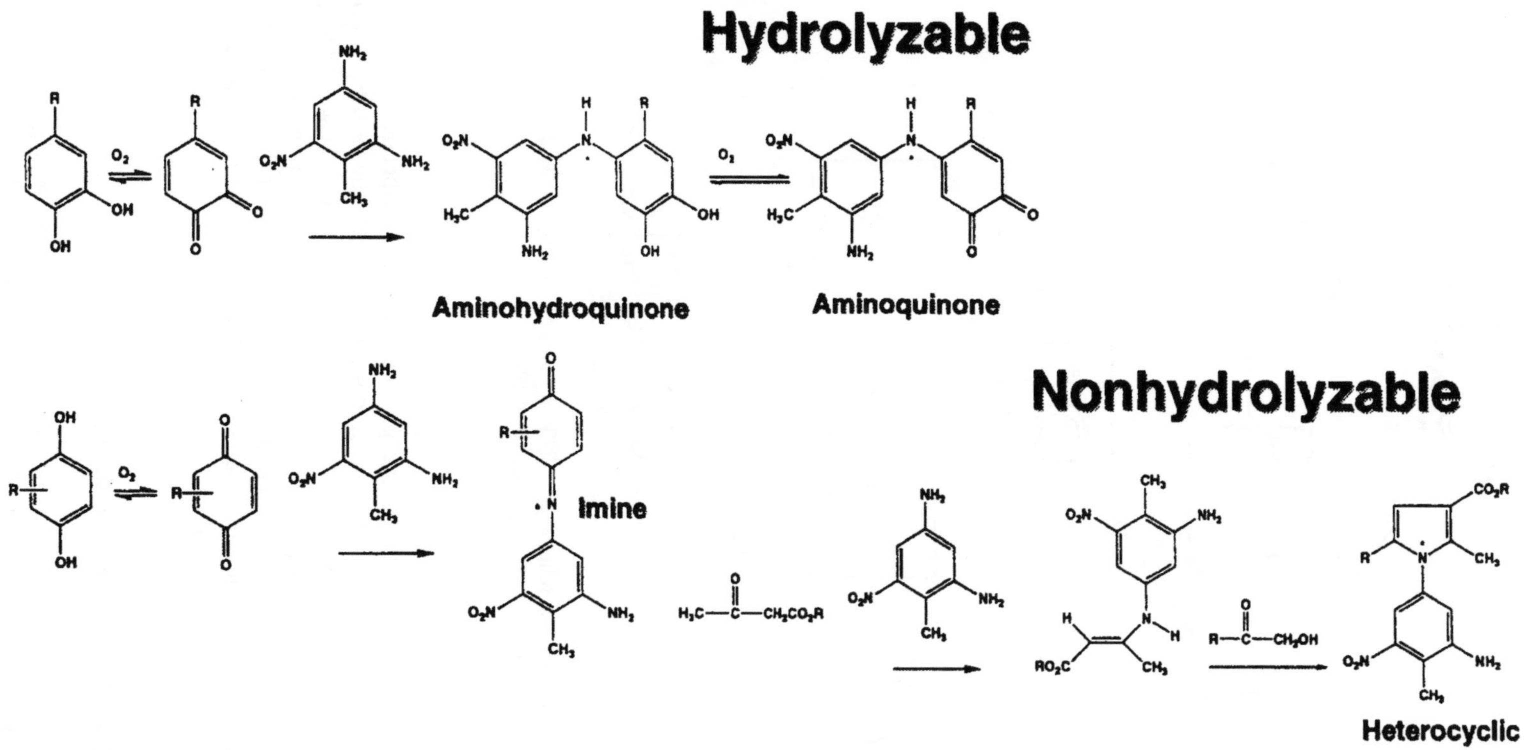

Figure 2 Irreversible (nonhydrolyzable) and moderately reversible (hydrolyzable or partially hydrolyzable) covalent bonds between amino substituents and surrogate humic acid functional groups (Kevin A. Thorn, personal communication, 1998, USGS, Arvada, CO).

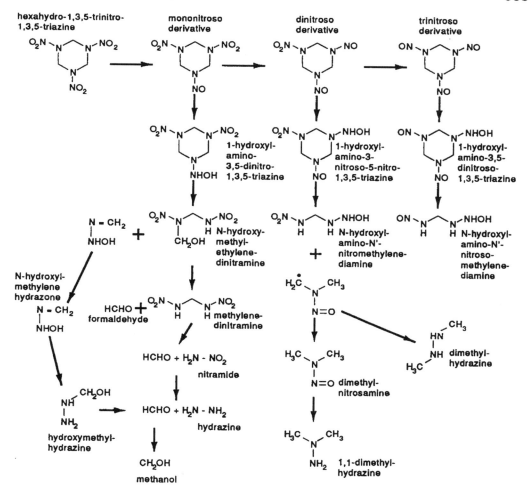

Figure 3 Anaerobic mineralization pathway for RDX (7).

olites leaching from the vadose into an underlying aquifer (32). Each column consisted of a stainless steel tube (46 cm length, 7.2 cm inside diam.) with stainless steel end plates and equipped with sampling ports 1, 2, and 3 at 5, 23, and 40 cm depth, respectively. The deeper port, no. 3, consisted of a hollow stainless steel tube (64 mm length, 6.3 mm O.D.) perforated along its length, and covered with a 106-µm stainless steel screen. A Teflon®-coated stainless steel plug valve was connected to each port through an adapter. Further details of the design of the columns can be found elsewhere (33). Columns were packed with aquifer material collected at the well-documented unconfined aquifer at Canadian Forces Base Borden. To obtain homogenous and reproducible results, successive 120-g portions of sand were transferred and compacted with a standard-weight plunger until the columns were filled (34). The aquifer material consisted of 96% sand with 2% of silt and 2% clay. The solid-phase organic carbon was 0.02%. The distribution or sorption coefficient (K_d) for the soil to nitroaromatics was obtained from batch equi-

Table 1 Freundlich Isotherm Parameters for Various Soils

	TNT		2ADNT		4ADNT		2ADNT		4ADNT[a]	
	K_d	n	K_d	n	K_d	n	K_d	n	K_d	n
Sand (Borden)	0.4	0.84	1	0.97	1	0.73				
Sand (4% clay)	1.3	0.84	18	0.94	16	0.75	42	0.65	59	0.98
Loam	1.6	0.96								
Kaolinite clay	1	0.87								
Montmorillonite	156	0.74								

[a] K_d values were obtained in the presence of both aminometabolites (2 ADNT + 4ADNT).

librium experiments. TNT adsorption to various sandy, loamy, and clay soils was shown to follow a Freundlich-type isotherm:

$$q = K_d C^{1/n}$$

where

q = amount of solute retained by the soil (μg/g)
C = solute concentration in solution (mu;g/mL)
K_d = distribution coefficient (cm^3/g)
n = dimensionless (typically $n < 1$)

Increasing the clay content of an aquifer material generally results in an increase in the distribution coefficients, as shown in Table 1. The two columns were equilibrated for 155 h, with a synthetic groundwater with composition as shown in Table 2.

Pore volumes of 541 mL and 563 mL for columns 1 and 2 were estimated from the breakthrough of the bromide tracer. The first column received 1.8 pore volumes of contaminated groundwater with concentrations of TNT, 2ADNT, and 4 ADNT of 16.4 ppm, 700 ppb, and 879 ppb, respectively, in a 1:1 ratio with methanol (control), while the second column received 1.6 pore volumes of the same contaminated groundwater mixed in a 1:1 ratio with synthetic groundwater (simulation column). To elute the TNT, column 1 and column 2 received 3.3 and 3.6 pore volumes of synthetic groundwater, respectively. The effective porosity ($\eta_e = 0.34$) and the dispersion coefficient ($D_x = 3.5$ cm^2/h) were obtained from the bromide tracer studies.

The retardation factors were derived from the column data. As shown in Fig. 4 (top), TNT and metabolites displayed sharp breakthrough curves, indicating no retardation in the control column. However, the percent recovery of 4ADNT and 2ADNT was 90% and 60%, respectively. In the simulation column, the percent recovery of 4ADNT and 2ADNT was 60% and 40%, respectively (Fig. 4, bottom). TNT and its aminometabolites were retarded with respect to bromide. Retardation factors were then calculated by dividing the mean arrival time of the organic solute by the mean arrival time of the nonreactive tracer (Table 3).

A. Biotransformation Studies

The columns were filled with nitroaromatic contaminated soil containing 20 ppm TNT and minor quantities of other nitroaromatics including 2,4-DNT and 2,6-DNT. Deionized water was recirculated at a flow rate of 1 mL/min. Column A was kept unsaturated, while

Table 2 Groundwater Composition

Component (in ppm unless stated)	Well 104	Well PZ3	Synthetic
Specific conductance (μS/cm)		845	1043
Dissolved oxygen (mg/L)	4.2	<0.5	
pH	6.7	8.8	7
Nitrate	134	8.8	440
Nitrite	107	0	
Sulfate	41.8	131	40
Ferric iron (Fe^{3+})	0.44	0	
Ferrous iron (Fe^{2+})	0.62	0.24	
Iron (Fe)	1.06	0.24	
Bicarbonate (HCO_3)	124	240	124
Calcium (Ca)	144	3.6	144
Mg (Mg)	15	23	10
Potassium (K)	2.4	76	2.6
Sodium (Na)	2.6	62	50
TNT	75.5	0.31	
2,4-DNT	104	0	
2,6-DNT	125	0	
2-NT	3.5	0	
3-NT	nd	0	
4-NT	0	0	
2A-DNT	0	0.14	
4A-DNT	0.09	0	

Table 3 Retardation Factors in Groundwater from a TNT-Contaminated Site

Compound	Retardation factor	Retention time (h)
Bromide	1	32
TNT	1.4	45
4ADNT	1.7	55
2ADNT	1.9	60

column B was saturated. This setup provided for the higher dissolved oxygen (DO) values of 8.5 mg/L in column A compared to column B (4.3 mg/L). Column B was kept saturated to promote anaerobic conditions.

As shown in Fig. 5, the TNT concentration decreased in the soil column without added substrate. The pseudo-first-order transformation rate constant (μ) was calculated as 1.4×10^{-3} h^{-1}, which is of the same order of magnitude as the rates found in the literature (35) but lower than the rates found in a subsequent study (36). It has been suggested that TNT transformation rates are strongly dependent on the redox potential (Eh), with lower Eh values resulting in higher rates (11,36,37).

As shown in Fig. 6, aminometabolites were produced in both columns. 4ADNT and 2ADNT seem to have reached stability after 60 days, with the exception of the 4ADNT in

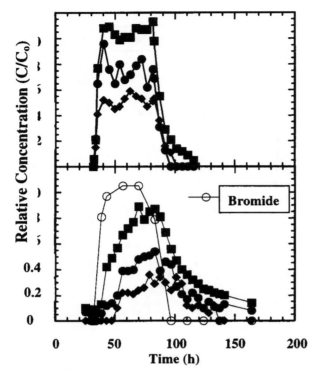

Figure 4 Transport of TNT and aminometabolites in contaminated groundwater from a TNT-contaminated site through aquifer material: (top) solution containing 50% methanol; (bottom) aqueous solution (bromide tracer, open circles; TNT, filled squares; 4ADNT, filled circles; 2ADNT, filled diamonds).

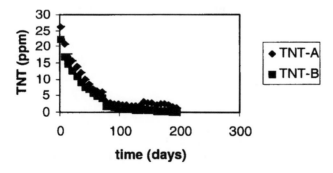

Figure 5 TNT concentration in recirculating "groundwater."

column B which is still rising (Fig. 6, left). The disappearance of the 2ADNT in column A after 30 days may be due to polymerization of this metabolite with the soil matrix under higher DO contents (Fig. 6, right). Kaplan had shown a greater tendency for production of 4ADNT over the 2ADNT in line with the results shown here (38,39).

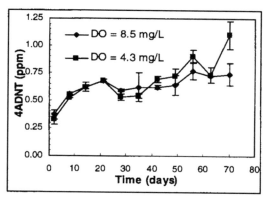
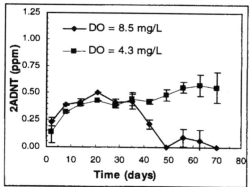

Figure 6 Production of aminometabolites 2ADNT and 4ADNT at different dissolved oxygen (DO) levels.

As shown in Fig. 7, both nitrite (NO_2) and nitrate (NO_3) were formed in the soil columns, likely as a result of biotransformations involving 2,4-DNT and the mononitroaromatics present in the soil. Microcosm data in Table 4 show significant decreases in 2,4-DNT and 2-NT after 21 days of incubation at 4°C as compared to TNT and 2,6-DNT.

Aerobic conditions are known to promote the formation of nitrite from mononitroaromatics and 2,4-DNT (26,40). However, the production of nitrite from TNT has not yet been reported. Preliminary site data have also indicated a significant production of nitrite and nitrate in the aerobic zone of the contaminated aquifer, likely as a result of mononitroaromatic and 2,4-DNT biodegradation.

As shown in Fig. 8, the large pool of nitrate/nitrite found in areas of high 2,4-DNT concentrations may be a first indication of biodegradation occurring in situ on a large scale.

In summary, laboratory data suggest that nitrate and nitrite release in soil columns and in the field is due to 2,4-DNT biodegradation. The other explosives present at high concentration, i.e., TNT and 2,6-DNT, do not show appreciable biodegradation. TNT biotransforms to monoamino and diamino intermediates, with prevalence of the 4-amino-DNT over the 2-amino-DNT under aerobic conditions. Sorption experiments in soil columns indicate that transformation to amino derivatives leads to compounds which are retarded with respect to TNT, hence biotransformation reactions result in an attenuation of the TNT plume. The toxicity of the biotransformation products compared to TNT will have to be taken into account for a full risk assessment of the contaminated site.

Table 4 Microcosms of Groundwater from Contaminated Well 104

Day	TNT	2,4-DNT	2,6-DNT	2A-DNA	4A-DNT	2-NT(**)	4-NT(**)	3-NT
1	69.6	104.240	124.782	0	0	3.500	0	0
21	66.5	86.400	104.463	0	0	1.120	0	0
97	62.0	2.440	97.780	0	0	0	0	0
123	60.6	0	—	—	—	—	—	—

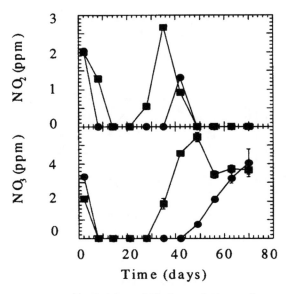

Figure 7 Production of nitrite and nitrate (squares = 8.4 ppm DO; circles = 4.3 ppm DO).

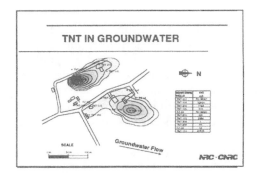 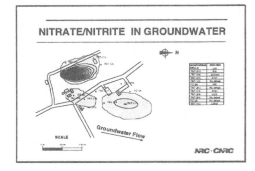

Figure 8 (Left) TNT plumes originating from well 104 and well 113 at a Canadian TNT-contaminated site. (Right) Nitrate/nitrite plumes centered on wells 104 and 113.

IV. IMPLEMENTATION

A. Site Characterization

Careful characterization of the explosive plume, in terms of the specific contaminants present, their concentrations and distribution, and also in terms of the hydrogeology of the aquifer, is essential. This requires installation of a sufficient number of groundwater monitoring wells within the plume to define the lateral and vertical extent of the contamination. Typically, wells will be arranged parallel to the groundwater flow direction, begin near the contaminant source if it is known, and extend beyond the down-gradient contaminant plume. Peripheral monitoring wells are needed to define the lateral extent of the plume and deep wells are needed to define the lower limits of the contamination. Wells are also needed between the contaminant plume and any potential receptors of concern, to serve as sentinel wells. Detection of contamination in sentinel wells suggests that MNA is not protective of receptors and will trigger consideration of contingency plans, which should already be prepared.

Provided that the substrate at the site is amenable to penetrations, a cone penetrometry (CPT) sampling event may assist in refinement of the plume conceptualization. The CPT can provide lithologic profiles at multiple locations and depths to refine the definition of the site hydrogeology and provide subsurface soils for further contaminant characterization and for any additional testing necessary, e.g., site attenuation capacity and adsorption properties, hydraulic conductivity determinations, and microbiological assessments.

B. Groundwater Monitoring

The groundwater monitoring protocol must be specific to generate data that are adequate for measuring trends in contaminant mass over time. To that end, all parties involved in sampling, handling, and analysis should follow a consistent plan. Deviations from established protocols can greatly impact the observation of small trends over short observation periods by increasing variability in the data. The following are suggestions for development of quality groundwater monitoring data.

1. To assure that highly contaminated well water cannot cross-contaminate less contaminated water, sampling should progress from low- to high-concentration wells, sample tubing should be dedicated to each well, and all sampling apparatus that comes into contact with the groundwater should be decontaminated between wells. To ensure that the decontamination procedure is effective, random samples of rinsate from decontamination should be submitted for explosives analyses.
2. Pumping with a low-flow pump to match removal rates to recharge (micropurge technique), minimizing the influence of removal on the contaminant plume, and removing at least three-well volumes before collection of the groundwater sample contribute to reliability of data. Oxygenation of the water in contact with air in the well before pumping is initiated has been observed to cause a decrease in explosives concentrations (30). Therefore, removal of this oxygenated water, which can introduce an artifact into the explosives concentration data, is essential. Additional studies showed that explosives concentrations stabilized when DO reading stabilized as water was pumped from the well using the micropurge technique. However, the micropurge technique cannot be used under conditions of low well volume.
3. Results of experiments conducted at LAAP indicated that acidification of groundwater samples to pH 2 with 1.5 g sodium bisulfate per liter of groundwater was an effective means of preserving the analytes of interest (30).
4. At least one of every 10 samples should be collected and analyzed in duplicate to assure precision in the data. At least two spiked blanks (distilled or reverse-osmosis water) and three spiked groundwater samples should be analyzed for each sampling event to assure accuracy. Spikes should represent both low and high analyte concentrations. Temporal representativeness of the data can be obtained by sampling with sufficient frequency to observe seasonal trends that may affect overall explosives concentration trends. Spatial representatives can be obtained by sampling from representative locations throughout the groundwater plume.

Table 5 Explosives and Related Analytes Monitored in Groundwater at LAAP and JAAP

Chemical name	Acronym
2,4,6-Trinitrotoluene	TNT
1,3,5-Trinitro-1,3,5-hexahydrotriazine	RDX
Octahydro-1,3,5,7-tetranitro-1,3,5,7-tetrazocine	HMX
N,2,4,6-Tetranitro-N-methylaniline	Tetryl
1,3,5-Trinitrobenzene	TNB
2,6-Dinitrotoluene	2,6DNT
2,4-Dinitrotoluene	2,4DNT
1,3-Dinitrobenzene	1,3DNB
Nitrobenzene	NB
4-Amino-2,6-dinitrotoluene	4ADNT
2-Amino-4,6-dinitrotoluene	2ADNT
3,5-Dinitroanaline	35DNA
2,4-Diamino-6-nitrotoluene	24DANT
2,6-Diamino-4-nitrotoluene	26DANT
2,2',4,4'-Tetranitro-6,6'-azoxytoluene	66'AZOXY
4,4',6,6'-Tetranitro-2,2'-azoxytoluene	22'-AZOXY
2,2',6,6'-Tetranitro-4,4'-azoxytoluene	44'AZOXY
Hexahydro-1-nitroso-3,5-dinitro-1,3,5-triazine	MNX
Hexahydro-1,3-dinitroso-5-dinitro-1,3,5-triazine	DNX
Hexahydro-1,3,5-trinitroso-1,3,5-triazine	TNX
Mononitroso-octahydro-1,3,5,7-tetranitro-1,3,5,7-tetrazocine	MN-HMX
Picric acid	

5. Sampling efficiency can be optimized by using two two-person field crews. While one crew monitors and demobilizes at one well, the other crew sets up and initiates monitoring at the next well. Using this "leapfrog" technique greatly reduces well sampling time.

Due to the several potential attenuation processes affecting explosives, analysis of a battery of parent compounds and transformation/degradation products is necessary, especially during initial characterization of the contaminant plume (Table 5). Some of the transformation products of TNT are as environmentally undesirable as the parent compound (41–44); therefore, these must be monitored to assure protection of potential receptors.

Monitoring of certain geochemical parameters can be useful (Table 6). Some of these are monitored in the field, while others are assayed in the laboratory. The use to which these data can be put is relatively site-specific. Attenuation processes are typically too slow for accumulation of sufficient degradation products (e.g., nitrates, ammonia, methane, and carbon dioxide) to be distinguished from background. However, if organic carbon is high at the site, e.g., if the contaminant source contains high-organic-carbon wastes of some kind monitoring for specific degradation products of explosives may be productive. Otherwise, the geochemical properties contribute to the weight of evidence that conditions in the site are suitable for attenuation processes. For example, reduced iron and low redox potential suggest an environment in which covalent bonding of amino transformation products of TNT to organic matter can occur. Initial characterization of geochemical parameter will indicate general water quality and tell whether the site

Table 6 Geochemical Parameters Monitored in Groundwater at LAAP and JAAP

	Assayed in the laboratory	
Monitored in the field	Aerobic collection	Anaerobic collection
pH	Total organic carbon	Reduced iron
Conductivity	Total iron	Methane
Dissolved oxygen	Calcium	
Temperature	Magnesium	
Redox potential	Manganese	
Turbidity	Nitrate/nitrite nitrogen	
Salinity	Sulfate	
	Chloride	

has any problem independent of the explosives contamination, e.g., excedance of drinking water standard for naturally occurring parameters such as total iron or manganese.

C. Groundwater Modeling

Site conceptualization will assist in determinating the potential effectiveness of MNA. A conceptual model combines available information on contaminant sources, what is known about distribution of the contaminants, site hydrogeology, and the location of potential receptors. The site conceptual model will assist in estimations of total contaminant mass and, subsequently, with determining estimates of mass reduction as attenuation progresses. Development of a numerical model is useful for predicting the future status of the contamination so that expected attenuation rates can be evaluated against clean-up goals and desired time frames. Many models are available that can be applied (5).

D. Case Study at Louisiana Army Ammunition Plant

1. Demonstration Site

A 2-year demonstration of MNA for explosives was conducted at the Louisiana Army Ammunition Plant (LAAP) in Minden, Louisiana, USA. The LAAP was selected because the source of contamination had already been remediated, a large body of contaminant data was available, and the site was well characterized with approximately 50 functional groundwater monitoring wells in place. The demonstration was conducted at Area P (Fig. 9), the site where wastes from loading, assembling, and packing of munitions had been disposed into 16 unlined lagoons. The lagoons had been pumped out and the sediment excavated and incinerate in the late 1980s. The area was capped with clay compacted to 90% of the standard proctor density for the clay used. The cap was covered with 4 in. of topsoil with a slope of at least 1% to facilitate drainage.

2. Site Geology and Contaminant History

The near-surface geology at LAAP has a complex stratigraphy of Pleistocene, terraced fluvial deposits (basal gravels fining upward to clays) unconformably overlying Eocene, nonmarine, massive sands, silty sands, silty clays, and occasional lignitic beds. An effectively impermeable boundary, the Cane River Formation, lies below the fluvial deposits. Matrix characteristics of this site that potentially affect attenuation include

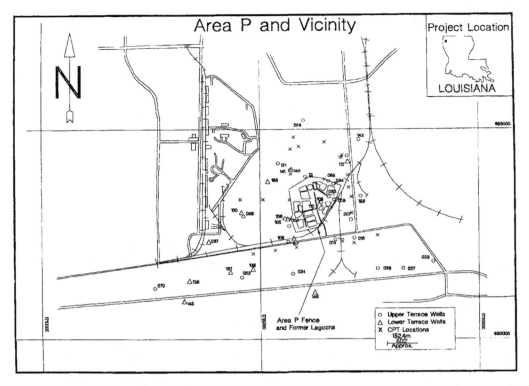

Figure 9 Area P at LAAP, showing former lagoons, capped area, monitoring wells, and CPT locations.

the following: clay lenses, silts, and low porosity, which can promote immobilization processes for transformation products of TNT and effectively reduce groundwater flow rates; and low organic carbon in aquifer soils, which decreases the likelihood for immobilization by interactions of TNT transformation products with organic carbon, but also decreases microbial degradation potential because of limited co-substrate. Data from the late 1980s indicated that the explosives contaminants from the Area P lagoons had entered the terrace aquifers. Groundwater plumes containing RDX, TNT, and TNB had been detected in excedance of drinking water Health Advisory Levels (45).

Concentrations detected in the upper terrace aquifer in 1994 were lower than concentrations detected in 1990, indicating an improvement in the groundwater quality at Area P since the removal of the lagoons. The groundwater contaminant plumes had not advanced very far laterally, suggesting either attenuation or very slow transport. Conceptualization of the lateral extent of the TNT contaminant plume in the upper terrace aquifer at the inception of the study (February 1996) is illustrated in Fig. 10.

The extreme variability of the historic contaminant concentration data for wells in Area P made determinations of trends in concentration over time difficult. Observed concentrations of some explosives spanned several orders of magnitude in the most highly contaminated area in both the upper and lower terraces. Most of the historical data were collected prior to source removal. Therefore, the implications of historical data trends for future contaminant trends were limited. Establishment of trends in attenuation required implementation of a new monitoring plan.

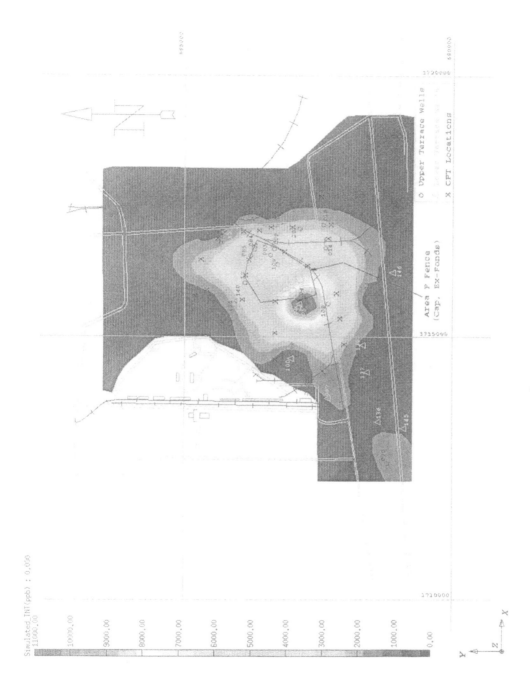

Figure 10 Conceptualization of the lateral distribution of TNT in the upper terrace groundwater at LAAP at the beginning of the demonstration.

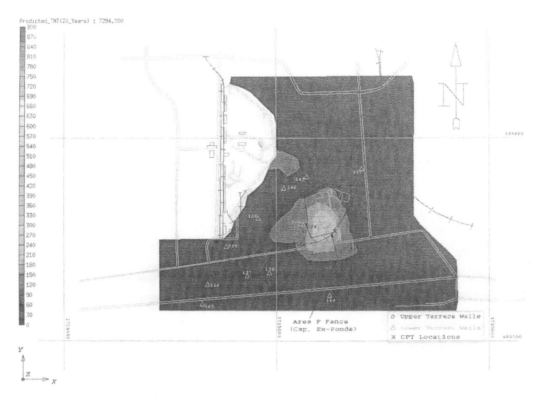

Figure 11 Predicted TNT contaminant plume (model at LAAP after 20 years).

3. Monitoring

Thirty monitoring wells were sampled using the micropurge technique and with care given to the above suggestions. Cone penetrometry (CPT) soil samples were collected along transects that bisected the plume from highest to lowest contaminant concentrations in four directions and vertically. Samples were collected using a split spoon sampler according to procedures described by Pennington et al. (30). A field test kit was used to determine the concentration of explosives immediately after CPT sample collection. These analyses guided the placement of the next penetration to minimize the cost of taking data outside the plume. Data precision, accuracy, representativeness, and comparability were assured by adopting field and laboratory quality control and quality assurance procedures (30).

Trends in contaminant concentrations over the 2-year study period were analyzed statistically for the 11 wells in which most analytes were consistently detected. Significant declines in contaminant concentrations occurred in 9 of the 11 wells (Fig. 12).

4. Modeling

A comprehensive computer graphical and integral modeling system, the Department of Defense Groundwater Modeling System (GMS) (46), was used. The GMS contains tools to facilitate site conceptualization, geostatistical computations, and postprocessing.

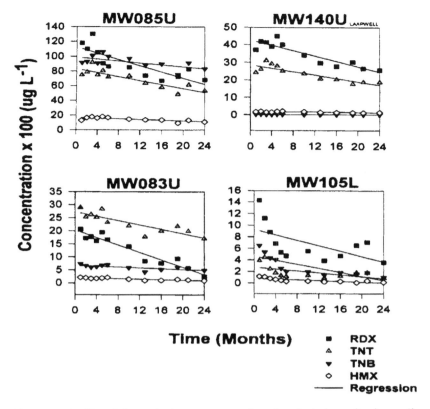

Figure 12 Trends in explosives concentrations in selected monitoring wells over the 2-year sampling period at LAAP.

The GMS links transport and water quality models to predict the fate and transport of contaminants. Sensitivity of the model simulations and predictions to input parameters was coupled with the desired level of accuracy to determine the level of detail required for field and laboratory measurements. Contaminant mass over time was also calculated using the measured and predicted explosives concentrations.

The GMS provided efficient numerical tools to integrate and translate the complex field data into simple graphic forms that could be used to determine fate and transport of the contaminant plumes. The measured and simulated flow data indicated slow subsurface flow at the LAAP site because of the low-permeability media and low-hydraulic gradients. The TNT and RDX plumes were virtually static. The simulated flow directions were consistent with the direction of explosives plume propagation. The simulated results indicated that the explosives at LAAP may be reduced naturally without posing any threat to offsite receptors. Even though the reduction process is very slow, the plume is confined to a limited area and is not moving significantly. The results of contaminant mass calculations indicated that the initial mass of TNT and RDX was reduced significantly during 20 years of simulation (Fig. 11). The sensitivity analysis suggested that the important model-input parameters are the adsorption rates and biodegradation rates. Contaminant mass calculations also indicated declining mass from 52 and 78 to 1.0 and 0.8 metric tons for TNT and RDX, respectively.

5. Site Capacity Estimates

Site capacity for adsorption and transformation can vary considerably across a site; therefore, accurate lithological profiles and determination of sorption and transformation parameters on a sufficient number of aquifer soil samples is essential. For LAAP, batch tests by shaking and/or in columns followed standard procedures [40CFR796.2750 (*http://frwebgate.access.gpo.gov/cgi-bin/multidb.cgi*) at 40CFR796.2750-Sec. 796.2750 Sediment and soil adsorption isotherm]. Aquifer soils from LAAP were generally high in sand, ranging from 65% to 92.5% sand. Silt and clay was present in all samples, although in lower levels. Total organic carbon was low, ranging from 0.015% to 0.162%. Cation-exchange capacity (CEC) was also low, ranging from 3.5 to 8.1 Meq/100 g. Soil pH was acidic and relatively consistent for all soil types (average of 5.55). Permeabilities of the soils ranged from 10^{-4} to $>10^{-9}$ cm/s. Most explosive compounds reached steady-state distribution between aquifer soils and water within 24 h (47), making equilibrium partitioning an appropriate assumption for sorption of most of the explosives at this site.

Adsorption of explosives from groundwater by the LAAP aquifer soils was limited. The measured values of K_d were below 1 L/kg for all soils and contaminants, ranging from no significant adsorption to a high value of 0.84 L/kg. The highest degree of sorption was associated with the soils highest in clay and CEC. The range of sorption coefficients between soils varied over an order of magnitude for explosive compounds. For modeling of contaminant transport at the LAAP site, the use of an average value of K_d to represent sorption was appropriate. Although these data indicate limited retardation, low permeability may exert a greater influence on movement of the plume than is exerted by sorption and transformation. These results suggest that mass transport limitations rather than site capacity restrict transport at LAAP.

6. Biomarker Analyses

Three sets of tests were evaluated to determine their usefulness in supporting site evaluation of natural attenuation. These included (a) soil mineralization radioassays, (b) lipid biomarker analyses, and (c) nucleic acid biomarker analyses. These tests were designed to evaluate of the microbial degradation potential of the site. The radioassays demonstrate the mineralization potential of the microflora present in aquifer soils; the lipid assays estimate the microbial biomass, and the nucleic acid assays assess the degradation potential of the site microflora by detection of genes specific to degradation processes. For the radioassays, soils from the aquifer receive radiolabeled acetate, TNT, or RDX in slurry reactors. Evolution of radiolabeled CO_2 from acetate indicates the general activity of the microflora, while evolution of radiolabeled CO_2 from the explosives indicates degradation potential and estimate rate. For the lipid assays, ester-linked phospholipid fatty acids (PLFA) are extracted from aquifer soil, fractionated on a silica gel column, and analyzed by gas chromatography. PLFA data are quantified by comparison with known internal and external standards and from patterns and types of PLFA profiles determined from a growing database. Lipid analyses provide quantitative data on the in-situ viable microbial biomass, community composition, and physiological status. For nucleic acid assays, DNA is isolated from the aquifer soils, the bacterial biodegradation genes targeted to nitroaromatic contaminants are amplified, and banding patterns, which are diagnostic of the presence of targeted genes, are determined by slab gel electrophoresis. The presence of the targeted catabolic genes is correlated with the radioassay results to establish a link between the presence of the requisite genes in the soil and the observed mineralization activity.

Rates of TNT and RDX mineralization were very low in LAAP soils. Rate constants for TNT ranged from less than 1×10^{-4} to 2.2×10^{-3}; for RDX the range was from less than 1×10^{-4} to 2.0×10^{-3}. Few significant correlations between geochemical parameters and biomarkers were found. Several nucleic acid probes correlated positively with mineralization rate, as did the following parameters determined by lipid biomarkers: biomass; abundance of gram-negative, sulfate-reducing, and iron-reducing bacteria; and a sulfite reductase. The potential for aerobic degradation of TNT in LAAP soils was suggested by the presence of two catechol oxygenase gene probes. Other observed genes supported potential for both anaerobic and aerobic metabolism of TNT.

7. Conclusions

At LAAP the low permeability and groundwater flow rate provide sufficient time for explosives attenuation processes to occur. Results of the groundwater monitoring over 2 years demonstrated declines in explosives concentration at specific points in the plume and reduction in total estimated contaminant mass. Site capacity data showed that mass transport limitations rather than site capacity for attenuation restrict transport at LAAP. Results of biomarker assays demonstrated the potential for microbial degradation processes at the site and contributed to the weight of evidence for these processes. The numerical model predicted continuing declines in explosives mass over time. This information can be used to predict when site clean-up goals may be achieved. Even though attenuation processes are slow, the site hydrogeology provides sufficient time for attenuation processes to protect potential receptors.

V. ADVANTAGES AND LIMITATIONS

The greatest advantage of monitored natural attenuation over engineered remedial alternatives is the potential cost savings. In a cost comparison with two other technologies, in-situ bioremediation and activated carbon adsorption, estimated cost savings by use of monitored natural attenuation were significant (47). Monitored natural attenuation also generates less waste, reduces the risk of human and environmental exposure to contaminants during remediation, and is less intrusive than other remedial alternatives.

For explosives, the following technical limitations are imposed by the current state of the science.

1. Possibly the most important attenuation mechanism for TNT is immobilization by interactions between TNT transformation products and soil components. However, monitoring immobilization processes is restricted to measuring reductions in mass of TNT and its transformation products in groundwater over time. Removal and analysis of the complex, insoluble products of immobilization are not technically feasible at this time. Therefore, the current state of the science does not provide methods for following in-situ immobilization processes by direct biological or geochemical measurements.
2. Since microbial mineralization is primarily co-metabolic, degradation rates in natural systems are typically so slow that products (e.g., nitrates, ammonia, methane, and carbon dioxide) typically do not accumulate in sufficient quantities to be distinguished from background levels. Sites in which groundwater

is atypically high in organic carbon, e.g., sites receiving leachate from a feedlot or other high-carbon wastes, may accumulate measurable degradation products of the explosives.
3. Biomarker techniques, including laboratory radioassays, and lipid and nucleic acid analyses using site material, measure degradation *potential* in the aquifer, but do not demonstrate in-situ degradation. These tests are not currently widely conducted and they contribute to the weight of evidence only.

In spite of these limitations, MNA should be considered at sites where attenuation processes can be demonstrated to protect potential receptors and sites at which clean-up goals can be met within a reasonable amount of time.

VI. CONCLUSIONS

Monitored natural attenuation of explosives should be among the options considered for remediation of explosives-contaminated sites. Implementation of monitored natural attenuation requires sufficient characterization of the subsurface hydrogeology and contaminant distribution to develop conceptual and predictive models of the plume, initiation and careful execution of a long-term groundwater monitoring plan, periodic reevaluation of anticipated attenuation rates and plume migration, and development of contingency plans should MNA prove less effective than envisioned. Regulatory acceptance of MNA at a specific site will require assurance that the groundwater system is sufficiently well understood to make reliable predictions concerning the future fate of the contaminants, that remediation goals will be met within a reasonable time frame, and that potential human health and environmental receptors will be protected.

REFERENCES

1. Federal Register. 1990. *Fed. Reg.*, 55(46):March 8.
2. Weidemeier, T., Wilson, J. T., Kampbell, D. H., Miller, R. N., and Hansen, J. E. 1996. Technical protocol for implementing intrinsic remediation with long-term monitoring for natural attenuation of fuel contamination dissolved in groundwater, volume I. Air Force Center for Environmental Excellence, San Antonio, TX.
3. RTDF. 1996. *Guidance Handbook on Natural Attenuation of Chlorinated Solvents*. Remediation Technologies Development Forum, September 1996, available on worldwide web at [http://www.icubed.com/rtdf/].
4. Weidemeier, T. H., Swanson, M. A., Moutoux, D. E., Gordon, E. K., Wilson, J. T., Wilson, B. H., Kampbell, D. H., Hansen, J. E., Haas, P., and Chapelle, F. H. 1996. Technical protocol for evaluating natural attenuation of chlorinated solvents in groundwater. Air Force Center for Environmental Excellence, Brooks Air Force Base, San Antonio, TX.
5. Pennington, J. C., Bowen, R., Brannon, J. M., Gunnison, D., Harrelson, D. W., Zakikhani, M., Clarke, J., Mahannah, J., and Gnewuch, S. 1999. Draft protocol for evaluating, selecting and implementing monitored natural attenuation at explosives-contaminated sites. Tech. Rep., EL-99-10, U.S. Army Engineer Waterways Experiment Station, Vicksburg, MS.
6. U.S. EPA. 1999. *Use of Monitored Natural Attenuation at Superfund, RCRA Corrective Action, and Underground Storage Tank Sites*, Draft interim final (OSWER Directive 9200.4-17P), April 21, 1999, available on worldwide web at [http://www.epa.gov/swerustl/directiv/d9200417.htm].

7. Kaplan, D. L. 1993. Biotechnology and bioremediation for organic energetic compounds. In: Paul Marikas (ed.), *Organic Energetic Compounds*. Nova Science, New York.
8. Kaplan, D., Ross, E., Emerson, D. LeDoux, R., Mayer, J., and Kaplan, A. M. 1985. Effects of environmental factors on transformation of 2,4,6-trinitrotoluene in soils. Tech. Rep. Natick/TR-85/052, U.S. Army Natick Research and Development Center, Natick, MA.
9. Kaplan, D. L., and Kaplan, A. M. 1982. Thermophilic biotransformations of 2,4,6-trinitrotoluene under simulated composting conditions. *Appl. Environ. Microbiol.*, 44(3):757–760.
10. McCormick, N. G., Cornell, J. H., and Kaplan, A. M. 1981. Biodegradation of hexahydro-1,3,5-trinitro-1,3,5-triazine. *Appl. Environ. Microbiol.*, 42(5):817–823.
11. Brannon, J. M., Price, C. B., and Hayes, C. 1998. Abiotic transformation of TNT in montmorillonite and soil suspensions under reducing conditions. *Chemosphere*, 36(6):1453–1462.
12. Pennington, J. C., and Patrick, W. H., Jr. 1990. Adsorption and desorption of 2,4,6-trinitrotoluene by soils. *J. Environ. Qual.*, 19:559–567.
13. Sitzman, M. E. 1974. Chemical reduction of 2,4,6-trinitrotoluene—Initial products. *J. Chem. Eng. Data*, 19:179–181.
14. Pennington, J. C., Hayes, C. A., Myers, K. F., Ochmann, M., Gunnison, D., Felt, D. R., and McCormick, E. F. 1995. Fate of 2,4,6-trinitrotoluene in a simulated compost system. *Chemosphere*, 30(3):429–438.
15. Caton, J. E., Ho, C.-H., Williams, R. T., and Griest. W. H. 1994. Characterization of insoluble fractions of TNT transformed by composting. *J. Environ. Sci. Health*, A29(4):659–670.
16. Folsom, B. L., Jr., Pennington, J. C., Teeter, C. L., Barton, M. R., and Bright, J. A. 1988. Effects of soil pH and treatment level on persistence and plant uptake of 2,4,6-trinitrotoluene. Tech. Rep. EL-88-22, U.S. Army Engineer Waterways Experiment Station, Vicksburg, MS.
17. Thorn, K. A. 1997. Covalent binding of the reductive degradation products of TNT to humic substances examined by N-15 NMR. *Am. Chem. Soc. Abstr.*, 37:305–306.
18. Osmon, J. L., and Klausmeier, R. E. 1972. The microbial degradation of explosives. *Dev. Ind. Microbiol.*, 14:247–252.
19. Carpenter, D. F., McCormick, N. G., Cornell, J. H., and Kaplan, A. M. 1978. Microbial transformation of ^{14}C-labeled 2,4,6-trinitrotoluene in an activated-sludge system. *Appl. Environ. Microbiol.* 35(5):949–954.
20. Hallas, L. E., and Alexander, M. 1983. Microbial transformation of nitroaromatic compounds in sewage effluent. *Appl. Environ. Microbiol.*, 45:1234–1241.
21. Fernando, T., Bumpus, J. A., and Aust, S. D. 1990. Biodegradation of TNT (2,4,6-trinitrotoluene) by *Phanerochaete chrysosporium*. *Appl. Environ. Microbiol.*, 56:1666–1671.
22. Spanggord, R. J., Spain, J. C., Nishino, S. F., and Mortelmans, K. E. 1991. Biodegradation of 2,4-dinitrotoluene by a *Pseudomonas* sp. *Appl. Environ. Microbiol.*, 57:3200–3205.
23. Valli, K., Brock, B. J., Joshi, D. K., and Gold, M. H. 1992. Degradation of 2,4-dinitrotoluene by the lignin-degrading fungus. *Appl. Environ. Microbiol.*, 58:221–228.
24. Funk, S. B., Roberts, D. J., Crawford, D. L., and Crawford, R. L. 1993. Initial-phase optimization for bioremediation of munition compound-contaminated soils. *Appl. Environ. Microbiol.*, 59:2171–2177.
25. Boopathy, R., Wilson, M., and Kulpa, C. F. 1993. Metabolism of 2,4,6-trinitrotoluene (TNT) by *Desulfovibrio* sp. (B. strain). *Appl. Microbiol. Biotechnol.*, 39:270–275.
26. Bradley, P. M., Chapelle, F. H., Landmeyer, J. E., and Schumacher, J. G. 1994. Microbial transformation of nitroaromatics in surface soils and aquifer materials. *Appl. Environ. Microbiol.*, 60:2170–2175.
27. Bradley, P. M., Chapelle, F. H., Landmeyer, J. E., and Schumacher, J. G. 1997. Potential for intrinsic bioremediation of a DNT-contaminated aquifer. *Groundwater*, 35:12–17.
28. Duque, E., Haidour, A., Godoy, F., and Ramos, J. L. 1993. Construction of a *Pseudomonas* hybrid strain that mineralizes 2,4,6-trinitrotoluene. *J. Bacteriol.*, 175:2278–2283.

29. Preuss, A., Fimpel, J., and Diekert, G. 1993. Reduction of nitroaromatic compounds by anaerobic bacteria isolated from the human gastrointestinal tract. *Arch. Microbiol.*, 159:345–353.
30. Pennington, J. C., Gunnison, D., Harrelson, D. W., Brannon, J. M., Zakikhani, M., Jenkins, T. F., Clarke, J. U., Hayes, C. A., Myers, T., Perkins, E., Ringelberg, D., Townsend, D., Fredrickson, H., and May, J. H. 1999. Natural attenuation of explosives in soil and water systems at Department of Defense sites: Interim report. Tech. Rep., EL-99-8, U.S. Army Engineer Waterways Experiment Station, Vicksburg, MS.
31. McCormick, N. G., Cornell, J. H., and Kaplan, A. M. 1985. The anaerobic biotransformation of RDX, HMX, and their acetylated derivatives. Tech. Rep. NATICK/TR-85/007, U.S. Army Natick Research and Development Center, Natick, MA.
32. Cattaneo, M. V., Masson, C., Hawari, J., Sunahara, G., Greer, C. W., Thiboutot, S., and Ampleman, G. 1997. Natural attenuation of TNT in soil columns. In: B. Alleman and A. Leeson (eds.), *In Situ and On-Site Bioremediation*, IV Int. Symp., 2:3–8. Battelle Press.
33. Patrick, G. C., Ptacek, C. J., Gillham, R. W., Barker, J. F., Cherry, J. A., Major, D., Mayfield, C. I., and Dickhout R. D. 1985. The behaviour of soluble petroleum product derived hydrocarbons in groundwater. PACE Rep. 85-3.
34. Millette, D. 1995. Ph.D. thesis, University of Waterloo, p. 71.
35. Townsend, D. M., Myers, T. E., and Adrian, D. D. 1995. 2,4,6-trinitrotoluene (TNT) transformation/sorption in thin-disk soil columns. Tech. Rep. IRRP-95-4, U.S. Army Engineer Waterways Experiment Station, Vicksburg, MS.
36. Townsend, D. M., and Myers, T. E. 1996. Recent developments in formulating model descriptors for subsurface transformation and sorption of TNT, RDX and HMX. USACE Tech. Rep. IRRP-96-1.
37. Price, C. B., Brannon, J. M., and Hayes, C. A. 1997. Effect of redox potential and pH on TNT transformation in soil-water slurries. *J. Environ. Eng.*, October: 988–992.
38. Kaplan, D. L. 1990. Biotransformation pathways of hazardous energetic organo-nitro compounds. In: D. Kamely, A. Chakrabarty, and G. S. Omenn (eds.), *Biotechnology and Biodegradation*. Gulf Publ., Houston, TX.
39. Walker, J. E., and Kaplan, D. L. 1992. Biological degradation of explosives and chemical agents. *Biodegradation*, 3:369–385.
40. Haigler, W. E., and Spain, J. C. 1996. Degradation of nitroaromatic compounds by microbes. *SIM News*, 46:59–68.
41. Honeycutt, M. E., Jarvis, A. S., and McFarland, V. A. 1996. Cytotoxicity and mutagenicity of 2,4,6-trinitrotoluene and its metabolites. *Ecotoxicol. Environ. Safety*, 35:282–287.
42. Jarvis, A. S., McFarland, V. A., and Honeycutt, M. E. 1998. Assessment of the effectiveness of composting for the reduction of toxicity and mutagenicity of explosives-contaminated soil. *Ecotoxicol. Environ. Safety*, 39:131–135.
43. Tan, E. L., Ho, C.-H., Griest, W. H., and Tyndall, R. L. 1992. Mutagenicity of trinitrotoluene and its metabolites formed during composting. *J. Toxicol. Environ. Health*, 36:165–175.
44. Won, W. D., DiSalvo, L. H., and Ng, J. 1976. Toxicity and mutagenicity of 2,4,6-trinitrotoluene and its microbial metabolites. *Appl. Environ. Microbiol.*, 31:576–580.
45. Science Applications International Corporation. 1994. Five-year review report of interim remedial action at former Area P lagoons, Louisiana Army Ammunition Plant, Shreveport, LA. Prepared for U.S. Army Toxic and Hazardous Materials Agency Contract No. DAAA15-91-D0017, by Science Applications International Corporation, McLean, VA.
46. GMS. 1996. *Department of Defense Groundwater Modeling System, Reference Manual*, Version 2.0. Brigham Young University, Engineering Computer Graphic Laboratory, Provo, UT.
47. Pennington, J. C., Zakikhani, M., and Harrelson, D. W. 1999. Monitored natural attenuation of explosives in groundwater—ESTCP completion report. Tech. Rep., EL-99-7, U.S. Army Engineer Waterways Experiment Station, Vicksburg, MS.

41
Municipal Solid Waste Generation and Management in Caracas and Metropolitan Area, Venezuela

Adriana Diaz-Triana
Northeastern University, Boston, Massachusetts

I. INTRODUCTION

Like big cities around the world, Caracas is not exempt from the problem of municipal solid waste (MSW) management. The metropolitan area population has grown sharply in the last 10 years, due to expected growth and also due to migration of working population from other regions of the country. Current infrastructure and service for MSW collection, transport, and final disposal seems not to be sufficient. This chapter provides a general overview of this situation, including some figures that may give a better idea of this reality. This chapter is a comparison to a paper by Tin et al. (1), "Cost-Benefit Analysis of the Municipal Solid Waste Collection system in Yangon, Myanmar." In this study, the authors analyze the problem of MSW collection for a city in a developing country in Asia. There are many similarities between the problems encountered in Yangon and the problems of MSW management in Caracas and its metropolitan area (CMa), the capital city of Venezuela, a developing country in South America.

II. BRIEF DESCRIPTION OF CARACAS AND METROPOLITAN AREA

CMa is located in a valley surrounded by mountains. Its topography is varied, having flat and hilly areas. The metropolitan area has a population of 3,665,765 inhabitants, mainly in seven countries, with different characteristics. Inhabitants of poor areas account for 41% (1,502,964) of the total population in this area. Per-capita production of waste is approximately 0.6–1.8 kg/day (2). In terms of this chapter, the seven counties that provide information about MSW generation are located in two states: Miranda State and the Federal District (DF). Miranda State counties studied were Baruta, Chacao, El Hatillo, Plaza, and Sucre. DF counties studied were Libertador and Vargas. Table 1 presents information on population distribution for each country and for the two states, corresponding to year 1999. Table 2 contains the information on the amount of MSW generated by county per day. Detailed descriptions of the characteristics of wastes

Table 1 Population Distribution Among Counties Studied for Year 1999

State	County	Population
Miranda	Baruta (B)	312,855
Miranda	Chacao (C)	72,473
Miranda	El Hatillo (H)	59,645
Miranda	Plaza (P)	168,099
Miranda	Sucre (S)	768,927
DF	Libertador (L)	1,975,463
DF	Vargas (V)	308,303
Total		3,665,765

Source: Ref. 3.

Table 2 MSW Generation per County, on a Daily Basis, 1999

County	Wastes (tons/day)	Distribution (%)
Baruta (B)	395.36	11.20
Chacao (C)	214.27	6.07
El Hatillo (H)	105.25	3.01
Plaza (P)	13.06	0.37
Sucre (S)	769.19	21.79
Libertador (L)	1961.27	55.56
Vargas (V)	70.60	2.0
Total	3530	100

Source: Ref. 4.

for each county are not available. It is estimated that since most of the counties comprise residential areas, the majority of wastes are organic, easily degradable wastes. Some of the counties have important commercial and business activities that generate a different type of waste. Industrial wastes are limited due to the small industrial activity in this area.

III. EXISTING MSW MANAGEMENT SYSTEM IN CMa

Until 1994, MSW management in CMa was in the charge of a governmental institution (Municipal Urban Cleaning Institute, IMAU), but due to the privatization of this sector in 1995, three private companies have since managed the MSW system. These companies are Cotécnica C.A., Fospuca C.A., and Sabenpe C.A. These companies, each having different resources, serve the seven counties in CMa, according with the information presented in Table 3.

The MSW management companies have two types of services: regular and special. Regular service corresponds to collection, transport, and final disposal of wastes (from

Table 3 MSW Management Companies: Inventory of Resources, 1999

Company/Item	Cotécnica	Fospuca	Sabenpe	Total
Counties served	L, C	L, B	H, P, S, V	B, C, H, P, S, L, V
Wastes collected (tons/day)	1,160	1,500	870	3,530
Workers in street	1,000	1,231	693	2,924
Compactor trucks	90	159	88	337
Population served[a]	1,060,205	1,300,586	1,304,974	3,665,765
Population served/worker	1,060	1,057	1,883	Average: 1,333
Workers per 10,000 population	9.4	9.5	5.3	Average: 8.1
Wastes collected/work (kg/labor day)	1,160.0	1,218.52	1,255.41	Average: 1,211.3
Population served/truck	11,780	8,180	14,289	Average: 11,596
Trucks per 15,000 population	1.27	1.83	1.01	Average: 1.4

[a] For calculation of population served and corresponding to workforce and trucks; it was assumed that 50% of the population of Libertador county was served by Cotécnica C.A. and the other 50% was served by Fospuca C.A.

residential and commercial areas, health centers, hospitals, clinics, schools, markets, etc.), cleaning of streets, sidewalks, parks, beaches, and other public spaces with the exception of highways. Companies charge customers for this service, and this charge is included as a separate item in electricity bills. Special or fee-based service refers to collection of wastes and cleaning services for other private companies that request the service. These charges are not included in regular bills (4). The companies send the collected waste to La Bonanza landfill, the main site for final disposal of wastes. There are three other landfills, with less capacity, that sometimes serve counties from the area studied. These landfills are Vargas, Guaicaipuro, and Tejerías landfills, but for the purpose of this study the main landfill is La Bonanza. La Bonanza also receives wastes from other counties not included in this study. It is able to process 4000 tons/day of MSW. This site has 70 workers, but also has the inconvenience of 200 wastes removers, trying to get some benefits from wastes they collect. This landfill has inclined terraces where the compaction and burying of wastes take place. There are also operations for the separation of valuable wastes for recycling (glass, plastic, metal, paper and cardboard, and wood) (5).

Along with this landfill, the waste management infrastructure is completed with one transfer unit, called Las Mayas. Some counties send their wastes to this transfer unit, while others send their wastes directly to La Bonanza. This is determined by the distance from the county to the final disposal site, and the routes of collection for each company. For the transportation of solid wastes to the landfill, different types of vehicles are in use, the most common being rear-loading hydraulic compactor trucks, followed by high-sided open trucks. In poor areas, due to the difficult access, trucks for the transportation of wastes are smaller. In these areas, not incorporated in the formal waste collection and transport system, the alternative for temporary storage of wastes consists of open containers placed in streets and roads (communal storage dumps (2)).

Some commercial and residential counties, where most buildings, houses, stores, and offices are located, have compaction machines. Each machine is able to store 15 tons of waste in a hermetically closed camera. Animals and people cannot get in touch with wastes, eliminating the esthetic and health problems associated with exposed refuse. There is a frequency of cleaning for these machines, depending on the type of waste they store: every 3 days for organic putrefying wastes and every 7 days for nonputrefying wastes (6).

IV. PROBLEMS OF THE MSW MANAGEMENT SYSTEM IN CMA

The main problems that influence the waste handling process in CMa are:

- Areas of uncontrolled population growth that are not incorporated into the formal system of waste management
- Poor net of paved roads and streets
- Lack of resources and infrastructure to handle the amount of waste produced
- Lack of enforcement of laws for environment protection

These interrelated aspects are analyzed in the following discussion. Referring to the first two aspects, recall from previous discussion that 41% of the population of CMa lives in poor areas, also known as "barrios" (shantytowns). The inhabitants of poor neighborhoods live in severe poverty conditions, most of the time lacking basic services such as water and electricity. These areas have a deficient system of road and streets, because they are located on slopes of mountains or hills surrounding the city, and also because of the nature of construction. There is approximately 7% street and road access for every hectare of construction. Of the inhabitants of poor areas, 75% have to walk down the equivalent of 22 stories to take their garbage to the communal storage dumps; this means that a person employs between US$2 and $4 daily (based on the cost of labor hours) to take the wastes to the container. This explains why the population prefers to throw wastes on the slopes of the mountains where they live. This situation results in accumulation of wastes, bad odor, ugly appearance, and potential health problems. Another reality that contributes to this problem is that there are not enough containers for the amount of waste generated in these areas. Wastes are placed in open spaces and on slopes. In CMa there are 133 uncontrolled sites for waste disposal, of which 61% remain more than 15 days with wastes exposed to open air, and 99% remain more than a week in this condition (2).

Referring to the third item, approximately between 7–15% of each county's budget goes to waste management operations, and of this amount, 50% goes to transport and final disposal of wastes. Each MSW management company charges between US$2 and $15 (depending on the type of customer and service) for the collection, transport, and disposal of waste (4). At the same time, companies have to pay a certain amount to La Bonanza landfill for the final disposal of wastes. According to figures from 1998, companies owed US$2.95 million to the landfill. This critical situation determined that the landfill was not able to operate and improve its facilities, and, in fact, it was about to be closed indefinitely. Special funds from the regional government helped reopen the landfill, but it is still operating under budget, with loss of money due to lack of payment (7). This landfill has various technical problems that are worth mentioning. Some of the problems are lack of treatment facilities for liquid leachates from waste terraces, no control of foreign individuals entering installations to collect wastes, poor working conditions (workers are under the risk of suffering from skin and respiratory diseases), accumulation of wastes exposed to open air, lack of enough compaction machines (until 1998 this landfill had only one compaction machine), bad odor, and animals scavenging wastes (these could be potential vectors for diseases). Figure 1 shows wastes to be compacted and buried in La Bonanza landfill. Figure 2 shows a terrace where wastes are buried.

Figure 1 Waste accumulation in La Bonanza landfill, prior to compaction and burying (7).

Figure 2 Terrace for compaction and waste burying, La Bonaza landfill (7).

In Table 3, figures refer to physical resources that MSW management companies have, such as number of trucks and workers in the streets. The total workforce in the streets is 2924 workers. Between 5.3 and 9.4 workers serve a population of 10,000 (from 1060 to 1900 people are served daily by a waste collection worker). These figures are below the minimum standard of 20–50 collectors per 10,000 population. Regarding the number of trucks, between 1.2 and 1.8 trucks serve 15,000 inhabitants (approximately 8,000–15,000 people per truck). This vehicle force number is above the minimum standard of 1 truck per 15,000 population served (1). For the calculation of these figures it is assumed that the three companies have an even distribution of its resources among counties, and among areas within counties. This is certainly not exact, but there is not sufficient information on real distribution of resources. Poor areas have deficient waste collection services compared to other commercial and residential areas in the same county.

With regard to law enforcement for environment protection, the fourth aspect mentioned above, there is a significant number of uncontrolled dumping sites in CMa. It is difficult to punish those responsible, because entire communities dispose of their wastes in this manner. The environmental situation of these sites is alarming. The accumulation of waste in open spaces promotes the growth of rodents, insects, and pathogenic microorganisms that can cause health problems to populations living close to these areas. When there is excess accumulation of wastes, burning is a common practice to reduce their volume. Burning of wastes generates contaminant emissions in the form of vapor, particulate matter, and odor. Again, this practice is unlawful, but rarely are those responsible punished. The inappropriate use of land for waste disposal implies degradation of its quality, its value, and its appearance.

A problem often found in business and commercial areas is the accumulation of wastes on streets and sidewalks, because the schedules for waste collection are not observed. Again, violators are not punished, and this practice results in esthetic problems and complications in the transit of pedestrians and cars. Figures 3, 4, and 5 show different areas of the city where solid wastes accumulate in the streets and sidewalks.

V. SIMILARITIES BETWEEN THE PROBLEM OF MSW IN YANGON AND IN CMa

Although the information about CMa is very general and was taken mainly from articles in the newspaper *El Universal*, a general comparison of the MSW management systems in CMa and Yangon is possible. The main similarities of both MSW management systems are summarized in this section:

> In both cities, the majority of wastes are of domestic and commercial type, and there is not a significant amount of industrial waste. The domestic and commercial wastes have a high content of organic matter that degrades easily. This determines that the frequency for waste collection has to be high, and that wastes have to be safely handled, to avoid exposure to humans and to the environment.
> There are serious deficiencies of waste collection systems in areas of uncontrolled growth. The waste quantifying process is difficult and hence the estimation of resources and infrastructure required is not accurate. Due to the economic situation, the MSW management systems in both cities operate with loss of money. Public and private companies managing solid wastes are not able to charge customers the real value of waste collection and disposal services.
> There are difficulties associated with the topography and construction characteristic of counties in both cities that determine high operational costs, for example, the fact that both cities have many areas where access is limited and difficult. This particular vehicle use implies more frequent maintenance and change of spare parts, placing a burden on the budget of the company.
> The economic deficiencies determine that workforce and trucks for waste collection are not sufficient. The worker-per-10,000 population ratios in Yangon and CMa are below the minimum standard (20–50 collectors per 10,000 population). The vehicle-per-15,000 population ratio is below minimum standard in Yangon, but is above the standard in CMa (1 vehicle per 15,000 population served).
> Noncontrolled disposal of wastes is present in both cities. Open spaces serve as disposal sites for MSW. This situation generates other problems, such as proliferation of insects, rodents, and other animals, contamination of soil and water courses due to infiltration and runoff, exposure of people living close to the site, unpleasant sight, waste scavengers, and degradation in the use of land.

Figure 3 Accumulation of wastes on the slope of a mountain (2).

Figure 4 Wastes from a residential area (7).

Figure 5 Waste from a commercial area (7).

Another situation common to both cities is burning of wastes. This practice produces toxic gases, bad odor, and air pollution. Legislation in both cities makes the dropping of litter an offense, but this is not enforced and regional authorities take no action.

In both cities there are communal storage dumps that attempt to lower the costs of collection by facilitating access for collection vehicles, but often these containers are not enough for the amount of wastes generated in the community. In the case of seasonal variations in waste generation, the problem is even more acute. (In certain periods of the year waste generation increases, e.g., at Christmas, but MSW management infrastructure remains the same.)

The combination of different factors discussed before constrains the use of foreign technology to handle the problem of wastes in both cities. This determines the study, evaluation, and implementation of particular solutions and technologies, according to the needs of each city.

VI. PROPOSED ALTERNATIVES TO IMPROVE THE MSW MANAGEMENT SYSTEM IN CMa

Since the main cause of the problems in MSW management are almost impossible to solve (population growth and difficult economic situation), the alternatives must focus somehow on increasing and improving infrastructure for waste collection, transport, and final disposal. In this context, the following information is available about alternatives to improve the MSW management system in CMa.

Government officials from CMa are considering the construction of a modern transfer plant, four landfills, and the closure of La Bonanza landfill, at a cost of US$250 million. The transfer plant will also serve as a separation center for valuable recycling wastes: plastic, metal, paper, wood, and glass. Advanced technology for landfill construction will include geotextile membranes for ground lining and special systems for leachate treatment. No information was available on the possibility of increasing the number of trucks and workers in the streets.

Chacao County government is planning to increase the number of compaction machines in the streets, from 7 to 15, by the year 2000. These machines are a good alternative to maintain streets clean, and to keep wastes isolated from humans and animals.

Baruta County government officials are studying the construction of an incineration plant, which will be able to convert 95% of nonrecycled wastes of the county into ash and energy. Ashes will be used in mixtures for construction and paving materials. Burning wastes will generate 96 million kilowatts of electricity per year. The cost of the plant is estimated to be US$45 million.

Other innovative technologies under consideration are mixing of organic wastes with soil to produce a compost that could be used as fertilizer, and mixing and compaction of wastes to be used in construction.

VII. CONCLUSIONS

Any waste management program has to protect people and the environment from the hazard of wastes, by providing an effective and efficient system that minimizes, treats, and stores all wastes, disposing of them as soon as possible. Unfortunately, due to specific conditions of cities like Caracas and its metropolitan area, and Yangon, this is not always true. Important facts about the MSW management system in CMa are the following.

> The seven counties of Caracas and its metropolitan area studied produce 3530 tons of wastes daily. The final disposal site for these wastes is La Bonanza landfill, with a capacity of 4000 tons per day.
>
> In poor areas of the city the waste collection service is deficient, causing the accumulation of solid wastes and subsequent degradation of the environment and risks to human health.
>
> Communal storage dumps are not sufficient, and frequency of collection is determined by available access by paved roads and streets.
>
> Collection vehicles are mainly of the rear-loading hydraulic type. The number of vehicles gives an average ratio of 1.4 vehicles per 15,000 population served.
>
> The workforce in the streets comprises 2924 workers, for an average ratio of 8.1 workers per 10,000 population.
>
> The existing solid waste management system is poor, unsatisfactory, and inefficient due to factors as lack of required collection vehicles, lack of required workforce, and shortage of financial resources.
>
> Some alternatives for improvement of MSW management are under consideration. Among these are construction of a new transfer plant with a center for the separation of recycling goods, construction of four landfills, and the construction of an incineration plant.

REFERENCES

1. Tin et al. 1995. Cost-benefit analysis of the municipal solid waste collection system in Yangon, Myanmar. *Resources, Conservation and Recycling*, 14:103–131.
2. Canizales, M. 1998. La basura rueda por los cerros. *El Universal*, May 8.
3. Oficina Central De Estadistica e Informatica de Venezuela, OCEI (Venezuelan Central Office of Statistics and Information). Web page: http//www.ocei.gov.ve.
4. Servicios, cifras y sanciones. 1997. *El Universal*, November 12.
5. Cotecnica C.A. Web page: http//www.cotecnica.com.
6. Lando, M. 1996. Compactadoras de basura acaban con el negocio de los lateros. *El Universal*, November 5.
7. Marcano, R. 1997. Paralizada recoleccion de basura en la ciudad. *El Universal*, July 30.
8. Aguirre, M. 1996. Basura municipal puede ser entable. *El Universal*, June 26.
9. Gomez, B. 1998. Alcaldes conoceran plan resctor para tratamiento de la basura. *El Universal*, January 7.
10. Mollejas, C. 1998. Baruta quiere sacar luz de la basura. *El Universal*, September 14.

Index

Abelmoschos esculentus, 598
Abiotic attenuation (degradation)
    mechanisms, 776–779
Acacia, 598
Acinetobacter, 360–364, 770
Additives, and trace metal remediation, effect of, 511
Adequacy of hazardous landfilling, 204–209 (*see also* Evaluation of hazardous chemical site landfilling)
    excessive leaching, definition of, 207–209
    reliability of TCLP, 205–207
Adequacy of regulations, 194–195 (*see also* Policy)
Adsorption
    batch, 66–67
    and chromium-contaminated soil, 593–594
Aglaonema commutatum, 492, 496
Airborne pollutants, 140
Air emissions, (*see* Industrial trickling bed biofilters for abatement of VOCs from air emissions)
Air purification plants, 451
Air quality control, 265–266
Air sparging (AS), 269–274 (*see also* Low-permeability soils, and bioremediation)
    addressing groundwater contamination, 269–272
    monitoring AS/SVE, 266–267
        biological removal, 266–267
        physical removal, 266
    pneumatic and hydraulic fracturing, 273–274
    technical description of AS/SVE system, 264–266

[Air sparging (AS)]
    air quality control, 265–266
    construction, 264–265
Aldrin, 114
Alkaline pH, heavy metal uptake, 550–557
Amino groups, 950–951
Anolyte pH, enhancement of, 100
Antagonistic and synergistic interactions, 135–137
Amaranthus hybridus, 751
Anthurium andreanum, 492, 496
Applications
    of soil flushing, 618–619
    of soil respiration measurements, 581–582
    of soil washing, 632
Aqueous-phase ligand, addition of, 547–548, 550–557
Aqueous reactions, 162
Aqueous solvent removal of contaminants from soils, 617–638
    soil flushing, 617–631
        additional treatment requirements, 625–626
        background, 618
        case studies, 629–631
        contaminant types, 620–622
        design considerations, 626–629
        soil types, 619
        suitable applications, 618–619
        system equipment requirements, 622–625
        theory, 618
    soil washing, 631–637
        background, 632
        case studies, 636–637
        contaminant types, 633
        design considerations, 635–636

[Aqueous solvent removal of contaminants from soils]
    equipment requirements, 634–635
    soil types, 632–633
    solutions, 633–634
    suitable applications, 632
    theory, 632
Arsenic, 114, 115, 621, 637, 715–758, 830
    remedial action, 739–758
        conclusions, 752
        environmental biochemistry of As, 745–752
        introduction, 739–740
        sources of As in the environment, 740–742
    risk assessment, 715–737
        Austrian case study, 722–733
        Bangladesh case study, 719–722
        concentration in plants, 728–729
        conclusions, 733
        introduction, 715
        regions affected by unusual environmental levels, 717–719
        potential risks, 716–717
Arthorbacter, 294–295
Artificial soils construction, structural and functional aspects, 491–505
    conclusions, 504–505
    introduction, 491–492
    materials presented, 492–504
Aspergillus niger, 600, 606
Austrian case study, 722–733

Bacillus, 606, 770
Bacteriodes, 603
Bangladesh case study, 719–722
Barrier materials (*see* Contaminant migration in barrier materials)
Bayou Saint John, 133
Bayou Trepagnier samples, 127, 130–133
BCD reactor, 910 (*see also* Reactors)
Beets, 933–946
Beijerinckia, 601, 606
Bentonite, ion-selectivity, 549, 563–566
Benzene, 261, 621
Benzo(a)pyree, 114
Beta vulgaris, 496
Bioavailability, constraints to bioremediation, 217–241, 311–322 (*see also* Enhanced naphthalene bioavailability in a liquid-liquid biphasic system)
    characteristics of the microorganisms, 229–230

[Bioavailability, constraints to bioremediation]
    effect of an organic phase, 317–319
    effect of particle surface area on growth, 314–315
    effect of surfactants on naphthalene bioavailability, 316–317
    introduction, 217–219
    mass transfer, 225–229
    multiphase partitioning, 229
    overcoming problems, 230–235
        co-solvents, 233–234
        electrokinetics, 234–235
        hydraulic/pneumatic fracturing, 234, 273–274
        surfactants, 230–233
        thermal enhancements, 234
    and toxicity, 219–225
        bioluminescence reporter system, 220–221
        earthworm and invertebrates test, 221
        microalgal and cyanobacterial growth inhibition factors, 221
        microbial bioassays, 220
        plant tests, 224
        uptake of contaminants, 224–225
        vertebrate tests, 221–224
Biocatalyst development, 456–457
Biodegradation
    activity of RC1, 369–372
    of BTEX compounds, 769–775
    of chlorinated aliphatic hydrocarbons, 779
    of oil in soil, 286–287
    of oil in water, 284–286
Biofilters, (*see* Industrial trickling bed biofilters)
Biogeochemically based techniques of As remediation, 749–752
Biological degradation of organics, 916–920
Biological processes of remediation, 53–54, 263–264, 266–267, 439–440
Bioluminescence reporter system, 220–221
Biomarker analysis, 966–967
Biosensors, 145
Biosorption and reduction of chromium, 605
Biotransformation studies, 954–957
Biphasic systems, (*see* Enhanced naphthalene bioavailability in a liquid-liquid biphasic system)
Borden Canadian Air Forces Base site, 630, 953–954
    and explosives removal, 953–954
    TCE removal, 630
Boundary conditions, 165–167

Index

Brassica chinensis, 496
Brewery mash, 936
BTEX, 272, 274, 769–775, 781–784, 806, 830
  biodegradation of, 769–775

Cadmium, 114, 131, 513–515, 541, 596, 621, 636–637, 641–650, 698, 701, 703, 704, 705, 830
  extraction, 513–515
Calcium carbonate, 832–836, 860, 868
Calcium magnesium acetate, 682–683
Calculating soil respiration, 577–578
  simplified, 578
Canadian Forces Base Borden, (*see* Borden Canadian Air Forces Base site)
Candida fusarium, 936–946
Candida tropicalis, 770, 936–946
Candida parapsilosis, 770
Capsicum annum, 598
Carbohydrate-protein fodder, (*see* Vegetable raw material, processing of)
Carbon dioxide in remediation, 829–838
  conclusions, 836
  of contaminants, 830–836
  future research needs, 836–837
  introduction, 829–830
  methods, 830
Casuarina, 598
Cation-exchange, 67–68, 965
Catholyte neutralization, 99
Charge flux, 159–160
Chelating or complexing agents, 99–100
Chemical background, 859–865
Chemical concentration measurement, 115–118
Chemical detoxification processes, 906–911
Chemical reactions, (*see* Dispersing chemical reaction (DCR) remediation technology)
Chemical site landfilling (*see* Evaluation of hazardous chemical site landfilling)
Chlorinated aliphatic hydrocarbons, 779
  mechanisms of, 779–781
Chlorinated compounds, degradation of, 775–785, 806
  abiotic attenuation (degradation) mechanisms, 776–779
Chlorinated solvent plumes, overall evolution of, 781–784
Chromium, 115, 131, 396, 399, 403, 413, 583–615, 621, 631, 636–637, 641–650
  in groundwater, 588–592
  mobility of, 591–592

[Chromium]
  in soil, 587–588
Chromium-contaminated soil, due to tannery wastes disposal, 583–615
  characterization of tannery wastes, 586–587
  chromium contamination in the environment, 587–592
  in groundwater, 588–592
  in soil, 587–588
  pollution due to disposal of tannery wastes, 585–586
  remediation of, 592–603
    bioremediation, 599–603
    chemical in-situ treatment, 593–594
    physical in-situ treatment, 592–593
    phytoremediation, 594–599
  tanning processes, 584–585
    chrome tanning, 584–585
    vegetable tanning, 584
  technologies for minimization of chromium contamination, 603–606
    biological treatment of effluent, 603–604
    composting of tannery sludge, 606
    constructed wetlands, 606
    microbial reduction, 605–606
    reed bed, 604–605
CIM characteristics, 542–545
Cladosporium, 770
Clay and minerals, 594, 620, 743, 745, 954
Cleaning technique, of groundwater, 189
Clean-up, of site, 194
Cobalt, 596
Colorimetric and selective ion electrodes, 140–141
Colorimetry, 117–118
Column test protocols, 68–71
  details, 69–71
Compact reactor, 839–841,
  (*see also* Reactors)
Composting, 602–603, 606
  and other organic amendments, 602–603
  and tannery sludge, 606
Cone penetrometer technology, 183–191
  for groundwater monitoring, 184–189
    cleaning technique, 189
    fluorescence measurements, 186
    MicroWell, 188–189
    sample recovery, 186–188
    SCAPS system, 184–185, 190
    site characterization, 186

[Cone penetrometer technology]
    volatile organic compound sensors, 188, (*see also* Volatile organic compounds)
    observation method, 183–184
Conservation of mass and charge, 160–161
Construction materials, 890–894
Contaminant migration in barrier materials, 63–82
    experimental methods, 66–71
        batch adsorption, 66–67
        cation-exchange test protocols, 67–68
        column test protocols, 68–71
    field assessment, 72–80
        electrical methods, 77–79
        electrochemical methods, 79
        geophysical monitoring, 73–77
        in-situ monitoring, 73
        retrieval and testing of samples, 72–73
    numerical approaches, 63–66
Contaminant types and concentrations, 101–103, 114–115, 262–263, 633
    classes of, 262–263
Contaminated land, international perspectives, 1–62
    conclusions, 59–61
    policy issues, 2–48
        Australia, 2–3
        Austria, 3–5
        Belgium, 5–8
        Canada, 8–10
        Czech Republic, 10–12
        Denmark, 12–15
        Finland, 15–17
        France, 17–20
        Germany, 20–22
        Greece, 22–24
        Hungary, 24–27
        Netherlands, 28–29
        New Zealand, 29–32
        Norway, 32–34
        Slovenia, 34–35
        Sweden, 35–38
        Switzerland, 38–40
        Turkey, 40–43
        United Kingdom, 43–46
        United States, 46–48
    technologies and issues, 49–59
        biological processes, 53–54
        in-situ technologies, 49–51
        integration of technologies, 58–59
        physical-chemical treatment, 52–53
        solidification/stabilization, 56–58

[Contaminated land, international perspectives]
    thermal treatment, 54–56
Copper, 115, 131, 146, 396, 399, 413, 414, 415, 416, 541, 548–551, 555, 557, 561, 563–564, 596, 621, 636–637, 641–650, 830, 877
Corn, 933–946
Co-solvents, 233–234
Costs, 105–108
    electric energy cost, 105
    enhancement agent, 106
    fabrication and installation of electrodes, 105–106
    posttreatment, 106–107
    total, 107–108
    variable, 107
Cryptosporidium, 195
Cumulative risk assessment, uncertainties of, 762–769
Cyanides, 621
Cyanobacterial and microalgal growth inhibition factors, 221
Cyclic process
    to buffered sludge, 548
    for heavy metal extraction, 549
Cyclic voltammetry, 117, 123, 138–140, 141–147
    testing, 123, 138–140
Cynodon dactylon, 751
Cyperus corymbosus, 598

DDT, 114, 225, 228
DCR, (*see* Dispersing chemical reactions remediation technology)
Degree of immobilization, 281–284
Deposited sediments—recent muds, 389–394
Design considerations for hazardous waste landfills, 83–93, 635–636
    construction, operation, and postclosure, 91–92
    cover configuration, 90–91
    cut, fill, and air space, 87
    example case, 85–86
    leachate collection system, 89–90
    liner configuration, 88–89
    objectives, 84–86
    permit application, 92
    settlement and stability analysis, 90
    site characterization, 87–88
    waste characterization, 87
Desmanthus vergatus, 598
Desorption, 545, 561–566

Index

Detoxification of hazardous compounds, 905–920
  biological degradation of organics, 916–920
  chemical detoxification processes, 906–911
  DCR detoxification, 911–916
  organochlorines, 905–906
Dieldrin, 114
Diffenbachia picta, 492, 496
Dimethylformamide (DMF), 139
Dioxins, 914–916
Direct-current resistivity method, 78–79
Dispersing chemical reactions (DCR) remediation technology, 849–929
  chemical background, 859–865
  detoxification of hazardous compounds, 905–920
    biological degradation of organics, 916–920
    chemical detoxification processes, 906–911
    DCR detoxification, 911–916
    organochlorines, 905–906
  evaluation methods, 920–927
  hazardous waste problems, 849–859
    appropriate strategy, 853–859
    potential and endangering value, 850–853
  heavy metals, 865–879
    fixation by functionalized macromolecules, 879
    immobilization by a trapping matrix, 873–879
    through precipitation, 865–873
  oily phases, 879–897
    converting to dry pulverulent products, 879–882
    organic contaminants, 882–884
    in soil, 884–897
  production of solid fuels, 897–899
  removal of oily phases, 899–904
    on oil-polluted solids, 900–904
    on water, 899–900
  summary, 927–928
Distribution of heavy metals by depth, 401–408
  spatial, 394–401
Distribution of organic micropollutants by depth, 408–413
Drilling cuttings, (*see* Mineral oil-contaminated drilling cuttings)

Earth construction materials, 890–894
Earthworm and invertebrates test, 221
Elbe river, hydrological regime of, 378–379

Electrical methods, for detecting contamination, 77–79
Electrical neutrality, 164–165
Electric energy cost, 105
Electrochemical methods, 79
Electrodes
  configurations and time requirements, 103–104
  fabrication and installation, costs of, 105–106
  optimum spacings, 108–109
Electrokinetic remediation, 95–111, 120, 155–171, 234–235
  costs, 105–108
    electric energy cost, 105
    enhancement agent, 106
    fabrication and installation of electrodes, 105–106
    posttreatment, 106–107
    total, 107–108
    variable, 107
  electrolysis reactions, 98
  electrolyte enhancement, 99–100
    catholyte neutralization, 99
    chelating or complexing agents, 99–100
    enhancement of anolyte pH, 100
    ion-selective membranes, 99
  ion migration, 97–98
  models, 167–169
  optimum electrode spacings, 108–109
  phenomena in soils, 95–97
  practical considerations, 100–105
    contaminant types and concentrations, 101–103
    electrode configurations and time requirements, 103–104
    energy expenditure, 104–105
    soil type, 100–101
    voltage and current levels, 103
  soil pH and geochemical reactions, 98
  theoretical simulation, 155–171
    boundary conditions, 165–167
    charge flux, 159–160
    conservation of mass and charge, 160–161
    electrical neutrality, 164–165
    electrokinetic models, 167–169
    fluid flux, 156–157
    geochemical reactions, 161–163
    mass flux, 157–159
    modeling transport, 163–164
    numerical strategies, 167
Electrolysis reactions, 98

Electrolyte enhancement, 99–100
  catholyte neutralization, 99
  chelating or complexing agents, 99–100
  enhancement of anolyte pH, 100
  ion-selective membranes, 99
Electron acceptor reactions, 779–780
Electron donor reactions, 780–781
Electrophoretical mobility, 281–283
EMP
  cell experiments, 550
  process, 566–570
Energy expenditure, 104–105
Endomycopsis fibuligera, 938–946
Engineering techniques of As remediation, 746–749
Enhanced naphthalene bioavailability in a liquid-liquid biphasic system, 311–322
  conclusions, 319–320
  effect of an organic phase on bioavailability, 317–319
  effect of particle surface area on growth, 314–315
  effect of surfactants on naphthalene bioavailability, 316–317
  growth and degradation in mineral medium, 313–314
  introduction, 311–312
  isolation of a PAH-degrading culture, 312–313
Enhancement agents, cost of, 106
Environmental biochemistry of As, 745–752
Environmental information system (EIS) and holistic macroengineering approaches to biodegradation, 759–827
  of BTEX compounds, 769–775
  case study, 803–817
    conclusions, 815–817
    groundwater site conditions, 806–808
    present contamination description, 804–805
    simulation results, 808–813
    site description, 803
    soil profile, 806
    summary, 813–815
  of chlorinated aliphatic hydrocarbons, 779
  conclusions, 818–826
  degradation of chlorinated compounds, 775–785
    abiotic attenuation (degradation) mechanisms, 776–779
  introduction, 759–760

[Environmental information system (EIS) and holistic macroengineering approaches to biodegradation]
  mechanisms of chlorinated aliphatic hydrocarbon biodegradation, 779–781
  mixed evolution, 784–785
  overall evolution of chlorinated solvent plumes, 781–784
  risk-based approach, 760–769
    uncertainties of cumulative risk assessment, 762–769
  theoretical framework of EIS modeling, 786–803
    mass balance, 789–792
    natural processes influencing a representative volume, 789
    notation and definitions, 787–789
    solution of the governing equations, 793–799
    uncertainty analysis, 799–803
Equation for the liquid phase, 792, (*see also* Modeling; Numerical approaches; Theoretical framework of EIS modeling)
  solution of, 793–799
Equilibrium isotherms of free metal (II) cations, 547–548, (*see also* Heavy metals)
Equipment configurations, 124–125
Escherichia coli, 220
Ethanol, 136
Ethyl acetate, 467
Eucalyptus, 598
Eudrilus euginae, 606
Evaluation of hazardous chemical site landfilling, 193–215
  adequacy of hazard classification, 204–209
    excessive leaching, definition of, 207–209
    reliability of TCLP, 205–207
  adequacy of regulations, 194–195
  inadequate approaches, 196–200
  landfill and waste management area covers, 201–203
  liability for site clean-up, 194
  monitoring of capped waste management units, 204
  natural attenuation, 210–213
  time of waste as a threat, 200–201
Expanded reactor, 841–843, (*see also* Reactors)
Explosives, natural attenuation of, 114, 631, 949–970

Index

[Explosives, natural attenuation of]
  advantages and limitations, 967–968
  conclusions, 968
  implementation, 958–967
    case study at LAAP, 961
    groundwater modeling, 961
    groundwater monitoring, 959–961
    site characterization, 958–959
  introduction, 949–950
  process of, 950–951
  transport case study, 951–957
    biotransformation studies, 954–957
Extraction, 120
Extractive, ex-situ soil washing, 245–249 (*see also* Modeling extractive washing of HOC-contaminated soils; Soil washing)

Fabrication and installation of electrodes, costs of, 105–106
Fibrous components of artificial soils, 493, 497
Field techniques for sampling and measurement of soil gas constituents, 657–692
  background, 658
  case studies, 678–682
    site descriptions, 678
    typical soil gas results, 678–682
  introduction, 651
  measurement of soil gas constituents, 672–676
    direct read versus sample collection, 674–675
    instruments, 672–673
    serial dilution, 675–676
  quality control, 676
  sampling issues, 683–690
    direct-read versus Tedlar bag sampling, 687
    purge volume prior to sampling, 683–687
    short-circuiting, 688–690
  soil gas sampling and measurement systems, 660–672
    equipment, 661
    methods, 660–661
    permanent sampling points, 669–672
    rapid-deployment samplers, 661–669
  state-of-practice, 658–660
    gas sampling with probes, 659–660
    headspace measurements, 659
  summary, 690
  temperature, 676–678
Field testing, 121–128
Finger millet, 598

First-order decay model solution, 794–795
Flower crops, 597–598
Flow rate
  effect on soil texture, 69
  effect on sorption, 68–69
Fluid flux, 156–157
Flush filter probe with surface sampling, 661–664
  with downhole oxygen sensor, 668–669
Fluorescence measurements, of groundwater, 186
Fodder, 598, 933–946
  grass, 598
Formaldehyde, 468–471

Gaseous bioreduction, 605–606
Gasoline, 270
Gas plant wastes, manufactured, 847
Gas sampling, 659–660, (*see also* Field techniques for sampling and measurement of soil gas constituents)
Geochemical reactions, 161–163
Geophysical monitoring techniques, 73–77
Geotrichum candidum, 360–372
Goals, 173–174
Ground-permeating radar, 75–77
Groundwater (*see also* Cone penetrometer technology; Low-permeability soils; Policy)
  addressing contamination, 269–272
  modeling, 961, 964–965
  monitoring, 184–189, 959–961
    cleaning technique, 189
    fluorescence measurements, 186
    MicroWell, 188–189
    sample recovery, 186–188
    SCAPS system, 184–185
    site characterization, 186
    volatile organic compound sensors, 188
  in regional perspective, 174–180
    action plan, 179–180
    background of Emmen, 175
    framework under the new policy, 177–179
    situation under former policy, 175–177
Grouting and cleaning techniques, 189
Growth and degradation in mineral medium, 313–314
GTP-binding proteins, 298–301

Halogenated compounds, 806, 811–813
Harbor sediment, 844–847
Hazardous compounds, detoxification of, 905–920

[Hazardous compounds, detoxification of]
  biological degradation of organics, 916–920
  chemical detoxification processes, 906–911
  DCR detoxification, 911–916
  organochlorines, 905–906
Hazardous waste landfilling (*see* Design considerations for hazardous waste landfilling; Evaluation of hazardous chemical site landfilling; Wastes)
Hazardous waste problems, 849–859
  appropriate strategy, 853–859
  potential and endangering value, 850–853
  removal of, 853–854
HCB, 136
Headspace measurements, 659
Heavy metals, 114, 138, 234, 394–401, 401–408, 541–572, 639–655, 693–713, 806, 865–879
  distribution of heavy metals by depth, 401–408
  fixation by functionalized macromolecules, 879
  immobilization by a trapping matrix, 873–879
  ion-exchange process, new, 541–572
  -laden sludges and soils, decontamination of, 541–572
    CIM characteristics, 542–545
    conclusions, 570–571
    effect of background composition, 546–547
    experimental section, 547–550
    introduction, 541–542
    process configuration, 545–546
    results and discussion, 550–570
  mobility and the presence of organic wastes, 639–655
    concluding remarks, 651
    effects on metal mobility in soil columns, 649–650
    effects on the metal sorbing properties of soils, 646–648
    effects on sequentially extracted fractions of soil metals, 650–651
    introduction, 639
    metal complexes with organic matter in residues, 644–646
    metal-mobilizing action of olive mill wastewater, 642–644
    organic residues as metal-mobilizing agents, 641–642
    twofold interaction, 640–641
  precipitation, 865–873

[Heavy metals]
  spatial distribution of heavy metals, 394–401
  speciation of, 693–713
    availability and immobilization, 704–705
    introduction, 693–694
    particulate-bound, 698–704
    solution, 695–698
    summary and outlook, 705–706
Hexachlorobenzen (HCB), 114
Hexadecane, labeled, into CO_2 and the mycelia, 296–298
Hexavalent chromium, 114, (*see also* Chromium; Chromium-contaminated soil)
Hill Air Force Base site, 630–631
HDPE, 196, 202
HOC-contaminated soils (*see* Modeling extractive washing of HOC-contaminated soils)
Hormodendrum, 770
Hydraulic gradient
  effect on soil texture, 69
  and flow rate, 68–69
Hydraulic/pneumatic fracturing, 234, 273–274
Hydrocarbon
  contamination, 279–289 (*see also* Immobilized microorganisms)
  uptake and utilization by streptomyces strains, 291–310
    GTP-binding proteins, 298–301
    labeled hexadecane into CO_2 and the mycelia, 296–298
    monitoring the kinetics, 302–307
    oil utilization strains, 292–294
    uptake and transformation of *n*-alkanes, 295–296
Hydrological regime of the River Elbe, 378–379

Identification of toxic organic compounds in sediments, 417–424
Immobilizing hazardous waste, (*see also* Dispersing chemical reactions; Hazardous wastes; Waste)
  by technical means, 856–857
  through DCR, 857–859
Immobilized microorganisms, and removal of hydrocarbon contamination, 279–289
  conclusions, 287–288
  degree of, 281–289
  experimental program, 280–287
    biodegradation of oil in soil, 286–287
    biodegradation of oil in water, 284–286

Index

[Immobilized microorganisms, and removal of hydrocarbon contamination]
  degree of immobilization, 281–284
  methods and materials, 280–281
  introduction, 279–280
Industrial microorganisms, 454 (*see also* Microorganisms)
Industrial trickling bed biofilters for abatement of VOCs from air emissions, 449–473, (*see also* Volatile organic compounds)
  conclusions, 471
  introduction, 449–452
  results and discussion, 452–471
    biocatalyst development, 456–457
    industrial plants, 464–471
    isolation and selection of microorganisms, 452–456
    laboratory microreactors, 457
    pilot plants, 457–464
In-situ technologies, 49–51, 73
Instantaneous reaction model solution, 795–796
Instruments used in measuring soil gas constituents, 672–673
Integration of technologies, 58–59
Invertebrates and earthworm test, 221
Ion-exchange process, new, 541–573 (*see also* Heavy metals)
Ion migration, 97–98
Ion-selectivity
  of bentonite, 549, 563–566
  membranes, 99
Iron, and oxides, 621, 642, 694, 700, 702, 704, 722, 743–744, 749–750, 775
Isolation of oil-degrading microorganisms, 360–364
Isolation of a PAH-degrading culture, 312–313
Isolation and selection of microorganisms, 452–456, (*see also* Microorganisms)

Jasminum grandiflorum, 597
Jerusalem artichoke, 933–946

Kinetics, monitoring, 302–307
King of Prussia superfund site, 636–637
Korai grass, 599

Laboratory microreactors, 457
Lambito mourtii, 606
Landfill and waste management area covers, 201–203
Leachate collection system, 89–90
Leaching, 205–207
  excessive, 207–209
Lead, 114, 131, 146, 396, 413, 414, 416, 541, 549, 557, 596, 621, 636–637, 641–650, 806, 830, 860
  oxide, 114
Liability for site clean-up, 194
Light nonaqueous-phase liquid (LNAPL), 657–658
Liner configuration, 88–89
Liquid-liquid biphasic system, (*see* Enhanced naphthalene bioavailability in a liquid-liquid biphasic system)
Louisiana Army Ammunition Plant, 961 (*see also* Explosives, natural attenuation of)
Low-permeability soils, and bioremediation, 251–278
  AS/SVE studies, 269–274
    addressing groundwater contamination, 269–272
    pneumatic and hydraulic fracturing, 234, 273–274
  conclusion, 274
  impact efficiencies, 267–269
    continuous versus pulsed, 268–269
    removal, 267–268
  introduction, 261–264
    biological remediation, 263–264
    classes of contaminants, 262–263
    definitions, 261–262
  monitoring AS/SVE, 266–267
    biological removal, 266–267
    physical removal, 266
  technical description of AS/SVE system, 264–266
    air quality control, 265–266
    construction, 264–265

Maize, 598, 603
  effect of amendments on yield, 603
Malathion, 139
Manufactured gas plant wastes, 847
Mass balance, 789–792
Mass flux, 157–159
Mass transfer, 225–229
Measurement of soil gas constituents, (*see also* Field techniques for sampling and measurement of soil gas constituents)
  direct reed versus sample collection, 674–675
  instruments, 672–673
  serial dilution, 675–676
Mercury, 114, 541, 621, 641–650, 860

Metal remediation using solid additives, 507–521
  conclusions, 519–521
  introduction, 507–510
  materials and methods, 510–512
    analytical methods, 511
    effect of additives, 511
    rapid characterization of, 510
    soil characteristics, 510
    thermodynamic simulations, 511–512
  results and discussion, 512–519
    experimental study, 512–516
    thermodynamic simulations, 516–519
Methanol, 468–471
Methoxychlor, 114
Methyobacterium, 454, 470
Methylmercury chloride, 114
Microalgal and cyanobacterial growth inhibition factors, 221
Microbial bioassays, 220
Microbial reduction, 605–606
Microorganisms, characteristics of, 229–230, 279–289, 452–456, 599–602 (*see also* Immobilized microorganisms)
  isolation and selection of, 452–456
  use of, 599–602
Microreactors, laboratory, 457
Mineral medium, 313–314
Mineral oil-contaminated drilling cuttings, 323–329
  conclusions, 328–329
  experimental details, 323–325
    drilling mud, 323–324
    experimentation, 324–325
  results and discussion, 325–328
    experiments at 4.5-L scale, 325–327
    experiments at 65-L scale, 327–328
Mirex, 114
Mobility and immobilization, 855–859
Modeling extractive washing of HOC-contaminated soils, 243–260
  conclusions, 258–259
  extractive, ex-situ soil washing, 245–249
  groundwater, 964–965
  introduction, 243–245
  results of simulations, 255–258
  staged, extractive washing model, 249–255
    countercurrent separations, 253–255
    multiple-fraction equilibrium washing, 249
    multisize washing, 251–253
    physical separation processes, 253

[Modeling extractive washing of HOC-contaminated soils]
    rate-limited extraction, 249–251
Modeling, instantaneous reaction solution, 795–796
Modeling transport, 163–164, 167–169, 964–965
Monitoring, 184–189, 204, 266–267, 302–307, 963–964 (*see also* Groundwater monitoring)
  AS/SVE, 266–267
    biological removal, 266–267
    physical removal, 266
  of capped waste management units, 204
  kinetics, 302–307
Monod kinetics model solution, 796–799
Muds, recent, 389–394
Multiphase partitioning, 229
Municipal solid waste generation and management, 971–979
  alternatives, 978
  Caracas and metro area, 971–972
  conclusions, 978–979
  existing system, 972–974
  introduction, 971
  problems of, 974–976
  similarities in Yangon and in Cma, 976–978

n-alkanes, uptake and transformation, 295–296
Naphthalene, 136
NAPLs, 229
Natural attenuation, 210–213
Natural processes influencing a representative volume, 789
Nerium oleander, 597
Nickel, 396, 399, 403, 407, 541, 549, 596, 621, 636–637
Nitrates, 114
Nonextractable ^{14}C-PAH residues in contaminated soils, 429–447
  fate and stability of, 437–445
    impact of biological treatments, 439–440
    impact of chemical treatments, 441–445
    impact of physical treatments, 440–441
  formation of, 431–437
  introduction, 429–431
Norcardia, 454, 464
Numerical approaches, (*see also* Equation for the liquid phase; Modeling; Theoretical framework of EIS modeling)
  of contaminant migration, 63–66
  of electrokinetic remediation, 167

Index

Odor control, 464
Oil-contaminated land, recultivation of, 884–490
Oil-degrading microorganisms, 292–294, 360–364
Oily phases, 879–897, 899–904
  converting to dry pulverulent products, 879–882
  organic contaminants, 882–884
  removal of, 899–904
    on oil-polluted solids, 900–904
    on water, 899–900
  in soil, 884–897
Olive mill wastewater, 642–644
Organic and inorganic contaminants in a river floodplain, 375–428, 882–884, (*see also* Heavy metals)
  conclusions, 424–425
  hydrological regime of the River Elbe, 378–379
  identification of toxic organic compounds in sediments, 417–424
  introduction, 375–377
  soil characterization, 379–383
  soil hydrological regime, 383–385
  soil organic matter (SOM), 382–383
  soil pollution, 385–413
    deposited sediments—recent muds, 389–394
    distribution of heavy metals by depth, 401–408
    distribution of organic micropollutants by depth, 408–413
    input of suspended particulate matter, 386–389
    spatial distribution of heavy metals, 394–401
  transfer of heavy metals to plants, 413–417
Organic wastes, 639–655, (*see also* Heavy metals; Wastes)
  concluding remarks, 651
  effects on metal mobility in soil columns, 649–650
  effects on the metal sorbing properties of soils, 646–648
  effects on sequentially extracted fractions of soil metals, 650–651
  introduction, 639
  metal complexes with organic matter in residues, 644–646
  metal-mobilizing action of olive mill wastewater, 642–644

[Organic wastes]
  organic residues as metal-mobilizing agents, 641–642
  twofold interaction, 640–641
Organochlorines, 905–906
Orthic luvisol, 579
Oryza sativa, 598

Particle surface area, effect on growth, 314–315
Particulate matter, input of suspended, 386–389
PCBs, (*see* Polychlorinated biphenyl)
Pea (grain), 933
Pencillium, 606
Pentachlorophenol (PCP), 114
Permanent sampling points, 669–672
Permeameter chamber, 69–71
Pesticides, 114
Petroleum-related compounds, 114
pH, soil, and geochemical reactions, 98
Phenols, 114, 136, 806
Phenomena in soils, 95–97
Phosphates, 114
Photinus pyralis, 220
Photobacterium fischerii, 221
Photobacterium phosphoreum, 220, 221
Photolysis, enhanced, and PCBs, 523–540
  of PCBs in surfactant soil washing solutions, 526, 531–538
    identification and quantification, 531–534
    rates and quantum yields, 529–531
  of 2,4,4′ CB in nonionic surfactant [POL(10)]solutions, 526–531
  of 2,4,4′ CB in surfactant solutions, 524–526
Photooxidation treatment, of soil, 53
Physical-chemical treatment of soil, 52–53, 266–267, 440–441, 441–445, 592–594
Phytoremediation, 121, 594–599, 751–752
  and tannery wastes, 594–599
Pilot plants, 457–464
Plant tests, 224
Pleurotus ostreatus, 431–437, 440
Pneumatic/hydraulic fracturing, 234, 273–274
Poa, 415
Polarographic techniques, 116–117
Polianthus tuberosa, 597
Policy issues, 2–48, 173–181
  Australia, 2–3
  Austria, 3–5
  Belgium, 5–8
  Canada, 8–10
  Czech Republic, 10–12

[Policy issues]
  Denmark, 12–15
  Finland, 15–17
  France, 17–20
  Germany, 20–22
  Greece, 22–24
  Hungary, 24–27
  Netherlands, 28–29, 173
  new policy, and techniques, 173–181
    groundwater in regional perspective, 174–180
  New Zealand, 29–32
  Norway, 32–34
  Slovenia, 34–35
  Sweden, 35–38
  Switzerland, 38–40
  Turkey, 40–43
  United Kingdom, 43–46
  United States, 46–48
Pollutant distribution, toxicity, and remediation techniques, 113–153
  background, 114–115
    contaminants of interest, 114–115
    speciation and risk assessment issues, 115
  discussion, 141–146
    future, 141–142
    implications, 142–143
    new approaches, 143–146
    status, 141
  materials and methods, 115–130
    chemical concentration measurement, 115–118
    cyclic voltammetry, 117
    field testing, 121–128
    remediation techniques, 119–121
    theoretical concerns, 128–130
    toxicity assessment, 118–119
  results, 130–141
    Bayou Saint John, 133
    Bayou Trepagnier samples, 130–133
    colorimetric and selective ion electrodes, 140–141
    cyclic voltammetry testing, 138–140
    sequential extraction, 134–135
    synergistic and antagonistic interactions, 135–137
Pollution due to tannery wastes, 585–586 (*see also* Tannery wastes)
Polychlorinated biphenyl (PCB), 243–248, 524–538, 621, 633, 806, 905–914
  in surfactant soil washing solutions, 526, 531–538, 621, 633
    identification and quantification, 531–534

[Polychlorinated biphenyl (PCB)]
  of 2,4,4′ CB in nonionic surfactant [POL(10)]solutions, 526–531
  of 2,4,4′ CB in surfactant solutions, 524–526
Polycyclic aromatic hydrocarbons (PAHs), 114, 129, 136, 146, 225–227, 233, 311–322, 429–447, 806, 844–847
  ^{14}C-PAH residues in contaminated soils, 429–447
  isolation of a PAH-degrading culture, 312–313
Polygunum sachalinese, 751
Polynuclear aromatics, 621, 633
Powder processing,
  fine, 844
  pilot, 844
  pilot-scale evaluation, 844–847
    harbor sediment, 844–847
    manufactured gas plant wastes, 847
Practical considerations, 100–105
  contaminant types and concentrations, 101–103
  electrode configurations and time requirements, 103–104
  energy expenditure, 104–105
  soil type, 100–101
  voltage and current levels, 103
Precipitation/dissolution reactions, 162–163
Pressure sensors in unsaturated media, 79
Principle responsible particles (PRPs), 193, 194
Process reactors, 839–848, (*see also* Reactors)
  characteristics, 843
  conclusions, 847–848
  fine powder processing, 844
  fuel injection, 843–844
  particle residence time distribution, 843
  pilot powder processing, 844
  pilot-scale evaluation, 844–847
    harbor sediment, 844–847
    manufactured gas plant wastes, 847
  technologies, 839–843
    compact reactor, 839–841
    expanded reactor, 841–843
Pseudomonas aeruginosa, 601, 606
Pseudomonas cepacia, 360–364
Pseudomonas fluorescens, 220–221, 360–364, 454, 467–468
Pseudomonas oleovorans, 454, 460–463, 467
Pseudomonas putida-36, 280–288, 360–372, 770
Pteris vittata, 751

Quality control, 676

Index

Quantum yields, and photolysis, 529–531

Rapid-deployment samplers, 661–669
    flush filter probe, 661–664
        with downhole oxygen sensor, 668–669
        recessed filter probe, 665–666
        retractable-tip probe, 666–668
RC1 consortium for soil decontamination, 359–374
    conclusion, 372–373
    introduction, 359–360
    isolation of oil-degrading microorganisms, 360–364
    possibilities in degradation of oil products, 368–369
    technology of RC1 production, 364–368
    verification of biodegradation activity, 369–372
RDX, 949–970, (*see also* Explosives, natural attenuation of)
Reactors
    BCD, 910
    compact, 839–841
    expanded, 841–843
    process, 839–848,
        characteristics, 843
        conclusions, 847–848
        fine powder processing, 844
        fuel injection, 843–844
        particle residence time distribution, 843
        pilot powder processing, 844
        pilot-scale evaluation, 844–847
        technologies, 839–843
Recessed filter probe with surface sampling, 665–666
Retardation factors, calculation of, 954–956
Retractable-tip probe with surface sampling, 666–668
Reed bed, 604–605
Remediation of arsenic-contaminated soils, 739–758, (*see also* Arsenic)
Remediation of chromium-contaminated soil, 592–603 (*see also* Chromium; Chromium-contaminated soil, due to tannery wastes disposal)
    bioremediation, 599–603
    chemical in-situ treatment, 593–594
    physical in-situ treatment, 592–593
    phytoremediation, 594–599
Remediation techniques, 119–121, 173–181
    set by new policy, 173–181
        groundwater in regional perspective, 174–180

Resource Conservation Recovery Act (RCRA), 196, 201–203, 205, 213
Retrieval and testing of samples, 72–73
Rhodococcus, 294–295, 439–440, 454, 464
Rhodosporidium, 770
Rice, 598
Risk assessment issues, and speciation, 115
River Elbe, 375–428
Road construction, and DCR products, 894–896

Saccharam officinalis, 599
Sampling, 186–188, 576, 660–672, 683–690
    (*see also* Field techniques for sampling and measurement of soil gas constituents)
    equipment, 661
    issues, 683–690
        direct read versus Tedlar bag sampling, 687
        meter limitations, 687–688
        purge volume prior to sampling, 683–687
        short-circuiting, 688–690
    methods, 660–661
        rapid deployment, 661–669
        permanent sampling points, 669–672
    and preparation, 576
    recovery, 186–188
SCAPS system, 184–185, 190
Sediments—recent muds, 389–394
Selenium, 596, 621, 830
Semivolatile organic compounds, 114
Sequential extraction, 134–135, 701–704
Shoe-making factory, 467
Single chemical extractants, 700–701
Site characterization, 186, 510, 678, 958–959
Site, liability of clean-up, 194
Solanum melongena, 598
Solvents, mixed, 461–463
Soil characterization, 379–383
Soil construction, (*see* Artificial soils construction, structural and functional aspects)
Soil flushing, 617–631, (*see also* Soil washing)
    additional treatment requirements, 625–626
    background, 618
    case studies, 629–631
    contaminant types, 620–622
    design considerations, 626–629
    soil types, 619
    suitable applications, 618–619
    system equipment requirements, 622–625
    theory, 618

Soil gas constituents, (see Field techniques for sampling and measurement of soil gas constituents)
Soil hydrological regime, 383–385
Soil-like materials, DCR treatment of, 896–897
Soil organic matter (SOM), 382–383
Soil pH and geochemical reactions, 98
Soil pollution, 385–413
  deposited sediments—recent muds, 389–394
  distribution of heavy metals by depth, 401–408, (see also Heavy metals)
  distribution of organic micropollutants by depth, 408–413
  input of suspended particulate matter, 386–389
  spatial distribution of heavy metals, 394–401
Soil respiration, determination of, for scientific and routine analysis, 573–582
  applications, 581–582
  calculating soil respiration, 577–578
    simplified, 578
  calibration, 578
  carrying out measurements, 577
  documenting and assessment, 577
  introduction, 573–574
  measurements, 574, 575–576, 578–579
    comparative, 578–579
    principle of, 574
    range, 575–576
    system, 575
  orthic luvisol, 579
  potential and limits, 581
  reproducibility, 580
  sampling and preparation, 576
  volume determination, 577
  woodland soil, 580
Soil substitutes, 491–505
  fibrous components, 493, 497
Soil types, 100–101, 619, 632–633
Soil vapor extraction (SVE), 269–274 (see also Low-permeability soils, and bioremediation)
  addressing groundwater contamination, 269–272
  monitoring AS/SVE, 266–267
    biological removal, 266–267
    physical removal, 266
  pneumatic and hydraulic fracturing, 273–274
  technical description of AS/SVE system, 264–266
    air quality control, 265–266

[Soil vapor extraction (SVE)]
    construction, 264–265
Soil washing, 52, 592, 631–637, (see also Soil flushing)
  background, 632
  case studies, 636–637
  and chromium-contaminated soils, 592, (see also Chromium; Chromium-contaminated soil)
  combined with other technologies, 52
  contaminant types, 633
  design considerations, 635–636
  equipment requirements, 634–635
  soil types, 632–633
  suitable applications, 632
  theory, 632
  typical, 52
  washing solutions, 633–634
Solid additives, (see Metal remediation using solid additives)
Solid fuels, production of, 897–899
Solidification/stabilization, 56–58
Solid waste, (see Municipal solid waste generation and management)
Solvent removal, (see Aqueous solvent removal of contaminants from soils)
Sorption reaction, 161–162, 545, 549, 557–561
Spatial distribution of heavy metals, 394–401, (see also Heavy metals)
Speciation
  and heavy metals, 693–713, (see also Heavy metals)
  and risk assessment issues, 115
Specimen preparation, 125–126
Streptomyces strains, uptake and utilization by, 291–310
  GTP-binding proteins, 298–301
  labeled hexadecane into CO_2 and the mycelia, 296–298
  monitoring the kinetics, 302–307
  oil utilization strains, 292–294
  uptake and transformation of n-alkanes, 295–296
Styrene, 467–468
Sugar cane, 599, 933–946
Sulfuric acid, 114
Superfund sites, 617, 621, 630, 631, 636–637
Surfactants, 230–233, 523–540
  effect on naphthalene bioavailability, 316–317
  -enhanced photolysis of PCBs, 523–540
    conclusions, 538
    introduction, 523–524

Index

[Surfactants]
    materials and methods, 524–526
    results and discussion, 526–528
Suspended particulate matter (SPM), 385–389, 401, 402 (*see also* Particulate matter)
Synergistic and antagonistic interactions, 135–137
Synthetic compounds, and soils, 510 (*see also* Artificial soils)

Tannery wastes, 583–615, (*see also* Chromium contaminated soil, due to tannery wastes disposal)
    characterization of, 586–587
    pollution due to disposal of, 585–586
    remediation of, 592–603
        bioremediation, 599–603
        chemical in-situ treatment, 593–594
        physical in-situ treatment, 592–593
        phytoremediation, 594–599
    tanning processes, 584–585
        chrome tanning, 584–585
        vegetable tanning, 584
Technologies and issues, 49–59
    biological processes, 53–54
    in-situ technologies, 49–51
    integration of technologies, 58–59
    physical-chemical treatment, 52–53
    solidification/stabilization, 56–58
    thermal treatment, 54–56
Tedlar bag sampling, 687
Temperature, 676–678
Tetrachlorodibenzodioxin (TCDD), 114
Tetrandine, 114
Theoretical concerns, 128–130, 155–171, 786–803
Theoretical framework of EIS modeling, 786–803, (*see also* Equation for the liquid phase; Modeling; Numerical approaches)
    mass balance, 789–792
    natural processes influencing a representative volume, 789
    notation and definitions, 787–789
    solution of the governing equations, 793–799
    uncertainty analysis, 799–803
Thermal enhancements, 234
Thermal treatment, 54–56
Thermodynamic simulations, 511–512, 516–519, 780–781
Thiobacillus ferrooxidans, 867
Time of waste as a threat, 200–201

TNT, 949–970, (*see also* Explosives, natural attenuation of)
Tobacco, 114
TORBED process reactor, 839–848, (*see also* Reactors)
    characteristics, 843
    conclusions, 847–848
    fine powder processing, 844
    fuel injection, 843–844
    particle residence time distribution, 843
    pilot powder processing, 844
    pilot-scale evaluation, 844–847
        harbor sediment, 844–847
        manufactured gas plant wastes, 847
    technologies, 839–843
        compact reactor, 839–841
        expanded reactor, 841–843
Toxicity, 118–119, 126–127, 219–225, 854–855
    assessment, 118–119, 126–127
    bioluminescence reporter system, 220–221
    earthworm and invertebrates test, 221
    microalgal and cyanobacterial growth inhibition factors, 221
    microbial bioassays, 220
    plant tests, 224
    tests, 222–223
    uptake of contaminants, 224–225
    vertebrate tests, 221–224
Toxicity characteristic leaching procedures (TCLP), 205–207, 208–209, 213, 541
Trace metals, remediation of, (*see* Metal remediation using solid additives)
Trametes versicolor, 439–440
Triticum vulgare, 496
2,4,4′ CB
    in nonionic surfactant [POL(10)]solutions, 526–531
    in surfactant solutions, 524–526

Uncertainty analysis, 799–803
Underground storage tanks (UST), 261
United Chrome Products site, 629–630
Uptake of contaminants, 224–225

Vanadium, 114, 621
Vegetable raw material, processing of, to obtain carbohydrate-protein fodder, 931–948
    body of material, 932–946
    conclusion, 947
    introduction, 931–932
    for tanning, 584, (*see also* Tannery wastes)

Verification of biodegradation activity, 369–372
Vertebrate tests, 221–224
Vibrio fischeri, 423
Volatile organic compounds (VOCs), 114, 188, 449–452, 631, 686–687, 760, 806 (*see also* Industrial trickling bed biofilters for abatement of VOCs from air emissions)
Voltage and current levels, 103
Voltammetry, cyclic, 117, 123, 138–140, 141–147
  testing, 138–140

Washing, of HOC-contaminated soil, 243–260 (*see also* Modeling extractive washing of HOC-contaminated soils)
  difficulties, 258
Wastes, 83–93, 201–203, 510, 639–655, 847, 849–859, 931–948, 971–979, (*see also* Design considerations for hazardous waste landfills; Evaluation of hazardous chemical site landfilling; Organic and inorganic contaminants)
  characterization of, 87

[Wastes]
  design considerations, 83–93
  hazardous waste problems, 849–859
    appropriate strategy, 853–859
    potential and endangering value, 850–853
  landfill and waste management area covers, 201–203
  manufactured gas plant, 847
  organic, 639–655
  solid, generation and management, 971–979
  tannery, 583–615
  vegetable raw material, 931–948
Wet deposition, and arsenic, 730, 740
Wetlands, constructed, 606
Woodland soil, 580

Xylene, 460–461

Yarrowia lipolytica, 360–372

Zea mays, 598
Zero-valent metals, 120–121
Zinc, 396–398, 402, 403, 407, 413–416, 513–515, 596, 621, 636–637, 641–650, 701, 703, 704, 705, 806, 830
  extraction, 513–515